50 Years of the Renormalization Group

Dedicated to the Memory of Michael E Fisher

Other World Scientific Titles by Michael E. Fisher

Excursions in the Land of Statistical Physics
ISBN: 978-981-314-489-7
ISBN: 978-981-314-490-3 (pbk)

50 Years of the Renormalization Group

Dedicated to the Memory of Michael E Fisher

Editors

Amnon Aharony
Tel Aviv University, Israel

Ora Entin-Wohlman
Tel Aviv University, Israel

David A Huse
Princeton University, USA

Leo Radzihovsky
University of Colorado, USA

World Scientific

NEW JERSEY • LONDON • SINGAPORE • BEIJING • SHANGHAI • HONG KONG • TAIPEI • CHENNAI • TOKYO

Published by

World Scientific Publishing Co. Pte. Ltd.
5 Toh Tuck Link, Singapore 596224
USA office: 27 Warren Street, Suite 401-402, Hackensack, NJ 07601
UK office: 57 Shelton Street, Covent Garden, London WC2H 9HE

British Library Cataloguing-in-Publication Data
A catalogue record for this book is available from the British Library.

50 YEARS OF THE RENORMALIZATION GROUP
Dedicated to the Memory of Michael E Fisher

ISBN 978-981-12-8237-9 (hardcover)
ISBN 978-981-12-8238-6 (ebook for institutions)
ISBN 978-981-12-8239-3 (ebook for individuals)

For any available supplementary material, please visit
https://www.worldscientific.com/worldscibooks/10.1142/13571#t=suppl

Desk Editor: Joseph Ang

Typeset by Stallion Press
Email: enquiries@stallionpress.com

https://doi.org/10.1142/9789811282386_fmatter

Preface

This book has two strongly overlapping themes: it is a memorial to Michael E. Fisher and his scientific legacy, and it is a review of the modern era of the renormalization group, which was initiated by Wilson and Fisher five decades ago and was one of the scientific topics where Michael Fisher made major and seminal contributions.

The 1972 Wilson–Fisher paper, on critical phenomena in 3.99 dimensions, started the modern era in the study of critical phenomena using the renormalization group. This *Physical Review Letter* (PRL), included in PRL's "Letters from the Past — A PRL Retrospective" (https://journals.aps.org/prl/50years/milestones), has become a major tool for such analytic calculations, opening the way to the study of many physical systems. After 50 years, we find it appropriate to review the history of the field over the last five decades and its current status. Indeed, as this book shows, the field has a fascinating history of development and it continues to be very active, applying RG methods in ever more diverse ways and fields of research.

On May 21, 2023, at 16:00 Israel time, Google found 417000 papers and books, which included "renormalization group" in their titles. This number grew by the hour. The topic has been covered in many textbooks and is included in practically all advanced courses on statistical physics and quantum field theory. Examples of books specifically on the renormalization group include those by Ma, Toulouse and Pfeuty, Cardy, Goldenfeld, Zinn-Justin, Binney, Dowrick, Fisher and Newman, Vasil'ev, Uzunov, Creswick, Farach, and Poole, Amit and Martin-Mayor, and many more. This book complements these books by reviewing more recent developments.

The major contributions of Kenneth G. Wilson have already been celebrated after his death in *Ken Wilson Memorial Volume: Renormalization, Lattice Gauge Theory, The Operator Product Expansion and Quantum Fields*, edited by Belal E. Baaquie, Kerson Huang, Michael E. Peskin, and K. K. Phua (World Scientific, 2015). After 1974, Wilson moved on to apply the renormalization group to other topics, while Fisher continued research in critical phenomena. This book is dedicated to the memory of Michael E. Fisher, who died on 26 November 2021 (50 years after the Wilson–Fisher paper was received by PRL on 11 October 1971). Fisher was the direct and indirect teacher for all of us and practically influenced the careers of many statistical physics researchers throughout his long career.

The modern history of critical phenomena would not have been the same without the leadership of Michael E. Fisher. The book *Excursions in the Land of Statistical Physics* (World Scientific, 2017) contains a collection of reviews by Fisher, including his review on the history of the renormalization group theory, the epsilon expansion and Ken Wilson. Some of these reviews are mentioned in our book and can be seen as a follow-up of Fisher's many contributions.

Many of the contributors to this book were Fisher's students, postdocs and colleagues (due to obvious constraints, we could not invite all of Fisher's followers and we apologize for that). Their contributions contain vivid memories of their personal and scientific encounters with him. Some of them also include reviews and new recent results related to Fisher's legacy. Other (younger) authors did not interact directly with Fisher, but their contributions present recent important studies on topics that reflect the current trends in the field. We thank all the authors for their enthusiastic acceptance of our invitations, and also for helping to referee other author's contributions.

We hope that this book will stimulate more activity in the field and look forward to the next 50 years.

Amnon Aharony
Ora Entin-Wohlman
David A. Huse
Leo Radzihovsky

https://doi.org/10.1142/9789811282386_fmatter

Contents

Preface v

Michael Fisher Memories 1

Chapter 1. Personal History with MEF and Some Related Topics **3**
H. Au-Yang and J.H.H. Perk

Chapter 2. Michael Fisher and Critical Phenomena in the 1970s **19**
J.M. Kosterlitz

Chapter 3. My Memories of Michael Fisher **31**
J.S. Langer

Chapter 4. Some Memories from a 60-Year Friendship with Michael E. Fisher **33**
J.L. Lebowitz

Chapter 5. My Life with Michael E. Fisher **39**
N.D. Mermin

Chapter 6. Michael E. Fisher: Teacher, Mentor, Colleague and Friend: A (Very) Personal Account **47**
E. Domany

Chapter 7. Michael Fisher as a Role Model 61
B.I. Halperin

Chapter 8. Some Reminiscences of Michael E. Fisher 65
A.J. Liu

Phase Transitions and Critical Phenomena 67

Chapter 9. Fluctuation-Dominated Phase Ordering 69
M. Barma

Chapter 10. Mixed-Order Phase Transitions 89
D. Mukamel

Chapter 11. Bounds on the Superconducting Transition Temperature 103
M. Randeria

Chapter 12. Critical Fluctuations in Polymer Solutions: Crossover from Criticality to Tricriticality 119
M.A. Anisimov, T.J. Longo, and J.V. Sengers

Chapter 13. Series Expansions in Statistical Physics: Large Dimensionality, Quenched Randomness and Griffiths Singularities 147
R.R.P. Singh

Chapter 14. Deconfined Quantum Critical Points: A Review 169
T. Senthil

Chapter 15. Experimental Studies of the Critical Casimir Effect 197
M.H.W. Chan

Renormalization Group 213

Chapter 16. Fisher Renormalization and Fisher Reminiscences 215
D.R. Nelson

Chapter 17. 50 Years of Correlations with Michael Fisher and the Renormalization Group 245
Amnon Aharony

Chapter 18. The Yang–Lee Edge Singularity and Related Problems 281
J. Cardy

Chapter 19. There's Plenty of Room in the Middle: The Unsung Revolution of the Renormalization Group 303
Nigel Goldenfeld

Chapter 20. Knots from a Random Matrix Theory with Replicas 327
S. Hikami

Chapter 21. Wilson-Fisher Fixed Points in the Presence of Dirac Fermions 347
I.F. Herbut

Chapter 22. Strong-Randomness Renormalization Groups 363
David A. Huse

Chapter 23. Line Defects and the Renormalization Group 373
Z. Komargodski

Chapter 24. Theories and Mysteries of the Normal-Metal-to-Superconductor and Nematic-to-Smectic-A Phase Transitions 385
T. Lubensky

Chapter 25. Beyond the Bulk: An Appreciation of the Enduring Influence of Michael E. Fisher 405

D. Jasnow and J. Rudnick

Chapter 26. Normal Forms, Universal Scaling Functions, and Extending the Validity of the RG 415

J.P. Sethna, D. Hathcock, J. Kent-Dobias, and A. Raju

Chapter 27. Scaling Behaviors at Quantum and Classical First-Order Transitions 437

A. Pelissetto and E. Vicari

Chapter 28. Michael Fisher and Localization 477

F.J. Wegner

Chapter 29. Fermi Surface Renormalization Group 503

R. Shankar

Chapter 30. Critical Matter 513

L. Radzihovsky

Finite Size Scaling, Monte Carlo 549

Chapter 31. Multiparameter Universality and Intrinsic Diversity in Weakly Anisotropic Bulk and Confined Systems 551

V. Dohm

Chapter 32. Michael Fisher's Relationship to Monte Carlo 587

D.P. Landau and W. Selke

Chapter 33. Finite-Size Scaling 599

A.P. Young

Field Theory 617

Chapter 34. Should We Worry about Renormalons in the ϵ-expansion? 619

E. Brézin

Chapter 35. Field-Theoretic Renormalization Group in Models of Critical Behavior, Turbulence and Non-equilibrium Dynamics 631

L. Adzhemyan, N. Antonov, M. Hnatič, J. Honkonen, P. Kakin, D. Kazakov, M. Komarova, M. Kompaniets, A. Kudlis, T. Lučivjanský, M. Nalimov

Chapter 36. Critical Phenomena and the Bootstrap 675

D. Poland and D. Simmons-Duffin

Chapter 37. Quantum Statistical Mechanics of the Sachdev–Ye–Kitaev Model and Charged Black Holes 693

S. Sachdev

Soft Matter, Complex Matter and Biological Physics 719

Chapter 38. Topological Transformations of Biomembranes and Vesicles *via* Fission and Fusion 721

R. Lipowsky

Chapter 39. Universal Scaling Function Ansatz for Finite Temperature Jamming 769

S.A. Ridout, J.P. Sethna, and A.J. Liu

Chapter 40. How I Met Michael Fisher and Started Working in Biophysics 785

A.B. Kolomeisky

Chapter 41. Topological Defects in Living Matter **795**

J.M. Yeomans

Chapter 42. Critical Casimir Forces in Soft Matter **805**

A. Gambassi and S. Dietrich

Michael E. Fisher — Publications Until 2017 855

Index 891

Michael Fisher Memories

https://doi.org/10.1142/9789811282386_0001

Chapter 1

Personal History with MEF and Some Related Topics

H. Au-Yang* and J.H.H. Perk†

Department of Physics, Oklahoma State University,
145 Physical Sciences, Stillwater, OK 74078-3072, USA
†*perk@okstate.edu*

We present our personal histories with Michael Fisher. We describe how each one of us first came to Cornell University. We also discuss our many subsequent interactions and successful collaborations with him on various physics projects.

1. From Shanghai to Cornell University

Born under the name Ou Yang Yee Sun in Shanghai and after having suffered through much of the great famine of 1959–1962, Helen got official permission to join her parents in Hong Kong. It was quite a shock, not knowing Cantonese and English, as she only spoke Shanghainese, Mandarin and some Russian that she had learned in high school. Fortunately, the written Chinese is universal and Helen found a poster announcing a competition for two fellowships at Chu Hai College, Kowloon. Even though Helen had not yet finished high school, she won one of them and was admitted under the new name Helen Au-Yang. After graduating in 1965, she was admitted to San Diego State University.

In 1968, Helen entered graduate school at the State University of New York in Stony Brook. Here, she generalized the work of McCoy and Wu[1] on the Ising model on the half-plane with boundary magnetic field to the case with boundary coupling different from the coupling in the bulk. Thus, she provided exact results[2] related

*Helen's part of the story is told by her co-author and husband of 40 years, as, at the time of writing, she was not (yet) sufficiently recovered from a stroke.

to the series, Monte Carlo and mean field results of Binder and Hohenberg[3,4] on 2D and 3D Ising and Heisenberg models with a boundary.

Helen continued with the main topic of her PhD thesis[5]: layered Ising models with period n, deriving exact results for the specific heat[6] and for the pair correlation in a row parallel to the layering.[7] That last calculation led to a block Toeplitz determinant with a 2×2 matrix generating function $a(\xi)$. After reviewing the then known theorems for such determinants, especially for those for which $a(\xi)$ can be properly factorized, Helen gave $n = 2$ results for the $T < T_c$ spontaneous magnetization and for the leading long-distance behavior of the pair correlation for $T < T_c$ and for $T > T_c$.

In 1973, Helen graduated with a Ph.D. in theoretical physics, with her thesis signed by Professors B.M. McCoy, C.N. Yang, T.T. Wu and others.[a] She became a postdoc with Michael Fisher at Cornell University.

2. Helen at Cornell University

With Helen in his group, Michael Fisher continued the work on his project "Bounded and Inhomogeneous Ising Models" started with Arthur Ferdinand[9,10] many years earlier. In 1969, they had published finite-size effects on the specific heat of an $m \times n$ Ising lattice, asymptotically for large n with $\xi = m/n$ fixed.[10]

Helen and Michael first studied the specific-heat scaling function for an infinitely long Ising strip of width n in the limit $n \to \infty$.[11] After this, they studied the finite-size effects of regularly spaced point defects,[12–14] also incorporating some earlier calculations of Ferdinand.

Michael was very happy to have Helen around, as he liked things to be done exactly if they could be done exactly. One time a student solved a cubic equation numerically by computer. Michael told Helen, "Helen, you show him how that is done exactly!"

Michael wanted to keep Helen for a longer time by creating a special research position for her. He asked C.N. Yang to take her back as a postdoc for a year, while he created such a new position at Cornell. At Stony Brook, Helen calculated Ising multispin correlations along the diagonal[15] to check the operator reduction formulae of Kadanoff. She also calculated a new scaling form for the four-spin correlation along the diagonal.[16]

Back at Cornell University, Helen got involved in a generalization of earlier work,[11] namely the study of an $n \times \infty$ Ising strip with field h_1 on the first infinite

[a]The publication of the thesis[5] and the papers[6,7] got delayed, as her advisor Barry McCoy was involved with the massive project on Painlevé III scaling functions in 2D Ising correlations.[8]

layer and field h_n on the last (nth) layer. One motivation for this work was to investigate the validity of the *ad hoc* local free-energy functional of Fisher and de Gennes.[17] In spite of serious limitations of that theory, their most striking predictions were confirmed and finite-size scaling forms were derived in the limit $n \to \infty$.[18,19] A more detailed summary of this research is given in Section 4.3 of Fisher's "Simple Ising Models Still Thrive!" lecture.[20]

At the same time period, much numerical work was done to study inhomogeneous differential approximants for power series of one or more variables.[21] Nowadays, with Maple and Mathematica available, this is a lot easier, but then it was not. At that time, using Fortran, one had to prepare a stack of punchcards for each example with up to 80 characters per card and hand that in for processing overnight, carefully keeping the stacks of cards in the right order. This work was very labor-intensive, as not only existing high- and low-temperature series coefficients and activity (high-field) expansions of various well-known spin models but also many test examples derived from mathematical functions with a given singularity and various background noise contributions were examined.

One case was studied in detail: universality tests at Heisenberg bicritical points.[22] The experimentally determined amplitude ratio $Q_{\text{fit}} \approx 1.6 \pm 0.35$ found in MnF_2 was much smaller than theory seemed to predict, $Q_{\text{th}} \approx 2.39$. Noise contributions in the theory could bring that value about 10% down, but that is not enough. Fisher *et al.* then note[22] that shifts in the critical lines compatible with experimental uncertainties could raise Q_{fit} enough to bring theory and experiment in agreement. In other words, Q_{fit} should have been reported with significantly larger error bars.

During a visit back to Stony Brook, Helen calculated the scaling form for the four-spin correlation function of two parallel nonaligned pairs in the planar Ising model.[23]

Some time in 1980, Helen's oldest sister and her family were released from labor camp in Western China to settle in Hongkong with the rest of the family there. Helen then abruptly quit her position at Cornell in order to help tutor her two nieces into the Hongkong school system, using the earlier experience that she had had moving from Shanghai to Hongkong in 1962. In 1982, Helen wanted to return to the USA, and Barry McCoy and Joel Lebowitz organized a special invitation to the December 1982 Rutgers Stat. Mech. Meeting.

3. Jacques' Visits to Cornell and Rutgers

In 1978, I (Jacques) received the Royal Dutch Shell Oil Travel Prize awarded at Leiden University for my thesis research. This award allowed me to visit research groups at several universities and research institutes in the USA. Thus, I came to also visit Professors Fisher and Widom at Cornell University, giving a seminar on (in)stability of critical behavior,[24,25] as that seemed more suitable than my thesis work on time-dependent correlations in XY chains. Students and postdocs had beforehand told me that Fisher had "destroyed" the speakers in the seminar the previous several weeks, but that did not happen this time. I had seen Fisher in action at earlier conferences and I was prepared. I got a few questions about how things were proved that I could answer.

Noteworthy was the party at Fisher's home. It started with a box of Chinese metal wire puzzles put in front of me. As I had about the same collection, I knew these well. I quickly took a few apart and put them together again, upon which Fisher removed the box, causing a sigh of relief among the others. Fisher then got his Spanish guitar out. He was quite accomplished and entertaining. This and the special treats Mrs. Fisher had prepared were the high points of the party.

At the December 1979 Stat. Mech. meeting at Rutgers, I asked Leo Kadanoff if he knew that there are continuously varying critical exponents if one varies the coupling constants within one line in the bulk of the square-lattice Ising model.[26–28] With Michael Fisher and others standing by, Kadanoff said that that is impossible, upon which I said to expect a preprint soon. Fisher and Kadanoff followed that up with some further remarks on the phenomenon.[29,30]

At this point, we note that, like many others, both of us were very surprised that Fisher and Kadanoff did not share the 1982 Nobel Prize in Physics with Kenneth Wilson.

4. Helen and Jacques Married

It all started with Barry McCoy asking Jacques to pick up Helen, who had visited Michael Fisher at Cornell, at the Greyhound bus terminal in Manhattan on New Year's eve of 1982 and to bring her to the McCoy home. Jacques and Helen did not know that Barry and Martha McCoy conspired and had a detailed plan designed to get both of us married. All the scheming worked and six days later, we were engaged to be married, with the wedding taking place on January 22, beginning a very happy marriage lasting more than 40 years so far.

Soon after Helen asked, "Jacques, have you ever done anything with your quadratic difference equations for 2D Ising correlations?" This led to the first of many joint papers following, namely a letter with fairly complete results to calculate critical Ising pair correlations[31] and even a result for the monomer–monomer correlation in the square dimer problem. Many years later, students in the statistical mechanics class of a colleague told us that this letter was cited in Pathria's textbook.[32]

Having seen several papers on chiral clock models written at Cornell University during her time there, Helen was led to extend the search for integrable Potts models to chiral ones, discovering the genus 10 curve condition for solving the star-triangle equations for the three-state non-selfdual chiral Potts model.[33] The full parametrization of the N-state case was found during a visit to Canberra[34] and the history of these discoveries is described in a *Topical Review* celebrating Baxter's 75th birthday and implicitly Helen's 70th.[35]

5. Later Collaborations with Michael Fisher

Michael Fisher had been very apprehensive about our marriage at the beginning, but he noted that Helen was very happy, seeing her at several of Lebowitz's Stat. Mech. meetings at Rutgers.

Fig. 1. Helen and Michael at December 2013 Stat. Mech. Meeting at Rutgers University.

Then, at the end of 2008, Fisher called me (Jacques) that he wanted a calculation done and that Helen was the only one he trusted to do it correctly. I told him that he was free to ask Helen and let her decide what to do. Fisher then emailed a letter in his typical style, which is reproduced in the appendix. Fisher wanted some exact calculations done on the specific heat to qualitatively understand some features seen in the experiments of Gasparini's group in Buffalo.[36,37]

There was some delay, but in the end, two papers were published on 2D Ising models with alternating layers with weaker and stronger interactions,[38,39] the second with the details left out in the first one. Fisher was very happy that I helped with some of the LaTeX coding and cleaning up the figures.

Later, Helen thought of a better way to compare with an array of 3D cubes of helium covered with a 2D film. Correspondingly, she studied 2D Ising layers connected by layers of 1D Ising strings, with the strings 1, 2, or 3 units apart. First, she did a 60-page calculation of the free energy in her usual fine print handwriting using Hamm's method,[40] followed by a 20-page calculation, reproducing the results using another method.[13] When she showed me the results, we could guess the general answer for Ising strings N units apart and next prove it.[41,42] We also obtained exact results for the spontaneous magnetization.[41,43]

Finally, Fisher contacted me, as he had to do something about the 3D Ising deceptions of Zhang Zhidong. He asked me if I had any follow-up on my original comment. I sent him what I had, including my last comment.[44] Fisher also got very upset that his friend of long ago, Norman March, was fooled by the deceptions. Together we wrote another comment,[45] in which we also introduced the statistical mechanics community to the new bootstrap results for 3D Ising critical exponents. Fisher was particularly happy, when I pointed out that the result for the correction to scaling exponent $\Delta = \omega\nu = 0.5231(12)$ is very close to the $\frac{1}{2}$ proposed by Andrea Liu and him.[46]

In conclusion, it has been a great pleasure to work with Michael Fisher. In some sense, Michael was the conscience of the statistical mechanics community, keeping people honest with probing questions during and after their talks and often making good suggestions. He will be dearly missed.

Appendix

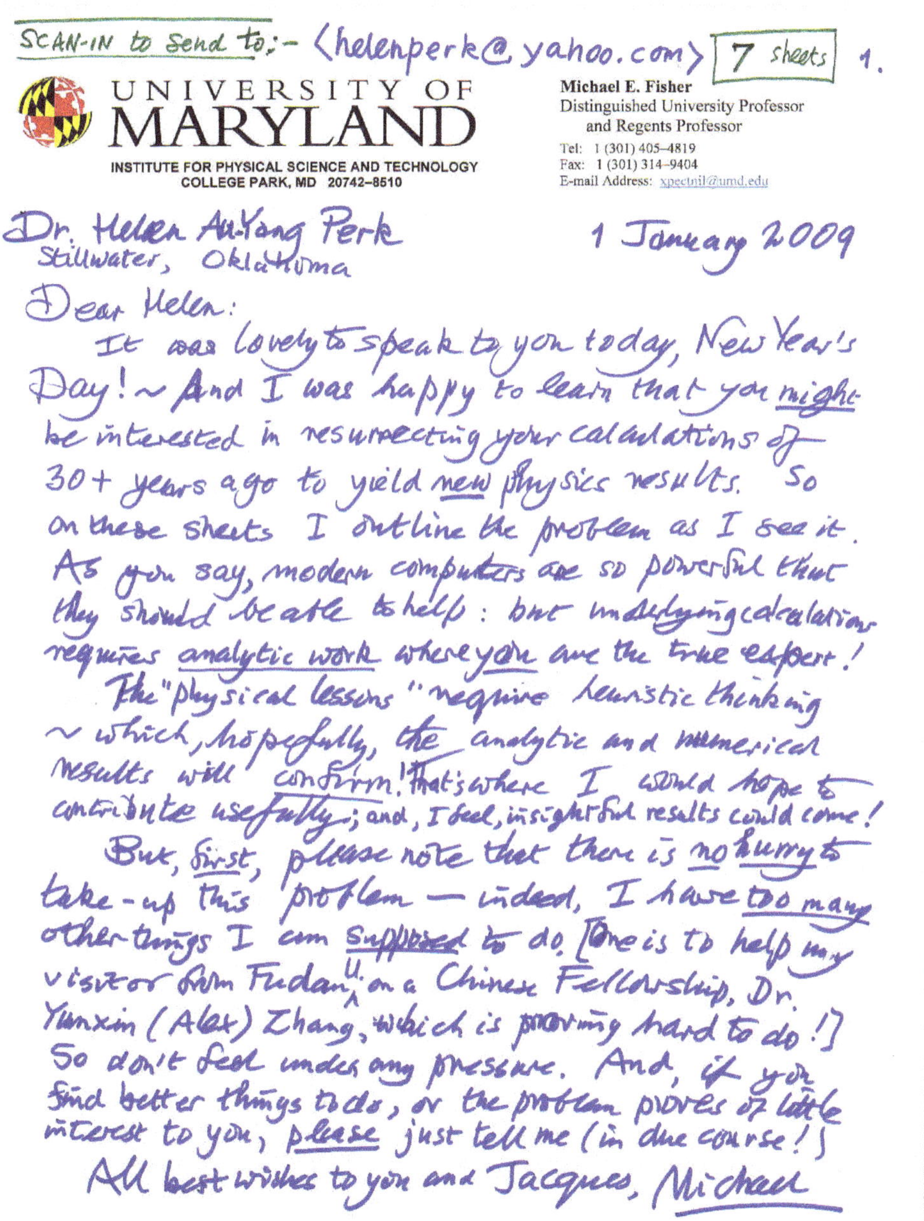

SCAN-IN to send to:- <helenperk@yahoo.com> 7 sheets 1.

UNIVERSITY OF MARYLAND
INSTITUTE FOR PHYSICAL SCIENCE AND TECHNOLOGY
COLLEGE PARK, MD 20742-8510

Michael E. Fisher
Distinguished University Professor
and Regents Professor
Tel: 1 (301) 405-4819
Fax: 1 (301) 314-9404
E-mail Address: xpectnil@umd.edu

Dr. Helen Au-Yang Perk
Stillwater, Oklahoma

1 January 2009

Dear Helen:

It was lovely to speak to you today, New Year's Day! ~ And I was happy to learn that you <u>might</u> be interested in resurrecting your calculations of 30+ years ago to yield <u>new</u> physics results. So on these sheets I outline the problem as I see it. As you say, modern computers are so powerful that they should be able to help: but underlying calculations requires <u>analytic work</u> where <u>you</u> are the true expert!

The "physical lessons" require heuristic thinking ~ which, hopefully, <u>the</u> analytic and numerical results will <u>confirm</u>! That's where I would hope to contribute <u>usefully</u>; and, I feel, insightful results could come!

But, <u>first</u>, please note that there is <u>no hurry</u> to take-up this problem — indeed, I have <u>too many</u> other things I am <u>supposed</u> to do. [One is to help my visitor from Fudan U. on a Chinese Fellowship, Dr. Yunxin (Alex) Zhang, which is proving hard to do!] So don't feel under any pressure. And, if you find better things to do, or the problem proves of little interest to you, <u>please</u> just tell me (in due course!)

All best wishes to you and Jacques, Michael

Fig. A1. Page 1 of Fisher's January 2009 email to Helen.

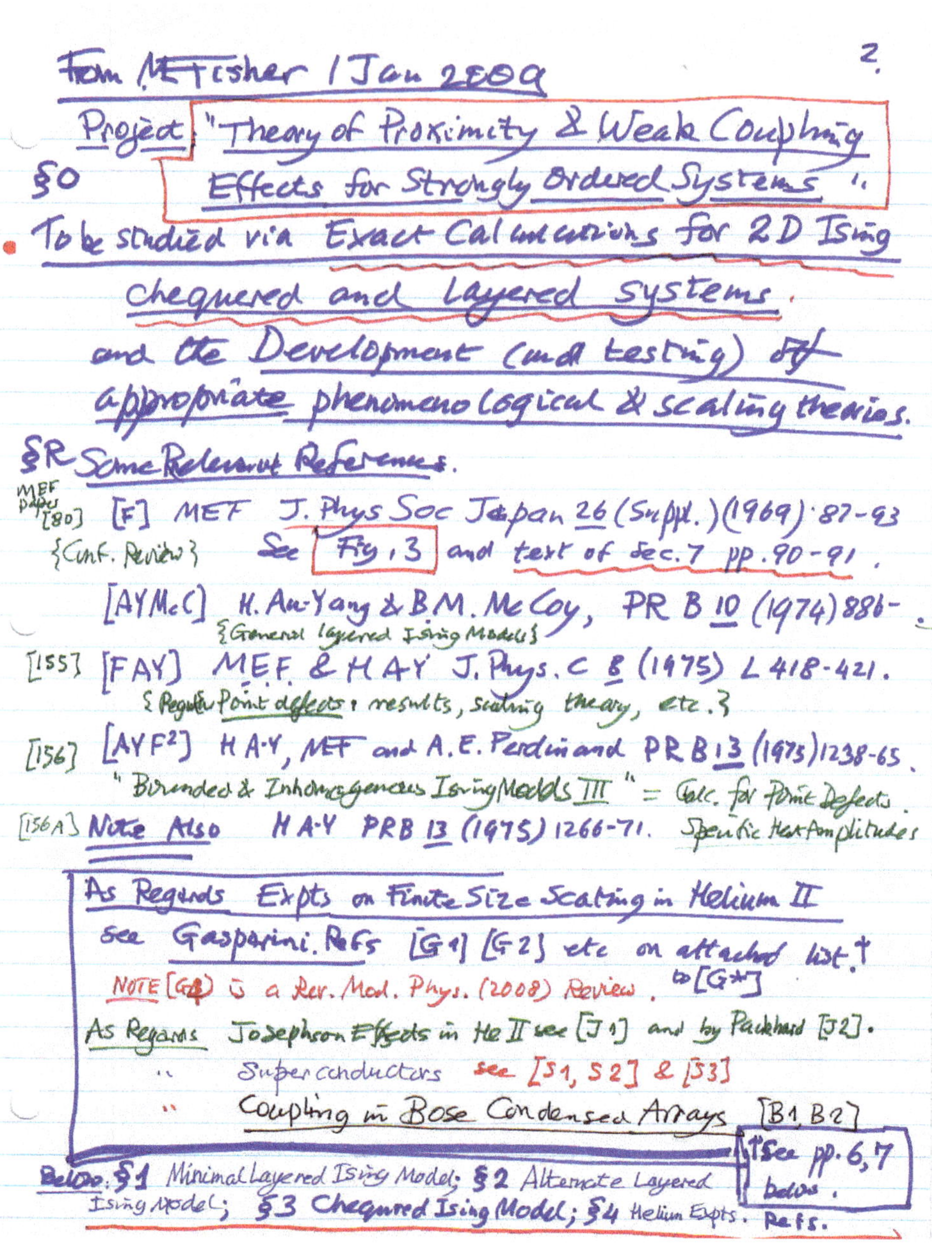

From MEFisher 1 Jan 2009 2.

Project "Theory of Proximity & Weak Coupling Effects for Strongly Ordered Systems".

§0

• To be studied via Exact Calculations for 2D Ising chequered and layered systems. and the Development (and testing) of appropriate phenomenological & scaling theories.

§R Some Relevant References.

MEF paper [80] [F] MEF J. Phys Soc Japan 26 (Suppl.) (1969) 87–93 {Conf. Review} See Fig. 3 and text of Sec. 7 pp. 90–91.

[AYMcC] H. Au-Yang & B.M. McCoy, PR B 10 (1974) 886–. {General layered Ising Models}

[155] [FAY] MEF & H AY J. Phys. C 8 (1975) L 418–421. {Regular Point defects: results, scaling theory, etc.}

[156] [AYF²] H A-Y, MEF and A.E. Ferdinand PR B 13 (1975) 1238–65. "Bounded & Inhomogeneous Ising Models III" = Calc. for Point Defects.

[156A] Note Also H A-Y PRB 13 (1975) 1266–71. Specific Heat Amplitudes

As Regards Expts on Finite Size Scaling in Helium II see Gasparini Refs [G1] [G2] etc on attached list.†

NOTE [G*] is a Rev. Mod. Phys. (2008) Review. to [G*]

As Regards Josephson Effects in He II see [J1] and by Packard [J2].

" Superconductors see [S1, S2] & [S3]

" Coupling in Bose Condensed Arrays [B1, B2]

†See pp. 6,7 below. Refs.

Below: §1 Minimal Layered Ising Model; §2 Alternate Layered Ising Model; §3 Chequered Ising Model; §4 Helium Expts.

Fig. A2. Page 2 of Fisher's January 2009 email to Helen.

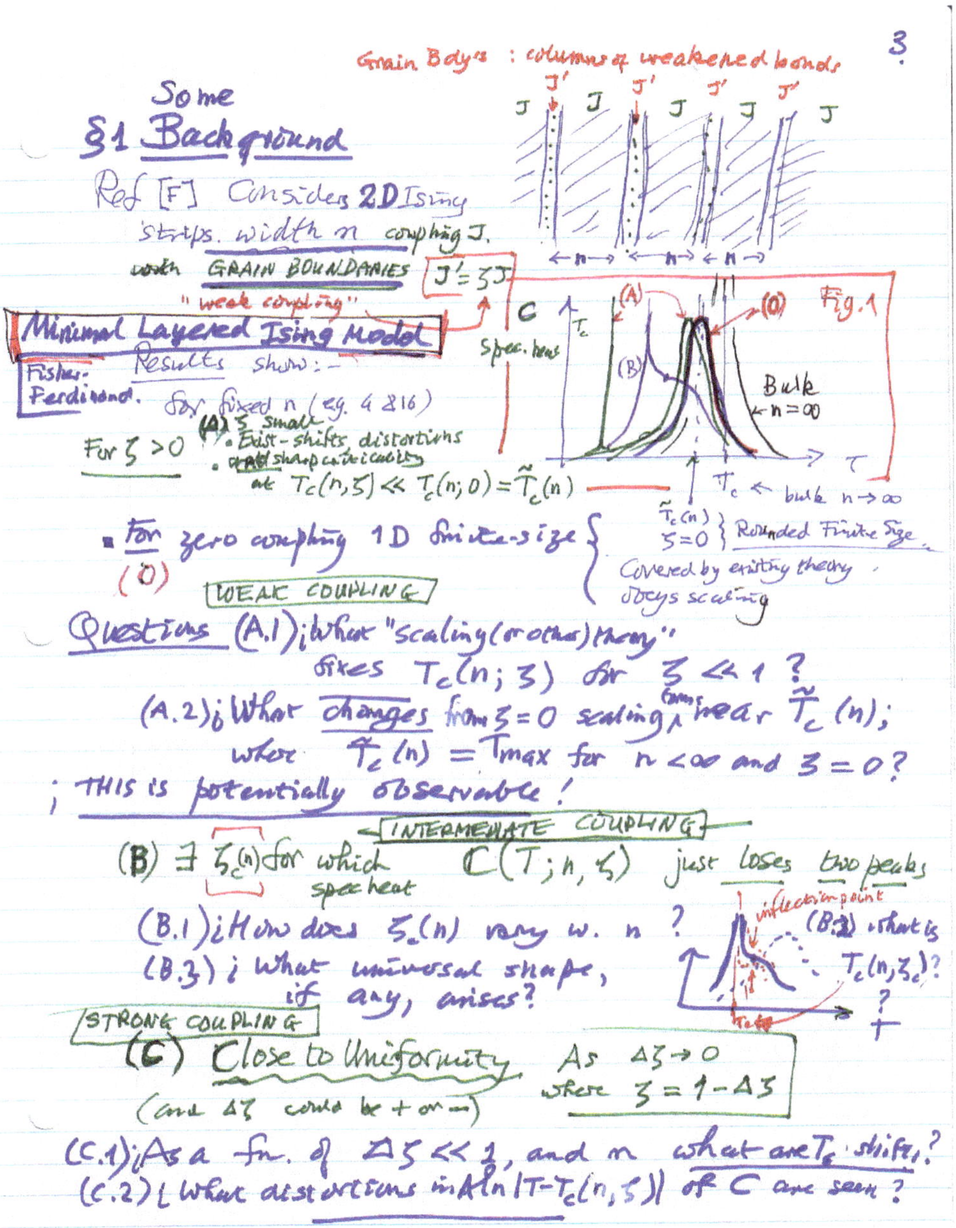
3

Grain Bdy's : columns of weakened bonds

J J' J J' J J' J J' J

← n → ← n → ← n →

Some

§1 Background

Ref [F] Considers 2D Ising strips, width n coupling J, with GRAIN BOUNDARIES $J' = \zeta J$ "weak coupling"

Minimal Layered Ising Model

Fisher-Ferdinand. Results show:–

for fixed n (eg. 4 & 16)

For $\zeta > 0$ (A) ζ small
- Exist – shifts, distortions
- sharp criticality

at $T_c(n, \zeta) \ll T_c(n; 0) = \tilde{T}_c(n)$

C, Spec. heat, T_c, (A), (0), (B), Fig. 1, Bulk $n = \infty$, T, T_c ← bulk $n \to \infty$

■ For zero coupling 1D finite-size (0) { $\tilde{T}_c(n)$, $\zeta = 0$ } Rounded Finite Size. Covered by existing theory, obeys scaling

WEAK COUPLING

Questions (A.1) ¿What "scaling (or other) theory" fixes $T_c(n; \zeta)$ for $\zeta \ll 1$?

(A.2) ¿ What changes from $\zeta = 0$ scaling forms near $\tilde{T}_c(n)$; where $\tilde{T}_c(n) = T_{max}$ for $n < \infty$ and $\zeta = 0$?

; THIS is potentially observable!

INTERMEDIATE COUPLING

(B) ∃ $\zeta_c(n)$ for which $C(T; n, \zeta)$ just loses two peaks (spec heat)

(B.1) ¿How does $\zeta_c(n)$ vary w. n?

(B.2) what is $T_c(n, \zeta_c)$?

(B.3) ¿ What universal shape, if any, arises?

inflection point

STRONG COUPLING

(C) Close to Uniformity As $\Delta\zeta \to 0$ where $\zeta = 1 - \Delta\zeta$

(and $\Delta\zeta$ could be + or –)

(C.1) ¿ As a fn. of $\Delta\zeta \ll 1$, and n what are T_c shifts?

(C.2) ¿ What distortions in $A \ln|T - T_c(n, \zeta)|$ of C are seen?

Fig. A3. Page 3 of Fisher's January 2009 email to Helen.

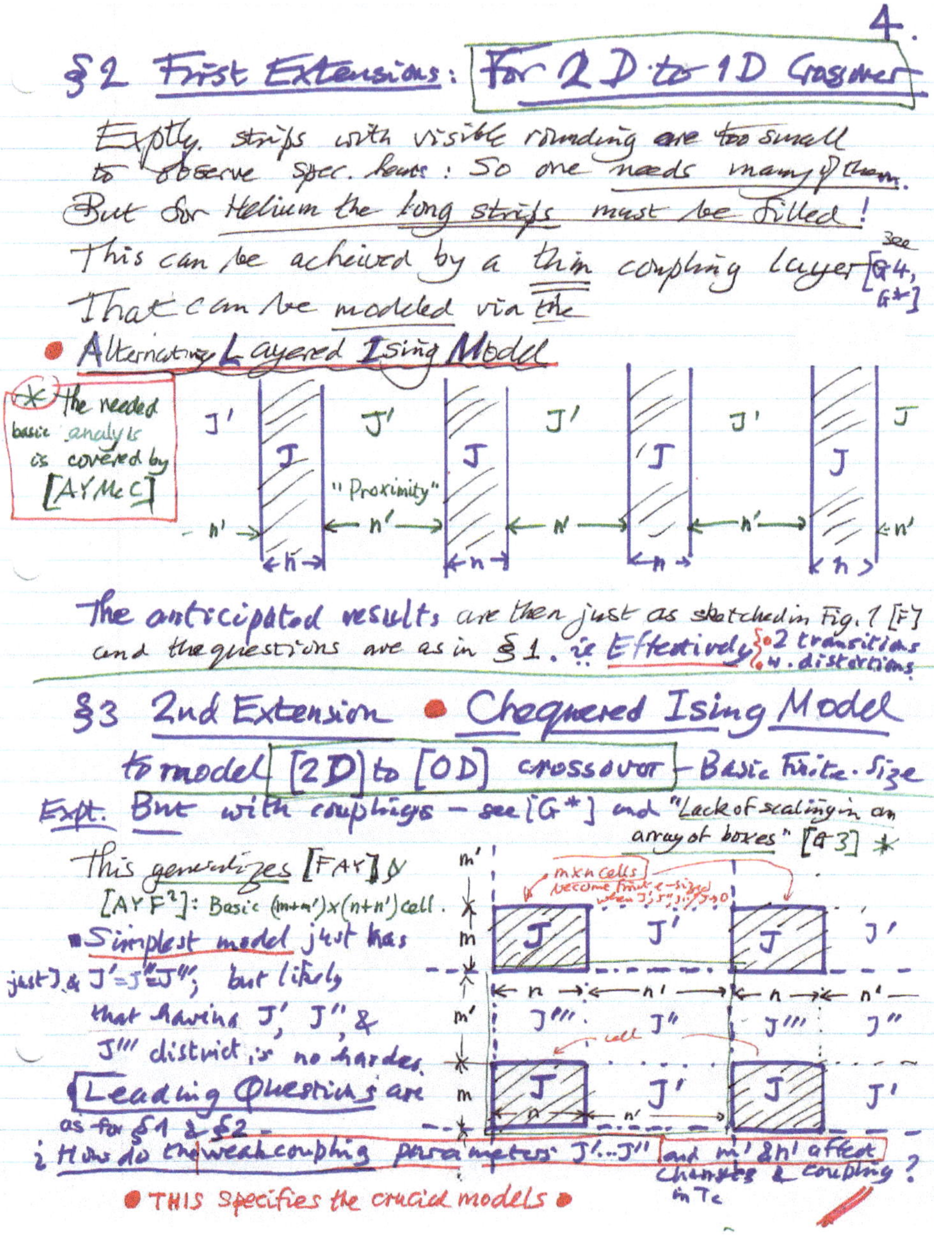

4.

§2 First Extensions: For 2D to 1D Crossover

Exptly. strips with visible rounding are too small to observe spec. heats: So one needs many of them. But for Helium the long strips must be filled! This can be acheived by a thin coupling layer [see G4, G*] That can be modeled via the

• Alternating Layered Ising Model

The anticipated results are then just as sketched in Fig. 1 [F] and the questions are as in §1. i.e. Effectively {• 2 transitions, • ∼. distortions

§3 2nd Extension • Chequered Ising Model

to model [2D] to [0D] crossover — Basic Finite-Size Expt. But with couplings — see [G*] and "Lack of scaling in an array of boxes" [G3] *

This generalizes [FAY] & [AYF²]: Basic (m+m′)×(n+n′) cell.

■ Simplest model just has just J & J′=J″=J‴; but likely that having J′, J″ & J‴ distinct is no harder.

Leading Questions are as for §1 & §2

¿ How do the weak coupling parameters J′...J‴ and m′ & n′ affect changes & coupling in T_c ?

• THIS specifies the crucial models •

Fig. A4. Page 4 of Fisher's January 2009 email to Helen.

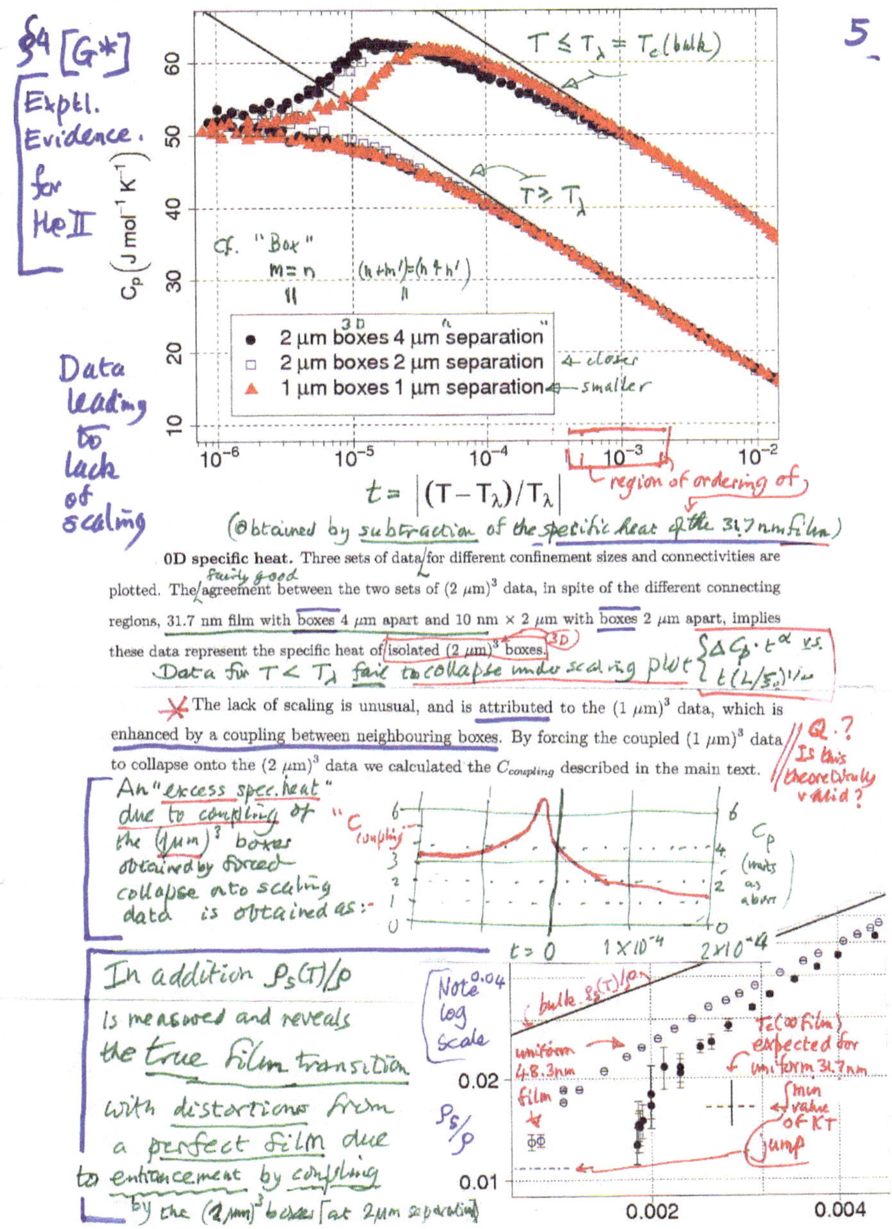

0D specific heat. Three sets of data for different confinement sizes and connectivities are plotted. The agreement between the two sets of $(2\ \mu m)^3$ data, in spite of the different connecting regions, 31.7 nm film with boxes 4 μm apart and 10 nm × 2 μm with boxes 2 μm apart, implies these data represent the specific heat of isolated $(2\ \mu m)^3$ boxes.

Data for T < T$_\lambda$ fail to collapse under scaling plot {$\Delta C_p \cdot t^\alpha$ vs. $t(L/\xi_0)^{1/\nu}$}

The lack of scaling is unusual, and is attributed to the $(1\ \mu m)^3$ data, which is enhanced by a coupling between neighbouring boxes. By forcing the coupled $(1\ \mu m)^3$ data to collapse onto the $(2\ \mu m)^3$ data we calculated the $C_{coupling}$ described in the main text.

Q.? Is this theoretically valid?

Fig. A5. Page 5 of Fisher's January 2009 email to Helen.

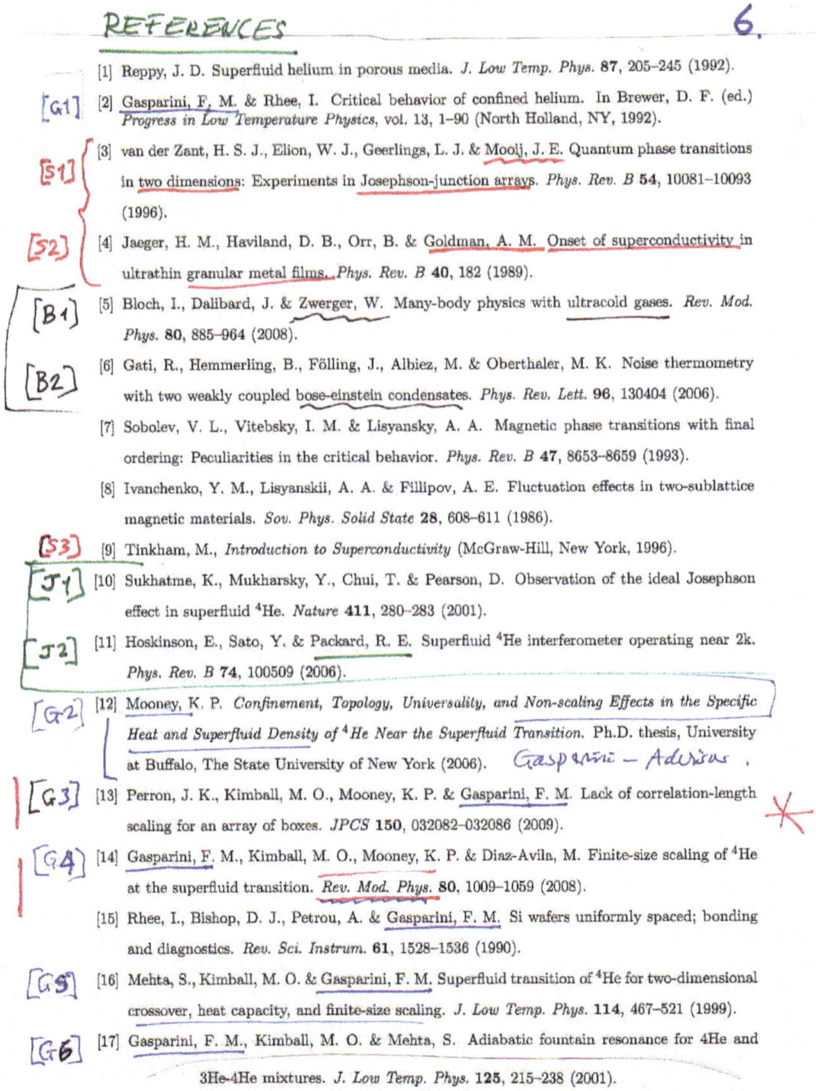

REFERENCES 6.

[1] Reppy, J. D. Superfluid helium in porous media. *J. Low Temp. Phys.* **87**, 205–245 (1992).

[G1] [2] Gasparini, F. M. & Rhee, I. Critical behavior of confined helium. In Brewer, D. F. (ed.) *Progress in Low Temperature Physics*, vol. 13, 1–90 (North Holland, NY, 1992).

[S1] [3] van der Zant, H. S. J., Elion, W. J., Geerlings, L. J. & Mooij, J. E. Quantum phase transitions in two dimensions: Experiments in Josephson-junction arrays. *Phys. Rev. B* **54**, 10081–10093 (1996).

[S2] [4] Jaeger, H. M., Haviland, D. B., Orr, B. & Goldman, A. M. Onset of superconductivity in ultrathin granular metal films. *Phys. Rev. B* **40**, 182 (1989).

[B1] [5] Bloch, I., Dalibard, J. & Zwerger, W. Many-body physics with ultracold gases. *Rev. Mod. Phys.* **80**, 885–964 (2008).

[B2] [6] Gati, R., Hemmerling, B., Fölling, J., Albiez, M. & Oberthaler, M. K. Noise thermometry with two weakly coupled bose-einstein condensates. *Phys. Rev. Lett.* **96**, 130404 (2006).

[7] Sobolev, V. L., Vitebsky, I. M. & Lisyansky, A. A. Magnetic phase transitions with final ordering: Peculiarities in the critical behavior. *Phys. Rev. B* **47**, 8653–8659 (1993).

[8] Ivanchenko, Y. M., Lisyanskii, A. A. & Fillipov, A. E. Fluctuation effects in two-sublattice magnetic materials. *Sov. Phys. Solid State* **28**, 608–611 (1986).

[S3] [9] Tinkham, M., *Introduction to Superconductivity* (McGraw-Hill, New York, 1996).

[J1] [10] Sukhatme, K., Mukharsky, Y., Chui, T. & Pearson, D. Observation of the ideal Josephson effect in superfluid ^{4}He. *Nature* **411**, 280–283 (2001).

[J2] [11] Hoskinson, E., Sato, Y. & Packard, R. E. Superfluid ^{4}He interferometer operating near 2k. *Phys. Rev. B* **74**, 100509 (2006).

[G2] [12] Mooney, K. P. *Confinement, Topology, Universality, and Non-scaling Effects in the Specific Heat and Superfluid Density of ^{4}He Near the Superfluid Transition.* Ph.D. thesis, University at Buffalo, The State University of New York (2006). Gasparini – Advisor.

[G3] [13] Perron, J. K., Kimball, M. O., Mooney, K. P. & Gasparini, F. M. Lack of correlation-length scaling for an array of boxes. *JPCS* **150**, 032082–032086 (2009).

[G4] [14] Gasparini, F. M., Kimball, M. O., Mooney, K. P. & Diaz-Avila, M. Finite-size scaling of ^{4}He at the superfluid transition. *Rev. Mod. Phys.* **80**, 1009–1059 (2008).

[15] Rhee, I., Bishop, D. J., Petrou, A. & Gasparini, F. M. Si wafers uniformly spaced; bonding and diagnostics. *Rev. Sci. Instrum.* **61**, 1528–1536 (1990).

[G5] [16] Mehta, S., Kimball, M. O. & Gasparini, F. M. Superfluid transition of ^{4}He for two-dimensional crossover, heat capacity, and finite-size scaling. *J. Low Temp. Phys.* **114**, 467–521 (1999).

[G6] [17] Gasparini, F. M., Kimball, M. O. & Mehta, S. Adiabatic fountain resonance for 4He and 3He-4He mixtures. *J. Low Temp. Phys.* **125**, 215–238 (2001).

Fig. A6. Page 6 of Fisher's January 2009 email to Helen.

7.

References (contd).

[18] Fisher, M. E. The theory of critical point singularities. In Green, M. (ed.) *Critical Phenomenon Proc. 51st "Enrico Fermi" Summer School, Varenna, Italy* (Academic Press, NY, 1971).

[19] Costa, R. M., Ferreira, L. M., Vieira, V. N., Pureur, P. & Schaf, J. Coherence transition in granular $YBa_2Cu_3O_{7-\delta}$, $YBa_2Cu_{2.95}Zn_{0.05}O_{7-\delta}$, and $YBa_{1.75}Sr_{0.25}Cu_3O_{7-\delta}$ superconductors. *European Physical Journal B* **58**, 107–113 (2007).

[G7] [20] Mehta, S. & Gasparini, F. M. Specific heat and scaling of ^4He confined in a planar geometry. *Phys. Rev. Lett.* **78**, 2596–2599 (1997).

[G8] [21] Kimball, M. O. & Gasparini, F. M. Superfluid fraction of ^{3}He-^{4}He mixtures confined at 0.0483 μm between silicon wafers. *Phys. Rev. Lett.* **86**, 1558–1561 (2001).

[22] Greywall, D. S. & Ahlers, G. Second-sound velocity, scaling, and universality in He-II under pressure near the superfluid transition. *Phys. Rev. Lett.* **28**, 1251–1254 (1972).

[23] Kosterlitz, J. M. & Thouless, D. J. Ordering, metastability and phase transitions in two-dimensional systems. *J. Phys. C* **6**, 1181–1203 (1973).

[24] Nelson, D. R. & Kosterlitz, J. M. Universal jump in the superfluid density of two-dimensional superfluids. *Phys. Rev. Lett.* **39**, 1201–1205 (1977).

[25] Mamaladze, Y. G. & Cheishvili, O. D. Flow of a superfluid liquid in porous media. *Soviet Physics JETP* **23**, 112–117 (1967).

[G9] [26] Kimball, M. O., Mooney, K. P. & Gasparini, F. M. Three-dimensional critical behavior with 2D, 1D, and 0D dimensionality crossover: Surface and edge specific heats. *Phys. Rev. Lett.* **92**, 115301 (2004).

[G10] [27] Kimball, M. O. *Helium Confined to a Restricted Geometry – A Study of Lower Dimensionality Crossover and Universality.* Ph.D. thesis, University at Buffalo, The State University of New York (2005). under Gasparini

List Extracted from UNDER REVIEW article:

[G*] **Superfluid transition in granular ^{4}He: Coupling and proximity effects**

UNPUBLISHED

Justin K. Perron,* Mark O. Kimball,† Kevin P. Mooney,‡ and Francis M. Gasparini§

Department of Physics, University at Buffalo,

the State University of New York, Buffalo, NY 14260

*Electronic address: jkperron@buffalo.edu

†Electronic address: mok2@enthalpy.physics.buffalo.edu

‡Electronic address: mooney.kp@gmail.com

§Electronic address: fmg@buffalo.edu

Fig. A7. Page 7 of Fisher's January 2009 email to Helen.

References

1. B. M. McCoy and T. T. Wu, Theory of Toeplitz determinants and the spin correlations of the two-dimensional Ising model. IV. *Phys. Rev.* **162**(2), 436–475 (1967).
2. H. Au-Yang, Thermodynamics of an anisotropic boundary of a two-dimensional Ising model. *J. Math. Phys.* **14**(7), 937–946 (1973).
3. K. Binder and P. C. Hohenberg, Phase transitions and static spin correlations in Ising models with free surfaces. *Phys. Rev. B* **6**(9), 3461–3487 (1972).
4. K. Binder and P. C. Hohenberg, Surface effects on magnetic phase transitions. *Phys. Rev. B* **9**(5), 2194–2214 (1974).
5. H. Au-Yang, A study on layered Ising model. Ph. D. Thesis, State University of New York at Stony Brook, 1974, pp. i–v, 1–94.
6. H. Au-Yang and B. M. McCoy, Theory of layered Ising models: Thermodynamics. *Phys. Rev. B* **10**(3), 886–891 (1974).
7. H. Au-Yang and B. McCoy, Theory of layered Ising models. II. Spin correlation functions parallel to the layering. *Phys. Rev. B* **10**(9), 3885–3905 (1974).
8. T. T. Wu, B. M. McCoy, C. A. Tracy and E. Barouch, Spin-spin correlation functions for the two dimensional Ising model: Exact theory in the scaling region. *Phys. Rev. B* **13**(1), 315–374 (1976).
9. M. E. Fisher and A. E. Ferdinand, Interfacial, boundary, and size effects at critical points. *Phys. Rev. Lett.* **19**(4), 169–172 (1967).
10. A. E. Ferdinand and M. E. Fisher, Bounded and inhomogeneous Ising models. I. Specific-heat anomaly of a finite lattice. *Phys. Rev.* **185**(2), 832–846 (1969).
11. H. Au-Yang and M. E. Fisher, Bounded and inhomogeneous Ising models. II. Specific-heat scaling function for a strip. *Phys. Rev. B* **11**(5), 3469–3487 (1975).
12. M. E. Fisher and H. Au-Yang, Regularly spaced defects in Ising models. *J. Phys. C* **8**(20), L418–L421 (1975).
13. H. Au-Yang, M. E. Fisher and A. E. Ferdinand, Bounded and inhomogeneous Ising models. III. Regularly spaced point defects. *Phys. Rev. B* **13**(3), 1238–1265 (1976).
14. H. Au-Yang, Bounded and inhomogeneous Ising models. IV. Specific-heat amplitude for regular defects. *Phys. Rev. B* **13**(3), 1266–1271 (1976).
15. H. Au-Yang, Multispin correlation functions for Ising models. *Phys. Rev. B* **15**(5), 2704–2707 (1977).
16. H. Au-Yang, Four-spin correlation function along the diagonal. *Phys. Rev. B* **16**(11), 5016–5031 (1977).
17. M. E. Fisher and P.-G. de Gennes, Wall phenomena in a critical binary mixture. *C. R. Acad. Sc. B* **287**(8), 207–209 (1978).
18. M. E. Fisher and H. Au-Yang, Critical wall perturbations and a local free energy functional. *Physica A* **101**(1), 255–264 (1980).
19. H. Au-Yang and M. E. Fisher, Wall effects in critical systems: Scaling in Ising model strips. *Phys. Rev. B* **21**(9), 3956–3970 (1980).
20. M. E. Fisher, Simple Ising models still thrive! A review of some recent progress. *Physica A* **106**(1–2), 28–47 (1981).[b]
21. M. E. Fisher and H. Au-Yang, Inhomogeneous differential approximants for power series. *J. Phys. A* **12**(10), 1677–1692 (1979); **13**(4), 1517 (1980) (Corrigenda).

[b]Note that most of Helen's papers up to that time are cited in this work. Helen is specially mentioned in the Acknowledgements.

22. M. E. Fisher, J.-H. Chen and H. Au-Yang, Universality tests at Heisenberg bicritical points. *J. Phys. C* **13**(18), L459–L464 (1980).
23. H. Au-Yang, Four-spin correlation function of two parallel nonaligned pairs in the planar Ising model. *Phys. Rev. B* **23**(9), 4749–4754 (1981).
24. H. W. Capel, J. H. H. Perk and L. W. J. Den Ouden, Critical-exponent renormalization and first-order transitions. *Phys. Lett. A* **66**(6), 437–439 (1978).
25. H. W. Capel, L. W. J. Den Ouden and J. H. H. Perk, Stability of critical behaviour, critical-exponent renormalization and first-order transitions. *Physica A* **95**(3), 371–416 (1979).
26. R. Z. Bariev, Effect of linear defects on the local magnetization of a plane lsing lattice. *Zh. Eksp. Teor. Fiz.* **77**(3), 1217–1229 (1979) [English translation: *Sov. Phys. JETP* **50**(3), 613–619 (1979)].
27. B. M. McCoy and J. H. H. Perk, Two-spin correlation functions of an Ising model with continuous exponents. *Phys. Rev. Lett.* **44**(13), 840–844 (1980).
28. B. M. McCoy and J. H. H. Perk, *Continuous Exponents of Spin Correlation Functions of Inhomogeneous Layered Ising Models*, *Lecture Notes in Mathematics*, Vol. 925 pp. 12–27 (Springer, Berlin, 1982).
29. Section 5 of Simple Ising Models Still Thrive![20]
30. L. P. Kadanoff, Many-point correlation functions in a modified Ising model. *Phys. Rev. B* **24**(9), 5382—5383 (1981).
31. H. Au-Yang and J. H. H. Perk, Ising correlations at the critical temperature. *Phys. Lett. A* **104**(3), 131–134 (1984).
32. R. K. Pathria and P. D. Beale, Statistical Mechanics, (3rd edn.) (Elsevier, Oxford UK, 2011), Section 13.4.
33. H. Au-Yang, B. M. McCoy, J. H. H. Perk, S. Tang and M.-L. Yan, Commuting transfer matrices in the chiral Potts models: Solutions of the star-triangle equations with genus greater than 1. *Phys. Lett. A* **123**(5), 219–223 (1987).
34. R. J. Baxter, J. H. H. Perk and H. Au-Yang, New solutions of the star-triangle relations for the chiral Potts model. *Phys. Lett. A* **128**(3–4), 138–142 (1988).
35. J. H. H. Perk, The early history of the integrable chiral Potts model and the odd-even problem. *J. Phys. A: Math. Theor.* **49**(15), 153001 (2016) (20 pp.).
36. F. M. Gasparini, M. O. Kimball, K. P. Mooney and M. Diaz-Avila, Finite-size scaling of ^{4}He at the superfluid transition. *Rev. Mod. Phys.* **80**(3), 1009–1059 (2008).
37. J. K. Perron and F. M. Gasparini, Critical point coupling and proximity effects in ^{4}He at the superfluid transition. *Phys. Rev. Lett.* **109**(3), 035302 (2012) (4 pp.).
38. H. Au-Yang and M. E. Fisher, Criticality in alternating layered Ising models: I. Effects of connectivity and proximity. *Phys. Rev. E* **88**(3), 032147 (2013) (12 pp.).
39. H. Au-Yang, Criticality in alternating layered Ising models: II. Exact scaling theory. *Phys. Rev. E* **88**(3), 032148 (2013) (8 pp.).
40. J. R. Hamm, Regularly spaced blocks of impurities in the Ising model: Critical temperature and specific heat. *Phys. Rev. B* **15**(11), 5391–5411 (1977).
41. H. Au-Yang and J. H. H. Perk, Ising models with holes: Crossover behavior. arXiv:1806.00873.
42. H. Au-Yang and J. H. H. Perk, Specific heat of Ising models with holes: Mathematical details using dimer approaches. arXiv:1808.07525.
43. H. Au-Yang and J. H. H. Perk, Spin-spin correlations in central rows of Ising models with holes. arXiv:1907.04748.

44. J. H. H. Perk, Comment on 'Mathematical structure of the three-dimensional (3D) Ising model'. *Chinese Physics B* **22**(8), 080508 (2013) (5 pp.).
45. M. E. Fisher and J. H. H. Perk, Comments concerning the Ising model and two letters by N.H. March. *Phys. Lett. A* **380**(13), 1339–1340 (2016).
46. A. J. Liu and M. E. Fisher, On the corrections to scaling in three-dimensional Ising models. *J. Stat. Phys.* **58**(3/4), 431–442 (1990).

https://doi.org/10.1142/9789811282386_0002

Chapter 2

Michael Fisher and Critical Phenomena in the 1970s

J.M. Kosterlitz
Department of Physics, Brown University, Providence, RI 02912, USA

This chapter is about my memories of being a young postdoc from the UK coming to the center of activity on a very new and successful method for dealing with the hitherto impenetrable topic of phase transitions and critical phenomena, the Renormalization Group and the ϵ-expansion of Kenneth G. Wilson and Michael E. Fisher.

In 1970, I was a high-energy postdoc in Torino, Italy and intended to go next to CERN in Geneva. Unfortunately, I had a bad habit of delaying things I really did not want to do until the pressure became too high to wait. I finally did send my application to CERN, but the response was "Sorry, all the space is taken for this year, please reapply next year." Unemployment was imminent, but my new wife said a few choice words, walked to the railway station where one could buy English language newspapers, returned with several and sat me at the kitchen table with the academic job advertisements and ordered me to apply. I duly applied for vacant postdoc positions and was offered a two to three year postdoc position in the Department of Mathematical Physics at Birmingham University in England. Although this was the last place I really wanted to go to because Birmingham was a large and dirty industrial city in the flat middle of England without a mountain in sight. At that time, I was a very enthusiastic rock climber, who was actually quite good at the sport, so Birmingham was very low on my list of desirable places to live.

For the first year, between my climbing adventures, I continued working on the so-called dual resonance models in high-energy physics but found myself in direct competition with a group of physicists at Berkeley, California. I had finished a long tedious calculation and was in the process of writing a preprint when a paper arrived on my desk which contained precisely what I was doing. I cursed a bit, shrugged my shoulders, thought "these things happen, do not take it too seriously" and began the next tedious calculation. I was in the process of writing up those calculations when exactly the same thing happened with the same group from Berkeley. Needless to say, I was very upset and decided that, since I was on my own in a small office in Birmingham, I just could not compete with the Berkeley group, so I should look for something else to work on. I went from office to office asking everyone, "Do you have a problem I could look at?" The replies were invariably negative until I came to David Thouless' office. I was rather apprehensive about approaching the great man because he had a reputation in the department of "not suffering fools gladly" and, compared to David, I certainly was a fool.

After a slightly rocky start, to my surprise and pleasure, David and I worked well together mostly because David had the physical ideas and I was the ignorant upstart to whom David's heretical ideas made perfect sense. We wrote two papers together on phase transitions in two dimensions and I wrote a third paper which laid out the details of a renormalization-group treatment of the vortex-driven transition in two-dimensional films of ^{4}He. I was later very surprised to find that this work was part of the reason for our share of the 2016 Physics Nobel Prize, although many years earlier, David did confide to me that he thought the work was good enough for such a prize. When he learned that, for several years, I had wanted to work in the USA, he volunteered to use his contacts at Cornell to try to arrange a postdoc position there so I could learn more about the very hot topic of the Wilson renormalization group and its applications to critical phenomena at the hub of activity on phase transitions. Also, I really wanted to understand the theoretical basis for the remarkably successful scaling hypothesis.[1] Needless to say, I was very pleased and excited when the offer of a one-year position at Cornell University arrived, which I immediately and enthusiastically accepted.

I arrived at Clark Hall, Cornell University anxious to learn and work on phase transitions and critical phenomena from the experts[6] but was a bit disappointed to discover that the main interests of the physicists were in superfluid ^{3}He. This was very new and interesting but could be explained by generalizations of the BCS theory of superconductivity. As far as I was concerned, this was just an elaborate version of mean-field theory which, although important, did not address the question of strongly interacting fluctuations. The seminars at Cornell were fascinating mostly because of one faculty member who would ask extremely penetrating questions and the speaker would be torn apart if he or she was unable to answer appropriately and knowledgeably.

This particular questioner was Michael Fisher, Professor of Chemistry, Physics and Mathematics, who seemed to know everything about everything. My impression was that he was a more outgoing, very friendly and politely aggressive version of David Thouless. Michael had a great interest in critical phenomena since around 1958 when he wrote some of his very early papers on the subject.[2–6] As a new postdoc, I was expected to give a seminar on my own work to this audience who were eager to ask very hard questions. I think it was either Michael Fisher or Ken

Wilson who asked about irrelevant variables and I was completely unable to answer. I had no idea what a relevant or irrelevant operator was. All I knew was that one or both of the terms in my free-energy functional decreased or increased when the effective lattice spacing was increased. I had no idea that my work was actually a version of a renormalization-group analysis of the two-dimensional XY model.

As an aside, I found a very relevant article[7] about Michael written by David Mermin, a colleague for many years of Michael Fisher at Cornell University. It is an after-dinner talk at the 70th Birthday Conference at Rutgers in December 2001 and this should be required reading for anyone interested in Michael Fisher. There is a quote in that article by David Mermin, which I cannot resist repeating:

> "I would like to thank God for arranging our lives so I could spend over two decades with Michael Fisher at Cornell, and His servant, the National Science Foundation, for supporting this investigation through Grant PHY9022796."

After this seminar, I also realized that of all the experts there, only one person had actually understood my work in detail and this was one of Michael Fisher's brilliant young graduate students, David Nelson, who was doing an accelerated six-year undergraduate to Ph.D. program. I also listened to wonderful talks by Michael Fisher on scaling in the critical region complemented by his amazing hand-drawn sketches on the chalk board and realized that my major interest at Cornell would be in trying to understand some beautiful experimental data on phase transitions by using the new ϵ-expansion of Wilson and Fisher.[8] Of course, first I had to understand what was the Wilson version of the renormalization group which is a formal mathematical formulation of Leo Kadanoff's scaling idea[10,11] of mapping a system with a short-distance cut-off of a into one with cut-off ae^l.

I was based in the Physics Department in Clark Hall and, because I was on my own from Britain with nowhere to live, I was given a few days lodging in an apartment on the top of the building while apartment hunting. This was a bit peculiar, but I did like the setup because, should I be awake at any time of the night, I could always go down to the basement where I would find a group of John Reppy's students and postdocs trying to repair their apparatus where a superconducting contact had broken when the experiment was being cooled down to about 1°K. They would be only too happy to take a break from repairing the broken contact and chat over a beer. Occasionally, the cooling would be successful and the experiment would be run all night because a successful cooling without breaking a contact had to be used to the maximum. I learned a lot about the disadvantages of experimental physics and the advantages of being a theorist. Elsewhere in the basement, was a group of David Lee, Bob Richardson and Doug Osheroff working at even lower temperatures doing experiments with ^{3}He. In 1973, Clark Hall at Cornell University was an amazing place with four or five future Nobel laureates in the same building.

Although I was not one of Michael Fisher's official postdocs, I drifted into his group which was working on the field of phase transitions and critical phenomena. This was a remarkable set of young physicists and chemists including Ben Widom, Amnon Aharony, Alastair Bruce, Monique Combescot, David Nelson and myself, all of whom are now well-known for their contributions to the field. Of course, in the background always available for advice and consultation was the ultimate authority on the renormalization group and the epsilon expansion, Ken Wilson whom David Nelson and I frequently asked for advice when we collaborated on a complicated crossover problem in critical phenomena. This group, composed and led by Michael Fisher, was probably the most impressive group of young experts in one institution I have ever come across. All of us benefitted greatly from Michael's insistence on understanding experimental data, fitting data to a theory and comparing theoretical predictions with the final authority of experiment.

The first important lesson is that the properties of a system in the critical region can be understood in terms of a free-energy functional defining the partition function

$$\begin{aligned} Z(T,H) &= \int \mathcal{D}\mathbf{s}(\mathbf{r})\exp\left(-\int d^d r\mathcal{H}[\mathbf{s}(\mathbf{r})]\right), \\ \mathcal{H}[\mathbf{s}] &= \frac{1}{2}(\nabla\mathbf{s}\cdot\nabla\mathbf{s} + r_0\mathbf{s}\cdot\mathbf{s}) + u(\mathbf{s}\cdot\mathbf{s})^2 + \cdots + \mathbf{h}\cdot\mathbf{s}, \end{aligned} \tag{1}$$

where the spin $\mathbf{s}(\mathbf{r}) = (s_1(\mathbf{r}), s_2(\mathbf{r}), \ldots, s_n(\mathbf{r}))$ has n components and the Hamiltonian $\mathcal{H}$ is invariant under $O(n)$ rotations in spin space or whatever symmetry group is appropriate. The parameters $r, u, \ldots$ depend on temperature and other fields with $r_0(T) = a(T - T_0)$ and $u = u(q_1, \ldots, q_4)$ but independent of T. For various reasons, we can take the parameters in Eq. (1) to be constants whose values are unimportant but their signs are.[8] In the early 1970s, the subject of critical phenomena was not well understood from experimental or theoretical points of view. There is an interesting review article by Kadanoff *et al.*[12] discussing the state of critical phenomena when little was known except for the ideas of scaling based on the assumption of homogeneity of the dimensionless free-energy density $f(t,h) = V^{-1}F(T,H)/k_BT$. Here, $t = (T - T_c)/T_c$ is the fractional deviation from the critical temperature T_c and $h = H/k_BT$ is the applied ordering field in units of k_BT_c.

For convenience, the theory is stated in terms of a uniform ferromagnetic system under a change of scale of the short distance cut-off or lattice spacing, $a \to ae^l$ so that

$$f(t,h) = e^{-ld}f(t(l),h(l)) \quad \text{where } t(l) = e^{\lambda_t l}t, \quad h(l) = e^{\lambda_h l}h.$$

The relations between critical exponents result from this scaling assumption and all critical exponents $\alpha, \beta, \gamma, \delta$ etc. are given in terms of the two exponents λ_t and λ_h. All that one needed was some theory which could cope with strongly interacting fluctuations near the critical point from which one could calculate the values of λ_t, λ_h and also derive the scaling form of the free energy in the critical region. This was finally realized in the early 1970s by the renormalization group of K.G. Wilson and the development of the ϵ expansion by Wilson and Fisher[8] which made calculations of measurable quantities possible.

The Wilson renormalization group is most readily understood in terms of a Ginsburg–Landau–Wilson free-energy functional $\mathcal{H}(\mathbf{s})$ so that, in the critical region, the partition function Z has the form of Eq. (1) with the functional $\mathcal{H}$ a phenomenological free-energy functional of the order parameter field called $\mathbf{s}(\mathbf{r})$. This is constructed as the well-known Landau free energy which is written as a sum of terms consistent with the symmetry of the system and whose minimum yields the mean-field transition. The various coefficients are unknown functions of the microscopic parameters of the system. For example, for an isotropic ferromagnet with n-component spins $\mathbf{s}(\mathbf{r})$, the GLW functional is composed of several $O(n)$ rotationally invariant terms in spin space but only the signs of the coefficients matter, not the values of the coefficients, because the critical behavior is determined by their fixed point values as $l \to \infty$ and the system will reach the fixed point provided the initial parameter values are in the domain of attraction[9]

$$\mathcal{H}[\mathbf{s}] = \int d^d r \left(\frac{1}{2} \sum_{\alpha=1}^{n} (\nabla s_\alpha(\mathbf{r}) \cdot \nabla s_\alpha(\mathbf{r}) + r_0(T) s_\alpha(\mathbf{r}) s_\alpha(\mathbf{r})) \right.$$
$$\left. + u \sum_{\alpha,\beta} s_\alpha^2(\mathbf{r}) s_\beta^2(\mathbf{r}) - \sum_\alpha h_\alpha s_\alpha(\mathbf{r}) \right), \tag{2}$$

where $r_0(T) = a(T - T_0)/T_0$ is chosen to change sign at T_0 and depends linearly on $T - T_0$. The coefficient of the quartic term in $\mathcal{H}$ is taken to be a positive constant. The coefficients do depend on q^2, but this is not relevant for our purposes. We do not consider higher-order terms in $\mathcal{H}$, but an s^6 term is necessary for stability when $u \leq 0$, which happens at a tricritical point at $u = 0$.

The object is to compute the partition function

$$Z(r, u, h, \ldots) = \int \mathcal{D}\mathbf{s} \exp\left(-\mathcal{H}[\mathbf{s}]\right), \tag{3}$$

where the functional integral just means the sum over all possible configurations of the spins $\mathbf{s}(\mathbf{r})$. The symbol $\mathbf{r}$ denotes the sites of some underlying crystal lattice and, provided the spins $\mathbf{s}(\mathbf{r})$ on neighboring sites are almost parallel, the original sum over the lattice sites can be approximated by the continuum and the true free-energy functional can be approximated by Eq. (2), especially in the critical region. The big problem of the early 1970s was to understand the essential collective behavior of a large number of interacting microscopic systems in the critical region near a continuous phase transition. Of course, many aspects were understood phenomenologically, but the underlying principles were a mystery.

Michael Fisher was one of the leading figures of that period and he wrote a very influential review[6] which summarized the understanding of critical phenomena just before Wilson wrote his seminal papers[13,14] on the renormalization group and its application to critical phenomena. These papers form the essential framework of the Renormalization Group (RG) as applied to critical phenomena but were not easy to understand or to use in applications, but, in the following year, Michael Fisher teamed up with Ken Wilson and wrote the vital paper[8] which introduced the physics world to a more systematic renormalization-group transformation.

Despite all the important and fascinating physics being performed in the Clark Hall basement on superfluid ^{3}He by a group of three future Nobel Laureates, I was interested only in the renormalization group and critical phenomena, finally meeting Michael Fisher who had gathered a group of very bright young postdocs and graduate students working on applying the Wilson–Fisher renormalization-group and ϵ expansion methods to the ordered low-temperature phase with a finite value of the order parameter[25] by his brilliant young graduate student, David Nelson. They explained the Wilson–Fisher $\epsilon = 4 - d$ expansion and how it worked. Michael then suggested to a group of three physics postdocs, Monique Combescot, Michel Droz and myself,[27] that they should calculate the two-point correlation function to $\mathcal{O}(\epsilon^2)$ for finite magnetization for Ising spins

$$G(\mathbf{q}, T, m) = \sum_{\mathbf{R}} e^{i\mathbf{q}\cdot\mathbf{R}} \langle [(\sigma(\mathbf{R}) - m)(\sigma(\mathbf{0}) - m)\rangle.$$

Fisher and Aharony[26] had previously studied the critical scattering problem for zero field to $\mathcal{O}(\epsilon^2)$ and he suggested that we extend their treatment to include general fields and temperatures to the same order in ϵ. Eventually, we succeeded with Michael's help, but I, at least, did not find this problem easy. When I look back at our calculation,[27] I am amazed that we were able to get the horrific calculations correct or, at least, consistent. Evaluating the necessary Feynman diagrams involved inventing a new way of calculating them because, again at Michael's suggestion, we worked in position instead of momentum space. I have a sneaking suspicion that Michael gave us this project to check that this group of three European physicists were capable of correctly carrying out the calculations necessary for making progress on the ϵ expansion approach to the renormalization group.

The Wilson version of the RG is simple in principle and involves performing a few operations on Eq. (1). The first step is to perform a partial statistical average by integrating out the short-distance degrees of freedom. The idea behind this is that, because the important degrees of freedom near a critical point are the very long wavelength modes, these should be kept and the rapidly varying modes averaged over in some approximate way with their effects absorbed into a suitable redefinition of parameters in the action. This is most easily done by first going from real space to Fourier space by writing

$$\begin{gathered} \mathbf{s}(\mathbf{r}) = \mathbf{s}^{>}(\mathbf{r}) + \mathbf{s}^{<}(\mathbf{r}) = \int_q^{>} e^{i\mathbf{q}\cdot\mathbf{r}} \mathbf{s}^{>}(\mathbf{q}) + \int_q^{<} e^{i\mathbf{q}\cdot\mathbf{r}} \mathbf{s}^{<}(\mathbf{q}), \\ \mathbf{s}^{>}(\mathbf{q}) = \mathbf{s}(\mathbf{q}), \quad 1 < q = |\mathbf{q}| < 1/b \quad \text{and} \quad \mathbf{s}^{<}(\mathbf{q}) = \mathbf{s}(\mathbf{q}), \quad 0 < q < 1/b, \end{gathered} \tag{4}$$

where the parameter $b > 1$ is arbitrary and $\int_{\mathbf{q}} \equiv \int \frac{d^d q}{(2\pi)^d}$. The first step in the RG procedure is to perform the statistical average in Eq. (3) over the $\mathbf{s}^{>}(\mathbf{q})$ degrees of freedom which is exactly Wilson's idea of eliminating the short-wavelength degrees of freedom, resulting in an effective description of the partition function in terms of fewer degrees of freedom. The remaining steps are designed to rescale the variables in such a way that the new or renormalized Hamiltonian $\mathcal{H}(\mathbf{s})$ is of the same functional form as the original one but with renormalized parameters, $r(b)$, $u(b)$, $h(b)$. The scaling factor $\mathbf{s}(\mathbf{q}) \to \zeta_b \mathbf{s}(\mathbf{q})$ is chosen so that the coefficient of the $(\nabla \mathbf{s})^2$ term remains fixed under the whole rescaling operation. This is for convenience and ease of interpretation of the renormalized system, which leads to $\zeta_b = b^{(d+2-\eta)/2}$. The RG equations are with $b = e^{\delta l}$ become particularly simple because we have to do an expansion in powers of the quartic and higher coefficients of $\mathbf{s}^{2n}$, and when the integrated modes are in an infinitesmal spherical shell with $1 > q > 1 - \delta l$, we have

$$\begin{aligned} \frac{dr}{dl} &= 2r + 4(n+2)\frac{u}{1+r}, \\ \frac{du}{dl} &= \epsilon u - 4(n+8)\frac{u^2}{(1+r)^2}. \end{aligned} \tag{5}$$

Note that we have absorbed some numerical factors into a redefinition of the fourth-order term $uK_d \to u$ and $\epsilon = 4 - d$. Equation (5) has two fixed points where $dr/dl = 0 = du/dl$. As explained by Wilson and Fisher,[8,13,14] a fixed point of the recursion relations corresponds to a special point in the corresponding physical phase diagram. There are two fixed points: (i) $r^* = u^* = 0$ which is the Gaussian fixed point and (ii) the important fixed point at $u^* = \epsilon/4(n+8), r^* = -2(n+2)u^*$ corresponding to a critical point (T_c, h_c). Since both quantities are assumed to be $\mathcal{O}(\epsilon)$ in the critical region, one can linearize Eq. (5) and define a temperature variable $t(l) = r(l) + 2(n+2)u(l)$, which obey

$$\begin{aligned} \frac{dt}{dl} &= (2 - 4(n+2)u(l))t(l) + \mathcal{O}(\epsilon^2), \\ \frac{du}{dl} &= \epsilon u - 4(n+8)u^2 + \mathcal{O}(\epsilon^3), \\ \frac{dh}{dl} &= \frac{1}{2}(d+2-\eta)h. \end{aligned}$$

The gaussian fixed point $r^* = u^* = 0$ has two unstable eigenvalues for $d \leq 4$, $r(l) = r(0)e^{2l}$ and $u(l) = u(0)e^{\epsilon l}$ which can describe a tricritical point that requires fixing two fields to access while the other fixed point has one positive and one negative eigenvalue, $(u(l) - u^*) = (u(0) - u^*)\exp(-\epsilon l)$ and $t(l) = t(0)\exp[2 - \epsilon(n+2)/(n+8)]l$, which means that the correlation length exponent $\nu = 1/2 + \epsilon(n+2)/4(n+8)$.

The values of the exponents can be formally derived as a power series in the parameter $\epsilon = 4 - d$, but the numerical values compare very poorly with other methods such as high-temperature expansions at which Fisher was an expert. Also, one can compute other universal quantities such as critical amplitude ratios and the ϵ expansion values compare poorly with other methods. At this point, one can ask about the point of performing low-order ϵ expansions, particularly to $\mathcal{O}(\epsilon)$ which is relatively easy but very difficult to extend to higher orders. Also, it is very doubtful that such higher-order calculations done at the cost of much greater effort can give any extra physical information. Our major conclusion was that calculations to lowest order in ϵ gave useful qualitative information about what phases can exist and the shape of the corresponding phase diagram, but one should not take numerical values seriously.

The problem which fascinated me was a uniaxial anisotropic antiferromagnet in a uniform magnetic field with n-component spins which was known to have a spin flop transition from an Ising ordered phase to an XY-ordered phase, as was already known from a simple mean-field calculation. This system has a special multicritical point where several phases meet and this point was of great interest at that time, so it seemed to Kosterlitz, Nelson and Fisher that this was an ideal candidate for a renormalization-group study, which resulted in at least three papers.[16,21,22] Again, Michael Fisher initiated our study of systems which displayed the new sort of multicritical points where several phases met and the typical system chosen was the uniaxially anisotropic antiferromaget. In view of Ken Wilson's argument about universality in critical phenomena, the aim was to write the spin Hamiltonian as some Ginsburg–Landau free-energy functional suitable for a RG analysis. The Hamiltonian is in terms of n-component spins $\mathbf{s}(\mathbf{r}) = (s_{||}(\mathbf{r}), \mathbf{s}_\perp(\mathbf{r}))$ at the sites $\mathbf{r}$ of a d-dimensional hypercubic lattice

$$\begin{aligned}\mathcal{H}_{\text{int}} = &-\sum_{\mathbf{r},\mathbf{r}'}[J(\mathbf{r}-\mathbf{r}')\mathbf{s}(\mathbf{r})\cdot\mathbf{s}(\mathbf{r}') + D(\mathbf{r}-\mathbf{r}')s_{||}(\mathbf{r})s_{||}(\mathbf{r}')]\\ &-\sum_{\mathbf{r}}[H_{||}s_{||}(\mathbf{r}) + \mathbf{H}_\perp\cdot\mathbf{s}_\perp(\mathbf{r}) + e^{i\mathbf{k}_0\cdot\mathbf{r}}\mathbf{H}^\dagger\cdot\mathbf{s}(\mathbf{r})].\end{aligned}$$

After some tedious algebra,[21,22] we finally arrive at the effective Hamiltonian in a uniform magnetic field suitable for a RG analysis. Since we are dealing with an antiferromagnetic system with two sublattices, to diagonalize the quadratic part of the Hamiltonian, we must introduce transformed spin variables some of which can just be integrated out, which eventually yields a reduced Hamiltonian for an effective three-component spin $(\mathbf{s}_{||}, \mathbf{s}_\perp) = (\sigma, \mathbf{s})$ with $n_{||} = 1$, $n_\perp = 2$.

$$\mathcal{H}_{\text{red}} = -\int d\mathbf{R}\left[\frac{1}{2}\left((\nabla\sigma)^2 + r_{||}\sigma^2 + (\nabla\mathbf{s})^2 + r_\perp|\mathbf{s}|^2\right) + u\sigma^4 + 2w\sigma^2|\mathbf{s}|^2 + v|\mathbf{s}|^4\right].$$

The one-loop recursion relations are as follows:

$$\begin{aligned}\frac{dr_{||}}{dl} &= (2-4(n_{||}+2))r_{||} - 2n_\perp r_\perp + 4(n_{||}+2)u + 2n_\perp w + \mathcal{O}(\epsilon^2),\\ \frac{dr_\perp}{dl} &= (2-4(n_\perp+2)v)r_\perp - 2n_{||}wr_{||} + 4(n_\perp+2)v + 2n_{||}w + \mathcal{O}(\epsilon^2),\end{aligned}$$

$$\frac{du}{dl} = \epsilon u - 4(n_{||} + 8)u^2 - 4n_{\perp}w^2 + \mathcal{O}(\epsilon^3),$$

$$\frac{dv}{dl} = \epsilon v - 4(n_{\perp} + 8)v^2 - 4n_{||}w^2 + \mathcal{O}(\epsilon^3),$$

$$\frac{dw}{dl} = \epsilon w - 16w^2 - 4(n_{||} + 2)uw - 4(n_{\perp} + 2)vw + \mathcal{O}(\epsilon^3).$$

Our analysis[22] is limited to $\mathcal{O}(\epsilon)$, but it does show some remarkable phenomena. The last three equations determine the fixed points, four of these having $w^* = 0$ representing Hamiltonians with decoupled $\mathbf{s}_{||}$ and $\mathbf{s}_{\perp}$. These Hamiltonians all satisfy the mean-field condition for a tetracritical point[24] The RG eigenvalues which take the system out of the $w = 0$ plane are ϵ, $6\epsilon/(n_{||} + 8)$, $6\epsilon/(n_{\perp} + 8)$, $\epsilon(32 - 2n_{||} - 2n_{\perp} - n_{||}n_{\perp})/(n_{||} + 8)(n_{\perp} + 8)$.

The first three fixed points are clearly unstable to w perturbations for all $n_{||}$ and $n_{\perp} > -8$. The last fixed point is, however, *stable* against w perturbations when $n_{||}n_{\perp} + 2n_{||} + 2n_{\perp} > 32 + \mathcal{O}(\epsilon)$ and will determine the critical behavior. For the case when $n_{||} = 1$ and $n_{\perp} = n - 1$, the decoupled fixed point is stable only when $n > 11$, which is difficult to realize physically.

The two remaining fixed points are at $w^* \neq 0$ and the first is the isotropic Heisenberg fixed point at

$$u^* = v^* = w^* = \frac{\epsilon}{4(n_{||} + n_{\perp} + 8)},$$

$$r^*_{||} = r^*_{\perp} = -\frac{\epsilon(n_{||} + n_{\perp} + 2)}{2(n_{||} + n_{\perp} + 8)}.$$

The interaction parameters u, v, w at this fixed point, which we expect to dominate for $n \leq 3$ in $d = 3$, satisfies the mean-field condition, $(w^*)^2 > u^*v^*$, for bicritical behavior, and we can conclude that this fixed point is fully stable and describes the critical behavior for $n_{||} + n_{\perp} \leq 4 + \mathcal{O}(\epsilon)$. The domain of stability of the last fixed point with $n_{||}n_{\perp} + 2(n_{||} + n_{\perp}) > 32 + \mathcal{O}(\epsilon)$ and $n_{||} + n_{\perp} \geq 4 + \mathcal{O}(\epsilon)$ is called *biconical.*[22] To the best of my knowledge, the existing experimental data is consistent with this prediction.[17–20]

For an anisotropic $n = 3$ line antiferromagnet with $n_{||} = 1$ and $n_{\perp} = 2$ in a uniform field $H_{||}$ parallel to the anisotropy axis, there is a critical line of Ising transitions at $T_c^{||}$ which terminates in a bicritical point at a critical value of temperature and field. Then, as the field is increased beyond the bicritical point, the transition line continues as an $n = 2$ transition with different exponents. One of the main conclusions of the low-order RG study is the shape of the phase diagram, as shown in Figure I of Ref. 22. In particular, note that the Ising and XY phase boundaries approach the bicritical Heisenberg point tangentially because the crossover exponent $\phi > 1$. The number of components $n = n_{||} + n_{\perp} < n^X(d) = 4 - 2\epsilon$ and our best estimate is that $n^X > 3$ in $d = 3$, so we expect the bicritical point to be

Heisenberg-like in $d = 3$. Also, the $T_c^I(g)$, $T_c^H(g)$ and the spin flop line separating the two ordered phases are predicted to merge tangentially at the bicritical point as $T_c(H_{||}) - T_b \sim (H_{||} - H_b)^{\phi}$ where the crossover exponent $\phi > 1$.[16]

Michael Fisher was a remarkable man who was also a remarkable physicist, chemist and mathematician. He was born in 1931, got his BSc from King's College, London in 1951, earned his PhD in 1957, was appointed as Lecturer in 1958 and as full professor in 1965. The following year, he went to Cornell University as Professor of Chemistry, Physics and Mathematics, and in 1971, he was one of the first faculty at Cornell to be elected as a Horace White Professor and also became Secretary to the Cornell University Senate. In 1983, Fisher was elected as a foreign associate of the chemistry section of the US National Academy of Sciences because he remained a UK citizen. He finally retired from the University of Maryland College of Computer, Mathematical and Natural Sciences in 2012 and died in November 2021 at the ripe old age of 90.

His list of honors includes the Wolf Prize in 1980 with Ken Wilson and Leo Kadanoff, for which he was credited for bringing together and teaching a common language to chemists and physicists working on diverse problems of phase transitions. Three years later in 1983, he was awarded the Boltzmann Medal "for his many illuminating contributions to phase transitions during the past 25 years," and in 1995, he won the Lars Onsager Prize "for his numerous and seminal contributions to statistical mechanics, including the theory of phase transitions and critical phenomena, scaling laws, critical exponents, finite size effects and the application of the renormalization group to many of these problems" (official laudatio). In 2005, he got the Royal Medal from the Royal Society "for his seminal contributions to wetting transitions, dislocation melting and criticality of ionic solutions and many other topics in Statistical Mechanics."

This amazing list of well-deserved honors leaves one a bit breathless and wondering how could one man achieve so much in a short lifetime. Another of his important achievements was an amazing ability to bring together a group of overachieving scientists to work on topics which interested him. It helped very much that he and Sorrel, his wife of many years, were so hospitable and invited all sorts of disreputable graduate students and postdocs into their home and made them feel as if they belonged there. I, for one, will miss Michael's presence at various physics conferences and his inimitable way of bringing a speaker back down to reality with one or two penetrating questions but with a sense of humor. I will also miss those conversations about physics and life in general I used to have when I was a lowly postdoc at Cornell. The only minor regret that I still have is that Michael never won a Nobel Prize, which is still the ultimate prize in physics.

References

1. H. E. Stanley, *Introduction to Phase Transitions and Critical Phenomena* (Oxford University Press, Oxford and New York, 1971).
2. M. F. Sykes and M. E. Fisher, Susceptibility of the Ising model of an antiferromagnet. *Phys. Rev. Lett.* **1**, 321–322 (1958).
3. M. E. Fisher, The susceptibility of the plane Ising model. *Physica* **25**, 521–524 (1959).
4. M. E. and M. F. Sykes, Excluded-volume problem and the Ising model of ferromagnetism. *Phys. Rev.* **114**, 45–58 (1959).
5. M. E. Fisher, Transformations of Ising models. *Phys. Rev.* **113**, 969–981 (1959).
6. M. E. Fisher, The theory of equilibrium critical phenomena. *Rep. Prog. Phys.* **30**, 615–730 (1967).
7. N. David Mermin, My life with Fisher, *J. Stat. Phys.* **110**, 467–473 (2003).
8. Kenneth G. Wilson and Michael E. Fisher, Critical exponents in 3.99 dimensions. *Phys. Rev. Lett.* **28**(4), 240–243 (1972).
9. K. G. Wilson and J. Kogut, The renormalization group and the ϵ expansion. *Phys. Rep.* **12**, 78–199 (1974).
10. L. P. Kadanoff, *Physics* **2**, 263 (1966).
11. L. P. Kadanoff, Spin-spin correlations in the two-dimensional Ising model. *Nuovo Cimento* **44**, 276–305 (1966).
12. L. P. Kadanoff, W. G\`otze, D. Hamblen, R. Hecht, E. A. S. Lewis, V. V. Palciauskas, M. Rayl, J. Swift, D. Aspnes and J. Kane, Static phenomena near critical points: Theory and experiment. *Rev. Mod. Phys.* **39**(2), 395–431 (1967).
13. K. G. Wilson, Renormalization group and critical phenomena. I. Renormalization group and the Kadanoff Scaling picture. *Phys. Rev. B* **4**, 3174–3183 (1971).
14. K. G. Wilson, Renormalization group and critical phenomena II. Phase-space analysis of critical behavior. *Phys. Rev. B* **4**, 3184–3205 (1971).
15. M. E. Fisher, Renormalization group theory: Its basis and formulation in statistical physics. *Rev. Mod. Phys.* **70**, 653–681 (1998).
16. M. E. Fisher and David R. Nelson, Spin flop, supersolids, and bicritical and tetracritical points. *Phys. Rev. Lett.* **32**, 1350–1353 (1974).
17. L. S. Jacobs, Spin-flopping in MnF_2 by high magnetic fields. *J. Appl. Phys.* **32** Suppl., 61S–62S (1961).
18. V. A. Schmidt and S. A. Friedberg, Low-temperature magnetic phase transitions in $CoCl_2{\cdot}6H_2O$ and $MnBr_2{\cdot}4H_2O$. *J. Appl. Phys.* **38**, 5319–5326 (1967).
19. Y. Shapira, S. Foner and A. Misetich, Manetic-phase diagram of MnF_2 from ulttrasonic and differential magnetization measurements. *Phys. Rev. Lett.* **23**, 98–101 (1969).
20. J. H. Schelling and S. A. Friedberg, Thermal behavior of the antiferromagnet $MnBr_2.4H_2O$ in applied magnetic fields. *Phys. Rev.* **185**, 728–734 (1969).
21. David R. Nelson, J. M. Kosterlitz and Michael E. Fisher, Renormalization-group analysis of bicritical and tetracritical points. *Phys. Rev. Lett.* **33**, 813–817 (1974).
22. J. M. Kosterlitz, David R. Nelson and Michael E. Fisher, Bicritical and tetracritical points in anisotropic antiferromagnetic systems. *Phys. Rev. B* **13**, 412–432 (1976).
23. M. E. Fisher, The renormalization group in the theory of critical behavior. *Rev. Mod. Phys.* **46**, 597–616 (1974).

24. K.-S. Liu and M. E. Fisher, Quantum lattice gas and the existence of a supersolid. *J. Low Temp. Phys.* **10**, 655–683 (1973).
25. D. R. Nelson, Coexistence-curve singularities in isotropic ferromagnets. *Phys. Rev. B* **13**, 2222–2230 (1976).
26. M. E. Fisher and A. Aharony, Scaling function for critical scattering. *Phys. Rev. Lett.* **31**, 1238–1241 (1973).
27. M. Combescot, M. Droz and J. M. Kosterlitz, Two-point correlation function for general fields and temperatures in the critical region. *Phys. Rev. B* **11**, 4461–73 (1975).

https://doi.org/10.1142/9789811282386_0003

Chapter 3

My Memories of Michael Fisher

J.S. Langer

In the fall of 1966, I had known Michael Fisher for several years. I don't remember where or when we first met. But we needed no introduction when he arrived at Cornell as the new Horace White Professor of Chemistry, Physics and Mathematics, and I came there on a one-year leave from Carnegie Tech. In the back of my mind is an image of Michael with Sorrel and young versions of their children arriving on campus not long after Elly and I and our even younger children came there. I was looking forward to working with Michael on a wide range of ideas in theories of phase transitions, and I wasn't disappointed.

The biggest of these new ideas were in critical phenomena. Ken Wilson was at Cornell. Leo Kadanoff had introduced his block-spin analysis of the Ising model. One day that fall, Michael and I covered the blackboard in my office with block-spin equations, but we didn't yet know what to do with them. It was Ken who figured that out. Michael's work quickly moved in that direction, developing renormalization-group methods for studying critical phenomena. He made some wonderful contributions to this field, which is the theme of this book of essays. From the beginning, he insisted that Ken Wilson deserved the primary credit. This was the message of the talk he gave, at my invitation, at Carnegie Tech in 1971. It was the message of his Loeb Lecture at Harvard in 2017. I've never been sure that he was right, or that this question of priority should have been important for the scientific community. But it tells us something about Michael.

Michael was intensely interested in an unusually broad range of subjects. That's why it was especially appropriate for him to be a Professor of "Chemistry, Physics, and Mathematics" at Cornell. The academic year 1966–1967 was a very active one there, and Michael seemed to be at the center of everything that was happening. He, Vinay Ambegaokar and I used some of my later-to-be-called "instanton" ideas to study critical fluctuations in systems such as superfluids and superconductors, all while trying to keep up with and — especially in Michael's case — contribute to the renormalization-group developments.

In the following decades, Michael and I never directly collaborated on other research projects, but we kept in touch with each other regularly and crossed paths at various kinds of scientific events. He never liked email. I have in my files just a few recent messages from him at "xpectnil@umd.edu," some of them discussions about whether or not we could come to each other's birthday celebrations or retirement events.

There was a non-scientific dimension to Michael's remarkable range of talents. He was a skilled guitar player, specializing in Flamenco music. I remember that once, when a small group of us at a statistical physics meeting went out to dinner at a local restaurant, Michael politely borrowed the guitar from an in-house musician and played a lovely recital. It was a special evening.

I too have had a non-scientific dimension to my range of interests. I studied art very seriously as a teenager and always loved to draw pictures of people, trying to capture the expressions on their faces. In the 1980s and later, as I became increasingly occupied with science-related public affairs, I couldn't resist quietly sketching people across the table from me at various kinds of meetings. Here's one of Michael, dated 1988. I have no record of where this was drawn. It's obviously on the back of some sheet of instructions. It's a typical Michael Fisher expression meaning "I don't believe this. Prove it to me."

https://doi.org/10.1142/9789811282386_0004

Chapter 4

Some Memories from a 60-Year Friendship with Michael E. Fisher

J.L. Lebowitz
Departments of Mathematics and Physics,
Rutgers University, Piscataway, NJ 08854-8019, USA
lebowitz@math.rutgers.edu

1. Introduction

I first met Michael in the summer of 1961. I was in London for a few days, on my way to a conference in Vienna, visiting Oliver Penrose who was then at Imperial College. While there I went to Kings College to see Cyril Domb. Cyril took me to lunch in a kosher restaurant. After returning from lunch, Cyril introduced me to his group of young researchers. These included Martin Sykes and Michael Fisher, who told me something about their work. I cannot remember anymore of what Michael was working on, but I remember (or think I do) that his office was down a long corridor and had no windows. This must have been before Michael's promotion to a Reader at Kings College.

My next meeting with Michael would have been at a conference in Brown University in the spring of 1962. The conference was in honor of Lars Onsager on the 30th anniversary of the publication, while at Brown, of his paper on reciprocal relations (which got him the Nobel Prize). Michael was an invited speaker at that conference which marked him as an upcoming star in statistical mechanics. It led to his invitation by Marc Kac to visit Rockefeller University for the upcoming academic year. I do not remember talking with Michael at that conference, but we must have connected somehow because my next memory of Michael is having dinner at his house while he was a visitor at Rockefeller. The house was located on the New Jersey side of the George Washington Bridge spanning the Hudson River. It was at that dinner that I first met Sorrel, Michael's wife. Sorrel and Michael were excellent hosts and the dinner was very pleasant. I do not think any of the children were at the dinner, so they must have had their meal earlier.

Also present at that dinner was Shneior Lifson, a well-known physical chemist from the Weizmann Institute. Lifson was a great teller of jokes and stories, some of which I still remember. This made the dinner very jolly.

During Michael's visit at Rockefeller, he and I became good friends and when Michael moved to Cornell, he would come down to Yeshiva University in New York where I organized semiannual meetings, beginning in the fall of 1959, to discuss developments in statistical mechanics.

These occasions became a regular event for meeting with Michael. He usually came down with Ben Widom and their students and postdocs in a large station wagon. I still remember vaguely Amnon Aharony, then a postdoc of Michael's, talking about the renormalization group for systems with long-range potentials. The renormalization group was of course the big topic of the meeting in the early 1970s. Michael was, next to Wilson, the leader and chief spokesman for the field. The meetings became quite popular and at some point speakers were restricted to five minutes later to one minute unless their intensity quotient (IQ) for talking was high. The meeting lasted just one day. They were announced by letter and the respondents who wanted to present their works were asked to indicate their IQ. That changed sometimes in the 1970s when the meetings were extended to two and then to three days with longer "invited" talks of which Michael gave many (probably more than anyone else).

After the meetings, some of us used to go for dinner at some restaurant. I remember one time going to a Chinese restaurant with Michael and about a dozen others. As usual, in such places, we ordered a large variety of dishes, some of them doubles. Michael kept careful track of the order and when the dishes came reminded the waiter forcefully that some of the doubles we ordered came as singles. That was Michael's style. He did not let mistakes go by unnoticed.

Going back a bit in time to the fall of 1968, Michael and I participated in a small conference on rigorous results in statistical mechanics at the University of California, in Irvine, California. We got adjacent seats on the plane flying back from California to New York. During the flight, we discussed the problem of proving the existence of the thermodynamic limit for systems with periodic boundary conditions. Michael and David Ruelle had proven the existence of this limit, under suitable conditions on the interparticle potentials, for systems in regular boxes tending to infinity, but their proof did not extend immediately to systems with periodic boundary conditions. Since periodic boundaries are frequently used in analyzing physical systems, we thought that proving the existence of this limit was worth doing. We exchanged some ideas about possible strategies. We finished this work during a visit I made to Ithaca sometime in 1969 or the spring of 1970. I stayed in Michael's house and we solved this problem. This led to a joint publication "Asymptotic Free Energy of a System with Periodic Boundary Conditions", *Communication Mathematical Physics* 19, 251–272, 1970.

Fig. 1. Picture taken by Predrag Cvitanović at the 100th Statistical Mechanics Conference. From left to right: Pierre Hohenberg, Michael Fisher, (not sure), John Cardy, David Chandler.

This reminds me of the case when I received, as the Editor of the *Journal of Statistical Physics*, a manuscript claiming to prove the non-existence of the thermodynamic limit for a fluid with Lennard Jones potentials. I realized immediately that it was wrong, but to give the author some feedback I sent it to Michael for refereeing. Michael took it as a personal insult that someone dared question his proof and wrote back a fierce letter saying he would break off all relations with me, or the *Journal* (I can't remember which) if I published that paper. Of course, I did not publish it.

When I moved from Yeshiva to Rutgers in the summer of 1977, Michael first objected to my also moving the statistical mechanics conference to Rutgers. New Brunswick, where Rutgers is located, is farther away from Ithaca than New York and less convenient to get to. Michael, however, kept participating. This was especially so after he moved from Cornell to Maryland. Getting to Rutgers was now simpler than getting to New York and Michael would usually drive to the meeting. In more recent years, he would be given a ride by Eugene (Genya) Kolomeisky from the University of Virginia. Until the last few years, Michael would always sit in the front row, make notes on his program and scare the speakers with his questions. He would also be a strong voice for human rights at the session devoted to that topic at the meeting.

During one of the meetings, when Michael was still driving, there was a heavy snow fall on the day the meeting ended. Michael stayed over at my house in Princeton. I served some sherry before dinner in a "port" glass and got a lesson from Michael on the proper glasses for serving sherry and port.

Another time Michael stayed over after the conference, we had dinner at the house of Wendy, my daughter, who also lives in Princeton. She has a tree house in the garden left over from the previous owner of her house. She remembers Michael as the only adult who climbed into that house. Michael must have been past 80 at that time.

The last time I saw Michael in good shape was at the conference organized for his retirement from the University of Maryland. I had been on a visiting committee for the university a few years earlier and had written in my report how important a part Michael played in the intellectual life of the university. At the banquet of that conference, I was seated next to one of Michael's first students, Arthur Ferdinand, who was then Tax Commissioner of Atlanta. So, Michael's influence went far.

I will conclude these reminiscences of Michael, whom I miss very much, by giving some "poems" I wrote about Michael.

The first poem, *Homage à Michael*, was written (I think) for Michael's 70th birthday which we celebrated at the banquet of the 86th Statistical Mechanics Conference in December 2001.

Homage à Michael

Michael has long been our teacher,
And also our very dear friend.
At our meetings he is the main feature
And it's he who determines the trend.

He listens very carefully,
Sitting in the front row,
And then asks very tactfully,
"Is that really so?"

In matters of exponents critical
Michael is the ruler supreme,
For they are creations alphabetical
And obey his power law scheme.

His honors are many,
His achievements amazing;
His insights are canny,
His ideas trailblazing.

So let's cheer for Michael
And for dear Sorrel, too!
May the Goddess of Fortune
Shine on both of you!

The next two were written on the occasion of the 106th Statistical Mechanics Conference which took place at Rutgers on December 18–20, of the year 2011, when Michael, Jerry Percus and Ben Widom were guests of honor.

The Guests of Honor

Tonight we honor three great men:
Our good friends, Michael, Jerry and Ben!

Each one has been a trusted guide
On topics ranging far and wide.
They've been our leaders all the way,
So now it's time to say "Hurray!"
The big advances they have made
Are now the basics of our trade.
They're innovative, polymath
They've opened many a new path.

So now let's praise them once again —
Here's to Michael, Jerry and Ben —
They have earned our rousing cheers;
May they flourish for many years!

Michael Fisher

Michael Fisher's *raw-ther* British
And he can be *raw-ther* skittish!
He insists on sheer perfection
And allows us no defection.
In stat-mech Michael is a star;
He also plays a mean guitar
While Sorrel dances a flamenco
      Enchanting all, both near and far.
Michael's work is truly seminal,
Be it physical, biological or chemical;
      Of course, it's also mathematical,
With theorems precise and very practical.
His exponents range both far and wide,
      Getting renormalized on each side;
And show their universality
Whenever there's criticality.
Here's a toast to say we wish a
Long happy life to Michael Fisher!
And to his lovely Sorrel, too.
Let us make that toast times 2!

Finally, I reproduce a self-portrait by Michael which he gave me as a birthday present for my 70th birthday.

To Joel, for his 70th Birthday

With love from the artist
as a young musician!

Michael
L.F.

https://doi.org/10.1142/9789811282386_0005

Chapter 5

My Life with Michael E. Fisher*

N.D. Mermin

I first heard of Michael Fisher in 1963 at the start of my postdoctoral year with Walter Kohn at La Jolla. I had been introduced to another young postdoc, Bob Griffiths. He told me that he was currently working on a proof that the free energy of a spin system exists. "That it *what?*" I asked. "That it *exists*," said Bob firmly. "I'm using some ideas I got from Michael Fisher." Well, I thought, this Fisher must be a man with deep philosophical interests — a sort of Plato of thermodynamics.

I next heard of Michael a year later, shortly after I began my career at Cornell. Ben Widom told me that somebody named Michael Fisher would be coming for a visit. Remembering Griffiths telling me about his mentor, I looked forward to meeting a quiet, contemplative man. Michael's Cornell visit lasted over 20 years. It was the most unquiet and exciting period of my professional life.

The earliest record of his impact on me is a 1966 acknowledgement in the paper where Herbert Wagner and I give our version of Hohenberg's theorem. Herbert and I had tried to explain to Michael how a still unpublished argument of Pierre Hohenberg about superfluids could also be used to prove that there could not be spontaneous magnetization in the two-dimensional Heisenberg model. I hadn't known Michael very long, and one of the first things I learned was that you should think twice before claiming to *prove* something to the man who encouraged a postdoc to prove that the free energy of a spin system *exists*.

He questioned everything. Spectral functions, indeed! How did we know those frequency integrals converged? It became evident that we were dealing with a man who knew nothing about quantum field theory, didn't care that he didn't, and was convinced that we ourselves would be better off to forget it. Immediately.

So, we worked backwards in the face of this unexpected onslaught, unbundling our result from the conceptual wrappings in which it had been enshrouded by some

*This memoir evolved from after-dinner talks that I gave at Michael's 60th and 70th birthday celebrations. It is revised and expanded from the version that appears as "My Life with Fisher" in my 2016 collection of essays, *Why Quark Rhymes with Pork*. I have also drawn from "My Life with Wilson" in the same volume.

of the great thinkers of the previous decade. We peeled off layer after layer, day after day, in the face of his unrelenting skepticism. Finally, we boiled it down to an elementary statement about finite-dimensional matrices.

Then an astonishing change took place. "Publish!", he practically shouted. "It's very important!" Having experienced what it was like to be at the end of a Michael Fisher attack, I learned what it was like to have him on your side. Freeman Dyson came to Ithaca. Michael introduced us. "Mermin and Wagner have just proved that there's no spontaneous magnetization in the two-dimensional Heisenberg model," Michael proudly informed him, as Herbert and I basked in his admiration. "Of course there isn't," Dyson responded. "But they have *proved* that there isn't" Michael insisted. One Dyson eyebrow may have moved up a millimeter in response. No matter. I was hooked on arguing with Michael Fisher. My life would never be the same.

The best of the many Cornell scientific gatherings from the mid-1960s to the mid-1980s were the weekly lunch-time "Chemical Physics Seminars" run by Ben and Michael. There was an ironclad rule: blackboard only. And nothing could be written on the blackboard in advance. Nothing could be projected on a screen. No prepared pages could be distributed without Michael's prior approval. Were the seminars held today, PowerPoint and its variants would certainly have been prohibited.

The talks began at 1:00. You could bring lunch. They generally ended when Michael ran out of questions and remarks, usually by 3:00. The room held over a 100 and was typically at least half full. It was on the top floor of Clark Hall and had a spectacular view of Cayuga Lake, stretching to the North, and the distant hills beyond its western shore. Talks were usually given by Cornell people, but there were also speakers from outside Cornell, and people visiting for a few weeks or more were often invited to speak. The audience was encouraged to interrupt throughout the lecture with questions and pertinent remarks, following the example of Michael himself. This was where the Wilson–Fisher–Widom–Kadanoff theory of phase transitions was born in the mid-1960s and fully developed in the early 1970s.

Ben Widom put forth his famous scaling hypothesis in 1965 right at the start. In 1966, I received a preprint from Leo Kadanoff, whom I knew from both college and graduate school. I volunteered to give a report on Leo's manuscript. I remember my exact title: "My Early Impressions of Kadanoff's Partial Justification of Widom's Conjecture on the Homogeneity of the Equation of State Near the Critical Point."

I didn't think much of Leo's partial justification. It struck me as a sloppy piece of work. He started with a well-defined problem and then proceeded to butcher it, replacing detailed local configurations by gross averages, making unbelievably crude approximations to force those averages to look like the original problem at a different length scale and then repeating the whole messy business over and over again.

To my surprise, Ken Wilson, already a famous Cornell particle-physics theorist, showed up for my talk. Even more to my surprise, he seemed to take Kadanoff seriously. About a week later, he dropped by my office, ostensibly to talk about Kadanoff's paper. But what he actually had to say was that the crucial thing was to think about a ball rolling up a hill with just enough energy to get exactly to the top. I humored him. "Yes Ken, ... Very nice Ken," I said as I ushered him out the door.

I had known Wilson since he was 16 and I was 17. We were both freshmen at Harvard in the same German class. I knew even then that he was a mathematical wizard. But now I thought he was losing his wits. There was all this sloppy stuff in momentum space. He didn't even write proper integral signs.

I was away from Cornell on sabbatical during the culminating 1970–1971 revolution in our understanding of phase transitions. But Michael told me all about it when I returned to Ithaca. I was amazed to find that Michael — a man who was interested in whether the free energy *existed* — had waded right in to Ken's obscure remarks and was able to understand what he was getting at. He, who sneered at quantum field theory, had even learned about Feynman diagrams. Michael lucidly explained exactly what Ken had been trying to tell me, along with what the two of them had accomplished during my year abroad.

Many of my later enlightening interactions with Michael are recorded in other acknowledgments:

1967: "The analysis given here was constructed at the suggestion and with the vigorous assistance of M.E. Fisher."

1968: "M.E. Fisher's insistence on the difficulty of specifying a criterion for crystalline ordering led me to discard several earlier versions of the argument."

1976: "We are indebted to M.E. Fisher for lending us what seems to be the only copy of de Gennes' book now in Ithaca."

1977: "The importance of these considerations was brought home to me by a ferocious lunchtime discussion with M.E. Fisher."

1979: "It was M.E. Fisher who first suggested and repeatedly insisted that I should publish my lecture notes, but I am not sure he deserves thanks for this."

Most importantly, in our 1976 book, *Solid State Physics*, Neil Ashcroft and I, after thanking 47 alphabetically arranged colleagues, devote an entire paragraph to the 48th:

> One person, however, has influenced almost every chapter. Michael E. Fisher, Horace White Professor of Chemistry, Physics, *and* Mathematics, friend and neighbor, gadfly and troubadour, began to read the manuscript six years ago and has followed ever since, hard upon our tracks, through chapter, and, on occasion, through revision and re-revision, pouncing on

> obscurities, condemning dishonesties, decrying omissions, labeling axes, correcting misspellings, redrawing figures, and often making our lives very much more difficult by his unrelenting insistence that we could be more literate, accurate, intelligible, and thorough. We hope he will be pleased at how many of his illegible red marginalia have found their way into our text and expect to be hearing from him about those that have not.

We called Michael a troubadour to recognize the pleasure he took traveling with his Flamenco guitar. And it was only after getting to know Michael that I fully understood what the Athenians had meant when they called Socrates a gadfly. Shortly after that, I also started to understand why they had made him drink the hemlock. I think many of our readers knew that Neil and I had Socrates in mind when we used the term "gadfly". But when our book was translated into Japanese, it was Michael himself who reported to me, with only a tinge of acidity, that a Japanese friend had nervously asked him why our preface called him a "small, but loud and annoying insect." The Russian translator changed "gadfly" into "pedant." I did not understand what Michael was called in the Polish translation, until I sought help from Wojciech Zurek:

> "In our book Neil Ashcroft and I refer to Michael Fisher as "gadfly and troubadour". In the Polish edition I believe "gadfly" has become *ciety jak osa* and troubadour, *wesoly jak trubadur.* Am I right? Can you give me a translation of these phrases? I have to give an after-dinner speech at a banquet in Fisher's honor."

Here are some excerpts from Zurek's reply:

> "The translation is not bad, though it does change the meaning of the original phrase a bit: *ciety jak osa* means "ready to bite like a wasp". You could also say *giez* (which is literal for "gadfly"), but you would not say this about anyone in an after-dinner speech in his honor. On the other hand, *wesoly jak trubadur* (literally "gay as a troubadour") probably changes the intent."

The translators of the French edition in 2002, like the Russians several decades earlier, refused to try, replacing "gadfly and troubadour" with *rigoriste et poète.* But the Portuguese translation of 2011 has *moscardo e trovador* which, as far as I can tell, gets it exactly right.

Neil and I had not planned to ask Michael for a line-by-line critique of *Ashcroft and Mermin.* But early in our project, he asked what we were up to and we gave him our draft of Chapter 1. A few days later, he returned it, bristling with corrections, queries, complaints, and suggestions. Our publishers had been paying two physicists to comment on our draft chapters, forwarding to us their pleasant notes of approval. We urged them to drop those reviewers and offer both jobs to Michael. To our delight, he accepted their offer. Our subsequent writings were haunted by the specter of Michael, peering over our shoulders, ready to pounce. And pounce he did, as we commemorate in our acknowledgment.

Michael once gave a wonderful Cornell physics colloquium on what mathematical physics was all about. This is a hard thing to talk about in a colloquium, but he managed to make it gripping. I coincidentally had just come up with my own definition of the difference between mathematical physics and theoretical physics. I was planning to state it in a colloquium I was to give at Princeton the following week, so I tried it out on him after his lecture. The distinction, I told him, was to be found not in the physics, but in the sociology of physics: theoretical physics was done by physicists who lacked the necessary skills to do real experiments; mathematical physics was done by mathematicians who lacked the necessary skills to do real mathematics.

Michael was not amused. "I'd advise you not to say that at Princeton," he snarled. I said it anyway. It set off a riot. He was right.

But one of the wonderful things about Michael was that he was always ready to give you advice about anything whatsoever, and if you didn't take his advice, he didn't hold it against you. He never forgot, of course, that you didn't, and he was glad to remind you of his advice very sympathetically, when you get into trouble because you didn't. The reason he was so good at giving advice was that he thought very seriously about everything and always sought out the best advice himself.

He once asked me how I would find out where to buy a typewriter in New York city. I said I really couldn't tell him because all I would do would be to ask my father-in-law. What's his name, he asked? The next time I spoke to my father-in-law he remarked that a strange thing had happened. A man with a very loud voice who said he knew me had phoned him in his law office and asked where to buy a typewriter. "What did you do?" I asked. "I told him, of course," said my father-in-law impatiently. He was like Michael in some ways.

Michael always had strong opinions about everything, What fascinated me about his opinions was that although they were among the strongest and most forcibly argued I had ever encountered, I rarely could predict in advance what direction they would point in. Closely related to this was the most profound unwillingness that I have ever run across to settle for things the way they were.

What did Michael do when he checked into a hotel room for a night? He'd rearrange the furniture. He'd rotate the bed 90°, put the TV in the closet to make more room on the desk, carry the desk over to the window to get more light. He was renting the room for the night, and for that night, it was *his* room.

Often, I found it valuable to ask myself at difficult moments what Michael would do. Many years ago, I was at a meeting of the Danish Physical Society at a conference center south of Copenhagen. Each conferee had a little apartment with a tiny attic. Downstairs was a living room and bathroom. Up a narrow ladder was a built-in bed in a room with no light. Since one used the apartment only at night, this was an irritating arrangement. It was clear what Michael would have done. So, I dragged the mattress and bedding down the ladder, remade the bed on the living room floor, and never climbed up to the attic again. This solution only came to me because I had asked myself "What would Michael do?" The next day various Danish conferees complained about the arrangement. Ah, I said, under such trying circumstances, you should ask yourself what Michael Fisher would do. That night the air was filled with mattresses hurtling down ladders.

Sometimes, the answer to "What would Michael do?" is clear, but one lacks the courage to do it. Michael and I were once returning to Ithaca on a flight that stopped in London. After we had reboarded the plane and the door had shut, the plane was slow to leave the gate. As time went on, it began to look more and more like we would miss the flight from New York to Ithaca. When the likelihood began to approach certainty, Michael, muttering that there was no reason to spend the night on a bench at Kennedy when he had a brother-in-law with an apartment in London, rose from his seat and announced to the flight attendant that he was getting off. "You can't," she said. "Yes I can," said he. "We're about to depart," she said. "You've been saying that for an hour and a half," said he. "Michael, sit down," I said. "Shut up," said he. And he strode past her toward the closed door. "Open the door and let me out," he ordered. "Your baggage is on board," said they. "Hold it for me in New York, I'll pick it up tomorrow," said he.

And then something happened that I wouldn't have believed. The door opened, a ramp appeared, and off he strode, shouting back to me (who had for some time been pretending he was a complete stranger), "See you tomorrow in Ithaca!" Immediately thereafter, the door closed and the plane took off. It landed in New York just in time for me to catch the Ithaca flight which had, as usual, been delayed.

I conclude this story of my life with Michael E. Fisher with the tale of how Dorothy and I came to own our first microwave oven. Many years ago, before microwaves were common, I agreed to spend three months in Leiden as Lorentz Professor. My immediate predecessor in that position had been Michael E. Fisher. I remarked to Pierre Hohenberg that Michael would be a tough act to follow. No, Pierre said, on the contrary, following Michael had to be the easiest way to be a Lorentz Professor because "nothing you ask of them will seem unreasonable."

When we were first shown the Lorentz Professors' apartment, I was surprised to see a microwave oven in the kitchen. I remarked on what a well-appointed kitchen it was. "Yes," our host said, "the microwave is new. We just got it last year." Apparently Michael, on first being shown the apartment, had looked it over and said, "What, no microwave?!" So, for three months, we made good use of the Fisher microwave. When we got home, I looked around our kitchen and said, "What, no microwave?!" We immediately acquired one.

The Lorentz Professor sits at Lorentz's old desk. Attached to it is a brass plaque stating that between 1878 and 1912, the desk was used by H.A. Lorentz. At Lorentz's desk was a chair. Attached to it, I found another brass plaque stating that in 1994, the chair was used by M.E. Fisher. Whatever Michael thought of Lorentz, he did not like his desk chair. As a result, I sat very comfortably at the Lorentz desk in the Fisher chair. There cannot be many who for three whole months had been made *more* comfortable by Michael. Gadflies do not make people more comfortable.

Michael lived just down the street from me. There were many physicists in our neighborhood. As you walked down the street looking at the mailboxes, you would read Berkelman, Mermin, Widom, **F I S H E R**, Webb. Life in Ithaca, without such flashes of excitement, was a shade quieter than it had been before Michael left town.

But life in Maryland heated up a bit when Michael left us. After he had moved there and bought a house, I asked how things were going. "Not well," he said. "Why?" I asked. "We decided to move the walls out three feet", he said. "Which walls?" I asked. "All of them," he said.

Were there a Heaven — which he categorically denied — things there would have got a lot bumpier with the arrival of Michael E. Fisher.

https://doi.org/10.1142/9789811282386_0006

Chapter 6

Michael E. Fisher: Teacher, Mentor, Colleague and Friend

A (Very) Personal Account

E. Domany

Department of Physics of Complex Systems,
Weizmann Institute of Science, Rehovot 7630031, Israel
eytan.domany@weizmann.ac.il

The only rational way of educating is to be an example. If one can't help it, a warning example.

— Albert Einstein

I had the good fortune and privilege of having Michael Fisher as my teacher, supervisor, mentor and friend. During my years as a scientist, teacher and supervisor of about 100 students and postdocs, I found myself innumerable times realizing that I am following or at least trying to follow Michael's example. These pages are my attempt to convey recollections of my association with Michael, focusing on how he served as an example for me.

1. Introduction

The most difficult part of writing a memorial for someone close to you is to get started — to decide how to go about it. The natural way is to put things in temporal order, but perhaps a more attractive alternative is a random walk in time and space, allowing associations to dictate the sequence in which events come to mind. The second aspect in terms of difficulty is to keep in mind that you are not writing about yourself, but about Michael, and this is really hard since the recollections you record are personal and naturally you are a central player in whatever you reminisce about.

I decided to be flexible on both accounts — while trying to keep chronological order, allowing my mind to take leaps when they pop up, and on the personal front — to shrug off sanctimonious reservations and criticism of self-promotion and talk about myself whenever I feel like it.

2. Pre-Ph.D.: First Year at Cornell

My first encounter with Fisher was in early 1972. I was working at a research center in Israel, finishing my M.Sc. After lunch, we used to pass through the library and I remember distinctly one of my older colleagues (he was probably about half my current age) picking up the issue of the *Physical Review Letters* that just arrived, opening it and exclaiming: there is really no limit to what people of leisure will publish! He then read aloud the title of a letter: "Critical Exponents in 3.99 Dimensions," by Wilson and Fisher.[1] I was intrigued and took a course on critical phenomena that briefly mentioned the Renormalization Group.

For a variety of personal reasons, I decided, some months later, to do a Ph.D. in the US. My M.Sc. advisor, Michael Kirson, recommended to me Cornell, where he did his Ph.D. with Hans Bethe. I applied and was accepted, and we (Pnina, my recently married wife and I) found ourselves in Ithaca in late August 1973. I was immediately immersed in the small community of Israeli graduate students, mainly from Business Administration, Political Science and the like, who were in fierce competition to get their degrees as fast as possible, so during the first semester, I took an unreasonably heavy workload of courses. One of these was an advanced statistical physics seminar given by Geoffrey Chester, in which students were supposed to propose and complete some research project. When the time came, I mentioned that I heard that something interesting was happening in critical phenomena, was curious to find out what it was, and that I would also like to do a Monte Carlo simulation. Chester said, "If so, let's go and talk to Fisher!"[a] So a meeting was set up, and Chester, K.K. Mon (my partner in the project) and I walked over from Clark Hall to see Fisher in his office in Baker.

Michael was very kind, friendly and straight. He told us that he believed that a good Monte Carlo was a dead Monte Carlo, but if we insisted, he was willing to give the idea a try. He agreed to propose a question, which on the one hand he did find interesting, but on the other, he had a good idea what the answer should be, so that if we generated complete nonsense, he would be able to tell. The problem he suggested was simulations of the eight-vertex model, solved exactly by Baxter,[2] which appeared as a special "multicritical" case in a much broader space of models he wanted to study. I dwell on this project and the resulting publication[3] because it highlights several attributes of Michael, which had a profound impact on my own approach and attitude to science, lasting throughout my career.

Michael's dead serious attitude to his science *and research was my first striking takeaway lesson.* Once he set out on a problem, even if it was a mere course project of two first-year graduate students, he treated it with full, uncompromising attention to every detail. The problem was embedded in the most general setting, the research questions were clearly and precisely stated, no stone was left unturned

[a]Forty years later, Michael wrote to me that he first met Chester when he started graduate work in 1952. In 1974, I was unaware of their long-standing connection.

and no corners were ever cut. As part of our project, background material was thoroughly read and reviewed — K.K. and I read papers on the Baxter model,[2] its spin representation,[4] marginal operators, scaling hypothesis[4]; we learned finite-size scaling theory[5] and how to do Monte Carlo simulations.[6] We started by simulating Ising models to demonstrate that we did it right, deriving known results from finite-size scaling and only then attacking the full problem.[b]

Michael met us and reviewed our progress periodically, ever suspicious, and insisted on testing and retesting every finding. When he was satisfied with the results, agreement of the numerically derived continuous variation of the specific heat exponent on the Baxter line with the known answer, confirmation of the various scaling hypotheses and the effect of symmetry breaking, he told me that I should now write a first draft of a paper.

I was quite literate in English and took pride in my writing skills and hence was stunned when my attempt came back completely rewritten in red ink between my lines. There was, however, a comment on top, "Not bad at all for a first draft!" followed by an immortal Fisher line, "But why do you use the *blunt end of a charred broomstick* to draw your figures?"

Michael's approach to writing his papers *was my next lesson.* The student who worked on the project always wrote the first draft, as part of his training, even if rewriting it was far more time-consuming than just writing it. Michael revised our paper several times, devoting to it immense attention and energy. Since his name was on the paper, it had to meet his standards of clarity, scientific and linguistic precision, and style. The number of future readers of the paper is completely irrelevant; there is only one person whose standards and quality control matter, and that person is Michael, and for all my future papers — me. This statement holds for the present manuscript too!

3. Ph.D.: First Year

At the end of the summer of 1974, having passed my qualifying exam, I went to see Michael to check the possibility of doing a Ph.D. under his supervision. "Going to see him" calls for some explaining about the setup at Baker.

The main entry to Michael's office was via a door that led to a tiny entrance space. To the right, there was a windowless cubicle where his Senior Graduate Student, David Nelson, sat. Straight ahead was a room occupied by Lynn, the secretary he shared with Ben Widom. From this room opened the doors to Ben's office on

---

[b]Our computer program was punched onto cards that we took down to the basement of Clark and fed them into a card reader connected to a CDC 7600 machine at Berkeley. At late hours, our only competitor at the reader was Ken Wilson, who ran his Exact Renormalization Group programs on the same machine. Between runs, we competed with Wilson also on a computer game, landing a simulator of a moon landing on a small green screen, which was part of a huge computer that controlled low-temperature experiments of Richardson, Lee and Reppy.

the left and Michael's on the right. You asked Lynn to set up an appointment with Michael, and she told you over the phone the day and time. When the front door of this complex was locked (e.g. after office hours) — you entered a tiny dark space directly from the corridor and waited there till Michael opened a door to his office. In this small space, there were several black-and-white photographs of some fluid in motion — taken presumably by Michael when he did his Ph.D. thesis, solving some differential equations. It was through this anteroom that I came to see him. The prime exhibit in his office was a photograph of young Michael, confident and self-assured, facing Lars Onsager.

As always, *Michael was direct, honest and to the point.* He told me that I had to wait since he took only one Ph.D. student at a time, there was another enquiry on line, and the other candidate was preferred. It was Mike Peskin, a fellow student and friend, who asked also Ken Wilson about a Ph.D. When some time later Ken agreed to be his advisor, I got a call to come back and see Michael, who informed me that I was accepted, but first he wanted to discuss with me two issues.

The first, he said with a measure of unease, was that "Israelis were known to be brusque" and he hoped that this would not be the case with me. To which I responded, "Professor Fisher, would you mind explaining to me what is 'brusque'?" which he did, and I assured him that I would be the last person his secretary would ever complain about. I did not enquire what gave him the notion that Israelis were brusque — he had an Israeli postdoc at the time, one Amnon Aharony, who was some six years earlier the Teaching Assistant of my undergraduate quantum mechanics course at Tel Aviv University and became a friend and collaborator years after Cornell.[c] Amnon just finished two years with Michael, during which he wrote more than 30 papers, and I could imagine the stress he caused among the secretaries who had to type them (on typewriters, mind you!).

The second issue Michael brought up also caused him some discomfort. At Cornell, once you started working on your Ph.D., your advisor paid from a grant your tuition, fees and fellowship as a Research Assistant (RA). As an RA, you worked full time on your thesis research and did not have to spend about 10 hours a week as a TA, teaching or assisting in a lab.[d] Michael informed me that his current grants did not have funds for an RA and I will have to continue as a TA. Now, it was my turn to be embarrassed; I apologized for not bringing this up before, but I did intend to ask him whether it would be possible to work with him without taking a position as his RA. He was very surprised, and I explained that a few months before I found a summer job as RA of Professor Jim Krumhansl. I calculated effective

[c]To do justice to Amnon, at some later time I did ask Michael about this and he mentioned two other very prominent Israeli scientists who passed through Cornell before Amnon. Michael was well aware of the dangers of inference based on a single observation.

[d]Actually, as an RA, you were expected to contribute to teaching as a grader of some course. I had the good luck to grade the last Quantum Mechanics course given by Hans Bethe prior to his retirement. It was one of my most amazing experiences at Cornell.

elastic constants of some composite materials, and following the request of Jim Gubernatis, the postdoc in charge of the project, Krumhansl asked me to stay on as RA on a huge DARPA grant he just landed (on ultrasonic detection of defects in the engine of the just approved B1 bomber aircraft). He emphasized that this was "a bread and butter job," i.e. with no Ph.D. prospect down the line, and he expected me to look for someone else as my thesis advisor. I found out many years later that Michael was profoundly surprised and deeply impressed by my willingness and daring to do a Ph.D. with him and work in parallel on some other research project. When a year later Michael told me that now he did have funding and I could become his RA, I again asked him if it would be okay with him if I stayed on with Jim Krumhansl. Gubernatis (who became a close friend) got a position at Los Alamos and was leaving Cornell, a new postdoc was expected to arrive several months later and would need time to get into the project, so Jim really depended on me to run his huge DARPA grant. Michael of course agreed and I mention all this for two reasons: to show ***how open-minded and non-possessive Michael was*** and to sketch the special and unique atmosphere at Cornell at the time. The advanced students and postdocs talked to one another all the time, and small *ad hoc* teams continuously formed and dissolved discussing problems and writing papers. Postdocs in particular were roaming freely around. People like Mike Kosterlitz, Alistair Bruce, the Combescots, Mike Peskin, Serge Aubry, Alan Bishop, Steve Teitel and many others were enthusiastically discussing whatever excited them at any moment. I never encountered any barrier due to "I am working with Prof. X and you with Y"; on the contrary, you had to make an effort to filter out stuff that was flowing around you.

So, in the Fall of 1974, I started my Ph.D. with Fisher. This statement also needs some qualifying; actually, I started working with David Nelson. Now, I am fully aware that David is alive and well, and it is not his obituary that I am writing, but it is impossible to describe my years at Cornell with Michael without talking about David. He was some years younger than I and on a six-year Ph.D. program. Clearly one of the sharpest and smartest guys around, he already published seminal work using the epsilon expansion to analyze critical behavior near multicritical points; he just wrote a paper with Rudnick[7] on calculating free energy and tricritical crossover scaling functions by integrating various quantities along RG trajectories. He proposed to Michael using trajectory integrals to calculate crossover scaling functions near bi-critical points, and I was assigned to work on this problem. To say that working with David was challenging is a severe understatement, but we did get along well and remained friends. Some years ago, I visited him at Harvard and gave a seminar — he introduced me saying that we were graduate students together and we both had to put up with Fisher, to which I responded — you had to put up with Fisher, I had to put up with Fisher and with you! He was filling spiral notebooks with error-free analytic calculations extremely fast, with his writing speed being the rate-limiting factor. Our project progressed well and we calculated the scaling functions in the disordered phase, with relatively little involvement from Michael, who was not a co-author on this paper[8] (first of a series of three).

This was another important lesson I learned from Michael: ***let your students initiate projects, and work on them independently, teaming up with other students. Never "sign" papers to which you did not make some significant contribution.***[e] While at Cornell, I wrote papers alone and with several collaborators. Michael was always supportive but never ever made me feel that as my thesis advisor, he should be a co-author. This does not imply that he was indifferent, on the contrary: his door and mind were open to ideas. His curiosity was boundless and if you came up with an idea that he found interesting and challenging, his eyes would glitter; at times, he would get involved and work on it enthusiastically (see the following example). He always found the right balance between letting you work at your own pace, shifting to the problems that excited you most, and acting, at the same time, as advisor and collaborator.

One aspect of Michael's attitude to science and students, which to my regret I did not adopt, was ***attending and taking students to conferences***. He went to every major meeting, many times giving keynote invited talks, but when he was interested in learning something new (e.g. biology), he had no ego problem registering and giving a poster presentation. He took me along to the major conferences in the field; the first one, which I remember most vividly was a Yeshiva meeting in New York in December 1974. I even gave a one-minute "talk" standing in the back of a large dark auditorium and screaming something, with the formidable Joel Lebowitz on the stage, shutting up everyone who went over time just by standing up and looking at the speaker with a big smile, without saying a word.

4. Ph.D.: Second Year

When after my first year David Nelson left for Harvard, Michael and I switched to a closer mode of collaboration. I used to see him on Tuesdays at the "Fisher seminar" (see the following section) and every Friday for one hour alone in his office. Around that time, he became Chair of the Department of Chemistry, so he was very busy, but my allocated hour was always there. When I did not make much progress and did not have anything of substance to discuss and report, I would call and say so and we canceled our meeting. At a Magnetism Conference to which we went together, we met two guys from IBM Zurich, Schneider and Stoll, who were considering sending their student, Stefan Sarbach, to Cornell on a Swiss postdoctoral fellowship. Michael warned them that he was very busy and advised them to talk to me to find out how it was to work with him. I told them that

[e]This got me into trouble with my biologist colleagues: several times a paper I had never seen landed in my inbox with my name on it as a co-author (mostly because one of my students helped with some analysis). When I removed my name, my collaborator was offended — is this paper not good enough for you?

I had Michael's full attention for an hour every Friday, and I recall Toni Schneider asking, "Is one hour a week with your advisor enough?" I could not help reverting to brusque Israeli behavior and told him that it was far better to spend one hour a week with a wise man than eight hours a day with a fool. I am not sure whether it was this deep insight that convinced them, but Sarbach did come to Cornell. Another person who came, around the same time (Fall 1975), was an old friend of mine, David Mukamel — he and I shared an apartment and a boss in the late sixties. David finished a postdoc at Brookhaven and came for a second one at Cornell. During my next (final) year, while finishing the second[9] and third[10] papers on scaling functions, I collaborated with David and Michael on some other problems, *which serve as the example promised above, of **Michael being "turned on" by an idea to which he got exposed.***

While at Brookhaven, David identified physical systems whose phase transitions were characterized by order parameters with $n \geq 4$ components.[11] Using Landau theory, he wrote down their Landau–Ginzburg Hamiltonians,[f] and using the epsilon expansion, he discovered that some of them did not have stable fixed points of the RG.[12] Since the corresponding physical systems had first-order transitions, he proposed that the lack of stable RG fixed points indicates that a transition of the corresponding system must be first order. This was an extension of the symmetry-based Landau rules[13] that excluded symmetry breaking via a continuous transition.

Michael did not like Landau's theory: he viewed it as an embellishment of mean-field theory by fancy group-theoretical terminology (in which he was not well versed and hence suspicious of its relevance). He did take a deep interest, however, in first-order transitions. In his course on phase transitions and critical phenomena, which he gave in late 1974 to a huge class of 120 students, Michael emphasized that we do not understand first-order transitions. I dared raise my hand and mentioned that there were elaborate Landau rules based on symmetry that excluded the existence of continuous transitions and hence identified situations in which symmetry breaking must be first order.[13] He gave me a look and said, "The trouble with Israelis is that they take the Russians too seriously!" 119 students burst out laughing.

David did succeed in convincing Michael that the Landau framework, based on symmetry, was the natural one to construct the correct Landau–Ginzburg Hamiltonians for physical systems undergoing a phase transition. Whatever you do next, e.g. how you approximate the ensuing functional integral, is your business — be it Mean Field, Monte Carlo, or your favorite form of RG.

Interest in first-order transitions fueled two studies. The first one addressed the question of what happens when a field is applied to a system for which the transition is first order due either to the absence of a stable fixed point or it being inaccessible

[f] I always wondered why was Wilson's name added to this concept?

under the RG flows. In some cases, the field can make a stable fixed point accessible, and we investigated how the transition turns from first order to continuous.[14]

I was intrigued by the fact that if you look at a cube down the (111) axis, it has three-fold symmetry. David and I thought about an ordered cubic ferromagnet with the six (100) directions being the easy axes. In a strong magnetic field pointing along (1,1,1), the spins will be aligned with the field. As the field is reduced, the magnetization will flip toward one of the three equivalent directions. David showed that this transition is in the three-state Potts universality class; this system provided a physical realization of the three-state Potts model in $d = 3$ dimensions! There was some evidence indicating that the transition was first order — whereas in $d = 2$, it was known to be continuous.[15] Michael got immersed in the problem — it was generalized to external fields in an arbitrary direction, the full phase diagram was mapped out, in which the Potts transition was a special multicritical point. A quick Letter was written,[16] which turned the spotlight to yet another specialty of

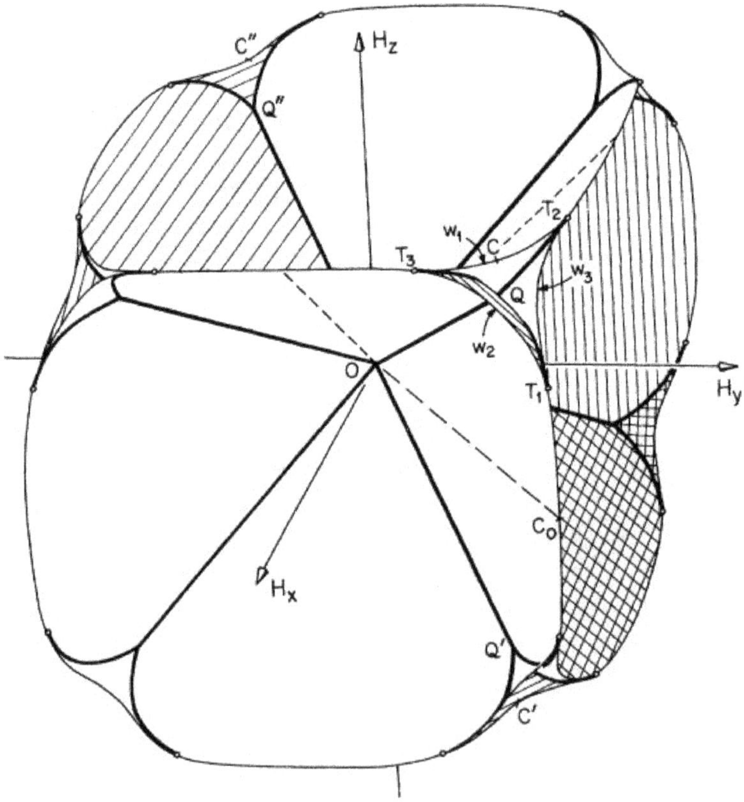

Fig. 1. Art in Science: Schematic mean-field phase diagram of a cubic ferromagnet in an external field. The three-state Potts transitions are at point Q (and at symmetry-related points Q′, Q″). See Ref. 16.

Michael: *his love for drawing and his talent at it.* The published phase diagrams, for first-order and continuous Potts transitions, are remarkable examples of combining art and aesthetics in scientific drawings. Many years later, in our apartment in Rehovot, my six-year-old son Gil was fascinated by Michael's drawings of nudes — in particular, how he was able to bring out their three-dimensional curvature. Michael, ever willing to explain and teach, spent hours showing Gil how to draw female bosoms.

5. Fisher Seminar, Teaching, Super-Referee

I remember admiring the drawings he made on Styrofoam cups while listening to seminars. The only "formal" requirement that came with being associated with Michael was attending the "Fisher seminar" every Tuesday afternoon. The audience sat around tables; blackboard and chalks were provided to speakers, but transparencies were strictly forbidden.[g] If the speaker wanted to show figures (which would take too long to draw on the blackboard), a large number of copies were made and distributed to the audience. Michael felt that transparencies convey the illusion that whatever was on them was correct, while text written on the blackboard was more inviting to discussion and criticism. Besides, writing on the blackboard slowed the speaker down, giving the participants time to think and take notes. Michael was notorious for mercilessly grilling speakers with his questions. The foremost reason was his genuine boundless intellectual curiosity; if he found the topic interesting enough to have a seminar on it, he really wanted to understand everything. He had no tolerance for ambiguity and weak unfounded arguments and did not hide his aversion to them. While he considered it his privilege and even duty to interrupt the speaker whenever he felt like it, he was much less patient with questions from other people (e.g. me), expecting them to think carefully before asking. *Here is another lesson I learned:* ***the importance of an exciting informal weekly seminar series (or group meetings).***

Michael was a wonderful teacher and lecturer. ***The pride he took in his teaching and the time he devoted to planning and preparing his lectures — classes as well as scientific seminars*** *— were yet another aspect in which he served as an example for me.* In those prehistoric, pre-PowerPoint times, overhead projectors were used to show handwritten transparencies. Michael insisted on using two projectors simultaneously, leaving a transparency on while the next one was introduced, making the transition between them smoother and easier to follow. Clearly, the two projectors had to be close enough so that he did not

[g] There were exceptions. Roald Hoffmann from the Chemistry Department was allowed transparencies, and when I asked Michael, "How come?" he answered that people who got or will get the Nobel Prize were exempt from the no transparencies rule. I have not heard yet of the Woodward–Hoffmann rules, but sure enough, Hoffmann received the Nobel Prize some years later. Interestingly, neither Ken Wilson nor Mike Kosterlitz were exempt.

have to run a long distance between them, but far enough to keep the projections well separated (on two screens, of course). His transparencies were meticulously prepared and he liked artistic effects and surprises. When talking about critical exponents, he compared the increasing accuracy with which they were calculated and measured with our improving knowledge of π. He heard that somewhere in the Bible the value $\pi = 3$ appeared and asked me whether I knew about this. I asked around and informed him that indeed, there is mention of a large circular copper basin cast for Solomon's Temple, "a molten sea, ten cubits from the one brim to the other: it was round all about... and a line of thirty cubits did compass it round about."[17] He was delighted and asked me to make a copy of the original Hebrew page, which he showed proudly for years.

Returning to the opening Einstein quote, I recall one instance where Michael served for me as a "warning example." One day I told him of an idea I had — to combine the RG and Monte Carlo. He gave me a dismissive look and said, "You will combine the evil of the two methods." I did not pursue the idea, and a few months later Sheng Ma[18] published a letter on this, followed two years afterward by the wonderful work of Bob Swendsen.[19] As a result, whenever one of my students announced that she (or he) had an idea, no matter how far out it sounded, I always expressed surprise and pleasure, and encouraged her to work on it for a few weeks — and then either drop it or come see me again about continuing. Someone having an idea is a cause for celebration! Beware of extinguishing the quest for new ideas.

One other role Michael took, which of course I could not adopt, was that of *a* ***"super-referee" of the field****. Anyone who published a paper on phase transitions and critical phenomena had to be prepared to receive a letter from Fisher, complimenting his work and expressing criticism of parts that he did not agree with.* Everyone, especially people who worked with him in the past, always had this super-referee in mind, asking ourselves: "I wonder, what will Fisher think about this?" His involvement gave rise, every now and then, to amusing incidents; I will mention two, one of which I only heard second hand, while about the second I am confident, having seen the correspondence.

The first story concerns a letter on exchange effects in bcc ^{3}He by Hetherington and Willard.[20] Michael did not like something about it, assumed that Willard was the senior author, and rushed off a "Dear Prof. Willard" letter. As it turned out, it was originally a single-author paper by Hetherington who used the plural form all over. When his letter was accepted, the copy editor requested to change "we" to "I" everywhere, which meant retyping a technically difficult paper. Hetherington preferred to add an author, F.D.C. Willard,[h] his cat, who now had to correspond with Fisher!

The second story concerns S. Shtrikman, a very prominent Israeli physicist whom I knew and respected, and was surprised when Michael complained about him.

[h] F.D.C. stood for Felix Domesticus Chester.

When I asked him why, he sent me a letter he wrote to Shtrikman, reprimanding him about a serious oversight: identifying[21] the measured (finite) peak of the susceptibility of an antiferromagnet as the Neel temperature, whereas Michael has shown that T_N was, in fact, at a temperature close but below that of the peak, where the derivative of the function diverged.[22] He was offended by Shtrikman not bothering to answer his letter. I looked at the letter and asked him, "Have you noticed the date? May 30, 1967." "So?" I told him that this was a week before the six-day war started and even if the letter ever arrived, Shtrikman was probably on reserve duty and had other things on his mind. Not a man who concedes easily, Michael alerted me to the fact that Hornreich, the junior co-author of the paper was at the time on sabbatical at Yale and could have answered.

No account of Michael would be complete without mention of his family. I remember fondly the huge house in Ithaca, presided over by Sorrel, the kindest, funniest and warmest person you would ever have met. The Christmas lunch morphed into dinner, with the strange Christmas pudding, a round object made of undefined ingredients that had been buried for six months before being dug up and consumed — I have no idea whether this was true or invented to check my naivety. The house was full of annoying puzzles that were stuck in your hand with the challenge of taking them apart, putting them together, or removing them from the buttonhole of your jacket. The flamenco guitar, which had a seat of its own on flights to Israel, and the Fisher kids, of whom he was enormously proud also deserve special mention.

6. Wrapping Up: Post Cornell

In the spring of 1976, my wife informed me that "I do not know about you, but this past winter was *my* last one in Ithaca." The original plan was to stay for four years at Cornell, i.e. till Fall 77, so I told Michael that I would have to leave sooner. It was time to write a Ph.D. thesis, and the way this happened brought to fore *two aspects of Michael which I also tried to adopt*: ***foresight about theses and dealing with administration firmly but without confrontation.***

When I wrote my first crossover paper, the last thing I had on my mind was my thesis. Michael told me to make sure that the final version was typed precisely in the format of a Cornell thesis. The same was true for the second paper. Now, when the time came for writing the thesis, I already had the main part of it done and ready! I wrote a few brief introductory chapters and was all set to put everything together when disaster struck. When the second paper was ready for typing, Michael's secretary Lynn was either ill or overloaded, and it was farmed out to an excellent professional typist in Ithaca, who did a superb job. Her IBM typewriter had, however, a "head" with a font slightly different font from Lynn's. When I discovered this, I rushed to warn Michael that the secretary in charge of quality control of theses had a reputation of being quite difficult and the rules were very explicit about using the same typewriter throughout the entire thesis. The departure date for my postdoc was already set and time was starting to run out. Michael looked with disbelief at the typescripts of the two papers and exclaimed: "But they are practically identical!" I insisted — if I noticed the difference, no doubt the secretary would also discover it. I was curious to see how Michael would fight this out and was impressed watching him in action. He wrote an extremely polite lengthy letter to the Dean in charge, explaining to her the problem I was working on, how technically difficult it was to type this thesis, the problems with Lynn, how hard we tried to match typewriters and alas, one of the chapters is in a barely noticeably different font. Would she be so kind as to allow this, just for once? When two weeks passed and there was no answer, Michael wrote a second letter, thanking the Dean profusely for her flexibility and agreement to his request of having one chapter in a slightly different font. When I went, shaking, to see the secretary whose approval was essential for submission, she immediately caught the two distinct typewriters. I informed her that Professor Fisher had discussed the matter extensively with the Dean, an agreement had been reached, and produced a copy of Michael's letter of thanks. She was shaken, but went on to check the thesis. Her eyes lit up when she discovered, to my horror, that in the "Chapter of the Second Typewriter" a typo, about which I completely forgot, had been corrected using another typewriter! She kicked me out, saying "The Dean may have agreed to two typewriters, but not to three!" Michael phoned the lady who typed it for us and begged her to drop whatever she was doing and retype the single offensive page. I rushed over and sat with her while she did it, rushed back to the secretary, had the thesis approved, photocopied everything on paper of the right weight and submitted it a few hours before our plane for my postdoc at Seattle took off, three years and four months after my arrival at Cornell.

From teacher and student, we progressed over the years to friendship. We both liked to write, so many letters were exchanged. He came a lot to Israel, as a member of the Weizmann Board (see Fig. 2) and various scientific and academic advisory committees, driving physicists, mathematicians and chemists crazy with his questions, critique and advice. He always spoke his mind in a clear, loud and direct fashion, but was also aware of the politics of the place. As a young just tenured Associate Professor, I campaigned to stop the construction of a huge solar tower, a showy, wasteful and crazy large-scale engineering project driven by some donor's whim, and told Michael about my fight. He listened and asked me a single question: "Who is the Weizmann scientist pushing for this?" I explained to him who it was and Michael summarized the situation in one sentence; "I see, so he is way out of your league," advising me to stop a futile fight. Over the years, whenever I started to feel important, I walked to a window to look at the solar tower to place things in proper perspective.

After his move to Maryland, I tried to pass through whenever I had a chance. I was touched when during my last visit to the fancy establishment to which Michael

Fig. 2. Michael responding to the award of Ph.D. Honoris Causa from the Weizmann Institute, Nov. 2009.

and Sorrel moved he gave me as a gift a Sheffield steel hunting knife that belonged to his father. The knife is on my desk, and the "gadfly and troubadour"[23] who gave it to me is carved into my heart.

References

1. K. G. Wilson and M. E. Fisher, Critical exponents in 3.99 dimensions. *Phys. Rev. Lett.* **28**(4), 240–243 (1972).
2. R. J. Baxter, Eight-vertex model in lattice statistics. *Phys. Rev. Lett.* **26**(14), 832–835 (1971).
3. E. Domany, K. K. Mon, G. V. Chester and M. E. Fisher, Monte Carlo study of multicriticality in finite Baxter models. *Phys. Rev. B* **12**(11), 5025–5033 (1975).
4. L. P. Kadanoff and F. J. Wegner, Some critical properties of the eight-vertex model. *Phys. Rev. B* **4**(11), 3989–3993 (1971).
5. M. E. Fisher and M. N. Barber, Scaling theory for finite-size effects in the critical region. *Phys. Rev. Lett.* **28**(23), 1516–1519 (1972).
6. J. M. Hammersley and D. C. Handscomb, *Monte Carlo Methods*, Methuen & Co. London (1964).
7. D. R. Nelson and J. Rudnick, Equations of state for tricritical points. *Phys. Rev. Lett.* **35**(3), 178–181 (1975).
8. D. R. Nelson and E. Domany, Equations of state for bicritical points. I. Calculations in the disordered phase. *Phys. Rev. B* **13**(1), 236–250 (1976).
9. E. Domany, D. R. Nelson and M. E. Fisher, Equations of state for bicritical points. II. Ising-like ordered phases. *Phys. Rev. B* **15**(15), 3493–3509 (1977).

10. E. Domany and M. E. Fisher, Equations of state for bicritical points. III. Cubic anisotropy and tetracriticality. *Phys. Rev. B* **15**(7), 3510–3521 (1977).
11. D. Mukamel, Physical realizations of $n \geq 4$ vector models. *Phys. Rev. Lett.* **34**(8), 481–484 (1975).
12. P. Bak, S. Krinsky and D. Mukamel, First-order transitions, symmetry, and the ε expansion. *Phys. Rev. Lett.* **36**(1), 52–55 (1976).
13. L. D. Landau and E. M. Lifshitz, *Statistical Physics*, Pergamon, New York (1968), 2nd ed., Chap. XIV.
14. E. Domany, D. Mukamel and M. E. Fisher, Destruction of first-order transitions by symmetry-breaking fields. *Phys. Rev. B* **15**(11), 5432–5441 (1977).
15. R. J. Baxter, Potts model at the critical temperature. *J. Phys. C: Solid State Phys.* **6**(23), L445–L448 (1973).
16. D. Mukamel, M. E. Fisher and E. Domany, Magnetization of cubic ferromagnets and the three-component potts model. *Phys. Rev. Lett.* **37**(10), 565–568 (1976).
17. 1 Kings Ch. 7, pp. 23-26, King James Version (400 B.C.)
18. S. K. Ma, Renormalization group by Monte Carlo methods. *Phys. Rev. Lett.* **37**(8), 461–464 (1976).
19. R. H. Swendsen, Monte Carlo renormalization group. *Phys. Rev. Lett.* **42**(14), 859–861 (1979).
20. J. H. Hetherington and F. D. C. Willard, Two-, three-, and four-atom exchange effects in bcc ^{3}He. *Phys. Rev. Lett.* **35**(21), 1442–1444 (1975).
21. R. Hornreich and S. Shtrikman, Susceptibility-derived sublattice magnetization in antiferromagnets. *Phys. Rev.* **159**(2), 408–410 (1967).
22. M. E. Fisher, Relation between the specific heat and susceptibility of an antiferromagnet. *Phil. Mag.* **7**(82), 1731–1743 (1962).
23. N. W. Ashcroft and N. D. Mermin, *Solid State Physics*, Holt, Rinehart and Winston, New York (1976) Preface p. x.

https://doi.org/10.1142/9789811282386_0007

Chapter 7

Michael Fisher as a Role Model

B.I. Halperin
Harvard University, Massachusetts Hall,
Cambridge, MA 02138, USA

Unfortunately, I never had the privilege of collaborating directly with Michael Fisher on a research project, and we were never colleagues at the same institution. My interactions with him were therefore largely limited to encounters at scientific conferences and some brief visits to each other's home institutes. Nevertheless, Michael was an important role model for me, and he had a major influence on my work. Beyond this, I came to consider him as a very good friend.

At the scientific level, I was most immediately influenced by his ideas of scaling in the description of classical critical phenomena. When, together with Pierre Hohenberg, I became involved in the problem of dynamical properties at or near a critical point, it was natural for us to try to generalize ideas of scaling to include these properties.

Following Michael's focus on critical exponents and the relations between them, Pierre and I defined a critical exponent, which we denoted z, relating characteristic frequencies ω to characteristic wavevectors k in the critical region, as $\omega \propto k^z$.[1] We discovered afterwards that we had inadvertently violated one of Michael's cardinal rules: that critical exponents should be denoted by Greek letters, not Latin ones. Michael took the offense in good spirits, however, and it never interfered with our friendship.

When Pierre, Shang-keng Ma, and I came up with our first application of renormalization-group methods to dynamic critical phenomena,[2,3] we were heavily influenced by the famous paper of Wilson and Fisher on critical phenomena in $d = 3.99$ dimensions.[4] Though our first paper and others written later with Eric Siggia leaned heavily on Wilson's subsequent development of a formal series expansion for critical exponents in powers of $\varepsilon = 4 - d$, and though we calculated exponents to second order in ε, our work was fundamentally motivated by the simpler, more physical ideas in the Wilson–Fisher paper, which formulated the renormalization group in terms of recursion relations.[5–9] Although the Wilson–Fisher recursion analysis was only valid to first order in ε, by generalizing their approach we could more easily set up the problem and understand what the simplest model would be for analyzing dynamics when we considered physical systems with different sets of symmetries and commutation rules. It helped us to understand the importance of dividing systems into universality classes and to understand when two systems would or would not ultimately wind up in the same dynamic universality class. In most cases, the distinction between universality classes was already evident at first order in ε.

Although Michael is best known for his work on static critical phenomena, he also made direct personal contributions to the analysis of dynamic critical phenomena. I cite particularly the 2006 *Physical Review Letter* on critical dynamics in a model of a binary fluid that Michael co-authored with Subir Das, Jan Sengers, Jürgen Horbach and Kurt Binder.[10]

Apart from my indebtedness to Michael's ideas about scaling and the renormalization group on an intellectual level, I was at least a second-hand beneficiary of his mentorship in the 1970s. Amnon Aharony spent a year at Bell Laboratories, fresh from his pioneering work with Michael, in the early days of applications of the renormalization group and ε expansions to critical phenomena. During that period, I had many stimulating interactions with Amnon, which included particularly discussions of universal ratios between the amplitudes of various singular quantities near a critical point. Amnon and I published two papers on this subject, one of which was co-authored with Pierre Hohenberg and Eric Siggia, which considered the three-dimensional dipolar Ising model as well as various systems with short-range interactions in 4 and $(4 - \varepsilon)$ dimensions.[11,12] This work was strongly motivated by the work of Stauffer, Ferrer and Wortis on two-sale factor universality,[13] but of course, it also made heavy use of the ε expansion.

On a personal level, Michael was a role model for how a faculty member or a researcher should behave in relation to other researchers and to the profession more generally. Michael was a faithful participant in conferences on statistical mechanics. I remember particularly seeing him at the semiannual meetings organized by Joel Lebowitz, first at Yeshiva University and later at Rutgers. Michael would sit in the first row of the auditorium, listening carefully to every talk, and very often raising polite inciteful questions at the end. Of course, Michael's own lectures were a model of clarity, precision, and elegance.

Michael was famous for the degree to which he took seriously his obligations as a professor. As an example, I remember hearing from David Bishop about the detailed comments on his Ph.D. thesis that he received from Michael, who was not his advisor but was a member of his thesis committee. David was amazed at the amount of time and effort Michael had devoted to the task, when, David said only half in jest, most other professors would have just picked up the thesis and said, "Okay, it weighs enough, let's grant him the degree." During my own career as a professor, I did try to make detailed comments about the presentation in my reviews of theses that I was assigned to read, because I felt this was an important opportunity to make the student aware of the importance of clear and careful exposition. However, I doubt that I came close to Michael's example in this regard.

In 1976, Michael gave me one of the highest compliments I have received in my career. At the time, I was moving from Bell Laboratories to Harvard, and Michael encouraged his oldest son, Daniel, to enroll in Harvard as a graduate student, with the assumption that he would work with me. Indeed, Daniel came to Harvard, and I was his thesis advisor, although he also worked extensively with David Nelson, who was a member of Harvard's Society of Fellows at the time. Of course, Daniel proved to be one of the most outstanding students I would ever have, but I am not sure that I taught him much that he did not already know. I only wrote two papers with Daniel when he was a graduate student, but at least one of them stands out in my mind as one which has always given me particular satisfaction to recall. That paper, written together with Phil Platzman at Bell Laboratories, showed how resonances observed by Grimes and Adams in an experiment with electrons forming a Wigner crystal on the surface liquid helium could be understood as arising from a coupling of the lateral motion of the electrons to ripplons on the helium surface, and it showed how the frequencies of their frequencies could be used to obtain direct information on the lattice constant and the triangular structure of the electron crystal.[14,15]

I frequently recall one piece of personal advice that Michael gave me later in life, roughly at the time when he and Sorrel left their house and moved to a senior citizens community. He urged me to follow his example, to downsize and move to a supportive community sooner, rather than later, as, he said, the older you are, the harder such a move will become. So far, I have resisted this advice, as I think that my wife and I are still too young for this, but we shall see whether I come to regret my obstinance.

References

1. B. I. Halperin and P. C. Hohenberg, Scaling laws for dynamic critical phenomena. *Phys. Rev.* **177**, 952 (1969).
2. B. I. Halperin, P. C. Hohenberg and S. Ma, Calculation of dynamic critical properties using Wilson's expansion methods. *Phys. Rev. Lett.* **29**, 1548 (1972).
3. B. I. Halperin, P. C. Hohenberg and S. Ma, Renormalization group methods for critical dynamics; I. Recursion relations and effects of energy conservation. *Phys. Rev. B* **10**, 139 (1974).
4. K. G. Wilson and M. E. Fisher, Critical exponents in 3.99 dimensions. *Phys. Rev. Lett.* **28**, 240 (1972).
5. B. I. Halperin, P. C. Hohenberg and E. D. Siggia, Renormalization group calculations of divergent transport coefficients of critical points. *Phys. Rev. Lett.* **32**, 1289 (1974).
6. B. I. Halperin, P. C. Hohenberg and E. D. Siggia, Renormalization group treatment of the critical dynamics of superfluid helium, the isotropic antiferromagnet, and the easy-plane ferromagnet. *Phys. Rev. B* **13**, 1299 (1976).

7. E. D. Siggia, B. I. Halperin and P. C. Hohenberg, Renormalization-group treatment of the critical dynamics of the binary-fluid and gas-liquid transitions. *Phys. Rev. B* **13**, 2110 (1976).
8. B. I. Halperin, P. C. Hohenberg and S. K. Ma, Renormalization-group and methods for critical dynamics: II detailed analysis of the relaxational models. *Phys. Rev. B* **13**, 4119 (1976).
9. P. C. Hohenberg and B. I. Halperin, Theory of dynamic critical phenomena. *Rev. Mod. Phys.* **49**, 435 (1977).
10. S. K. Das, M. E. Fisher, J. V. Sengers, J. Horbach and K. Binder, Critical dynamics in a binary fluid: Simulations and finite-size scaling. *Phys. Rev. Lett.* **97**, 025702 (2006).
11. A. Aharony and B. I. Halperin, Exact relations among amplitudes at critical points of marginal dimensionality. *Phys. Rev. Lett.* **35**, 1308 (1975).
12. P. C. Hohenberg, A. Aharony, B. I. Halperin and E. D. Siggia, Two scale factor universality and the renormalization group. *Phys. Rev. B* **13**, 2986 (1976).
13. D. Stauffer, M. Ferrer and M. Wortis, Universality of second-order phase transitions: The scale factor for the correlation length. *Phys. Rev. Lett.* **29**, 345 (1972).
14. D. S. Fisher, P. M. Platzman and B. I. Halperin, Phonon-ripplon coupling and the two-dimensional electron solid on a liquid helium surface. *Phys. Rev. Lett.* **42**, 798 (1979).
15. C. C. Grimes and G. Adams, Evidence for a liquid-to-crystal phase transition in a classical, two-dimensional sheet of electrons. *Phys. Rev. Lett.* **42**, 795 (1979).

https://doi.org/10.1142/9789811282386_0008

Chapter 8

Some Reminiscences of Michael E. Fisher

A.J. Liu
Department of Physics and Astronomy, University of Pennsylvania, Philadelphia, PA 19104, USA

Working as a graduate student with Michael Fisher was one of the defining experiences of my life. It is hard to imagine how not only my career but also my character would be different if I had worked with someone else.

It is more than a little shocking to realize that I am older now than Michael was when I was in his group. To a beginning graduate student, the breadth and depth of his knowledge and understanding and his clarity of thought and expression were overwhelming. Above all, Michael was offended by sloppy thinking or incorrect statements. He would dress down anyone from a student to a seminar speaker for showing a plot without proper labels or with error bars that allowed for violations of thermodynamics. I remember an occasion when I had a meeting with him after a talk that ran late and wasted his time. In his annoyance, he decimated me and I felt utterly helpless — I simply could not think quickly enough and did not understand the physics well enough to defend myself. It was not easy to maintain confidence, let alone self-respect, under those conditions.

What saved me from quitting physics was that in the most important regard, he truly was gender-blind. He treated me exactly the same way he would treat a male student with the same level of ability. He only cared about the physics and getting it right.

However, he was, even for that time, somewhat old-fashioned in his beliefs about gender roles. A year before I planned to finish, I told him that I was engaged to marry a fellow Cornell physics graduate student, Doug Durian. His immediate response was, "Oh — then I've wasted all my time with you." I assured him that I was not planning to quit physics and fully intended to seek a postdoc position. To my great relief and to his credit, he did not treat me differently after this — he continued to spend as much time with me as he had before. However, when Doug and I started looking for postdoc positions, trouble arose. One day Michael came to my office to tell me that a former postdoc of his was interested in my application and had just called him up to discuss the possibility of offering me a postdoctoral fellowship. Michael informed me that he had declined on my behalf, since Doug did not have a postdoc offer at that university. I responded that it was my decision to make and that I would consider working at a different institution from Doug if that would increase our probability of finding positions together at a later stage. I asked him to call back. To his great credit he did, and eventually Doug and I both received postdoc offers there.

Shortly after this, I told Michael that Doug was offered a postdoc position at Harvard. Michael immediately offered to call David Nelson and ask him to give me a desk. This sounded a little strange to me, so I asked if he meant that he would be recommending me for a postdoc position. When I heard that he literally meant just a desk, I told him that I did not want to be a volunteer — that I had worked hard and wanted to be a professional and to be paid. He responded, "Every woman who has ever worked for me has worked at some point for free. I don't understand why you wouldn't. I love physics so much I would happily work for free." At this point, I lost control of my temper and blurted out, "You are the most highly paid Professor at the University of Maryland, and you can say that to me?" He turned bright red with anger. I apologized and left, and after a couple of uneasy days, we both pretended that we forgot this exchange.

I hope it is clear from these stories that Michael was willing to listen and reconsider. He not only respected but also supported what I wanted for my career and life, when I was his student and during the following decades. He was a complex and difficult person with extremely strong opinions, but he also maintained a remarkable flexibility of mind and could exhibit considerable caring and charm. He set a standard for attention to detail and thoroughness that was invaluable to learn, and above all, he set a standard for integrity as a person and a scientist that I will always remember.

Phase Transitions and Critical Phenomena

https://doi.org/10.1142/9789811282386_0009

Chapter 9

Fluctuation-Dominated Phase Ordering

M. Barma
TIFR Centre for Interdisciplinary Sciences,
Tata Institute of Fundamental Research,
Hyderabad 500046, India
barma23@gmail.com

Fluctuation-dominated phase ordering refers to a steady state in which the magnitude of long-range order varies strongly owing to fluctuations and to the associated coarsening phenomena during the approach to steady state. Strong fluctuations can lead to a number of interesting phenomena, including a cusp singularity in the scaled correlation function, implying the breakdown of the Porod Law. First identified in a nonequilibrium system of passively sliding particles on a fluctuating surface, fluctuation-dominated order also occurs in several other systems, including an equilibrium Ising model with long-range interactions. This chapter discusses these systems and others where clustering effects are stronger.

1. Meeting and Working with Michael Fisher

As a graduate student at Stony Brook in the early 1970s, I first heard about Michael Fisher from Helen Au-Yang, who was then a senior student working with Barry McCoy. She recommended strongly that I read Michael's 1967 Reports on Progress in Physics Review of Critical Phenomena.[1] This was my introduction to the subject, and I was captivated by Michael's description of singularities, scaling and universality, and the diversity of phenomena they describe.

My first encounter with Michael in person happened a few years later. I attended a summer school on critical phenomena and the renormalization group organized by Donald Betts in Banff, where Michael was one of the lecturers. Younger participants were asked to give short talks on their work, and I talked about the isostructural first-order phase transition in samarium sulphide, using Ising variables to represent the two possible valencies of the samarium ions. A prominent experimentalist criticized me strongly, saying the model was not realistic enough. But before I could say anything in reply, Michael spoke out strongly about the importance of simple models in clarifying complex phenomena and saved me from having to respond!

Then in the early 1980s, I had an opportunity to work with Michael when I spent a couple of years at Cornell as a research associate, on leave from the Tata Institute of Fundamental Research (TIFR) in Mumbai (then Bombay). We worked on the application of partial differential approximants to study singular corrections to scaling in two-dimensional Ising-like systems and the Gaussian to Ising crossover which showed some unusual features. It was a rewarding experience, and I learned a lot. Michael took every bit of science very seriously and was able to get the best out of his collaborators, no matter what their backgrounds or skills. While preparing a manuscript, he was meticulous. Every word had to be just right, and the figures I prepared would invariably be redrawn by him before being sent to the drafting section. He was extremely caring about the welfare of his students and postdocs. During summers, I was asked to interact with students who were yet to start on doctoral research. I wrote a paper with one of them, Steve White, which we requested Michael to read. It came back covered in red, with remarks all over and arrows to mark transpositions, along with the admonition, "Don't put the cart before the horse!"

I learned a lot from him, physicswise and otherwise, both abstract and practical. He was extremely organized, and instructed me on the virtues of colored separators in files (hard-bound, not soft). A discussion I remember well had to do with loyalty. Michael remarked that in addition to loyalty to the family and to the country, it is very important to be loyal to one's profession. In this connection, one should never refuse invitations to go and give lectures, including at colleges and schools. Over the years, that bit of advice has stayed with me and I have passed it on to my students, along with many other points I learned from him.

In 2007, he visited India and delivered a special commemorative lecture at the Tata Institute. The subject was phase transitions (Fig. 1). The lecture was preceded by a good deal of preparation by the auditorium staff, who had been asked to ensure there was a table with a minimum specified length and a pointer which was long enough. Michael later told me that he found the arrangements at TIFR among the best he had seen anywhere. But the arrangements at the hotel he stayed at with Sorrel and her sister Jacinta were not quite to his satisfaction. As was his wont, Michael moved the furniture in the hotel room around until he felt it was right. This sort of rearrangement was not unexpected, as it was an unusual but endearing habit that he was well known for.[2]

2. Introduction to FDPO

Michael's work encompassed the subject of phase transitions and critical phenomena in a wide variety of systems, ranging from multicritical points, where several critical loci meet, to multiphase points, from which emanate an infinite number of phases. As emphasized by him, criticality defines a new state of matter.[3] Different sorts of

Fig. 1. Michael Fisher delivering the Homi Bhabha lecture at the Tata Institute of Fundamental Research, Mumbai in 2007.

Source: Tata Institute of Fundamental Research. This photograph first appeared in *Current Science*, **123**, 817 (2022).

critical points represent different sorts of states, and each has its own particularness and peculiarities.

It is in this vein that the current chapter discusses *fluctuation-dominated phase ordering* (FDPO). Like the critical points discussed above, FDPO often arises at the separatrix between ordered and disordered phases in a number of equilibrium and non-equilibrium systems. But it differs from a normal critical state in a fundamental way. As explained in the sections that follow, the state is characterized by long-range order whose magnitude varies strongly owing to fluctuations; hence the name "fluctuation-dominated order."

In general, there are two properties which characterize FDPO. The first is that the distribution of the order parameter remains broad even in the thermodynamic limit. The second is that the two-point spin–spin or density–density correlation function depends on separation scaled by system size, and the scaling function exhibits a cusp singularity. These points are discussed in the following sections. First identified in a non-equilibrium system of passively sliding particles on a fluctuating surface,[4,5] FDPO has been found in several other non-equilibrium systems as well as an equilibrium Ising model with long-range interactions.[6]

2.1. *Broad distribution of the order parameter*

In a typical phase ordering system which exhibits two phases, a large finite system in steady state resides in one phase for a long time before switching over to the other. For instance, in an Ising ferromagnet, if s_j is the spin at site j, then the order parameter with non-conserved dynamics is $m = \frac{1}{L}\sum_{j=1}^{L} s_j$. The probability distribution of m is sharply peaked around the two values m_s or $-m_s$, with very infrequent transitions between the two. The peak widths around $\pm m_s$ approach zero in the thermodynamic limit $L \to \infty$.

By contrast, in the FDPO steady state, the system shows large fluctuations without losing macroscopic order. Thus, the order parameter distribution remains broad:

$$Lim_{L\to\infty} P(m) = f(m) \qquad \text{[FDPO]}. \tag{1}$$

The limiting distribution $f(m)$ does not narrow down, and has support in a broad range of m, even in the thermodynamic limit.

If the order parameter is conserved, the magnetization is constant in both ordered and disordered phases. Quantities which are sensitive to the difference between phases are the long-wavelength Fourier modes $Q_k = \frac{1}{L}\sum_{j=1}^{L} |e^{iqj}(1+s_j)/2|$, with $q = \frac{2\pi k}{L}$ and k a small integer. For a given conserved value of magnetization m, the mode Q_1 is an appropriate order parameter. In an Ising ferromagnet, the probability distribution of Q_1 is sharply peaked at a single value. But in the FDPO state with a conserved order parameter, the distribution of Q_1 remains broad in the thermodynamic limit, as do the distributions of $Q_2, Q_3, \ldots$[7]

Another important characteristic of the order is obtained from the asymptotic value of the two-point spin–spin correlation function $C(r) = \langle s_0 s_r \rangle$ in the absence of a field. At large separations r, the correlation function is expected to decouple

$$Lim_{r\to\infty} Lim_{L\to\infty} \langle s_0 s_{0+r} \rangle = \langle s_0 \rangle \langle s_{0+r} \rangle = {m_c}^2. \tag{2}$$

A non-zero value of m_c indicates that the system has long-range order. In a finite system, C is a function of the scaled variable r/L for large enough values of r and L. The value ${m_c}^2$ can be read off as the intercept in a plot of C versus r/L. While m_c and m_s coincide in systems such as the ferromagnetic Ising model, this is typically not the case with FDPO.

2.2. *Singularity in the scaled correlation function*

As discussed above, FDPO often arises at the separatrix between ordered and disordered phases. Normal critical behavior is characterized by a power-law decay $C(r) \sim r^{d-2+\eta}$, where η is the Fisher exponent; by contrast, in the FDPO state, the correlation function is a function of r/L rather than r alone. It shares some features with phase ordering systems (scaling as a function of r/L) and other features with critical systems (strong fluctuations, power law decays, singular scaling functions).

To appreciate the significance of the cusp singularity in C, first let us recall some facts about phase ordering in a system with regular ferromagnetic Ising order. When the order parameter is conserved, the low-T state is phase separated, with each phase occupying a macroscopic region. An interface of width W separates the phases, where W is of the order of correlation length ξ. Thus, on length scales $\xi \ll r \ll L$, the interface is structureless and sharp. If further we consider $r/L \ll 1$, then $C(r)$ takes the form $C(r) \approx {m_c}^2(1 - 2r/L)$. The linear fall can be understood as follows: spatial averaging of $s_o s_r$ produces $+{m_c}^2$ with probability $(1 - r/L)$ if points 0 and r lie within a phase and $-{m_c}^2$ if they lie across an interface. Further, the linear drop in r/L implies that the structure factor $S(q)$ (the Fourier transform of $C(r)$) is given for large-scaled wavevectors $qL \gg 1$ by $S(q)/L^d \sim 1/(qL)^{d+1}$. This form of the decay of $S(q)$ is known as the Porod law. The Porod law is well established in scalar order parameter systems with a sharp interface and is widely observed.[8,9]

A prominent signature, found in most systems that exhibit FDPO, is the singular behavior of the two-point correlation function $C(r, L)$ between points separated by a distance r in the steady state of a system of size L. As for a regular ordered state, the data show a scaling collapse when plotted against the ratio r/L, and the steady state scaling function C_{ss} has a finite intercept at the origin (Fig. 2). The intercept gives the value of long-range order in the system, as the ratio r/L is driven to zero as $L \to \infty$ for every value of the separation r. The key distinguishing feature of FDPO is the 'cuspy' approach to $r/L = 0$, as seen in Fig. 2. The steady-state scaling function shows a cusp singularity:

$$C_{ss} \approx {m_0}^2 - b_{ss}\left(\frac{r}{L}\right)^{\alpha} \qquad (r/L \to 0) \qquad \text{[FDPO]}, \tag{3}$$

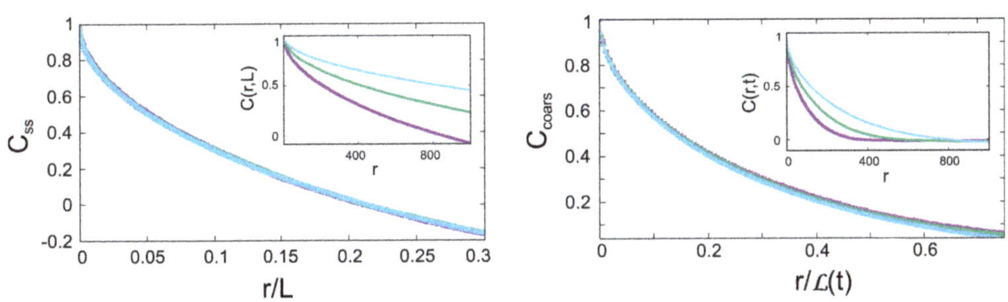

Fig. 2. In most systems which exhibit FDPO, the two-point correlation function is a singular function of the scaled separation. The data shown are for the coarse-grained depth (CD) model, which is defined in Section 3. In the left panel, the separation r is scaled by the system size L in steady state (with $L = 4000, 8000, 16000$), while the right panel shows scaling by a time-dependent length scale $\mathcal{L}(t)$ while coarsening (with $\mathcal{L}(t) = 68, 100, 147$). In both cases, the scaling function exhibits a cusp singularity at the origin, indicating that the Porod law beaks down. The unscaled data are shown in the insets.

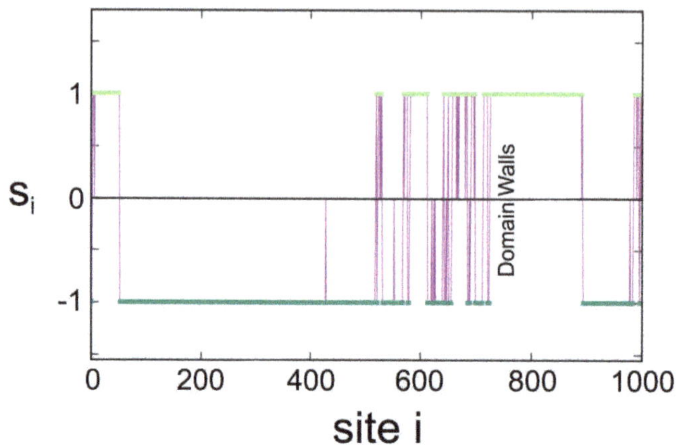

Fig. 3. A typical configuration in the steady state of the coarse-grained depth model defined in Section 3. The horizontal lines depict regions of like spins, while the vertical lines are domain walls. Macroscopic stretches of like spins are responsible for long-range order, while spins are scrambled in disordered regions with many domain walls. Large disordered regions are responsible for the breakdown of the Porod law.

where the cusp exponent $\alpha < 1$. This feature implies the breakdown of the Porod law discussed above, according to which the intercept is approached linearly ($\alpha = 1$). As argued above, the Porod law relies only on the existence of a well-defined interface separating coexisting phases, with the interface width being finite. That the law no longer holds implies that in FDPO, the region between coexisting ordered phases is also proportional to L and occupies a fluctuating but finite fraction of the system. This is illustrated in Fig. 3 which shows a typical configuration of the CD model in steady state. Two features stand out: (i) the occurrence of some very long (horizontal) stretches of like spins; (ii) these long stretches are separated from each other by a large disordered region where spins are scrambled up owing to a conglomeration of close-by domain walls. The longest stretches of like spins are of order system size L, consistent with the occurrence of long-range order. On length scales set by L, the scrambled-spin regions are disordered, and the largest of these are associated with the departure from the Porod law.

A similar situation also arises while the system is coarsening. Consider an infinite system evolving in time, starting from a completely disordered state. As known from studies of phase ordering dynamics,[9] the system reaches a local quasi-steady state on length scales smaller than $\mathcal{L}(t) \sim t^{1/z}$, where z is the dynamic exponent. In particular, we expect the analogue of Eq. (1) to hold, with system size L replaced by the length scale $\mathcal{L}(t)$. The right panel of Fig. 2 shows that this is indeed the case, with the coarsening data showing that the scaling function C_{coars} has a cusp as a function of $r/\mathcal{L}(t)$.

2.3. *Systems which exhibit FDPO*

Most systems which show FDPO show both a broad order parameter distribution and a singular scaled correlation function. In exceptional cases, one of these may be suppressed,[10] but we will not consider such cases in the following. We summarize the content and principal results of the sections that follow.

Three simple models which exhibit FDPO are discussed in Section 3. The first is a model of particles sliding on a fluctuating surface, where numerical studies establish FDPO in the system. Next, motivated by the extreme adiabatic limit of this model, we study the coarse-grained depth (CD) model of the fluctuating surface. Both the broad order parameter distribution and the singularity in the scaling function can be established analytically in this case. Finally, we study the truncated inverse distance squared Ising (TIDSI) model, an Ising model with long-range interactions that act within a cluster of like spins. Along a part of the critical locus in this model, the system shows FDPO, and the cusp exponent can be determined analytically. Interestingly, it is found to depend continuously on the coupling constants. This model is also discussed in the article on mixed-order phase transitions in this volume.[11]

In Section 4, several issues related to FDPO are discussed, including the imprint of FDPO on time-dependent properties. The autocorrelation function in the steady state of the model shows the analogue (in the time domain) of the cuspy behavior discussed above for the equal-time two-point correlation function. Further, while the system is coarsening, it shows an interesting aging behavior which is captured by a two-time correlation function.

Also discussed in Section 4 is *fluctuation-dominated clustering* (FDC), which generalizes FDPO to describe situations with much more intense clustering. FDC is accompanied by even larger fluctuations, and the cusp singularity of Eq. (3) is replaced by a divergence in this case. Finally, several systems which display FDPO and FDC are discussed, including passive particles carried by a compressible fluid, lipid clustering in cell membranes, coarsening in the Vicsek model, active nematics, vibrated rods and inelastically colliding particles.

3. Models of FDPO

3.1. *Particles sliding on a fluctuating surface*

Consider a system of particles which lie on a fluctuating surface and slide downward on it (Fig. 4). The time evolution of the surface is assumed to be governed by either Edwards Wilkinson (EW) or Kardar Parisi Zhang (KPZ) dynamics.[12,13] The particles are guided downward by the surface gradient, but are passive in that they do not influence the time evolution of the surface. While the sliding particle (SP) model can be defined and studied in any dimension, many of the most interesting features are seen in one dimension. Results for the two-dimensional case are discussed at the end of this section.

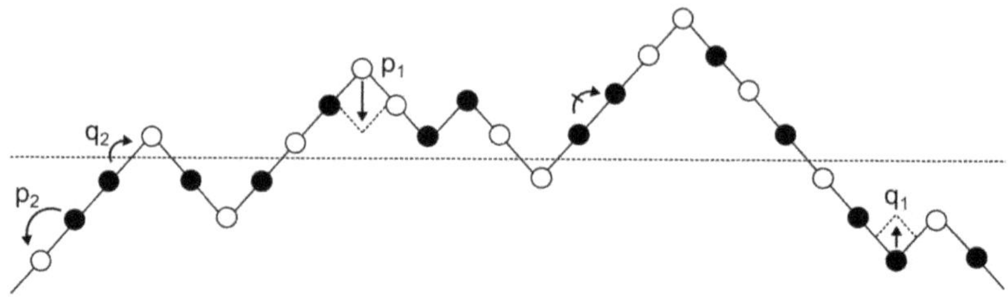

Fig. 4. The sliding particle (SP) model describes passive particles sliding on a fluctuating surface. The surface moves involve a local hill becoming a local valley, and vice versa, while a particle can hop onto a neighboring vacant site, with downward hops occurring at a greater rate. Also shown is a horizontal cut, required for the definition of the CD model, in which the state on site are decided by the height of that site with respect to the cut.

In a lattice representation of the one-dimensional model, particles (which are taken to have a hard-core interaction which enforces exclusion) occupy the sites of a one-dimensional lattice with periodic boundary conditions whose bonds represent the local slopes of the surface. The surface dynamics is that of the single-step model in which corners are flipped stochastically,[14] i.e. $\wedge \to \vee$ with rate p_1, while the reverse transition $\vee \to \wedge$ occurs with rate $q_1 \leq p_1$ (Fig. 4). Equal rates $q_1 = p_1$ correspond to EW dynamics, while unequal rates lead to KPZ surface dynamics. The rate of particle–hole exchanges on nearest-neighbor sites depends on the state of the bond in between. If the exchange across a bond results in a lower location of the particle, then the move occurs with rate p_2, while the reverse transition occurs with rate $q_2 < p_2$.

Surface fluctuations lead to particle clustering. The reason is that particles easily slide down a surface fluctuation which creates a valley; but, should the reverse surface fluctuation occur, it is very difficult for particles to climb up a hill as the process involves the slow process of activation against a field. Given this propensity to promote particle clustering, it is interesting to ask about the large-scale behavior as time increases.

Starting from an initially random placement of particles on the surface, the evolution of the local density can be tracked by numerical simulation. In a system with N_P particles, particularly revealing is the evolution of the two-point density–density correlation function $C(r,t) \equiv \langle n_i n_{i+r} \rangle - (N_P/L)^2$. It is found that the data at different times collapse, on scaling the separation by a growing length scale $\mathcal{L}(t) \sim t^{1/z}$, much as in the right panel of Fig. 2, which refers to the coarse-grained depth (CD) model defined in the following section. This is expected, as catchment areas for particles are provided by fresh valleys which form when coarse-grained hills overturn, and the bases of such hills grow as $t^{1/z}$, where z is the dynamic exponent which governs surface fluctuations ($z = 2$ for EW evolution, $z = 3/2$ for KPZ evolution). In the scaling limit, the data collapse to a single function $C_{\text{coars}}(y = r/\mathcal{L}(t))$. The intercept indicates the value of long-range order ${m_c}^2$. The approach to the intercept exhibits a cusp singularity $C_{\text{coars}}(y) = m_c{}^2 - by^{\alpha}$, which signals the breakdown of the Porod law, according to which C_{coars} should fall linearly.

In a finite system of length L, the system reaches the steady state in a time of order L^z. In steady state, the two-point correlation function $C(r, L)$ varies with separation r on the scale of system size L, and the data for different sizes L show a collapse when plotted as a function of r/L in the scaling limit $r \to \infty, L \to \infty$ with r/L held fixed (as in the left panel of Fig. 2, which refers to the CD model). The scaling function C_{ss} for steady-state correlations shows a cusp singularity as $y = r/L$ approaches zero, $C_{\text{ss}}(y) = m_0{}^2 - by^{\alpha}$, pointing to FDPO. This is to be contrasted with customary phase ordering, where $C(r) = m_c^2(1-2r/L)$ for $r/L \ll 1$, on length scales r much larger than the correlation length ξ, which determines the width of the interface.

An interesting point is that the SP model actually lies on the phase boundary separating a fully ordered state from a disordered state. With EW driving, this is the phase boundary of the Lahiri–Ramaswamy model, a lattice model of sedimenting colloidal crystals,[15] while with KPZ driving, this is the phase boundary separating disordered and ordered states of the light–heavy (LH) model.[16,17] In both cases, the SP model shows FDPO, while the order parameter shows a first-order 0–1 jump from the disordered to the ordered phase. This is consistent with the transition being of mixed-order, but in order to clinch the mixed- order character (see Section 3.3), the divergence of the correlation length on approaching the critical locus from the disordered side needs to be shown for the SP model.

In two dimensions, a study of hard-core particles sliding down a surface evolving through KPZ dynamics shows that particles cluster together on a length scale which grows in time.[18] The final fluctuation-dominated phase-separated steady state is characterized by a power-law distribution of cluster sizes of particles and holes, though the two are not identical, unlike in 1D. The scaled correlation function shows a cusp singularity and the order parameter distribution remains broad in the large-size limit. We conclude that FDPO is present in this case as well.

3.2. *The coarse-grained depth model*

Michael Fisher always emphasized the importance of simple models, preferably those that can be analyzed analytically, to shed light on the phenomenon of interest. In the context of FDPO, this role is played by the CD model of a fluctuating surface.[4,5,22]

The CD model is motivated by considering the extreme adiabatic limit of the sliding particle (SP) model considered above. In this limit, particles tend toward the minimum energy configuration for a particular height profile. Since the particles have hard-core interactions, this state corresponds to filling the height profile, bottom upwards; the topmost particles define a horizontal level below which all sites are occupied by particles.

We define the CD model by taking a horizontal cut on a fluctuating surface with zero net slope (Fig. 4). For a given configuration of the surface, if the cut is at height $H(t)$, we define a spin variable at site i as $s_i(t) = -sgn[h_i(t) - H(t)]$. Evidently $s_i(t)$ can take values $+1$, -1 or 0 depending on whether the height at the ith site is below, above or at the cut. It provides a coarse-grained description of the surface height. The time evolution of s_i is induced by the dynamics of the underlying surface. Invoking the correspondence with the adiabatic SP model, sites with $s_i = 1$ would correspond to occupied sites. However, the correspondence is not exact, as $\sum s_i$ is not conserved by the dynamics.

Different choices of the reference height $H(t)$ correspond to different variants of the CD model.[5,22] Two useful choices are as follows: (i) the *CD2 model*, in which the reference level is chosen to be at the instantaneous height of the site $i = 0$, i.e. $H(t) = h_{i=0}(t)$. This choice leads to a useful mapping to random walks. (ii) The *CD3 model*, in which the reference level is chosen to be at the instantaneous mean height $H(t) = \langle h(t)\rangle = \frac{1}{L}\sum h_i(t)$.

There is a simple mapping between each surface configuration in the steady state of the CD2 model and a random walk (RW) trajectory, with a site label in the former representing time in the latter, and the local surface slope $+1$ or -1 representing a rightward or leftward RW step. Several results follow immediately from the mapping: (a) The length of a cluster of like spins maps to the time between successive returns to the origin, implying that the distribution of cluster lengths $P(l)$ in steady state $\sim l^{-3/2}$ for $1 \ll l < L$. (b) For the CD2 model with periodic boundary conditions, the problem is isomorphic to the Brownian bridge.[19] (c) As the longest return time for a T-step RW is of order T, it follows that the length $l_{\max}$ of the longest cluster in the CD2 model is of order L. The distribution of the longest cluster $P(l_{\max})$ is a function of the scaled variable $l_{\max}/L$.[19,20] Thus, both the mean and standard deviation of $l_{\max}$ are of $O(L)$. (d) The order parameter $\tilde{m} = \frac{1}{L}\sum s_i$ maps to the excess time a random walker spends on one side of the origin. This distribution is known to be uniform,[21] implying that the order parameter distribution in Eq. (1) is independent of $\tilde{m}$. (e) Knowing $P(l)$ allows the two-point correlation function to be calculated within the independent interval approximation.[4,5] The result is that $C(r, L)$ is a scaling function of $y = r/L$, which, for small argument, reduces to Eq. (2) with the cusp exponent $\alpha = 1/2$.[5] We conclude that the CD2 model exhibits the hallmarks of FDPO: a broad distribution of the order parameter and a cusp singularity in the scaled correlation function. These features are confirmed by numerical simulation for the CD3 model as well.[22] For instance, Fig. 2 shows that $C(r, L)$ is a scaling function of r/L with a cusp singularity for small argument.

We expect these steady-state results for the CD models to generalize to the coarsening regime, with the coarsening length scale $\mathcal{L}(t)$ taking the place of L. Indeed, the cluster size distribution shows a power-law decay $P(l) \sim l^{-3/2}$ with an exponential cut-off at $l \sim \mathcal{L}(t)$.[22] Further, as shown in the right panel of Fig. 2, the two-point correlation function of the CD3 model during coarsening is a scaling function of $r/\mathcal{L}(t)$, with a cusp singularity at small argument.

Finally, we remark that the CD model can be generalized to higher dimensions by taking a cut of an evolving self-affine surface. On a 2D KPZ surface, there is a cusp singularity in the scaled correlation function and a broad distribution of the order parameter, pointing to FDPO.[18]

3.3. *FDPO at a mixed order transition: Ising model with long-range interactions*

Fluctuation-dominated order also occurs in an equilibrium system, namely a 1D Ising model with truncated power-law decaying interactions.[6] This model shows a *mixed-order transition*, characterized by a discontinuity of the order parameter across a critical locus, along with a divergent correlation length from the disordered side; FDPO occurs along part of the critical locus (Fig. 4).

In the truncated inverse-square decay Ising (TIDSI) model, a nearest-neighbor interaction $-J_{NN}\sigma_j\sigma_{j+1}$ is present, along with a long-range interaction term $-J(i-j)\sigma_i\sigma_j I(i\ j)$, where $I(i\ j) = 1$ if sites i and j belong to the same domain of all up or all down spins.[23] The coupling $J(r)$ is taken to be $J(r) \approx J_{LR}/r^2$. The Hamiltonian is thus

$$\mathcal{H} = -J_{NN}\sum_{i=1}^{L}\sigma_i\sigma_{i+1} - \sum_{i<j} J(i-j)\sigma_i\sigma_j \prod_{k=i}^{j}\frac{1+\sigma_k\sigma_{k+1}}{2}. \tag{4}$$

The analysis of the model is described in the article on mixed-order transitions which appears in this volume,[11] and thus, we give only a brief account in the following. It is convenient to go to a domain representation, in which a configuration is described by a set of domains $(l_1, l_2, \ldots, l_N)$ with each domain being a stretch of parallel spins. The number of domains N can vary from one configuration to another. In the domain representation, the Hamiltonian is rewritten in terms of $\{l_i\}$, and the partition function can be expressed as

$$Z_y(L) = \sum_{N=1}^{\infty} y_0^N \sum_{l_1=1}^{\infty} \cdots \sum_{l_N=1}^{\infty} \prod_{n=1}^{N} \frac{1}{l_n^c}\ \delta_{\sum_{n=1}^{N} l_n,\, L}. \tag{5}$$

where $y_0 = \exp -\beta\Delta$ is the fugacity of a domain wall and $c = \beta J_{LR}$. Here, β is the inverse temperature, while $\Delta = 2J_{NN} + J_{LR} + b_0$. The analysis of the partition function as a function of the coupling constants c and y_0 leads to the phase diagram shown in Fig. 5, with the critical locus given by the equation $y_0 = 1/\zeta(c)$, where ζ is the Riemann zeta function. In the following, we discuss the behavior in the disordered and ordered phases and along the critical locus.[6]

In the disordered phase, the net magnetization $m_0 = \frac{1}{L}\sum\langle s_i\rangle$ vanishes. The probability distribution of domain lengths has an exponential cutoff, $p(l) \sim exp(-l/\xi)$ which serves to define the correlation length ξ. As the critical line

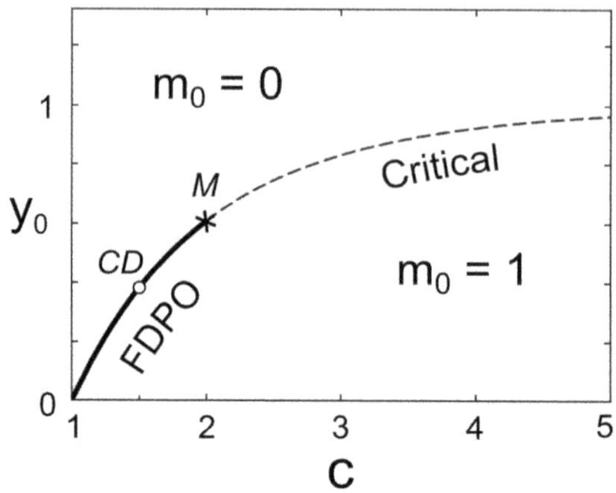

Fig. 5. Phase diagram of the TIDSI model in the $(c - y_0)$ plane, with $c = \beta J_{LR} > 1$. The critical line separates the paramagnetic $m_0 = 0$ phase from the ferromagnetic $m_0 = 1$ phase. The portion of the critical line with $1 < c < 2$ exhibits FDPO. The multicritical point M marks the change from FDPO to normal critical behavior, while at the point marked CD, the large-distance behavior is the same as that of the CD model. (Adapted from Ref. 6).

is approached, the domain wall density approaches zero and $\xi \sim t^{-\nu}$ where $t = (T - T_c)/T_c$ is the reduced temperature; the value of the critical exponent is $\nu = 1$ for $c > 2$ and $\nu = 1/(c-1)$ for $1 < c < 2$.

In the ordered phase, the value of m_0 is 1, as the fraction of sites in the longest cluster approaches unity in the limit $L \to \infty$;[24] the state is one with near-perfect order. Thus, m_0 shows a 0–1 jump, while $\xi \to \infty$ in the disordered phase, implying a mixed-order phase transition across the critical line.

Along the critical locus, the system exhibits two distinct behaviors.[6] For $c > 2$, the system shows normal critical behaviour, whereas it exhibits FDPO in the range $1 < c < 2$. In the former case, the mean cluster size $\langle l \rangle$ is finite, and the mean size of the largest cluster is a vanishing fraction of the system size as $L \to \infty$. The correlation function $C(r, L)$ decays asymptotically as a simple power law $\sim r^{-(c-2)}$. On the other hand, for $1 < c < 2$, the mean size of the largest cluster $\sim L$.[24] The correlation function can be found from the cluster size distribution within the independent interval approximation. The result is $C(r, L) \approx 1 - (c-1)^{-1} \left(\frac{r}{L}\right)^{2-c}$ which is of the FDPO form of Eq. (1). Interestingly, the cusp exponent α depends continuously on the parameter $c = \beta J_{LR}$.

These results for the steady state suggest that a similar form would hold for the TIDSI model while coarsening, with system size L replaced by the coarsening length $\mathcal{L}(t)$. This has been checked by numerical simulation using single spin-flip (Glauber) moves for the dynamics.[25]

The domain representation of the probability of a configuration brings out an interesting connection between the TIDSI and CD models, namely that the steady-state properties of the CD model correspond to those of the critical TIDSI model with $c = 3/2$.[22] However, this correspondence for static quantities does not extend to the dynamics, as the interface-driven dynamics of the CD model is quite different from the single-spin flip dynamics in the TIDSI model.

4. Discussion and Conclusion

In the following, we discuss several issues concerning FDPO and related states. In turn, we discuss the distinctive imprint of FDPO on dynamic properties; a generalization of FDPO in which clustering is much more intense, namely fluctuation-dominated clustering (FDC); the occurrence of non-Porod behaviour in systems with quenched disorder; and finally, examples of several physical phenomena and models in which fluctuation-dominated states are key.

4.1. *Dynamic manifestations of FDPO*

Besides its effect on spatial correlation functions, FDPO has a strong effect on dynamics. This is true both for the autocorrelation function in steady state and for aging correlation functions during coarsening for both SP and CD models.[26] Interestingly, the cuspy behavior of scaled spatial correlation functions is inherited by dynamic correlations.

We define the autocorrelation function as $A(t) = \langle s_i(0)s_i(t)\rangle$. It is found that $A(t, L)$ is a scaling function of t/L^z, where z is the dynamic exponent governing the surface dynamics. For $t/L^z \ll 1$, we find

$$A(t, L) \approx {m_c}^2 - a\left(\frac{t}{L^z}\right)^{\beta'}. \tag{6}$$

In the cases considered, we find that the exponent β' is smaller than 1, implying that cusp singularities in the scaling function are to be found in dynamics as well. In the case of the CD model, $A(t, L)$ can be calculated analytically for the CD3 model driven by an EW surface.[26] One finds that Eq. (6) holds, with ${m_c}^2 = 1$, $\beta' = 1/4$, $z = 2$. With KPZ evolution for the surface, Monte Carlo simulation results for $A_{CD}(t)$ show that Eq. (6) holds with $z = 3/2$ and the cusp exponent $\beta' \simeq 0.31$. For the SP models too, numerical simulations confirm the validity of Eq. (6). Finally, the two-time correlation function which characterizes aging dynamics in the coarsening regime can be calculated analytically in all regimes for the CD3 model with EW driving and determined numerically in other cases.[26]

4.2. *More intense clustering: Fluctuation-dominated clustering (FDC)*

In FDPO, the two-point correlation function exhibits scaling when separations are scaled by L in steady state (or by $\mathcal{L}(t)$ during coarsening). In either case, the scaling function exhibits a cusp singularity in the limit that the scaled separation $y = r/L$ (or $r/\mathcal{L}(t)$) approaches zero.

However, a cusp is not the only possibility.[27] In a number of systems, $C(r, L)$ shows a stronger singularity, namely a divergence:

$$C(r, L) \approx L^{-\mu} Y\left(\frac{r}{L}\right), \tag{7}$$

with

$$Y(y) \sim y^{-\nu} \qquad (y \to 0) \qquad [FDC] \tag{8}$$

The divergence indicates much stronger clustering. It embodies two effects: (i) the formation of clusters, each made of a large number of particles; (ii) the clustering of clusters close to each other. The clusters are dynamic entities which keep changing in time, as they make and break. Correspondingly, the state is accompanied by large fluctuations and exhibits spatial and temporal intermittency, i.e. it shows *fluctuation-dominated clustering* (FDC).

In the context of particles sliding on a fluctuating surface, FDC arises when the particles are noninteracting. Numerical simulations show that particles move toward valley bottoms, and since there is no restriction on the number n_i of particles on each site i, the degree of clustering is much larger than in FDPO.[28] Parenthetically, we note that the "strong clustering states" discussed in this reference are identical to the FDC states under discussion here. The divergence in Eq. (8) is strongest in the case of a KPZ surface, in which case $\nu = 3/2$ and $\mu = 1/2$. With driving by an EW surface, Eqs. (8) and (9) hold with $\mu = 0$ and $\nu \simeq 2/3$. In both cases, there are strong signatures of intermittency in space and time.[29]

The scaling function can be found analytically in the adiabatic limit when the surface profile is stationary; the problem reduces to the Sinai problem of random walkers in a random medium, which itself is the trajectory of a random walk. In an analytic treatment of the steady state of the model, the sitewise distribution of particle numbers was derived, as well as the two-point correlation function which was shown to be of the FDC scaling form (Eqs. (7) and (8)).[28,30] The results agree well with numerical results for non-interacting particles sliding down a KPZ surface[30] and are consistent with earlier work on random walkers in a Sinai landscape.[31]

We note an interesting point regarding the connection between giant number fluctuations and the form of the scaling function.[32] While singularities in the small $y \equiv r/L$ behavior (cusps for FDPO, divergences for FDC) indicate a departure from the Porod Law, the correlation function may, in some cases, exhibit a different power-law decay for large values of y. Now, fluctuations of the total number of particles in a given volume involve a sum over correlation functions in that region, and this sum is dominated by the contribution from large separations. In such cases, it is the large-y power law in the scaling function which is pertinent to determining number fluctuations.[32]

4.3. *Disordered systems which show non-Porod behavior*

Correlation functions in some systems with quenched disorder show strong deviations from the Porod law. It is useful to distinguish between two classes of quenched systems. In the first class, we are interested directly in the correlation functions of the quenched degrees of freedom, while in the second class, the quenched disorder provides a background for other degrees of freedom, and we are interested in the correlations in the latter set.

In the first class, the degrees of freedom are frozen. An important example of this class is porous materials, in which two-point correlation functions have been studied extensively by scattering techniques, and non-Porod behavior has been established.[33,34] In a different context, deviations from Porod behavior are also seen in the microstructure due to mineralization in mammalian cortical bone.[35]

A well-studied example in the second class is the Ising model in a quenched random field. The degrees of freedom in this case are spins which evolve in time in a quenched random background. Monte Carlo simulations performed to study ordering dynamics with conserved order parameter reveal that the correlation function exhibits scaling and shows a cusp singularity at small argument.[36]

In a broad sense, the second class of systems has some points of resemblance with a limiting case of fluctuation-dominated order discussed in earlier sections, even though the microscopic dynamics is completely different. In particular, recall the adiabatic limit of particles sliding down fluctuating surfaces, in which particles move on a frozen surface configuration, with the average over configurations being done later. The earlier-described treatments of the CD model (in the case of interacting particles) and random walkers in a frozen Sinai potential (for non-interacting particles) may then be viewed as dealing with problems of quenched disorder.

An interesting point brought out by studies on non-Porod behaviour in disordered systems is that the cusp exponent is related to fractal structure.[33,34,37] It would be interesting to establish a similar correspondence in FDPO states by studying the fractal structure of the disordered regions between long-ordered stretches, as shown in Fig. 3.

4.4. *Other systems which exhibit FDPO and FDC*

Besides the examples of sliding particles on fluctuating surfaces and long-range Ising models that we considered earlier, fluctuation-dominated states of the FDPO and FDC variety arise in diverse situations. Some of these are discussed in the following.

Passive particles carried by a compressible fluid: A highly compressible fluid described by a noisy vortex-free Burgers equation can be mapped to a surface evolving through KPZ dynamics.[38] This enables results for FDPO and FDC for passive sliders on fluctuating surfaces to be translated to describe clustering of passive particles in the Burgers fluid.

Lipid clustering in cell membranes: The collective dynamics of active polar filaments near the surface of a cell can lead to the formation of asters, which form and disappear at a certain rate, but survive long enough to bring about lipid clustering.[39] The mechanism by which lipids are drawn into the center of the aster resembles particles falling into valleys in sliding particle models. A large region in parameter space shows fluctuation-dominated states, with a violation of the Porod law and macroscopic fluctuations in steady state.

Coarsening in the Vicsek model: Studies of the coarsening dynamics of the Vicsek model of bird flocking show that there are two distinct values of the dynamic exponent z for the velocity and density fields and, correspondingly, two distinct scaling functions.[40] Interestingly, the scaling function for density exhibits FDPO with a cusp singularity, while the scaling function for velocity shows a divergence, characteristic of FDC.

Active nematics: A model of passive particles advected by a nematic field shows FDPO.[41] For a model of apolar rods whose movement depends on the orientation, Monte Carlo studies suggest that the density of rods shows FDPO,[32] though a hydrodynamic calculation concludes that active currents can destroy phase separation in two dimensions.[42]

Vibrated rods: Experiments on rods vibrated in a tray show giant fluctuations of the density.[43] A subsequent analysis[32] shows that the data are consistent with standard deviation growing proportionally to the mean, with sub-leading corrections coming from FDC-like corrections in the density.

Inelastically colliding particles: The coefficient of restitution for inelastically colliding particles depends on the relative velocity of approach. In a model which incorporates this effect, the free collapse of a 1D system of particles results in a state in which the structure functions of density and velocity both depart from the Porod law.[44]

4.5. *Conclusion*

FDPO embodies the idea of a phase with long-range order, whose dynamics is dominated by large fluctuations of the order of the mean, resulting in a broad distribution of the order parameter. In most systems which exhibit FDPO, the two-point correlation function is a function of the distance scaled by system size or coarsening length, and the scaling function is singular — showing a cusp for FDPO, and a divergence for FDC, which describes systems with more intense clustering. These themes — systems with strong fluctuations, resulting in singularities which may be cusps or divergences — recall the life and legacy of Michael Fisher.

Acknowledgments

I gratefully acknowledge collaborations over the years on FDPO and related matters, with Dibyendu Das, Satya Majumdar, Manoj Gopalakrishnan, Sakuntala Chatterjee, Apoorva Nagar, Rajeev Kapri, Malay Bandyopadhyay, Shauri Chakraborti, David Mukamel, Tapas Singha, Samvit Mahapatra, Kabir Ramola, Souvik Sadhukhan, Saroj Nandi and Arghya Das. I am also grateful to Anusheela Chatterjee for the useful inputs. This work was supported by the Department of Atomic Energy, India through the DAE Homi Bhabha Chair Professorship and intramural funds at TIFR Hyderabad.

References

1. M. E. Fisher, The theory of equilibrium critical phenomena. *Rep. Prog. Phys.* **30**, 615 (1967).
2. N. D. Mermin, My life with Fisher. *J. Stat. Phys.* **110**, 467 (2003).
3. M. E. Fisher, Proceedings of the Robert A. Welch Conferences on Chemical Research XXIII, Modern Structural Methods, Houston, Texas (1979).
4. D. Das and M. Barma, Particles sliding on a fluctuating surface: Phase separation and power laws. *Phys. Rev. Lett.* **85**, 1602 (2000).
5. D. Das, M. Barma and S. N. Majumdar, Fluctuation-dominated phase ordering driven by stochastically evolving surfaces: Depth models and sliding particles. *Phys. Rev. E* **64**, 046126 (2001).
6. M. Barma, S. N. Majumdar and D. Mukamel, Fluctuation-dominated phase ordering at a mixed order transition. *J. Phys. A: Math. Theor.* **52**, 254001 (2019).
7. R. Kapri, M. Bandyopadhyay and M. Barma, Order-parameter scaling in fluctuation-dominated phase ordering. *Phys. Rev. E* **93**, 012117 (2016).
8. G. Porod, *Small Angle X-Ray Scattering*, (edited by O. Glatter and L. Kratky) (Academic Press, New York, 1983).
9. A. J. Bray, Theory of phase-ordering kinetics. *Adv. Phys.* **43**, 357 (1994).
10. C. Iyer, A. Das and M. Barma, Coarsening, condensates, and extremes in aggregation-fragmentation models. *Phys. Rev. E* **107**, 014122 (2023).
11. D. Mukamel, Mixed order phase transitions. this volume (2023).
12. S. F. Edwards and D. R. Wilkinson, The surface statistics of a granular aggregate. *Proc. R. Soc. Lon. Ser. A* **381**, 17 (1982).
13. M. Kardar, G. Parisi and Y.C. Zhang, Dynamic scaling of growing interfaces. *Phys. Rev. Lett.* **62**, 89 (1986).
14. M. Plischke, Z. Rácz and D. Liu, Time-reversal invariance and universality of two-dimensional growth models. *Phys. Rev. B* **35**, 3485 (1987).
15. R. Lahiri, M. Barma and S. Ramaswamy, Strong phase separation in a model of sedimenting lattices. *Phys. Rev. E* **61**, 1648 (2000).
16. S. Chakraborty, S. Chatterjee and M. Barma, Ordered phases in coupled nonequilibrium systems: Static properties. *Phys. Rev. E* **96**, 022127 (2017).
17. S. Mahapatra, K. Ramola K and M. Barma, Light and heavy particles on a fluctuating surface: Bunchwise balance irreducible sequences and local density-height correlation. *Phys. Rev. Res.* **2**, 043279 (2020).
18. G. Manoj and M. Barma, Phase separation driven by a flucuating two-dimensional self-affine potential field. *J. Stat. Phys.* **110**, 1305 (2003).

19. C. Godrèche, Two-time correlation and occupation time for the Brownian bridge and tied-down renewal process. *J. Stat. Mech.* 073205 (2017).
20. L. Frachebourg, I. Ispolatov and P. L. Krapivsky, Extremal properties of random systems. *Phys. Rev. E* **52**, R5727.
21. W. Feller, *An Introduction to Probability Theory and its Applications* Vol-I (John Wiley & Sons).
22. A. Das and M. Barma, Fluctuation-dominated phase ordering in coarse-grained depth models: Domain wall structures, extreme values and coarsening. *J. Stat. Mech.* 043203 (2023).
23. A. Bar and D. Mukamel, Mixed-order phase transition in a one-dimensional model. *Phys. Rev. Lett.* **112**, 015701 (2014).
24. A. Bar, S. N. Majumdar, G. Schehr and D. Mukamel, Exact extreme-value statistics at mixed-order transitions. *Phys. Rev. E* **93**, 052130 (2016).
25. S. Sadhukhan and S. Nandi, private communication.
26. S. Chatterjee and M. Barma, Dynamics of fluctuation-dominated phase ordering: Hard-core passive sliders on a fluctuating surface. *Phys. Rev. E* **73**, 011107 (2006).
27. M. Barma, Singular Scaling Functions in Clustering Phenomena. *Eur. Phys. J. B* **64**, 387 (2008).
28. A. Nagar, M. Barma and S. N. Majumdar, Passive sliders on fluctuating surfaces: strong- clustering states. *Phys. Rev. Lett.* **94**, 240601 (2005).
29. T. Singha and M. Barma, Clustering, intermittency, and scaling for passive particles on fluctuating surfaces. *Phys. Rev. E* **98**, 052148 (2018).
30. A. Nagar, S. N. Majumdar and M. Barma, Strong clustering of non-interacting, passive sliders driven by a Kardar-Parisi-Zhang surface. *Phys. Rev. E* **74**, 021124 (2006).
31. A. Comtet and C. Texier, Supersymmetry and Integrable Models. In Proceedings, Chicago, IL, edited by H. Aratyn, T. Imbo, W. Y. Keung, and U. Sukhatme (Springer, Berlin, 1998).
32. S. Dey, D. Das and R. Rajesh, Spatial structures and giant number fluctuations in models of active matter. *Phys. Rev. Lett.* **108**, 238001 (2012).
33. H. D. Bale and P. Schmidt, Small-angle X-Ray-scattering investigation of submicroscopic porosity with fractal properties. *Phys. Rev. Lett.* **53**, 596 (1984).
34. P. Wong, Scattering by inhomogeneous systems with rough internal surfaces: Porous solids and random-field Ising systems. *Phys. Rev. B* **32**, 7417 (1985).
35. A. Mayya, A. Banerjee and R. Rajesh, Mammalian cortical bone in tension is non-Haversian. *Sci. Rep.* **3**, 2533 (2013). DOI:10.1038/srep02533.
36. M. Kumar, V. Banerjee and S. Puri, Random field Ising model with conserved kinetics: Super-universality violation, logarithmic growth law and the generalized Tomita sum rule. *Europhy. Lett.* **117**, 10012 (2017).
37. G. P. Shrivastav, V. Banerjee, and S. Puri, Non-Porod behavior in systems with rough morphologies. *Eur. Phys. J. E* **37**, 98 (2014).
38. E. Medina, T. Hwa and M. Kardar, Yi-cheng Zhang, Burgers equation with correlated noise: Renormalization-group analysis and applications to directed polymers and interface growth. *Phys. Rev. A* **39**, 3053 (1989).
39. A. Das, A. Polley and M. Rao, Phase segregation of passive advective particles in an active medium. *Phys. Rev. Lett.* **116**, 068306 (2016).
40. N. Katyal, S. Dey, D. Das and S. Puri, Coarsening dynamics in the Vicsek model of active matter. *Eur. Phys. J. E* **43**, 10 (2020).
41. S. Mishra and S. Ramaswamy, Active nematics are intrinsically phase separated. *Phys. Rev. Lett.* **97**, 090602 (2006).

42. S. Shankar, S. Ramaswamy and M. C. Marchetti, Low-noise phase of a two-dimensional active nematic system. *Phys. Rev E* **97**, 012707 (2018).
43. V. Narayan, S. Ramaswamy and N. Menon, Long-lived giant number fluctuations in a swarming granular nematic. *Science* **317**, 105 (2007).
44. M. Shinde, D. Das and R. Rajesh, Violation of the Porod law in a freely cooling granular gas in one dimension. *Phys. Rev. Lett.* **99**, 234505 (2007).

https://doi.org/10.1142/9789811282386_0010

Chapter 10

Mixed-Order Phase Transitions

D. Mukamel
Department of Physics of Complex Systems, Weizmann Institute of Science
Rehovor 7610001, Israel

Mixed-order phase transitions are transitions which have common features with both first-order and second-order transitions. I review some results obtained in the context of one of the prototypical models of mixed-order transitions, the one-dimensional Ising model with long-range coupling that decays as truncated inverse square distance between spins. The correspondence between this model and the Poland Scheraga model of DNA denaturation, a subject to which Michael Fisher made substantial contributions, is then outlined.

1. Joining Michael Fisher as a Postdoc at Cornell

I was a postdoctoral fellow of Michael Fisher at Cornell in the years 1975–1977. Before arriving in Cornell, I was a research fellow at Brookhaven National Lab where I applied symmetry considerations of the Landau theory to argue that n-vector spin models with $n \geq 4$ are in fact realized in a large variety of physical systems. Renormalization-group analysis of these models thus provides their classification into universality classes. Furthermore, it suggests that when an accessible stable fixed point is lacking, they should display a first-order transition.

At the time Michael was skeptical about the use of the Landau theory, beyond it being perhaps a useful but uncontrolled approximation. So, upon my arrival in Cornell, he asked me to present my work in one of his famous Tuesday seminars. The seminar turned into a tutorial series of three lengthy and grueling lectures where Michael kept posing questions and making infinitely many comments and suggestions. After the third seminar, he asked me to come to his office for a chat in which at some point he said, "You know the Landau theory is underappreciated in the West," and then he added candidly "...and maybe I am to blame". We ended up collaborating on a number of studies combining Landau's symmetry considerations with renormalization-group calculations together with my old friend and his then student Eytan Domany. We have, for example, suggested a physical realization of the three-state Potts model[a] and demonstrated how a symmetry-breaking field can turn a first-order transition into a continuous one.

Cornell at the time was an exciting and intellectually stimulating place to be, and working with Michael was a unique experience. I have learned a lot from Michael's non-compromising approach to science, how to try to select and pursue meaningful scientific problems and how to conduct research. My stay with Michael has deeply influenced my research throughout my career, and I am grateful for that.

2. Introduction

In this chapter, I briefly review some results on mixed-order phase transitions, a subject to which Michael has made important contributions. Mixed-order phase transitions, also known as first-order critical transitions or hybrid transitions, particularly in the percolation community, are phase transitions which have common features with both first- and second-order phase transitions. In particular, they display a discontinuity of the order parameter as is common in first-order phase transitions and a diverging correlation length as is typical in second-order phase transitions. They have been observed theoretically in a variety of models and found experimentally in some physical systems. Examples include the denaturation transition of DNA,[1–5] jamming transitions,[6–10] k-core percolation,[11–14] non-enclave percolation,[15] models of wetting,[16,17] networks dynamics,[18] models of moving condensates,[19] the one-dimensional Ising model with long-range interactions that decay as $1/r^2$ with the distance between the interacting spins[20–27] and a truncated interaction modification of this model.[28–30]

While in some of the models, the correlation length ξ was found to diverge exponentially rapidly with the deviation, ϵ, from criticality as $\xi \sim e^{\epsilon^{-\mu}}$, in other models, the divergence follows a power law, $\epsilon^{-\nu}$, with μ and ν some positive exponents. The former class includes models of jamming transitions[7] and the 1D Ising model with $1/r^2$ interaction,[24,26] while the latter includes models of DNA denaturation,[2–5] percolation transitions[12,15] and a truncated version of the $1d$ Ising model with long-range interactions.[28–30]

One of the prototypical models which exhibits a mixed-order transition is the $1d$ Ising model with long-range interactions which decay as $\sim 1/r^2$ at large distance r. The model, later termed the inverse distance square Ising model (IDSI), was introduced by Thouless,[20] who argued that it possesses a phase transition with a jump in the magnetization at the transition. This magnetization discontinuity became known as the "Thouless effect." A closely related hierarchical model was then introduced and solved by Dyson[21] who demonstrated that magnetization discontinuity indeed takes place. An exact proof of the discontinuous nature of the transition

[a]For an impressive piece of art generated by Michael for this project where he displays his artistic drawing skills, see *PRL* 37, 565 (1976) and the contribution of Eytan Domany in this volume, arXiv:2301:11086.

in the IDSI model was eventually given by Aizenman *et al.*[27] In addition, scaling arguments[24] and renormalization-group analysis[26] have shown that the correlation length ξ diverges at the transition with essential singularity, $\xi \sim e^{b/\sqrt{T-T_c}}$, where b is a constant and T_c is the critical temperature. Although a lot is known about the thermodynamic behavior of this model, it is not exactly solvable. In $d = 2$ and higher dimensions, the model with $1/r^2$ interaction results in energy which scales super-linearly with the volume. When properly scaled with the volume, the model exhibits an ordinary continuous transition.

More recently, a modified version of the IDSI model was introduced.[28–30] This model is exactly solvable and was shown to display a mixed-order transition which is closely related to that of IDSI although with some differences. In what follows, this model is introduced and its main propertied are reviewed. I will then comment on the relation of this model to the Poland Scheraga model of DNA denaturation. I will also highlight the contributions of Michael to the understanding of the nature of mixed-order transitions in general and to DNA denaturation in particular.

3. The Truncated Inverse Distance Square Ising (TIDSI) Model

The TIDSI model is a truncated version of the 1D IDSI model. Here, as in IDSI, the long-range interaction between spins at distance r apart decay as $1/r^2$ with the distance. However, the interaction is confined to spins which lie within the same domain of either all up or all down spins. The Hamiltonian of the model is

$$\mathcal{H} = -\Sigma_{i<j} J(i-j)\sigma_i\sigma_j I(i \sim j) - J_{NN}\Sigma_i \sigma_i \sigma_{i+1}, \tag{1}$$

where $\sigma_i = \pm 1$, $I(i \sim j) = 1$ as long as sites i and j are in the same domain of either all up or all down spins and $I(i \sim j) = 0$ otherwise. The long-range coupling is taken to be

$$J(r) \approx \frac{C}{r^2} \quad \text{for } r \gg 1, \tag{2}$$

and, for simplicity, one may take periodic boundary conditions $\sigma_i = \sigma_{i+L}$ in a lattice of length L. Note that the indicator function $I(i \sim j)$ may be expressed in terms of the spin variables as

$$I(i \sim j) = \prod_{k=i}^{j-1} \delta_{\sigma_k \sigma_{k+1}} = \prod_{k=i}^{j-1} \frac{1 + \sigma_k \sigma_{k+1}}{2}, \tag{3}$$

where $\delta_{\sigma_k \sigma_{k+1}}$ is the Kronecker delta function. The Hamiltonian (1) is thus not of the usual binary spin interaction form, but rather it involves multi-spin interactions.

It is convenient to express the Hamiltonian in terms of the domain length representation, where a domain is defined as a stretch of successive parallel spins (see Fig. 1). The long-range interaction in the first term operates only between pairs of spins which belong to the same domain, while the nearest-neighbor interaction in the second term results in an energy cost for each domain wall. A typical

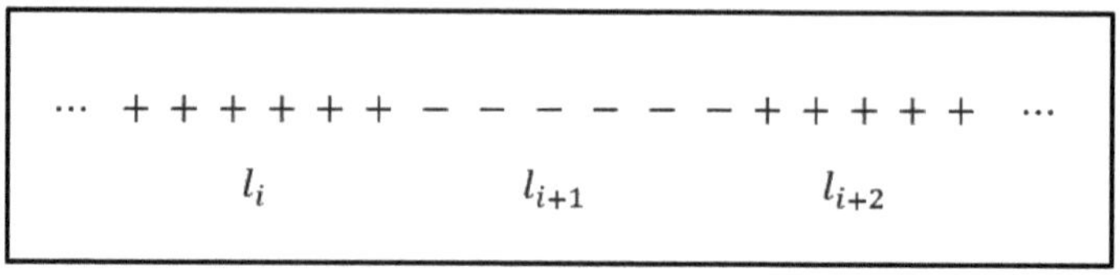

Fig. 1. Microscopic configuration of the Ising model with the domain lengths indicated.

configuration $\mathcal{C}$ is thus described by a set of domains with lengths $\{l_1, l_2, \cdots, l_N\}$, where the number of domains N can vary from one configuration to another. In terms of these variables, the Hamiltonian can be expressed as

$$\mathcal{H} = \sum_{n=1}^{N} \mathcal{H}_n, \tag{4}$$

where

$$\mathcal{H}_n = -J_{NN}(l_n - 2) - \sum_{r=1}^{l_n} (l_n - r)J(r). \tag{5}$$

The variables l_n are constrained by the system size L to satisfy

$$\sum_{n=1}^{N} l_n = L. \tag{6}$$

Using the form of $J(r)$ in Eq. (2), one can estimate the sum by replacing it with an integral as

$$\begin{aligned} \sum J(r) &\approx a_0 - \frac{C}{l_n}, \\ \sum rJ(r) &\approx b_0 + C \ln l_n, \end{aligned} \tag{7}$$

where we have assumed that l_n is large and kept the two leading order terms for large l_n. This is justified since we are interested in phenomena close to the critical line where domains are typically large. Dropping an overall unimportant constant, one obtains the effective Hamiltonian

$$\mathcal{H} = C \sum_{n} \ln l_n + \Delta N, \tag{8}$$

where the constant C is the amplitude of the long-range interaction and $\Delta = 2J_{NN} + C + b_0$ acts as a chemical potential for the number of domains. The resulting Hamiltonian is that of a Coulomb gas of alternating sign charges in 1D, where the interaction is confined to nearest-neighbor charges only, and it is logarithmic in the distance, as is the case of charges in 2D. In this representation, the charges correspond to the domain boundaries, and the domain lengths, l_n, are the distance between nearest-neighbor charges.

The free energy and the resulting thermodynamic properties of the model have been calculated in recent studies and the mixed-order nature of the transition has been clarified.[28,29] In addition, a discussion of this model within the context of a different class of models exhibiting fluctuation-dominated phase ordering has been given by Barma *et al.*[30] and in the contribution of Mustansir Barma in the present volume. Let us briefly outline this analysis. The probability of a configuration $\mathcal{C}$, defined by the domain lengths $(l_1, \ldots, l_N)$ and their number N is given by its Boltzmann weight

$$P(l_1, l_2, \ldots, l_N, N|L) = \frac{y^N}{Z_y(L)} \prod_{n=1}^{N} \frac{1}{l_n^c} \delta_{\sum_{n=1}^{N} l_n, L}, \tag{9}$$

where

$$c = \beta C, \quad y = e^{-\beta\Delta}, \tag{10}$$

and $\delta_{i,j}$ is the Kronecker delta function that enforces the sum rule. The normalization constant $Z_y(L)$ is the partition function given by

$$Z_y(L) = \sum_{N=1}^{\infty} y^N \sum_{l_1=1}^{\infty} \cdots \sum_{l_N=1}^{\infty} \prod_{n=1}^{N} \frac{1}{l_n^c} \delta_{\sum_{n=1}^{N} l_n, L}. \tag{11}$$

Here, we have ignored a geometrical factor which arises from the number of ways of placing the sequence of domains on the lattice. This factor results in a sub-leading contribution to the partition sum and hence it does not affect the thermodynamic properties of the model.

In order to analyze the behavior of the model and particularly the phase diagram and the probability distribution of the domains length, we proceed by considering the grand canonical ensemble. The grand partition sum, $Q_y(z)$ is given by

$$Q_y(z) = \sum_{L=1}^{\infty} z^L Z_y(L), \tag{12}$$

where z is the fugacity. Carrying out the sums in Eqs. (11) and (12), one obtains

$$Q_y(z) = \frac{y\phi_c(z)}{1 - y\phi_c(z)}, \tag{13}$$

with

$$\phi_c(z) = \sum_{l=1}^{\infty} \frac{z^l}{l^c} \equiv \mathrm{Li}_c(z), \tag{14}$$

where $\mathrm{Li}_c(z)$ is the polylogarithmic function. The sum in Eq. (14) converges for $z < 1$, diverges for $z > 1$ and it exhibits a singularity at $z = 1$.

The phase diagram and the nature of the phase transition can be deduced by direct inspection of $Q_y(z)$. The partition sum of the model, $Z_y(L)$, is determined by the pole of $Q_y(z)$ when it exists. The pole is obtained at z^* which satisfies

$$y\phi_c(z^*) = 1, \tag{15}$$

and it exists as long as $z^* < 1$. In this case, the partition sum is given by

$$Z_y(L) \propto z^{*-L}. \tag{16}$$

A phase transition takes place in the model at $z^* = 1$, namely at y and c satisfying

$$y\phi_c(1) = 1. \tag{17}$$

The nature of the phases and the phase transition are simply found by calculating the distribution function, $P(l)$, of a single domain. This is done by summing over the lengths of all other domains in (9), yielding

$$P(l) = \frac{Z_y(L-l)}{Z_y(L)} \frac{y}{l^c}, \tag{18}$$

which, given (16), results in

$$P(l) = \frac{yz^{*l}}{l^c} \propto \frac{e^{-l/\xi}}{l^c}, \tag{19}$$

as long as $z^* < 1$. Here, $\xi \equiv -1/\ln z^*$ is a length scale which controls the typical domain size. For finite ξ (namely, for $z^* < 1$), the equilibrium domains are of finite typical length controlled by ξ, and their average number N scales with the system's length L. As a result, the domain's density $n = N/L$, is finite. As $z^* \to 1$, the correlation length ξ diverges and the density of domains (or charges) vanishes. One can think of two order parameters involved in this transition: the magnetization m and the average number of domains $\langle n \rangle$. The two order parameters yield somewhat different results as to the nature of the phase transition, as is outlined in the following.

In the spin representation, a typical configuration in the paramagnetic phase is composed of a finite density of domains of alternating up and down spins, so that the overall magnetization vanishes. At the transition point, $z^* = 1$, the domain's density vanishes, leading to a phase which is composed of a single domain of length L with some small domains whose total length is sub-leading in L. This is a ferromagnetic phase with average magnetization per spin $m = \pm 1$ for any $c > 1$. The transition from the paramagnetic to this ferromagnetic phase was thus termed "extreme Thouless effect". The fact that the discontinuous order parameter is accompanied by a diverging correlation length makes the transition of mixed-order nature for any $c > 1$.

In order to explore the nature of the phase transition using the $\langle n \rangle$ order parameter, let us consider the domain size distribution at the transition, $P(l) \propto 1/l^c$, which exists as a normalizable distribution for $c > 1$. For $1 < c \leq 2$, the average domain length, $\langle l \rangle$, diverges, and thus, the average density of domains, $\langle n \rangle$, vanishes. This indicates that the order parameter of the transition, $\langle n \rangle$, goes continuously to zero at the transition. The transition is thus a usual second-order transition with diverging correlation length and continuously vanishing order parameter. On the other hand, for $c > 2$, $\langle n \rangle$ is finite at the transition, indicating that the transition into the

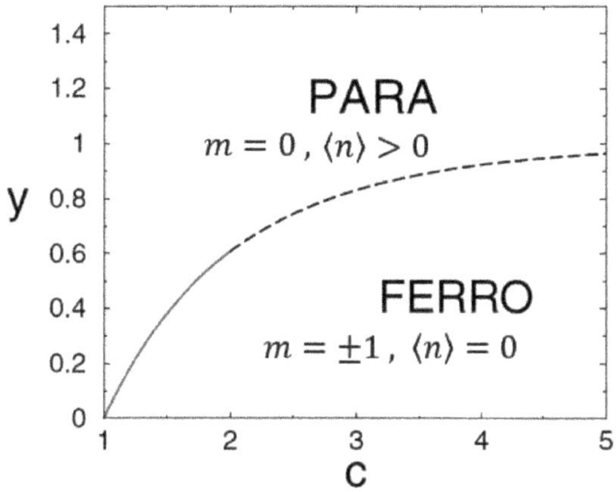

Fig. 2. The (c, y) phase diagram of the TIDSI model. The red solid line represents a continuous transition with $\langle n \rangle$ vanishing continuously at the transition. The dashed line represents a mixed-order transition with a discontinuous order parameter $\langle n \rangle$ and diverging length. Note that the m order parameter is discontinuous at the transition for any $c > 1$.

ordered phase, at which $\langle n \rangle$ vanishes, involves a discontinuous change of the order parameter. Thus, for $c > 2$, the model is of a mixed-order nature, with a diverging length ξ and a discontinuous order parameter $\langle n \rangle$. Note that the parameter c of the model, given in (10), is temperature-dependent, and the nature of the transition is determined by the value of c at the transition temperature, namely $c = \beta_c C$. The phase diagram of the model in the y–c plane is given in Figure 2.

The critical behavior of the model can be extracted from the singularity of the polylogarithmic function at $z = 1$. In the disordered phase, where the domain density $\langle n \rangle$ is non-vanishing and $z < 1$, the fugacity is related to temperature variable y ($\equiv \beta\Delta$) via $y\phi_c(z) = 1$. Close to the transition point, and for $1 < c < 2$, the polylogarithmic function satisfies $\phi_c(1 - \delta z) - \phi_c(1) \propto \delta z^{c-1}$. Thus, close to the transition, where $y = y_{crit} + \delta y$, one has $\delta y \propto \delta z^{c-1}$. The correlation length $\xi = -1/\ln(\delta z)$ thus satisfies $\xi \propto \delta z^{-1}$, or $\xi \propto \delta y^{-\frac{1}{c-1}}$. Since y is basically the temperature variable, δy is proportional to the reduced temperature $t = (T - T_c)/T_c$, resulting in

$$\xi \propto t^{-\nu} \quad \text{with} \quad \nu = 1/(c-1). \tag{20}$$

For $c > 2$, the singular term in $\phi_c(z)$ is sub-leading with $\phi_c(1 - \delta z) - \phi_c(1) \propto z$, which yields $\nu = 1$.

The order parameter of the transition is $\langle n \rangle = 1/\langle l \rangle$. Using the domain length distribution (19), one finds that for $1 < c < 2$,

$$\langle n \rangle = 1/\xi^{2-c} \propto t^{\frac{2-c}{c-1}}. \tag{21}$$

For $c > 2$, the average domain length $\langle l \rangle$ is finite and the order parameter is discontinuous at the transition.

As mentioned above, in the IDSI model, the correlation length diverges at the transition with an essential singularity rather than a power law. An insight into this behavior is gained by considering the model within the framework of the charge configurations. Here, as in the TIDSI model, a microscopic configuration is expressed as a sequence of alternating charges marking the domain boundaries. However, unlike TSIDI, the logarithmic interaction between the charges is not restricted to nearest-neighbor charges only but is rather all to all. This makes the transition of a Kosterlitz–Thouless type, with its characteristic essential singularity. This has first been studied by Anderson *et al.*[24] using scaling arguments and later elaborated by more detailed renormalization-group approach by Cardy.[26]

An interesting question is what is the effect of an ordering field H on a mixed-order phase transition. In ordinary second-order phase transition, an ordering field removes the transition, as, for example, is the case for the Ising model. On the other hand, first-order transitions are retained in the presence of an ordering field, resulting in a fork shape phase diagram in the (T, H) plane where three lines of first-order transitions meet at a triple point (see Fig. 3). In the case of the IDSI model, it was first conjectured by Anderson *et al.*[24] that the (T, H) phase diagram of this mixed-order transition is like that of an ordinary second-order transition (see Fig. 3(a)). Fisher, who has always been intrigued by first-order transitions, developed together with Berker a scaling approach to first-order transitions and argued that this is indeed the case and termed the transition "first-order critical point."[25] The phase diagram of the TIDSI model is quite different. In Figure 4, a schematic phase diagram of a model with $c > 2$ is illustrated. In this figure, the $H = 0$ mixed-order transition survives the ordering field. The phase diagram has a fork shape with two lines of mixed-order phase transitions meeting the $H = 0$ first-order transition line. The mixed-order lines are characterized by a continuously varying parameter $c(H)$. This parameter decreases with H until it reaches the value $c = 2$, where the transition becomes continuous.

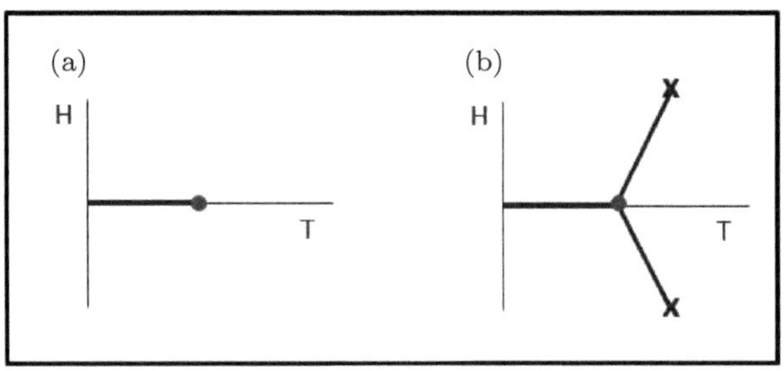

Fig. 3. Typical schematic (T, H) phase diagram for (a) second-order transition at $H = 0$ and (b) first-order transition at $H = 0$ with three first-order lines intersecting at a triple point. The two fork branches terminate at critical points marked x in the figure. Scaling arguments suggest that the phase diagram of the IDSI model is of the form of (a).

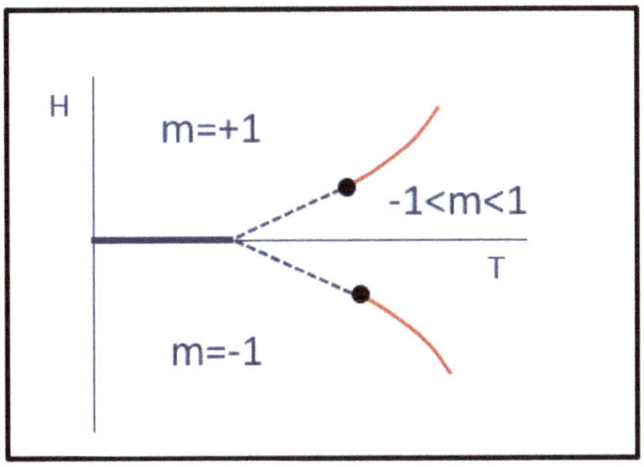

Fig. 4. Schematic (T, H) phase diagram of the TIDSI model with $c > 2$. Dashed lines represent lines of mixed-order phase transition of the TIDSI type with H-dependent parameter $c(H) > 2$, while red thin lines represent continuous transitions with $1 < c(H) < 2$. On the thin lines, the magnetization m approaches $m = \pm 1$ continuously, while on the dashed lines, it exhibits a discontinuous jump. The thick solid line at $H = 0$ is a first-order transition line on which the magnetization changes discontinuously from $+1$ to -1.

On the mixed-order sections of the branches, the magnetization changes discontinuously from some value in the paramagnetic phase to $|m| = 1$ in the ferromagnetic phase. However, on the continuous section of the branches, the magnetization changes continuously to its $|m| = 1$ value in the ferromagnetic phase (for details, see Ref. 29).

4. Comment on DNA Denaturation

DNA denaturation corresponds to melting of double-stranded DNA, a phase transition at which the two bound strands of the molecule become separated. This transition has been extensively studied experimentally and theoretically for over 60 years by now. The canonical model used to study the transition is the Poland–Scheraga (PS) model introduced in the 1960s.[2] Within this model, a DNA molecule is schematically viewed at an alternating sequence of segments of bound strands and denatured loops of unbound single strands (see Fig. 5). The statistical weight of a microscopic configuration, which is characterized by the number of segments and length of each one of them is determined by the following simplifying assumptions: each bound pair carries a (negative) binding energy u so that the statistical weight of a bound segment of length l is $\omega = e^{-\beta u l}$. For simplicity, all bound pairs carry the same energy and the bound segment is non-flexible so that it carries no configurational entropy. A segment of length l of unbound pairs constitutes a loop of length $2l$. The loop carries no energy, however it is viewed as composed of a flexible string, and thus, it possesses configurational entropy which has to be taken into account as calculating the statistical weight of the DNA configuration. In the PS model, as introduced by Poland and Scheraga, loops are considered as fully flexible and they are modeled as non-interacting random loops. In the large loop limit, the

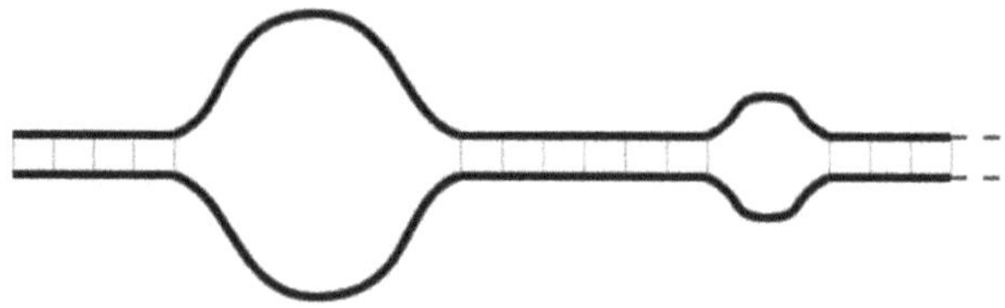

Fig. 5. Schematic configuration of DNA molecule with alternating bound segments and open loops.

statistical weight of a loop of length l in d dimensions is

$$\Omega(l) \propto \frac{s^l}{l^c}, \tag{22}$$

where s is some d-dependent constant and $c = d/2$. Here, $1/l^c$ represents the probability of a random walk in dimension d to return to the starting point after l steps.

Equation (22) implies that as in the TIDSI model, the statistical weight of a loop may be expressed by two opposite charges interacting logarithmically as $c \ln l$, leading to the same type of transition as in TIDSI. In particular, the loop length distribution takes the form of Eq. (19), leading to no transition for $c \leq 1$, continuous transition for $1 < c < 2$ and a mixed-order transition for $c \geq 2$. A major distinction between the two models is that while the parameter c in TIDSI is temperature-dependent, and thus non-universal, in the PS model, c is universal. It is independent of temperature and of any type of short-range interactions that may exist between the two strands. The prediction of the PS model for DNA denaturation is that $c = 3/2$ in $d = 3$, implying a continuous transition. This result disagrees with the experimental observations which suggest that the transition is discontinuous.

An important step in resolving this discrepancy was made by Michael Fisher in 1966,[3] who observed that exclusion interaction between the strands in a loop, which was neglected in the PS analysis, should be taken into account. This interaction is long-ranged and can thus affect the value of c. Noting that the end-to-end distance R of a self avoiding random walk of length l scales as $R \sim l^\nu$, the probability of the walk to return to the origin becomes $1/l^{d\nu}$, making $c = d\nu$. Given that in $d = 3$, the Flory exponent is estimated to be $\nu \approx 0.583$, the parameter c increases from 1.5 in the non-interacting model to $\approx$1.75 when exclusion is taken into account. While the predicted transition is still continuous, the inclusion of the exclusion interaction is a step in the right direction.

This approach was later taken a step forward by Kafri *et al.*,[4,5] who noted that the exclusion interaction between a loop and the rest of the chain modifies c even further. Modeling the rest of the chain as a linear flexible polymer, one can apply a renormalization-group approach developed for evaluating the entropy of a general polymer network[31–33] to study the length distribution of a self-avoiding loop embedded in a chain. It was found that the loop size distribution is of the same form as that of an isolated loop but with a modified parameter c which takes the value $c \approx 2.11$ in $d = 3$ dimensions. This makes the denaturation transition of mixed-order type with a discontinuous order parameter in line with what experiments suggest.

5. Summary

Mixed-order phase transitions have been suggested to exist in a variety of theoretical models of physical systems. In this paper, some properties of this intriguing class of phase transitions are reviewed and their derivation is outlined in the framework of the TIDSI model. This model is exactly solvable and thus provides useful insights into the mechanism leading to a transition with discontinuous order parameter together with a diverging correlation length. The model can be expressed in terms of charges of alternating sign in $1d$ whose interaction is restricted to the nearest neighbor only. This feature is found to lead to a power-law divergence of the correlation length. When the restriction is removed and the interaction becomes all-to-all, the model represents the IDSI model, which displays a Kosterlitz–Thouless type mixed-order transition with exponential divergence of the correlation length. It would be of great interest to explore the relevance of this picture of interacting charges to models in higher dimensions, such as models of jamming transitions.

Acknowledgments

I thank Amir Bar, Mustansir Barma, Yariv Kafri, Satya N Majumdar, Luca Peliti and Gregory Schehr with whom I have collaborated over the years on studies of properties and peculiarities of mixed-order transitions. I also thank Eytan Domany and Martin Evans for the comments on the manuscript. This work was supported by a research grant from the Center for Scientific Excellence at the Weizmann Institute of Science.

References

1. R. M. Wartell and A. S. Benight, Thermal denaturation of DNA molecules: a comparison of theory with experiment. *Phys. Rep.* **126** (2), 67–107 (1985).
2. D. Poland and H. A. Scheraga, Occurrence of a phase transition in nucleic acid models. *J. Chem. Phys.* **45** (5), 1464–1469 (1966).
3. M. E. Fisher, Effect of excluded volume on phase transitions in biopolymers. *J. Chem. Phys.* **45** (5), 1469–1473 (1966).
4. Y. Kafri, D. Mukamel and L. Peliti, Why is the DNA denaturation transition first order? *Phys. Rev. Lett.* **85** (23), 4988 (2000).
5. Y. Kafri, D. Mukamel and L. Peliti, Melting and unzipping of DNA. *Eur. Phys. J. B-Condens. Matt. Complex Syst.* **27** (1), 135–146 (2002).
6. D. J. Gross, I. Kanter and H. Sompolinsky, Mean-field theory of the Potts glass. *Phys. Rev. Lett.* **55**, 304–307 (Jul, 1985).

7. C. Toninelli, G. Biroli and D. S. Fisher, Jamming percolation and glass transitions in lattice models. *Phys. Rev. Lett.* **96**, 035702 (Jan, 2006).
8. M. Jeng and J. Schwarz, Comment on "jamming percolation and glass transitions in lattice models". *Phys. Rev. lett.* **98** (12), 129601 (2007).
9. C. Toninelli, G. Biroli and D. Fisher, Toninelli, Biroli and Fisher reply. *Phys. Rev. Lett.* **98** (12), 129602 (2007).
10. J. M. Schwarz, A. J. Liu and L. Chayes, The onset of jamming as the sudden emergence of an infinite k-core cluster. *EPL (Europhys. Lett.)* **73** (4), 560 (2006).
11. J. Chalupa, P. L. Leath and G. R. Reich, Bootstrap percolation on a bethe lattice. *J. Phys. C: Solid State Phys.* **12** (1), L31 (1979).
12. S. N. Dorogovtsev, A. V. Goltsev and J. F. F. Mendes, K-core organization of complex networks. *Phys. Rev. lett.* **96** (4), 040601 (2006).
13. J. Park, S. Yi and B. Kahng, Hysteresis and criticality in hybrid percolation transitions. *Chaos: Interdiscipl. J. Nonlinear Sci.* **30** (5), 051102 (2020).
14. D. Lee, W. Choi, J. Kértesz and B. Kahng, Universal mechanism for hybrid percolation transitions. *Sci. Rep.* **7** (1), 1–7 (2017).
15. M. Sheinman, A. Sharma, J. Alvarado, G. Koenderink and F. MacKintosh, Anomalous discontinuity at the percolation critical point of active gels. *Phys. Rev. Lett.* **114** (9), 098104 (2015).
16. M. E. Fisher, Walks, walls, wetting and melting. *J. Stat. Phys.* **34** (5), 667–729 (1984).
17. R. Blossey and J. Indekeu, Diverging length scales at first-order wetting transitions. *Phys. Rev. E.* **52** (1), 1223 (1995).
18. K. Bassler, W. Liu, B. Schmittmann and R. Zia, Extreme thouless effect in a minimal model of dynamic social networks. *Phys. Rev. E.* **91** (4), 042102 (2015).
19. J. Whitehouse, A. Costa, R. A. Blythe and M. R. Evans, Maintenance of order in a moving strong condensate. *J. Stat. Mech.: Theory Exp.* **2014** (11), P11029 (2014).
20. D. Thouless, Long-range order in one-dimensional Ising systems. *Phys. Rev.* **187** (2), 732 (1969).
21. F. J. Dyson, An ising ferromagnet with discontinuous long-range order. *Commun. Math. Phys.* **21** (4), 269–283 (1971).
22. P. W. Anderson and G. Yuval, Exact results in the kondo problem: Equivalence to a classical one-dimensional coulomb gas. *Phys. Rev. Lett.* **23** (2), 89 (1969).
23. P. W. Anderson, G. Yuval and D. Hamann, Exact results in the Kondo problem. ii. scaling theory, qualitatively correct solution, and some new results on one-dimensional classical statistical models. *Phys. Rev. B.* **1** (11), 4464 (1970).
24. P. Anderson and G. Yuval, Some numerical results on the Kondo problem and the inverse square one-dimensional Ising model. *J. Phys. C: Solid State Phys.* **4** (5), 607 (1971).
25. M. E. Fisher and A. N. Berker, Scaling for first-order phase transitions in thermodynamic and finite systems. *Phys. Rev. B.* **26** (5), 2507 (1982).
26. J. L. Cardy, One-dimensional models with 1/r*2 interactions. *J. Phys. A: Math. Gen.* **14** (6), 1407 (1981).
27. M. Aizenman, J. Chayes, L. Chayes and C. Newman, Discontinuity of the magnetization in one-dimensional $1/_{|}$ x- $y_{|}^{|}$*2 Ising and Potts models. *J. Stat. Phys.* **50** (1), 1–40 (1988).
28. A. Bar and D. Mukamel, Mixed-order phase transition in a one-dimensional model. *Phys. Rev. Lett.* **112** (1), 015701 (2014).
29. A. Bar and D. Mukamel, Mixed order transition and condensation in an exactly soluble one dimensional spin model. *J. Stat. Mech.: Theory Exp.* **2014** (11), P11001 (2014).

30. M. Barma, S. N. Majumdar and D. Mukamel, Fluctuation-dominated phase ordering at a mixed order transition. *J. Phys. A: Math. Theor.* **52** (25), 254001 (2019).
31. B. Duplantier, Polymer network of fixed topology: Renormalization, exact critical exponent γ in two dimensions, and d = 4-ε. *Phys. Rev. Lett.* **57** (8), 941 (1986).
32. B. Duplantier, Statistical mechanics of polymer networks of any topology. *J. Stat. Phys.* **54** (3), 581–680 (1989).
33. L. Schäfer, C. von Ferber, U. Lehr and B. Duplantier, Renormalization of polymer networks and stars. *Nucl. Phys. B.* **374** (3), 473–495 (1992).

https://doi.org/10.1142/9789811282386_0011

Chapter 11

Bounds on the Superconducting Transition Temperature

M. Randeria
Department of Physics, The Ohio State University,
Columbus, OH 43210, USA
randeria.1@osu.edu

I summarize recent progress on obtaining rigorous upper bounds on superconducting transition temperature T_c in two dimensions independent of pairing mechanism or interaction strength. These results are derived by finding a general upper bound for the superfluid stiffness for a multi-band system with arbitrary interactions, with the only assumption that the external vector potential couples to the kinetic energy and not to the interactions. This bound is then combined with the universal relation between the superfluid stiffness and the Berezinskii–Kosterlitz–Thouless T_c in 2D. For parabolic dispersion, one obtains the simple result that $k_B T_c \leq E_F/8$, which has been tested in recent experiments. More generally, the bounds are expressed in terms of the optical spectral weight and lead to stringent constraints for the T_c of low-density, strongly correlated superconductors. Results for T_c bounds for models of flat-band superconductors, where the kinetic energy vanishes and the vector potential must couple to interactions, are briefly summarized. Upper bounds on T_c in 3D remains an open problem, and I describe how questions of universality underlie the challenges in 3D.

1. Introduction

This chapter summarizes recent works[1–3] on the question of ultimate limits on the superconducting transition temperature T_c, where we have obtained sharp results in two dimensions (2D), with connections to many interesting experiments.[4–6]

This is a contribution to a volume on 50 years of the renormalization group (RG), dedicated to the memory of Michael E. Fisher, and it may seem ironic that I have chosen to focus on that most *non*-universal of quantities, the transition temperature T_c, even though it is a quantity of great general interest for superconductivity! There are, however, many aspects of the analysis presented in the following that

I am sure Michael would have approved of. First, these are rigorous[a] bounds on T_c, which do not make any mean-field or other uncontrolled approximations. Second, the 2D bounds use in an essential way universal results from the RG analysis of the Berezinskii–Kosterlitz–Thouless (BKT) transition. Finally, universality and quantum critical phenomena play an important role in understanding the unresolved challenges that arise in attempting to prove similar bounds in 3D.

Let me note at the outset that there is no upper bound on the superconducting T_c measured in Kelvin. Superconductivity at 250 K has been established in high-pressure experiments on hydrogen-based materials like LaH.[7] It is widely believed that superfluidity of paired neutrons occurs at $T_c \simeq 10^9 K$ in neutron stars, but then all energy scales here are huge compared to terrestrial quantum materials, e.g. the Fermi energy $E_F \simeq 10 - 100\,\text{MeV}$.

An insightful way to organize data on diverse superconductors is the Uemura plot[8] of $\log T_c$ versus $\log E_F$ in Figure 1. There is considerable ambiguity in defining "the Fermi energy" E_F in strongly correlated and multi-band materials, nevertheless, the data strongly suggest that all known superconductors have a T_c that scales at most like a constant times E_F. The systems that come closest to a presumed bound are magic-angle twisted bilayer graphene,[5] monolayer FeSe/STO,[6] and ultracold Fermi gases ^{40}K and ^{6}Li in the BCS–BEC crossover,[9] which have little in common except that they are all strongly correlated superconductors or paired superfluids.

These data motivated us to see if one could prove such a bound. The rest of the chapter summarizes the progress we have made in 2D, recent experiments on Li:ZrNCl[4] that came after our theoretical prediction, applications of our bound to various models and materials, the question of flat-band superconductivity, and, finally, the unsolved problem in 3D.

2. Models

We consider a system of spin-1/2 fermions described by a general multi-band Hamiltonian of the form

$$\mathcal{H} = \mathcal{H}_K + \mathcal{H}_{\text{int}}; \quad \mathcal{H}_K = \sum_{\mathbf{k},m,\sigma} \epsilon_m(\mathbf{k}) c^\dagger_{\mathbf{k}m\sigma} c_{\mathbf{k}m\sigma} \tag{1}$$

where $\mathbf{k}$ is crystal momentum, m is a band label, and σ the spin. $\mathcal{H}_K$ describes the kinetic energy of the "low-energy" bands that are impacted by superconductivity; we ignore bands far from the chemical potential that are either fully occupied or completely empty. The external vector potential $\mathbf{A}$ enters $\mathcal{H}$ through the

[a]'Rigorous' is used here in the sense of theoretical physics, i.e., an exact result; however, we do not address subtleties of the thermodynamics limit that mathematical physicists would care about.

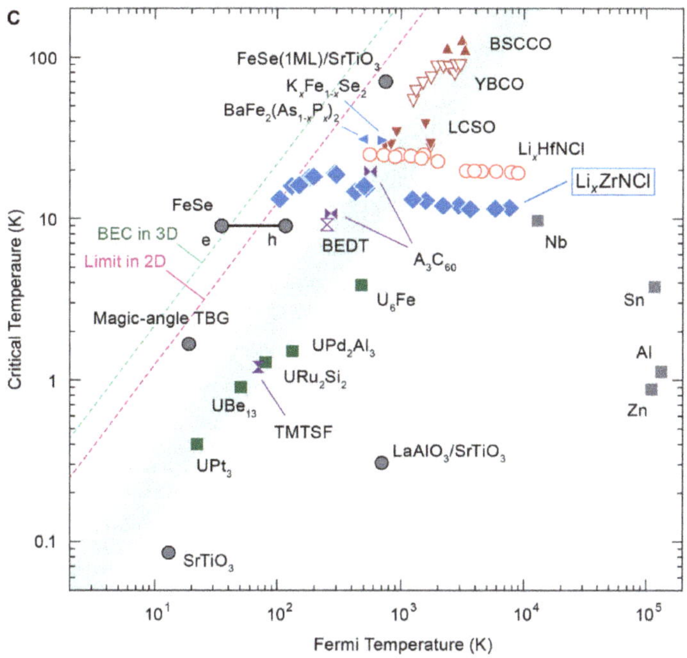

Fig. 1. Uemura plot of critical temperature T_c versus Fermi temperature $T_F = E_F/k_B$ for various superconductors; see text for more details. There are many conventional BCS superconductors in the lower right region (below the green shaded band) that are not shown, however there are no materials in the upper left region. The "limit in 2D" line $T_c = T_F/8$ is the 2D bound derived in Ref. 1 for a parabolic dispersion. This bound, together with the Li_xZrNCl data (filled diamonds) from Ref. 4, are discussed in Section 6. T_c bounds for arbitrary dispersion and multi-band superconductors are described in later sections. We discuss in Section 11 why the 3D BEC temperature $T_c = 0.218\, T_F$ is *not* a bound on the superconducting T_c in three-dimensional systems. This figure from Ref. 4 is reprinted with permission from AAAS.

Peierl's substitution in the tight-binding representation of $\mathcal{H}_K$ or minimal subtraction in a continuum model. We consider an arbitrary interaction term $\mathcal{H}_{\rm int}$ that leads to pairing and superconductivity, with the single assumption that $\mathbf{A}$ does *not couple* to $\mathcal{H}_{\rm int}$. These are very natural assumptions in most theoretical models of superconductivity, including the usual electron–phonon interaction, electron–electron interactions like in the Hubbard model, and super-exchange in the tJ model. The assumptions are also valid for the Feshbach resonance that leads to pairing of ultracold Fermi atoms, with rotation playing the role of the magnetic field.

We postpone until Section 10 the discussion of cases in which these assumptions are not valid, and the analysis then becomes more involved.[2] We discuss in that section T_c bounds for models of flat-band superconductivity, where $\mathcal{H}_K \equiv 0$, as well as topological bands, where the Wannier functions are not exponentially localized, and thus, tight-binding and Peierl's substitution cannot be used straightforwardly.

For now, we focus on the class of Hamiltonians described in the preceding paragraph, which after all have very broad applicability.

3. Bounds on Superfluid Stiffness

There are two distinct ways in which superconductivity can be destroyed: (1) by suppressing the amplitude of the order parameter or (2) by destroying phase coherence, even though the amplitude is robust. BCS-Eliashberg mean-field theory falls into the first category, with the transition temperature $T_c \propto |\Delta|$, the superconducting gap. The BKT theory of vortex–antivortex unbinding in 2D falls into the second category, as do the ideas of Uemura[8] and of Emery–Kivelson[10] in 3D. When phase fluctuations control the destruction of superconductivity $T_c \propto D_s$, the superfluid stiffness, and we will focus on bounding D_s.

The superfluid stiffness controls the free energy cost $\frac{D_s}{2}\int d^d\mathbf{r}|\nabla\theta|^2$ of distorting the phase of the SC order parameter $|\Delta|e^{i\theta}$ and is related to the London penetration depth in 3D via $1/\lambda_L^2 = (4\mu_0 e^2/\hbar^2)D_s$. The stiffness is essentially the same as the helicity modulus Υ of Michael Fisher and coworkers.[11]

Microscopically, D_s can be calculated using linear response theory[12,13]:

$$D_s = \widetilde{D} - \frac{\hbar^2}{4e^2}\chi^{\perp}_{j_x j_x}(\mathbf{q}\to 0, \omega = 0), \tag{2}$$

where $\widetilde{D}$ is the "diamagnetic term", of central interest in this work, and $\chi^{\perp}_{j_x j_x}$ is the transverse paramagnetic current–current correlation function. We use the Lehmann representation to obtain $\chi^{\perp}_{j_x j_x}(\mathbf{q}\to 0, \omega = 0) \geq 0$, as can be seen from

$$\chi^{\perp}_{j_x j_x} = \frac{1}{\mathcal{Z}}\sum_{ij}\left[\frac{e^{-\beta E_i} - e^{-\beta E_j}}{E_j - E_i}\right]|\langle i|j_x^P(\mathbf{q})|j\rangle|^2, \tag{3}$$

using $(e^{-x} - e^{-y})/(y-x) \geq 0$. Here, $|i\rangle$'s are the exact eigenstates of $\mathcal{H}$ of Eq. (1) with eigenvalues E_i and the partition function $\mathcal{Z} = \mathrm{Tr}[e^{-\beta\mathcal{H}}]$, with $\beta = 1/k_B T$. We thus obtain a rigorous upper bound

$$D_s(T) \leq \widetilde{D}(T). \tag{4}$$

4. Optical Spectral Weight

Before turning to the question how one can calculate $\widetilde{D}$, let us develop some intuition for what it means and why it gives an upper bound on the superfluid stiffness. Using the Kubo formula for the frequency-dependent conductivity $\sigma(\omega)$ and the Kramers–Krönig relation, it is straightforward to derive the optical sum rule

$$\int_0^{\infty} d\omega\ \mathrm{Re}\ \sigma(\omega) = \frac{2\pi e^2}{\hbar^2}\widetilde{D}, \tag{5}$$

where the frequency integration includes all the "low-energy" bands (but excludes the fully occupied or completely empty bands that were not part of Eq. (1)). By identifying $\widetilde{D}$ with the *optical spectral weight*, we see that it is gauge-invariant. Moreover, the second law of thermodynamics implies that the dissipative response $\text{Re}\ \sigma(\omega) \geq 0$, and thus, so is $\widetilde{D}$.

Now, in any superconductor $\text{Re}\ \sigma(\omega) = (4\pi e^2/\hbar^2)D_s\delta(\omega) + \text{Re}\ \sigma_{\text{reg}}(\omega)$, where the singular term describes the infinite d.c. conductivity of the condensate and the "regular" term is a non-negative contribution arising from excitations. Using the sum rule (5), it is easy to understand the inequality of Eq. (4).

Next, we turn to computing the optical spectral weight for our model. From the linear response calculation that leads to Eq. (2), we see that $\widetilde{D}$ is essentially $\langle \partial^2\mathcal{H}/\partial A_x^2 \rangle$, and thus, we find

$$\widetilde{D}(T) = \frac{\hbar^2}{4V} \sum_{\mathbf{k},mm',\sigma} M^{-1}_{mm'}(\mathbf{k})\, \langle c^\dagger_{\mathbf{k}m\sigma} c_{\mathbf{k}m'\sigma} \rangle. \tag{6}$$

For a simple parabolic band with dispersion $\epsilon(\mathbf{k}) = \hbar^2 k^2/2m$, the optical spectral weight $\widetilde{D} = \hbar^2 n/4m$ is proportional to the the plasma frequency. For non-parabolic dispersion or multiple bands, however, one cannot uniquely decompose $\widetilde{D}$ into the ratio of a density and a mass. In the general result of Eq. (6), $1/m$ is replaced by an inverse mass matrix $M^{-1}_{mm'}(\mathbf{k})$ and the density n is replaced by the density matrix (V is the volume of the system).

The inverse mass matrix is given by

$$M^{-1}_{mm'}(\mathbf{k}) = \sum_{\alpha\beta} U^*_{m,\alpha}(\mathbf{k})\, \frac{\partial^2 t_{\alpha\beta}(\mathbf{k})}{\partial(\hbar k_a)^2}\, U_{\beta,m'}(\mathbf{k}), \tag{7}$$

where α, β label orbitals/sites within a unit cell (labelled by i) of a Bravais lattice, $t_{\alpha\beta}(\mathbf{k})$ is the Fourier transform of the hopping matrix element $t_{\alpha\beta}(\mathbf{r}_{i\alpha} - \mathbf{r}_{j\alpha})$, and $U_{\alpha,m}(\mathbf{k})$ is the unitary transformation that diagonalizes $t_{\alpha\beta}(\mathbf{k})$ to obtain the band structure: $U^*_{m,\alpha}(\mathbf{k})t_{\alpha\beta}(\mathbf{k})U_{\beta,m}(\mathbf{k}) = \epsilon_m(\mathbf{k})\delta_{m,m'}$. Note that M^{-1} depends only on band dispersion and Bloch wave functions, and not on interactions or temperature. The density matrix $\langle c^\dagger_{\mathbf{k}m\sigma} c_{\mathbf{k}m'\sigma} \rangle$ in Eq. (6), on the other hand, depends in general on both interactions and temperature, since the equilibrium expectation value is computed using $\exp(-\mathcal{H}/k_BT)/\mathcal{Z}$.

5. T_c bounds in Two Dimensions

The optical spectral weight bound on the superfluid stiffness is valid in any dimension. Now, we restrict attention to *two dimensions* (until Section 11), since it is only in 2D that there is a universal relation between the BKT transition temperature and the superfluid stiffness. $D_s(T)$ exhibits a jump discontinuity at T_c and the RG analysis[14] of Nelson and Kosterlitz leads to the celebrated result $k_BT_c = \pi D_s\left(T_c^-\right)/2$. Remarkably, the 'amplitude ratio' $\pi/2$ is universal; it is related to value of the critical exponent $\eta(T_c^-) = 1/4$.

Combining this result with Eq. (4), we obtain a bound on T_c, namely

$$k_B T_c \leq \frac{\pi}{2} \widetilde{D}(T_c), \tag{8}$$

with $\widetilde{D}$ given by Eq. (6). Clearly, this is an exact result, independent of pairing mechanism or strength, but is it useful? In the rest of this chapter, I show that there are many situations where this bound leads to insights into the T_c of strongly correlated superconductors with low-density and narrow bands, systems where the standard BCS-Eliashberg mean-field approach fails. Conversely, in regimes where mean-field theory works, the bound is still valid but not useful, since it gives a huge overestimate.

6. Parabolic Bands and 2D BCS–BEC Crossover

Consider the problem of superconductivity in a 2D parabolic band with dispersion $\hbar^2 k^2/2m$, density n, or Fermi energy $E_F = \pi\hbar^2/n/m$, and arbitrary interaction that leads to pairing. We can then write the stiffness $D_s(T) = \hbar^2 n_s(T)/4m$ in terms of the *superfluid density* $n_s(T)$, and our bound of Eq. (4) reduces to the simple statement that the superfluid density $n_s(T) \leq n$ the total density. Equation (8) then leads to the inequality

$$k_B T_c \leq \frac{1}{8} E_F. \tag{9}$$

Plotting the line $k_B T_c = E_F/8$ on the Uemura plot looks very nice and gives a general scale for the limits on T_c, as seen in Figure 1. But we should caution that only few of the systems on that plot fulfill the condition of parabolic bands in 2D. In that sense, this is like a "Drude model" result that gives a sense of what should happen, but if one looks at multi-band systems with complicated band structures, only the optical spectral weight bound is justified.

This raises the question: Can one experimentally test this bound for parabolic bands? This can definitely be tested in the BCS–BEC crossover in 2D (which we discuss in the following) in experiments and in quantum Monte Carlo simulations of the ultracold Fermi gases. But, surprisingly, the first experimental test has been in a quantum material. To investigate the BCS–BEC crossover, one needs to tune Δ/E_F, the ratio of the gap to Fermi energy. Unlike cold atoms, where the Feshbach resonance is used to control Δ, there is no way in general to tune the attractive interactions between electrons in a solid. However, through a combination of intercalation and ionic liquid gating, the Iwasa group was able to change the density, and thus the Fermi energy, by *two orders of magnitude* and investigate the BCS–BEC crossover in Li_xZrNCl.[4] Electronic structure calculations[15] show that this 2D material has a parabolic band.

The experiments measure the electron density using the Hall effect and probe superconducting properties, like the pair size ξ_0 (from the upper critical field) and

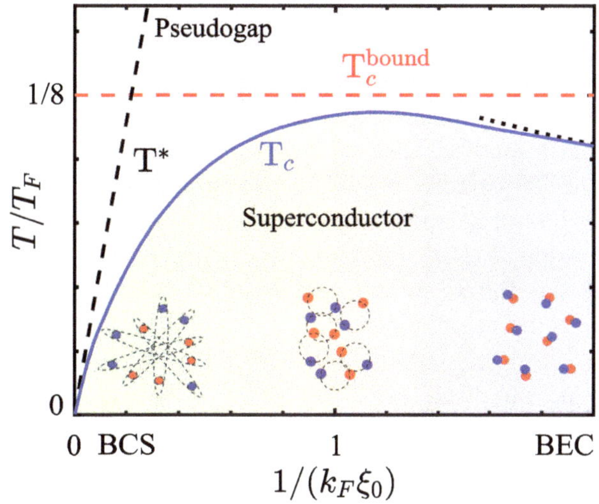

Fig. 2. Schematic phase diagram of the BCS–BEC crossover in a 2D Fermi gas with short-range attractive interactions. The ratio of the inter-particle separation $1/k_F$ to the pair size ξ_0 shows how the attraction between fermions increases from the BCS regime of large pairs to the BEC of tightly bound pairs. The blue curve is the schematic transition temperature T_c (in units of the Fermi temperature $T_F = E_F/k_B$), which interpolates between known results in the BCS and BEC limits; see text. T_c vanishes in the extreme BEC limit in 2D, but does so very slowly (double logarithm), as indicated by the dotted asymptote; see text. The black dashed line is the pseudogap temperature T^* below which pairs form, while superconductivity develops only below the BKT T_c. The red dashed line is the theoretical limit $Tc \leq T_F/8$ for parabolic dispersion.

energy gap Δ (from tunneling). As the density is decreased by gating, the measured values of $1/(k_F\xi_0)$ and Δ/E_F show clear evidence of tuning into the strongly interacting BCS–BEC crossover regime, as well as the characteristic normal state pairing pseudogap seen in tunneling.

Remarkably, the measured values of k_BT_c/E_F increase upon approaching the crossover regime and then saturate at a value of 0.12, very close to the predicted 1/8, as can be seen in Figure 1.

We next discuss in more detail the BCS–BEC crossover in 2D dilute Fermi gases; see Figure 2. These systems are dilute in the sense that the range of the interaction is much smaller than $1/k_F$. The 2D crossover for s-wave pairing is parametrized by the dimensionless interaction $\log(E_b/E_F)$, where E_b is the binding energy of the two-body bound state in vacuum.[16] In the weak-coupling BCS limit $E_b/E_F \ll 1$, the pair size $\xi_0 \gg 1/k_F$. The mean field T_c has a BCS essential singularity and is exponentially smaller than our bound. In the opposite BEC limit $E_b/E_F \gg 1$, the bosons of size $\xi_0 \ll 1/k_F$ have mass $2m$, density $n/2$ and an inter-boson scattering length a_b, which has been computed[17] in terms of E_b. Here too, our bound is larger than $k_BT_c \sim E_F/\log\log(\text{const.}\, E_b/E_F)$,[18,19] valid in the regime $\log\log \gg 1$. Thus, $T_c \to 0$ in the extreme BEC limit, as it must for non-interacting bosons, but does so extremely slowly. These results are shown in a schematic phase diagram in Figure 1, where we see that $E_F/8$ is likely to be saturated in the crossover regime with pair size ξ_0 or order the the interparticle spacing $1/k_F$.

To conclude this section, we note that a transition temperature of $E_F/8$ was seen in various approximate calculations (using mean-field theory plus fluctuations) for 2D models for cold atoms[20,21] and for quantum materials,[22] with parabolic band structures and specific choices of pairing interactions. But the generality of the result as an upper bound independent of the nature of the interactions was not recognized. In fact, some of these approximate calculations[20] obtained the erroneous result that $E_F/8$ was the T_c in the extreme BEC limit, where, in fact the T_c is known to vanish, as discussed above.

7. Non-parabolic Dispersion

For a single band with arbitrary dispersion $\epsilon(\mathbf{k})$, Eq. (6) simplifies to

$$\widetilde{D} = \frac{1}{4V}\sum_{\mathbf{k}\sigma}\frac{\partial^2\epsilon(\mathbf{k})}{\partial k_x^2}\, n(\mathbf{k}), \tag{10}$$

where $n(\mathbf{k}) = \langle c^\dagger_{\mathbf{k}\sigma}c_{\mathbf{k}\sigma}\rangle$ is the momentum distribution. For the specific case of nearest neighbor hopping on a square lattice $\epsilon(\mathbf{k}) = -2t\,[\cos(k_x a) + \cos(k_y a)]$ and $\widetilde{D}$ is proportional to the thermal expectation value of the kinetic energy $\langle\mathrm{KE}\rangle$. The optical spectral weight in lattice systems is often just called "kinetic energy." This identification is true neither for a single band with arbitrary dispersion nor for a multi-band system. (In the latter case, Eq. (6) depends not only on the dispersion but also on the single-particle wave functions through the U-matrix.)

It is instructive to compare our bound for the single band $\widetilde{D}$ with sign-problem-free Quantum Monte Carlo (QMC) simulations[23] of the attractive Hubbard model on a 2D square lattice, with $\mathcal{H}_{\mathrm{int}} = -|U|\sum_i (n_{i\uparrow} - 1/2)\,(n_{i\downarrow} - 1/2)$. For a density $n \neq 1$, the system has a superconducting ground state, exhibiting a crossover from a weak coupling BCS state ($|U|/t \ll 1$) to a BEC of hard-core on-site bosons ($|U|/t \gg 1$). The QMC result[23] for T_c is a non-monotonic function of $|U|/t$ at a fixed density n. The BCS mean-field theory correctly describes the weak coupling $T_c \sim t\exp(-t/|U|)$, but for $|U|/t \gtrsim 2$, the mean field "T_c" represents a pseudogap[24,25] crossover temperature T^* below which pairs form and lies well above the QMC T_c, at which phase coherence is established. The two scales exhibit quantitatively different behavior at large $|U|/t$, where $T^* \sim |U|$, while $T_c \sim t^2/|U|$ the effective boson hopping.

Using the triangle inequality and $n(\mathbf{k}) \leq 1$, it is easy to see that we get $k_B T_c \leq \pi\widetilde{D}/2 \leq (\pi/4V)\sum_{\mathbf{k},\sigma}\left|\partial^2\epsilon/\partial k_x^2\right|$. This U-independent bound is obviously better than the MFT "T_c" that diverges as $|U|$ at strong coupling. The use of $n(\mathbf{k}) \leq 1$ is too crude, however, to capture the $t^2/|U|$ asymptotic behavior of T_c. We can use $n(\mathbf{k})$ as obtained from the Leggett crossover theory of the BCS–BEC crossover[9] in Eq. (10) to get an *estimate* (rather an exact bound) on T_c. This leads to an "approximate bound" that goes to a constant at small $|U|$ and crosses over to $t^2/|U|$ behavior at large $|U|$. In summary, we see that at weak coupling, the $\widetilde{D}$ bound greatly overestimates T_c, but is useful at intermediate and strong coupling for the attractive Hubbard model. The reader is referred to Figure 2 and related discussion in Ref. 1 for details.

8. Multi-band Systems

In the general multi-band case, we need to deal with the density matrix of the interacting system about which we have little *a priori* knowledge. We can, however, proceed as follows. $\widetilde{D} \geq 0$ from the optical sum rule, and we use the triangle inequality to obtain $\widetilde{D} \leq \frac{\hbar^2}{4V} \sum_{\mathbf{k},mm',\sigma} |M^{-1}_{mm'}(\mathbf{k})|\,|\langle c^\dagger_{\mathbf{k}m\sigma} c_{\mathbf{k}m'\sigma}\rangle|$. Next, we define an inner product in the space of operators $(A, B) = \langle A^\dagger B\rangle$ in terms of a thermal expectation value. This allows us to use the Cauchy–Schwarz inequality $|\langle c^\dagger_{\mathbf{k}m\sigma} c_{\mathbf{k}m'\sigma}\rangle| \leq [n_m(\mathbf{k})\, n_{m'}(\mathbf{k})]^{1/2}$, where $n_m(\mathbf{k}) = \langle c^\dagger_{\mathbf{k}m\sigma} c_{\mathbf{k}m\sigma}\rangle$ is the momentum distribution of the mth band. Now, for any Fermi system, $n_m(\mathbf{k}) \leq 1$, and we obtain by this sequence of inequalities that $\widetilde{D} \leq \frac{\hbar^2}{4V} \sum_{\mathbf{k},mm',\sigma} |M^{-1}_{mm'}(\mathbf{k})|$.

In fact, a better bound is obtained by first transforming the sum on right-hand side of Eq. (6) to a basis in which $M^{-1}_{mm'}(\mathbf{k})$ is diagonalized and then using this sequence of inequalities. We then obtain

$$k_B T_c \leq \frac{\pi}{2}\,\widetilde{D}(T_c) \leq \frac{\pi}{2}\frac{\hbar^2}{4V} \sum_{\mathbf{k},a,\sigma} |\lambda_a(\mathbf{k})|, \tag{11}$$

where $\lambda_a(\mathbf{k})$'s are the eigenvalues of $M^{-1}_{mm'}(\mathbf{k})$. We have greatly degraded the bound, so the final result is likely to be a large overestimate, nevertheless, we have arrived at a general multi-band bound that depends only on the band structure and is independent of the mechanism of superconductivity.

9. 2D Materials

In this section, we apply our bound to various strongly correlated 2D superconductors. We have already discussed the gate-controlled BCS–BEC crossover in Li:ZrNCl[4] in Section 6. Here, we briefly discuss Monolayer FeSe/STO and magic angle twisted bilayer graphene.

Monolayer FeSe/STO: A monolayer of FeSe grown on a $SrTiO_3$ substrate exhibits T_c's around 65 K, with some reports of T_c as high as 109 K; for a review, see Ref. 6. This is remarkable since bulk FeSe has a T_c around 10 K and free-standing monolayers are not superconducting even at temperatures as low as 2 K. There is no consensus on the pairing mechanism and the role of the substrate in increasing T_c. Our bound is useful here because it is agnostic to the pairing mechanism, symmetry of the order parameter, and other issues under debate.

We use an effective two-orbital model consistent with $\mathbf{k} \cdot \mathbf{p}$ perturbation theory that describes the two electron pockets around the M-point of the Brillouin zone,[27,28] and fit the parameters of the electronic structure to ARPES data.[29] The orbitals derive from the same two-dimensional irreducible representation and the Fermi surfaces are small so we can truncate the $\mathbf{k}\cdot\mathbf{p}$ dispersion at quadratic order in k_x and k_y. We thus find that the inverse mass $M^{-1}_{mm'}(\mathbf{k})$ turns out to be proportional to $\delta_{m,m'}$ and $\mathbf{k}$-independent. Thus, the $\mathbf{k}$-sum in Eq. (6) is simply proportional to the Fermi energy, and we find $T_c \leq 164$ K; see Ref. 26.

Magic angle twisted bilayer graphene (MA-TBG): There has been an explosion of interest in MA-TBG, where superconductivity was first reported[5] in the vicinity of correlation-induced insulating state. This is not the place to review the large, and ever-growing, experimental and theoretical literature on this topic; however, it is safe to say that there is at present no consensus on the pairing mechanism (phonons versus correlations) or the symmetry of superconducting order parameter.

The electronic band structure of MA-TBG is complex. There are 8 ($= 4 \times 2$ for spin degeneracy) narrow bands,[30] with a total bandwidth $\lesssim 10\,\text{meV}$, and these bands have non-trivial topology.[31,32] There is much discussion in the literature about whether inclusion of additional bands may help deal with both the topology and the interactions. There is also the issue of whether one can write down a tight-binding model for the low-energy bands and have an effective interaction such that the vector potential couples only to the kinetic energy. We address these questions in Section 10, where we obtain T_c bounds for simple models of flat-band superconductivity.

For now, we simply focus on models of MA-TBG that meet the conditions spelled out at the outset in Section 2. We use the tight-binding model of Ref. 33, which is a multi-parameter fit to the continuum model of Ref. 30, to calculate $M^{-1}_{m,m'}(\mathbf{k})$ of Eq. (7). We could equally well use other tight-binding models[34] for MA-TBG. We compute the inverse mass matrix $M^{-1}_{mm'}(\mathbf{k})$ from the tight-binding model and then use Eq. (11) to find that $T_c \leq 48\,\text{K}$. This is a large overestimate because we have degraded the inequality in Section 8 as explained above.

10. Flat-Band Superconductors

In recent years, there has been interest in the effect of correlations in flat bands that has only intensified following the discovery of superconductivity in magic-angle-twisted bilayer graphene (MA-TBG). There has been important progress on the *mean-field theory* (MFT) of flat-band superconductivity[35–39] in attractive (negative U) Hubbard models. The gap scale was found to be proportional to $|U|$ and the superfluid stiffness proportional to $|U|$ times the trace of the quantum metric $\sum_{\mathbf{k}} \text{Tr} g(\mathbf{k})$ for the flat-band wave functions. The *quantum metric* $g_{ab}(\mathbf{k})$ defined by $ds^2 = 1 - |\langle\psi(\mathbf{k}+d\mathbf{k})|\psi(\mathbf{k})\rangle|^2 = g_{ab}(\mathbf{k}) dk_a dk_b$ characterizes the distance between states in Hilbert space.[40,41]

But to what extent should one trust MFT in a problem where the attractive interaction $|U|$ is much larger in the bandwidth, in fact, infinitely larger in the flat-band limit? Our goal was to find an exact bound on D_s, for which we need to face up to multiple challenges. First, if we use the multi-band bound derived above, we find that it is far from optimal, since it includes optical spectral weight from inter-band transitions, while we would like to focus only on intra-band spectral weight to get a tight bound. Second, if one projects down to the flat band to focus on the intra-band spectral weight, there is no kinetic energy term, and thus one cannot couple the vector potential $\mathbf{A}$ to flat-band KE as we had done above. Finally, there is a lot of interest in topological flat bands, motivated by MA-TBG. Non-trivial band topology, however, acts as an obstruction to finding exponentially localized Wannier functions obeying all the symmetries of the problem, and one cannot write down a tight-binding Hamiltonian describing the topological bands alone.

All of these challenges have been addressed for both non-topological and topological flat bands with an attractive Hubbard interaction in Ref. 2. Omitting all technical details of the derivations, we summarize the main results. Note that flat bands cannot exist "in isolation" and arise in models, such as the Lieb lattice, with multiple orbitals in each unit cell. We work in a regime where the chemical potential lies in the flat band, and the (possibly zero) bandwidth $w \ll |U| \ll E_0$, the gap between the flat band and other bands. We project onto the flat band and focus on the low-energy (intra-band) optical spectral weight $\widetilde{D}_{\text{low}} = \left(\hbar^2/2\pi e^2\right) \int_0^\Lambda d\omega \; \text{Re}\sigma(\omega)$, where the upper cutoff satisfies $|U| \ll \Lambda \lesssim E_0$.

For the Lieb lattice, one finds an exactly flat band at zero energy and two p–h symmetric dispersive bands, using "staggered hopping" to open up the gap E_0. We can then derive an upper bound $\widetilde{D}_{\text{low}} \leq \widetilde{n}|U|\Omega/2$. Quite naturally, $|U|$ sets the scale of the optical weight, since there is no transport in a flat band in the absence of interactions. The dependence on the density n comes in via $\widetilde{n} = \min(n, 2-n)$, which vanishes, as expected for a completely empty/full flat band. Finally, Ω is the Marzari–Vanderbilt[41] spread of the flat-band Wannier function (WF). The optimal bound on $\widetilde{D}_{\text{low}}$ is thus obtained by choosing the most highly localized WFs. The quantum metric prevents us from reducing this as much as we please, since $\Omega \geq \sum_{\mathbf{k}} \text{Tr}g(\mathbf{k})$. The Lieb lattice has non-topological flat bands, and thus, the system can be continuously deformed into a trivial insulator where individual unit cells (each with three sites) become decoupled from each other. In this limit, $\Omega \to 0$ so that the optical spectral weight vanishes and that there is no transport.

Using $D_s \leq \widetilde{D}_{\text{low}}$ and the BKT relation between T_c and D_s, we obtain $k_B T_c \leq \pi\,\widetilde{n}|U|\min\Omega/4$. We note that for the Lieb lattice with its particle–hole symmetric band structure, it is natural to say that in the flat-band regime, $E_F = 0$. This re-emphasizes the point that there is in general no upper bound on T_c/E_F and the only general bound on T_c is in terms of the optical spectral weight.

Many authors have emphasized lower bounds on D_s arising from a mean-field approximation.[36–39] Using the well-known inequality,[41] $\text{Tr}g(\mathbf{k}) \geq \mathcal{C}$, it is easy to see that the mean field D_s, proportional to the trace of the quantum metric, is bounded below by the Chern number $\mathcal{C}$, We emphasize that these mean field 'bounds' are not exact lower bounds, since fluctuations ignored in mean field theory would be expected to reduce the superfluid stiffness.

We can also obtain an *exact lower bound* on $\widetilde{D}_{\rm low}$ at $T = 0$ by exploiting rigorous mathematical physics results[43] on the off-diagonal long-range order (LRO) in the ground state of the attractive Hubbard model on the Lieb lattice. (This result is the $U < 0$ counterpart of Lieb's theorem on ferromagnetic LRO in flat bands with $U > 0$; see Ref. 43). We can show that[2] $\widetilde{D}_{\rm low} \geq \widetilde{n} \sum_{\mathbf{k}} {\rm Tr} g(\mathbf{k})/72$.

Finally, we briefly summarize our bounds on D_s in *topological* flat bands. The tight-binding model of Ref. 44 has flat bands with $\mathbb{Z}_2$ odd topology, and one cannot obtain localized WFs with a finite spread Ω. There are two degenerate, (nearly) flat bands $|\mathbf{k}\uparrow\rangle$ and $|\mathbf{k}\downarrow\rangle$ with spin Chern numbers $\mathcal{C}_{\uparrow,\downarrow} = \pm 1$. Using the methodology developed in Refs. 45 and 46, we "unwind the topology" and find two linear combinations $|\mathbf{k}\,1\rangle$ and $|\mathbf{k}\,2\rangle$, both of which have zero Chern number. The price we pay is that the corresponding exponentially localized WFs are not a Kramers pair and the states have a non-local representation of time-reversal symmetry. Nevertheless, this allows us to proceed with our analysis, and the resulting upper bound on $\widetilde{D}_{\rm low}$ is proportional to $|U|$, to $\widetilde{n} = \min(n, 2-n)$, and to a generalized second moment of the WFs. The bound is also found to compare favorably with D_s computed using sign-problem-free QMC simulations[47] for the same model; see Ref. 2 for details.

11. T_c in Three Dimensions

Given the success in bounding T_c in 2D, it is natural to ask: What about 3D? Experiments seem to suggest at first sight that there might well be such a bound on T_c in 3D systems; see the Uemura plot in Figure 1. For example, the BCS–BEC crossover in 3D experiments and quantum Monte Carlo calculations suggest that $\max T_c \simeq 0.22 E_F/k_B$; see Ref. 9 for a more detailed discussion and references. But there are no rigorous bounds available on T_c in 3D even though the optical spectral weight bound on the superfluid stiffness of Eq. (4) is valid in *any* dimension. We now discuss the challenges for deriving such bounds and whether they even exist.

First, there is a widespread misconception that the non-interacting BEC temperature $k_B T_{\rm BEC} = [n/2\zeta(3/2)]^{2/3} \pi\hbar^2/m \simeq 0.218 E_f$ of tightly bound pairs (bosons with density $n/2$ and parabolic dispersion with mass $2m$) provides an "upper bound" to the transition temperature of a strongly coupled superconductor. However, it is now rigorously established in the mathematical physics literature that the T_c of the dilute Bose gas in 3D is *higher* than the non-interacting BEC temperature. The reader is directed to Ref. 48 for the proof, as well as a summary of the decades-long history of incorrect statements on this issue by a long list of luminaries!

Next, let us look more carefully at the relation between the stiffness and T_c. From the free energy cost for phase distortions $\mathcal{F} = \frac{D_s}{2} \int d^d\mathbf{r} |\nabla\theta|^2$, we see that

D_s has dimensions of energy *only* in 2D, and this underlies the universal relation between T_c and D_s in the BKT theory. If thermal phase fluctuations govern the 3D transition, then following Ref. 10, $k_B T_c = \mathcal{C}\ D_s(0)\,\overline{a}$, where $\overline{a}$ is a "cutoff" length scale, the scale up to which one has to coarse-grain the underlying fermionic model to derive an effective XY model, and $\mathcal{C}$ is a *non-universal* constant. Given a cutoff $\overline{a}$, one would need to bound $\mathcal{C}$ to get a bound on T_c. See Ref. 1 for further discussion.

From a practical point of view, one is interested in learning how high T_c can be in a class of models or materials. But, if a general bound were to exist, it should be equally valid in situations where both T_c and D_s are driven to zero near a quantum critical point (QCP). Quantum Josephson scaling[49] implies $D_s(0) \sim \delta^{(z+d-2)\nu}$ in the vicinity of a QCP tuned by the dimensionless parameter $\delta \to 0^+$, where ν is the correlation length exponent and z the dynamical exponent in d spatial dimensions. Using $T_c \sim \delta^{z\nu}$, one obtains $T_c \sim D_s^{z/(z+d-2)}$. Thus, in 2D, one finds $T_c \sim D_s$, independent of z. However, in 3D, one obtains $T_c \sim D_s^{z/(z+1)}$, a sub-linear scaling sufficiently close to the QCP which necessarily violates any proposed upper bound on T_c that scales linearly with D_s. Such a sub-linear scaling is indeed observed in experiments in highly underdoped[50,51] and highly overdoped[52,53] cuprates, with $T_c \sim D_s^{1/2}$ consistent with $z = 1$. Thus, the linear scaling between T_c and D_s, central to the 2D bounds presented above, is necessarily violated near a 3D QCP.

12. Conclusions

In conclusion, I take this opportunity to record my gratitude to Michael Fisher for taking me under his wing when I first moved to Cornell as a graduate student, even though his sabbatical plans led me to eventually do my thesis research under Jim Sethna's supervision. I am grateful for much that I learned from Michael while working on the two papers[54,55] that I wrote with him (and one that we never finished) and for his amazing lectures on critical phenomena and RG, whose notes I still refer back to after so many years.

Acknowledgments

The results described here were part of the Ph.D. thesis of Tamaghna Hazra (now at Rutgers) and of Nishchhal Verma (now at Columbia), whose contributions are gratefully acknowledged. I acknowledge the support from the NSF Materials Research Science and Engineering Center Grant DMR-2011876.

References

1. T. Hazra, N. Verma and M. Randeria, Bounds on the superconducting transition temperature: Applications to twisted bilayer graphene and cold atoms. *Phys. Rev. X* **9**, 031049 (2019).

2. N. Verma, T. Hazra and M. Randeria, Optical spectral weight, phase stiffness and T_c bounds for trivial and topological flat band superconductors. *Proc. Nat. Acad. Sci.* **118**, e2106744118 (2021).
3. M. Randeria, Limits on superconductivity in flatland. *Science* **372**, 132 (2021).
4. Y. Nakagawa, *et al.*, Gate-controlled BCS-BEC crossover in a two-dimensional superconductor. *Science* **372**, 190 (2021).
5. Y. Cao, *et al.* Unconventional superconductivity in magic-angle graphene superlattices. *Nature* **556**, 43 (2018).
6. D. Huang and J. E. Hoffman, Monolayer FeSe on $SrTiO_3$. *Ann. Rev. Cond. Mat. Phys.* **8**, 311 (2017).
7. A. P. Drozdov, *et al.*, Superconductivity at 250 K in lanthanum hydride under high pressures. *Nature* **569**, 528 (2019).
8. Y. J. Uemura, *et al.* Universal Correlations between T_c and n_s/m^* (Carrier Density over Effective Mass) in High- T_c Cuprate Superconductors. *Phys. Rev. Lett.* **62**, 2317 (1989).
9. M. Randeria and E. Taylor, BCS-BEC Crossover and the unitary Fermi gas. *Ann. Rev. of Cond. Mat. Phys.* **5**, 209 (2014).
10. V. J. Emery and S. A. Kivelson, Importance of phase fluctuations in superconductors with small superfluid density. *Nature* **374**, 434 (1995).
11. M. E. Fisher, M. N. Barber and D. Jasnow, Helicity Modulus, Superfluidity, and scaling in isotropic systems. *Phys. Rev. A* **8**, 1111 (1973).
12. G. Baym, The microscopic description of superfluidity. In *Mathematical Methods in Solid State and Superfluid Theory*, edited by R. C. Clark and G. H. Derrick (Springer, New York, 1968).
13. D. J. Scalapino, S. R. White and S. Zhang, Insulator, metal, or superconductor: The criteria. *Phys. Rev. B* **47**, 7995 (1993).
14. D. R. Nelson and J. M. Kosterlitz, Universal jump in the superfluid density of two-dimensional superfluids. *Phys. Rev. Lett.* **39**, 1201 (1977).
15. M. Calandra, P. Zoccante and F. Mauri, Universal increase in the superconducting critical temperature of two-dimensional semiconductors at low doping by the electron-electron interaction. *Phys. Rev. Lett.* **114**, 077001 (2015).
16. M. Randeria, J.-M. Duan and L.-Y. Shieh, Bound states, Cooper pairing, and Bose condensation in two dimensions. *Phys. Rev. Lett.* **62**, 981 (1989).
17. D. S. Petrov, M. A. Baranov and G. V. Shlyapnikov, Superfluid transition in quasi-two-dimensional Fermi gases. *Phys. Rev. A* **67**, 031601 (2003).
18. V. N. Popov, On the theory of the superfluidity of two- and one-dimensional Bose systems. *Theo. Math. Phys.* **11**, 565 (1972).
19. D. S. Fisher and P. C. Hohenberg, Dilute Bose gas in two dimensions. *Phys. Rev. B* **37**, 4936 (1988).
20. S. S. Botelho and C. A. R. Sa de Melo, Vortex-Antivortex Lattice in Ultracold Fermionic Gases. *Phys. Rev. Lett.* **96**, 040404 (2006).
21. G. Bighin and L. Salasnich, Finite-temperature quantum fluctuations in two-dimensional Fermi superfluids. *Phys. Rev. B* **93**, 014519 (2016).
22. A. V. Chubukov, I. Eremin and D. V. Efremov, Superconductivity versus bound-state formation in a two-band superconductor with small Fermi energy: Applications to Fe pnictides/chalcogenides and doped $SrTiO_3$. *Phys. Rev. B* **93**, 174516 (2016).
23. T. Paiva, R. Scalettar, M. Randeria and N. Trivedi, Fermions in 2D Optical lattices: Temperature and entropy scales for observing antiferromagnetism and superfluidity. *Phys. Rev. Lett.* **104**, 066406 (2010).

24. M. Randeria, N. Trivedi, A. Moreo and R. T. Scalettar, Pairing and spin gap in the normal state of short coherence length superconductors. *Phys. Rev. Lett.* **69**, 2001, (1992).
25. N. Trivedi and M. Randeria, Deviations from Fermi liquid behavior above Tc in 2D short coherence length superconductors. *Phys. Rev. Lett.* **75**, 312 (1995).
26. T. Hazra, N. Verma and M. Randeria, (unpublished).
27. P. M. Eugenio and O. Vafek, Classification of symmetry derived pairing at the M point in FeSe. *Phys. Rev. B* **98**, 014503 (2018).
28. D. F. Agterberg, *et al.*, Resilient nodeless d-wave superconductivity in monolayer FeSe. *Phys. Rev. Lett.* **119**, 267001 (2017).
29. Y. Zhang, *et al.*, Superconducting gap anisotropy in monolayer FeSe thin film. *Phys. Rev. Lett.* **117**, 117001 (2016)
30. R. Bistritzer and A. H. MacDonald, Moire Bands in twisted double-layer graphene, *Proc. Natl. Acad. Sci.* **108**, 12233 (2011).
31. H.C. Po, L. Zou, T. Senthil and A. Vishwanath, Faithful tight-binding models and fragile topology of magic-angle bilayer graphene. *Phys. Rev. B* **99**, 195455 (2019).
32. Z. Song, *et al.*, All magic angles in twisted bilayer graphene are topological. *Phys. Rev. Lett.* **123**, 036401 (2019).
33. M. Koshino, *et al.*, Maximally localized Wannier orbitals and the extended Hubbard Model for twisted bilayer graphene. *Phys. Rev.* X **8**, 031087 (2018).
34. J. Kang and O. Vafek, Symmetry, Maximally localized Wannier states, and a low-energy model for twisted bilayer graphene narrow bands. *Phys. Rev.* X **8**, 031088 (2018).
35. N. B. Kopnin, T. T. Heikkila, G. E. Volovik, High-temperature surface superconductivity in topological flat-band systems. *Phys. Rev. B* **83**, 220503 (2011).
36. S. Peotta, P. Torma, Superfluidity in topologically nontrivial flat bands. *Nat. Commun.* **6**, 8944 (2015).
37. A. Julku, S. Peotta, T. I. Vanhala, D. H. Kim and P. Torma, Geometric origin of superfluidity in the Lieb-lattice flat band. *Phys. Rev. Lett.* **117**, 045303 (2016).
38. X. Hu, T. Hyart, D. I. Pikulin and E. Rossi, Geometric and conventional contribution to the superfluid weight in twisted bilayer graphene. *Phys. Rev. Lett.* **123**, 237002 (2019).
39. F. Xie, Z. Song, B. Lian and B. A. Bernevig, Topology-bounded superfluid weight in twisted bilayer graphene. *Phys. Rev. Lett.* **124**, 167002 (2020).
40. J. P. Provost and G. Vallee, Riemannian structure on manifolds of quantum states. *Commun.Math. Phys.* **76**, 289 (1980).
41. N. Marzari, D. Vanderbilt, Maximally localized generalized Wannier functions for composite energy bands. *Phys. Rev. B* **56**, 12847 (1997).
42. C. Brouder, G. Panati, M. Calandra, C. Mourougane and N. Marzari, Exponential localization of Wannier functions in insulators. *Phys. Rev. Lett.* **98**, 046402 (2007).
43. H. Tasaki, *Physics and Mathematics of Quantum Many-Body Systems* (Springer, 2020).
44. T. Neupert, L. Santos, C. Chamon and C. Mudry, Fractional quantum Hall states at zero magnetic field. *Phys. Rev. Lett.* **106**, 236804 (2011).
45. A. A. Soluyanov and D. Vanderbilt, Wannier representation of Z2 topological insulators. *Phys. Rev. B* **83**, 035108 (2011).
46. A. A. Soluyanov, D. Vanderbilt, Smooth gauge for topological insulators. *Phys. Rev. B* **85**, 115415 (2012).
47. J. S. Hofmann, E. Berg and D. Chowdhury, Superconductivity, pseudogap, and phase separation in topological flat bands. *Phys. Rev. B* **102**, 201112 (2020).

48. R. Seiringer and D. Ueltschi, Rigorous upper bound on the critical temperature of dilute Bose gases. *Phys. Rev. B* **80**, 014502 (2009).
49. M. P. A. Fisher, P. B. Weichman, G. Grinstein and D. S. Fisher, Boson localization and the superfluid-insulator transition. *Phys. Rev. B* **40**, 546 (1989).
50. I. Hetel, T. R. Lemberger, and M. Randeria, Quantum critical behavior in the superfluid density of strongly underdoped ultrathin cuprate films. *Nat. Phys.* **3**, 700 (2007).
51. D. M. Broun, *et al.* Superfluid density in a highly underdoped $YBa_2Cu_3O_{6+y}$ Superconductor. *Phys. Rev. Lett.* **99**, 237003 (2007).
52. T. R. Lemberger, I. Hetel, A. Tsukada, M. Naito and M. Randeria, Superconductor-to-metal quantum phase transition in overdoped LaSrCuO. *Phys. Rev. B* **83**, 140507(R) (2011).
53. I. Božović, *et al.* Dependence of the critical temperature in overdoped copper oxides on superfluid density. *Nature* **536**, 309 (2016).
54. M. E. Fisher and M. Randeria, Location of renormalization-group fixed points. *Phys. Rev. Lett.* **56**, 2332 (1986).
55. M. Randeria and M. E. Fisher, Multisingularity and scaling in partial differential approximants I. *Proc. Roy. Soc. A* **419**, 181 (1988).

https://doi.org/10.1142/9789811282386_0012

Chapter 12

Critical Fluctuations in Polymer Solutions: Crossover from Criticality to Tricriticality

M.A. Anisimov, T.J. Longo, and J.V. Sengers
Institute for Physical Science and Technology
and Department of Chemical and Biomolecular Engineering,
University of Maryland, College Park, MD 20742, USA

Critical fluctuations in fluids and fluid mixtures yield a non-analytic asymptotic Ising-like critical thermodynamic behavior in terms of power laws with universal exponents. In polymer solutions, the amplitudes of these power laws depend on the degree of polymerization. Non-asymptotic behavior (upon the departure from the critical point) is particularly interesting in the case of polymer solutions, where it is governed by a competition between the correlation length of the critical fluctuations and the radius of gyration of the polymer molecules. If the correlation length is the dominant length scale, Ising-like critical behavior is observed. If, however, the radius of gyration exceeds the correlation length, tricritical behavior with mean-field critical exponents is observed. The Ising-like critical region shrinks with the increase of the polymer molecular weight. In the limit of an infinite degree of polymerization, the Ising-like critical region vanishes, yielding to theta-point tricriticality.

1. Introduction

Almost 30 years ago, Michael E. Fisher published a review on phase transitions in ionic fluids,[1] in which he discussed a remarkable analogy between the demixing of polymer solutions at an infinite degree of polymerization and a tricritical phase transition, as observed in a variety of systems, such as He^3–He^4 mixtures[2,3] and metamagnets,[4,5] with coupled order parameters.[6–8,76] We illustrate this analogy in Fig. 1 by comparing the phase behavior in solutions of polystyrene in cyclohexane with a ferromagnet containing a non-magnetic impurity. In this paper, we describe the physics that governs the crossover between Ising-like asymptotic critical behavior and tricritical theta-point behavior in high-molecular-weight polymer solutions.

Thermodynamic behavior of substances asymptotically close to their critical points (second-order phase transitions) follows the principle of critical-point universality: the microscopic details become unimportant if the correlation length of the order-parameter fluctuations exceeds the range of intermolecular interactions.[14–20]

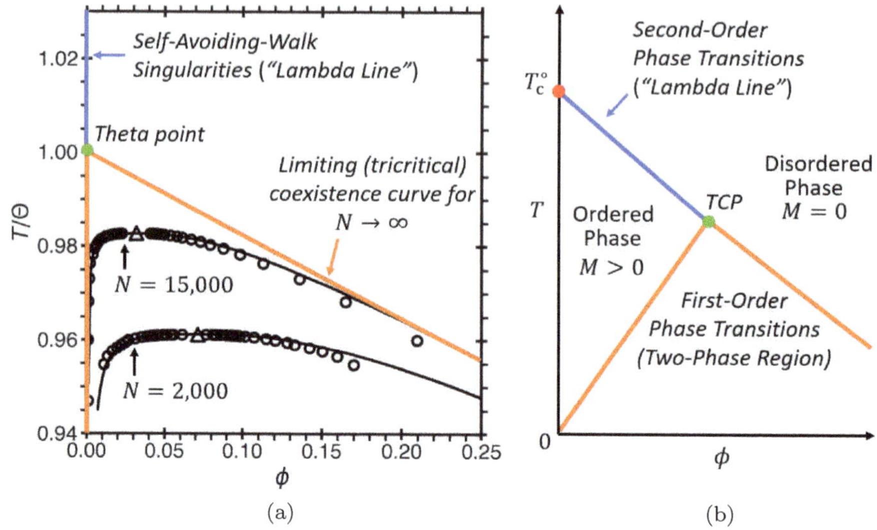

Fig. 1. Comparison of the theta-point phase behavior of a polymer solution in the limit of an infinite degree of polymerization (N) with a ferromagnet containing a non-magnetic impurity. (a) Phase behavior of polymer solutions of polystyrene in cyclohexane. As the polymer chain becomes longer, the curved shape of the liquid–liquid coexistence transforms into the "angle-like" shape near the theta point, where the two branches of the phase coexistence become pure solvent and a phase containing the polymer (volume fraction, $\phi > 0$).[9] The open circles are experimental data,[10–12] the triangles are the critical points of demixing, and the solid black curves are the theoretical prediction.[13] (b) Schematic phase diagram for a ferromagnet (with magnetization, M) containing a non-magnetic impurity (with concentration, ϕ). T_c° (red circle) is the Curie temperature of a pure ferromagnet. The tricritical point (TCP) separates the line of second-order phase transitions from the first-order phase transitions, accompanied by phase separation.

All fluids and fluid mixtures, simple and complex, belong to the three-dimensional Ising-model universality class.[16,19] Asymptotically, their thermodynamic properties are described by power laws with universal Ising-model critical exponents and system-dependent amplitudes (which are interrelated by universal relations).[20] However, there is a famous quotation of Peter Debye in response to the critical power laws (proposed by Michael Fisher) at the first international conference on critical phenomena[21]:

> "I would like that the theoretical people tell me when I am so and so far away from the critical point, then my curve should look so and so."

This question received the answer in the theory of crossover critical phenomena developed in the 1980s and 1990s.[9,22–27] The crossover between asymptotic Ising critical behavior and mean-field behavior away from the critical point is governed by a ratio of the distance between the molecules and the range of molecular interactions, known as the Ginzburg number.[19,28] Since, in ordinary fluids, the range of molecular interactions does not exceed the intermolecular distance, there is no region where the critical anomalies can be fully described by the mean-field ("Landau") theory.[29] The only possibility to observe complete crossover from Ising criticality to mean-field criticality in simple systems is to consider an Ising-like model with a tunable range of interactions. An accurate computational study of such a model[27] has perfectly confirmed the crossover theory.

An alternative kind of crossover from Ising criticality to mean-field-like behavior, the subject of this review, is a striking phenomenon observed in solutions of high-molecular-weight polymers in a low-molecular-weight solvent. At a finite degree of polymerization, asymptotically close to the demixing critical point, one observes universal Ising criticality where the amplitudes of the power laws depend on the degree of polymerization.[30–32] Upon an increase in the degree of polymerization, the Ising-like critical region shrinks, yielding to theta-point tricritical behavior.[9,13,33–36] The crossover to theta-point tricriticality is governed by a competition between the correlation length of the critical fluctuations and the radius of gyration of the polymer molecules. If the correlation length is the dominant length scale, Ising-like critical behavior is observed. If, however, the radius of gyration exceeds the correlation length, tricritical behavior is observed.[35,36] Theta-point tricritical behavior is characterized by mean-field critical exponents,[7,8] but the tricritical amplitudes are affected by fluctuation-induced logarithmic corrections.[13] This picture of global non-asymptotic critical behavior in polymer solutions has been confirmed by light-scattering experiments in our laboratory.[34,35,37,38]

This chapter is organized as follows. In Section 2, we discuss the Ising-like critical phenomena in polymer solutions at finite degrees of polymerization. In particular, we show that accurate experiments validate the de Gennes-Sanchez scaling for the critical amplitudes.[30–32] In Section 3, we review the thermodynamics of the theta point in polymer solutions and show that the theta point is a tricritical point. In Section 4, we demonstrate the crossover between Ising criticality and theta-point tricriticality in the shape of the liquid–liquid coexistence curve and the values of the critical exponents. In Section 5, we discuss the Ising-like critical phenomena in polymer blends. Lastly, in Section 6, we present some concluding remarks.

2. Ising-Like Critical Phenomena in Polymer Solutions

The thermodynamic behavior of polymer solutions near their critical demixing transitions, like of all fluids and fluid mixtures, belongs to the three-dimensional Ising-model universality class. Asymptotically close to the liquid–liquid critical point, the thermodynamic properties of polymer solutions (e.g., phase coexistence, osmotic susceptibility, correlation length, etc.) are described by power laws with universal critical exponents, as given in Table 1. This scaling behavior has been confirmed in a variety of systems, including polystyrene in cyclohexane,[10–13,34–38] polystyrene in methylcyclohexane,[39] and in polymethylmethacrylate in 3-octanone.[40,41] In this section, we focus on light-scattering measurements in solutions of polystyrene in

Table 1. Asymptotic power laws of thermodynamic properties with the reduced distance to the critical temperature, $\Delta\hat{T} = (T - T_c)/T_c$, and their critical exponents.[44]

Property	Scaling Law	Mean-Field Critical Exponents	3D Ising-Model Critical Exponents
Heat Capacity	$C = A_0\lvert\Delta\hat{T}\rvert^{-\alpha}$	$\alpha = 0.0$	$\alpha = 0.110 \pm 0.003$
Order Parameter	$\varphi = \pm B_0\lvert\Delta\hat{T}\rvert^{\beta}$	$\beta = 0.5$	$\beta = 0.326 \pm 0.002$
Susceptibility	$\hat{\chi} = \Gamma_0\lvert\Delta\hat{T}\rvert^{-\gamma}$	$\gamma = 1.0$	$\gamma = 1.239 \pm 0.002$
Correlation Length	$\xi = \xi_0\lvert\Delta\hat{T}\rvert^{-\nu}$	$\nu = 0.5$	$\nu = 0.630 \pm 0.002$
Interfacial Tension	$\sigma = \sigma_0\lvert\Delta\hat{T}\rvert^{\mu}$	$\mu = 1.5$	$\mu = 1.260 \pm 0.004$

Note: The asymptotic critical amplitudes of each property (A_0, B_0, Γ_0, ξ_0, and σ_0) are system-dependent. In the text, we use an upper bar to differentiate between the mean-field and scaling theory prediction for the amplitude.

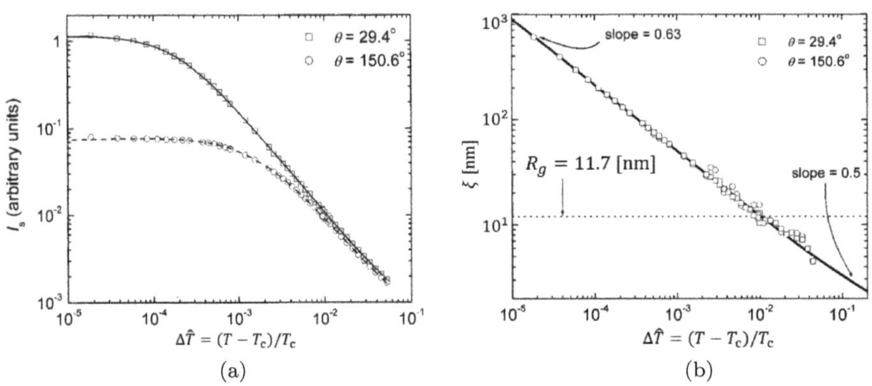

Fig. 2. Demonstration of Ising-model scaling behavior in polymer solutions from (a) Light-scattering intensity for a solution of polystyrene (M_w = 195,900 g/mol) in cyclohexane as a function of the distance to the critical temperature, (b) Correlation length for the same system. The symbols represent the experimental data,[37] while the solid curves represents the theoretical prediction.[35,37] Adapted with permission from Ref. 37 © The Optical Society.

cyclohexane. For instance, as shown in Fig. 2, accurate light-scattering measurements in a solution of polystyrene (M_w = 195,900 g/mol) in cyclohexane, performed by Jacob *et al.*,[37] have confirmed the universal Ising-model behavior of the correlation length, ξ, and of the osmotic susceptibility, $\hat{\chi}$.

In light-scattering experiments on near-critical mixtures, the intensity is proportional to the spatially dependent ("local") osmotic susceptibility, $\hat{\chi}(q\xi)$, and can be written as[34,35]

$$I_s = I_0\hat{\chi}(q\xi) + I_b, \tag{1}$$

where $q = 4\pi n\lambda_0^{-1}\sin(\theta/2)$ is the light-scattering wave number, in which n is the refractive index, λ_0 is the wavelength of incident light, and θ is the scattering angle, while I_0 is an instrumental constant and I_b is a regular "background" scattering intensity (not associated with critical fluctuations). In the mean-field approximation and in the first-epsilon expansion of the renormalization-group (RG) theory,[42,43] the susceptibility is described through the Ornstein–Zernike correlation function[29]

$$\hat{\chi} = \frac{\hat{\chi}_{q=0}}{1 + q^2\xi^2}, \tag{2}$$

where $\hat{\chi}_{q=0}$ is the "bulk" (thermodynamic) susceptibility. In the mean-field approximation and in the first-epsilon RG expansion, the correlation length is related to the thermodynamic susceptibility as $\xi = (c_0\hat{\chi}_{q=0})^{1/2}$, where c_0 is of the order of the square of the range of molecular interactions.[29]

In the scaling theory, the spatially dependent susceptibility is best described by using the following interpolation based on the Fisher–Burford approximation[45,46]:

$$\hat{\chi} \simeq \frac{\hat{\chi}_{q=0}}{(1 + q^2\xi^2)^{1-\eta/2}}. \tag{3}$$

In the scaling theory, the thermodynamic susceptibility is related to the correlation length by $\hat{\chi}_{q=0} \sim \xi^{2-\eta}$. Thus, the critical exponent of the susceptibility, γ, is universally connected to that of the correlation length, ν, through the relation $\gamma = \nu(2 - \eta)$.[16–20] The difference between Eqs. (2) and (3) is almost negligible because the critical exponent, η, in the Ising-model universality class is a very small number, $\eta = 0.033$.[47–49] The two limits of the susceptibility, $\hat{\chi}(q\xi \ll 1) \sim \xi^{2-\eta}$ and $\hat{\chi}(q\xi \gg 1) \sim q^{\eta-2}$, have been observed in the light-scattering measurements of Jacob *et al.*,[37] presented in Fig. 2a, and in Anisimov *et al.*[35] In addition, it was observed that the correlation length, as shown in Fig. 2b, crosses over from asymptotic scaling behavior toward mean-field behavior far away from the critical point. This occurs when the correlation length becomes smaller than the radius of gyration. Due to the two scaling relations between critical exponents, namely $d\nu = 2 - \alpha$, where d is the dimensionality of space, and $2 - \alpha = 2\beta + \gamma$,[16–20] one can see that by measuring the critical exponents of any two thermodynamic properties, all other exponents become known.

A further confirmation of the scaling behavior of polymer solutions has been observed through the measurements of the mutual diffusion coefficient, D, by dynamic light scattering.[37,38] The diffusion coefficient may be expressed as a sum of a critical contribution, D_c, and a background contribution, D_b, as $D = D_c + D_b$. The mode-coupling theory of critical dynamics predicts that, asymptotically close to the critical point, D_c, should satisfy a modified Stokes–Einstein–Kawasaki relation of the form[38,47,50,51]

$$D_c = \frac{R_D k_B T}{6\pi\eta_{app}\xi} K(q\xi) \left[1 + \left(\frac{q\xi}{2}\right)^2\right]^{z_\eta/2} \Omega(\xi/\xi_D), \tag{4}$$

where η_{app} is the apparent mesoscopic viscosity, k_B is Boltzmann's constant, and R_D is a universal dynamic amplitude ratio, typically, of the order of unity.[38,50]

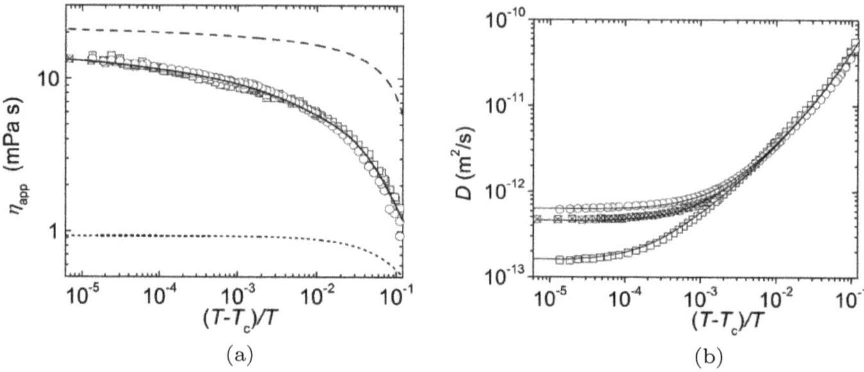

Fig. 3. (a) Apparent (mesoscopic) viscosity of a solution of polystyrene ($M_w = 195,900$ g/mol) in cyclohexane. The data correspond to three light-scattering angles: $\theta = 30°$ (open squares), $\theta = 90°$ (crossed squares), and $\theta = 150°$ (open circles). The solid curve is the prediction from the mode-coupling theory, the dashed curve represents the viscosity of the solution, and the dotted curve represents the viscosity of the solvent. (b) Diffusion coefficient for the same system. The curves represent the fit of the mode-coupling theory to the experimental data using the mesoscopic viscosity presented in (a).[38] Reprinted with permission from Ref. 38.

Asymptotically close to the critical temperature, the actual shear viscosity of the solution, η (not to be confused with the critical exponent of the correlation function), diverges as $\eta = \eta_b(Q_0\xi)^{z_\eta}$, in which Q_0 is a system-dependent amplitude, $z_\eta = 0.068$,[52] and η_b is the non-critical background viscosity.[53] Figure 3(a) depicts the behavior of the viscosity in a solution of polystyrene in cyclohexane. In Eq. (4), the factor in square brackets provides the necessary crossover from the diverging critical viscosity to the constant background viscosity far away from the critical point, $q\xi \gg 1$. Meanwhile, the Kawasaki function, $K(q\xi) \equiv K(x) = [3/(4x^2)](1 + x^2 + [x^3 - x^{-1}]\arctan x)$, satisfies the limits: $K(q\xi \ll 1) = 1$ and $K(q\xi \gg 1) = q\xi$.[50] Lastly, the factor $\Omega(\xi/\xi_D) = (2/\pi)\arctan(\xi/\xi_D)$ accounts for deviations from asymptotic dynamic critical behavior due to a finite cutoff length scale, ξ_D, that is of the order of the radius of gyration. It follows from Eq. (4) that the diffusion coefficient vanishes asymptotically as $D_c \sim \xi^{-(1+z_\eta)}$ at the critical point. The background contribution to the diffusion coefficient, D_b, can be estimated as[38]

$$D_b \approx \frac{k_B T}{16\eta_b\xi}\left[\frac{1+q^2\xi^2}{q_C\xi}\right], \tag{5}$$

where the wavenumber, q_C, can be approximated as ξ_D^{-1}.[38] From Eqs. (4) and (5), the behavior of the diffusion coefficient has the following two limits. In the hydrodynamic regime ($q\xi \ll 1$), the diffusion coefficient becomes independent of q varying with the distance to the critical temperature as $D \sim |\Delta\hat{T}|^{\nu-z_\eta}$,[54,55] while in the non-hydrodynamic limit ($q\xi \gg 1$), $D(q)$ becomes independent of the temperature. Dynamic light-scattering measurements of Kostko *et al.*,[38] presented in Fig. 3(b), have confirmed this behavior of the diffusion coefficient.

Table 2. de Gennes-Sanchez scaling relations for various thermodynamic properties in polymer solutions.[30–32]

Property	Scaling Law	de Gennes–Sanchez Relations	Mean-Field Amplitude	Ising-Model Amplitude
Order Parameter	$\varphi \sim \pm N^{-b}\lvert\Delta\hat{T}\rvert^{\beta}$	$b = (1-\beta)/2$	$b = 0.25$	$b = 0.34$
Heat Capacity	$C \sim N^{-a}\lvert\Delta\hat{T}\rvert^{-\alpha}$	$a = (1+\alpha)/2$	$a = 0.5$	$a = 0.555$
Susceptibility	$\hat{\chi} \sim N^{g}\lvert\Delta\hat{T}\rvert^{-\gamma}$	$g = (1-\gamma)/2$	$g = 0$	$g = -0.12$
Correlation Length	$\xi \sim N^{n}\lvert\Delta\hat{T}\rvert^{-\nu}$	$n = (1-\nu)/2$	$n = 0.25$	$n = 0.185$
Interfacial tension	$\sigma \sim N^{-m}\lvert\Delta\hat{T}\rvert^{\mu}$	$m = (2-\mu)/2$	$m = 0.25$	$m = 0.37$

Note: The amplitude exponents are given by their mean-field and three-dimensional Ising-model forms.

While the critical exponents in polymer solutions, like in all fluids and fluid mixtures, belong to the Ising-model universality class, the critical amplitudes, unique to polymer solutions, diverge or vanish with the degree of polymerization, N, as given in Table 2. Static light-scattering measurements of solutions of polystyrene in cyclohexane from $M_w = 195.900\,\text{g/mol}$ to $M_w = 11.4$ million g/mol, performed in our laboratory,[35] have confirmed that the amplitude of the correlation length and the amplitude of the susceptibility satisfy the predictions of de Gennes and Sanchez,[30–32] both in the vicinity of the critical point and in the mean-field regime far away from the critical point (see Fig. 4). The N-dependent nature of the amplitudes raises a fundamental question about the critical behavior of extremely high-molecular-weight polymer solutions close to the limit of an infinite degree of polymerization.

Typically, the critical behavior of simple fluids is controlled by a single mesoscopic length scale, the correlation length of the concentration fluctuations, ξ.[16,18–20] However, in polymer solutions, there is a second competing length scale, namely the size of the polymer coil (as defined through the radius of gyration, R_g) that depends on the degree of polymerization, N. With increasing values of N, upon approaching the limit $N \to \infty$, the critical concentration (volume fraction), ϕ_c, decreases to zero, while the critical temperature, T_c, approaches the theta temperature (see Fig. 1). In the asymptotic vicinity of the critical temperature, where the correlation length is larger than the radius of gyration, the range of Ising-like critical behavior decreases as R_g increases. The radius of gyration in the polystyrene–cyclohexane system, obtained by neutron-scattering experiments,[56,57] obeys the scaling behavior $R_g \sim N^{1/2}$ along the critical line, see Fig. 4(b). Dynamic light-scattering experiments reveal a second correlation length,[38] associated with the entanglements and disentanglements of the polymer chain (a "viscoelastic" length scale[58]), which we identify with ξ_ψ. Figure 4(b) verifies the theoretical prediction $\xi_\psi \propto R_g$ (see the following section for details). Thus, the asymptotic amplitude of the correlation length scales with the degree of polymerization as $\xi_0 \sim N^{0.185}$ in the Ising-like

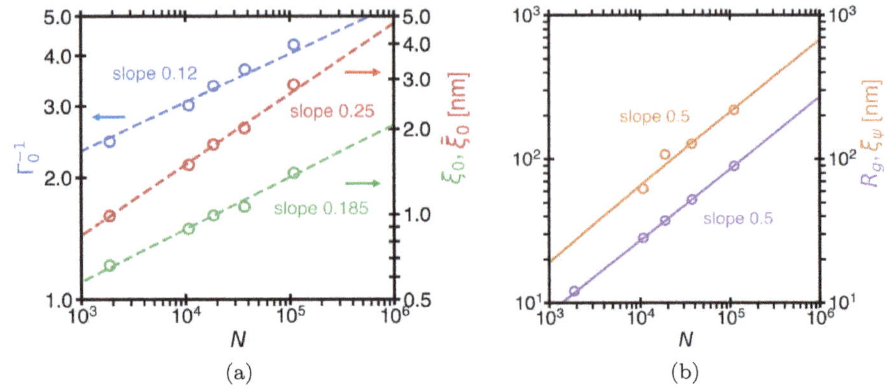

Fig. 4. (a) Experimental confirmation of the de Gennes–Sanchez scaling relations (Table 2). The left axis presents the amplitude of the susceptibility, Γ_0 (blue), while the right axis presents the amplitudes of the correlation length of concentration fluctuations, in the critical region, ξ_0 (blue), and away from the critical point, $\bar{\xi}_0$, in the mean-field region (orange).[35] The slopes of the dashed lines correspond to the values predicted by de Gennes and Sanchez, as given in Table 2. (b) The radius of gyration R_g, as a function of the degree of polymerization N, from neutron-scattering measurements.[56,57] The purple line corresponds to the scaling law, $R_g \sim N^{1/2}$, along the critical line. Also depicted is the correlation length associated with the polymer-chain fluctuations, ξ_ψ, obtained from the coupling between diffusive and viscoelastic dynamic modes.[38] The orange line corresponds to the theoretical prediction $\xi_\psi \propto R_g$ for large N — more details provided in Section 4, see Eq. (22).

critical region (where $\xi > \xi_\psi$), while it scales as $\bar{\xi}_0 \sim N^{0.25}$ in the mean-field region (where $\xi < \xi_\psi$), see Fig. 4(b). Such shrinking of the scaling critical regime is also observed in the behavior of the coexistence curve (see Fig. 1), which leads to the "angle-like" shape in the limit of an infinite degree of polymerization. In the following section, we prove thermodynamically that in the limit $N \to \infty$, the Ising-critical regime disappears completely and the polymer solution exhibits theta-point tricritical behavior.

3. Tricritical Behavior near the Theta-Point

The phase separation behavior of a polymer solution below the critical point of demixing was originally explained in the mean-field theory by Flory and Huggins.[59–61] The Flory–Huggins theory, based on a lattice model, in which the monomers link to form linear polymer chains and are assumed to be the same size as the solvent molecules. According to the Flory–Huggins theory,[61] the Gibbs free energy of mixing per monomer unit is given by

$$\frac{\Delta G}{nk_B T} \equiv \hat{G} = (1-\phi)\ln(1-\phi) + \frac{\phi}{N}\ln\phi + \varpi\phi(1-\phi), \tag{6}$$

where n is the total number of solvent and monomer molecules, ϕ is the volume fraction (the monomer–solvent number ratio), and $\varpi = \Theta/2T$ is the Flory interaction parameter (where Θ is known as the "theta" temperature). We consider the case when the mixture is incompressible and Θ is independent of pressure. The theta temperature is the limiting value of the critical temperature of phase separation in the limit of an infinite degree of polymerization.[62–64]

The critical demixing point is found from the thermodynamic stability conditions, $(\partial^2\hat{G}/\partial\phi^2)|_{T,N} = 0$ and $(\partial^3\hat{G}/\partial\phi^3)|_{T,N} = 0$. In the Flory–Huggins theory, the critical temperature, T_c, and critical volume fraction, ϕ_c, of demixing depend on the degree of polymerization, N, as[61]

$$T_\mathrm{c} = \frac{\Theta}{(1+1/\sqrt{N})^2} \qquad \text{and} \qquad \phi_\mathrm{c} = \frac{1}{1+\sqrt{N}}. \tag{7}$$

In the limit of $N \to \infty$ in the Flory–Huggins theory, the strong asymmetry in the contribution to the Gibbs energy from the entropy of mixing causes an "entropy-driven" phase transition[65] to occur at the theta point, where $T_\mathrm{c} \to \Theta$ and $\phi_\mathrm{c} \to 0$.

Using the relations in Eq. (7), we define the order parameter as $\varphi = \phi - \phi_\mathrm{c}$ and the reduced distance to the critical temperature as $\Delta\hat{T} = (T - T_\mathrm{c})/T_\mathrm{c}$. Expanding the critical part of the Gibbs free energy in the vicinity of the liquid–liquid critical point and in the large N limit gives a typical truncated Landau expansion of the form[29,66]

$$\Delta\hat{G} = \hat{G} - \hat{G}_\mathrm{b} = \frac{1}{2!}\Delta\hat{T}\varphi^2 + \frac{1}{4!}u_0\varphi^4, \tag{8}$$

where the coefficient of the fourth-order term is $u_0 = 2\sqrt{N}$ in the large N limit.[9] The Landau expansion represents mean-field critical behavior. The background ("regular") part of the Gibbs free energy is given by

$$\hat{G}_\mathrm{b} = \hat{\mu}_\mathrm{s}(\phi_\mathrm{c}) + \phi\hat{\mu}_\mathrm{ms}(\phi_\mathrm{c}, \varpi), \tag{9}$$

where the chemical potential of the solvent molecules, $\hat{\mu}_\mathrm{s}$, and the monomer–solvent exchange chemical potential, $\hat{\mu}_\mathrm{ms}$, are defined from appropriate derivatives of the Gibbs free energy: $\hat{\mu}_\mathrm{s} = \hat{G} - \phi\hat{\mu}_\mathrm{ms}$ and $\hat{\mu}_\mathrm{ms} = (\partial\hat{G}/\partial\phi)|_{T,N}$. This part of the free energy does not affect either the shape of the phase boundary or the osmotic susceptibility.

The effects of fluctuations of the order parameter can be incorporated in the mean-field approximation as well as in the first-order RG expansion through the "local" (spatial-dependent) Gibbs free-energy density (per unit volume, V), known as the Landau–Ginzburg functional,[29]

$$\frac{\mathrm{d}(\Delta\hat{G})}{\mathrm{d}V} = \frac{1}{2}\Delta\hat{T}[\varphi(\boldsymbol{r})]^2 + \frac{1}{4!}u_0[\varphi(\boldsymbol{r})]^4 + \frac{1}{2}c_0|\nabla\varphi(\boldsymbol{r})|^2, \tag{10}$$

where $\boldsymbol{r}$ is the spatial coordinate, the first two terms denote the "bulk" thermodynamic potential, given by the Landau expansion, Eq. (8), and the last term is the energy penalty associated with the fluctuation-induced inhomogeneous mixing of the solute and solvent. The coefficient, c_0, is a length scale on the order of the square of the range of molecular interactions. For polymer solutions, c_0 is N-dependent, scaling as $c_0 \sim \sqrt{N}$.[30,35,67–69] From the Landau–Ginzburg functional, the correlation length is related to the thermodynamic susceptibility through $\xi = \sqrt{c_0 \hat{\chi}_{q=0}}$.[29] Since, in the mean-field approximation, the susceptibility is independent of N, scaling as $\hat{\chi}_{q=0} = (\partial^2 \Delta\hat{G}/\partial\varphi^2)|_{T,N} \sim 1/\Delta\hat{T}$ from Eq. (8), then the N-dependence of the correlation length emerges from c_0, such that $\xi \sim \sqrt{c_0/\Delta\hat{T}} = N^{1/4}|\Delta\hat{T}|^{-1/2}$ (see Table 2).

The Landau expansion of the Flory–Huggins free-energy density reveals two problems. First, asymptotically close to the critical point, the Flory–Huggins theory does not capture the Ising-like critical behavior of real solutions (see Table 1). For example, the mean-field Flory–Huggins theory predicts that the volume fraction along coexistence, found from $(\partial\Delta\hat{G}/\partial\varphi)|_{T,N} = 0$, scales as $\varphi = \pm\sqrt{3}N^{-1/4}|\Delta\hat{T}|^{1/2}$, while asymptotically close to the critical point, $\varphi \sim |\Delta\hat{T}|^{\beta}$, with $\beta = 0.326$. Second, Eq. (8) does not describe the phase behavior in the vicinity of the theta point, where $N \to \infty$. In this limit, the coefficient of the fourth-order term diverges at the theta point, and the Landau expansion no longer remains valid. Therefore, the range of applicability of Eq. (8) shrinks with increasing degree of polymerization. To solve these problems, in this section, we present a phenomenological thermodynamic theory to describe the theta point as a tricritical point, and, in the following section, we present a crossover from the asymptotic Ising-like critical behavior to the mean-field-like tricritical behavior near the theta point.

First, we consider the consequences of the Flory–Huggins theory in the case of very high degree of polymerization. In this case, the critical composition becomes a small physical parameter, and the appropriate expansion of the free energy could be given through an expansion of the osmotic pressure. The osmotic pressure, Π, is expressed through the difference of the chemical potential of the pure solvent, $\hat{\mu}_s^\circ$, and its value in solution, $\hat{\mu}_s$, as $\Pi = \rho(\hat{\mu}_s^\circ - \hat{\mu}_s)$, where ρ is the density. The osmotic pressure (expressed in a reduced form as $\hat{\Pi} = \Pi/\rho k_B T$) is related to the Gibbs energy through a Legendre transform, $\hat{\Pi} = \phi\hat{\mu}_{ms} - \hat{G}$, and may be represented as a virial expansion around the theta point as[64,70]

$$\hat{\Pi} \approx \frac{\phi}{N} + B\phi^2 + C\phi^3, \tag{11}$$

where the second-order virial coefficient is related to the distance to the theta temperature by $B = (1/2)(T - \Theta)/T$, while the third-order coefficient is $C = 1/3$.

In the scaling theory, the osmotic pressure, like ordinary pressure, is a field-dependent free energy. This is in contrast to the Gibbs energy, which is a "density"-dependent free energy. One may interconvert between field-dependent and density-dependent free energies through Legendre transforms. For instance, in single-component substances, the density of the grand canonical free energy, Ω, is related to both the pressure, P, of the system and the Helmholtz free energy, A, by a Legendre transform of the form $\Omega/V = -P = \rho A - \rho\mu$, where ρ is the density and the field variable is the chemical potential, μ.

To describe the theta-point thermodynamics in terms of an appropriate density-dependent free energy, $\hat{\Phi}$, we first need to introduce an appropriate order parameter for such an energy. De Gennes,[31,71] based on the ideas of Edwards,[72] defined the order parameter characterizing the long polymer chain as a vector, $\psi(\boldsymbol{r})$, associated with the statistical behavior of the polymer-chain ends.[73–75] De Gennes proved that by integrating over the length of the polymer chain, $\psi(\boldsymbol{r})$ is related to the volume fraction by $|\psi|^2 = \phi$.[31] Assuming that the polymer-chain free energy, $\hat{\Phi}$, may be expanded in powers of the polymer-chain order parameter, we write the expansion truncated at the sixth-order term as

$$\hat{\Phi} = \frac{1}{2}h_2|\psi|^2 + \frac{1}{4}u_\psi|\psi|^4 + \frac{1}{6}g_\psi|\psi|^6, \tag{12}$$

where the coefficient g_ψ is independent of T and ϕ, while u_ψ vanishes at the theta point, as shown in the following. From Eq. (12), we define the "ordering" field $h_1(\boldsymbol{r})$, conjugate to the order parameter, as $h_1(\boldsymbol{r}) = \partial\hat{\Phi}/\partial[\psi(\boldsymbol{r})]$, which, like the order parameter, is also a vector. The second field, h_2, conjugate to $|\psi|^2 = \phi$, is expected to play a similar role as the reduced distance to the critical temperature, $\Delta\hat{T}$, in Eq. (8).

To connect the expansion coefficients (u_ψ and g_ψ) with physical properties of the polymer chain (e.g., ϕ, T, and N), we must relate the polymer-chain free energy, $\hat{\Phi}$, to the osmotic pressure, $\hat{\Pi}$. This can be done through a Legendre transform, $\hat{\Pi}(T, h_1) = \psi(\boldsymbol{r}) \cdot h_1(\boldsymbol{r}) - \hat{\Phi}(T, |\psi|)$ — analogous to the Legendre transform between the pressure and the Helmholtz free-energy density. Performing this Legendre transform, with $\hat{\Phi}$ given by Eq. (12), we find

$$\hat{\Pi} = \frac{1}{2}h_2|\psi|^2 + \frac{3}{4}u_\psi|\psi|^4 + \frac{5}{6}g_\psi|\psi|^6. \tag{13}$$

By definition, Eq. (13) is the reduced osmotic pressure, such that the coefficients of the Landau expansion of the polymer-chain free energy in Eq. (13) must be related to the coefficients of the virial expansion in Eq. (11). Substituting $\phi = |\psi|^2$ in Eq. (11) and matching the corresponding terms, we obtain the following relationships. From the coefficients of the $|\psi|^2$ term,

$$h_2 = \frac{2}{N} - 4B|\psi|^2 - 3C|\psi|^4, \tag{14}$$

while from the coefficients of the $|\psi|^4$ and $|\psi|^6$ terms

$$u_\psi = 4B \quad \text{and} \quad g_\psi = 3C. \tag{15}$$

With use of Eqs. (14) and (15), the expansion of the polymer-chain free energy can be expressed through the virial expansion coefficients as

$$\hat{\Phi} = -\frac{1}{N}|\psi|^2 + B|\psi|^4 + C|\psi|^6. \tag{16}$$

This expression describes the behavior of the polymer solution in the close vicinity of the theta point.

One can observe that the polymer-chain free energy, Eq. (16), contains all of the features present in the Landau theory of tricritical phase transitions.[29] Phenomenologically, a tricritical point separates lines of second-order phase transitions from first-order phase transitions, as shown in Fig. 1. In the Landau theory, a tricritical point exists where both the coefficients of the second- and fourth-order terms in Eq. (16) are zero. The renormalization of the fourth-order term may happen due to the coupling between order parameters (a scalar, φ, and a vector, ψ), described by a coupling term in the Landau expansion in the form of $\lambda\varphi|\psi|^2$.[6–8] This coupling term originally exists in the Flory–Huggins Gibbs energy, Eq. (6), as $\varpi\phi^2 = \varpi\phi|\psi|^2$. Therefore, the Flory–Huggins interaction parameter, ϖ, plays the same role as the coupling constant λ in the theory of tricriticality. The renormalization of the fourth-order term in the polymer-chain free energy, Eq. (16), occurs upon the approach to the theta temperature as $B = 1/2 - \varpi = 1/2(1 - \Theta/T)$ in accordance with the Flory–Huggins theory.

As shown in Eq. (16), this occurs in polymer solutions when $N \to \infty$ and $T \to \Theta$, which indicates that the theta point is a tricritical point. In polymer solutions, states above the theta temperature, along the line of zero polymer volume fraction $\phi = 0$, correspond to the "critical states," characterized as self-avoiding walk singularities.[1,31] While no real second-order phase transition occurs for these states, the line (along $\phi = 0$ with $N \to \infty$) is analogous to the critical (*lambda*) line in other systems with tricritical phase transitions, such as in mixtures of He^3–He^4.[2,3]

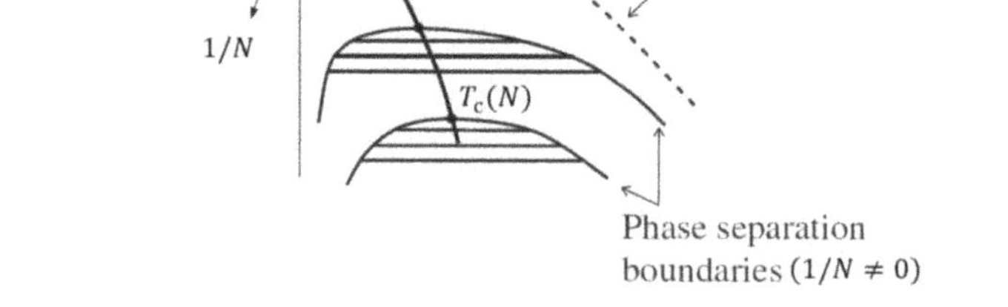

Fig. 5. Polymer-chain scaling fields and phase separation of polymer solution in the vicinity of the theta point (tricritical point). For temperatures $T > \Theta$, the self-avoiding walk singularity line ("lambda line") for a polymer solution in the limit $N \to \infty$ is analogous to a line of second-order phase transitions ($h_1 = h_2 = 0$). For temperatures $T < \Theta$, the lambda-line transforms into a first-order phase transition line. The line splits into two branches: the polymer branch (dashed curve) and pure solvent (vertical line), where the critical line of demixing at finite N, is shown by the solid black curve.

For temperatures below the theta temperature, the system is phase separated, in which the polymer chain entirely collapses into one phase. This collapse of the polymer chain is accompanied by the emergence of a finite polymer volume fraction in this phase and thus can be regarded as a first-order phase transition. In Eq. (16), this behavior is observed depending on the sign of the coefficient of the fourth-order term, B. When $B > 0$, there is no phase separation, while when $B < 0$ the phase separation is first order.

The value for the polymer-chain order parameter, $\psi_0(\boldsymbol{r})$, which lies along coexistence for $T < \Theta$ and along the critical composition for $T > \Theta$, is given from the minimization of the polymer-chain free energy, $\partial\hat{\Phi}/\partial[\psi(\boldsymbol{r})] = 0$, as

$$|\psi_0|^2 = \begin{cases} 0 & \text{for } T > \Theta, \\ -\dfrac{B}{2C} \pm \sqrt{\left(\dfrac{B}{2C}\right)^2 + \dfrac{1}{N}} & \text{for } T \leq \Theta. \end{cases} \tag{17}$$

Thus, the "angle-like" behavior of the coexistence curve is described in the limit $N \to \infty$ (for $T \leq \Theta$) as

$$|\psi_0(N \to \infty)|^2 = \varphi_0 = \frac{3}{2}\left(\frac{\Theta - T}{T}\right). \tag{18}$$

We note that the same result for the phase behavior is predicted from the Flory–Huggins theory by taking the excess chemical potential of the solvent to be zero in the osmotic pressure expansion, Eq. (11), when $N \to \infty$. Similarly, in the limit $T \to \Theta$, the spontaneous polymer-chain order parameter is given by

$$|\psi_0(T \to \Theta)|^2 = \varphi_0 = \sqrt{1/N}. \tag{19}$$

The fact that when $B = 0$, the spontaneous value of the order parameter scales with the leading coefficient to the power $1/4$ is another feature common to tricritical phase transitions.[29]

When the degree of polymerization is large, but still finite, the behavior of the polymer chain is affected by the degree of polymerization. From the polymer-chain free energy, we can define the ordering field for the polymer chain, $h_1(\boldsymbol{r})$, as

$$h_1(\boldsymbol{r}) = \left.\frac{\partial\hat{\Phi}}{\partial[\psi(\boldsymbol{r})]}\right|_{h_2} = \psi(\boldsymbol{r})\left[h_2 + 4B|\psi|^2 + 3C|\psi|^4\right] = \frac{2}{N}\psi(\boldsymbol{r}). \tag{20}$$

Physically, the ordering field in the polymer chain can be related to the energy difference between the chain ends. The closer the polymer chain ends are to each other, the more the chain is affected by this ordering field.

The response of the polymer chain to the ordering field is given by the polymer-chain susceptibility defined as $\hat{\chi}_\psi = (\partial^2\hat{\Phi}/\partial[\psi(\boldsymbol{r})]^2)^{-1}$. Along the spontaneous value of the order parameter, $|\psi_0|$, and in the limit $N \gg B$, the inverse susceptibility is given by

$$\hat{\chi}_{\psi}^{-1} \propto \frac{1}{N} + B^2. \tag{21}$$

The susceptibility scales with the degree of polymerization at the theta temperature, as $\hat{\chi}_{\psi}(T \to \Theta) \propto N$, while in limit $N \to \infty$, the susceptibility is inversely proportional to the distance to the theta point, as $\hat{\chi}_{\psi}(N \to \infty) \propto B^{-2}$. Moreover, the behavior of the correlation length of the polymer-chain order parameter, ξ_{ψ}, in the vicinity of the theta point can be determined through a Landau–Ginzburg functional, similar to Eq. (10), such that

$$\xi_{\psi} = \sqrt{c_{\psi}\hat{\chi}_{\psi}} \propto \frac{\sqrt{N}}{1 + \sqrt{N}B}, \tag{22}$$

where it is important to note that c_{ψ} only depends on the size of the monomer. Equation (22) has two key limits: at the theta temperature, $\xi_{\psi}(T \to \Theta) \propto \sqrt{N}$ and for an infinite degree of polymerization, $\xi_{\psi}(N \to \infty) \propto B^{-1} \sim (\Theta - T)^{-1}$.

A remarkable feature of tricriticality is the fact that in three dimensions, the system near tricritical points demonstrates essentially mean-field behavior. For Ising-like systems, the marginal dimension that separates mean-field criticality and scaling criticality is $d = 4$. For tricritical systems, the marginal dimension is $d = 3$.[13,35,76] Therefore, the crossover between criticality and theta-point tricriticality is essentially a crossover between asymptotic Ising-like critical behavior and mean-field tricritical behavior, which is discussed in detail in the following section.

4. Crossover between Ising Criticality and Theta-Point Tricriticality

As first demonstrated by Widom,[77] the phase coexistence in the mean-field Flory–Huggins theory may be represented in a universal way by rescaling the order parameter, $\varphi = \phi - \phi_c$, and the reduced distance to the critical temperature, $\Delta\hat{T} = (T - T_c)/T_c$, by $\sqrt{N}$. A schematic diagram of the universal mean-field-scaled behavior is illustrated in Fig. 6. Following Widom,[77] we define a "Widom variable" as $x = \sqrt{N}|\Delta\hat{T}|$, and since the critical volume fraction scales as $\phi_c \approx 1/\sqrt{N}$ for large N, we define the rescaled order parameter as $\hat{\varphi} = \varphi/\phi_c \approx \sqrt{N}\varphi$.

Widom demonstrated that there are two key limits of such scaled behavior.[77] First, asymptotically close to the critical point, where $\Delta\hat{T} \gg \sqrt{N}$ and $x \ll 1$, the order parameter along coexistence is $\hat{\varphi} = \sqrt{3x}$. This result also corresponds to the behavior of the Landau expansion of the Flory–Huggins theory. However, Widom's approach also elucidates the behavior of the polymer solution in the alternative limit, where $\Delta\hat{T} \ll \sqrt{N}$ and $x \gg 1$. In this limit, the order parameter along coexistence adopts two solutions: for the polymer-rich phase, $\hat{\varphi} = (3/2)x$, while for the solvent-rich phase, $\hat{\varphi} = -1$, which corresponds to the "angle-like" shape of the coexistence observed in the theta-point (tricritical) limit, Eq. (18).

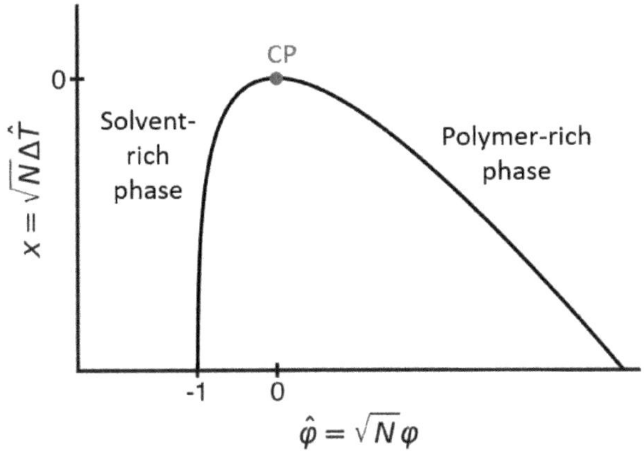

Fig. 6. A universal phase diagram in Flory–Huggins theory, as suggested by Widom,[77] is achieved by rescaling the volume-fraction order parameter, φ, and the reduced distance to the critical temperature, $\Delta\hat{T}$, by the square root of the degree of polymerization.

As already discussed, the actual (fluctuation-induced) behavior of the phase coexistence asymptotically close to the critical point follows the Ising-model universality class. Thus, in the region where $x \ll 1$, according to the scaling theory, $\hat{\varphi} \simeq x^{\beta}$, where $\beta = 0.326$ (see Table 1). The region where the mean-field approximation becomes insufficient, and the effects of critical fluctuations must be taken into account, is given by the "Ginzburg criterion,"[19,28] $|\Delta\hat{T}| \ll |\Delta\hat{T}|_{\times} = N_{\mathrm{G}}$, where N_{G} is the Ginzburg number and $|\Delta\hat{T}|_{\times} = (T_{\times} - T_{\mathrm{c}})/T_{\mathrm{c}}$. The Ginzburg number can be estimated by the following expression[19,28]:

$$|\Delta\hat{T}|_{\times} \approx u_0^2 \left(\frac{v_0}{\bar{\xi}_0^3}\right)^2, \tag{23}$$

where u_0 is the fourth-order coefficient in the Landau expansion and $v_0^{1/3}$ is of the order of the range of interactions, ξ_0. The Ginzburg number indicates the crossover temperature from mean-field behavior away from the critical point to Ising-like behavior in the asymptotic vicinity of the critical point. For ordinary second-order phase transitions, the Ginzburg number is controlled by the range of interactions, ξ_0, while u_0 and v_0 are effectively constant.[27] For polymer solutions, $u_0 = 2\sqrt{N}$, as given in Eq. (8), v_0 is the volume of the monomer, and $\bar{\xi}_0 \approx v_0^{1/3}N^{1/4}$ is the amplitude of the correlation length in the mean-field approximation. Thus, Eq. (23) can be simplified as $|\Delta\hat{T}|_{\times} = c/\sqrt{N}$, where c is a system-dependent constant of the order of unity. Analogously, at $x = \sqrt{N}|\Delta\hat{T}| = 1$, the Widom crossover indicates the crossover temperature from mean-field tricritical behavior in the vicinity of the theta point to the behavior (either mean-field or Ising-like) in the vicinity of the critical-demixing point.

The Widom procedure for the crossover between the mean-field critical regime and the theta-point tricritical regime in polymer solutions was generalized by

Anisimov and co-workers to include the effects of critical fluctuations in the vicinity of the critical-demixing point.[9,13,34–36,43] We refer to this approach as the "generalized Widom crossover," and in this review, we present a simplified version of this crossover, emphasizing the essential physics of this phenomenon.

First, in the vicinity of the theta point, there are logarithmic corrections to the mean-field behavior, induced by fluctuations, since the marginal dimension for tricriticality is $d = 3$, that must be accounted for.[13,35,76] In this case, the classical chain-length dependence of the critical parameters, ϕ_c and T_c, as given by Eq. (7), are modified by tricritical fluctuations. The modified critical parameters are[13]

$$\phi_c = \frac{1}{1 + a\sqrt{N}(1 + \ln N)^{-1/2}} \quad \text{and} \quad \frac{\Theta - T_c}{\Theta} \approx \frac{1}{(\ln N)^{3/22}\sqrt{N}}, \tag{24}$$

where the coefficient a is of the order of unity. Note that there is a significant effect of logarithmic fluctuations on the critical volume fraction, as shown in Fig. 7, as compared with the prediction of the Flory–Huggins theory, Eq. (7). Meanwhile, the effect of logarithmic corrections on the critical temperature is relatively small, but still consequential.

In the generalized Widom crossover, as a result of the logarithmic correction to the theta-point tricritical behavior, a more accurate representation for the Widom variable, x, is $\tilde{x} = (T - T_c)/(\Theta - T_c)$, where the critical temperature is measured experimentally and the theta temperature is obtained from extrapolation.[13,34–36] In the Flory–Huggins theory, $\tilde{x} = x$, as $(\Theta - T_c)/T_c \approx 1/\sqrt{N} = \phi_c$ for large N. Moreover, the reduced order parameter, $\hat{\varphi} = (\phi - \phi_c)/\phi_c$, is rescaled with a critical volume fraction determined from experiment. From here, a simple interpolation

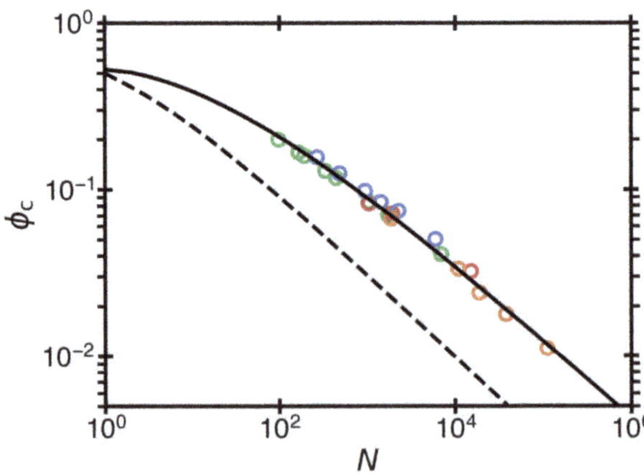

Fig. 7. Critical volume fraction, ϕ_c, of solutions of polystyrene in cyclohexane measured as a function of the degree of polymerization, N, by Kojima and Nakata *et al.*[10–12] (red), polystyrene in cyclohexane by Anisimov *et al.*[35] (orange), polystyrene in methylcyclohexane by Dobashi *et al.*[39] (green), and PMMA in 3-octanone by Xia *et al.*[40,41] (blue). The solid curve represents Eq. (24), with coefficient $a = 0.95$, while the dashed curve represents the prediction from the Flory–Huggins theory, Eq. (7).

function, with a crossover parameter given through the Widom variable, is utilized to modify the universal variables ($\tilde{x}$ and $\hat{\varphi}$) to match the scaling limits in the asymptotic Ising-like critical regime. The crossover parameter responsible for the transition from Ising-like criticality to theta-point tricriticality is given as the ratio $z = \xi/R_g$.[9,13,33–36] We note that in the Flory–Huggins theory, the crossover variable reduces to the Widom variable as $z = \tilde{x}^{-1/2}$. Therefore, the crossover behavior of the phase coexistence in polymer solutions can be represented by a simple generalization of the Widom crossover as

$$\hat{\varphi} \approx \hat{\varphi}_{\mathrm{MF}} \left(1 + z^2\right)^{(1-2\beta)/4\nu}, \tag{25}$$

where the amplitude function, $\hat{\varphi}_{\mathrm{MF}}$, is predicted from the mean-field Flory–Huggins-like theory. The Ising critical exponent is $\beta = 0.326$ (see Table 1). Equation (25) has the limits: $\hat{\varphi}(z \ll 1) \to \hat{\varphi}_{\mathrm{MF}}$ and $\hat{\varphi}(z \gg 1) \to B_0|\Delta\hat{T}|^{\beta}$, where $B_0 \sim N^{-0.337}$ (see Table 2).

In a form similar to the phase coexistence behavior Eq. (25), the crossover expressions for the susceptibility and correlation length are given by[35,36]

$$\hat{\chi}^{-1} \approx \hat{\chi}_{\mathrm{MF}}^{-1}(1 + z^2)^{(\gamma-1)/2\nu}, \tag{26}$$

$$\xi \approx \xi_{\mathrm{MF}}(1 + z^2)^{(2\nu-1)/4\nu}, \tag{27}$$

where the amplitude functions are the susceptibility and correlation length predicted by the mean-field Flory–Huggins-like theory, $\hat{\chi}_{\mathrm{MF}}^{-1} = \bar{\Gamma}_0|\Delta\hat{T}|$ and $\xi_{\mathrm{MF}} = \bar{\xi}_0|\Delta\hat{T}|^{-0.5}$ (where $\bar{\Gamma}_0 \sim N^0$ and $\bar{\xi}_0 \sim N^{-0.25}$ are the mean-field amplitudes), see Eq. (10). Asymptotically close to the critical point (for $z \gg 1$), Eqs. (26) and (27) satisfy the critical power laws, $\hat{\chi} = \Gamma_0|\Delta\hat{T}|^{-\gamma}$ and $\xi = \xi_0|\Delta\hat{T}|^{-\nu}$ (see Table 1). For example, consider the phase behavior of the eight different solutions of polystyrene in methylcyclohexane with different degrees of polymerization[13,39] presented in Fig. 8(a). The universal crossover behavior of this system is shown in Fig. 8(b).

By rescaling the thermodynamic properties, we may represent the crossover phenomenon in a universal form as

$$\tilde{\varphi} \equiv \frac{\hat{\varphi}|\Delta\hat{T}|_{\times}^{-\beta}}{B_0} \approx \frac{(1 + z^2)^{(1-2\beta)/4\nu}}{(\Delta\hat{T}_{\times}/\Delta\hat{T})^{1/2}}, \tag{28}$$

$$\tilde{\chi} \equiv \frac{\chi|\Delta\hat{T}|_{\times}^{\gamma}}{\Gamma_0} \approx \frac{(1 + z^2)^{(1-\gamma)/2\nu}}{\Delta\hat{T}/\Delta\hat{T}_{\times}}, \tag{29}$$

$$\tilde{\xi} \equiv \frac{\xi|\Delta\hat{T}|_{\times}^{\nu}}{\xi_0} \approx \frac{(1 + z^2)^{(2\nu-1)/4\nu}}{(\Delta\hat{T}/\Delta\hat{T}_{\times})^{1/2}}. \tag{30}$$

Figure 9 shows the universal susceptibility (a) and correlation length (b) for five different degrees of polymerization in a polystyrene-cyclohexane solution.[10–12,35,39–41] The inset of Fig. 9(a) shows the deviation of the susceptibility from the asymptotic Ising-like behavior in the vicinity of the theta point. In their rescaled forms,

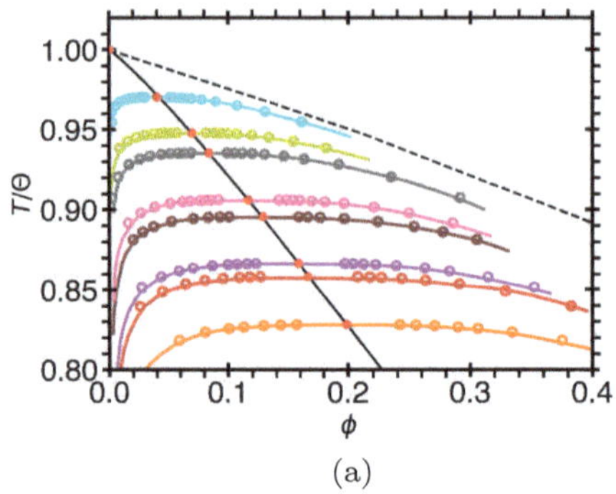

Fig. 8. Phase diagram and universal behavior of a solution of polystyrene in methylcyclohexane. (a) Phase behavior of the polymer solution. The open circles denote experimental data,[39] while the curves denote the theoretical prediction.[33] (b) Rescaling the volume fraction and temperature generates a universal phase diagram. As shown, the polymer-rich phase exhibits a crossover from the asymptotic Ising-like critical region to the tricritical theta-point region, while the solvent-rich phase exhibits a crossover from the asymptotic Ising-like region to the limit of pure solvent.

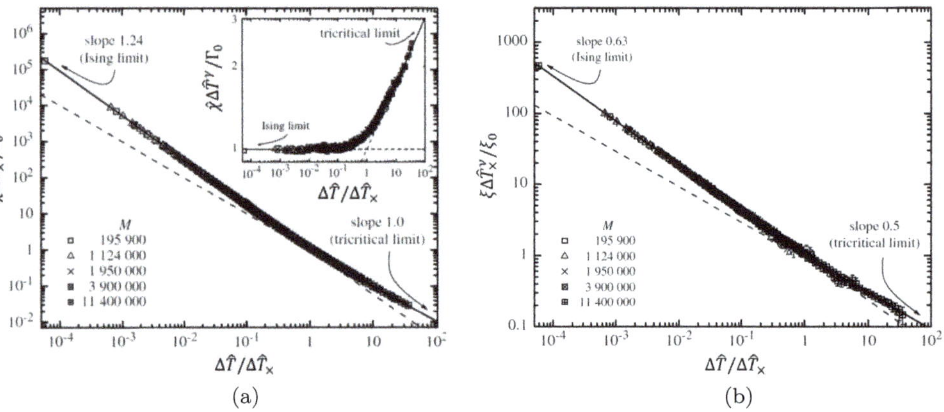

Fig. 9. Universal scaling behavior of thermodynamic properties. (a) Scaled susceptibility and deviation of the susceptibility from Ising critical behavior (shown as inset) as a function of the normalized distance from the critical temperature, $\Delta\hat{T}/\Delta\hat{T}_\times$, for polystyrene–cyclohexane solutions with five different molecular weights of polystyrene. (b) The scaled correlation length for the same data. The symbols indicate experimental data, while the solid curve represents the crossover theory. The dashed lines represent the two limiting behaviors: Ising asymptotic behavior and theta-point (tricritical) behavior. Adapted with permission from Ref. 35.

the thermodynamic properties exhibit the middle of their crossover at the Widom crossover point, $x = \Delta\hat{T}/\Delta\hat{T}_\times = 1$.

Another phenomenon associated with the crossover to theta-point tricriticality is the dynamic coupling of the diffusive relaxation of the critical concentration fluctuations discussed in Section 2, and entanglements/disentanglements of long polymer chains (commonly referred to in the literature as a "viscoelastic mode"[58]).

Experimental verification of the dynamic coupling between these two modes was demonstrated independently by Tanaka *et al.*[58] and Kostko *et al.*[35,78] According to the theory of de Gennes and Brochard,[79,80] the normalized time-dependent intensity correlation function, $g_2(t)$, is given in the form

$$g_2(t) - 1 = \left[f_+ e^{-\omega_+ t} + f_- e^{-\omega_- t}\right]^2, \tag{31}$$

with decay frequencies given by

$$\omega_\pm = \frac{\omega_\psi}{2}\left[1 + q^2\xi_\psi^2 + \frac{\omega_q}{\omega_\psi} \pm \sqrt{\left(1 + q^2\xi_\psi^2 + \frac{\omega_q}{\omega_\psi}\right)^2 - 4\frac{\omega_q}{\omega_\psi}}\right] \tag{32}$$

and corresponding amplitudes f_+ and f_- given by

$$f_\pm = \pm\frac{\omega_\pm - \omega_\psi(1 + q^2\xi_\psi^2)}{\omega_+ - \omega_-}, \tag{33}$$

where $\omega_\pm$ are the coupled modes, in which the relaxation of the faster mode is designated as, ω_+, and consequently, the slower mode is designated as ω_-. In previous works, ξ_ψ was defined as a mesoscopic viscoelastic length scale, which we identify with the correlation length of the polymer chain, given by Eq. (22). In Eqs. (32) and (33), ω_q represents the uncoupled q-dependent diffusion relaxation mode, as specified through the diffusion equations given by Eqs. (4) and (5), and ω_ψ is an uncoupled q-independent viscoelastic relaxation mode associated with the polymer chain and assumed to be inversely proportional to the viscosity of the solution, which diverges in the limit $N \to \infty$, as $\omega_\psi \propto \eta(T)^{-1}$.[38] The diffusive relaxation mode, ω_q, is associated with the conserved order parameter, $\varphi = \phi - \phi_c$, while the viscoelastic mode is associated with the non-conserved polymer-chain order parameter $|\psi|$.[38,58,78] The coupling between the two dynamic modes reflects the coupling between these two order parameters that is responsible for the emergence of the theta-point tricriticality.

The effect of coupling between the two dynamic modes is most clearly observed in polymer solutions with a high degree of polymerization (closer vicinity to the theta point). For instance, Fig. 10 depicts the dynamic coupling in a solution of polystyrene in cyclohexane with molecular weight of 11.4 million g/mol (N = 110,000).[38] The avoided crossing in the behavior of the fast and slow frequency modes, $\omega_\pm$, is reminiscent of the Landau-Zener effect in the transition dynamics of a two-state quantum system.[81,82] As illustrated in Fig. 10, in the vicinity of the critical point, both the uncoupled viscoelastic and diffusive modes saturate. The diffusive relaxation time saturates at a finite wave number, controlled by the scattering angle, while the viscoelastic relaxation time saturates because the degree of polymerization is finite and ξ_ψ reaches a finite value. In this region, the characteristic frequency of the faster coupled mode scales with the correlation length of the polymer chain as $\omega_+ = \omega_\psi(1 + q^2\xi_\psi^2)$, while the frequency of the slower coupled mode scales as $\omega_- = \omega_q(1 + q^2\xi_\psi^2)^{-1}$. In the region far away from

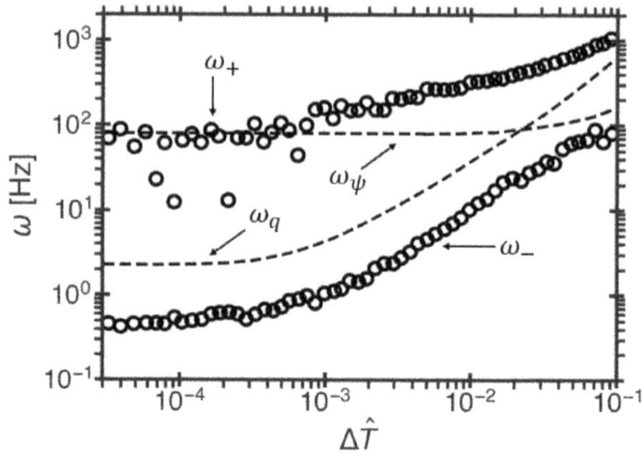

Fig. 10. Decay frequencies, $\omega_{\pm}$, for a solution of polystyrene in cyclohexane, with $N = 110,000$, measured at scattering angle, $\theta = 30°$.[38] The symbols represent the values for the two coupled modes extracted from the fit of the correlation-function data, Eq. (31). The points correspond to the faster mode, ω_+ (upper), and the slower mode, ω_- (lower). The dashed curves represent the theoretical calculations of the uncoupled viscoelastic mode, ω_ψ, and the uncoupled diffusive mode, ω_q.

the critical temperature, the two modes uncouple as $\omega_- \approx \omega_\psi$ and $\omega_+ \approx \omega_q$.[38,58,78] We conclude that dynamic light-scattering experiments enable one to obtain the polymer-chain correlation length, ξ_ψ.

5. Critical Phenomena in a Polymer Blend

Our treatment of the critical behavior in polymer solutions may be generalized to a mixture of two incompressible polymer species, A and B, with degrees of polymerization $N_{\rm A}$ and $N_{\rm B}$, respectively. Modifying the Flory–Huggins theory to account for the lengths of each species, the Gibbs energy of mixing of the blend, $\hat{G}_{\rm bd}$, is given by[59–61,83–85]

$$\hat{G}_{\rm bd} = \left[\frac{\phi}{N_{\rm A}}\ln\phi + \frac{(1-\phi)}{N_{\rm B}}\ln(1-\phi)\right] + \varpi\phi(1-\phi), \tag{34}$$

where the Flory interaction parameter is approximated as $\varpi = \Theta/2T$, see Eq. (6). Equation (34) has the following two limits: first, in the case when one species is much longer than the other, $N_{\rm A} \equiv N \gg N_{\rm B}$ and $N_{\rm B} \approx 1$, then Eq. (34) reduces to the Gibbs energy of mixing in the Flory–Huggins theory, Eq. (6). Hereafter, we refer to this case as the asymmetric limit. Second, in the case when each species is of the same size, $N_{\rm A} = N_{\rm B} = N$, then the contribution from the entropy of mixing, the term in brackets in Eq. (34), scales with $1/N$. We refer to this case as the symmetric limit.

The critical volume fraction and critical temperature are determined from the thermodynamic stability conditions, $(\partial^2\hat{G}_{\rm bd}/\partial\phi^2)|_{T,N_{\rm A},N_{\rm B}} = 0$ and

$(\partial^3 \hat{G}_{\text{bd}}/\partial\phi^3)|_{T,N_\text{A},N_\text{B}} = 0$, yielding

$$\phi_\text{c} = \frac{\sqrt{N_\text{B}}}{\sqrt{N_\text{A}} + \sqrt{N_\text{B}}} \quad \text{and} \quad T_\text{c} = \Theta\left[\frac{1}{\sqrt{N_\text{A}}} + \frac{1}{\sqrt{N_\text{B}}}\right]^{-2}. \tag{35}$$

In the asymmetric limit, the critical parameters given by Eq. (35) reduce to those given by Eq. (7). In the symmetric limit, $\phi_\text{c} = 1/2$ and $T_\text{c} = \Theta N/2$, and, when both polymer chains are sufficiently long, such that $N \gg 1$, then $T_\text{c} \to \infty$, and the polymer blend would not mix for any degree of non-ideality.

Just as given by Eq. (8), the truncated Landau expansion may be represented in the form

$$\Delta\hat{G}_{\text{bd}} = \frac{1}{2}\tilde{a}_0 \Delta\hat{T}\varphi^2 + \frac{1}{4!}\tilde{u}_0\varphi^4, \tag{36}$$

where the expansion coefficients for the blend are given by

$$\tilde{a}_0 = \left(\frac{1}{\sqrt{N_\text{A}}} + \frac{1}{\sqrt{N_\text{B}}}\right)^2 \quad \text{and} \quad \tilde{u}_0 = \sqrt{N_\text{A}}\sqrt{N_\text{B}}\left(\frac{1}{\sqrt{N_\text{A}}} + \frac{1}{\sqrt{N_\text{B}}}\right)^4. \tag{37}$$

The Landau expansion of the Flory–Huggins Gibbs free energy, Eq. (8), is recovered in the asymmetric limit. In the symmetric limit, the second- and fourth-order coefficients of Eq. (37) are given by $a_0 = 4/N$ and $u_0 = 32/N$. In the symmetric blend, any degree of non-ideality will produce phase separation in the limit $N \to \infty$.

The osmotic pressure for the polymer blend is related to the Gibbs energy through a Legendre transform, $\hat{\Pi}_{\text{bd}} = \phi(\partial\hat{G}_{\text{bd}}/\partial\phi)|_{T,N_\text{A},N_\text{B}} - \hat{G}_{\text{bd}}$, and may be represented as a virial expansion around the theta point as[84]

$$\hat{\Pi}_{\text{bd}} = \frac{\phi}{N_\text{A}} + \frac{1}{2}\left(\frac{1}{N_\text{B}} - \frac{\Theta}{T}\right)\phi^2 + \frac{1}{3N_\text{B}}\phi^3. \tag{38}$$

Comparing with the asymmetric limit for osmotic pressure, Eq. (11), we see that the second and third virial coefficients depend on N_B as

$$B = \frac{1}{2}\left(\frac{1}{N_\text{B}} - \frac{\Theta}{T}\right) \quad \text{and} \quad C = \frac{1}{3N_\text{B}}. \tag{39}$$

This dependence on N_B has important implications for the polymer-chain free energy, Eq. (16). Following the derivation in Section 4, the polymer-chain free energy of the blend is given by

$$\hat{\Phi}_{\text{bd}} = -\frac{1}{N_\text{A}}|\psi|^2 + \frac{1}{2}\left(\frac{1}{N_\text{B}} - \frac{\Theta}{T}\right)|\psi|^4 + \frac{1}{3N_\text{B}}|\psi|^6. \tag{40}$$

The condition for theta-point tricriticality is only observed in the asymmetric limit, where $N_\text{A} \gg N_\text{B} \to \infty$. However, in the symmetric limit, the polymer-chain free energy becomes undefined at the theta point as $T \to \Theta$ and $N \to \infty$, while $\psi(\boldsymbol{r}) \to 0$, implying that there is no theta-point tricriticality in a symmetric blend.

It is important to note that, for finite N, the polymer blend exhibits Ising-like critical behavior in the vicinity of the liquid–liquid critical point,[86,87] just like all fluid mixtures.[16,19] The size of the critical region is determined through the Ginzburg criterion, Eq. (23), which scales with the degree of polymerization of each species as[87–90]

$$N_G \sim \frac{1}{\sqrt{N_A N_B}}. \tag{41}$$

Equation (41) has the following two limits: In the asymmetric limit, $N_G \sim 1/\sqrt{N}$, just as discussed in Section 4, while in the symmetric limit, $N_G \sim 1/N$. Thus, a symmetric blend, with a large degree of polymerization, has a much smaller critical region than a polymer blend in the asymmetric limit. This fact, combined with the consequences of Eq. (40), means that the symmetric polymer blend would only exhibit one type of crossover, namely the crossover from Ising-like critical behavior in the vicinity of the critical point to mean-field behavior away from the critical point. As the polymer blend becomes asymmetric, a tricritical region would emerge as discussed in Section 4.

6. Conclusion

We have shown that the nature of critical fluctuations in polymer solutions, including the crossover from Ising-like critical behavior near the critical point of mixing to tricritical behavior near the theta point, has been resolved both theoretically and experimentally. Specifically, this crossover is governed by a competition between two characteristic length scales, namely the correlation length of concentration fluctuations, ξ, related to the osmotic susceptibility, which diverges at the critical point, and the polymer-chain correlation length, ξ_ψ, related to the radius of gyration, R_g. This competition causes a coupling between the two order parameters: the conserved order parameter associated with the polymer concentration, ϕ, and the non-conserved order parameter associated with the fluctuations of the long polymer chains, $\psi(\boldsymbol{r})$, which accounts for the observed static and dynamic phenomena in polymer solutions.

Acknowledgments

J.V.S. and M.A.A. acknowledge a long-lasting friendship and fruitful interactions with Michael E. Fisher and Benjamin Widom. M.A.A. and T.J.L. also thank Michael Rubinstein, Ralph H. Colby, and Nicholas Tang for the recent stimulating discussions on the nature of the crossover from Ising-like criticality to theta-point tricriticality. The research of M.A.A. and T.J.L. was supported by the NSF Award No. 1856479.

References

1. M. E. Fisher, The story of coulombic criticality. *J. Stat. Phys.* **75**, 1–36 (1994).
2. D. Vollhardt and P. Wölfle, *The Superfluid Phases of Helium 3* (Taylor and Francis, London, 1990).
3. A. Schmitt, *Introduction to Superfluidity.* vol. 888. *Lecture Notes in Physics* (Springer, New York, 2015).
4. I. S. Jacobs and P. E. Lawrence, Metamagnetic phase transitions and hysteresis in $FeCl_2$. *Phys. Rev.* **164**, 866–878 (1967).
5. M. E. Fisher and D. R. Nelson, Universality of magnetic tricritical points, *Phys. Rev. B.* **12**, 263–266 (1975).
6. M. A. Anisimov, E. E. Gorodetskiĭ and V. M. Zaprudskiĭ, Phase transitions with coupled order parameters. *Sov. Phys. Usp.* **24**, 57–75 (1981).
7. C. M. Knobler and R. L. Scott, Multicritical points in fluid mixtures: Experimental studies. In C. Domb and J. L. Lebowitz (eds.), *Phase Transitions and Critical Phenomena*, Vol. 9 (Academic, New York, 1984).
8. M. A. Anisimov, *Critical Phenomena in Liquids and Liquid Crystals.* (Gordon and Breach, Philadelphia, 1991).
9. A. A. Povodyrev, M. A. Anisimov and J. V. Sengers, Crossover flory model for phase separation in polymer solutions. *Physica A.* **264**, 345–369 (1999).
10. J. Kojima, N. Kuwahara and M. Kaneko, Light scattering and pseudospinodal curve of the system polystyrene — cyclohexane in the critical region. *J. Chem. Phys.* **63**, 333 (1975).
11. M. Nakata, N. Kuwahara and M. Kaneko, Coexistence curve for polystyrene — cyclohexane near the critical point. *J. Chem. Phys.* **62**, 4278 (1975).
12. M. Nakata, T. Dobashi, N. Kuwahara, M. Kaneko and B. Chu, Coexistence curve and diameter of polystyrene in cyclohexane. *Phys. Rev. A.* **18**, 2683–2688 (1978).
13. J. S. Hager, M. A. Anisimov, J. V. Sengers and E. E. Gorodetskiĭ, Scaling of demixing curves and crossover from critical to tricritical behavior in polymer solutions. *J. Chem. Phys.* **117**, 5940 (2002).
14. H. E. Stanley, *Introduction to Phase Transitions and Critical Phenomena* (University, Oxford, 1971).
15. K. G. Wilson, The renormalization group — introduction. In C. Domb and M. S. Green (eds.), *Phase Transitions and Critical Phenomena*, Vol. 6 (Academic, New York, 1976).
16. M. E. Fisher, Scaling, universality and renormalization group theory. In F. J. W. Hahne (eds.), *Critical Phenomena*, Lecture Notes in Physics, pp. 1–139 (Springer, Berlin, Heidelberg, 1983).
17. K. G. Wilson, The renormalization group and critical phenomena. *Rev. Mod. Phys.* **55**, 583–600 (1983).
18. N. Goldenfeld, *Lectures on Phase Transitions and the Renormalization Group* (Addison-Wesley, New York, 1992).
19. M. A. Anisimov and J. V. Sengers, *Critical Region*, In J. V. Sengers, R. F. Kayser, C. J. Peters and H. J. White, Jr. (eds.), *Equations of State for Fluids and Fluid Mixtures, Part I*, chapter 11, pp. 381–433 (Elsevier, Amsterdam, 2000).
20. A. Pelissetto and E. Vicari, Critical phenomena and renormalization-group theory. *Phys. Rep.* **368**, 549–727 (2002).

21. M. S. Green and J. V. Sengers (eds.), *Critical Phenomena: Proceedings of a Conference Held in Washington, D. C., April 1965* (National Bureau of Standards Miscellaneous Publication, 1966 p. 273.
22. J. F. Nicoll and P. C. Albright, Crossover functions by renormalization-group matching: Three-loop results. *Phys. Rev. B.* **31**, 4576–4589 (1985).
23. Z. Y. Chen, P. C. Albright and J. V. Sengers, Crossover from singular critical to regular classical thermodynamic behavior of fluids. *Phys. Rev. A.* **41**, 3161–3177 (1990).
24. Z. Y. Chen, A. Abbaci, S. Tang and J. V. Sengers, Global thermodynamic behavior of fluids in the critical region. *Phys. Rev. A.* **42**, 4470–4484 (1990).
25. S. Tang, J. V. Sengers, and Z. Y. Chen, Nonasymptotic critical thermodynamical behavior of fluids. *Physica A.* **179**, 344–377 (1991).
26. A. Kostrowicka Wyczalkowska, M. A. Anisimov, and J. V. Sengers, Global crossover equation of state of a van der Waals fluid. *Fluid Phase Equilibria.* **158–160**, 523–535 (1999).
27. Y. C. Kim, M. A. Anisimov, J. V. Sengers and E. Luijten, Crossover critical behavior in the three-dimensional Ising model. *J. Stat. Phys.* **110**, 591–609 (2003).
28. M. A. Anisimov, S. B. Kiselev, J. V. Sengers and S. Tang, Crossover approach to global critical phenomena in fluids. *Physica A.* **188**, 487–525 (1992).
29. L. D. Landau and E. M. Lifshitz. *Statistical Physics Part 1.* Vol. 5. *Course of Theoretical Physics* (3rd edn.), (Pergamon, Oxford, 1980).
30. P.-G. de Gennes, Critical opalescence of macromolecular solutions. *Phys. Lett.* **26A**, 313–314 (1968).
31. P.-G. de Gennes, *Scaling Concepts in Polymer Physics* (Cornell University, Ithaca, 1979).
32. I. C. Sanchez, Critical amplitude scaling laws for polymer solutions. *J. Phys. Chem.* **93**, 6983–6991 (1989).
33. M. A. Anisimov, V. A. Agayan and E. E. Gorodetskiǐ, Scaling and crossover to tricriticality in polymer solutions. *Jetp Lett.* **72**, 578–582 (2000).
34. M. A. Anisimov, A. F. Kostko and J. V. Sengers, Competition of mesoscales and crossover to tricriticality in polymer solutions. *Phys. Rev. E.* **65**, 051805 (2002).
35. M. A. Anisimov, A. F. Kostko, J. V. Sengers and I. K. Yudin, Competition of mesoscales and crossover to theta-point tricriticality in near-critical polymer solutions. *J. Chem. Phys.* **123**, 164901 (2005).
36. M. A. Anisimov and J. V. Sengers, Scaling, tricriticality, and crossover in polymer solutions. *Mol. Phys.* **103**, 3061–3070 (2005).
37. J. Jacob, M. A. Anisimov, J. V. Sengers, V. Dechabo, I. K. Yudin and R. W. Gammon, Light scattering and crossover critical phenomena in polymer solutions. *Appl. Opt.* **40**, 4160–4169 (2001).
38. A. F. Kostko, M. A. Anisimov and J. V. Sengers, Dynamics of critical fluctuations in polymer solutions. *Phys. Rev. E.* **76**, 021804 (2007).
39. T. Dobashi, M. Nakata and M. Kaneko, Coexistence curve of polystyrene in methylcyclohexane. I. Range of simple scaling and critical exponents. *J. Chem. Phys.* **72**, 6685 (1980).
40. K. Xia, Interfacial tensions of phase-separated polymer solutions. *J. Chem. Phys.* **97**, 1446 (1992).
41. K. Xia, X. An and W. Shen, Measured coexistence curves of phase-separated polymer solutions. *J. Chem. Phys.* **105**, 6018 (1996).
42. K. G. Wilson and M. E. Fisher, Critical exponents in 3.99 dimensions. *Phys. Rev. Lett.* **28**, 240–243 (1972).

43. V. A. Agayan, M. A. Anisimov and J. V. Sengers, Crossover parametric equation of state for Ising-like systems. *Phys. Rev. E.* **64**, 026125 (2001).
44. H. Behnejad, J. V. Sengers and M. A. Anisimov. Thermodynamics behaviour of fluids near critical points. In A. R. Goodwin, C. J. Peters and J. V. Sengers (eds.), *Applied Thermodynamics of Fluids*, Chapter 10, pp. 321–367 (Royal Society of Chemistry, Cambridge, 2010).
45. M. E. Fisher, Correlation functions and the critical region of simple fluids. *J. Math. Phys.* **5**, 944 (1964).
46. M. E. Fisher and R. J. Burford, Theory of critical-point scattering and correlations. I. the Ising model. *Phys. Rev.* **156**, 583–622 (1967).
47. B. I. Halperin, P. C. Hohenberg and E. D. Siggia, Renormalization-group calculations of divergent transport coefficients at critical points. *Phys. Rev. Lett.* **32**, 1289–1292 (1974).
48. E. D. Siggia, B. I. Halperin and P. C. Hohenberg, Renormalization-group treatment of the critical dynamics of the binary-fluid and gas-liquid transitions. *Phys. Rev. B.* **13**, 2110–2123 (1976).
49. J. V. Sengers, Transport properties of fluids near critical points. *Int. J. Thermophys.* **6**, 203 (1985).
50. K. Kawasaki, Mode coupling and critical dynamics. In C. Domb and M. S. Green (eds.), *Phase transitions and Critical Phenomena*, vol. 5A. Academic, New York (1976).
51. H. C. Burstyn, J. V. Sengers, J. K. Bhattacharjee, and R. A. Ferrell, Dynamic scaling function for critical fluctuations in classical fluids. *Phys. Rev. A.* **28**, 1567–1578 (1983).
52. H. Hao, R. A. Ferrell, and J. K. Bhattacharjee, Critical viscosity exponent for classical fluids. *Phys. Rev. E.* **71**, 021201 (2005).
53. T. Ohta, Multiplicative renormalization of the anomalous shear viscosity in classical liquids. *J. Phys. C: Solid State Phys.* **10**, 791 (1977).
54. A. Onuki, *Phase Transition Dynamics* (Cambridge University, Cambridge, 2002).
55. S. K. Das, J. Horbach, K. Binder, M. E. Fisher and J. V. Sengers, Static and dynamic critical behavior of a symmetrical binary fluid: A computer simulation. *J. Chem. Phys.* **125**, 024506 (2006).
56. Y. B. Melnichenko and G. D. Wignall, Dimensions of polymer chains in critical semidilute solutions. *Phys. Rev. Lett.* **78**, 686–688 (1997).
57. G. D. Wignall and Y. B. Melnichenko, Recent applications of small-angle neutron scattering in strongly interacting soft condensed matter. *Rep. Prog. Phys.* **68**, 1761 (2005).
58. H. Tanaka, Y. Nakanishi and N. Takubo, Nonuniversal nature of dynamic critical anomaly in polymer solutions. *Phys. Rev. E.* **65**, 021802 (2002).
59. P. J. Flory, Thermodynamics of high polymer solutions. *J. Chem. Phys.* **9**, 660 (1941).
60. M. L. Huggins, Solutions of long chain compounds. *J. Chem. Phys.* **9**, 440 (1941).
61. P. J. Flory. *Principles of Polymer Chemistry* (Cornell University, Ithaca, 1953).
62. R. H. Colby and M. Rubinstein, Two-parameter scaling for polymers in θ solvents. *Macromol.* **23**, 2753–2757 (1990).
63. J. Teixeira Azevedo, M. J. E. de Mello Cardoso and E. C. da Silva, Theoretical determination of the theta temperature for solvent–polymer systems. *J. Mol. Struct: Theochem.* **464**, 95–105 (1999).
64. M. Rubinstein and R. H. Colby. *Polymer Physics* (University, Oxford, 2003).
65. D. Frenkel, Entropy-driven phase transitions. *Phy. A.* **263**, 26–38 (1999).

66. E. M. Lifshitz and L. P. Pitaevskii. *Statistical Physics: Part 2*. Vol. 9. *Landau and Lifshitz Course of Theoretical Physics* (2nd edn.) (Pergamon, Oxford, 1981).
67. K. Shinozaki and T. Nose, Molecular weight dependence of critical amplitudes in polymer solutions. *Polym. J.* **13**, 1119–1123 (1981).
68. T. Nose, K. Shinozaki, T. Van Tan, and Y. Saito. Interfacial tension of demixed polymer solutions near critical solution temperature. In W. Bailey and T. Tsuruta (eds.), *Contemporary Topics in Polymer Science*, Chapter 52, pp. 789–805 (Springer, New York, 1984).
69. S. Z. Mirzaev, T. Heimburg and U. Kaatze, Critical behavior of polystyrene-cyclohexane: Heat capacity and mass density. *Phys. Rev. E.* **82**, 061502 (2010).
70. I. M. Withers, A. V. Dobrynin, M. L. Berkowitz and M. Rubinstein, Monte Carlo simulation of homopolymer chains. I. second virial coefficient. *J. Chem. Phys.* **118**, 4721 (2003).
71. P.-G. de Gennes, Collapse of a polymer chain in poor solvents. *J. Phys. (Paris) Lett.* **36**, L–55 (1975).
72. S. F. Edwards, The statistical mechanics of polymers with excluded volume. *Proc. Phys. Soc.* **85**, 613 (1965).
73. J. des Cloizeaux, The lagrangian theory of polymer solutions at intermediate concentrations. *J. Phys. France.* **36**, 281–291 (1975).
74. M. A. Moore, Theory of semi-dilute polymer solutions. *J. Phys. France.* **38**, 265–271 (1977).
75. L. Schäfer and T. A. Witten, Renormalized field theory of polymer solutions. I. scaling laws. *J. Chem. Phys.* **66**, 2121, (1977).
76. I. D. Lawrie and S. Sarbach, Theory of tricritical points. In C. Domb and J. L. Lebowitz (eds.), *Phase Transitions and Critical Phenomena*, Vol. 9 (Academic, New York, 1984).
77. B. Widom, Phase separation in polymer solutions. *Physica A.* **194**, 532–541 (1993).
78. A. F. Kostko, M. A. Anisimov, and J. V. Sengers, Dynamic crossover to tricriticality and anomalous slowdown of critical fluctuations by entanglements in polymer solutions. *Phys. Rev. E.* **66**, 020803 (2002).
79. F. Brochard and P. de Gennes, Dynamical scaling for polymers in theta solvents. *Macromol.* **10**, 1157 (1977).
80. F. Brochard, Gel-like modes of polymer solutions in $\ll \theta \gg$ solvents. *J. Phys. France.* **44**, 39–43 (1983).
81. L. Landau, On the theory of transfer of energy at collisions II. *Phys. Z. Sowj.* **2**, 46 (1932).
82. C. Zener, Non-adiabatic crossing of energy levels. *Proc. R. Soc. Lond. A.* **137**, 696–702 (1932).
83. I. C. Sanchez, Bulk and interface thermodynamics of polymer alloys. *Annu. Rev. Mater. Sci.* **13**, 387–412 (1983).
84. K. Binder, Collective diffusion, nucleation, and spinodal decomposition in polymer mixtures. *J. Chem. Phys.* **79**, 6387 (1983).
85. C. G. Panayiotou, Thermodynamics of polymer mixtures: Factors affecting miscibility in polymer blends, In P. J. Lemstra and L. A. Kleintjens (eds.), *Integration of Fundamental Polymer Science and Technology — 3*, pp. 105–110 (Springer, Dordrecht, 1989).
86. G. Meier, B. Momper and E. W. Fischer, Critical behavior in a binary polymer blend as studied by static and dynamic light scattering. *J. Chem. Phys.* **97**, 5884 (1992).

87. D. W. Hair, E. K. Hobbie, A. I. Nakatani and C. C. Han, Deviation from mean-field behavior in a low molecular weight critical polymer blend. *J. Chem. Phys.* **96**, 9133 (1992).
88. F. S. Bates, J. H. Rosedale, P. Stepanek, T. P. Lodge, P. Wiltzius, G. H. Fredrickson and R. P. Hjelm, Static and dynamic crossover in a critical polymer mixture. *Phys. Rev. Lett.* **65**, 1893–1896 (1990).
89. D. Schwahn, T. Schmackers and K. Mortensen, Ginzburg criterion for the mean-field to three-dimensional Ising crossover in polymer blends. *Phys. Rev. E.* **52**, R1288–R1291 (1995).
90. Z.-G. Wang, Concentration fluctuation in binary polymer blends: χ parameter, spinodal and Ginzburg Criterion. *J. Chem. Phys.* **117**, 481 (2002).

https://doi.org/10.1142/9789811282386_0013

Chapter 13

Series Expansions in Statistical Physics: Large Dimensionality, Quenched Randomness and Griffiths Singularities

R.R.P. Singh

Department of Physics and Astronomy,
University of California, Davis, CA 95616, USA
rrpsingh@ucdavis.edu

We discuss the role of series expansions in studies of critical phenomena. We review some general aspects of graph enumeration and counting and present the basis for using dimensionlity d as a continuous variable in series expansion studies. We discuss the series analysis method of differential approximants and partial differential approximants and emphasize the importance of embedding a particular model in a wider class of models to obtain better numerical estimates and a more comprehensive description of the critical behavior. We consider series studies of quenched random systems, especially classical and quantum Ising spin glasses. We argue that in classical spin glasses, Griffiths singularities hardly affect series extrapolations. In contrast, for quantum spin glasses, series expansions provide a powerful method for exploring Griffiths–McCoy singularities.

1. Introduction

Series expansion methods for the study of critical phenomena were pioneered by the Kings College group in London led by Cyril Domb.[1] They succeeded in showing that even modest-length high-temperature expansions of 6–20 terms for a variety of lattice statistical models could be used to build the concept of universality and get numerical estimates of critical properties. Early computational successes included the demonstration of power-law singularities for susceptibility and heat capacity of Ising models near the critical point, the notion that critical exponents are independent of lattice geometry in a given dimension and the validation of some scaling relations. Since then, advancements in methodologies for the calculation of series coefficients and their extensions to many other lattice statistical models as well as the development of series extrapolation/analysis techniques using Pade approximants and their generalizations[2] have greatly widened the scope and applicability of these methods. Michael Fisher greatly admired Cyril Domb and treasured his days at King's College (See Ref. 3 for a detailed exposition of Michael's tenure at the King's College), and series expansion methods were always dear to Michael's heart.

Michael's influence is well known over the entire field of phase transitions and critical phenomena and it is especially true for series expansion studies. His focus was not only on addressing new physical problems and understanding the fundamentals of universality in critical behavior using series expansions but also on developing the basic formalism and methodology for such studies. He contributed extensively to series generation and graph-counting methods as well as to series analysis and extrapolation techniques.

The work of Essam and Fisher[4] compiled systematic notations and definitions for graph-theoretic concepts needed for series expansion studies of lattice statistical models. It is unfortunate that such systematic definitions and lists and counts of graphs to avoid duplication of efforts and make the methods universally available have not always been followed. Even though there is an excellent recent text on series expansion methods by Oitmaa *et al.*,[5] there is no systematic database for graphs to this day. This leads to limited number of researchers using these methods and many researchers having to repeat the same calculations and even invent their own terminology and notation for graph counting.

Michael was one of the pioneers in tailoring series analysis to suit the singular behavior of a system rather than having a one-method-fits-all approach, which could be blindly applied to all problems. He was particularly fond of the method of partial differential approximants[6–8] and held the view that these methods were by far the most suited to studies of confluent singularities and corrections to scaling. In a single-variable analysis of the properties of a given model, it is difficult to disentangle uncertainties in the dominant exponent from corrections to scaling. Thus, the end result is an effective value for all the exponents. While analysis methods (such as higher-order differential approximants[9]) can allow for the determination of multiple scaling exponents, in practice the values of different exponents will be strongly correlated with each other and the analysis could be numerically unstable to biasing. In contrast, partial differential approximant analysis of a two-variable series naturally allows for multiple scaling variables and, thus, for two confluent singularities to arise at a critical point.

In addition to contributing extensively to series generation and analysis methods, the range of physical systems for which Michael employed the series method was vast. It included self-avoiding walks and percolation,[10,11] criticality and correlations in Ising and n-vector models,[12,13] crossover in XXZ models,[14] random ferromagnets,[15] spin glasses,[16,17] and even the Hubbard model spectra.[18]

In this chapter, we focus on selected topics related to series expansions in the study of critical phenomena that were of particular interest to Michael with a few recent additions. In Section 2, we discuss general aspects of series expansions and their relation to singular behavior of functions. In Section 3, we focus on graph counting and series expansions for lattice statistical models defined on d-dimensional hypercubic lattices as a function of d, a work pioneered by Fisher and Gaunt.[11] In Section 4, we discuss some aspects of series extrapolation methods using differential approximants and ways to incorporate corrections to scaling and confluent singularities using partial differential approximants. The rest of the chapter is focused on series expansion studies of disordered systems. In Section 5, we discuss classical systems with quenched randomness and Griffiths singularities.[19] We argue that in classical systems, Griffiths singularities, which arise due to rare regions in disorder configurations, are extremely weak and nearly unobservable in a numerical study. They do not significantly affect numerical extrapolations of the series. In Section 6, we discuss $T = 0$ properties of random quantum systems. In this case, the Griffiths–McCoy (GM) singularities[20] can be quite strong, especially in any time-dependent or time-integrated properties. We discuss a recent study of Griffiths–McCoy (GM) singularities in quantum spin glasses done in collaboration with Peter Young,[21,22] where the presence of such GM singularities can be clearly demonstrated. Finally in Section 7, we present our summary and conclusions.

2. Series Expansions and Singular Behavior of Functions

Consider a thermodynamic function, such as order-parameter susceptibility, χ, expanded as a power series in the inverse temperature β

$$\chi(\beta) = \sum_n a_n \beta^n. \tag{1}$$

We learn in undergraduate courses that the radius of convergence of a series and the singular behavior of the function $\chi(\beta)$ is contained in the asymptotic or large-n behavior of the power series coefficients. Let us assume that the physical singularity controls the radius of convergence of the series and denote the ratio of successive terms by

$$r_n = a_n / a_{n-1}. \tag{2}$$

Then the critical value of β is determined by the large-n limit

$$\beta_c = \lim_{n\to\infty} \frac{1}{r_n}. \tag{3}$$

Furthermore, let the function $\chi(\beta)$ have a power-law singularity at the critical point with exponent γ, i.e. near T_c, the susceptibility behaves as

$$\chi \sim \frac{A}{|T - T_c|^\gamma}.$$

Then, the critical exponent γ can be obtained from the asymptotic slope of the ratio function r_n for large n:

$$r_n = \frac{1}{\beta_c}\left(1 + \frac{\gamma - 1}{n} + o(1/n)\right). \tag{4}$$

What is not evident from these equations is how large n has to be to obtain good estimates of critical parameters such as β_c and γ to be able to distinguish different universality classes. In simple statistical models, such as the Ising model, the large-n behavior of the coefficients already manifests itself at relatively small n (of order $n = 10$), and this forms the basis of the series expansion–extrapolation method for the calculation of critical properties. While very few statistical models are tractable for an exact solution, a few terms in the series expansion can be obtained for many thermodynamic properties of interest for a variety of models.

The challenge for the series expansion method is twofold:

(i) how to obtain long enough series to obtain meaningful estimates of critical parameters;
(ii) how to get the most accurate information from the series by analyzing it in ways that maximizes the available information.

We will touch briefly upon both these issues in this chapter. Much early work in this field can be found in the third volume of Domb and Green edited series on *Phase Transitions and Critical Phenomena*,[1] which is entirely dedicated to series expansions. In terms of length of series coefficients that can be calculated, currently the highest-order series for three-dimensional classical models are known to be approximately 25 terms.[23,24] Ten times longer series can be obtained for some two-dimensional models.[25]

3. Graph Enumeration and Series Expansions in General Dimensions

While there are a variety of methods for obtaining the series expansion coefficients for thermodynamic properties of lattice statistical models, one aspect that is common to most methods is counting of graphs embedded in a lattice. For example, in a linked cluster method, an extensive property P on a large translationally invariant lattice of N sites is expressed as[5]

$$\frac{P}{N} = \sum_c L(c) \times W(c), \tag{5}$$

where the sum is over all distinct linked clusters or graphs that can be embedded in the lattice. The lattice constant of the cluster $L(c)$ is the number of ways the cluster c can be embedded in the lattice per lattice site. Translational symmetry ensures that the number of embeddings of each linked cluster is proportional to the number of sites in the lattice, with strict proportionality requiring periodic boundary conditions. The weight of the cluster $W(c)$ depends entirely on the property P evaluated for the cluster c and its subclusters and can also be used to develop a purely numerical linked cluster expansion without reference to a small parameter.[26]

The calculation of series expansion coefficients as an analytic function of dimensionality d was pioneered by Fisher and Gaunt.[11] Dimensionality d is an important control parameter in studies of critical phenomena. Such an expansion provides a way to vary d continuously and to take the $d \to \infty$ limit, which often results in a mean-field theory. One can also calculate systematic $1/d$ corrections to properties like the critical temperature.[11,16,17,21] In this section, we discuss enumeration and counting of graphs on d-dimensional hypercubic lattices needed for such expansions.

Clusters or graphs that can be embedded in the lattice are characterized by a vertex set V and an edge set E associated with the vertex set.[4] The vertex set is the set of vertices or points in the cluster, where the degrees of freedom of the model are defined. The edge set is the set of edges or bonds connecting the vertices and we associate the set of interactions between these degrees of freedom with the bonds. For thermodynamic properties, clusters with the same Hamiltonian or connectivity which can be defined by the adjacency matrix of the cluster make identical contributions. Thus, the goal of graph counting is to obtain a list of all such distinct clusters and obtain their lattice constants.

We focus on series expansions where the expansion variables are associated with bonds of the lattice (such as in high-temperature series expansions). In this case, the weight $W(c)$ of a cluster c with n bonds is non-zero only in nth and higher orders of the expansion. Thus, to obtain the the expansion coefficients for the property P to order n, it suffices to restrict the sum in Eq. (5) to clusters with n or less bonds. Thus, an important part of the calculation is to enumerate all distinct clusters with up to n bonds and find their lattice constants. Such bond-based counts have also been called weak embeddings in contrast to site-based graph enumeration and counts which are called strong embeddings.[1,5,27]

3.1. *Graph counts on d-dimensional hypercubic lattices*

We consider the d-dimensional hypercubic lattices, whose vertices can be labelled by d integer coordinates. In two dimensions, this reduces to the square lattice. In three dimensions, it becomes the simple cubic lattice. Nearest neighbors on the lattice are two vertices which differ by one in one of the coordinates and all other coordinates are identical. We will consider interactions associated with nearest-neighbor bonds only, which greatly simplifies the counting problem. We are interested in enumerating a list of all linked clusters up to some order n that can be embedded in this family of lattices and find their lattice constants $L_d(c)$ as an analytic function of dimensionality d.

Graph counting in low dimensions such as $d = 2$ or $d = 3$ can be done by a direct enumeration method, i.e., by explicitly counting the number of embeddings per site for each distinct graph using an algorithm that enumerates every embedding modulo translations. All graphs of length $n + 1$ can be obtained by adding a bond to some graph of length n. This idea can be used to develop automated computer programs that will start with a single bond graph and one by one generate embeddings of all graphs up to some order.[1,5] But, lattice constants of graphs increase rapidly with coordination number and dimensionality. Thus, with increasing dimensionality, such a brute force counting becomes highly inefficient. Furthermore, this does not give us lattice constants as an analytic function of d. The key to obtaining $L_d(c)$ is to realize that each embedding of a cluster in d dimensions extends in a certain number of dimensions less than or equal to d, which can be called the embedding dimension. The total lattice constant is the sum of counts for all embedding dimensions multiplied by appropriate combinatoric factors, i.e., the lattice constant of a cluster in d dimensions can be expressed as

$$L_d(c) = \sum_{i=d_1}^{d_2} C_i^d \, \tilde{L}_i(c), \tag{6}$$

where the combinatoric factor is

$$C_i^d = \frac{d!}{i!(d-i)!}, \tag{7}$$

and $\tilde{L}_i(c)$ is the count of embeddings of cluster c that extend in exactly i dimensions. The latter are embeddings absent in dimensions less than i but present in i dimensions. As example, we show in Fig. 1 the embeddings of a chain graph that embed in one, two, and three dimensions, respectively. In Eq. (6), we have denoted the minimum number of dimensions in which the cluster c can be embedded as d_1, while d_2 is the smaller of d or $d_{\max}$ the largest number of dimensions in which the cluster c can extend. Since each bond extends in exactly one dimension, a cluster with n bonds can extend in at most n dimensions, so that $d_{\max} \leq n$. Thus, once the embeddings of a cluster are known up to dimensionality $d = d_{\max}$, they are known analytically in all higher dimensions as well. To obtain the lattice constants of various clusters, one can count their embeddings $L_d(c)$ explicitly in dimensions

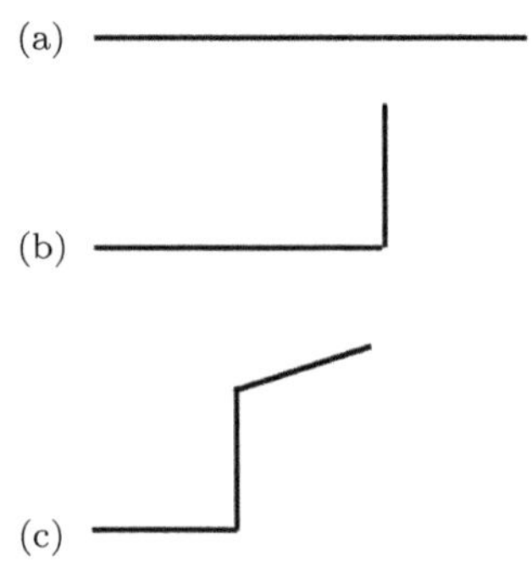

Fig. 1. Embeddings of a linear chain graph that extend in exactly (a) one, (b) two and (c) three dimensions, respectively.

d_1 through $d_{\max}$ and then use the Eq. (6) to obtain $\tilde{L}_i(c)$. Once $\tilde{L}_i(c)$ are known, the lattice constants are known as a function of dimensionality d.

3.2. *Organizing graphs by cyclomatic numbers*

The valency of a vertex in a cluster is defined as the number of bonds that meet at that vertex. A vertex with valency one is called a 'free end'. An open chain has two 'free ends'. In a graph with no 'free ends,' there must be at least one closed loop. The cyclomatic number $I(c)$ of a cluster is defined as

$$I(c) = B(c) - N(c) + 1,$$

where $N(c)$ is number of vertices and $B(c)$ is number of bonds in a cluster. An open chain, or any graph with no closed loops, has cyclomatic number zero, whereas a single closed loop has cyclomatic number one. Cyclomatic number is the number of independent cycles or loops in a cluster. It is easier to count embeddings of clusters with fewer 'free-ends' and more loops simply because the existence of closed loops significantly reduces the number of distinct embeddings.

An open chain with n bonds can extend in n-dimensions, whereas a closed loop of n bonds can extend in at most $n/2$ dimensions as they must return back to the point of origin. In general, graphs with higher cyclomatic numbers have much fewer counts and extend in fewer dimensions and are thus much easier to count by direct enumeration. It is useful to organize clusters in order of increasing cyclomatic number.

Fisher and Gaunt introduced another important concept that can greatly help in counting lattice embeddings. One can write down simple linear relations between lattice constants of graphs of order n and cyclomatic number I in terms of graphs with either lower n or higher cyclomatic number. Thus, if count of all clusters of order $n-1$ and all clusters of order n and cyclomatic numbers higher than I are known, those at order n and cyclomatic number I can be obtained through these relations without any further enumeration. Thus, in each order, at worst, only the clusters with no free ends need to be counted by direct enumeration.

An example of such a relation is given by looking at graphs in Fig. 2. When a bond is added to one of the ends of a linear chain graph of length n, three things

Fig. 2. On a hypercubic lattice, adding a bond at the end of a chain of length five shown in (a) results in three distinct graphs: (b) a chain of length six, (c) a closed loop of length six and (d) a tadpole graph with a loop of length four with a protruding external line segment of length two.

can happen: (i) a linear chain graph of length $n+1$ can result, (ii) one can obtain a closed loop graph of length $n+1$, or (iii) one can obtain a tadpole graph with a closed loop of length $m < n+1$ and a protruding linear segment of length $n+1-m$. On a hypercubic lattice, loops must be of even length and the smallest loop is of length four. Hence, only one tadpole graph is present in this relation when the linear chain is of length five, as shown in the figure. The extra bond can be added on either end and in d dimensions at each end, it can be placed in $(2d-1)$ positions so as to not overlap with an existing bond. Graph (b) could have arisen in two ways from either of two free-end bonds added to some chain of length n. Each member of graph (c) can arise in 12 ways, any of its six bonds could be the last bond added to close the loop and the bond could have been added to either end of an existing graph. Finally, for graph (d), the last bond added could be either of the two bonds inside the loop that are adjacent to the protruding line segment. This leads to the relation between lattice constants of graphs (a)–(d) in Fig. 2 as

$$2(2d-1)L_d(a) = 2L_d(b) + 2 \times 6L_d(c) + 2L_d(d). \tag{8}$$

Since the arising clusters in the Eq. (8) and various factors multiplying the lattice constants depend on the topological properties of the clusters, various such relations can also be generated by an automated computer program.[22]

Using these methods, Fisher and Gaunt obtained a complete list of a class of graphs called star graphs that can be embedded in d-dimensional hypercubic lattice to 11th order. Star graphs are multiply connected graphs which are not disconnected when a single bond is cut. There are a total of only 26 distinct star graphs to 11th order that can be embedded in hypercubic lattices and the complete list was published by Ditzian and Kadanoff,[28] who used them to calculate free-energy series for Ising spin glasses in d-dimensions to order w^{11}, where $w = \tanh^2 \beta J$. This list was later very helpful to us as a check when Sudip Chakravarty and I obtained series expansions for spin-glass susceptibility of Ising spin glasses using star graph expansions to higher orders.[29–31]

More recently, Peter Young and I used these methods[21,22] to enumerate and obtain lattice constants for all linked clusters to order 10 in arbitrary dimensions and used that to study the problem of Ising spin glasses in a transverse field in general dimensions. There were 933 distinct linked clusters to 10th order.

4. Series Analysis Using Differential Approximants

Among the many developments in series extrapolation methods, the use of d-log Pade approximants represented a key step forward.[2] It is a method well suited to estimating critical temperature and critical exponents from series expansions, especially in cases where the physical singularity does not control the radius of convergence of the series and the ratio method breaks down.

Given the series coefficients for a function $f(x)$ in powers of x to order L

$$f(x) = \sum_{i=0}^{L} a_i x^i, \tag{9}$$

one represents $d\ln f(x)/dx$ by a Pade approximant or a ratio of polynomials

$$\frac{1}{f(x)}\frac{df(x)}{dx} = \frac{P_M(x)}{Q_N(x)}, \tag{10}$$

where $P_M(x)$ and $Q_N(x)$ are polynomials of order M and N, respectively, with $Q_N(x = 0) = 1$. The polynomials are obtained by demanding that the series expansion for both sides of Eq. (10) be exactly matched to order $M+N$. Given the order L up to which series coefficients are known, one most choose $M+N+1 \leq L$. Once the differential equation has been constructed, one can show that the differential equation Eq. (10), has power-law singularities, $f(x) \sim A(x_c - x)^{-\gamma}$ near $x = x_c$ where the critical point x_c is given by the x value where the polynomial in the denominator $Q_N(x)$ vanishes. The critical exponent γ can be determined by the relation

$$\gamma = -\frac{P_M(x_c)}{Q'_N(x_c)}, \tag{11}$$

where $Q'_N(x)$ denotes the derivative of $Q_N(x)$.

This method is easily generalized to an inhomogeneous differential approximant, where the function $f(x)$ is assumed to satisfy the slightly more general differential equation:

$$Q_N(x)\frac{df(x)}{dx} - P_M(x)f(x) = U_J(x), \tag{12}$$

where we have added a third polynomial $U_J(x)$. If the latter is set to zero, one is left with a d-log Pade approximant. The polynomial $U_J(x)$ allows a slowly varying background term near the critical point. Otherwise, the expressions for x_c and γ are not changed.

4.1. *Partial differential approximants*

A powerful extension of these methods is that of partial differential approximants (PDAs),[6–8,32] which is particularly important for incorporating corrections to scaling in the description of the critical behavior. Near the critical point, a thermodynamic function such as the zero-field order-parameter susceptibility χ of a generic model has corrections to dominant scaling behavior giving it the form

$$\chi(t) = A\, t^{-\gamma}\, (1 \;+\; B\, t^{\theta}), \tag{13}$$

where $t = |T - T_c|/T_c$ is the reduced temperature and $\theta > 0$ is called the correction to scaling exponent.

PDAs require the study of a model Hamiltonian with two or more parameters. There are several ways in which two or more parameters can be introduced to study zero-field critical behavior of an Ising system. One example is the double Gaussian model.[7,8,32] The partition function is defined as

$$Z = \int \prod_i ds_i \; W(s_i; y) \exp\{-\mathcal{H}/k_B T\}, \tag{14}$$

where

$$\mathcal{H} = -\sum_{\langle i,j \rangle} J_{ij} s_i s_j \tag{15}$$

is the usual Ising Hamiltonian and the weight function $W(s_i; y)$ is given by

$$W(s_i; y) \propto b(e^{(-s_i - \sqrt{y})^2/2g} + e^{(-s_i + \sqrt{y})^2/2g}), \tag{16}$$

with $g = 1 - y$ and $b = 1/2(1 - y)$. Note that for $y = 0$, this model reduces to a Gaussian model which is exactly soluble and has free-field behavior, whereas for $y = 1$, the weights reduce to two delta functions at $s_i = \pm 1$ and the model becomes the ordinary Ising model. Thus, this model interpolates between a purely Gaussian model and an Ising model.

For such a model, one can develop a two-variable series expansion

$$f(x, y) = \sum_i \sum_j a_{i,j} x^i y^j, \tag{17}$$

where $x = J/k_B T$ and y serves as an auxiliary variable.

To use PDAs, one represents the function $f(x, y)$ by a partial differential equation

$$\begin{aligned} Q_N(x, y) \frac{\partial f(x, y)}{\partial x} &+ R_K(x, y) \frac{\partial f(x, y)}{\partial y} \\ &= U_J(x, y) + P_M(x, y) f(x, y), \end{aligned} \tag{18}$$

The two-variable polynomials $Q_N(x, y)$, $R_K(x, y)$, $U_J(x, y)$ and $P_M(x, y)$ are determined by matching coefficients on both sides of Eq. (18).

PDAs are well suited to capturing the behavior at multicritical points with two relevant scaling behavior. They are equally useful for studying corrections to scaling at an ordinary critical point with only one relevant variable, where the corrections to scaling come from an irrelevant variable. The solution to the differential equation, Eq. (18), has singularities when $Q_N(x, y)$ or $R_K(x, y)$ vanishes. However, the important fixed points controlling the singularities are where the zeros of $Q_N(x, y)$ and $R_K(x, y)$ coincide. Let us call such points (x_c, y_c). In the vicinity of such points, the solution to the partial differential equation has a scaling behavior:

$$f(x,y) \approx |\Delta\tilde{x}|^{-\gamma}\, Z\left(\frac{\Delta\tilde{y}}{|\Delta\tilde{x}|^{\phi}}\right), \tag{19}$$

with

$$\Delta\tilde{x} = (x - x_c) - (y - y_c)/e_2 \tag{20}$$

and

$$\Delta\tilde{y} = (y - y_c) - e_1(x - x_c). \tag{21}$$

All the critical parameters, x_c, y_c, e_1, e_2, γ, ϕ and the scaling function $Z(z)$ including its possible singular behavior can be determined from the partial differential equation.[6–8,32] A multicritical point corresponds to $\phi > 0$ and then the scaling function describes a crossover from one critical behavior to another.[14] An ordinary critical point will have $\phi < 0$ and will instead give a correction to scaling[7] with exponent $\theta = -\phi$.

Thus, in order to get a more comprehensive view of the critical behavior and a more accurate estimate of the critical parameters, it is important to numerically study not just an isolated model but rather a family of models. Behavior of such models can also be analyzed one parameter set at a time. But, such an analysis will always show critical exponents varying in a continuous manner with parameters. A two-variable series analyzed by partial differential approximants (PDA) can correctly capture the expected scaling in two parameters. Michael Fisher was particularly keen on the use of the PDA method. Unfortunately, given the technical challenges in both generating multi-variable series expansions and their analysis, this method remains underexplored.

In addition, Michael also advocated incorporating any additional information into series extrapolations to improve convergence, such as using exact location of critical temperature in estimating exponents, or knowledge of the singularity caused by spin waves in estimating the function,[33–36] or using knowledge of low-temperature properties to improve extrapolation of high-temperature series expansions which was done particularly effectively by Bernu and Misguich for low-dimensional quantum spin systems.[37]

5. Disordered Systems

Series expansions have been used to study macroscopic properties of many disordered systems, i.e., systems with quenched randomness. As is well known, for such systems, disorder average must be done not for the partition function but for the logarithm of the partition function or other extensive properties as only these properties self-average. For many physical problems with quenched disorder, expansion coefficients for such quenched average properties can be calculated exactly up to some order, using methods similar to those used for non-random systems. Once the averaging over randomness is performed, the system becomes translationally invariant and the usual graph counting methods apply.

The simplest case of such an expansion is where the disorder arises only in the variable whose powers the expansion is defined in. In this case, disorder averaging is particularly simple and series expansions can be calculated for an arbitrary random distribution. We consider here the problem of random-bond Ising model.[15,29–31,38–40] The Hamiltonian is

$$\mathcal{H} = \sum_{\langle i,j \rangle} J_{i,j} S_i S_j, \tag{22}$$

where the sum runs over all nearest-neighbor bonds of the lattice. The Ising variables S_i take values ± 1, while the exchange constants J_{ij} are quenched random variables. We will assume that J_{ij} are independent on every bond and identically distributed with some probability distribution $P(J_{ij})$. In a multi-variable high-temperature expansion, each term is of the form

$$c \times \prod_b (\beta J_b)^{n_b}, \tag{23}$$

where the product is over all bonds of the lattice and n_b are non-negative integers. Averaging such an expression is straightforward for many simple $P(J_b)$, including Gaussian, box and $\pm J$ distributions. The $\pm J$ distribution makes calculations particularly easy, where upon disorder averaging each term either becomes zero or $(\beta J)^n$, where $n = \sum_b n_b$. Thus, the longest series have been obtained for the $\pm J$ distribution.[29]

Computing series expansion coefficients for disorder-averaged quantities is much harder when the quenched random variable is part of the unperturbed Hamiltonian and enters the coefficients in more complex ways such as being part of energy denominators. In this case, numerical linked cluster expansion has proven an effective method.[41,42]

5.1. *Griffiths singularities and series extrapolations*

While obtaining series expansion coefficients is straightforward, series expansion-based studies of critical phenomena in quenched random systems face a different conceptual challenge. Quenched disorder reduces the critical temperature relative to that of the pure system or increases the critical value of β. But, the existence of rare regions in the disorder configurations where the system is nearly uniform or non-random invariably leads to additional singularities in the free energy of the system at the ordering temperature of the pure system that is well above the ordering system of the random system. This was first demonstrated rigorously in a diluted Ising model by Griffiths[19] and such singularities are known as Griffiths singularities.

The question arises: Do Griffiths singularities invalidate the use of high-temperature series expansions to study phase transitions and critical phenomena in random spin systems? The argument supporting this proposition goes as follows: Series expansion methods rely on analytic continuity between the parameter region of critical phenomena of interest (behavior near T_c) and the starting point of the expansions ($T = \infty$) and this analytic continuity is lost once Griffiths singularities intervene in the middle. The free energy is non-analytic at the highest temperature where the Griffiths singularities set in (typically at the ordering temperature of the pure system). Hence, any analytic continuation beyond this point is invalid.

Michael Fisher's firm view was that series extrapolations can provide valid insights into such phase transitions despite Griffiths singularities. Series extrapolation with a modest number of terms (typically 10–20) should be regarded as a numerical extrapolation rather than an analytic continuation in the strict mathematical sense. Any analytic continuation of a function from a finite number of terms in a power series, in the strict mathematical sense, is ill-posed anyway. A numerical extrapolation of a finite series is typically dominated by the strong singularities of the function rather than the weak sometimes only essential singularities which might arise at the highest temperature. Numerical results bear this out.[29–31,43–45] The series extrapolation estimates of critical temperatures and exponents in various dimensions are consistent with what was obtained from Monte Carlo simulations, which do not face such an analytic continuation barrier.

This issue is separate from the limitation in series extrapolations coming from a short series and the possible role of other factors such as corrections to scaling that could complicate a numerical analysis. In principle, these limitations are also present for the pure Ising model. One should regard exponents determined from the analysis of a short series as effective exponents and the value of such exponents are known to be correlated with estimates of the critical temperatures. This issue may be particularly severe for three-dimensional Ising spin-glass models. In more recent Monte Carlo work (for a compilation of several estimates, see Ref. 46), the estimated value of the critical temperature is about 10% lower than that in earlier studies and this changes the estimates for critical exponents by almost a factor of two. This reflects the fact that the transition behavior in the 3D Ising spin-glass model is more complex, perhaps with a strong correction to scaling or the system being close to the lower critical dimension. In four dimensions, no such problems arise and critical behavior first calculated by series expantion[29] was later confirmed by Monte Carlo simulations.[47] In fact, for four and higher dimensions, critical temperatures can be estimated quite well by a large-d expansion derived from the high-temperature series expansions.[17]

On the contrary, one might ask if such series expansion coefficients have any validity regarding the behavior of the function beyond the highest temperature where the function becomes non-analytic? The answer is an emphatic *yes* for the following reasons: Non-analyticities arise only in the thermodynamic limit, whereas finite numbers of terms in a series expansion of any property of a model are identical for the infinite system and a finite but large system with periodic boundary conditions (typically for the series to be identical with the infinite system, the linear dimension of the finite system must be equal to the number of terms in the series or larger). For a finite system, there are no non-analyticities and thus no formal barrier to analytic continuation at any temperature. In principle, a finite but large number of terms in the series expansion for the finite system should be able to capture numerically exact behavior of the finite system at all temperatures. This idea is exploited in Sandvik's stochastic series expansion Monte Carlo method.[48] Thus, a numerical extrapolation of a finite but large number of terms in the series should capture the behavior near the critical point in spite of Griffiths singularities.

Griffiths singularities in finite temperature thermodynamic behavior of classical spin models are known to be weak essential singularities, where the function and all its derivatives remain finite. In a numerical extrapolation of a finite series, past such a temperature, the function is hardly affected by such weak singularities. For similar reasons, such singularities have virtually no visible signatures in Monte Carlo studies of finite systems. For the same reasons, they are almost invisible in series extrapolations. Michael's mental picture of the effect of Griffiths singularities was of a gentle prick as one brushed some leaves in walking past them in a forest and no more. The function is hardly altered by their presence. Indeed, series expansions have been used to study critical phenomena in spin glasses,[29] random-bond ferromagnets,[15] random anisotropy magnets,[49] diluted spin systems,[39] and many other quenched random systems.[50–57]

6. Griffiths–McCoy Singularities in Random Quantum Systems

Griffiths-like singularities in ground-state properties of quantum systems tell another story.[58] Here, they can be quite strong, even divergent, and thus cannot escape a series extrapolation study. The strength of these singularities depends crucially on the property being studied. An equal-time property may be largely unaffected by such singularities as they are expected to be exponentially weak even in quantum ground states, whereas the singularities can dominate time-dependent or time-integrated properties. The reason the Griffith singularities are much stronger in the time-dependent properties can be most clearly understood by the Trotter expansion of the model in imaginary time. Disorder exists only in the spatial direction and not in the time direction. Thus, relatively small ordered clusters in the spatial direction correspond to infinite clusters in spacetime. The problem of quenched disorder that is infinitely correlated along one axis was first studied in the context of two-dimensional Ising Models by McCoy, and hence, such singularities have been called Griffiths–McCoy (GM) singularities.

For concreteness, we consider the ground-state properties of an Ising spin-glass model in a transverse-field on d-dimensional hypercybic lattices.[21,59–65] The Hamiltonian for the system is

$$\mathcal{H} = \sum_{\langle i,j \rangle} J_{ij}\sigma_i^z\sigma_j^z - h_T \sum_i \sigma_i^x - \sum_i h_L^i \sigma_i^z, \tag{24}$$

where h_T is the transverse field and the nearest-neighbor exchange interactions J_{ij} are independently distributed with the probability distribution

$$P(J_{ij}) = \frac{1}{2}\ [\delta(J_{ij} - J) + \delta(J_{ij} + J)]. \tag{25}$$

To study the phase behavior of the model, we have also added local longitudinal fields h_L^i and properties are studied in the limit that all the h_L^i go to zero. One can calculate[21,22] ground-state properties by a series expansion in

$$x = (J/h_T)^2.$$

The following properties were calculated:

(a) The equal time spin-glass structure factor is

$$SF_{SG} = \frac{1}{N}\sum_{i,j} \left[\langle 0|\sigma_i^z\sigma_j^z|0\rangle^2\right]_{av}, \tag{26}$$

here and in all subsequent expressions, $|0>$ denotes the ground state of the system for a given configuration of exchange constants and $[\cdot]_{av}$ denotes an average over the distribution of exchange constants.

(b) The static spin-glass susceptibility is defined as

$$\chi_{SG} = \frac{1}{N}\sum_{i,j} [\chi_{ij}^2]_{av}, \tag{27}$$

where the susceptibility χ_{ij} is the magnetization at site i due to infinitesimal longitudinal field at site j and can be calculated from the ground-state energy expanded to quadratic order in local longitudinal fields

$$E(h_L^i) = E(0) - \frac{1}{2}\sum_{i,j} \chi_{ij} h_L^i h_L^j. \tag{28}$$

(c) Various moments of the local susceptibility distribution arising from the bond randomness are defined as

$$\chi_m = \frac{1}{N}\sum_i [\chi_{ii}^m]_{av}. \tag{29}$$

A key result of the study[21] was that in the large-dimensionality limit, the model maps term by term on to the quantum Sherrington–Kirkpatrick (SK) model.[66,67] We further found that from $d = \infty$ down to $d = 6$, the behavior captured by the series expansion study remained very close to that of the SK model not only for the critical exponents associated with SF_{SG} and χ_{SG} but also for various log corrections associated with these quantities.[21,59–65] There were no visible signatures of GM singularities in high dimensions.

In contrast, for $d = 4$ and lower, the GM singularities become important and grow stronger as dimensionality is lowered down toward two. The first dramatic signature of GM singularities is that the equal-time spin-glass structure factor and the zero-frequency spin-glass susceptibility show divergences at very different values of the transverse field h. The separation between the location of divergence of the two increases as the dimensionality is lowered. In $d = 3$, χ_{SG} diverges at $x = 0.102$, whereas SF_{SG} diverges at $x = 0.106$. In $d = 2$, χ_{SG} diverges at $x = 0.198$, where as SF_{SG} diverges at $x = 0.221$.

The fact that zero frequency susceptibility has a divergence at a smaller value of the expansion parameter $x = J^2/h^2$ compared to the equal-time structure factor can be understood from the following consideration: Since each term inside the summation in Eq. (26) is bounded by unity, divergence of the equal-time structure factor requires equal-time correlation length to diverge. Thus, its divergence can be taken as the onset of a thermodynamic phase transition to the spin-glass phase. In contrast, dynamical properties can diverge while the equilibrium spatial correlation length is still finite as χ_{ii} itself could grow in an unbounded manner due to energy denominators.

In the Griffiths–McCoy phase, one expects a tail in the distribution of local (and non-local) susceptibilities. Spins that are surrounded by a more uniform set of couplings should have large local susceptibilities once the J/h ratio exceeds the critical value of the pure Ising model, since one is locally in an ordered phase. Spin operators will have matrix elements between states with very small energy differences as droplet excitations can become soft. The energy denominators can be exponentially small in the size of the droplet compensating for the exponentially small probability of the rare region and causing the susceptibilities to become large. Another way to think about the susceptibility divergence is in terms of the integration over imaginary time. Since susceptibilities involve a time integration, correlation times can become large in the GM phase and cause the susceptibilities to diverge.

Moments of the susceptibility distribution provide sensitive probes of the tails of the distribution. The higher the powers of the susceptibility, the more the number of energy denominators or time integrations and hence the stronger the divergence. Hence, we calculated various moments of the local susceptibility distribution. Indeed, we found that the higher moments diverge at larger values of the transverse-field, i.e., at smaller values of the expansion parameter.

In Figs. 3 and 4, we plot the critical values of $x_c = (J/h)^2_c$ where the different moments of the local susceptibility distribution diverge in three and two dimensions, respectively. The location of the spin-glass transition deduced from the divergence of the structure factor is shown by a horizontal red line, while the critical coupling where the pure system is known to diverge is shown by a horizontal blue line.

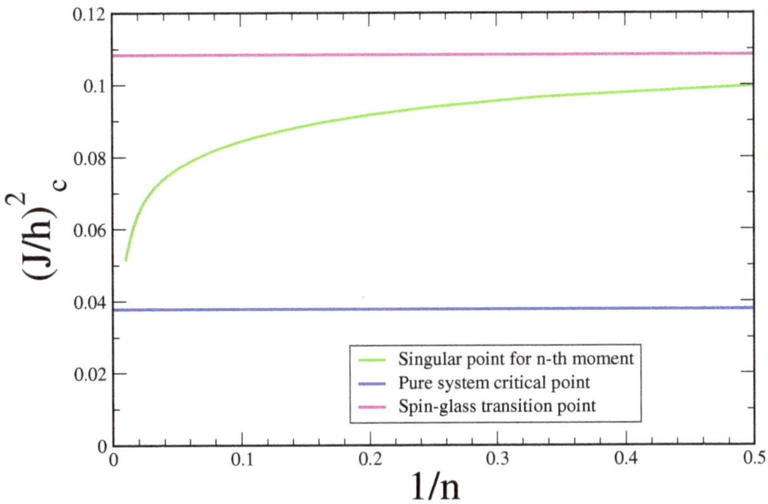

Fig. 3. Critical values of the expansion parameter $(J/h)^2_c$, where the nth moment of the local susceptibility distribution diverges in three dimensions, are shown as a function of $1/n$ by a green curve. The critical point of the non-random Ising model is shown by the horizontal blue line, while the spin-glass transition temperature deduced from the divergence of the equal-time spin-glass structure factor is shown by a horizontal red line.

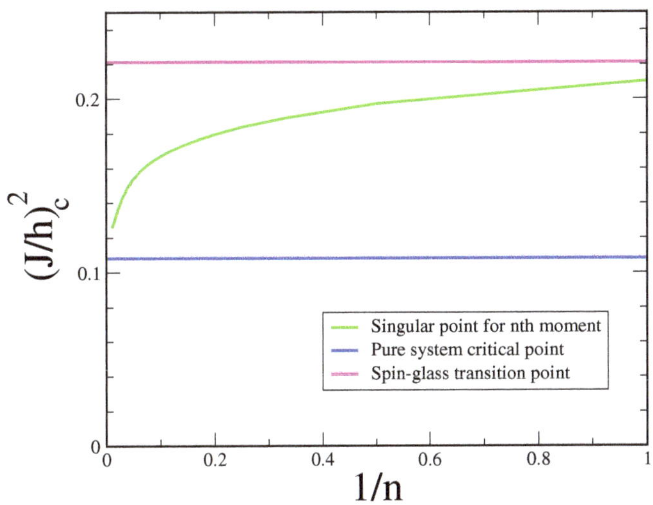

Fig. 4. Critical values of the expansion parameter $(J/h)^2_c$, where the nth moment of the local susceptibility distribution diverges in two dimensions, are shown as a function of $1/n$ by a green curve. The critical point of the non-random Ising model is shown by the horizontal blue line, while the spin-glass transition temperature deduced from the divergence of the equal-time spin-glass structure factor is shown by a horizontal red line.

As noted in Ref. 21, the spin-glass susceptibility which has two powers of the susceptibility appears to diverge at the same point as the second moment with two powers of the susceptibility. This correspondence is clearer in two dimensions and suggests that the divergence is dominated by an enhancement of the local susceptibility.

Another striking feature of Figs. 3 and 4 is that as the power of the susceptibility n becomes large, the critical couplings appear to approach the critical coupling of the pure system. This is suggestive of the expected behavior that GM singularities extend all the way to the critical coupling of the pure system. We see that one needs a high enough power of the local susceptibility to see this. The full distribution of susceptibilities can be studied by exact diagonalization of small clusters,[68] where one can also see the development of long tails in the distribution although it is difficult to get estimates of critical points or exponents from the study of small systems.

7. Summary and Conclusions

In this chapter, we have discussed a few selected topics associated with series expansion studies of critical phenomena. We discussed graph counting on d-dimensional hypercubic lattices as a function of dimensionality. We also discussed series analysis methods using differential approximants and partial differential approximants. We argued that the partial differential approximant analysis of a two-variable series is particularly effective in studying multicritical behavior and corrections to scaling.

We also discussed series expansion studies of disordered systems, including spin glasses and quantum spin glasses. In classical spin glasses, we argued that Griffiths singularities arising due to rare regions in the disorder configuration are so weak that they do not significantly affect the numerical analysis of the series. However, in quantum systems such as ground states of Ising spin glasses in a transverse field, Griffiths–McCoy singularities can be quite strong and series expansion methods can provide an effective way to elucidate their detailed nature.

Acknowledgements

It is a great honor for me to contribute a chapter to this volume in memory of Michael Fisher. I had the good fortune to work with Michael as a postdoctoral associate at Cornell University and at the University of Maryland while he was transitioning between the two schools. Parts of this chapter are based on my memory of the conversations with Michael. Michael considered calculation of series expansion coefficients and getting them correct as a character-building exercise, since even the smallest of errors could completely impair the entire analysis. I would also like to thank Sudip Chakravarty, David Huse, Jaan Oitmaa and Peter Young for their mentoring and collaborations on topics related to series expansions.

References

1. C. Domb and M. S. Green, *Phase Transitions and Critical Phenomena Vol. 3* (Academic Press, London, New York, San Francisco, 1974).
2. G. A. Baker, *Essentials of Pade' Approximants* (Academic Press, New York, San Francisco, London, 1975).
3. C. Domb, Michael Fisher at King's College, London. *Physica A.* **177**, 1, (1991).
4. J. W. Essam and M. E. Fisher, Some basic definitions in graph theory. *Rev. Mod. Phys.* **42**, 272–288 (April, 1970), doi: 10.1103/RevModPhys.42.272.
5. J. Oitmaa, C. Hamer, and W. Zheng, *Series Expansion Methods for Strongly Interacting Lattice Models* (Cambridge University Press, Cambridge, UK, 2006).
6. M. E. Fisher and R. M. Kerr, Partial differential approximants for multicritical singularities. *Phys. Rev. Lett.* **39**, 667–670 (September, 1977), doi: 10.1103/PhysRevLett.39.667.
7. J.-H. Chen, M. E. Fisher and B. G. Nickel, Unbiased estimation of corrections to scaling by partial differential approximants. *Phys. Rev. Lett.* **48**, 630–634 (March, 1982), doi: 10.1103/PhysRevLett.48.630.
8. M. Barma and M. E. Fisher, Corrections to scaling and crossover in two-dimensional Ising and scalar-spin systems. *Phys. Rev. Lett.* **53**, 1935–1938 (November, 1984), doi: 10.1103/PhysRevLett.53.1935.
9. A. J. Guttmann, Asymptotic analysis of power-series expansions. In C. Domb and J. Lebowitz (eds.), *Phase Transitions and Critical Phenomena* (Academic Press, London, New York, San Francisco, 1989).
10. M. E. Fisher and M. F. Sykes, Excluded-volume problem and the Ising model of ferromagnetism. *Phys. Rev.* **114**, 45–58 (April, 1959), doi: 10.1103/PhysRev.114.45.
11. M. E. Fisher and D. S. Gaunt, Ising model and self-avoiding walks on hypercubical lattices and "high-density" expansions. *Phys. Rev.* **133**, A224–A239 (January, 1964), doi: 10.1103/PhysRev.133.A224.
12. D. S. Ritchie and M. E. Fisher, Theory of critical-point scattering and correlations. II. Heisenberg models. *Phys. Rev. B.* **5**, 2668–2692 (April, 1972), doi: 10.1103/PhysRevB.5.2668.
13. P. R. Gerber and M. E. Fisher, Critical temperatures of classical n-vector models on hypercubic lattices. *Phys. Rev. B.* **10**, 4697–4703 (December, 1974), doi: 10.1103/PhysRevB.10.4697.
14. P. Pfeuty, D. Jasnow and M. E. Fisher, Crossover scaling functions for exchange anisotropy. *Phys. Rev. B.* **10**, 2088–2112 (September, 1974), doi: 10.1103/PhysRevB.10.2088.
15. R. R. P. Singh and M. E. Fisher, Random coupling crossover in Ising ferromagnets. *Phys. Rev. B.* **37**, 1980–1985 (February, 1988), doi: 10.1103/PhysRevB.37.1980.
16. R. R. P. Singh and M. E. Fisher, Short range Ising spin glasses in general dimensions. *J. Appl. Phys.* **63**, 3994–3996, (1988), doi: 10.1063/1.340575.
17. M. E. Fisher and R. R. P. Singh, Critical points, large dimensionality expansions, and the Ising spin-glass. In C. R. Grimmett and D. J. A. Welsh (eds.), *Disorder in Physical Systems*, pp. 87–111 (Clarendon Press, Oxford, 1990).
18. M. E. Fisher and W. J. Camp, Estimation of spectra from moments — application to the hubbard model. *Phys. Rev. B.* **5**, 3730–3737 (May, 1972), doi: 10.1103/PhysRevB.5.3730.
19. R. B. Griffiths, Nonanalytic behavior above the critical point in a random Ising ferromagnet. *Phys. Rev. Lett.* **23**, 17–19 (July, 1969), doi: 10.1103/PhysRevLett.23.17.

20. B. M. McCoy, Theory of a two-dimensional Ising model with random impurities. IV. Generalizations. *Phys. Rev. B.* **2**, 2795–2803 (October, 1970), doi: 10.1103/PhysRevB.2.2795.
21. R. R. P. Singh and A. P. Young, Critical and Griffiths-Mccoy singularities in quantum Ising spin glasses on d-dimensional hypercubic lattices: A series expansion study. *Phys. Rev. E.* **96**, 022139 (August, 2017), doi: 10.1103/PhysRevE.96.022139.
22. R. R. P. Singh and A. P. Young, Efficient generation of series expansions for $\pm j$ Ising spin glasses in a classical or a quantum field. *Phys. Rev. B.* **96**, 214210 (December, 2017), doi: 10.1103/PhysRevB.96.214210.
23. P. Butera and M. Comi, Extension to order β^{23} of the high-temperature expansions for the spin-$\frac{1}{2}$ Ising model on simple cubic and body-centered cubic lattices. *Phys. Rev. B.* **62**, 14837–14843 (December, 2000), doi: 10.1103/PhysRevB.62.14837.
24. P. Butera and M. Comi, An on-line library of extended high-temperature expansions of basic observables for the spin-s Ising models on two- and three-dimensional lattices. *J. Stat. Phys.* **109**, 311–315 (2002).
25. W. P. Orrick, B. G. Nickel, A. J. Guttmann and J. H. H. Perk, Critical behavior of the two-dimensional Ising susceptibility. *Phys. Rev. Lett.* **86**, 4120–4123 (April, 2001), doi: 10.1103/PhysRevLett.86.4120.
26. M. Rigol, T. Bryant and R. R. P. Singh, Numerical linked-cluster approach to quantum lattice models. *Phys. Rev. Lett.* **97**, 187202 (November, 2006), doi: 10.1103/PhysRevLett.97.187202.
27. M. P. Gelfand and R. R. P. Singh, High-order convergent expansions for quantum many particle systems. *Adv. Physics.* **49**, 93–140 (2000).
28. R. V. Ditzian and L. P. Kadanoff, High-temperature expansion methods for Ising systems with quenched impurities. *Phys. Rev. B.* **19**, 4631–4645 (May, 1979), doi: 10.1103/PhysRevB.19.4631.
29. R. R. P. Singh and S. Chakravarty, Critical behavior of an Ising spin-glass. *Phys. Rev. Lett.* **57**, 245–248 (July, 1986), doi: 10.1103/PhysRevLett.57.245.
30. R. R. P. Singh and S. Chakravarty, High-temperature series expansion for spin glasses. II. Analysis of the series. *Phys. Rev. B.* **36**, 559–566 (July, 1987), doi: 10.1103/PhysRevB.36.559.
31. R. R. P. Singh and S. Chakravarty, High-temperature series expansion for spin glasses. I. Derivation of the series. *Phys. Rev. B.* **36**, 546–558 (July, 1987), doi: 10.1103/PhysRevB.36.546.
32. M. Barma and M. E. Fisher, Two-dimensional Ising-like systems: Corrections to scaling in the klauder and double-gaussian models. *Phys. Rev. B.* **31**, 5954–5975 (May, 1985), doi: 10.1103/PhysRevB.31.5954.
33. D. A. Huse, Ground-state staggered magnetization of two-dimensional quantum Heisenberg antiferromagnets. *Phys. Rev. B.* **37**, 2380–2382 (February, 1988), doi: 10.1103/Phys.Rev.B.37.2380.
34. R. R. P. Singh, M. P. Gelfand and D. A. Huse, Ground states of low-dimensional quantum antiferromagnets. *Phys. Rev. Lett.* **61**, 2484–2487 (November, 1988), doi: 10.1103/PhysRevLett.61.2484.
35. R. R. P. Singh, Thermodynamic parameters of the $t = 0$, spin-1/2 square-lattice Heisenberg antiferromagnet. *Phys. Rev. B.* **39**, 9760–9763 (May, 1989), doi: 10.1103/PhysRevB.39.9760.
36. Z. Weihong, J. Oitmaa and C. J. Hamer, Square-lattice Heisenberg antiferromagnet at $T = 0$. *Phys. Rev. B.* **43**, 8321–8330 (April, 1991), doi: 10.1103/PhysRevB.43.8321.

37. B. Bernu and G. Misguich, Specific heat and high-temperature series of lattice models: Interpolation scheme and examples on quantum spin systems in one and two dimensions. *Phys. Rev. B.* **63**, 134409 (March, 2001), doi: 10.1103/PhysRevB.63.134409.
38. R. Fisch and A. B. Harris, Series study of a spin-glass model in continuous dimensionality. *Phys. Rev. Lett.* **38**, 785–787 (April, 1977), doi: 10.1103/PhysRevLett.38.785.
39. L. Klein, J. Adler, A. Aharony, A. B. Harris and Y. Meir, Dilute spin glass at zero temperature in general dimension. *Phys. Rev. B.* **40**, 4824–4832 (September, 1989), doi: 10.1103/PhysRevB.40.4824.
40. L. Klein, J. Adler, A. Aharony, A. B. Harris and Y. Meir. Series expansions for the Ising spin glass in general dimension. *Phys. Rev. B.* **43**, 11249–11273 (May, 1991), doi: 10.1103/PhysRevB.43.11249.
41. T. Devakul and R. R. P. Singh, Early breakdown of area-law entanglement at the many-body delocalization transition. *Phys. Rev. Lett.* **115**, 187201 (October, 2015), doi: 10.1103/PhysRevLett.115.187201.
42. M. D. Mulanix, D. Almada and E. Khatami, Numerical linked-cluster expansions for disordered lattice models. *Phys. Rev. B.* **99**, 205113 (May, 2019), doi: 10.1103/PhysRevB.99.205113.
43. K. Binder and A. P. Young, Spin glasses: Experimental facts, theoretical concepts, and open questions. *Rev. Mod. Phys.* **58**, 801–976 (October, 1986), doi: 10.1103/RevModPhys.58.801.
44. A. T. Ogielski and I. Morgenstern, Critical behavior of three-dimensional Ising spin-glass model. *Phys. Rev. Lett.* **54**, 928–931 (March, 1985), doi: 10.1103/PhysRevLett.54.928.
45. R. N. Bhatt and A. P. Young, Search for a transition in the three-dimensional j Ising spin-glass. *Phys. Rev. Lett.* **54**, 924–927 (March, 1985), doi: 10.1103/PhysRevLett.54.924.
46. H. G. Katzgraber, M. Körner and A. P. Young, Universality in three-dimensional Ising spin glasses: A monte carlo study. *Phys. Rev. B.* **73**, 224432 (June, 2006), doi: 10.1103/PhysRevB.73.224432.
47. R. N. Bhatt, J. D. Reger and A. P. Young, The four dimensional Ising spin glass: A monte carlo study. *J. Appl. Phys.* **69**, 5219 (1991), doi: 10.1063/1.348085.
48. A. W. Sandvik, Stochastic series expansion method with operator-loop update. *Phys. Rev. B.* **59**, R14157–R14160 (June, 1999), doi: 10.1103/PhysRevB.59.R14157.
49. R. Fisch and A. B. Harris, High-temperature series for random-anisotropy magnets. *Phys. Rev. B.* **41**, 11305–11313 (June, 1990), doi: 10.1103/PhysRevB.41.11305.
50. J. Adler, Y. Meir, A. Aharony and A. B. Harris, Series study of percolation moments in general dimension. *Phys. Rev. B.* **41**, 9183–9206 (May, 1990), doi: 10.1103/PhysRevB.41.9183.
51. A. Aharony, A. Brooks Harris and Y. Meir, Dilute random-field Ising models and uniform-field antiferromagnets. *Phys. Rev. B.* **32**, 3203–3213 (September, 1985), doi: 10.1103/PhysRevB.32.3203.
52. R. Fisch and A. B. Harris, Critical behavior of random resistor networks near the percolation threshold. *Phys. Rev. B.* **18**, 416–420 (July, 1978), doi: 10.1103/PhysRevB.18.416.
53. M. Gofman, J. Adler, A. Aharony, A. B. Harris and M. Schwartz, Evidence for two exponent scaling in the random field Ising model. *Phys. Rev. Lett.* **71**, 1569–1572 (September, 1993), doi: 10.1103/PhysRevLett.71.1569.
54. M. Gofman, J. Adler, A. Aharony, A. B. Harris and M. Schwartz, Critical behavior of the random-field Ising model. *Phys. Rev. B.* **53**, 6362–6384 (March, 1996), doi: 10.1103/PhysRevB.53.6362.

55. A. B. Harris and R. Fisch, Critical behavior of random resistor networks. *Phys. Rev. Lett.* **38**, 796–799 (April, 1977), doi: 10.1103/PhysRevLett.38.796.
56. Y. Meir, R. Blumenfeld, A. Aharony and A. B. Harris, Series analysis of randomly diluted nonlinear resistor networks. *Phys. Rev. B.* **34**, 3424–3428 (September, 1986), doi: 10.1103/PhysRevB.34.3424.
57. Y. Shapir, A. Aharony and A. B. Harris, Localization and quantum percolation. *Phys. Rev. Lett.* **49**, 486–489 (August, 1982), doi: 10.1103/PhysRevLett.49.486.
58. T. Vojta, Quantum Griffiths effects and smeared phase transitions in metals: Theory and experiment. *J Low Temp Phys.* **161**, 299–323, (2010), doi: 10.1007/s10909-010-0205-4.
59. J. Ye, S. Sachdev and N. Read, Solvable spin glass of quantum rotors. *Phys. Rev. Lett.* **70**, 4011–4014 (June, 1993), doi: 10.1103/PhysRevLett.70.4011.
60. J. Miller and D. A. Huse, Zero-temperature critical behavior of the infinite-range quantum Ising spin glass. *Phys. Rev. Lett.* **70**, 3147–3150 (May, 1993), doi: 10.1103/PhysRevLett.70.3147.
61. N. Read, S. Sachdev and J. Ye, Landau theory of quantum spin glasses of rotors and Ising spins. *Phys. Rev. B.* **52**, 384–410 (July, 1995), doi: 10.1103/PhysRevB.52.384.
62. H. Rieger and A. P. Young, Zero-temperature quantum phase transition of a two-dimensional Ising spin glass. *Phys. Rev. Lett.* **72**, 4141–4144 (June, 1994), doi: 10.1103/PhysRevLett.72.4141.
63. M. Guo, R. N. Bhatt and D. A. Huse, Quantum critical behavior of a three-dimensional Ising spin glass in a transverse magnetic field, *Phys. Rev. Lett.* **72**, 4137–4140 (June, 1994), doi: 10.1103/PhysRevLett.72.4137.
64. M. Guo, R. N. Bhatt and D. A. Huse, Quantum Griffiths singularities in the transverse-field Ising spin glass. *Phys. Rev. B.* **54**, 3336–3342 (August, 1996), doi: 10.1103/PhysRevB.54.3336.
65. A. P. Young and H. Rieger, Numerical study of the random transverse-field Ising spin chain. *Phys. Rev. B.* **53**, 8486–8498 (April, 1996), doi: 10.1103/PhysRevB.53.8486.
66. D. Sherrington and S. Kirkpatrick, Solvable model of a spin-glass. *Phys. Rev. Lett.* **35**, 1792–1796 (December, 1975), doi: 10.1103/PhysRevLett.35.1792.
67. A. P. Young, Stability of the quantum sherrington-kirkpatrick spin glass model. *Phys. Rev. E.* **96**, 032112 (September, 2017), doi: 10.1103/PhysRevE.96.032112.
68. T. You and R. R. P. Singh, Distribution of local susceptibilities in quantum Ising spin-glasses. *To be published* (2023).

https://doi.org/10.1142/9789811282386_0014

Chapter 14

Deconfined Quantum Critical Points: A Review

T. Senthil
Department of Physics, Massachusetts Institute of Technology, Cambridge, MA, 02139
senthil@mit.edu

Continuous phase transitions in equilibrium statistical mechanics were successfully described 50 years ago with the development of the renormalization group framework. This framework was initially developed in the context of phase transitions whose universal properties are captured by the long wavelength (and long time) fluctuations of a Landau order parameter field. Subsequent developments include a straightforward generalization to a class of $T = 0$ phase transitions driven by quantum fluctuations. In the last 2 decades it has become clear that there is a vast landscape of quantum phase transitions where the physics is not always usefully (or sometimes cannot be) formulated in terms of fluctuations of a Landau order parameter field. A wide class of such phase transitions — dubbed deconfined quantum critical points — involve the emergence of fractionalized degrees of freedom coupled to emergent gauge fields. Here I review some salient aspects of these deconfined critical points.

To be published in "50 years of the renormalization group", dedicated to the memory of Michael E. Fisher, edited by Amnon Aharony, Ora Entin-Wohlman, David Huse, and Leo Radzihovsky, World Scientific.

1. Introduction: the Landau-Ginzburg-Wilson-Fisher paradigm

The development of the theory of phase transitions in critical phenomena in the 1960s and 1970s is one of the towering achievements in modern physics. Michael Fisher played a leading role in this development. The resulting understanding, associated primarily with the names of Landau, Ginzburg, Wilson, and Fisher, has come to be known as the LGWF theory (or sometimes the LGWF paradigm). This theory — which we briefly review below — provided a conceptual framework to discuss equilibrium phase transition phenomena. Combined with renormalization group methods and the dimensionality expansion introduced by Wilson and Fisher, the theory enables computations of critical exponents and other universal properties in a systematic approximation. The LGWF theory was generalized in many different directions, eg, to deal with dynamical critical phenomena, and more pertinently to this chapter, to quantum phase transitions at $T = 0$.

Phase transition theory begins by first asking how to distinguish distinct phases of matter. A crucial idea (due to Landau) is to focus on the symmetries of the microscopic Hamiltonian describing the many body system. The equilibrium state in a range of parameters (or the ground state if, at $T = 0$) may not have the symmetries of the microscopic Hamiltonian, *i.e* the symmetry is spontaneously broken. Then this state is sharply distinct from a symmetry preserving state that may exist at other parameter values. The evolution from the broken symmetry phase to one where the symmetry is restored must then involve (at least one) phase transition. Landau also introduced the notion of an order parameter to quantify the extent of broken symmetry. The Landau order parameter appears as a new thermodynamic variable needed to fully characterize the macroscopic state of the symmetry broken phase. The universal properties of the phase transition are then associated with the long wavelength (and long time, if dealing with quantum or dynamical critical phenomena) fluctuations of the order parameter field. The energy (or action, for quantum critical points) is written as a spatial integral of sums of an energy density that is expanded in powers of the order parameter field and its gradients. This is a coarse grained continuum description (an 'effective field theory') of the original microscopic model. The statistical mechanics of the resulting action (the Ginzburg-Landau action) then provides a description of the universal physics of the critical point. In practice, this can be analysed within the framework of the renormalization group with the Wilson-Fisher ϵ expansion (or other expansions) as a tool to access the fixed points.

These beautiful set of ideas are sufficient to describe phase transitions in a large number of systems. The classic examples are the classical Ising or $O(N)$ magnets. However, in the last few decades, it has become clear that they turn out to not be the full story in a number of situations, particularly at quantum critical points. In this chapter we will survey a few different examples of how the LGWF paradigm breaks down. Nevertheless the framework of effective field theory and the renormalization group will remain central. What is different at these 'beyond LGWF' transitions is that the effective field theory will not be the naive one written in terms of just the order parameter field.

As we will see, there are a number of distinct routes for 'beyond LGWF' phase transitions. First, there are phases of matter which cannot be simply characterized in terms of the usual broken symmetry paradigm and hence do not admit a description in terms of a Landau order parameter. Obviously, their phase transitions will not be described by the LGWF paradigm. More surprisingly, we can have Landau-forbidden quantum critical points between phases that themselves are described as broken symmetry phases with a Landau order parameter. Suppose the microscopic symmetry of some system is described by a group G, and consider two phases of matter where the symmetry G is broken to a subgroup H in one phase, and to a different subgroup H' in another phase. In standard Landau theory. a second order phase transition between the two phases is possible only if one of the two groups H, H' is a subgroup of the other. Remarkably, we have learnt that this rule can be broken at a variety of quantum crtical points.[1,2] The resulting critical theory is most usefully described in terms of a continuum field theory with emergent gauge fields coupled to matter fields carrying fractional quantum numbers of the microscopic global symmetry. These emergent fractionalized fields and associated gauge fields are not simply associated with quasiparticle excitations of either phase of matter; rather they rear their head at the critical point as useful variables to access the critical fixed point. Thus they have been dubbed 'deconfined quantum critical points'.

There are a number of examples of deconfined quantum critical points and related phenomena that have been studied in recent years. For instance, examples have been found where the same phase transition allows multiple universality classes.[3] These include situations[4] in which a Landau-allowed phase transition has both the conventional universality class as well as an unconventional one that is not described by the LGWF theory. Thus even when the phase transition occurs between a trivial gapped phase and a broken symmetry one described by a Landau order parameter, the effective field theory of the critical point may not be captured by the LGWF action. Even more striking are "unnecessary" critical points[3] which do not separate distinct phases of matter but rather live suspended within a single phase of matter. It is possible to find a path in parameter space which avoids the critical point completely.

Together these results underscore the idea that we have very little intuition for whether a phase transition is allowed to be second order, and if so, what its field theoretic description will be. As such they embolden us to contemplate the possibility that many mysterious quantum critical phenomena observed in metallic systems might find explanations in a form that is beyond a conventional description (known as the Hertz-Millis paradigm) in terms of electronic quasiparticles coupled to a critically fluctuating bosonic order parameter field.

1.1. *Phases of matter beyond Landau*

For orientation, let us consider the different kinds of non-Landau phases of matter that are known to exist. The focus of this article will be on $T = 0$ quantum phases of matter, *i.e*, ground states of quantum many body systems. A familiar example of phases whose distinction is not captured by the concept of broken symmetry are electronic band insulators as compared to band metals. The discovery of the integer and fractional quantum hall phases in the 1980s — which clearly do not break any microscopic symmetries — emphasized the possibility of new kinds of quantum ground states which fall under the rubric of 'topological quantum order'[a]. Since then, a large variety of 'beyond-Landau' phases have been shown to be possible theoretically, and some have been discovered in experiment. These include other topologically ordered states (such as gapped quantum spin liquids in insulating magnets), symmetry protected topological states (such as topological band insulators and the Haldane spin-1 chain), and various gapless phases of matter where the gaplessness is not due to Goldstone fluctuations associated with a broken symmetry. The most famous example is the Landau fermi liquid itself. Other variants include Dirac materials such as graphene. Most novel are phases of matter with no quasiparticle excitations at all: examples include some gapless quantum spin liquids, and metallic non-fermi liquid phases.

Obviously, if the phase itself is not described simply by specifying a Landau order parameter, the phase transition out of it will not be described by an LGWF action in terms of any fluctuating order parameter field. In Fig. 1 we sketch the variety of quantum phase transitions that accomodate this variety of phases of matter. The LGWF paradigm in its original form is only applicable to the situation where the symmetry preserving phase is gapped and trivial, and the symmetry broken phase is trivial apart from the broken symmetry. In all other cases, we should expect to find other descriptions of the phase transition.

1.2. *Landau ordering transitions: a brief review*

We begin with a quick review of Landau ordering quantum phase transitions. The canonical example is the transverse field Ising model on a d-dimensional lattice witha global Z_2 symmetry. This has a disordered phase that preserves the global Z_2 symmetry, and an ordered phase that spontaneously breaks the symmetry. The quantum phase transition between these two phases is second order, and is in the universality class of the $d+1$-dimensional classical Ising model. The universal critical singularities are captured by an LGWF action of the form

$$\mathcal{S} = \int d^d x d\tau \left(\frac{\partial \phi}{\partial \tau}\right)^2 + (\nabla \phi)^2 + r\phi^2 + u\phi^4 \tag{1}$$

[a]In recent years, it has become popular to say that some such phases can actually be captured by the Landau paradigm if one generalizes the notion of symmetry. For a clear presentation of this point of view, see Ref. 5. For instance topologically ordered states of matter are considered to spontaneously break 'higher-form' symmetries rather than ordinary (0-form) symmetries. This point of view is interesting and useful. In applying it to the systems typically of interest in condensed matter physics, we must recognize that these higher form (or other generalized) symmetries are not present in the microscopic system. They are emergent symmetries in particular phases of matter that may then be spontaneously broken. This is different from the conventional Landau paradigm which dictates that we distinguish phases by asking whether *microscopic* symmetries are present/absent in the equilibrium (ground) state. Thus, in my view, incorporating topological ordered or other "non-Landau" phases into the Landau paradigm requires not only generalizing the notion of symmetry but also generalizing what is meant by the Landau paradigm itself. Thus, in this article, I will refer to such phases as non-Landau as has been common for years.

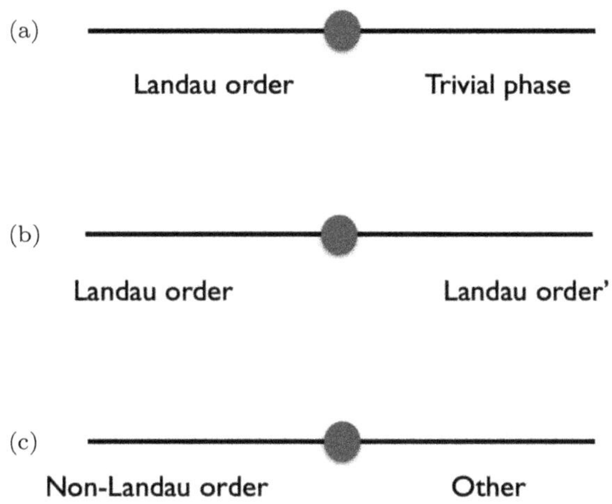

Fig. 1. Varieties of quantum critical points. (a) The only case where the traditional LGWF paradigm is usually applicable. (b) Unless the unbroken symmetry group of one phase is a subgroup of the unbroken symmetry of the other phase, the transition is Landau-forbidden (c) Non-Landau order, if present in at least one of the two phases guarantees a transition beyond the LGWF paradigm.

where ϕ is the coarse-grained Ising order parameter field; τ is the imaginary-time coordinate, and $\mathbf{x}$ is the spatial coordinate.

A simple generalization is to models with a global $O(N)$ symmetry which (for $d > 1$) have a second order phase transition between a symmetry preserving trivial gapped phase, and a symmetry broken phase where the global $O(N)$ symmetry is broken to the subgroup $O(N-1)$. The transition is described by an action of the same form as in eqn. 1 but with ϕ replaced by an N-component order parameter field $\mathbf{N}$ that transforms as a vector under the global $O(N)$ symmetry.

When might we expect the LGWF continuum field theory to correctly describe the phase transition out of the broken symmetry phase? Note that the LGWF theory should really be understood as an expansion about the disordered symmetry preserving phase which is implicitly assumed to be trivial and gapped. There are quasiparticle excitations charged under the global $O(N)$ symmetry. The most elementary such excitations transform under the vector representation of $O(N)$. As the transition is approached, the gap for these excitations closes.

Thus we should expect LGWF actions of the form of Eqn. 1 to fail to describe the transition if the phase where the symmetry is restored is non-trivial in some way. For instance it may not be a trivial gapped phase and may have topological or other quantum order. Alternately it may break some other symmetry spontaneously. Clearly such cases are beyond the purview of the standard LGWF theory.

An important and interesting class of examples arises in situations where a trivial gapped symmetry preserving phase is forbidden on general grounds. A classic example is in quantum many body systems of interacting particles with conserved particle number (corresponding to a global $U(1)$ symmetry) on a translation invariant lattice. If the lattice filling ν is not an integer, then a trivial symmetric gapped ground state is not possible through a (generalization[6,7] of) theorem by Lieb, Schultz, and Mattis (LSM). Such LSM restrictions apply very generally also to a number of translation-invariant models on a lattice where the degrees of freedom at each unit cell transform projectively under the global internal symmetry. For instance, in a spin system with global $SO(3)$ symmetry, if we have a spin-1/2 (or other half-integer spin) moment at each unit cell, then a trivial gapped ground state is forbidden by the LSM restriction.

It follows that phase transitions in systems obeying LSM restrictions are never to be expected to be described by the LGWF theory.

A closely related situation arises for phase transitions that occur at the boundary of a higher dimensional Symmetry Protected Topological[8] (SPT) phase[b]. An SPT phase has a ground state that cannot be adiabatically deformed through any path that preserves the global symmetries into a completely trivial ground state. Examples of SPT phases include topological band insulators in diverse dimensions and the spin-1 Haldane spin chain. In the last decade a great deal has been understood about the classification and physical properties of such SPT phases. A defining feature is that it is not possible for the boundary of such an SPT phase to be trivially gapped. Thus phase transitions at SPT boundaries are also not expected to be described by the LGWF theory. Indeed the connection[10] of the theory of deconfined quantum criticality to SPT boundaries has been very effective in understanding the latter.

2. LGWF* Phase Transitions

We begin by describing a simple example of a non-Landau transition, one that involves a critical point between a symmetry broken phase and one where the symmetry is restored but there is topological quantum order. Consider models of interacting bosons on a lattice with spatial dimension $d = 2$. The total boson number is conserved and there is a corresponding global $U(1)$ symmetry. At every lattice filling, a superfluid phase where the global $U(1)$ symmetry is spontaneously broken is clearly possible. As is well known, if the average number of bosons per unit cell is an integer, a Mott insulating phase is also possible where both the global $U(1)$ and translation symmetry are preserved. A physical picture is that in the Mott insulator, there are an integer number of bosons localized to each unit cell of the lattice. This Mott insulator is a symmetric trivial gapped phase. The corresponding superfluid-Mott phase transition is described by LGWF theory, and is in the universality class of the classical XY model in 3 dimensions.

[b]As emphasized in Ref. 9 we can regard a system satisying the LSM restriction itself as the boundary of a specific case of an STP phase in one higher dimension, one involving lattice translation as well as internal symmetries.

Next consider the situation when the number of bosons per unit cell is 1/2. Then a trivially gapped Mott insulator is forbidden. Nevertheless a symmetry preserving gapped ground state can exist so long as it has topological order and associated fractionally charged excitations. This state may be dubbed a fractionalized Mott insulator. The simplest example is when the topological order is that of a deconfined Z_2 gauge theory. These can also be thought of as a quantum spin liquid ground state of a spin-1/2 magnet; for reviews, see Ref. 11,12. This theory has gapped quasiparticles in distinct topological superselection sectors denoted $1, e, m$ and ϵ. The corresponding e and m quasiparticles have bosonic self-statistics, and a mutual braiding statistics of π (*i.e*, when one of e, m is taken on a loop around the other, there is a phase of π). They may be identified with the Z_2 gauge charge and the Z_2 gauge flux of the Z_2 gauge theory[c]. Though the e and m are both good quasiparticles in this state of matter, their non-local statistical interaction implies that neither can be created locally, *i.e* by acting with local operators in the physical Hilbert space. This allows them to carry fractional quantum numbers of the global symmetry. Though a single e or m particle is not a local excitation, e^2 or m^2 (where a pair of e or m particles are created) are local. Thus we can create, say, a pair of e particles locally. The non-trivial statement is that the individual e-particles can be moved apart with a finite energy cost. Thus e^2, m^2 cannot carry fractional quantum numbers of the microscopic symmetries. It follows that e can have a quantum number 1/2 under the global $U(1)$ symmetry. (In the presence of time reversal, it can be shown[10,13,14] that the m particle cannot also have fractional $U(1)$ quantum numbers but will fractionalize lattice symmetries.[15,16]) Indeed it is possible to construct concrete models where precisely such 1/2-charged e particles, and neutral m particles exist.[17–19] Now if the microscopic boson model is at a lattice filling of 1/2 a boson per unit cell, we can contemplate a fractionalized Mott insulator where there is a single e particle that is localized in each unit cell.

The phase transition between the superfluid and this fractionalized Mott insulator is conveniently described as a condensation of the e particles while the gap to m stays non-zero. At energy scales well below the m gap, we can describe this transition through the action

$$\mathcal{S} = \int d^d x d\tau \left|\frac{\partial \phi}{\partial \tau}\right|^2 + |\nabla \phi|^2 + r|\phi|^2 + u|\phi|^4 \tag{2}$$

where ϕ is a coarse-grained complex field whose quanta (in the disordered phase) describe the e particles. This looks just the same as the standard LGWF action for the superfluid-Mott transition at integer filling but there is an important subtlety. As the ϕ field creates the e-particle, it should not be directly identified with the superfluid order parameter, or indeed with any local observable in the original boson

[c]Their bound state ϵ is a fermion, and has π braiding statistics with both e and with m; the superselection sector 1 is simply associated with excitations that can be created by the action of local operators.

model. Relatedly this action does not know about the gapped m-sector which is part of the universal physics of both the fractionalized Mott insulator and the critical point. These subtleties can be handled by explicitly coupling the ϕ field to a Z_2 gauge field whose flux represents the m-particles. In that description, the ϕ field is not gauge invariant under Z_2 gauge transformations (though ϕ^2 is).

The action in Eqn. 2 is nevertheless useful to read off many universal critical properties. We simply note that it describes the gapless sector as the $3DXY$ universality class except that operators which are not Z_2 gauge invariant are thrown out. The correlation length exponent ν (which is determined by the scaling dimension of $|\phi|^2$ which is gauge invariant) is the same as in the classical $3D$ XY model. However the scaling dimension of the physical superfluid order parameter ψ is different. This is because we can identify $\psi = \phi^2$. It thus has the anomalous exponent (introduced by Michael Fisher) $\eta_\Psi \approx 1.49$. This large anomalous exponent must be contrasted with that of the usual superfluid-Mott transition at integer filling where $\eta \approx 0.03$. For numerical studies of this transition, see Ref. 20.

Thus the superfluid-fractionalized Mott insulator transition is beyond the standard LGWF paradigm, albeit with a simple relation to the standard LGWF theory. For these reasons, this kind of transition has come to be known as the LGWF* theory. For a sample of discussions of other LGWF* transitions, see Ref. 21–23.

Theories with a different structure also arise at certain phase transitions between topologically ordered and symmetry broken phases, for instance between fractional quantum Hall and superfluid phases of bosons on a periodic lattice. Unlike in the example above, there is usually no anyon in the fractional quantum Hall state that can be directly condensed to both break the global $U(1)$ symmetry and simultaneously destroy the topological order. Nevertheless, as argued in Ref. 24, a continuous transition is possible and can be formulated in terms of emergent fermionic degrees of freedom coupled to a $U(1)$ gauge field. These fermions can be roughly thought of as composite fermions used to great effect in describing quantum Hall phenomena.

3. Landau-forbidden Quantum Critical Points Between Landau Allowed Phases

We now turn to phase transitions between two phases which can both be characterized in terms of Landau order parameters corresponding to distinct broken symmetries. We will show that, contrary to naive expectations, such a phase transition can be second order. The primary example — which we review below — involves phase transitions in two dimensional insulating square lattice quantum antiferromagnets. These are described by models of systems of interacting quantum spins on the lattice. A number of other examples in two space dimension have also been presented. (We postpone the discussion of other dimensions to later sections). An interesting example involves a phase transition in a system of interacting fermions, as we also review below.

3.1. *Neel-VBS transition on the square lattice*

Consider a system of $SU(2)$ spins on the square lattice with short range interactions. More precisely we will take the spin symmetry to be $SO(3) = \frac{SU(2)}{Z_2}$. and place spins that transform in the spin-1/2 (projective) representation of $SO(3)$ at each site. We consider situations where the microscopic Hamiltonian has global $SU(2)$, time reversal, and square lattice space group symmetries.

The standard ground state of such a magnet is the Néel state that breaks spin $SO(3)$ and some lattice symmetries. Specifically it preserves lattice rotation symmetry, and a combination of unit lattice translation and time reversal. A different symmetry breaking state is a Valence Bond Solid (VBS) that preserves spin $SO(3)$, and time reversal but breaks unit lattice translations in one direction, and lattice rotational symmetry. Clearly these phases break distinct symmetries, and the remnant symmetry in neither phase is a subgroup of the other. Naively we might expect the transition between these two phases to either be first order, or go through a region where both broken symmetries coexist, or go through an intermediate phase with no broken symmetry. Note that LSM restrictions applied to these spin models tell us that the last possibility is only allowed if such an intermediate symmetric phase has topological or some other non-trivial quantum order. In particular a trivially gapped symmetric phase is not possible. Thus, as discussed above, a theory for the destruction of either the Néel or the VBS order will necessarily be 'beyond LGWF'.

In Ref. 1,2 it was argued that a direct second order transition is potentially possible and is described by the field theory

$$\mathcal{L}_0 = \sum_{\alpha=1,2} |D_b z_\alpha|^2 + \left(|z_1|^2 + |z_2|^2\right)^2. \tag{3}$$

Here z_α ($\alpha = 1, 2$) are scalar fields (known as 'bosonic spinons') that are coupled to a dynamical $U(1)$ gauge field b, and $D_{b,\mu} = \partial_\mu - ib_\mu$ is the covariant derivative. (This

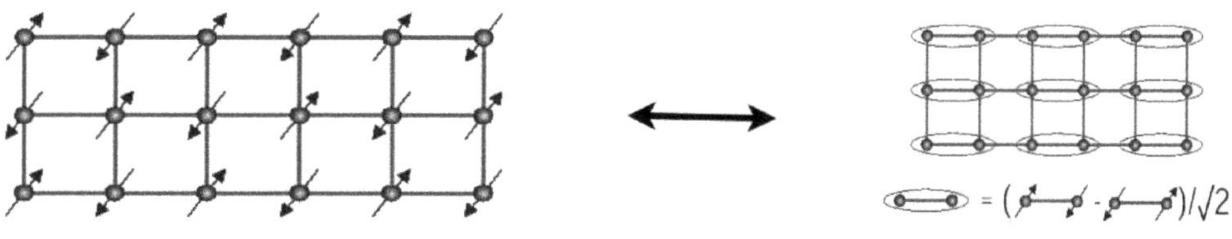

Fig. 2. The Neel and VBS states of a square lattice spin-1/2 magnet.

Euclidean action and subsequent similar actions are short-hands for the appropriate strongly coupled Wilson-Fisher critical theory where a background gauge field has been promoted to a dynamical field.) The model has a global $SO(3)$ symmetry under which z_α transforms as a spinor.[d] In the microsopic lattice spin model, this corresponds to the $SO(3)$ spin rotation. It also has a global $U(1)$ symmetry associated with the conservation of the flux of b. In the microsopic lattice spin model, this is not an exact symmetry. Consequently monopole operators (which pick up a phase under a $U(1)$ rotation) must be added to the Lagrangian. However, it is known that lattice symmetries ensure that the minimal allowed monopole operator (with continuum angular momentum $\ell = 0$) has strength 4.

Analytic arguments[1,2] and numerical calculations[25–27] strongly support the possibility that these monopoles are irrelevant at the putative critical fixed point of Eq. (3). The Néel phase is obtained when z_α is condensed, and the VBS phase when z_α is gapped. The Néel phase breaks $SO(3)$ to a $U(1)$ subgroup while the VBS phase breaks the $U(1)$ flux conservation symmetry. The Néel order parameter is simply $\mathbf{N} = z^\dagger \sigma z$ (σ are Pauli matrices), and the VBS order parameter is the strength 1 monopole operator $\mathcal{M}_b$ which creates 2π flux of b.

The field theory in Eqn. 3 in the absence of monopoles is known as the Non-Compact CP^1 ($NCCP^1$) model. It describes[28] the phase transition in $O(3)$ models where hedgehog topological defects have been suppressed by hand. These hedgehog defects correspond precisely to monopoles in the CP^1 formulation of the $O(3)$ non-linear sigma model.

A heroic body of numerical work on specific quantum magnets and related systems[25–27,29–42] shows a striking apparently continuous transition[e] between the N'eel and VBS phases with properties consistent with expectations based on the field theory in Eqn. 3.

What makes such a second order transition possible? The key physics is the properties of topological defects of the order parameters of either symmetry broken phase. To see this most simply, consider starting in the VBS phase which breaks the discrete symmetries of lattice translation and lattice rotation. There are four degenerate VBS ground states which are related to each other by Z_4 lattice rotations about a site. Naively we might then expect a phase transition associated with the disordering of the VBS order to be in the universality class of the phase transition of a Z_4 clock model in $2+1-D$. As is well known, this is the same as the $2+1-D$ XY

[d]Though z_α transforms as a spinor under $SU(2)$, rotations in the center of $SU(2)$ can be compensated by a $U(1)$ gauge transformation so that the spin rotation symmetry of the model is $\frac{SU(2)}{Z_2} = SO(3)$.

[e]It is however not yet clear whether the transition is truly second order or whether it displays only 'quasiuniversal' behavior up to a very large but finite length scale.[27,31–34,37,43–46] We will discuss this below.

universality class: four-fold anisotropy of the XY order parameter is *dangerously irrelevant*[f] at the $3D$ XY fixed point.

This naive expectation is in fact not correct due to non-trivial structure of the topological defects of the VBS order parameter. Before exploring this further, it is useful to first recall the physics of topological defects at the usual $2+1-D$ XY phase transition. For an XY order parameter, the topological defects are of course vortices. The transition out of the XY ordered phase requires the proliferation of these vortices. The most famous example is in the BKT transition of the *classical* $2d$ XY model but it remains true for the quantum $2+1-D$ XY transition as well. Note that the $2+1-D$ XY transition is described by the Wilson-Fisher fixed point with an action of the form in Eqn. 2. Equivalently we can describe it with a 'dual' field theory in terms of a complex scalar field ϕ_v (identified physically with the vortex field) coupled to a dynamical $U(1)$ gauge field:

$$\mathcal{L}_v = |D_b\phi_v|^2 + \left(|\phi_v|^2\right)^2. \tag{4}$$

This the celebrated particle-vortex duality.[47–49] The disordered phase of the XY model is identified with the phase in which ϕ_v is condensed. This gaps out the gauge field b through the Anderson-Higgs mechanism and we get a trivial gapped phase. The ordered phase of the XY model is identified with the phase in which ϕ_v is gapped. At low energies, we integrate out ϕ_v and get the free Maxwell action for the gauge field b. The corresponding 'photon' is then identified with the Goldstone mode of the broken XY symmetry. The XY order parameter itself is identified with the monopole operator $\mathcal{M}_b$. If the microscopic system has four-fold anisotropy on the XY order parameter, then we are allowed to add the operator $\mathcal{M}_b^4 + c.c$ to the dual action.

Armed with this understanding, let us now return to the task of disordering the VBS order in the square lattice quantum magnet. For more detail, please see Ref. 50. Naturally the topological defects are domain walls. Four such elementary domain walls can end at a vortex which is a point defect. It is easy to see that these VBS vortices necessarily involve a site of the lattice that is not paired with any other sites through a valence bond. Thus this site carries an unpaired spin-1/2 moment. This is a key difference with an ordinary Z_4 clock model. Now consider quantum disordering the VBS order by proliferating (*i.e condensing*) these vortices. The spin-1/2 moment carried by these vortices will then lead to the breaking of spin rotation symmetry. It can be argued that the resulting state is precisely the Néel state. Indeed, following the principles of particle-vortex duality, the action in Eqn. 3 can be very simply understood as an effective field theory for the VBS vortices which are identified with the z_α field[g].

---

[f]Another concept first introduced by Michael Fisher.

[g]From a modern point of view, the structure of the topological defects implies that the $NCCP^1$ theory has a mixed ' t Hooft anomaly between the global $U(1)$ and $SO(3)$ symmetries.[51] Specifically, if we introduce a background gauge field A for the global $U(1)$ symmetry, the monopole

An alternate perspective is obtained by starting from the Néel state. The Néel order parameter **N** transforms in the vector representation of $SO(3)$. It thus admits skyrmion defects corresponding to $\Pi_2(S^2) = Z$. In the original lattice system, the conservation of skyrmions can potentially be violated by 'hedgehog' events in spacetime. However for spin-1/2 magnets on the square lattice, it can be shown that single skyrmions transform non-trivially under lattice symmetries.[1,2,55,56] Thus a single skyrmion cannot be created or destroyed while preserving microscopic symmetries. In fact, the symmetry transformation of the skyrmion creation operator is identical to that of the VBS order parameter. If we now disorder the Néel order by proliferating skyrmions, the resulting paramagnetic phase will have VBS order.

The structure of the topological defects also reveals a problem with a naive Ginzburg-Landau action to describe the competition between the Néel and VBS orders. Indeed, a description of this competition will focus on the Néel order parameter **N** and the VBS order parameter ψ, and write an action for their fluctuations. To that end it is convenient to introduce a 5-component unit vector n^a ($a = 1, \ldots, 5$) such that $n^{3,4,5}$ correspond to the three components of the Néel vector **N**, and $n^{1,2}$ to the two real components of the VBS order parameter ψ. A naive Ginzburg-Landau action will then take the form of a non-linear sigma model

$$\mathcal{S}_{naive} = \frac{1}{2g} \int d^3x \, (\partial n^a)^2 + \lambda \left(\mathbf{N}^2 - |\psi|^2\right) + \ldots.. \tag{5}$$

The first term is fully $O(5)$ symmetric under rotations of n^a; the second term introduces anisotropy between the 5 components of n^a so that the symmetry is reduced to $O(3) \times O(2)$. This term can also be used to tune between Néel and VBS ordered states. The ellipses include terms that further break these symmetries to those of the lattice spin model. This kind of Ginzburg-Landau action would usually suggest that the Néel-VBS transition is either first order, or goes through an intermediate phase with neither order. Further in this naive sigma model, this intermediate phase will be a trivially gapped phase[h] However since such a phase is forbidden in the lattice spin-1/2 magnet, clearly the naive sigma model needs to be modified in some way.

The crucial point is that the naive Ginzburg-Landau model does not capture the structure of the topological defects described above. Consider for instance the VBS ordered state where n^a points in the n^4, n^5 plane, and construct a vortex configuration of $\psi = n^4 + in^5$. At the core of the vortex, $|\psi| = 0$ and so the unit

operators in A will transform in the spinor representation of $SO(3)$ which is a reflection of the anomaly. Similarly, the hypothesized emergent $SO(5)$ symmetry discussed below of the low energy theory has its own 't Hooft anomaly which reduces to the mixed $U(1)$ — $SO(3)$ anomaly upon restricting to that subgroup.[51] The anomaly perspective provides a powerful tool to obtain some non-perturbative constraints on the RG flows of the theory. For a discussion on the relationship between the anomalies of the low energy theory and LSM constraints of the lattice spin model, see Ref. 52–54.

[h] A direct second-order Neel-VBS transition can then only be expected at a multicritical point.

vector n^a will point along the Néel direction. The dynamics of the Néel vector in the vortex core will be described by an $SO(3)$ rotor model and will have quantum states transforming in integer representations of $SO(3)$. However we saw above that the VBS vortex in our physical setting must transform as spin-1/2 under $SO(3)$. This leads to a 'quantum intertwinement' of the two orders that must be incorporated into the Ginzburg-Landau theory. The correct action[57,58] turns out to require inclusion of a Wess-Zumino-Witten (WZW) term at level-1 for the dynamics of the $SO(5)$ vector n^a:

$$S = \frac{1}{2g} \int d^3x \, (\partial n^a)^2 + 2\pi\Gamma \, [n^a] \, . \tag{6}$$

The WZW term Γ is defined in the standard way: the field n^a defines a map from spacetime S^3 to the target space S^4, and Γ is the ratio of the volume in S^4 traced out by n_a to the total volume of S^4. If $n^a(x,u)$ is any smooth extension of $n^a(x)$ such that $n^a(x,0) = (0,0,0,0,1)$ and $n^a(x,1) = n^a(x)$, then

$$\Gamma = \frac{\epsilon_{abcde}}{\text{Area}(S^4)} \int_0^1 du \int d^3x \, n^a \partial_x n^b \partial_y n^c \partial_t n^d \partial_u n^e. \tag{7}$$

The WZW term correctly captures the non-trivial quantum numbers of the topological defects and is responsible for the non-Landau physics. Repeating the calculation of the structure of the VBS vortex, it is easy to see that the WZW term reduces — at the vortex core — to the standard spin Berry phase as appropriate for a spin-1/2 degree of freedom deescribed by $\mathbf{N}$.

The sigma model is well-defined as a continuum field theory only in the weak coupling limit, where it is ordered. Here there is a clear semiclassical picture for the effect of the WZW term on the topological defects in the ordered state. For the transition itself — driven by the λ term — this ordered state corresponds to a first order phase transition. In particular the sigma model formulation does not by itself make it at all natural that the Néel-VBS transition can be second order. To directly say anything about the putative second-order Neel-VBS phase transition in this formulation requires extending the sigma model to strong coupling, and looking for a symmetry preserving critical phase (see below for discussion of a recent novel numerical study[59]). At strong coupling the sigma model theory is non-renormalizable and requires an alternative formulation as a continuum field theory. Physically, disordered phases of the sigma model (defined with an explicit UV cutoff) correspond to phases where topological defects of the order parameter have proliferated. Thus a modification of the topological defects leads to a modification of the corresponding disordered phases. The sigma model formulation thus exposes the seed, in the ordered phase, of the impending non-Landau physics of the disordered critical regime.

The field theory in Eqn. 3 thus raises the possibility that the Neel-VBS transition may be second order. Whether it is actually so or not depends on the IR fate of the $NCCP^1$ model. This is however hard to study. To get a handle, it is extremely useful to consider the general case of interacting $SU(N)$ quantum spins. The proposed field theory for the non-Landau transition can be studied analytically in the limit of large-N through a $1/N$ expansion. A second order transition described by a conformal field theory is obtained. Thus the lattice spin model for sufficiently large-N provides a solvable and, hence important, example of a non-Landau deconfined quantum critical point. Extrapolating to small values of N, the critical fixed point has large values of the anomalous exponent η for both the N'eel and VBS order parameters. The four-fold monopoles are irrelevant at the critical fixed point, but of course they are important in the VBS ordered phase where they pin the orientation of the VBS order parameter to one of four degenerate values. This dangerous irrelevance of four-fold monopoles leads to the presence of two divergent length scales: one is the spin correlation length ξ, and the other — which may be identified as a scale associated with the pinning of the VBS order — ξ_{VBS} diverges as a power of ξ.

The field theory for the $N = 2$ model can only be studied numerically, and its precise fate remains murky. In particular, many studies[27,31–34,37,43–45,60,61] notice vexing problems (such as drifts with increasing system size in some critical properties).

Further studies[35] show the emergence of an $SO(5)$ symmetry that rotates the Néel and VBS order parameters into one another. Such a symmetry is of course suggested by the sigma model formulation of the intertwinement betweeen Neel and VBS orders, which however is not directly suited to describe the QCP as discussed above. This symmetry is however not at all apparent in the $NCCP^1$ formulation. Ref. 51 provided an understanding of the emergence of the $SO(5)$ symmetry from many points of view. The field theory admits multiple ('dual') descriptions. In any one of these descriptions only a subgroup of the full symmetry is apparent but it is a different subgroup in different descriptions. Thus by combining the information from these multiple descriptions we can get an understanding of the full emergent $SO(5)$ symmetry.

A manifestly $SO(5)$ invariant effective field theory[51] of the intertwinement of the Néel and VBS orders is obtained by starting with a fermionic spinon description of the square lattice spin-1/2 magnet. This naturally leads to a low energy theory of two flavors of massless Dirac fermions coupled to a dynamical $SU(2)$ gauge field — a theory denoted $N_f = 2$ QCD_3. It is expected[i] that this theory shares the same universal infrared physics as the $NCCP^1$ theory.

The possibility that the putative fixed point theory may have full $SO(5)$ symmetry enables another check on its existence as a conformal field theory using conformal bootstrap methods. Unfortunately an $SO(5)$ symmetric CFT with exponents in the range expected from numerical calculations of lattice spin models is not found.[45] Combined with the drifts in critical exponents mentioned above, it appears possible

[i]It is a bosonic theory with global $SO(5)$ symmetry, and the same anomaly.

that for $N = 2$, the Neel-VBS transition is weakly first order but with a very long correlation length leading to the appearance of a continuous transition.

A possible explanation of these facts (an unnaturally weak first order transition together with the emergence of the $SO(5)$ symmetry) is provided by the idea that the transition in the $N = 2$ model is 'pseudocritical'.[51,62] Pseudocriticality refers to a situation where the transition is second order 'close' to the physical model (in a generalized space including spatial dimensionality d, number of components N, etc) but the fixed points move slightly away from the real axis for the physical model. Then along the renormalization group flow the coupling constants linger ('walk') for a long time when they are closest to the fixed point but never really reach it, and ultimately flow away. Pseudocriticality commonly occurs when a stable fixed point merges with an unstable one, and they annihilate each other. A famous example,[63] is the 5-state classical Potts model in two dimensions. For other examples and a general discussion, see Ref. 64. For the Néel-VBS transition, the pseudocritical scenario finds support[65,66] in an epsilon expansion near space dimension $d = 1$ of the WZW non-linear sigma model.

An alternate suggestion[67–71] is that at $N = 2$, there is (in a parameter range) a quantum spin liquid intervening between the N'eel and VBS phases. As the parameters are varied, the quantum spin liquid region shrinks and is eventually replaced by a first order N'eel-VBS transition. There is an associated multicritical point when the spin liquid terminates and the line of first order transitions begins, and this is identified[70] with the deconfined quantum critical point.

Very recently, a beautiful study[59] provides strong support for the emergence of approximate conformal invariance and pseudocriticality at this phase transition. This work studied the $SO(5)$ WZW model regulated as a fermion model on a "fuzzy sphere" (a sphere with a magnetic monopole at the center, which leads to the formation of Landau levels for the fermions). As shown in a previous work[72] on the $2+1-D$ Ising transition, such a fuzzy sphere regularization enables examining the emergence of conformal symmetry by studying the spectrum of eigenstates through exact diagonalization. For conformal field theories, these energies are related to operator scaling dimensions, and the presence of conformal symmetry leads to many restrictions on the spectrum. For the regulated $SO(5)$ WZW model, the fuzzy sphere calculation finds a spectrum nicely consistent with the emergence of conformal symmetry. However with increasing system size, an $SO(5)$ singlet operator becomes weakly relevant. As we emphasized above, for the Néel-VBS transition, it is necessary that the $SO(5)$ WZW model at strong coupling have a critical phase (the phase transition is then tuned by an anisotropy term that explicitly breaks $SO(5)$ to $SO(3) \times O(2)$). The presence of a weakly relevant $SO(5)$ singlet is thus nicely consistent with the conjectured pseudocriticality at the Néel-VBS transition.

Clearly further clarifying the precise fate of the $N = 2$ model is an important goal for the near future. Nevertheless we emphasize that the existence of a second order Néel-VBS transition for large enough N establishes the matter-of-principle question on whether such transitions can exist at all.

3.2. *Other realizations*

A different realization of a deconfined critical point is a phase transition in a system of interacting electrons. The continuous global internal symmetry is $U(2)$ corresponding to conservation of the total electronic charge and spin. In additon there is time reversal symmetry under which the electrons transform as Kramers doublets. Consider a phase where $SU(2)$ spin rotation symmetry is broken spontaneously to a $U(1)$ subgroup but the charge $U(1)$ and time reversal are preserved. The corresponding order parameter is a three component field that transforms in the spin-1 representation of the $SU(2)$ spin rotation symmetry. Now let us further specialize to a situation where — at the mean field level — there is a quantum spin Hall effect.[73,74] In the bulk of such a system, the electronic spectrum is gapped but there will be gapless edge states at a boundary to the vacuum.

It was shown in Ref. 75 that fat skyrmion defects of the order parameter carry electrical charge-2. Restoring the broken $SO(3)$ symmetry by condensing skyrmion defects then leads to superconducting order[j]. *i.e*, to a spontaneous breakdown of the global $U(1)$ associated with charge conservation. The phase transition between these two states is described at low energies by the $NCCP^1$ field theory, and hence is another realization of the deconfined quantum critical point. In the $NCCP^1$ formulation, the flux of the $U(1)$ gauge field corresponds to the physical charge density, and hence is conserved exactly. Thus monopole operators of the $U(1)$ gauge field correspond to Cooper pair operators that change the charge by 2. Obviously these monopole operators are not allowed in the action (unlike in the quantum magnetism realization).[k]

The absence of monopoles of any strength implies that (in contrast to the Neel-VBS example), here there is only a single diverging length scale at the transition. Fermionic lattice models that realize this transition have been written down

[j]This 'skyrmion condensation' mechanism of superconductivity has, in recent years, been proposed to be realized in twisted bilayer graphene.[76,77] It has been demonstrated numerically both on lattice models,[78] and on continuum models similar to those appropriate for twisted bilayer graphene.[79]

[k]The astute reader will recall that the low energy $NCCP^1$ field theory has a mixed t' Hooft anomaly between the global $U(1)$ and the global $SO(3)$ symmetries. While we are certainly used to emergent symmetries having 't Hooft anomalies, it may be surprising that there is such an anomaly in a system where these symmetries are well-defined as 'on-site' symmetries of a microscopic lattice model. The resolution is that the anomaly is only present if the theory is regarded as a bosonic theory. In the presence of the (gapped) electronic degrees of freedom, the anomaly is trivialized, and this allows an on-site realization of the $SO(3) \times U(1)$ symmetry. This has the consequence[80,81] that in a system with boundaries, the fermion will be gapless at the edge even in the presence of the gapless critical bulk.

and are amenable to large scale Monte Carlo studies[78,82] which find evidence for a second order transition.

Deconfined critical field theories also arise in the context of phase transitions between trivial and SPT phases.[10,83,84] These are often related by duality transformations to the theories of the Landau-forbidden transitions.[51]

4. DQCP in 1d

In the last few years, there have been a number of interesting studies[85–90] on one dimensional lattice quantum models that possess deconfined quantum critical points. In contrast to their 2-dimensional counterparts, these one dimensional models can be studied reliably with field theoretic and numerical methods. Models have even been constructed which are exactly solvable and show a deconfined quantum critical point.

Consider a $1d$ lattice model[85] of spin-1/2 moments with the Hamiltonian

$$\mathcal{H} = -\sum_j \left(J_x \sigma_j^x \sigma_{j+1}^x + J_z \sigma_j^z \sigma_{j+1}^z\right) + \sum_j \left(K_x \sigma_j^x \sigma_{j+2}^x + K_z \sigma_j^z \sigma_{j+2}^z\right) \tag{8}$$

The model has two global Z_2 symmetries corresponding to π rotations of all the spins about either the x or the z-axis. These are denoted Z_2^z and Z_2^x respectively. It also has lattice translation symmetry, and the (anti-unitary) time reversal symmetry. By tuning the various parameters $J_{x,z}, K_{x,z}$, a number of phases can be accessed. For instance, a $z - FM$ phase spontaneously breaks the Z_2^x and time reversal symmetries but preserves Z_2^z and translation. A distinct VBS phase preserves both Z_2 symmetries and time reversal but breaks unit lattice translation symmetry. The phase transition between these two phases can be shown to be second order. The structure of the resulting theory has many similarities with that of the Neel-VBS transition in two space dimensions.[85] The Landau-forbidden phase transition is enabled by the structure of the topological defects. In the VBS phase, a domain wall leaves behind an unpaired site with a single spin-1/2 moment. This unpaired spin transforms projectively under the $Z_2^x \times Z_2^y$ symmetry. Destroying the Neel order by proliferating these domain walls then necessarily breaks $Z_2^x \times Z_2^y$ symmetry.

Ref. 88 constructed an exactly solvable $1d$ model with global $Z_2 \times Z_2$ symmetry which features a continuous phase transition between two phases which break the two different Z_2 symmetries. The symmetry action is however anomalous (one of the Z_2 symmetries is not an on-site symmetry). Thus this system is usefully viewed as living at the boundary of a $2d$ Symmetry Protected Topological phase of spins with the same $Z_2 \times Z_2$ symmetry. The model could be exactly mapped to a Z_4 clock model coupled to a Z_2 gauge field. In $1d$ the presence of the Z_2 gauge field simply implies global selection rules on physical states when the system is placed on a ring. The Landau-forbidden transition can then be related to the ordering transition of the Z_4 model which is continuous.

5. DQCP in 3d: Solvable Examples

A fruitful approach to generating a wealth of examples of deconfined criitcal points whose universal IR physics is solvable is described in Ref. 3. The idea is to study deconfined critical theories that are free in the IR but become strongly coupled and confine upon moving away from criticality. To illustrate this, consider a quantum field theory in $3+1$ spacetime dimensions described by N_f massless Dirac fermions coupled to, say, an $SU(2)$ gauge field. The one-loop beta function for the gauge coupling g is well known, and takes the form

$$\beta(g^2) = \frac{dg^2}{dl} = \frac{g^3}{24\pi^2}(N_f - 11) \tag{9}$$

The theory is thus IR-free for $N_f > 11$. The gauge coupling g is irrelevant at the IR fixed point. The low energy physics is then given by a theory of free massless Dirac fermions ('quarks'), and a separate set of free bosons ('gluons'). Now consider perturbing the theory by adding a mass m for the Dirac fermions. If we demand time reversal symmetry, then there is a unique (real) mass term that is allowed. However once the fermions are massive, they can be integrated out and the low energy theory is that of pure $SU(2)$ gauge theory; in this phase, g flows to strong coupling, and the theory is confining. The RG flows are shown in Fig. 3. The massless point $m = 0$ can be interpreted as a quantum critical point in the phase diagram. The nature of the two phases depends on the value of N_f. For even N_f, the $m = 0$ point describes a phase transition between a trivial gapped phase and an SPT phase[1] (we discuss odd N_f below). Though this model may seem esoteric to condensed matter physics, it has the advantage that it realizes a particularly simple and solvable (in the IR) example of deconfined quantum criticality.

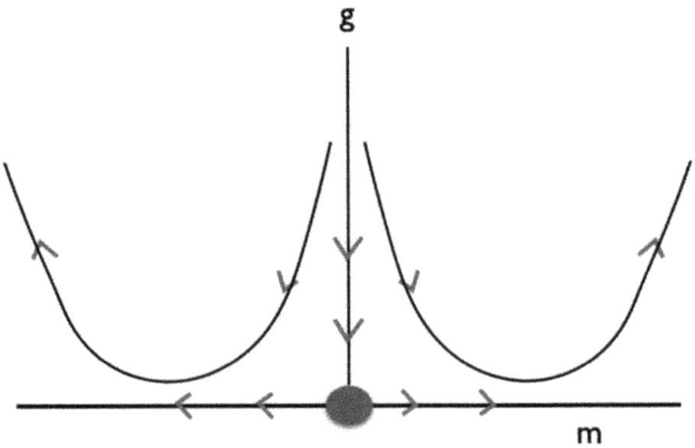

Fig. 3. Renormalization group flows of the IR-free non-abelian gauge theory.

[1]This field theory should be regarded as describing a microscopic system of interacting bosons: all local, *i.e* gauge invariant, operators in the theory are bosons. See Ref. 3 for identification of the protecting symmetries and further details.

6. Other Novel QCPs

In this section, we briefly discuss work in the last few years demonstrating a number of interesting critical phenomena. These were first found in the context of IR-free deconfined critical theories of the kind described above (by considering different gauge groups and different matter content). More natural models (from a condensed matter perspective) that realize some of these phenomena have since been found.

6.1. *Multiple universality classes*

In traditional phase transition theory, it is common to assume that, given two phases that are separated by a second order transition, there is a unique universality class for the phase transition. This is not generally true. For instance in disordered systems (*eg*, the $\pm J$ Ising spin glass), it has been known for a long time that there may be multiple universality classes for the same phase transition. It is perhaps to be expected that this phenomenon of multiple universality classes also occurs in clean systems but the present author had not seen examples till recently. The IR-free non-abelian gauge theories naturally lead to many examples of multiple (deconfined critical) universality classes for the same trivial-SPT phase transition. For a simpler example in a $d = 1$ spin systems, see Ref. 91.

6.2. *Landau beyond Landau*

The examples of multi-universality mentioned above were for transitions between gapped symmetry preserving phases. These of course are not to be described within the LGWF paradigm. It is thus interesting that it also possible to find examples where a Landau ordering phase transition that admits a universality class within the LGWF paradigm also allows a distinct universality class that is beyond LGWF. An example of this phenomenon is found in the model of the IR-free non-abelian gauge theory. Specifically, the $SU(2)$ gauge theory with N_f massless fermions, with N_f odd and sufficiently large, is a (deconfined) critical point between a trivial gapped phase and a phase that spontaneously breaks time reversal symmetry but is otherwise gapped. The transition is driven by changing the sign of the fermion mass. This transition can clearly also be in the usual Ising universality class. The deconfined critical universality class has (many) more gapless degrees of freedom which further transform non-trivially under some global symmetries[m] — other than time reversal — present in this system. There is thus no formulation of this new universality class in terms of a theory written just in terms of the Ising order parameter field; it thus represents a dramatic breakdown of the LGWF paradigm.

[m]These global symmetries are unbroken on either side of the phase transition; in the usual Ising universality class, excitations charged under these other symmetries will be gapped. The deconfined universality class however has gapless excitations that are charged under these other global symmetries.

At the time this article was written, there are no other examples of this 'Landau beyond Landau' phenomenon in simpler models in condensed matter physics, and finding one such is a fascinating future endeavor.

6.3. *"Unnecessary" critical points*

It is often stated — in discussions on quantum materials — that the most fundamental question about a quantum critical point is to ask about what distinguishes the phases on either side. However, Ref. 3 described an interesting phenomenon where a second order critical line lives "inside" a single phase of matter. This was dubbed an "unnecessary" quantum critical point. In this case there is no sharp distinction between the two sides of the transition and it is possible to go around it smoothly. It is analogous to the liquid-gas transition except that it is second order. Furthermore in the initial example described in Ref. 3, this unnecessary critical point had emergent deconfined gapless degrees of freedom though the phase it is submerged in is a trivial gapped phase. This example was once again constructed by studying the phase diagram of perturbed IR-free non-abelian gauge theories.

Subsequent work has found a number of examples in simpler systems. These include spin chains[80,91] in $1d$, and free fermion systems[3,92] in $2d$ or $3d$. It appears that unnecessary criticality is not an uncommon phenomenon.

7. Discussion

As reviewed here, there are a number of phase transitions where the strategy of focusing on long wavelength, long time fluctuations of a Landau order parameter field is not adequate. Nevertheless the framework of effective field theory and the renormalization group remain crucial. This is arguably one of the most important insights of the theory of critical phenomena developed in the 1960s and 1970s. At the deconfined critical points described in this review, useful effective field theories can be constructed in terms of emergent gapless fields that are charged under an emergent gauge field.

There are many further developments and applications that we have not discussed. For instance, recent work[93] studies the dynamics of the Neel-VBS transition in a driven $2d$ quantum magnet through numerical simulations. The deconfinement at the putative critical point impacts the annealing process, and leads to dynamics distinct from that described by the usual Kibble-Zurek scaling of conventional phase transitions.

An outgrowth of the theory of deconfined criticality is the understanding that there can be stable gapless 'deconfined' phases in two dimensional quantum magnets whose low energy theories involve fermions coupled to gauge fields.[94] An interesting example is a phase known as an algebraic (or Dirac) spin liquid where there are emergent gapless fermions coupled to a dynamical $U(1)$ gauge field. Interesting generalizations of these states have been proposed[95] where it is not clear that there is *any* Lagrangian-based field theoretic description at all. If such states do exist, they may pose a challenge even to the paradigm of writing down an action for an effective field theory to describe the low energy physics.

At the time of writing, there are relatively few experimental realizations of deconfined quantum critical points. For the Neel-VBS transition, very recent experiments on the material $SrCu_2(BO_3)_2$ have been interpreted in terms of a deconfined critical point.[96] A provocative recent paper suggests[97] that a deconfined critical point may underlie the onset of superconductivity in the monolayer material WTe_2. A realization of a $1d$ deconfined quantum critical point in Rydberg atom simulators has been proposed.[98]

One of the prime motivations for the original development of the theory of deconfined criticality in Ref. 1,2 was the observation of non-fermi liquid physics at the transition between two fermi liquid phases in the heavy fermion materials.[99,100] These critical points are usually associated with the onset of magnetic order out of the heavy fermi liquid but there may be a concomitant destruction[101,102] of the Kondo screening responsible for the formation of the heavy fermi liquid in the first place. There is of course a qualitative resemblance to the simpler deconfined critical points discussed above. However a theory of these kinds of heavy fermion critical points must deal with hard questions on the fate of the fermi surfaces in either phase. Despite years of effort, these critical points remain mysterious. Nevertheless it is encouraging that the study of simpler models (without fermi surfaces) already demonstrates that Landau order parameter fluctuations may not be the full story at phase transitions, and, even worse, may distract from the essential physics of some phase transitions.

This work was supported by NSF grant DMR-2206305, and partially through a Simons Investigator Award from the Simons Foundation. Part of this article was written at the KITP and supported in part by the National Science Foundation under Grant No. NSF PHY-1748958. I am honored to be part of an academic lineage which places me as Michael Fisher's academic 'great grandson' (through David Nelson and Subir Sachdev). A different connection is through my postdoctoral advisor (Matthew Fisher) — one of Michael's physicist sons. Reading an article on critical phenomena and the renormalization group by Michael was one of the reasons I got attracted to condensed matter physics as an undergraduate. It is thus a pleasure to contribute to this memorial volume.

References

1. T. Senthil, A. Vishwanath, L. Balents, S. Sachdev, and M. P. A. Fisher, Deconfined quantum critical points, *Science.* **303**, 1490, (2004).
2. T. Senthil, L. Balents, S. Sachdev, A. Vishwanath, and M. P. A. Fisher, Quantum criticality beyond the landau-ginzburg-wilson paradigm, *Phys. Rev. B.* **70**, 144407 (Oct, 2004). doi: 10.1103/PhysRevB.70.144407. http://link.aps.org/doi/10.1103/PhysRevB.70.144407.

3. Z. Bi and T. Senthil, Adventure in topological phase transitions in 3+ 1-d: non-abelian deconfined quantum criticalities and a possible duality, *Physical Review X.* **9**(2), 021034, (2019).
4. Z. Bi, E. Lake, and T. Senthil, Landau ordering phase transitions beyond the landau paradigm, *Physical Review Research.* **2**(2), 023031, (2020).
5. J. McGreevy, Generalized symmetries in condensed matter, *Annual Review of Condensed Matter Physics.* **14**, 57–82, (2023).
6. M. Oshikawa, Commensurability, excitation gap, and topology in quantum many-particle systems on a periodic lattice, *Physical review letters.* **84**(7), 1535, (2000).
7. M. B. Hastings, Lieb-schultz-mattis in higher dimensions, *Phys. Rev. B.* **69**, 104431 (Mar, 2004). doi: 10.1103/PhysRevB.69.104431. https://link.aps.org/doi/10.1103/PhysRevB.69.104431.
8. T. Senthil, Symmetry-protected topological phases of quantum matter, *Annu. Rev. Condens. Matter Phys.* **6**(1), 299–324, (2015).
9. M. Cheng, M. Zaletel, M. Barkeshli, A. Vishwanath, and P. Bonderson, Translational symmetry and microscopic constraints on symmetry-enriched topological phases: A view from the surface, *Physical Review X.* **6**(4), 041068, (2016).
10. A. Vishwanath and T. Senthil, Physics of Three-Dimensional Bosonic Topological Insulators: Surface-Deconfined Criticality and Quantized Magnetoelectric Effect, *Physical Review X.* 3(1):011016 (Jan., 2013). doi: 10.1103/PhysRevX.3.011016.
11. L. Savary and L. Balents, Quantum spin liquids: a review, *Reports on Progress in Physics.* **80**(1), 016502, (2016).
12. C. Broholm, R. Cava, S. Kivelson, D. Nocera, M. Norman, and T. Senthil, Quantum spin liquids, *Science.* **367**(6475), eaay0668, (2020).
13. C. Wang and T. Senthil, Boson topological insulators: A window into highly entangled quantum phases, *Phys. Rev. B.* **87**, 235122 (Jun, 2013). doi: 10.1103/PhysRevB.87.235122. http://link.aps.org/doi/10.1103/PhysRevB.87.235122.
14. M. A. Metlitski, C. L. Kane, and M. P. A. Fisher, Bosonic topological insulator in three dimensions and the statistical witten effect, *Phys. Rev. B.* **88**, 035131 (Jul, 2013). doi: 10.1103/PhysRevB.88.035131. http://link.aps.org/doi/10.1103/PhysRevB.88.035131.
15. R. A. Jalabert and S. Sachdev, Spontaneous alignment of frustrated bonds in an anisotropic, three-dimensional ising model, *Physical Review B.* **44**(2), 686, (1991).
16. T. Senthil and M. P. Fisher, Z 2 gauge theory of electron fractionalization in strongly correlated systems, *Physical Review B.* **62**(12), 7850, (2000).
17. L. Balents, M. P. Fisher, and S. M. Girvin, Fractionalization in an easy-axis kagome antiferromagnet, *Physical Review B.* **65**(22), 224412, (2002).
18. T. Senthil and O. Motrunich, Microscopic models for fractionalized phases in strongly correlated systems, *Physical Review B.* **66**(20), 205104, (2002).
19. O. Motrunich and T. Senthil, Exotic order in simple models of bosonic systems, *Physical review letters.* **89**(27), 277004, (2002).
20. S. V. Isakov, R. G. Melko, and M. B. Hastings, Universal signatures of fractionalized quantum critical points, *Science.* **335**(6065), 193–195, (2012).
21. A. V. Chubukov, T. Senthil, and S. Sachdev, Universal magnetic properties of frustrated quantum antiferromagnets in two dimensions, *Physical review letters.* **72**(13), 2089, (1994).
22. Y.-C. Wang, M. Cheng, W. Witczak-Krempa, and Z. Y. Meng, Fractionalized conductivity and emergent self-duality near topological phase transitions, *Nature Communications.* **12**(1), 5347, (2021).

23. Y.-H. Zhang, Z. Zhu, and A. Vishwanath, Xy* transition and extraordinary boundary criticality from fractional exciton condensation in quantum hall bilayer, *arXiv preprint arXiv:2302.03703.* (2023).
24. M. Barkeshli and J. McGreevy, Continuous transition between fractional quantum hall and superfluid states, *Physical Review B.* **89**(23), 235116, (2014).
25. A. W. Sandvik, Evidence for deconfined quantum criticality in a two-dimensional heisenberg model with four-spin interactions, *Phys. Rev. Lett.* **98**, 227202 (Jun, 2007). doi: 10.1103/PhysRevLett.98.227202. http://link.aps.org/doi/10.1103/PhysRevLett.98.227202.
26. J. Lou, A. W. Sandvik, and N. Kawashima, Antiferromagnetic to valence-bond-solid transitions in two-dimensional SU(n) heisenberg models with multispin interactions, *Phys. Rev. B.* **80**, 180414 (Nov, 2009). doi: 10.1103/PhysRevB.80.180414. http://link.aps.org/doi/10.1103/PhysRevB.80.180414.
27. A. Nahum, J. T. Chalker, P. Serna, M. Ortuño, and A. M. Somoza, Deconfined quantum criticality, scaling violations, and classical loop models, *Phys. Rev. X.* **5**, 041048 (Dec, 2015). doi: 10.1103/PhysRevX.5.041048. http://link.aps.org/doi/10.1103/PhysRevX.5.041048.
28. O. I. Motrunich and A. Vishwanath, Emergent photons and transitions in the o (3) sigma model with hedgehog suppression, *Physical Review B.* **70**(7), 075104, (2004).
29. R. G. Melko and R. K. Kaul, Scaling in the fan of an unconventional quantum critical point, *Phys. Rev. Lett.* **100**, 017203 (Jan, 2008). doi: 10.1103/PhysRevLett.100.017203. http://link.aps.org/doi/10.1103/PhysRevLett.100.017203.
30. A. Banerjee, K. Damle, and F. Alet, Impurity spin texture at a deconfined quantum critical point, *Phys. Rev. B.* **82**, 155139 (Oct, 2010). doi: 10.1103/PhysRevB.82.155139. http://link.aps.org/doi/10.1103/PhysRevB.82.155139.
31. A. W. Sandvik, Continuous quantum phase transition between an antiferromagnet and a valence-bond solid in two dimensions: Evidence for logarithmic corrections to scaling, *Phys. Rev. Lett.* **104**, 177201 (Apr, 2010). doi: 10.1103/PhysRevLett.104.177201. http://link.aps.org/doi/10.1103/PhysRevLett.104.177201.
32. K. Harada, T. Suzuki, T. Okubo, H. Matsuo, J. Lou, H. Watanabe, S. Todo, and N. Kawashima, Possibility of deconfined criticality in su(n) heisenberg models at small n, *Phys. Rev. B.* **88**, 220408 (Dec, 2013). doi: 10.1103/PhysRevB.88.220408. http://link.aps.org/doi/10.1103/PhysRevB.88.220408.
33. F.-J. Jiang, M. Nyfeler, S. Chandrasekharan, and U.-J. Wiese, From an antiferromagnet to a valence bond solid: evidence for a first-order phase transition, *Journal of Statistical Mechanics: Theory and Experiment.* **2008**(02), P02009, (2008). http://stacks.iop.org/1742-5468/2008/i=02/a=P02009.
34. K. Chen, Y. Huang, Y. Deng, A. B. Kuklov, N. V. Prokof'ev, and B. V. Svistunov, Deconfined criticality flow in the heisenberg model with ring-exchange interactions, *Phys. Rev. Lett.* **110**, 185701 (May, 2013). doi: 10.1103/PhysRevLett.110.185701. http://link.aps.org/doi/10.1103/PhysRevLett.110.185701.
35. A. Nahum, P. Serna, J. T. Chalker, M. Ortuño, and A. M. Somoza, Emergent so(5) symmetry at the néel to valence-bond-solid transition, *Phys. Rev. Lett.* **115**, 267203 (Dec, 2015). doi: 10.1103/PhysRevLett.115.267203. http://link.aps.org/doi/10.1103/PhysRevLett.115.267203.
36. O. I. Motrunich and A. Vishwanath, Comparative study of Higgs transition in one-component and two-component lattice superconductor models, *ArXiv e-prints* (May. 2008).

37. A. B. Kuklov, M. Matsumoto, N. V. Prokof'ev, B. V. Svistunov, and M. Troyer, Deconfined criticality: Generic first-order transition in the su(2) symmetry case, *Phys. Rev. Lett.* **101**, 050405 (Aug, 2008). doi: 10.1103/PhysRevLett.101.050405. http://link.aps.org/doi/10.1103/PhysRevLett.101.050405.
38. L. Bartosch, Corrections to scaling in the critical theory of deconfined criticality, *Phys. Rev. B.* **88**, 195140 (Nov, 2013). doi: 10.1103/PhysRevB.88.195140. http://link.aps.org/doi/10.1103/PhysRevB.88.195140.
39. D. Charrier, F. Alet, and P. Pujol, Gauge theory picture of an ordering transition in a dimer model, *Phys. Rev. Lett.* **101**, 167205 (Oct, 2008). doi: 10.1103/PhysRevLett.101.167205. http://link.aps.org/doi/10.1103/PhysRevLett.101.167205.
40. G. Chen, J. Gukelberger, S. Trebst, F. Alet, and L. Balents, Coulomb gas transitions in three-dimensional classical dimer models, *Phys. Rev. B.* **80**, 045112 (Jul, 2009). doi: 10.1103/PhysRevB.80.045112. http://link.aps.org/doi/10.1103/PhysRevB.80.045112.
41. D. Charrier and F. Alet, Phase diagram of an extended classical dimer model, *Phys. Rev. B.* **82**, 014429 (Jul, 2010). doi: 10.1103/PhysRevB.82.014429. http://link.aps.org/doi/10.1103/PhysRevB.82.014429.
42. G. J. Sreejith and S. Powell, Scaling dimensions of higher-charge monopoles at deconfined critical points, *Phys. Rev. B.* **92**, 184413 (Nov, 2015). doi: 10.1103/PhysRevB.92.184413. http://link.aps.org/doi/10.1103/PhysRevB.92.184413.
43. H. Shao, W. Guo, and A. W. Sandvik, Quantum criticality with two length scales, *Science.* **352**(6282), 213–216, (2016). ISSN 0036-8075. doi: 10.1126/science.aad5007. http://science.sciencemag.org/content/352/6282/213.
44. D. Simmons-Duffin. unpublished.
45. Y. Nakayama and T. Ohtsuki, Necessary condition for emergent symmetry from the conformal bootstrap, *Phys. Rev. Lett.* **117**, 131601 (Sep, 2016). doi: 10.1103/PhysRevLett.117.131601. http://link.aps.org/doi/10.1103/PhysRevLett.117.131601.
46. Z. Wang, M. P. Zaletel, R. S. Mong, and F. F. Assaad, Phases of the (2+ 1) dimensional so (5) nonlinear sigma model with topological term, *Physical Review Letters.* **126**(4), 045701, (2021).
47. M. E. Peskin, Mandelstam-'t hooft duality in abelian lattice models, *Annals of Physics.* **113**(1), 122–152, (1978).
48. C. Dasgupta and B. I. Halperin, Phase transition in a lattice model of superconductivity, *Phys. Rev. Lett.* **47**, 1556–1560 (Nov, 1981). doi: 10.1103/PhysRevLett.47.1556. http://link.aps.org/doi/10.1103/PhysRevLett.47.1556.
49. M. P. A. Fisher and D. H. Lee, Correspondence between two-dimensional bosons and a bulk superconductor in a magnetic field, *Phys. Rev. B.* **39**, 2756–2759 (Feb, 1989). doi: 10.1103/PhysRevB.39.2756. http://link.aps.org/doi/10.1103/PhysRevB.39.2756.
50. M. Levin and T. Senthil, Deconfined quantum criticality and Néel order via dimer disorder, *Phys. Rev. B.* 70(22):220403 (Dec., 2004). doi: 10.1103/PhysRevB.70.220403.
51. C. Wang, A. Nahum, M. A. Metlitski, C. Xu, and T. Senthil, Deconfined quantum critical points: symmetries and dualities, *Physical Review X.* **7**(3), 031051, (2017).
52. M. A. Metlitski and R. Thorngren, Intrinsic and emergent anomalies at deconfined critical points, *Physical Review B.* **98**(8), 085140, (2018).
53. Z. Komargodski, T. Sulejmanpasic, and M. Ünsal, Walls, anomalies, and deconfinement in quantum antiferromagnets, *Physical Review B.* **97**(5), 054418, (2018).

54. W. Ye, M. Guo, Y.-C. He, C. Wang, and L. Zou, Topological characterization of lieb-schultz-mattis constraints and applications to symmetry-enriched quantum criticality, *SciPost Physics.* **13**(3), 066, (2022).
55. F. D. M. Haldane, O(3) nonlinear σ model and the topological distinction between integer- and half-integer-spin antiferromagnets in two dimensions, *Phys. Rev. Lett.* **61**, 1029–1032 (Aug, 1988). doi: 10.1103/PhysRevLett.61.1029. http://link.aps.org/doi/10.1103/PhysRevLett.61.1029.
56. N. Read and S. Sachdev, Spin-peierls, valence-bond solid, and néel ground states of low-dimensional quantum antiferromagnets, *Phys. Rev. B.* **42**, 4568–4589 (Sep, 1990). doi: 10.1103/PhysRevB.42.4568. http://link.aps.org/doi/10.1103/PhysRevB.42.4568.
57. A. Tanaka and X. Hu, Many-body spin berry phases emerging from the π-flux state: Competition between antiferromagnetism and the valence-bond-solid state, *Phys. Rev. Lett.* **95**, 036402 (Jul, 2005). doi: 10.1103/PhysRevLett.95.036402. http://link.aps.org/doi/10.1103/PhysRevLett.95.036402.
58. T. Senthil and M. P. A. Fisher, Competing orders, nonlinear sigma models, and topological terms in quantum magnets, *Phys. Rev. B.* **74**, 064405 (Aug, 2006). doi: 10.1103/PhysRevB.74.064405. http://link.aps.org/doi/10.1103/PhysRevB.74.064405.
59. Z. Zhou, L. Hu, W. Zhu, and Y.-C. He, The so(5) deconfined phase transition under the fuzzy sphere microscope: Approximate conformal symmetry, pseudo-criticality, and operator spectrum, *arXiv preprint arXiv:2306.16435.* (2023).
60. J. Zhao, Y.-C. Wang, Z. Yan, M. Cheng, and Z. Y. Meng, Scaling of entanglement entropy at deconfined quantum criticality, *Physical Review Letters.* **128**(1), 010601, (2022).
61. Y. Da Liao, G. Pan, W. Jiang, Y. Qi, and Z. Y. Meng, The teaching from entanglement: 2d deconfined quantum critical points are not conformal, *arXiv preprint arXiv:2302.11742.* (2023).
62. A. Nahum, J. Chalker, P. Serna, M. Ortuño, and A. Somoza, Deconfined quantum criticality, scaling violations, and classical loop models, *Physical Review X.* **5**(4), 041048, (2015).
63. B. Nienhuis, A. N. Berker, E. K. Riedel, and M. Schick, First- and second-order phase transitions in potts models: Renormalization-group solution, *Phys. Rev. Lett.* **43**, 737–740 (Sep, 1979). doi: 10.1103/PhysRevLett.43.737. http://link.aps.org/doi/10.1103/PhysRevLett.43.737.
64. V. Gorbenko, S. Rychkov, and B. Zan, Walking, weak first-order transitions, and complex cfts, *Journal of High Energy Physics.* **2018**(10), 1–49, (2018).
65. R. Ma and C. Wang, Theory of deconfined pseudocriticality, *Physical Review B.* **102**(2), 020407, (2020).
66. A. Nahum, Note on wess-zumino-witten models and quasiuniversality in 2+ 1 dimensions, *Physical Review B.* **102**(20), 201116, (2020).
67. B. Zhao, J. Takahashi, and A. W. Sandvik, Multicritical deconfined quantum criticality and lifshitz point of a helical valence-bond phase, *Physical Review Letters.* **125**(25), 257204, (2020).
68. W.-Y. Liu, J. Hasik, S.-S. Gong, D. Poilblanc, W.-Q. Chen, and Z.-C. Gu, Emergence of gapless quantum spin liquid from deconfined quantum critical point, *Physical Review X.* **12**(3), 031039, (2022).
69. D.-C. Lu, C. Xu, and Y.-Z. You, Self-duality protected multicriticality in deconfined quantum phase transitions, *Physical Review B.* **104**(20), 205142, (2021).

70. J. Yang, A. W. Sandvik, and L. Wang, Quantum criticality and spin liquid phase in the shastry-sutherland model, *Physical Review B.* **105**(6), L060409, (2022).
71. H. Shackleton, A. Thomson, and S. Sachdev, Deconfined criticality and a gapless z 2 spin liquid in the square-lattice antiferromagnet, *Physical Review B.* **104**(4), 045110, (2021).
72. W. Zhu, C. Han, E. Huffman, J. S. Hofmann, and Y.-C. He, Uncovering conformal symmetry in the 3d ising transition: State-operator correspondence from a quantum fuzzy sphere regularization, *Phys. Rev. X.* **13**, 021009 (Apr, 2023). doi: 10.1103/PhysRevX.13.021009. https://link.aps.org/doi/10.1103/PhysRevX.13.021009.
73. C. L. Kane and E. J. Mele, Quantum spin hall effect in graphene, *Physical Review Letter.* **95**, 226801, (2005).
74. C. L. Kane and E. J. Mele, Z_2 topological order and the quantum spin hall effect, *Physical Review Letter.* **95**, 146802, (2005).
75. T. Grover and T. Senthil, Topological spin hall states, charged skyrmions, and superconductivity in two dimensions, *Physical review letters.* **100**(15), 156804, (2008).
76. E. Khalaf, S. Chatterjee, N. Bultinck, M. P. Zaletel, and A. Vishwanath, Charged skyrmions and topological origin of superconductivity in magic-angle graphene, *Science advances.* **7**(19), eabf5299, (2021).
77. M. Christos, S. Sachdev, and M. S. Scheurer, Superconductivity, correlated insulators, and wess–zumino–witten terms in twisted bilayer graphene, *Proceedings of the National Academy of Sciences.* **117**(47), 29543–29554, (2020).
78. Y. Liu, Z. Wang, T. Sato, M. Hohenadler, C. Wang, W. Guo, and F. F. Assaad, Superconductivity from the condensation of topological defects in a quantum spin-hall insulator, *Nature Communications.* **10**(1), 2658, (2019).
79. S. Chatterjee, M. Ippoliti, and M. P. Zaletel, Skyrmion superconductivity: Dmrg evidence for a topological route to superconductivity, *Physical Review B.* **106**(3), 035421, (2022).
80. R. Thorngren, A. Vishwanath, and R. Verresen, Intrinsically gapless topological phases, *Physical Review B.* **104**(7), 075132, (2021).
81. R. Ma, L. Zou, and C. Wang, Edge physics at the deconfined transition between a quantum spin hall insulator and a superconductor, *SciPost Physics.* **12**(6), 196, (2022).
82. M. Hohenadler, Y. Liu, T. Sato, Z. Wang, W. Guo, and F. F. Assaad, Thermodynamic and dynamical signatures of a quantum spin hall insulator to superconductor transition, *Physical Review B.* **106**(2), 024509, (2022).
83. K. Slagle, Y.-Z. You, and C. Xu, Exotic quantum phase transitions of strongly interacting topological insulators, *Phys. Rev. B.* **91**(11):115121 (Mar., 2015). doi: 10.1103/PhysRevB.91.115121.
84. Y.-Y. He, H.-Q. Wu, Y.-Z. You, C. Xu, Z. Y. Meng, and Z.-Y. Lu, Bona fide interaction-driven topological phase transition in correlated symmetry-protected topological states, *Phys. Rev. B.* **93**(11):115150 (Mar., 2016). doi: 10.1103/PhysRevB.93.115150.
85. S. Jiang and O. Motrunich, Ising ferromagnet to valence bond solid transition in a one-dimensional spin chain: Analogies to deconfined quantum critical points, *Physical Review B.* **99**(7), 075103, (2019).
86. B. Roberts, S. Jiang, and O. I. Motrunich, Deconfined quantum critical point in one dimension, *Physical Review B.* **99**(16), 165143, (2019).

87. R.-Z. Huang, D.-C. Lu, Y.-Z. You, Z. Y. Meng, and T. Xiang, Emergent symmetry and conserved current at a one-dimensional incarnation of deconfined quantum critical point, *Physical Review B*. **100**(12), 125137, (2019).
88. C. Zhang and M. Levin, Exactly solvable model for a deconfined quantum critical point in 1d, *Physical Review Letters*. **130**(2), 026801, (2023).
89. C. Mudry, A. Furusaki, T. Morimoto, and T. Hikihara, Quantum phase transitions beyond landau-ginzburg theory in one-dimensional space revisited, *Physical Review B*. **99**(20), 205153, (2019).
90. L. Su, Boundary criticality via gauging finite subgroups: a case study on the clock model, *arXiv preprint arXiv:2306.02976*. (2023).
91. A. Prakash, M. Fava, and S. Parameswaran, Multiversality and unnecessary criticality in one dimension, *arXiv preprint arXiv:2209.00037*. (2022).
92. C.-M. Jian and C. Xu, Generic unnecessary quantum critical points with minimal degrees of freedom, *Physical Review B*. **101**(3), 035118, (2020).
93. Y.-R. Shu, S.-K. Jian, A. W. Sandvik, and S. Yin, Equilibration of topological defects at the deconfined quantum critical point, *arXiv preprint arXiv:2305.04771*. (2023).
94. M. Hermele, T. Senthil, M. P. Fisher, P. A. Lee, N. Nagaosa, and X.-G. Wen, Stability of u (1) spin liquids in two dimensions, *Physical Review B*. **70**(21), 214437, (2004).
95. L. Zou, Y.-C. He, and C. Wang, Stiefel liquids: Possible non-lagrangian quantum criticality from intertwined orders, *Physical Review X*. **11**(3), 031043, (2021).
96. Y. Cui, L. Liu, H. Lin, K.-H. Wu, W. Hong, X. Liu, C. Li, Z. Hu, N. Xi, S. Li, *et al.*, Proximate deconfined quantum critical point in srcu2 (bo3) 2, *Science*. **380**(6647), eadc9487, (2023).
97. T. Song, Y. Jia, G. Yu, Y. Tang, P. Wang, R. Singha, X. Gui, A. J. Uzan, M. Onyszczak, K. Watanabe, T. Taniguchi, R. J. Cava, L. M. Schoop, N. P. Ong, and S. Wu. Unconventional superconducting quantum criticality in monolayer wte2, (2023).
98. J. Y. Lee, J. Ramette, M. A. Metlitski, V. Vuletic, W. W. Ho, and S. Choi, Landau-forbidden quantum criticality in rydberg quantum simulators, *arXiv preprint arXiv:2207.08829*. (2022).
99. P. Coleman, C. Pépin, Q. Si, and R. Ramazashvili, Topical review: How do fermi liquids get heavy and die?, *Journal of Physics Condensed Matter*. **13**(35), R723–R738, (2001).
100. H. v. Löhneysen, A. Rosch, M. Vojta, and P. Wölfle, Fermi-liquid instabilities at magnetic quantum phase transitions, *Reviews of Modern Physics*. **79**(3), 1015, (2007).
101. Q. Si, S. Rabello, K. Ingersent, and J. L. Smith, Locally critical quantum phase transitions in strongly correlated metals, *Nature*. **413**(6858), 804–808, (2001).
102. T. Senthil, M. Vojta, and S. Sachdev, Weak magnetism and non-fermi liquids near heavy-fermion critical points, *Physical Review B*. **69**(3), 035111, (2004).

https://doi.org/10.1142/9789811282386_0015

Chapter 15

Experimental Studies of the Critical Casimir Effect

M.H.W. Chan

Department of Physics, Penn State University,
University Park, PA 16802, USA

Fisher and de Gennes showed that the cutoff of the order-parameter fluctuations in a liquid film of finite thickness near its critical point leads to a thickness-dependent free energy. This article presents a brief review of experimental studies of this Casimir-like effect in helium films near the superfluid transition and the tricritical point and in binary mixture liquid films near their critical points.

1. Introduction

The thermodynamic properties of a bulk system near its critical point are dominated by the diverging correlation length ξ, which measures the spatial extent of the order-parameter fluctuations. For a finite system such as a liquid film confined between two walls or adsorbed on a solid substrate, the fluctuations are limited to the thickness of the film. In 1978, Fisher and de Gennes[1] considered this effect on a critical binary mixture film and concluded that this cutoff of the order-parameter fluctuations translates to a thickness-dependent free energy. This phenomenon is analogous to the Casimir effect[2] arises from the cutoff of electromagnetic zero-point fluctuations when two conducting surfaces are brought close to each other. The critical point Casimir effect is not limited to binary mixtures but applicable to all confined liquid films undergoing critical phase transitions. This chapter reviews briefly the experiments spawned by the Fisher and de Gennes paper.

The first evidence of the critical Casimir effect surfaced in 1989[3] in an experiment studying the wetting of superfluid helium films on silver. Dionne and Hallock[3] used a parallel plate capacitor with silver electrodes evaporated on glass to examine the wetting properties of liquid helium through the superfluid transition. They found the helium film wets silver below as well as above the superfluid or the lambda transition at T_λ. The measured capacitance was found to increase smoothly and nearly linearly with temperature, reflecting the increase in the density of helium vapor, through T_λ. Interestingly, the data also show a dip, roughly twice the scatter of the data near 2 mK below T_λ. This dip was interpreted as a thinning of the adsorbed liquid helium film. Such a thinning of the liquid film is consistent with an attractive critical Casimir force exerted by the liquid helium–silver and the liquid helium–vapor interfaces on the film. An attractive Casimir force is consistent with theoretical expectations.[4–8]

The superfluid transition, belonging to the three-dimensional (3D) X–Y universality class, is particularly well suited for the study of the critical Casimir and other finite size-related effects. Liquid helium is chemically inert and wets nearly all solid surfaces except for Cs.[9] The confinement of superfluid helium films between two solid walls or the adsorption of such a film on a solid wall with a liquid–vapor interface provides an ideal experimental configuration to investigate critical fluctuation-related confinement effects.[10–12] Next to a solid substrate, the first one or two atomic layers of the adsorbed liquid helium film are compressed by the van-der-Waal potential of the substrate into an 'inert' solid layer. The thickness of this inert layer, owing to the much stronger helium–substrate attraction than that between helium atoms, remains constant from absolute zero to temperatures well above T_λ at 2.1768 K. The order parameter of the transition, the amplitude of the superfluid density of the liquid film vanishes next to this inert layer and at the liquid–vapor interface, thus assuming symmetrical Dirichlet–Dirichlet boundary conditions at both interfaces. In addition, the superfluid transition of bulk liquid helium is a model system of continuous phase transition with the most precise experimental information. The specific heat exponent $\alpha(-0.0127 \pm 0.0003)$[13] and the superfluid density exponent $\nu(0.6705 \pm 0.0006)$,[14] obtained from measurements that extend from 10^{-2} to 10^{-9} (specific heat) and 10^{-7} (superfluid density) in reduced temperature, $t = T/T_\lambda - 1$, are in complete agreement with the values deduced from renormalization-group calculations and simulations.[15–17] These reasons motivated Garcia and Chan[18] to investigate systematically the dip observed by Dionne and Hallock.[3]

2. Critical Casimir-Induced Film Thinning Near the Superfluid Transition, Measurements with Copper Parallel Plate Capacitors

In the Garcia and Chan experiment[18] six coin-shape Copper (Cu) disks were stacked to form five parallel plate capacitors as shown schematically in Fig. 1. The capacitors were suspended above a small reservoir of bulk liquid helium inside an oxygen-free high thermal conductivity Cu sample cell. The diameter and thickness of the six Cu electrodes are respectively ~1.7 and 0.25 cm and the width of the gaps between the electrodes is 0.2 mm. The Cu disks were encased in annular Cu guard rings to minimize error due to fringing electric field during measurements. The surfaces of the encased electrodes were polished to mirror finish using nominal 1000 Å diamond powder. However, under optical microscope, micron size scratches and micron size dust particles are visible on the Cu surface. The thicknesses of the helium films adsorbed on the Cu electrodes of the five parallel plate capacitors are progressively thinner with increasing heights above the bulk liquid. In the absence of critical

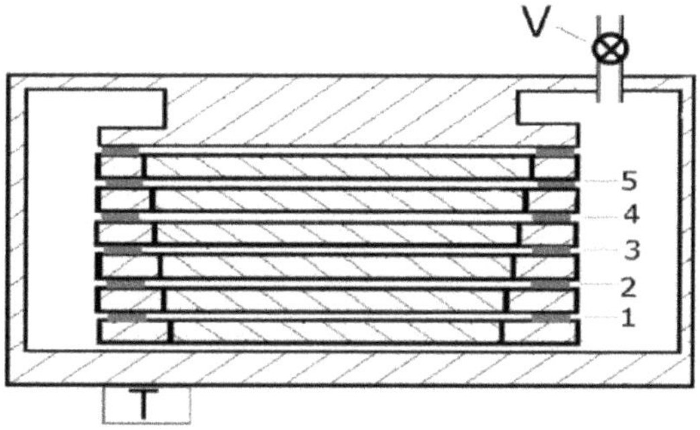

Fig. 1. Schematics of the sample cell, showing six Cu electrodes forming five parallel plate capacitors. A germanium thermometer, T, is attached to the cell bottom and the cell is sealed by needle valve, V, during the measurement. Figure adapted from Ref. 18.

fluctuation or Casimir effect, the equilibrium film thickness d of an adsorbed film at a height h above the bulk liquid can be written as

$$\left(\frac{\gamma_0}{d^3}\right)\left(1+\frac{d}{d_{1/2}}\right)^{-1} = mgh. \tag{1}$$

This equation, a simplified version[19,20] of the theory of Dzyaloshinskii, Lifshitz, and Pitaevskii (DLP),[21] describes the competition between gravitational potential energy, mgh, and the van-der-Waals attraction of the substrate $\frac{\gamma_0}{d^3}$ (= 2600K Å^3 for Cu) on liquid helium. The factor $\left(1+\frac{d}{d_{1/2}}\right)^{-1}$ corrects for the effect of the finite speed of light[19] and the value of $d_{1/2}$ is substrate-dependent, for Cu, it is 193 Å.[19,20] The DLP theory was verified by Sabisky and Anderson in 1972[22] in an experiment measuring the thicknesses of saturated helium films adsorbed on cleaved surfaces of alkaline-fluoride crystals at 1.38 K.

Near the superfluid transition, the cutoff of the order-parameter fluctuation introduces an additional term[1,4–8] to Eq. (1)

$$\left(\frac{\gamma_0}{d^3}\right)\left(1+\frac{d}{d_{1/2}}\right)^{-1} + \frac{k_B T_\lambda V}{d^3}\vartheta = mgh. \tag{2}$$

In Eq. (2), k_{B} and $V = 45.81$ Å^3/atom are, respectively, the Boltzmann constant and the specific volume of liquid ^{4}He. ϑ is a scaling function whose argument according to finite-size scaling,[23–25] is the ratio d/ξ. Since $\xi = \xi_0(\mathrm{t})^{-\nu}$, ϑ can be expressed with a scaling variable that is experimentally more readily assessable, specifically, $\mathrm{x} = (\xi_0.d/\ \xi)^{1/\nu} = \mathrm{t}d^{1/\nu}$. Experimentally, if ϑ is found to be > 0, this critical fluctuation term adds as an enhancement to γ_0/d^3 and leads to a thickening of the film. If ϑ is negative, the film thins. As noted above, the sign and functional form of ϑ are sensitive to the universality class of the transition and to the boundary conditions of the order parameter.[5–8] Under symmetric Dirichlet–Dirichlet boundary conditions at the two interfaces, ϑ is predicted to be negative.[5–8] The measurement of adsorbed films of five different thicknesses allows a test of whether the critical Casimir-induced thinning of the film is described by a universal scaling function ϑ consistent with the premise of finite-size scaling.[23–25]

The thicknesses d of the films adsorbed on the Cu electrodes in the five capacitors from bottom to top, based on their heights above the bulk liquid helium surface, estimated from the geometry of the sample cell and the quantity of liquid helium within, are deduced from Eq. (1) to be 423, 339, 299, 275, and 257 Å. The thickness of the bulk liquid at the bottom of the sample cell is estimated to be 2.5 ± 0.1 mm. The uncertainty in the heights of the capacitors above the liquid and hence the values of the thickness of the adsorbed films is estimated to be ~2%. This estimate assumes no systematic error due to the uncertainty in the value of $\frac{\gamma_0}{d^3}$. The thicknesses of the liquid films adsorbed on the two Cu electrodes comprising each parallel plate capacitor are within 0.5%.[18] The vertical dimension of the cell limits the maximum h and thus the minimum film thickness. The maximum film thickness is limited by the onset of capillary condensation in the 0.20 mm gap in the bottom capacitor. Once a desired quantity of helium was introduced, the sample cell was shut-off from the filling capillary with a mechanical needle valve. This keeps the amount of liquid in the cell and the thicknesses of the absorbed films constant and eliminates any superfluid heat flush effect from the liquid helium column in the filling capillary. This is essential to achieve good temperature homogeneity within the sample cell. Two concentric thermal shields were installed to enclose the sample cell. The outer thermal shield is thermally connected, but weakly, to an evaporating liquid helium reservoir that serves as the cooling source. These thermal shields are designed to intercept any stray heat leaks into the sample cell. With this arrangement, the temperature of the sample cell was regulated to within 2 μK and the temperature homogeneity throughout the sample cell was found to be the same. The empty cell (without helium inside the sample cell) capacitances C_0 of each of the five capacitors were found to be constant over the temperature range of the experiment. The capacitances were measured, by means of an ac bridge circuit, against a reference capacitor of similar construction thermally anchored outside the sample cell. The precision for the measured dielectric constant $\varepsilon\ (T) = C/C_0$ is better than five parts in 10^8.[26] Figure 2 shows the measured dielectric constant of the bottom or capacitor 1. It increases nearly linearly from ~1.000467 to 1.000510 between 2.15 and 2.195 K. The vapor contributes about 4.4×10^{-4} to ε and the increase reflects the density increase of the saturated helium vapor. The adsorbed film contributes only 5×10^{-5} to ε. A dip in ε is found 2.6 mK below T_λ, confirming the findings of Ref. 3. Similar dips were found in $\varepsilon\ (T)$ measured with the other capacitors. The position of the dips is found to move to lower temperatures with decreasing film thickness d. The dips are reproducible in size and temperature upon warming and cooling of the sample cell and are independent of the frequency and excitation voltage used in the capacitance bridge. The dip cannot be accounted for by any known anomaly in the helium vapor.[27] It is 40 times sharper

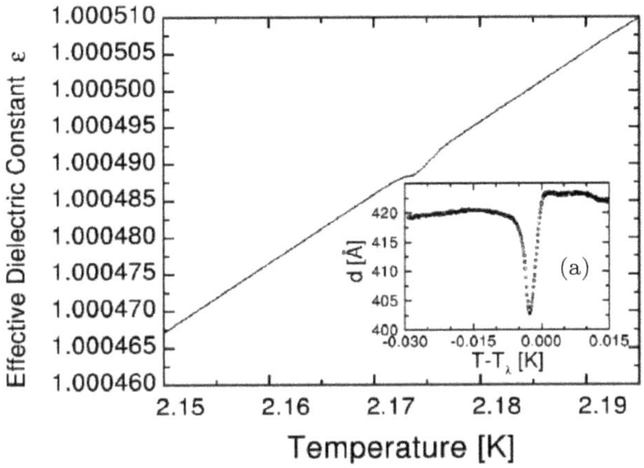

Fig. 2. Effective dielectric constant for capacitor 1 as a function of temperature. Inset (a) shows the film thickness calculated using Eq. (3). A dip is found 2.6 mK below T_λ. Figure adapted from Ref. 18.

and 30 times larger than that can be attributed to lambda point anomalies in the liquid density[28] or surface tension at the liquid–vapor interface.[29] To make sure the dip is not related to superfluid film flow driven by a minute thermal gradient within the sample cell, e.g., between the bulk liquid and the adsorbed film, a liquid sample with 10% ^{3}He was studied. In such a sample, the thermal conductivity of the superfluid film is reduced 1000-fold.[28] There is no significant effect on the size, shape, or position relative to the new superfluid transition temperature of the observed dip.

Since the capacitance of the liquid film and the vapor inside the gap are added in series, the thickness of the liquid film d can be calculated from the measured dielectric constant $\varepsilon(T)$ according to Eq. (3), where G is the gap of the capacitor:

$$d = \frac{G}{2}\left(\frac{1}{\varepsilon_{\text{vapor}}} - \frac{1}{\varepsilon(T)}\right) \Big/ \left(\frac{1}{\varepsilon_{\text{vapor}}} - \frac{1}{\varepsilon_{\text{film}}}\right). \tag{3}$$

The dielectric constant of the liquid film $\varepsilon_{\text{film}}$ can be assumed to be the same as bulk liquid, $\varepsilon_{\text{liquid}} \sim 1.0576 \pm 0.00005$.[30] The 'uncertainty' quoted here corresponds to the variation of $\varepsilon_{\text{liquid}}$ between 2.14 and 2.2 K.[28] $\varepsilon_{\text{vapor}}$, the dielectric constant of helium vapor is calculated by means of the Clausius–Mossotti equation, $(\varepsilon - 1)/(\varepsilon + 2) = 4\pi\rho\chi/3M$, where ρ is the density of helium vapor, and χ and M are, respectively, the molar polarizability ($0.123296 \pm 0.000030\,\text{cm}^3/\text{mole}$)[31] and the molar mass of helium. The density of helium vapor is estimated using a modified second virial coefficient that makes the film thicknesses outside the critical region to be independent of temperature.[18] The film thicknesses calculated from Eq. (3) for the five capacitors from the bottom for $T > T_\lambda$ are, respectively, 574, 514, 588, 341 and 298 Å, or 36%, 52%, 97%, 42% and 16% thicker than the values given by

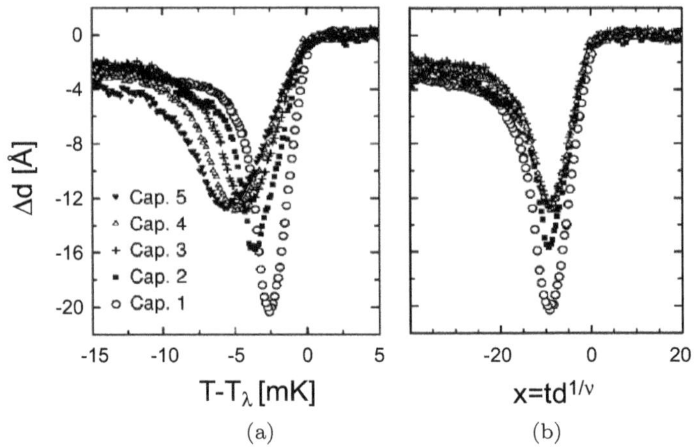

Fig. 3. (a) Fluctuation-induced film thinning Δd vs. temperature, Cap. 1 is closest to the bottom of the sample cell. (b) The same data plotted vs. the scaling variable x, showing the minimum of all five films occur at $x = -9.2 \pm 0.2$ Å$^{1/\nu}$. For clarity, only every fourth point is shown. Figure adapted from Ref. 18.

the DLP theory from Eq. (1). The large discrepancies are most likely due to capillary condensation of liquid helium around the micron and submicron scratches, and micron size dust particles on the polished surfaces of the copper electrodes. Since $\varepsilon_{\text{liquid}}$ is significantly larger than $\varepsilon_{\text{vapor}}$, these liquid 'puddles' increase the measured ε and give rise to an apparent and erroneous thicker liquid film. These considerations prompted Garcia and Chan to subtract a temperature-independent constant from the result of Eq. (3) so that $d = d_0$ for normal fluid film away from the transition region. d_0 is the film thickness predicted by Eq. (1) according to the DLP theory. The inset (a) of Fig. 2 shows the liquid film thickness adsorbed on the bottom (shown as Cap. 1 in Fig. 3) capacitor as a function of temperature calculated from Eq. (3) and adjusted to be consistent with the DLP theory for $T > T_\lambda$. It shows a dip in the film thickness that reaches ~20 Å at 2.6 mK below T_λ. In addition, the superfluid film at temperature below the dip is ~3 Å thinner than that for $T > T_\lambda$ in the normal fluid film.

Figure 3(a) shows the fluctuation-induced film thinning $\Delta d = (d - d_0)$ versus temperature for the adsorbed films measured by the five capacitors. Δd, as noted above is adjusted to be zero for normal fluid films away from the transition region. For all five films, the thinning appears to commence near 2 mK above T_λ and accelerates for $T < T_\lambda$. The magnitude and the temperature of maximum Δd decrease systematically for the five capacitors with decreasing film thickness. According to Eq. (2), the maximum film thinning should coincide with the minimum of the scaling function $\vartheta(x)$ and therefore should occur at a specific $x = td^{1/\nu}$ for all films. This is in fact the case, as shown in Figure 3(b). The minimum in $\vartheta(x)$ is found at $x = -9.2 \pm 0.2$ Å$^{1/\nu}$. This result supports the assumption that the DLP theory

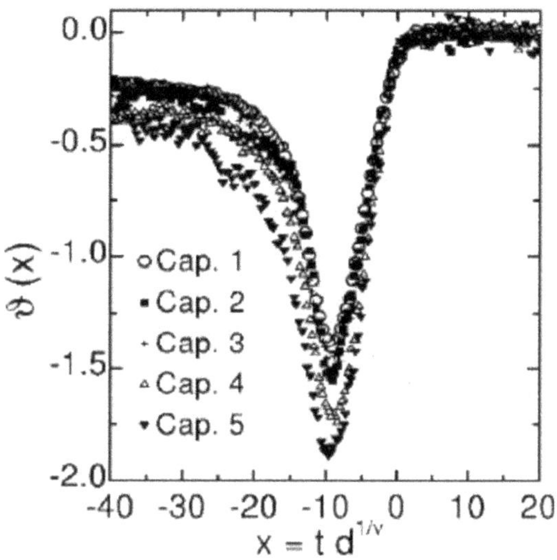

Fig. 4. The scaling functions of the five adsorbed films do not collapse into a universal curve. Figure adapted from Ref. 16.

from Eq. (1) gives the correct thicknesses for the normal fluid films. While capillary condensation induced by scratches and dust particles on the Cu surfaces increase substantially the total amount of liquid helium adsorbed on the Cu electrodes, the great fraction of the Cu electrode surface is covered by a liquid helium film with a thickness consistent with the DLP theory. According to finite-size scaling, the ϑ calculated from Eq. (2) from the five capacitors should collapse into one universal curve. This collapse, as shown in Figure 4, is not seen.

A quartz crystal microbalance (QCM) experiment[32] was carried out to confirm the film thinning and determine the superfluid onset temperature of the adsorbed helium film. The 0.5-mm-thick QCM, oscillating at 5 MHz, was placed horizontally with the bottom and top gold electrodes, respectively, 1.5 and 2 mm above bulk liquid helium in the sample cell. The thicknesses of the normal fluid films adsorbed on the top and bottom gold electrodes are deduced from Eq. (1) to be 420 and 460 Å, respectively. An increase in the resonant frequency of the QCM, consistent with film thinning that reaches 20 Å is found centering at 2.2 mK below T_λ. This agrees with the capacitor measurements. The QCM also shows a superfluid onset temperature at 3.2 mK below T_λ. (Reference 32 mis-identified the superfluid onset temperature at 4.2 mK below T_λ). This means the adsorbed films shown in Figs. 3 and 4 become superfluid 1 mK below the Casimir dip.

Gasparini and his collaborators[10–12] carried out extensive and systematic measurements of the specific heat of liquid helium near the superfluid transition confined between two SiO_2/Si wafers with different spacings. With increasing confinement, the specific heat peak becomes more rounded and shifts to lower temperatures. Finite-size-scaling analysis of the specific heat of confined 2D helium films over a wide range thicknesses (from 483 to 6918 Å) shows the peaks center at a common value of $x = td^{1/\nu} = 9 \pm 1$ Å$^{1/\nu}$. However, the scaling functions of the specific heat peaks do not collapse into a universal curve. There is no explanation for this non-universal behavior.[10–12] The fact that the maximum thinning of the adsorbed films and the maximum specific heat peaks are found at the same x (within experimental precision) indicates that $x = -9$ Å$^{1/\nu}$ locates the 'position' of maximum critical fluctuations.[12]

3. Measurements of Superfluid Film Thinning with Silicon Wafer Capacitor, Confirmation of Finite-Size Scaling

In order to determine whether the lack of scaling shown in Figure 4 is real or an experimental artifact due to capillary condensation on the less-than-perfect surfaces of the Cu electrodes, the Penn State group carried out a new experiment[33] using two single-crystal silicon wafers as the electrodes for the measurement capacitor. The silicon wafers used are heavily doped with phosphorous, chosen for their high electrical and thermal conductivity at low temperatures. Tapping-mode atomic force microscopy scans over five 0.01 μm^2 areas of the surface yields, rms roughness of 1 Å and scans over areas larger than 1 μm^2 regions give 8 Å due to occasional scratches. Figure 5 shows a schematic diagram of the silicon wafer sample cell. To minimize scratches and dust particles on the electrodes, the components of the sample cell were washed, dried, assembled and then sealed inside a class 10 clean room at the Penn State Nanofabrication Facility. The top (1 inch diameter) and bottom (2 inch diameter) wafers are secured into Cu guard rings with rubber cement. The thickness of the wafers is 0.4 mm. To minimize errors due to fringe fields,

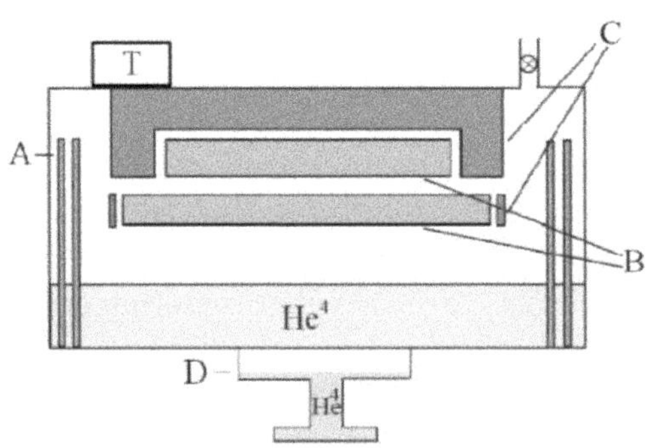

Fig. 5. The schematics of the silicon wafer capacitor sample cell. The height h of the Si wafer electrodes (B) above bulk liquid helium is determined by an annular capacitor (A). The Si electrodes are attached inside electrically grounded Cu guard rings (C) by rubber cement. A germanium thermometer (T) calibrated vs. ^{3}He vapor pressure and a lambda fixed point device (D) are anchored at the bottom of the cell. Figure adapted from Ref. 31. The sample cell is enclosed in two concentric thermal shields, S1 and S2. The outer shield S1 is weakly connected to a cooling source, an evaporating liquid helium reservoir.

the top and bottom guard rings are electrically grounded and held apart by 0.2-mm-thick Cu spacers. The gap between the two silicon wafer electrodes at low temperature was determined to be 0.235 mm. The 1-inch wafer is placed at virtual ground in the ac bridge circuit[26] for capacitance measurements. To determine T_λ, a lambda point device is anchored at the bottom of the sample cell. Like the sample cell used by Garcia and Chan,[18] the sample cell used here is also enclosed in two concentric thermal shields, S1 and S2. There is only one parallel plate capacitor in this sample cell. In order to increase the thickness d of the adsorbed films on the electrodes, additional liquid helium is added into the sample cell to reduce the height h between the silicon wafer capacitor and the bulk liquid helium at the bottom of the sample cell. The height h is measured by an annular capacitor inside the sample cell. The critical Casimir effect were measured on liquid helium film with d_0 equal to 238, 285 and 340 ± 10 Å.

Figure 6 shows the thinning of the adsorbed films, Δd, as a function of temperature for the three adsorbed films with thicknesses of 238, 285 and 340 Å from 2 mK above T_λ down to 50 mK below T_λ. $\Delta d = (d_0 - d)$ where d_0 is the film thickness calculated from Eq. (1). The temperature-dependent film thickness, $d(T)$ is calculated from Eq. (3) following the same procedure used by Garcia and Chan.[18] The silicon wafer empty cell capacitance C_0 show a linear increase with temperature. This temperature effect may be due to the differential thermal contraction among the various materials that make up the capacitor plates. Using $C_0(T) = C_0(T_\lambda)(1 - 3.5 \times 10^{-5}(T_\lambda - T))$, the calculated d for all three films are found to be temperature-independent above T_λ and also for $T < (T_\lambda - 15\text{ mK})$ in the superfluid phase. The opening and closing of the low-temperature needle valve used to seal off the sample cell and the refilling of liquid helium to the cryostat often

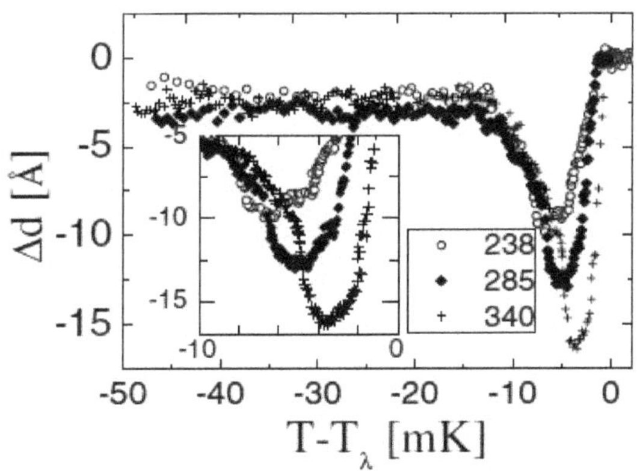

Fig. 6. Film thinning vs. temperature from 2 mK above to 50 mK below T_λ. The inset shows a blowup near the minimum. There is a kink in each of the three the datasets at temperature just below the minimum. These kinks locate the superfluid transition temperature of these three films. Figure adapted from Ref. 33.

results in a shift in the measured capacitance on the order of 50 ppm. To correct for this, $C_0(T_\lambda)$ is adjusted so that the calculated d at temperature $T > (T_\lambda + 2$ mK) is equal to d_0. The most efficient method to obtain reliable and reproducible data is by controlling the outer thermal shield S1 at a fixed value, typically ~0.1 K below T_λ and add a controlled amount of heat to the inner thermal shield S2 to allow the temperature of the sample cell to drift slowly up or down. A drift rate of 70–300 μK/hour was used at temperature more than 15 mK below T_λ and 10–40 μK/hour closer to and above T_λ. At these rates, the temperature of the sample cell appears to be in excellent thermal equilibrium and the capacitance (with precision on the order of 1 part in 5×10^8) and temperature (with precision better than 2 μK) readings are found to be reproducible upon warming and cooling. These slow rates mean it takes more than two weeks to complete the measurements for each adsorbed film.

Similar to that seen by Garcia and Chan,[18] a systematic trend in both the magnitude and position of the dip is seen (Fig. 6), specifically, the thickest film show the largest dip closest to T_λ. The thickness of the superfluid films at temperatures below the dip is a constant down to 50 mK below T_λ. The inset of Fig. 6 expands the data near the position of maximum thinning. Each dataset shows a kink or change of slope on the low-temperature side of the dip. Gasparini and colleagues have determined the onset temperature of superfluidity as a function of helium film thickness,[12] the temperatures of the kinks shown in the inset of Fig. 6 are found to track the onset of superfluidity in these three films.[12] As noted above, the QCM measurements also found a superfluid onset temperature that is just ~1 mK below the position of maximum film thinning.[32] These observations allow one to conclude that the adsorbed films are superfluid at temperatures below the Casimir dip. Figures 3, 4 and 6 show that the superfluid films well below the Casimir effect dip is 2.5±0.8 Å thinner in comparison with the normal fluid film. In the Garcia and Chan experiment, the film thinning due to the Casimir effect appears to commence at 2 mK above T_λ and accelerates for $T < T_\lambda$. Figure 6, however, shows no evidence of any film thinning above T_λ.

Figure 7 shows that the scaling functions ϑ of the three adsorbed films, calculated from Eq. (2), collapse into one single curve within experimental scatter. The minimum of ϑ at -1.3 ± 0.03 is found at $x = td^{1/\nu} = -9.7 \pm 0.8$ Å$^{1/\nu}$. The position of the minimum agrees with that found by Garcia and Chan[18] at $x = -9.2 \pm 0.2$ Å$^{1/\nu}$ and the position of maximum in the scaling function of the specific heat at $x = -9 \pm 1$ Å$^{1/\nu}$.[10–12] As noted above, this concurrence indicates x at -9 Å$^{1/\nu}$ is the position of maximum fluctuation of the order parameter of confined films. The collapse of the scaling functions of critical Casimir film thinning into a single curve confirms the prediction of finite-size scaling.[23–25] Finite-size scaling in the context of critical Casimir effect has been studied in considerable detail.[6,7,34] As noted above, the scaling functions of the specific heat maxima of confined helium films do not collapse into a universal curve. It would be

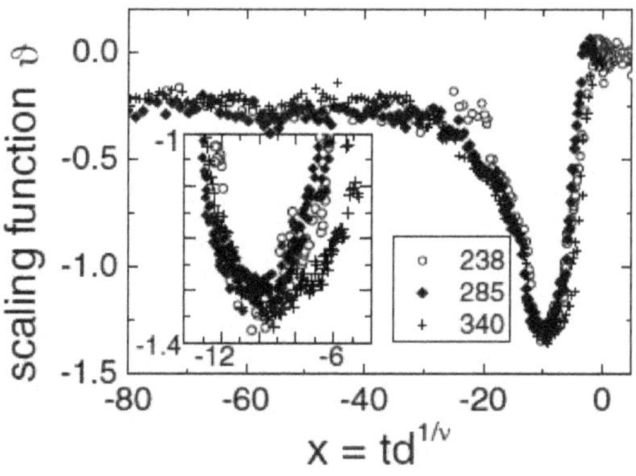

Fig. 7. The scaling function ϑ vs. the scaling variable x. The scaling functions for all three films collapse to a single curve with common minimum at $x = -9.7 \pm 0.8$ A$^{1/\nu}$.

interesting to obtain Casimir thinning data over a wider range, e.g., more than one decade in film thickness, like the specific heat experiments, to confirm the robustness of the finite-size scaling result. However, this is not easy to implement since the thickness of an adsorbed film d is constraint on the thick film limit by possible capillary condensation inside the capacitor and on the thin film limit by the height, h, of the sample cell, since, as shown in Eq. (1), h scales as $1/d^3$.

Krech and Dietrich[5] carried out field theoretic calculations of the scaling function for $T > T_\lambda$ and found a small negative ϑ that appears to agree with that shown in Fig. 4. However, the apparent small film thinning observed by Garcia and Chan above T_λ is of the order of the scatter of the data.[18] As shown in Figs. 6 and 7, this small film thinning is not seen in the experiment measured with the silicon capacitor. Zandi and collaborators[35] carried out a mean-field calculation for $T < T_\lambda$ and found a scaling function in qualitative agreement with that shown in Fig. 7. Biswas and collaborators[36] carried out an independent mean-field calculation for temperature above and below T_λ and found agreement with Refs. 5 and 35. Hucht,[37] Vasilyev *et al.*[38,39] and Hasenbusch[40] independently carried out Monte Carlo simulations of the Casimir force over the entire critical region and found scaling functions that are in quantitative agreement with that shown in Fig. 7. Zandi, Rudnick and Kardar[41] showed that the thinning of the superfluid films below the Casimir effect dip originates from phase fluctuation within the superfluid film, i.e., Goldstone mode and surface fluctuations at the superfluid–vapor interface. The sum of these two effects appears to agree with the thinning shown in Fig. 7.[41]

4. Critical Casimir Effects in ^{3}He–^{4}He Mixture Films Near the Tricritical Point

The tricritical point in ^{3}He–^{4}He mixtures provides another experimental system for the study of the critical Casimir effect. The superfluid transition of pure ^{4}He decreases continuously in temperature upon the dilution of ^{3}He until the mixture phase separates into ^{3}He and ^{4}He-rich regions. The lambda line is found to intersect the phase separation region at the peak of the boundary at the tricritical point (TP) at $X_3 = 0.672$ and 0.870 K.[42] X_3 is the molar ^{3}He concentration of mixture. The critical behavior near the TP is well described by the Blume–Emery–Griffiths (BEG) model.[43] Garcia and Chan[44] also studied the critical Casimir effect near the TP. In the tricritical point experiment, two capacitors were used. The lower capacitor is submerged in the liquid to determine the isotopic concentration of the mixture and pinpoint the tricritical point. The upper capacitor is placed in the vapor phase to measure the thickness of the adsorbed film on the electrodes. Due to the different zero-point energies of the two isotopes, the adsorbed film has a concentration gradient with a ^{4}He-rich superfluid layer residing next to the capacitor plate. The concentration of the film near the vapor surface, on the other hand, is the same as the bulk mixture. In other words, the adsorbed film has opposite or non-symmetrical boundary conditions at the two interfaces. Near the tricritical point a thickening of the film, indicative of a repulsive Casimir force with a positive scaling function, is found. The scaling function for mixtures with X_3 close to the tricritical point concentration show a peak value of 13 at the scaling variable $x = td = 14$ Å (near TP, $\nu = 1$). For the mixture at the TP, the scaling function is found to be 8.7 ± 1.7. The uncertainty in the scaling function is due primarily to the uncertainty of the absolute thickness of the film. These experimental findings agree with theoretical calculations.[45–47]

5. Critical Casimir Effect in Binary Mixture Films

The critical Casmir effect has also been confirmed in 3D Ising systems, specifically in binary liquid mixtures.[48–50] An optical ellipsometry[48,49] experiment was carried out in 1999 to measure the thickness of methanol–hexane mixture and 2-methoxyethanol–methylcyclohexane mixture films adsorbed on Si wafer near the critical points of these binary fluid systems. Thickening of the films, consistent with that expected for systems with opposite boundary conditions at the substrate and vapor interfaces, were observed. However, the magnitude of the scaling functions was found to be two orders smaller than expected.[48,49] It was suggested that the discrepancy in the magnitude of the scaling function may be due to the incomplete wetting of the adsorbed films.[50] An X-ray reflectivity measurement in 2005[50] showed an enhancement in the thickness of the wetting binary mixture film of methylcyclohexane and perfluoro-methylcyclohexane on a SiO_2/Si substrate, also with opposite boundary conditions at the two interfaces, near the critical point. The magnitude of enhancement in this experiment is consistent with theoretical expectations.[5–8,39,51] A third experiment in 2008[52,53] used total internal reflection microscopy to directly measure, with femto-newton resolution, the Casimir force between a single colloidal sphere and a flat silica surface immersed in lutidine–water liquid mixture near the critical point. Depending on whether the adsorption preferences of the colloidal sphere and the silica surface are identical or opposite, attractive or repulsive forces, respectively, are found. The sign and the magnitude of the force and a large part of the scaling function was found to agree with theoretical predictions.[52,53]

Acknowledgments

The author is grateful and fortunate to have Michael Fisher as one of his mentors. The mentor–mentee relation started when the author was a graduate student of John Reppy at Cornell. Michael showed keen interest in nearly all the experiments, including the critical Casmir experiments, conducted by the author and his students. Michael's penetrating questions often served as the guiding light for us. The experiments at Penn State described here were carried out with extraordinary dedication by Andriy Ganshin, Sarah Scheidemantel, Steve Jordon, John Lazzaretti and, particularly, Rafael Garcia. Rafael also played a key role in the design of the experiments and the analysis of the data. We have benefitted from discussions with many colleagues besides Michael over the course of the experiments. They include Michael Krech, Siegfried Dietrich, Milton Cole, Mehran Kardar, Klaus Knorr, Horst Meyer, and Frank Gasparini. The experiments at Penn State were supported by the Division of Material Research of NSF and the Office of Biological and Physical Research of NASA.

References

1. M. Fisher and P. G. de Gennes, Wall phenomena in a critical binary mixture. *C. R. Acad. Sci. Ser. B* **287**, 207–209 (1978).
2. H. B. G. Casimir, On the attraction between two perfectly conducting plates. *Proc. K. Ned. Akad. B* **51**, 793–795 (1948).
3. R. J. Dionne and R. B. Hallock, Wetting of helium films to silver substrates. *AIP Conf. Proc.* **194**, 199–200 (1989).
4. M. P. Nightingale and J. O. Indekeu, Effect of criticality on wetting layers. *Phys. Rev. Lett.* **54**, 1824–1827 (1985).
5. M. Krech and S. Dietrich, Finite-size-scaling for critical films. *Phys. Rev. Lett.* **66**, 345–348 (1991).
6. M. Krech and S. Dietrich, Specific heat of critical films, the Casimir force, and wetting films near critical end points. *Phys. Rev. A* **46**, 1992–1940 (1992).
7. M. Krech and S. Dietrich, Free energy and specific heat of critical films and surfaces. *Phys. Rev. A* **46**, 1886–1921 (1992).
8. M. Krech, Fluctuation-induced force in critical fluids. *J. Phys.: Cond. Matt.* **11**, R391–R412 (1999).
9. P. J. Nacher and J. Dupont-Roc, Experimental evidence for non-wetting with superfluid film. *Phys. Rev. Lett.* **67**, 2966–2969 (1991).
10. S. Mehta and F. M. Gasparini, Specific heat and scaling of ^{4}He confined in a planar geometry. *Phys. Rev. Lett.* **78**, 2596–2599 (1997).

11. S. Mehta, M. O. Kimball and F. M. Gasparini, Superfluid transition of ^{4}He for two-dimensional crossover, heat capacity, and finite-size scaling. *J. Low Temp. Phys.* **114**, 467–521 (1999).
12. F. M. Gasparini, M. O. Kimball, K. D. Mooney and M. Diaz-Avilla, Finite-size-scaling of ^{4}He at the superfluid transition. *Rev. of Mod. Phys.* **80**, 1019–1059 (2008).
13. J. A. Lipa *et al.*, Specific heat of liquid helium in zero gravity very near the lambda point. *Phys. Rev. B* **68**, 174518 (2003).
14. L. S. Goldner and G. Ahlers, Superfluid fraction of ^{4}He very close to T_λ. *Phys. Rev. B* **45**, 13129–13132 (1992).
15. H. Kleinert, Theory and satellite experiment for critical exponent α of λ-transition in superfluid helium. *Phys. Lett. A* **277**, 205–211 (2000).
16. M. Campostrini, *et al.*, Critical behavior for the three-dimensional X-Y universality class. *Phys. Rev. B* **63**, 214503 (2001).
17. A. Pelissetto and E. Vicari, Critical phenomena and renormalization-group theory. *Phys. Rep.* **368**, 549–727 (2002).
18. R. Garcia and M. H. W. Chan, Critical fluctuation-induced thinning of ^{4}He films near the superfluid transition. *Phys. Rev. Lett.* **83**, 1187–1190 (1999).
19. E. Cheng and M. W. Cole, Retardation and many body effects in multilayer-film adsorption. *Phys. Rev. B* **38**, 987–995 (1988).
20. G. Vidali, C. Ihm, H. Y. Kim and M. W. Cole, Potentials of physical adsorption. *Surf. Sci. Rep.* **12**, 135–181 (1991).
21. I. E. Dzyaloshinskii, E. M. Liftshitz and L. P. Pitaevskii, The general theory of Van der Waals forces. *Adv. Phys.* **10**, 165–209 (1961).
22. E. S. Sabisky and C. H. Anderson, Verification of the Liftshitz theory of the van der Waals potential using liquid-helium films. *Phys. Rev. A* **7**, 790–805 (1973).
23. M. E. Fisher and A. E. Ferdinand, Interfacial, boundary, and size effects at critical points. *Phys. Rev. Lett.* **19**, 169–172 (1967).
24. M. E. Fisher and M. N. Barber, Scaling theory for finite-size effects in the critical region. *Phys. Rev. Lett.* **28**, 1516–1519 (1972).
25. V. Privman, Finite-size scaling theory. In V. Privman (ed.), *Finite-size Scaling and Numerical Simulations of Statistical Systems*. pp. 1–98 (World Scientific, Singapore).
26. M. W. Pestak and M. H. W. Chan, Equation of state of N_2 and Ne near their critical points, scaling, correction to scaling, and amplitude ratios. *Phys. Rev. B* **30**, 274–288 (1984).
27. M. E. Fisher and P. J. Upton, Universality and interface at critical end points. *Phys. Rev. Lett.* **65**, 2402–2405 (1990).
28. J. Wilks (Clarendon Press, Oxford 1967), *Properties of Liquid and Solid Helium*, 247–295.
29. J. H. Magerlein and T. M. Sanders, Surface tension of ^{4}He near T_λ. *Phys. Rev. Lett.* **36**, 258–261 (1976).
30. C. E. Chase, E. Maxwell and W. E. Millet, The dielectric constant of liquid helium. *Physica* **27**, 1129–1145 (1961).
31. R. F. Harris and K. A. Smee, Thermal expansion of liquid helium II. *Phys. Rev. A* **2**, 158–163 (1970).
32. R. Garcia, S. Jordan, J. Lazzaretti and M. H. W. Chan, Quartz microbalance study of thick ^{4}He films near the superfluid transition. *J. Low Temp. Phys.* **134**, 527–533 (2004).
33. A. Ganshin, S. Scheidemantel, R. Garcia and M. H. W. Chan, Critical Casimir force on ^{4}He films: Confirmation of finite-size scaling. *Phys. Rev. Lett.* **97**, 075301 (2006).

34. A. Maciolek, A. Gambassi and S. Dietrich, Critical Casimir effect in superfluid wetting films. *Phys. Rev. E* **76**, 031124 (2007).
35. R. Zandi, *et al.*, Thinning of superfluid films below the critical point. *Phys. Rev. E* **76**, 030601(R) (2007).
36. S. Biswas, *et al.*, The critical Casimir force in the superfluid phase: effect of fluctuations. *New J. of Phys.* **12**, 063039 (2010).
37. A. Hucht, Thermodynamic Casimir effect in ^{4}He films near T_λ : Monte Carlo results. *Phys. Rev. Lett.* **99**, 185301 (2007).
38. O. Vasilyev, A. Gambassi, A. Maciolek and S. Dietrich, Monte Carlo simulation results for critical Casmir forces. *Europhys. Lett.* **80**, 60009 (2007).
39. O. Vasilyev, A. Gambassi, A. Maciolek and S. Dietrich, Universal scaling functions of critical Casimir forces obtained by Monte Carlo simulations. *Phys. Rev. E* **79**, 041142 (2009).
40. M. Hasenbusch, The thermodynamic Casimir effect in the neighbourhood of the λ-transition: A Monte-Carlo study of an improved three-dimensional lattice model. *J. Stat. Mech: Theory Exp.* P07031 (2009).
41. R. Zandi, J. Rudnick and M. Kardar, Casimir forces, surface fluctuations, and thinning of superfluid film. *Phys. Rev. Lett.* **93**, 155302 (2004).
42. E. H. Graf, D. M. Lee and J. D. Reppy, phase separation and the superfluid transition in liquid ^{3}He-^{4}He mixtures. *Phys. Rev. Lett.* **19**, 417–419 (1967).
43. M. Blume, V. J. Emery and R. B. Griffith, Ising model for the λ transition and phase separation in ^{3}He-^{4}He mixtures. *Phys. Rev. A* **4**, 1071–1077 (1971).
44. R. Garcia and M. H. W. Chan, Critical Casimir effect near the ^{3}He-^{4}He tricritical point. *Phys. Rev. Lett.* **88**, 086101 (2002).
45. A. Maciolek and S. Dietrich, Critical Casimir effect in ^{3}He-^{4}He films. *Europhys. Lett.* **74**, 22–28 (2006).
46. N. F. Bafi, A. Maciolek and S. Dietrich, Phase diagram of fluid phases in ^{3}He-^{4}He mixtures. *Phys. Rev. E* **91**, 022138 (2015).
47. N. F. Bafi, A. Maciolek and S. Dietrich, Tricritical Casimir forces and order parameter profile in wetting films of ^{3}He-^{4}He mixtures. *Phys. Rev. E* **95**, 032802 (2017).
48. A. Mukhopadhyay and B. M. Law, Critical Casimir effect in binary wetting films. *Phys. Rev. Lett.* **83**, 772–775 (1999).
49. A. Mukhopadhyay and B. M. Law, Casimir effect in critical films of binary liquid mixtures. *Phys. Rev. E* **62**, 5201–5215 (2000).
50. M. Fukuto, Y. F. Yano and P. S. Pershan, Critical Casimir effect in three-dimensional Ising system: Measurements in binary wetting films. *Phys. Rev. Lett.* **94**, 135702 (2005).
51. M. Hasenbusch, Thermodynamic Casimir effect for films in the three-dimensional Ising universality class: Symmetry-breaking boundary conditions. *Phys. Rev. B* **82**, 104425 (2010).
52. C. Hertlein, L. Helden, A. Gambassi, S. Dietrich and C. Bechinger, Direct measurement of critical Casimir forces. *Nature* **451**, 172–175 (2008).
53. A. Gambassi, *et al.*, Critical Casimir effect in classical binary liquid mixtures. *Phys. Rev. E* **80**, 061143 (2009).

Renormalization Group

https://doi.org/10.1142/9789811282386_0016

Chapter 16

Fisher Renormalization and Fisher Reminiscences

D.R. Nelson

Lyman Laboratory, Harvard University, Cambridge, MA 02138, USA

This chapter covers my personal remembrances in three Sections. Section 1 contains a brief personal perspective on Michael E. Fisher's contributions to science. Section 2 tells how I came to work with Michael and describes events while I was under his supervision during my graduate years at Cornell University. Section 3 summarizes recent work, done in collaboration with Suraj Shankar, on thermalized buckling of isotopically compressed thin (perhaps atomically thin) sheets of materials, such as graphene or MoS_2. These investigations were inspired by Michael's beautiful work on the effect of constraints at critical points, with fluctuations at all length scales, which leads to "Fisher renormalization" of critical exponents.[1] Thin fluctuating sheets embedded in three dimensions, when they are tensionless as in a cantilever or "diving board" geometry, are automatically at a critical point everywhere in a low-temperature flat phase. However, when we consider thin sheets supported on multiple sides in various ways, Fisher's ideas lead to the inequivalence of isotensional and isometric thermodynamic ensembles, which triggers dramatic differences in some of the critical exponents associated with the two types of boundary conditions.[2] Readers not interested in my experiences as a student at Cornell University may only want to read Sections 1 and 3. Section 4 provides some concluding remarks.

1. Remarks on Michael Fisher's Contributions to Science

Most readers of this book will know that Michael Fisher, who was University Professor Emeritus and Regents Professor at the University of Maryland and Horace White Professor Emeritus of Chemistry, Physics, and Mathematics at Cornell University, was a giant of statistical mechanics. Michael was in fact a towering influence for at least 60 years, in *many* fields, including condensed matter physics, statistical mechanics, applied mathematics, and physical chemistry. With Ben Widom, Leo Kadanoff, and Ken Wilson, he helped usher in the modern age of critical phenomena and condensed matter physics, with anomalous dimensions, scaling, and the epsilon expansion. He was a member of numerous honorary societies, including the Brazilian Academy of Sciences, the Indian Academy of Sciences, and the

Royal Norwegian Society of Sciences and Letters, to name just the more exotic organizations. His awards included the Irving Langmuir Prize in Chemical Physics, the Guthrie Medal, the Michelson–Morley Award, the Boltzmann Medal, and the Onsager Prize. In 1980, Michael, Leo Kadanoff, and Ken Wilson shared the Wolf Prize in Physics for their seminal contributions to understanding phase transitions and critical phenomena.

I would like to provide my personal perspective about Michael's striking contributions to science over the years. As many readers will know, during the final two decades of his scientific career, Michael did beautiful work on physical biology, which you will learn about elsewhere in this volume. Remarkably, physical biology *also* benefited considerably from his pioneering work on statistical mechanics in the 1960s. For example, Fisher demonstrated, via fugacity expansions, the existence of non-classical polymer critical exponents;[3] when applied to compute the bubble entropy for the classic Poland–Scheraga model of the helix coil transition,[4] one could finally understand the double-stranded DNA melting which underlies the polymerase chain reaction. Another of Michael's forays into statistical mechanics — the exact solution of the infinite spin one-dimensional Heisenberg model[5] was later adapted by John Marko and Eric Siggia to predict the force extension curves in some beautiful single molecule DNA stretching experiments, carried out by the group of Carlos Bustamante, in the 1990s.[6]

In addition to these biophysics contributions, going back at least to the early 1960s and moving onwards, Fisher's work on critical point correlations laid the foundations, along with Widom and Kadanoff, for the scaling theory of critical phenomena. The citation for his 1980 Wolf Prize (shared equally with Ken Wilson and Leo Kadanoff) reads in part as follows:

> "*Professor Michael E. Fisher has been an extraordinarily productive scientist, and one ... at the height of his powers and creativity. Fisher's major contributions have been in equilibrium statistical mechanics, and have spanned the full range of that subject. He was mainly responsible for bringing together, and teaching a common language to, chemists and physicists working on diverse problems of phase transitions.*"

Fisher is uniquely responsible for the prediction of the anomalous critical exponent η (Michael once owned a boat that sailed on Lake Cayuga in Ithaca, New York, named the Eta), which controls the decay of order parameter correlations at the critical temperature. This exponent also plays a key role in quantum field theories, where it is closely related to anomalous scaling dimensions. As Michael himself once said, the exponent η is numerically small, but nevertheless quite important.

Perhaps, Michael's most brilliant work was his invention, with Ken Wilson, of the ε expansion.[7] Nothing less than perturbation theory in the deviation of the dimensionality from four, and indirectly related to the concept of dimensional regularization and field theory, the ε-expansion led, almost overnight, to a huge number of essentially exact solutions to previously intractable problems in static and dynamic critical phenomena. The relative stability of various fixed points explained the existence of universality classes; multi-critical points were introduced and classified; scaling functions were calculated; the dependence of critical exponents on dimensionality and symmetry was finally understood and so on. Although this revolution required the invention by Wilson of a new kind of renormalization group, Fisher's crucial idea of critical exponents in 3.99 dimensions is what brought this revolution to life.

It's interesting to recall how close Ken Wilson came to the ε-expansion before his pivotal collaboration with Michael. In his second paper laying down the foundations of his renormalization group ideas,[8] Wilson constructed an "approximate recursion formula" for the quartic Landau potential describing coarse-grained Ising spins after ℓ renormalization-group transformations of the form

$$Q_\ell(y) = r_\ell y^2 + \lambda_\ell y^4. \tag{1}$$

With the help of an inspired wave-packet analysis that thinned out the degrees of freedom with each renormalization group iteration by a factor 2 (Ken was very fond of powers of 2), he produced a renormalization group which led to recursion relations in d dimensions for the coupling constants r_ℓ and λ_ℓ after ℓ iterations that read

$$\begin{aligned} r_{\ell+1} &= 4(r_\ell + 3q_\ell\lambda_\ell - 9q_\ell^3\lambda_\ell^2), \\ \lambda_{\ell+1} &= 16 \times 2^{-d}(\lambda_\ell - 9q_\ell^2\lambda_\ell^2), \end{aligned} \tag{2}$$

with $q_\ell = 1/(1 + r_\ell)$. This transformation represents a thinning of the degrees of freedom corresponding to "Kadanoff blocks" of size $b = 2$. After an uncontrolled truncation, Ken then proceeded to solve his approximate recursion relation numerically on a PDP-10 computer at Cornell's Wilson Synchrotron Laboratory and estimate critical exponents directly in three dimensions. However, in 20–20 hindsight, he could instead have noticed that the second entry in Eq. (2) can be rewritten, for a small quartic coupling constant λ_ℓ, as

$$\lambda_{\ell+1} \approx b^{4-d}\lambda_\ell = b^\varepsilon\lambda_\ell, \quad \varepsilon = 4 - d. \tag{3}$$

Thus, the nonlinear coupling constant shrinks with iteration for $d > 4$, but grows and approaches a non-trivial fixed point (now known as the "Wilson–Fisher fixed point") when $d < 4$! In 2017, I hosted Michael Fisher's David M. Lee Historical Lecture at Harvard, entitled "Ken Wilson, As I Knew Him", with Dave Lee in attendance, which Michael told me was likely the last public talk he would ever give, see Fig. 1.

In this lecture, Fisher described in vivid terms how he and Ken were walking over to the Statler Faculty Club at Cornell University to have lunch after a seminar one day in 1971, when he pointed out that the physics of self-avoiding polymers simplifies to produce non-interacting random-walk exponents above four dimensions and suggested the idea of a systematic perturbation theory in powers of $\varepsilon = 4 - d$ for

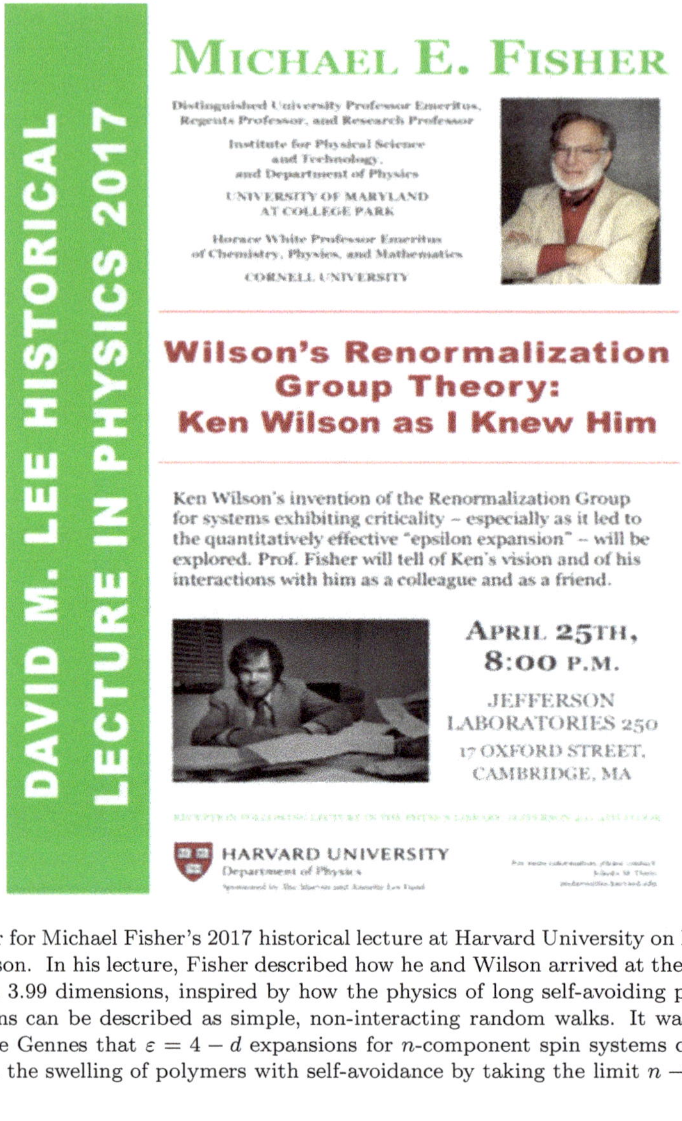

Fig. 1. Poster for Michael Fisher's 2017 historical lecture at Harvard University on his connections with Ken Wilson. In his lecture, Fisher described how he and Wilson arrived at the idea of critical phenomena in 3.99 dimensions, inspired by how the physics of long self-avoiding polymers above four dimensions can be described as simple, non-interacting random walks. It was later pointed out by P.G. de Gennes that $\varepsilon = 4 - d$ expansions for n-component spin systems can be adapted to understand the swelling of polymers with self-avoidance by taking the limit $n \to 0$.

critical phenomena more generally. Ken Wilson produced the rough draft of their famous paper together a few days later. The final version of this paper[7] included not only the first epsilon expansion formula for the Ising model susceptibility critical exponent γ,

$$\gamma = 1 + \varepsilon/6 + O(\varepsilon^2)(\text{Ising model}), \tag{4}$$

but also a beautiful analysis of two *coupled* Ising models, a kind of reformulation of Rodney Baxter's eight-vertex model.[9] Their analysis revealed that thermal fluctuations lead a *spontaneously generated* XY rotational symmetry at a critical point with the susceptibility exponent

$$\gamma = 1 + \varepsilon/5 + O(\varepsilon^2)(XY\,\text{model}). \tag{5}$$

Thus, the Wilson–Fisher ε-expansion not only illustrated the universality of critical exponents but also revealed their dependence on the symmetry of the order parameter.

Michael's larger-than-life personality, a feature shared by all members of his family, led to pugnacious responses to various challenges. He was famous for standing up after talks at Joel Lebowitz's influential Statistical Mechanics Conference Series (then at Yeshiva University; the 124th version took place at Rutgers University in May of 2023) and delivering blunt evaluations of talks he didn't like. (He could also deliver praise when he thought appropriate.) At a Cornell University talk by an experimentalist who claimed to have measured a violation of an exponent inequality near a critical point, Michael is alleged to have asked the hapless speaker if he really meant to overthrow the second law of thermodynamics! Thorough cross-examinations of speakers at the stimulating weekly lunchtime seminars (blackboard only, no transparencies allowed) he organized with Ben Widom were an important feature of the education of graduate students and postdocs. Participating in these seminars encouraged (and probably expanded) my natural inclination to ask questions. Once, when I questioned a distinguished theorist from Canada at a neighboring seminar on superfluidity in Clark Hall, the speaker said he didn't answer questions from graduate students. There was a chorus of disapproval from the audience... Michael actively encouraged questions from everyone, including graduate students! Michael Fisher told me that he had a passion for raising the "level of hygiene" of the fields he entered. Those of us who learned the field of critical phenomena in the early 1970s were fortunate in at least two ways: (1) We had a new tool embodied in the renormalization group, a kind of calculus of functional integrals that few others knew, which enabled us to understand previously unsolved problems; (2) We entered a field in which the issues and assumptions were clearly and precisely defined. Many of us discovered the importance of a high "level of hygiene" when we moved on to work in other areas of science without the benefit of gadflies like Michael to clean up wrong ideas, sloppy thinking, and misconceptions.

Michael Fisher's relentless challenging of authority often revealed itself when we met over the years. He insisted we ignore the warning signs, get out of a car driven by Eytan Domany and move quite close to a fierce-looking free-ranging buffalo, when we were together at a summer school in 1976 near the small Rocky Mountain town of Banff, Canada. When I hosted Michael's 2017 Historical Lecture at Harvard, I took Michael to our faculty club for lunch. True to form, he ignored the velvet ropes deflecting us away from a fancy grand piano and took a straight-line path to the relevant dining room. This action resembled his response to the velvet ropes blocking the stairs to the roof of the Church of Holy Sepulcher in Jerusalem, which he, my wife Pat and I visited during a Jerusalem Winter School in 1988.

This pushing past a barrier of velvet ropes took place during the first Palestinian uprising... Luckily, we weren't arrested; in fact, we got to meet the interesting Coptic Christians from Ethiopia who had been camping out on the roof for many years.

I now turn to a more personal account of my time during the early years of the renormalization-group revolution at Cornell University, heavily influenced of course, by the presence of Michael.

2. Interactions with Michael Fisher at Cornell University

It was my privilege to have Michael as my thesis advisor and postdoctoral mentor from the fall of 1972 to the summer of 1975. However, I almost didn't get to work with him.

How this happened is somewhat involved. I was in the fourth and final class of Cornell's Six Year Ph.D. program, an ill-fated experiment in elitist education, funded by the Ford Foundation, designed to take promising high-school graduates and help them get a Ph.D. at Cornell in only six years. I say "ill-fated" because many members of this program who entered Cornell, some as young as 15 or 16, were more concerned with learning how to drive or going on their first date than with dealing with the rigors of an accelerated academic program. As I recall, some had to leave Cornell before graduating and only 3/40 of the members of my class of "Phuds" got degrees from Cornell in the allotted time. Fortunately, I had already turned 18 (with some dating and driving experience...) when I arrived at Cornell in July of 1969 (we Phuds watched the first Apollo moon landing together in the basement of our dormitory), and this program turned out to be perfect for me. I was already in love with physics and mathematics, and the Six Year Ph.D. Program allowed me to construct a double-major in these subjects, with no complaints when I skipped various courses along the way. In addition to completing the requirements for a math major, I managed to take virtually every graduate physics course at Cornell in my first three years, with amazing teachers such as Ken Wilson, Hans Bethe, Bob Silsbee, N. David Mermin, Tung-Mow Yan and Peter Carruthers. When I took graduate statistical mechanics with Mark Nelkin, he told the class of an amazing Professor of Chemistry, Michael E. Fisher, who would teach a similar course with a complementary perspective out of our Chemistry Department in a future year. Little did I know, I would eventually become a TA for this course.

As an undergraduate, I worked in the Clark Hall Science Library most evenings from 6 pm to midnight. While wandering around this library in early 1972, I stumbled across the just-published *Physical Review Letter* by Ken Wilson and Michael mentioned above,[7] entitled "Critical Phenomena in 3.99 Dimensions." I was hooked and resolved to finish my final three years in the Six Year Ph.D. Program as Michael's graduate student. However, first, I had to be formally admitted to Cornell's graduate physics program. I had done well in virtually every graduate physics course Cornell offered and had overall grades that would allow me to become the salutatorian of my undergraduate class. My expenses as a graduate student, if accepted in the physics program, were guaranteed for three years by the Ford Foundation. I was thus somewhat disheartened to receive a polite letter from Cornell turning me down (*We receive many talented applicants and cannot accept everyone; we wish you all the best, etc.*). I showed my rejection letter to John Wilkins, my undergraduate physics advisor, who placed it on the desk of Bob Silsbee, the Director of graduate admissions in Physics. I then left town on Friday for a three-day weekend on a dubious trip to Kentucky (unrelated to my turndown) with my college roommate to purchase a bottle of "Rebel Yell" bourbon, a product only sold "South of the Mason–Dixon Line," due to unresolved issues associated with the US Civil War. Fortunately for me, in addition to acquiring a bottle of Rebel Yell, things were straightened out about my admission by the following Monday: my rejection letter was intended for another David Nelson, who lived in Arizona.[10] I was formally admitted to graduate school in physics, decided to stay at Cornell, and was delighted when that fall Michael Fisher became my thesis advisor.

When I entered graduate school at Cornell in the fall of 1972, there were other exciting options. John Wilkins pointed out that the critical points of interest to Ken and Michael were merely a "set of measure zero" in the pressure–temperature phase diagram of most substances and urged me to explore instead the beautiful experiments of Doug Osheroff, Bob Richardson and Dave Lee, who had recently uncovered the remarkable superfluid phases of He^3 fermions in the basement of Clark Hall. I was in the lab with Doug Osheroff when some of their amazing data were coming in. I also vividly remember Bob Richardson presenting their latest results at an informal condensed matter physics lunch, where he revealed for the first time glitches in the time evolution of the pressure in their Pomeranchuk cell during compression and subsequent decompression. Bob eloquently replied to theorists insisting they had discovered the "Balian–Wertheimer phase" or the "Anderson–Brinkman–Morel" state, with something like "no, that feature is a reproducible pressure jump on our chart recorder whose explanation remains to be seen." This view of the primacy of experimental data over potential theoretical interpretations, however beautiful, made a deep impression on me. It was probably better for me that I *didn't* enter this exciting field. There are in fact *two* distinct superfluid phases of superfluid He^3, and it took quite some time to sort out all the subtle theoretical details, with significant contributions by brilliant experts like Tony Leggett[11]

Then there was of course Ken Wilson himself, who was already a legendary theorist in high-energy physics. My graduate school roommate (who worked in high-energy experiment) claimed that Ken's letter of recommendation for a Cornell faculty position from his Cal Tech professor Murray Gell-Mann read, "Ken Wilson is the smartest person at Cal Tech except Feynman!" Feynman's letter supposedly said, "Ken is the smartest person at Cal Tech except me!" Ken had a self-effacing sense of humor — at an afternoon talk in the Laboratory of Nuclear Physics, there was a tremendous clap of thunder — he looked up at the sky outside and said calmly, "Did I say something wrong?" I wouldn't recommend trying this today, but Ken was rumored to have gotten tenure at Cornell without publishing a single paper while he was an assistant professor, on the very strong recommendation of Hans Bethe. However, I was already in love with statistical mechanics and, despite its close connections with quantum field theory (especially when formulated via path integrals), Michael Fisher seemed a better fit for me. In any case, I was extremely fortunate that my academic program allowed me to get to the frontiers of physics at Cornell already in 1972 — two future Nobel Laureates (Bob Richardson and Ken Wilson) were on my thesis committee. Dave Lee and Doug Osheroff, who were to share a Nobel Prize with Bob, were doing beautiful science in the basement of Clark Hall. Many of us felt that Michael Fisher and Leo Kadanoff (then at Brown University) should have shared the Nobel Prize with Ken[12] — I was thus surrounded by no less than five likely future Nobel laureates when I entered graduate school.

When I formally entered graduate school in 1972, Ken had recently returned from a stay at the Institute of Advanced Study at Princeton, where he gave a series of lectures on the renormalization group and the ε-expansion. These were published in Physics Reports with John Kogut.[13] Fortunately for me, Ken gave a course based on these lectures that fall, attended by students, postdocs and professors at Cornell. I was one of the two graduate students who took Wilson's course for credit and was again the beneficiary (as I was when he taught me graduate-level quantum mechanics) of his beautiful, inspirational lecture style. Since I was ready to do research, Michael suggested that I, in parallel, apply similar methods to the exactly soluble one dimensional Ising model, where I could test fancy ideas about relevant and irrelevant variables, corrections to scaling and non-linear scaling fields[14] in a simple context. Michael was particularly intrigued about how the exact 1D renormalization-group method could be adapted to handle antiferromagnetic interactions, a subject that would figure prominently in the rest of my thesis. In my investigations, I showed that integrating out every other spin mapped a 1D antiferromagnet onto a 1D ferromagnet after one iteration, but that integrating out two out of every three spins preserved antiferromagnetism. I presented this work that spring of 1973 at a seminar, where several luminaries were present. I remain grateful to this day for the encouragement I received from David Mermin after that talk.

As I'm sure others will relate as well, Michael looked out for his students. In late May of 1973, Mel Green and others organized a wonderful conference to review the progress in understanding critical phenomena (and its implications for high energy physics) over the previous decade. The meeting was held in a Temple University Conference Center in rural Pennsylvania.[15] Michael used his persuasive powers (he was one of the organizers...) to arrange for me to attend, even though I was just a first-year graduate student.[16] My roommate at that conference was Eric Siggia who came down from Harvard, where he was in the process getting a Ph.D. in only *five* years supervised by Paul Martin, without the benefit of a fancy program! This conference was an amazing opportunity for me, as I was able to hear, for the first time, a constellation of giants in the field expound on the ongoing revolution in statistical mechanics and quantum field theory, inspired by renormalization-group ideas and the ε-expansion. Experiments were represented by J.M.H. (Anneke) Levelt-Sengers from the National Bureau of Standards (now the National Institute of Standards and Technology) and also by Guenter Ahlers from Bell Labs. I remember Michael going out of his way to suggest that Guenter's talk on phase transitions in superfluid He^4 was likely to be one of the best talks by an experimenter I would ever hear. (It was. . ..) He also strongly recommended the lucid, forthright talk of Michael Wortis on results from high-temperature series expansions. I was introduced in addition to my future collaborator and postdoctoral mentor Bert Halperin by hearing him explain beautiful work on renormalization groups for dynamic critical phenomena carried out with Pierre Hohenberg and Shang-keng Ma, and remember Bert wondering whether we really understood the concept of a "metric" in the space of Hamiltonians when acted upon by Wilson's renormalization procedure. It was a pleasure to meet or hear for the first time talks from those outside my Cornell acquaintances who would eventually become close personal friends and/or admired colleagues, such as Halperin, Shang Ma, Paul Martin, Leo Kadanoff, Franz Wegner, Eberhard Riedel, David Bergman, Daniel Amit, Jean Zinn-Justin, Masuo Suzuki, Edouard Brezin and Georgio Parisi. I still remember Edouard eloquently holding forth about the potential impact of renormalization groups on multiple fields of physics at one end of a long table at lunch, while Ken sat quietly eating at the other end.

Leo Kadanoff's talk had a particular impact on me because he illustrated his seminal "Kadanoff block" renormalization-group ideas by iteratively thinning the degrees of freedom of a 1D Ising model, thus "decimating" out every other spin in a long chain and scooping me on my first year of graduate work. Michael spoke on my behalf to Leo, for whom I'm sure exact renormalization-group transformations on 1D Ising models were a relatively minor development. Leo graciously encouraged Michael and me to go ahead and publish, although we did cite his Temple University talk in our eventual publication.[17] I was deeply impressed with Kadanoff and ultimately had the privilege (after I left Cornell to become a postdoc) of collaborating with him, while he was still at Brown University, on an investigation with Jorge Jose and Scott Kirkpatrick on two-dimensional XY models with symmetry-breaking perturbations that ultimately became one of the most cited papers on our publication lists.[18]

My first year as a graduate student was not without other mishaps which required the assistance of Michael Fisher. When I took the oral exam that would allow me to officially start my thesis work, Bob Richardson, Ken Wilson and Michael were on the committee. I thought I could answer most questions they would throw at me and was reasonably confident. However, Bob Richardson, probably because he thought theorists needed a healthy dose of experimental reality, was out to get me. I *was* able to answer Ken's question about estimating the ground-state energy of the hydrogen atom using the uncertainty principle. But then, after I explained Leo Kadanoff's ideas about the operator product expansion (independently invented by Ken in the context of quantum field theory), Bob asked me how a television worked. Well, I had wanted to become an electrical engineer during most of my high-school years — although I ended up at Cornell, I applied to and was accepted by MIT for undergraduate studies for this reason. I had replaced the vacuum tubes on our home TV and I *knew* how a television worked... Undaunted by my correct answer, Bob tripped me up when he then asked me to compute the capacitance of a metal sphere... I panicked and failed miserably because I had no idea where to put the other plate. Ken Wilson kept saying "what's the only thing you can write down." The answer in MKS units is $C = 4\pi\varepsilon_0 R$, where R is the sphere radius. Perhaps, this was only thing that *Ken* could have written down. However, there were *many* things I could have written down, especially because I didn't know that the other plate was at infinity![19]

My committee decided that I would pass the exam, but with two important conditions, guided I'm sure by Michael: I had to serve as a teaching assistant for the section meetings of a freshman physics course in electricity and magnetism in the fall of 1973 (taught by Bobby Pohl and Boris Batterman), and I had to work for one semester in the laboratory of Watt Webb every Friday afternoon for a semester on an experiment on superfluid He^3–He^4 mixtures with an exceptional postdoc named Paul Leiderer. Teaching section lectures on electricity and magnetism forced me to really learn this subject and taught me about teaching more generally. (Although most of the students liked my section lectures in their evaluations, one student asserted that "jumping up and down excitedly and waving my hands was a poor substitute for actually transmitting information.") The time in the laboratory studying dynamical critical phenomena near tricritcal points in He^3–He^4 mixtures with Paul Leiderer and Watt Webb led to my first *Physical Review Letter*[20] and was a truly transformative experience. Paul did most of the real experimental work with the background whine of a leak checker while hovering over a cryostat, and I helped mostly with aligning the apparatus and with the theory. Paul and Watt became lifelong friends; I connected with Watt's lab again when I returned to Cornell on sabbatical in 2004, which led to a biophysics theory paper on phase separation of lipid bilayers in giant unilamellar vesicles.[21]

My other indoctrination into experimental reality was a graduate course in experimental physics, required of theorists as well as experimenters at Cornell. (Harvard University has a similar requirement, instituted I believe by John van Vleck, who felt that theorists should not be on a "vitamin-free diet."). I still remember the thrill of finding a sharp singularity in the specific heat of a sample of beta brass, a random alloy of copper and zinc. At a 50–50 concentration, beta-brass orders like a two-sublattice antiferromagnet below an order–disorder transition. The head of that course, Paul Hartman, was very helpful and encouraging. He later wrote an engaging history of the Cornell Physics Department. I vividly remember learning two interesting things from Professor Hartman: (1) In the late 1930s, when he was a Cornell graduate student, he was worried about *his* oral exam. He went to Hans Bethe and said he knew most of the material but was worried because he hadn't really learned fluid mechanics. Bethe told him not to worry because the physics faculty was getting rid of this subject to make room for quantum mechanics! Sadly, fluid mechanics was long gone from US physics curricula by the time I entered school — I had to learn this subject on my own from a now battered copy of the text by L.D. Landau and E.M. Lifshitz. My copy still has squashed bugs from reading this text with the help of a kerosene lamp on a camping trip with my girlfriend, Patricia Schneider. (2) Paul Hartman also said the best Cornell student ever to take his legendary graduate course in graduate experimental physics was the famous theoretical physicist, Freeman Dyson. I was inspired by this remark to try some simple experiments later in my career. Reactions when I tried to get involved with experiments while on sabbatical at Brandeis with Bob Meyer and at Rockefeller University with Albert Libchaber ranged from lukewarm acquiescence to active discouragement! I finally succeeded in doing range expansion experiments on fluorescently labelled bacteria and Baker's yeast years later, while on leave from teaching at Harvard in the laboratory of Sharad Ramanathan (a former student of Michael's son, Daniel Fisher), in collaboration with Oskar Hallatschek, another theorist with a taste for experiments.[22] I shall always be grateful to Michael Fisher and Bob Richardson for insisting that I engage with experiment and avoid becoming a "vitamin-free" theorist.

As mentioned above, some of the "Cornell crowd" were at that Temple Conference in the spring of 1973. Not surprisingly, given the recent breakthroughs in understanding superfluids and renormalization groups in Ithaca, this group included some exceptional postdocs! My friend, Amnon Aharony, was one of the Cornell postdocs at this meeting; I believe he reported on his beautiful work on phase transitions in spin systems with long-range dipole–dipole interactions. Other postdocs from around the world (Amnon was from Israel) I knew at Cornell were Tim Padmore (USA), Joe Sak (Czechoslovakia), Roland Combescot (France), Monique Combescot (France), Michel Droz (Switzerland), Pierre Pfeuty (France), Bernie Nickel (Canada) and Alastair Bruce (Scotland). Some of them took me under their wing during the summer of 1972, when I shared a windowless office on the fifth floor of Clark Hall. They were generous in helping me deal with the Matsubara frequency sums in a Green's function calculation suggested by John Wilkins, before I signed on with Michael Fisher. But those were difficult times in condensed matter theory — the previous year there was only a single tenure-track job opening in this still-developing field in all of the United States, claimed by Joseph Straley (a former postdoc of Michael Fisher) at the University of Kentucky; at least one well-meaning postdoc asked me if I really wanted to go into this challenging field.

J. Michael Kostlerlitz showed up as a postdoc during my second year as a graduate student and talked about his beautiful paper[23] implementing ideas about topological phase transitions from his earlier work with David J. Thouless[24] and independent work by Berezinski.[25] We knew Mike was going to talk about his renormalization-group solution of the famous two-dimensional XY model, but it wasn't "our" renormalization group, i.e., the Wilsonian formulation involving local field theories, momentum shells and the ε-expansion. Kosterlitz adapted a renormalization method developed for long-range interactions in the Kondo problem by Anderson and Yuval[26] to understand virtually all important equilibrium properties of XY models in two dimensions. Mike's talk was my first encounter with phase transitions that had a *line* of fixed points, controlling an entire low-temperature phase. Similar lines of fixed points (in fact, a whole surface of fixed points) played an important role in the two-dimensional melting transitions, with an intermediate fourth hexatic phase of matter, as I found when subsequently producing a theory of melting with Bert Halperin while a postdoc at Harvard. When Mike Kosterlitz and I were both at Cornell, we began to compete a bit to understand the then fashionable subject of phase transitions with bicritical and tetracritical points. Michael Fisher stepped in and wisely insisted that we collaborate, which led to several papers together.[27] This important early connection with Mike helped us to collaborate effectively later (when I was a postdoc) on a paper that proposed a universal jump discontinuity in superfluid Helium films,[28] a prediction that played a role, via beautiful experiments by Dave Bishop and John Reppy at Cornell,[29] in convincing people that 2–3 competing theories were wrong, and Kosterlitz and Thouless were correct with their ideas about two-dimensional superfluids and XY models of magnetism. We used renormalization-group methods coupled with Josephson's scaling relation for superfluids in two dimensions to show that Mike's prediction of the critical exponent $\eta = 1/4$ describing the decay of correlations exactly at the phase transition implied a universal jump in the superfluid density, a prediction subsequently confirmed in a variety of experiments by Bishop, Reppy, Isadore Rudnick and others.[30]

My interactions with graduate students at Cornell were regrettably less extensive. Most of them were initially engaged with coursework, while I was trying to get started in research, although we did bond over on intramural games of softball and touch football on a team called the "Physics Zeroes." I remember breaking the nose of the captain of our football team in an unfortunate collision and later playing shortstop in a softball game next to an open can of beer (as was also the case for other infielders and outfielders) in a so-called "beer game" against a competing physics team. However, I did make a close friend in H.R. Krishnamurthy, who arrived from India to enter in my year, to write an important thesis with John Wilkins and Ken Wilson, inspired by Ken's numerical approach to a renormalization-group solution to the Kondo problem.[31] One day, Krish walked into my office, concerned about a dispute between his two mentors, Wilkins and Wilson. John Wilkins always wanted more out of Ken and said he wasn't satisfied with just the static susceptibility for the Kondo model, he wanted dynamics! To which Ken Wilson replied with an obscene gesture. At least, I *think* it was an obscene gesture. When Kirsh tried to reproduce it, he used his index finger. I suggested that some other finger was likely involved. I also had enjoyable interactions with Eytan Domany, a tough Israeli graduate student who came to work with Michael shortly after me. Although Michael Fisher was most certainly Eytan's primary thesis advisor, I had the pleasure of working with Eytan somewhat in the capacity of a mentor during my final year. Eytan and I used a renormalization-group trajectory integral method I had helped develop to calculate equations of state for bicritical points.[32,33] As I remember our interactions, Domany initially thought most of the things I said were wrong... I was forced to work hard to explain my claims properly to him, a very valuable experience. Eytan became a close friend and was later the sponsor of numerous sightseeing adventures for me in Israel. This experience helped me when I later mentored graduate students while a postdoc at Harvard University, where I had the pleasure of working a bit with Bob Pelcovits and Daniel Fisher, when they were officially the students of their true thesis advisor, Bert Halperin.

The method Eytan and I used in our work together was an outgrowth of an accidental encounter I had over the telephone with Joseph Rudnick (the son of the UCLA experimenter Isadore Rudnick, mentioned above). While relatively isolated from the latest renormalization-group developments as a postdoc at the Technion in Israel, Joe had done some remarkable diagrammatic renormalization-group calculations.[34] He was now back in the US and was phoning Michael Fisher, who was never in his office when Joe called. My tiny office (again, windowless) was then in the Baker Laboratory of Chemistry, next to Michael. Finally, with Joe again on the phone, Michael's frustrated assistant, Lynn Rabenstein, handed me the phone and said, "Here, YOU talk to him!" Joe and I hit it off, discovered we had very similar interests and quickly wrote a paper together,[35] before ever meeting in person. I knew Joe was exceptional when I showed our draft paper to Ken Wilson, who said "Joe Rudnick, there's a smart guy." Ken had been a referee for one of Joe's papers. Joe eventually came to Cornell to give a talk, which led to our second paper together.[36]

In addition to our work on exactly soluble renormalization models in one dimension, my thesis with Michael Fisher concerned layered antiferromagnets called metamagnets,[37] as well as phase transitions at bicritical and tetracritical points.[38] We briefly thought that our metamagnet calculations could be adapted to demonstrate that Ising magnets and liquid–gas phase transitions were in different universality classes, until Ken Wilson pointed out to me that a crucial auxiliary field could be integrated out to show that the two theories were in fact equivalent.

(The *dynamics* near these two types of critical points is indeed different.) Michael and I were once talking about such problems in his office (which had a picture of Winston Churchill with an inspiring wartime quote), when he got an irate phone call from a colleague who claimed he had written an unfavorable letter of recommendation. Michael proceeded to pull out a copy of his (favorable) letter and read it out loud over the telephone! Thanks to Michael's support, I managed to get my Ph.D. in 5.5 instead of 6 years. He employed me as a postdoc for the final semester of my last year, although I also had to help as a teaching assistant of his popular course on graduate statistical mechanics, offered by the Cornell Chemistry Department. Michael graciously gave me a lot of independence during my final year. Six of the papers originating from research I was involved in that year did *not* have Michael as a co-author.

During my time at Cornell, Michael and Sorrel Fisher would have wonderful dinner gatherings at their home on the Parkway in Ithaca. (Ben Widom's house was just across the street.) I remember meeting his sons Daniel and Matthew, who went on to become famous physicists themselves. I recall Matthew being very good at table tennis and persistently trying to get me to solve various mechanical puzzles. Daniel eventually came to me for advice on what physics courses he could skip when he entered Cornell at age 15. He would eventually become a valued colleague at Harvard, where he, Leo Radzihovsky and I played squash together, before Daniel was stolen away by Stanford. I brought Patricia Schneider (by then my fiancé) to the evenings at the Fisher home and was pleased to see that Michael clearly approved of her. Pat and I were married at Sage Chapel in Ithaca, New York (where we met singing in the choir), on a snowy day shortly after Christmas of 1975. Michael came to our wedding reception, where he explained the rules of English darts to my father, with a detailed follow-up letter. Our wedding present from Sorrel and Michael was a lovely chiming clock. This clock still sits today on the mantle above the fireplace in our living room. For 35 years or so after I became a faculty member myself, inspired by the generous hospitality of the Fisher family, Pat and I would invite all the stray postdocs and graduate students we could find, mostly international students from far away, to Thanksgiving dinners with our growing family at our home in Lexington, Massachusetts.

Thanks I'm sure to recommendations from Michael Fisher and Ken Wilson, I was selected as Harvard Junior Fellow, a three-year postdoctoral appointment that began in the summer 1975, a position that enabled me to work with Michael J. Stephen (then commuting to Harvard weekly from Rutgers University to be with his wife Johanna, on the faculty at the Harvard Medical School) to apply dynamical renormalization-group methods to understand the large-distance and long-time properties of a randomly stirred fluid. Thus, I was able to capitalize on learning fluid mechanics during graduate student camping trips near Ithaca, New York, and eventually to work with two other important mentors besides Michael, first Leo Kadanoff and then Bert Halperin.

I now turn to some recent work inspired by Michael Fisher.

3. Fisher Renormalization and Thermalized Buckling of Isotopically Compressed Thin Sheets

One of Michael Fisher's many contributions to critical phenomena prior to the invention of the ε-expansion was his 1968 paper[1] on the renormalization of critical exponents by hidden variables. What Fisher had in mind was, for example, an annealed density of mobile impurities in a background Ising model whose local density $c(\vec{r})$ can fluctuate, but whose overall number,

$$\int d^d r c(\vec{r}) = N_{\text{imp}}, \tag{6}$$

is conserved, thus providing a constraint on the statistical mechanics. Fisher assumed that the impurity density led to an additional contribution to the Hamiltonian H of an Ising-like system with an "energy" coupling, say, to the square of the Ising order parameter. The constraint of course drops out if there are no impurities, $N_{\text{imp}} = 0$. In fact, such constraints will in general merely shift the location of the critical point a bit if the ensemble becomes grand canonical in the impurities, so that there is an impurity chemical potential μ_{imp}, which plays the role of a Lagrange multiplier enforcing the constraint on average in the statistical mechanics,[39]

$$H \to H - \mu_{\text{imp}} \int d^d r c(\vec{r}), \tag{7}$$

Here, H already includes the coupling between the impurities and the Ising degrees of freedom. Fisher was able to show that an ensemble imposing a rigid constraint of a non-zero impurity concentration *changed* the path of approach to the critical point, so that the effective critical exponents close to T_c such as α, β and γ of the specific heat, magnetization and susceptibility, respectively, would change (in a slightly different notation from that used by Fisher) to[1]

$$\alpha \to \alpha' = -\alpha/(1-\alpha), \quad \beta \to \beta' = \beta/(1-\alpha), \quad \gamma \to \gamma' = \gamma/(1-\alpha). \tag{8}$$

Michael's paper on renormalized critical exponents, which has received over 1000 citations, is extremely important conceptually. Normally, we expect statistical mechanical averages to be *independent* of the precise statistical ensemble in the large number of degrees of freedom ($N \to \infty$) thermodynamic limit of large system size. However, the standard proofs of ensemble equivalence[40] assume linear system sizes in d dimensions $L \sim N^{1/d}$ much greater than the correlation length ξ, $L \gg \xi$. Michael realized that this assumption fails near a critical point, where there are fluctuations at all scales up to a diverging correlation length. Fisher's conclusions were subsequently confirmed via a renormalization group calculation in $d = 4 - \varepsilon$ dimensions by Rudnick *et al.*,[41] who found the exponent changes are reflected in a crossover from an unstable Wilson–Fisher Ising fixed point to a new stable fixed point with Ising-like Fisher renormalized critical exponents. (A related Ising-to-Ising fixed point crossover occurs in metamagnets.[37])

Although conceptually important, Fisher renormalization typically produces only small changes in actual exponents, since the specific heat exponent α appearing in Eq. (8) is almost always very small.[42] Recently, however, Suraj Shankar and I studied related Fisher renormalization effects for critical exponents near the finite temperature buckling transition of isotopically compressed thin sheets of materials, such as graphene or MoS_2.[2] Here, the inequivalence of isotensional (laser tweezers apply a fixed force at the edge) and isometric ensembles (here, a given hole size enforces a fixed strain) for thin supported membranes leads to dramatic differences between these two different boundary conditions for membranes, see Fig. 2.

Although experiments with atomically thin monolayers like graphene typically suspend the elastic sheets across fixed-size holes,[43,44] it is useful to imagine a *variable* aperture size that could in principle be tuned by a camera shutter mechanism to change the strain isotropically and possibly adjust it to zero. An external symmetry-breaking electric field E perpendicular to average plane of the sheet could also be applied in either ensemble to bias the direction of buckling for charged membranes. In the absence of an external electric field, and in situations where

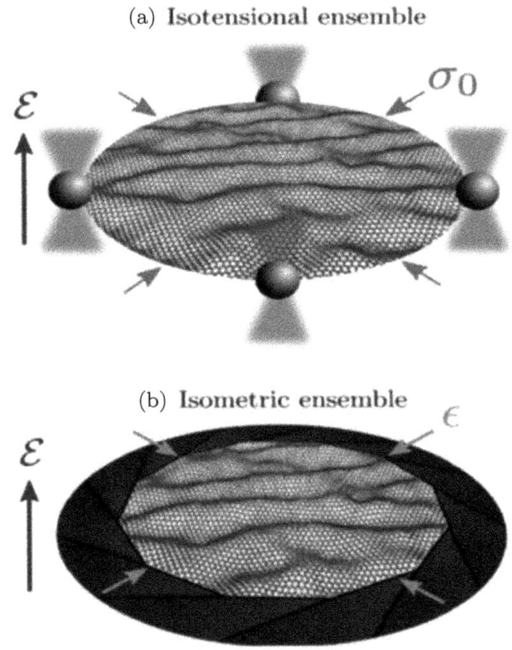

Fig. 2. A possible realization of two mechanical ensembles relevant, for example, to an atomically thin sheet of free-standing graphene, and not, as is typically the case, plastered to a 3D solid substrate that kills off flexural phonons by giving them a mass. (a) In the isotensional ensemble, multiple laser tweezers enforce a constant external stress applied to the membrane edge (in the spirit of single molecule biophysics experiments), while the edge boundary displacement fluctuates. (b) The isometric ensemble instead implements a clamped boundary condition used more often in experiments, with the external load imposed via a strain imposed on, say, a graphene membrane suspended across a hole of a fixed, but tunable size. Figure taken from Ref. 2, to which the reader is referred for more details.

boundary constraints allow easy relaxation to the tensionless state, such systems are *automatically* at a critical point at any point in the low-temperature flat phase.[45] For example, when ribbons are in a cantilever geometry,[46] one end is free and the geometry resembles an extremely thin diving board where macroscopic strains can easily relax. Such systems have fluctuations at all length scales without the need for fine-tuning, like long self-avoiding polymer chains in a good solvent,[47] and a collection of out-of-equilibrium problems often called "self-organized criticality."[48]

To see how striking differences between isotensional and isometric ensembles can arise for thin, supported membranes, consider the long-wavelength free energy associated with a distorted elastic sheet characterized by in-plane phonon fields $u_1(x_1, x_2)$ and $u_2(x_1, x_2)$, as well as an out-of-plane "flexural" phonon field $f(x_1, x_2)$, see Fig. 3.

Points in the original flat reference state, $\vec{r}_0 = x_1\vec{e}_1 + x_2\vec{e}_2$, are now mapped into three dimensions via

$$\vec{r}_0 = x_1\vec{e}_1 + x_2\vec{e}_2 \to \vec{r}(x_1, x_2) = \vec{r}_0 + \begin{pmatrix} u_1(x_1, x_2) \\ u_2(x_1, x_2) \\ f(x_1, x_2) \end{pmatrix}. \tag{9}$$

The effect of this mapping is to distort every line element $d\vec{r}_0 \to d\vec{r}$ with a new length given by

$$\begin{aligned} dr^2 &= dr_0^2 + 2u_{ij}dx_idx_j \equiv g_{ij}dx_idx_j, \\ g_{ij} &= \delta_{ij} + 2u_{ij}, \end{aligned} \tag{10}$$

where deviations from the identity δ_{ij} of the metric tensor g_{ij} associated with the distortion is given by the non-linear strain matrix

$$u_{ij}(\vec{x}) = \frac{1}{2}\left[\frac{\partial u_i(\vec{x})}{\partial x_j} + \frac{\partial u_j(\vec{x})}{\partial x_i} + \frac{\partial f(\vec{x})}{\partial x_i}\frac{\partial f(\vec{x})}{\partial x_j}\right]. \tag{11}$$

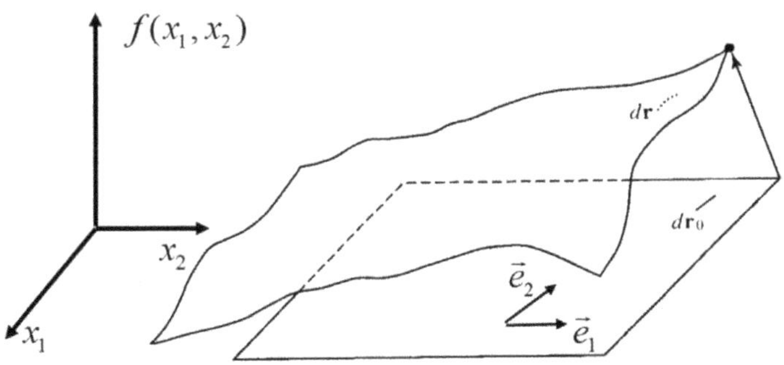

Fig. 3. Three-dimensional displacement vector from the upper right corner of a flat membrane reference state to a distorted state in three dimensions induced by, say, thermal fluctuations with both conventional in-plane phonon displacements and an out-of-plane flexural phonon field. See text for details.

(There are also higher-order terms in the in-plane strains $\partial_i u_j$, but these turn out to be irrelevant compared to the terms kept here.) In the presence of edge forces and a symmetry-breaking field, the elastic energy of the distorted state is given by[49]

$$E = \frac{1}{2}\int d^2x[\kappa(\nabla^2 f(\vec{x}))^2 + 2\mu u_{ij}^2(\vec{x}) + \lambda u_{kk}^2(\vec{x}) - Ef(\vec{x})] - \oint_C d\ell \hat{\nu}_i \sigma_{ij}^{ext} u_j. \quad (12)$$

Here, the Lamé elastic parameters parametrizing the stretching energy are μ and λ, and κ is a bending rigidity that penalizes deviations from the flat state. The final boundary integral is the work done by an external stress σ_{ij}^{ext} with $\hat{\nu}_i$ being the outward unit normal (within the plane) to the boundary curve C. The final term in square brackets represents a potential energy proportional to an external out-of-plane electric field E which couples directly to the height of a charged membrane. Note that the non-linear term in Eq. (11) for the strain can be viewed as a matrix vector potential created by the flexural (out of plane) phonon field,

$$A_{ij}(\vec{x}) = \frac{\partial f(\vec{x})}{\partial x_i}\frac{\partial f(x)}{\partial x_j}. \quad (13)$$

Since the symmetric 2×2 matrix $A_{ij}(\vec{x})$ has in general three independent components, whereas the in-phonon field $\vec{u}(\vec{x})$ in Eq. (11) (at finite wavevectors) has only two components, flexural phonon distortions lead to geometrical frustration in the statistical mechanics whenever there is a non-zero Gaussian curvature.[50]

To see why the statistical membranes of thin elastic membranes with a shear modulus is like a system at its critical point (or a massless field theory), it is helpful to first set $E = 0$ and $\sigma_{ij}^{\text{ext}} = 0$ in Eq. (12) integrate out the in-plane phonon degrees of freedom $u_1(\vec{x})$ and $u_2(\vec{x})$ in the partition function. These quantities only occur quadratically in the membrane energy function. The resulting Gaussian functional integrals lead to an effective free energy F_{eff}, where[45]

$$\begin{aligned} F_{\text{eff}} &= -k_BT\ln\left(\int D\{u_1(\vec{x})\}\int D\{u_2(\vec{x})\}e^{-E/k_BT}\right) \\ &= \frac{1}{2}\kappa\int d^2x[(\nabla^2 f)^2] + \frac{1}{4}Y\int d^2x[P_{ij}^T(\partial_i f\partial_j f)]^2 \equiv F_0 + F_1. \end{aligned} \quad (14)$$

Here, $P_{ij}^T = \delta_{ij} - \frac{\partial_i\partial_j}{\nabla^2}$ is the transverse projection operator and $Y = \frac{4\mu(\mu+\lambda)}{2\mu+\lambda}$ is the twodimensional Young's modulus of the membrane. To better understand Eq. (14), note that if we represent the membrane position in space via the Monge form as $(x_1, x_2, f(\vec{x}))$, the unit normal is given by

$$\hat{n}(\vec{x}) = \frac{1}{\sqrt{1+\left|\vec{\nabla} f\right|^2}}\begin{pmatrix} -\partial_1 f(\vec{x}) \\ -\partial_2 f(\vec{x}) \\ 1 \end{pmatrix}. \quad (15)$$

The physics of a two-dimensional membrane embedded in three dimensions in its low temperature flat phase in fact resembles somewhat a collection of interacting classical Heisenberg spins in two dimensions, with, however, genuine long-range order in the normal field $\hat{n}(\vec{x})$, which plays the role of a three-component spin vector.[45] See Refs. 51 and 52 for discussions of how and why elastic membranes are able to evade the famous Hohenberg–Mermin–Wagner theorem, prohibiting a continuous broken symmetry and long-range order in two dimensions. The first term of the bottom line of the effective free energy in Eq. (14), $F_0 = \frac{1}{2}\kappa \int d^2x[(\nabla^2 f)^2]$, describes interactions that tend to align neighboring tipping vectors $\delta\vec{n}(\vec{x}) \sim -\vec{\nabla} f(\vec{x})$ of the normal away from the direction of long-range order, which we take to be the z-axis. A Heisenberg-like exchange coupling κ favors alignment. Thus, Eq. (14) is like a massless field theory in the tipping vector $\delta\vec{n}(\vec{x})$, with, however, an unusual non-linear term $F_1 = \frac{1}{4}Y \int d^2x[P^T_{ij}(\partial_i f \partial_j f)]^2$ quartic in the tipping vector. It can be shown that

$$-\nabla^2 \left(\frac{1}{2} P^T_{ij}(\partial_i f \partial_j f)\right) = \det\left(\frac{\partial^2 f}{\partial x_i \partial x_j}\right) = G(\vec{x}), \tag{16}$$

where $G(\vec{x})$ is the Gaussian curvature of the membrane at position $\vec{x}$. This observation leads to a long-range interaction proportional to the Young's modulus Y between distant membrane Gaussian curvatures,[50] which is one way of understanding how elastic membranes evade the Hohenberg–Mermin–Wagner prohibition against a continuous broken symmetries in two dimensions.

The profound effect of the non-linear term in Eq. (14) is already evident in perturbation theory. A straightforward low-temperature spin-wave-like calculation in perturbation theory to lowest order in Y gives for the renormalized wave-vector-dependent bending rigidity[45]

$$\kappa_R(q) = \kappa + k_B T Y \int \frac{d^2k}{(2\pi)^2} \frac{[\hat{q}_i P^T_{ij}(\vec{k})\hat{q}_j]^2}{\kappa|\vec{q}+\vec{k}|^4} + \cdots \tag{17}$$

For a membrane of linear size L in the limit of small external wavevectors q, we have

$$\lim_{q\to 0} \kappa_R(q) \approx \kappa[1 + 3(vK)k_B T/(32\pi^3\kappa) + \cdots]. \tag{18}$$

Here, the Foeppl–von Karman number is $vK = YL^2/\kappa$, a dimensionless measure of the strength of the non-linear coupling. At first sight, the perturbative result, Eq. (18), looks reassuring: For atomically or molecularly thin membranes, the dimensionless factor $k_B T/\kappa$ is typically quite small (of order 1/40) even at room temperature.[45] The numerical factor $3/32\pi^3$ is also rather small. Unfortunately for this perturbation theory, however, the factor vK in the correction term *diverges* with system size! Upon substituting the elastic constants for a 8.5 inch × 11 inch piece of writing paper, one finds $vK \sim 10^7$, while for a relatively small $10\mu \times 10\mu$ square of graphene $vK \sim 10^{12}$. Thus, ordinary perturbation theory simply will not work for elastic membranes of any appreciable size. Such infrared divergences are tailor-made for graphical resummations of the most diverging terms in perturbation theory,[50] or alternatively for Ken Wilson's renormalization group. Such methods applied to the renormalized bending rigidity of elastic membranes lead to a scale-dependent elastic parameter characterized by a universal critical exponent η:[53,54]

$$\kappa_R(l) \approx \kappa(l/l_{th})^{\eta}, \quad \eta \approx 0.82, \tag{19}$$

where ℓ is a length scale such as the system size. The renormalized Young's modulus Y_R also becomes scale-dependent and is characterized by η_u, another universal critical exponent,[53,54]

$$Y_R(l) \approx Y(l_{th}/l)^{\eta_u} \quad \eta_u \approx 0.36. \tag{20}$$

Note that the elastic "constants" $\kappa_R(\ell)$ and $Y_R(\ell)$ are not in fact constant but in fact depend on the length scale ℓ whenever $\ell \geq \ell_{th}$, where ℓ_{th} is a thermal length scale that can be calculated (in the the spirit of the Ginzburg criterion for when to expect non-classical exponents near a critical point) by asking at what length scale does the correction in Eq. (18) become comparable to the quantity it is correcting. The result is[55,56]

$$\ell_{th} = \sqrt{\frac{16\pi^3\kappa^2}{3k_BTY}}. \tag{21}$$

The theory sketched above, although fairly well understood already by the late 1980s, received new life with the pioneering 2015 graphene cantilever experiments in the group of Paul McEuen at Cornell University, as illustrated schematically in Fig. 4.[46]

These authors pointed out that, using standard unrenormalized graphene elastic parameters, the thermal length at room temperature was only ~0.2 nm. *Thus, all graphene experiments in materials above this tiny size will have strongly scale-dependent elastic parameters.* Although Blees *et al.* did not systematically vary the *width* of their free-standing graphene ribbons to demonstrate a scale dependence,

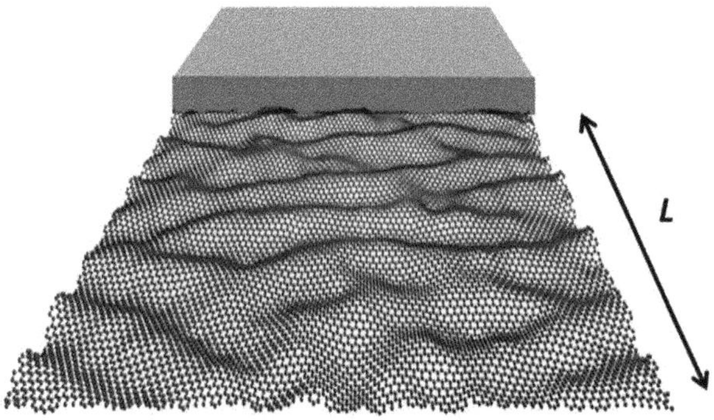

Fig. 4. Schematic diagram of a thermally fluctuating free-standing atomically thin graphene ribbon in a cantilever geometry, attached only at one end to a gold pad. Adapted with permission from Ref. 46.

they did change the length and were able to measure (both by measuring deflections of a metal pad and by measuring thermal fluctuations) a thermally renormalized bending rigidity at room temperature that was approximately *4000* times larger than that expected from measurements and quantum density functional theory near $T = 0$. We stress that relaxed graphene cantilevers in the flat phase like those shown in Fig. 4 are *automatically* at a critical point, with universal critical exponents and thermal fluctuations at all lengths up to the system size.

What happens when we apply isotensional or isometric constraints to this critical system, i.e., what is the effect of going from experimental setups like Fig. 4 to setups as in Fig. 2? Just as there are two ways to move away from a conventional critical point (temperature-like and magnetic field-like perturbations), there are two ways to destroy the criticality of free-standing elastic membranes. Upon returning to the full energy expression of Eq. (12), these turn out to be an external electric field E applied normal to a charged membrane (a field-like perturbation) and an external strain or stress (σ_{ij}^{ext}) applied to the edge of the membrane (a temperature-like perturbation). For the isotensional ensemble, we assume an isotropic external edge force[2]

$$\sigma_{ij}^{\text{ext}} = \sigma_0 \delta_{ij}, \tag{22}$$

with no further constraints on the zero modes of the displacements or strain. The in-plane boundary displacements can adjust freely when $\sigma_0 \neq 0$, where with our conventions $\sigma_0 > 0$ corresponds to a tensile stress that smooths out thermal fluctuations and $\sigma_0 < 0$ represent a compressive stress that (when large enough) can cause a buckling transition, see the following. For the isometric ensemble (likely to be easier to achieve experimentally), suspending a membrane across a circular hole of radius R is assumed to lead to a fixed radial boundary displacement Δ_C along the contour C:[2]

$$\vec{u} = \Delta_C \vec{\nu} \equiv \varepsilon R/2, \tag{23}$$

where ε (not to be confused with $4 - d$) is the fixed strain associated with the isometric ensemble. The displacement at the boundary (measured relative to the relaxed state) in this fixed-strain ensemble is given. For $\varepsilon > 0$, thermal fluctuations will be suppressed, while for $\varepsilon < 0$, they can be enhanced due to a buckling transition.

With two relevant perturbations to the tensionless state introduced, we can now define critical exponents. We only mention here three such exponents, two of which have large ensemble-dependent changes. It has been known since the mid-18th century and the time of Euler that thin plates like those shown in Fig. 2 will undergo a buckling transition for σ_0 or ε sufficiently large and negative. Although Euler originally studied the buckling of rods, very similar phenomena arise when one compresses thin circular plates.[57] As shown in Fig. 5, for $E = 0$, there is a symmetry-breaking buckling phase transition beyond a critical value of the stress

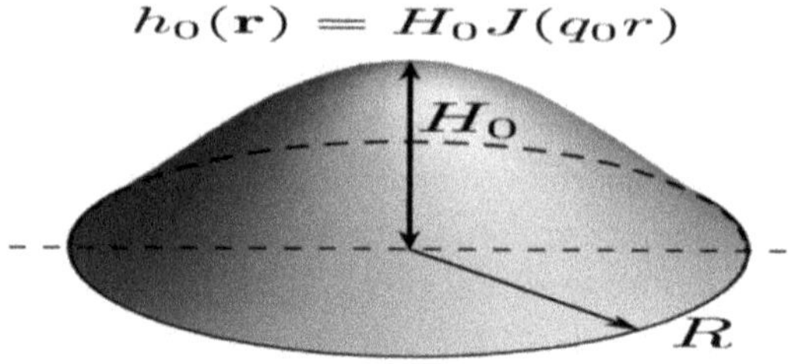

Fig. 5. Sketch of the first buckling mode with boundary conditions such that the height at the rim vanishes for a circular plate of average radius R. There is an overall Bessel-function shape, and amplitude of the mode at the center of the circular frame acts like an order parameter corresponding to a small wave vector. Hinged boundary conditions are assumed at $r = R$ in this example for simplicity. In the absence of a symmetry-breaking field, there is a broken symmetry, and states with positive or negative heights at the center are equally likely.

σ_c (isotensional ensemble) or a critical value of the strain ε_c (isometric ensemble), such that the maximum height H_0 of the thermalized structure rises continuously near the buckling transition, much like an order parameter,[58]

$$\begin{aligned} H_0 &\sim |\sigma_0 - \sigma_c|^{\beta} \text{ (isotensional)}, \\ H_0 &\sim |\varepsilon - \varepsilon_c|^{\beta'} \text{ (isometric)}. \end{aligned} \tag{24}$$

The response of the H_0 to the external field is given by

$$\begin{aligned} \chi &= \left.\frac{\partial H_0}{\partial E}\right|_{E=0} \sim |\sigma_0 - \sigma_c|^{-\gamma} \text{ (isotensional)}, \\ \chi &= \left.\frac{\partial H_0}{\partial E}\right|_{E=0} \sim |\varepsilon - \varepsilon_c|^{-\gamma'} \text{ (isometric)}. \end{aligned} \tag{25}$$

We expect these critical exponents γ and γ' to be the same on both sides of the transition.

In fact, here, unprimed exponents refer to the isotensional ensemble and primed exponents refer to the isometric ensemble. Our final critical exponents refer to the non-linear response of H_0 to E precisely at the buckling transition,

$$\begin{aligned} H_0 &\sim E^{1/\delta} \text{ (isotensional)}, \\ H_0 &\sim E^{1/\delta'} \text{ (isometric)}. \end{aligned} \tag{26}$$

At zero temperature, these mean field Euler exponents are just,[58] $\beta = \beta' = \frac{1}{2}$, $\gamma = \gamma' = 1$ and $\delta = \delta' = 3$. However, as shown using the Wilson–Fisher renormalization group (as well via the self-consistent screening approximation[54]) in Ref. 2, they can be dramatically different when one studies buckling transitions at finite temperature. The best current estimates are $\beta = \frac{1}{2}, \beta' = 0.718$ and $\gamma = 1, \gamma' = 1.436$, although there is no ensemble difference for δ: $\delta = \delta' = 3$. (There is also no ensemble difference for the η-exponents associated with the running elastic constants: $\eta = \eta' = 0.821$, $\eta_u = \eta_u' = 0.358$.) Although the exponents β and γ are unchanged by thermal fluctuations in the isotensional ensemble, there are large differences in these quantities when one switches to the more experimentally accessible isometric ensemble. As mentioned above, the critical buckling thresholds in both ensembles are negative: $\sigma_c < 0$ and $\varepsilon_c < 0$. In fact, both these thresholds vanish in the thermodynamic limit, with $\sigma_c, \varepsilon \sim -1/R^2$, where R is the membrane radius, as was known to Euler.[57] Thus, in the thermodynamic limit of infinite membrane size ($R \to \infty$), classical buckling is a thresholdless long-wavelength ($q_0 \sim 1/R \to 0$) instability, even though the buckling amplitude itself remains macroscopic in this limit. See Ref. 2 for a detailed exposition of the theory sketched above.

4. Concluding Remarks

In this tribute to Michael E. Fisher, I attempted three things:

First, I touched upon his massive contributions to science and highlighted his larger-than-life, somewhat pugnacious personality. Michael's contrarian instincts were in full display in a provocative 1985 talk he gave at a symposium in honor of Niels Bohr at the American Academy of Arts and Sciences organized by Herman Feshbach, entitled "Condensed Matter Physics: Does Quantum Mechanics Matter?" Although I did not attend this meeting, Fisher explained to me in person the thesis in his talk that quantum mechanics was basically a side issue for most problems in the forefront of finite-temperature condensed matter physics at the time. He maintained that if the giants of 19th century physics, such as Boltzmann or Gibbs or Rayleigh, could be brought back to life in 1985, they could all be doing front-line research in about a week without taking time off to learn quantum mechanics. After all, to paraphrase Michael, in a Feynman path integral picture, say for massive bosons, most aspects of quantum mechanics can be captured just adding another dimension, of size $\beta\hbar$ ($\beta = 1/k_B T$) in an imaginary time direction arising from the quantum partition function.

With a pointer from Amnon Aharony, I was able to review Michael's contribution to the proceedings of this symposium.[59] Feshbach's original charge to Fisher asked for an "emphasis on any major unsolved problems (in condensed matter physics) and comments on any overlap with Bohr's ideas regarding the fundamentals of quantum mechanics." Feshbach's request clearly aroused Michael Fisher, much like waving a red cape at a bull! Although Michael provided a wonderful review of the hot topics in condensed matter physics in the mid-1980s, such as the physics of polymer solutions, critical points in low dimensions, materials with quenched random disorder, quasicrystals, etc., he argues, whenever possible, for the irrelevance of quantum mechanical details, at least in the limit where finite temperature correlation lengths become large compared to a suitably defined length in the imaginary time dimension. Fisher of course accepted the importance of quantum mechanics for the collective behavior of most fermion problems, and he acknowledged that it underpins at a more microscopic level most of the problems that interest him. However, I doubt Michael's ideas were well received at the Niels Bohr Symposium:

His 50-page paper, in which he must have invested considerable effort, has only been cited only nine times in 35 years...

Of course, all physicists (including Michael) love the beauty of quantum mechanics. Two of my favorite courses at Cornell involved graduate quantum mechanics presented (in quite different ways) by Ken Wilson and Hans Bethe. I myself have taught the standard two-semester sequence of graduate non-relativistic quantum mechanics frequently over the past 45 years. Nevertheless, I agree with Fisher that there can be advantages to stripping away some of the "quantum mysticism" by recasting quantum statistical mechanics as Feynman path integrals in imaginary time,[60] full of classical world lines and commuting c-numbers. In the same vein, systematic coherent state path integral methods[61] allow us to describe He^4 superfluids at low temperatures in terms of a single macroscopic complex-valued wave function $\psi(\vec{r})$ that behaves classically upon coarse-graining. A middle road might be to slide back and forth between the two perspectives as needed. To me, with a background in soft matter physics as well as quantum condensed matter, the thermally excited (and possibly melted) Abrikosov flux lattices of wiggling quantized flux lines that appear in high-temperature superconductors in a magnetic field[62] look like directed polymer melts! It can nevertheless be very helpful to pass via a transfer matrix technique to the equivalent quantum problem, which looks like a two-dimensional quantum superfluid at low temperature, albeit with a somewhat unusual ideal Bose gas boundary condition in imaginary time.[63] For a system of interacting, thermally excited vortex lines in 2+1 dimensions piercing a set of parallel CuO_2 planes in a high-temperature superconductor, it can be shown that the analog of the long-sought-after quantum supersolid phase in two dimensions is all but *inevitable* at high magnetic fields.[64] Thick superconducting slabs, with microscopic cuprate layers ticking off discrete imaginary time intervals, map onto low temperatures in the equivalent 2D quantum system, with the physical temperature of the cuprate playing the role of Planck's constant. In a similar vein, it can be shown in this 2+1-dimensional world that the analog of a quantum *hexatic* phase is also quite likely,[65] an intermediate phase of matter with long-range bond orientational order that seems to have been observed with diffraction off directed assemblies, not of actual Abrikosov flux lines in a superconductor, but of aligned, charged DNA polymers in $2 + 1$ dimensions at room temperature, which have quite similar interactions.[66] Which is more interesting: *genuine* 2d quantum supersolids or quantum hexatics near $T = 0$, or analog systems of long-directed polymer-like lines at finite temperatures in 2+1 dimensions? The answer seems to me to be a matter of taste. Another realm where the quantum $\leftrightarrow$ classical analogy typically goes the other way is quantum antiferromagnets, where a hard quantum problem in d dimensions is mapped onto a better understood classical Heisenberg spin problem in a d+1 dimensional slab of thickness $\beta\hbar$.[67] Although this connection helped launch the exciting field of quantum phase transitions,[68] it *can* overlook subtleties like a Berry phase term in the quantum action.[69] Overall, however, I think Michael Fisher would have *approved* of sliding back and forth between the quantum problems in d dimensions and classical analogs in $d+1$ dimensions, as needed, to solve important problems. After all, Michael's favorite classical system, the two-dimensional Ising model, was solved by Onsager via a brilliant mapping of its transfer matrix onto the Hamiltonian of one-dimensional system of quantum fermions.[70] It is worth noting that the Hamiltonians associated with transfer matrices for classical statistical physics problems can easily be *non-Hermitian*, as it happens for the asymmetric six-vertex model.[71] Hence, the palette of Hamiltonians arising naturally from transfer matrices generated by classical statistical mechanics problems in one higher dimension (such as ice models in a tilted electric field) is richer than the Hermitian problems associated with conventional quantum mechanics. Similar ideas led to a study of one-dimensional localization transitions in "non-Hermitian quantum mechanics," generated by tight-binding models with asymmetric hopping matrix elements and diagonal disorder.[72] More recently, an imaginary time analog of the quantum Thouless pump has been studied for 2D directed polymers with repulsive interactions and interacting with a set of grooves in a 2D substrate.[73]

In the second section of this chapter, I tried to give a flavor of what it was like to work with Michael and be a student at Cornell in the early 1970s, an exceptionally exciting place and time.[74] I learned an amazing amount from the people around

Fig. 6. Image created to advertise a college production of the Shakespeare play *Twelfth Night* by Michael E. Fisher. Gift of Michael for the 60th birthday of the author. Michael may have chosen to send this because the quote "Out hyperbolical fiend, how vexest thou this man!" conjures up work on geometrical frustration and the glass transition in spaces of constant negative Gaussian curvature.

me not only about science but also from exhortations such as "Drop everything when you have an opportunity to work with experimentalists!" or "*Physical Review Letters* is mere advertising, document your work in long papers in more humble journals!" via Michael. I also made lifelong friends, such as Amnon Aharony, Eytan Domany, H. R. Krishnamurthy, Paul Leiderer and J. Michael Kosterlitz.[75]

In the third section, I gave a brief summary of recent work on the different critical exponents exihbited by isotensional and isometric ensembles for atomically thin elastic membranes. In this work, Michael's prescient ideas about "Fisher renormalized critical exponents" from 1968[1] live on in modern context over a half century later.[2] (For a treatment of a related problem involving thermal buckling, see Ref. 76). I hope Michael Fisher will also live on in Fig. 6.

The original came from an illustration by Michael for a college performance of the Shakespearean play *Twelfth Night*. To me, the image of the jester evokes Michael's catalytic role in science, whether it be defining and understanding critical exponents, inventing ε-expansions, questioning orthodox perspectives about the importance of quantum theory or just generally raising the "level of hygiene" for us all.

Acknowledgments

I am pleased to acknowledge the fruitful collaboration with Suraj Shankar that led to Ref. 2 and Section III. I am also grateful for the inputs on the drafts of this chapter from David Andelman, Amnon Aharony, Eytan Domany, H.R. Krishnamurthy and Leo Radzihovsky. Any errors and possible omissions are, of course, my responsibility.

References

1. M. E. Fisher, Renormalization of critical exponents by hidden variables. *Phys. Rev.* **176**, 257 (1968).
2. S. Shankar and D. R. Nelson, Thermalized buckling of isotropically compressed thin sheets. *Phys. Rev. E* **104**, 054141 (2021).
3. M. E. Fisher, Effect of excluded volume on phase transitions in biopolymers. *J. Chem. Phys.* **45**, 1469–1473 (1966).
4. D. Poland and H. A. Scheraga, Occurrence of a phase transition in nucleic acid models. *J. Chem. Phys.* **45**, 1464–1469 (1966).
5. M. E. Fisher, Magnetism in one-dimensional systems — the Heisenberg model for infinite spin. *Am. J. Phys.* **32**, 343–346 (1964).
6. J. F. Marko and E. D. Siggia, Stretching dna. *Macromolecules* **28**, 8759–8770 (1995).
7. K. G. Wilson and M. E. Fisher, Critical exponents in 3.99 dimensions. *Phys. Rev. Lett.* **28**, 240–243 (1972).
8. K. G. Wilson, Renormalization group and critical phenomena. II. Phase-space cell analysis of critical behavior. *Phys. Rev. B* **4**(9), 3184 (1971).
9. R. J. Baxter, Eight-vertex model in lattice statistics. *Phys. Rev. Lett.* **26**(14), 832 (1971).
10. I received a gracious letter of apology from the Department Secretary, a wonderful woman who had made a perfectly understandable mistake (the letter of acceptance

to the "wrong" David Nelson had not yet gone out.) The name "David Nelson" is quite frequent in the US. I discovered this again in the aftermath of the 9/11 terrorist attack in 2001, when my very common name was placed on a terrorist watch list — a 'bad' David Nelson seems to have misbehaved while attempting to board a flight in Los Angeles. As a result, I (and approximately 10,000 other US David Nelsons) received extra scrutiny while boarding both domestic and international plane flights for the following 5 years.

11. D. Vollhardt and P. Wolfle, *The Superfluid Phases of Helium 3* (Courier Corporation, 2013.)
12. Years later, at a Stanford celebration honoring Michael Fisher's physicist son Daniel at Stanford University, I learned from Michael's wife Sorrel, a story from 1982, when Ken Wilson was notified of his award in an early morning phone call from the Nobel Prize Committee. Ken was surprised that Michael and Leo Kadanoff were not included in the award, as had been the case when they shared the Wolf Prize in Physics. After the Committee insisted that he receive the award on his own or not at all, he accepted and eventually called the Fisher home on the Parkway in Ithaca to let Michael know what had happened. Sorrel took the early morning call, but told Ken in no uncertain terms that she wasn't going to wake her husband from a sound sleep to tell him that he had *not* won the Nobel Prize! (Michael and Ken did talk later that morning, and all was well.) Leo Kadanoff's reaction on that day was to come to work at the University of Chicago in a full tuxedo with tails. When asked why, he replied "it's not every day that you *don't* win the Nobel Prize!"
13. K. G. Wilson and J. Kogut, The renormalization group and the ϵ expansion. *Phys. Rep.* **12**, 75–199 (1974).
14. F. J. Wegner, Corrections to scaling laws. *Phys. Rev. B* **5**, 4529 (1972).
15. J. D. Gunton and M. S. Green, (Eds.), Renormalization Group in Critical Phenomena and Quantum Field Theory: Proceedings of a Conference, held at Chestnut Hill, Pennsylvania, 29–31 May 1973 (Temple University, Philadelphia, 1974).
16. Michael drove some of us down from Ithaca, New York, to the Temple University site outside of Philadelphia. One of the *very* few times I ever saw him back down from authority is when he was stopped for speeding by a member of the Pennsylvania State Highway Patrol.
17. D. R. Nelson and M. E. Fisher, Soluble renormalization groups and scaling fields for low-dimensional Ising systems. *Ann. Phys.* **91**, 226–274 (1975).
18. J. V. José, L. P. Kadanoff, S. Kirkpatrick and D. R. Nelson. Renormalization, vortices, and symmetry-breaking perturbations in the two-dimensional planar model. *Phys. Rev. B* **16**, 1217 (1977). This work was an investigation of XY models of magnetism with various symmetry-breaking fields, using a combination of duality transformations, renormalization group transformations and numerical work. My collaboration with Leo began in the spring of 1976 when he was still at Brown University, in Providence, Rhode Island and I was a postdoc in Cambridge, Massachusetts. I would drive down to Providence to see Leo, and he would drive up to Cambridge to see me. One day when working with Leo down at Brown, I suggested that we look up an important paper by J. Michael Kosterlitz, see Ref. 23. In those days, you had to search for papers in physical bound back issues of journals in the library. I was astonished to learn that Leo, after a decade on the faculty at Brown, had no idea where Brown's science library was located! We eventually found both the library and Mike's seminal paper, which built on earlier work with David J. Thouless. Only someone with Kadanoff's extraordinary creativity and ability to derive things from first principles

could have functioned for 10 years at a major university without ever visiting the library.

19. Needless to say, I never forgot the formula for the capacitance of a sphere. This formula is also important for physical biology, via a lovely analogy between electrostatics and nutrient uptake of a cell following from a steady-state solution of the diffusion equation. As exploited in a famous paper by H. Berg and E. M Purcell, Physics of chemoreception. *Biophy. J.* **20**, 193–219 (1977), the nutrient concentration $c(\vec{r})$ in the steady state plays the role of the electrostatic potential, and nutrient current $\vec{J}(\vec{r}) = -D\vec{\nabla}c(\vec{r})$ taken up by a perfectly absorbing cell is analogous to the electric field.

20. P. Leiderer, D. R. Nelson, D. R. Watts and W. W. Webb, Tricritical Slowing Down of Superfluid Dynamics in He 3-He 4 Mixtures. *Phys. Rev. Lett.* **34**, 1080 (1975).
21. K. S. Korolev, and D. R. Nelson, Defect-mediated emulsification in two dimensions. *Phys. Rev. E* **77**, 051702 (2008).
22. O. Hallatschek, P. Hersen, S. Ramanathan and D. R. Nelson, Genetic drift at expanding frontiers promotes gene segregation. *Proc. Nat. Acad. Sci.* **104**, 19926 (2007).
23. J. M. Kosterlitz, The critical properties of the two-dimensional xy model. *J. Phys. C: Solid State Phys.* **7**, 1046 (1974).
24. J. M. Kosterlitz and D. J. Thouless, Ordering, metastability and phase transitions in two-dimensional systems. *J. Phys. C: Solid State Phys.* **6**, 1181 (1973).
25. V. L. Berezinskii, Destruction of long-range order in one-dimensional and two-dimensional systems having a continuous symmetry group I. Classical systems. *Sov. Phys. JETP* **32**, 493–500 (1971).
26. G. Yuval and P. W. Anderson, Exact results for the Kondo problem: One-body theory and extension to finite temperature. *Phys. Rev. B* **1**, 1522 (1970).
27. D. R. Nelson, J. M. Kosterlitz and M. E. Fisher, Renormalization-group analysis of bicritical and tetracritical points. *Phys. Rev. Lett.* **33**, 813 (1974); J. M. Kosterlitz, D. R. Nelson and M. E. Fisher. Bicritical and tetracritical points in anisotropic antiferromagnetic systems. *Phys. Rev. B* **13**, 412 (1976).
28. D. R. Nelson and J. M. Kosterlitz, Universal jump in the superfluid density of two-dimensional superfluids. *Phys. Rev. Lett.* **39**, 1201 (1977).
29. D. J. Bishop and J. D. Reppy, Study of the superfluid transition in two-dimensional He 4 films. *Phys. Rev. Lett.* **40**, 1727 (1978).
30. J. D. Reppy, In J. G. Dash and J. Ruvalds (eds.), *Phase Transitions in Surface Films*, p. 233 (Plenum Press, New York, 1980).
31. H. R. Krishna-Murthy, K. G. Wilson and J. W. Wilkins, Temperature-dependent susceptibility of the symmetric Anderson model: Connection to the Kondo model. *Phys. Rev. Lett.* **35**(16), 1101 (1975).
32. D. R. Nelson and E. Domany, Equations of state for bicritical points. I. Calculations in the disordered phase. *Phys. Rev. B* **13**, 236 (1976).
33. E. Domany, D. R. Nelson and M. E. Fisher. Equations of state for bicritical points. II. Ising-like ordered phases. *Phys. Rev. B* **15**, 3493 (1977).
34. J. Rudnick, Diagrammatic renormalization-group method. *Phys. Rev. B* **11**, 363 (1975).
35. D. R. Nelson and J. Rudnick, Equations of state for tricritical points. *Phys. Rev. Lett.* **35**, 178 (1975).
36. J. Rudnick and D. R. Nelson, Equations of state and renormalization-group recursion relations. *Phys. Rev. B* **13**, 2208 (1976).

37. D. R. Nelson and M. E. Fisher, Renormalization-group analysis of metamagnetic tricritical behavior. *Phys. Rev. B* **11**, 1030 (1975).
38. M. E. Fisher and D. R. Nelson, Spin flop, supersolids, and bicritical and tetracritical points. *Phys. Rev. Lett.* **32**, 1350 (1974).
39. A grand canonical description would be natural if a 2d Ising model with impurities coexisted with a 3d vapor phase with which it could exchange impurities.
40. L.D. Landau and E. M. Lifshitz, *Statistical Physics* Vol. 5. (Elsevier, 2013).
41. J. Rudnick, D. J. Bergman and Y. Imry, Renormalization group analysis of a constrained Ising model. *Phys. Lett. A* **46**, 449–450 (1974).
42. M. E Fisher, The renormalization group in the theory of critical behavior. *Rev. Mod. Phys.* **46**(4) 597 (1974).
43. R. J. T. Nicholl, H. J. Conley, N. V. Lavrik, I. Vlassiouk, Y. S. Puzyrev, V. P. Sreenivas, S. T. Pantelides and K. I. Bolotin, The effect of intrinsic crumpling on the mechanics of free-standing graphene. *Nat. Commun.* **6**, 8789 (2015).
44. R. J. T. Nicholl, N. V. Lavrik, I. Vlassiouk, B. R. Srijanto and K. I. Bolotin, Hidden area and mechanical nonlinearities in freestanding graphene. *Phys. Rev. Lett.* **118**, 266101 (2017).
45. D. Nelson, T. Piran and S. Weinberg (eds.) *Statistical Mechanics of Membranes and Surfaces* (World Scientific, 2004).
46. M. K. Blees, A. W. Barnard, P. A. Rose, S. P. Roberts, K. L. McGill, P. Y. Huang, A. R. Ruyack, et al, Graphene kirigami. *Nature* **524**, 204–207 (2015).
47. P. G. De Gennes, *Scaling Concepts in Polymer Physics* (Cornell University Press, 1979).
48. P. Bak, C. Tang and K. Wiesenfeld, Self-organized criticality. *Phys. Rev. A* **38**, 364 (1988).
49. L. D. Landau, E. M. Lifshitz, E. M. Kosevich, A. M. and L. P. Pitaevskii, *Theory of Elasticity,* Vol. 7 (Elsevier, 1986).
50. D. R. Nelson and L. Peliti, Fluctuations in membranes with crystalline and hexatic order. *J. Phys.* **48**(7), 1085–1092 (1987).
51. M. E. H. Bahri and Y. Sinai, Statistical Mechanics of Freely Fluctuating Two-Dimensional Elastic Crystals. *J. Stat. Phys.* **180**(1–6) 739–748 (2020).
52. B. I. Halperin, On the Hohenberg–Mermin–Wagner theorem and its limitations. *J. Stati. Phys.* **175**(3–4) 521–529 (2019).
53. J. A. Aronovitz and T. C. Lubensky, Fluctuations of solid membranes. *Phys. Rev. Lett.* **60**(25) 2634 (1988).
54. P. Le Doussal and L. Radzihovsky, Anomalous elasticity, fluctuations and disorder in elastic membranes. *Ann. Phys.* **392**, 340–410 (2018).
55. Y. Kantor and D. R. Nelson, Phase transitions in flexible polymeric surfaces. *Phys. Rev. A* **36**(8), 4020 (1987).
56. A. Košmrlj and D. R. Nelson, Response of thermalized ribbons to pulling and bending. *Phys. Rev. B* **93**(12), 125431 (2016).
57. S. P. Timoshenko and J. M. Gere, *Theory of Elastic Stability* (Courier Corporation, 2009).
58. P. Z. Hanakata, S. S. Bhabesh, M. J. Bowick, D. R. Nelson and D. Yllanes, Thermal buckling and symmetry breaking in thin ribbons under compression. *Extreme Mech. Lett.* **44**, 101270 (2021).
59. M. E. Fisher, Condensed matter physics: does quantum mechanics matter. *Niels Bohr: Phys. World* (1988), pp. 65–115.

60. R. P. Feynman, A. R. Hibbs and D. F. Styer, *Quantum Mechanics and Path Integrals* (Courier Corporation, 2010).
61. J. W. Negele and J. Orland, *Quantum Many Particle Systems* (Addison-Wesley, New York, 1988).
62. G. W. Crabtree and D. R. Nelson, Vortex physics in high-temperature superconductors. *Phys. Today* **50**(4), 38–45 (1997).
63. D. R. Nelson and H. S. Seung, Theory of melted flux liquids. *Phys. Rev. B* **39**(13), 9153 (1989).
64. E. Frey, D. R. Nelson and D. S. Fisher, Interstitials, vacancies, and supersolid order in vortex crystals. *Phys. Rev. B* **49**(14), 9723 (1994).
65. M. C. Marchetti and D. R. Nelson, Dislocation loops and bond-orientational order in the Abrikosov flux-line lattice. *Phys. Rev. B* **41**(4), 1910 (1990).
66. H. H. Strey, J. Wang, R. Podgornik, A. Rupprecht, L. Yu, V. A. Parsegian and E. B. Sirota, Refusing to twist: demonstration of a line hexatic phase in DNA liquid crystals. *Phys. Rev. Lett.* **84**(14), 3105 (2000).
67. S. Chakravarty, B. I. Halperin and D. R. Nelson, Two-dimensional quantum Heisenberg antiferromagnet at low temperatures. *Phys. Rev. B* **39**(4), 2344 (1989).
68. S. Sachdev, Quantum phase transitions. *Phys. World* **12**(4), 33 (1999).
69. T. Senthil, A. Vishwanath, L. Balents, S. Sachdev and M. P. A. Fisher, Deconfined quantum critical points. *Science* **303**(5663), 1490–1494 (2004).
70. L. Onsager, Crystal statistics. I. A two-dimensional model with an order-disorder transition. *Phys. Rev.* **65**(3–4), 117 (1944).
71. B. M. McCoy and T. T. Wu, Hydrogen-bonded crystals and the anisotropic Heisenberg chain. *Il Nuovo Cimento B (1965–1970)* **56**(2), 311–315 (1968).
72. N. Hatano and D. R. Nelson, Localization transitions in non-Hermitian quantum mechanics. *Phys. Rev. Lett.* **77**(3), 570 (1996).
73. R. P. Pedro, J. Paulose, A. Souslov, M. Dresselhaus and V. Vitelli, Topological protection can arise from thermal fluctuations and interactions. *Phys. Rev. Lett.* **122**(11), 118001 (2019).
74. Cornell's amazing physics faculty included theorists like Michael Fisher, Ken Wilson, Vinay Ambegaokar, John Wilkins, N. David Mermin and Neil Ashcroft and experimenters like John Reppy, Watt Webb, Dave Lee and Bob Richardson. Their presence resulted from a wonderful burst of junior hires by Cornell University in the mid-1960s. But as John Wilkins said in 1972, "we can now all watch ourselves grow old together!"
75. In 1976, while we were working on our paper about the universal jump the superfluid density of Helium films,[28] Mike Kosterlitz visited Harvard and pointed out that Richard Feynman had his own ideas the energy–entropy balance involved for dislocations in two-dimensional melting. As described in a paper by Elgin and Goodstein: "Thermodynamic study of the He^4 monolayer adsorbed on Grafoil. *Physical Review A* **9**, 2657, (1974) Feynman had, in unpublished work, computed the coefficient of the logarithmically diverging energy of an isolated dislocation in terms of the Lame elastic constants and had gotten a different answer from the one in the textbooks. This was Feynman, so we worked very hard to convince ourselves that the textbook result was in fact correct. This exercise was helpful when Bert Halperin and I later developed a full theory of 2D melting, which predicted a fourth hexatic phase of matter. Several years later, I had the privilege of giving a physics colloquium at Cal Tech about our theory, hosted by David Goodstein, with Feynman in the front row.
76. P. Le Doussal and L. Radzihovsky, Thermal buckling transition of crystalline membranes in a field. *Phys. Rev. Lett.* **127**, 015702 (2021).

https://doi.org/10.1142/9789811282386_0017

Chapter 17

50 Years of Correlations with Michael Fisher and the Renormalization Group

Amnon Aharony

School of Physics and Astronomy, Tel Aviv University,
Ramat Aviv, Tel Aviv 6997801, Israel
aaharonyaa@gmail.com

I start with a review of my personal and scientific interactions with Michael E. Fisher, who was my post-doc mentor in 1972-1974. I then describe several recent renormalization group studies, which started during those years, and still raise some open issues. These include the magnets with dipole-dipole interactions, the puzzle of the bicritical points and the random field Ising model.

1. My Life with Michael E. Fisher

1.1. *Personal*

I worked on my Ph.D. thesis with the late Yuval Ne'eman at Tel Aviv University, in parallel to my military service in the Israeli army. Expecting to be slow relative to the level of activity required in more competitive topics, I chose to work on the arrows of time, combining recent developments on the breaking of time-reversal symmetry in the decay of the K-meson with the second law of thermodynamics. I thus studied both high energy physics and statistical physics, but knew very little about critical phenomena. When I wrote my thesis (while on duty on the Golan heights in September 1971), I started looking for a place for my postdoctoral training, and I almost accepted a position where I would continue my work on irreversible thermodynamics.

However, two crucial events changed my life. First, the late Joe Imry just returned from his postdoc at Cornell, and he told me about his good experience there. Second, the late Pierre Hohenberg visited Tel Aviv, and told us about Wilson's renormalization group (RG). Both became close friends and collaborators for many years. As a result, I wrote to Michael Fisher, and applied for a postdoc position in his group. In spite of my lack of sufficient background in phase transitions, Fisher soon offered me a postdoc position. In particular, he asked me to respond within two weeks! Being very precise, he indicated that I shall have only two duties: do research and attend the weekly Widom-Fisher seminar. I immediately cancelled my alternative possibilities and accepted Fisher's offer.

My wife, my two-year-old son Ofer (now a high energy physicist at the Weizmann Institute, studying string theory — Fisher won: he has two sons in physics...) and I used my Fulbright travel grant,[a] arrived in Ithaca in July 1972, and were told that we missed the summer, which occurred on the day before our arrival (Ithaca is cold, and summers are short.) We had already rented an apartment (for which Michael had kindly loaned us the first rent) and thus we quickly settled down. During the following two years, we were often invited to the Fisher home, where we enjoyed the warm hospitality of both Sorrel and Michael. As on many later occasions, in October 1973 Michael drove me in his car to my first magnetism conference in Boston, where I talked about our joint work on dipolar magnets (see below). We returned to Ithaca a few days before my daughter Tamar was born (in parallel to worrying about the Yom Kippur war in Israel). Kindly, the Fishers held a party in their home, for exhibiting Tamar to the local community. I last met Sorrel in 2009, when Michael was awarded an honorary doctorate at the Weizmann Institute (Fig. 1), and I was very sad to hear about her death in 2016. It was difficult for Michael to survive without Sorrel. He reacted emotionally when I sent him the photos in Fig. 1.

After 1974, Michael and I met many times at conferences (e.g., Yeshiva and Rutgers Statistical Mechanics meetings,[b] IUPAP conferences on Statistical Physics and on Magnetism, Gordon Conferences, and more). He also visited Tel Aviv in

Fig. 1. Michael and Sorrel Fisher in 2009, when he received an honorary PhD from the WIS. Left: with David Mukamel, Joe Imry, AA. Right: with Sir Michael Berry and AA.

[a]Most of the postdocs at Cornell came with some external fellowships.

[b]My first Yeshiva meeting, in December 1972, was a special experience. In particular I met there my life-long friends and collaborators, the late Dietrich Stauffer and (long live) Sushanta Dattagupta, and many other players in the field.

1980 — on the occasion of receiving the Wolf prize[c] (with Wilson and Kadanoff), in 1983 — when he was honored with a Sackler distinguished visiting scholarship at Tel Aviv university,[d,e] in 1988, when he participated in "Frontiers of Physics," Landau Memorial Conference,[f] and in 1992 — when he was awarded an honorary doctorate from Tel Aviv University.[g] As I describe below, we collaborated on many topics, and I learned a lot from him — on how to use elegant mathematics, how to write papers, how to fight with referees and with editors (especially in PRL — both Michael and I refrained for a while from publishing as well as refereeing there) and how to present talks (with two parallel transparency projectors!). As other authors in this book describe, I also "suffered" from his many red comments on my paper drafts and enjoyed seeing his art of drawing figures for our papers, both of which taught me a lot. In addition, I enjoyed participating in Michael's 60th,[1] 70th[2] and 80th[3] birthdays, see e.g. Figs. 2, 3, which were great reunions of the large group of Michael's students and postdocs. Michael also attended my 60th birthday conference in Eilat, in 2004,[h] see Fig. 4. In 1982, Michael Stephen, Michael Fisher and I joined others to teach a course on critical phenomena in South Africa.[4,i] Also, I visited Michael in Maryland in 2002. When I arrived at Dulles airport, the person at customs asked me where I am going, and I told him I am giving a colloquium on quantum mechanics at the University of Maryland. He immediately asked when we shall get rid of airports and use teleportation, and I told him this will not happen in my lifetime. I mentioned this conversation in my colloquium, and Michael immediately reacted: "This must have been our student"! Note: my talk was about quantum interference in mesoscopic devices, and at that time Fisher raised questions about the relevance of quantum mechanics to condensed matter physics,[5] but we did not debate on this. Finally, I must have done something right: Michael also

[c]A typical Fisher story: The Wolf prize was young. Documents were prepared in Hebrew and then translated into English. In this translation, "phase transitions" became "transient phases", and this appeared on the certificate given to Michael by Israel's president at the ceremony. I caught this, and asked the Wolf people to be more careful in the future. However, when they asked Michael to return the certificate for correction he refused: "It will look much more impressive on my office wall as it is". I do not know how this ended.

[d]https://ias.tau.ac.il/IAS-Distinguished-Scholars; this was before the Sacklers' involvement in the opioid crisis became known; since then I no longer use the Sackler name when I refer to my School and Faculty at TAU.

[e]On that visit Michael stayed for a month in an apartment with an appropriately placed desk and room for playing the guitar — being a perfectionist in everything, Michael always insisted on proper arrangements in his rooms and offices.

[f]"Frontiers of Physics," Proc. Landau Memorial Conf., Eds. E. Gotsman, Y. Ne'eman and A. Voronel (Pergamon Press, Oxford, 1990).

[g]http://www3.tau.ac.il/honorary#phd. Michael did not bring a formal suit for the ceremony, so we rushed downtown to buy one, and I saw him wear that suit also on later occasions; even when Michael was not in Tel Aviv, Tel Aviv stayed with him...

[h]https://www.tau.ac.il/~aharony/public/eilatposter.jpg.

[i]Both kindly drove on the wrong side of the road, given their prior experience.

Fig. 2. Fisher and his students/postdocs on his 60h birthday, NAS, Washington DC 1991. Top: Mustansir Barma, AA, Shmuel Fishman, Joseph Straley, Pierre Pfeuty, David Nelson, David Jasnow, Ubo Felderhof, Masuo Suzuki, Michael Barber, Gunduz Caginalp, Arthur Ferdinand. Bottom: Vladimir Privman, MEF, Lev Mikheev.

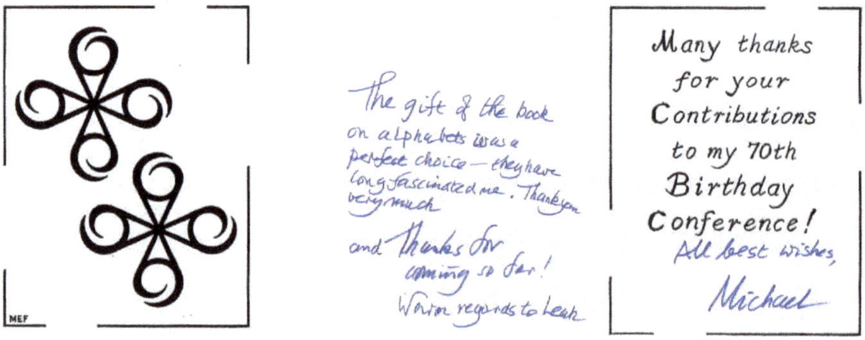

Fig. 3. Card drawn and written by MEF to AA after MEF's 70th birthday. Note the typical drawings and calligraphy.

Fig. 4. AA's 60th birthday, Eylat, 2004. Top: Fisher lecturing with transparencies; AA, MEF, Joe Imry, Édouard Brézin and Eytan Domany. Bottom: MEF with Sir Sam Edwards; Yuval Ne'eman, AA and MEF (Peter Fulde and the Mukamels are behind).

invited my first doctoral student, the late Shmuel Fishman, to be a postdoc in his group (first at Cornell and then at Maryland; Fig. 2).

1.2. *Scientific*

My time at Cornell was unique.[j] The Wilson-Fisher paper on the ϵ−expansion[6] had just appeared, and the authors were there to help with its applications to phase transitions in a variety of physical systems. That paper, in which they considered the dimensionality $d = 4 - \epsilon$ as a continuous variable, allowed simple analytic calculations (instead of the rather more complicated numerics proposed by Wilson earlier). This opened the field to many researchers, and led to the flourishing of critical phenomena ever after.[k] Half a century later, this is one of the main reasons to celebrate Michael Fisher and the RG in this book.

[j]In a talk celebrating Wilson, David Mermin called these years "the Valhalla years of Cornell".

[k]Since I came into the field in 1972, I do not feel competent to review its earlier history, accept for noting that Wilson and Fisher followed the scaling ideas set earlier by Ben Widom and Leo Kadanoff, and Widom was also at Cornell — open for discussions and help. Fisher himself reviewed this history in Ref. 7.

I overlapped at Cornell with many postdocs and students, both in the Baker laboratory of Chemistry, where Fisher was centered (with Ben Widom, who told me proudly that he got Fisher there, but was unable to contribute a chapter to this book), and in the Clark Hall of Physics. The list included Pierre Pfeuty and Helen Au-Yang, with whom I shared an office, David Jasnow, Howard Tarko, Alastair Bruce, Michel Droz, Michael Kosterlitz, David Nelson, Eytan Domany, Paul Gerber, Joe Sak and many more, not to mention Widom's group (e.g., Jim Bartis) and Wilson's group (e.g. H. R. Krishnamurthy). Also, Bob Griffiths was there on sabbatical. Among others, Pfeuty talked about his work with Elliot on the Ising model in a transverse field, and Kosterlitz talked about his work with Thouless on the XY model in two dimensions. The atmosphere was friendly, and Michael encouraged joint discussions and collaborations.[1]

Ken Wilson gave a course on the RG, which he also gave in Princeton (later published by Wilson and Kogut[8]). Most of us worked on the RG, and often the course gave us the hints we needed for making progress in our research. Many visitors came through, and we heard their seminars both on the blackboard in the Fisher-Widom seminar and in the more conventional physics seminar. I brought a pile of slides based on my doctoral thesis (transparencies only began being used then), but did not use them — being forced to use only the blackboard. Away from this seminar, Michael did use transparencies; he kept using them all his years, avoiding the more modern use of PowerPoint. In particular, I liked his way of using two projectors in parallel, so that one could compare theory and experiments, or keep track of definitions. I adopted this method for many years, until I did switch to PowerPoint. Several papers in this volume also mention Michael's questions to seminar and conference speakers. Although sometimes looking sleepy, he usually understood much of the talks and posed important questions for which the speakers were not always prepared. In the following years, I tried to imitate Michael also in this respect, which may have given me the reputation of being "brusque".

Immediately after my arrival, Michael suggested that I study the dependence of the critical temperature on the many physical parameters in specific systems. He later returned to that problem in his work on non-linear fields, and I return to it in Sec. 2.1 below. However, shortly after that I attended a physics seminar in Clark Hall, discussing some aspects of dipole-dipole interactions. Already during the talk I realized that the propagator related to the two-spin correlation function of such interactions is similar to the propagator of the photon (which vanishes for longitudinal fluctuations), which I knew from my field theory courses at Tel Aviv. This propagator projects out longitudinal correlations. I immediately spoke to Michael about this, and we agreed that the critical behavior of dipolar systems is a better direction to follow. Indeed, this project soon ended up with six papers,[9–14] with

[1]We also had the good fortune to read the preprints of chapters from Ashcroft and Mermin's "Solid State Physics" (I still have a pile of these heavy typed booklets), and be inspired by both authors.

the first appearing already in March 1973. However, only two of these were co-authored by Michael! The 'Heisenberg' dipolar interactions involve scalar products like $(\mathbf{S}(\mathbf{R}_i) \cdot \mathbf{R}_{ij})$, involving both the coordinate vector $\mathbf{R}_{ij}$ and the spin vector $\mathbf{S}(\mathbf{R}_i)$, and therefore we used $n = d = 4 - \epsilon$, where n is the number of spin components.[m] These scalar products also required angular integrations in continuous d dimensions. This connection between space and spin also led me to generalize the Wilson-Fisher paper (which had $n = 2$ coupled Ising spins) to systems with cubic symmetry, i.e., coupled n component spins,[15,16] which raised issues of competing fixed points (FPs), still being investigated today (see below). After we published these two papers, Michael allowed me to proceed on my own, taking advantage of Wilson's course and including Wilson's direct diagrammatic way of calculation,[11] antiferromagnets,[12] anisotropies (in which $n < d$),[13] $1/n$−expansions (in which n was also treated as a continuous variable) and more. Still at Cornell, I also wrote papers on dipolar systems with Alastair Bruce.[17] Needless to say, all my papers from Cornell acknowledge both Fisher and Wilson for their advice.

In particular, in Ref. 14 I was able to use the Fisher-Wilson approach to show that the upper critical dimension for the uniaxial dipolar problem ($n = 1$) is 3, with logarithmic corrections,[14] confirming the earlier Larkin-Khmelnitzkii calculations (Larkin then visited Cornell, before he knew English, and Khmelnitzkii later became a friend) and correcting the simple Landau theory approach (without the logarithmic corrections) used earlier for structural phase transitions (about which I learned from Alastair Bruce, who was a student of Roger Cowley). Guenter Ahlers (whom I also met through Michael, and later collaborated with[18]) later received at a Gordon conference (Fig. 5[n]) a prize of 1/3 of a log (of wood) for confirming these corrections experimentally.[19] Unfortunately, Guenter was unable to contribute to this book. I return to dipolar systems below, in Sec. 2.1.

During my years at Cornell, many experimentalists were attracted to visit and understand the implications of the RG on their work. I later enjoyed very much collaborations with many of these experimentalists, including, e.g., Ahlers, Birgeneau, Cowley, Garland, Litster, and Rohrer. One special example was the visit of Alex Müller, who worked on structural phase transitions and on mulicritical points in perovskites.[o] This visit resulted in many collaborations with Alex and with Alastair Bruce.[20,21,23] Fortunately, after a visit in Tel Aviv Alex ignored my

[m]The equality $n = d$ arises in many physical systems. Generally, one can extrapolate to the physical limit $n = d = 3$ along many trajectories, e.g., at fixed $n = 3$ and expand in $\epsilon = 4 - d$, expand in ϵ along $n = d = 4 - \epsilon$, at fixed $d = 3$ and expand in $1/n$, etc. Similar questions arise for $n = d = 2 + \epsilon$ and $n = d = 1 + \epsilon$. At the end, all these expansions and extrapolations should approach the same limit at $n = d = 3$. It will be useful to compare all these with existing modern tools.

[n]David Nelson reminds me that the conference was announced in Science magazine as "solids and fruits", to which Seb Doniach proposed an improvement: "salads and fruits".

[o]The history of my collaborations with Alex is reviewed in: Amnon Aharony, My encounters with Alex Müller and the perovskites, Physica C: Superconductivity and its applications, Special issue commemorating Alex Müller, **613**, 1354336 (2023).

Fig. 5. Color version of the original black and white photo of the participants in the Gordon conference on Dynamics of quantum solids and fluids: cooperative phenomena in solids and fluids, July 1965. David Nelson, Bob Birgeneau and AA so far identified Aharony, Ahlers, Als Nielsen, Bak, Baker, Berker, Birgeneau, Betts, Bishop, Bruce, Chang, Ditzian (now Kadanoff), Doniach, Emery, Fisher, Grest, Greytak, Griffiths, Hertz, Hohenberg, Imry, Kadanoff, Kitchens, Kosterlitz, Krishnamurthy, Luther, Martin, Mandelbrot, Nelson, Nickel, Passell, Pindak, Pynn, Rice, Rohrer, Rudnick, Siggia, Stephen, van Leeuwen, Widom, Wolf and Wortis. Many of these are also mentioned in the text. I leave it for the readers to connect names with pictures.

enthusiastic recommendation to stay with structural transitions, and he switched to working on high temperature superconductivity, for which he was awarded the Nobel prize. Similarly, Heini Rohrer ignored my advice to stay in magnetism (we collaborated on the random field multicritical antiferromagnets, see below), and went on to receive the Nobel prize for the scanning tunneling microscope. Apparently, doing experiments on phase transitions and talking to theorists are good preparations for Nobel prizes... Multicritical points were also studied then by Kosterlitz, Nelson, and Fisher,[24] and later by Domany, Mukamel and Fisher.[25] Both of these papers contain examples of Fisher's special talent for drawing figures. Both Domany and Mukamel later became friends and collaborators. I return to this topic, in Sec. 2.2.

In May 1973, less than a year after my arrival at Cornell, Mel Green and Jim Gunton organized a meeting at Temple University, entitled "Renormalization Group in Critical Phenomena and Quantum Field Theory". Kindly, Fisher took me along to that meeting, and gave me a chance to meet all the great people who were working on that title. I talked on the dipolar systems and met Paul Martin (who hosted me at Harvard in the fall of 1974, where I learned about random systems from Alan Luther and Geoff Grinstein), Shang-keng Ma (who hosted me at UCSD in the spring of 1975, starting a long collaboration on random systems), and Bert Halperin and Pierre Hohenberg (who hosted me at Bell Labs in the summer of 1975, starting a long collaboration on universal amplitude ratios and starting my long collaboration with Bob Birgeneau). I also met some of the experimentalists mentioned above, and people like Franz Wegner, Eberhard Riedel and Édouard Brézin. In the following years, all of these people worked on the same topics as myself, we corresponded a lot, and became good friends. The 'RG community' of those early years was stimulating and friendly, and I have been grateful to Michael for helping me become part of it. Another result of that meeting was that Green invited me to write Ref. 16, joining a select group of the above list.[p]

As I mentioned, Wilson also gave the same course at Princeton, where it was attended by Édouard Brézin and David Wallace. The three of them then used the RG to derive the universal scaling function of the equation of state.[26] (Brézin and I continued to read and write about each other's papers for many years). As soon as we saw that paper, Michael and I realized that a similar universal scaling function should also exist for the two-spin correlation function. This resulted in the next two joint papers on the $\epsilon-$expansion of this function,[27,28] and by my own paper on its $1/n-$expansion.[29] The $\epsilon-$expansion demonstrated for me Michael's excellent mathematical insights. In Fourier space, this function depends on the 'momentum' q, and the perturbative expansion (in the quartic spin terms) only contributed to this $q-$dependence at second order, apparently yielding contributions only at second order in ϵ. However, the short range correlations were known to have terms which behave like the energy,[30] involving, e.g., the specific heat exponent α, which starts at order ϵ. Michael soon realized that the terms of first order in ϵ cancel out, so that our expansion of the correlation function fully agreed with the known scaling forms, at both short and long distances. This work, and Michael's insights into the universality of scaling functions, later led me to work on universal amplitude ratios.[31] In fact, these topics came up again during Michael's visits to Tel Aviv, in 1980 and 1983, resulting with two papers[32,33] on analytic and non-analytic corrections to scaling.[q] In these papers we may have accomplished Michael's first assignment to me, back in 1972.

I cannot end this historical introduction without mentioning another 'interaction' with Michael. In 1989, he spoke at the Gibbs symposium, and emphasized the Gibbs rule on the convexity of geometrical surfaces in thermodynamic phase diagrams. The summary of that talk[35] ends with the following: "The striking phase diagrams observed in doped crystals displaying high-T_c superconductivity [12]

[p]That series of 20 books, first edited by Cyril Domb (Michael's mentor at King's college) and Green and then by Domb and Joel Lebowitz (so far leader of 124 Yeshiva→ Rutgers meetings), contains many great reviews on phase transitions and critical phenomena.

[q]After this paper appeared on the arXiv, Jacques Perk kindly reminded me that higher order series showed that these papers lacked higher order log terms (which could not be seen by the short series we had at the time).[34]

suggest the intriguing Gibbsian question: What, if anything, replaces the convexity of the $U(S, V)$ surface when there is a variable density of frozen or quenched random impurities?" Ref. [12] refers to a paper,[36] in which some of us tried to predict details of the phase diagrams of the recently discovered high-temperature supercoductors by Müller and Bednorz (mentioned above). That phase diagram was drawn only schematically, and some lines apparently did not seem to obey the Gibbs rules. Michael noticed that, and he would not miss any occasion to tell me that he keeps watching me, following the biblical rule "spare the rod — spoil the child".[37] OK, Michael, I am proud to have been your scientific child!

2. Recent Developments

The wave of econophysics, sociophysics and the like 'pushed' me away from the 'modern' statistical physics community. Therefore, my interests shifted to random systems, magnetism, quantum mesoscopics and spintronics. However, many statistical physicists stayed within real physics, and I kept following the literature on critical phenomena. I often returned to that field, and wrote on topics which in many respects followed Fisher's legacy and/or addressed issues which remained open. Examples include Ref. 38, related to Fisher's interest in random systems,[35,39] and Ref. 40, related to multicritical points.[24]

Below I mention a few issues which I think are still open, even after 50 years. Celebrating the 50th anniversary of the ϵ−expansion, it may be appropriate to remember that this expansion treated the dimensionality of space as an algebraic abstract variable, without any geometrical picture of a real system with 3.99 dimensions. One attempt to find such systems came up when I met Benoit Mandelbrot at Harvard in 1980. Mandelbrot's fractals are geometrical constructions with non-integer dimensions, which are not translationally invariant but in which the 'mass' M within a volume of linear size L scales as $M \sim L^D$, identifying D as the (non-integer) fractal dimensionality. It turned out that real space RG, in the original spirit of Kadanoff, becomes exact on many fractals,[41] and the ease of such calculations started a large wave of publications. However, it turned out that the physics of fractals required much more than just the fractal dimensionality,[42] and this path of research did not find a geometrical description of the Wilson-Fisher non-integer d. However, fractal models have turned out to be very useful to model non-translationally invariant physical systems, e.g., percolation clusters. Hence my long personal involvement with percolation.[43,r]

A major development in recent years was the entrance of field theorists into the field of critical phenomena, many years after Wilson set the example (not to mention my own background in high energy). They not only rediscovered many of the

[r]In fact, Michael worked on percolation clusters much before I met him.[44] He also discussed fractal vesicles in a conference which I organized with Jens Feder.[45]

physical systems studied in the 1970'ies, but they also developed and applied many novel and very accurate methods to study them, including re-summed high-order ϵ−expansions,[46] fixed-dimensions RG calculations,[47] conformal bootstrap calculations,[48] and accurate Monte Carlo simulations.[49] (To save space, I give only recent references, exemplifying the large current activity of field theorists on critical phenomena). Specifically, these calculations identified unambiguously the stable FPs of the RG, and gave consistent and accurate values for the critical exponents, both at critical and at multicritical points. The availability of such tools (and the 'pushing' of Slava Rychkov) led me recently back to the puzzle of the bicritical points[50–52] discussed in more detail in the following.[s] Below I also present selected other topics, in which Fisher was interested already 50 years ago, and demonstrate that even half a century later, the RG is still an active and exciting field.

2.1. *The dipole-dipole interactions*

Fifty years after my early papers on the dipolar systems, Refs. 9–14, sometimes people ask me if it was my father that wrote them. Well, it was me, and I am still alive and kicking. (The issue may become more tricky in 30 years, when they will ask my son Ofer the same question...). Occasionally, some of these results are rediscovered "independently", and I send E-mails to those rediscoverers.

In any case, every material with magnetic ions has dipole-dipole interactions, and therefore it is always necessary to check their relevance to experimental measurements. Originally, the ϵ−expansion for the $n = d = 4 - \epsilon$ isotropic dipolar case[9,10] seemed numerically close to the short-range Heisenberg case ($2\nu_{dip} = 1+9\epsilon/34+\cdots$ versus $2\nu_{iso} = 1 + \epsilon/4 + \cdots$). Similar closeness arose for $1 << d$.[13] However, these numbers should be improved, and I still await extensions to higher order expansions and other accurate high energy methods, like those mentioned above. (As Slava Rychkov tells me,[53] the dipolar model is not conformal, hence bootstrap is not applicable. Other methods should still work.)

As mentioned, the uniaxial Ising case is particularly interesting. The upper critical dimension is $d_u = 3$, and one could prove analytically that at this dimension, e.g., the susceptibility diverges near criticality as $\chi \sim |t|^{-1}|\log|t||^{1/3}$, where t is the scaling field related to the temperature distance from the critical point (this is the one third of a log mentioned above), similar to the behavior of the short range Ising model at $d = 4$. However, the next correction, of order $\log|\log|t||/\log|t|$, has different amplitudes in these two cases.[54] Note: this dimensionality shift from d up to $d + 1$ dimensions differs from that of the random field

[s]Assuming that Michael would have liked this recent work, I tried to communicate it to him, but it was too late. His E-mail address, xpectnil@umd.edu, no longer responded. (The name of this address reflected Michael's attitude to E-mail, but he did use it more after he retired. More interestingly, Michael kept copies of every letter he ever sent or received. I am sure a historian of science will find many jewels in that archive. I am told that UMD is keeping it.)

shift from d to $d+2$ dimensions, discussed below;[55] the latter results from supersymmetry,[56] and therefore *all* the terms in the $\epsilon-$expansions map from d to $d+2$. Interestingly, unlike at $d < d_u^{dip}$, universal amplitude ratios *at* $d = d_u^{dip}$ have logarithmic corrections, e.g. the free energy per correlation volume[57]

$$\xi^2 \xi_{\|} C t^2 / k_B = (3/32\pi)|\log|t||, \tag{1}$$

where C is the singular specific heat, ξ is the correlation length, $\xi_{\|} = g_d^{1/2}\xi^2$, g_d measures the strength of the dipolar interactions and k_B is the Boltzmann factor. This implies logarithmic corrections to the hyperscaling relation, $d\nu = 2-\alpha$, which do not arise at $d<3$, where the free energy per correlation volume is a universal temperature-independent constant.[31,t]

The situation is even more interesting when randomness is included. In that case the expansion in $\epsilon = 3-d$ starts at order $\epsilon^{1/2}$, and the logarithmic corrections for the susceptibility at $d = d_c = 3$ are replaced by $\chi \sim |t|^{-1}\exp[(D|\log|t||)^{1/2}]$, with the universal $D = 9/[8\log(4/3)+53]$, different form $D = 6/53$ for the short range Ising model in $d = 4$.[58] I found one reference that claims to have seen a crossover to this behavior,[59] but the evidence there is somewhat indirect.

My recent interest in the dipolar system arose when my colleague at Ben Gurion University Moshe Schechter told me about his paper on $LiHo_xY_{1-x}F_4$.[60,u] This is a dilute uniaxially anisotropic quantum dipolar ferromagnet. At $x=1$, an external transverse field B on this non-random ferromagnet causes a decrease of the critical temperature from $T_c(B=0) \approx 1.53$K all the way down to a zero temperature quantum transition at a field of $B \approx 4.9$T. T_c also decreases with decreasing x (the Y ion is not magnetic), approaching a spin glass phase at low x. Shechter *et al.* review various numerical calculations of the dependence of T_c on both B and x, including their own — which emphasized the quantum fluctuations due to off-diagonal spin terms which come from the dipolar interactions.

Since fluctuations are the "bread and butter" of the RG, I now propose to use the RG for an alternative approach to this system. As I show, $LiHo_xY_{1-x}F_4$ practically contains all the topics studied by RG during the recent decades: crossover from short range interactions to isotropic dipolar interactions, then to anisotropic (uniaxial) dipolar interactions, then to random (dilute) uniaxial Ising behavior (both random exchange and random fields), and finally adding quantum effects. The order of these crossovers may change, depending on the relative strengths of the relevant parameters. Below I extend many of my old papers on some of these topics, and suggest how one can derive the critical temperature as function of the short/long interaction strengths, the spin anisotropy, the dilution and the low temperature

[t]To show the close community in those days, the acknowledgements in that paper include G. Ahlers, J. Als-Nielsen, R. Birgeneau, P. C. Hohenberg, J. M. Kosterlitz, D. R. Nelson and E. Siggia. Unfortunately the first 3 experimentalists were unable to contribute to this volume.

[u]This seems to be a relative of $LiTbF_4$, where Ahlers originally measure the 1/3 log correction.[19]

quantum effects. This discussion is rather technical, and the unmotivated reader can skip it.

At first, I ignore quantum fluctuations. These can be added later, by adding another dimension of imaginary times of a width that grows with decreasing temperature.[61–64] Classically, we start with the isotropic Heisenberg short range system. Pure $LiTbF_4$ then requires adding the isotropic dipole-dipole interactions, with strength g_d, as treated in Ref. 10, and the spin anisotropy, with strength g_a, as treated in Ref. 13. We start at $B = 0$ and $x = 0$. The classical (rescaled) Ginzburg-Landau-Wilson Hamiltonian of the $n-$component spin model in momentum space is[v]

$$\bar{\mathcal{H}}_0(\{\mathbf{S}\}) = -\sum_{\alpha\beta}\left[\frac{1}{2}\int_{\mathbf{q}} U_2^{\alpha\beta}(\mathbf{q}) S^\alpha_{\mathbf{q}} S^\beta_{-\mathbf{q}} + u_{\alpha\beta}\int_{\mathbf{q}}\int_{\mathbf{q}'}\int_{\mathbf{q}''} S^\alpha_{\mathbf{q}} S^\alpha_{\mathbf{q}'} S^\beta_{\mathbf{q}''} S^\beta_{-\mathbf{q}-\mathbf{q}'-\mathbf{q}''}\right], \tag{2}$$

where the integrals are over a spherical Brillouin zone of radius 1, and[10]

$$\begin{aligned} U_2^{\alpha\beta}(\mathbf{q}) &= [r_\alpha + q^2]\delta_{\alpha\beta} + g_d q^\alpha q^\beta/q^2, \quad \mathbf{q}\neq 0,\\ &= r_\alpha\delta_{\alpha\beta} + g_d\mathcal{D}^{\alpha\beta}, \quad \mathbf{q}=0, \end{aligned} \tag{3}$$

with $r_\alpha = (T - T^{MF}_\alpha)/T^{MF}_\alpha$, and $\mathcal{D}^{\alpha\beta}$ being the (shape-dependent) demagnetization tensor. The quartic coefficients $u_{\alpha\beta}$ will be chosen below (for the simplest isotropic case, $u_{\alpha\beta} \equiv u$.) Our final aim is to find the shift of the critical temperature T_c from its mean field value, T^{MF}_α, for the highest 'initial' T^{MF}_α, which we assume arises for $\alpha = n = \parallel$. For simplicity we ignored in $U_2^{\alpha\beta}$ the terms $q^\alpha q^\beta$ and $(q^\alpha)^2$.[10,65] In pure $LiHoF_4$ there is also a uniaxial anisotropy, $g_a = r_\perp - r_\parallel$, so that

$$r_n = r_\parallel = r - (1-1/n)g_a, \quad r_\perp = r_1 = \cdots = r_{n-1} = r + g_a/n. \tag{4}$$

Adding a magnetic field will add more shifts in the r's, see below.

Near the isotropic short range FP, the RG iterations for the three main scaling fields give $t(\ell) = t(0)e^{\ell/\nu}$, $g_d(\ell) = g_d(0)e^{(2-\eta)\ell}$, $g_a(\ell)e^{\ell\phi/\nu}$. Here, t is the true scaling field for the temperature, which vanishes at the critical temperature T_c, and ϕ is the crossover exponent for the anisotropy.[66] Eventually, for $g_d(\ell)$ and $g_a(\ell)$ sufficiently large, the system will cross over to the uniaxial dipolar problem, treated in Ref. 14. At that point, the scaling field $t(\ell)$, associated with the single critical temperature parameter $r_\parallel$, will be proportional to the true $T - T_c$, and therefore it will yield the temperature shift $T^{MF}_c - T_c$, as measured experimentally.

The $\epsilon-$expansion is based on iteratively integrating out of the partition function the $S_\alpha(\mathbf{q})$'s in the spherical shell $1 - \delta\ell < |\mathbf{q}| < 1$ and then rescaling the spin and

[v]Here and below we denote $\bar{\mathcal{H}} = -\mathcal{H}/(k_B T) - \sum_{\mathbf{R}}\left[|\mathbf{S}_{\mathbf{R}}|^2/2 + u[|\mathbf{S}_{\mathbf{R}}|^4\right]$.[16]

momenta so that the momentum cutoff and the coefficient of q^2 return to 1.[8] At $d = 4 - \epsilon$ the recursion relation for r_α is[66]

$$dr_\alpha/d\ell = 2r_\alpha + 12u_{\alpha\alpha}A^{\alpha\alpha} + 4\sum_{\alpha\neq\beta} u_{\alpha\beta}A^{\beta\beta} + \cdots\,, \tag{5}$$

where the dots indicate higher orders in the u's,

$$A^{\alpha\beta} = \int^{>} G^{\alpha\beta}(\mathbf{q})/(\delta\ell), \tag{6}$$

the integrals are over a spherical shell in momentum space, with $1 - \delta\ell < |\mathbf{q}| < 1$, and the matrix $G \equiv [U_2]^{-1}$ (for $\mathbf{q} \neq 0$) is[13]

$$G^{\alpha\beta}(\mathbf{q}) = \frac{1}{r_\alpha + q^2}\left[\delta_{\alpha\beta} - \frac{g_d}{q^2 + g_d Q^2}\frac{q^\alpha q^\beta}{r_\beta + q^2}\right], \tag{7}$$

with $Q^2 = \sum_\gamma (q^\gamma)^2/[r_\gamma + q^2]$. For $g_d = 0$ this reduces to $A^{\alpha\beta} = K_d\delta_{\alpha\beta}/(r_\alpha + 1)$, where K_d is the area of a $d-$dimensional unity radius sphere. In the dipolar case $G^{\alpha\alpha}$ depends on the direction of $\mathbf{q}$, and the integrals, which depend on g_d, are more complicated.[10,13,14]

Since $g_d(\ell) = g_d(0)e^{(2-\eta)\ell}$ is exact [and we ignore the $\ell-$depedence of $\eta = O(\epsilon^2)$], we can substitute this $\ell-$dependence into Eqs. (5) and find $r_{||}(\ell)$ and $r_\perp(\ell)$. Now we can follow the example of Nelson and Domany:[67] iterate until $r_\perp(\ell_1) = 1$, obtain $r_{||}(\ell_1)$, integrate the transverse spins out of the partition function and then proceed with the recursion relations for the uniaxial case with only $S^{||}$. Note: the propagator at that stage may still have the form

$$G^{||||} = \frac{1}{r_{||} + q^2}\left[1 - \frac{g_d}{q^2 + g_d Q^2}\frac{(q^{||})^2}{r_{||} + q^2}\right], \tag{8}$$

with $Q^2 = (q^{||})^2/[r_{||} + q^2]$. It will approach

$$G^{\alpha\beta} = \delta_{\alpha,n}\delta_{\beta,n}\frac{1}{r_n + q^2 + g_d\cos^2\theta}, \tag{9}$$

where θ is the angle between the unit vector $\hat{\mathbf{q}}$ and the $n-$axis, only when $g_d(\ell) \gg 1$. Although some calculations of the crossover in the presence of both g_a and g_d were done previously,[68,69] I think the above prescription is more straight forward.

Adding a magnetic field B (in energy units — including the factor $g\mu_B$) on the spin component S^1, adds a term $-BS^1_{q=0}$ to $\mathcal{H}_0$. It is then convenient to replace

$S^1_{\mathbf{q}} \Rightarrow S^1_{\mathbf{q}} + M\delta(\mathbf{q})$. This shifts the coefficients in $U_2^{\alpha\beta}$,[17]

$$\begin{aligned} r_1 &\to \bar{r}_1 = r_\perp + 12uM^2 = \bar{r}_\perp + 8uM^2, \\ r_\perp &\to \bar{r}_\perp = r_\perp + 4uM^2, \quad \text{applies for } \beta = 2, \ldots, n-1, \\ r_n &\to \bar{r}_\parallel = r_\parallel + 4uM^2, \end{aligned} \tag{10}$$

where, for simplicity, we set $u_{\alpha\beta} \equiv u$. The value of M is determined by requiring that the new coefficient of S^1 vanishes,[17]

$$(r_\perp + g_d\mathcal{D}^{11} + 4uM^2)M = B, \tag{11}$$

and at small M we can use $M \approx B/(r_\perp + g_d\mathcal{D}^{11})$. In addition, this shift also generates a cubic spin term, $4uM\sum_\alpha S^1_{\mathbf{q}} S^\alpha_{\mathbf{q}'} S^\alpha_{-\mathbf{q}-\mathbf{q}'}$.[17]

Equations (10) imply

$$\bar{r}_1 > \bar{r}_\perp > \bar{r}_\parallel, \tag{12}$$

implying that $S^n = S^\parallel$ will be the main order parameter. As we iterate the recursion relations (5), all the $\bar{r}$'s grow with ℓ, but $\bar{r}_1(\ell)$ will reach the value 1 first, at $\ell = \ell_1$. At that point the correlations in $S^\parallel$ have been eliminated, and we can use mean field theory to integrate $S^\parallel$ out of the partition function, leaving us with an effective Hamiltonian in the remaining $n-1$ spin components. We can next iterate until ℓ_2, when $\bar{r}_\perp(\ell_2) = 1$, eliminate the $(n-2)$ transverse components with $\beta = 2, \ldots, n-1$ and remain with the single temperature variable $\bar{r}_\parallel(\ell_2)$. The scaling field for that variable finally yields

$$T_c(B) = T_\parallel^{MF}[\bar{r}_\perp(\ell_2) - \bar{r}_\parallel(\ell_2) + 8u(\ell_2)M(\ell_2)^2]. \tag{13}$$

The dilution is easily introduced via replicas.[58,70–72] Each spin component S^α is replaced by m replicas, $S^{\alpha i}$, $i = 1, \ldots, m$, and at the end one takes the limit $m \to 0$. In real space, the Hamiltonian of a specific realization of the occupations, $\{p_{\mathbf{R}}\}$, is

$$\begin{aligned} \mathcal{H}(\{p_{\mathbf{R}}\}, \{\mathbf{S}_{\mathbf{R}}\}) = -\Bigg[&\frac{1}{2} \sum_{\mathbf{R}\neq\mathbf{R}'} \sum_{\alpha\beta}^{n} p_{\mathbf{R}} p_{\mathbf{R}'} W^{\alpha\beta}(\mathbf{R}-\mathbf{R}') S^\alpha_{\mathbf{R}} S^\beta_{\mathbf{R}'} \\ &+ \sum_{\mathbf{R}} p_{\mathbf{R}} \Big(\sum_\alpha (r_\alpha - r)(S^\alpha_{\mathbf{R}})^2 - BS^1_{\mathbf{R}} + u|\mathbf{S}_{\mathbf{R}}|^4 \Big) \Bigg], \end{aligned} \tag{14}$$

where $W^{\alpha\beta}(\mathbf{R}-\mathbf{R}')$ contains all the interactions and $p_{\mathbf{R}} = 1$ or 0 with probabilities $1-x$ or x. The free energy for this realization is $F = -\log Z(\{p_{\mathbf{R}}\})$, where $Z(\{p_{\mathbf{R}}\}) = \mathrm{Tr}_{\{\mathbf{S}_{\mathbf{R}}\}} \exp[-\beta\mathcal{H}(\{p_{\mathbf{R}}\}, \{\mathbf{S}_{\mathbf{R}}\})]$.

The quenched average free energy requires averaging of the free energy over the distribution of the $p_{\mathbf{R}}$'s,

$$-\beta\overline{F} = \overline{\log Z} = \overline{\lim_{m\to 0}[Z^m - 1]/m} = \overline{\lim_{m\to 0}[Z_m - 1]/m}, \tag{15}$$

where the overline denotes averaging over the $p_{\mathbf{R}}$'s, and

$$\overline{Z_m} = \overline{\mathrm{Tr}_{\{\mathbf{S_R}\}} \exp\Big[-\beta \sum_{i=1}^{m} \mathcal{H}(\{p_{\mathbf{R}}\}, \{\mathbf{S}^i_{\mathbf{R}}\})\Big]}. \tag{16}$$

The exponential on the RHS contains m replicas of the original Hamiltonian, each with its own spins $\mathbf{S}^i_{\mathbf{R}}$'s.

The average of the free energy over the quenched random variables is achieved using the identity[71,72]

$$\overline{e^{\bar{\mathcal{H}}}} = 1 + \overline{\bar{\mathcal{H}}} + \overline{\bar{\mathcal{H}}^2}/2 + \cdots = e^{\bar{\mathcal{H}}_{ran}}, \tag{17}$$

where

$$\bar{\mathcal{H}}_{ran} = \overline{\bar{\mathcal{H}}} + \overline{[\bar{\mathcal{H}}^2]_{cum}/2 + \cdots}, \tag{18}$$

and "cum" means the cumulant of the quenched average. Averaging over cumulants of products of the $p_{\mathbf{R}}$'s without the magnetic field we end up with

$$\bar{\mathcal{H}}_{ran} = -\sum_{i=1}^{m} \frac{1}{2}\int_{\mathbf{q}} U_2^{\alpha\beta}(\mathbf{q}) S^{\alpha i}_{\mathbf{q}} S^{\beta i}_{-\mathbf{q}} - \sum_{i,j=1}^{m} (v + u\delta_{ij}) \int_{\mathbf{q}}\int_{\mathbf{q}'}\int_{\mathbf{q}''} S^{\alpha i}_{\mathbf{q}} S^{\alpha i}_{,\mathbf{q}'} S^{\beta j}_{\mathbf{q}''} S^{\beta j}_{-\mathbf{q}-\mathbf{q}'-\mathbf{q}''}, \tag{19}$$

with

$$T^{MF}_{\alpha} \to (1-x)^2 T^{MF}_{\alpha}, \quad g_a \to (1-x) g_a, \quad r_d \to (1-x)^2 g_d,$$
$$B \to (1-x)B, \quad u \to (1-x)u \tag{20}$$

and — to leading order in x, $v = x\Delta^2$, where Δ^2 is the mean square deviation of the 'bare' transition temperature.[72] One can then follow Ref. 71, derive recursion relations for u and v, send m to zero and follow the three steps that led to Eq. (13). Generally, I have shown that T_c is lowered by fluctuations in all three types of spins.

The term with the magnetic field adds a quadratic coupling between the replicas,

$$\overline{\mathcal{H}}_{RF} = x(1-x)B^2 \sum_{\mathbf{R}} \sum_{i,j} S^{1i}_{\mathbf{R}} S^{1j}_{\mathbf{R}} + \cdots, \tag{21}$$

where the dots indicate higher orders. This is the same term generated by the random field Ising model, see Sec. 2.3, but for the non-ordering spin component S^1. This will further lower T_c, but the details still need detailed analysis. It should also be noted that, even without randomness, the off-diagonal terms in $\mathcal{H}$, e.g.

$\sum_{\mathbf{R},\mathbf{R}'} S^{\alpha}_{\mathbf{R}} S^{\beta}_{\mathbf{R}'}$, also generate effective-field-like effects, which may turn the problem onto the random field model (see Sec. 2.3).[73–75] As discussed in Sec. 2.3 below, this problem is still open, and may be responsible for the loss of ferromagnetic order at low temperatures in the preset case.

As stated, quantum fluctuations can be added for low temperatures by adding an imaginary time axis. At sufficiently low temperatures, these fluctuations imply mapping of d to $d+1$,[w] thus decreasing all the effects discussed above. This may explain the flattening of the curves $T_c(B, x)$ at low temperatures (observed experimentally.[60])

So far, we only emphasized the shift in the transition temperature. After eliminating all the fluctuations, the expected critical behavior should be that of the uniaxial dipolar system in a transverse field, with (at finite temperatures) logarithmic corrections to mean field theory. The simulations in Ref. 60 and the experiments do not yet seem to have reached this limit, so they probably observe effective exponents in the middle of the crossovers towards it.

Unfortunately, the editors of this book imposed a deadline, preventing me from finishing all these calculations. I invite everybody to do that, and — more importantly — extend these calculations to higher orders in ϵ or using other, more accurate methods to solve this RG project. Notably, this material seems to involve all the outstanding issues in critical phenomena, and it certainly deserves more research.

2.2. *Multicritical points*

As mentioned, the modern history of bicritical and tetracritical points, using the RG, started with Ref. 24, which followed an earlier mean field paper by Liu and Fisher.[76] The phase diagram of a system with two order parameters, having n_1 and n_2 components, respectively, contains two phases, in which these order parameters are (separately) non-zero. Fig. 6 exhibits three possible phase diagrams for the special case of the XXZ antiferromagnet (a Heisenberg $n = 3$ antiferromagnet, with an anisotropy which prefers ordering along the $Z-$axis.) In that example, the longitudinal magnetic field $H_{\parallel}$ modifies the spin anisotropy, causing the transition from the longitudinally ordered phase ($n_1 = 1$) to the 'flopped' phase ($n_2 = 2$). Experimentally and numerically, these phases are often separated by a first-order "flop" line, which ends at a *bicritical* point, Fig. 6(a). Alternatively, the two phases can be separated via second order transition lines by a "mixed" intermediate phase, and these lines meet the two transition lines between the ordered phases and the disordered phase at a *tetracritical* point, Fig. 6(b).[20,24] A third possibility, which

[w]This dimensionality shift is based on the relation $\tau \sim \xi^z$, where τ and ξ are the temporal and spacial correlation lengths. Here we follow Ref. 62 and adopt $z = 1$, as also found for the quantum one dimensional Ising model in a transverse field.[61] The role of fluctuations will decrease for any $z > 0$, e.g., $z = 2$ for $n > 1$.

will be the main topic of the present subsection, is shown in Fig. 6(c): the bicritical point of Fig. 6(a) is replaced by a *triple* point, at which three first order lines meet.[25] In this case, the transition from the disordered phase into the "mixed" phase on the "flop" line, along the 'isotropic' line $g = 0$, is also first order.

Before presenting the detailed RG arguments, here is a summary: At the multicritical point, and along the line $g = 0$ in Fig. 6, the symmetry between the n spin components is maintained. For a while, it was thought that this point is described by the isotropic (rotationally invariant) Heisenberg fixed point, implying a bicritical point. However, many modern accurate techniques show that for $n = d = 3$ this fixed point is slightly *unstable*, and the RG trajectories flow slowly either to a biconical fixed point, implying a tetracritical point, or towards a first order transition, implying a triple point. So far, neither the experiments nor the Monte Carlo simulations show either of these asymptotic predictions, hence the long-time puzzle. Below I present a novel recent way to calculate the RG trajectories, expanding around the isotropic fixed point. Since the experiments and simulations show an apparent bicritical phase diagram, we conclude that they follow the second scenario, but do not reach the triple point because of the slow flow, which requires a very large correlation length. However, close to the latter limit the effective exponents change significantly, and could be measured for checking the theory.

The nature of this multicritical point in $d = 3$ has been a matter of dispute for many years. The critical properties at this point are determined by the RG analysis along the line $g = 0$ in Fig. 6. This analysis starts with a general Ginzburg-Landau-Wilson Hamiltonian (in momentum space),

$$\overline{\mathcal{H}} = -\left[\frac{1}{2}\int_{\mathbf{q}}\sum_{\alpha}(r_\alpha + q^2)S^\alpha_{\mathbf{q}}S^\alpha_{-\mathbf{q}} + \sum_{\alpha\beta\gamma\delta} u_{\alpha\beta\gamma\delta}\int_{\mathbf{q}}\int_{\mathbf{q}'}\int_{\mathbf{q}''} S^\alpha_{\mathbf{q}}S^\beta_{\mathbf{q}'}S^\gamma_{\mathbf{q}''}S^\delta_{-\mathbf{q}-\mathbf{q}'-\mathbf{q}''}\right], \tag{22}$$

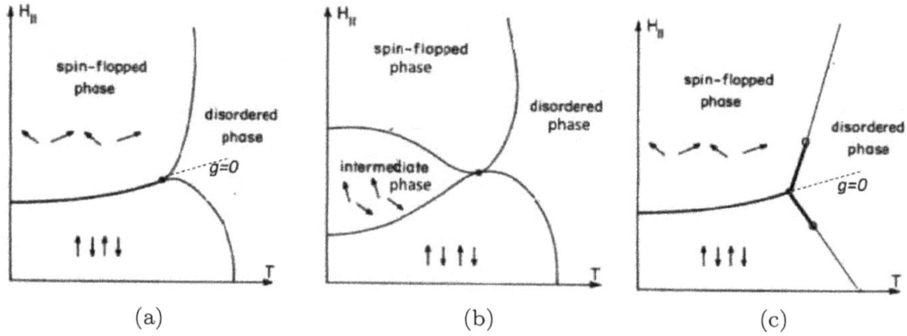

Fig. 6. Possible phase-diagrams for the XXZ antiferromagnet in a longitudinal magnetic field. (a) Bicritical phase diagram. (b) Tetracritical phase diagram. (c) Diagram with a triple point. Thick lines — first-order transitions. Thin lines — second-order transitions. The first-order transition lines between the ordered phases and the disordered paramagnetic phase end at tricritical points (small empty circles). After Refs. 20, 25.

where the integrals are over a spherical Brillouin zone of radius 1. Generalizing Eq. (4) to the case of $n_1 + n_2 = n$ components, we have

$$r_\alpha = r_\parallel = r - (n_2/n)g, \quad 1 \leq \alpha \leq n_1;$$
$$r_\alpha = r_\perp = r + (n_1/n)g, \quad n_1 + 1 \leq \alpha \leq n. \tag{23}$$

The quadratic anisotropy g causes the crossover from the multicritical point to the n_1- and n_2-component critical lines below and above it.[66] From now on we set $g = 0$.

All the previous work started at the Gaussian FP, $u^{*,G}_{\alpha\beta\gamma\delta} = 0$, and expanded the free energy in powers of the quartic coefficients $\{u\}$. The recursion relation $\partial u_{\alpha\beta\gamma\delta}/\partial\ell = \epsilon u_{\alpha\beta\gamma\delta} - O(u^2)$ then gave non-trivial FPs, with $u = O(\epsilon)$. Depending on details, one always found four FPs: the Gaussian one, the isotropic FP, at $u_{\alpha\beta\gamma\delta} \equiv u^{*,I(n)} = 8\pi^2\epsilon/(n+8)+O(\epsilon^2)$, and the decoupled FP, with $u_{\alpha\beta\gamma\delta} \equiv u^{*,I(n_1)}$ for $\alpha, \beta, \gamma, \delta \leq n_1$ and $u_{\alpha\beta\gamma\delta} \equiv u^{*,I(n_2)}$ for $\alpha, \beta, \gamma, \delta > n_1$. The fourth FP depends on the details: Kostelitz, Nelson and Fisher[24] found the biconical FP, and in the cubic case $u_{\alpha\alpha\beta\beta} \equiv u + v\delta_{\alpha\beta}$ Ref. 15 found the cubic FP. Only one of the four FP's was stable (in the RG sense: at criticality ($T = T_c$) there is a region in the $u-$space which flows under the RG iterations to that FP). The isotropic $n-$component FP was always stable (in the critical space spanned by the u's) for n below a threshold $n_c(d)$. Based on early low order $\epsilon-$expansions,[77] it was thought that $n_c(3) > 3$, and therefore the multicritical point is associated with the isotropic FP. However, as mentioned, higher order expansions (up to ϵ^6) and modern re-summations of these divergent (but Borel summable) series, as well as Monte Carlo simulations yielded $2.85 < n_c(3) < 3$. Therefore, the isotropic FP is unstable at $d = 3$: As n crosses $n_c(d)$, an operator which was irrelevant at the isotropic FP for $n < n_c(d)$ becomes relevant, and the RG trajectory related to it flows away. If this flow goes to an alternative, stable FP (biconical, cubic or decoupled), the transition at the multicritical point is continuous, and is then identified as a tetracritical point (at these points the minimum of the quartic Hamiltonian below the multicritical point has a mixed state[76]). Below we return to what happens when the flow does not reach a stable FP.

Until now, no experiments nor simulations observed any sign of a first order transition at the multicritical point. If indeed the transition is second order, it must be associated with one of the above stable FP's, hence the tetracritical scenario. This conclusion generated Ref. 40, commenting on a Monte Carlo study[78] of Zhang's $SO(5)$ model[79] for the high temperature superconductors, which included the $n_1 = 3-$component antiferromagnet and the $n_2 = 2-$component superconductor. In that case there exists an exact proof that the stable FP is the decoupled FP, and my comment wondered why the $SO(5)$ did not find that, and why they seemed to observe a bicritical point with exponents associated with the isotropic FP. Obviously, the late Shoucheng Zhang was not happy with that comment. However, I did end the comment saying: "All the above statements assume that the initial Hamiltonian is in the region of attraction of the decoupled FP. Alternatively, one should expect a first order transition... The apparent experimental observation of isotropic exponents may still indicate that the initial Hamiltonian is close to the isotropic FP." Indeed, the calculation below confirms these last comments: if one has a finite sample of linear size L then after a finite number of RG iterations $\ell = \ln L$ the RG flow stays near the isotropic FP, and does not reach the triple point. In that case one would observe an apparent bicritical phase diagram.

In many earlier papers, when the RG trajectories flowed away from an unstable point, without reaching any alternative stable FP, it has been generally expected that the renormalized effective free energy, which can be solved after all the fluctuations have been removed, has a first order transition. If the transition at the multicritical point (along the line $g = 0$ in Fig. 6) becomes first order, then (by continuity) the two transition lines from the disordered phase also become first order in its vicinity, and the multicritical point then becomes a *triple point*.[25] These first order regions then end at *tricritical* points, Fig. 6(c). However, the experiments and simulations have not yet exhibited such triple points. Also, there did not exist accurate estimates on how close to criticality should the triple point arise. Hence the "puzzle of the bicritical points", which I studied recently with my long-time colleague Ora Entin-Wohlman,[52] based on calculations with Andrey Kudlis,[50,51,80] who was a major player in the re-summed high order $\epsilon-$expansions.[46] Since these references give full details of our calculations, I give here only a brief summary.

Generally, the recursion relation near a FP for a parameter g_i has the form

$$\partial g_i/\partial \ell = \overline{\beta}_i[g_1,\ g_2, \ldots]. \tag{24}$$

In our case, the $\overline{\beta}-$functions are generated as sixth order expansions in the $u's$, and people solved them for the FP's. However, we need more that just the FP's. To follow the RG trajectories in the $u-$space we need the actual variation of $\overline{\beta}_i$ on its $\ell-$dependent arguments, and (like the $\epsilon-$expansion) the series in these arguments do not converge! For $n = 3$ we know that the isotropic and biconical (or cubic) FP's are close to each other, and therefore we decided to construct recursion relations in the vicinity of the isotropic FP, rather than near the Gaussian FP. Thus, we expanded $\overline{\beta}_i$ in powers of $\delta g_i = g_i - g_i^{*,I(3)}$ around the isotropic FP. The coefficients in these expansions are derivatives of $\overline{\beta}_i$ with respect to the various g_j's, at the isotropic FP, which are known series in the $g_i^{*,I(3)}$. The latter series can be written as series in ϵ, and can then be resummed to give good estimates of their accurate values at $d = 3$. Close to the isotropic FP it was sufficient to use only second order expansions in the deviations δg_i's; the resulting values of the cubic or of the biconical FPs were quite accurate.

We now turn to our main task: what happens if we start with a Hamiltonian out of the region of attraction of the stable (biconical or cubic) FP? Group theory[49,81–83] shows that there exist only three independent stability exponents near the isotropic

n–component FP. The quartic terms in Eq. (22) split into three subgroups of the rotationally invariant $O(n)$ group. For even functions in the spins, the quartic terms can always be written as

$$\mathcal{H}_4 = (u^{*,I(n)} + p_0)\mathcal{P}_0 + p_2\mathcal{P}_2 + p_4\mathcal{P}_4, \tag{25}$$

where the p_i's are scaling fields which flow under the RG iterations, with (for $d = 3$) stability exponents (agreed by all the extrapolations)[48]

$$\lambda_0 \approx -0.78, \quad \lambda_2 \approx -0.55, \quad \lambda_4 \approx 0.01. \tag{26}$$

The operator $\mathcal{P}_0 \equiv |\mathbf{S}|^4$ is the single fully isotropic quartic term, which characterizes the flow to the isotropic FP. The other operators break rotational symmetry. The operator $\mathcal{P}_2$ is a linear combination of operators which 'prefer' ordering of one of the competing order parameters, e.g., $\mathcal{P}_2 = |\mathbf{S}|^2(n_2|\mathbf{S}_{\parallel}|^2 - n_1|\mathbf{S}_{\perp}|^2)$. For $n = 3$ there are five such operators. As we said, the multicritical point requires 'isotropy', hence $g = p_2 = 0$. The remaining nine operators form $\mathcal{P}_4$, which (with no cubic terms, $v = 0$) has the form

$$\mathcal{P}_4 \equiv | \mathbf{S} |^4 \left[\frac{n_1 n_2}{(n+2)(n+4)} + x(1-x) - \frac{n_1(1-x) + n_2 x}{n+4}\right], \quad x = \frac{|\mathbf{S}_{\parallel}|^2}{|\mathbf{S}|^2}. \tag{27}$$

It is easy to express the original u's in terms of the three p's, and thus find the recursion relations for the p's.

Since we set $p_2 = 0$, we are left with the coupled recursion relations for p_0 and p_4. Since $|\lambda_0|$ is large, $p_0(\ell) = p_4(0)e^{\lambda_0 \ell}$ decays fast, and can be neglected after ℓ_1 iterations. Beyond that point, the quadratic recursion relation for p_4 can be solved analytically, yielding[50,80]

$$p_4(\ell) = \frac{p_4(\ell_1)e^{\lambda_4(\ell-\ell_1)}}{1 + Bp_4(\ell_1)\left[e^{\lambda_4(\ell-\ell_1)} - 1\right]/\lambda_4}, \tag{28}$$

where the biconical (or cubic) FP is at $p_4^{*,B} = \lambda_4/B > 0$. Since $\lambda_4 > 0$ is small, the initial flow is very slow, and the trajectory stays near the isotropic FP for many iterations. This may explain why Hu's Monte Carlo simulations of Zhang's $O(5)$ model gave an apparent bicritical diagram with isotropic exponents. For $p_4(0) > 0$, the trajectory goes to the biconical (or cubic) FP, but then one should immediately see tetracritical behavior.

The situation changes for $p_4(0) < 0$. In that case the trajectory again remains close to the isotropic FP, but at some point the denominator in (28) vanishes, and $|p_4|$ diverges. Although our quadratic approximation in the recursion relation may no longer be valid, it is clear that p_4 varies strongly before this divergence. In Refs. 50–52, 80 we show that if the iterations stop at ℓ_2, when $r(\ell) \sim 1$, then we can use mean field theory and identify when the transition becomes first order, ending us with the triple point of Fig. 6(c).

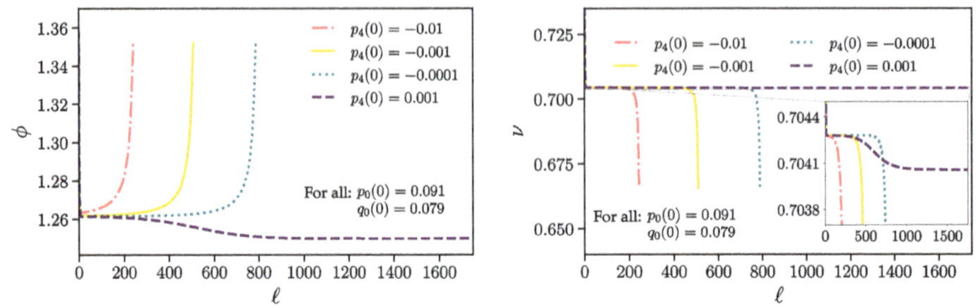

Fig. 7. Dependence of the effective critical exponents ν and ϕ on the RG flow parameter ℓ. Different lines correspond to different initial values $p_4(0)$, for a fixed $p_0(0)$. In this plot, the trajectories are shown only at $p_4(\ell) > -.8$, where our quadratic approximation is reasonable. From Ref. 80.

Once we have the $\ell-$dependent p's, we can also substitute them into other recursion relations, e.g. for r and g, and derive the exponents for the associated scaling fields, $t(\ell) = e^{\ell/\nu(\ell)}t(0)$ and $g(\ell) = e^{\ell\phi(\ell)/\nu(\ell)}g(0)$. Figure 7 shows these effective exponents, for the case $n = 1{+}2$ without cubic symmetry, and demonstrates their fast variation as the bicritical point approaches the triple point.

All these calculations should be extended to higher orders in the δg_i's. In our papers we also suggest how one may change the parameters, in order to approach the triple point and demonstrate Fig. 6(c) by experiments or simulations.

2.3. *Random fields Ising model (RFIM)*

Quenched random fields exist in many physical systems.[84–86] Experiments are difficult, due to equilibration problems,[87–89] but still very desired. Note that some of these experiments study the drastic effects of the random fields on bicritical phase diagrams (e.g., destroying XY order below the random field lower critical dimension $d_l = 4$). This model was first solved for the $n \to \infty$ limit (random sources and sinks of superfluid particles) by Lacour-Gayet and Toulouse,[90] who found the new upper and lower critical dimensions $d_u = 6$, $d_l = 4$. Imry and Ma[91] then gave qualitative arguments that showed that below a lower critical dimension d_l, random magnetic fields cause a breakdown of the system into domains with opposite spins, hence a destruction of the magnetic long range order. The size L of the domains is determined by the competition between the gain in bulk energy ($\pm h_0 L^{d/2}$, h_0 is the typical random field) and the loss in boundary energy (JL^{d-1} for $n = 1$ or JL^{d-2} for $n > 1$; J is the exchange energy). They found $d_l = 4$ for the rotationally invariant models (with $n \geq 2$) and $d_l = 2$ for the Ising case ($n = 1$). In both cases, the upper critical dimension was found to be $d_u = 6$. As experimental realizations, Imry and Ma suggested antiferromagnetic magnetic impurities in ferromagnets, or lattice distortions. Following Ref. 84, many experiments were done with uniform magnetic fields on dilute antiferromagnets.

At dimensions $d > d_l = 2$, there exists an ordered ferromagnetic phase below $T_c(B)$ for the dilute antiferromagnet in a field B, but the Imry-Ma domains still exist above the transition, with typical size $L \sim (h_0/J)^{2/(d-2)}$, representing metastable states which take a long time to equilibrate (via avalanches of flopping domains) for large h_0. This typical size decreases with h_0, eventually allowing long-range order. Indeed, Ref. 92 identified a temperature $T_i(B) > T_c(B)$, below which these domains cause slow dynamics. This may explain the difficulties of experiments to equilibrate.[93] As far as I know, there is not yet a complete consistent theory to describe this scenario.

Looking at the diagrammatic expansions in the quartic spin terms and in the random fields, it was shown that each of the leading diagrams in the random fields model in d dimensions maps exactly onto a similar diagram for the non-random model in $d-2$ dimensions.[55,94] It was thus concluded that to all orders, the ϵ−expansion of the random field model in $\epsilon = 6 - d$ is the same as for the non-random model with $\epsilon = 4 - d$.[55] This dimensionality shift by 2 was later found to be an exact result of supersymmetry.[56] Indeed, this dimensionality shift is consistent with the Imry-Ma argument for $n \geq 2$, where the lower critical dimension without the random fields is equal to 2.

However, the rule $d \to d-2$ does not agree with the Ising case, where Imry and Ma predicted $d_l = 2 \neq 1 + 2$ (1 is the lower critical dimension for the Ising model without the random fields). Also, Imbrie[95] and Brickmont and Kupiainen[96] showed that — unlike the non-random Ising model at $d = 1$ — there is long range order for the RFIM at $d = 3$ $(= 1+2)$. This disagreement with dimensionality shift by 2 has been a source of much research in the last 45 years, suggesting a variety of explanations. Some of these are reviewed below, but may questions still remain open for future research.

In addition to the problem with the dimensionality shift, there were two other (related) questions: (1) Do all the random field distributions belong to the same universality class? and (2) Is the phase transition of the RFIM always continuous? The literature studied several distributions of the site random fields, e.g., the Gaussian, $P(h_i) \sim \exp[-(h_i/h_0)^2/2]$ and the bimodal, $P(h_i) \sim [\delta(h_i+h_0)+\delta(h_i-h_0)]/2$. Mean field theory[97,98] showed that the transition can be continuous at all temperatures for a distribution with a maximum at $h_i = 0$, like the Gaussian one, but it must become of first order at low temperatures (and high random fields) for distributions with a minimum at $h_i = 0$, like the bimodal one.

To test the issue of the dimensionality shift for the RFIM, in 1993 we returned to the old and good method of high temperature series.[99] As reviewed in this book by Singh, Michael Fisher played a crucial role in the development of this method (see also his list of scientific achievements at the end of this book). He is also acknowledged in this paper.[x] Series for the susceptibility were developed for

[x]The other acknowledgement was to Brooks Harris, with whom I have been fortunate to publish many papers, both on series and on the RG.

$d = 2, 3, .., 8$. The results confirmed mean field exponents for $d \geq 6$ (logarithmic corrections could not be seen), and it was difficult to observe transitions at $d = 2$ and $d = 3$. It was emphasized that when the random fields are switched on, one has a crossover from the non-random to the RFIM behavior (without random exchange, the crossover exponent is equal to the non-random susceptibility exponent γ.[100] Random exchange causes small corrections). At large random fields and low temperatures the analysis is erratic, probably hinting at a first order transition, and this leaves only a intermediate range of fields for extracting the RFIM exponents. At $d = 5$ the result was $\gamma \approx 1.18$, but it was difficult to give error bars. This value is smaller than the $d = 3 = 5 - 2$ non-random $\gamma(d = 3) \approx 1.241$, but the uncertainties led the authors to say that dimensionality shift may hold at $d = 5$. Results were not clear at $d = 3$.

To clarify this we produced much better series in 1996.[101] After stating that "series expansions have a great advantage over Monte Carlo simulations, since (like the real experiments) they do not suffer from metastability and equilibration problems" [see Michael Fisher's comments on Monte Carlo simulations in Domany's paper, Chapter 6 of this book[y]], Ref. 101 presented new high temperature series for the RFIM in dimensions $d = 3, 4, 5, 8$. Results at $d = 8$ gave mean-field exponents, as expected for $d_u = 6$. Results at $d < 6$ confirmed the two-exponent (and not three-exponent, see below) scenario, but the numerical values of the exponents and their error bars did not seem consistent with the $d \to d - 2$ rule. For example, in $d = 4,\ 5$ we found the susceptibility exponent for *both* field distributions in the ranges $\gamma = 1.45 \pm 0.05,\ 1.13 \pm 0.02$, distinct from the non-random values $\gamma = 1.75,\ \sim 1.241$ for the non-random Ising model at $d = 2,\ 3$. In addition, the range of temperature for which the series were stable was shorter for the bimodal distribution, which may have hinted at an approach to a tricritical point and a first order transition at low temperatures (presumably including $T = 0$). We conclude that dimensionality shift indeed breaks down somewhere near $d \sim 5$, and that the critical exponents seem to come from another, supersymmetry breaking, fixed point.

Many theories tried to generalize the $d \to d - 2$ rule in the hyperscaling relations (e.g., $d\nu = 2 - \alpha$) by a new rule, $d \to d - \theta$, breaking supersymmetry and the dimensionality shift. This procedure introduced a new independent exponent θ (presumably related to the dimensionality of the roughened domain boundaries[103]). These models thus require three (instead of the usual two) independent critical exponents. Some of these models, which involve zero temperature fixed points, were reviewed e.g. in Ref. 101. The non-random fixed point associated with the transition as function of T at $T_c(0)$, is strongly unstable against the random fields, and the RG trajectories flow from that fixed point to the RFIM fixed point, associated with the transition as function of h_0 at $T = 0$, see RHS of Fig. 8. Bray and Moore[104] use simple arguments to identify the recursion relations near that fixed point, and argue

[y] However, Fisher withdrew that "slip of the tongue" in 1999.[102]

that it is stable with respect to the temperature. Therefore, thermal fluctuations are irrelevant, and it is sufficient to find the ground state as function of the random fields at $T = 0$. Such a scenario would replace the supersymmetric fixed point at low dimensions. However, it remained to find how the supersymmetric fixed point turns into this new fixed point. Also, this scenario, which assumes a continuous transition, cannot be valid for the binary field distributions, which should yield a first order transition (as predicted by mean field theory; fluctuations may narrow the range of such a transition, but not eliminate it completely). As explained in Ref. 97, for the binary distributions there must exist an ustable tricritical point on the critical line of the RFIM, between the non-random fixed point at $T_c(h = 0)$ and the one at $T = 0$. This would require a stable RFIM fixed point at a finite temperature between the non-random fixed point and the tricritical fixed point, as plotted on the LHS of Fig. 8. This phase diagram differs significantly from that of Bray and Moore. at temperatures below the tricritical fixed point, the RG trajectories would have to flow to a zero-temperature fixed point which represents a first order transition,[105] which is presumably different from the Bray-Moore RFIM fixed point. The apparent universality found in both series and Monte Carlo must imply that all distributions are described by this finite$-T$ RFIM fixed point, and not by the zero temperature fixed point. Although some candidates for this fixed point have been proposed (see below), the full details of such an RG scenario have not yet been investigated.

Some of the above questions were already addressed in 1997 by Swift *et al.*[106] They used a finite size scaling analysis of the exact ground states of small RFIM

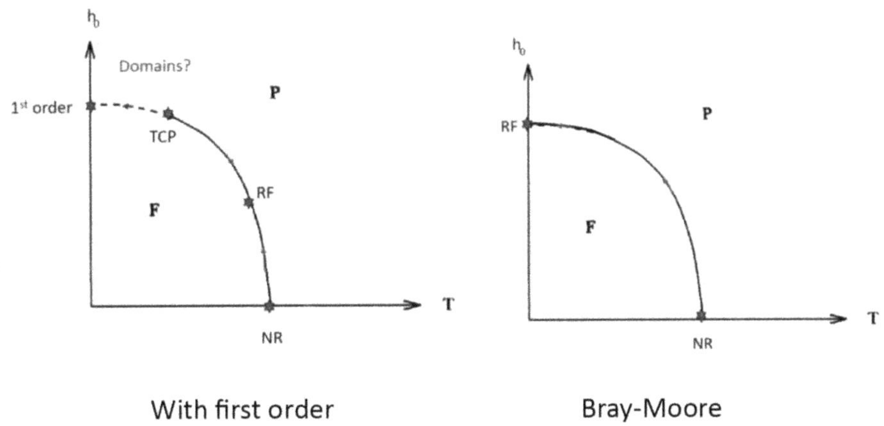

Fig. 8. Schematic phase diagrams in the temperature-random field $T - h_0$ plane for $d = 3$. Right: The random field fixed point is at zero temperature. Left: conjectured diagram for the binary and similar field distribution. P=paramagnetic (but with Imry-Ma domains), F=ferromagnetic (or any other ordered Ising phase). NR=non-random, RF=random field, TCP=tricritical point. h_0^2 is the mean square average of the field distribution. The blue arrows indicate the renormalization flow at criticality.

samples in three and four dimensions, and found a difference between the Gaussian and the bimodal field distributions in four dimensions, with the latter apparently having a discontinuous jump in the magnetization. The two distributions generated very different distributions of the magnetization. Although their analysis was restricted to zero temperature, they did not rule out a possible RFIM fixed point at a finite temperature. A similar study at $T = 0$ by Sourlas,[107] at $d = 3$, indicated that the Gaussian and the bimodal distributions do not belong to the same universality class. He noticed important differences between individual realizations (which showed discontinuities) and their averages, but still concluded that all the distributions give continuous transitions.

The last point raises the important issue of self-averaging. Away from criticality, the central limit theorem predicts that the width of the distribution of any measured quantity decays with system size L as $(\xi/L)^d$, where ξ is the correlation length. However, near a random (exchange or field) fixed point this width approaches a constant.[108] In particular, for the RFIM fixed point near six dimensions this width is proportional to $(h_0^2 u)^*$ at the random field fixed point, and independent of L, where h_0^2 is the second cumulant of the field distribution and u is the coefficient of the non-random quadratic spin term.[108] Since for large L this width becomes finite and independent of L, simulations must be very careful with averages over small numbers of realizations. Except for a few examples (below), I am not aware of systematic studies of self-averaging for the general RFIM problem. Such studies might explain the different results found by different groups.

One exception is the 2002 work of Parisi and Sourlas.[109] They studied the correlation functions of the three dimensional RFIM, and found that the correlation length (at finite L) is not self-averaging. They blamed this result on the nonperturbative formation of a bound state in the underlying replicated field theory. This nonperturbative result may replace the supersymmetric fixed point at low dimensions, but it is not clear what are the predictions for the critical behavior.

Assuming the Bray-Moore scenario, of a continuous transition at all T, Ref. 110 performed Monte Carlo simulations at zero temperature. Apparently, such simulations avoid some of the difficulties in previous simulations. At $d > 5$ these simulations agreed with the supersymmetry predictions. However, for $d < 5$ they found deviations from supersymmetry. Like our series results, these authors also found the same exponents from both distributions. However, they found the zero temperature transition continuous for all the distributions — in contrast to the mean field theory for the bimodal distributions. As stated, such distributions require more research. A possible way out may come from the effects of random quenched fluctuations on first order transitions. Indeed, Imry and Wortis[112] presented a domain argument similar to that of Imry and Ma, to show that randomness (e.g., random interactions or random fields) can turn a first order transition into a continuous one (the system breaks into domains of the two competing phases). This argument was supported by various other arguments.[113,114] However, all of these arguments still required some finite temperature below which the argument fails, and there should still be a first order transition at low temperature. Hopefully, the above story will stimulate theoreticians (and experimentalists!) to look into this. These authors also checked self-averaging,[111] and found that the widths of the susceptibility distribution approaches finite $L-$independent values, as predicted in Ref. 108. However, they find different values for different distributions, which may indicate different universality classes.

The absence of self-averaging implied that the average description may differ significantly from the behavior of a single realization. Interestingly, Kumar *et al.*[115] simulated the zero temperature configurations of RFIM with a uniform ordering magnetic field at $d = 3$, for specific field realizations. The domain shapes seemed to depend on the field distributions. However, they measured the dependence of the magnetization M on the uniform field H, and identified the universal (distribution-independent) critical exponents β and δ [$M \sim (T - T_c)^\beta$ at $H = 0$, $M \sim H^{1/\delta}$ at $T = T_c$], finding $\beta = .025 \pm .005$ and $\delta = 72 \pm 4$. Remembering that a first order transition would have $\beta = 0$, $\delta = \infty$, I wonder whether this allows for the phase diagram on the LHS of Fig. 8, alas for specific realizations.[116] It is not clear yet how to relate such realizations with the averages used in all the theories.

Apart from the above numerics, there have also been some analytic calculations, within the field theory and the Wilson-Fisher renormalization group approach. In 1998, Brézin and De Dominicis[117] noted that the breakdown of supersymmetry should arise due to additional operators, which break supersymmetry, to the Parisi-Sourlas Lagrangian.[118] If the supersymmetric fixed point becomes unstable with respect to such operators, this would explain the breaking of supersymmetry. However, they found a singular dependence on the replica number n, ending up with a runaway from the supersymmetric fixed point, without any other, stable, fixed point.

In 2004, Tarjus and Tissier[119] started a long series of papers, in which they reformulated the functional nonperturbative RG in a superfield formalism, generalizing the Parisi-Sourlas supersymmetric theory. They concluded that the failure of dimensionality reduction and standard perturbation theory below $d \simeq 5.1$ is due to the nonanalytic nature of the zero-temperature 'cuspy' fixed point, which is associated with the existence of many metastable states and the avalanches between them. The cusps appear in the cumulants of the renormalized field distributions. Near the zero temperature fixed point, the temperature is a dangerous irrelevant field, related to the appearace of droplets, which cause rounding of the cusps.

Another exciting result came from a recent theoretical study, which combined nonperturbative conformal field theory with perturbative RG.[120] These authors also added interactions which break supersymmetry, but are irrelevant near the supersymmetic fixed point near $d = 6$. Examples include quartic spin terms, which may become relevat below $d = 4$. They found that two such operators become relevant below $d_c \approx 4.2 - 4.7$, indicating a crossover to a new, stable, fixed point, at which supersymmetry is broken. As these authors say, this situation is similar to that discussed in the previous subsection, where the $n = 3$ isotropic fixed point becomes stable at a dimension slightly below 3, replacing the biconical (or cubic) fixed point (which are stable at higher dimensions). Since this calculation uses the expansion near $d = 6$ down to the vicinity of $d = 4$, it still requires support from other approximations. In any case, how this new fixed point develops at lower dimensions, how it relates to the phase diagrams in Fig. 8, and does it disappear at the Imry-Ma $d_l = 2$, remain open questions.

In summary, here are some of the remaining open questions:

(a) Which of all the suggested fixed points replaces the supersymmetric fixed point, and breaks supersymmetry? Are essential singularities important in this context?
(b) Is that fixed point at zero or at a finite temperature?
(c) If the LHS of Fig. 8 is correct, what are the details of the tricritical point and of the first-order fixed point?
(d) How does that fixed point relate to the Imry-Ma scenario?
(e) Is that fixed point universal, for all field distributions?
(f) Does that fixed point confirm the breakdown of self-averaging? If so, how does it relate to observations for single field realizations?
(g) How can the slow dynamics and avalanches of domain flipping be incorporated into the renormalization group treatment?
(h) All the above was for symmetric field distributions, $P(h) = P(-h)$. There were claims that non-symmetric distributions, e.g., for a binary mixture in a porous media, may yield a different behavior,[121] but this issue deserves further research.

Surely, more work will discover may more questions. There is much more to do on the RFIM.

3. Conclusions

I hope that my story has convinced you that Michael E. Fisher was a great scientist, a great leader of the statistical physics community, and a great mentor and teacher. I owe him a lot.

I also hope that I convinced you that both the life of Michael Fisher and the golden jubilee of the $\epsilon-$expansion deserve a celebration. Although the life of Fisher ended, the $\epsilon-$expansion and other RG methods are still very active, as I hope I demonstrated by my few examples. There is much more to do!

Acknowledgements

I thank Ora Entin-Wohlman and Andrey Kudlis for collaboration, and Slava Rychkov and Nikos Fytas for very useful discussions. Eytan Domany and David Nelson kindly helped checking the history and identifying people in Fig. 2, Jacques Perk caught many misprints and added some references, and Leo Radzihovsky corrected some crucial errors in an earlier version. I apologize to the many people whose references were not included, only due to space and time constraints.

References

1. Current problems in statistical mechanics. E. Domany and D. Jasnow, *Physica A* **177**(1–3), (1991).
2. *J. Stat. Phys.* **110**, Special issue in honor of Michael E. Fisher's 70th birthday, (edited by Joel Lebowitz), Numbers 3–6 (March 2003).
3. 106th Rutgers Statistical Physics Conference, December 2011: https://cmsr.rutgers.edu/docman-lister/cmsr/events-cmsr/statistical-mechanics-conference/106/1070-smm106-finalprogram/file.
4. **Critical Phenomena, Proc. Stellenbosch, S. A. 1982**, F. J. W. Hahne, ed., (Springer, Berlin, 1983).
5. M. E. Fisher, *Condensed Matter Physics: Does Quantum Mechanics Matter?* Excursions in the Land of Statistical Physics, pp. 207–253 (World Scientific 2016). Reprinted from "Niels Bohr: Physics and the World," Eds. H. Feshbach, T. Matsui and A. Oleson (Harwood Academic Publisher, Chur, 1988) pp. 65–115.
6. K. G. Wilson and M. E. Fisher, Critical exponents in 3.99 dimensions. *Phys. Rev. Lett.* **28**, 240 (1972).
7. Michael E. Fisher, Renormalization group theory: Its basis and formulation in statistical physics. *Rev. Mod. Phys.* **70**, 653 (1998). See Also, Michael E. Fisher, Renormalization group theory, the epsilon expansion and Ken Wilson as I knew him. *Int. Jn Mod. Phys. B* **29**, 1530006 (2015). Selected reviews by Fisher appeared in **Excursions in the land of statistical physics** (World Scientific, 2017); see also Cyril Domb's paper there.
8. K. G. Wilson and J. Kogut, The renormalization group and the ϵ expansion. *Phys. Repts.* **12**, 75 (1974).
9. M. E. Fisher and A. Aharony, Dipolar interactions at ferromagnetic critical points. *Phys. Rev. Lett.* **30**, 559 (1973).
10. A. Aharony and M. E. Fisher, Critical behavior of magnets with dipolar interactions, I. Renormalization group near four dimensions. *Phys. Rev. B* **8**, 3323 (1973).
11. A. Aharony, Critical behavior of magnets with dipolar interactions, II. Feynman graph expansion for ferromagnets near 4 dimensions. *Phys. Rev. B* **8**, 3342 (1973).
12. A. Aharony, Critical behavior of magnets with dipolar interactions, III. Antiferromagnets. *Phys. Rev. B* **8**, 3349 (1973).
13. A. Aharony, Critical behavior of magnets with dipolar interactions, IV. Anisotropy. *Phys. Rev. B* **8**, 3358 (1973).
14. A. Aharony, Critical behavior of magnets with dipolar interactions, V. Uniaxial magnets in d dimensions. *Phys. Rev. B* **8**, 3363 (1973). (*Erratum Phys Rev. B* **9**, 3946 (1974)).

15. A. Aharony, Critical behavior of anisotropic cubic systems. *Phys. Rev.* **8**, 4270 (1973).
16. A. Aharony, Dependence of universal critical behavior on symmetry and range of interaction. In C. Domb and M. S. Green, (eds.), Phase Transitions and Critical Phenomena, Vol. 6, pp. 357–424 (Academic Press, NY, 1976).
17. A. Aharony and A. D. Bruce, Equation of state and scaling relations for isotropic ferromagnets with dipolar interactions. *Phys. Rev. B* **10**, 2973 (1974).
18. A. Aharony and G. Ahlers, Universal ratios among corrections to scaling amplitudes and effective critical exponents. *Phys. Rev. Lett.* **44**, 782 (1980).
19. G. Ahlers, A. Kornblit and H. J. Guggenheim, Logarithmic corrections to the Landau specific heat near the Curie temperature of the dipolar Ising ferromagnet $LiTbF_4$. *Phys. Rev. Lett.* **34**, 1227 (1975).
20. A. D. Bruce and A. Aharony, Coupled order parameters, symmetry breaking irrelevant scaling fields and tetracritical points. *Phys. Rev. B* **11**, 478 (1975).
21. A. Aharony, K. A. Müller and W. Berlinger, Trigonal-to-tetragonal transition in stressed $SrTiO_3$: A realization of the 3-state Potts model. *Phys. Rev. Lett.* **38**, 33 (1977). This paper was directly triggered by Ref. 22.
22. D. Mukamel, M. E. Fisher and E. Domany, Magnetization of cubic ferromagnets and the three-component potts model. *Phys. Rev. Lett.* **37**, 565 (1976).
23. The following 2 papers, describing theory and experiments, appeared back to back: A. Aharony and A. D. Bruce, Lifshitz point, critical and tricritical behavior in anisotropically-stressed Perovskites, *Phys. Rev. Lett.* **42**, 462 (1979); J. Y. Buzaré, J. C. Fayet, W. Berlinger and K. A. Müller, Tricritical behavior in uniaxially stressed $RbCaF_3$. *Phys. Rev. Lett.* **42**, 465 (1979).
24. J. M. Kosterlitz, D. R. Nelson and M. E. Fisher, Bicritical and tetracritical points in anisotropic antiferromagnetic systems. *Phys. Rev. B* **13**, 412 (1976).
25. E. Domany, D. Mukamel and M. E. Fisher, Destruction of first-order transitions by symmetry-breaking fields. *Phys. Rev. B* **15**, 5432 (1977).
26. E. Brézin, D. J. Wallace and Kenneth G. Wilson, Feynman-graph expansion for the equation of state near the critical point (Ising-like Case). *Phys. Rev. Lett.* **29**, 591 (1972).
27. M. E. Fisher and A. Aharony, Scaling function for critical scattering. *Phys. Rev. Lett.* **31**, 1238 (1973).
28. M. E. Fisher and A. Aharony, Scaling function for two point correlations, I. Epsilon expansion near 4 dimensions. *Phys. Rev. B* **10**, 2818 (1974).
29. A. Aharony, Scaling function for two-point correlations. II. Expansion to order $1/n$. *Phys. Rev. B* **10**, 2834 (1974). Interestingly, these results were rediscovered independently by Andrey Katanin, in 2021! After our correspondence, which followed the ArXiv version of his paper, he found a misprint in my paper, and the corrected results appeared: A. Katanin, Nonanalytic momentum dependence of spin susceptibility for Heisenberg magnets in the paramagnetic phase and its effect on critical exponents. *Phys. Rev. B* **103**, 054415 (2021).
30. M. E. Fisher and J. S. Langer, Resistive anomalies at magnetic critical points. *Phys. Rev. Lett.* **20**, 665 (1968).
31. V. Privman, P. C. Hohenberg and A. Aharony, Universal critical-point amplitude relations. In C. Domb and J. L. Lebowitz, (eds.), Phase Transitions and Critical Phenomena, Vol. 14, pp. 1–134, 364–367 (Academic, NY, 1991).
32. A. Aharony and M. E. Fisher, Universality in analytic corrections to scaling for planar Ising models. *Phys. Rev. Lett.* **45**, 679 (1980).
33. A. Aharony and M. E. Fisher, Nonlinear scaling fields and corrections to scaling near criticality. *Phys. Rev. B* **27**, 4394 (1983).

34. Y. Chan, A. J. Guttmann, B. G. Nickel and J. H. H. Perk, The Ising susceptibility scaling function. *J. Stat. Phys.* **145** 549 (2011) and references therein.
35. Michael E. Fisher, Phases and Phase Diagrams — Gibbs' Legacy Today. *Physica A* **163**, 15 (1990).
36. A. Aharony, R. J. Birgeneau, A. Coniglio, M. A. Kastner and H. E. Stanley, Magnetic phases and magnetic pairing in doped La_2CuO_4. *Phys. Rev. Lett.* **60**, 1330 (1988).
37. Proverbs **13**, 20.
38. A. Aharony and A. B. Harris, Absence of self-averaging and universal fluctuations in random systems near critical points. *Phys. Rev. Lett.* **77**, 3700 (1996).
39. P. B. Weichman and M. E. Fisher, Helium in Vycor, constrained randomness and the Harris criterion. *Phys. Rev. B* **34**, 7652 (1986); R. R. P. Singh and M. E. Fisher, Random coupling crossover in Ising ferromagnets. *Phys. Rev. B* **37**, 1980 (1988); R. P. Singh and Michael E. Fisher, Disordered systems which escape the bound $\nu \geq 2/d$. *Phys. Rev. Lett.* **60**, 548 (1988).
40. A. Aharony, Comment on "Bicritical and tetracritical phenomena and scaling properties of the SO(5) theory". *Phys. Rev. Lett.* **88**, 059703 (2002); A. Aharony, *Old and new results on multicritical points*, in Ref. 2, dedicated to Fisher's 70th birthday, p. 659.
41. Y. Gefen, B. B. Mandelbrot and A. Aharony, Critical phenomena on fractal lattices. *Phys. Rev. Lett.* **45**, 855 (1980).
42. Y. Gefen, A. Aharony, B. B. Mandelbrot and S. Kirkpatrick, Solvable fractal family, and its possible relation to the backbone at percolation. *Phys. Rev. Lett.* **47**, 1771 (1981).
43. D. Stauffer and A. Aharony, Introduction to Percolation Theory (Taylor and Francis, London, 1992); revised 2nd edition (1994).[z]
44. M. E. Fisher, Critical probabilities for cluster size and percolation problems. *J. Math. Phys.* **2**, 620 (1960).
45. M. E. Fisher, Fractal and nonfractal shapes in two-dimensional vesicles. *Physica D* **38**, 112 (1989); Proc. Conf. "Fractals in Physics," 1-4 October 1989, Vence, France, Eds. A. Aharony and J. Feder (North Holland Publ. Co., 1989).
46. E.g. L. T. Adzhemyan, E. V. Ivanova, M. V. Kompaniets, A. Kudlis and A. I. Sokolov, Six-loop ϵ expansion study of three-dimensional $n-$vector model with cubic anisotropy. *Nucl. Phys. B* **940**, 332 (2019) and references therein. See also chapter by Adzhemyan *et al.* in this book.
47. J. M. Carmona, A. Pelissato and E. Vicari, $N-$component Ginzburg-Landau Hamiltonians with cubic anisotropy: A six-loop study. *Phys. Rev. B* **61**, 15136 (2000) and references therein.
48. D. Poland, S. Rychkov and A. Vichi, The conformal bootstrap: Theory, numerical techniques, and applications. *Rev. Mod. Phys.* **91**, 015002 (2019); S. M. Chester, W. Landry, J. Liu, D. Poland, D. Simmons-Duffin, N. Su and A. Vichi, Bootstrapping Heisenberg magnets and their cubic anisotropy. *Phys. Rev. D* **104**, 105013 (2021) and references therein. See also chapter by Poland and Simmons-Duffin in this book.

[z]This book was written while Saddam Hussain sent missiles from Iraq to Israel. The universities were closed, and I had time to work in the bomb shelter. However, no journal editor allowed to acknowledge Hussain for helping my research — so here it is. At that time, Michael wrote to me about his experiences during the German Blitz bombing of London and invited me and my family to come and stay in safe Maryland. We did not come, and stayed alive, but remain grateful for the kind intention.

49. M. Hasenbusch and E. Vicari, Anisotropic perturbations in three-dimensional $O(N)-$symmetric vector models. *Phys. Rev. B* **84**, 125136 (2011); M. Hasenbusch, Cubic fixed point in three dimensions: Monte Carlo simulations of the φ^4 model on the simple cubic lattice. *Phys. Rev. B* **107**, 024409 (2023) and references therein.
50. A. Aharony, O. Entin-Wohlman and A. Kudlis, Different critical behaviors in perovskites with a structural phase transition from cubic-to-trigonal and cubic-to-tetragonal symmetry. *Phys. Rev. B* **105**, 104101 (2022).
51. A. Aharony, O. Entin-Wohlman and A. Kudlis, Bi- and tetracritical phase diagrams in three dimensions. *Low Temperature Phys.* **48**, 483 (2022) [Fiz. Niz. Temp. (Kharkhov) **48**, 542 (2022)].
52. A. Aharony and O. Entin-Wohlman, The puzzle of bicriticality in the XXZ antiferromagnet. *Phys. Rev. B* **106**, 094424 (2022).
53. S. Rychkov, Numerical conformal bootstrap: targets. Talk at the Simons Collaboration on the Nonperturbative Bootstrap Annual Meeting, Simons Foundation, New York, November 8–9 (2018). Recording is available: https://www.simonsfoundation.org/event/simons-collaboration-on-the-nonperturbative-bootstrap-annual-meeting-2018/ Also, Aleix Gimenez-Grau, Yu Nakayama, Slava Rychkov, *Scale without Conformal Invariance in Dipolar Ferromagnets*, arXiv:2309.02514.
54. E. Brézin and J. Zinn-Justin, Critical behavior of uniaxial systems with strong dipolar interactions. *Phys. Rev. B* **13**, 251 (1976).
55. A. Aharony, Y. Imry and S.-k. Ma, Lowering of dimensionality in phase transitions with random fields. *Phys. Rev. Lett.* **37**, 1364 (1976).
56. G. Parisi and N. Sourlas, Random magnetic fields, supersymmetry, and negative dimensions. *Phys. Rev. Lett.* **43**, 744 (1979).
57. A. Aharony and B. I. Halperin, Exact relations among amplitudes at critical points of marginal dimensionality. *Phys. Rev. Lett.* **35**, 1308 (1975).
58. A. Aharony, New singularities in the critical behavior of random Ising models at marginal dimensionalities. *Phys. Rev. B* **13**, 2092 (1976).
59. A. Ferdinand, A.-C. Probst, A. Michels, R. Birringer and S. N. Kaul, Critical behaviour of nanocrystalline gadolinium: Evidence for random uniaxial dipolar universality class. *J. Phys.: Condens. Matt.* **26**, 056003 (2014).
60. T. Dollberg, J. C. Andresen and M. Schechter, Effect of intrinsic quantum fluctuations on the phase diagram of anisotropic dipolar magnets. *Phys. Rev. B* **105**, L180413 (2022).
61. R. J. Elliott, P. Pfeuty and C. Wood, Ising model with a transverse field. *Phys. Rev. Lett.* **25**, 443 (1970).
62. M. Suzuki, Relationship between $d-$dimensional quantal spin systems and $(d+1)$-dimensional Ising systems. *Prog. Theor. Phys.* **56**, 1454 (1976).
63. S. Sachdev, Quantum Phase Transitions (Cambridge University Press, 2011).
64. D. Andelman and A. Aharony, Critical behavior with axially correlated random bonds. *Phys. Rev. B* **31**, 4305 (1985) and references therein.
65. Cubic symmetry is reflected by terms like $hq_\alpha^2\delta_{\alpha\beta}$ in Eq. (3).[10] In parallel to our work, Thomas Nattermann sat isolated in East Germany, and impressively found independently many of our dipolar results (produced in the supporting Cornell atmosphere). In particular, he concentrated on these cubic terms: e.g., T. Nattermann and S. Trimper, Critical behaviour and cubic anisotropy. *J. Phys. A: Math. Gen.* **8**, 2000 (1975). Here I ignore these terms for simplicity. They (as well as the terms with $q^\alpha q^\beta$ also seem to be irrelevant, but only at order ϵ^2. In those years Nattermann and I corresponded a lot by snail mail (this was before the electronic mail!), and met

many times after he escaped from East Germany and joined the university of Köln in West Germany.
66. M. E. Fisher and P. Pfeuty, Critical behavior of the anisotropic $n-$vector model. *Phys. Rev. B* **6**, 1889 (1972).
67. D. R. Nelson and E. Domany, Equations of state for bicritical points. I. Calculations in the disordered phase. *Phys. Rev. B* **13**, 236 (1976).
68. K. Reid, Y. Millev, M. Fähnle and H. Kronmüller, Phase transitions in ferromagnets with dipolar interactions and uniaxial anisotropy. *Phys. Rev. B* **51**, 15229 (1995).
69. S. N. Kaul, Critical behaviour of heisenberg ferromagnets with dipolar interactions and uniaxial anisotropy. *Lect. Notes Phys.* **678**, 11 (2005).
70. S. Grinstein and A. Luther, Application of the renormalization group to phase transitions in disordered systems. *Phys. Rev. B* **13**, 1329 (1976).
71. A. Aharony, Critical properties of random and constrained dipolar magnets. *Phys. Rev. B* **12**, 1049 (1975).
72. A. Aharony, Y. Imry and S. Ma, Comments on the critical behavior of random systems. *Phys. Rev. B* **13**, 466 (1976). This reference averaged over the local transition temperature, while here we average over the site occupation.
73. A. Aharony, Absence of ferromagnetic long range order in random isotropic dipolar magnets and in similar systems. *Solid State Comm.* **28**, 667 (1978).
74. M. Schechter, $LiHo_xY_{1-x}F_4$ as a random-field Ising ferromagnet. *Phys Rev. B* **77**, 020401(R) (2008).
75. D. M. Silevitch, D. Bitko, J. Brooke, S. Ghosh, G. Aeppli and T. F. Rosenbaum, A ferromagnet in a continuously tunable random field. *Nature* **448**, 567 (2007).
76. K. S. Liu and M. E. Fisher, Quantum lattice gas and the existence of a supersolid. *J. Low Temp. Phys.* **10**, 655 (1973).
77. I. J. Ketley and D. J. Wallace, A modified epsilon expansion for a Hamiltonian with cubic point-group symmetry. *J. Phys. A: Math. Nucl. Gen.* **6** 1667 (1973).
78. X. Hu, Bicritical and tetracritical phenomena and scaling properties of the SO(5) theory. *Phys. Rev. Lett.* **87**, 057004 (2001).
79. E. Demler, W. Hanke and S.-C. Zhang, $SO(5)$ theory of antiferromagnetism and superconductivity. *Rev. Mod. Phys.* **76**, 909 (2004).
80. A. Kudlis, A. Aharony and O. Entin-Wohlman, Effective exponents near bicritical points. ArXiv:2304.08265. *Eur. Phys. J. Spec. Top.* (https://doi.org/10.1140/epjs/s11734-023-00971-w).
81. F. J. Wegner, Critical exponents in isotropic spin systems. *Phys. Rev. B* **6**, 1891 (1972).
82. A. Codello, M. Safari, G. P. Vacca and O. Zanusso, Critical models with $n \leqq 4$ scalars in $d = 4 - \epsilon$. *Phys. Rev. D* **102**, 065017 (2020) and references therein.
83. P. Calabrese, A. Pelissetto and E. Vicari, Multicritical phenomena in $O(n_1) \bigoplus O(n_2)$-symmetric theories. *Phys. Rev. B* **67**, 054505 (2003).
84. S. Fishman and A. Aharony, Random field effects in disordered anisotropic antiferromagnets. *J. Phys. C: Solid State Phys.* **12**, L729 (1979).
85. J. L. Cardy, Random-field effects in site disordered Ising antiferromagnets. *Phys. Rev. B* **29**, 505 (1984).
86. P. G. de Gennes, Liquid-liquid demixing inside a rigid network. Qualitative features. *J. Phys. Chem.* **88**, 6469 (1984).
87. H. Rohrer, A. Aharony and S. Fishman, Critical and multicritical properties of random antiferromagnets. *J. Mag. Mag. Mat.* **15-18**, 396 (1980).

88. E.g., J. P. Hill, T. R. Thurston, R. W. Erwin, M. J. Ramstad and R. J. Birgeneau, *Transition to long-range order in the three-dimensional random-field Ising model. Phys. Rev. Lett.* **66**, 3281 (1991).
89. R. A. Cowley, A. Aharony, R. J. Birgeneau, R. A. Pelcovits, G. Shirane and T. R. Thurston, The bicritical phase diagram of two-dimensional antiferromagnets with and without random fields. *Zeitschrift für Physik B Condensed Matter* **93**, 5 (1993); R. J. Birgeneau, A. Aharony, R. A. Cowley, J. P. Hill, R. A. Pelcovits, G. Shirane and T. R. Thurston, Effects of random fields on bicritical phase diagrams in two and three dimensions. in Ref. 2, dedicated to Fisher's 70th birthday, p. 58.
90. P. Lacour-Gayet et G. Toulouse, Ideal Bose Einstein condensation and disorder effects. *J. Phys. France* **35**, 425 (1974).
91. Y. Imry and S.-k. Ma, Random-fields instability of the ordered state of continuous symmetry. *Phys. Rev. Lett.* **35**, 1399 (1975).
92. U. Nowak *et al.*, Dyamic state model for exchage bias. I. Thory. *Phys. Rev. B* **66**, 014430 (2002).
93. R. J. Birgeeau, Random fields ad phase transitions in model magnetic systems. *J. Mag. Mag. Mat.* **177-178**, 1 (1998).
94. See also P. A. Young, On the lowering of dimensionality in phase trasitions with random fields. *J Physics C* **10**, L257 (1977).
95. John Z. Imbrie, Lower critical dimension of the random-field Ising model. *Phys. Rev. Lett.* **53**, 1747 (1984).
96. J. Brickmont and A. Kupiainen, Lower critical dimension for the random-field Ising model. *Phys. Rev. Lett.* **59**, 1829 (1987).
97. A. Aharony, Tricritical points in systems with random fields. *Phys. Rev. B* **18**, 3318 (1978).
98. D. Andelman, First- and second-order phase transitions with random fields at low temperatures. *Phys. Rev. B* **27**, 3079 (1983).
99. Y. Shapir and A. Aharony, High-temperature series and exact relations for the Ising model in a random field. *J. Phys. C: Solid State Phys.* **15**, 1361 (1982).
100. Y. Shapir and A. Aharony, Rigorous relations for the cross-over behaviour due to random perturbations. *J. Phys. C: Solid State Phys.* **14**, L905 (1981).
101. M. Gofman, J. Adler, A. Aharony, A. B. Harris and M. Schwartz, Evidence for two exponent scaling in the random field Ising model. *Phys. Rev. Lett.* **71**, 1569 (1993); Critical behavior of the random-field Ising model. *Phys. Rev. B* **53**, 6362 (1996).
102. M. E. Fisher, Some views from forty years as a statistical mechanician. *Proc. STATPHYS20, Physica A* **263**, 554 (1999).
103. e.g., G. Grinstein and S.-k. Ma, Roughening and lower critical dimension in the random-field Ising model. *Phys. Rev. Lett.* **49**, 685 (1982); D. S. Fisher, Interface fluctuations in disordered systems: $5-\epsilon$ expansion and failure of dimensional reduction. *Phys. Rev. Lett.* **56**, 1964 (1986).
104. A. J. Bray and M. A. Moore, Scaling theory of the random-field Ising model. *J. Phys. C: Solid State Phys.* **18**, L927 (1985).
105. M. E. Fisher and A. N. Berker, Scaling for first-order transitions in thermodynamic and finite systems. *Phys. Rev. B* **26**, 2507 (1982).
106. M. R. Swift, A. J. Bray, A. Marita, M. Cieplak and J. R. Baavar, Scaling of the random-field Ising model at zero temperature. *EuroPhys. Lett.* **38**, 273 (1997).
107. N. Sourlas, Universality in random systems: the case of the 3D random field Ising model. *Computer Phys. Comm.* **121-122**, 183 (1999).

108. A. Aharony and A. B. Harris, Absece of self-averagig and uiversal fluctuatios i random systems near critical poits. *Phys. Rev. Lett.* **77**, 3700 (1996).
109. G. Parisi and N. Sourlas, Scale invariance in disordered systems: The example of the random-field Ising model. *Phys. Rev. Lett.* **89**, 257204 (2002).
110. Nikolaos G. Fytas and Víctor Martín-Mayor, Universality in the three-dimensional random-field ising model. *Phys. Rev. Lett.* **110**, 227201 (2013); Nikolaos G. Fytas, V. Martín-Mayor, M. Picco, and N. Sourlas, Phase transitions in disordered systems: The example of the random-field ising model in four dimensions. *Phys. Rev. Lett.* **116**, 227201 (2016); N. G. Fytas, V. Martín-Mayor, G. Parisi, M. Picco, and N. Sourlas, Evidence for supersymmetry in the random-field Ising model at $D = 5$. *Phys. Rev. Lett.* **122**, 240603 (2019).
111. Nikolaos G. Fytas and Víctor Martín-Mayor, Efficient numerical methods for the random-field Ising model: Finite-size scaling, reweighting extrapolation, and computation of response functions. *Phys. Rev. E* **93**, 063308 (2016).
112. Y. Imry and M. Wortis, Influence of quenched impurities on first-order phase transitions. *Phys. Rev. B* **19**, 3580 (1979).
113. M. Aizenman and J. Wehr, Rounding of first-order phase trasitions in systems with quenched disorder. *Phys. Rev. Lett.* **62**, 2503 (1989).
114. K. Hui and A. N. Berker, Random-field mechanism in random-bond multicritical systems. *Phys. Rev. Lett.* **62**, 2507 (1989).
115. M. Kumar, V. Banerjee and S. Puri, Random field Ising model in a uniform magnetic field: Ground states, pinned clusters and scaling laws. *Eur. Phys. J. E* **40**, 96 (2017).
116. Similar apparent first order trasitions for sigle realizations were also indicated by L. Hernández ad H. Ceva, Wang-Landau study of the critical behavior of the bimodal 3D random field Ising model. *Physica A* **387**, 2793 (2008) and by N. G. Fytas and A. Malakis, Phase diagram of the 3D bimodal random-field Ising model. *Eur. Phys. J. B* **61**, 111 (2008).
117. E. Brézin and C. De Dominicis, New phenomena in the random field Ising model. *Europhys. Lett.* **44**, 13 (1998).
118. In fact, such a possibility was already mentioned by G. Parisi, *An introduction to the statistical mechanics of amorphous systems*, in **Recent advances in field theory and statistical mechanics**, ed. J.-B. Zuber and R. Stora (Elsevier, 1984).
119. G. Tarjus and M. Tissier, Random-field Ising model and $O(N)$ models: theoretical description through the functional renormalization group. *Eur. Phys. J. B* **93**, 50 (2020) and references therein.
120. A. Kaviraj, S. Rychkov and E. Trevisani, Parisi-Sourlas supersymmetry in random field models. *Phys. Rev. Lett.* **129**, 045701 (2022); Random field Ising model and Parisi-Sourlas supersymmetry I. Supersymmetric CFT. *JHEP* **04**, 090 (2020); Random field Ising model and Parisi-Sourlas supersymmetry. Part II. Renormalization group. *J. High Energ. Phys.* **2021**, 219 (2021); S. Rychkov, Lectures on the Random Field Ising Model; From Parisi-Sourlas Supersymmetry to Dimensional Reduction, Springer 2023.
121. A. Maritan *et al.*, Orderig and phase trasitions in random-field Ising model. *Phys. Rev. Lett.* **67**, 1821 (1991).

https://doi.org/10.1142/9789811282386_0018

Chapter 18

The Yang–Lee Edge Singularity and Related Problems

J. Cardy
All Souls College, University of Oxford,
Oxford OX1 4AL, UK

The Yang–Lee edge singularity is a prototypical example of the application of renormalization-group ideas to critical behavior and one to which Michael Fisher made several important contributions. Moreover, it has connections to several other problems such as the statistics of branched polymers, and its scaling limit in two dimensions provides a simple example of integrable field theory. This chapter aims to give a pedagogical introduction to these matters, with a few new ideas thrown in.

1. Introduction

In 1952, T.D. Lee and C.N. Yang published two remarkable back-to-back papers in *The Physical Review.*[1,2,a] Exploiting the equivalence between the ferromagnetic Ising model in an external magnetic field h and a lattice gas with activity $z = e^{h/kT}$, they proved a theorem which states that, on any finite graph of degree N, the partition function Z, which is proportional to a polynomial in z of degree N, has its zeros only on the unit circle $|z| = 1$, corresponding to purely imaginary h. Moreover, since Z is a sum of positive terms for real h, these zeros are bounded away from the real axis. They argued that this should persist at sufficiently high temperatures in the thermodynamic limit $N \to \infty$ on suitably regular lattices with finite-range interactions and that the zeros should become dense along the imaginary axis for $|\mathrm{Im}\, h|$ greater than some critical value h_c. This limit point is known as the Yang–Lee edge. They argued that mechanism for the free energy per site $\lim_{N\to\infty}(-N^{-1}) \log Z$ in zero field becoming singular as T approaches the critical temperature T_c from above is that the Yang–Lee edge approaches the real axis, i.e., $h_c \to 0$.

[a]Since the author order is reversed between the two papers, in the literature, this work is referred to interchangeably as the Yang–Lee theory or the Lee–Yang theory. In this chapter, I follow Fisher and use the first option.

At the time, this must have seemed a promising idea. Very little was understood theoretically about the critical behavior beyond the Curie–Weiss theory, Onsager's result for the free energy of the two-dimensional Ising model in zero field,[3] and Yang's expression for the spontaneous magnetization.[4] However, although locating Yang–Lee zeros remains a useful numerical tool for many varied problems, its analytic power has turned out to be less impactful than the understanding gained from the spectrum of the transfer matrix in integrable models. Compared with renormalization-group ideas, it gives little insight into the origin of anomalous scaling behavior.

Almost 20 years later, Kortman and Griffiths[6] argued, on the basis of series expansions, that the density of zeros has a power-law behavior near the edge, namely $\rho(h') \propto (h' - h_c)^\sigma$ as the imaginary part $h' \downarrow h_c$. They estimated $\sigma \approx -0.1$ for the square lattice and $\approx +0.1$ for the three-dimensional diamond lattice. Yang and Lee had already showed that $\sigma = -\frac{1}{2}$ for a one-dimensional model (see Section 2.2), while $\sigma = \frac{1}{2}$ in the mean-field theory.

However, it was Fisher[7] who recognized that the Yang–Lee edge singularity is a *bona fide* example of critical behavior, with a diverging correlation length and universal critical exponents which depend only on the dimensionality d of space. The only feature that makes it unusual is that, because of the imaginary magnetic field, the Boltzmann weights are no longer real and non-negative, so do not give a probability measure. This makes some properties, for example, the arguments for the existence of the thermodynamic limit, questionable. However, this has not stopped theoretical physicists from embracing it as one of the simplest examples of a critical point with a diverging correlation length and unleashing the arsenal of modern theoretical methods upon it. It is the aim of this chapter to describe some of these. It turns out that the problem is an ideal classroom for the understanding of some of the fundamental ideas of the last 50 years of theoretical statistical physics, many of which bear Michael Fisher's mark.

2. Yang–Lee Edge

2.1. *Yang–Lee theorem*

In this section, we give a brief account of the Yang–Lee theorem and its proof, since the original papers are not easy to follow for a modern reader, and some of the accounts given in the literature are overly complicated or incomplete. In fact, there are at least two different lines of approach, algebraic or analytic, with different generalizations. We give a simplified version of the latter due to Ruelle.[8]

Suppose we have a finite graph $\mathcal{G}$ with vertices labelled by $i = 1, \ldots, N$ and edges by (jk) with $j < k$. At each vertex i, there is an Ising spin s_i taking the values ± 1. The partition function is

$$Z^{\mathcal{G}}(h_1, \ldots, h_N) = \sum_{\{s_j\}} e^{\beta \sum_{jk} V_{jk} s_j s_k + \beta \sum_j h_j s_j}, \tag{1}$$

where we have allowed for a different external magnetic field h_j at each vertex. This may be mapped to a lattice gas by defining occupation numbers $n_j = \frac{1}{2}(s_j + 1)$ taking the values 0 or 1. Restricting $n \leq 1$ then models an on-site repulsion, while taking $V_{jk} \geq 0$ corresponds to a general attractive two-body potential (not necessarily nearest neighbor) for the lattice gas or a ferromagnetic interaction in the Ising model. In the gas picture, we may think of Z as being proportional to the grand partition function

$$\Xi^{\mathcal{G}}(z_1, \ldots, z_N) = \sum_{\{n_j=0,1\}} \prod_j z_j^{n_j} e^{-4\beta \sum_{jk} V_{jk}(n_j - n_k)^2}, \tag{2}$$

where $z_j = e^{2\beta h_j}$ is now interpreted as the local activity of the gas. We have written this in a form where it is a multinomial of degree one in each of its arguments z_j. Under the Ising symmetry $(s_j \to -s_j, h_j \to -h_j)$, which translates into $(n_j \to 1 - n_j, z_j \to z_j^{-1})$,

$$\Xi^{\mathcal{G}}(z_1, \ldots, z_N) = \left(\prod_{j=1}^{N} z_j \right) \Xi^{\mathcal{G}}(1/z_1, \ldots, 1/z_N). \tag{3}$$

Note also that, as long as V_{jk} is real,

$$\Xi^{\mathcal{G}}(\{z_j\}) = \overline{\Xi^{\mathcal{G}}}(\{\bar{z}_j\}). \tag{4}$$

The first part of the proof is to show the following.

Proposition 1. *$\Xi^{\mathcal{G}}(z_1, \ldots, z_N)$ is non-zero in $\cap_{j=1}^{N}\{|z_j| < 1\}$.*

Proof. The proof proceeds by induction on N. Suppose it is true for all graphs of degree $\leq N-1$. Considering the sum over n_1 in (2), the term with $n_1 = 0$ just gives the partition function on $\mathcal{G} \setminus \{1\}$, while that with $n_1 = 1$ modifies the activities at the remaining vertices of $\mathcal{G} \setminus \{1\}$. Explicitly, setting $a_{jk} \equiv e^{-\beta V_{jk}} \leq 1$,[b]

$$\begin{aligned} \Xi^{\mathcal{G}}(z_1, z_2, z_3, \ldots) &= \Xi^{\mathcal{G}\setminus\{1\}}(a_{12}z_2, a_{13}z_3, \ldots) \\ &\quad + z_1 a_{12} \ldots a_{1n} \Xi^{\mathcal{G}\setminus\{1\}}(z_2/a_{12}, z_3/a_{13}, \ldots). \end{aligned} \tag{5}$$

The first term is non-zero by hypothesis, and, using (3) and (4), the second may be written as

$$z_1 z_2 \ldots z_n \overline{\Xi^{\mathcal{G}\setminus\{1\}}}(a_{12}/\bar{z}_2, a_{13}/\bar{z}_3, \ldots), \tag{6}$$

[b] Although if V_{jk} is a nearest-neighbor interaction, the repulsive case $V_{jk} < 0$ on a bipartite lattice may be mapped to the ferromagnetic case by reversing the sign of s_j on one sublattice, this is special and the Yang–Lee theorem does not generally apply to repulsive interactions.

so we need to show that this is smaller in modulus than the first. A sufficient condition is

$$\sup_{|a_{1j}|\leq 1}\sup_{|z_j|<1}\left|\frac{z_1z_2\ldots z_n\Xi^{\mathcal{G}\setminus\{1\}}(a_{12}/\bar{z}_2,a_{13}/\bar{z}_3,\ldots)}{\Xi^{\mathcal{G}\setminus\{1\}}(a_{12}z_2,a_{13}z_3,\ldots)}\right|\leq 1. \tag{7}$$

By continuity, the left-hand side does not change if we strengthen the conditions $|a_{1j}|\leq 1$ to the open set $|a_{1j}|<1$. Then, by the maximum modulus principle, which states that the supremum is attained on the boundary, the left-hand side does not change if we replace the condition $|z_j|<1$ by $|z_j|=1$. However, with these modified conditions, the expression to be maximized is a constant 1, and the bound is satisfied trivially. The case $N=1$ is trivial, and, for $N=2$, $\Xi^{\mathcal{G}}(z_1,z_2)=1+a_{12}(z_1+z_2)+z_1z_2$, which vanishes when $z_2=-(1+a_{12}z_1)/(a_{12}+z_1)$. This maps $|z_1|<1$ into $|z_2|>1$, so the proposition holds. This completes the inductive proof. □

We note in passing that the inhomogeneous case has application to the problem of a lattice gas in a random medium, equivalent to the Ising model in a random field. However, for the present, we specialize to the case when all the z_j are equal to z, and $\Xi^{\mathcal{G}}$ is a polynomial of degree N in z. We then have the following.

Theorem 1 (Yang–Lee). *The zeros of the* $\Xi^{\mathcal{G}}(z)$ *all lie on the unit circle* $|z|=1$.

Proof. From Prop. 1 the interior of the unit circle is free of zeros. But from (3) the exterior of the unit circle is also zero-free. This completes the proof. □

Since $\Xi(z)$ is a polynomial of degree N, it is determined up to a constant by its zeros $\{e^{i\theta_j}\}$. If, for example, we are interested in the mean density (magnetization),

$$N\langle n\rangle = z\partial_z\log\Xi = z\partial_z\sum_{j=1}^{N}\log(1-e^{-i\theta_j}z)=\sum_{j=1}^{N}\frac{z}{z-e^{i\theta_j}}. \tag{8}$$

It is remarkable the density of Yang–Lee zeros contains in principle all the information about the equation of state of the Ising model in a magnetic field. As Fisher was fond of pointing out, all the existing approximations to the equation of state failed to take the Yang–Lee singularity into account. More recently, however, Xu and Zamolodchikov[5] have made progress in this direction.

2.2. *One dimension*

A nice feature of the Yang–Lee singularity is that in one dimension, it has nontrivial behavior and is easily solvable as an exercise. Suppose we have a lattice gas with nearest-neighbor attraction, so that

$$\Xi=\sum_{\{n_i=0,1\}}\prod_{j=0}^{L-1}z^{n_j}e^{-v(n-n_{j+1})^2}, \tag{9}$$

where, for convenience, we may impose periodic boundary conditions $n_L = n_0$. This may then be written as $\mathrm{Tr}\,\mathcal{T}^L$, where $\mathcal{T}$ is the transfer matrix

$$\mathcal{T} = \begin{pmatrix} 1 & w \\ zw & z \end{pmatrix}, \tag{10}$$

where $w = e^{-v} < 1$, with eigenvalues

$$\lambda_\pm = \frac{1}{2}[1 + z \pm \sqrt{(1-z)^2 + 4w^2 z}], \tag{11}$$

so that $\Xi(z) = \lambda_+^L + \lambda_-^L$. The zeros are where $\lambda_+/\lambda_- = e^{(2j+1)\pi i/L}$, with $j = 0, \ldots, L-1$. After some algebra, one finds their location to be $z = e^{i\theta_j}$, where

$$\cos\theta_j = (1 - w^2)\cos[(2j+1)\pi/L] - w^2. \tag{12}$$

We see that at high temperature ($w \to 1$), the zeros cluster around $\theta = \pi$, while at low temperatures ($w \to 0$), they are more or less uniformly spaced around the unit circle.

The Yang–Lee edge singularity is where the correlation length $\xi \sim \log(\lambda_+/\lambda_-)$ diverges, i.e., $\cos\theta_c = 1 - 2w^2$. Near the edge,

$$\theta_j - \theta_c \propto [(2j+1)\pi/L]^2, \tag{13}$$

so that in the thermodynamic limit, the density of zeros diverges as $(\theta - \theta_c)^{-1/2}$.

For general complex w, the locus of Yang–Lee singularities is

$$z = 1 - 2w^2 \pm 2w(w^2 - 1)^{1/2}. \tag{14}$$

In the repulsive case $w > 1$, therefore, there are singular points on the negative real z-axis, the one closer to the origin moving from $z = -1+$ to $z = 0-$ as w increases from $1+$ to ∞. This is an example of the *repulsive gas singularity*. Since, in this case, it was found by analytic continuation from the Yang–Lee edge singularity and has the same critical behavior. This, as we shall see, should hold more generally.

Although above we considered the simple case of the spin-$\frac{1}{2}$ Ising model, in one dimension the transfer matrix method may be applied to higher spin and also to models with continuous symmetries like $\mathrm{O}(n)$. Fisher[9] gave convincing arguments that the Yang–Lee singularity is always the result of the meeting and annihilation of the two largest eigenvalues and therefore has the same critical exponents.

3. Renormalization Group Analysis

We now come to Fisher's most celebrated contribution[7] to this subject: the realization that the scaling behavior of the Yang–Lee edge singularity is given by a scalar field theory with a ϕ^3 interaction and a purely imaginary coupling, and the application of the renormalization-group ideas, including the ϵ-expansion, that he and Wilson had developed for the Ising critical point (ϕ^4 theory) six years previously.[10] I shall describe his approach and then set this in the context of renormalized-field theory.

3.1. *Field theory formulation*

The most systematic way of arriving at a field theory from a lattice Ising model is through a Hubbard–Stratonovich transformation, which is simply a Gaussian integral. Starting from the partition function on a regular lattice, whose vertices are labeled by a d-dimensional vector r, we write

$$Z = \mathrm{Tr}_{\{s(r)=\pm 1\}} e^{\frac{1}{2}\sum_{r,r'} s(r)V(r-r')s(r')+h\sum_r s(r)} \tag{15}$$

$$\propto \mathrm{Tr}_{\{s(r)=\pm 1\}} \int [d\phi(r)] e^{-\frac{1}{2}\sum_{r,r'} \phi(r)V^{-1}(r-r')\phi(r')+\sum_r (h+\phi(r))s(r)} \tag{16}$$

$$= \int [d\phi(r)] e^{-\frac{1}{2}\sum_{r,r'} \phi(r)V^{-1}(r-r')\phi(r')+\sum_r \log\cosh(h+\phi(r))}, \tag{17}$$

where V^{-1} is the inverse of the matrix with rows and columns labeled by r and r' and elements $V(r-r')$. Shifting $\phi \to \phi - h$ removes the h-dependence in the last term and generates a linear coupling to ϕ. If h is purely imaginary, so is this coupling.

The inverse is easily taken in terms of the Fourier transforms V_q and ϕ_q. So far, this is exact, but now we start expanding in powers of q and the fluctuations about the minimum of the potential in ϕ. Denoting the shifted field by ψ, this leads to terms of $O(\psi^2, \psi^3, \ldots)$. In real space, the resulting effective free energy functional may be written as

$$\mathcal{H} = \int \left[\frac{1}{2}((\nabla\psi)^2 + r\psi^2) - iw\psi^3\right] d^d r. \tag{18}$$

Although the details of the dependence of r and w on the original couplings are tedious, they are analytic and only the general form of (18) is important. In Fourier space,[7]

$$\mathcal{H} = h\hat{\psi}_0 - \frac{1}{2}\int_q (r+q^2)\hat{\psi}_q\hat{\psi}_{-q} - iw\int_q\int_{q'} \hat{\psi}_q\hat{\psi}_{q'}\hat{\psi}_{-q-q'}. \tag{19}$$

In principle, the q integrals run over the first Brillouin zone. However, anticipating that rotational invariance emerges in the scaling limit near the critical point, this is replaced by $|q| < \Lambda = O(a^{-1})$, where a is the lattice spacing. The neglect of all the terms discarded in reaching this point is justified *a posteriori* by their irrelevance (coefficients flowing to zero) under the RG to be described.

3.2. *Momentum shell RG*

Wilson's momentum-shell RG[11,12] then involves (i) integrating out the wavelengths $b^{-1}\Lambda < |q| < \Lambda$ where $b > 1$ and (ii) rescaling $q' = bq$, so the cutoff is again $|q'| < \Lambda$. Consider first the rescaling. In order to keep the coefficient of $q^2 = 1$, we need to rescale $\psi_q = b^{d/2+1}\psi'_{q'}$, so then $h' = b^{d/2+1}h$, $r' = b^2 r$ and $w' = b^{\epsilon/2}w$, where $\epsilon = 6-d$.

This gives the upper critical dimension for this theory, above which the RG flows go into the Gaussian fixed point with $w = 0$, as $d_u = 6$, as compared with 4 for the ϕ^4 theory. Note that the ψ^3 theory has a peculiarity which is absent from, or rather unimportant for, the ϕ^4 case, in that, as long as $w \neq 0$, one may shift the field $\psi_{q=0}$ either to set the $r' = 0$ or $h' = 0$. This is an example of the presence of a *redundant* field, given by the variation of the free energy with respect to $\psi_{q=0}$. Such redundant fields do not usually affect the calculation, but here one is forced to make a choice. It turns out that maintaining the coefficient r or the less relevant field gives less singular, if mathematically equivalent, RG flows.

More importantly, it implies that the RG eigenvalues of h and r are not independent, as is the case with the Wilson–Fisher fixed point: the Yang–Lee edge singularity has only one independent critical exponent (apart from corrections to scaling from irrelevant fields), all others determined by scaling relations. If the RG eigenvalue of h is y_h, then standard scaling arguments[13] show that the correlation length ξ diverges as $h \to h_c$ like $|h - h_c|^{-\nu}$, where $\nu = 1/y_h$, the scaling dimension[c] of ψ is $x_h = d - y_h$, so the two-point function at the critical point decays as $|r|^{-(d-2+\eta)}$, where $d - 2 + \eta = 2x_h$, and the magnetization $M \sim \langle\psi\rangle \sim \xi^{-(d-y_h)} \sim |h - h_c|^{d-y_h)/y_h}$. In particular, this gives the density of Yang–Lee zeros behaving as $(h - h_c)^\sigma$ where $\sigma = (d - y_h)/y_h = (d - 2 + \eta)/(d + 2 - \eta)$. This is Fisher's scaling relation.[7]

To return to Fisher's actual calculation, the next step is the elimination of the degrees of freedom with $b^{-1}\Lambda < |q| < \Lambda$. This is most easily carried out perturbatively in w, and in the end it involves the same one-loop Feynman diagrams we shall encounter below in the renormalized field-theoretic approach, so we do not give details here. It was, however, Fisher's crucial observations that (a) these diagrams depend meromorphically on d, and (b) the emergence, for the ϕ^4 case, of a weak-coupling fixed point at a value of the coupling $O(4 - d)$. These were the crucial ingredient in he and Wilson's discovery of the ϵ-expansion as the first controlled approach to a RG calculation for a non-trivial fixed point.[10] In the 1978 paper[7] it must have been relatively straightforward for him to apply these methods to the ψ^3 theory, although by then other authors had considered field theories with cubic interactions.

3.3. *RG in renormalized-field theory*

The renormalized-field theory approach gives a systematic way of deriving the ϵ-expansion, which is perhaps less intuitive than the momentum shell approach, but is much more efficient at higher orders, especially when combined with minimal subtraction, to be described in the following.

It is worth separating the principle of the argument from the details of the calculation. In general, an interacting statistical field theory, formally defined by

[c]Modern approaches using conformal field theory regard the scaling dimensions x of scaling fields as the fundamental data, all other critical exponents being derived from these by scaling relations.

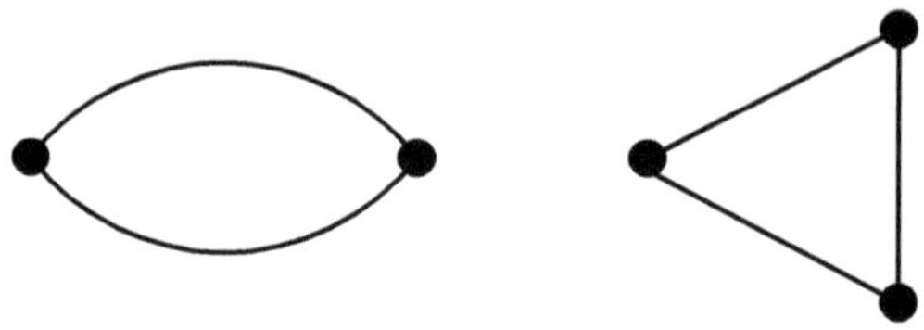

Fig. 1. one-loop primitively divergent graphs for $\Gamma^{(2)}$ and $\Gamma^{(3)}$ in ψ^3 theory.

a functional integral $\int[d\psi]e^{-\mathcal{H}[\psi]}$, is in fact ill defined even in perturbation theory due to ultraviolet divergences. The process of renormalization makes sense of the theory. We regard the integration field ψ and the parameters (r, w) in (18) as 'bare' quantities (ψ_0, r_0, w_0), The first step is to regulate the divergences by, for example, imposing a momentum cutoff $|q| < \Lambda$, or by dimensional regularization, that is continuing in d in such a way that all Feynman integrals converge, at least apart from special kinematic points. The next step is to identify by dimensional analysis which parts of which diagrams depend on the regulator. These are called the primitively divergent irreducible vertex parts. For the ψ^3 theory, these are shown in Fig. 1. The *statement of renormalizability*, at least for this ψ^3 theory, says that *if*, for $d \leq 6$, we do

- shift the bare field so that at each order $\langle\psi_R\rangle = 0$;
- *mass renormalization*: tune r_0 as a function of w_0 so that the renormalized mass vanishes, i.e., $\Gamma_R^{(2)}(q) = \langle\psi_q^R\psi_{-q}^R\rangle^{-1}$ vanishes at $q = 0$ (these first two conditions serve to put the theory on its critical manifold, and remove the worst, power-law, divergences);
- *field renormalization*: $\phi_0 = Z_\psi^{1/2}\psi_R$ such that $\partial_{q^2}\Gamma_R^{(2)}(q) = 1$, evaluated at some fixed but arbitrary normalization point $q^2 = \mu^2$; this cannot be at $q^2 = 0$ because of infrared divergences for $d \leq 6$;
- *coupling constant renormalization*: $w_R = Z_w w_0 = i\Gamma_R^{(3)}(q, q', -q-q')$ evaluated once again at an infrared finite normalization point dependent on μ^2;

then all the correlation functions

$$G_R^{(N)} = \langle\psi_R(r_1)\ldots\psi_R(r_N)\rangle = Z_\psi\langle\psi_0(r_1)\ldots\psi_0(r_N)\rangle, \tag{20}$$

when considered at a function of w_R, have a finite limit (at non-coincident points) as the regulator is removed, i.e., $\Lambda \to \infty$ or $d \to 6$. In the latter case, the normalization constants Z_ψ and Z_w have poles at $d = 6$ (as well as other values of d which are not important). In minimal subtraction, we retain only their divergent pieces, since the rest corresponds to some finite redefinition of the renormalized theory.

The above works only if $d \leq 6$: For $d < 6$, there are primitive divergences in only a finite number of subdiagrams, and the theory is *super-renormalizable*. However, there is then a proliferation of infrared divergences in the massless theory, i.e., at the critical point. The ϵ-expansion effectively resumes these into something sensible.

At $d = 6$, there are an infinite number of primitively divergent subgraphs, but they occur only in $\Gamma^{(2)}$ and $\Gamma^{(3)}$, where they are logarithmic. The theory is *exactly* renormalizable. For $d > 6$, divergences proliferate and the theory is *non-renormalizable*. However, the infrared behavior becomes trivial, that of the Gaussian fixed point, so $d = 6$ is called the upper critical dimension from the point of view of statistical physics.

Historically, in particle physics, the property of renormalizability was taken to be essential: The renormalized correlation functions are the physical ones. More recently, this point of view has altered: all field theories may be considered as *effective*, valid over some large but finite range of scales. However, since we want the predictions of the theory to be insensitive to the particular value of the regulator, this is more or less equivalent to demanding renormalizability.

In statistical physics, however, the physical quantities are the bare, cutoff correlation functions. Why then does renormalizability matter? From the insensitivity of the renormalized theory to the cutoff, we can say that

$$\Lambda\frac{\partial}{\partial\Lambda}\bigg|_{w_R} G_R^{(N)} = \Lambda\frac{\partial}{\partial\Lambda}\bigg|_{w_R}\left(Z_\psi^{-N/2}\, G_0^{(N)}\right) = 0. \tag{21}$$

Defining the bare dimensionless coupling, $g_0 = w_0\Lambda^{(6-d)/2}$ so that $G_0^{(N)}$ depends on g_0 and Λ, we can expand out (21) to get

$$\left[\Lambda\frac{\partial}{\partial\Lambda} + \beta(g_0)\frac{\partial}{\partial g_0} - \frac{N}{2}\gamma(g_0)\right] G_0^{(N)}(\{q_i\}) = 0, \tag{22}$$

where $\beta(g_0) = \Lambda(\partial/\partial\Lambda)|_{w_R} g_0$ and $\gamma(g_0) = \Lambda(\partial/\partial\Lambda)|_{w_R}\log Z_\psi$. The first term acts on the explicit Λ dependence of $G_0^{(N)}$ and the other terms on its implicit dependence on keeping w_R fixed.

These are examples of RG flow equations. In the language of Wilson's RG, if we send $\Lambda \to b^{-1}\Lambda$, with $b = 1 + \delta\ell$, then to keep the same long-distance physics, we have $dg_0/d\ell = -\beta(g_0)$.

It remains to compute $\beta(g_0)$ and $\gamma(g_0)$, at least to one-loop order. The first diagram in Fig. 1 gives

$$\Gamma_0^{(2)}(p) = p^2 + \frac{(6!g_0)^2}{2}\int^\Lambda \frac{d^6q/(2\pi)^6}{q^2(q-p)^2}, \tag{23}$$

where the denominator factor may be written, using Feynman parameters, $\int_0^1[xq^2 + (1-x)(q-p)^2]^{-2}dx$, which, on shifting the q integration, becomes $\int_0^1[q^2 + x(1-x)p^2]^{-2}dx$. In this form, it is straightforward to take the derivative with respect to p^2 to get an expression $\propto \log\Lambda$, so that $\gamma(g_0) = -6K_6g_0^2 + \cdots$, where K_6 is the area of a unit sphere in $d = 6$. Similarly,

$$\Gamma_0^{(3)}(p,p') = w_0 - w_0(6!g_0)^2\int^\Lambda \frac{d^6q/(2\pi)^6}{q^2(q-p)^2(q-p')^2}, \tag{24}$$

again $\propto \log \Lambda$. Multiplying by $Z_\psi^{-3/2}$ gives w_R. We then have to invert this to find g_0 as a function of w_R and Λ. Finally, we find

$$\beta(g_0) = -(\epsilon/2)g_0 + 56K_6 g_0^3 + \cdots . \tag{25}$$

The important feature of this is the appearance of a zero, corresponding to an RG fixed point, at $g_0^2 = g_0^{*2} = O(\epsilon)$. We see that the RG flows go toward this fixed point for $\epsilon = 6 - d > 0$. If we had taken the coupling constant to be real, this fixed point would not have been accessible. At the fixed point, $[\Lambda(\partial/\partial\Lambda) - \gamma^*]G_0^{(2)}(q) = 0$. To get physics out of this, we need simple dimensional analysis: $G_0^{(2)}(q) = q^{-2}$ times a function of $|q|/\Lambda$. Finally, we find that, at the fixed point, $G_0^{(2)}(q) \propto |q|^{-2+\eta}$, where $\eta = -\epsilon/9 + O(\epsilon^2)$. That η can be negative may happen in a non-unitary theory. The ϵ-expansion for the Yang–Lee universality class has been taken to five loops,[14] and, when suitably resummed, extrapolates all the way down to $d = 2$, where it agrees well with exactly known results, as well as more recent data in $d = 3$ from the conformal bootstrap.[15]

3.4. *Discussion*

The success of RG theory from the point of view of renormalized perturbative Euclidean quantum field theory led some theoreticians to assert that this is its natural home. Anyone who believes this to be the case is invited to read Fisher's masterly deconstruction of the idea in the published version of a talk he gave in 1996 to a group of quantum field theorists and philosophers of physics.[16] Among other things, he cites many of the successes of RG theory that are not simply encapsulated in renormalized-field theory, in particular the whole explanation of universality, which depends on the idea that RG flows actually take place in a large-dimensional space of possible coupling constants, not just one or two; crossover theory between nearby fixed points, a subject to which Fisher made many important contributions; other RG methods, such as Monte Carlo, functional RG, and so on, which often succeed where perturbative methods fail.

4. The Universal Repulsive Gas and Yang–Lee Singularities

We now turn to the problem of the repulsive gas singularity, and its relation to the Yang–Lee edge singularity, which Fisher and collaborators investigated in depth in two papers.[17,18]

Consider a classical gas with purely repulsive two-body interactions $V(r) \geq 0$. In that case, the Mayer function $e^{-V} - 1$ is negative, and it follows that the expansion of the logarithm of the grand partition function (the pressure times the volume), in powers of the activity z, alternates in sign with real coefficients. Since it is known to converge for small enough $|z|$, it follows that if it has a finite radius of convergence, this is given by a singularity on the negative real z-axis.

Moreover, this singularity appears to be universal in the sense of the RG, i.e., while its location is model-dependent, the critical exponents and other critical amplitude ratios are not. This was first proposed by Poland[19] on the basis of cluster expansions of lattice and continuum fluids and further confirmed by Baram and Luban.[20] The relevant exponent is defined by the singular part of the pressure $p(z)_{\text{sing}} \propto (z - z_c)^{\phi}$, where ϕ was estimated to take the values $\frac{1}{2}, \frac{5}{6}, \approx 1.06, \approx 1.2, \approx 1.3$ for $d = 1, 2, 3, 4, 5$, respectively. The first comes from the simple calculation outlined in Section 2.2 and the second from the highly non-trivial result from Baxter's solution of the hard hexagon model,[21] which agrees with extensive recent series analysis by Jensen[22] of hard squares.

Fisher and Lai[17] suggested that the origin of this universality lies in its equivalence to another problem which also exhibits a wide degree of universality: the problem of the Yang–Lee edge which Fisher had studied 17 years previously.[7] On this basis, they conjectured that the exponent $\phi(d)$ of the universal repulsive singularity is related to $\sigma(d)$, giving the behavior of the density of Yang–Lee zeros by $\phi(d) = \sigma(d) + 1$. The agreement, backed up by further numerical analysis of different models, was impressive (see Table 1 of their paper for an example of Fisher's thoroughness.)

However, in this paper, they were not able to devise an analytic argument for the correspondence between these problems. This gap was quickly filled in a paper with Park,[18] where they showed that both of these critical points appear in the same theory and are described by a ψ^3 field theory with purely imaginary coupling. Their arguments make for rather heavy algebra however, and in the following section, we give a more straightforward derivation.

4.1. *Unified field theory of the critical points*

In this section, we give a unified treatment of the Ising, Yang–Lee and repulsive critical points. We start from the Hubbard–Stratonovich representation for the partition function of the Ising model in a uniform magnetic field

$$Z = \int [d\phi] e^{\int [-\frac{1}{2}\phi \circ V^{-1} \circ \phi + \log\cosh(\phi+h)] d^d r}, \tag{26}$$

where V^{-1} is the positive-definite matrix inverse of the interaction, and $\circ$ denotes a convolution.

The first step in a perturbative RG analysis is to identify the critical points of the Landua–Ginzberg–Wilson action above. The first variation gives

$$V^{-1} \circ \phi = \tanh(\phi + h), \tag{27}$$

which of course is the mean-field equation, identifying ϕ as the internal magnetic field. Assuming translational invariance,

$$\phi = V_{q=0} \cdot \tanh(\phi + h). \tag{28}$$

At a critical point, the Hessian of second-order variations vanishes. In q-space, this is simple since it is diagonal, and assuming only ferromagnetic ordering, only the $q = 0$ components are important, and we get

$$1 = V_0 \operatorname{sech}^2(\phi + h) = V_0(1 - V_0^{-2}\phi^2). \tag{29}$$

Solving for ϕ, we find after a little algebra

$$V_0 = \cosh^2 u, \quad \phi = (1/2)\sinh(2u), \quad h = u - (1/2)\sinh(2u), \tag{30}$$

where $u = \phi + h$.

This gives a regular parametrization (uniformization) of the mean-field critical surface in terms of the complex parameter u which lives on the cylinder $|\mathrm{Im}\, u| \leq \pi/2$. The physical cases of interest arise where this manifold intersects the real V_0 axis:

- u purely imaginary, so $0 < V_0 < 1$, h pure imaginary: Ising model above critical temperature in purely imaginary magnetic field = Yang–Lee edge;
- $\mathrm{Im}\, u = \pm i\pi/2$, $V_0 < 0$, $z = e^{2h} < 0$ with two solutions, one with $|z| < 1$ = repulsive gas singularity;
- u real, $V_0 > 1$, $h \neq 0$ = the spinodal line in a metastable low-temperature phase;
- $u = 0$: the Ising critical point(in this context a multicritical point).

In all these cases, the cubic term in the fluctuations has a coefficient $2V_0^{-2}\phi$. However, as we move around the manifold, we need to rotate the contour in ϕ so that the V_0 term remains positive, thus $\phi \to V_0^{-1/2}\phi = \sinh u$. This is purely imaginary at the Yang–Lee and repulsive singularities, thus confirming their identification near their critical points. Elsewhere on the critical manifold, we need to choose the correct phase for h so as to generate a purely imaginary cubic coupling and flow to the Fisher fixed point. It vanishes as expected at the Ising critical point.

5. Branched Polymers and Dimensional Reduction

One of the most remarkable features of the Yang–Lee/universal repulsive singularity is that it also describes critical phenomena in higher dimensions, related to the counting of self-avoiding geometrical objects on the lattice or in the continuum. Specifically, these have the form of trees or, more physically, branched polymers, although, if loops are included, they may be argued to be irrelevant in the RG sense. That the counting of such objects in d dimensions should be related to the Yang–Lee singularity in $d - 2$ dimensions was first suggested by Parisi and Sourlas[23] on the basis of field-theoretic formulations of each problem. They observed an emergent supersymmetry in the leading diagrams, whereby two anticommuting coordinates canceled two of the usual commuting Euclidean coordinates and led to the aforesaid correspondence. There is ample numerical evidence for this, see, for example, Table 1 of Fisher and Lai.[17] However, it is based on perturbative RG ideas and as such could be questioned, especially when a similar earlier argument by Parisi and Sourlas[24] for the random-field Ising model fails in low enough dimensions.[25,d]

However, in 2001–2002, Brydges and Imbrie[27,28] devised a class of models for which the connection between the repulsive singularity and branched polymers in two higher dimensions is both rigorous and appealing in its physical description. Although their arguments ultimately appeal to supersymmetry, we shall give a simpler version based on Kirchhoff's matrix-tree theorem of 1847.[29]

Consider a simple classical gas in d dimensions with repulsive two-body interactions. The grand partition function is

$$\Xi(z) = \sum_{N=0}^{\infty} \frac{z^N}{N!} \prod_{i=1}^{N} \int d^d r_i \prod_{1 \leq j < k \leq N} e^{-V(r_j - r_k)} \sim e^{p(z)|\Omega|}, \tag{31}$$

where z is the activity, p is the pressure and $|\Omega|$ is the total volume, assumed large.

The cluster expansion proceeds by writing $e^{-V} = 1 + f$ and expanding in powers of the Mayer function f. This gives a sum over subgraphs which span the complete graph with N vertices, each edge of the subgraph carrying a factor $f(r_j - r_k)$. Taking the logarithm of this expansion gives a sum over connected diagrams, and the mean density $\langle n \rangle = z(d/dz)\log \Xi$ is given by a sum over *rooted* connected spanning subgraphs C of the complete graph K_N:

$$\langle n \rangle_d = \sum_{N=1}^{\infty} \frac{z^N}{(N-1)!} \sum_{C} \prod_{i=}^{N} \int d^d r_i \delta^d(r_1) \prod_{jk \in e(C)} f(r_j - r_k), \tag{32}$$

where $e(C)$ denotes the set of edges of C.

Now, suppose for simplicity that $f(r_j - r_k) = e^{-\alpha(r_j - r_k)^2}$ (a more general dependence may be written as a superposition of such terms). Then

$$\prod_{jk \in e(C)} f(r_j - r_k) = e^{-\sum_{\mu=1}^{d} r_j^\mu A_{jk} r_k^\mu}, \tag{33}$$

where $A_{jk} = A_{kj} = -\alpha$ if $(kj) \in e(C)$, and $A_{jj} = -\sum_{k \neq j} A_{jk}$, i.e., A is the laplacian on C. For a fixed C, the Gaussian integral over $r_2, \ldots, r_N$ gives

$$\pi^{(N-1)d/2}(\det A')^{-d/2} = \pi^{-(N-1)}(\det A') \times \pi^{(N-1)(d+2)/2}(\det A')^{-(d+2)/2}, \tag{34}$$

where A' is the matrix formed from A by striking out the first row and column. According to the matrix-tree theorem,[29] $\det A'$ can be written as a sum over rooted spanning trees T

$$\det A' = \sum_{T \subseteq C} \prod_{jk \in T} A_{jk}, \tag{35}$$

where in this case $A_{jk} = -\alpha$.

[d]Various explanations have been offered for this failure. However, it may be accounted for by observing that dimensional reduction applies only to the tree diagrams, which describe the long-range limit of the random-field Ising model.[26]

The second factor in (34) is the weight of the cluster diagram in $d+2$ dimensions. This may be written back in terms of $e^{-V(r_j-r_k)} = 1 + e^{-\alpha(r_j-r_k)^2}$. The outcome is that

$$\langle n(z)\rangle_d = z\sum_{N=1}^{\infty}\frac{(-\alpha z/\pi)^{N-1}}{(N-1)!}C_N\prod_i\int d^{d+2}r_i\delta^{d+2}(r_1)\prod_{1\leq j<k\leq N}(1+e^{-\alpha(r_j-r_k)^2}), \tag{36}$$

where $C_N = N^{N-1}$ is the number of rooted spanning trees on K_N. This may be viewed as the partition function, or generating function, for branched polymers in $d+2$ dimensions, with a repulsive weight for each pair of monomers, or vertices of the graph. More generally, we have[27]

$$\langle n(z)\rangle_d = z\sum_{N=1}^{\infty}\frac{(-z/\pi)^{N-1}}{(N-1)!}\prod_i\int d^{d+2}r_i\delta^{d+2}(r_1)\sum_T\prod_{jk\notin e(T)}P(r_{jk}^2)\prod_{jk\in e(T)}Q(r_{jk}^2), \tag{37}$$

where $P(r^2) = e^{-V(r)}$ and $Q(r^2) = P'(r^2)$. For a reasonable potential V, $P<1$ and $P\to 1$ as $r\to\infty$, corresponding to a short-range repulsive weight for pairs of monomers which are not neighbors on the tree, and $Q>0$ with $Q\to 0$ at infinity, giving a weight which is peaked at some value of r^2, corresponding to finite-distance attraction and short-distance steric repulsion between neighboring monomers.

Equation (37) may be integrated with respect to z to give

$$p(z)_d = -\pi\sum_{N=1}^{\infty}(-z/\pi)^N\mathcal{Z}(N)_{d+2}, \tag{38}$$

where now $\mathcal{Z}(N)_{d+2}$ is the partition function for unrooted branched polymers with N monomers. This is the main result of Brydges and Imbrie.[27] It may be extended to correlation functions.[28] The change of sign of the activity is important because it implies that the repulsive gas singularity at $z=-z_c$ controls the large N behavior of $\mathcal{Z}(N)_{d+2}$. Explicitly, if $p(z)_d$ has a singular piece $\propto (z+z_c)^{\phi(d)}$, then

$$\mathcal{Z}(N)_{d+2} \sim N^{-1-\phi(d)}(\pi z_c)^{-N}. \tag{39}$$

5.1. *Directed branched polymers*

Similar reasoning may be applied to the problem of directed branched polymers or trees, in which each edge is preferentially oriented in a particular spatial direction, labeled by a coordinate, say t. In this case, the correspondence is between the branched polymer problem and the relaxational dynamics of the repulsive gas. This was first conjectured using a field theory formulation,[30] and Dhar[31] showed an exact correspondence between a discrete dynamical model and hard squares and hexagons. Subsequently, Imbrie,[32] using methods similar to that in Ref. 27, constructed a model for which the correspondence is exact. Here, we present a more refined version of Ref. 30.

Let us begin with the sine-Gordon transformed version of the repulsive lattice gas partition function

$$\Xi = \int [d\phi] e^{-\mathcal{H}} = \int [d\phi] e^{-\frac{1}{2}\sum_{r,r'}\phi(r)V^{-1}(r,r')\phi(r')+z\sum_r e^{i\phi(r)}}, \tag{40}$$

where V^{-1} is the matrix inverse of the two-body potential. The derivation of this is similar to that of (17), except that we sum over occupation numbers $n(r) = 0, 1, 2, \ldots$, and the factor i necessary to make the gaussian integral converge.

Now, we take $\mathcal{H}$ as an effective Landau–Ginzberg–Wilson free energy, and consider its relaxational 'model A' dynamics,[e] defined by the stochastic equation[f]

$$\frac{\partial\phi(r)}{\partial t} = -\Gamma\frac{\partial\mathcal{H}}{\partial\phi(r)} + \eta(r,t), \tag{41}$$

where Γ is a frictional coefficient and η is a Brownian noise with covariance $\langle\eta(r,t)\eta(r',t')\rangle = 2D\delta_{r,r'}\delta(t-t')$. From the corresponding Fokker–Planck equation, one may show that as long as $D = \Gamma kT$, the joint probability distribution of $\{\phi(r)\}$ relaxes to a thermal one $\propto e^{-\mathcal{H}/kT}$.

To proceed, we use the response function formalism,[33] writing a delta function of (41) as

$$\int [d\bar{\phi}] e^{\int dt\sum_r \bar{\phi}(r,t)[\partial_t\phi(r,t)+\Gamma\partial\mathcal{H}/\partial\phi(r)-\eta(r,t)]}. \tag{42}$$

In this form, it is straightforward to average over the noise to get a term $D\bar{\phi}^2$ in the action. Physical observables are then defined by adding source terms.

In our case, the response functional is

$$S = \int dt \sum_r \left(\bar{\phi}(r,t)[\partial_t\phi(r,t) + \Gamma\sum_{r'} V^{-1}(r,r')\phi(r') - ize^{i\phi(r,t)}] - D\bar{\phi}(r,t)^2\right). \tag{43}$$

Let us now compare with a field theory derived from an explicit lattice model of directed branched polymers. We use a variant of the sine-Gordon transformation first deployed in Ref. 30. On each site (r,t) of a directed lattice introduce a commuting ring consisting of complex linear combinations of pseudospins $(a(r,t), \bar{a}(r,t))$, satisfying the algebra

$$a^2 = a, \quad \bar{a}^2 = 0, \tag{44}$$

---

[e] Although this comes from a fluid whose density is conserved and which therefore should satisfy model B dynamics,[33] the field ϕ is essentially the local chemical potential and is not conserved.

[f] This is the physicist's crude way of writing $d\phi_t(r) = \cdots + D^{1/2}dB_t(r)$, where $B_t(r)$ are independent Brownian motions.

together with a linear mapping Tr to the complex numbers defined by

$$\mathrm{Tr}\,1 = 1, \quad \mathrm{Tr}\,a = 0, \quad \mathrm{Tr}\,\bar{a} = 1, \quad \mathrm{Tr}\,\bar{a}a = 1. \tag{45}$$

Consider the expression

$$\mathrm{Tr}\,\bar{a}(0,0)\prod_{r,r'}\prod_{t'>t}\big(1 + xV(r,t;r',t')\bar{a}(r',t')a(r,t)\big). \tag{46}$$

Expanding this out in powers of x, each term is associated with a subgraph of the lattice, and using the above rules the operation Tr projects onto directed trees rooted at $(0,0)$, so gives the generating function for these. Since $\bar{a}^2 = 0$, this can be exponentiated and written as a gaussian integral

$$\exp\left(x\sum_{r,r',t,t'}V(r,t;r',t')\bar{a}(r',t')a(r,t)\right) = \int[d\bar{\phi}][d\phi]e^{\sum_{r,t}x^{-1}\bar{\phi}V^{-1}\phi + i\bar{a}\bar{\phi} + ia\phi}. \tag{47}$$

The operation Tr may now be performed to give a functional

$$S' = \int dt\sum_r\Big(\bar{\phi}(r,t)[\partial_t\phi(r,t) + \Gamma\sum_{r'}V^{-1}(r,r')\phi(r') + \log(1 + ix\bar{\phi}\,e^{i\phi})\Big), \tag{48}$$

where we have rescaled $\bar{\phi}$ and assumed that V is short-ranged in t, so we may make a derivative expansion in ∂_t. Now, expanding the logarithm and keeping relevant terms, we get (43) with negative activity $z = -x$.

Some of the critical exponents of directed branched polymers may now be inferred from the theory of dynamic critical behavior.[33] The Fourier transform $G(\omega, q)$ of the response function $\langle\phi(r,t)\bar{\phi}(0,0)\rangle$, which gives the local density of the branched polymer at (r,t), when evaluated at $\omega = 0$, should equal the static correlation function of the repulsive singularity. Thus, $G(0,0) \sim (x - x_c)^{-\gamma}$ and $\xi_\perp \sim (x - x_c)^{-\nu}$, where $\xi_\perp$ is the correlation length in the r directions, and γ, ν are the exponents of the repulsive gas singularity, or, equivalently, the Yang–Lee singularity, in d dimensions. By our earlier arguments in Section 3.2, $\nu = 1/y_h$ and $\gamma = (2y_h - d)/y_h$. By hyperscaling, the singularity in the free energy is $(x - x_c)^{-1+\sigma(d)}$, and therefore, the number of directed branched polymers of mass N in $d+1$ dimensions grows like $N^{1+\sigma(d)}x_c^{-N}$. Note that these arguments give no immediate information about the correlation length ξ_t, as this requires knowledge of the dynamic exponent of the Yang–Lee problem.

6. Exact Results in Two Dimensions

The scaling limit of the Yang–Lee/universal repulsive gas singularity at or near the critical point has been a fertile paradigm for theoretical investigations of two-dimensional critical behavior because of the following: (i) its simplicity as it has only one relevant scaling field, making it simpler in some ways than the Ising model, although it is not a free theory in any sector, as is the Ising model; (ii) it gives the simplest example of a non-unitary theory, albeit not exhibiting all the features of which more general such theories are capable. Unfortunately, we have space for only a brief description and refer the reader to the literature for more details.

6.1. *Conformal field theory*

At a rotationally invariant critical point with short-range interactions, scale invariance is usually enlarged to conformal symmetry, which may loosely be considered as invariance (or covariance, as Fisher always insisted upon saying) under non-uniform scaling. In two dimensions, it is very powerful, as any complex analytic function generates a locally conformal mapping of the complex plane and even in higher dimensions. Once the kinematic clutter was tidied away, the conformal bootstrap (described elsewhere in this volume[15]) has become the most effective numerically and analytically. The conformal field theory of the Yang–Lee edge singularity in two dimensions was identified[34] early on, shortly after Belavin, Polyakov and Zamolodchikov[35] formulated their classification of the so-called *minimal* models, for which the number of irrelevant fields grows less fast than generically, owing to an abundance of redundant fields, or null fields, which might be expected to occur in polynomial field theories like $i\psi^3$.

The minimal models are characterized by two co-prime positive integers (p, p'). The scaling fields are organized into representations of the Virasoro algebra, labeled by integers (r, s), within which the scaling dimensions are spaced by integers. The most relevant field $\phi_{r,s}$ in each representation has scaling dimension

$$x_{r,s} = \frac{(rp - sp')^2 - (p - p')^2}{2pp'} = x_{p'-r,p-s}, \tag{49}$$

where $1 \le r \le p' - 1$, $1 \le s \le p - 1$. This dual appearance of each field and the existence of null fields restricts the allowed structure of the operator product expansion to be

$$\phi_{r,s} \cdot \phi_{r',s'} = \sum_{|r-r'| \le r'' \le r+r'-1} \; \sum_{|s-s'| \le s'' \le s+s'-1} \phi_{r'',s''}. \tag{50}$$

Assuming that the Yang–Lee conformal field theory is a minimal model, it must satisfy the following: (a) there is only one relevant field, i.e., with $x_{r,s} < 2$ (apart from the identity $\phi_{1,1}$); (b) the three-point function is non-zero, i.e., the OPE $\phi_{r,s} \cdot \phi_{r,s}$ should contain $\phi_{r,s}$ itself or its image $\phi_{p'-r,p-s}$. It turns out that the latter can hold with $(r, s) = (1, 2)$, with

$$\phi_{1,2} \cdot \phi_{1,2} = \phi_{1,1} + \phi_{1,3} = \phi_{1,1} + \phi_{2-1,5-2}, \tag{51}$$

as long as $(p, p') = (2, 5)$. Using (49), this gives $x = -\frac{2}{5}$, and using Fisher's scaling relation, $\sigma = -\frac{1}{6}$, in agreement with Baxter's exact result $\phi = \frac{5}{6}$ for hard hexagons.[21]

This was the first successful application of CFT methods to a non-unitary theory. It predicts not only the leading exponent but also the full finite-size scaling spectrum when the theory is compactified on a cylinder. Moreover, since as was argued by Belavin *et al.*, the existence of a null field in the representation (r, s) implies that the correlators satisfy linear partial differential equations of order $r \cdot s$, the four-point function may be computed exactly (using crossing, which is a main ingredient of the conformal bootstrap) and the universal coefficient of the three-point function extracted, the result being

$$\langle \phi(r_1)\phi(r_2)\phi(r_3)\rangle = iC^{1/2}|r_{12}r_{23}r_{31}|^{2/5}, \tag{52}$$

where the normalization $\langle \phi(r_1)\phi(r_2)\rangle = |r_{12}|^{4/5}$ and

$$C = \frac{\Gamma(\frac{6}{5})^2\Gamma(\frac{1}{5})\Gamma(\frac{2}{5})}{\Gamma(\frac{3}{5})\Gamma(\frac{4}{5})^3}. \tag{53}$$

Note that the factor of i in the three-point function confirms the imaginary cubic coupling in the field theory. This has been verified numerically.[36] We quote it in full to show just how non-trivial these types of CFT result are.

Although the Yang–Lee CFT is non-unitary, its spectrum is real. This is a consequence of the symmetry of the associated quantum Hamiltonian under simultaneous time reversal $H \to H^*$ and $\phi \to -\phi$.

6.2. *The simplest S-matrix*

Just as the universal properties at a critical point are encoded in a conformal field theory, those of the neighboring scaling region are generally encoded in a massive field theory. To be precise, this scaling limit involves taking the short-distance cutoff, e.g., the lattice spacing, $a \to 0$, while keeping the correlation length ξ, as measured in laboratory units, fixed. Since in general $\xi \sim ag^{-1/y}$, where g is a relevant coupling with RG eigenvalue $y > 0$, this limit involves letting $g \to 0 \propto a^y$. The mass scale of the limiting field theory is then given by the inverse correlation length ξ^{-1}. Note that making an RG transformation $a \to ba$ simply rescales the constant of proportionality above. Thus, all points on the relevant outflow trajectory from the fixed point correspond to the *same* massive continuum theory, when expressed in terms of the physical masses.[g]

A massive field theory is characterized by its particle spectrum and S-matrix. In general, the spectrum may contain particles of different masses and quantum numbers, but, since the Yang–Lee theory has only one type of particle, we shall assume this. A single particle of mass m with momentum p^1 has energy $p^0 = \sqrt{m^2 + {p^1}^2}$. It is convenient to parametrize $(p^0, p^1) = (m\cosh\theta, m\sinh\theta)$ in terms of the rapidity θ.

[g]Fisher was very fond of pointing out what is often termed the so-called RG in renormalized-field theory, which ignores the infinitude of other couplings, is a tautology.

A convenient basis is then that of the asymptotic states $|\theta_1, \theta_2, \ldots\rangle$, which may be thought of as a multiparticle state before the interaction is switched on. This is turned on for some long but finite time, after which the system will be in some linear superposition of such states which, because of the quantum evolution being unitary, may be written $S^{-1}|\theta_1, \theta_2, \ldots\rangle$ where S is a unitary operator. As usual, scattering processes conserve total energy and momentum, which are integrals of components of the conserved energy–momentum tensor $T_{\mu\nu}$ over space. In a CFT, however, there are extra conservation laws, which are more easily understood in light-cone coordinates $x^{\pm} = x^0 \pm x^1$, where the conservation laws read

$$\partial_- T_{++} + \partial_+ T_{+-} = 0, \quad \partial_+ T_{--} + \partial_- T_{-+} = 0. \tag{54}$$

In a 2D CFT, $T_{+-} = T_{-+} = 0$, so T_{++} and T_{--} are separately conserved. Moreover, so are any powers T^p_{++} and T^p_{--}.

Although this is not generic, it may happen that when a relevant coupling g is switched on, at least some of these currents remain conserved. Zamolodchikov[37] argued that, for certain deformations of minimal CFTs, this is the case. The reasoning is simply that if one asks what can lie on the right-hand side of the equation $\partial_- T^p_{++} =?$, the only possibilities are derivatives ∂_+ of some other local field. His argument applies to the deformation of the Yang–Lee CFT by its (only) relevant field.[38]

The consequences for the scattering are profound. It implies that not only are $\sum_j e^{\pm\theta_j}$ the same in the initial and final states, but also $\sum_j e^{\pm p\theta_j}$ for an infinite set of values of p. This can only happen if $\{\theta'_j\}$ in the final state are some permutation of those in the initial state. Moreover, it implies factorization of the S matrix: the n particle $\to n$ particle S matrix can be written as a product of $2 \to 2$ matrices. Thus, we need focus on only $S_{2\to 2}(\theta_1, \theta_2)$, which may be written $S(\theta_1 - \theta_2)$ by Lorentz invariance.

The usual constraints of analyticity, unitarity and crossing are simple in the rapidity basis:

- $S(\theta)$ is analytic in $0 < \operatorname{Im}\theta < \pi$ apart from possible poles on the imaginary axis corresponding to bound states;
- unitarity $S(\theta)S(-\theta) = 1$;
- crossing symmetry $S(\theta) = S(i\pi - \theta)$.

Moreover, we want 3 three-point coupling to be non-zero, i.e., the particle should occur as a bound state in the two-particle spectrum. By considering $3 \to 3$ scattering, we may go to the pole in, say the (12) channel and find its residue. this gives a non-linear relation for the $2 \to 2$ S matrix, called the *bootstrap* condition[h] $S(\theta) = S(\theta - \frac{1}{3}i\pi)S(\theta + \frac{1}{3}i\pi)$.

[h]This is the original meaning of the term, not to be confused with the conformal bootstrap.

The minimal solution to these conditions is then[38]

$$S(\theta) = \frac{(e^\theta - e^{-2\pi i/3})(e^\theta - e^{-i\pi/3})}{(e^\theta - e^{2\pi i/3})(e^\theta - e^{i\pi/3})}. \tag{55}$$

This simple expression encodes all the properties of the scaling limit away from the critical point. For example, it may used to construct the form factors of local operators $\langle 0|\Phi|\theta_1, \theta_2, \ldots\rangle$, which may then be sewn together using unitarity to find highly convergent expressions for the two-point functions. It may also be used as input to the thermodynamic Bethe ansatz to extract the CFT data and check the consistency of the whole circular argument.

7. Conclusions

The Yang–Lee edge problem, in addition to being an important way of understanding criticality in Ising and other systems, is in itself a wonderful playground to explore criticality in general, and in particular, systems with complex weights or equivalently non-unitary field theories, which arise in many areas of condensed matter physics. The connections with polymer problems in higher dimensions are a bonus. It is not surprising that Michael Fisher found it fascinating, returning to it more than once.

This work was prepared while the author was attached to the University of California, Berkeley.

References

1. C. N. Yang and T. D. Lee, Statistical theory of equations of state and phase transitions I. Theory of condensation. *Phys. Rev.* **87**(3), 404–409 (1952).
2. T. D. Lee and C. N. Yang, Statistical theory of equations of state and phase transitions II. Lattice gas and Ising model. *Phys. Rev.* **87**(3), 410–419 (1952).
3. L. Onsager, Crystal statistics. I. A two-dimensional model with an order-disorder transition. *Phys. Rev.* **65**, 117–149 (1944).
4. C. N. Yang, The spontaneous magnetization of a two-dimensional Ising model. *Phys. Rev.* **85**, 808–816 (1952).
5. H.-L. Xu and A. B. Zamolodchikov, 2D Ising field theory in a magnetic field: The Yang-Lee singularity. *J. High Energy Phys.* **8**, 57 (2022).
6. P. J. Kortman and R. B. Griffiths, Density of zeros on the Lee-Yang circle for two Ising ferroinagnets. *Phys. Rev. Lett.* **27**(21), 1439–1442 (1971).
7. M. E. Fisher, Yang-Lee edge singularity and ϕ^3 field theory. *Phys. Rev. Lett.* **40**(25), 1610–1613 (1978).
8. D. Ruelle, Characterization of Lee-Yang polynomials. *Ann. Math.* **171**, 589–603 (2010).
9. M. E. Fisher, Yang-Lee edge behavior in one-dimensional systems. *Prog. Theor. Phys. Supp.* **69**, 14–29 (1980).
10. K. G. Wilson and M. E. Fisher, Critical exponents in 3.99 dimensions. *Phys. Rev. Lett.* **28**(4), 240–243 (1972).

11. K. G. Wilson, Renormalization group and critical phenomena. I. Renormalization group and the Kadanoff scaling picture. *Phys. Rev. B* **4**(9), 3174–3183 (1971).
12. K. G. Wilson, Renormalization group and critical phenomena. II. Phase-space cell analysis of critical behavior. *Phys. Rev. B* **4**(9), 3184–3205 (1971).
13. J. Cardy, *Scaling and Renormalization in Statistical Physics* (Cambridge University Press, Cambridge, 1996), pp. 45–55.
14. M. Borinsky, J. A. Gracey, M. V. Kompaniets and O. Schnetz, Five loop renormalization of ϕ^3 theory with applications to the Lee-Yang edge singularity and percolation theory. *Phys. Rev. D* **103**, 116024 (2021).
15. S. Hikami, Conformal bootstrap analysis for Yang-Lee edge singularity. *Prog. Theor. Exp. Phys.* **2018**(5), 053101 (2018).
16. M. E. Fisher, Renormalization group theory: Its basis and formulation in statistical physics. *Rev. Mod. Phys.* **70**(2), 653–681 (1998).
17. S.-N. Lai and M. E. Fisher, The universal repulsive-core singularity and Yang-Lee edge criticality. *J. Chem. Phys.* **103**, 8144–8155 (1995).
18. Y. Park and M. E. Fisher, Identity of the universal repulsive-core singularity with Yang-Lee edge criticality. *Phys. Rev. E* **60**(6), 6323–6328 (1999).
19. D. Poland, On the universality of the nonphase transition singularity in hard-particle systems. *J. Stat. Phys.* **35**, 341–353 (1984).
20. A. Baram and M. Luban, Universality of the cluster integrals of repulsive systems. *Phys. Rev. A* **36**, 760–765 (1987).
21. R. J. Baxter, Hard hexagons: Exact solution. *J. Phys. A: Math. Gen.* **13**, L61 (1980).
22. I. Jensen, Comment on Series expansions from the corner transfer matrix renormalization group method: The hard-squares model. *J. Phys. A: Math. Gen.* **45**, 508001 (5pp.), (2012).
23. G. Parisi and N. Sourlas, Critical behavior of branched polymers and the Lee-Yang edge singularity. *Phys. Rev. Lett.* **46**, 871–874 (1981).
24. G. Parisi and N. Sourlas, Random magnetic fields, supersymmetry, and negative dimensions. *Phys. Rev. Lett.* **43**, 744–745 (1979).
25. J. Z. Imbrie, Lower critical dimension of the random-field Ising model. *Phys.Rev. Lett.* **53**, 1747–1750 (1984).
26. J. Cardy, The liquid-gas critical point in a random medium and dimensional reduction, in preparation.
27. D. C. Brydges and J. Z. Imbrie, branched polymers and dimensional reduction. *Ann. Math.* **158**, 1019–1039 (2003).
28. D. C. Brydges and J. Z. Imbrie, Dimensional reduction formulas for branched polymer correlation functions. *J. Stat. Phys.* **110**, 503–518 (2003).
29. G. Kirchhoff, Ueber die Auflijsung der Gleichungen, auf welche man bei der Untersuchung der linearen Vertheilung Galvanischer Ströme geführt wird. *Ann. Phys. Chem.* **72**, 497–508 (1847).
30. J. Cardy, Directed lattice animals and the Lee-Yang edge singularity. *J. Phys. A: Math. Gen.* **15**, L593–595 (1982).
31. D. Dhar, Exact solution of a directed-site animals-enumeration problem in three dimensions. *Phys. Rev. Lett.* **51**, 853–856 (1983); Erratum *Phys. Rev. Lett.* **51**, 1499 (1983).
32. J. Z. Imbrie, Dimensional reduction for directed branched polymers. *J. Phys. A: Math. Gen.* **37**, L137–L142 (2004).
33. U. C. Täuber, *Critical Dynamics* (Cambridge University Press, Cambridge & New York, 2014).

34. J. Cardy, Conformal invariance and the Yang-Lee edge singularity in two dimensions. *Phys. Rev. Lett.* **54**(13), 1354–1356 (1985).
35. A. A. Belavin, A. M. Polyakov and A. B. Zamolodchikov, Infinite conformal symmetry of critical fluctuations in two dimensions. *J. Stat. Phys.* **34**, 763 (1984).
36. T. Wydro and J. F. McCabe, Tests of conformal field theory at the Yang-Lee singularity. *AIP Conference Proceedings* **1198**, 216 (2009).
37. A. B. Zamolodchikov, Integrable field theory from conformal field theory (Proc. Taniguchi Symp. (Kyoto, 1988)). *Adv. Stud. Pure Math.* **19**, 641–674 (1989).
38. J. Cardy and G. Mussardo, S-matrix of the Yang-Lee edge singularity in two dimensions. *Phys. Lett. B* **225**(3) (1989).

https://doi.org/10.1142/9789811282386_0019

Chapter 19

There's Plenty of Room in the Middle: The Unsung Revolution of the Renormalization Group

Nigel Goldenfeld

Department of Physics,
University of California San Diego,
La Jolla, CA 92093, USA
nigelg@ucsd.edu

The remarkable technical contributions of Michael E. Fisher to statistical physics and the development of the renormalization group are widely known and deeply influential. But less well known is his early and profound appreciation of the way in which renormalization group created a revolution in our understanding of how physics — in fact, all science — is practiced and the concomitant adjustment that needs to be made to our conception of the purpose and philosophy of science. In this chapter, I attempt to redress this imbalance, with examples from Fisher's writings and my own work. It is my hope that this tribute will help remove some of the confusion that surrounds the scientific usage of minimal models and renormalization group concepts, as well as their limitations, in the ongoing effort to understand emergence in complex systems.

1. Introduction

In 1988, Michael E. Fisher authored what must surely be a candidate for his least cited paper. Entitled "Condensed Matter Physics: Does Quantum Mechanics Matter?"[1,2] and bestowed with all of nine citations (according to Google Scholar), it poses an outrageous question that in Fisher's hands is developed and answered with his characteristic brilliance, clarity and originality. Fisher was writing at the behest of Herman Feshbach to review the state of condensed matter physics as it stood on the 100th anniversary of the birth of Niels Bohr, with a special remit to comment on the overlap with Bohr's ideas on the fundamentals of quantum mechanics. The timing could not have been more propitious. The previous two years had seen the discovery of the high-temperature superconductors LBCO[3] and YBCO[4]; the discovery of quasicrystals had been made four years previously[5]; and the integer[6] and fractional quantum hall effects[7] had been discovered and largely explained during the previous eight years.[8] Five major discoveries in condensed matter physics in less than a decade, ultimately garnering five Nobel Prizes and each a splendid manifestation of quantum mechanics.

Nevertheless, Fisher evidently did not regard his role as to provide a self-congratulatory pat on the back to condensed matter physics. Instead, he took it upon himself to explain his thoughts on "... what condensed matter physics does, or should be doing, and what defines condensed matter physics, and thence to approach the question 'Does quantum mechanics matter?'" The viewpoint that he expounded was at the time still rather unorthodox in many areas of physics and science, in particular his perspective that "the question of connecting the models with fundamental principles is *not* a very relevant issue or central enterprise." This would come as news to those in the physics community engaged in the reductionist program, including to some extent condensed matter physicists but to a far greater extent those working in what at the time was sometimes called elementary particle physics or fundamental physics and what is now generally called high-energy physics, or in some areas, with less hubris and reflecting the progress in accelerator physics and budget, medium-energy physics. This dichotomy within physics was certainly pervasive at the time and important enough that Fisher chose to make it the focus of his article.

Fisher undoubtedly hoped to use the opportunity provided by Feshbach to pop the bubble of the reductionist approach to condensed matter physics, and this he did with gusto, and one assumes, with the twinkle in his eye and suave[9] cravat by which I will always remember him.

Over the last 35 years since Fisher's article, physics and its philosophical roots have undergone a significant shift. This shift influences the way we do physics, the way we interpret what we do, and the way we use physics to explore interdisciplinary scientific questions. Accordingly, I have chosen for my title a deliberately ambiguous and evocative phrase. "There is plenty of room in the middle" is of course a parody of Feynman's famous lecture "There is plenty of room at the bottom," which is often credited with foreseeing the nanotechnology revolution.[10] In my case, I am drawing attention away from the microscopic scale to focus instead on the middle scale or level of description, which depending on the problem might be the scale intermediate between the lattice spacing and the correlation length (in the case of critical phenomena) or the scale larger than the dissipation scale and the scale of energy input (in the case of fluid turbulence). In the first of these, one can make this scale arbitrarily large by tuning the temperature close to the critical temperature, and in the second, by making the Reynolds number large. This middle scale was where many of Fisher's interests lay, understanding the universal, cooperative phenomena that arise there. And of course that is why he chose his provocative theme for his article.

But I have a second motivation for my title. On more than one occasion, I remember Michael saying, "You can't have interdisciplinary research without first having disciplines" (for my personal comments, I will lapse into first name terms). Thus, the "middle" in my title also refers to the space between disciplines, where interdisciplinary research happens and transdisciplinary research begins. I happen to believe that this is where the greatest scientific excitement can often be generated. Fisher himself was one of the shining examples of an interdisciplinary scientist, holding for many years at Cornell University the Horace White Professorship of Chemistry, Physics, and Mathematics, each of which he illuminated with his scholarship and unique contributions. Thus, I propose to use this chapter to pay tribute to Fisher's deep understanding of the way in which his most important work on renormalization group theory led to a transformation in our understanding of physics and its explanatory purpose and to extend the scope of his examples and comments, especially with regard to non-equilibrium statistical physics. I will spend very little time on critical phenomena and exponents because I believe that although this was the problem that Fisher and the statistical mechanics community set out to solve, the impact of the renormalization group is equally important away from criticality. One of the ways that I will argue this is through the connections between universality, renormalization group, levels of description and asymptotics. I also want to argue that the perspective he espoused is especially relevant today to applications outside of physics, where there has arguably been less self-consciousness about the way we choose to construct theories.

2. Condensed Matter Physics: Does Quantum Mechanics Matter?

To appreciate Fisher's subversive intentions, it is worth reminding the reader of the intellectual background to his article. It is no exaggeration to say that physics underwent a remarkable transformation sometime around the middle of the 20th century. During the previous decades, the reductionist program had been profoundly successful, with the discovery of the electron, neutron, proton; the theoretical predictions of the neutrino (by Pauli[11] in 1930) and the meson (by Yukawa[12] in 1934); the detection of the former by Cowan, Reines, Harrison, Kruse and McGuire[13] in 1956 and of the latter in 1936 by Neddermeyer and Anderson[14] — although they actually discovered what we now call the muon, a lepton whose mass was close to Yukawa's prediction for the nuclear force carrier mass; the actual discovery in 1947 of Yukawa's particle, the pion, in cosmic rays[15]; and accelerating with the discovery of a multitude of mesons during the 1950s and the subsequent decades. Yet, with the birth of solid-state physics around 1940 (reflected by the establishment of the American Physical Society Division of Solid State Physics (DSSP) in 1947), the initial focus on one-electron properties of metals and semiconductors turned in the 1950s to the collective properties of matter through many-body theory and the application of non-relativistic quantum field theory. In due course, the field became known as condensed matter physics, reflecting the increasing focus on all matter, not just solids, and especially on collective phenomena[16] (and marked by the American Physical Society renaming the Division of Solid State Physics as the Division of Condensed Matter Physics in 1978).

It quickly became clear that novel macroscopic phenomena transcended the additive properties of single degrees of freedom, and a renewed focus on collective properties was seen by many as the defining aspect of condensed matter physics, particularly through the field's visionary leadership by P.W. Anderson.[17,18] Indeed, Fisher is very clear where he stands on this point[1,2]: "The basic problem which underlies the subject is to understand the many, varied manifestations of ordinary matter in its condensed states and to elucidate the ways in which the properties of the 'units' affect the overall, many-variable systems." In other words, condensed matter physics is about interactions primarily.

For present purposes, I would argue that the pivotal developments in the field of condensed matter theory were two iconically quantum mechanical theories: Bogoliubov's 1947 theory of superfluidity for the weakly interacting Bose gas,[19] and the Bardeen–Cooper–Schrieffer 1957 theory of superconductivity.[20,21] I choose these problems because they are relevant to Fisher's conception of the task of condensed matter physics. Again, he is very explicit about the questions that animate the task of addressing the basic problem underlying condensed matter physics: "What are the states of matter?... What is their nature?... How do the various states transform into one another?" As we will see, inextricably linked with these questions are two concepts that are subtle and took a long time to be appreciated: the notions of "minimal models" and "levels of description". In fact, there are several features of superfluidity and superconductivity theory to which I especially want to draw attention.

2.1. *Minimal models*

First, their starting point. Both these works showed the surprising explanatory power of a ludicrously simple model of interacting bosons or electrons, respectively, that is clearly a brutal idealization of the real complexities of atomic or electron–phonon interactions. Bogoliubov assumed that the interactions between bosons were weak in the sense that "corresponds to a neglection of the finiteness of molecular radius, since we do not take into account the intensive increase of [the potential] for small r, which causes the impenetrability of molecules."[19] At the end of his paper, he argued that the approximation could also be extended to include the finite "molecular" radius by replacing the Fourier transform of the potential by the amplitude of the binary collision probability in the Born scattering approximation (an observation for which he thanks Landau). This amounts to replacing the real interaction with a contact interaction $U\delta(\mathbf{r})$. Bogoliubov's explicit recognition of the fact that his approximation was valid at long wavelengths where the atomic (molecular) size was far smaller than the mean particle separation was prescient, but it took several decades for this insight to be translated into a renormalization group description.[22–25] Bardeen, Cooper and Schrieffer (BCS) took the Bardeen–Pines Hamiltonian for electron–phonon interactions including Coulomb effects[26] and replaced the complex interaction with one in which "each state is connected to n other states by the same matrix element $-V$."[20] They explained that this idealization applied to a subset of states that were paired with equal and opposite spin and momentum and that this was sufficient to capture the formation of the condensed state for arbitrary weak matrix element/potential V.

The models that both Bogoliubov and BCS used in their work are examples of what are sometimes known as "minimal models,"[27–31] "model[s] that most economically caricature the essential physics."[28] However, what this actually means is rather subtle and sometimes misunderstood; in order to elaborate, we need to discuss the role of asymptotics, in particular the process by which the minimal models were analyzed.

2.2. *Asymptotics and universality*

One reason that both Bogoliubov and BCS were successful where others had failed was because they used non-perturbative methods to extract their physical conclusions. BCS used a variational method, while Bogoliubov invented a canonical transformation applied after the naïve perturbation theory in the weak potential had been performed. Shortly after the publication of the BCS letter announcing their results,[20] Bogoliubov's method was applied to the BCS model.[32,33] I emphasize the non-perturbative nature of the theories because in both cases, the results are non-analytic; for example, the depletion of the condensate in the Bose gas problem at zero temperature due to interactions is proportional to $U^{3/2}$, while in the BCS problem, the energy gap depends on the matrix element V and the density of states at the Fermi level $N(0)$ through an essential singularity of the form $\exp(-1/N(0)V)$. Such results are, of course, beyond the scope of simple, regular, finite order perturbation theory and, due to the way they are obtained, are believed to be valid asymptotically as leading-order approximations for $U, V \to 0$.

The necessity of using non-perturbative methods to obtain non-analytic results that describe the emergent states of superfluidity and superconductivity is not a result of any "exotic" character of these states. Phase boundaries are by definition the loci where the partition function is non-analytic, and the emergent states are physically not simple perturbations around the normal state, possessing generalized rigidity and elasticity as a result of the symmetry breaking associated with the transition.[18,28] The non-analytic nature of phase transitions means that any mathematical expansion about the transition point cannot be convergent, and so, our results are at best asymptotic in the minimal model interaction parameter. As a result, there will be sub-leading corrections to our predictions, in addition to other regular terms that may arise from the details ignored in constructing the minimal model in the first place. I also want to stress that in both the examples given here, the non-analytic results arise even at mean-field level, and proper treatment of the interaction of fluctuations adds another level of non-analyticity.

For example, in the case of the Bogoliubov Bose gas, the minimal model itself is constructed to be an asymptotic long-wavelength limit when the potential range is negligible compared to the mean separation of the particles. Corrections to this approximation arising from including a finite potential range or a non-vanishing particle density will decorate the leading-order asymptotic result in a complicated way that needs careful consideration of the various approximations. Nevertheless, the notion of a minimal model as being an economical caricature of the essential physics must be understood in this asymptotic context. The two cannot be separated. Thus, it is not simply that a minimal model is an inaccurate representation of a physical system, a poor approximation that can be improved by embellishment. Instead, the point is that the explanatory requirement of a minimal model is to account for the phases and phase transitions of matter, just as Fisher emphasized, and also the mathematical structure of the description of the latter, which necessarily involves asymptotic methods. Gratuitously realistic details of the starting model will not only complicate the technical task of extracting the asymptotic structure, but also will not impact the leading order behavior. Thus, a minimal model represents a universality class of models; whether or not one uses the minimal model or a decoration of it, the asymptotic outcome of a non-perturbative calculation will not be different.

Overall, we are led inexorably to the conclusion that there is a strong connection between asymptotics and universality. Indeed, in my own work, I have shown how renormalization group methods can be used to treat singular mathematical problems, such as those arising in differential equation theory[28,34] and low Reynolds number fluid dynamics.[35] In these problems, there are no fluctuations at all: The non-analyticity arises at mean field level.

2.3. *Minimal models in action: The case of superconductivity*

This narrative of the nature of minimal model explanation is not a subjective philosophical interpretation that is open to debate, although it has been challenged[36] and defended[37] in the philosophical literature.[29,31,38] It is what physicists actually do. In the case of BCS theory, we have a very important demonstration of the validity and utility of this perspective. The BCS minimal model of superconductivity leaves out a number of embellishments that are desirable to include if one wants to calculate accurately observables such as the transition temperature or energy gap at zero temperature for materials where details of the actual electron–phonon interactions are measurable. Such an elaboration of the BCS theory was initiated by Eliashberg,[39] who accounted for the detailed form of the electron–phonon frequency spectrum even in the strong coupling limit, using methods developed by Migdal for the normal state,[40] and by McMillan,[41] who provided a synthesis that included electron–electron interactions and band structure effects. In Eliashberg's theory, the effects of phonons lead to integral equations accounting for the phonon excitation spectrum and so the theory becomes intrinsically non-local. In fact, the interaction between electrons, mediated by phonons, is inherently a retarded interaction because of the mass difference between the lattice ions contributing to the phonon modes and the electrons themselves. Distortions of the crystal lattice, described by the phonons, inevitably relax on a much slower time scale than that of the electrons, and basically, it is the fact that electrons experience the local deformations that causes them to interact and be attracted to one another, in terms of their kinematics — hence the BCS coupling of electrons together in time-reversed momentum states. In McMillan's main result[41] (Eq. (18)), the structure of the BCS formula, with its essential singularity is modified to $\exp(-1.04(1+\lambda)/(\lambda-\mu^*(1+0.62\lambda))$, where λ and μ^* are constants that are, respectively, calculable from the Eliashberg theory and the effects of Coulomb interactions.[42] For small values of the parameter λ (i.e., the weak coupling limit), the formula reduces to the BCS formula, with $\lambda - \mu^*$ being identified as $N(0)V$. Thus, we see explicitly how the minimal model can be embellished by including greater levels of realism in order to compare with experimental details, but staying within its universality class and mathematical structure.

2.4. *Levels of description and emergence*

Bogoliubov's work showed how an interacting system can, within a first approximation, be treated as a collection of non-interacting quasiparticles, each of which involves all the original "real" atoms in the system. The quasiparticles were recognized right at the outset as forming an explicitly long-wavelength description of the system, the details of which were later developed in 1957 by Lee and Yang, and others.[22–25,43–45] The many-body ground state and the excitations of the interacting Bose system depend on a single parameter U characterizing the strength of interactions of the original, "real" bosons; both the ground state and the excitations are collective properties of the system. In the case of superconductivity, the additional complication is the formation of composite bosons — Cooper pairs — which simultaneously undergo a similar ordering as in the Bogoliubov gas. The upshot of these developments is that for many purposes, a superfluid or superconductor has a universal description that effectively hides certain details of the underlying constituents and their interactions. In the case of the weakly interacting bosons, the interaction parameter U depends on the s-wave scattering length derived from the collision of two bosons, and the theory is valid when this length is much shorter than the mean separation of the bosons. Thus, it can be said that U relates two different *levels of description*; the microscopic world inhabited by the minimal model of interacting bosons and the macroscopic world of the condensate with its superfluid properties and excitation spectrum for the quasiparticles that have emerged. It would be possible in principle to infer the existence of the quasiparticles phenomenologically from thermodynamic measurements, as Landau famously did,[46] without knowledge of the properties, or even the existence, of the atomic level of description.

Although much of the impact of the Bogoliubov and especially the BCS theories derived from their ability to bridge levels of description and enable quantitative comparisons of the predictions of minimal models with experiment at the microscopic level of description (e.g., the work of McMillan[41] already mentioned), Fisher is emphatic that this is not their main importance. We already cited his views about connecting "the models with fundamental principles," and elsewhere in the same article, he writes, "The basic problem which underlies the subject is to understand the many, varied manifestations of ordinary matter in its condensed states... an important role is, and, indeed, should be played by various special 'models.' ...One might as a theorist argue that more attention should be paid to the connection between models and the fundamental principles as embodied in quantum mechanics."[1,2] However, he argues, this is misguided for the following reason: "I stress again, that in any science worth the name, the important point is to gain understanding. The language in which the understanding is best expressed cannot be dictated ahead of time, but must, rather, be determined by the subject as it develops. Accordingly, it really would not be a 'great success' to derive the Ising model from atomic theory or quantum mechanics. Indeed, for which physical system would one be 'deriving' it?"[1,2] His point is that a minimal model such as the Ising model is a minimal model for a ferromagnet, a liquid–gas system, an alloy, a ferroelectric material and, I would add, many other systems, including examples in biology and ecology, ranging from the neurobiology of retinal ganglion cells[47] to the annual nut distribution of pistachio trees.[48] Thus, although "...one does want to understand what the important variables actually are... in many ways, it is more important to understand how these variables interact together and what their 'cooperative' results will be."[1,2]

To close this important section on levels of description, I want to comment again, in Fisher's prescient words, of the important role played by asymptotics: "Personally, I find the elucidation of the connections between various disciplinary levels a topic of great interest. Typically, delicate matters of understanding appropriate asymptotic limits, both physical and mathematical, are entailed. The issues involved are often very subtle; nevertheless, one must admit, they seldom add significantly to one's understanding of either the 'fundamental' starting theory or of the target discipline. Sad but true!"[1,2]

To this, I would add a point that is not made frequently enough: technical derivations of higher levels of description from more microscopic ones have a limited regime of validity. For example, it is possible to derive the Navier–Stokes equations from Boltzmann's kinetic equation. This is a long and complicated calculation that I often show the physics students in my statistical mechanics classes. After I have done that, I show them a phenomenological motivation that of course is very brief and closer in spirit to the derivations usually given in fluid mechanics texts.

Most students, when asked, find the former derivation more convincing. They are surprised to learn that I do not share this view and are somewhat shocked to realize that the seemingly more careful, technical derivation is only valid at very low density, whereas we know that the Navier–Stokes equations are an excellent description of fluid phenomena up to very high densities.

2.5. *Universal and asymptotic scaling laws in non-critical matter*

Fisher structured his article first by describing his perspective on the nature of condensed matter physics and second by giving multiple examples of forms of matter to illustrate his main thesis. The first form of condensed matter that he chose was polymeric matter, focusing first of all on the excluded volume problem for a single polymer,[49,50] and then the problem of the statistical thermodynamics of dilute polymer solutions.[27,51] Fisher chose this example and not the obvious example of critical phenomena scaling. Why?

I conjecture that he wanted to make two points. First, that there are other ways in which matter can enter asymptotic realms other than just being close to a critical point. Fisher was intimately acquainted with asymptotic methods and understood that an example of scaling which has nothing to do with a critical point would be consciousness-raising for some of his audience. This is elaborated on with numerous "critical phase" examples in Chapter 30 by Leo Radzihovsky. Second, that he wanted to show the relevance of condensed matter physics to a discipline other than physics. He writes,[1,2] "...other challenging problems remain beyond the size of a single isolated molecule. One might say, 'Well, aren't polymers for chemists?' The answer is that many of their properties have proved too difficult for traditionally trained chemists!"

At this time, Fisher was in fact a Professor of Chemistry — among other things — and had in the early part of his career at King's College, London, worked extensively on the statistics of self-avoiding walks and Ising models, developing methods to enumerate configurations, extract exponents using series expansions, Padé approximants and so on. In an especially interesting paper from this period,[52] he and Gaunt estimated the behavior of self-avoiding walks and Ising models in dimensions $d > 4$, finding that the deviations from mean-field theory seemed to become very small above four dimensions, a result that must have been on his mind during the development with Wilson of the $\epsilon = 4 - d$ expansion.[53] One of the earliest applications of renormalization group theory, and in particular the ϵ-expansion, was to the conformations of polymers.

In 1965, Edwards had[50] developed a field-theoretic approach to the statistics of a self-avoiding polymer chain, based on Bogoliubov's idea of the pseudopotential,[19] and used a functional integral approach to derive a self-consistent theory that showed how in the limit of an asymptotically long chain, the radius of gyration of a single polymer scaled with length L as $L^{3/5}$. This theory, like Bogoliubov's, had a regime of validity associated with a description at long wavelengths compared to the range of interparticle forces, in this case being that L was much greater than the monomer scale. Edwards' result is, like Bogoliubov's, a mean-field theory approximation of course and ignores fluctuations which add still another level of non-analyticity to the problem. The much more complicated problem of semi-dilute polymer solutions was treated the following year by the same methods[54] subject to the same limitations of course.

In the early 1980s, Ohta and Oono[27,51] developed a conformation-space renormalization group method which was not based on de Gennes' analogy to the n-component magnet[55] and thus was able to account for arbitrary molecular weight distributions of the polymers. Ohta and Oono used their method to go far beyond computing simple scaling laws, not only correcting the Edwards mean-field results to include fluctuations but also calculating the universal scaling functions for the statistical thermodynamics of polymer solutions as a function of concentration. Fisher could not hide his excitement in reporting their results, especially the complete comparison with experiment[56] over several decades for the functional form of the osmotic compressibility and other variables. In particular, Fisher enthused: "Thus we see in polymeric matter new, subtle and universal behavior which we have succeeded in understanding theoretically. But quantum mechanics has had essentially nothing to say about the problem! Indeed, one feels that if some of the giants of the past, like Boltzmann or Gibbs or Rayleigh, were able to rejoin us today, they would be able to engage in research at the cutting edges of condensed matter physics without taking time off to study quantum mechanics first!"[1,2]

3. Turbulence: Does Fluid Mechanics Matter?

I want to turn now to a topic that did not play a large role in Fisher's article or his research interests: systems far from equilibrium. At the time of his article, condensed matter physicists were beginning to address problems in non-equilibrium pattern formation, phase transition kinetics, and there is a long story to tell there about asymptotic scaling laws, since this is the field where I started my own career. That story is for a future occasion. Here, I want to mention briefly two vignettes from my own research on turbulent fluids. The first of these I know he enjoyed because I got to tell him about it during a visit to Cornell. Unfortunately, I do not remember if I had the chance to tell him about the second one.

3.1. *A turbulent analogue of Widom scaling*

The classic problem of fluid dynamics is that fluids appear to be scale-invariant when strongly turbulent.[57,58] Specifically, in a fluid geometry with characteristic scale D (which might be the diameter of a pipe), characteristic velocity U (which might be the average mean flow velocity along a pipe) and kinematic viscosity ν, we define the Reynolds number as $Re = UD/\nu$. In addition, there is the energy dissipation rate ϵ, which is in steady state equal to the energy input rate. The so-called energy spectrum — the kinetic energy per unit mass of fluid per unit wavenumber k — was argued by Kolmogorov,[59] in a paper universally known as K41, to vary as $E(k) = \epsilon^{2/3}k^{-5/3}$ at large Re, and this is broadly in agreement with experimental data going back more than 50 years.[60] The classical measurements and theories only concerned themselves with the lowest-order moments of the velocity fluctuation probability density, but today, there is strong evidence for multifractal scaling of higher-order moments.[58] Kolmogorov further showed that the range in k-space over which this scaling occurred was intermediate between the scale of forcing D and the Kolmogorov scale of dissipation $\eta_K = (\nu^3/\epsilon)^{1/4}$, in other words $2\pi/D \ll k \ll 2\pi/\eta_K$. As $Re \to \infty$, this range of scaling increases because the Kolmogorov scale gets smaller and smaller, but can be shown to always be above the mean free path, so that the fluid is always in the hydrodynamic limit. In fact, K41 is not quite correct, even for the second-order correlation function, and there is a so-called large-scale intermittency correction, η, which changes the scaling result to $E(k) \sim k^{-5/3+\eta}$.[61,62]

To a statistical mechanic, this scaling behavior is reminiscent of what happens in a magnet near a critical point. In general, correlation functions decay as a power law within a range of wavenumber that is intermediate between the lattice spacing a and the correlation length ξ. As the temperature T approaches its critical value T_c, the correlation length diverges to infinity, and at any scale accessible to experiment, the correlations will be in the power-law scaling range, ultimately scaling as $G(k) \sim k^{-2+\eta}$ at the critical temperature itself. Here, η is a correction to the mean-field result, sometimes called Fisher's exponent, and now understood to be the anomalous scaling dimension of the magnetization M, which grows with a power law β as $T \to T_c^-$.

Statistical mechanicians also know that these results are only valid when there is no external field H. When $H \neq 0$, there are more complicated scaling laws. For example, in general, we expect that the magnetization is proportional to the external field, but this linear response law breaks down exactly at the critical temperature: $M(H, T_c) \sim H^{1/\delta}$, where δ is another critical exponent that can of course be calculated by renormalization group. In the critical region $H \sim 0$ and $t \to 0$ (where $t \equiv (T - T_c)/T_c$), Widom showed[63] that the behavior of the magnetization, ostensibly a function of both H and t is actually a function of a combined variable, predicting a data collapse of $M(H, T)$ when plotted appropriately, similar to the data collapse that we talked about with polymer solution theory. This scaling was independently discovered by Kadanoff and explained in a famous paper the following year,[64] and this discovery by Widom and Kadanoff was instrumental in the development of the renormalization group.

Thus, it is a natural question to ask: Is there a counterpart to the Widom scaling in turbulence?

I approached this question by asking what are the analogous quantities to t and H in turbulence. With Greg Eyink, I had long ago argued that it is natural for $1/t$ to be analogous to Re because both control the size of the intermediate regime in k space where there is power-law scaling as $1/t$ and Re, respectively, go to infinity.[65] But what is the turbulent analogue of H? My reasoning was that H is a variable that couples to and induces magnetization. In a pipe, it turns out that the laminar flow is linearly stable, but if it has rough walls, the roughness will excite turbulence. Thus, my guess was that the roughness scale s could be analogous to H. Using the same sort of scaling arguments that Kadanoff and Widom had used, and assuming that turbulence was in fact a non-equilibrium steady state with its own critical point at $Re \to \infty$, $s \to 0$, it was possible to predict the analogue of Widom's scaling law.[66]

Fortunately, there were experimental data on this very question! In 1933, Nikuradse had measured the friction experienced by a turbulent fluid as it transited a pipe with rough walls.[67] Importantly, Nikuradse had systematically varied both Re and the roughness scale s. His data collapsed satisfactorily when plotted according to the formula.[66] However, I had only used the mean-field exponents in my collapse, the ones found in K41. Mehrafarin and Pourtolami extended my calculation to include the intermittency correction.[68] Even though the correction is very small for the second-order correlation function, they were able to show convincingly that they could extract it by improving the data collapse fit. Their result was consistent with known estimates for the intermittency exponent obtained by direct measurements of velocity fluctuations.

This result is extraordinary for the following reason. In phase transition theory, it is known that there is a connection between purely thermodynamic critical exponents and those associated with spatial correlations.[28] It turns out that it is possible to deduce Fisher's exponent for correlations, the anomalous dimension of the order parameter, directly from the thermodynamic exponent δ that we mentioned above quantifies the breakdown of linear response theory at the critical point. In my treatment of the turbulent scaling problem, the analogue of the thermodynamic exponent is one that concerns the scaling of the pressure drop along the pipe with Re. So, using the data collapse scaling law, if it had been known in 1933, Nikuradse could have measured indirectly the intermittency exponent of turbulence characterizing the strong spatial fluctuations! In other words, the fluctuations of turbulence are directly related to the pressure drop along a pipe, i.e., the dissipation experienced. This implies that there are non-equilibrium fluctuation–dissipation relations in turbulence, but at the present time, we do not know how to extract them.

There were other developments that are important to mention. My Illinois colleagues Pinaki Chakraborty and Gustavo Gioia came up with an ingenious heuristic mean-field argument that predicted the various exponents that are visible in different regimes of the Nikuradse data,[69,70] at least at mean-field level.[71] With Nicholas Guttenberg, I worked out the scaling laws and simulated the flow in a rough two-dimensional pipe (i.e., a soap film suspended between two wires),[72] where the point was that in two dimensions, a different scaling of the velocity correlations is possible than those arising from the K41 theory for the energy cascade, because of the flow of angular momentum fluctuations (enstrophy). This fact meant that we could in principle construct flows with different velocity correlation scaling from the K41 one, and observe the effect on the pressure drop, i.e., the friction, and thus test the fluctuation–dissipation relation that had been discovered. Our predictions were fully confirmed in experiments that we performed with Walter Goldburg at Pittsburgh and Hamid Kellay in Bordeaux.[73–75]

Overall, these findings suggest that fully developed turbulence is controlled by a non-equilibrium critical point, with strong connections to a far from equilibrium statistical mechanics through an unknown fluctuation–dissipation relation.

However, this was not the only surprise in treating turbulence as a problem in statistical mechanics.[76]

3.2. *Fluids become turbulent through a non-equilibrium phase transition*

Up to now, we have talked about the large *Re* behavior of turbulence. But how do fluids transition from predictable, smooth, laminar flows to unpredictable, fluctuating turbulent flows? In certain shear flows where the laminar–turbulence transition is sub-critical, including pipe flow, we now have compelling theoretical,[77–83] computational[84,85] and experimental[86–89] evidence that this transition is a non-equilibrium phase transition in the universality class of directed percolation.[90] A synthesis is beginning to emerge,[76,91,92] and most workers in this small field tend to agree that the one-dimensional problem is understood in varying levels of detail, but the problem of the transition to turbulence in two dimensions is still not understood theoretically, even though the experimental data are very compelling.[88]

I believe that Fisher would have been excited by these developments because it is fascinating to see phase transition physics, which he calls "my own love," emerge unexpectedly in a field such as fluid dynamics, where there is no explicit stochasticity, no partition function, and no obvious connection to directed percolation. This is an intriguing example of universality, to be sure. But I think Fisher would have been equally delighted by the way that this prediction was made. When my students and I set out to tackle this problem, our perspective was that the worst starting point for understanding the transition to turbulence was the Navier–Stokes equations. Instead, our goal was to construct the level of description that corresponded to what would be Landau theory for an equilibrium transition. To this end, we used direct numerical simulations to identify the weak long-wavelength modes that are relevant at a putative continuous laminar–turbulence transition, and once we had identified these, constructed the generic description, which turned out to be a stochastic predator–prey or activator–inhibitor model.[81] In other words, we wanted to identify the variables that defined the universality class of the laminar–turbulence transition. This model led to directed percolation in a straightforward way using known techniques,[93] and as we showed was able to reproduce the universal aspects of the phenomenology at the laminar–turbulence transition.[76,81,83]

We did not set out to derive the directed percolation transition from the Navier–Stokes equations because we already knew that this sort of derivation from low levels of description cannot be done systematically, let alone rigorously, even for the simplest systems, such as equilibrium magnets. We also did not attempt to predict the critical Reynolds number of the laminar–turbulence transition because we know from the renormalization group in equilibrium statistical mechanics that this is not universal.

In a later paper reviewing the statistical mechanics approach to turbulence,[76] we did give a heuristic motivation for the predator–prey description, at least at mean-field level, but I believe that it is fair to say that the majority of fluid mechanicians do not yet share Fisher's perspective that it is more important to understand the cooperative interactions of the emergent degrees of freedom at the appropriate level of description than to systematically derive these degrees of freedom. The reason for his perspective, and the reason why it is so counterintuitive, is that the renormalization group approach relies on a heuristic calculation or guess for the emergent level of description, and then calculates the relevant degrees of freedom at the appropriate fixed points. This approach obviates the need to calculate quantities that will neither affect the universal properties nor the final goal of qualitiative understanding.

4. Concluding Remarks

It is an unavoidable temptation to imagine the reaction when Fisher's manuscript arrived on Feshbach's desk. Not only does its very title threaten to undermine the rationale for the whole enterprise, but Fisher doubles down, arguing at one point that "...it is a mistake to view complementarity as merely a two-terminal black box ...a fully reductionist philosophy, while tenable purely as philosophy, is the wrong way to practice real science!" For many editors and organizers of a symposium on the legacy of Niels Bohr, Fisher's article might have seemed like an unwarranted provocation. But I suspect that in Feshbach's case, he appreciated the article as a masterly contribution that both recalled the extraordinary collision between physics and philosophy of a previous age, and heralded the dawn of a new era where this confrontation would be renewed, but on fresh ground, whose boundaries had been demarcated by Fisher's warning shot across the bow.

Surprisingly, it would be over a decade later before the first skirmishes took place. I would place their date to be the 2001 publication of a book by R.W. Batterman, entitled *The Devil in the Details: Asymptotic Reasoning in Explanation, Reduction, and Emergence.*[29] Drawing on sources such as the works of Michael Berry on caustics and asymptotics,[94,95] Leo Kadanoff,[64] Michael Fisher[96] and Ken Wilson[97] on critical phenomena and renormalization group, and my own work on asymptotics and renormalization group,[98,99] as summarized especially in Ref. 28, Batterman weaves together the concepts of minimal models, asymptotics, renormalization group theory and emergence in a thesis that explicitly recognizes how the modern way of doing science is neither Popperian nor Kuhnian, partly because of a more sophisticated notion of falsifiability entailed by concepts such as universality. This work and its extensions[31,38] have been increasingly influential during the last 20 years, and the battles have been fought mostly by philosophers of science. Most recently, Batterman has used hydrodynamic linear response theory as an example to expound further on these ideas, echoing many of the themes that Fisher advocated and which I have described here.[38]

I believe that the philosophical debates about minimal models are especially important for complex systems such as biology and turbulent fluid mechanics, because the choices one makes in modeling of these fields are so much more difficult than in physics. I discussed above how the minimal model and renormalization group approach to modeling at the appropriate level of description was successful in understanding quantitatively the laminar–turbulence transition in certain shear flows. The idea of making effective models of turbulence is, however, not new to fluid mechanics. Indeed, there is a long tradition of making approximate models, and these are discussed in the textbooks, especially Pope's excellent monograph on turbulent flow, see Section 8.3 in Ref. 57. These are not minimal models because the "ultimate objective is to obtain a tractable quantitative theory or model that can be used to calculate quantities of interest and practical relevance."[57] This means using models as approximations for computer simulations or simple analytical calculations whose utility is assessed by factors such as cost and ease of use, range of applicability and accuracy.[57] In general, it is not the goal to devise universal scaling functions, for example, in the way that minimal modeling was used in the semi-dilute polymer problem.[27,51]

That being said, it must not be forgotten that Kolmogorov's program of analyzing turbulence began with two similarity hypotheses about universality and this is to my knowledge the first example of where the renormalization group perspective was articulated clearly[59] and followed up in many seminal works by Barenblatt (Kolmogorov's last student!).[100,101] The first of these is that the energy spectrum can be written in the middle or inertial range of scales as $E(k) = (\nu^2/\eta_K)F(k\eta_K)$, where the limit $kL \to \infty$ has been assumed to exist and been taken, and the scaling function $F(z)$ must be the same for all cases of locally isotropic turbulence.[59] This hypothesis does *not* require the large Reynolds number limit to have been taken, and indeed, it has been verified to a good approximation even for transitional turbulent flows.[102] Only when a second similarity hypothesis is taken, that the $Re \to \infty$ limit exists, does the K41 scaling emerge. In fact, the connection with critical phenomena is very close: the first similarity hypothesis is not quite correct. The $kL \to \infty$ limit does not exist, and the asymptotics exhibit incomplete similarity[100,101] leading to the existence of a new exponent, the intermittency exponent that we described earlier.

In the renormalization group-informed perspective presented throughout this chapter, minimal models are in the universality class of the transition and thus give a quantitatively accurate account of the transition. They are not approximations in the sense that they have inadequate realism. As we have discussed above, the predictions of minimal models are quantitatively accurate, despite the fact that the minimal models are seemingly lacking in realism. In biology, there are many more levels of description than in physical systems because of the intertwining of molecular sequence, three-dimensional structure, large-scale molecular motions, elasticity, gene expression, metabolism and energy flow, regulation, signaling, cell division, tissue mechanics, electrical activity, . . ., all the way up to ecology and evolutionary processes. Choosing the right questions and the right levels of description is critically important because there is no other way forward. As Fisher wrote: "Schrodinger's equation really does have *almost no* relevance to what one actually does in vast areas of chemistry."[1,2] I believe that the analysis of biological modeling from the perspective of philosophy of science would be a very interesting project, one which might bring a broader range of topics into the field and which might ultimately help biophysicists and biologists in their efforts to understand these complex systems by being aware of the necessity to choose the right level of description. In my own work, there are a number of examples where this has been attempted, in topics ranging from the evolution of the genetic code,[103] the open-ended growth of biological complexity,[104] to the topological scaling laws that have recently been uncovered in phylogenetic trees.[105]

For physicists, it was all over by the early 1990s, as even elementary particle physicists came to the clear realization that their subject was about effective field theory and not about fundamental physical law. This was a remarkable 180° turn for a community which by and large had regarded renormalizability as a necessary but not sufficient condition for fundamental quantum field theories and which still adhered to the program of first understanding the basic constituents of matter, so that they could then be combined to understand the low-energy world to which we have direct access. But within a few years of Fisher's article, this view was superseded by the general acceptance of effective field theory, the fascinating history of which has been beautifully described from a personal perspective by Weinberg[106] and summarized intellectually by Georgi.[107] I want to end by briefly talking about effective field theory, through the lens of Georgi's review, because it is truly remarkable to see the convergence between the modern perspective of condensed matter, as championed by Fisher, and the modern perspective of high-energy physics, as embodied in effective field theory.

4.1. *High-energy physics: Does high energy matter?*

The starting point of effective field theory is the recognition that the goal has changed. The goal is not to create a theory of everything but, according to Georgi, "to isolate a set of phenomena from all the rest, so that we can describe it without having to understand everything. . . Fortunately this is often possible. We can divide the parameter space of the world into different regions, in each of which there is a different appropriate description of the important physics."[107] This is the pragmatic rationale for doing effective theory, but historically, this is not how it developed. If we had a theory of everything, it would be awfully unwieldy to calculate something specific and so for convenience you "use the effective theory... it makes calculations easier because you are forced to concentrate on the important physics."[107] In practice, this means doing an expansion in scale, shrinking to zero size the features of the physics that are smaller than the scale of interest. As Georgi writes,[107] "This gives a useful and simple picture of the important physics. The finite size effects that you have ignored are all small and can be included as perturbations." The attentive reader will have noticed that this passage is uncannily reminiscent of the modeling strategy articulated earlier in this article by Bogoliubov in 1947.[19] In high-energy physics, this strategy is equivalent to removing massive particles, but the price paid for this massive oversimplification (sorry, I could not resist!) is that the ultraviolet regularization is now non-trivial because of the way that coupling constants vary with scale. Integrating out the high-energy degrees of freedom results in a non-local theory because of virtual exchange processes with the neglected massive particles. The program of effective theory replaces these non-local interactions with local interactions that are[107] "...constructed to give the same physics at low energies... Thus the domain of utility of an effective theory is necessarily bounded from above in energy scale." This is of course reminiscent of Bogoliubov (prompted by Landau) connecting his minimal model to reality by replacing his point interaction amplitude with the amplitude of the binary collision probability in the Born scattering approximation and of course to the Eliashberg extension of the BCS model of superconductivity.

I pause to note another important connection to condensed matter physics. The scale dependence of interactions was discovered in the context of critical phenomena by Kadanoff[64] and Patashinski and Pokrovsky[108] in 1966 but identified earlier during the development of the renormalization group,[109,110] and used by the Landau[111] school in the context of quantum electrodynamics and the famous "Moscow Zero"[112–116] — the divergence of the coupling constant that renders the field theory ill defined at high energies. I refer the interested reader to a thoughtful modern perspective on the Moscow Zero in the context of condensed matter physics[117] and the remarkable experimental observation of the scale dependence of the coupling constant in graphene.[118]

The effective field theory is also bounded below in energy. This is because the effective field theory will generate its own massive particles, and on energy scales smaller than these masses, these particles can again be eliminated to generate a new effective theory at a lower-energy scale too. Thus, an effective field theory lives in an intermediate or "middle" scale of energy.

Effective field theory is in practice looked at in a different way because we do not have any information about the high-energy theory that we invoked above. So, effective field theory is used to describe the physics of an energy scale of interest up to a certain level of accuracy, with a small or finite number of parameters that "parameterize our ignorance in a useful way."[107] This is not the same as the old-fashioned view of renormalizability in field theory because it is expected that with increasing energy, the non-renormalizable interactions get replaced with a new effective theory.

Weinberg recounts[106] how his perspective started to change in 1976, when he learned about Wilson's approach to critical phenomena by integrating out the ultraviolet degrees of freedom and using the renormalization group equation to ensure that physical quantities are cutoff independent. He realised that this entails introducing "every possible interaction renormalizable or not, to keep physics strictly cutoff independent. From this point of view, it doesn't make much difference whether the underlying theory is renormalizable or not... Non-renormalizable theories, I realized, are just as renormalizable as renormalizable theories."[106]

Georgi describes this change in perspective in a remarkable comment that sounds the death knell for the fundamental viewpoint of elementary particle physics[107]: "How does this process end? It is possible, I suppose, that at some very large energy scale... the theory is simply renormalizable in the old sense. This seems unlikely... It may even be possible that there is no end, simply more and more scales as one goes to higher and higher energy. Who knows? Who cares?"

Michael Fisher would agree.

Acknowledgments

First and foremost, I want to acknowledge the many inspirations, insights and perspectives that I learned from Michael's writings and personal discussions in our scientific interactions. One of my own struggles with renormalization group was to figure out how to separate it from the context of statistical and quantum field theory. The way Michael presented his own work, as well as his pedagogical introductions to the field, aided my understanding of the field that he had helped invent. Indeed, Michael cared deeply about the presentation of science as well as its content, and in a pre-powerpoint world, he was ahead of the curve in bringing his talks to life by writing on his slides in real time, filling in the gaps he had purposefully left. In this way, he managed to combine the immediacy of a blackboard talk with the opportunity to create a skillful layout that aided the presentation. I felt that I had got to know him properly when he shared with me his secret: the slides were written in indelible marker, but the in-lecture annotations were written in water-soluble marker and were easily washed away after the talk, ready for the next one. To the extent that I am successful as an educator and communicator of science, it is partly because of Michael's example, along with those of Jim Langer and Sir James Lighthill.

I also want to use this opportunity to acknowledge with thanks my friend and collaborator Yoshi Oono, whose brilliant and sometimes iconoclastic perspectives uniquely influenced my understanding of science, even going back to the time when I was a graduate student. It is no accident that Michael had the breadth and intellectual taste to choose to begin the "forms of matter" section of his article with Yoshi's largely overlooked work on the conformational space renormalization group of polymeric matter. I remember that Yoshi was excited that Michael had highlighted his work in his article and that is how I learned of Michael's article.

I also wish to thank my other collaborators and students, whose work was mentioned here: Gregory Eyink, Hong-Yan Shih, Tsung-Lin Hsieh, Chi Xue, Zhiru Liu, Xueying Wang, Lin-Yuan Chen, Nicholas Guttenberg, Tuan Tran, Alisia Prescott, Pinak Chakraborty, Hamid Kellay, Gustavo Gioia, Walter Goldburg, Maksim Sipos, Olivier Martin, Fong Liu, Kalin Vetsigian, Carl Woese, John Veysey.

I wish to thank Robert Batterman for many discussions on the philosophy of science, and for his skillful articulation and observations about the way modern statistical physicists approach the task of practicing science.

Lastly, I thank Amnon Aharony and Leo Radzhihovsky for inviting me to contribute this essay, for their helpful suggestions that improved the manuscript, and for their patience during the unforeseen delays.

This work was partially supported by the Simons Foundation through Targeted Grant "Revisiting the Turbulence Problem Using Statistical Mechanics" (Grant No. 662985(N.G.)).

References

1. M. E. Fisher, Condensed matter physics: Does quantum mechanics matter? In H. Feshbach, T. Matsui and A. Oleson (eds.), *Niels Bohr: Physics and the World* (Gordon and Breach, New York, 1988), pp. 65–115.
2. M. E. Fisher, Condensed matter physics: Does quantum mechanics matter? In F. Michael E. (ed.), *Excursions in the Land of Statistical Physics* (World Scientific Publishing Co. Pte. Ltd., 2017), pp. 207–253.
3. J. G. Bednorz and K. A. Müller, Possible high Tc superconductivity in the Ba-La-Cu-O system. *Zeitschrift für Physik B Condensed Matter* **64**(2), 189–193 (1986).
4. M. Wu, J. Ashburn, C. Torng, P. Hor and R. Meng, Superconductivity at 93 K in a new mixed-phase Y-Ba-Cu-O compound system at ambient pressure. *Phys. Rev. Lett.* **58**(9), 908–910 (1987).
5. D. Shechtman, I. Blech, D. Gratias and J. Cahn, Metallic phase with long-range orientational order and no translational symmetry. *Phys. Rev. Lett.* **53**(20), 1951–1953 (1984).
6. K. Klitzing, G. Dorda and M. Pepper, New method for high-accuracy determination of the fine-structure constant based on quantized hall resistance. *Phys. Rev. Lett.* **45**(6), 494–497 (1980).
7. D. Tsui, H. Stormer and A. Gossard, Two-dimensional magnetotransport in the extreme quantum limit. *Phys. Rev. Lett.* **48**(22), 1559–1562 (1982).
8. R. Laughlin, Anomalous quantum hall effect: An incompressible quantum fluid with fractionally charged excitations. *Phys. Rev. Lett.* **50**(18), 1395–1398 (1983).
9. By the sartorial standards of physicists, at least.
10. Feynman's lecture "There's plenty of room at the bottom" is available online at: https://www.zyvex.com/nanotech/feynman.html.

11. W. Pauli, Pauli's proposal was recorded in a letter to Lisa Meitner (1930), available online at CERN: http://cds.cern.ch/record/83282/files/meitner.
12. H. Yukawa, On the interaction of elementary particles. i. *Proc. Physico-Mathematical Soc. Jpn. 3rd Ser.* **17**, 48–57 (1935).
13. F. Harrison, H. Kruse and A. McGuire, Detection of the free neutrino: A confirmation. *Science* **124**(3212), 103–104 (1956).
14. S. H. Neddermeyer and C. D. Anderson, Note on the nature of cosmic-ray particles. *Phys. Rev.* **51**(10), 884–886 (1937).
15. C. M. Lattes, H. Muirhead, G. P. Occhialini and C. F. Powell, Processes involving charged mesons. *Nature* **159**(4047), 694–697 (1947).
16. A graph of the relative usage during the years 1960–2019 in books of the terms "solid state physics" and "condensed matter physics" can be seen online at: https://tinyurl.com/ hfjebn68.
17. P. W. Anderson, More is different: Broken symmetry and the nature of the hierarchical structure of science. *Science* **177**(4047), 393–396 (1972).
18. P. W. Anderson, *Basic Notions of Condensed Matter Physics* (CRC Press, 2018).
19. N. Bogoliubov, On the theory of superfluidity. *J. Phys.* **11**(1), 23–32 (1947).
20. J. Bardeen, L. Cooper and J. Schrieffer, Microscopic theory of superconductivity. *Phys. Rev.* **106**(1), 162–164 (1957).
21. J. Bardeen, L. Cooper and J. Schrieffer, Theory of superconductivity. *Phys. Rev.* **108**(5), 1175–1204 (1957).
22. E. B. Kolomeisky and J. P. Straley, Renormalization-group analysis of the ground-state properties of dilute bose systems in d spatial dimensions. *Phys. Rev. B* **46**(18), 11749–11756 (1992).
23. D. S. Fisher and P. Hohenberg, Dilute bose gas in two dimensions. *Phys. Rev. B* **37**(10), 4936–4943 (1988).
24. M. Bijlsma and H. Stoof, Renormalization group theory of the three-dimensional dilute bose gas. *Phys. Rev. A* **54**(6), 5085–5103 (1996).
25. J. O. Andersen, Theory of the weakly interacting bose gas. *Rev. Mod. Phys.* **76**(2), 599–639 (2004).
26. J. Bardeen and D. Pines, Electron-phonon interaction in metals. *Phys. Rev.* **99**(4), 1140–1150 (1955).
27. Y. Oono, Statistical physics of polymer solutions: Conformation-space renormalization-group approach. In *Advances in Chemical Physics*. pp. 301–437 (1985).
28. N. Goldenfeld, *Lectures On Phase Transitions and the Renormalization Group* (Addison-Wesley Reading, MA, 1992).
29. R. W. Batterman, *The Devil in the Details: Asymptotic Reasoning in Explanation, Reduction, And Emergence* (Oxford University Press, 2001).
30. R. W. Batterman, Asymptotics and the role of minimal models. *Br. J. Philos. Sci.* **53**(1), 21–38 (2002).
31. R. W. Batterman and C. C. Rice, Minimal model explanations. *Philos. Sci.* **81**(3), 349–376 (2014).
32. N. N. Bogoliubov, A new method in the theory of superconductivity. 1. *Sov. Phys. JETP* **7**(1), 41–46 (1958).
33. J. Valatin, Comments on the theory of superconductivity. *Il Nuovo Cimento.* **7**(6), 843–857 (1958).
34. L.-Y. Chen, N. Goldenfeld and Y. Oono, Renormalization group and singular perturbations: Multiple scales, boundary layers, and reductive perturbation theory. *Phys. Rev. E* **54**(1), 376–394 (1996).
35. J. Veysey and N. Goldenfeld, Simple viscous flows: From boundary layers to the renormalization group. *Rev. Mod. Phys.* **79**(3), 883–927 (2007).

36. M. Lange, On minimal model explanations: A reply to batterman and rice. *Philos. Sci.* **82**(2), 292–305 (2015).
37. T. McKenna, Lange on minimal model explanations: A defense of batterman and rice. *Philos. Sci.* **88**(4), 731–741 (2021).
38. R. W. Batterman, *A Middle Way: A Non-fundamental Approach to Many-body Physics* (Oxford University Press, 2021).
39. G. Eliashberg, Interactions between electrons and lattice vibrations in a superconductor. *Sov. Phys. JETP* **11**(3), 696–702 (1960).
40. A. Migdal, Interaction between electrons and lattice vibrations in a normal metal. *Sov. Phys. JETP* **7**(6), 996–1001 (1958).
41. W. McMillan, Transition temperature of strong-coupled superconductors. *Phys. Rev.* **167**(2), 331–344 (1968).
42. P. Morel and P. Anderson, Calculation of the superconducting state parameters with retarded electron-phonon interaction. *Phys. Rev.* **125**(4), 1263–1271 (1962).
43. T. Lee and C. Yang, Many-body problem in quantum mechanics and quantum statistical mechanics. *Phys. Rev.* **105**(3), 1119–1120 (1957).
44. T. Lee, K. Huang and C. Yang, Eigenvalues and eigenfunctions of a bose system of hard spheres and its low-temperature properties. *Phys. Rev.* **106**(6), 1135–1145 (1957).
45. K. Brueckner and K. Sawada, Bose-Einstein gas with repulsive interactions: Hard spheres at high density. *Phys. Rev.* **106**(6), 1128–1135 (1957).
46. L. Landau, Two-fluid model of liquid helium II. *J. Phys. Ussr.* **5**(1), 71–90 (1941).
47. S. Cocco, S. Leibler and R. Monasson, Neuronal couplings between retinal ganglion cells inferred by efficient inverse statistical physics methods. *Proc. Natl. Acad. Sci.* **106**(33), 14058–14062 (2009).
48. A. E. Noble, T. S. Rosenstock, P. H. Brown, J. Machta and A. Hastings, Spatial patterns of tree yield explained by endogenous forces through a correspondence between the Ising model and ecology. *Proc. Natl. Acad. Sci.* **115**(8), 1825–1830 (2018).
49. P. J. Flory, The configuration of real polymer chains. *J. Chem. Phys.* **17**(3), 303–310 (1949).
50. S. Edwards, The statistical mechanics of polymers with excluded volume. *Proc. Phys. Soc.* **85**(4), 613–624 (1965).
51. T. Ohta and Y. Oono, Conformation space renormalization theory of semidilute polymer solutions. *Phys. Lett. A* **89**(9), 460–464 (1982).
52. M. E. Fisher and D. S. Gaunt, Ising model and self-avoiding walks on hypercubical lattices and "high-density" expansions. *Phys. Rev.* **133**(1A), 224–239 (1964).
53. K. G. Wilson and M. E. Fisher, Critical exponents in 3.99 dimensions. *Phys. Rev. Lett.* **28**(4), 240–243 (1972).
54. S. Edwards, The theory of polymer solutions at intermediate concentration. *Proc. Phys. Soc.* **88**(2), 265–280 (1966).
55. P. de Gennes, Exponents for the excluded volume problem as derived by the wilson method. *Phys. Lett. A* **38**(5), 339–340 (1972).
56. P. Wiltzius, H. R. Haller, D. S. Cannell and D. W. Schaefer, Dynamics of long-wavelength concentration fluctuations in solutions of linear polymers. *Phys. Rev. Lett.* **53**(8), 834–837 (1984).
57. S. Pope, *Turbulent Flows* (Cambridge University Press, Cambridge, UK, 2000).
58. K. Sreenivasan and R. Antonia, The phenomenology of small-scale turbulence. *Ann. Rev. Fluid Mech.* **29**, 435–472 (1997).
59. A. N. Kolmogorov, Local structure of turbulence in incompressible fluid at a very high reynolds number. *Dokl. Akad. Nauk. SSSR* **30**, 301–305 (1941). [English translation in *Proc. R. Soc. London Ser. A* **434**, 9–13 (1991)].

60. H. L. Grant, R. W. Stewart and A. Moilliet, Turbulence spectra in a tidal channel. *J. Fluid Mech.* **12**, 241–268 (1962).
61. A. Kolmogorov, A refinement of previous hypotheses concerning the local structure of turbulence in a viscous incompressible fluid at high reynolds number. *J. Fluid Mech.* **13**, 82–85 (1962).
62. K. Sreenivasan and P. Kailasnath, An update on the intermittency exponent in turbulence. *Phys. Fluids A* **5**(2), 512–514 (1993).
63. B. Widom, Equation of state in the neighbourhood of the critical point. *J. Chem. Phys.* **43**, 3898–3905 (1965).
64. L. P. Kadanoff, Scaling laws for ising models near tc. *Physics* **2**, 263–272 (1966).
65. G. Eyink and N. Goldenfeld, Analogies between scaling in turbulence, field theory and critical phenomena. *Phys. Rev. E* **50**, 4679–4683 (1994).
66. N. Goldenfeld, Roughness-induced critical phenomena in a turbulent flow. *Phys. Rev. Lett.* **96**, 044503 (2006).
67. J. Nikuradze, Stromungsgesetze in rauhen Rohren. *VDI Forschungsheft.* **361**(1) (1933). [English translation available as National Advisory Committee for Aeronautics, Tech. Memo. 1292 (1950). Online at: http://hdl.handle.net/2060/19930093938].
68. M. Mehrafarin and N. Pourtolami, Intermittency and rough-pipe turbulence. *Phys. Rev. E* **77**, 055304 (2008).
69. H. Blasius, Das Ähnlichkeitsgesetz bei Reibungsvorgängen in Flüssigkeiten. *Forschg. Arb. Ing. — Wes.* **134** (1913).
70. A. Strickler, Beitrage zur frage der geschwindigkeitsformel und der rauhigkeitszahlen fur strome, kanale und geschlossene leitungen. Mitteilungen des Eidgenössischen Amtes für Wasserwirtschaft 16, Bern, Switzerland. Translated as "Contributions to the question of a velocity formula and roughness data for streams, channels and closed pipelines." by T. Roesgan and W. R. Brownie, Translation T-10, W. M. Keck Lab of Hydraulics and Water Resources, Calif. Inst. Tech., Pasadena, Calif. January 1981 (1923).
71. G. Gioia and P. Chakraborty, Turbulent friction in rough pipes and the energy spectrum of the phenomenological theory. *Phys. Rev. Lett.* **96**, 044502 (2006).
72. N. Guttenberg and N. Goldenfeld, Friction factor of two-dimensional rough-boundary turbulent soap film flows. *Phys. Rev. E* **79**(6), 65306 (2009).
73. T. Tran, P. Chakraborty, N. Guttenberg, A. Prescott, H. Kellay, W. Goldburg, N. Goldenfeld and G. Gioia, Macroscopic effects of the spectral structure in turbulent flows. *Nature Phys.* **4**, 438–441 (2010).
74. H. Kellay, T. Tran, W. Goldburg, N. Goldenfeld, G. Gioia and P. Chakraborty, Testing a missing spectral link in turbulence. *Phys. Rev. Lett.* **109**(25), 254502 (2012).
75. A. Vilquin, J. Jagielka, S. Djambov, H. Herouard, P. Fischer, C.-H. Bruneau, P. Chakraborty, G. Gioia and H. Kellay, Asymptotic turbulent friction in 2d rough-walled flows. *Sci. Adv.* **7**(5), eabc6234 (2021).
76. N. Goldenfeld and H.-Y. Shih, Turbulence as a problem in non-equilibrium statistical mechanics. *J. Stat. Phys.* **167**(3–4), 575–594 (2017).
77. Y. Pomeau, Front motion, metastability and subcritical bifurcations in hydrodynamics. *Physica* **23D**, 3–11 (1986).
78. D. Barkley, Simplifying the complexity of pipe flow. *Phys. Rev. E* **84**(1), 016309 (2011).
79. N. Goldenfeld, N. Guttenberg and G. Gioia, Extreme fluctuations and the finite lifetime of the turbulent state. *Phys. Rev. E* **81**(3), 035304 (2010).

80. M. Sipos and N. Goldenfeld, Directed percolation describes lifetime and growth of turbulent puffs and slugs. *Phys. Rev. E* **84**(3), 035304 (2011).
81. H.-Y. Shih, T.-L. Hsieh and N. Goldenfeld, Ecological collapse and the emergence of travelling waves at the onset of shear turbulence. *Nat. Phys.* **12**(3), 245–248 (2016).
82. D. Barkley, Theoretical perspective on the route to turbulence in a pipe. *J. Fluid Mech.* **803**(1) (2016).
83. X. Wang, H.-Y. Shih and N. Goldenfeld, Stochastic model for quasi-one-dimensional transitional turbulence with streamwise shear interactions. *Phys. Rev. Lett.* **129**(3), 034501 (2022).
84. M. Chantry, L. S. Tuckerman and D. Barkley, Universal continuous transition to turbulence in a planar shear flow. *J. Fluid Mech.* **824** (2017).
85. S. Gomé, L. S. Tuckerman and D. Barkley, Statistical transition to turbulence in plane channel flow. *Phys. Rev. Fluids* **5**(8), 083905 (2020).
86. G. Lemoult, L. Shi, K. Avila, S. V. Jalikop, M. Avila and B. Hof, Directed percolation phase transition to sustained turbulence in couette flow. *Nat. Phys.* **12**(3), 254–258 (2016).
87. V. Mukund and B. Hof, The critical point of the transition to turbulence in pipe flow. *J. Fluid Mech.* **839**, 76–94 (2018).
88. L. Klotz, G. Lemoult, K. Avila and B. Hof, Phase transition to turbulence in spatially extended shear flows. *Phys. Rev. Lett.* **128**(1), 014502 (2022).
89. G. Lemoult, V. Mukund, J. M. Lopez, H.-Y. Shih, G. Linga, J. Mathiesen, N. Goldenfeld and B. Hof, Directed percolation and puff jamming near the transition to pipe turbulence (2023). (unpublished).
90. H. Hinrichsen, Non-equilibrium critical phenomena and phase transitions into absorbing states. *Adv. Phys.* **49**(7), 815–958 (2000).
91. B. Hof, Directed percolation and the transition to turbulence. *Nat. Rev. Phys.* **5**(1), 62–72 (2023).
92. M. Avila, D. Barkley and B. Hof, Transition to turbulence in pipe flow. *Ann. Rev. Fluid Mech.* **55**, 575–602 (2023).
93. M. Mobilia, I. T. Georgiev and U. C. Täuber, Phase transitions and spatio-temporal fluctuations in stochastic lattice Lotka-Volterra models. *J. Stat. Phys.* **128**(1–2), 447–483 (2007).
94. M. Berry, Beyond rainbows. *Curr. Sci. (Bangalore)* **59**(21–22), 1175–1191 (1990).
95. M. Berry, Asymptotics, singularities and the reduction of theories. In *Studies in Logic and the Foundations of Mathematics*, Vol. 134 (Elsevier, 1995), pp. 597–607.
96. M. E. Fisher, Scaling, universality and renormalization group theory. In F. J. W. Hahne (ed.), *Critical Phenomena* (Springer Berlin Heidelberg, Berlin, Heidelberg, 1983), pp. 1–139.
97. K. G. Wilson and J. Kogut, The renormalization group and the epsilon expansion. *Phys. Rep.* **12**(2), 75–199 (1974).
98. N. Goldenfeld, O. Martin and Y. Oono, Intermediate asymptotics and renormalization group theory. *J. Sci. Comput.* **4**, 355–372 (1989).
99. N. Goldenfeld, O. Martin, Y. Oono and F. Liu, Anomalous dimensions and the renormalization group in a nonlinear diffusion process. *Phys. Rev. Lett.* **64**(12), 1361–1364 (1990).
100. G. Barenblatt and Y. B. Zel'Dovich, Self-similar solutions as intermediate asymptotics. *Ann. Rev. Fluid Mech.* **4**(1), 285–312 (1972).
101. G. Barenblatt, *Scaling, Self-similarity, and Intermediate Asymptotics* (Cambridge University Press, 1996).

102. R. T. Cerbus, C.-C. Liu, G. Gioia and P. Chakraborty, Small-scale universality in the spectral structure of transitional pipe flows. *Sci. Adv.* **6**(4), eaaw6256 (2020).
103. K. Vetsigian, C. Woese and N. Goldenfeld, Collective evolution and the genetic code. *Proc. Natl. Acad. Sci.* **103**(28), 10696–10701 (2006).
104. N. Guttenberg and N. Goldenfeld, Cascade of complexity in evolving predator-prey dynamics. *Phys. Rev. Lett.* **100**(5), 058102 (2008).
105. C. Xue, Z. Liu and N. Goldenfeld, Scale-invariant topology and bursty branching of evolutionary trees emerge from niche construction. *Proc. Natl. Acad. Sci.* **117**(14), 7879–7887 (2020).
106. S. Weinberg, Effective field theory, past and future. *Int. J. Mod. Phys. A* **31**(06), 1630007 (2016).
107. H. Georgi, Effective field theory. *Ann. Rev. Nucl. Part. Sci.* **43**(1), 209–252 (1993).
108. A. Patashinskii and V. Pokrovskii, Behavior of ordered systems near the transition point. *Sov. Phys. JETP* **23**(2), 292–297 (1966).
109. E. C. G. Stueckelberg von Breidenbach and A. Petermann, La normalisation des constantes dans la theorie des quanta. *Helvetica Physica Acta* **26**(5), 499–520 (1953).
110. M. Gell-Mann and F. Low, Quantum electrodynamics at small distances. *Phys. Rev.* **95**(5), 1300–1312 (1954).
111. D. Ter Haar (ed.), Collected papers of LD Landau (1965). Pergamon. available online at https://www.sciencedirect.com/book/9780080105864/collected-papers-of-l-d-landau.
112. L. Landau, A. Abrikosov and I. Khalatnikov, On the removal of infinities in quantum electrodynamics. *russ.) Dokl. Akad. Nauk SSSR* **95**, 497–500 (1954).
113. L. Landau, A. Abrikosov and I. Khalatnikov, Asymptotic expression for green's function of electrons in quantum electrodynamics. *Dokl. Akad. Nauk SSSR* **95**, 773 (1954).
114. L. Landau, A. Abrikosov and I. Khalatnikov, An asymptotic expression for the photon green function in quantum electrodynamics. *Dokl. Akad. Nauk SSSR* **95**(6), 1177 (1954).
115. L. Landau, A. Abrikosov and I. Khalatnikov, The electron mass in quantum electrodynamics. *Dokl. Akad. Nauk SSSR* **96**, 261 (1954).
116. L. Landau, On the quantum field theory. In W. Pauli (ed.), *Niels Bohr and the Development of Physics* (Pergamon, London, 1955), p. 52.
117. S.-K. Jian, E. Barnes and S. Das Sarma, Landau poles in condensed matter systems. *Phys. Rev. Res.* **2**(2), 023310 (2020).
118. J. P. Reed, B. Uchoa, Y. I. Joe, Y. Gan, D. Casa, E. Fradkin and P. Abbamonte, The effective fine-structure constant of freestanding graphene measured in graphite. *Science* **330**(6005), 805–808 (2010).

https://doi.org/10.1142/9789811282386_0020

Chapter 20

Knots from a Random Matrix Theory with Replicas

S. Hikami
Okinawa Institute of Science and Technology Graduate University,
1919-1 Tancha, Okinawa 904-0495, Japan
hikami@oist.jp

A classical knot is described by a one-stroke trajectory of a string with entanglements. The replica method is a powerful tool in statistical mechanics for dealing with string-like objects like polymers or self-avoiding walks. We consider here the ($N \to 0$) replica limit for Gaussian means of products of traces of $N \times N$ Hermitian matrices, which correspond to one-stroke graphs for knots. The Seifert surfaces of knots and links are thus related to a random matrix model. The zeros of Alexander polynomials on the unit circle are discussed for the case of n-vertices in analogy with the Yang–Lee edge singularity. The extension of one-matrix models to higher dimensional knots is considered, and also to the half-integral level k in a Chern–Simons gauge theory.

1. Introduction

The replica method is a powerful tool in statistical mechanics to deal with disordered systems. The critical behavior of a single polymer or equivalently the geometric properties of self-avoiding walks are given by a zero-replica limit ($N \to 0$) of the $O(N)$ vector model,[1] allowing the use of Wilson–Fisher-ϵ expansion.[2] The Anderson localization in a random potential is known to be the zero replica limit of a Grassmannian non-linear σ model, opening the way to a renormalization-group analysis.[3] The intersection numbers of p-spin curves for the moduli space of Riemann surfaces[5,13] have also been investigated by a replica method using a random matrix theory with an external source.[6–8] The replica limits ($N \to 0$) appears as one-stroke trajectories for Wilson lines in a quantum field theory.

Generally, those problems are related to one-stroke trajectories of strings. We examine in this chapter how the replica method for a Gaussian random matrix theory corresponds to the theory of knots studied intensively in mathematics and applied in statistical mechanics (for instance, see the textbook of knots[9]). This classical theory has bee extended to higher-dimensional knots, such as 2-knots,[10] which may also be put in correspondence with matrix models.

2. Knot Graphs and Matrix Models

Let us recall first some earlier results on Gaussian matrix integrals. We consider Gaussian averages as in[8] of the form

$$\langle X(M)\rangle = \frac{1}{Z}\int dM X(M) e^{-\frac{1}{2}\mathrm{tr}M^2}, \tag{1}$$

where $X(M)$ is a product of multi-traces of $\mathrm{tr}M^n$, i.e., $(\prod_i \mathrm{tr}M^{n_i})$. The matrices M are Hermitian $N \times N$ and Z is a normalization, which ensures that $\langle 1\rangle = 1$.

For multi-traces such as $\langle \frac{1}{N}\mathrm{tr}(M^{n_1}\cdots \mathrm{tr}M^{n_k})\rangle$, the application of Wick's theorem produces a sum of Feynman graphs with double lines between the k vertices. In the $N \to 0$ limit, the only graphs which remain consist of a single stroke line going along the propagators from vertex to vertex. From our previous work on intersection numbers of the moduli space of curves on a Riemann surface,[6] starting with the Konsevitch model,[13] we have found an explicit formula for the replica limit of the generating function of multi-traces. If

$$U(\sigma_1,\ldots,\sigma_n) = \frac{1}{N}\left\langle \prod_{i=1}^{n} \mathrm{tr}e^{\sigma_i M}\right\rangle \tag{2}$$

denotes this generating function, its zero-replica limit is given by

$$\lim_{N\to 0} U(\sigma_1,\ldots,\sigma_n) = \frac{2^n}{\chi^2}\prod_{i=1}^{n}\sinh\left(\frac{\chi\sigma_i}{2}\right), \tag{3}$$

with $\chi = \sum_{i=1}^{n}\sigma_i$. The expansion of the generating function (3) in powers of σ_i leads easily to explicit results such as selection rules, for instance,

$$\lim_{N\to 0}\frac{1}{N}\left\langle (\mathrm{tr}M^3)^k\right\rangle = 0 \tag{4}$$

unless $k = 2$ (mod 4). For other powers (1), one finds, for instance,

$$\lim_{N\to 0}\frac{1}{N}\langle (\mathrm{tr}M^3)^{4g-2}\rangle = \frac{3^{3g-2}2^{-2g}(6g-4)!(4g-2)!}{g!(3g-2)!}. \tag{5}$$

In our earlier work, g was identified with the genus of the Riemann surface whose intersection was the r.h.s. of (5). The Gaussian mean of the above case in the replica limit has a number, which counts the different contraction of legs. For instance, $g = 1$ case, this number becomes 3, which is three possibilities of contraction of six legs of the two vertices in the replica limit. Thus, such numbers are related to the numbers of different knots and circles.

On the other hand, mathematicians describe knots with the help of Seifert surfaces, whose edge provides a directed knot.[9] A Seifert surface is made of disks each connected to one another by twisted bands (Seifert band). One establishes a correspondence between such a surface with Feynman diagrams, in which a disk corresponds to a vertex and a connected Seifert band between disks as a propagator. From the point of view of knots, this replica limit in the matrix side involves both knots and unknotted circles: the correspondence between knots and matrix averages is not one-to-one. The matrix model ignores the two ways of the over or down trails in the Seifert band. The matrix model belongs to two dimensions, while knots belong to three dimensions, embedded into S^3. The description of directed trails over and down is characterized by a skein relation.[9] If one endows the vertex of a matrix graph with a sign, like the head and tail of a coin, the connection of two vertices (head and tail) produces a Seifert band with the two choices of up or down crossing.

The standard catalogue of knots (Rolfsen catalogue) lists the knots with their signature, such as $3_1, 4_1, 5_1, 5_2, \ldots$ The knot of 3_1, named trefoil, 4_1 is a figure 8 knot, etc. The correspondence with matrix averages in the zero-replica limit is built as follows. For instance, a trefoil knot (3_1) corresponds to $\lim_{N\to 0} \frac{1}{N}\langle(\mathrm{tr}M^3)^2\rangle$ in (1). The trivalent vertex is described by $\mathrm{tr}M^3$. The average is Gaussian average, but for the Seifert surface, the Seifert band is taken as a connector (propagator). It is obtained by flipping half of the vertices and connecting all vertices by Seifert bands.

For the 5_1 knot, the corresponding Gaussian average is $\langle(\mathrm{tr}M^5)^2\rangle$. (From now on, a factor $1/N$ and a limit $N \to 0$ is meant). The Gaussian average of this is 165, which is the sum of three different graphs, and the sum of the numbers of the contractions of these graphs becomes 165. The knot of 5_1 has a contraction number as 5. These different graphs lead to 5_1 and circles. For the 7_1 knot, the corresponding term is $\langle(\mathrm{tr}M^7)^2\rangle$. In general, the term of $\langle(\mathrm{tr}M^n)^2\rangle$ gives the knot, named as n_1, where n is odd.

The 4_1 (figure-eight) knot corresponds to the average $\langle(\mathrm{tr}M^4)(\mathrm{tr}M^2)^2\rangle$. It is easy to find the Gaussian means of the product of matrices for each Seifert graphs of knots by the trails which take the opposite direction path at the crossing points of knot graphs. Gaussian means of (1) include unknotted circles, therefore we need some selection rule which separate knots graphs from unknotted circles. The separation of the unknotted circles is made through a skein relation.[9] Table 1 lists the correspondence between the Rolfsen numbers and the Gaussian averages.

The same Gaussian means appear for different knots. The difference is due to the different crossing ways in a Seifert band. For instance, the knots of $8_9, 8_{17}, 8_{18}, 8_{19}, 8_{20}, 8_{21}$ have the same Gaussian means. They correspond indeed to several different configurations of the crossings in the Seifert band. The contractions of 8_9, 8_{18} and 8_{17} are all different (Feynman diagrams are different). Starting from the alternating knot of 8_{17}, by the changes of the paired crossing of over or

Table 1. Standard knot notation (left) and replica limit of Gaussian mean (right) with Seifert band contractions denoted by s.

3_1	$\langle(\mathrm{tr}M^3)^2\rangle_s$
4_1	$\langle(\mathrm{tr}M^4)(\mathrm{tr}M^2)^2\rangle_s$
5_1	$\langle(\mathrm{tr}M^5)^2\rangle_s$
5_2	$\langle(\mathrm{tr}M^3)^2(\mathrm{tr}M^2)^2\rangle_s$
6_1	$\langle(\mathrm{tr}M^4)(\mathrm{tr}M^2)^4\rangle_s$
6_2	$\langle(\mathrm{tr}M^6)(\mathrm{tr}M^4)(\mathrm{tr}M^2)\rangle_s$
6_3	$\langle(\mathrm{tr}M^6)(\mathrm{tr}M^3)^2\rangle_s$
7_1	$\langle(\mathrm{tr}M^7)^2\rangle_s$
7_2	$\langle(\mathrm{tr}M^3)^2(\mathrm{tr}M^2)^4\rangle_s$
7_3	$\langle(\mathrm{tr}M^5)^2(\mathrm{tr}M^2)^2\rangle_s$
7_4	$\langle(\mathrm{tr}M^3)^2(\mathrm{tr}M^2)^4\rangle_s$
7_5	$\langle(\mathrm{tr}M^5)(\mathrm{tr}M^4)(\mathrm{tr}M^3)(\mathrm{tr}M^2)\rangle_s$
7_6	$\langle(\mathrm{tr}M^7)(\mathrm{tr}M^3)(\mathrm{tr}M^2)^2\rangle_s$
7_7	$\langle(\mathrm{tr}M^6)(\mathrm{tr}M^3)^2(\mathrm{tr}M^2)\rangle_s$

down of these Seifert bands, other knots of non-alternating knots 8_{19}, 8_{20} and 8_{21} are obtained.

Some cases appear with the same Gaussian mean but different Seifert graphs. The cases of 7_2 and 7_4 have the same Gaussian mean $\langle(\mathrm{tr}M^3)^2(\mathrm{tr}M^2)^4\rangle_s$, but the Seifert graphs are different.

The Gaussian mean is evaluated by Wick contraction, which is a commutative algebraic procedure. However, if one writes a Feynman diagram in the zero-replica limit, using for the drawing double-line propagators, the order chosen for writing the diagram matters. According to the order chosen for the double-line propagators, sometimes one generates knotted diagrams, but some ordering of lines may lead to different knots or even unknotted circles. For instance, in the case of $\langle(\mathrm{tr}M^3)^2\rangle$, the maximum crossing diagram gives the trefoil knot 3_1 or a circle, depending on the choices for drawing a maximum crossing Feyman diagram. This is the same in a Seifert band, which gives a knot (3_1) or a circle, depending upon the choices of which line goes over or down at a crossing.

If we use a standard Feynman diagram, instead of twisted Seifert graphs, there appear several graphs which are different, depending on how we write the propagators in a time order on a paper, i.e., the way of the crossing of the ribbons. For instance, in the case of $\langle(\mathrm{tr}M^3)^2\rangle$, the maximum crossing graphs give one-stroke graphs. If the change of order of ribbons crossing, we have 3_1 and also 4_1. We have mentioned that the head and tail connected vertex of Seifert surface gives 3_1 or a circle. Thus, there is a difference between a standard Feynman diagram for matrices and twisted Seifert graphs. We shall discuss the way of the distinctions for different knots by the numbering in the following section.

3. Chern–Simons Gauge Theory and Random Matrix

This Seifert band is similar to the propagator of following Chern–Simons action of gauge field[11]:

$$\mathcal{L} = \frac{k}{4\pi}\int_W \mathrm{tr}(A \wedge dA + \frac{2}{3} A \wedge A \wedge A), \tag{6}$$

where k is integer, which corresponds to a level of Lie group and W is a three-dimensional manifold.

Near the crossing point of Seifert band, two edges are twisted, and it means the two edges are bound to each other. The two bounded edges are analogous to two bounded merons, which form an instanton. The merons were discussed in QCD,[16] and they are also found in a superconductor as a vortex, which pairs condensate into two-dimensional superconductor.[17,18] These bound states are described by an Abelian or non-Abelian gauge theory.

The knot should be related to a Chern–Simons gauge theory, since the knot-invariant polynomial (Jones polynomial)[27] can be found from the Chern–Simons gauge theory with a quantum parameter $q = e^{\frac{2\pi i}{p}}$ with $p = k+2$.[11] This parameter p has a meaning of a spin p, and p-spin curve on the Riemann surface has a topological invariant quantity of an intersection number of the moduli space.[5,8]

The boundary of the three dimensional Chern–Simons theory, which is turned to two dimensions, is described by Wess–Zumino–Witten (WZW) non-linear sigma model. Thus, the matrix model in the replica limit should be connected to the WZW or Chern–Simons gauge theory.

The Seifert surface is interpreted as a matrix model with two-type head and tail vertices. The Seifert band shows the crossing point which has over or down two choices. This crossing can be interpreted as a twist operator of Pauli matrices σ_x and σ_y, which are elements of $su(2)$ Lie algebra. The vertex of tail can be defined by the vertex multiplied with twist operator σ. Then, the standard Wick contraction with replica limit gives knots. The WZW model for $SU(2)$ Lie group naturally appears by taking this $su(2)$ Lie algebra with k-valence vertices, where k is a level of $SU(2)_k$. The knot polynomial of Jones can be evaluated from the skein relation with a quantum parameter $q = e^{\frac{2\pi i}{k+2}}$.[11] The twist operator, which is coupled to vertices of tail, is similar to the external source, which is deterministic in a random matrix theory.[8]

In this respect, the super matrix $SU(n|m)$[19] may be useful. The replica limit is $n \to m$. When $n = 2$, this supermatrix becomes $SU(2|2)$ and the previous Seifert band with the twist operator σ will appear.

4. Numbering

The correspondence between matrix models, in the zero-replica limit, and knots can be made one-to-one by numbering the edges. We assign numbers to the edge lines, numbering the floor to which this line belongs. The floors are connected to over or down floors by two types of Seifert bands (staircase or escalator). In the case of an over-connection from nth to $(n+1)$th floor, the number increases of one bit from n to $n+1$. For a downward connection, n becomes $n-1$. At the crossing point of a Seifert band, the total number is conserved. For incoming numbers n and m, the sum of outgoing numbers is also $n+m$.

For the trefoil knot, the edges of two vertices carrying $\mathrm{tr}(M^3)$ are assigned numberings 1 and 2. The Seifert band of this case is $1 \to 2$ and $2 \to 1$, according to over and down twisted connections. This assignment is consistent with the fact that the trefoil is an alternating knot. Along the directed path, the trefoil is written as the sequence of numbers (2121212), which is expressed as the paired sequence of the crossings

$$\begin{pmatrix} 2 & 1 & 2 & 1 \\ 1 & 2 & 1 & 2 \end{pmatrix}, \tag{7}$$

where 3, the sum of the numbers of the first and second row in the same column, is conserved. This matrix is a sequence of 2 by 2 scattering matrices. Trivalent vertices can also provide unknotted cases if one Seifert band can be changed from over to down and vice versa. Then we have

$$\begin{pmatrix} 2 & 3 & 2 & 1 \\ 1 & 0 & 1 & 2 \end{pmatrix}. \tag{8}$$

Although the number conservation, the sum of incoming and outgoing, holds, the total sum of the first row is 8, whereas the sum of the second row is 4. It shows that this case reduces to an unknotted circle.

The number n introduced here may be transformed to a phase $e^{i\theta n}$ with an angle θ. The increase of the number n to $n+1$ leads to a change $e^{i\theta n} \to e^{i\theta(n+1)}$. This interpretation of phase is more close to a gauge-field picture. This phase is related to parameter of Jones polynomial $t^n = e^{\frac{2i\pi n}{p}}$ if we take θ as pure imaginary.

There are many non-alternating knots, which start from the eight crossing points of the Rolsen table as $8_{19}, 8_{20}$ and 8_{21}. The knot 8_{17} is an alternating knot. These four knots have the same graphs in the replica limit of matrix model, $\langle \mathrm{tr} M^8 (\mathrm{tr} M^4)^2 \rangle_s$. The non-alternating knots have repeated connections of over or down crossings. Thus, the numbers are increasing or decreasing in several digits. So, the numbers are needed to be greater than 2. For the alternating knot of 8_{17}, the numbers of sequence are (1212121212121212), while the non-alternating knot 8_{19} has a sequence (1232123212323212). These two knots have the same graph except the over or down connections. The following graph represents the knots of $8_{17}, 8_{19}, 8_{20}$ and 8_{21}. The three strands should be connected as the upper points to the lowest points. The path has upper to down direction. Then three directed circles are obtained which are connected by the eight transverse rungs. The center strand which has eight rungs represents $\mathrm{tr}(M^8)$, and the other both sides column represents tr(M^4). The rung means the Seifert band, which has two choices of over or down directed path. For the case 8_{17}, all segments of the center strand are assigned by 2, and 1 assigned for the vertical segments of strands of both sides. This alternating knot corresponds to the sequence (1212121212121212). The graph is called traditionally as "Amida-kuji" (Kuji means a lottery).

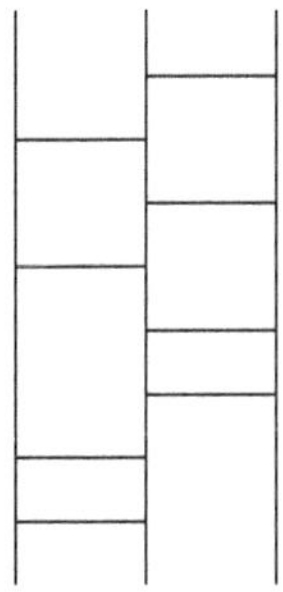

This numbering on *amida-kuji* can also distinguish the unknotted circle from the linked one. For the link of two circles, the path becomes $2 \to 1 \to 2$ or $1 \to 2 \to 1$. The starting number is the same as the ending number. For the unlinked circles, if they are placed in an overlap configuration, the numbers become $1 \to 2 \to 3$. The starting number 1 becomes different from the ending number 3. Thus, this case is easily eliminated for the knots and links.

When the average of Gaussian mean is given, and the crossing points are n, it may be not the minimal number of the crossing points. The example is $\langle(\mathrm{tr}M^4)^2\mathrm{tr}M^8\rangle$, which has eight crossing points. By the choice of up and down crossing of eight Seifert bands, one reduces to knot 5_2, which has five crossing points. This example appears in Teneva's transformation of a knot 5_2,[12] and it has four coloring. The previous numbering of floor height becomes a sequence of 12101234323212321, which involves a sequence of the increasing numbers 1234. After Reidemeister transformation, it reduces to 5_2 knot.

The numbering, which is shown here, is similar to introduction of the time. The increase of the height of floor means the increase of time. The standard treatment of knot invariants is based on the Knitzhnik–Zamolodchikov equation with the Cauchy repeated integration,[13] which uses the integration of time. We find the one-to-one correspondence of the replica matrix model by the numberings to the different knot configurations. This may be the same as Seifert surface, but the replica method has an advantage for generating all knot configurations.

We consider the numbering of the height of floors. This can be generalized to any numbers, such as a complex number of the root of unity. The general skein relation is[9]

$$xL_+ - yL_- = wL_0, \tag{9}$$

where x, y and w are arbitrary. One may take these numbers as $x = t^{-1}$, $y = t$ and $w = (\sqrt{t} - \frac{1}{\sqrt{t}})$, which leads to Jones polynomials.[27]

The skein relation is similar to the algebra of $su(2)$, which has Pauli matrices σ_x, σ_y and σ_z with $[\sigma_x, \sigma_y] = i\sigma_z$. The operator σ_z corresponds to eraser of the rung of *amida-kuji*, and $\sigma_x\sigma_y$ and $\sigma_y\sigma_x$ correspond to the over and down crossing of the Seifert band. The introduction of the operator $\hat{\sigma}$ for the vertices of the matrix model is an alternative method to the numbering for the knot construction. This leads to $SU(2)_k$ Wess–Zumino–Witten non-linear σ model in two dimensions, which is equivalent Chern–Simon gauge action in three dimensions.

The numbering of the segments of the vertices of the matrix in the replica limit gives the Chern–Simons gauge field A in (6). The n point function $U(\sigma_1, \ldots, \sigma_n)$ in (2) is similar to the multi-loop invariants of the Chern–Simons theory,

$$\hat{\mu}(\sigma_i, \ldots \sigma_n) = \int_A \prod_{i=1}^{n} \mathrm{tr}\left(e^{\int_{\sigma_i} A}\right) d\mu(A). \tag{10}$$

The matrix M with a numbering turns to a gauge-invariant field, which has $SU(2)_k$ symmetry. The physical interpretation may be that the one-stroke trajectory knot is a string (vortex flux ring) which is a three-dimensional gauge-invariant field.

5. Derivation of Replica Formula

For the case of two vertices, we have from (3)

$$\begin{aligned}\lim_{N\to 0} U(\sigma_1, \sigma_2) &= \sigma_1\sigma_2 + \frac{1}{3}\left(\sigma_1^5\sigma_2 + 2\sigma_1^4\sigma_2^2 + 2\sigma_1^3\sigma_2^3 + \sigma_1^2\sigma_2^4 + \sigma_1\sigma_2^5\right) + O\left(\sigma^{10}\right)\\ &= \sigma_1\sigma_2\langle(\mathrm{tr}M)^2\rangle + \frac{\sigma_1^5\sigma_2}{5!}\langle \mathrm{tr}M^5\mathrm{tr}M\rangle + \frac{\sigma_1^4\sigma_2^2}{4!2!}\langle \mathrm{tr}M^4\mathrm{tr}M^2\rangle\\ &\quad + \frac{\sigma_1^3\sigma_2^3}{3!3!}\langle(\mathrm{tr}M^3)^2\rangle + \frac{\sigma_1^2\sigma_2^4}{2!4!}\langle \mathrm{tr}M^4\mathrm{tr}M^2\rangle + \frac{\sigma_1\sigma_2}{5!}\langle \mathrm{tr}M\mathrm{tr}M^5\rangle + \cdots\end{aligned} \tag{11}$$

The coefficient of $\sigma_1^3\sigma_2^3$ gives the knot of 3_1, which is Seifert graph of $\langle(\mathrm{tr}M^3)^2\rangle$. Other terms are circle graphs, which are unknotted. From Table 1, the knot of two vertices of $U(\sigma_1, \sigma_2)$ are $3_1, 5_1, 7_1, \ldots$, which are torus knots, and they are represented in the replica limit of the average of product of two traces $\langle(\mathrm{tr}M^{2n+1})^2\rangle$ $(n = 0, 1, 2, \ldots)$. The number of the trace is the number of Seifert disks, and it is equal to the number of the vertices of Seifert graphs.

For three-point vertices, the formula of replica limit (3) gives, for the order of σ^{16},

$$\begin{aligned}\lim_{N\to 0} U(\sigma_1, \sigma_2, \sigma_3)|_{\sigma^{16}} &= \frac{11}{1152}\sigma_1^8\sigma_2^4\sigma_3^4 + \frac{11}{1440}\sigma_1^8\sigma_2^5\sigma_3^3 + \frac{47}{11520}\sigma_1^8\sigma_2^6\sigma_3^2\\ &\quad + \frac{53}{11520}\sigma_1^7\sigma_2^7\sigma_3^2 + \frac{347}{34560}\sigma_1^7\sigma_2^6\sigma_3^3 + \frac{89}{5760}\sigma_1^7\sigma_2^5\sigma_3^4\\ &\quad + \frac{623}{34560}\sigma_1^6\sigma_2^6\sigma_3^4 + \frac{511}{23040}\sigma_1^6\sigma_2^5\sigma_3^5 + \text{symmetric terms},\end{aligned} \tag{12}$$

which corresponds to the knots of $(8_8, 8_{17}, 8_{18}, 8_{19}, 8_{20}, 8_{21})$ for $\sigma_1^8\sigma_2^4\sigma_3^4$. The term of $\sigma_1^8\sigma_2^5\sigma_3^3$ corresponds to $(8_{16}, 8_{10}, 8_7)$. The term of $\sigma_1^8\sigma_2^6\sigma_3^2$ corresponds to $(8_5, 8_2)$ knots.

There is a relation of the knots and the intersection numbers of the curves. The intersection numbers of the curve is a special case of Gromov–Witten invariants which maps from the manifold to a point. The Kontsevich Airy matrix model has an integral representation as[13]

$$Z = \int dM e^{\frac{1}{3}\mathrm{tr}M^3 + \mathrm{tr}M^2\Lambda}. \tag{13}$$

Expanding the term $\mathrm{tr}M^3$, we obtain the power of the vertex $\mathrm{tr}M^3$ for the Gaussian means as $\langle(\mathrm{tr}M^3)^n\rangle$ (n is even number). We found they correspond to 3_1, $9_{23}, \ldots$ knots in the previous section. The numerical values of the Gaussian means are interpreted as intersection numbers, which are denoted as $\langle\tau_n\rangle$. In the case of Kontsevich model, the intersection numbers for one marked point are

$$\langle\tau_1\rangle = \frac{1}{24}, \quad \langle\tau_{3g-2}\rangle = \frac{1}{(24)^g g!}. \tag{14}$$

The Gaussian means are obtained from the derivatives of the logarithm of the partition function Z for the matrix model.

The figure-eight 4_1 knot has an expression of the Gaussian mean $\langle(\mathrm{tr}M^4)(\mathrm{tr}M^2)^2\rangle$ in Table 1. This is obtained by the derivatives of c_4 and c_2 for the partition function Z, where coefficients of $\mathrm{tr}M^j$ are c_j,

$$Z = \int dM e^{c_4 trM^4 + c_2\mathrm{tr}M^2} \tag{15}$$

$$\langle(\mathrm{tr}M^4)(\mathrm{tr}M^2)^2\rangle|_{c_4=0, c_2=-\frac{1}{2}} = \frac{\partial^3}{\partial c_4\partial^2 c_2}\mathrm{logZ}. \tag{16}$$

We consider the k-point function,

$$U(\sigma_1, \ldots, \sigma_k) = \frac{1}{N}\langle\mathrm{tr}e^{\sigma_1 M}\cdots\mathrm{tr}e^{\sigma_k M}\rangle, \tag{17}$$

where M is $N \times N$ Hermitian random matrix. There is an integral formula for k-point function[6]

$$U(\sigma_1, \ldots, \sigma_k) = (-1)^{\frac{k(k-1)}{2}} e^{\sum\frac{1}{2}\sigma_i^2} \oint \prod_{i=1}^k \frac{du_i}{2i\pi} e^{\sum u_i\sigma_i} \prod_1^k (1 + \frac{\sigma_i}{u_i})^N \det\frac{1}{u_i + \sigma_i - u_j}. \tag{18}$$

In the limit $N \to 0$, we have logarithmic terms,

$$\lim_{N\to 0}\frac{1}{N}U(\sigma_1, \ldots, \sigma_k) = (-1)^{\frac{k(k-1)}{2}} e^{\sum\frac{1}{2}\sigma_i^2} \oint \prod_{i=1}^k \frac{du_i}{2i\pi} e^{\sum u_i\sigma_i}$$

$$\left(\sum_{i=1}^k \log(1 + \frac{\sigma_i}{u_i})\right) \det\frac{1}{u_i + \sigma_i - u_j}. \tag{19}$$

When $k = 1$ one-point function, this integral is evaluated by the cut of logarithm. The integral of u becomes a line integral $-\sigma < u < 0$. (For $k = 1$, the determinant is just $\frac{1}{\sigma}$).

$$\lim_{N\to 0} U(\sigma) = \frac{1}{\sigma} e^{\frac{1}{2}\sigma^2} \oint \frac{du}{2i\pi} e^{u\sigma} \log\left(1 + \frac{\sigma}{u}\right)$$
$$= \frac{1}{\sigma} e^{\frac{1}{2}\sigma^2} \int_{-\sigma}^{0} du e^{u\sigma} = \frac{\sinh\left(\frac{\sigma^2}{2}\right)}{\left(\frac{\sigma^2}{2}\right)}. \tag{20}$$

This result is also obtained by the contour integral for the pole $u = 0$ by the expansion of the logarithmic term,

$$U(\sigma) = \frac{1}{\sigma} e^{\frac{\sigma^2}{2}} \oint \frac{du}{2i\pi} e^{u\sigma} \left(\sum_{n=1}^{\infty} (-1)^{n-1} \frac{1}{n} \left(\frac{\sigma}{u}\right)^n \right)$$
$$= 1 + \frac{1}{24}\sigma^4 + \frac{1}{1920}\sigma^8 + O(\sigma^{12}), \tag{21}$$

which leads to (20).

As a categorification of knot theory, Kontsevich loop expansion has been studied, which provides Vassiliev invariants as coefficients.[13,14] The contribution of one-loop order is given by $\sinh(\frac{x}{2})/(\frac{x}{2})$, and the logarithm of this one-loop term becomes a series expansion of x^2 which coefficients $B_n/(4n \cdot (2n)!)$, where B_n is Bernoulli number:

$$\sum b_{2n} x^{2n} = \frac{1}{2} \log\left(\frac{\sinh(\frac{x}{2})}{\frac{x}{2}} \right). \tag{22}$$

This Knotsevich one-loop contribution coincides with (20). Vassiliev invariant for knot K of order j (i.e., number of double points in knot K) is denoted by $v_j(K)$, and Vassiliev invariants has a generating function $\sum_j v_j(K) x^j$. These invariants are related to Jones polynomial for knot K, $V_K(t)$, with $t = e^x$. The expansion of Jones polynomial in power of x provides Vassiliev invariants $v_j(K)$ as[15]

$$\sum_j v_j(K) x^j = V_K(e^x). \tag{23}$$

The Vassiliev invariant $v_2(K)$ is equal to $(-3)\times$ [z^2 coefficient in Conway polynomial $P(z)$]. For instance, $v_2(5_2) = -6$ and $P(z) = 1 + 2z^2$, where $z = \sqrt{t} - \frac{1}{\sqrt{t}}$. The replica formula of $U(\sigma_1, \ldots, \sigma_n)$, which is used for the intersection theory,[8] is interesting in the connections of Kontsevich loop expansion and Vassiliev invariants.

For two-point function ($k = 2$), we have

$$\lim_{N\to 0} U(\sigma_1, \sigma_2) = -e^{\frac{1}{2}(\sigma_1^2 + \sigma_2^2)} \oint \frac{du_1 du_2}{(2i\pi)^2} e^{u_1\sigma_1 + u_2\sigma_2}$$
$$\times \left[\left(\log\left(1 + \frac{\sigma_1}{u_1}\right) + \left(\log\left(1 + \frac{\sigma_2}{u_2}\right) \right) \right] \frac{1}{(u_1 - u_2 + \sigma_1)(u_2 - u_1 + \sigma_2)}. \tag{24}$$

If we consider the contour of u_1, there is no pole for the contour of the logarithmic term. Thus, it vanishes. Then the integral becomes the first term,

$$\lim_{N\to 0} U(\sigma_1,\sigma_2) = -e^{\frac{1}{2}(\sigma_1^2+\sigma_2^2)} \oint \frac{du_1 du_2}{(2i\pi)^2} e^{u_1\sigma_1+u_2\sigma_2} \times \left(\log\left(1+\frac{\sigma_1}{u_1}\right)\right)\frac{1}{(u_1-u_2+\sigma_1)(u_2-u_1+\sigma_2)}. \quad (25)$$

By taking the pole of $u_2 = u_1 - \sigma_2$ and $u_2 = u_1 + \sigma_1$, we have

$$e^{\frac{1}{2}(\sigma_1^2+\sigma_2^2)}\frac{1}{\chi}(e^{-\sigma_2^2} - e^{\sigma_1\sigma_2}) \oint \frac{du_1}{2i\pi} e^{u_1\chi}\log\left(1+\frac{\sigma_1}{u_1}\right) = \frac{4}{\chi^2}\sinh\frac{\chi\sigma_1}{2}\sinh\frac{\chi\sigma_2}{2}, \quad (26)$$

where $\chi = \sigma_1 + \sigma_2$. Repeating this process, we obtain the replica formula of (3) for $k > 2$.[6]

5.1. *Intersection numbers*

Kontsevich matrix model is given by

$$Z = \int dM e^{\frac{i}{3}\mathrm{tr}M^3 - \lambda \mathrm{tr}M^2}, \quad (27)$$

and the intersection numbers for the one marked point $\langle\tau_n\rangle_g$ is[13]

$$\langle\tau_n\rangle_g = \sum \langle\tau_n\rangle_g t_n. \quad (28)$$

The values of the intersection numbers for trivalent matrix model (Airy matrix model) is

$$\langle\tau_n\rangle_g = \frac{1}{(24)^g g!}, \quad (29)$$

where n is determined by Riemann–Roch relation of Riemann surface,

$$3g - 2 = n. \quad (30)$$

Thus, we find the direct relation of trefoil knot to the intersection numbers of $\langle\tau_1\rangle_{g=1}$. The knots of trivalent Seifert vertices are all alternating knots, and therefore, the Kontsevich Airy matrix model corresponds to knots $3_1, 9_{23}, \ldots$ one to one. Thus, the intersection theory of one marked point gives the knot of trivalent vertices.

5.2. *Links*

The $N \to 0$ limit of the matrix expectation values give diagrams that can be traced on a sheet of paper without lifting a pen: single-stroke diagrams. The next order in N describes two entangled knots, in which the entanglement is due to a double layout of two continuous intricate lines. There again one can obtain the number of entangled knots from the same matrix result. For the one-point function, the linking is obtained from the terms of order N^2 of $N \times N$ Hermitian matrix.

$$
\begin{aligned}
U(\sigma) &= \frac{e^{\frac{\sigma^2}{2}}}{N\sigma} \oint \frac{du}{2i\pi} e^{\sigma u} e^{N\log(1+\frac{\sigma}{u})} \\
&= \frac{e^{\frac{\sigma^2}{2}}}{\sigma} \oint \frac{du}{2i\pi} e^{\sigma u} \left[\log\left(1 + \frac{\sigma}{u}\right) + \frac{N}{2}\left[\log\left(1+\frac{\sigma}{u}\right)\right]^2 + O(N^2). \right. \qquad (31)
\end{aligned}
$$

Thus, the one-point one link expectations are obtained from the term of order N in (31), describing the diagrams made of two-stroke trails. The contour integral in (31) provides

$$
U(\sigma) = e^{\frac{\sigma^2}{2}} \sum_{1} \frac{(N-1)(N-2)\cdots(N-k+1)}{k!(k-1)!} \sigma^{2k-2}, \qquad (32)
$$

and in the $N \to 0$-limit,

$$
\begin{aligned}
&(N-1)(N-2)\cdots(N-k+1) = (-1)^{k-1}(k-1)! \\
&+N(-1)^k(k-1)!\left[1 + \frac{1}{2} + \cdots + \frac{1}{k-1}\right] + O(N^2). \qquad (33)
\end{aligned}
$$

Expressing the sum

$$
1 + \frac{1}{2} + \cdots + \frac{1}{k-1} = \int_0^1 dx \frac{1-x^{k-1}}{1-x}, \qquad (34)
$$

one finds the terms of order N in terms of the finite integral[22]

$$
\begin{aligned}
U(\sigma)|_N &= \frac{e^{\sigma^2/2}}{\sigma^2} \int_0^1 \frac{dx}{1-x} [(e^{-\sigma^2} - 1) - \frac{1}{x}(e^{-x\sigma^2} - 1)] \\
&= \frac{1}{2}\sigma^2 + \frac{1}{72}\sigma^6 + O(\sigma^8). \qquad (35)
\end{aligned}
$$

The one-point function $U(\sigma)$ for arbitrary N is expressed as

$$
\begin{aligned}
U(\sigma) &= \frac{1}{N\sigma} \oint \frac{du}{2i\pi} e^{\sigma u} \left(\frac{1 + \frac{\sigma}{2u}}{1 - \frac{\sigma}{2u}} \right)^N \\
&= \frac{1}{\sigma} \sum_{k=1}^{\infty} \frac{N^{k-1}}{k!} \oint \frac{du}{2i\pi} e^{\sigma u} \left[\log\left(\frac{1 + \frac{\sigma}{2u}}{1 - \frac{\sigma}{2u}} \right) \right]^k. \qquad (36)
\end{aligned}
$$

For small σ expansion becomes

$$
\begin{aligned}
U(\sigma) = \sum_{k=1}^{\infty} \frac{N^{k-1}}{k!} \Bigg[& \frac{1}{(k-1)!}\sigma^{2k-2} + \frac{1}{12(k+1)!}\sigma^{2k+2} \\
& + \left(\frac{k}{80} + \frac{k(k-1)}{288} \right) \frac{1}{(k+3)!}\sigma^{2k+6} \\
& + \left(\frac{k}{448} + \frac{k(k-1)}{960} + \frac{k(k-1)(k-2)}{10368} \right) \frac{1}{(k+5)!}\sigma^{2k+10} \\
& + \left(\frac{k}{2304} + \frac{71k(k-1)}{268800} + \frac{k(k-1)(k-2)}{23040} \right. \\
& \left. + \frac{k(k-1)(k-2)(k-3)}{497664} \right) \frac{1}{(k+7)!}\sigma^{2k+14} + O(\sigma^{2k+18}) \Bigg]. \qquad (37)
\end{aligned}
$$

The above expansion is consistent with (20) for $k = 1$ and also with (35) for $k = 2$. When the scaling $\sigma^2 \to \sigma^2/N$ is taken, the expansion of the large N is obtained. The leading order of N is expressed as

$$
\lim_{N\to\infty} U(\sigma) = \frac{1}{t} J_1(2t), \qquad (38)
$$

where $\sigma = -it$ and $J_1(x)$ is a Bessel function. The next order of $\frac{1}{N^2}$ is also expressed by the Bessel function of J_2. Thus, the scaling $\sigma \to \sigma/\sqrt{N}$ in (37) makes an interesting relation which connects the large N and the replica limit $N \to 0$.

For the supermatrices in the previous section, one-point function is given with external source eigenvalues r_i and ρ_j,[19]

$$
U(\sigma) = \frac{1}{\sigma} \oint \frac{du}{2i\pi} e^{\sigma u} \prod_{i=1}^{n} \left(\frac{u - r_i + \frac{\sigma}{2}}{u - r_i - \frac{\sigma}{2}} \right) \prod_{j=1}^{m} \left(\frac{u - \rho_i - \frac{\sigma}{2}}{u - \rho_j + \frac{\sigma}{2}} \right). \qquad (39)
$$

When the external source r_i and ρ_j are put to zero, this reduces to (20) in the limit $n \to m$.

The link graphs are obtained from two-point function $u(\sigma_1, \sigma_2)$ in order of N. For instance, from $\langle \mathrm{tr} M^2 \mathrm{tr} M^2 \rangle \sigma_1^2 \sigma_2^2$, with Seifert band connections, there appear a link and two circles, depending on the choice of crossings. The construction of the Gaussian means are the same as knots. There is a correspondence similar to Table 1 between Gaussian means and links. The contractions of propagators are given by twisted Seifert bands. The link of two knots are described by the diagram of order N.

6. Characteristic Polynomial for Trivalent Vertices

When Seifert disk (vertex) has trivalent connector, the corresponding Gaussian mean average becomes $\langle (\mathrm{tr} M^3)^n \rangle$. The non-vanishing one is the case of $n = 4m - 2$,

$m = 1, 2, 3, \ldots$ from (5). The number m is genus g of the Riemann surface. This provides a series of genus g intersection numbers for trivalent vertices (Kontsevich model) and a series of knots, $3_1, 9_{23}, \ldots$

These knot is classified into 2-bridge knot,[9] and Seifert matrix $\tilde{M}$ is made of the diagonal elements plus one line of the same element 1. From the linking of the Seifert bands of this type of knots, it becomes easy to get Alexander polynomials through the characteristic polynomials,

$$\Delta(t) = \det(t\tilde{M} - \tilde{M}^T), \tag{40}$$

where $\tilde{M}$ is Seifert matrix and $\tilde{M}^T$ is a transpose of $\tilde{M}$. For the knot of the series of the trivalent vertices, Seifert matrix $\tilde{M}$ has diagonal elements of 2, except the $(1,1)$ element and (n,n) element, which are 1. For instance, in the case of trefoil 3_1 and knot of 9_{23}, Seifert matrix $\tilde{M}$ become

$$\tilde{M}_{3_1} = \begin{pmatrix} 1 & 1 \\ 0 & 1 \end{pmatrix}, \tag{41}$$

$$\tilde{M}_{9_{23}} = \begin{pmatrix} 1 & 1 & 0 & 0 \\ 0 & 2 & 1 & 0 \\ 0 & 0 & 2 & 1 \\ 0 & 0 & 0 & 1 \end{pmatrix}. \tag{42}$$

Alexander polynomial $\Delta(t)$, which is a characteristic polynomial defined by (40), has zeros in a complex t-plane. In the trivalent vertices of the previous section, the determinant is similar to Toeplitz determinant and has an interesting zero locus.

It is well known that Ising model with a magnetic field has unit circle zero locus, and there is Yang–Lee edge singularity studied by Fisher.[20] The critical exponent σ for the magnetization $m \sim (h - h_c)^\sigma$ is $\sigma = (d-2+\eta)/(d+2-\eta) = \frac{\Delta_\phi}{d-\Delta_\phi}$. The scale dimension $\Delta_\phi = (d-2+\eta)/2$. The exact value of $\Delta_\phi = -\frac{2}{5}$ in Yang–Lee edge singularity is known for two dimensions. This model is given by the Lagrangian $\mathcal{L} = \frac{1}{2}(\partial\phi)^2 + i(h - h_c)\phi + ig\phi^3$, and $\epsilon = 6 - d$ expansion can be obtained. The density of characteristic polynomial of trivalent knots have a similarity to edge singularity of ϕ^3 theory.

The locus of the zero of the characteristic polynomial (Alexander polynomial) is on the arc of the unit circle $x^2 + y^2 = 1$ in the region $-\sqrt{\frac{3}{4}} < y < \sqrt{\frac{3}{4}}$ and $\frac{1}{2} < x < 1$, as shown in Fig. 1. The density of the zeros increases in the approach to the edge point in the complex plane $(x,y) = (\frac{1}{2}, \sqrt{\frac{3}{4}})$. This point is analogous to the edge point of the zeros of Yang–Lee Ising model in a magnetic field. The distribution of zeros for trivalent vertex knot shows the square root singularity as the edge singularity. This behavior is originated from the relation of cusp singularity $y^2 = x^3$ to the trefoil knot. This is also related to Landau–Ginzburg potential for primary fields, which are obtained from matrix models.[5,21]

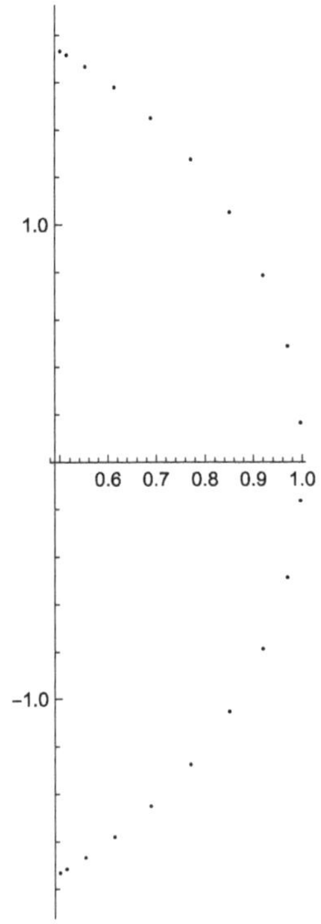

Fig. 1. Zeros of Alexander polynomial for trivalent vertex, which are on an arc of the unit circle. The edge points of spectrum are at $(x, y) = (\frac{1}{2}, \pm\sqrt{\frac{3}{4}})$ in a complex coordinate.

The knot 5_1 has a vertex with five external legs. This is related to p-spin curve with $p = 4$. In this case, the same ladder structure for higher crossing knots made of vertices with 5-legs only are expressed as $\langle(\mathrm{tr}M^5)^n\rangle$, $n = 3, 6, \ldots$ The zeros of Alexander polynomials are located on the unit circle similar to the trivalent vertices. The region of the zeros on the unit circle is limited to the region $(-1)^{\frac{3}{5}} < x$, where x is the real part of the zeros. The zeros at $(-1)^{\frac{1}{5}}$ and $(-1)^{\frac{3}{5}}$, for positive imaginary region, appear independent of n, and they are the end points of edges of a spectrum.

7. Higher-Dimensional Knots and Singularities

The classical knot theory concerns with the embedding of S^1 into S^3. The extension of this knot to higher-dimensional case has been studied with covering spaces of the classical knots.[9,10] n-dimensional knot (N, M, k) is embedding of n-dimensional manifold N^n in $(n+2)$-dimensional manifold M^{n+2} $(k : N^n \subset M^{n+2})$. The embedding k:$N^n \subset M^m$ with codimension $m-n \geq 3$ leads to only unknotting. So, we need the codimension 2 for knots. A classical knot is 1-knot, $n = 1$. By spinning classical knots, the surface knot (2-knots) is obtained, which is embedded in 4-manifold $(S^2 \subset S^4)$).

For 2-knot, the Seifert surface is constructed by the spinning of a knot around a certain axis, which leads to generation of surfaces.[23] Natural way of 2-knot is obtained from 1-knot by taking the movement in a time direction, i.e., movie picture.[10]

The replica limit of (3) is applied for the time-dependent Gaussian matrix model.[21] By adding $\mathrm{tr}(dM/dt)^2$ term to action, the matrix describes the movement on time. This time-dependent matrix model reduces to two matrix model by a path integral formulation.[21]

When 1-knot is embedded in to four dimensions, the knot becomes resolved due to the additional dimension of time. However, the pair of strings, which makes a surface becomes knotted. The axis of the spun knot correspond to the interaction of two matrices M_1 and M_2 coupled as $(c\mathrm{tr}M_1M_2)^2$.

The 2-knot has singularities such as double point and triple point, which are generated by the degeneracies of Reidemeister relation in the moving picture.[10] The Reidemeister relations for 1-knot are generalized to Roseman relations. Over and down distinction in 1-knot with previous numberings does not make sense for 2-knot. The coloring (numbering) may be useful instead.[12] We consider the replica limit of the time-dependent matrix model for the description of the surface knot.

The two-point correlation function $U(\sigma_1, \sigma_2)$ is

$$U(\sigma_1, \sigma_2) = \frac{1}{Z}\int dM_1 dM_2 (\mathrm{tr}e^{\sigma_1 M_1}\mathrm{tr}e^{\sigma_2 M_2})e^{-\frac{1}{2}\mathrm{tr}(M_1^2+M_2^2-2cM_1M_2)}. \tag{43}$$

The parameter c is a coupling constant of two matrices, which is equal to e^{-t} with a time t. In the limit $t \to \infty$, c becomes vanishing, and we obtain two non-interacting matrices.

For the picture of the spinning, we take order the terms of c^2, which represent the axis of the spinning. The n-point correlation function for $c \neq 0$ can be formulated in a contour integrals.[21] The replica limit for this case is obtained by taking $N \to 0$ in the contour integral. For instance, the two-point case is expressed as

$$\begin{aligned} U(\sigma_1, \sigma_2) = &-\frac{1}{N}\oint \frac{dudv}{(2i\pi)^2}\left(1+\frac{\sigma_1}{u}\right)^N \left(1+\frac{\sigma_2}{cv}\right)^N \\ &\times \frac{1}{(u-v+\frac{\sigma_1}{N})(u-v+\frac{\sigma_2}{N})} e^{\frac{\sigma_1 u}{1-c^2}+\frac{\sigma_2 v}{1-c^2}-\frac{\sigma_1^2}{2N(1-c^2)}+\frac{\sigma_2^2}{2N(1-c^2)}} \end{aligned} \tag{44}$$

8. Extension to Half-Integer Spins

Recently, the extension of the p spin curve of integral value of p to half integer p has been considered by Brézin and myself.[24–26] The skein relation is expressed with a parameter $z = \sqrt{t} - \frac{1}{\sqrt{t}}$. This parameter t appears in Alexander polynomials and Jones polynomials.[27] This parameter t can be a root of unity, $t = e^{\frac{2\pi i}{p}}$. When p is half-integer, the skein relation provides a new relation of knots. There appears also a factor $\tilde{z} = \sqrt{t} + \frac{1}{\sqrt{t}}$, which represents a trivial link. It is interesting to note that this factor appears in the contour integral with a change variable from u to y as $u = \frac{i}{2}(y^2 - \frac{1}{y^2})$, which gives a Laurent series in y.[24–26] In the expression of the integrand for correlation function, there appears a factor $\tilde{z} = \sqrt{t} + \frac{1}{\sqrt{t}}$ with $y = \sqrt{t}$. The μ component trivial link is expressed as $V_{O_\mu}(t) = (-1)^{\mu-1}(\sqrt{t} + \frac{1}{\sqrt{t}})^{\mu-1}$, which appears in the construction of Jones polynomials.[9]

This half-integer spin leads to Fermionic case. The case $p = \frac{1}{2}$ is Dirac spin and $p = \frac{3}{2}$ is Rarita–Schwinger case. Such half-integer p and half-integer level k ($p = k + 2$) has new properties for knots. Since k is a coefficient of Chern–Simons action, this fractional level leads to a fractional charge and a new conformal field theory, which may have application on topological semimetals of spin $\frac{3}{2}$, such as half-Heusler alloy (PdBiSe, etc.).

9. Discussion

In this chapter, we point out explicitly that the expectation values of vertices for the Gaussian Hermitian matrix model correspond to knots in the zero-replica limit. The diagrammatic expansions show an explicit correspondence between Seifert graphs and Gaussian means after the introduction of a height function. By the tuning of an external matrix source, the Hermitian matrix model can be chosen to generate p-spin curves.[8] We found the correspondence of the n-vertex Gaussian means to the singularity of Landau–Ginzburg potentials. This correspondence appears in the distribution of zeros of the corresponding Alexander polynomial, which is related to Yang–Lee edge singularity.

The replica limit of $O(N)$ vector model has been used for polymers as a way to implement the self-avoiding behavior in d dimensions, where $1 < d < 4$. Recently, polymers have been studied by a conformal bootstrap method for general space dimensions d.[28,29] The Yang–Lee edge singularity has been studied by the conformal bootstrap in a determinant method[30,31] for dimensions $2 \leq d \leq 6$. The value $\Delta_\phi \sim -\frac{2}{5}$ has been found for two dimensions. The d-dimensional branched polymer is equivalent to Yang–Lee edge singularity with the dimensional reduction $d \to d - 2$, which can be seen explicitly in Wilson–Fisher-ϵ expansion.[2,29] It is interesting to investigate the dimensional reduction[32] for the knots of higher dimensions. This will be an interesting future work.

Recently, in a four-dimensional $\mathcal{N} = 4$ super Yang–Mills theory, similar replica formula of (3) appears (see footnote 15 in Ref. 33). This may be an indication that a gauge theory appears from the random matrix theory with a replica.

The spin glass in a disordered system has been discussed by a replica method as a replica symmetry breaking,[34] which may be equivalent to p-adic theory.[35] We have seen the knot or unknotted circle appear in the random matrix theory with replica limit. The unknotted circles provide a large entropy, and a glass transition may be considered as a transition from the knotted system to unknotted (circle) transitions. It may be interesting to apply the present study of the replica limit of the matrix model to such glass transition and gauge glass transition as a vortex lattice melting in a superconductor.[36,37]

Acknowledgments

The author thanks Edouard Brézin for the many discussions about replica method and fractional p spin curves. He also thanks Andreani Petrou for the help in the evaluation of knots. This work is supported by JSPS KAKENHI 19H01813.

References

1. P.-G. De Gennes, Exponents for the excluded volume problem as derived by the Wilson method. *Phys. Lett. A* **38**, 339 (1972).
2. K. G. Wilson and M. E. Fisher, Critical exponents in 3.99 dimensions. *Phys. Rev. Lett.* **28**, 240 (1972).
3. S. Hikami, Anderson localization in a non-linear σ model representation. *Phys. Rev. B* **24**, 2671 (1981).
4. M. Kontsevich, Intersection theory on the moduli space of curves and matrix Airy function. *Commun. Math. Phys.* **147**, 1 (1992).
5. E. Witten, Algebraic geometry associated with matrix models of two dimensional gravity. In *Topological Methods in Modern Mathematics, Publish or Perish, Houston* (1993), p. 235.
6. E. Brézin and S. Hikami, Intersection theory from duality and replica. *Commun. Math. Phys.* **283**, 507 (2008).
7. E. Brézin and S. Hikami, Intersection numbers of Riemann surfaces from Gaussian matrix models. *JHEP* **10**, 096 (2007).
8. E. Brézin and S. Hikami, *Random Matrix Theory with an External Source, Springer Briefs in Mathematical Physics*, Vol. 19 (Springer, 2016).
9. K. Murasugi, *Knot Theory and Its Application* (Birkhaeuser, 1996).
10. S. Kamada, *Surface-knots in 4-space, Springer Monographs in Mathematics* (Springer, 2017).
11. E. Witten, Quantum field theory and the Jones polynomial. *Commun. Math. Phys.* **121**, 351 (1989).
12. M. Saito, Minimal numbers of Fox colors and quandle cocycle invariants of knots. *J. Knot Theory Its Ramificat.* **19**, 1449 (2010).
13. M. Kontsevich, Vassiliev's knot invariants. *Adv. Sov. Math.* **16**, 137 (1993).
14. A. Kricker, The lines of Kontsevich integral and Rozansky's rationality conjecture. arxiv: math.GT/0005284.
15. J. S. Birman and X.-S. Lin, Knot polynomials and Vassiliev's invariants. *Invent. Math.* **111**, 225 (1993).
16. C. G. Callan, R. Dashen and D. J. Gross, A mechanism for quark confinement. *Phys. Lett. B* **66**, 375 (1977).
17. J. M. Kosteritz and D. J. Thouless, *J. Phys. C* **6**, 1181 (1973).
18. S. Hikami and T. Tsuneto, Phase transition of quasi-two dimesional planar system. *Prog. Theor. Phys.* **63**, 387 (1980).
19. E. Brézin and S. Hikami, Random supermatrices with an external source. *JHEP* **8**, 086 (2018).

20. M. E. Fisher, Yang–Lee singularity and ϕ^3 field theory. *Phys. Rev. Lett.* **46**, 1610 (1978).
21. E. Brézin and S. Hikami, Computing topological invariants with one and two-matrix models. *JHEP* **4**, 110 (2009).
22. E. Brézin, *A Private Communication* (2023).
23. E. C. Zeeman, Twisting spun knots. *Trans. Am. Math. Soc.* **115**, 471 (1965).
24. E. Brézin and S. Hikami, Punctures and p-spin curves from matrix models. *J. Stat. Phys.* **180**, 1031 (2020).
25. S. Hikami and E. Brézin, Punctures and p-spin curves from matrix models II. *J. Stat. Phys.* **183**, 36 (2021).
26. S. Hikami, Punctures and p-spin curves from matrix models III, D_l types and logarithmic potential. *J. Stat. Phys.* **188**, 20 (2022).
27. V. F. R. Jones, Hecke algebra representations of braid groups and link polynomials. *Ann. Math.* **126**, 335 (1987).
28. H. Shimada and S. Hikami, Fractal dimensions of self-avoiding walks and Ising high-temperature graphs in 3D conformal bootstrap. *J. Stat. Phys.* **165**, 1006 (2016).
29. S. Hikami, Conformal bootstrap analysis for single and branched polymers. *Prog. Theor. Exp. Phys.* **2018**, 123I01 (2018).
30. F. Gliozzi and A. Rago, Critical exponents of the 3d Ising and related models from conformal bootstrap. *JHEP* **2014**, 42 (2014).
31. S. Hikami, Conformal bootstrap analysis for the Yang–Lee edge singularity. *Prog. Theor. Exp. Phys.* **2018**, 053I01 (2018).
32. S. Hikami, Dimensional reduction by conformal bootstrap. *Prog. Theor. Exp. Phys.* **2019**, 083A03 (2019).
33. T. Bargheer, F. Coronado and P. Vieira, Octagons I: Combinatorics and non-planar resummations. *JHEP* **8**, 162 (2019).
34. G. Parisi, Infinite number of order parameters for spin-glaces. *Phys. Rev. Lett.* **43**, 1754 (1979).
35. G. Parisi and N. Sourlas, P-adic numbers and replica symmetry breaking. *Eur. Phys. J. B* **14**, 535 (2000).
36. E. Brézin, A. Fujita and S. Hikami, Large-order behavior of the perturbation theory for superconductors. *Phys. Rev. Lett.* **65**, 1949 (1990).
37. S. Hikami, A. Fujita and A. I. Larkin, Magnetic-flux-lattice melting in a strong magnetic field. *Phys. Rev. B* **44**, 10400(R) (1991).
38. A. Fujita and S. Hikami, Matrix model approach to the flux lattice melting in two dimensional superconductors. *Phys. Rev. B* **51**, 16259 (1995).

https://doi.org/10.1142/9789811282386_0021

Chapter 21

Wilson-Fisher Fixed Points in the Presence of Dirac Fermions

I.F. Herbut

Department of Physics, Simon Fraser University, Burnaby, British Columbia, Canada V5A 1S6

Wilson–Fisher expansion near upper critical dimension has proven to be an invaluable conceptual and computational tool in our understanding of the universal critical behavior in the ϕ^4 field theories that describe low-energy physics of the canonical models, such as Ising, XY, and Heisenberg. Here, I review its application to a class of the Gross–Neveu–Yukawa (GNY) field theories, which emerge as possible universal description of a number of quantum phase transitions in electronic two-dimensional systems such as graphene and d-wave superconductors. GNY field theories may be viewed as minimal modifications of the ϕ^4 field theories in which the order parameter is coupled to relativistic Dirac fermions through Yukawa term and which still exhibit critical fixed points in the suitably formulated Wilson–Fisher ϵ-expansion. I discuss the unified GNY field theory for a set of different symmetry-breaking patterns, with focus on the semimetal-Néel-ordered-Mott insulator quantum phase transition in the half-filled Hubbard model on the honeycomb lattice, for which a comparison between the state-of-the-art ϵ-expansion, quantum Monte Carlo, large N, and functional renormalization-group calculations can be made.

1. Introduction

Critical behavior of the $O(N)$ models for $N = 1, 2, 3$ in the physical three ($d = 3$) dimensions is an inherently strong-coupling problem for which no obvious small paramater exists. Anybody who tried some real-space decimation procedure for two-dimensional Ising model, for example, has certainly experienced a feeling of frustration in having to terminate the generated series of ever-further-neighbors-couplings in a more or less *ad hoc* manner.[1] The Wilson–Fisher (WF) expansion[2] in powers of the parameter $\epsilon = 4 - d$ was therefore a conceptual, and as it turns out, even a computational breakthrough par excellence. I can still recall my own sense of wonder when I first learned of this approach as an uninitiated undergraduate in the course on phase transitions. It seemed to me to be a truly imaginative idea to take an integer quantity such as dimension of space and consider it in a mathematically consistent and apparently useful way as a real number. At the same time, however, I could not see how a procedure that relied on the parameter such as ϵ being small could be expected to be sensible even for its value as large as unity! Little did I know that I would spend a fair part of my professional life wrestling with precisely these two issues.

Today, the significance of the WF ϵ-expansion around the upper critical dimension for general studies of critical phenomena cannot be overstated. At the cost of entertaining non-integer values of system's dimensionality d, it allows one to directly follow the emergence of the non-trivial critical WF fixed point as the upper critical dimension is crossed from above and to monitor for the relevance of all couplings at the critical point with the increase of ϵ. As long as the evolution of the WF fixed point is smooth, one can hope to rely on the perturbation theory to extract the desired critical exponents in powers of ϵ. The unpleasant fact is that the series is certainly not convergent, but it is thought to be asymptotic. The first few terms often already provide a decent estimate of the universal quantities in $d = 3$, and even in $d = 2$, for the Ising model, for example, where the exact Onsager's solution can be used for comparison. Elaborate procedures[3] for resummation of the series exist nowadays, which yield the most accurate values of the critical exponents for various values of the parameter N, when the expansion is pushed to higher order, the sixth order being the highest at the time of writing.[4] Even for the Ising model where a more accurate conformal bootstrap[5] amounts to an essentially exact solution in $d = 3$, such a resummed ϵ-expansion is still competitive in accuracy.

The success of the ϵ-expansion for the $O(N)$ ϕ^4 field theories is rooted in the fact that there is a single self-interaction coupling constant that becomes relevant at the non-interacting Gaussian fixed point as the upper critical dimension is crossed from above. One therefore needs only to track the evolution of the WF fixed point along a well-defined line in the coupling space, and provided there are no other non-perturbative fixed points along the same line,[6] the WF critical fixed point continues to exist at all values of ϵ. Such a smooth evolution is no longer guaranteed when there is more than one coupling in the theory. First, it is in principle possible that different couplings have their canonical dimensions vanish in different physical dimensions, in which case there would be no well-defined upper critical dimension around which to expand. Second, even if there is an upper critical dimension in the problem in the standard sense, the number of fixed points below it may depend on some fixed parameter of the system, such as the number of the field's components N, for instance. An important early example of this was provided by the original Ginzburg–Landau theory for the complex scalar field, i.e. the superconducting order parameter, coupled to the fluctuating electromagnetic gauge field, also known as the scalar electrodynamics.[7,8] In this canonical field theory,[9] both the self-interaction coupling and the electromagnetic charge become relevant at the Gaussian fixed point below the upper critical dimension of $d = 4$, and for a number of complex fields n larger than the critical value n_c, there is indeed a critical fixed point of the renormalization-group (RG) flow, and the concomitant universal critical behavior. As $n \to n_c+$, however, this critical point is approached in the coupling space by another, bicritical fixed point, until they coincide at $n = n_c$. Both fixed points become complex for $n < n_c$, when there is no longer a real-valued critical fixed point left, and only a runaway flow remains. Interestingly, the critical number of complex components n_c can itself be computed in the ϵ-expansion,[10] and one finds $n_c = 182.95(1 - 1.752\epsilon + 0.798\epsilon^2 + 0.362\epsilon^3) + O(\epsilon^4)$ in the four-loop computation.[11] The series is obviously badly behaved, but using additional information about the behavior of the scalar electrodynamics near $d = 2$, the value of n_c can be estimated to be around 12, with significant uncertainty in this number. The main point is that the fixed-point structure of the RG flow in this case depends crucially on the value of the parameter n, with its critical value n_c itself being rather strongly dependent on ϵ, and consequently poorly known for $\epsilon = 1$. For different values of n, the RG flow at small and large values of ϵ therefore may or may not be smoothly connected.

In this contribution, I discuss a set of field theories where the $O(N)$ order parameter is also coupled to soft modes such as the gauge field as in the above example of scalar electrodynamics, except that the modes are being fermionic instead of bosonic. These field theories are believed to describe low-dimensional condensed matter systems which at low energies feature Dirac fermions, such as graphene or d-wave superconductors. The electronic Fermi surface collapses to a set of (Dirac) points, and the energy spectrum of fermionic quasiparticles becomes effectively relativistic, which facilitates a controlled field-theoretic treatment of quantum phase transitions that ensue with an increase of electron–electron interactions. A paradigmatic example is provided by the standard Hubbard model on the honeycomb lattice at filling one-half and at zero temperature. For weak on-site repulsive interaction U, the electronic system is a paramagnetic semimetal and presumably in the same ground state as in the real graphene layer. At high U, on the other hand, the ground state is an antiferromagnetic Mott insulator, with a finite value of the three-component Néel order parameter. There is strong accumulated evidence that there exists a single critical value of the interaction U_c, so that for $U > U_c$, Dirac fermions acquire a relativistic mass gap and that the Néel order simultaneously develops. The resulting semimetal–insulator quantum critical point should be described by a model closely related to the Gross–Neveu model in $2 + 1$ spacetime dimensions.[12] Variants with Ising ($N = 1$) and XY ($N = 2$) order parameters also have realizations on the honeycomb lattice with fermion–fermion interactions suitably modified to include nearest- and the next-nearest-neighbor terms.

Gross–Neveu-like models in $2 + 1$ dimensions can be treated in $1/N_\psi$ expansion, with N_ψ as the number of Dirac fermions. Alternatively, one can explicitly include the bosonic order parameter Yukawa-coupled to the Dirac fermions so that the theory features two coupling constants: the order-parameter self-interaction, and the Yukawa coupling, both marginally irrelevant in $3+1$ spacetime dimensions in the infrared.[13] A possible advantage of this formulation is that one can attempt the standard Wilson–Fisher expansion around the upper critical spatial dimension of three. The algebraic structure of the RG β-functions is similar to those in the scalar electrodynamics, with one important difference: fermionic statistics of Dirac fermions reverses the signs of the analogous terms in the scalar electrodynamics, so that the problematic collisions of the fixed points described above are avoided. This allows one to push the ϵ-expansion to higher orders and to try to extract some quantitative information about the quantum critical points in the Hubbard, and the Hubbard-like models on honeycomb lattice. Some of these models can also be independently studied by sign-problem-free quantum Monte Carlo calculations, which can then be compared with the analytic results.

The rest of the paper is organized as follows. In Section 2, I describe the construction of the GNY theory for various order parameters on honeycomb lattice. In Section 3, the WF ϵ-expansion for general GNY theory is discussed, and one-loop results for the critical exponents are given. Section 4 gives a review of the higher-order results for the chiral-Heisenberg universality class relevant for the Hubbard model on the honeycomb lattice, and compares them with the results of other analytical and numerical techniques. Further discussion and extensions of the GNY theory to other patterns of symmetry-breaking are provided in Sectino 5. Summary is given in Section 6.

2. GNY Field Theories for Graphene

The motion of non-interacting electrons on the graphene's honeycomb lattice can be described by the simple tight-binding Hamiltonian

$$H_0 = -t \sum_{\vec{R},i,\sigma} \left[u_\sigma^\dagger(\vec{R}) v_\sigma(\vec{R}+\vec{\delta}_i) + \text{h.c.} \right], \qquad (1)$$

with nearest-neighbor hopping amplitude t.[14] Here, u and v are the electron annihilation operators at the two triangular sublattices of the honeycomb lattice, and the sum runs over the sites $\vec{R}$ of the first triangular sublattice with position vectors $\vec{R}_1 = a\,(\sqrt{3}/2, -1/2)$ and $\vec{R}_2 = a\,(0,1)$. The lattice spacing a is set to $a=1$ and the three nearest-neighbor vectors $\vec{\delta}_i$ read $\vec{\delta}_1 = \left(1/(2\sqrt{3}), 1/2\right)$, $\vec{\delta}_2 = \left(1/(2\sqrt{3}), -1/2\right)$ and $\vec{\delta}_3 = \left(-1/\sqrt{3}, 0\right)$. $\sigma = \pm$ labels the third projection of the electron spin.

The diagonalization of the Hamiltonian H_0 yields the spectrum with two degenerate energy bands with the dispersion $\epsilon_{\vec{k}} = \pm t|\sum_{i=1}^{3} \exp(i\vec{k}\cdot\vec{\delta}_i)|$. At the corners of the Brillouin zone, given by the two points $\vec{K} = \pm(2\pi/\sqrt{3}, 2\pi/3)$, the two energy bands touch linearly and isotropically, and give rise to two inequivalent Dirac points. Retaining only the Fourier modes near the Dirac points, the continuum low-energy effective theory for H_0 can be written down in terms of the free Dirac Lagrangian

$$L_\psi = \psi^\dagger(x)(1_2 \otimes 1_2 \otimes (\partial_\tau - i\sigma_1\partial_1 - i\sigma_2\partial_2) + O(\partial^2))\psi(x), \tag{2}$$

where σ_i are the conventional Pauli matrices, 1_2 is a two-dimensional unit matrix, and the eight-component Dirac field $\Psi^T = (\Psi_+^T, \Psi_-^T)$, with $\psi_\sigma(x) = \int d^D q e^{iqx}\psi_\sigma(q)$ given by $\psi_\sigma^\dagger(q) = \left[u_\sigma^\dagger(K+q), v_\sigma^\dagger(K+q), iv_\sigma^\dagger(-K+q), -iu_\sigma^\dagger(-K+q)\right]$. The $D = 2+1$-energy–momentum vector $q = (\omega, \vec{q})$ collects together the Matsubara frequency ω and the wavevector $\vec{q}$, $K = (0, \vec{K})$, and τ represents the imaginary time. The reference frame is chosen so that $q_x = \vec{q}\cdot\vec{K}/|\vec{K}|$ and $q_y = (\vec{K}\times\vec{q})\times\vec{K}/|\vec{K}|^2$.[12] We have also set the Fermi velocity $v_F = t\sqrt{3}/2$ to unity.

With the above definition of the four-component Dirac fermions, it is evident that the leading term in the low-energy Lagrangian is invariant under a global unitary transformation

$$\psi(x) \to (U \otimes 1_2)\psi(x), \tag{3}$$

with the unitary matrix $U \in SU(4)$. The Lagrangian is also symmetric under an arbitrary global change of the phase of the Dirac field, which of course implies the familiar particle number conservation. Hereafter, I will for the reasons of economy of presentation assume that the particle-number $U(1)$ symmetry is always preserved and will not consider possible superconducting states.[15,16]

The general relativistic "mass term," which if simply added by hand to L_ψ would gap out the Dirac fermions, is then

$$L_{\phi\psi} = \phi_i\psi^\dagger(x)(H_i \otimes \sigma_3)\psi(x), \tag{4}$$

where ϕ_i are real constants ("masses") and H_i are either the 15 Hermitian generators of $SU(4)$, when $i = 1, 2, \ldots, 15$ or the unit matrix when $i = 0$. Summation over the repeated index is assumed. When $\phi_i \neq 0$ for some $i \neq 0$, $L_\psi + L_{\phi\psi}$ has the symmetry reduced from $SU(4)$ to $U(1) \times SO(4)$. The masses ϕ_i $i = 1, 2, \ldots, 15$ transform as the adjoint representation under $SU(4)$, whereas ϕ_0 transforms as a scalar.

The preserved particle-number $U(1)$ symmetry implies that $L_\psi + L_{\phi\psi}$ would describe the low-energy spectrum of an insulator. We may further discern the following broken symmetry states:

(1) $H = 1_2 \otimes 1_2$ corresponds to the quantum anomalous Hall state,[17] which violates only the time-reversal symmetry and otherwise preserves the entire $SU(4)$,
(2) $H = 1_2 \otimes \sigma_i$, $i = 1, 2, 3$, correspond to the charge-density wave[14] and the two Kekule bond-density waves,[18] which break the valley-rotation (sometimes also called "chiral") $SO(3)$ symmetry, but preserve the spin-rotation $SO(3)$ and the time-reversal symmetry,
(3) $H = \sigma_i \otimes 1_2$, $i = 1, 2, 3$, correspond to the anomalous spin-Hall state,[19] which breaks the spin-rotation $SO(3)$ while preserving the valley-rotation $SO(3)$ and the time reversal, and finally,

(4) $H = \sigma_i \otimes \sigma_j$, $i, j = 1, 2, 3$, correspond to the spin-density wave[12] and the triplet versions of the two Kekule bond-density waves, which break both the valley-rotation $SO(3)$ and the spin-rotation $SO(3)$ symmetry, as well as the time reversal.

The antiunitary time-reversal operator in the above representation is given by $T = (\sigma_2 \otimes \sigma_2 \otimes \sigma_2)C$, where C stands formally for the complex conjugation. In terms of the $SO(4) \simeq SO(3) \times SO(3)$ subgroup of the $SU(4)$ symmetry group, with the two $SO(3)$ groups as the spin-rotation and the valley-rotation symmetries, the above matrices transform as $(0, 0)$, $(0, 1)$, $(1, 0)$, and $(1, 1)$ irreducible representations, respectively.

The large $SU(4)$ symmetry of the low-energy Dirac Hamiltonian for electrons in graphene is an artifact of the linearization of the energy dispersion, and the $O(\partial^2)$ term in Eq. (2) already reduces it. Explicitly, it reads

$$O(\partial^2) = 1_2 \otimes \sigma_3 \otimes (\sigma_1(\partial_1^2 - \partial_2^2) - \sigma_2 \partial_1 \partial_2), \tag{5}$$

so the $SU(4)$ symmetry is reduced to $SO(3) \otimes SO(2)$, which are the spin rotations, and the translation symmetry in disguise,[20] the latter generated by $1_2 \otimes \sigma_3$. The electron–electron interaction terms which derive from the Coulomb repulsion can also be expected to respect only the same reduced symmetry. One is therefore led to consider the expectation values of the following fermion bilinears to be possibly dynamically generated at strong interactions:

(1) $\phi_{\text{cdw}} = \langle \psi^\dagger(x)(1 \otimes \sigma_3 \otimes \sigma_3)\psi(x)\rangle$, which would preserve the group $SO(3) \times SO(2)$, but break the discrete (Ising) sublattice symmetry $\psi \to (1_2 \otimes 1_2 \otimes \sigma_1)\psi$, $\partial_2 \to -\partial_2$, which exchanges the two triangular sublattices of the honeycomb lattice. Generation of such a finite bilinear average is favored, for example, by a sufficiently strong nearest-neighbor repulsion.[12]
(2) $\phi_{\text{kek},1} = \langle \psi^\dagger(x)(1 \otimes \sigma_1 \otimes \sigma_3)\psi(x)\rangle$ and $\phi_{\text{kek},2} = \langle \psi(x)^\dagger(1 \otimes \sigma_2 \otimes \sigma_3)\psi(x)\rangle$, which preserve spin-rotation $SO(3)$, but break the translation $SO(2)$ subgroup of the valley-rotation symmetry.[18] This order parameter is dynamically induced by sufficiently strong nearest-neighbor and next-nearest-neighbor repulsions, when they are of comparable strength.[21]
(3) $\phi_{\text{sdw},i} = \langle \psi^\dagger(x)(\sigma_i \otimes \sigma_3 \otimes \sigma_3)\psi(x)\rangle$, which breaks the spin-rotation $SO(3)$, preserves translation $SO(2)$, and breaks sublattice symmetry. A finite vector Néel order parameter $\vec{\phi}_{sdw}$ is induced by sufficiently strong on-site Hubbard repulsion.[12]

Note that the factor in the charge-density-wave mass matrix $1 \otimes \sigma_3$ can be transformed as $U_1(1 \otimes \sigma_3)U_1^\dagger = \sigma_3 \otimes \sigma_3$, with some unitary $U_1 \in SU(4)$. Similarly, the factors in the two Kekule bond-density-wave mass matrices, $1 \otimes \sigma_i$, $i = 1, 2$, can be transformed as $U_2(1 \otimes \sigma_i)U_2^\dagger = \sigma_i \otimes \sigma_3$, with a different unitary transformation $U_2 \in SU(4)$. Both transformations belong to the $SU(4)$, the group of symmetry of the Dirac Lagrangian L_ψ. One can therefore study all three above symmetry-breaking quantum phase transitions which would be induced by increasing different components of electron–electron interactions by considering the single GNY field theory in the following form:

$$L = L_\psi + L_{\phi\psi} + L_\phi, \tag{6}$$

with

$$L_{\phi\psi} = g\phi_i(x)\psi^\dagger(x)(\sigma_i \otimes \sigma_3 \otimes \sigma_3)\psi(x), \tag{7}$$

and

$$L_\phi = \frac{1}{2}((\partial_\mu \phi_i(x))^2 + m^2\phi_i(x)\phi_i(x)) + \lambda(\phi_i(x)\phi_i(x))^2 \tag{8}$$

by restricting the index i to take the values $i = 1$ (charge-density wave), $i = 1, 2$ (Kekule bond-density wave), and $i = 1, 2, 3$ (Néel). Index $\mu = 0, 1, 2$ goes over imaginary time and space dimensions. The tuning parameter for the transition is $m^2 \sim (V_c - V)$, where V is the strength of the interaction relevant to the particular phase transition and V_c is its (non-universal) critical value. Coupling λ is the order parameter's self-interaction. The form of the Yukawa coupling of the bosonic order parameter $\phi_i(x)$ to Dirac fermions implies that

$$\langle\phi_i(x)\rangle = -\frac{g}{m^2}\langle\psi^\dagger(x)(\sigma_i \otimes \sigma_3 \otimes \sigma_3)\psi(x)\rangle + O(\lambda\langle\phi_j\phi_j\phi_i\rangle, \partial^2\langle\phi_i\rangle), \tag{9}$$

so that the system becomes a broken-symmetry gapped Mott insulator when $V > V_c$, i.e. when $m^2 < 0$. Anticipating some of the results that follow, we have also set the velocity of the bosonic order parameter to be the same as the velocity of Dirac fermions, that is, to unity.

3. ϵ-Expansion for GNY

By comparing the Dirac and the Yukawa terms in the GNY field theory, one finds that in terms of their canonical dimensions,

$$g\phi \sim L^{-1}, \tag{10}$$

where L is a length. Comparing the derivative and the self-interaction terms, on the other hand,

$$\lambda\phi^2 \sim L^{-2}. \tag{11}$$

Eliminating the order-parameter field yields therefore that in terms of their canonical dimensions

$$\lambda \sim g^2. \tag{12}$$

The canonical dimensions of the self-interaction λ and of the square of the Yukawa coupling g are the same, and $\lambda \sim g^2 \sim L^{d-3}$, where d is the number of spatial dimensions. In the physical case, $d = 2$, and they are both infrared-relevant couplings at the Gaussian fixed point. Extending d to real values and in particular following Wilson and Fisher[2] and assuming it to be near and below $d = 3$ would thus bring both canonical dimensions to be small and positive. This opens up the possibility for the ϵ-expansion for the GNY theory of the order parameter coupled to the Dirac fermions.[13]

Standard one-loop computation then leads to the RG flow,[22]

$$\beta_\lambda = \frac{d\lambda}{d\ln b} = \epsilon\lambda - 4N_\psi y\lambda - 4(N+8)\lambda^2 + N_\psi y^2, \tag{13}$$

$$\beta_y = \frac{dy}{d\ln b} = \epsilon y - (2N_\psi + 4 - N)y^2, \tag{14}$$

with the elimination of both the order parameter's and Dirac field's modes in the momentum shell $\Lambda/b < q < \Lambda$, with $\Lambda \ll |\vec{K}|$ as the high-energy cutoff in the theory. We left N_ψ as the general number of Dirac-fermions, with $N_\psi = 2$ in graphene, and N as the number of order-parameter components: $N = 1, 2, 3$, for Ising (charge-density wave), XY (Kekule), and Heisenberg (Néel) order parameters, respectively. We have also redefined the coupling constants as $\lambda/(8\pi^2\Lambda^\epsilon) \to \lambda$ and $y = g^2/(8\pi^2\Lambda^\epsilon)$.

Since the one-loop function β_y is independent of the self-interaction λ, the Yukawa coupling y is equally relevant at the standard $O(N)$ WF fixed point at $y = 0$ and $\lambda = \epsilon/4(N+8)$ as it is at the Gaussian fixed point $y = \lambda = 0$. Starting anywhere at $y > 0$ and $\lambda > 0$, the RG flow in the critical plane $m^2 = 0$ is attracted to the new critical fixed point where both $y = y^* = O(\epsilon)$ and $\lambda = \lambda^* = O(\epsilon)$. At this fixed point and for $N_\psi = 2$, the order parameter's anomalous dimension is

$$\eta_\phi = \frac{4\epsilon}{8-N}, \tag{15}$$

and thus of the order $O(\epsilon)$, in contrast to its $O(\epsilon^2)$ value at the standard WF fixed point.[9] One may therefore expect η_ϕ at semimetal–insulator quantum phase transitions in graphene not to be particularly small, in contrast to the usual $O(N)$ universality classes.

The correlation-length critical exponent may also be evaluated, and to the leading order, it equals

$$\nu = \frac{1}{2} + \frac{3(4+N)}{(8-N)(8+N)}\epsilon. \tag{16}$$

The Lorentz invariance of the GNY theory implies that the dynamical critical exponent is exactly

$$z = 1. \tag{17}$$

The hyperscaling is expected to hold, and therefore the remaining critical exponents are given by the usual scaling laws.[9]

The fermion propagator also acquires an anomalous dimension; at the critical point, it behaves as $G_f^{-1} \sim (\omega^2 + k^2)^{(1-\eta_\psi)/2}$, with η_ψ as the fermion's anomalous dimension.[12] To the leading order in ϵ, one finds it equals

$$\eta_\psi = \frac{3\epsilon}{2(8-N)} \tag{18}$$

and thus to be comparable to the order parameter's anomalous dimension. The scaling implies that the residue of the quasiparticle pole on the semimetallic side vanishes as a power law[12]

$$Z \sim (m^2)^{\nu\eta_\psi}. \tag{19}$$

Similarly, the velocity of the Dirac fermions scales as

$$v_F \sim (m^2)^{\nu(z-1)}. \tag{20}$$

The Lorentz invariance of the GNY theory thus implies that the Dirac velocity remains finite at the transition, while the residue of the Dirac quasiparticle's pole vanishes continuously as the critical point is approached from the semimetal side.

The systematic expansion in ϵ has been pursued to higher order,[23–25] the highest at the moment of writing being the fourth order in λ and y. For the "chiral-Ising" ($N = 1$) and the "chiral-Heisenberg" ($N = 3$) GNY theories, for example, the four-loop computation entails summing up 31671 Feynman diagrams. Such a computationally intensive calculation is possible only because of the recent breakthroughs in automatization of high-order perturbative calculations designed for the standard model of particles physics.[25]

4. Chiral-Heisenberg Universality Class

We focus next on the chiral-Heisenberg universality class, i.e. the GNY theory with $N = 3$, which is supposed to describe the quantum phase transition between the Dirac semimetal and the Néel-ordered Mott insulator in the canonical Hubbard model on honeycomb lattice at half-filling. The fourth-order ϵ-expansion yields the critical exponents[25]

$$\nu^{-1} = 2 - 1.527\epsilon + 0.4076\epsilon^2 - 0.8144\epsilon^3 + 2.001\epsilon^4, \tag{21}$$

$$\eta_\phi = 0.8\epsilon + 0.1593\epsilon^2 + 0.02381\epsilon^3 + 0.2103\epsilon^4, \tag{22}$$

$$\eta_\psi = 0.3\epsilon - 0.0576\epsilon^2 - 0.1184\epsilon^3 + 0.04388\epsilon^4 \tag{23}$$

$$\omega = \epsilon - 0.4830\epsilon^2 + 0.9863\epsilon^3 - 2.627\epsilon^4, \tag{24}$$

where we included the leading-correction-to-scaling-exponent ω as well.

One may immediately observe the usual poor convergence properties of the series: for the physical value of $\epsilon = 1$, the ϵ^3 terms become larger than the preceding ϵ^2 terms in three out of the four displayed series. Possibly useful estimates may be obtained therefore by simply terminating the series at the order $O(\epsilon^2)$. This leads to $\nu = 1.13$, $\eta_\phi = 0.96$, $\eta_\psi = 0.24$, and $\omega = 0.52$. (Expanding ν and terminating again at the second order in ϵ would, for example, lead to a similar value of $\nu = 1.07$.) The crudeness of the approximation notwithstanding, the results are in the same ballpark as the results of the more elaborate summation using Padé approximants; although the series are probably too short to give stable results, [3/1] Padé approximant, for example, yields $\nu = 1.2352$, $\eta_\phi = 0.9563$, and $\eta_\psi = 0.1560$.[25]

The Hubbard model on the honeycomb lattice at the filling one-half can also be studied directly by the auxiliary-field quantum Monte Carlo method, as the calculation does not suffer from the sign problem. Large-scale calculations[26–28] support the overall picture provided by the GNY theory: (1) there is a direct continuous quantum phase transition between the semimetallic and the insulating antiferromagnetic phases, (2) the Néel order parameter scales the same way with the size of the system and the deviation from the critical point as the fermion single-particle gap, (3) the values of the critical exponents are distinctly unconventional, with both the correlation length exponent ν and the order parameter's anomalous dimension η_ϕ close to unity, (4) the residue of the Dirac quasiparticle pole is reduced continuously as the critical point is approached from the semimetallic side, while the Fermi velocity remains finite. While in broad agreement, different Monte Carlo calculations still mutually disagree on the precise values of the critical exponents, which also differ somewhat from the field-theoretic estimates based on the GNY theory. For example, Ref. 28 finds $\nu = 1.02(1)$, $\eta_\psi = 0.20(2)$, whereas Ref. 27 finds $\nu = 0.84(4)$, $\eta_\phi = 0.70(15)$, and Ref. 29 gives $\nu = 1.185(43)$ and $\eta_\phi = 0.71(5)$. It is encouraging, on the other hand, that the results seem to be independent of the details of the microscopic model and to depend only on the broken symmetry and the number of Dirac fermions, just as the GNY field theory would imply. This way the Hubbard model on the honeycomb and the staggered-flux square lattice, which both feature two Dirac fermions and the Néel-ordered phase, but have very different critical values of the interaction, for example, show numerically identical finite-size scaling functions and the critical exponents.[27] Even starting from an entirely different single-particle Hamiltonian, such as d-wave Cooper-paired electrons at half-filled square lattice, which lacks particle-number $U(1)$ symmetry but does have the same number of Dirac fermions, seems to lead to the quantum phase transition in the same chiral-Heisenberg universality class with an increase of Hubbard on-site repulsion U: The values of the critical exponents are $\nu = 1.05(5)$, $\eta_\phi = 0.75(4)$, $\eta_\psi = 0.23(4)$.[30]

5. Discussion

While we have postulated Lorentz invariance of the GNY field theory from the outset, and both the velocities of the order parameter and the Dirac fermions have been set to unity, one may also assume the two velocities to be different. Within the

ϵ-expansion, they are then found to flow to the same value in the infrared, both if their difference is initially small[22] or even large.[31] In fact, the relativistic invariance in the GNY-like theories becomes restored in the infrared under very general conditions, with and without couplings of the order parameter and the Dirac fermions to the fluctuating gauge-field, with the gauge field having yet another different bare velocity, in 3+1 dimensions and below it.[31,32] It thus seems safe to assume the breaking of relativistic invariance to be an irrelevant perturbation at the critical point in the physical 2+1 dimensions. Similarly, the long-range $\sim 1/r$ tail of the Coulomb interaction between electrons also represents an irrelevant perturbation,[31,33] although the detailed interplay between the Hubbard on-site interaction U and the long-range part may be quite intricate.[33] The effect of Coulomb interaction's long-range tail on the GNY criticality is similar as at the standard $O(N)$ WF quantum critical points without Dirac fermions.[9,34,35]

The Gross–Neveu model and in particular the chiral-Heisenberg universality class has been studied also in the large-N_ψ limit.[36–38] The correlation-length critical exponent and the order-parameter anomalous dimensions have been computed[38] to the order $O(1/N_\psi^2)$; in $2+1$ dimensions and for $N_\psi = 2$, one finds $\nu = 1.182$ and $\eta_\phi = 1.184$. The fermion's anomalous dimension is found to the order $O(1/N_\psi^3)$ and for the same parameters $\eta_\psi = 0.105$. Finally, the functional renormalization group has also been brought to bear[39]: the most elaborate computation to date yields $\nu = 1.26$, $\eta_\phi = 1.032$, and $\eta_\psi = 0.071$.[40]

Within the last decade, conformal bootstrap has led to the most accurate values of the critical exponents for the Ising model,[5] and has become competitive with the high-order ϵ-expansion for the XY and Heisenberg. It therefore seems natural to attempt to extend it to the GNY field theories. While this has not been done at the time of writing for the chiral-Heisenberg model, it has been done for a close cousin of the chiral-Ising field theory, which in the context of graphene, for example, describes the quantum phase transition into the quantum anomalous Hall state.[41] This GNY theory would correspond to the Ising ($N = 1$) order parameter in Eq. (8) coupled to the fermion bilinear as in

$$L_{\phi\psi} = g\phi(x)\psi^\dagger(x)(1_2 \otimes 1_2 \otimes \sigma_3)\psi(x). \tag{25}$$

Both this and the chiral-Ising theory describe the spontaneous symmetry-breaking of the Ising sublattice symmetry, but with the order parameter coupled to different fermion bilinears; a finite $\langle \psi^\dagger(1_2 \otimes 1_2 \otimes \sigma_3)\psi \rangle$ would violate the time-reversal symmetry as well. It has been argued[41] that the two GNY Ising theories differ at higher order in the $1/N_\psi$ expansion and therefore should not be expected to have the identical critical behavior; on the other hand, the actual difference in the exponents could be expected to be small. Indeed, the exponents extracted from the four-loop ϵ-expansion for the chiral-Ising model[42] agree within their error bars with the bootstrap values for the anomalous Hall transition, and even the difference with the quantum Monte Carlo calculations[43] is of the order of few percent.

Other quantum phase transitions have also been addressed within the framework of the GNY field theory. The quantum phase transition from the Dirac semimetal into the quantum spin Hall state on the honeycomb lattice also exhibits breaking of spin-rotational symmetry, but with the vector order parameter coupled to a different Dirac bilinear $\sim \psi^\dagger(\sigma_i \otimes \sigma_3 \otimes \sigma_3)\psi$.[44] The transition into the nematic state that breaks rotational symmetry but remains gapless has also been studied both numerically and analytically.[45–47] Both phase transitions appear to be continuous and to be described by an $O(\epsilon)$ fixed point of the RG flow in the corresponding GNY theories. Lattice models that circumvent the Nielsen–Ninomiya[48] fermion-doubling theorem and display the transitions involving a single two-component Dirac fermion, with or without spin, have also been put forward[49,50] and studied by Monte Carlo methods. They corroborate and extend further the physical picture implied by the GNY field theory and discussed above. The GNY phase transitions in the presence of quenched disorder[51] or cubic terms that could render the transition discontinuous have also been addressed.[52–55] Multicritical behavior in the presence of Dirac fermions has been studied as well.[56,57] Surprisingly, emergence of larger symmetries at the criticality induced by Dirac fermions has been found within ϵ-expansion.[58]

Although not discussed here, one can also formulate a GNY-type field theory for the transition into the s-wave superconducting state.[59,60] Finally, GNY-like field theories with fermions with quadratic instead of linear Dirac energy dispersion have also been considered and their $O(\epsilon)$ fixed points identified.[61,62]

6. Summary

In conclusion, we reviewed the construction and applications of the Gross–Neveu–Yukawa field theories for bosonic order parameters coupled to Dirac fermions, primarily as they arise in the system of interacting electrons on honeycomb lattice at the filling one half. These field theories generically exhibit critical fixed points points that are not of the standard $O(N)$ variety, but which nevertheless can be identified and systematically studied using the time-honored expansion around the upper critical dimension proposed by Kenneth Wilson and Michael Fisher more than 50 years ago. This extends the relevance of the method of the ϵ-expansion to the domain of quantum many-body systems and to the fundamental electronic models such as the Hubbard model, where the hope is that it could prove just as fertile as it has been in the classical statistical physics.

Acknowledgments

The author is grateful to Shaffique Adam, Fakher Assaad, Igor Boettcher, Laura Classen, John Gracey, Martin Hohenadler, Lukas Janssen, Vladimir Juričić, Bitan Roy, Michael Scherer, Francesco Parisen Toldin, and Oskar Vafek for the many useful discussions and collaborations on the subject of this review and especially to

Michael Scherer for also reading the manuscript. This work has been supported by the NSERC of Canada.

References

1. K. G. Wilson, *Nobel Lectures in Physics*, G. Ekspong (eds.), (World Scientific, Singapore, 1997).
2. K. G. Wilson and M. E. Fisher, Critical exponents in 3.99 dimensions. *Phys. Rev. Lett.* **28**, 240 (1972).
3. H. Kleinert and Schulte-Frohlinde, *Critical Properties of* ϕ^4 — *Theories* (World Scientific, Singapore, 2001).
4. M. V. Kompaniets and E. Panzer, Minimally subtracted six loop renormalization of $O(n)$-symmetric ϕ^4 theory and critical exponents. *Phys. Rev. D* **96**, 036016 (2017).
5. S. El-Showk, M. F. Paulos, D. Poland, S. Rychkov, D. Simmons-Duffin and A. Vichi, Solving the 3d Ising model with the conformal bootstrap II. c-minimization and precise critical exponents. *J. Stat. Phys.* **157**, 869 (2014).
6. M. Aizenman, Proof of the triviality of ϕ^4_d field theory and some mean-field features of Ising models for $d > 4$. *Phys. Rev. Lett.* **47**, 1 (1981); J. Fröhlich, On the triviality of ϕ^4_d theories and the approach to the critical point in $d(-) > 4$ dimensions. *Nucl. Phys. B* **200**, 281 (1982).
7. B. I. Halperin, T. C. Lubensky and S.-K. Ma, First-order phase transitions in superconductors and smectic-A liquid crystals. *Phys. Rev. Lett.* **32**, 292 (1974).
8. See also: T. C. Lubensky, Fluctuations and Gauges in Superconductors and Liquid Crystals, contribution to this volume.
9. I. Herbut, *A Modern Approach to Critical Phenomena* (Cambridge University Press, Cambridge, England, 2007).
10. I. F. Herbut and Z. Tešanović, Critical fluctuations in superconductors and the magnetic field penetration depth. *Phys. Rev. Lett.* **76**, 4588 (1996); Herbut and Tešanović Reply. *Phys. Rev. Lett.* **78**, 980 (1997).
11. B. Ihrig, N. Zerf, P. Marquard, I. F. Herbut and M. M. Scherer, Abelian Higgs model at four loops, fixed-point collision, and deconfined criticality. *Phys. Rev. B* **100**, 134507 (2019).
12. I. F. Herbut, Interactions and phase transitions on graphene's honeycomb lattice. *Phys. Rev. Lett.* **97**, 146401 (2006).
13. J. Zinn-Justin, Four-fermion interaction near four dimensions. *Nucl. Phys. B* **367**, 105 (1991).
14. G. Semenoff, Condensed-matter simulation of a three-dimensional anomaly. *Phys. Rev. Lett.* **53**, 2449 (1984).
15. C. Honerkamp, Density waves and Cooper pairing on the honeycomb lattice. *Phys. Rev. Lett.* **100**, 146404 (2008).
16. B. Roy and I. F. Herbut, Unconventional superconductivity on honeycomb lattice: Theory of Kekule order parameter. *Phys. Rev. B* **82**, 035429 (2010).
17. F. D. M. Haldane, Model for a quantum Hall effect without Landau levels: Condensed-matter realization of the "parity anomaly". *Phys. Rev. Lett.* **61**, 2015 (1988).
18. C.-Y. Hou, C. Chamon and C. Mudry, Electron fractionalization in two-dimensional graphene-like structures. *Phys. Rev. Lett.* **98**, 186809 (2007).
19. C. L. Kane and E. J. Mele, Quantum spin Hall effect in graphene. *Phys. Rev. Lett.* **95**, 226801 (2005).

20. I. F. Herbut, V. Juričić and B. Roy, Theory of interacting electrons on the honeycomb lattice. *Phys. Rev. B* **79**, 085116 (2009).
21. C. Weeks and M. Franz, Interaction-driven instabilities of a Dirac semimetal. *Phys. Rev. B* **81**, 085105 (2010).
22. I. F. Herbut, V. Juričić and O. Vafek, Relativistic Mott criticality in graphene. *Phys. Rev. B* **80**, 075432 (2009).
23. B. Rosenstein, H.-L. Lu and A. Kovner, Critical exponents of new universality classes. *Phys. Lett.* **314**, 381 (1993).
24. L. N. Mihaila, N. Zerf, B. Ihrig, I. F. Herbut and M. M. Scherer, Gross–Neveu–Yukawa model at three loops and Ising critical behavior of Dirac systems. *Phys. Rev. B* **96**, 165133 (2017).
25. N. Zerf, L. N. Mihaila, P. Marquard, I. F. Herbut and M. M. Scherer, Four-loop critical exponents for the Gross–Neveu–Yukawa models. *Phys. Rev. D* **96**, 096010 (2017).
26. F. F. Assaad and I. F. Herbut, Pinning the order: The nature of quantum criticality in the Hubbard Model on honeycomb lattice. *Phys. Rev. X* **3**, 031010 (2013).
27. F. Parisen Toldin, M. Hohenadler, F. F. Assaad and I. F. Herbut, Fermionic quantum criticality in honeycomb and π-flux Hubbard models: Finite-size scaling of renormalization-group-invariant observables from quantum Monte Carlo. *Phys. Rev. B* **91**, 165108 (2015).
28. Y. Otsuka, S. Yunoki and S. Sorella, Universal quantum criticality in the metal-insulator transition of two-dimensional interacting Dirac electrons. *Phys. Rev.* **6**, 011029 (2016).
29. J. Ostmeyer, E. Berkowitz, S. Krieg, T. A. Lähde, T. Luu and C. Urbach, Semimetal Mott insulator quantum phase transition of the Hubbard model on the honeycomb lattice, *Phys. Rev. B* **102**, 245105 (2020).
30. Y. Otsuka, K. Seki, S. Sorella and S. Yunoki, Dirac electrons in the square-lattice Hubbard model with a d-wave pairing field: The chiral Heisenberg universality class revisited. *Phys. Rev. B* **102**, 235105 (2020).
31. B. Roy, V. Juričić and I. F. Herbut, Emergent Lorentz symmetry near quantum critical points in two and three dimensions. *J. High Energ. Phys.* **2016**, 18 (2016).
32. M. M. Anber and J. F. Donoghue, Emergence of universal limiting speed. *Phys. Rev. D* **83**, 105027 (2011).
33. H.-K. Tang, J. N. Leaw, J. N. B. Rodrigues, I. F. Herbut, P. Sengupta, F. F. Assaad and S. Adam, The role of electron-electron interactions in two-dimensional Dirac fermions. *Science* **361**, 570 (2018).
34. M. P. A. Fisher and G. Grinstein, Quantum critical phenomena in charged superconductors. *Phys. Rev. Lett.* **60**, 208 (1988).
35. I. F. Herbut, Quantum critical points with the Coulomb interaction and the dynamical exponent: when and why z = 1. *Phys. Rev. Lett.* **87**, 137004 (2001).
36. A. N. Vasilev, S. É. Derkachev, N. A. Kivel' and A. S. Stepanenko, The $1/n$ expansion in the Gross–Neveu model: Conformal bootstrap calculations of the index η in order $1/n^3$. *Theor. Math. Phys.* (Engl. Transl.) **94**, 127 (1993).
37. J. A. Gracey, Computation of the critical exponent η at $O(1/n^3)$ in the four-Fermi model in arbitrary dimension. *Int. J. Mod. Phys. A* **9**, 727 (1994).
38. J. A. Gracey, Large N critical exponents for the chiral Heisenberg Gross–Neveu universality class. *Phys. Rev. D* **97**, 105009 (2018).
39. L. Janssen and I. F. Herbut, Antiferromagnetic critical point on graphene's honeycomb lattice: A functional renormalization group approach. *Phys. Rev. B* **89**, 205403 (2014).

40. B. Knorr, Critical chiral Heisenberg model with the functional renormalization group. *Phys. Rev. B* **97**, 075129 (2018).
41. R. S. Erramilli, L. V. Iliesiu, P. Kravchuk, A. Liu, D. Poland and D. Simmons-Duffin, The Gross-Neveu-Yukawa Archipelago. *J. High Energ. Phys.* **2023**, 36 (2023).
42. B. Ihrig, L. N. Mihaila and M. M. Scherer, Critical behavior of Dirac fermions from perturbative renormalization. *Phys. Rev. B* **98**, 125109 (2018).
43. Y. Liu, W. Wang, K. Sun and Z. Y. Meng, Designer Monte Carlo simulation for the Gross–Neveu–Yukawa transition. *Phys. Rev. B* **101**, 064308 (2020).
44. Y. Liu, Z. Wang, T. Sato, W. Guo and F. F. Assaad, Gross-Neveu Heisenberg criticality: Dynamical generation of quantum spin Hall masses. *Phys. Rev. B* 104, 035107 (2021).
45. M. Vojta, Y. Zhang and S. Sachdev, Renormalization group analysis of quantum critical points in d-wave superconductors. *Int. J. Mod. Phys.* **14**, 3719 (2000).
46. Y. Huh and S. Sachdev, Renormalization group theory of nematic ordering in d-wave superconductors. *Phys. Rev. B* **78**, 064512 (2008).
47. J. Schwab, L. Janssen, K. Sun, Z.-Y. Meng, I. F. Herbut, M. Vojta and F. F. Assaad, Nematic quantum criticality in Dirac systems. *Phys. Rev. Lett.* **128**, 157203 (2022).
48. H. B. Nielsen and M. Ninomiya, Absence of neutrinos on a lattice: (I). Proof by homotopy theory. *Nuc. Phys. B* **185**, 20 (1981).
49. T. C. Lang and A. M. Läuchli, Quantum Monte Carlo simulation of the chiral Heisenberg Gross–Neveu–Yukawa phase transition with a single Dirac cone. *Phys. Rev. Lett.* **123**, 137602 (2019).
50. S. M. Tabatabaei, A.-R. Negari, J. Maciejko and A. Vaezi, Chiral Ising Gross–Neveu criticality of a single Dirac cone: A quantum Monte Carlo study. *Phys. Rev. Lett.* **128**, 225701 (2022).
51. H. Yerzhakov and J. Maciejko, Random-mass disorder in the critical Gross-Neveu-Yukawa models. *Nucl. Phys. B* **962**, 115241 (2021).
52. Z.-X. Li, Y.-F. Yang, S.-K. Jian and H. Yao, Fermion-induced quantum critical point. *Nat. Commun.* **8**, 314 (2017).
53. M. M. Scherer and I. F. Herbut, Gauge-field-assisted Kekule quantum criticality. *Phys. Rev. B* **94**, 205136 (2016).
54. S.-K. Jian and H. Yao, Fermion-induced quantum critical points in two-dimensional Dirac semimetals. *Phys. Rev. B* **96**, 195162 (2017).
55. E. Torres, L. Classen, I. F. Herbut and M. M. Scherer, Fermion-induced quantum criticality with two length scales in Dirac systems. *Phys. Rev. B* **97**, 125137 (2018).
56. L. Classen, L. Janssen, I. F. Herbut and M. M. Scherer, Mott multicritcality of Dirac electrons in graphene. *Phys. Rev. B* **92**, 035429 (2015); Competition of density waves and quantum multicritical behavior in Dirac materials from functional renormalization. *Phys. Rev. B* **93**, 125119 (2016).
57. I. F. Herbut and M. M. Scherer, SO(4) multicriticality of two-dimensional Dirac fermions. *Phys. Rev. B* **106**, 115136 (2022), and references therein.
58. L. Janssen, I. F. Herbut and M. M. Scherer, Compatible orders and fermion-induced emergent symmetry in Dirac systems. *Phys. Rev. B* **97**, 041117 (2018).
59. S.-S. Lee, Emergence of supersymmetry at a critical point of a lattice model. *Phys. Rev. B* **76**, 075103 (2007).
60. B. Roy, V. Juričić and I. F. Herbut, Quantum superconducting criticality in graphene and topological insulators. *Phys. Rev. B.* **87**, 041401(R) (2013).

61. L. Janssen and I. F. Herbut, Nematic quantum criticality in three-dimensional Fermi system with quadratic band touching. *Phys. Rev. B* **92**, 045117 (2015); Phase diagram of electronic systems with quadratic Fermi nodes in $2 < d < 4$: $2 + \epsilon$ expansion, $4 - \epsilon$ expansion, and functional renormalization group, *Phys. Rev. B* **95**, 075101 (2017).
62. I. Boettcher and I. F. Herbut, Superconducting quantum criticality in three-dimensional Luttinger semimetals. *Phys. Rev. B* **93**, 205138 (2016).

https://doi.org/10.1142/9789811282386_0022

Chapter 22

Strong-Randomness Renormalization Groups

David A. Huse
Department of Physics, Princeton University, Princeton, NJ 08544, USA

Strong-randomness renormalization groups were first developed to treat various quantum critical ground states, especially in one-dimensional systems. After briefly examining some of the earlier works with these methods, the recent application of this approach to the many-body localization phase transition is reviewed.

1. Introduction

Renormalization-group (RG) approaches to systems in critical states have been developed for a wide variety of systems. As a consequence of this variety, there are many different types of RG calculations. What do they all have in common? Almost all RGs coarse-grain a system of many degrees of freedom by some type of "integrating out" of the highest energy or in some sense "stiffest" degrees of freedom. Good analytical control of the RG usually requires some level of simplicity of the couplings between the degrees of freedom that are integrated out and the remaining degrees of freedom.

For many translationally invariant systems, the RG is implemented in momentum space, as in the seminal Wilson–Fisher paper.[1,2] In many cases, the highest-energy (or stiffest) degrees of freedom are the remaining modes with the highest momenta, and these are integrated out, thus decreasing the ultraviolet momentum cutoff. For systems of many degenerate fermions, "highest momenta" may be replaced with momenta farthest in energy from the Fermi energy (see, e.g., Ref. 3).

Some other RGs work instead in real space, one notable example being the Kosterlitz–Thouless RG for two-dimensional superfluids, which can be formulated with a real-space cutoff on the distance between a vortex and an antivortex.[4,5] Vortex–antivortex pairs at this cutoff distance are integrated out, thus increasing the cutoff distance. The good analytic control of the Kosterlitz–Thouless RG is due to the fixed line governing the superfluid phase and its critical point being at zero density of vortices and antivortices, so the pairs that are being integrated out are dilute when we are near the fixed line. A similar real-space RG for some one-dimensional spin systems, where one integrates out pairs of domain walls at the cutoff distance, was developed earlier in Ref. 6.

For systems with strong quenched spatial randomness, the highest-energy (or stiffest) degrees of freedom are typically at some particular location in real space where the couplings are the strongest. This naturally leads to a real-space RG, as discussed in the following, and as has been reviewed thoroughly by Iglói and Monthus,[7,8] and more briefly by Refael and Altman.[9] Such strong-randomness RGs are the focus of this very brief review, which begins with some discussion of their use for quantum critical ground states and finishes with more recent generalizations that address the many-body localization (MBL) phase transition that can occur in highly excited quantum states.

2. Random-Singlet Ground State

The first strong-randomness RG was developed by Ma, Dasgupta and Hu (1979) for the low-temperature properties of random quantum antiferromagnets.[10–13] The system they considered is an antiferromagnetic Heisenberg spin-1/2 chain, with quenched random nearest-neighbor spin interactions. The Hamiltonian is

$$H = \sum_n J_n \boldsymbol{\sigma}_n \cdot \boldsymbol{\sigma}_{n+1}, \tag{1}$$

where the couplings $J_n > 0$ are drawn from a probability distribution $P(\log J)$ that is broad and $\boldsymbol{\sigma}_n$ is the vector of Pauli operators for the spin-1/2 at site n. The ground state of this spin chain is a so-called random-singlet state.[10–13] This random-singlet ground state consists, to first approximation, of pairs of spins that are in their total spin zero singlet state, with the two spins in each such pair being at various distances along the chain determined by local details of that sample's Hamiltonian.

The strong-randomness RG treatment of this random-singlet ground state proceeds as follows: Each coarse-graining step of the RG "integrates out" the two spins, n and $(n+1)$, that are coupled by the strongest remaining renormalized coupling J_n. The order in which the spins are integrated out is dictated by energy and so proceeds in a sequence that is specific to each particular sample. Since these pairs of strongly coupled spins are each located at some position in real space, this RG is often called a real-space RG. But really it more fundamentally works in "energy space". The two spins to be integrated out are also coupled to their other neighbors via the couplings J_{n-1} and J_{n+1}, which are both weaker than J_n. The small parameters that allow good analytic control of this RG are the ratios J_{n-1}/J_n and J_{n+1}/J_n. When the distribution $P(\log J)$ is broad, this makes these ratios typically small, so this RG approach is a controlled approximation. Assuming that these two ratios are small, those weaker couplings can be treated in low-order perturbation theory.

Thus, we have the three terms in the local Hamiltonian:

$$H = \cdots + J_{n-1}\boldsymbol{\sigma}_{n-1} \cdot \boldsymbol{\sigma}_n + J_n\boldsymbol{\sigma}_n \cdot \boldsymbol{\sigma}_{n+1} + J_{n+1}\boldsymbol{\sigma}_{n+1} \cdot \boldsymbol{\sigma}_{n+2} + \cdots, \tag{2}$$

with the middle term the strongest. The lowest-order approximation to the ground state is as follows: spins n and $(n+1)$ are in their total spin zero singlet state. If we stop at that very lowest perturbative order, the chain is cut, so this does not address what are the ground-state correlations between, for example, spins $(n-1)$ and $(n+2)$. The leading term that produces an interaction across the two spins that we are "integrating out" is produced at second order in perturbation theory in the weaker couplings, resulting in the renormalized Hamiltonian:

$$H' = \cdots + \delta E_0 + J'\boldsymbol{\sigma}_{n-1} \cdot \boldsymbol{\sigma}_{n+2} + \cdots, \tag{3}$$

with renormalized coupling $J' = J_{n-1}J_{n+1}/(2J_n) + \cdots$ as well as a contribution $\delta E_0 = -3J_n + \cdots$ to the ground- state energy. The ...'s refer to higher-order perturbative effects, which are asymptotically RG-irrelevant. So, this RG step puts spins n and $(n+1)$ in the local ground state, thus removing those degrees of freedom, and introduces a new renormalized coupling between spins $(n-1)$ and $(n+2)$, producing a renormalized spin chain that is two spins shorter in length.

As the initial authors understood,[10–12] and as was analyzed in much more detail by Daniel Fisher,[13] this approach results in a *functional* RG for the probability distribution $P(\log J)$ of the nearest-neighbor couplings. Under the RG flow, this distribution broadens without limit due to the very weak new renormalized couplings that are produced. This produces an RG flow to an *infinite-randomness fixed point*, where the approximations $J_{n\pm1} \ll J_n$ used in the RG become asymptotically exact, so this RG gives the correct low-energy description of this system. Since the three couplings that entered in setting the renormalized coupling J' are all removed from the renormalized Hamiltonian H', no correlations between different couplings J_m are generated by the leading-order RG. Thus, although one could have a joint probability distribution $P(\{J_m\})$ with short-range correlations between the couplings on different bonds, such correlations are RG-irrelevant: the RG flow reduces these correlations and the fixed-point distribution is asymptotically uncorrelated, which facilitates the analysis.[13]

This random-singlet ground state is a quantum critical state. Any spin-1/2 chain of this form, with random nearest-neighbor-only antiferromagnetic couplings drawn from a continuous joint probability distribution with only short-range correlations in the couplings, has a ground state governed by this same infinite-randomness fixed point. Asymptotically, when the remaining spins are at typical distance r bare lattice spacings, the typical renormalized couplings J scale as $-\log J \sim r^{1/2}$ for large r. Thus, there is an exponential dynamical scaling (dynamical critical exponent $z \to \infty$), with the energy scales decreasing as the exponential of a power of the length scale. Such exponential dynamical scaling is typical of infinite-randomness quantum critical points. The equal-time spin–spin correlations in this ground state are very broadly distributed, with the mean correlation falling off with distance as $\sim r^{-2}$ due to rare strongly correlated spins, while the typical correlations fall off with distance exponentially in $r^{1/2}$, as do the renormalized couplings.[13]

3. Other Infinite-Randomness Fixed Points

There are (infinitely) many other one-dimensional models, both quantum and classical, whose asymptotic ground state and/or low-frequency properties can be systematically and correctly obtained with a strong-randomness RG approach.[7–9] One notable example is the quantum critical point of the transverse-field Ising spin chain, whose infinite-randomness fixed point is very closely related to that of the random-singlet state.[14,15] An infinite series of discretely different infinite-randomness fixed points was also found. These govern the quantum critical ground states of certain spin chains with larger spin S, as well as chains of interacting non-Abelian anyons.[7,8,16,17]

For many of these one-dimensional models, the strong-randomness RG can be solved analytically. When the method is generalized to systems in more than one dimension ($d > 1$), parts of the calculations must instead be implemented numerically. It is found that, unlike in one dimension, for $d > 1$ and most models the RG flow near infinite randomness is instead toward weaker randomness, so such systems are not governed by infinite-randomness fixed points. The principal exception to this is the quantum critical point of the random transverse-field Ising model, which appears to remain governed by infinite-randomness fixed points for all dimensions d.[7,8]

4. Many-Body Localization

One recent development in the use of strong-randomness renormalization groups is in the study of many-body localization (MBL). Some of this RG work was reviewed in Refs. 8 and 9, so here I will focus on the more recent work.

MBL is the Anderson localization for many interacting particles or spins in highly excited states at thermodynamic conditions that correspond to non-zero entropy density. The system is assumed to be isolated from and not interacting with any other system or environment. The key question that is asked is whether or not this isolated many-body system "thermalizes": Does it successfully serve as a thermal "bath" for itself and, under its own unitary quantum dynamics, bring all of its subsystems to thermal equilibrium with each other: The MBL (localized) phase is the part of this system's phase diagram where the system remains (Anderson) localized near any non-thermal initial state, so it fails to bring itself to thermal equilibrium: It fails to thermalize. Some recent reviews about MBL include.[18–21]

Most uses of the strong-randomness RG for many-body quantum systems are for the study of the system's ground-state and low-lying excited states. In the study of MBL, on the other hand, the focus is on highly excited states, often typical states that at thermal equilibrium would correspond to infinite temperature. One main question that we want to address for MBL systems is whether or not the eigenstates of the system's dynamics, as well as its long-time dynamical states, are at thermal equilibrium. The MBL phase is the regime where the system does not go to thermal equilibrium nor are the eigenstates of its dynamics at thermal equilibrium, even in the limits of large systems and long times. It appears that a true MBL phase that remains localized in the standard thermodynamic limit and in the infinite-time limit is a possibility only for one-dimensional systems with short enough range interactions,[22,23] so those are the systems the remainder of this review considers.

The dynamics of an MBL system may be due to a time-independent Hamiltonian or due to a Floquet unitary operator that produces the dynamics for one period of a Hamiltonian that is periodic in time. MBL does not require randomness, for example, it may occur due to nonrandom quasiperiodicity of the system,[24] but here we only consider MBL due to quenched randomness. An example Hamiltonian is a Heisenberg spin-1/2 chain with a random field:

$$H = \sum_n [h_n S_{n,z} + \mathbf{S}_n \cdot \mathbf{S}_{n+1}], \tag{4}$$

with the quenched random fields $\{h_n\}$ at each site n drawn independently and uniformly from $[-W, +W]$. This is one of the most studied MBL models,[25] it is sometimes called the "standard model" for MBL, but only because it is highly studied and not because it is the best choice for a model to study MBL. The MBL phase is at large W (strong random fields) for this model. The precise location of the phase boundary is still not clear, but it appears to be at $W_c > 15$.[26,27]

The MBL phase is a gapless quantum-critical dynamic phase, and the strong-randomness RG has been used within the MBL phase by various authors; much of this work is reviewed in Ref. 8. Here, I instead review some of the RG studies of the dynamic quantum phase transition between the MBL phase, where the system does not thermalize, and the thermal phase, where the system does thermalize in the long time and large system limits. Early numerical work on this MBL phase transition in one-dimensional systems found behavior consistent with exponential dynamical scaling (dynamic critical exponent $z \to \infty$), suggesting that it might be governed by an infinite-randomness fixed point.[25] Since then, there has been a series of publications steadily exploring and developing strong-randomness RG treatments of this MBL phase transition in one-dimensional systems with short-range interactions and/or hoppings.[28–38]

When one uses a strong-randomness RG to study a many-body quantum ground state, this is usually done working directly with the microscopic degrees of freedom of the system, and the highest-energy remaining local excitations are usually integrated out, leaving a renormalized system with fewer degrees of freedom, as in most renormalization groups. For the MBL phase transition, on the other hand, a direct fully controlled RG calculation has not been found, so the RG is instead phenomenological and/or approximate, and in most cases both. The number of degrees of freedom sets the density of states (many-body states per unit energy), and thus, the thermodynamic entropy, which is a key ingredient in the physics of the MBL phase transition, so no true "integrating out" that removes degrees of freedom is done. Thus, the RG renormalizes (coarse-grains) the dynamics, scaling to longer times and larger length scales, but without the usual reduction of the number of degrees of freedom. As a result, the RG in most cases does not work directly with the microscopic degrees of freedom, but instead with more coarse-grained measures of the local dynamics.

In strong-randomness RG treatments of the MBL phase transition, the one-dimensional system is to some extent coarse-grained in to "blocks", which are segments of the chain with certain coarse-grained properties.[28–38] In many of these works, the blocks are assumed to be of only two types: locally thermalizing blocks that are locally less random and/or more interacting, so these thermal blocks do become strongly entangled; locally MBL blocks that are locally more random and/or less interacting, so these locally MBL blocks have localized states that remain area-law entangled unless thermalized by a nearby thermal block. Such a "binary" classification of the local properties is only approximate when applied at small length scales. However, since the resulting RG does exhibit a flow to infinite randomness in the MBL phase and at the phase transition, this binary description then becomes self-consistent and thus potentially asymptotically correct. Multiple assumptions are used in developing these RGs, so there certainly remains the possibility that some important and relevant physics has been left out.

Near the fixed point governing the MBL phase transition, the locally thermalizing blocks are rare, occupying an asymptotically vanishing fraction of the length of the one-dimensional system as one renormalizes to the limit of low-energy scales. These widely spaced thermal blocks are randomly placed, and their lengths have a probability distribution. This strong-randomness RG is, as usual, a *functional* RG, with this distribution of thermal block lengths as the function that is being renormalized.

Each thermal block is locally thermalizing the nearby locally MBL regions. As we renormalize to lower-energy scales in this RG, the locally thermal blocks thermalize the spins in the adjacent MBL blocks out to a distance that grows logarithmically with the (inverse) energy scale. There are two "events" that happen as this RG flow runs: (i) A single thermal block may reach the limit of how many spins in the nearby MBL regions it can thermalize. Once we renormalize below that energy scale, that thermal block is found to instead be localized (no longer able to spread more thermalization), so it is removed and that region is instead localized at this lower-energy scale. (ii) If two nearby thermal blocks manage to thermalize the entire MBL region that lies in between them, then they merge with each other and with this intermediate region that they have thermalized to make a new, much longer renormalized thermal block.[35,36] The latter process broadens the distribution of the lengths of the thermal blocks. The energy scale of a thermal block is set by its many-body energy-level spacing which is exponentially small in its length, so this length distribution becoming broader means a flow to infinite randomness.

A nice analogy with the Kosterlitz–Thouless (KT) RG and transition has been found and developed:[32–36] In KT, if vortices and antivortices are forbidden, then the two-dimensional superfluid remains stable. This critical KT superfluid phase is governed by a RG fixed line, with the superfluid stiffness divided by the temperature being the dimensionless parameter that varies along this fixed line. The KT phase transition occurs where the RG flow near this fixed line changes from being stable to being unstable to allowing vortices and antivortices.[4,5]

For the MBL transition in one-dimensional systems with quenched randomness, the role of vortices and antivortices is instead played by locally thermalizing rare regions (thermal blocks).[18,22,35,36] In this approach, the MBL phase is governed by a fixed line of the RG, with the dimensionless parameter that varies along the fixed line being the ratio of two lengths: one being the length per bit of thermodynamic entropy and the other length being a decay length for the exponential dependence of the relaxation rate of spins on their distance from the nearest thermal block.[22,35,36] The MBL phase is governed by the part of this fixed line that is stable to allowing locally thermalizing rare regions (thermal blocks); in the MBL phase, these rare regions are RG-irrelevant, so the RG flow goes to this fixed line. In the MBL phase, these thermal blocks only manage to locally thermalize a limited number of spins in the adjacent MBL blocks before the energy scale is reached where this thermalization stops and the thermal block becomes instead localized. The MBL-to-thermal phase transition is then governed by the point on this fixed line where these locally thermalizing rare regions (added thermal blocks) become RG-relevant, so the fixed line becomes unstable in the thermal phase and the RG flow heads off toward thermalization. This occurs when, under the RG flow, enough of the thermal blocks can merge and produce much longer thermal blocks, so the fraction of the system occupied by the thermal blocks instead increases under the RG flow.

Although there is this strong *qualitative* analogy to the KT RG, the strong-randomness RG flow for the MBL transition is mathematically distinct.[36] For KT, the RG is a two-parameter flow, with the vortex fugacity and the reduced superfluid stiffness being the two parameters. For MBL, on the other hand, it is instead a *functional* RG, with the function being the probability distribution of the lengths of the locally thermalizing rare regions. When two of these rare regions (thermal blocks) are close enough together to thermalize the locally MBL typical region in between them, then, under the RG flow, they merge in to a much longer rare region, extending the distribution function to those longer lengths.[35,36] This produces a rather direct connection within the functional RG flow between two very different length scales, a feature that is not present in the simpler KT RG flow.

Recent numerical work to quantitatively estimate where this asymptotic MBL-to-thermal phase transition occurs in the phase diagram of microscopically defined spin-chain models has shown that it is actually very deep in the regime where the behavior of those models appears to be strongly localized for sample sizes and times accessible to standard numerical and experimental methods.[26,27] Thus, it currently appears that the physics of the asymptotic MBL phase transition as found in the strong-randomness RG may only apply on rather large length scales and thus only for extremely low-energy scales (equivalently, extremely long times). One way to put a "positive spin" on our present understanding of this situation is to note that the strong-randomness RG method allows one to develop what appears to be a controlled asymptotic low-energy theory of this phase transition, even though other approaches to studying this novel and theoretically challenging phase transition have not been able to reach those very low-energy scales.

5. Conclusion

The strong-randomness RG methods are versions of the RG that are useful for treating certain systems with quenched randomness, particularly those whose low-energy behavior is governed by an infinite-randomness fixed point. These methods have wide applicability in certain one-dimensional quantum systems as well as in various other systems, as reviewed in Refs. 7–9. More recently, such strong-randomness RG methods have been generalized to study the dynamic phase transition between many-body localization (MBL) and thermalization, as I have briefly reviewed here.

Acknowledgments

I thank Michael Fisher for being a wonderful teacher, graduate advisor, and collaborator, and for all that he taught me about critical phenomena, the renormalization group, and many other things. I am deeply indebted to him for his contributions to my education and development as a scientist.

References

1. K. G. Wilson and M. E. Fisher, Critical exponents in 3.99 dimensions. *Phys. Rev. Lett.* **28**, 240 (1972).
2. M. E. Fisher, The renormalization group in the theory of critical behavior. *Rev. Mod. Phys.* **46**, 597 (1974).
3. R. Shankar, Renormalization-group approach to interacting fermions. *Rev. Mod. Phys.* **66**, 129 (1994).
4. J. M. Kosterlitz and D. J. Thouless, Ordering, metastability and phase transitions in two-dimensional systems. *J. Phys. C* **6**, 1181 (1973).
5. J. M. Kosterlitz, The critical properties of the two-dimensional XY model. *J. Phys. C* **7**, 1046 (1974).

6. P. W. Anderson, G. Yuval and D. R. Hamann, Exact results in the Kondo problem: II. Scaling theory, qualitatively correct solution, and some new results on one-dimensional classical statistical models. *Phys. Rev. B* **1**, 4464 (1970).
7. F. Iglói and C. Monthus, Strong disorder RG approach of random systems. *Phys. Rep.* **412**, 277 (2005).
8. F. Iglói and C. Monthus, Strong disorder RG approach — A short review of recent developments. *Eur. Phys. J. B* **91**, 290 (2018).
9. G. Refael and E. Altman, Strong disorder renormalization group primer and the superfluid-insulator transition. *Comptes Rendus Physique* **14**, 725–739 (2013).
10. S.-k. Ma, C. Dasgupta and C.-K. Hu, Random antiferromagnetic chain. *Phys. Rev. Lett.* **43**, 1434 (1979).
11. C. Dasgupta and S.-K. Ma, Low-temperature properties of the random Heisenberg antiferromagnetic chain. *Phys. Rev. B* **22**, 1305 (1980).
12. R. N. Bhatt and P. A. Lee, Scaling studies of highly disordered spin-1/2 antiferromagnetic systems. *Phys. Rev. Lett.* **48**, 344 (1982).
13. D. S. Fisher, Random antiferromagnetic quantum spin chains. *Phys. Rev. B* **50**, 3799 (1994).
14. D. S. Fisher, Random transverse field Ising spin chains. *Phys. Rev. Lett.* **69**, 534 (1992).
15. D. S. Fisher, Critical behavior of random transverse-field Ising spin chains. *Phys. Rev. B* **51**, 6411 (1995).
16. K. Damle and D. A. Huse, Permutation-symmetric multicritical points in random antiferromagnetic spin chains. *Phys. Rev. Lett.* **89**, 277203 (2002).
17. L. Fidkowski, H.-H. Lin, P. Titum and G. Refael, *Permutation-symmetric critical phases in disordered non-Abelian anyonic chains. Phys. Rev. B* **79**, 155120 (2009).
18. W. De Roeck and J. Z. Imbrie, Many-body localization: Stability and instability. *Phil. Trans. R. Soc. A* **375**, 20160422 (2017).
19. F. Alet and N. Laflorencie, Many-body localization: An introduction and selected topics. *Comptes Rendus Physique* **19**, 498 (2018).
20. D. A. Abanin, E. Altman, I. Bloch and M. Serbyn, Colloquium: Many-body localization, thermalization, and entanglement. *Rev. Mod. Phys.* **91**, 021001 (2019).
21. S. Gopalakrishnan and S. A. Parameswaran, Dynamics and transport at the threshold of many-body localization. *Phys. Reps.* **862**, 1 (2020).
22. W. De Roeck and F. Huveneers, Stability and instability towards delocalization in many-body localization systems. *Phys. Rev. B* **95**, 155129 (2017).
23. S. Gopalakrishnan and D. A. Huse, Instability of many-body localized systems as a phase transition in a nonstandard thermodynamic limit. *Phys. Rev. B* **99**, 134305 (2019).
24. S. Iyer, V. Oganesyan, G. Refael and D. A. Huse, Many-body localization in a quasiperiodic system. *Phys. Rev. B* **87**, 134202 (2013).
25. A. Pal and D. A. Huse, The many-body localization phase transition. *Phys. Rev. B* **82**, 174411 (2010).
26. A. Morningstar, L. Colmenarez, V. Khemani, D. J. Luitz and D. A. Huse, Avalanches and many-body resonances in many-body localized systems. *Phys. Rev. B* **105**, 174205 (2022).
27. D. Sels, Bath-induced delocalization in interacting disordered spin chains. *Phys. Rev. B* **106**, L020202 (2022).
28. R. Vosk, D. A. Huse and E. Altman, Theory of the many-body localization transition in one dimensional systems. *Phys. Rev. X* **5**, 031032 (2015).

29. A. C. Potter, R. Vasseur and S. A. Parameswaran, Universal properties of many-body delocalization transitions. *Phys. Rev. X* **5**, 031033 (2015).
30. L. Zhang, B. Zhao, T. Devakul and D. A. Huse, Many-body localization phase transition: A simplified strong-randomness approximate renormalization group. *Phys. Rev. B* **93**, 224201 (2016).
31. P. T. Dumitrescu, R. Vasseur and A. C. Potter, Scaling theory of entanglement at the many-body localization transition. *Phys. Rev. Lett.* **119**, 110604 (2017).
32. T. Thiery, F. Huveneers, M. Müller and W. De Roeck, Many-body delocalization as a quantum avalanche. *Phys. Rev. Lett.* **121**, 140601 (2018).
33. A. Goremykina, R. Vasseur and M. Serbyn, Analytically solvable renormalization group for the many-body localization transition. *Phys. Rev. Lett.* **122**, 040601 (2019).
34. P. T. Dumitrescu, A. Goremykina, S. A. Parameswaran, M. Serbyn and R. Vasseur, Kosterlitz–Thouless scaling at many-body localization phase transitions. *Phys. Rev. B* **99**, 094205 (2019).
35. A. Morningstar and D. A. Huse, Renormalization-group study of the many-body localization transition in one dimension. *Phys. Rev. B* **99**, 224205 (2019).
36. A. Morningstar, D. A. Huse and J. Z. Imbrie, Many-body localization near the critical point. *Phys. Rev. B* **102**, 125134 (2020).
37. U. Agrawal, S. Gopalakrishnan and R. Vasseur, Universality and quantum criticality in quasiperiodic spin chains. *Nature Comm.* **11**, 2225 (2020).
38. Z. D. Shi, V. Khemani, R. Vasseur and S. Gopalakrishnan, Many-body localization transition with correlated disorder. *Phys. Rev. B* **106**, 144201 (2022).

https://doi.org/10.1142/9789811282386_0023

Chapter 23

Line Defects and the Renormalization Group

Z. Komargodski
Simons Center for Geometry and Physics
Stony Brook University, Brookhaven, NY, USA
zkomargo@gmail.com

We define a notion of zero-temperature entropy for impurities (line defects) in $d+1$ space–time dimensions. We show that this entropy obeys a simple evolution equation under the renormalization group. We apply this result for impurities in magnets and in gauge theories. We find new critical impurities and phase transitions.

1. General Introduction

Conformal Field Theories (CFTs) in $d+1$ space–time dimensions are continuum theories describing second-order phase transitions. They arise in a variety of physical constructions, for instance,

- Certain zero-temperature quantum systems in d space dimensions at loci of the coupling constants corresponding to a diverging correlation length. Here the relativistic invariance in $d+1$ space–time dimensions and the emerging light-cone hold only near the phase transition.
- Classical statistical systems in thermal equillibrium in $d+1$ Euclidean dimensions at a point with a diverging correlation length.
- Quantum systems in $d+1$ space dimensions, where one space dimension is taken to be a circle whose radius is tuned to a point with a diverging correlation length.

There are many other instances in which CFTs are encountered.

The symmetry algebra of the continuum theory is $SO(d+1,2)$, which includes space–time rotations and boosts, translations, dilation (scaling) symmetry $x^\mu \to \lambda x^\mu$ (with $x^\mu \equiv (t, x^i)$) and special conformal transformations, which are generated by combining an inversion $x^\mu \to x^\mu/x^2$ with a translation in some direction b^μ followed by another inversion: $\frac{x'^\mu}{x'^2} = \frac{x^\mu}{x^2} + b^\mu$.

A central problem about CFTs is to understand the space of local operators $\mathcal{O}_i(x)$, which can be chosen to carry a well-defined scaling dimension Δ_i and spin J_i.

The scaling dimensions have played a significant role in the history of the subject due to the relation between the scaling dimensions of the low-lying operators and the critical exponents near the phase transition.[1]

A less trodden path is the study of extended operators in CFTs. Relativistic invariance and the Wick rotation allow us to think of extended operators as either nonlocal operators acting at a given time, or as a modification of the Hamiltonian by an insertion of an impurity in some region of space. The simplest example is line operators, which can be viewed as either extended operators localized on a line acting on the usual Hilbert space of the theory or, equivalently, as a localized point-like impurity in space, say at $x^i = 0$. In the latter point of view, the impurity potentially changes the Hilbert space, the space of local operators acting on it, and a new ground state must be found in the presence of the impurity.

Even though the theory away from the impurity is at a fixed point of the Renormalization Group (RG), the impurity may or may not be at a fixed point. Generally, we expect that in the infrared the impurity flows to a fixed point. The system with an impurity at a fixed point would preserve a

$$SO(1,2) \subset SO(d+1,2)$$

subgroup of the conformal symmetry. (For defects invariant under transverse rotations, including all the examples appearing here we also have an unbroken $SO(d)$.) The $SO(1,2) \simeq SL(2,\mathbb{R})$ consists of the conformal transformations that leave the defect at $x^i = 0$, namely, translations in time, scaling symmetry, and the special conformal transformation where $b^i = 0$. On the defect worldline $x^i = 0$, these transformations act by $t \to \frac{at+b}{ct+d}$ with $ad - bc = 1$.

We denote defect operators with a hat as $\hat{O}_I(t)$, which are distinguished from bulk local operators $O_i(t,x)$. Bulk operators can be expanded in terms of defect operators via the bulk-defect operator product expansion, which takes the following

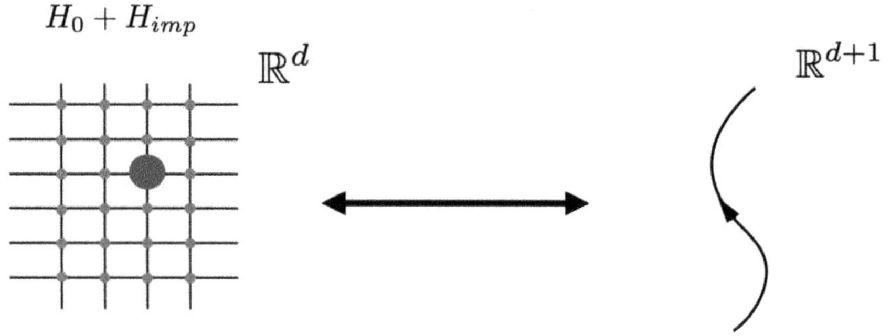

Fig. 1. Two ways of thinking about line defects: either as a modification of the Hamiltonian due to a point-like impurity in $\mathbb{R}^d$ or as a line in space–time $\mathbb{R}^{d+1}$ (which could be, in particular, space-like and act on the Hilbert space).

schematic form (ignoring transverse spin indices):

$$O_i(t,x) = \sum_I \frac{A_i^I}{|x|^{\Delta_i - \Delta_I}} \hat{O}_I(t).$$

This subject was reviewed recently.[2] The contribution corresponding to the unit operator on the right-handside (r.h.s.), for which we define $A_i^1 \equiv a_i$, is particularly interesting, since it leads to one-point functions for bulk operators

$$\langle O_i(t,x) \rangle = \frac{a_i}{|x|^{\Delta_i}}, \tag{1}$$

which are perhaps some of the simplest observables in the presence of a conformal impurity.

If all the $a_i = 0$ (except for the unit operator in the bulk), then the impurity is screened. Indeed, in the absence of an impurity, all one-point functions, other than that of the unit operator, vanish. (In fact, line operators with $a_i = 0$ could be either completely screened or topologically nontrivial but screened. We will not dwell on this distinction since the examples discussed in the following do not admit topologically nontrivial screened line operators.) In that case, the bulk-defect operator product expansion becomes simply Taylor's expansion around $x^i = 0$ and the defect operators are simply the original bulk operators and their derivatives. For nontrivial conformal defects, the space of defect operators and their scaling dimensions need to be determined and cannot be simply obtained from the data about the bulk operators.

To close, we would like to discuss a somewhat abstract observable, which will however turn out to be quite useful. Rotating to Euclidean signature, we can consider the expectation value of the circular loop lying in a plane in our $d+1$ Euclidean dimensions (normalized by the vacuum partition function without the circular loop).

As often happens for partition functions in the continuum limit, such an observable is slightly ambiguous due to local counter-terms. In $d > 1$, there is only one counter-term, the cosmological constant term on the worldline, which is essentially the mass of the impurity $\frac{1}{2\pi}\Lambda \int ds$. For a circular loop of radius R, this leads to a contribution to the partition function $\log Z \sim \Lambda R$. In $d = 1$, there is another possible counter-term due to the extrinsic curvature $\Lambda' \int \kappa$, with Λ' a purely imaginary dimensionless constant. This counter-term is not a local functional of the data on the curve in $d > 1$. We see that in any d the real part of $(1 - R\frac{d}{dR}) \log Z$ is a scheme-independent (finite) observable. In general, this is a complicated function of R and as we increase R, we probe the RG flow on the impurity. For a conformal line operator, $(1 - R\frac{d}{dR}) \log Z$ is just a constant, which is often denoted by $\log g$, where g represents the "quantum dimension" of the impurity. A simple screened impurity has $\log g = 0$, where "simple" means that it cannot be written as a direct sum of two other impurities.

If the bulk is a topological theory in $d = 2$, then it is well known that $\log g \geq 0$ and the terminology "quantum dimension" originates from the fact that if k impurities are inserted on S^2, then the dimension of the Hilbert space scales like g^k for large k. If the bulk is a conformal theory, then we will see that $\log g$ can be either negative or positive. Intuitively, the quantum dimension can be smaller than 1 since the impurity may expel the degrees of freedom around it, which cannot happen in topological theories, since there are no local bulk degrees of freedom to begin with.

Unlike the space of bulk local operators $O_i(t, x)$, about which a lot is known, it is not known how many nontrivial (nonscreened, simple[a]) conformal line operators exist even in the most familiar nontrivial fixed points. Several constructions are known for interesting line operators: magnetic defects, spin defects (as in the Kondo problem[3]), and Wilson and 't Hooft lines, which correspond to external electrically or magnetically charged impurities. Whether such defects flow to nontrivial defect fixed points is the subject of the following sections, after we discuss in some detail the subject of RG flows on the impurity.

2. RG Flows

Starting from a conformal defect, which may or may not be trivial, we can trigger an RG flow if we deform the defect action by integrating a defect operator with $\Delta < 1$ (or a marginally relevant defect operator with $\Delta = 1$), as follows:

$$\delta S = \mu^{1-\Delta_{\hat{O}}} \int dt \hat{O}(t),$$

with μ the physical scale in the problem. This drives the theory on the impurity to an infrared fixed point (which may be trivial).

Conformal invariance is violated throughout the RG flow, and is only restored at very short or very long distances. Therefore, unlike (1) (which is valid at the defect fixed points), the formula for the following one-point function is significantly less constrained:

$$\langle O_i(t, x) \rangle = \frac{g(\mu|x|)}{|x|^{\Delta_i}},$$

with g a function that needs to approach the constants a_i^{uv} and a_i^{ir} for small $|x|$ and large $|x|$, respectively.

A more abstract observable that follows the RG flow is the partition function of the circular loop in Eucildean signature. As we have seen, $(1 - R\frac{d}{dR}) \log Z$ is an R independent constant, $\log g$, for a conformal line defect. If there is an RG flow,

[a]For a nonsimple line operator, the Hilbert space is a direct sum, and it is thus sufficient to restrict to simple line operators without loss of generality.

then it is useful to define a function of R, the defect entropy function

$$s(R) = \left(1 - R\frac{d}{dR}\right) \log Z,$$

which interpolates between $\log g_{uv}$ and $\log g_{ir}$ for small and large R, respectively.

This quantity obeys the following interesting identity[4]:

$$\frac{ds}{dR} = -R^2 \int d\phi_1 d\phi_2 \langle \hat{T}(\phi_1)\hat{T}(\phi_2)\rangle_c (1 - \cos(\phi_1 - \phi_2)) \leq 0.$$

The operator $\hat{T}$ measures the energy density on the defect. It identically vanishes for conformally invariant impurities (the energy cannot be localized on a conformally invariant impurity as it tends to spread to the bulk, but for impurities that are not conformally invariant, the energy can be localized), consistent with that $ds/dR = 0$ at fixed points. This identity shows that in nontrivial RG flows, we must have

$$\log g_{uv} > \log g_{ir} \tag{2}$$

and that $\log g$ is constant on conformal manifolds (i.e. it is constant under exactly marginal deformations). Intuitively, it means that as we coarse grain, the number of effective degrees of freedom near the defect decreases (and correspondingly, intuitively, the number of defect operators should decrease as less degrees of freedom remain near the impurity).

Equation (2) shows that the space of conformal impurities is foliated by $\log g$ and that defect RG flows are irreversible. In the next section, we will see some applications of this general result. Our discussion is valid in arbitrary d, generalizing the classic results[5–7] in $d = 1$.

3. Impurities in Magnets

In this section, we will study some defects in the Wilson–Fisher fixed points[8] in $2+1$ dimensions. We will assume that the bulk fields are at the infrared fixed point of the following action[b]:

$$S = \int d^2x dt \left(\frac{1}{2}\partial\vec{\phi}\cdot\partial\vec{\phi} - \frac{\lambda_*}{4!}(\vec{\phi}\cdot\vec{\phi})^2\right),$$

with $\vec{\phi} = (\phi_1, \ldots, \phi_N)$. This fixed point is nontrivial for all $N \geq 1$ and has $O(N)$ global symmetry. For what follows, we need to recall a few of the salient properties of this family of fixed points:

- The dimension of the order parameter $\Delta(\vec{\phi})$ approaches 0.5 from above at large N and its N dependence is rather weak. For $N = 1$, $\Delta(\phi) = 0.518\ldots$.
- The scaling dimension $\Delta(\vec{\phi}\cdot\vec{\phi})$ approaches 2 from below at large N and it again has comparatively weak N dependence and for $N = 1$, $\Delta(\phi^2) = 1.412\ldots$.

[b]Of course, where the fixed point lies in coupling space is a scheme-dependent question, which will not concern us here.

3.1. *Pinning fields*

An example of a pinning field defect is the line operator

$$e^{-h\int dt\phi_1(t,x=0)}. \tag{3}$$

The physical meaning of this line operator is that we have turned on a localized magnetic field in the 1 direction of $O(N)$ at the location $x = 0$. This is a point-like impurity corresponding to turning on an external field at a point. Such perturbations are sometimes referred to as pinning field defects (we usually study the response of the system at the phase transition to turning a uniform magnetic field, by contrast, here we turn it on only around a few lattice sites).

The theory with this defect now only preserves $O(N-1)$ symmetry. The ultraviolet fixed point of the defect is at $h = 0$, which is a trivial screened defect. Since $\Delta(\phi_1) < 1$ for all N, the coupling constant h is relevant at the ultraviolet fixed point $h = 0$. The pinning field defect therefore describes an RG flow from the trivial line defect in the ultraviolet to some infrared defect fixed point. Since the ultraviolet $h = 0$ fixed point is trivial and the deformation is relevant, we immediately learn from (2) that $g_{ir} < 1$, which implies that the infrared limit of the defect cannot be trivial! This means that the localized magnetic field is not screened in the infrared, which is a nice application of the monotonicity theorem.

We can think about the line defect as an insertion into the following action:

$$S = \int d^2x dt \left(\frac{1}{2}\partial\vec{\phi}\cdot\partial\vec{\phi} - \frac{\lambda_*}{4!}(\vec{\phi}\cdot\vec{\phi})^2\right) - h\int dt\ \phi_1(t, x = 0).$$

As with the Wilson–Fisher fixed point, the infrared limit of the line defect (3) can be understood analytically in the ϵ-expansion, and, in addition, it turns out that an exact large N solution exists. There is some evidence[9] from the ϵ expansion for a mild N dependence for certain observables with the pinning field defect (as with more familiar observables in the Wilson–Fisher fixed point). Therefore, let us simply quote results from the large N solution in $2+1$ dimensions. The large N solution is obtained by identifying a $SO(1,2)$ invariant saddle point corresponding to the pinning field.[9] It may be possible to solve the entire RG flow for the pinning field parameter h at large N, but at present, only results at the infrared (and trivial ultraviolet) fixed point are available. The saddle point at large N is quite complicated (but explicit) and when we quote the following predictions, we retain a few digits after the decimal point. For the infrared fixed point, we find the following g function:

$$\log g_{ir} = -0.153673N,$$

which shows that, unlike defects in topological theories, the quantum dimension of conformal defects can be smaller than 1 (and arbitrarily close to 0). More easily accessible quantities are the defect scaling dimensions and expectation values of bulk operators. Consider the bulk-order parameter operator $\phi_1(t, x)$, which has scaling dimension 1/2 in the large N limit. Due to the pinning field defect it has a decaying expectation value as in (1). We find from the large N solution

$$\frac{\langle\phi_1(0,x)\rangle}{\sqrt{\langle\phi_1(0,\infty)\phi_1(0,0)\rangle_{\text{no defect}}}} = \frac{a_\phi}{|x|^{1/2}}, \quad a_\phi^2 = 0.55813N. \tag{4}$$

Finally, the order parameter can be brought close to the defect and in addition to the unit operator we have the defect operator $\hat{\phi}_1(t)$, which measures the response of the infrared fixed point to changing the value of the magnetic field away from criticality. We find

$$\Delta(\hat{\phi}_1) = 1.542. \tag{5}$$

Note that in the ultraviolet $h = 0$ fixed point, $\Delta(\hat{\phi}_1) = 0.5$, which shows strong renormalization of the defect operator dimensions due to the RG on the impurity. While we reported here results from the large N limit only, they are in rough agreement with what is seen in simulations at small N.[10]

3.2. *Spin impurities*

The pinning field defect originates from turning on a localized external field. A more general class of constructions is to allow for new degrees of freedom at the defect. Here we will focus on the $O(3)$ Wilson–Fisher fixed point in $d = 2+1$. While the bulk operators transform under the orthogonal group, we will now add a defect in a $2s+1$-dimensional representation of the $so(3)$ algebra, coupled to the bulk. For half integer spin s, this cannot be completely screened by the bulk degrees of freedom, but it can still flow to a trivial nonsimple defect (e.g. a decoupled qubit).

The defect Lagrangian, before coupling to the bulk, is

$$S_{QM} = i\int dt\bar{z}\dot{z}, \quad \bar{z}z = 2s.$$

$z = (z_1, z_2)$ are two complex variables on the line subject to the constraint $\bar{z}z = 2s$ and subject to the gauge transformation $z \to e^{i\alpha(t)}z$. The target space of this QM model is therefore S^2 and the Hilbert space consists of $2s+1$ degenerate states in the spin s representation of $so(3)$.[c] The $so(3)$ operators $S^a = \frac{1}{2}\bar{z}\sigma^a z$ can be coupled to the bulk degrees of freedom via

$$S = S_{O(3)} + S_{QM} - \gamma\int dt\sum_{a=1}^{3} S^a\phi_a,$$

where γ is a coupling constant. For $\gamma = 0$, we have a decoupled defect with $2s+1$ states and $\log g_{uv} = \log(2s+1)$. In the infrared we expect to obtain a fixed point, but whether this fixed point is decoupled or not may depend on s. Simulations

[c] One can think about this QM model as the low-energy approximation of a charged particle moving around a magnetic monopole of charge 2s.

suggest that there is an infrared fixed point all the way to $s = 1/2$.[11,12] This defect, unlike the pinning field defect, preserves the $SO(N)$ symmetry of the model and hence, the one-point function of the order parameter vanishes (in contrast to (4)). However, the one-point function of singlet operators, such as $\vec{\phi}^2$, does not vanish.

At large enough s, there is a nontrivial infrared fixed point. Furthermore, there are several exact results at large s. The main idea is that there is an expansion in $1/s$, where at $s = \infty$ there are reliable saddle points with small fluctuations of S^a.[13] We need to sum over all these saddle points. This allows to obtain several exact results that we quote as follows. An interesting defect operator is S^a itself. In the ultraviolet fixed point, it has dimension 0. But in the nontrivial infrared fixed point, at large s, we find $\Delta(S) \sim 1/s^2$. The coefficient of the $1/s^2$ term is not yet known exactly (in the ϵ expansion, one finds $\Delta(S) = \frac{11}{4\pi^2 s^2}$ for sufficiently large $s \gg \frac{1}{\sqrt{\epsilon}}$). Another interesting defect operator is $\hat{\phi}_a$, for which we know the large s scaling dimension:

$$\Delta(\hat{\phi}_a) = 1 + \mathcal{O}(1/s).$$

Another interesting fact is that the one-point function of $\vec{\phi}^2$ in the large s limit exactly coincides with the one-point function of $\vec{\phi}^2$ in the pinning field defect, up to $\mathcal{O}(1/s)$ corrections. Finally, it is possible to show that in the large s limit the defect entropy at the infrared fixed point is smaller than $\log(2s+1)$ by the (negative) $\log g$ of the $N = 3$ pinning field defect. Here we discussed only spin impurities in the $O(3)$ model, other possible models can be found in Ref. 14.

4. Charged Impurities in Deconfined Critical Points

In $2+1$ dimensions, there exist fixed points with deconfined (fluctuating) gauge fields. A class of interesting defects are external charged particles (charged impurities). These are described by Wilson lines. We will only make remarks here about $2+1$-dimensional QED$_3$ with $2N_f$ Dirac fermions of charge 1:

$$L = \frac{-1}{4e_*^2} F^2 + i \sum_{a=1}^{2N_f} \bar{\Psi}_a \gamma^\mu (\partial_\mu - iA_\mu) \Psi_a.$$

By e_* we refer to the fact that we flow to the infrared fixed point of this bulk deconfined theory. At large N_f, this fixed point can be understood in a $1/N_f$ expansion.[15]

Inserting a charged particle of charge q corresponds to deforming the action by $\Delta S = q \int dt A_0(t, x = 0)$. This is the usual charge q Abelian Wilson line. The influence of this charged particle on the bulk can be understood in two steps. First, we need to determine the strength of the electric field that this probe causes and then, in the second step, we have to check for self-consistency, i.e. whether the bulk Ψ_a fermions can propagate in a healthy fashion in such an electric field.

To find the electric field, we integrate out Ψ_a and obtain (in Euclidean signature)

$$S = -2N_f \mathrm{Tr}\log(\gamma^\mu\partial_\mu - i\gamma^\mu A_\mu) + iq\int d\tau A_0. \tag{6}$$

The saddle point, which is reliable at large N_f, allows to determine the electric field to leading order

$$F_{\tau i} = iE\frac{x^i}{|x|^3}, \quad E = \frac{4q}{\pi N_f} + \mathcal{O}\left(\frac{q^2}{N_f^2}\right). \tag{7}$$

For $q \sim N_f$, at large N_f, the formula for the electric field is more complicated, but it can be obtained by a numerical solution of the saddle point of (6).[16] For concreteness, we provide a plot of $E(q/N_f)$ for large N_f in Figure 2.

Now let us ask if the fermions Ψ_a can self-consistently propagate in such a background with an electric field. For this we consider fermion bilinears $\hat{\bar{\Psi}}\hat{\Psi}$ and compute their scaling dimensions as defect operators (in the bulk, at large N_f, their scaling dimension is close to 2, but there are large corrections to the scaling dimension near the defect). We find[16] that at leading order in N_f

$$\Delta(\hat{\bar{\Psi}}\hat{\Psi}) = 1 + \sqrt{1-4E^2}.$$

(In the absence of a defect, $E = q = 0$, the scaling dimension of the fermion bilinear of course agrees with the scaling dimension of the bulk primary.) When the scaling dimension turns complex, the fermions are expected to condense, and integrating them out naively as we did before (6) is not self-consistent. Therefore, we must

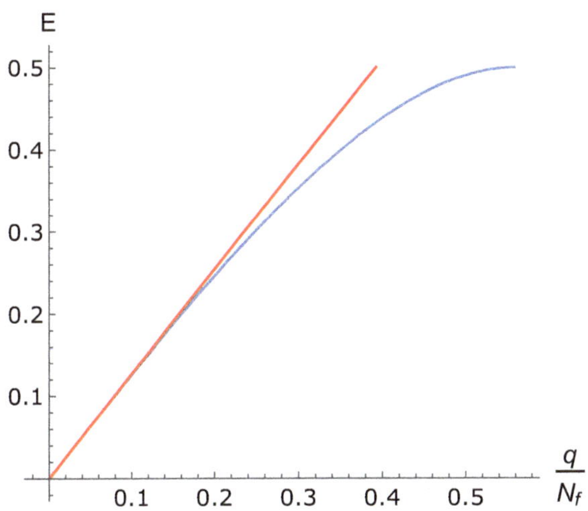

Fig. 2. We plot $E(q/N_f)$ for large N_f. The red line represents the linear approximation $E = \frac{4q}{\pi N_f}$, while the blue line is the correct value of the electric field, taking into account the nonlinear nature of the saddle point equation that is obtained from the action (6).

have $|E| \leq 1/2$ for stability of the vacuum with the Coulomb field.[d] If $|E| \leq 1/2$, then the infrared is a conformal defect with the electric field one-point function (7). If $|E| > 1/2$, then conjecturally there is an RG flow on the defect that reduces q until $|E| = 1/2$. The RG flow on the defect can be understood as a flow of the defect bilinear operator $\bar{\hat{\Psi}}\hat{\Psi}$.

Numerically, from Figure 2, we find that $|E| = 1/2$ corresponds to $|q_c| \sim 0.56 N_f$ (for large N_f). Therefore, Wilson lines with $|q| > |q_c|$ are partially screened and in the infrared flow to the conformal defect with $|q| = |q_c|$. While the Wilson lines with $|q| < |q_c|$ correspond to the conformal defects with $E = E(q, N_f)$ obtained from the the saddle point of (6).

A curious point is that the massive phases of QED$_3$ are $U(1)_{\pm N_f}$ Chern–Simons theory. (And the critical point is a non-Landau–Ginzburg transition between these topological theories.) These massive phases are obtained by deforming the theory with the $SU(2N_f)$ symmetric mass term for the fermions that can be taken to be either large and positive or large and negative, resulting in the topological theories $U(1)_{\pm N_f}$, respectively. Each of these topological theories admits N_f lines with nontrivial mutual braiding. Since $|q_c| \sim 0.56 N_f$, we have "slightly" more than N_f distinct conformal lines at the fixed point. We therefore have just enough distinct conformal lines to provide us with the nontrivial lines in the topological phases.

References

1. J. L. Cardy, Scaling and renormalization in statistical physics.
2. M. Billò, V. Gonçalves, E. Lauria and M. Meineri, Defects in conformal field theory. *JHEP* **4**, 91 (2016). doi:10.1007/JHEP04(2016)091 [arXiv:1601.02883 [hep-th]].
3. I. Affleck, Conformal field theory approach to the Kondo effect. *Acta Phys. Polon. B* **26**, 1869–1932 (1995). [arXiv:cond-mat/9512099 [cond-mat]].
4. G. Cuomo, Z. Komargodski and A. Raviv-Moshe, Renormalization group flows on line defects. *Phys. Rev. Lett.* **128**(2), 021603 (2022). doi:10.1103/PhysRevLett.128.021603 [arXiv:2108.01117 [hep-th]].
5. I. Affleck and A. W. W. Ludwig, Universal noninteger 'ground state degeneracy' in critical quantum systems. *Phys. Rev. Lett.* **67**, 161–164 (1991). doi:10.1103/PhysRevLett.67.161.
6. D. Friedan and A. Konechny, On the boundary entropy of one-dimensional quantum systems at low temperature. *Phys. Rev. Lett.* **93**, 030402 (2004). doi:10.1103/PhysRevLett.93.030402 [arXiv:hep-th/0312197 [hep-th]].
7. H. Casini, I. Salazar Landea and G. Torroba, The g-theorem and quantum information theory. *JHEP* **10**, 140 (2016). doi:10.1007/JHEP10(2016)140 [arXiv:1607.00390 [hep-th]].
8. K. G. Wilson and M. E. Fisher, Critical exponents in 3.99 dimensions. *Phys. Rev. Lett.* **28**, 240–243 (1972). doi:10.1103/PhysRevLett.28.240.

[d] This phenomenon of vacuum instability with $|E| > 1/2$ was already observed in Graphene.[17]

9. G. Cuomo, Z. Komargodski and M. Mezei, Localized magnetic field in the O(N) model. *JHEP* **2**, 134 (2022). doi:10.1007/JHEP02(2022)134 [arXiv:2112.10634 [hep-th]].
10. F. Parisen Toldin, F. F. Assaad and S. Wessel, Critical behavior in the presence of an order-parameter pinning field. *Phys. Rev. B* **95**(1), 014401 (2017). doi:10.1103/PhysRevB.95.014401 [arXiv:1607.04270 [cond-mat.stat-mech]].
11. K. H. Höglund, A. W. Sandvik and S. Sachdev, Impurity induced spin texture in quantum critical 2D antiferromagnets. *Phys. Rev. Lett.* **98**(8), 087203 (2007).
12. K. H. Höglund and A. W. Sandvik, Anomalous Curie response of impurities in quantum-critical spin-1/2 Heisenberg antiferromagnets. *Phys. Rev. Lett.* **99**(2), 027205 (2007).
13. G. Cuomo, Z. Komargodski, M. Mezei and A. Raviv-Moshe, Spin impurities, Wilson lines and semiclassics. *JHEP* **6**, 112 (2022). doi:10.1007/JHEP06(2022)112 [arXiv:2202.00040 [hep-th]].
14. S. Liu, H. Shapourian, A. Vishwanath and M. A. Metlitski, Magnetic impurities at quantum critical points: Large-N expansion and connections to symmetry-protected topological states. *Phys. Rev. B* **104**(10), 104201 (2021). doi:10.1103/PhysRevB.104.104201 [arXiv:2104.15026 [cond-mat.str-el]].
15. T. Appelquist, D. Nash and L. C. R. Wijewardhana, Critical Behavior in $(2+1)$-Dimensional QED. *Phys. Rev. Lett.* **60**, 2575 (1988). doi:10.1103/PhysRevLett.60.2575.
16. O. Aharony, G. Cuomo, Z. Komargodski, M. Mezei and A. Raviv-Moshe, Phases of Wilson lines in conformal field theories. [arXiv:2211.11775 [hep-th]].
17. Y. Wang *et al.*, Observing atomic collapse resonances in artificial nuclei on graphene. *Science* **340**(6133), 734–737 (2013).

https://doi.org/10.1142/9789811282386_0024

Chapter 24

Theories and Mysteries of the Normal-Metal-to-Superconductor and Nematic-to-Smectic-A Phase Transitions

T. Lubensky

University of Pennsylvania, Philadelphia, PA 19104, USA
tom@physics.upenn.edu

Unlike many of the authors who are contributing to this book, I was neither a student nor a postdoc with Michael Fisher, nor did I ever collaborate with him. Nevertheless, his impact on my work and career has been enormous. In my first encounter with him, however, his outstanding research and leadership in the general field of statistical physics took second place to his other great prowess as a flamenco guitarist.

From the summer of 1969 to the summer of 1970, I was a postdoc at the Orsay branch of the University of Paris, mostly attached to the group of Pierre Gilles de Gennes. In the summer of 1970, my wife and I drove from Paris to Cappadocia in Turkey and back, in our little green Simca, stopping at the great historical sites — Venice, Diocletian's Palace, Istanbul and Hagia Sofia, Greek ruins like Ephesus, the Bulgarian mountains, and more. On the return segment of our journey, we stopped at Lake Como in Italy where the Lake Como School of Advanced Studies was hosting a summer course in statistical physics at which Michael was lecturing on critical phenomena at a time just before the Wilson–Fisher enlightenment regarding physics in 3.99 dimensions. Though I had finished my degree at Harvard in 1969, I knew almost nothing about critical phenomena, and I found the few lectures I attended to be quite exciting. In truth, however, I was more impressed with Michael's performance as a flamenco guitarist. The story is that he worked his way through graduate school in London playing in nightclubs and the like — He was a professional, and it showed. I play (or play at) classical guitar myself, and Michael and I immediately developed a close rapport.

I spent the 1970–1971 academic year in Leo Kadanoff's group at Brown before, ⇒ before joining the physics faculty at the University of Pennsylvania. The publication of the paper on critical points in 3.99 dimensions in January 1972 opened a new frontier, and the statistical-mechanics community rushed to be a part of it. I obtained a copy of Ken Wilson notes through Bert Halperin, and because Penn was doing surface physics, along with a colleague Mort Rubin, I wrote my first paper using the renormalization group in a study of phase transitions in systems with a boundary. Shortly thereafter, in summer 1973 at the Aspen Center for

Physics, Bert Halperin, Shang-keng Ma and I began a collaboration on first-order phase transitions in superconductors and liquid crystals that led to the now well-known "HLM" paper, which is the subject of this chapter. Michael was there in Aspen to advise us.

I found the manuscripts coming out of Michael's group at Cornell, with its string of graduate students and postdocs, who themselves quickly became leaders in the field, to be among the most informative and inspiring of the steady string of new work that began with the publication of the 3.99 dimension paper.

Michael always acted as a friend, taking seriously his responsibility of nurturing and including younger scientists. One of my greatest pleasures were dinners with Michael and Sorrel, and other members of the statistical-physics community at the formal National Academy balls. He was a giant among us, and his presence is and will be missed.

1. Preliminaries

This chapter deals almost exclusively with phase transitions and critical properties of two "gauge" models: (1) the Landau–Ginzburg model for superconductors (SCs) and the normal-metal-to-SC (M–SC) transition[1,2] and (2) the de Gennes model[3,4] for the nematic (N) and smectic-A (SA) liquid crystal (LC) phases and the N–SA transition between them. Both models have a "gauge" field and a complex order parameter ψ with $n/2$ complex components. The field ψ is the order parameter for both the SC and SA phases, related to the SC energy gap in the former and equal to a its complex one-dimensional mass-density amplitude in the latter.

The gauge field $\mathcal{A}$ of the SC is the standard electromagnetic vector potential $\mathbf{A}$ while that of the liquid crystal is the unit-length director field, $\mathbf{n}$, specifying the direction of molecular alignment in the N and SA phases, or more precisely the deviation $\delta\mathbf{n} = \mathbf{n} - \mathbf{n}_0$, where $\mathbf{n}_0$ is the equilibrium director and $\delta\mathbf{n} \cdot \mathbf{n}_0 = 0$. The free energies of both the SC and the LC phases can be expressed as a sum of a part F_g depending only on the square of spatial derivatives of the gauge field $\mathcal{A}$ and a part $F_c[\psi, A]$ consisting of a part depending only on ψ and a part coupling ψ to $\mathcal{A}$:

$$F_g[\psi, \mathcal{A}] = \int d^d x \, (\mathcal{F}_c[\psi, \mathcal{A}] + \mathcal{F}_g[\mathcal{A}]), \tag{1}$$

where $\mathcal{F}_g[\psi, \mathcal{A}]$ and $\mathcal{F}_c[\mathcal{A}]$ are energy densities.

This is a review of theoretical work done mostly 45–50 years ago, as published in Refs. 5–9. As such, it will cite references that have a direct relationship to the text and make no attempt to cite the many many publications published in later periods.

2. Transitions in Superconductors

This section outlines three approaches to the study of the M–SC transition: (a) a modified mean-field theory that takes into account the interaction between the gauge field **A** and the order parameter ψ and that predicts all transitions are surprisingly first order;[5] (b) an ϵ-expansion about four dimensions with "runaway" trajectories that predict a first-order transition;[7] (c) Monte Carlo simulation of a lattice Villain model[10] that makes a strong case for a second-order transition in three dimensions.[11] It will not discuss in any detail self-consistent field theories[12,13] that have produced results consistent with at least some experiments.[14]

The energies controlling superconducting properties are

$$\mathcal{F}_c[\psi, \mathbf{A}] = a|\psi|^2 + \frac{1}{2}b|\psi|^4 + \gamma|(\nabla - iq_0\mathcal{A})\psi|^2, \tag{2a}$$

$$\mathcal{F}_g[\mathbf{A}] = \frac{1}{8\pi\mu_0}\sum_{i<j}(\nabla_j A_i - \nabla_i A_j)^2, \tag{2b}$$

where μ_0 is the magnetic permeability and

$$a = a'(T - T^*)/T^* \equiv a'\,t, \tag{3}$$

where T^* is temperature at the limit of metastability on cooling from the M phase. At a second-order transition, T^* becomes the transition temperature T_c. The parameter q_0 is equal to $2e/\hbar c$ in c.g.s. units.

2.1. *SC-MF*

Standard mean-field theory neglects **A** and predicts a second-order xy-phase transition at $T = T^*$ with order parameter, $\psi = \sqrt{-a/b}$. **A** is, however, coupled to ψ, and its fluctuations at non-zero temperature modify the equation of state. To see this, we restrict our attention to three dimensions (and $n = 2$) and develop an effective energy $F_{\text{eff}}[|\psi|]$ by integrating over **A**: $F_{\text{eff}}[|\psi|]/T = -\log[\int d\{\mathbf{A}\}\exp\{-F[\psi, \mathbf{A}]/T\}$. Since F is gauge-invariant, we choose to use the Coulomb gauge, $\nabla \cdot \mathbf{A} = 0$, in which the only coupling between $|\psi|$ and **A** is $\gamma q_0^2|\psi|^2|\mathbf{A}|^2$. Differentiation with respect to $|\psi|$ produces the equation of state,

$$\frac{1}{2}\frac{\partial\mathcal{F}_{\text{eff}}}{\partial|\psi|} = a|\psi| + b|\psi|^3 + q_0^2\gamma|\psi|\langle A^2\rangle_\psi = 0. \tag{4}$$

The gauge-field fluctuation $\langle A^2\rangle$ is the sum of equal contributions from its two independent components:

$$\langle A^2\rangle_\psi = \frac{T^*\mu_0}{8\pi^3}\int_{|\mathbf{q}|<\Lambda} d^3q\frac{8\pi}{q^2 + k_s^2}, \tag{5}$$

with Λ the upper cutoff and

$$k_s^2 = 8\pi\gamma q_0^2\mu_0|\psi|^2 = \lambda^{-2} = 8\pi\mu_0\gamma q_0^2|a|/b \tag{6}$$

where λ is the London penetration depth.[1] For sufficiently small $|\psi|$, $k_s \ll \Lambda$ and $\langle A^2\rangle_\psi$ can be expanded in powers of $|\psi|$.

$$\langle A^2\rangle = 4\mu_0 T^*\Lambda\pi^{-1} - 3w\psi, \tag{7}$$

where

$$w^2 = \frac{128}{9}\pi(\gamma q_0^2\mu_0)^3(T^*)^2. \tag{8}$$

The negative linear term in $|\psi|$ translates to a negative third-order term in $|\psi|$ in

$$\mathcal{F}_{\text{eff}}[|\psi|] = a|\psi|^2 - w|\psi|^3 + \frac{1}{2}b|\psi|^4. \tag{9}$$

This is a standard model for systems that undergo a first-order transition,[15] such as the isotropic-to-nematic (I–N) transition, between two different states. It predicts a first-order transition from the normal-metal to the superconducting state at a temperature T_c and a limit of metastability of the superconducting state on heating at temperature $T^{**} > T_c$:

$$T_c - T^* = \frac{w^2}{2a'b}T^*, \quad T^{**} - T_c = \frac{w^2}{16a'b}T^*. \tag{10}$$

In reality the right-hand sides of these equations are linear in T rather than T^*. This approximation simplifies the algebra and is correct to lowest order in $w^2/(a'b)$.

The difference $T^{**} - T^*$ provides a measure of the *size* of the first-order transition. When $\kappa = \lambda/\xi = (8\pi\mu_0\gamma q_0^2/b)^{-1/2}$, the ratio of the London penetration depth to the coherence length $\xi = \sqrt{\gamma/|a(T)|}$, is sufficiently large,

$$\Delta T \equiv T^{**} - T^* = \frac{9w^2}{16a'b}T^* = \frac{1}{2}\varepsilon_c T^*\kappa^{-6}, \tag{11}$$

where $\varepsilon_c = b^2(T^*)^2/(32\pi^2 a\gamma^3)$ is a Ginzburg parameter[16] indicating the approximate value of $t = (T - T^*)/T^*$ below which fluctuations become important and mean-field theory becomes inaccurate. ΔT is estimated to be of order of a few μK for Aluminum, which has a transition temperature of 1.19 K.

Though this generalized mean-field theory provides a more nuanced description of the phase transition to the SC state than the standard mean-field theory, it is still basically a mean-field theory that becomes less and less accurate as κ decreases and $\Delta T/T^*$ becomes less than ϵ_c, indicating that fluctuations associated with the xy transition at $q_0 = 0$ cannot be neglected. Thus, Eq. (11) applies to type-I SCs ($\kappa < 1/\sqrt{2}$) but not type-II systems ($\kappa > 1/\sqrt{2}$) for which the expected expression[6,7,17] is $\Delta T \sim \varepsilon_c T^*\kappa^{-2/\epsilon\nu}$, where $\epsilon = 4 - d$ and ν is the correlation length exponent of the xy model.

2.2. ϵ-*expansion*

This section reviews renormalization-group ϵ-expansion calculations and discusses their results. The ϵ-expansion[15,18–20] provides detailed information about phase transitions near an upper critical dimension, which in this case is $d_c = 4$.

The important variables for the M–SC theory are the reduced temperature $a = a't$, where $t = (T - T^*)/T^*$ is the reduced temperature with T^* the limit of metastability of the NM phase on cooling; the coefficient $b/2$ of the fourth-order term in $|\psi|$ of the field theory for the xy model; and the variable $f = 4\pi\mu_0 q_0^2/C_d$, where $C_d = 2^{d-1}\pi^{d/2}\Gamma(d/2)$ is an integration constant, coming from the new gauge field. The critical dimension for both b and f is $d_c = 4$, and fixed points for the theory exist in the b–f plane. Near a critical point, the ψ and $\mathbf{A}$ correlation functions obey scaling functions:

$$G(\mathbf{q}, a, u, f) = \langle|\psi(\mathbf{q})|^2\rangle = e^{(2-\eta)l}G(e^l\mathbf{q}, a(l), u(l), f(l)), \tag{12a}$$

$$\tilde{D}_{ij}(\mathbf{q}, a, u, f) = \langle A_i(-\mathbf{q})A_j(\mathbf{q})\rangle = e^{(2-\eta_A)l}\tilde{D}_{ij}(e^l\mathbf{q}, a(l), u(l), f(l)). \tag{12b}$$

Renormalization proceeds as usual by removal of degrees of freedom with wavenumber $\mathbf{q}$ with magnitude between $e^{-\ell}\Lambda$ and Λ and rescaling fields via $\mathbf{q} \to e^{-\ell}\mathbf{q}$, $\psi \to e^{-[(d-2-\eta)/2]\ell}\psi$, and $\mathbf{A} \to e^{-[(d+2-\eta_A)/2]\eta}\mathbf{A}$. The flow equations for the parameters of the energy defined in Eqs. (1)–(2b) are as follows[6,7]:

$$\frac{da}{dl} = [2 - \eta(l)]a(l) + \frac{1}{2}\frac{(n+2)C_d b(l)}{1+a(l)} + 3f(l)C_d, \tag{13a}$$

$$\frac{db(l)}{dl} = [\epsilon - 2\eta(l)]b(l) - \frac{1}{2}\frac{(n+8)b^2(l)C_d}{[1+a(l)]^2} - 6nf(l)C_d, \tag{13b}$$

$$\frac{d\mu^{-1}}{dl} = -\eta_A(l)\mu^{-1}(l) + \frac{1}{6}n[4\pi\mu q^2(l)], \tag{13c}$$

$$\frac{dq_0(l)}{dl} = \frac{1}{2}[\epsilon - \eta_A(l)]q_0(l), \tag{13d}$$

$$\eta(l) = -3\frac{f(l)}{(1+a(l))}, \tag{13e}$$

$$\frac{df(l)}{dl} = \epsilon f(l) - \frac{1}{6}nf^2(l). \tag{13f}$$

Note that η_A does not appear in the equations for $f(l)$ and $b(l)$. Thus, the fixed point structure is independent of η_A, and η_A can take on any real value. We choose $\eta_A = \epsilon$, so the charge q_0 remains unchanged under renormalization. Near a fixed point, each field g has a characteristic scaling exponent λ_g such that $g(l) = e^{\lambda_g l}g$.

The fixed points are as follows:

Gaussian: $b^* = f^* = 0$; $\lambda_b = \lambda_f = \epsilon$, $\lambda_r = 2$. This fixed point describes mean-field behavior with $\nu = 1/2$. It is unstable with respect to both u and f.

Heisenberg: $f^* = 0$, $b^* = [2/(n+8)][\epsilon/C_d]$; $\lambda_g = -\epsilon$; $\lambda_t = 2 - [(n+2)/(n+8)]\epsilon$, where $g = b - [2/[(n+8)][\epsilon/C_d] - [6/(n+8)]f$. When $n = 2$, this fixed point describes the λ-transition in helium.

Superconducting: In general, there are two new fixed points (denoted by subscripts "+" and "−") associated with a finite value of the charge. The fixed point values are

$$f^* = \frac{6\epsilon}{nC_d}, \tag{14}$$

$$b^*_\pm = \frac{\epsilon}{(n+8)C_d}\left[\left(1+\frac{36}{n}\right) \pm \frac{\Delta}{n}\right], \tag{15}$$

and their associated stability exponents are

$$\lambda_{r\pm} = 2 + \frac{1}{n}\epsilon - \frac{1}{2}(n+2)b^*_\pm C_d, \tag{16}$$

$$\lambda_{g\pm} = \mp\frac{\Delta}{n}, \tag{17}$$

$$\lambda_f = -\epsilon, \tag{18}$$

where $\Delta = (n^2 - 360n - 2160)^{1/2}$ and

$$g_\pm = \left(1 \pm \frac{\Delta}{n}\right)\left(\frac{u^*}{8}\right) - \left(\frac{33n + 180 \mp 3\Delta}{4n(n+8)}\right) f^* \tag{19}$$

are the eigen-directions associated with the superconducting fixed points $(b^*_\pm, f^*)$ and stability exponents $\pm\Delta/n$ above.

Δ is imaginary for $n < n_c = 365.9$, and b is complex for both the "superconducting" fixed points, which are physically inaccessible. Thus, the only accessible fixed points for $n < n_c$ are the Gaussian and Heisenberg points, both of which are unstable with respect to f (i.e., with respect to turning on charge). There is, therefore, a "runaway" that usually signals a first-order transition and that is already expected from the generalized mean-field analysis. When $n > n_c$, there are two fixed points, other than the Gaussian and Heisenberg points, with real b and non-zero f. As shown in Fig. 1, both are stable with respect to flows along a line at an angle to the b-axis, but the left fixed point [at (b^*_-, f^*)] is unstable and the right [at (b^*_+, f^*)] is stable for flows parallel to the b-axis. At $n = n_c$, the two fixed points join to a point that is a fixed point for flow along the slanted axis but at which flows are to the left along the b-axis and uninterrupted. The parameters of this point are as follows:

$$b^* = \frac{\epsilon}{(n_c+8)C_d}\left(1+\frac{36}{n_c}\right), \quad f^* = \frac{6\epsilon}{n_c C_d}, \tag{20a}$$

$$\lambda_g = 0, \quad \lambda_f = -\epsilon \tag{20b}$$

$$\lambda_t = 2 + \frac{18}{n_c}\epsilon - \frac{1}{2}\frac{n_c+2}{n_c+8}\left(1+\frac{36}{n_c}\right)\epsilon. \tag{20c}$$

A full analysis of Eqs. (13) provides a derivation of the results $\Delta T \sim \kappa^{-2/\epsilon\nu}$.[7]

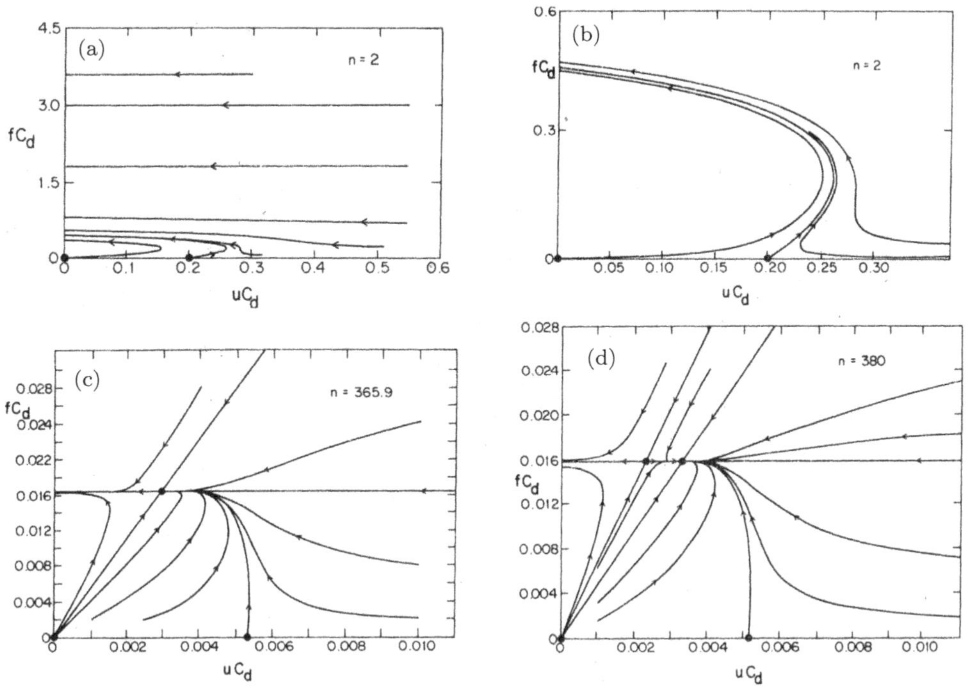

Fig. 1. Fixed points and flows for (a) $n = 2$ showing the Heisenberg fixed point at $f = 0$, $C_d b = 0.2$. Note the runaway to negative b; (b) same as (a) but at higher magnification; (c) fixed point and flow at the critical $n = n_c = 365.9$ showing the merging of the two fixed points for $n > n_c$ in (d).[7]

2.3. *Second-order transition from a dual theory*

Wolfgang Helfrich was the first to point out[21] the possible importance of dislocations to the N–SA transition. Following observations by Kosterlitz and Thouless[22] that the transition from the high-temperature disordered phase to the low-temperature phase should occur when the energetic contribution of a single vortex equals that of its entropic energy, Helfrich argued that the N–SA transition should occur at a temperature when the energy of a long dislocation equals its entropic free energy. Dasgupta and Halperin[11] took Helfrich's idea a step further by carrying out Monte Carlo simulations of what they call a "lattice superconductor model (LSM)," which should be in the same universality as that of a type-I superconductor. They relate this model to a generalized Villain model[10] that is a function of temperature and charge e and show that the point $(T, e^2 = 0)$ maps onto the point $(T = 0, e^2 = 4\pi^2/T)$. Thus, the xy transition at $(T = T_c, e^2 = 0)$ maps onto a loop transition at $(T = 0, e^2 = 4\pi^2/T_c)$. Since e^2 is proportional to $1/T$, the transition at $T = 0$ as a function of charge will have amplitudes that are inverted relative to those of xy transition at $e^2 = 0$. Thus, the specific-heat exponent α is very nearly zero for the

3D xy model, and the specific heat satisfies

$$c_V = -A \ln |t| - \frac{1}{2} DA \text{ sgn } t + B. \tag{21}$$

The sign of D will change in going from the xy to the loop model.

The loop transition is stable with respect to turning on T so that there should be an inverted xy transition in the vicinity of $(T = 0,\ e^2 = 4\pi^2/T_c)$. Dasgupta and Halperin then studied the transition at an intermediate value of e^2 and found strong evidence for an inverted xy transition. The situation in the vicinity of $(T_c,\ e^2 = 0)$ is complicated because of first-order effects, but it seems reasonable that the transition will remain second-order inverted-xy-like all the way down to the xy critical point. These calculations strongly support a second order in superconductors in $d = 3$.

3. The N–SA Transition

As mentioned in the introductory paragraphs, both SCs and SAs have complex order parameters. In addition, the SA has the director field $\delta\mathbf{n} = \mathbf{n} - \mathbf{n}_0$ that behaves much as the gauge field. As a result, the Landau–de Gennes free energy for the N and SA phases are very similar to those of the MN and SC phases. This section is based on Refs. 5, 8, 9.

3.1. *The LC energies and correlation functions*

Both the $\mathcal{F}_c$ and $\mathcal{F}_g$ energies reflect the anisotropy of the LC states:

$$\mathcal{F}_c[\psi, \delta\mathbf{n}] = a|\psi|^2 + \frac{1}{2} b|\psi|^4 + \gamma_{||}|\nabla_{||}\psi|^2 + \gamma_{\perp}|(\nabla_{\perp} - iq_0\delta\mathbf{n})\psi|^2, \tag{22a}$$

$$\mathcal{F}_g[\delta\mathbf{n}] = \frac{1}{2}\left[K_1(\nabla_i\delta n_i)^2 + K_2\sum_{i>j}|\nabla_i\delta n_j - \nabla_j\delta n_i|^2 + K_3\sum_j(\nabla_i\delta n_j)^2\right], \tag{22b}$$

where K_1, K_2, and K_3 are the Frank elastic constants.[3] In three dimensions, the director deviation $\delta\mathbf{n}$ lies in the plane perpendicular to $\mathbf{n}_0$. The unit vectors $\mathbf{e}_\perp$ and $\mathbf{e}_t$, respectively, lie in the plane defined by $\mathbf{n}_0$ and the wave vector $\mathbf{q}$ and along the normal to that plane, as shown in Fig. (2). In three dimensions, the director energy to harmonic order in $\delta\mathbf{n}$ expressed in Fourier space is then

$$F_g(\delta\mathbf{n}) = \frac{1}{2}\int\frac{d^3q}{(2\pi)^3}\left[(K_3q_{||}^2 + K_1q_\perp^2)|\delta\mathbf{n}_\perp|^2 + (K_3q_{||}^2 + K_2q_\perp^2)|\delta\mathbf{n}_t|^2\right], \tag{23}$$

where $q_{||}$ and $q_\perp$ are the components of $\mathbf{q}$ parallel and perpendicular to $\mathbf{n}_0$, respectively. An immediate consequence of the anisotropy of $\mathcal{F}_c$ is the existence of two correlation lengths $\xi_{||} = \sqrt{\gamma_{||}/a}$ and $\xi_\perp = \sqrt{\gamma_\perp/a}$,[23] associated with ψ.

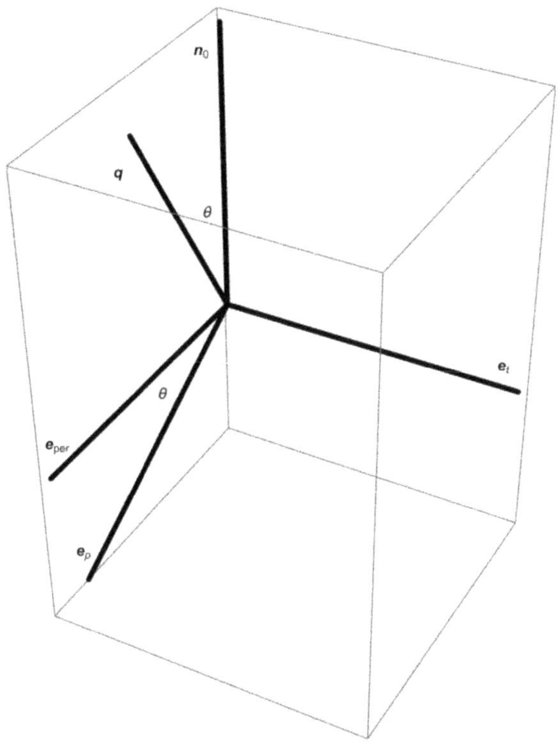

Fig. 2. Schematic diagram of the various directions that are important parts of the LC gauge energies: The vertical axis is along $\mathbf{n}_0$; The $\mathbf{e}_\perp$ and $\mathbf{e}_t$ are both perpendicular to $\mathbf{n}_0$ and to each other. In d-dimensions, there are $(d-2)$ $\mathbf{e}_t$ directions; The $\mathbf{q}$-direction is parallel to the wavevector $\mathbf{q}$ and $\mathbf{e}_p$ is perpendicular to $\mathbf{q}$ in the $\mathbf{n}_0 - \mathbf{e}_\perp$.[8]

The SC and LC models are very similar, and they not only have some common features, but also some that are quite different. The most important differences are as follows: (1) the energy for the LC is strongly anisotropic rather than isotropic and (2) the splay energy, $K_1 q_\perp^2 |\nabla_\perp \delta n_\perp|^2$, breaks the gauge invariance protected in the SC. As a result, all physically measurable quantities are independent of gauge in the SC model, whereas there is a single preferred gauge, the *physical* gauge, that describes physically measurable director fluctuations in the LC model.

The most obvious manifestation of an unusual property of the LC is that of the singular elastic energy of the SA phase. It follows directly from Eqs. (22b), (23), and an assumption that $\psi = |\psi| e^{iq_0 u}$ with $|\psi|$ is non-zero and constant throughout the sample.[a]

$$F_{\rm el} = \frac{1}{2} \int d^3x \left[B(\nabla_{||} u)^2 + D(\nabla_\perp u - \delta \mathbf{n})^2 \right] + F_g(\delta \mathbf{n}), \tag{24}$$

[a] $|\psi|$ is just a number without a phase. It is assumed to be constant. This constraint does does not produce long-range phase order.

where $B = 2\gamma_{||}q_0^2|\psi|^2$ and $D = 2\gamma_\perp q_0^2|\psi|^2$. Relaxing $\delta\mathbf{n}$ to equal $\nabla_\perp u$ and keeping only the leading terms in gradients in F_{el} yields the long-wavelength elastic energy:

$$F'_{\text{el}} = \int d^3x[B(\nabla_{||}u)^2 + K_1(\nabla_\perp^2 u)^2]. \tag{25}$$

Note that the presence of K_1, which breaks gauge invariance, and the essential role it plays in stabilizing the smectic phase. But this stabilization is not sufficient to prevent the destruction of long-range order as can be seen from the correlation function for ψ that it predicts:

$$G(\mathbf{x}, 0) = \langle\psi^*(\mathbf{x})\psi(0)\rangle = |\psi|^2 e^{-g(\mathbf{x})}, \tag{26a}$$

$$g(\mathbf{x}) = \frac{1}{2}q_0^2\langle[u(\mathbf{x}) - u(\mathbf{0})]^2\rangle = q_0^2 \int \frac{d^3q}{(2\pi)^3}\frac{T}{Bq_{||}^2 + K_1 q_\perp^4}\left(1 - e^{i\mathbf{q}\cdot\mathbf{x}}\right) \tag{26b}$$

$$= \eta_c \sim \begin{cases} \ln|x_{||}| & \text{if } \mathbf{x}_\perp = 0, \\ \ln x_\perp^2 & \text{if } x_{||} = 0, \end{cases} \tag{26c}$$

where

$$\eta_c = \frac{q_0^2 T}{8\pi\sqrt{K_1 B}}, \tag{27}$$

implying power-law decay rather than long-range order of

$$G(\mathbf{x}) \sim \begin{cases} |x_{||}|^{-\eta_c} & \text{if } \mathbf{x}_\perp = 0, \\ |\mathbf{x}_\perp|^{-2\eta_c} & \text{if } x_{||} = 0 \end{cases} \tag{28}$$

throughout the SA phase. This behavior, which has been measured in X-ray scattering experiments,[24,25] is a direct consequence of the anisotropy of the physical LC gauge that forces a coupling of the order parameter and gauge fields, a situation that is avoided in the Coulomb gauge.

In the N phase, the $\psi - \delta\mathbf{n}$ coupling leads to critical enhancement[23,26] of the Frank constants K_2 and K_3:

$$\delta K_2 \propto \xi_{||}^{-1}\xi_\perp^{\epsilon+1} \xrightarrow{d\to 3} \xi_\perp^2/\xi_{||}, \quad \delta K_3 \propto \xi_{||}\xi_\perp^{\epsilon-1} \xrightarrow{d\to 3} \xi_{||}, \tag{29}$$

where $\epsilon = 4 - d$. In three-dimensions, these relationships become $\delta K_2 \propto \xi_\perp^2\xi_{||}$ and $\delta K_3 \propto \xi_{||}$. In most cases, $\xi_\perp \propto \xi_{||}$ and both K_2 and K_3 are proportional to $\xi_{||} \propto |t|^{-\nu}$.

3.2. *A gauge transformation and the first-order transition*

Importantly, the splay constant K_1 does not undergo any changes arising from the coupling of ψ and $\delta\mathbf{n}$. This is an unusual situation. As will be shown in Section 4.1, the ϵ-expansion equation for K_1 would seem to indicate that K_1 has the same singularity as K_2, but it does not. It does not change as the N–SA critical point is approached and passed. Couplings to other variables, such as temperature or density, could induce changes in K_1, but it is unlikely that they would induce critical behavior near the critical point. Thus, K_1 is a kind of marginal variable[27] that can produce continuous changes in exponents or amplitudes as its magnitude varies in response to changes such as temperature, pressure, composition or other variables not included in the de Gennes model.

Though measurements are necessarily a reflection of the physical gauge, gauge changes are permitted in intermediate steps in calculations. In particular, changing to the Coulomb gauge is straightforward through a modification of the phase of ψ via $\psi = \Psi e^{iq_0 L}$. This amounts to the introduction of a new gauge field,[5,8]

$$\mathbf{A} = \delta\mathbf{n} - \nabla L. \tag{30}$$

The goal is to choose L so that $\nabla \cdot \mathbf{A} = 0$, but in order for the prescription to work, it is first necessary to introduce changes of length scales to create a rescaled isotropic version of $\mathcal{F}_c$:

$$x_{||} = s^{2/3} x'_{||}, \quad x_\perp = s^{-1/3} x'_\perp, \quad \delta\mathbf{n} \to s^{1/3}\delta\mathbf{n}'. \tag{31}$$

This rescaling preserves volume, $x_{||}x_\perp^2 \to s^{2/3}s^{-2/3}x_{||}x_\perp^2 = x_{||}x_\perp^2$, and spawns the following rescalings:

$$\begin{aligned} &\gamma_{||} \to s^{-4/3}\gamma, && \gamma_\perp \to s^{2/3}\gamma, \quad \gamma_{||}\gamma_\perp^2 = \gamma^3 \quad s^2 = \gamma_{||}/\gamma_\perp \\ &K_3 \to s^{-1/3}\mathsf{K}_3, && K_{1,2} \to s^{2/3}\mathsf{K}_{1,2}, \quad K_2/K_3 \to s\mathsf{K}_2/\mathsf{K}_3, \end{aligned} \tag{32}$$

where γ and K_α, $\alpha = 1, 2, 3$, are rescaled parameters.

With this rescaling, the gauge-coupling term adopts its isotropic form, $\gamma|(\nabla - iq_0\delta\mathbf{n}')\psi|^2$, and a gauge transformation implemented via Eq. (30) to produce a model in which $\mathbf{A}$ satisfies the Coulomb constraint $\nabla \cdot \mathbf{A} = 0$,

$$L = \frac{\nabla \cdot \delta\mathbf{n}}{\nabla^2} = \frac{\nabla_\perp \cdot \delta\mathbf{n}}{\nabla^2}, \tag{33}$$

and that $\mathbf{A}$ is the projection of $\delta\mathbf{n}$ on to the plane perpendicular to the gradient operator (or, equivalently, the wave number $\mathbf{q}$).

As already discussed, $\delta\mathbf{n}$ lives in the $(d-1)$-dimensional space perpendicular to $\mathbf{n}_0$. This space is spanned by a basis vector $\mathbf{e}_\perp$ lying in the plane defined by $\mathbf{n}_0$ and $\mathbf{q}$ and by $d-2$ orthogonal basis vectors $\mathbf{e}_t$ in the space perpendicular to the $\mathbf{q} - \mathbf{e}_\perp$ plane. On the other hand, $\mathbf{A}$ is restricted to the plane perpendicular to the vector $\hat{\mathbf{q}} = \cos\theta\mathbf{n}_0 + \sin\theta\mathbf{e}_\perp$ spanned by the vectors $\mathbf{e}_p = -\sin\theta\mathbf{n}_0 + \cos\theta\mathbf{e}_\perp$ and $\mathbf{e}_t$'s. Thus, $\delta\mathbf{n} = \delta n_\perp\mathbf{e}_\perp + \sum_t \delta n_t\mathbf{e}_t$, and $\mathbf{A} = A_p\mathbf{e}_p + \sum_t A_t\mathbf{e}_t$.

These constraints along with Eq. (34b) provide the relations between $\mathbf{A}$ and $\delta\mathbf{n}$:

$$A_i = (\delta_{ij} - \hat{q}_i\hat{q}_j)\delta n_j, \tag{34a}$$

$$A_p = e_{pi}(\delta_{ij} - \hat{q}_i\hat{q}_j)\, e_{\perp_j}\, \delta n_{\perp j} = \cos\theta\, \delta n_\perp, \quad \delta n_t = A_t. \tag{34b}$$

This finally leads to

$$G_{A,p}(\mathbf{q}) = \langle A_p(-\mathbf{q})A_p(\mathbf{q})\rangle = \frac{q_{||}^2}{q^2}\frac{T}{\mathsf{K}_3 q_{||}^2 + \mathsf{K}_1 q_\perp^2} = \frac{T}{\mathsf{K}_3 q^2 + \mathsf{K}_1 (q_\perp^2/q_{||}^2) q^2}, \tag{35}$$

$$G_{A,t}(\mathbf{q}) = \langle A_t(-\mathbf{q})A_t(\mathbf{q})\rangle = \frac{T}{\mathsf{K}_3 q_{||}^2 + \mathsf{K}_2 q_\perp^2} = D_t(\mathbf{q}), \tag{36}$$

and to

$$w_\alpha = \frac{T}{12\pi}\left(\frac{2\gamma q_0^2}{\mathsf{K}_3}\right)^{3/2} \zeta_\alpha, \tag{37}$$

where

$$\zeta_t = \frac{\mathsf{K}_3}{\mathsf{K}_1} \qquad \zeta_\perp = \left(\frac{\sqrt{\mathsf{K}_3}}{\sqrt{\mathsf{K}_3} + \sqrt{\mathsf{K}_1}}\right)^2. \tag{38}$$

Finally, we obtain $w = w_t + w_\perp$ and

$$\Delta T = \frac{1}{2}\epsilon_c \kappa^{-6}, \tag{39}$$

where

$$\epsilon_c = \frac{1}{32\pi^2}\frac{b^2}{a'}\frac{T}{\gamma_{||}\gamma_\perp^2} \tag{40}$$

and

$$\begin{aligned}\kappa^{-6} &= \frac{2\gamma^6 q_0^6}{b^3 \mathsf{K}_3^3}\left[\frac{\mathsf{K}_3}{\mathsf{K}_2} + \left(\frac{\sqrt{\mathsf{K}_3}}{\sqrt{\mathsf{K}_3} + \sqrt{\mathsf{K}_1}}\right)^2\right]^2 \\ &= \frac{2\gamma_{||}^2\gamma_\perp^4 q_0^6}{b^3 K_3^3}\left[\frac{K_3}{K_2} + \left(\frac{\sqrt{\gamma_{||}K_3}}{\sqrt{\gamma_\perp K_3} + \sqrt{\gamma_{||}K_1}}\right)^2\right]^2.\end{aligned} \tag{41}$$

Note that all of the above functions that depend on K_1 do not have the strong singularity that $D_\perp$ has when $\mathsf{K}_1 = 0$. Furthermore, as required, ΔT and κ^{-6} are identical to the results in the SC case when $K_1 = 0$ and $K_2 = K_3 = K$. As in the SC case, Eq. (41) applies to small-κ "type I" systems. In large-κ "type II" systems, ΔT becomes very small, and fluctuations of ψ become important. In this case, ΔT is expected to be proportional to $\kappa^{-2/\nu}$.

4. Physical Correlation Functions

As already noted, the physical order parameter ψ suffers from strong fluctuations that kill long-range order (in dimensions $d \leq 3$). On the other hand, the "gauged" order parameter Ψ is perfectly well behaved even when $K_1 = 0$, in which case the model is identical to that of the superconductor with long-range order in Ψ (which by definition is in the Coulomb gauge). The field L restores the singularities of ψ:

$$G_\psi(\mathbf{x}) = \langle \psi^*(\mathbf{x}), \psi(0)\rangle = \langle \Psi^*(\mathbf{x})\Psi(0)e^{iq_0(L(\mathbf{x})-L(0))}\rangle \approx G_\Psi(\mathbf{x})e^{-R(\mathbf{x})}, \tag{42}$$

where

$$R(\mathbf{x}) = \frac{1}{2}q_0^2\langle[L(\mathbf{x}) - L(0)]^2\rangle. \tag{43}$$

It is important to remember that L is a property of the physical model, and its correlation functions are to be calculated in the original physical gauge.

In the nematic phase, the Ψ correlation functions have the traditional decay lengths of the superconductivity model in the Coulomb gauge, but the physical order parameter has additional contributions from $R(\mathbf{x})$, which substantially modifies the nature of the decay of order. $G_\psi(\mathbf{x})$ decays with correlation lengths $\xi_{||}$ and $\xi_\perp$. Strikingly, $\exp(-R(\mathbf{x}))$ also decays exponentially but with different decay lengths[8]:

$$R(\mathbf{x}) = \frac{q_0^2 T}{2(2\pi)^3}\int d^3q \frac{q_\perp^2}{q^4}\frac{1}{K_3 q_{||}^2 + K_1 q_\perp^2}(1 - e^{\mathbf{q}\cdot\mathbf{x}}) \tag{44a}$$

$$= \begin{cases} \ell_{||}^{-1}|x_{||}| & \text{if } \mathbf{x}_\perp = 0, \\ \ell_\perp^{-1}|\mathbf{x}_\perp & \text{if } \mathbf{x}_{||} = 0, \end{cases} \tag{44b}$$

where

$$\ell_{||}^{-1} = \frac{q_0^2}{8\pi K_1}\frac{1 - (K_3/K_1) + (K_3/K_1)\ln(K_3/K_1)}{[1 - (K_3/K_1)]^2} \sim \xi_{||}^{-1}\ln\xi_{||}, \tag{45a}$$

$$\ell_\perp^{-1} = \frac{q_0^2}{8\pi\sqrt{K_1K_3}}\frac{2 + \sqrt{K_1/K_3}}{(1 + \sqrt{K_1/K_3})^2} \sim \xi_{||}^{-1/2}. \tag{45b}$$

Note the factor of $q_\perp^2/q^4$ in the expression for R. It causes the integrand to diverge as q^{-4} at small q, indicating that the integral itself should diverge as $|\mathbf{x}|$ at large $|\mathbf{x}|$ as Eqs. (44) and (45) verify. As will be discussed in the following section, K_1 is not only not critical, within the confines of the model at hand, it also does not change at all; it is a type of marginal field in the renormalization-group sense.[27] K_3 on the other hand is proportional to $\xi_{||}$ in three dimensions near a critical point, Eq. (29). The final expressions use this result and apply to situations in which $K_3 \gg K_1$.

Combining Eqs. (42) and (45) leads to decay lengths combining $\xi_{||,\perp}$ with $\ell_{||,\perp}$ and measurable by X-scattering:

$$(\xi_{||}^X)^{-1} = \xi_{||}^{-1} + \ell_{||}^{-1}, \tag{46}$$

$$(\xi_{\perp}^X)^{-1} = \xi_{\perp}^{-1} + \ell_{\perp}^{-1}, \tag{47}$$

whose dependence on temperature can vary considerably depending on the value of K_1 and on the ratios $\xi_{||}/\ell_{||}$ and $\xi_\perp/\ell_\perp$. Importantly, the experimental correlation lengths are anisotropic even if the universality class of the gauge theory is isotropic. The $\ln \xi_{||}$ factor in $\ell_{||}^{-1}$ is a relatively small correction to $\ell_{||}^{-1}$, and $(\xi_{||}^X)^{-1}$ is effectively proportional to $t^{\nu_{||}}$ for both isotropic and anisotropic universality classes. $(\xi_\perp^X)^{-1}$ is the sum of two terms proportional, respectively, to $t^{\nu_\perp}$ and $t^{\nu_{||}/2}$. Thus, sufficiently close to a critical point, $\nu_\perp^X = \nu_{||}/2$ regardless of the universality class. This is an important result whose impact has not received the attention it should have. There is ample evidence in experiments[14] that at least some samples exhibit this behavior, and there are at least two theoretical models, one based on dislocation melting[28,29] and the other on a self-consistent theory in the physical gauge,[12,13] that do as well.

In the ordered phase, Ψ (unlike ψ) has a well-defined long-range average $\langle\Psi\rangle$ and phase $iq_0u(\mathbf{x})$. The long-wavelength elastic excitations are described by Eqs. (24) and (25). Most importantly, because $2q_0\gamma_\perp\psi^*\nabla_\perp\psi$ is non-zero, interactions between $\delta\mathbf{n}$ and u cannot be ignored. It is clear from Eq. (24) that to lowest order in gradients $\delta\mathbf{n} = \nabla_\perp u(\mathbf{x})$ and, in the long length scales of interest,

$$R(\mathbf{x}) = q_0^2 T \int \frac{d^3q}{(2\pi)^3} \frac{q_\perp^4}{q^4[Bq_{||}^2 + K_1 q_\perp^4]}(1 - e^{i\mathbf{q}\cdot\mathbf{x}}). \tag{48}$$

The long-distance behavior of $R(\mathbf{x})$ matches that of $g(\mathbf{x})$ in Eq. (26c). The physical correlation function is thus

$$G(\mathbf{x}, 0) = |\langle\Psi\rangle|^2 e^{-R(\mathbf{x})}, \tag{49}$$

which has the same long-distance behavior as $G(\mathbf{x})$ in Eq. (26c), as required. Right at a second-order critical point (if it exists), $R(\mathbf{x}_{||}, 0) \propto |x_{||}|^{1/3}$ when $R(0, \mathbf{x}_\perp) \propto |\mathbf{x}_\perp|^{1/2}$.[8]

4.1. ϵ-*Expansion*

The ϵ-expansion provides useful insights into the nature of the N–SA transitions. This section reviews both how this theory is set up and what predictions it makes. First, to keep the presentation as simple as possible, the focus will be on the rescaled, "isotropic gauge-transformed model" discussed in Section 3.2, though references to the physical gauge will be made when appropriate.

The starting point is to set up a formalism that includes the possibility of anisotropic scaling. This is done by introducing exponents that rescale wavevectors parallel and perpendicular to $\mathbf{n}_0$ differently: $q_{||} \to e^{(1+\mu_{||})\ell} q_{||}$ and $\mathbf{q}_\perp \to e^{\ell}\mathbf{q}_\perp$. This scaling implies the existence of two correlation lengths $\xi_\perp \propto |t|^{-\nu_\perp}$ and $\xi_{||} \propto |t|^{-\nu_{||}}$, where $\nu_{||} = (1 + \mu_{||})\nu_\perp$. Incidentally, this relation reduces to $\nu_{||} = (5 - d)\nu_\perp$ or $\nu_{||} = 2\nu_\perp$ when $d = 3$.

The fields $\Psi(\mathbf{q})$ and $\delta\mathbf{n}(\mathbf{q})$ scale as

$$\Psi(e^{-\ell}\mathbf{q}) \to e^{[(d+2+\mu_{||}-\eta_\perp)/2]\ell}\Psi(\mathbf{q}), \tag{50a}$$

$$\delta\mathbf{n}(e^{-\ell}\mathbf{q}) \to e^{[(d+2+\mu_{||}-\eta_n)/2]\ell}\delta\mathbf{n}(\mathbf{q}), \tag{50b}$$

and correlation functions scale as

$$G(\mathbf{q}) = e^{(2-\eta_\perp)\ell} G(a(\ell), \{\mathsf{K}(\ell)\}, e^{\ell} q_\perp, e^{(1+\mu_{||})\ell}\mathbf{q}_\perp), \tag{51a}$$

$$D_{ij}(\mathbf{q}) = e^{(2-\eta_n)\ell} D_{ij}(a(\ell), \{\mathsf{K}(\ell)\}, e^{\ell} q_\perp, e^{(1+\mu_{||})\ell}\mathbf{q}_\perp). \tag{51b}$$

The scaling relations of the various parameters implied by Eqs. (22a), (23), (50b), and (51) include $\gamma_\perp \to \gamma_\perp e^{-\eta_\perp \ell}$, $\gamma_{||} \to \gamma_{||} e^{-\eta_{||}\ell}$, $q_0^2 \to q_0^2\, e^{(\epsilon-\mu_{||}-\eta_n)\ell}$. When interactions are included, the exponents $\eta_\perp$, η_n and $\mu_{||}$ can be chosen so that $\gamma_\perp = \gamma_{||} = 1$ and q_0 remains constant under changes in ℓ. Of particular importance is a constant q_0 implies $\eta_n = \epsilon - \mu_{||}$.

The recursion relations for the Frank constants are

$$\frac{d\mathsf{K}_1}{dl} = -(\epsilon - \mu_{||})\mathsf{K}_1, \tag{52a}$$

$$\frac{d\mathsf{K}_2}{dl} = -(\epsilon - \mu_{||})\mathsf{K}_2 + \frac{1}{6} n C_d q_0^2, \to -\left(\epsilon - \mu_{||} - (n/6) f_2\right)\mathsf{K}_2, \tag{52b}$$

$$\frac{d\mathsf{K}_3}{dl} = -(\epsilon + \mu_{||})\mathsf{K}_3 + \frac{1}{6} n C_d q_0^2 \to -\left(\epsilon + \mu_{||} - (n/6) f_3\right)\mathsf{K}_3, \tag{52c}$$

where $\{\mathsf{K}(\ell)\}$ represents the set of three Frank elastic constants and $f_\alpha = C_d q_0^2/\mathsf{K}_\alpha$. These equations are easily solved: K_1 does not depend f_2 or f_3, and

$$\mathsf{K}_3(\ell) = e^{(\epsilon+\mu_{||}-(n/6)f_3)\ell}\,\mathsf{K}_3^0, \tag{53}$$

$$\mathsf{K}_2(\ell) = e^{(\epsilon-\mu_{||}-(n/6)f_2)\ell}\,\mathsf{K}_2^0, \tag{54}$$

$$\mathsf{K}_1(\ell) = e^{(\epsilon-\mu_{||})}\,\mathsf{K}_1^0. \tag{55}$$

Inserting this into Eq. (51b) yields

$$D_\perp(\ell) = T\left[e^{-(n/6)f_3\ell}\mathsf{K}_3^0 q_{||}^2 + \mathsf{K}_1 q_\perp^2\right]^{-1}, \tag{56}$$

$$D_t(\ell) = T\left[e^{-(n/6)f_3\ell}\mathsf{K}_3^0 q_{||}^2 + e^{-(n/6)f_2\ell}\mathsf{K}_2 q_\perp^2\right]^{-1}. \tag{57}$$

This equation shows the physically measurable Frank constants. When $q_0 = 0$ (i.e. the Gaussian limit), the three K's end up independent of ℓ in spite of the presence of ℓ in the initial scaling expressions for $D_\perp$ and D_t. When $q_0 \neq 0$, interactions give K_2 and K_3 ℓ-dependence that eventually can be expressed in terms of t or $\mathbf{q}$. On the other hand, K_1 remains independent reflecting its independence of q_0. This result overrides the interpretation in Ref. 8 that the physical K_1 does vary with ℓ and has a flow toward or away from fixed points at $\mathsf{K}_1 = 0$, $\mathsf{K}_1 = \infty$, and possibly other points with $\epsilon = \mu_{||}$ at which $\nu_{||} = 2\nu_{\perp}$.

It turns out that flow in ℓ and fixed-point structure are better represented by $f_\alpha = C_d q_0^2/\mathsf{K}_\alpha$ rather than K_α. They obey the equations

$$\frac{df_1}{d\ell} = (\epsilon - \mu_{||}) f_1, \tag{58a}$$

$$\frac{df_2}{d\ell} = (\epsilon - \mu_{||}) f_2 - \frac{1}{6} n f_2^2, \tag{58b}$$

$$\frac{df_3}{d\ell} = (\epsilon + \mu_{||}) f_3 - \frac{1}{6} n f_3^2. \tag{58c}$$

When $\mu_{||} = 0$ and $K_1 = 0$, there is a single fixed point involving

$$f_2 = f_3 = n\epsilon/6, \tag{59}$$

from which it follows that $\mathsf{K}_2 = \mathsf{K}_3$.

4.2. $K_1 = 0$

When $K_1 = 0$, the values of these fields then determine the behavior of the SA order parameter, and the result is a system identical to that of the M–SC transition in superconductors: a first-order transition when $n < 365.9$ and a second-order transition when $n > 365.9$. To see how this scenario comes about, consider the expression of $\mu_{||}$:

$$\mu_{||} = -2C_d q_0^2 \left(\frac{2}{3} \frac{\sqrt{\mathsf{K}_3}}{(\sqrt{\mathsf{K}_1} + \sqrt{\mathsf{K}_3})^3} - \frac{4}{3} \frac{1}{\sqrt{\mathsf{K}_2}(\sqrt{\mathsf{K}_2} + \sqrt{\mathsf{K}_3})} \right), \tag{60a}$$

(60b)

which reduces when $K_1 = 0$ to

$$\mu_{||}(\mathsf{K}_1 = 0) = -\frac{4}{3} f_3 \frac{(1 + 2R_{32})(1 - R_{32})}{1 + R_{32}}, \tag{61}$$

where $R_{\alpha\beta} = \sqrt{\mathsf{K}_\alpha/\mathsf{K}_\beta} = \sqrt{f_\beta/f_\alpha}$. When $\mu_{||} = 0$, there is no symmetry difference between $q_{||}$ and $q_\perp$ and the system is in a sense isotropic. Clearly, $\mu_{||}$ is zero when $R_{32} = 1$. There is one more solution to $\mu_{||} = 0$, but it is negative and thus inaccessible. A final step is to show that solution $R_{32} = 1$ is compatible with the solutions for f_2 and f_3 in Eqs. (58b) and (58c). These two equations can be converted to an equation in terms of R_{32} alone:

$$\left[\frac{1}{6}n(1+R_{32})^2+\frac{8}{3}(1+2R_{32})\right](1-R_{32})=0, \tag{62}$$

which clearly vanishes at $R_{23}=1$. The arithmetic leading to a second-order transition for $n>365.9$ is identical to that for the SC and will not be repeated here. So far, only isotropic fixed points have been studied, and the question of whether or not there are anisotropic fixed points has not been answered. A particular case shows that there is and least one when $\mathsf{K}_1=\infty$. In this case,

$$\mu_{||}=\frac{8}{3}\frac{f_3}{(1+R_{32})}. \tag{63}$$

With this $\mu_{||}$,

$$f_2^*=\frac{6\epsilon(1+R_{32})}{16+n(1+R_{32})}, \tag{64a}$$

$$R_{32}^3+\left(1+\frac{32}{n}\right)R_{32}^2-R_{32}-1=0, \tag{64b}$$

where the second equation is the one constructed from Eqs. (52b) and (52c). The equations for the three f_α are totally independent of the parameters involving the xy part of the theory (i.e., a and b). The numerical values of the relevant parameters and critical exponents are easily calculated. Some results are shown in Table 1. The converse is not true: The recursion relations for a and b depend on the f_αs. It is the xy part of the theory, in particular the equation for b, that determines whether there is a first-order or a second-order transition:

$$C_d b^*=\frac{\epsilon}{n+8}\left\{\left(1+8\frac{f_2}{1+R_{32}}\right)\pm\left[\left(1+8\frac{f_2}{1+R_{32}}\right)^2-8f_3^2R_{32}^{-1}(n+8)\right]^{1/2}\right\}. \tag{65}$$

The argument of the square root in this equation is the difference between two positive functions, which with the aid of Eq. (64) can be expressed as a function of n, that passes through zero at $n=n_c=238.17$. Thus, the argument is negative for $n<n_c$ and positive for $n>n_c$, and as in the SC case, the fixed point for b is complex and unphysical for $n<n_c$ and real and positive for $n>n_c$.

Table 1. Numerical values of selected quantities in the anisotropic state for $n=2$.

f_2^*	f_3^*	R_{32}^*	$\mu_{\|\|}$	$\eta_\perp$
0.4114ϵ	5.9961ϵ	0.3713ϵ	0.8629ϵ	-1.7258ϵ

5. Some Experimental Data and Final Comments

Garland and Nounesis[14] have provided a comprehensive review of experimental work up to 1994 on the N–SA transition in 27 samples of different LCs. It is generally accepted that the MacMillan ratio, $\Omega_M = T_{NA}/T_{NI} \leq 1$, of the N–SA transition temperature to that of the nematic-to-isotropic (NI) transition temperature provides a crude but useful characterization of systems exhibiting an N–SA transition. As Ω_M decreases below one, the range of existence of the N phase grows and with it the nematic order parameter S, measuring the degree of molecular alignment, and the Frank elastic constants. Samples with the value of Ω_M greater than about 0.92 are near a tricritical point. Of particular interest is K_1, which grows with decreasing Ω_M via a mechanism that is not included in the de Gennes model but that can have a large influence on the magnitude of $\ell_{||}$ and $\ell_{\perp}$ (see Eq. (45)).

The Garland–Nounesis paper attributes deviation from classical xy behavior to (1) crossover from a second- to a first-order transition via a Gaussian tricritical point brought about by the coupling between S and ψ and (2) coupling between $\delta\mathbf{n}$ and Ψ which produces the HLM first-order transition and the anisotropic behavior seen both in experiments and theory. It analyzes the data it presents as though the N–SA transition is second order rather than as predicted by HLM.

Table I of the paper provides values of relevant critical exponents obtained from 27 samples with Ω_M, ranging from 0.660 to 0.995. The general trend of the data shows values of the specific-heat exponent α that is close to the xy value of -0.007 for Ω_M between 0.660 and 0.926 whereupon it becomes a little noisy until $\Omega_M = 0.954$ when it begins to grow monotonically reaching 0.5 at $\Omega_M = 0.994$. Though the value of α is close to that of the xy value (for $\Omega_M < 0.926$), its amplitude ratio has the usual sign, not the reversal of sign predicted by the the Dasgupta–Halperin dual model.[11] The $\nu_{||}$ grows more or less continuously from 0.669 (the xy value) at $\Omega_M = 0.660$ to 0.83 at $\Omega_M = 0.936$ at which it begins to decrease at $\Omega_M = 0.994$. Over the same range of Ω_M, $\nu_{\perp}$ starts at 0.669, decreases to 0.52 at $\Omega_M = 0.820$, rises to 0.68 at $\Omega_M = 0.936$, and then falls to 0.38 at $\Omega_M = 0.995$.

The most striking thing about this data is that critical exponents vary continuously with Ω_M. At first glance, this seems to be contrary of what we know about the renormalization group and its universality classes. The results, however strange, can be explained by the presence of K_1, which is a type of marginal variable[30] whose scaling exponent vanishes and and whose value does not change under renormalization flows. Typically, exponents and amplitude ratios for the standard, non-marginal fields can and do depend on the value of the marginal variable. This is the case for of the effective correlation lengths of Eq. (45) and for the Andereck–Patton theory.[12,13]

Acknowledgments

I am eternally grateful to have been at the same 1973 Aspen summer workshop as Bert Halperin and Shang-Ma and to have had the privilege to work closely with such outstanding scientists. I learned as much from the two of them in one summer and the slightly later interactions with Bert in our study of the liquid crystal part of the story as I did in whole semesters in graduate school.

I also thank Michael Fisher, who attended the same Aspen workshop, for his advice as Bert, Shang and I worked through the details of the HLM paper and for his being available to discuss research at any time since (when he was not playing guitar).

References

1. M. Tinkham, *Introduction to Superconductivity, 2nd. ed.* (Dover Publications, Mineola, NY, 2004).
2. P. de Gennes, *Superconductivity of Metals and Alloys* (Addison Wesley, Redwood City, CA, 1989).
3. P. de Gennes and J. Prost, *The Physics of Liquid Crystals* (Oxford University Press, Oxford, 1994).
4. P. G. de Gennes, Analogy between superconductors and smectics-a. *Solid State Commun.* **10**, 9, 753 (1972), doi:10.1016/0038-1098(72)90186-x.
5. B. I. Halperin and T. C. Lubensky, Analogy between smectic-a liquid crystals and superconductors. *Solid State Commun.* **14**, 10, 997–1001 (1974), doi:10.1016/0038-1098(74)90411-6.
6. B. I. Halperin, T. C. Lubensky and S. K. Ma, First-order phase transitions in superconductors and smectic-a liquid crystals. *Phys. Rev. Lett.* **32**, 6, 292–295 (1974), doi:10.1103/PhysRevLett.32.292.
7. J. H. Chen, T. C. Lubensky and D. R. Nelson, Crossover near fluctuation-induced 1st-order phase-transitions in superconductors. *Phys. Rev. B* **17**, 11, 4274–4286 (1978), doi:10.1103/PhysRevB.17.4274.
8. T. C. Lubensky, The nematic-to-smectic-a transition — a theoretical overview. *J. Chimie Phys. Phys.-Chimie Biol.* **80**, 1, 31–43 (1983).
9. T. C. Lubensky and J. H. Chen, Anisotropic critical properties of de gennes model for nematic-to-smectic-a transition. *Phys. Rev. B* **17**, 1, 366–376 (1978), doi:10.1103/PhysRevB.17.366.
10. J. Villain, Theory of one-dimensional and two-dimensional magnets with and easy magnetization plane.2. Planar, classical, 2-dimensional magnet. *J. Phys.* **36**, 6, 581–590 (1975), doi:10.1051/jphys:01975003606058100.
11. C. Dasgupta and B. I. Halperin, Phase-transition in a lattice model of superconductivity. *Phys. Rev. Lett.* **47**, 21, 1556–1560 (1981).
12. B. R. Patton and B. S. Andereck, Extended isotropic-to-anisotropic crossover above the nematic smectic phase transition. *Phys. Rev. Lett.* **69**, 10, 1556–1559 (1992), doi:10.1103/PhysRevLett.69.1556.

13. B. S. Andereck and B. R. Patton, Anisotropic renormalization of thermodynamic quantities above the nematic smectic phase transition. *Phys. Rev. E* **49**, 2, 1393–1403 (1994), doi:10.1103/PhysRevE.49.1393.
14. C. W. Garland and G. Nounesis, Critical behavior at nematic–smectic-a phase transitions. *Phys. Rev. E* **49**, 4, 2964–2971 (1994), doi:10.1103/PhysRevE.49.2964.
15. P. M. Chaikin and T. C. Lubensky, *Principles of Condensed Matter Physics* (Cambridge Univ. Press, Cambridge, 2000).
16. V. L. Ginzburg, Some remarks on phase transitions of the 2nd kind and the microscopic theory of ferroelectric materials. *Soviet Phys.-Solid State* **2**, 9, 1824–1834 (1961).
17. M. E. Fisher and P. Pfeuty, Critical behavior of anisotropic n-vector model. *Phys. Rev. B* **6**, 5, 1889 (1972), doi:10.1103/PhysRevB.6.1889.
18. K. G. Wilson and M. E. Fisher, Critical exponents in 3.99 dimensions. *Phys. Rev. Lett.* **28**, 4, 240 (1972), doi:10.1103/PhysRevLett.28.240.
19. N. Goldenfeld, *Lectures on Phase Transitions and the Renormalization Group* (CRC Press, Boca Raton, FL, 2019).
20. J. Cardy, *Scaling and Renormalization in Statistical Physics* (Cambridge Press, Cambridge, 1996).
21. W. Helfrich, Defect model of smectic a-nematic phase-transition. *J. Phys.* **39**, 11, 1199–1208 (1978).
22. J. M. Kosterlitz and D. J. Thouless, Ordering, metastability and phase-transitions in 2-dimensional systems. *J. Phys. C-Solid State Phys.* **6**, 7, 1181–1203 (1973).
23. F. Jahnig and F. Brochard, Critical elastics constants and viscosities above a nematic-smectic-a transition of second-order. *J. Phys.* **35**, 3, 301–313 (1974), doi:10.1051/jphys:01974003503030100.
24. C. R. Safinya, D. Roux, G. S. Smith, S. K. Sinha, P. Dimon, N. A. Clark and A. M. Bellocq, Steric interactions in a model multi-membrane system — A synchrotron x-ray study. *Phys. Rev. Lett.* **57**, 21, 2718–2721 (1986), doi:10.1103/PhysRevLett.57.2718.
25. D. Roux and C. R. Safinya, A syncrhotron x-ray study of competing undulation and electrostatic interlayer interactions in fluid multi-membrane lyotropic phases. *J. Phys.* **49**, 2, 307–318 (1988), doi:10.1051/jphys:01988004902030700.
26. F. Brochard, Dynamics of fluctuations near a 2nd-order smectic a-nematic phase-transition. *J. Phys.* **34**, 5–6, 411–422 (1973).
27. R. G. Leigh and M. J. Strassler, Exactly marginal operators amd duality in 4-dimensional n = 1 supersymmetric gauge-theory. *Nucl. Phys. B* **447**, 1, 95–133 (1995), doi:10.1016/0550-3213(95)00261-p.
28. J. Toner and D. Nelson, Smectic, cholesteric, and rayleigh-benard order in two dimensions. *Phys. Rev. B* **23**, 316 (1981).
29. J. Toner, Bond-orientationaly-ordered phases that scatter light strongly. *Phys. Rev. A* **27**, 2, 1157–1163 (1983).
30. M. C. Cetinkaya, S. Ustunel, H. Ozbek, S. Yildiz and J. Thoen, Convincing evidence for the halperin-lubensky-ma effect at the n-sma transition in alkyloxycyanobiphenyl binary mixtures via a high-resolution birefringence study. *Eur. Phys. J. E* **41**, 10 (2018), doi:10.1140/epje/i2018-11738-0.

https://doi.org/10.1142/9789811282386_0025

Chapter 25

Beyond the Bulk: An Appreciation of the Enduring Influence of Michael E. Fisher

D. Jasnow* and J. Rudnick†

*Department of Physics and Astronomy, University of Pittsburgh, Pittsburgh, PA 15260, USA

†Department of Physics and Astronomy, University of California, Los Angeles, CA 90095, USA

We discuss the research performed by us in the relatively early days of the renormalization group, focusing on "beyond the bulk" properties of systems at and near bulk criticality. In all instances, the topics, along with specific problem addressed, originate in whole or in part from the work of Michael E. Fisher.

1. Introduction

It is an honor and a delight to contribute to a collection celebrating the life and work of Michael E Fisher; he has meant so much to our respective careers and to our own collaboration and friendship extending nearly a half century. We begin with our brief, respective "origin stories," how our connections with Michael began and developed, and how our own collaboration began, seeded by renormalization-group developments and influenced considerably by thoughts and ideas largely originating from Michael. We then turn to some specifics and comment on some of our work on "beyond the bulk" physics, which we divide into three categories: (i) two-phase interfaces and helicity, (ii) finite-size scaling and (iii) critical wall perturbations.

2. Origin Stories

2.1. *DJ*

I met Michael the first time, I think, as a graduate student when he visited UIUC to deliver a seminar or colloquium. He gave an impromptu talk for Michael Wortis (my advisor) and me on extracting the correlation length exponent for the Ising model. I appreciated his keeping the presentation within reach. I wound up as a postdoc in Michael's group at Cornell in the late 1960s and had the opportunity to work closely with Michael and take some new directions, even trying my hand at rigorous statistical mechanics. During those years, I had contacts with the Fisher family and Michael's and Sorel's daughter Caricia baby-sat for my elder daughter (with Sorel as backup down the street). I returned the favor some years later "minding" Matthew (without his knowledge, I hope). I entered the real world taking up a position at the University of Pittsburgh, and I think the concept of a contribution to the Domb–Green–Lebowitz Series on Phase Transitions and Critical Phenomena had its beginnings. During the early 1970s, I had extended stays with the Fishers while working on the monograph. Alas, the renormalization group (RG) put an end to it.

It was a seismic shift; so many questions were answered in critical phenomena and contributions appeared at a prodigious rate. Michael was in on it from the beginning, and that was where his attention and the attention of his group was focussed. In 1975, if memory serves, Michael was the Chair of the Chemistry Department at Cornell, and I took a leave from my own department to stay in Ithaca for the most part to make one last attempt to finish off our monograph (which ultimately had an introductory chapter on RG). It was at Cornell at that time that I met my co-author on this essay. Joe visited Michael's group to finish up on some work with David Nelson, who tipped me off in advance with something like "you're gonna like this guy." David showed remarkable insight; Joe became a valued friend, colleague and collaborator over an interval stretching to the present day.

Times being what they were, we were going to employ RG methods, but we had to outflank the very energetic community investigating a range of systems and producing results at a rapid rate. We started out by turning the power of RG to "beyond the bulk physics," several examples of which will be summarized briefly here. All of that physics bears the fingerprints of Michael Fisher.

2.2. *JR*

My graduate education, and particularly my dissertation work, focused on the properties of the degenerate electron gas. I entered graduate school at the University of California, San Diego in 1965, a year or so after the publication of Pierre Hohenberg and Walter Kohn's article on density functionals — and the same year that Walter and Lu Sham's work on the Kohn–Sham approach to the ground-state properties of interacting electrons appeared in print. At that time of my arrival, Walter's interest had turned to inhomogeneous electron systems, among other things the surface properties of simple metals. After accepting me as an advisee Walter set me the task of working on problems relevant to the behavior of conduction electrons in the environment immediately outside a metal. That work formed the basis of my Ph.D. thesis, and I believe Walter and Norton Lang made use of it in their celebrated paper on the surface energy of simple metals. My first postdoc, with Ed Stern at the University of Washington, was also dominated by issues relating to the physics of metallic electrons.

While all this was going on, I became aware of work being done on critical phenomena, through seminars by Richard Ferrell, Pierre Hohenberg, Leo Kadanoff, Michael Wortis, Robert Buchler — and a particularly impressive invited talk by Michael Fisher at an APS March Meeting. However, I never thought seriously of working on topics in that field.

Then, in 1972, on my way to my second postdoc at the Technion in Haifa, Israel, I stopped by the Collège de France and met a member of the scientific staff there (I am embarrassed to admit I don't remember who he was). He showed me Kenneth Wilson's first paper on the renormalization group, which completely befuddled me. On my arrival at the Technion, I dropped by the office of Mike Revzen and asked what he would like me to do. He told me that it would be a service to the condensed matter group there if I read Ken Wilson's papers and reported on what they were about. This set the stage for the next two years of my professional life, punctuated by the 1973 Yom Kippur War and its aftermath.

I don't recall when and why I reached out to Michael. This was half a century ago, when there was no email or internet, and international phone calls were ridiculously expensive. For whatever reason, I wrote him a letter and introduced myself. To my delight and unending gratitude, he replied. My correspondence with him was, quite frankly, a lifeline.

On my return to the United states and my third postdoc at Case Western Reserve University in Cleveland Ohio, I placed a call to Michael's office at Cornell which, to my very good fortune, was shunted to David Nelson, Michael being unavailable. We immediately began a collaboration leading to a couple of papers on RG-generated equations of state that included corrections to scaling — and to an invitation to visit Cornell. There I first met David in person, along with Michael, also Kenneth Wilson — and David Jasnow, whom I first spotted walking down a hallway looking deeply distracted; I believe he was collaborating with Michael on their now-legendary unpublished manuscript on correlations in the critical regime. That was the beginning of a long collaboration and a warm friendship.

3. Helicity and Two-Phase Interfaces

3.1. *Helicity modulus aka superfluid density*

To our knowledge, the term "helicity modulus" was coined by Michael Fisher as a generalization to systems with continuous symmetry of an interfacial property like surface tension in a system with scalar order parameter. He realized it was a type of rigidity or stiffness involving imposing an order parameter twist in a system with continuous symmetry. It was part of a continuing exploration of order and correlations in systems with continuous symmetry and first appeared in 1973.[1] The analogue in the case of ^{4}He below the λ point is the superfluid density, ρ_s, but in that publication, a definition of helicity or superfluid density was presented involving only fundamental concepts of bulk, equilibrium statistical mechanics (along with familiar considerations of thermodynamic limit and short-ranged "wall potentials"). Analogously with various (supposedly equivalent) definitions of long-range order, alternative definitions of helicity were proposed using, e.g., proper thermodynamic limits and boundary conditions placed on the appropriate microscopic variables. Other methods for calculational "convenience," were introduced (see, e.g., Refs. 2 and 3), and we used one of those to incorporate the idea of helicity into a renormalization-group framework and along the way provide a proof for the Josephson exponent relation[4] using the renormalization group. We also commented on the relationship between our work and other results reported at the time.[5,6]

Michael played an important role in keeping us on the right track. One of us (JR) came up with what seemed to him to be an elegant proof of the Josephson exponent relation. When Michael was shown the details, he immediately pointed out a fatal flaw, which we were able to correct. On this occasion, we are pleased to publicly acknowledge his assistance.

3.2. *ϵ-expansion for an interfacial profile*

Continuing with the "beyond the bulk" theme, we turned to the manifestly non-translationally invariant system comprising two phases on the coexistence curve near but below T_c. There are some connections with the helicity modulus analysis since, as noted, it corresponds to the surface tension for an $n = 2$ system. However, preceding our work on a scalar order parameter, an imposing publication by Lubensky and Rubin[7] appeared, addressing a semi-infinite Ising-like system using a Wilson–Fisher-like ϵ-expansion. It was difficult to imagine addressing the interface similarly. Nevertheless, a successful approach was found. We gave some talks, but before we submitted our work for publication, Ohta and Kawasaki published an ϵ-expansion,[8] bearing some similarities to our own approach.[9] Takao Ohta became valued collaborator over the years and joined the two of us on an analysis of interfacial behavior near tricriticality.

Our approach at the time, based on an independent reformulation of the renormalization-group approach[10,11] appears somewhat archaic from the current vantage point. Within a few years, a more obviously field-theoretic approach would probably have been employed, closer to the renormalized perturbation theory of Ohta and Kawasaki,[8] (see also, e.g. Ref. 12), of course with some of the internal calculations looking similar. It might even include the introduction of an interface constraint, or collective coordinate, $Z(\vec{x}_{d-1})$, for a surface specifying the nominal center of the diffuse interface; see the following and, e.g., Ref. 13, with ultimately an average over $Z(\vec{x}_{d-1})$.

There was at the time available experimental evidence for the deviation of the interfacial profile from the Landau–Ginzburg or van der Waals hyperbolic tangent. Our order ϵ results with $\epsilon = 1$ showed reasonable agreement with empirical results due to Huang and Webb,[14] which eliminated the simple hyperbolic tangent form (scaled with the experimental correlation length).

Applying the results of our ϵ expansion, though numerically encouraging, faced a profound conceptual problem. In the absence of a stabilizing field (e.g., gravity coupling to the slight density difference between coexisting fluid phases or possibly finite-size effects precluding a true thermodynamic limit), a finite interfacial width does not exist in $d \leq 3$. The difficulty is the effect of long-wavelength fluctuations of the nominal interface position. To interpret the results of the ϵ expansion, one must consider the experiment stabilized by a weak z-dependent field coupling to the order parameter (as it is); the field is weak enough so as not to distort the interfacial profile significantly. In our follow-up work (on interfaces near tricriticality and criticality in $d = 3$), we were able to include the so-called "capillary wave fluctuations" in a more satisfactory fashion.[15]

3.3. *Interfacial behavior near a tricritical point*

We continued our explorations of interfacial phenomena by turning, in collaboration with Takao Ohta, to two-phase interfaces on the "triple line" (where three phases may coexist) near tricritical points.[16] For tricritical systems, $d = 3$ is the critcal dimension, so we worked in fixed dimension with analysis that includes logarithmic corrections. Furthermore, we addressed the question of capillary waves, whose effects are essentially the same as for a two-phase system on the vapor–pressure curve, to be summarized in the following. The main justification for doing the calculation for a general tricritical system is that the calculation is asymptotically exact. Furthermore, there were available light-scattering experiments for a reasonably direct comparison.

The quantity that corresponds most closely to the experimental normal specular reflectivity is the Fourier transform of the derivative of the interfacial profile,

$$\mathcal{R}(k) \propto \left| \int \left(\frac{ds}{dz} \right) \exp(2ikz)\, dz \right|^2 , \tag{1}$$

where $s(z)$ corresponds to the order parameter profile and k corresponds to the wavenumber of the light in a medium with the mean index of refraction. The derivative is proportional to the index-of-refraction gradient, and $\mathcal{R}(k)$ corresponds to the specular reflection of normally incident radiation on a medium translationally invariant in the x–y plane, the medium having a small, slowly varying index of refraction. Comparison with the available data was promising. The calculation for a *symmetric* tricritical system on the triple line has strong similarities with the critical case, to which we return for the incorporation of capillary wave fluctuations in three dimensions.

3.4. *Approximate interfacial profile in $d = 3$ near criticality*

The effect of capillary waves in three dimensions is quite similar on the triple line in a tricritical system and on the vapor pressure curve (or its generalization) for a system near a critical point. In the tricritical case, the renormalization-group mechanics are essentially exact in $d = 3$, with the appearance of logarithmic corrections to normal power-law singularities. For critical systems in fixed $d = 3$, one may be able to perform renormalization-group analysis — typically for homogeneous systems — if one has sufficient knowledge of the "structure" of the perturbation expansion around the free field.[17] This is a difficult proposition and not one that we attempted. Fortunately, to determine the contributions of critical and capillary wave fluctuations, one can do a reasonable approximation, namely one can perform a one-loop renormalized perturbation theory, which in other contexts might be called an $\eta = 0$ approximation. This can be done along with a proper inclusion of capillary wave fluctuations.[18]

We have been using the term "capillary waves" for long-wavelength fluctuations of an interface (not limited to fluid systems). Since such fluctuations are known to destabilize the two-phase interface in three-dimensional systems (actually $d \leq 3$), for a calculation of interfacial properties, it is necessary to integrate capillary wave fluctuations with ordinary critical critical fluctuations. This is done by implementing the following physical picture which is quite natural within the framework of renormalization-group analysis: Order parameter fluctuations with wavelength up to $O(\xi)$ contribute to a diffuse interfacial structure, which is distorted by long-wavelength capillary waves. This can be done by a method of constraints Ref. 13 and assigning a collective variable (auxiliary field) to the configuration of the nominal interface center, as noted above. Conceptually, to compute physical quantities, one must include a trace over the auxiliary field. Hence, one imagines running the renormalization-group process (in practice, running l^* iterations or integrating the RG equations up until $\xi(l^*) \sim \xi(0)\exp(-l^*)$), which starts large relative to a relevant microscopic length, is reduced to order unity and writing for the order parameter

$$\phi = \phi_0(l^*, z - Z(x, y)) + \sigma(x, y, z). \tag{2}$$

As noted above, the function Z plays the role of the nominal local interface position about which is centered the mean-field "intrinsic" profile. One finds at lowest order that the profile is given by

$$M(l^*, x, y, z) = \langle \phi_0(l^*, z - Z(x, y)) \rangle_Z, \tag{3}$$

where the probability distribution of interface shapes is controlled by free-field Gaussian statistics. This intuitive result is derived most simply from an identity due to Gervais and Sakita,[13] which was brought to our attention by a colleague R.D. Carlitz at the University of Pittsburgh. The end result for an experimental quantity like the normal reflectivity of an interface is that of a convolution of the intrinsic interface shape and a Gaussian factor analogous to a Debye–Waller factor in solid-state physics, as pointed out to us by B.I. Halperin.

4. Finite-Size Scaling

Michael Fisher was influential in the development of this subject from the start, perhaps dating back at least to 1967[19] and certainly to the work with Michael Barber.[20] To incorporate RG ideas and particularly the ϵ-expansion to deal with a d-dimensional hypercube required solving a problem first. The issue is that mean-field theories typically don't have dimensionality specified and typically will lead to a sharp transition in finite systems. By beginning formally from the Landau–Ginzburg free energy in a hypercubical d-dimensional domain with periodic boundary conditions, one writes for the order parameter field $s = m + \sigma$, where m is a constant (the thermodynamic order parameter) and σ representing fluctuations, which by construction have no $\vec{k} = 0$ component. At the level of mean-field theory, formally of order u^{-1}, where u is the usual fourth-order coupling constant, one finds

$$\log Z_{mf} = \log \left[\int dm \exp(L^d (rm^2/2 + um^4/4)) \right] \tag{4}$$

which doesn't show singularities as $r \to 0$ with $L < \infty$. (Normally, in constructing a mean-field approach, one does something like steepest descents now, but for finite-size effects, one takes the equation seriously.) It does, of course, produce a singular behavior for all d in an infinite system. Such common defect of mean-field approximations has to be lived with, but the approach allows building on a renormalization-group approach to the fluctuations and an $\epsilon = 4 - d$ expansion. The approach produces essentially exact results for the finite-size corrections in the critical dimension $d = 4$. Similar RG analysis was carried out by Brézin and Zinn-Justin[21] for other system shapes.

Additional finite-size calculations have been performed in hyper-cubic geometry using, for example, anti-periodic boundary conditions, and in other geometries.

5. Interactions Between Surfaces *at* The Critical Point

A collaboration between Michael and Pierre-Gilles de Gennes resulted in a foundational paper on fluctuation-mediated interactions between the two bounding surfaces of a substance with a slab-like geometry at a critical point.[22] Their approach was based on a phenomenological effective field theory that satisfies bulk scaling and incorporates a position-dependent correlation length. A key result in this paper was a prediction for the influence of one wall in a system with a slab-like geometry on the far wall in that system. According to the analysis of Fisher and de Gennes, at the critical point, the influence should depend on the distance, D, between the two walls as D^{-d}, where d is the spatial dimension of the system. This behavior will be familiar to anyone who has looked into the critical Casimir force. In that case, the specific power law follows directly from finite-size scaling.

However, Fisher and de Gennes were interested in the properties of the order parameter and the way in which order parameter correlations between points spanning the system behave.

The effective field theory leading to their prediction is open to the criticism that it is *local* — except for the gradient term, which is short-ranged — and that it introduces unwanted singularities when applied to a one-dimensional system of finite extent. Furthermore, variations of that effective field theory that preserve the properties of the form proposed by Fisher and de Gennes lead to different predictions for the exponent of D above. We worked to fill that gap making use of previously obtained results for infinite and semi-finite systems at and near bulk criticality along with new analysis and ϵ-expansions[23] to argue that Fisher and de Gennes were correct in their conclusion for a scalar order parameter — and remarkably prescient in their choice of an effective field theory (which we characterized as "uncommonly good"). We also found that in an n-vector system, the influence of one component of the order parameter near a wall on an orthogonal component near the other wall goes as $D^{-(d-1)}$. Subsequently, John Cardy elegantly demonstrated that the Fisher–de Gennes prediction for a scalar order parameter follows from conformal invariance at criticality.[24]

6. Concluding Remarks

It has been a pleasure to provide these recollections in honor of Michael Fisher, who has had a profound influence on our professional and personal lives, and on the development of the field in which we worked. Going over some of the old papers was an experience. We apologize if this essay comes across as recitation of long-past "glory days," but perhaps we can be forgiven after half a century. We also take responsibility for omissions and lapses resulting from our fading memories. Those years of development after the introduction of renormalization-group ideas and techniques were something to behold — and Michael played a major role in much of it.*

References

1. M. E. Fisher, M. N. Barber and D. Jasnow, Helicity modulus, superfluidity, and scaling in isotropic systems. *Phys. Rev. A (General Physics)* **8**(2), 1111–1124 (1973).

**Note added in proofs*: Michael Wortis recalls that DJ first met MEF when all three met for lunch while attending an APS meeting in Chicago, occurring after DJ had applied to MEF for a post-doctoral position. DJ certainly recalls the occasion. Evidently, the meeting in MW's office, recounted above, occurred somewhat later while DJ was still at UIUC. We thank Michael Wortis for the clarification.

2. M. N. Barber and M. E. Fisher, Critical phenomena in systems of finite thickness .1. spherical model. *Ann. Phys.* **77**(1-2), 1–78 (1973). Doi: 10.1016/0003-4916(73) 90409-0.
3. M. N. Barber and M. E. Fisher, Critical phenomena in systems of finite thickness .3. specific-heat of an ideal boson film. *Phys. Rev. A* **8**(2), 1124–1135 (1973). Doi: 10.1103/PhysRevA.8.1124.
4. B. D. Josephson, Relation between superfluid density and order parameter for superfluid he near t_c, *Phys. Lett.* **21**(6), 608 (1966).
5. C. Bervillier, Universal relations among critical amplitude — calculations up to order ϵ^2 for systems with continuous symmetry. *Phys. Rev. B* **14**(11), 4964–4975 (1976). Doi: 10.1103/PhysRevB.14.4964.
6. B. I. Halperin, P. C. Hohenberg and E. D. Siggia, Renormalization-group treatment of critical dynamics of superfluid-helium, isotropic antiferromagnet, and easy-plane ferromagnet. *Phys. Rev. B* **13**(3), 1299–1328 (1976). Doi: 10.1103/PhysRevB.13.1299.
7. T. C. Lubensky and M. H. Rubin, Critical phenomena in semi-infinite systems .1. Epsilon expansion for positive extrapolation length. *Phys. Rev. B* **11**(11), 4533–4546 (1975). Doi: 10.1103/PhysRevB.11.4533.
8. T. Ohta and K. Kawasaki, Renormalization group approach to interfacial order parameter profile near critical-point. *Prog. of Theor. Phys.* **58**(2), 467–481 (1977). Doi: 10.1143/ptp.58.467.
9. J. Rudnick and D. Jasnow, Epsilon-expansion for interfacial profile, *Phys. Rev. B* **17**(3), 1351–1354 (1978).
10. J. Rudnick, Diagrammatic renormalization-group method. *Phys. Rev. B* **11**(1), 363–376 (1975). Doi: 10.1103/PhysRevB.11.363.
11. J. Rudnick and D. R. Nelson, Equations of state and renormalization-group recursion relations. *Phys. Rev. B* **13**(5), 2208–2221 (1976). Doi: 10.1103/PhysRevB.13.2208.
12. E. Brezin, J. C. Le Guilloue and J. Zinn-Justin, *Field Theoretical Approach*, In eds. C. Domb and M. S. Green, *Phase Transitions and Critical Phenomena*, Vol. 6 (New York, 1976).
13. J. L. Gervais and B. Sakita, Extended particles in quantum field-theories. *Phys. Rev. D* **11**(10), 2943–2945 (1975). Doi: 10.1103/PhysRevD.11.2943.
14. J. S. Huang and W. W. Webb, Diffuse interface in a critical fluid mixture. *J. Chem. Phy.* **50**(9), 3677 (1969). Doi: 10.1063/1.1671613.
15. F. P. Buff, R. A. Lovett and F. H. Stillinger, Interfacial density profile for fluids in critical region. *Phys. Rev. Lett.* **15**(15), 621 (1965). Doi: 10.1103/PhysRevLett.15.621.
16. D. Jasnow, T. Ohta and J. Rudnick, Interfacial profile near tricritical points. *Phys. Rev. B* **20**(7), 2774–2784 (1979).
17. J. Baker, G. A., B. G. Nickel and D. I. Meiron, Critical indices from perturbation analysis of the callan-symanzik equation. *Phys. Rev. B (Solid State)* **17**(3), 1365–74 (1978).
18. D. Jasnow and J. Rudnick, Interfacial profile in 3 dimensions. *Phys. Rev. Lett.* **41**(10), 698–701 (1978).
19. M. E. Fisher and A. E. Ferdinand, Interfacial boundary and size effects at critical points. *Phys. Rev. Lett.* **19**(4), 169 (1967). Doi: 10.1103/PhysRevLett.19.169.
20. M. E. Fisher and M. N. Barber, Scaling theory for finite-size effects in critical region. *Phys. Rev. Lett.* **28**(23), 1516 (1972). Doi: 10.1103/PhysRevLett.28.1516.

21. E. Brezin and J. Zinn-Justin, Finite size effects in phase transitions. *Nucl. Phys. B* **257**, 867–893 (1985).
22. M. E. Fisher and P. G. D. Gennes, Wall phenomena in a critical binary mixture. *Comptes Rendus Hebdomadaires Des Seances De L Academie Des Sciences Serie B.* **287**(8), 207–209 (1978).
23. J. Rudnick and D. Jasnow, Critical wall perturbations — Scaling and renormalization-group. *Phys. Rev. Lett.* **49**(22), 1595–1598 (1982).
24. J. L. Cardy, Universal critical-point amplitudes in parallel-plate geometries. *Phys. Rev. Lett.* **65**(12), 1443–1445 (1990). Doi: 10.1103/PhysRevLett.65.1443.

https://doi.org/10.1142/9789811282386_0026

Chapter 26

Normal Forms, Universal Scaling Functions, and Extending the Validity of the RG

J.P. Sethna*, D. Hathcock†, J. Kent-Dobias‡, and A. Raju§

**LASSP, Cornell University, Ithaca, NY 14850, USA*
†*IBM T. J. Watson Research Center, Yorktown Heights, NY 10598, USA*
‡*INFN, Unitá di Roma 1, 00185 Rome, Italy*
§*Simons Centre for the Study of Living Machines, National Centre for Biological Sciences, Tata Institute of Fundamental Research, Bengaluru 560065, India*

Our community has a deep and sophisticated understanding of phase transitions and their universal scaling functions. We outline and advocate an ambitious program to use this understanding as an anchor for describing the surrounding phases. We explain how to use normal-form theory to write universal scaling functions in systems where the renormalization-group flows cannot be linearized. We use the 2D Ising model to demonstrate how to calculate high-precision implementations of universal scaling functions and how to extend them into a complete description of the surrounding phases. We discuss prospects and challenges involved in extending these early successes to the many other systems where the RG has successfully described emergent scale invariance, making them invaluable tools for engineers, biologists, and social scientists studying complex systems.

1. Introduction

Half a century ago, Michael Fisher, together with Wilson and Kadanoff, introduced the renormalization group to analyze systems with emergent, fractal scale invariance. For five decades, physicists have applied these techniques to equilibrium phase transitions, avalanche models, glasses and disordered systems, the onset of chaos, plastic flow in crystals, surface morphologies, etc. But these tools have not made a substantial impact on engineering and biology. We believe it is our duty to make these tools accessible to the broader science community.

• We need to provide them with tools that allow them to describe not only the critical point but also properties of systems that exhibit incipient scale-invariant fluctuations yet are far from the critical region. These demand that we understand *corrections to scaling*, which become more important farther from the critical point. In Section 3, we build on Michael's work on analytic corrections to scaling with

Aharony[1,2] and his work on the complex analytic properties of the Onsager solution[3] to extend our understanding of the 2D Ising critical point to a full description of the ferromagnetic and paramagnetic phases.[4]

• We need not only universal power laws but also a complete description of all behavior of the material. These demand convenient access to accurate *universal scaling functions* that govern the behavior involving more than two quantities at a time. Michael taught us about these powerful tools in his pedagogical reviews and lectures. So, magnetization as a function of temperature goes as $M \sim t^{\beta}$, but magnetization as a function of temperature and external field goes as $M \sim t^{\beta}\mathcal{M}(h/t^{\beta\delta})$, where $\mathcal{M}(X)$ is a universal function, in principle predicted from the renormalization group. In Section 2, we discuss the correct way[5] of writing universal scaling functions for systems where logarithms and exponentials are found in addition to power laws (e.g., in the upper and lower critical dimensions). In Section 4, we present a userfriendly solution[6] for $\mathcal{M}(X)$ in the 2D Ising model, with a systematic expansion that captures the correct singular behaviors with seven-digit accuracy.

• We need rapidly converging methods that can connect our deep understanding of the singularities in phases and in universal scaling functions to quantitative predictions which make the best use of limited information away from the critical points. In modern numerical analysis, one can integrate or approximate analytic functions of one variable with exponential accuracy [7, Sections 4.6 and 5.8.1] so long as one knows the singularities at the end points. Chebyshev, Gauss, and Romberg methods have superseded Simpson's rule and its generalizations for approximating and integrating analytic functions and can be adapted to capture known singularities. We know that all properties are analytic in phases, and our mission for five decades has been to understand the singularities between phases. Can we explain the phases using the critical points? The exponential convergence in Sections 3 and 4 demonstrate that the normal-form theory can do so.

Focusing on the 2D Ising model allows us to show proof of principle, but can we aspire to similar progress for other less well-studied critical points? We suggest this as a key task for the next stage of research in critical phenomena and emergent scale invariance. In Section 5, we discuss progress by those using nonperturbative functional renormalization-group (NPFRG) methods,[8] which have found broad application not only in equilibrium thermodynamic systems but also in avalanche models, quantum systems, turbulence, etc. They explicitly coarse-grain and renormalize a system in fixed dimension, implicitly calculating the universal scaling functions and the system-specific behavior far from criticality. They have mostly been used to extract high-precision critical exponents, amplitude ratios, and proofs of success. If we can organize these calculations into user-friendly universal scaling functions, we may provide experimentalists, simulators, and theorists in a variety of fields with tools to describe matter beyond systems tuned to a single point on the phase diagram.

In Section 6, we embark into deep issues in the renormalization group and the prospects and challenges they provide to our mission to make the theory of critical phenomena an organizing principle for much of science.

2. Normal-Form Theory, RG Flows, and Logs

The renormalization group takes an enormous leap of abstraction — studying emergent scale invariance as a flow under coarse-graining in the space of all possible systems. This reduces the problem to the study of fixed points for differential equations in an infinite-dimensional space (e.g., free energy f, temperature $t = (T - Tc)$, field h, other parameters $u, \ldots$). Near the critical point for the Ising model, coarse-graining and rescaling the free energy f by a factor e^ℓ is described by the differential equations

$$\begin{aligned} df/d\ell &= df + At^2 + \text{other non-linear}\ldots, \\ dt/d\ell &= \lambda_t t + \text{non-linear}\ldots, \\ dh/d\ell &= \lambda_h h + \text{non-linear}\ldots, \\ du/d\ell &= \lambda_u u + \text{non-linear}\ldots \end{aligned} \tag{1}$$

The first step in most treatments of these RG flows is to linearize the flow near the fixed point. Positive eigenvalues λ_t and λ_h correspond to relevant operators like field h and temperature t; negative eigenvalues λ_u correspond to 'irrelevant' perturbations like u that provide *singular corrections to scaling* that are subdominant near the critical point. Using this linearization, this treatment then argues for universal power laws for things like the correlation length $\xi(t) \sim t^\nu = t^{1/\lambda_t}$ at $h = 0$, and universal power laws times universal functions

$$f(t,h,u) \sim t^{d\nu}\mathcal{F}(h/t^{\beta\delta}, ut^{\omega\nu}) = t^{d/\lambda_t}\mathcal{F}(h/t^{\lambda_h/\lambda_t}, ut^{-\lambda_u/\lambda_t}), \tag{2}$$

$$M(t,h,u) \sim t^{\beta}\mathcal{M}(h/t^{\beta\delta}, ut^{\omega\nu}) = t^{(d-\lambda_h)/\lambda_t}\mathcal{M}(h/t^{\lambda_h/\lambda_t}, ut^{-\lambda_u/\lambda_t}) \tag{3}$$

for things like the field-dependent free energy and magnetization, which describe the relations between more than two quantities.

This linearization is not useful in many cases (e.g., the 1D, 2D, and 4D Ising models and all models in their upper and lower critical dimensions). How to systematically formulate universal scaling functions in these cases has hitherto been mysterious. In this section, we describe the use of normal-form methods from dynamical systems theory by Raju *et al.*[5] to understand when this linearization is possible and how to modify the invariant arguments to the universal scaling functions functions when it is not.

Wegner and co-workers[9,10] in the early days justified this linearization by changing variables to *non-linear scaling fields* which transform linearly under the renormalization group. Cardy[11] denotes these new variables u_t and u_h, but we shall use tildes $\widetilde{t}$ and $\widetilde{h}$. By choosing a suitable Taylor expansion of the change of variables,

$$t(\widetilde{t}, \widetilde{h}, \widetilde{u}, \ldots) = \widetilde{t} + a_{tu}\widetilde{t}\widetilde{u} + a_{th^2}\widetilde{t}\widetilde{h}^2 + \cdots,$$
$$h(\widetilde{t}, \widetilde{h}, \widetilde{u}, \ldots) = \widetilde{h} + b_{hu}\widetilde{h}\widetilde{u} + b_{ht}\widetilde{h}\widetilde{t} + \cdots, \tag{4}$$
$$\ldots$$

the equations simplify to $d\widetilde{f}/d\ell = d\widetilde{f}$, $d\widetilde{t}/d\ell = \lambda_t \widetilde{t}$, $d\widetilde{u}/d\ell = \lambda_u \widetilde{u}$, etc. Aharony and Fisher[1,2] noted 10 years later that this change of variables leads to what we call *analytic corrections to scaling*, again subdominant near the critical point. The analytic corrections to scaling have power laws that involve integers and combinations of existing critical exponents β, ν, δ, and include an analytic background to the free energy; these corrections can be written in terms of derivatives of the universal scaling function. The aforementioned singular corrections to scaling introduce new critical exponents and become independent variables in the universal scaling functions.

Dynamical systems theory[5,12] tells us that linearizing the flow can only be done for what are called hyperbolic fixed points. The change of variables $f, t, h, u \to \widetilde{f}, \widetilde{t}, \widetilde{h}, \widetilde{u}$ is calculated one polynomial order at a time, and its radius of convergence can be a subtle mathematical issue.[13] In Section 3, we take on the ambitious task of attempting to describe the entire surrounding paramagnetic and ferromagnetic phases for the 2D Ising model. There, we shall see that convergence is tricky, but a good choice of coordinates can yield a radius of convergence that appears to converge precisely in the range from zero to infinite temperature.

Even for the two-dimensional Ising model, the RG fixed point cannot be linearized. The specific heat has a logarithmic singularity: often deemed $\alpha = 0(\log)$, but incompatible with a linearized flow. This arises because in 2D no polynomial change can remove the At^2 term in the flow of f in Eq. (1) (although rescaling the relative magnitudes of f and t can change the value of A to one). Somewhat messy algebra can confirm that this is due to a integer *resonance* between the two linear eigenvalues $\lambda_f = d = 2 = 2/\nu = 2 * \lambda_t$ for the 2D Ising model. The simplest normal form is thus $d\widetilde{f}/d\ell = 2\widetilde{f} - \widetilde{t}^2$, $d\widetilde{t}/d\ell = \widetilde{t}$, etc. This results in a singularity in the free energy of the form $t^2 \log(t^2)$ which will play an important role in Sections 3 and 4.

For the 4D Ising model, the leading irrelevant operator u becomes marginal, with a zero eigenvalue λ_u. There, it has always been clear that one cannot linearize the RG flow. In the dynamical systems nomenclature, the flow undergoes a transcritical bifurcation in $d = 4$ (together with the same resonance $d = 2/\nu$ seen in the 2D Ising above). Our analysis[5] shows that the normal form for the RG flows is[a]

$$d\widetilde{f}/d\ell = 4\widetilde{f} - \widetilde{t}^2, \quad d\widetilde{t}/d\ell = 2\widetilde{t} - A\widetilde{u}\widetilde{t}, \quad d\widetilde{h}/d\ell = 3\widetilde{h}, \quad d\widetilde{u}/d\ell = -\widetilde{u}^2 + D\widetilde{u}^3. \tag{5}$$

[a]The non-linear term proportional to $\widetilde{u}\widetilde{h}$ in the equation for $d\widetilde{h}/d\ell$ needed in the general normal form is known to be zero for the 4D Ising model. Raju[14] has shown this is because there is a redundant variable proportional to the magnetization cubed in the renormalization group (see Section 6).

The universal scaling of the free energy in a system of length L does not take the usual scaling form $f(t,h,u,L) = L^{-d}\mathcal{F}(X,Y,Z) + f_a(t,h,u,L)$ with $X = \widetilde{t}L^2$, $Y = \widetilde{h}L^3$, and $Z = \widetilde{u}/L^{\omega\nu}$. First, we have quite unusual scaling variables

$$X = \widetilde{t}L^2(W(yL^{1/D})/(1/(D\widetilde{u}) - 1))^{-A}, Y = \widetilde{h}L^3, \quad \text{and} \quad Z = yL^{1/D}, \qquad (6)$$

Second, the free energy has a more complex form

$$f(t,h,u,L) = L^{-d}\mathcal{F}(X,Y,Z) - W(Z)^{-A}\left(\frac{W(Z)^{-A}}{1-A} - \frac{1}{A}\right) + f_a(t,h,u,L). \qquad (7)$$

where $W(x)$ is the Lambert W or product log function, $\widetilde{u}$ is the marginal quartic term in the Landau free energy, y is a messy known function of $\widetilde{u}$, A and D are the amplitudes of non-linear terms in the renormalization-group flow (Eq. (5)), and f_a is a non-singular, analytic function near the critical point.

Equations (6) and (7) capture the complete, correct singularity for the 4D Ising model; the traditional log and log–log corrections in the specific heat and the susceptibility arise from expansions of the Lambert W function for large arguments. At bifurcations like $d = 4$ and resonances as at $d = 2$, normal-form theory dictates the nature of the singularity at the critical point.

It *also* tells us that the free energy in the surrounding phase can be found by changing variables. So, for example, we know that the liquid–gas critical point at high pressures and temperatures is in the 3D Ising universality class. Hence, the liquid and gas phase properties should be given by

$$f(T,P) = \widetilde{t}(T,P)^{3\nu}\mathcal{F}_{3\text{DIsing}}(\widetilde{h}(T,P)/\widetilde{t}(T,P)^{\beta\delta}, \widetilde{u}(T,P)t^{\omega\nu},\dots) + f_a(T,P), \qquad (8)$$

where f_a, $\widetilde{t}$, $\widetilde{h}$, and $\widetilde{u}$ are analytic in temperature and pressure near the critical point. As usual, from the free energy (and a corresponding scaling form for the correlation function), one can derive all of the equilibrium and linear-response properties of the resulting phases. We shall implement a change of coordinates like this for the 2D Ising model in Section 3.

We highlight the use of our normal-form methods[5] to solve a 20-year-old puzzle — the unusual scaling of avalanches in the non-equilibrium 2D random-field Ising model[15] (see Fig. 1). Over 25 years ago, we used the random-field Ising model to study avalanches. We could understand the scaling in 3, 4, 5, 7, and 9 dimensions,[16] but two dimensions made no sense. Changing the arguments from ratios of power laws to the invariant functions determined from normal-form theory was one part of the puzzle.[b]

The distribution of avalanche sizes in the random-field Ising model is cut off at size $\Sigma(w)$ depending on the disorder w. In three dimensions and higher, it takes the traditional form $\Sigma(w) = w^{-d_f\nu}$. In two dimensions, normal-form theory predicts that $\Sigma = (B + 1/w)^{-Bd_f+C}\exp(d_f/w)$, with d_f a universal critical exponent, and B and C being universal constants associated with irremovable nonlinear terms

[b]Using random lattices to suppress faceting was the other obstacle.

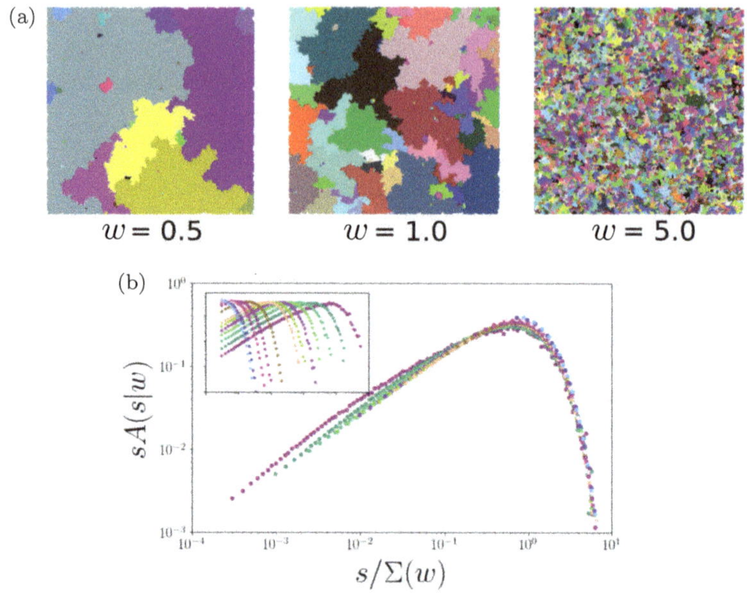

Fig. 1. Avalanche sizes for the 2D $T = 0$ random-field Ising model (from Ref. 15). (a) Avalanches for disorders $w = 0.5$, 1.0, and 5.0; each color is a separate avalanche. (b) Scaling collapse of the area-weighted avalanche size distribution. Note the factor of 10 range in disorder w, and the factor of 2000 range in typical avalanche size Σ. Here, the invariant scaling combination $\Sigma(w) = \Sigma_s(B + 1/w)^{-Bd_f+C} \exp(d_f/w)$ is not a ratio of power laws, but is derived directly from the non-linear normal form of the renormalization-group equations in the lower critical dimension (as in Eq. (5)). The area-weighted avalanche size distribution is thus $sA(s|w) = (s/\Sigma(w))^x \mathcal{A}((s/\Sigma(w))^y)$, with $\mathcal{A}$ a universal scaling function.

in the renormalization-group flow. As shown in Fig. 1, this does an excellent job explaining the behavior.

3. Changing Coordinates to Describe Phases: Matching Onsager

As foreshadowed in Section 2, normal-form theory tells us how to transform universal scaling forms to describe the entire phases surrounding a critical point. This section summarizes our recent results,[4] which implement this for the 2D Ising model in zero field, where Onsager's exact solution[17] for the free energy enables quantitative validation of the results.

Given the asymptotic scaling form for the free energy (or any other quantity) in normal-form coordinates, $\widetilde{f}(\widetilde{t}, \widetilde{h}, \widetilde{u})$, the free energy as a function of physical temperature, field, and irrelevant perturbations is $f(t, h, u) = \widetilde{f}(\widetilde{t}(t, h, u), \widetilde{h}(t, h, u), \widetilde{u}(t, h, u)) + f_a(t, h, u)$. The functions $\widetilde{t}(t, h, u)$ and $\widetilde{h}(t, h, u)$ are given by precisely the analytic change of variables that inverts Eq. (4), mapping the non-linear scaling variables back to their physical counterparts. Additionally, we must also add terms $f_a(t, h, u)$ accounting for the analytic background of the free energy. Note that the change of coordinates are linear at lowest order, e.g., $\tilde{t} \sim t$ and $\tilde{h} \sim h$. Therefore, $f(t, h, u) \sim \tilde{f}(t, h, u)$ near the critical point; in other words, the normal-form free energy $\tilde{f}$ is the asymptotic scaling form near the critical point, usually computed using RG and related techniques.

We have already seen that the 2D Ising model in zero field has a non-linear normal form due to the resonance between the free energy and temperature, $d\tilde{f}/d\ell = 2\tilde{f} - \tilde{t}^2$, $d\tilde{t}/d\ell = \tilde{t}$, and that these flow equations give rise to a logarithmic singularity, $\tilde{f} = -\tilde{t}^2 \log(\tilde{t}^2)$. Following the procedure outlined above, the free energy as a function of temperature is simply

$$f(t) = -\tilde{t}(t)^2 \log(\tilde{t}(t)^2) + f_a(t). \tag{9}$$

This result is consistent with Onsager's exact solution, which is known to have the form $a(t) \log t^2 + b(t)$ for some analytic functions a and b. Comparing these expressions for the free energy, we need to find the coordinate change $\tilde{t}(t)$ and analytic background $f_a(t)$ by solving

$$a(t) = -\tilde{t}(t)^2, \quad b(t) = \tilde{t}(t)^2 \log(\tilde{t}(t)^2/t^2) + f_a(t). \tag{10}$$

Note that because $\tilde{t}(t)$ is linear to lowest order in t, the term $\log\left(\tilde{t}(t)^2/t^2\right)$ is indeed analytic.

We recently[4] computed the free energy, Eq. (9), by perturbatively expanding Onsager's exact free energy around the critical point. A key question is the radius of convergence of the coordinate transformation $\tilde{t}(t)$ and $f_a(t)$ to the normal form. Unlike Taylor expansions about analytic points, the radius of convergence of this normal-form analytic expansion about a singular point is not simply the distance in the complex plane to the next-nearest singularity. One might hope that physics would govern the convergence — perhaps the distance to zero temperature, infinite temperature, or the nearest other phase transition. Indeed, in each of the expansions discussed in the following, the critical point is closest to zero temperature, with this distance determining the radius of convergence.

Our investigations showed the importance of the choice of variable used to parameterize the distance to the critical point (Fig. 2(a)). Expanding $\tilde{t}$ and f_a in temperature $t = (T - T_c)$, for example, converges in an estimated range $-T_c < t < T_c$ (or $0 < T < 2T_c$). This is not the only natural choice, however. The low-temperature expansions for the Ising model are expressed in powers of $X = \exp(-2/T)$. X is also a natural variable in that the zeros of the 2D Ising partition function in the complex X plane form a circle passing through X_c, as Fisher explained.[3] Using the Onsager solution to expand $\tilde{t}$ and f_a in terms of $x = X - X_c$ yields a radius of convergence that extends all the way from zero temperature to $X = 2X_c$ (corresponding to $T \approx 4.7\, T_c$), but fails to describe higher temperatures.

Here, we used special properties of the 2D Ising model to identify a new variable

$$V = \frac{5 - 3\sqrt{2} + X}{1 + \sqrt{2} + X} \quad v = V - V_c, \tag{11}$$

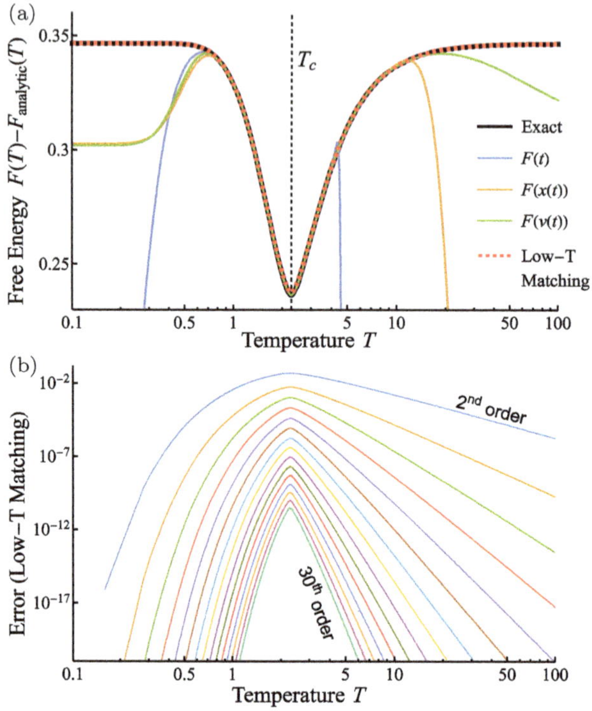

Fig. 2. Capturing the entire phase with analytic corrections, 2D Ising (from Ref. 4). (a) Radius of convergence depends on the expansion variable: Expanding Onsager's exact free energy (black line) in t, x, and v (colored solid lines) leads to increasing radii of convergence, with the convergence of the v-expansion estimated to cover the entire physical temperature range. Each expansion shown is 20th order. Determining analytic corrections in v by matching to the low-temperature expansion does even better, accurately reproducing the free energy across all temperatures even at low orders (red dashed line, sixth order). (b) Exponential convergence: Adding analytic corrections to the universal scaling function at T_c by fitting to the low-temperature expansion results in exponential convergence to the Onsager solution as we add more terms.

which allows our estimated radius of convergence to cover the full physical temperature range from zero to infinity (Fig. 2(a)). Our coordinate v, unlike t or x, respects the self-dual symmetry of the 2D Ising model. Fisher's circle of zeros in X (and thus x) breaks this self-dual symmetry. The linear fractional transformation in Eq. (11) precisely unwraps this circle of zeros into a straight line, extending the self-dual symmetry to the complex plane. The circle of zeros becomes the branch cut of the logarithm in the scaling function for the free energy (Eq. (9)).

Figure 2(a) compares the expansions $f(t)$, $f(x(t))$, and $f(v(t))$ to the exact free energy. In accordance with the discussion above, we see improved convergence for the x and v expansions. The expansion in v numerically appears to have a radius of convergence that extends to the entire physical range, zero to infinite temperature. Physics here does determine the range of convergence of normal-form theory.

The above results leave two questions: Can we improve our approximation for the free energy further, such that it converges, even at low orders, across all temperatures? Furthermore, can we approximate the coordinate change and analytic background without knowledge of the exact solution or the non-linear terms in the RG flows? To resolve these challenges, we again expand the free energy in v using the form given in Eq. (9), but fix the expansion coefficients of $\widetilde{t}$ and f_a by matching to low-temperature expansions of the free energy instead of expanding the exact solution. Importantly, this approach requires minimal knowledge about the critical point (only the asymptotic scaling form), with most of the information coming from deep within the low-temperature phase. Since the matching guarantees the correct low-temperature behavior and has the correct log singularity at the critical point built into the expansion, we see uniform convergence across all temperatures. For example, by sixth order, the approximation differs from the true free energy by at most 0.5% (red dashed line in Fig. 2(a)) and we see exponential convergence as we add additional terms (Fig. 2(b)).

Our approach for extending critical scaling forms to the neighboring phases should naturally generalize to unsolved statistical physics models and experimental systems. For example, low-temperature expansions are relatively easy to compute in all dimensions and for a variety of systems, and can be used to approximate the analytic corrections to established critical scaling forms. Candidates for this method include the 3D Ising model, where critical exponents are known to high precision from conformal bootstrap and the 2D Ising model in a field, whose scaling function is computed to high precision in Section 4. Finally, it is credible that the liquid phase, long known as being challenging because there is no 'small parameter,' could be described as a perturbation of the liquid–gas critical point, i.e. an Ising critical point plus analytic corrections (determined, for example, by matching to virial expansions).

4. 2D Ising Critical Point in a Field

Ever since Onsager solved the zero-field 2D Ising model, people have searched for a high-precision approximate solution for the 3D Ising model and occasionally also for the 2D Ising model in a field. The most successful of these attempts make use of parametric coordinates to interpolate behavior in temperature and field around the critical point in such a way that the singularity is naturally incorporated.[18,19] These coordinates, first introduced many decades ago,[20] are polar-like, with a radius-like component R that controls the proximity to the critical point and an angle-like coordinate θ that rotates around the temperature–field plane. When taken to parameterize the temperature–*magnetization* plane, such coordinates can be periodic in θ like true polar coordinates, something once studied by Michael and collaborators to describe the coexistence region.[21] However, when parameterizing the temperature–*field* plane, there is an inevitable cut resulting from the discontinuity between the up and down states at low temperatures, and θ is not a periodic coordinate but ends at the abrupt transition at some θ_0.

The power of adopting these coordinates is that, as a function of θ, the critical singularity does not appear, and one can reasonably well approximate the scaling function in this coordinate as a simple analytic function. Using this principle, Caselle *et al.* did a remarkably careful job of creating a function that satisfied lots of known properties and measured values (in yellow of Fig. 3). Their method was doomed to slow convergence at small external fields, though, because while the singularity of the critical point was indeed removed by the coordinate change, other more subtle singularities in the free energy remain, including a key essential singularity as one crosses the $h = 0$ line for $t < 0$. Figure 3 shows a comparison between the approximate form of Caselle *et al.* and that of our work including this singularity.

In order to rigorously realize the idea of a free energy scaling function broken into singular and analytic pieces, two singularities in the scaling function must be accounted for in addition to the critical one. One is ancient and well known to the experts: the Yang–Lee edge singularity for $T > T_c$ in the complex plane.[22,23] The other has only recently been realized to be relevant, but has a nice physical picture. As one crosses $h = 0$, the equilibrium magnetization jumps. But if you cross just a little bit, the up-spin magnetized state is metastable, lasting for a good while until a bubble of down spins forms and grows to flip the system. (Think of

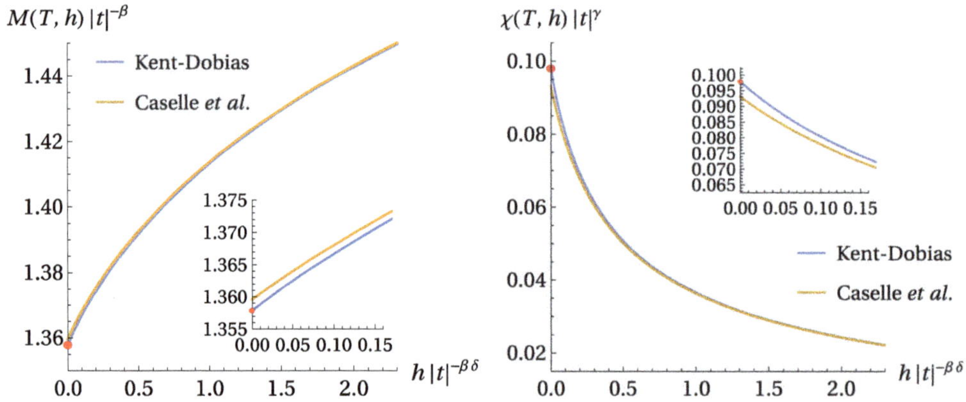

Fig. 3. Comparison between parametric approximations to the scaling functions with and without the appropriate singularities (from [6]). Scaling functions for the magnetization and susceptibility plotted as functions of the scaling invariant $h|t|^{-\beta\delta}$. The errors in the blue curves (Kent-Dobias) are estimated to be roughly 10^{-7}. Caselle's earlier work[19] that we build upon has significant errors near $h = 0$ (red point); these are due to the essential singularity at $h = 0$ and $t < 0$ that we address. These discrepancies grow larger with higher derivatives of the free energy, as shown in the right panel of Fig. 5.

supersaturated vapor — 101% humidity — and the nucleation of raindrops.) Here, the surface tension cost between the up and down regions for small droplets is bigger than the bulk free energy gain of aligning the interior with the external field. The energy barrier B to reach large sizes diverges as h vanishes, $B/k_BT = c/h$ for some constant c. Just like in quantum mechanics, where the lifetime of a state gives the energy an imaginary part, the free energy becomes complex for $h < 0$ with an essential singularity $\text{Im}f \sim e^{(c/h)}$. One can use a Kramers–Kronig transform to see that the real part also has an essential singularity, influencing f for $h > 0$ as well.

We can incorporate these two singularities into the universal scaling function by first generating a 'simplest' functional form that has the correct singularities and then changing variables $\theta \to \widetilde{\theta} = g(\theta)$ and adding an overall analytic function $\mathcal{F}_a(\theta) = G(\theta)$, in a precise analogy with the analytic corrections we introduce to match the universal scaling function to describe the surrounding phases (Section 3).

The 'simplest' form is taken from writing the most singular part of the imaginary part of the free energy and then taking advantage of a Kramers–Kronig-like relation to find the corresponding real part. The requisite contour integral in the complex-θ plane is shown in Fig. 4.

Though complicated, this procedure pays dividends. By incorporating these singularities and matching the power series terms in $g(\theta)$ and $G(\theta)$ (analogous to $\widetilde{t}(t)$ and $f_a(t)$ in section 3), we are able to achieve exponential convergence of the scaling function to exactly known values at $t = 0$ and also achieve exponential convergence of derivatives. The left-hand side of Fig. 5 shows this convergence in the free energy itself as a function of the scaling invariant $th^{-\beta\delta}$ by subtracting our best converged approximation $\mathcal{F}^{[6]}$ from the lower-order approximations $\mathcal{F}^{[n]}(\theta)$ for $n \in \{2, 3, 4, 5\}$. These provide evidence that our seven-digit convergence at $t = 0$ extends to the whole scaling function. On the right of this plot, we can see the origin of this good behavior: The series expansion for the free energy around the abrupt transition point has zero radius of convergence, but this is captured naturally by our approximate scaling form.

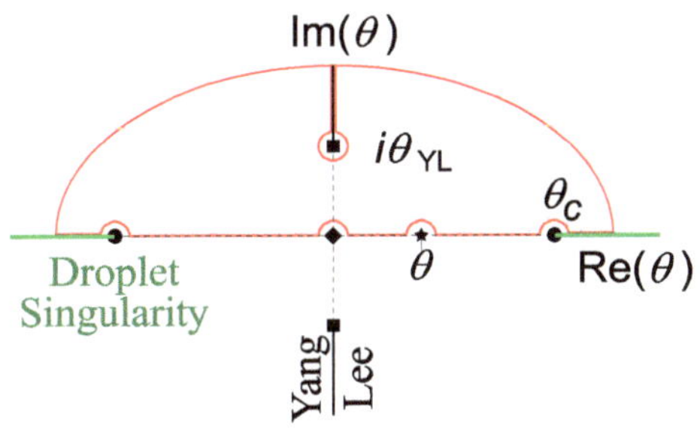

Fig. 4. Contour in the complex θ plane to generate a free energy scaling form with the correct singularities (from Ref. 6).

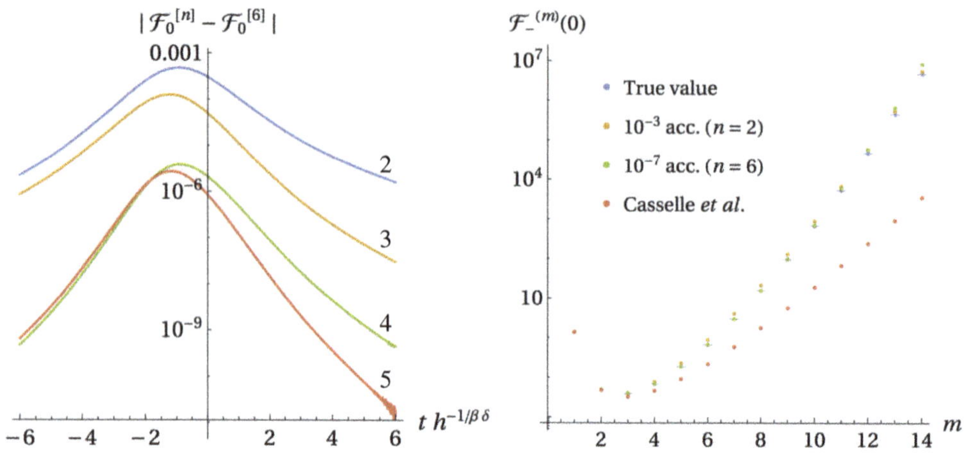

Fig. 5. Convergence of the universal scaling function (from Ref. 6). Left: The difference between the nth order approximation for the free energy scaling function in a field and the 6th order approximation as a function of the scaling invariant $th^{-1/\beta\delta}$. The 5th and 6th order approximations differ by at most 2×10^{-6}. Higher derivatives behave similarly. Right: The Taylor series coefficients $\mathcal{F}_-^{(m)}(0)$ of the free energy scaling function at the abrupt transition line (the location of the essential singularity) as a function of derivative m. By incorporating the essential singularity into the scaling function explicitly, the approximate forms have zero radius of convergence, matching numeric measurements of these coefficients.

5. Interpolating Scaling Functions Between Dimensions

Over five decades, the traditional thermodynamic critical points, illustrated in Fig. 6, have been extensively explored using ϵ-expansions, $1/N$ expansions, cluster expansions, etc. The universal critical exponents are known essentially exactly from conformal bootstrap methods.[c] Can we do the same for the universal scaling functions as functions of dimension d and the number of spin components N? And for (say) the random-field Ising model[24] or turbulence[25]?

Interpolation between dimensions could have several benefits. First, much is known analytically in two and four dimensions about the universal scaling functions that is only known numerically in three. Second, there are important features in scaling functions and their corrections to scaling that are a clear foreshadowing of properties in other dimensions. So, the leading correction to scaling in 3D is the echo of the marginal variable of the Wilson–Fisher transcritical bifurcation in 4D. And, in a more dramatic example, the universal scaling function for the avalanche size distribution in the 3D random-field Ising model has a striking feature (it grows

[c] In the same spirit, Kent-Dobias' solution[6] for the scaling function in 2D in Section 4 is also essentially exact: Both are well-defined algorithms that generate an exponentially converging approximation in a form useful for applications. After all, this is what we call the 'exact' solution for $\sin(x)$.

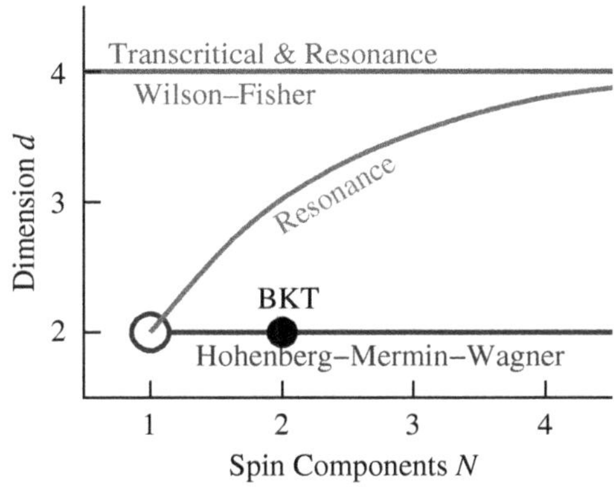

Fig. 6. Interpolating between dimensions: A schematic diagram showing the dimensions and spin components where traditional power-law scaling will break down. This leads to logarithms and exponentials replacing power laws and rich but complicated invariant scaling combinations as arguments for the universal scaling functions.[5] On the other hand, these complexities also promise to inform and accelerate convergence of the universal scaling functions for dimensions in between.

by a factor of 10 before cutting off exponentially[16]), which is clearly related to the unusual scaling of the avalanche sizes in two dimensions[15] (Fig. 1).

In ordinary differential equations, normal-form theory provides not only the behavior at the transition but also an unfolding of the behavior near the bifurcation (up to a smooth coordinate transformation). Here, we may expect to use this, not only to describe the phases near the critical points but also (following the lead of Wilson and Fisher) to describe the universal scaling functions as they evolve between dimensions.

How would we interpolate scaling functions, when even the arguments of the scaling functions vary with dimension (as in the W-functions we needed in Section 2 for the 4D Ising model)? Consider taking the equations for the 4D flow about the mean field fixed-point, Eq. (5), and keeping the unremovable non-linear terms to describe non-linear flows about the mean-field fixed point in d dimensions:

$$d\widetilde{f}/d\ell = d\widetilde{f} - \widetilde{t}^2, \quad d\widetilde{t}/d\ell = 2\widetilde{t} - A\widetilde{u}\widetilde{t}, \quad d\widetilde{u}/d\ell = (4-d)\widetilde{u} - \widetilde{u}^2 + D\widetilde{u}^3, \qquad (12)$$

Things to note are as follows: (1) The normal-form procedure of removing the non-linear terms allows us to set the non-linear terms in any way we wish. (2) We know the Wilson–Fisher fixed point for Eq. (12) at $(\widetilde{t}^*, \widetilde{u}^*)$ must have a linearization with eigenvalues $1/\nu$ and ω that, as a function of dimension, match the conformal bootstrap values (or the exact values in 2D). This constrains the two non-linear terms $A(d)$ and $D(d)$. Serendipitously, the $\widetilde{t}^2$ resonance term in the 4D free energy flow is also needed to get the logarithmic specific heat in 2D. A high-precision universal scaling function for the free energy using the non-linear scaling variables $\widetilde{f}$, $\widetilde{t}$ and $\widetilde{u}$ in Eq. (12) would describe the universal crossover scaling from mean-field to short-range magnetization in all dimensions between two and four.

One could then linearize that scaling function about the 3D Wilson–Fisher fixed point to extract the traditional scaling functions.

How would one calculate such scaling functions? Here, serendipity strikes in recent advances in non-perturbative functional renormalization-group (NPFRG) calculations,[8] whose critical exponents[26] are almost competitive with those of conformal bootstrap, but which, in the process, also compute a functional form for the coarse-grained free energy as a function of magnetization.

Matthew Tissier and colleagues at the Jussieu campus of the Sorbonne and the first author have been exploring how to calculate high-precision universal scaling functions by combining the strengths of the normal-form theory with the systematically improvable scaling functions of NPFRG. In initial work, we have found that the NPFRG for the 4D Ising model should indeed have a scaling function whose arguments are of the exotic form in Eq. (6). We are now learning to solve the partial differential equations for $\widetilde{t}(t,u)$ $\widetilde{u}(t,u)$ and $\widetilde{f}(t,u)$ in 3D to extract the universal scaling functions from the simplest of NPFRG models. We hope then to use the technology in Section 4 to demonstrate how to tabulate universal scaling functions in a form convenient for experimentalists and simulators.

The NPFRG method has had striking success in high-precision calculations of equilibrium thermodynamic systems, disordered thermodynamic and avalanche models, fully developed turbulence, quantum many-body systems, and in QCD and electroweak models.[8] Our hope and vision is to inspire our colleagues to gather and tabulate their results into universal scaling functions that can describe the behaviors both near and far from points of emergent scale invariance in a way accessible to engineers, biologists, and social scientists studying complex systems.

6. Prospects and Challenges for Future Work

We have summarized what we believe to be promising indications that we can indeed generate usable, high-precision universal scaling functions in a systematic way and can extend them systematically into high-precision descriptions of the surrounding phases. If this continues to work, one imagines our community will be able to provide quantitative theories of liquids, material plasticity and failure, fluctuations in biological systems, and potentially turbulence and glass behavior. In all these cases, we must mesh our understanding of the universal, emergent critical behavior and our system-dependent knowledge of properties away from criticality to describe the entire phases. Will this work in practice?

To show that our method works to high precision, we chose to use a problem where we knew a great deal about the answer: the 2D Ising model. The reader should legitimately be skeptical that this is the easiest case. In particular, the 2D Ising mode is special in that it (1) is self-dual, (2) is analytically solvable in zero field, and (3) has no singular corrections to scaling. Let us discuss causes for concern that our methods will become general tools and reasons for optimism:

(1) *The 2D Ising model is self-dual.* In Section 3, we found exponential convergence of the Ising free energy $f(t)$ in zero field using a power series of the Onsager solution about the critical point. To do so, we expanded in a variable $v(t)$ that was fine-tuned to the self-dual and complex analytic properties of the 2D Ising model. We then used the low-temperature expansion to avoid any use of Onsager's solution. Will we need to find the 'right' variable in 3D to proceed? It turns out that matching to properties outside the radius of convergence stabilizes the expansion. In the preliminary work, matching both high- and low-temperature series using the standard variable $x = X - X_c = e^{-2/T} - e^{-2/T_c}$ in 2D (ignoring duality), we found exponential convergence in the entire temperature region.[4]

(2) *The 2D Ising model in zero field has an exact solution* and also has an exact solution at T_c as a function of field.[d] But will the external field prevent us from finding an exponentially converging solution using analytic corrections to scaling $\widetilde{t}(t,h)$, $\widetilde{h}(t,h)$, and $f_a(t,h)$, by matching the 2D universal scaling function from Section 4 to the low-T/high-h and high-T/low-h cluster expansions, as we did for zero field in Section 3?

We have an exact solution for the magnetization — the first derivative with respect to field. This has allowed us to determine $\widetilde{t}$ and $\widetilde{h}$ to linear order in h. The susceptibility $\chi(T)$ at zero field is not known exactly, but a remarkable amount is known: there are formal expansions strongly indicating subdominant logarithmic corrections to scaling and a possible essential boundary in the complex plane at the Fisher circle of zeros in the complex temperature plane.[27–30] Log corrections to scaling are known to arise from resonances between relevant and irrelevant eigenvalues (Section 2); the observed logs could arise from irrelevant variables whose resonances contribute only in $O(h^2)$ to the free energy. It appears that we can use these results[27] not only to determine $\widetilde{t}$ and $\widetilde{h}$ to quadratic order in h, but also to extract information about irrelevant variables and their singular corrections to scaling in 2D (see below).

The universal scaling function in 3D presumably should succumb to the same kind of systematic approximation that we used in Section 4 for 2D. All the ingredients appear to be available.[e] The Yang–Lee singularity in 3D[33] has yielded to a high-precision NPFRG calculation. The essential singularity calculation depends only on the value of the surface tension near the critical point. In 2D, we used estimates of derivatives of the universal scaling function near $h = 0$ for $T > T_c$ and $T < T_c$. The latter (shown in Fig. 5(b)) are challenging to calculate using NPFRG methods (perhaps because of the essential singularity), but the former are now available.[34] Indeed, these authors used their estimates and the traditional

---

[d] It may seem in retrospect clear that the essential singularity at $h = 0$ would frustrate a search for an exact solution in a field below T_c.

[e] And, of course, high-precision NPFRG calculations of the critical exponents[26] implicitly have calculated the universal scaling function.

Schofield coordinates to estimate the scaling function; adding the proper droplet singularity and Yang–Lee singularities could be straightforward and hopefully will yield convincing exponential convergence.

Can we expect to extend a high-precision 3D Ising scaling function to describe the in-field behavior in the entire surrounding phases? Using analytic corrections to extend the critical singularity could break down at the roughening transition.[f] However, this subtle transition should in principle also impede the use of Dlog Padé methods using the low-temperature expansion of the magnetization to describe the critical-point behavior, which works well in practice. In any case, the roughening transition does not arise in isotropic systems, so these concerns will not interfere with finally extracting a quantitative theory of liquids by adiabatic continuation of our understanding of the 3D Ising critical point.

(3) *The Onsager solution exhibits no singular corrections to scaling.* First, this statement is more subtle than it seems. Onsager's solution has a pure logarithmic singularity, but conceivably there could be irrelevant eigenvalues with integer exponents that would lead to analytic terms in the free energy and magnetization. Also, there are irrelevant anisotropies in the correlation functions and the susceptibilities due to the square lattice, indeed with integer exponents. Even more subtle, these anisotropies are singular corrections to scaling in Wilson's RG formulation, but are clearly present at the fixed point of the real-space RG. Any coarse-graining procedure maintaining a square lattice will have a short-range square anisotropy.

Experts will remember that some of the irrelevant operators in the RG are *redundant*, and these redundant operators often involve flows between different fixed points in the same universality class.[g] The idea of redundant variables had been first explored by Wegner,[38] who pointed out that a RG transformation would always have certain operators that are redundant; they correspond to infinitesimal redefinitions of the field and do not contribute to the singular part of the free energy.[h]

---

[f]While one might argue that the behavior of an interface between spin-up and spin-down regions should not affect the bulk free energy, the (incredibly subtle) essential singularity as one crosses $h = 0$ (Section 4) depends on the surface energy of the critical droplet, which will itself have an essential singularity as a function of temperature as it develops facets at the roughening temperature.

[g]We have recently looked at this question[35] in the context of the period doubling onset of chaos in iterated maps $f(x)$. About half of the irrelevant eigenvalues are negative powers of the relevant critical exponent α and are redundant: One can set the amplitudes of their correction to scaling to zero using a coordinate change $x = \phi(y)$, so instead of $\alpha f(f(x/\alpha))$, one considers the renormalization group $\alpha\phi(f(f(\phi^{-1}(x)/\alpha))$. We called these *gauge irrelevant* as they depend on the choice of coordinate. A similar gauge invariance arises for the quasiperiodic onset of chaos[36,37] that we studied when the first author was a postdoc.

[h]How does one think of these redundant operators in the discrete theory? Murthy and Shankar had found a clever way to think about redundant operators for the discrete Ising model.[39] They essentially used a version of Kadanoff's RG projection operator without reducing the degrees of freedom. They found evidence that these redundant operators had observable consequences in simulations of the 2D Ising model.[40] Evidence for redundant operators had already been found in the 3D Ising model.[41]

The first author's first graduate student, Mohit Randeria, worked for Michael Fisher before shifting to his group. Randeria and Fisher wrote a comment on a paper by Swendsen[42] which claimed that one could modify the RG transformation to move the fixed point anywhere on the critical surface. They asserted that this could not in general be correct because the amplitudes of the singular corrections to scaling are zero at the fixed points. To quote them:

> "One fixed point may be mapped into another by a change of redundant variables and two RGs, say $\mathcal{R}$ and $\widehat{\mathcal{R}}$, may produce *formally* different fixed points by this mechanism. Nevertheless, a general point on a critical manifold *cannot* be transformed into a fixed point nor *vice versa*."

In the context of the 2D Ising model, there seem to be no singular corrections in the exact solution. Barma and Fisher had predicted that there should be logarithmic corrections to scaling coming from irrelevant variables.[43] They had also found evidence for an irrelevant operator with exponent $-4/3$.[44] Conformal field theory predicts a large number of irrelevant variables which come from descendant operators but does not predict the $-4/3$ exponent. It has been conjectured in the past that all of these descendent operators are redundant.[45] This may explain why they do not lead to any logarithmic corrections to scaling (which otherwise are generically expected[14]). We would consider all of these descendant operators as constituting 'gauge' corrections to scaling, which can be removed by an appropriate coordinate choice, whereas the $-4/3$ exponent would be a genuine singular correction to scaling.

In Section 3, we succeeded in solving for the free energy of the square-lattice Ising model at all temperatures by applying a normal-form change of variables to the universal scaling function for the critical point. Is that success due to the lack of (non-redundant) singular corrections to scaling for this particular universality class?[i] It is clear that in 3D, we shall need to incorporate the (genuine) singular corrections to scaling in order to accurately describe the behavior away from the fixed point. Indeed, these contributions are subdominant near the fixed point, meaning conversely that they grow faster as one leaves the fixed point than the relevant operators. Thus, we will need a universal scaling function that depends not only on $\widetilde{t}$ and $\widetilde{h}$ but also on potentially an infinite family of irrelevant variables $\widetilde{u}$. Can we nonetheless hope for exponential accuracy?

The NPFRG methods mentioned in Section 5, as it happens, precisely implement a flow with many irrelevant eigendirections at the fixed point. Applying our normal-form analysis, we have been able to linearize this flow in the simplest case, allowing us to extract a functional form for the leading correction subdominant as $\widetilde{t}^{\omega\nu}$. It

[i]It is believed that the powers of logarithms in the 2D Ising susceptibility vanish in isotropic systems. Since all irrelevant operators in 2D have rational eigenvalues, and the redundant ones cannot generate resonant logarithms, one suspects that all the isotropic corrections to scaling are redundant.

seems likely that one could generate exponential convergence for all properties of the phases by combining increasingly sophisticated NPFRG calculations, adding more irrelevant operators to our scaling functions, and matching to higher orders of low- and high-temperature expansions (or virial expansions, $1/N$ expansions, etc.).

Finally, there are intriguing hints of deep connections between normal-form theory, redundant operators, and the construction of universal scaling functions. We noted in Section 4 that the procedure we used to generate the universal scaling function recapitulated exactly the same steps as the procedure we used in Section 3 to generate the Onsager solution from the critical point. In both cases, we knew the singularities of the functions involved and fit parameters in two functions (an analytic change of variables and an additive analytic background) to match known properties measured separately. If this systematic procedure can be generally used as a best practice, rapid progress could be made.

There is also a striking analogy between redundant operators (removing some singular corrections to scaling by reparameterizing the space of predictions) and our normal-form theory (removing analytic corrections to scaling by reparameterizing the control parameters). Can we simultaneously remove both with a joint transformation? How are they related?

In thermodynamics, whether you write the free energies as a function of the external field and temperature or as a function of magnetization and temperature is considered a matter of choice. A Legendre transformation from control parameters (f, t, h) to (u, t, m) swaps the control parameter h with the prediction m: Does it swap normal-form transformations to redundant ones? (Indeed, the NPFRG methods coarse-grain the fixed-magnetization Gibbs free energy $u(t, m)$ rather than Wilson's free energy at fixed field.) When we use our normal-form theory to remove non-linear terms in the RG flows by changing h to $\widetilde{h}(t, h)$, we change the predicted magnetization. Is that change a redundant one?

Hankey and Stanley[46] showed that if you Legendre transform the generalized homogeneous functions we use for hyperbolic fixed points, you get another homogeneous function. Later, others, studying finite-size scaling of the Ising model[47,48] in the microcanonical ensemble predicted that it would be possible to speak of the entropy as a scaling function of the energy instead of the free energy as a function of the temperature.

Our preliminary work on this question ran into a difficulty, generalizing these ideas to the non-linear normal forms necessary in the upper and lower critical dimensions. We have found that it is possible to Legendre transform RG flows at the linear level, but non-linear flows generically lead to non-analyticities in the RG flow.[14,32,i] Whether this clarifies or confuses the correspondence between redundant and normal-form changes of variables needs to be explored.

---

[i]As noted in the footnote on page 418, even in four dimensions, the RG flow of the magnetization is linearizable. Presumably, this means that implementing an RG using flows in (f, t, h) and (u, t, m) will be possible even in four dimensions, while we find (f, t, h) to (S, E, h) leads to non-analytic RG flows.

As a long-time colleague of Michael Fisher, the lead author hopes that Fisher would have enjoyed this deep plunge into the complex history of the field and our ambitions for the future. He also hopes that Fisher's friends, colleagues, and collaborators who are contributing to this book, and of course our colleagues of the future who are reading it, will find this chapter more illuminating than obscure, and more useful than misleading or misguided.

Acknowledgments

JPS would like to thank Matthieu Tissier for his insights and patience explaining the NPFRG methods and for continuing collaborations in this work and Colin Clement for careful early work on the 2D Ising model. JPS benefited by funding from NSF DMR-1719490 and CNRS (the French National Center for Scientific Research). DH was partially supported by an NSF Graduate Research Fellowship Grant No. DGE-2139899. JK-D acknowledges Simons Foundation Grant No. 454943. AR acknowledges support from the Department of Atomic Energy, India under project No. RTI4006, and the Simons Foundation Grant No. 287975.

References

1. A. Aharony and M. E. Fisher, University in analytic corrections to scaling for planar Ising models. *Phys. Rev. Lett.* **45**, 679–682 (September, 1980). doi: 10.1103/PhysRevLett.45.679.
2. A. Aharony and M. E. Fisher, Nonlinear scaling fields and corrections to scaling near criticality. *Phys. Rev. B.* **27**, 4394–4400 (April, 1983). doi: 10.1103/PhysRevB.27.4394.
3. M. E. Fisher. The nature of critical points. In *Boulder Lectures in Theoretical Physics: Statistical Physics, Weak Interactions, Field Theory*, Vol. 7c, pp. 1–159 (University of Colorado Press, 1964).
4. D. Hathcock and J. P. Sethna, Phase transitions beyond criticality: extending Ising universal scaling functions to describe entire phases. https://arxiv.org/abs/2402.185 31, (2024).
5. A. Raju, C. B. Clement, L. X. Hayden, J. P. Kent-Dobias, D. B. Liarte, D. Z. Rocklin and J. P. Sethna, Normal form for renormalization groups. *Phys. Rev. X.* **9**, 021014 (April, 2019). doi: 10.1103/PhysRevX.9.021014.
6. J. Kent-Dobias and J. P. Sethna. Smooth and global Ising universal scaling functions. https://arxiv.org/abs/1707.03791 (2021).
7. W. H. Press, S. A. Teukolsky, W. T. Vetterling and B. P. Flannery, *Numerical Recipes 3rd Edition: The Art of Scientific Computing* (Cambridge University Press, 2007).
8. N. Dupuis, L. Canet, A. Eichhorn, W. Metzner, J. Pawlowski, M. Tissier and N. Wschebor, The nonperturbative functional renormalization group and its applications. *Phys. Rep.* **910**, 1–114 (2021). ISSN 0370-1573. https://doi.org/10.1016/j.physrep.2021.01.001. The nonperturbative functional renormalization group and its applications.
9. F. J. Wegner, Corrections to scaling laws. *Phys. Rev. B.* **5**, 4529 (1972).
10. F. J. Wegner and E. K. Riedel, Logarithmic corrections to the molecular-field behavior of critical and tricritical systems. *Phys. Rev. B.* **7**, 248 (1973).

11. J. Cardy, *Scaling and Renormalization in Statistical Physics*, Vol. 5 (Cambridge University Press, 1996).
12. S. Wiggins, *Introduction to Applied Nonlinear Dynamical Systems and Chaos*, Vol. 2, (Springer Science & Business Media, 2003).
13. J. Guckenheimer. (private communication).
14. A. Raju. Aspects of the renormalization group. PhD thesis, Cornell University, (201). https://sethna.lassp.cornell.edu/pubPDF/RajuPhD.pdf.
15. L. X. Hayden, A. Raju and J. P. Sethna, Unusual scaling for two-dimensional avalanches: Curing the faceting and scaling in the lower critical dimension. *Phys. Rev. Res.* **1**, 033060 (October, 2019). doi: 10.1103/PhysRevResearch.1.033060.
16. O. Perkovic, K. A. Dahmen and J. P. Sethna, Disorder-induced critical phenomena in hysteresis: Numerical scaling in three and higher dimensions. *Phys. Rev. B.* **59**, 6106–6119 (1999).
17. L. Onsager, Crystal statistics. I. A two-dimensional model with an order-disorder transition. *Phys. Rev.* **65**, 117–149 (February, 1944). doi: 10.1103/PhysRev.65.117.
18. R. Guida and J. Zinn-Justin, 3D Ising model: The scaling equation of state. *Nucl. Phys. B* **489**(3), 626–652 (1997).
19. M. Caselle, M. Hasenbusch, A. Pelissetto and E. Vicari, The critical equation of state of the two-dimensional Ising model. *J. Phys. A: Math. Gen.* **34**(14), 2923–2948 (March, 2001). doi: 10.1088/0305-4470/34/14/302.
20. P. Schofield, Parametric representation of the equation of state near a critical point. *Phys. Rev. Lett.* **22**(12), 606–608 (March, 1969). doi: 10.1103/physrevlett.22.606.
21. M. E. Fisher, S.-y. Zinn and P. J. Upton, Trigonometric models for scaling behavior near criticality. *Phys. Rev. B.* **59**(22), 14533–14545 (June, 1999). doi: 10.1103/physrevb.59.14533.
22. C. N. Yang and T. D. Lee, Statistical theory of equations of state and phase transitions. I: Theory of condensation. *Phys. Rev.* **87**(3), 404–409 (August, 1952). doi: 10.1103/physrev.87.404.
23. T. D. Lee and C. N. Yang, Statistical theory of equations of state and phase transitions. ii: Lattice gas and Ising model. *Phys. Rev.* **87**(3), 410–419 (August, 1952). doi: 10.1103/physrev.87.410.
24. I. Balog, G. Tarjus and M. Tissier, Dimensional reduction breakdown and correction to scaling in the random-field Ising model. *Phys. Rev. E.* **102**, 062154 (December, 2020). doi: 10.1103/PhysRevE.102.062154.
25. A. Gorbunova, G. Balarac, L. Canet, G. Eyink and V. Rossetto, Spatio-temporal correlations in three-dimensional homogeneous and isotropic turbulence. *Phys. Fluids* **33**(4), 045114 (2021). doi: 10.1063/5.0046677.
26. G. De Polsi, I. Balog, M. Tissier and N. Wschebor, Precision calculation of critical exponents in the O(N) universality classes with the nonperturbative renormalization group. *Phys. Rev. E.* **101**, 042113 (April, 2020). doi: 10.1103/PhysRevE.101.042113.
27. Y. Chan, A. J. Guttmann, B. Nickel and J. Perk, The Ising susceptibility scaling function, *Journal of Statistical Physics.* **145**, 549–590 (2011).
28. B. Nickel, On the singularity structure of the 2D Ising model susceptibility. *J. Phys. A: Math. Gen.* **32**(21), 3889–3906 (January, 1999). doi: 10.1088/0305-4470/32/21/303.
29. W. Orrick, B. Nickel, A. Guttmann and J. H. Perk, The susceptibility of the square lattice Ising model: new developments. *J. Stat. Phys.* **102**, 795–841 (2001).
30. M. Assis, J. L. Jacobsen, I. Jensen, J.-M. Maillard and B. M. McCoy, Analyticity of the Ising susceptibility. *J. Phys. A: Math. Theor.* **50**(36), 365203 (August, 2017). doi: 10.1088/1751-8121/aa8128.

31. M. Caselle, M. Hasenbusch, A. Pelissetto and E. Vicari, Irrelevant operators in the two-dimensional Ising model. *J. Phys. A: Math. Gen.* **35**(23), 4861 (May, 2002). doi: 10.1088/0305-4470/35/23/305.
32. C. B. Clement. Inference and Modeling in Physics: Image Reconstruction, Spin Glasses, and the Renormalization Group. PhD thesis, Cornell University, (2019). https://sethna.lassp.cornell.edu/pubPDF/ClementPhD.pdf.
33. A. Connelly, G. Johnson, F. Rennecke and V. V. Skokov, Universal location of the Yang-Lee edge singularity in O(N) theories. *Phys. Rev. Lett.* **125**, 191602 (November, 2020). doi: 10.1103/PhysRevLett.125.191602.
34. G. D. Polsi, G. Hernández-Chifflet and N. Wschebor, Precision calculation of universal amplitude ratios in o(n) universality classes: Derivative expansion results at order $\wr(\partial^4)$. *Phys Rev. E.* **104**(6), 064101 (12, 2021). doi: 10.1103/physreve.104.064101.
35. A. Raju and J. P. Sethna. Reexamining the renormalization group: Period doubling onset of chaos. https://arxiv.org/abs/1807.09517, (2018).
36. D. Rand, S. Ostlund, J. Sethna and E. D. Siggia, Universal transition from quasiperiodicity to chaos in dissipative systems. *Phys. Rev. Lett.* **49**, 132–135 (Jul, 1982). doi: 10.1103/PhysRevLett.49.132.
37. S. Ostlund, D. Rand, J. P. Sethna and E. D. Siggia, Universal properties of the transition from quasi-periodicity to chaos in dissipative systems. *Physica D.* **8**, 303–342 (1983).
38. F. Wegner, Some invariance properties of the renormalization group. *J. Phys. C: Solid State Phys.* **7**(12), 2098 (1974).
39. G. Murthy and R. Shankar, Redundant operators for Ising spins. *Phys. Rev. B.* **32** (9), 5851 (1985).
40. R. Shankar and R. Gupta, Clear evidence of redundant operators in monte carlo studies of the Ising model. *Phys. Rev. B.* **32**(9), 6084 (1985).
41. G. Pawley, R. Swendsen, D. Wallace and K. Wilson, Monte carlo renormalization-group calculations of critical behavior in the simple-cubic Ising model *Phys. Rev. B* **29**(7), 4030 (1984).
42. R. H. Swendsen, Optimization of real-space renormalization-group transformations, *Phys. Rev. Lett.* **52**(26), 2321 (1984).
43. M. Barma and M. E. Fisher, Corrections to scaling and crossover in two-dimensional Ising and scalar-spin systems. *Phys. Rev. Lett.* **53**(20), 1935 (1984).
44. M. Barma and M. E. Fisher, Two-dimensional Ising-like systems: Corrections to scaling in the klauder and double-gaussian models. *Phys. Rev. B.* **31**(9), 5954 (1985).
45. H. W. Blöte and M. P. den Nijs, Corrections to scaling at two-dimensional Ising transitions. *Phys. Rev. B.* **37**(4), 1766 (1988).
46. A. Hankey and H. E. Stanley, Systematic application of generalized homogeneous functions to static scaling, dynamic scaling, and universality. *Phys. Rev. B.* **6**(9), 3515 (1972).
47. R. C. Desai, D. W. Heermann and K. Binder, Finite-size scaling in a microcanonical ensemble. *J. Stat. Phys.* **53**, 795–823 (1988).
48. M. Kastner, M. Promberger and A. Hüller, Microcanonical finite-size scaling. *J. Stat. Phys.* **99**, 1251–1264 (2000).

https://doi.org/10.1142/9789811282386_0027

Chapter 27

Scaling Behaviors at Quantum and Classical First-Order Transitions

A. Pelissetto[*,‡] and E. Vicari[†,§]

[*] *Università di Roma Sapienza and INFN Sezione di Roma I, I-00185 Roma, Italy*

[†] *Università di Pisa and INFN, Largo Pontecorvo 3, I-56127 Pisa, Italy*

[‡] *andrea.pelissetto@roma1.infn.it*

[§] *ettore.vicari@unipi.it*

We consider quantum and classical first-order transitions, at equilibrium and under out-of-equilibrium conditions, mainly focusing on quench and adiabatic protocols. For these phenomena, we review the finite-size scaling theory appropriate to describe the general features of the large-scale (and long-time for dynamic phenomena) behavior of finite-size systems.

1. Introduction

Classical and quantum phase transitions (PTs) are phenomena of great interest in modern physics both theoretically and experimentally. Classical PTs are generally driven by thermal fluctuations,[1–8] while their quantum counterparts arise from quantum fluctuations in the limit of zero temperature.[9,10] Transitions separate qualitatively different phases of matter, each one of them being characterized by distinctive properties. In many cases, PTs are associated with the spontaneous breaking of a global symmetry and the emergence of corresponding ordering phenomena. Some transitions in systems with gauge symmetries are notable exceptions. In this case, the transitions are not characterized by a local order parameter associated with a global symmetry-breaking, but are driven by topological excitations.[11–14]

At PTs, in the infinite-volume thermodynamic limit, the free energy, or, in the case of zero-temperature quantum phase transitions (QPTs), the low-energy properties become non-analytic functions of the system parameters, such as the temperature, the coupling constants, etc.[2–10,15–23] Depending on the nature of the non-analyticity at the transition point, PTs are generally distinguished as first-order or continuous PTs. They are continuous when the bulk properties change continuously at the transition point, and the correlation functions develop a divergent length scale. They are of first order when the thermodynamic or ground-state properties in the infinite-volume (thermodynamic) limit are discontinuous across the transition point. See, for example, Ref. 20 for an introductory review of the main features of classical first-order PTs at finite temperature.

As already remarked in Ref. 20, the behaviors emerging at first-order PTs are more complex than those observed at continuous transitions. Critical phenomena at continuous PTs are essentially related to the presence of critical correlations, which decay as a power of the distance at the critical point, and of a diverging length scale ξ. When approaching the critical point, the long-distance behavior, i.e., on distances of the order of the scale ξ, shows universal features that only depend on few global properties of the microscopic short-distance interactions. The behavior at first-order transitions is instead more complex. Indeed, we typically observe the coexistence of one or more ordered phases, i.e., of phases in which the system variables are correlated on all length scales, apart from some short-range local fluctuations. The presence of coexisting phases gives rise to peculiar competing phenomena in the phase-coexistence region, such as metastability, nucleation, droplet formation, and coarsening.[20,24] Moreover, the observed bulk behavior crucially depends on the nature of the boundary conditions (BCs), at variance with what happens at continuous transitions, where BCs only affect some finite-size properties of the system, but are irrelevant in the thermodynamic limit.

In general, the singular behavior at PTs only occurs in the infinite-volume limit. If the size L of the system is finite, all properties are analytic as a function of the external parameter driving the transition. However, close to the transition point and for large system sizes, the finite-size behavior of thermodynamic quantities and long-distance properties shows some general scaling features. At continuous transitions, statistical systems develop finite-size scaling (FSS) behaviors when L is large and the length scale ξ of the critical modes is comparable with the size L of the system. More precisely, this occurs in the FSS limit, defined as the limit $L, \xi \to \infty$, keeping the ratio L/ξ fixed. This regime presents universal features, shared by all systems whose transition belongs to the same universality class.[22,23,25–29] Although originally formulated in the classical framework,[25] FSS also holds at continuous quantum transitions (CQTs).[9,23,29] Peculiar finite-size behaviors emerge at classical and quantum first-order transitions as well.[20,23,30–40] At variance with continuous transitions, the qualitative behavior varies with the BC, since boundaries drastically affect the bulk behavior of the system (for instance, they may induce domain walls). The sensitivity to the BCs represents one of the main qualitative differences between continuous and first-order PTs. Understanding finite-size effects in systems close to PTs is essential to correctly interpret experimental or numerical data, when phase transitions are investigated in relatively small systems,[20,22,23,25–29,39–41] or in particle systems trapped by external forces entailing an additional length scale, as in experiments with cold atoms confined by an external potential.[42–44]

In this chapter, we analyze several aspects characterizing quantum and classical first-order transitions (FOQT and FOCT, respectively) in equilibrium and out-of-equilibrium conditions. Our presentation is meant to give a unified picture of quantum and classical first-order transitions, not only emphasizing the deep connections between them, which can be understood using the so-called quantum-to-classical mapping,[10,23,29] but also stressing their crucial differences, which are due to the different nature of the fluctuations driving the transitions. We discuss the finite-size behavior at FOQTs and FOCTs, focusing on the interplay between the thermal and/or quantum fluctuations driving the PT and the size of the system, and on the role of the BC. Somewhat unconventionally, we begin our presentation by reviewing results for FOQTs, as in this case the FSS behavior emerges more clearly. Then, we extend the discussion to FOCTs. In particular, we use the general quantum-to-classical mapping of d-dimensional quantum systems onto D-dimensional ($D = d + 1$) classical systems in an anisotropic geometry to relate the quantum and classical results.

We stress that a good theoretical understanding of the behavior at first-order transitions is phenomenologically relevant, as they are ubiquitous. For example, the condensation of water, the melting of ice, etc. are crucially important FOCTs. FOQTs occur in quantum Hall systems,[45] itinerant ferromagnets,[46,47] heavy fermion metals,[48–50] SU(N) magnets,[51,52] quantum spin systems,[39,53,54] etc.

This chapter is organized as follows[a]:

• In Section 2, we discuss the equilibrium FSS behavior at the FOQTs occurring in the paradigmatic quantum Ising and Potts models on cubic-like lattices. We point out that intensive global quantities show a universal FSS behavior that can be parametrized in terms of the parameter driving the FOQT, the volume of the system and the energy gap (energy difference of the lowest states of the spectrum). Since the energy gap has a size dependence that is very sensitive to the BC (it can decrease exponentially or as a power of the size), the observed FSS behavior varies significantly with the BC, a specific feature of FOQTs.

• In Section 3, we show how the equilibrium FSS results can be extended to out-of-equilibrium phenomena at FOQTs. We define the FSS variables appropriate to describe the system behavior when the Hamiltonian parameters are suddenly changed (quench protocols), or when they are slowly (quasi-adiabatically) changed across the transition, as in Kibble–Zurek protocols.[55,56] The timescale of these out-of-equilibrium processes is also very sensitive to BC. In particular, if the BC are such that the spectrum is characterized by a gap that is exponentially small in the size of the system, the dynamic FSS functions can be obtained from a simpler quantum-mechanical model that only encodes the evolution of the lowest-energy states.

[a]Here, we provide the list of acronyms used in this chapter: PT (phase transition), CQT (continuous quantum transition), FOQT (first-order quantum transition), FOCT (first-order classical transition), RG (renormalization group), FSS (finite-size scaling), BC (boundary conditions), PBC (periodic BC), ABC (antiperiodic BC), OBC (open BC), FBC (fixed BC), OFBC (opposite FBC), EFBC (equal FBC).

• In Section 4, we discuss the quantum-to-classical mapping, which allows one to relate phenomena in d-dimensional quantum systems to D-dimensional ($D = d+1$) classical statistical systems at equilibrium.

• In Section 5, we outline the main features of the equilibrium FSS behavior at FOCTs, pointing out the analogies between the behavior of classical and quantum systems. Again, the sensitivity of the large-scale behavior to the nature of the BC emerges as a characterizing qualitative feature, distinguishing FOCTs from classical continuous transitions.

• In Section 6, we discuss some selected topics on the dynamics of classical systems close to FOCTs. We first discuss the equilibrium dynamics, focusing on the purely relaxational dynamics.[5,57] Then, we consider two different out-of-equilibrium dynamics, which are the classical analogue of those discussed in the quantum setting in Section 3. In particular, we present some recent results for the dynamics of the droplet formation in Potts systems using the Kibble–Zurek dynamics. We do not discuss coarsening phenomena, as they are already outlined in Ref. 24.

2. First-Order Quantum Transitions

Quantum PTs are striking signatures of many-body collective quantum behaviors.[9,10] They are continuous when the ground state of the system changes continuously at the transition point and correlation functions develop a divergent length scale. They are instead of first order when the ground-state properties are discontinuous across the transition point. FOQTs occur when the lowest-energy states cross in the infinite-volume limit. Since the ground state is different at the two sides of the transition, the physical properties change discontinuously. Note that, in the absence of conservation laws, a level crossing can only occur in the infinite-volume limit. In a finite system, the degeneracy is lifted and the level crossing is avoided.

At FOQTs, the low-energy properties are particularly sensitive to the BC, giving rise to a variety of behaviors, which are even more diverse than those occurring at CQTs.[9,10,23,29] Indeed, depending on the BC, for example, whether they are neutral or favor one of the phases, one can obtain qualitatively different static and dynamic properties.[39,40,58–62]

To understand the origin of the variety of behaviors observed at FOQTs, it is useful to discuss the size behavior of the gap Δ, which is the difference of the energies of the lowest Hamiltonian eigenstates, since it provides the relevant energy scale at the transition. At CQTs, Δ always scales as L^{-z}, where z is the universal dynamic exponent, independent of the BCs,[29] which only determine the amplitude of the asymptotic power law. On the other hand, at FOQTs, the size dependence of Δ varies significantly with the BCs.[39,59,63] For some BC the gap decays exponentially with increasing L, as $\Delta \sim e^{-cL^d}$. Instead, Δ scales as $L^{-\kappa}$ (where κ is a positive power that is usually an integer number), when BCs favor the presence of domain walls, which separate the space into distinct regions associated with different phases. The sensitivity of the large-size behavior of Δ and, in general, of all low-energy properties of the system to the BCs may be considered as the main difference between the behaviors of finite-size systems at CQTs and FOQTs.

In the following, we discuss the main features of the FSS behavior at FOQTs using the quantum Ising and Potts models as paradigmatic models. We first discuss FSS for Ising systems with neutral BCs, where one can predict the large-scale behavior using an effective model in which only the relevant low-energy states are considered. Then, we consider BCs that give rise to domain walls, obtaining a substantially different picture, and BCs that favor one of the two phases. Analogous issues are discussed in quantum Potts models.

2.1. *Quantum Ising model*

2.1.1. *Definition of the model*

To make our discussion concrete, we consider the paradigmatic quantum d-dimensional Ising model on a d-dimensional cubic-like lattice of linear size L. The Hamiltonian is

$$\hat{H} = -J \sum_{\langle \boldsymbol{x},\boldsymbol{y}\rangle} \hat{\sigma}^{(1)}_{\boldsymbol{x}} \hat{\sigma}^{(1)}_{\boldsymbol{y}} - g \sum_{\boldsymbol{x}} \hat{\sigma}^{(3)}_{\boldsymbol{x}} - h \sum_{\boldsymbol{x}} \hat{\sigma}^{(1)}_{\boldsymbol{x}}, \tag{1}$$

where $\hat{\sigma}^{(k)}$ are the usual spin-1/2 Pauli matrices ($k = 1, 2, 3$), the first sum is over all bonds connecting nearest-neighbour sites $\langle \boldsymbol{x}, \boldsymbol{y}\rangle$, and the other two sums are over all sites $\boldsymbol{x}$. The parameters g and h represent the external homogeneous transverse and longitudinal fields, respectively. In the following we set the Planck constant $\hbar = 1$. Without loss of generality, we assume $J = 1$, $g > 0$, and a lattice spacing $a = 1$. For $h = 0$, the infinite-volume Ising model is symmetric under the $\mathbb{Z}_2$ transformation $\sigma^{(i)}_{\boldsymbol{x}} \to U \sigma^{(i)}_{\boldsymbol{x}} U^\dagger$ with $U = \prod_{\boldsymbol{x}} \sigma^{(3)}_{\boldsymbol{x}}$, which maps $\sigma^{(1)}_{\boldsymbol{x}} \to -\sigma^{(1)}_{\boldsymbol{x}}$ and $\sigma^{(3)}_{\boldsymbol{x}} \to \sigma^{(3)}_{\boldsymbol{x}}$.

For $h = 0$, the model undergoes a CQT at a critical point $g = g_c$ ($g_c = 1$ in one dimension), separating a disordered phase ($g > g_c$) from an ordered one ($g < g_c$). The order parameter is the longitudinal magnetization $m_{\boldsymbol{x}} \equiv \langle \Psi_0(g,h)|\hat{\sigma}^{(1)}_{\boldsymbol{x}}|\Psi_0(g,h)\rangle$, where $|\Psi_0(g,h)\rangle$ is the ground state for the given g and h. The parameters $r \equiv g - g_c$ and h represent the (even and odd, respectively) relevant perturbations of the critical behavior at the CQT. For $g < g_c$, the longitudinal field h drives FOQTs, where the (average) longitudinal magnetization M becomes discontinuous in the infinite-volume limit:

$$\lim_{h\to 0^\pm} \lim_{L\to\infty} M = \pm m_0, \quad M \equiv L^{-d} \sum_{\boldsymbol{x}} m_{\boldsymbol{x}}, \quad m_{\boldsymbol{x}} \equiv \langle \Psi_0(g,h)|\hat{\sigma}^{(1)}_{\boldsymbol{x}}|\Psi_0(g,h)\rangle. \tag{2}$$

The FOQT separates two different phases characterized by opposite values of the magnetization m_0. The magnetization for $d = 1$ is known:[64] $m_0 = (1 - g^2)^{1/8}$.

Several BCs can be considered: (i) Periodic BCs (PBCs), for which $\hat{\sigma}^{(k)}_{\boldsymbol{x}} = \hat{\sigma}^{(k)}_{\boldsymbol{x}+L\vec{\mu}}$ ($\vec{\mu}$ indicates a generic lattice direction); (ii) Antiperiodic BCs (ABCs), for which $\hat{\sigma}_{\boldsymbol{x}}^{(k)} = -\hat{\sigma}_{\boldsymbol{x}+L\vec{\mu}}^{(k)}$; (iii) Open BCs (OBCs); (iv) Fixed BCs (FBCs), where the states of the spins on the lattice boundaries are fixed, typically choosing one of the eigenstates of the longitudinal spin operator $\hat{\sigma}^{(1)}$; (v) OBCs with boundary fields. For BCs of type (iv) and (v), we should additionally specify the behavior at the boundary. In one dimension, one can consider equal FBCs (EFBCs) — in this case, the spins on the two opposite boundaries are eigenstates of $\hat{\sigma}^{(1)}$ with the same eigenvalue — or opposite FBCs (OFBCs) — the boundary spins are eigenstates of $\hat{\sigma}^{(1)}$ with opposite eigenvalues. Analogously, we may consider systems with equal or opposite transverse boundary fields. Note that the finite system is translation invariant if PBCs and ABCs are used. On the other hand, in systems with OBCs, FBCs or with boundary fields, translation invariance is lost.

As we shall see, the FSS behavior depends on the behavior of the BCs with respect to the $\mathbb{Z}_2$ symmetry. We distinguish neutral BCs that do not break the $\mathbb{Z}_2$ symmetry and therefore do not favor any of the two phases and non-neutral BCs. In particular, PBCs, ABCs, and OBCs are neutral BCs along the FOQT line of Ising systems, while generic FBCs are not. In one dimension, OFBCs (or systems with opposite longitudinal boundary fields) are not $\mathbb{Z}_2$-invariant. However, a finite-size system defined on a lattice with $1 \leq x \leq L$ is still symmetric under combined $\mathbb{Z}_2$ and space-reflection transformations: Under a space reflection, $\sigma_x^{(i)} \to \sigma_{L-x+1}^{(i)}$. Thus, OFBCs can also be considered as neutral BCs.

2.1.2. *The finite-size gap at the quantum transitions of Ising systems*

Both at CQTs and FOQTs, the relevant energy scale is the gap Δ. At the CQT, i.e., for $g = g_c$ and $h = 0$, the gap $\Delta(L)$ for a system of size L scales as $\Delta(L) \sim L^{-z}$, with $z = 1$ for any spatial dimension d. In particular, the dynamic exponent z is independent of the BC. The amplitudes instead depend on the BC. For example, in one dimension, the gap $\Delta(L)$ at the critical point behaves as[29,59,65]

$$\Delta_{\rm PBC} = \frac{\pi}{2L} + O(L^{-2}), \quad \Delta_{\rm OBC} = \frac{\pi}{L} + O(L^{-2}), \quad \Delta_{\rm ABC} = \frac{3\pi}{2L} + O(L^{-2}), \tag{3}$$

for PBCs, OBCs, and ABCs, respectively. The behavior changes along the FOQT line $g < g_c$ and $h = 0$. Here, BCs play a crucial role in determining the FOQT behavior, as they control the bulk behavior of the system. We will distinguish three cases: (i) neutral BCs with a magnetized ground state; (ii) neutral BCs with domain walls; (iii) non-neutral BCs.

We consider first neutral BCs, such that the magnetized states $|+\rangle$ and $|-\rangle$ are the only relevant low-energy states for large values of L. This is the case of PBCs and OBCs. In this case, for $L \to \infty$ and $h = 0$, the magnetized states $|+\rangle$ and $|-\rangle$ cross at the transition and satisfy,

$$\langle +|\hat{\sigma}_{\boldsymbol{x}}^{(1)}|+\rangle = m_0, \quad \langle -|\hat{\sigma}_{\boldsymbol{x}}^{(1)}|-\rangle = -m_0, \tag{4}$$

independent of $\boldsymbol{x}$. The difference of the energies of the ground state and of the higher excited states remains finite in the infinite-volume limit. The degeneracy of the magnetized states is lifted by the longitudinal field h. Therefore, $h = 0$ is a FOQT point, where the longitudinal magnetization M becomes discontinuous, see Eq. (2). In a finite system of size L, the two lowest-energy states are superpositions of the two magnetized states. Due to tunneling effects, the gap $\Delta(L, h)$ for $h = 0$ is finite — the level crossing is avoided — but vanishes exponentially as L increases,[32,39]

$$\Delta(L, h = 0) \equiv E_1(L, h = 0) - E_0(L, h = 0) \sim e^{-cL^d}, \tag{5}$$

where c is a positive constant that depends on g (multiplicative powers of L may also be present). In particular, for a quantum Ising chain with $g < 1$ and $h = 0$, the gap at the transition point $h = 0$ scales as [64,66]

$$\Delta_{\rm OBC} = 2(1 - g^2)g^L[1 + O(g^{2L})], \quad \Delta_{\rm PBC} \approx 2\sqrt{\frac{1 - g^2}{\pi L}}\, g^L \tag{6}$$

for OBCs and PBCs, respectively. The differences $\Delta_n(L, h) \equiv E_n(L, h) - E_0(L, h)$ for the higher excited states $n > 1$ approach finite values for $L \to \infty$.

A second class of BCs are neutral BCs that induce the presence of domain walls. We consider first the one-dimensional case. This type of behavior can be observed for OFBCs, or for systems with sufficiently strong oppositely directed boundary fields. In this case, the magnetized states are no longer the relevant low-energy states. Instead, the lowest-energy states are domain walls (kinks), which, for small values of g, are simply characterized by the presence of nearest-neighbor pairs of antiparallel spins. Linear combinations of kink states can be considered as one-particle states with $O(L^{-1})$ momenta. Hence, there is an infinite number of excitations with a gap of order L^{-2} and, in the infinite-volume limit, the ground-state degeneracy is infinite. For the quantum Ising chain, the gap is exactly known for OFBCs at $h = 0$. We have [39,59,66]

$$\Delta_{\rm OFBC} = \frac{3g}{1 - g}\frac{\pi^2}{L^2} + O(L^{-3}). \tag{7}$$

An interesting limiting case is provided by systems with ABC or in the presence of oppositely directed boundary fields $\pm h_b$ for a particular value $h_b = h_{b,c}$ (the behavior for $h \approx h_{b,c}$ shows interesting scaling properties that are discussed in Refs. 59 and 63). In this case, both kink states and magnetized states are degenerate at the FOQT in the infinite-volume limit. The gap still behaves as L^{-2}, and, in particular, we have[39,59,66]

$$\Delta_{\rm ABC} = \frac{g}{1 - g}\frac{\pi^2}{L^2} + O(L^{-3}) \tag{8}$$

for ABCs. In more than one dimensions, this type of behavior can be obtained by using different BCs in the different directions. For instance, one could use ABCs or OFBCs in the x-direction and PBCs or OBCs in the remaining $(d - 1)$-directions. This choice would induce domain walls perpendicular to the x-axis.

Finally, we consider non-neutral BCs, for instance, equal and fixed BCs (EFBCs) which favor one of the two magnetized phases. They can be obtained by restricting the Hilbert space to states $|s\rangle$ such that $\hat{\sigma}_1^{(1)}|s\rangle = -|s\rangle$ and $\hat{\sigma}_L^{(1)}|s\rangle = -|s\rangle$ at the boundaries. The interplay between the size L and the bulk longitudinal field h turns out to be more complex than that observed with neutral BCs.[67] In the case of EFBCs, for small values of h, observables depend smoothly on h, up to $h_{\rm tr}(L) \approx c/L^d$ (c is a g-dependent constant), where a sharp transition to the oppositely magnetized phase occurs, see Fig. 1. The field $h_{\rm tr}(L)$ can be identified as the value of h where the gap $\Delta(L,h)$ between the two lowest states is the smallest. The minimum $\Delta_{\rm min}(L) = \Delta(L, h_{\rm tr})$ decreases exponentially with increasing L, i.e., $\Delta_{\rm min}(L) \sim e^{-bL}$. Note that the infinite-volume limit and the limit $h \to 0$ do not commute. Indeed, the gap $\Delta(L,0)$ is finite for $L \to \infty$.

2.2. *Equilibrium finite-size scaling for neutral boundary conditions*

At the FOQTs, the low-energy properties satisfy general FSS laws as a function of the temperature T, the field h, and the system size L. We will begin here by discussing the behavior for neutral BCs. In this case, for $h = 0$, the $\mathbb{Z}_2$ symmetry is not broken by the BC. If $h \neq 0$, the $\mathbb{Z}_2$ transformation maps $h \to -h$, and all

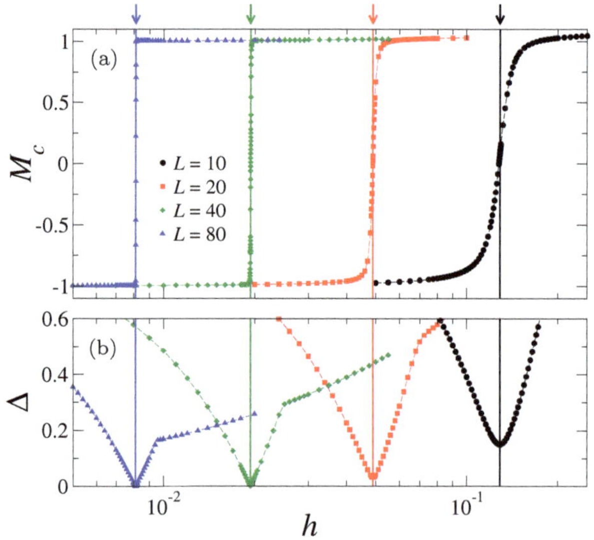

Fig. 1. The renormaized local magnetization in the middle of the system, $M_c \equiv m_{L/2}/m_0$ [top, panel (a)] and the energy gap Δ [bottom, panel (b)] in the quantum Ising chain with EFBC, as a function of the longitudinal field h, for a fixed transverse field $g = 0.8$, and several system sizes. The continuous vertical lines and arrows denote the magnetic fields $h_{\rm tr}(L)$ corresponding to the minimum of the energy gap. Results are taken from Ref. 67.

low-energy properties have well-defined properties under $h \to -h$. For instance, the gap Δ is even, while the local magnetization is odd under this transformation.

2.2.1. *General scaling behavior*

To describe the scaling behavior of the low-energy properties as h is turned on, we should identify the appropriate scaling variables. In a quantum system, the relevant energy scale is the gap Δ, and thus, we expect the interplay between h and the size L to be controlled by the ratio Φ between the energy variation δE_h due to addition of the longitudinal field h,

$$\delta E_h(L,h) \equiv E(L,h) - E(L,h=0), \tag{9}$$

and the gap $\Delta(L)$ at $h=0$.[39] Thus, we define the scaling variable

$$\Phi \equiv \delta E_h(L,h)/\Delta(L), \quad \Delta(L) \equiv \Delta(L,h=0). \tag{10}$$

If the BCs are such that the lowest-energy states are the magnetized states, we have[39]

$$\delta E_h(L,h) = 2m_0 h L^d \tag{11}$$

for sufficiently small h, where m_0 is the spontaneous magnetization, obtained approaching the transition point $h \to 0$ after the infinite-volume limit, see Eq. (2). Analogously, the scaling variable controlling the interplay between T and L is the ratio

$$\Xi \equiv T/\Delta(L). \tag{12}$$

The FSS limit corresponds to $L \to \infty$ and $h, T \to 0$, keeping Φ and Ξ fixed. In this limit, the gap Δ is expected to behave as[39]

$$\Delta(L,h) \approx \Delta(L)\mathcal{E}(\Phi). \tag{13}$$

By definition $\mathcal{E}(0) = 1$, and $\mathcal{E}(\Phi) \sim |\Phi|$ for $\Phi \to \pm\infty$, in order to reproduce the expected linear behavior $\Delta(L,h) \sim |h|L^d$ for sufficiently large $|h|$. Analogously, the local and global magnetization are expected to scale as

$$m_{\boldsymbol{x}}(L,T,h) \approx m_0 \mathcal{M}_x(\boldsymbol{x}/L,\Xi,\Phi), \quad M(L,T,h) \approx m_0 \mathcal{M}(\Xi,\Phi). \tag{14}$$

Due to the definition (2) of m_0, we have $\mathcal{M}(0,\Phi) \to \pm 1$ for $\Phi \to \pm\infty$. Moreover, $\mathcal{M}(\Xi,0)$ vanishes because of the $\mathbb{Z}_2$ invariance. For translation-invariant BCs (for instance, for PBCs), $m_{\boldsymbol{x}}(L,T,h) = M(L,T,h)$ and $\mathcal{M}_x(\boldsymbol{x}/L,\Xi,\Phi) = \mathcal{M}(\Xi,\Phi)$. The above FSS ansätze represent the simplest scaling behaviors compatible with the discontinuities arising in the thermodynamic limit.

The previous scaling relations apply generally to any type of neutral BCs, as they simply rely on the fact that the gap is the relevant energy scale. Therefore, also in the presence of domain walls, we expect Φ and Ξ, as defined in Eqs. (10) and (12), to be the relevant scaling variables. As an example, in Fig. 2, we report

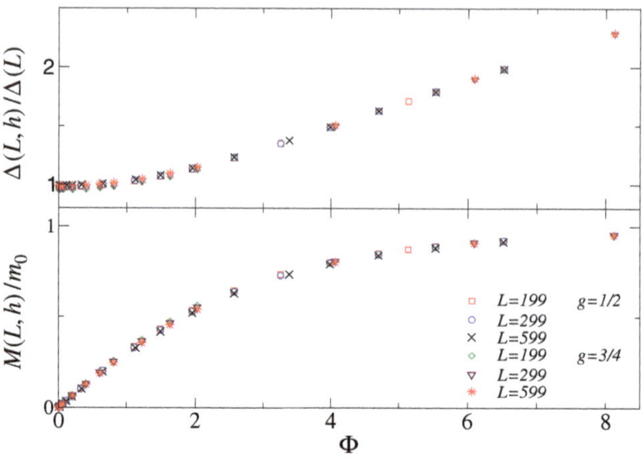

Fig. 2. Rescaled energy gap and magnetization for the quantum Ising chain with OFBCs at $g = 1/2$ and $g = 3/4$. The two panels show the ratios $\Delta(L,h)/\Delta(L)$ (top) and $M(L,h)/m_0$ (bottom) versus $\Phi = 2m_0hL/\Delta(L)$ (for OFBCs $\Delta(L) \sim 1/L^2$). The data approach non-trivial scaling curves with increasing L, which are independent of g, supporting Eqs. (13) and (14). Results are taken from Ref. 39.

numerical results for the quantum Ising chain with OFBCs for which $\Delta \sim L^{-2}$ at the transition point. In this case, the lowest-energy states are domain walls, see Section 2.1.2. Data satisfy the scaling relations (14) quite precisely.

We remark that the scaling behavior drastically depends on the nature of the BCs, if one expresses the scaling variables in terms of L, h, and T, given that the size behavior of $\Delta(L)$ at criticality depends crucially on the BCs. Indeed, the gap Δ may have both an exponential and a power-law dependence on L, as discussed in Section 2.1.2. For instance, for the quantum Ising chain, one has the following: (i) $\Phi \sim hLe^{cL}$ for OBC; (ii) $\Phi \sim hL^{3/2}e^{cL}$ for PBCs (c is a non-universal positive constant); (iii) $\Phi \sim hL^3$ for ABCs or OFBCs. This sensitivity to the BCs is one of the main features distinguishing FOQTs and CQTs in finite-size systems.

It is worth noting that the scaling behaviors (13) and (14) in terms of the scaling variables Φ and Ξ defined in Eqs. (10) and (12) also hold at the CQT, for example, at the critical endpoint of the FOQT line. At the CQT, the relevant parameters $r = g - g_c$ and h have RG dimensions $y_r = 1/\nu$ and $y_h = (d+z+2-\eta)/2$ (where η is the exponent associated with the power-law decay of the order-parameter two-point function at the critical point), respectively. If we fix $r = 0$ and vary the longitudinal field h only, the relevant FSS scaling variable is $\Phi_c = h\,L^{y_h}$, which, as we show in the following, can be identified with the scaling variable Φ defined in Eq. (10). Indeed, at the CQT, the energy change δE due to a perturbation $\hat{H}_h = -h\sum_{\boldsymbol{x}} \hat{P}_{\boldsymbol{x}}$, behaves as $\delta E_h(L,h) \sim h\,L^{d-y_p}$, where y_p is the RG critical dimension of the local operator $\hat{P}_{\boldsymbol{x}}$ at the RG fixed point. If $\hat{P}_{\boldsymbol{x}} = \hat{\sigma}^{(1)}_{\boldsymbol{x}}$, y_p satisfies the scaling relation[23,29] $y_h + y_p = d + z$, so that $\delta E_h(L,h) \sim h\,L^{y_h - z}$. Since, at criticality, we have $\Delta(L, h = 0) \sim L^{-z}$, we obtain $\Phi_c \sim L^z \delta E_h(L,h) \sim \Phi$. Analogously, the variable Ξ defined in Eq. (12) is also the relevant one at the CQT. Therefore, the FSS relations (13) and (14) in terms of the scaling variables (10) and (12) hold both to CQTs and FOQTs. Differences are encoded in the size dependence of Φ and Ξ and in the scaling functions.

The scaling behavior at FOQTs has also been investigated in the presence of an external local field $h_{\boldsymbol{x}}$.[39] The general scaling arguments reported above can also be applied here. If the external magnetic field is non-vanishing only at one lattice site x, the energy change is $\delta E \sim 2m_0 h_x$ and the appropriate scaling variable is $\Phi_x = 2m_0 h_x/\Delta(L)$. Numerical results for the Ising chain are in full agreement with these scaling predictions.[39] In the presence of inhomogeneous conditions, for example, when the external fields have a sufficiently smooth space dependence characterized by a further length scale ℓ, some scaling behaviors emerge as well, involving also the spatial position.[68]

2.2.2. *Scaling functions*

We now consider a class of neutral BCs (for instance, PBCs or OBCs), such that the magnetized states represent the lowest-energy excitations. As discussed in Section 2.1.2, at the transition point $\Delta(L) \sim \exp(-cL^d)$, while the energy differences $\Delta_n \equiv E_n - E_0$ associated with the higher excited states ($n > 1$) are finite (more generally, Δ_n/Δ decreases exponentially, with possible power corrections) in the infinite-volume limit.

For sufficiently large L, the low-energy properties in the crossover region $|h| \ll 1$ can be obtained by restricting the theory to the two lowest-energy states $|0\rangle$ and $|1\rangle$ or equivalently to the magnetized states $|+\rangle$ and $|-\rangle$.[39,40] In this restricted Hilbert space, we can consider an effective Hamiltonian given by

$$\hat{H}_2(h) = \varepsilon\hat{\sigma}^{(3)} + \zeta\hat{\sigma}^{(1)}. \tag{15}$$

Here, ε corresponds to the energy change due to the magnetic field h, while ζ is related to the small gap $\Delta(L)$ at $h = 0$,[39] i.e.,

$$\varepsilon = m_0 h L^d, \quad \zeta = \langle -|\hat{H}|+\rangle = \tfrac{1}{2}\Delta(L, h = 0). \tag{16}$$

The scaling variable Φ defined in Eq. (10) can be identified as $\Phi = \varepsilon/\zeta$. By diagonalizing the effective Hamiltonian $\hat{H}_2$, we obtain the ground state

$$|0\rangle = \sin(\alpha/2)|-\rangle + \cos(\alpha/2)|+\rangle, \quad \tan\alpha = \Phi^{-1}, \tag{17}$$

and the energy gap $\Delta(L, h)$ as a function of ε and ζ. Moreover, by taking the matrix element of the operator $\hat{\sigma}^{(3)}$, i.e., $\langle 0|\hat{\sigma}^{(3)}|0\rangle$, we obtain the magnetization of the quantum Ising model.

By matching these results with the scaling equations (13) and (14), one easily obtains the zero-temperature scaling behaviors:[39]

$$\mathcal{E}(\Phi) = \sqrt{1 + \Phi^2}, \quad \mathcal{M}(0, \Phi) = \Phi/\sqrt{1 + \Phi^2}. \tag{18}$$

Note that, for OBC, the above behavior for the local magnetization should hold for lattice sites $\boldsymbol{x}$ sufficiently far from the boundaries. For sufficiently small temperatures, i.e., for $T \sim \Delta$, the scaling function of the magnetization can be obtained by averaging the magnetization with the Gibbs weight, i.e., by considering

$$M = Z_2^{-1}\mathrm{Tr}\big[\hat{\sigma}^{(3)} e^{-\hat{H}_2/T}\big], \quad Z_2 = \mathrm{Tr}\big[e^{-\hat{H}_2/T}\big]. \tag{19}$$

The results reported above have been numerically confirmed for quantum Ising chains with OBC.[39] The convergence to the asymptotic two-level FSS behavior is generally fast, being roughly controlled by the ratio Δ/Δ_n between the gap $\Delta \sim e^{-cL^d}$ and the energy-level differences with the higher states, i.e., $\Delta_n \equiv E_n - E_0$ for $n > 1$, which are finite for $L \to \infty$. These results supposedly apply to any quantum system, in which the FOQT is driven by the competition of two different quantum states that are related by symmetry (the previous expression has been derived under the assumption that $\langle +|\hat{H}|+\rangle = \langle -|\hat{H}|-\rangle$, but this condition can be easily relaxed, see Ref. 67). Moreover, the above approach can be easily extended to systems in which a finite number of states are quasi-degenerate at the FOQT.

When systems at FOQTs develop domain walls, scaling functions cannot be obtained by performing a two-level truncation because the low-energy spectrum at the transition point presents a tower of excited states (the number of these states becomes infinite for $L \to \infty$) with $\Delta_n = E_n - E_0 = O(L^{-2})$. For one-dimensional chains, it was, however, possible to consider an effective model in terms of the relevant excitations (of order L) and determine the exact scaling functions, see Refs. 59 and 63 for details.

2.3. *Equilibrium finite-size scaling for non-neutral boundary conditions*

In the case of non-neutral BCs (in Ising chains, one can consider EFBCs), one should change the definition of the relevant scaling variables. Indeed, as discussed in Ref. 63, no transition is observed by taking the limit $L \to \infty$ at fixed $h = 0$. For instance, for EFBCs, the gap $\Delta(L, h = 0)$ is finite for $L \to \infty$. To observe the competition of the two different states, one should consider the limit $L \to \infty$ along the line $h = h_{\rm tr}(L)$ defined in Section 2.1.2. The corresponding gap $\Delta_{\rm min}(L) = \Delta(L, h_{\rm tr})$ vanishes exponentially. For $h \approx h_{\rm tr}(L)$, a universal FSS behavior emerges. The relevant scaling variable is

$$\Phi_{\rm tr} \equiv \frac{2m_0 L[h - h_{\rm tr}(L)]}{\Delta_{\rm min}(L)}. \tag{20}$$

The scaling variable $\Phi_{\rm tr}$ is the analogue of Φ defined in Eq. (10), with the essential difference that the finite-size pseudotransition now occurs at $h = h_{\rm tr}(L)$ and not at $h = 0$. Therefore, the relevant magnetic energy scale is the difference between the magnetic energy at h and that at $h_{\rm tr}(L)$, while the relevant gap is the one at

$h_{tr}(L)$. For $h \approx h_{tr}(L)$, observables are expected to satisfy FSS relations analogous to those reported before:

$$\Delta(L,h) \approx \Delta_{min}(L)\mathcal{E}_f(\Phi_{tr}), \quad m_c(L,h) \approx m_0\mathcal{M}_c(\Phi_{tr}), \tag{21}$$

where $m_c(L,h)$ is the magnetization at the center of the chain. An analogous relation holds for the average magnetization. These scaling behaviors have been confirmed by the numerical computations reported in Ref. 67. Since, also in this case, only two states are relevant for $L \to \infty$, scaling functions can be computed using a two-level truncation, as discussed in Section 2.2.2. Note that, at variance with what occurs in the PBC or OBC cases, the energy differences Δ_n with the higher states ($n > 1$) vanish as a power of L for $L \to \infty$. Nonetheless, since $\Delta_n/\Delta \to \infty$ for $L \to \infty$, asymptotically, we can still exploit the two-level truncation. The scaling functions for $\Delta(L,h)$ and $m_c(L,h)$ are the same as before, apart from a trivial normalization of the arguments. For the average magnetization $m(L,h)$, the expression differs because of the absence of $\mathbb{Z}_2$ symmetry, see Ref. 67 for details.

Note that the infinite-volume critical point $h = 0$ lies outside the region in which FSS holds. This implies that, in the definition of the scaling variable Φ_{tr}, the values of $h_{tr}(L)$ and $\Delta_{min}(L)$ cannot be replaced with their asymptotic behaviors. Analogous behaviors are expected in other FOQTs, for example, in higher dimensions, when BCs favor one of the two phases.

2.4. *The quantum Potts model*

2.4.1. *Definition of the model*

As a second paradigmatic example, we consider the one-dimensional quantum Potts chain,[69,70] which is the quantum analogue of the classical two-dimensional Potts model,[71–74] defined by the classical Hamiltonian

$$H = -J\sum_{\langle ij\rangle} \delta(s_i, s_j), \tag{22}$$

where the sum is over the nearest-neighbor sites of a square lattice, s_i are spin variables taking q integer values, i.e., $s_i = 1, ..., q$, and $\delta(m,n) = 1$ if $m = n$ and zero otherwise. The corresponding quantum Hamiltonian is derived from the continuous *time* limit of the transfer matrix.[69] In the quantum q-state Potts chain, one considers q states per site and the Hamiltonian[69,70]

$$\hat{H} = -J_q\sum_{j=1}^{L-1}\sum_{k=1}^{q-1} \Omega_j^k\Omega_{j+1}^{q-k} - g\sum_{j=1}^{L}\sum_{k=1}^{q-1} M_j^k - h\sum_{j}\sum_{k=1}^{q} \Omega_j^k, \tag{23}$$

where J_q, g, h are the model parameters, and Ω_j and M_j are $q\times q$ matrices: $\Omega_{mn} = \delta_{mn}e^{i2\pi(n-1)/q}$ and $M_{mn} = \delta_{\text{mod}(m,q)+1,n}$. These matrices commute on different sites and satisfy the algebra:[69] $\Omega_j^k\Omega_j^l = \Omega_j^{k+l}$, $M_j^kM_j^l = M_j^{k+l}$, $\Omega_j^q = M_j^q = \mathbb{I}$, $M_j^k\Omega_j^l = \omega^{kl}\Omega_j^lM_i^k$. For $q = 2$, the model is equivalent to the quantum Ising chain.

For $h = 0$ and fixed J_q, model (23) has a transition at $g = g_c$ that separates two phases: a disordered phase for $g > g_c$ and an ordered one for $g < g_c$. In the latter phase, the system magnetizes along one of the q *directions*. In the infinite-volume limit and for $h = 0$, the quantum Potts chain satisfies a self-duality relation[69] $H(g/J_q) = (g/J_q)H(J_q/g)$. This implies that the transition point g_c is located at $g_c = J_q$. In one dimension, the transition at $g = g_c$ is continuous for $q \leq 4$ and discontinuous for larger values of q. The FOQT at $g_c = J_q$ is somehow qualitatively different from those occurring in quantum Ising models, in which the energy density is continuous at the transition, and only the magnetization shows a discontinuity. Indeed, in the Potts chain, both quantities are discontinuous.

The ground-state magnetization

$$M = \langle \hat{\mu}_x \rangle, \quad \hat{\mu}_x = \frac{1}{q-1}\left(\sum_{k=1}^{q} \Omega_x^k - 1\right), \tag{24}$$

provides an order parameter. For $q > 4$, it vanishes in the disordered phase ($g > g_c$) and, as g is decreased, it jumps discontinuously to a non-zero value at the FOQT. In particular, $m_0 = \lim_{g \to g_c^-} \lim_{h\to 0} \lim_{L\to\infty} M$ is non-zero for $q > 4$.[75,76] The infinite-volume ground-state energy density E is also discontinuous across the FOQT. Indeed, the two limits $E_\pm = \lim_{g\to g_c^\pm} \lim_{L\to\infty} E$ differ for $q > 4$. Their difference $\Delta E \equiv E_+ - E_-$ is the analogue of the latent heat in FOCTs.

Beside the transition at $g_c = J_q$, there is also a line of FOQTs starting at g_c and ending at $g = 0$, driven by the magnetic field h of the Potts Hamiltonian (23). These magnetic transitions are analogous to those discussed in Ising systems.

2.4.2. *Finite-size scaling behavior*

As in quantum Ising models, the finite-size behavior of the gap at the FOQT at $g = g_c$ drastically depends on whether BCs are neutral or not. To specify boundary conditions, we define a boundary Hamiltonian[58,77]

$$\hat{H}_b = -J_q \left[h_1 \sum_{k=1}^{q-1} \Omega_1^k + h_L \sum_{k=1}^{q-1} \Omega_L^k \right], \tag{25}$$

which is added to the bulk Hamiltonian (23). OBCs and EFBCs correspond to considering $h_1 = h_L = 0$ and $h_1 = h_L = 1$. These two types of BCs are not neutral, as they favor the disordered and ordered phases, respectively. In both cases, the gap at $g = g_c$ approaches a constant as $L \to \infty$, so that it is necessary to consider a pseudotransition coupling $g_{\rm tr}(L)$ to observe a vanishing gap in the infinite-volume limit (see the discussion in Section 2.1.2). Neutral BCs are obtained by choosing $h_1 = 1$ and $h_L = 0$.[58,77] In this case, self-duality is preserved in a finite volume, and thus, the gap vanishes at g_c.

The available numerical estimates of the gap $\Delta(L)$ for $q = 10$ are compatible with the power-law behavior[58,68,77] $\Delta(L) \sim L^{-\kappa}$ with κ positive and of order 1.

For neutral BCs, equilibrium FSS is again controlled by the scaling variables Φ and Ξ defined in Eqs. (10) and (12). Thus, in the case at hand, $\Phi = (g - g_c)L/\Delta(L)$, where $\Delta(L)$ is the gap at the transition. In the zero-temperature limit, we expect

$$\Delta(L, g) \approx \Delta(L)\mathcal{D}(\Phi), \quad M(L, g) \approx m_0\mathcal{M}(\Phi). \tag{26}$$

Numerical evidence of the above behaviors was reported in Ref. 58. Similar scalings are expected in the non-neutral case, simply replacing $\Delta(L)$ with the gap at the pseudotransition point $g_{\rm tr}(L)$.

The behavior of the model along the line $g < g_c$, where the FOQTs are driven by the external field h, is analogous to that observed for the quantum Ising model. For $g < g_c$ and $h = 0$, in the infinite-size limit, the ground state is characterized by the degeneracy of q states. The degeneracy is lifted in a finite volume. The behavior of the gap for $h = 0$ depends on the nature of the BC — whether they are neutral or not — and on the nature of the ground states — whether domain walls are present or not. In the OBC case, the energy differences $E_n(h = 0, L) - E_0(h = 0, L)$ vanish exponentially as $L \to \infty$ for all $n < q$. Thus, not only can we predict the correct scaling variables, but we can also compute the scaling functions by performing a truncation of the spectrum: Only the lowest q levels need to be considered.[58] We finally mention that scaling phenomena at the FOQTs of the Potts model have also been investigated under inhomogeneous conditions.[68]

2.5. *The ground-state fidelity and the quantum Fisher information*

Quantum transitions in many-body systems are related to significant qualitative changes of the ground-state and low-excitation properties, induced by small variations of a driving parameter. A proper characterization of their main features may be obtained by using quantum-information concepts, such as the ground-state fidelity and the related fidelity susceptibility.[23,78,79] This quantity plays a central role in quantum estimation theory, as it is proportional to the quantum Fisher information, which, in turn, is related to the inverse of the smallest variance achievable in the estimation of the varying parameter. The fidelity quantifies the overlap between the ground states of quantum systems that share the same Hamiltonian, but with different values of the Hamiltonian parameters.[78,79] Its relevance can be traced back to the Anderson's orthogonality catastrophe:[80] the overlap of two many-body ground states corresponding to Hamiltonians differing by a small perturbation vanishes in the thermodynamic limit. At quantum transitions, the size behavior of the fidelity significantly differs from that observed in normal conditions. Indeed, the fidelity susceptibility is typically much larger in proximity of quantum transitions than under normal conditions, so that metrological performances are believed to drastically improve as a quantum transition is approached.[81,82]

In finite-size quantum Ising models at fixed g and varying h, the ground-state fidelity monitors the changes of the ground-state normalized wave function $|\Psi_0(L,h)\rangle$ when varying the control parameter h by δh. We define the fidelity as

$$A(L,h,\delta h) \equiv \big|\langle\Psi_0(L,h)|\Psi_0(L,h+\delta h)\rangle\big|. \tag{27}$$

Assuming δh small, one can expand Eq. (27) in powers of δh, obtaining[78]

$$A(L,g,h,\delta h) = 1 - \tfrac{1}{2}\delta h^2 \chi_A(L,g,h) + O(\delta h^3), \tag{28}$$

where χ_A is the fidelity susceptibility. The absence of the linear term in the expansion (28) is related to the normalization of the states and is obviously necessary to guarantee that $A \le 1$.

In normal conditions, for instance, in the disordered phase $g > g_c$, the fidelity susceptibility is expected to be proportional to the volume, $\chi_A \sim L^d$. However, at quantum transitions, the behavior may change. Indeed, at CQTs, there is a non-analytic contribution (it is obviously independent of the BCs) that behaves as $\chi_A \approx L^{2y_h}\mathcal{A}_2(hL^{y_h})$, where y_h is the RG dimension of the parameter h, see, e.g., Refs. 23,78,83 and references therein. If $d < 2y_h$, as it occurs at the CQT in Ising systems in any dimension, the singular contribution is the dominant one and the fidelity susceptibility diverges faster than under normal conditions. These results can be trivially extended to any CQT by replacing y_h with the RG dimension of the parameter that drives the transition.

At FOQTs, the fidelity susceptibility shows an even stronger divergence.[23,61] In terms of the scaling variables introduced in Section 2.2, the ground-state fidelity is conjectured to show the asymptotic FSS behavior

$$A(L,g,h,\delta h) \approx \mathcal{A}(\Phi,\delta\Phi), \quad \delta\Phi \equiv \delta E_h(L,\delta h)/\Delta(L), \tag{29}$$

close to the FOQT at $h=0$. This relation implies

$$\chi_A(L,h) \approx (\delta\Phi/\delta h)^2\,\mathcal{A}_2(\Phi) \sim L^{2d}\Delta(L)^{-2}\mathcal{A}_2(\Phi). \tag{30}$$

Therefore, at FOQTs, the fidelity susceptibility shows a divergent large-L behavior that depends on the BCs as a consequence of the different possible size behaviors of the gap $\Delta(L)$ discussed in Section 2.1.2. In particular, $\chi_A(L,h)$ may scale exponentially in the size (this is the case of, e.g., PBCs or EFBCs), or with a power of L. In the ABC case, for instance, since $\Delta(L) \sim L^{-2}$, the fidelity susceptibility diverges as L^6. As anticipated, these diverging behaviors are much stronger than at CQTs and under normal conditions.

If one considers, systems in which the magnetized states are the relevant low-energy excitations, the scaling function for the fidelity can be computed by performing a two-level truncation of the states. For neutral BCs such as OBCs or PBCs,

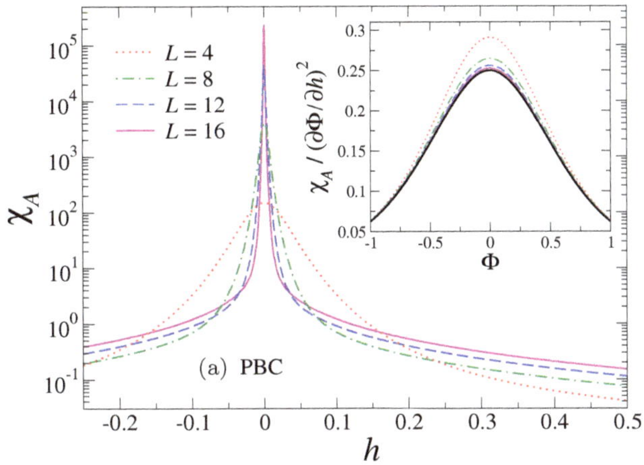

Fig. 3. Fidelity susceptibility $\chi_A(L, h)$ for the quantum Ising chain with PBC, at fixed $g = 0.9$, associated with changes of the longitudinal parameter h for some values of L. The inset displays $\chi_A/(\partial\Phi/\partial h)^2$, which clearly approaches the scaling function $\mathcal{A}_2(\Phi)$ (thick black line). Results are taken from Ref. 61.

one can use the effective Hamiltonian (15), obtaining[61]

$$\mathcal{A}_2(\Phi) = \frac{1}{4(1+\Phi^2)^2}. \tag{31}$$

This expression is expected to hold at any FOQT with only two symmetric relevant states in the large-L limit. The symmetry condition is not crucial, and $\mathcal{A}_2(\Phi)$ can also be computed for nonsymmetric states and therefore for non-neutral BCs.

These predictions have been confirmed by numerical computations for the quantum Ising chain.[61] In Figure 3 we report results for PBC in which $\Delta(L)$ decreases exponentially. Equation (29) with the scaling function (31) is clearly satisfied. The scaling predictions have also been verified for ABC,[61] for which $\Phi \approx hL^3$ and therefore $\chi_A \sim L^6$.

3. Dynamic Finite-Size Scaling at First-Order Quantum Transitions

In this section, we focus on the dynamical behavior of systems close to FOQTs. We show that the dynamic properties also obey general FSS laws that generalize and extend the static FSS relations outlined in Section 2. As it occurs for equilibrium properties, the out-of-equilibrium dynamic behavior at FOQTs is very sensitive to the BCs, at variance with what occurs at CQTs, where the power law of the time scale does not depend on the BC. We find that the size behavior of the long-time properties depends on the finite-size gap $\Delta(L)$ at the transition. If $\Delta(L)$ decreases exponentially — in Ising systems, this occurs when the magnetized states are the lowest-energy excitations — the relevant time scale of the dynamics increases exponentially with L. If, instead, $\Delta(L) \sim L^{-\kappa}$, as it occurs in the presence of domain walls, then the time scale increases as a power of L.

In the following, we consider dynamic protocols entailing instantaneous or slow changes of a Hamiltonian parameter. Specifically, the Hamiltonian is supposed to be of the form

$$\hat{H}[w(t)] = \hat{H}_0 + w(t)\hat{H}_p, \quad [\hat{H}_0, \hat{H}_p] \neq 0. \tag{32}$$

We assume that the Hamiltonian parameters of $\hat{H}_0$ are tuned at the FOQT and that $\hat{H}_p$ represents a perturbation that takes the system out of the FOQT. In the quantum Ising models, $\hat{H}$ is the Hamiltonian defined in Eq. (1): The Hamiltonian $\hat{H}_0$ corresponds to $\hat{H}$ with $h = 0$ and $g < g_c$, the parameter w corresponds to the magnetic field h and $\hat{H}_p = -\sum_{\boldsymbol{x}} \hat{\sigma}_{\boldsymbol{x}}^{(1)}$.

The equilibrium FSS theory outlined in Section 2 can be straightforwardly extended to the dynamic case by adding the time-dependent scaling variable

$$\Theta \equiv \Delta(L)\, t. \tag{33}$$

Note that Θ is also the relevant FSS variable for CQTs. In the following, we discuss dynamic FSS using the one-dimensional Ising chain, but these considerations are expected to also hold for more general systems (without continuous symmetries) and in larger dimensions.

It is worth mentioning that FOQTs have been studied in the context of adiabatic quantum computation as well.[53,84–86]

3.1. *Quantum quenches at quantum first-order transitions*

Instantaneous quench protocols are among the simplest protocols that can be used to investigate the out-of-equilibrium quantum dynamics of many-body systems with Hamiltonian (32). Initially, for $t < 0$, the system is in the ground state of $\hat{H}$ for a specific value w_i of the parameter w, i.e., in the eigenstate $|\Psi_0(w_i)\rangle$. Then, at $t = 0$, one suddenly changes the Hamiltonian parameter from w_i to $w \neq w_i$ and considers the quantum evolution defined by the Schrödinger equation

$$i\frac{d|\Psi(t)\rangle}{dt} = \hat{H}(w)|\Psi(t)\rangle, \quad |\Psi(t)\rangle = e^{-i\hat{H}(w)t}|\Psi_0(w_i)\rangle. \tag{34}$$

In the following, we consider quantum Ising systems and identify the parameters w_i and w with h_i and h.

The dynamic FSS relations for the different observables can be obtained by extending those that hold in the static case. First, we introduce the time scaling variable Θ defined in Eq. (33). Then, we define the scaling variables Φ_i and Φ. They are defined as in Eq. (10), using h_i and h, respectively. The dynamic FSS limit is defined as the limit $L \to \infty$, keeping the scaling variables Φ_i, Φ and Θ fixed. In this limit, the magnetization behaves as

$$M(t; L, h_i, h) \approx m_0 \mathcal{M}(\Theta, \Phi_i, \Phi), \tag{35}$$

where m_0 is the spontaneous magnetization defined in Eq. (4). This scaling behavior is expected to hold for any $g < 1$, and the scaling function $\mathcal{M}$ should be independent of g, apart from trivial normalizations of the arguments. As already discussed for the equilibrium FSS in Section 2, the size dependence of the scaling variables and the function $\mathcal{M}$ depend on the BC.

The above dynamic scaling relations can be straightforwardly extended to any FOQT, by identifying the scaling variable Φ as the ratio $\delta E_w(L)/\Delta(L)$, where $\delta E_w(L)$ is the energy variation associated with the perturbation $w\hat{H}_p$ and $\Delta(L)$ is the energy difference between the two lowest states at the transition point. The scaling variable Θ associated with time is always defined as in Eq. (33).

In the case of neutral BCs with magnetized low-energy states, for instance, for OBCs or PBCs, one may again exploit[40] the two-level truncation of the spectrum discussed in Section 2.2.2, which holds when $\delta E_w(L,w) = 2m_0|w|L^d = |\Phi|\,\Delta(L) \ll E_2(L)-E_0(L)$ (for systems with PBCs or OBCs, $E_2(L)-E_0(L) = O(1)$ in the large-L limit). This allows one to compute the scaling function $\mathcal{M}$ associated with the magnetization, extending the equilibrium computation reported in Section 2.2.2, see Ref. 40. For this purpose, we consider the unitary evolution of the reduced two-state system with Hamiltonian (15) and define ε_i and ε as the values of the parameter corresponding to h_i and h. The dynamics starts from the two-component ground state corresponding to the parameter ε_i. Simple calculations show that, apart from an irrelevant phase, the time-dependent state $|\Psi_2(t)\rangle$ evolves as[40]

$$|\Psi_2(t)\rangle = \cos\left(\frac{\alpha_i - \alpha}{2}\right)|0\rangle + e^{-i\Theta\sqrt{1+\Phi^2}}\sin\left(\frac{\alpha_i-\alpha}{2}\right)|1\rangle, \tag{36}$$

with $\tan\alpha_i = \Phi_i^{-1}$, $\tan\alpha = \Phi^{-1}$, and

$$|0\rangle = \sin(\alpha/2)|-\rangle + \cos(\alpha/2)|+\rangle, \quad |1\rangle = \cos(\alpha/2)|-\rangle - \sin(\alpha/2)|+\rangle, \tag{37}$$

$|\pm\rangle$ being the eigenstates of the operator $\hat{\sigma}^{(3)}$. The expectation value $\langle\Psi_2(t)|\hat{\sigma}^{(3)}|\Psi_2(t)\rangle$ gives the magnetization, from which we can infer the dynamic scaling function defined in Eq. (35):

$$\mathcal{M}(\Theta,\Phi_i,\Phi) = \cos(\alpha-\alpha_i)\cos\alpha + \cos\left(\Theta\sqrt{1+\Phi^2}\right)\sin(\alpha-\alpha_i)\sin\alpha. \tag{38}$$

Numerical results for the Ising chain with PBCs[40] confirm the validity of the scaling Ansatz (35) with the scaling function (38). The asymptotic FSS behavior is approached quite rapidly: Size corrections turn out to decrease exponentially in L. A similar analysis can be performed for the work fluctuations associated with the quench protocol.[87] The scaling predictions obtained by using the two-level truncation are nicely supported by the numerical results.

The results presented above are expected to be quite general. They should apply to any FOQT at which there are only two quasi-degenerate low-energy states which are related by symmetry. For instance, they should apply to quantum Ising models in any dimension along their FOQT line. It is trivial to extend the discussion to

FOQT driven by the competition of two different quantum states that are not related by symmetry or to systems (Potts models, for instance), in which there is a finite number of quasi-degenerate states.

We finally mention that, in the case of EFBC which favor a magnetized phase, see Section 2.3, one should take into account the shift in the pseudotransition point. As in the static case, the two-level truncation applies also to dynamic properties in the appropriate limit.[62]

The dynamic FSS behavior (35) also holds[62] in systems with ABC and OFBC, in which the gap behaves as $\Delta(L) \sim L^{-2}$. In this case, however, the low-energy behavior is controlled by a large number (of order L) of states, and thus, the dynamics cannot be computed by considering a system with a finite number of states. Analogous scaling behaviors are expected in higher dimensions, for example, in the presence of mixed boundaries, such as ABC or OFBC along one direction and PBC or OBC along the others.

We finally mention that out-of-equilibrium scaling behaviors arising from a quench at FOQTs have also been investigated in more complex composite models and, in particular, when a central spin is globally and homogenously coupled to a quantum many-body system.[88]

3.2. *Dynamic scaling: Kibble–Zurek dynamics at FOQT*

Another interesting out-of-equilibrium dynamics is obtained by slowly crossing a FOQT. In this case, the system parameters, for instance, $w(t)$ in Eq. (32), are changed with a very large time scale t_s, starting from an initial value $w_i < 0$ on one side of the FOQT, up to a final $w > 0$ on the other side of the transition. For instance, in the Kibble–Zurek protocol,[55,56] at time $t_i < 0$, the system is in the ground state of the Hamiltonian $H(w_i)$ with $w_i = w(t_i)$, and then it evolves unitarily with Hamiltonian $H[w(t)]$, where $w(t) = t/t_s$, up to a final time $t_f > 0$. Far from phase transitions, because of the adiabatic theorem, the system at time t is approximately (exactly in the limit $t_s \to \infty$) in the ground state of the Hamiltonian $H[w(t)]$. However, this adiabatic evolution is not generally realized when the system is driven across a phase transition because of the presence of low-energy modes that have a very long time scale that diverges when $L \to \infty$. Thus, in the infinite-volume limit, the large-scale modes do not equilibrate, however large the time scale t_s is. In a finite volume large-scale modes have a finite time scale $\tau(L)$, and thus, we can observe an out-of-equilibrium scaling behavior due to the interplay of the two (large) time scales t_s and $\tau(L)$. Thus, a slow (quasi-adiabatic) dynamics across critical points allows one to probe the universal features of the long-range modes emerging at the transition.

The nature of the out-of-equilibrium FSS behavior depends on the nature of the transition and on its universality class if it is continuous, see, e.g., Refs. 23,55,56, 89–91. To be specific, we now consider an Ising model with $g < g_c$ and a slowly varying longitudinal field $w(t) = h(t) = t/t_s$. At a time $t_i < 0$, the longitudinal field is $h_i = t_i/t_s$, and the system is in the ground state $|\Psi(t = t_i)\rangle \equiv |\Psi_0(h_i)\rangle$. Then, for $t > t_i$, the longitudinal field varies as $h(t) = t/t_s$ and the system evolves unitarily according to the Schrödinger equation

$$\frac{\mathrm{d}|\Psi(t)\rangle}{\mathrm{d}t} = -i\hat{H}[h(t)]|\Psi(t)\rangle, \quad |\Psi(t = t_i)\rangle = |\Psi_0(h_i)\rangle, \tag{39}$$

up to a time $t_f > 0$.

In the simultaneous limits $t_s \to \infty$ and $L \to \infty$, the system shows a universal FSS behavior. To simplify the discussion, we keep $h_i < 0$ fixed in the large-L and large-t_s limits. Since the equilibrium FSS behavior should be recovered when t_s is much larger than the time scale of the low-energy modes, one of the scaling variables should be identified with the corresponding equilibrium scaling variable. For neutral BCs, the relevant variable is given in Eq. (10) and therefore we define

$$\Phi_{\mathrm{KZ}} \equiv \frac{2m_0 h(t) L^d}{\Delta(L)} = \frac{2m_0 t L^d}{\Delta(L)\, t_s}. \tag{40}$$

The second scaling variable is $\Theta = \Delta(L)\, t$ defined in Eq. (33). Equivalently, we can define

$$\Upsilon \equiv \frac{\Theta}{\Phi_{\mathrm{KZ}}} = \frac{\Delta^2(L)\, t_s}{2m_0 L^d}, \tag{41}$$

which is independent of the time t. The dynamic FSS limit corresponds to t, t_s, $L \to \infty$, keeping the scaling variables Φ_{KZ} and Υ fixed. In this limit, the magnetization is expected to scale as

$$M(t, t_s, h_i, L) \approx m_0\, \mathcal{M}(\Upsilon, \Phi_{\mathrm{KZ}}), \tag{42}$$

independent of $h_i < 0$. In the adiabatic limit ($t, t_s \to \infty$ at fixed L and t/t_s), $\mathcal{M}(\Upsilon \to \infty, \Phi_{\mathrm{KZ}})$ must reproduce the equilibrium FSS function (14). The scaling functions are expected to be universal (i.e., independent of g along the FOQT line for a given class of BC). Note that the dynamic FSS behavior develops in a narrow range of magnetic fields. Indeed, since Φ_{KZ} is kept fixed in the dynamic FSS limit, the relevant scaling behavior develops in the interval $|h| \lesssim \Delta(L)/L^d$, which rapidly shrinks when increasing L. This implies that the dynamic FSS behavior is independent of the initial value $h_i < 0$ if it is kept fixed in the large-L limit.

The above scaling behaviors are expected to hold for generic neutral BCs, both in the presence of magnetized ground states and in the presence of domain walls. They can also be extended to nonneutral BCs, by replacing[62] $h(t)$ with $h(t) - h_{\mathrm{tr}}(L)$, where $h_{\mathrm{tr}}(L)$ is defined in Section 2.1.2. The previous scaling expressions are quite general and can be straightforwardly extended to any FOQT. The previous dynamic FSS relations have been numerically verified in Refs. 62 and 92 for Ising chains with global and local time-dependent perturbations and for several types of BCs.

If we consider neutral BCs with magnetized ground states, we can also compute the dynamic scaling functions associated with the Kibble–Zurek protocol.

Indeed, since the low-energy behavior is controlled by the two lowest-energy states, we can consider the effective Hamiltonian (15), which now becomes time-dependent:

$$\hat{H}_r = \varepsilon(t)\hat{\sigma}^{(3)} + \zeta\hat{\sigma}^{(1)}, \quad \varepsilon(t) = m_0 h(t) L^d = m_0 t L^d / t_s, \quad \zeta = \Delta(L)/2. \tag{43}$$

The system is thus equivalent to a two-level quantum mechanical system in which the energy separation of the two levels is a linear function of time. Its dynamics was first investigated by Landau and Zener[93,94] and solved exactly in Ref. 95. If $|\Psi_2(t)\rangle$ is the solution of the Schrödinger equation with the initial condition $|\Psi_2(t_i)\rangle = |+\rangle$ (where $|+\rangle$ is the eigenstate of $\hat{\sigma}^{(3)}$ with positive eigenvalue), the dynamic FSS function of the magnetization can be computed, obtaining[62,92]

$$\mathcal{M}(\Upsilon, \Phi_{\rm KZ}) = \langle\Psi_2(t)|\hat{\sigma}^{(3)}|\Psi_2(t)\rangle = 1 - \tfrac{1}{4}\Upsilon e^{-\frac{\pi\Upsilon}{16}} \left| D_{-1+i\frac{\Upsilon}{8}}(e^{i\frac{3\pi}{4}}\sqrt{2\Upsilon}\Phi_{\rm KZ})\right|^2, \tag{44}$$

where $D_\nu(x)$ is the parabolic cylinder function.[96] In this case, the deviations from the asymptotic dynamic FSS behavior are controlled by the ratio between $\Delta(L) \sim e^{-cL^d}$ and $\Delta_2 = O(1)$. Therefore, the approach to the asymptotic behavior is expected to be exponentially fast.[92]

The above results are expected to extend to generic FOQTs in which the universal features of the dynamics of the low-energy modes are effectively encoded in the dynamics of a two-level system. It is worth mentioning that an analogous dynamic scaling behavior is observed by considering time-dependent localized defects, such as a time-dependent magnetic field localized in one site of an Ising chain, essentially because the chain behaves rigidly along the FOQT line.

We point out that the dynamic scaling behaviors that have been outlined above have been numerically observed in relatively small systems. Therefore, given the need for high accuracy without necessarily reaching scalability to large sizes, the available technology for probing the coherent quantum dynamics of interacting systems, using, for instance, ultracold atoms in optical lattices,[97,98] trapped ions,[99–105] as well as Rydberg atoms in arrays of optical microtraps,[106,107] could offer possible playgrounds where the envisioned behaviors at FOQTs can be observed.

Metastability phenomena in Kibble–Zurek protocols in quantum Ising chains have also been investigated in Ref. 108, evidentiating quantized nucleation mechanisms. The out-of-equilibrium dynamics at FOQTs has been also investigated in the presence of dissipation[109,110] in the Lindblad framework.[111–114]

4. Quantum-to-Classical Mapping

In the previous sections, we have discussed FSS in quantum systems. However, historically, the basic concepts underlying our present understanding of critical and first-order transitions, were developed in a classical setting, in several seminal works by Kadanoff, Fisher, Wilson, among others (see, e.g., Refs. 3–6,17,18,115,116). Their extension to quantum systems is based on the quantum-to-classical mapping, which allows one to map a quantum system defined in a spatial volume V_s

onto a classical one defined in a box of volume $V_c = V_s \times L_T$, with $L_T = 1/T$ (using the appropriate units).[9,10,29] Under the quantum-to-classical mapping, the inverse temperature corresponds to the system size in an imaginary time direction. The BCs along the imaginary time are periodic or antiperiodic for bosonic and fermionic excitations, respectively. Thus, the universal FSS behavior observed at a quantum transition in d dimensions is analogous to the FSS behavior in a corresponding D-dimensional classical system with $D = d + 1$.

It is important to remark that the quantum-to-classical mapping does not generally lead to standard classical isotropic systems in thermal equilibrium. First, in many interesting instances, one obtains complex-valued Boltzmann weights (a well-known example is the Berry phase). Moreover, the corresponding classical systems are generally anisotropic. If the dynamic exponent z that controls the size behavior of the gap is equal to 1, as for the Ising CQT, the anisotropy is weak, as in the classical Ising model with direction-dependent couplings. In these cases, a straightforward rescaling of the imaginary time allows one to recover space-time rotationally invariant (relativistic) Φ^4 statistical field theories. However, there are also interesting transitions in which $z \neq 1$, such as the superfluid-to-vacuum and Mott transitions of lattice particle systems described by the Hubbard and Bose-Hubbard models.[10,117] In this case, we have $z = 2$ when the transitions are driven by the chemical potential. For CQTs with $z \neq 1$, the anisotropy is strong, i.e., correlations have different exponents in the spatial and thermal directions. Indeed, in the case of quantum systems of size L, under a RG rescaling by a factor b such that $\xi \to \xi/b$ and $L \to L/b$, the additional dimension related to the temperature rescales differently, as $L_T \to L_T/b^z$. However, an FSS theory has also been established in classical strongly anisotropic systems.[118]

5. Equilibrium Finite-Size Scaling at Classical First-Order Transitions

5.1. *General scaling results in discrete spin systems*

FSS for first-order classical transitions (FOCTs) was established in a series of seminal papers,[30–32,73] generalizing the ideas that had been previously developed in the context of critical transitions. Michael Fisher was actively involved in the development of a general FSS theory for first-order transitions. We do not plan to review here these very well-known issues, see, e.g., Ref. 20. We will only present some results, due to Michael Fisher and collaborators, that have a direct counterpart in quantum systems and directly relate to what we have been discussing in the quantum setting.

Reference 32 studied FSS for generic Ising systems with PBC, considering asymmetric D-dimensional classical systems of size $L^d \times L_\parallel$ with $D = d+1$. For "block" geometries in which the FSS limit is taken at fixed finite and non-vanishing ratio $L/L_\parallel$, the basic scaling variable turns out to be the dimensionless ratio of the total bulk ordering energy and the thermal energy. In Ising systems, along the low-temperature first-order transition line, the natural variable is therefore

$$\Phi_b = m_0 hV/T, \quad V = L^d L_\parallel = L^D \times (L_\parallel/L), \tag{45}$$

where T is the temperature, h the magnetic field, and m_0 the spontaneous magnetization. The free energy and the magnetization should therefore scale as

$$F(h,T) = A(T)\mathcal{F}(\Phi_b), \quad m(h,T) = m_0(T)\mathcal{M}(\Phi_b), \tag{46}$$

a result which was then proved rigorously, under very general conditions, in Ref. 35. It was also noted that this behavior does not hold when $L_\parallel \gg L$, the geometry that is relevant for quantum systems ($L_\parallel$ is the size of the imaginary-time direction that appears in the quantum-to-classical mapping). In this case, fluctuations break the system into successive regions of up and down magnetization with characteristic length

$$\xi_\parallel(T,L) \sim \exp[\sigma(T)L^d], \tag{47}$$

where $\sigma(T)$ is the reduced interfacial tension. Correspondingly, the relevant scaling variables are

$$\Phi = m_0 hL^d \xi_\parallel(T,L)/T, \quad \Xi = \xi_\parallel(T,L)/L_\parallel. \tag{48}$$

Under the quantum-to-classical mapping, the D-dimensional classical system with $L_\parallel \gg L$ is mapped onto a d-dimensional quantum sistem, $\xi_\parallel(T,L)/T$ and $1/L_\parallel$ playing the role of the energy gap and of the temperature T of the quantum system, respectively. Thus, for PBCs, Fisher and Privman had already identified the variables (10) and (12) as the correct scaling variables for FOQTs. Moreover, they postulated what they called "the two-eigenvalue dominance": Scaling functions can be computed by considering only the two largest eigenvalues of the transfer matrix. In the quantum context, this corresponds to the two-level truncation discussed in Section 2.2.2. Obviously, the quantum-to-classical mapping can also be used to extend quantum results to classical systems, i.e., we can use the results of Section 2.2 to predict the scaling behavior in cylinder-like systems with PBC in the *long* direction and different types of BC in the other ones. For instance, the results for the quantum Ising chain with ABC immediately predict the scaling behavior of a classical Ising model in a two-dimensional strip with ABC along the parallel boundaries at finite distance L. In this case, we would predict that the magnetization scales as $M(L,h) \approx m_0\mathcal{M}(hL^3)$, where L is the size of the strip. The same result would be obtained by considering sufficiently strong oppositely oriented boundary fields. For non-neutral BCs, the quantum analysis predicts a surface-induced shift of the pseudotransition $h_{\rm tr}(L)$ (note that, by applying the quantum-to-classical mapping, we expect $h_{\rm tr}(L) \sim L^{-d}$), a well-known phenomenon in the classical case.[31,119] Moreover, in the definition of the scaling variables, h should be replaced by $h - h_{\rm tr}(L)$, as verified explicitly in classical systems.[119]

We also mention that the effects of spatially inhomogeneous conditions, such as a temperature gradient, have been also analyzed at FOCTs,[120] showing the emergence of peculiar scaling behaviors, similar to those observed at FOQTs.

5.2. *General scaling results in continuous spin systems*

The analysis of the previous section can be extended to D-dimensional magnetic systems with a continuous $O(N)$-symmetry. Results for PBCs were obtained by Fisher and Privman in Ref. 33. Again, the behavior depends on the geometry of the system. In cubic systems, the modulus of the magnetization is essentially constant, while its direction is randomly distributed in the sphere, to recover the $O(N)$ invariance. The relevant scaling variable is Φ_b defined in Eq. (45). Corrections, such as those arising from spin-wave modes, are expected to decay as $1/L$.[33]

FSS substantially changes in anisotropic geometries. If we consider systems of size $L^d \times L_\parallel$ and $L_\parallel \gg L$, fluctuations give rise to spin waves in the "long" direction, with a finite length scale[33] $\xi_\parallel$ that scales as

$$\xi_\parallel \approx L_\parallel f_p(L_\parallel / L^2), \tag{49}$$

where $f_p(x)$ is a scaling function. The relevant scaling variables are again Φ and Ξ given in Eq. (48). These predictions have been confirmed by analytical and numerical computations for the three-dimensional $O(N)$ vector model.[121,122]

We can use the above-reported results and the quantum-to-classical mapping to conclude that the variables (10) and (12) are the relevant FSS variables also for a quantum system with continuous symmetry (at least when PBC are used). This result provides additional evidence that the only relevant system variable for FSS is the finite-size gap $\Delta(L)$ at the transition (or at the pseudocritical transition for non-neutral boundary conditions). Moreover, Eq. (49) implies that the gap scales as $1/L^2$, so that $\Phi \sim hL^{d+2}$ for a quantum d-dimensional system with continuous symmetry.

5.3. *Finite-size behavior of cumulants at first-order transitions*

First-order transitions are characterized by the discontinuity of thermodynamic quantities. In D-dimensional finite systems, they are smeared out, but they still give rise to easily identifiable properties. For instance, the specific heat and the magnetic susceptibility show a sharp maximum that diverges as a power of the volume and the same is true for the cumulants of the order parameter in magnetic transitions.

An FSS analysis of the crossover behavior in the coexistence region relies on some phenomenological approximations for the distribution of the quantity one is considering.[20,34,38,123] Let us consider a generic intensive scalar quantity A that is discontinuous at a transition that separates a single stable phase from q equivalent phases. For the magnetic low-temperature Ising transitions (or the equivalent nonsymmetric liquid–vapor transitions), q is 1. Instead, thermal first-order transitions in Ising-like or Potts models have $q \geq 2$. The basic assumption is that, close to the transition, A has a bimodal distribution,

$$P(A) = \frac{wL^{D/2}}{\sqrt{2\pi B_+}} \exp\left(-\frac{(A-A_+)^2L^D}{2B_+}\right) + \frac{(1-w)L^{D/2}}{\sqrt{2\pi B_-}} \exp\left(-\frac{(A-A_-)^2L^D}{2B_-}\right), \tag{50}$$

where $A_\pm$ and $B_\pm$ are regular functions of the parameter driving the transition. This dependence is irrelevant for the following discussion, and thus, $A_\pm$ and $B_\pm$ will represent the values of the corresponding functions at the transition point. We assume the "+" phase to be unique, while there are q equivalent "−" phases. The parameter w is related to the weight of the different phases. Since $\langle A \rangle = wA_+ + (1-w)A_-$, we immediately see that

$$\frac{w}{1-w} = \frac{1}{q}\exp(-\Delta F/T), \tag{51}$$

where ΔF is the free energy difference between the "+" phase and one of the "−" phases.[b] At the transition point, $\Delta F = 0$, and therefore, $w = 1/(q+1)$: The average value $\langle A \rangle$ is obtained as the average of A over the $(q+1)$ coexisting phases. If the transition is driven by varying a parameter λ coupled to A (i.e., the Hamiltonian contains a term $\lambda \sum_x A_x$), then we have $\Delta F = (\lambda - \lambda_c)(A_+ - A_-)L^D$, close to the transition point $\lambda = \lambda_c$, so that

$$w = \frac{e^{-\beta A_+ L^D \Delta\lambda}}{e^{-\beta A_+ L^D \Delta\lambda} + qe^{-\beta A_- L^D \Delta\lambda}}, \qquad \Delta\lambda = \lambda - \lambda_c. \tag{52}$$

The distribution (50) is motivated by phenomenological considerations, although many of the results we will summarize in the following have been proved rigorously for systems with PBCs.[35,124] It is important to stress that the Ansatz (50) is not expected to be true for any type of system geometry and boundary conditions. For instance, it should not hold for very elongated geometries or for systems with mixed boundary conditions that favor different phases, since, in these cases, stable domain walls are present in the system. In general, we expect the behavior to hold whenever interfaces have an exponentially small probability.

Distribution (50) allows us to predict the behavior of cumulants of A in the coexistence region. We first consider the susceptibility (if A is the energy, χ_A is proportional to the specific heat)

$$\chi_A = L^D(\langle A^2 \rangle - \langle A \rangle^2), \tag{53}$$

[b]Note that the expression (50) differs from that reported in Ref. 34, where the relative weights of the two Gaussians were computed using an equal-height prescription, but agrees with that reported in Ref. 20 (see Section 3 of Ref. 35 for a rigorous discussion).

which takes the values B_+ and B_- in the two phases. In the coexistence region, χ_A has a sharp maximum with

$$\chi_{A,\max} = \frac{1}{4}(A_+ - A_-)^2 L^D [1 + O(L^{-D})]. \tag{54}$$

For non-symmetric systems, the maximum does not occur at the critical point, but is shifted by a quantity proportional to L^{-2D} for $q = 1$ and to L^{-D} for $q \geq 2$ (see Ref. 35 for a critical discussion and proofs in the PBC case).

It is also interesting to consider the Binder parameter

$$U_A = \frac{\langle A^4 \rangle}{\langle A^2 \rangle^2}. \tag{55}$$

The behavior of U_A depends on the nature of the quantity that is considered. If both A_+ and A_- are non-vanishing at the transition, then U_A takes the value 1 at both sides of the transition and shows a maximum in between (the position of the maximum converges to the transition point with corrections of order L^{-D}) with

$$U_{A,\max} = \frac{(A_+^2 + A_-^2)^2}{4A_+^2 A_-^2} + O(L^{-D}). \tag{56}$$

At the transition point, the parameter is not trivial: It takes the value

$$U_{A,c} = \frac{(q+1)(A_+^4 + qA_-^4)}{(A_+^2 + qA_-^2)^2} + O(L^{-D}). \tag{57}$$

This is the expected generic behavior at thermal transitions where A is the energy density. In the Ising model along the low-temperature transition line, if A is the magnetization M, then $A_+ = -A_- = m_0$, where m_0 is the spontaneous magnetization. In this case, the Binder parameter behaves trivially as U_m is 1 in the whole coexistence region (it is trivial to verify that $U_{m,\max} = 1$).

The behavior drastically changes if $A_+ = 0$ or $A_- = 0$. Assuming for definiteness that $A_+ = 0$, U_A varies from 3 (in the "+" phase) to 1 and has a maximum for $w \approx 1 - B_+ L^{-D}/A_-^2$ — therefore in the "+" phase — with

$$U_{A,\max} = \frac{A_-^2}{4B_+} L^D. \tag{58}$$

At the transition point $w = 1/(q+1)$, we have instead $U_{A,c} = (1+q)/q + O(L^{-D})$. A scalar order parameter at a thermal FOCT is expected to have this type of behavior, since the its average vanishes in the disordered "+" phase. The same qualitative behavior is expected in the Potts models defined by the Hamiltonian (22), although, in this case, one should take into account the multicomponent nature of the order parameter.[38,123]

It should be noted that, in many cases, the definition (55) is somewhat arbitrary, as A may be defined modulo an additive constant (this is obviously the case if A is the energy). It is thus natural to define

$$U_A(A_0) = \frac{\langle (A - A_0)^4 \rangle}{\langle (A - A_0)^2 \rangle^2}, \tag{59}$$

where A_0 is an arbitrary constant. The maximum $U_{A,\max}(A_0)$ for a given value of A_0 shows a peak that becomes sharper and sharper as A_0 gets closer to A_+ or A_-. In particular, the maximum diverges as L^D, when A_0 is equal to A_+ or A_-.

The previous results can be used to distinguish first-order transitions from continuous ones. At FOCTs, the probability distributions of the energy and of the magnetization are expected to show a double peak for large values of L. Therefore, two peaks in the distributions are often taken as an indication of a first-order transition. However, as discussed, e.g., in Refs. 125–128 and references therein, the observation of two maxima in the distribution of the energy is not sufficient to conclude that the transition is an FOCT. For instance, in the two-dimensional Potts model with $q = 3$ and $q = 4$,[126,127] double-peak distributions are observed, even if the transition is known to be continuous. Analogously, in the 3D Ising model, the distribution of the magnetization at the critical point has two maxima.[129] In order to identify definitely a first-order transition, it is necessary to perform a more careful analysis of the large-L scaling behavior of the distributions.

A possible approach consists in considering the large-L behavior of the specific heat or of the susceptibility of the order parameter, which should have a maximum that diverges as L^D.[125–128] In practice, one should compute the maximum $\chi_{A,\max}(L)$ for each value of L and then fit the results to a power law aL^p. If one obtains $p \approx D$, the transition is of first order; otherwise, the transition is continuous, p being related to standard critical exponents. If A is the energy density, then χ_A is the specific heat that diverges as $L^{\alpha/\nu}$ (if α is positive), while if A is the magnetization, then $\chi_A \sim L^{2-\eta}$. This approach works nicely for strongly discontinuous transitions. In the opposite case, however, the asymptotic behavior $\chi_A \sim L^D$ may set in for values of L that are much larger than those considered in the simulations. Thus, data may show an effective scaling $\chi_{A,\max}(L) \sim L^p$, with p significantly smaller than D, effectively mimicking a continuous transition (the three-state Potts model in three dimensions is a good example, see Refs. 125–128). At order–disorder magnetic transitions, the Binder cumulant of the magnetization usually provides a better indicator.[38] Indeed, it diverges at first-order transitions, see Eq. (58), while it is smooth and finite at continuous transitions. Thus, the observation that $U_{m,\max}(L)$ increases with L is an evidence of the discontinuous nature of the transition, even if the maximum does not scale as L^D, as it should do asymptotically. This idea have been exploited to determine the nature of the transitions in several models, including systems with gauge symmetries, such as the Abelian Higgs model and scalar chromodynamics, see, e.g., Refs. 130–133.

Finally, note that we can exploit the behavior of the Binder parameter to estimate the transition point also at first-order transitions. For instance, at a thermal transition, the temperature $T_{\max}(L)$, where $U_A(L)$ attains its maximum, converges to T_c. One could also use the crossing method.[134] Since the value $U_{A,c}$ that the Binder parameter takes at the transition, Eq. (57), differs from the value it takes outside the coexistence region, the curves $U_A(L)$ for different values of L have an intersection point that converges to T_c as L becomes large. The energy cumulant U_E can also be used.[34] At continuous transitions, U_E is asymptotically approaching 1, while a peak should be observed at first-order transitions. Note that it is probably useful to consider the subtracted cumulant defined in Eq. (59) with E_0 close to the value that the energy takes in the transition region, as this would provide sharper maxima.

It is possible to generalize the previous discussion to systems with continuous symmetries. For the energy, nothing changes: All results also hold in this case, see, e.g., Ref. 135. For the order parameter, one should take into account its multi-component nature, which is only expected to give rise to quantitative changes with respect to the previous results.

6. Dynamic Scaling at Classical First-Order Transitions

As we have done for the quantum case, we will present here some general results for the equilibrium and out-of-equilibrium dynamics of classical systems, focusing again on the behavior close to FOCTs (see, e.g., Refs. 57, 136–138 for reviews on the critical dynamics at continuous transitions). At FOCTs, several interesting dynamic phenomena occur. Here, we will only consider some specific topics that can be somehow related to what we have discussed in the quantum case.

6.1. *Equilibrium dynamics*

Let us consider the dynamic behavior of a D-dimensional system in equilibrium, close to a FOCT, focusing on the long-time behavior of global observables. As it occurs at critical transitions,[57,138] scaling variables and scaling functions depend on some general features of the dynamics. The results presented here refer to the purely dissipative dynamics,[5,57] driven by one dissipative coupling without conservation laws (model A of the classification reported in Ref. 57), which can be easily realized in lattice models by using the Metropolis or the heat-bath algorithm.[139] As we shall see, in a finite volume, the FSS relations turn out to crucially depend on the BC and on the geometry of the system, similar to what happens at FOQTs.

As in the quantum case, the extension of the FSS theory to the dynamics is straightforward. Beside the scaling variable Φ defined in Section 5, it is enough to consider the additional scaling variable $\Theta = t/\tau(L)$, where $\tau(L)$ represents the time scale of the slowest modes of the system. In the dynamic FSS limit, i.e., at fixed Φ and Θ, the long-time behavior of the system shows a universal scaling behavior. For instance, in the dynamic FSS limit the time-dependent susceptibility $\chi(t-s) = L^D\langle M(t)M(s)\rangle$ ($M(t)$ is the instantaneous magnetization at time t and the average is over all possible dynamic evolutions of the system in equilibrium) is expected to scale as

$$\chi(h, L, t) = \chi(h, L, 0)\, \mathcal{X}(\Phi, \Theta). \tag{60}$$

The time scale $\tau(L)$ that appears in the definition of Θ crucially depends on the BC and geometry of the system. As an example, we consider an Ising system along the first-order low-temperature transition line with a small magnetic field h. We consider PBC and cubic systems of size L^D. As we discussed before, FSS occurs in a small interval of values of h, of the order of L^{-D}, and the relevant scaling variable is Φ_b defined in Eq. (45). At fixed Φ_b, the free-energy difference between the two magnetized phases is constant for $L \to \infty$, and typically, the system moves between the two magnetized phases, which are visited with a probability that depends on Φ_b only. In the coexistence region, the slowest dynamic mode is associated with the flip of the average magnetization, which occurs by generating configurations characterized by the presence of two coexisting phases separated by two approximately planar interfaces. Thus, the relevant time scale is expected to be

$$\tau(L) \approx L^{\alpha} e^{2\sigma L^{D-1}}, \tag{61}$$

where σ is the rescaled planar interface tension. We expect $\tau(L)$ to depend exponentially on L whenever interfaces are exponentially suppressed and indeed a similar result holds for OBC.[140] On the other hand, if domain walls are stabilized by appropriate BCs, we expect $\tau(L)$ to scale as a power of L, i.e., $\tau(L) \sim L^{\kappa}$. This issue was investigated in two-dimensional Ising and Potts models[140,141] using mixed boundary conditions. In the first direction, PBCs were used, while in the second direction, different conditions were used on the sides. In the Ising case, OFBCs were used, while in the Potts case, OBCs were used on one side and fixed magnetized BCs on the other one. The exponent z was estimated, obtaining $\kappa = 2.77(4)$ in the Ising case[140] (cubic systems of size L^2) and $\kappa = 3.0(1)$ in the Potts case[141] (anisotropic strip-like systems of size $L \times L_{\parallel}$, with $L_{\parallel} \gg L$).

The dynamic behavior of systems with O(N) symmetry was discussed in Refs. 121 and 122. In this case, because of the continuous symmetry, no interfaces are present and $\tau(L)$ always scales as L^{κ}. The power κ, however, depends on the BCs and geometry of the system. If one considers BCs that fix the direction of the magnetization, there are no long-range modes and one obtains $\kappa = 2$, as appropriate for a dissipative dynamics in an uncorrelated system. If instead, the BCs do not break the O(N) invariance, the slowest mode is represented by the dynamics of the average magnetization vector $\boldsymbol{M}(t)$, which makes a random walk on a sphere of radius $m_0(T)$. For cubic geometries, this implies[121] $\kappa = D$. For cylinder-like geometries, $L^{D-1} \times L_{\parallel}$ with $L_{\parallel} \gg L$, we should take into account the longitudinal correlation length $\xi_{\parallel}$, introduced in Section 5.2, which scales as L^2 in the FSS limit. In this case, one can argue[121] that $\tau(L) \sim L^{D-1}\xi_{\parallel} \sim L^{D+1}$.

6.2. *Out-of-equilibrium dynamics: Classical quench*

The results of the previous section can be used to predict the behavior of global observables under a soft classical quench. In particular, the two-level reduction we discussed in the quantum case also holds for classical systems in the absence of domain walls.

We consider an Ising system with $T < T_c$ in a cubic system L^D with PBCs, and a magnetized configuration with all spin equal to -1. Then, we consider the evolution of the system in the presence of a small positive magnetic field h. We can then define the average time-dependent magnetization $M(t)$, where the average is taken over the different realizations of the dynamics. In the FSS limit (keeping $\Phi_b = m_0 h V/T$ and $\Theta = t/\tau(L)$ fixed), this quantity satisfies the FSS relation[142]

$$M(t,h,L) = m_0 \, \mathcal{M}(\Phi_b, \Theta) \,. \tag{62}$$

The Θ dependence of this scaling function can be predicted by using a simpler coarse-grained dynamics, which represents the dynamic generalization of the two-level truncation we discussed in the quantum case. Since on time scales of the order of $\tau(L)$, the reversal of the sign of the magnetization $M(t)$ is essentially instantaneous, one can assume that $M(t)$ takes only two values, $\pm m_0$. Second, as the dynamics restricted within each free-energy minimum is rapidly mixing, one can assume that the coarse-grained dynamics is Markovian. Under these conditions, the dynamics is completely parameterized by the rates I_+ and I_- defined by

$$\begin{aligned} &P[M(t) = -m_0 \to M(t+dt) = +m_0] = I_+ dt, \\ &P[M(t) = +m_0 \to M(t+dt) = -m_0] = I_- dt, \end{aligned} \tag{63}$$

which satisfy $I_-/I_+ = e^{-2\Phi_b}$ to ensure the correct equilibrium distribution. Then, one obtains[143]

$$\mathcal{M}(\Phi_b, \Theta) = \tanh \Phi_b - (1 - \tanh \Phi_b) e^{-\Theta/T_s}, \tag{64}$$

where the time scale T_s is a function of Φ_b only. These results are confirmed by the numerical computations reported in Ref. 142. They also hold for the thermal transition in Potts models, apart from the necessary changes due to presence of q equivalent phases on the low-temperature side of the transition.[142]

6.3. *Out-of-equilibrium dynamics: Kibble–Zurek scaling*

We finally wish to discuss the Kibble–Zurek dynamics[55,56] in the classical setting, when the model parameters are slowly varied across the FOCT. For simplicity, let us consider again an Ising system along the low-temperature first-order transition line in the presence of a small magnetic field h. We consider the classical analogue of the dynamics discussed in Section 3.2. At $t_i < 0$, we consider an equilibrated configuration at magnetic field $h_i = t_i/t_s$, Then, we evolve the system using a diffusive dynamics (for instance, the heat-bath algorithm), varying h as $h(t) = t/t_s$ and follow the dynamics up to a time $t_f > 0$. As in the quantum case, we define an FSS regime in terms of $\Theta = t/\tau(L)$, the same dynamic variable we considered above, and of a time-dependent quantity $\Phi_{\rm KZ}(t)$, which is the time-dependent generalization of the FSS variable Φ defined in Section 5.1. We define

$$\Phi_{\rm KZ,cubic} = \frac{m_0}{T} L^D \frac{t}{t_s}, \quad \Phi_{\rm KZ,cylinder} = \frac{m_0}{T} L^{D-1} \xi_\parallel \frac{t}{t_s} \tag{65}$$

for cubic systems of size L^D and cylinder-shaped systems of size $L^{D-1} \times L_\parallel$, respectively. In analogy with what we have presented in the quantum case, it is also useful to define a scaling variable that is independent of t, such as

$$\Upsilon = \frac{\Theta}{\Phi_{\rm KZ}} = \frac{T}{m_0} \frac{t_s}{\tau(L) L^D}, \tag{66}$$

where the last equality holds for cubic L^D systems. The FSS limit is obtained by considering $t, t_s, L \to \infty$, keeping the previous scaling variables fixed. For simplicity, we keep the initial value h_i fixed with increasing L. Note that in the limit, we have $t_s \sim L^D \tau(L)$, i.e., t_s is much larger than the typical time scale of the system, so that the dynamics is effectively able to equilibrate the system at each time. In the FSS limit, the magnetization scales as

$$M(t, h, L) \approx m_0 \, \mathcal{M}(\Upsilon, \Phi_{\rm KZ}). \tag{67}$$

The behavior at thermal transitions is analogous. In this case, one varies the temperature as $\beta(t) = \beta_c(1 + t/t_s)$, so that $\Phi_{\rm KZ} = (\beta(t) - \beta_c)L^D = \beta_c t L^D/t_s$. These scaling behaviors have been verified in magnetic and thermal transitions.[121,141,143]

The Kibble–Zurek dynamics also allows one to investigate the dynamics of some coarsening phenomena by considering time scales that satisfy $t_s \ll \tau(L)$, effectively probing an out-of-equilibrium infinite-volume regime.[143,144] Let us briefly review the results of Ref. 143, for the two-dimensional Potts model (22) with $q = 10, 20$, in a finite volume with PBC. They considered a Kibble–Zurek dynamics. The dynamics starts from an equilibrated high-temperature disordered configuration which evolves under a purely relaxational heat-bath dynamics with time-dependent inverse temperature $\beta(t) = \beta_c(1 + t/t_s)$. They monitored the instantaneous energy at time t, computing its average (with respect to the thermal history and starting configuration) $E(t, t_s, L)$. It was first noted that, for fixed large t, t_s, the energy has a well-defined infinite-volume limit $E_\infty(t, t_s)$. If L is large, the most important dynamic phenomenon is the droplet formation. Since the time needed to create a droplet increases exponentially with its size R,[145] one can define a length scale $R(t) \sim \ln t$ that gives the typical size of droplets at time t. By analogy with the FSS scaling regime, one may define the scaling variable[143]

$$\Phi_{\rm dr} = \Delta\beta R^D = \frac{t}{t_s} \ln^D t \tag{68}$$

and consider the behavior of the system as $t, t_s \to \infty$ at fixed $\Phi_{\rm dr}$. A numerical analysis for the two-dimensional Potts model, see the left panel of Fig. 4, shows

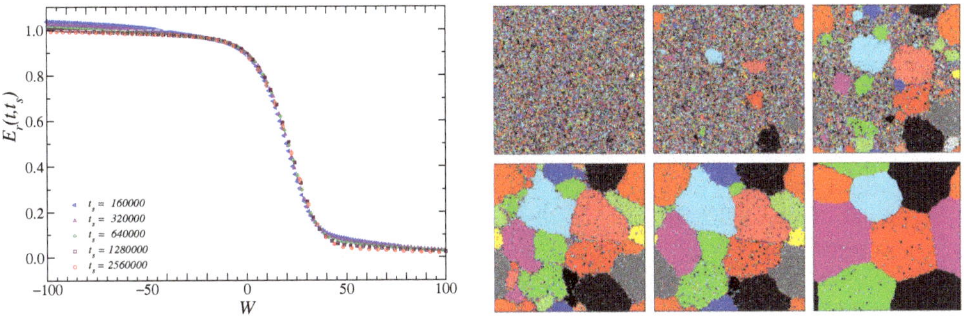

Fig. 4. Results for the two-dimensional Potts model, see Eq. (22), with $q = 20$ and PBC. Left: The renormalized energy density E_r versus W, see Eq. (69), for several large values of t_s. Right: Snapshots of a system of size $L = 512$ for $W = -20$, 0, 20, 40, 100, 1500 (from left to right, top to bottom) for $t_s = 640000$. We use different colors for each Potts state. Stable droplets start appearing for $W \approx 0$, See Ref. 143 for more details.

that the energy density obeys the large-t_s asymptotic scaling behavior[143]

$$E_r(t, t_s) \equiv \frac{E_\infty(t, t_s) - E_-}{E_+ - E_-} \approx f(W), \quad W = (\Phi_{\rm dr} - \Phi^*_{\rm dr})t^\theta, \tag{69}$$

where E_+ and E_- are the values of the transition-point energies associated with the the disordered and ordered phases (so that $0 \leq E_r \leq 1$ in the coexistence region), $\Phi^*_{\rm dr}$ is an appropriate critical value, and θ is an exponent characterizing the dynamics, which turns to be $\theta = 1/3$ within the accuracy of the computations. Figure 4 also shows snapshots of a configuration for different values of t; it shows droplets of increasing size as the scaling variable W increases. The scaling behavior can be interpreted as a dynamic transition whose position depends on t_s. Indeed, for each t_s, one can compute a time $t_{\rm dr}$ such that $(t_{\rm dr} \ln^d t_{\rm dr})/t_s = \Phi^*_{\rm dr}$ (this is roughly the time needed to create a stable large enough droplet), and a corresponding dynamical (pseudo)-transition inverse temperature $\beta_d = \beta_c(1+t_{\rm dr}/t_s)$. The singular behavior (69) resembles the one that occurs at the mean-field spinodal point,[20] but with β_d satisfying $\beta_d - \beta_c \sim (\ln t_s)^{-2} \to 0$. We mention that coarsening phenomena in Potts models with OBC were also investigated in Ref. 144. Similar results were obtained under different dynamic conditions in Refs. 146–152.

References

1. L. D. Landau and E. M. Lifshitz, Course of theoretical physics Vol. 5: *Statistical Physics* (Part 1), 3rd edition (Elsevier Butterworth-Heinemann, Oxford, (UK) — Burlingtoh MA, 1980).
2. M. E. Fisher, The nature of critical points In *Lecture Notes in Theoretical Physics*, Vol 7c, W. E. Brittin (ed.) pp. 1–159. (University of Colorado Press, Boulder, 1965).
3. K. G. Wilson and J. Kogut, The renormalization group and the ε expansion. *Phys. Rep.* **12**, 75 (1974).

4. M. E. Fisher, The renormalization group in the theory of critical behavior. *Rev. Mod. Phys.* **46**, 597 (1974).
5. S.-K. Ma, *Modern Theory of Critical Phenomena* (Routledge, New York, 2001).
6. K. G. Wilson, The renormalization group and critical phenomena. *Rev. Mod. Phys.* **55**, 583 (1983).
7. J. Zinn-Justin, *Quantum Field Theory and Critical Phenomena* (Clarendon Press, Oxford, 2002).
8. M. E. Fisher, *Excursions in the Land of Statistical Physics* (World Scientific, Singapore, 2016).
9. S. L. Sondhi, S. M. Girvin, J. P. Carini and D. Shahar, Continuous quantum phase transitions. *Rev. Mod. Phys.* **69**, 315 (1997).
10. S. Sachdev, *Quantum Phase Transitions* (Cambridge University, Cambridge, England, 1999).
11. F. J. Wegner, Duality in generalized ising models and phase transitions without local order parameters. *J. Math. Phys.* **12**, 2259 (1971).
12. X.-G. Wen, *Quantum Field Theory of Many-Body Systems* (Oxford University Press, New York, 2004).
13. S. Sachdev, Emergent gauge fields and the high-temperature superconductors. *Phil. Trans. R. Soc. A* **374**, 20150248 (2016).
14. S. Sachdev, Topological order, emergent gauge fields, and Fermi surface reconstruction. *Rep. Prog. Phys.* **82**, 014001 (2019).
15. D. Ruelle, Classical statistical mechanics of a system of particles. *Helv. Phys. Acta* **36**, 183 (1963); Statistical mechanics of quantum systems of particles. *Helv. Phys. Acta* **36**, 789 (1963).
16. M. E. Fisher, The free energy of a macroscopic system. *Arch. Ratl. Mech. Anal.* **11**, 377 (1964).
17. E. Brézin, J. C. Le Guillou and J. Zinn-Justin, Field theoretical approach to critical phenomena. In C. Domb and M. S. Green (eds.) *Phase Transitions and Critical Phenomena*, Vol. 6, p. 125, (Academic Press, London, 1976).
18. F. J. Wegner, The critical state, general aspects. In C. Domb and M. S. Green (eds.) *Phase Transitions and Critical Phenomena*, Vol. 6, p. 7, (Academic Press, London, 1976).
19. M. E. Fisher and G. W. Milton, Classifying first-order phase transitions. *Phys. A* **138**, 22 (1986).
20. K. Binder, Theory of first-order phase transitions. *Rep. Prog. Phys.* **50**, 783 (1987).
21. M. E. Fisher, Renormalization group theory: Its basis and formulation in statistical physics. *Rev. Mod. Phys.* **70**, 653 (1998).
22. A. Pelissetto and E. Vicari, Critical phenomena and renormalization group theory. *Phys. Rep.* **368**, 549 (2002).
23. D. Rossini and E. Vicari, Coherent and dissipative dynamics at quantum phase transitions. *Phys. Rep.* **936**, 1 (2021).
24. A. J. Bray, Theory of phase-ordering kinetics. *Adv. Phys.* **43**, 357 (1994).
25. M. E. Fisher and M. N. Barber, Scaling theory for finite-size effects in the critical region. *Phys. Rev. Lett.* **28**, 1516 (1972).
26. M. N. Barber, Finite-size scaling. In C. Domb and J. L. Lebowitz (eds.), *Phase Transitions and Critical Phenomena*, Vol. 8, p. 145, (Academic Press, London, 1983).
27. V. Privman (ed.), *Finite Size Scaling and Numerical Simulation of Statistical Systems* (World Scientific, Singapore, 1990).
28. J. Cardy (ed.) *Finite-Size Scaling* (North Holland, Amsterdam, 1988).

29. M. Campostrini, A. Pelissetto and E. Vicari, Finite-size scaling at quantum transitions. *Phys. Rev. B* **89**, 094516 (2014).
30. B. Nienhuis and M. Nauenberg, First-order phase transitions in renormalization-group theory. *Phys. Rev. Lett.* **35**, 477 (1975).
31. M. E. Fisher and A. N. Berker, Scaling for first-order phase transitions in thermodynamic and finite systems. *Phys. Rev. B* **26**, 2507 (1982).
32. V. Privman and M. E. Fisher, Finite-size effects at first-order transitions. *J. Stat. Phys.* **33**, 385 (1983).
33. M. E. Fisher and V. Privman, First-order transitions breaking O(n) symmetry: Finite-size Scaling. *Phys. Rev. B* **32**, 447 (1985).
34. M. S. S. Challa, D. P. Landau and K. Binder, Finite-size effects at temperature-driven first-order transitions. *Phys. Rev. B* **34**, 1841 (1986).
35. C. Borgs and R. Kotecky, A rigorous theory of finite-size scaling at first-order phase transitions. *J. Stat. Phys.* **61**, 79 (1990).
36. J. Lee and J. M. Kosterlitz, Finite-size scaling and Monte Carlo simulations of first-order phase transitions. *Phys. Rev. B* **43**, 3265 (1991).
37. C. Borgs and R. Kotecky, Finite-size effects at asymmetric first-order phase transitions. *Phys. Rev. Lett.* **68**, 1734 (1992).
38. K. Vollmayr, J. D. Reger, M. Scheucher and K. Binder, Finite size effects at thermally-driven first order phase transitions: A phenomenological theory of the order parameter distribution. *Z. Phys. B* **91**, 113 (1993).
39. M. Campostrini, J. Nespolo, A. Pelissetto and E. Vicari, Finite-size scaling at first-order quantum transitions. *Phys. Rev. Lett.* **113**, 070402 (2014).
40. A. Pelissetto, D. Rossini and E. Vicari, Dynamic finite-size scaling after a quench at quantum transitions. *Phys. Rev. E* **97**, 052148 (2018).
41. F. M. Gasparini, M. O. Kimball, K. P. Mooney and M. Diaz-Avila, Finite-size scaling of ^{4}He at the superfluid transition. *Rev. Mod. Phys.* **80**, 1009 (2008).
42. I. Bloch, J. Dalibard and W. Zwerger, Many-body physics with ultracold gases. *Rev. Mod. Phys.* **80**, 885 (2008).
43. M. Campostrini and E. Vicari, Critical behavior and scaling in trapped systems. *Phys. Rev. Lett.* **102**, 240601 (2009); (E) **103**, 269901 (2009).
44. M. Campostrini and E. Vicari, Trap-size scaling in confined particle systems at quantum transitions. *Phys. Rev. A* **81**, 023606 (2010).
45. V. Piazza, V. Pellegrini, F. Beltram, W. Wegscheider, T. Jungwirth and A. H. MacDonald, First-order phase transitions in a quantum Hall ferromagnet, *Nature London* **402**, 638 (1999).
46. T. Vojta, D. Belitz, T. R. Kirkpatrick and R. Narayanan, Quantum critical behavior of itinerant ferromagnets. *Ann. Phys. (Leipzig)* **8**, 593 (1999).
47. D. Belitz, T. R. Kirkpatrick and T. Vojta, First order transitions and multicritical points in weak itinerant ferromagnets. *Phys. Rev. Lett.* **82**, 4707 (1999).
48. M. Uhlarz, C. Pfleiderer and S. M. Hayden, Quantum phase transitions in the itinerant ferromagnet $ZrZn_2$. *Phys. Rev. Lett.* **93**, 256404 (2004).
49. C. Pfleiderer, Why first order quantum phase transitions are interesting. *J. Phys. Condens. Matter* **17**, S987 (2005).
50. W. Knafo, S. Raymond, P. Lejay and J. Flouquet, Antiferromagnetic criticality at a heavy-fermion quantum phase transition. *Nat. Phys.* **5**, 753 (2009).
51. J. D'Emidio and R. K. Kaul, First-order superfluid to valence-bond solid phase transitions in easy-plane SU(N) magnets for small N. *Phys. Rev. B* **93**, 054406 (2016).

52. N. Desai and R. K. Kaul, First-order phase transitions in the square-lattice easy-plane J-Q model. *Phys. Rev. B* **102**, 195135 (2020).
53. C. R. Laumann, R. Moessner, A. Scardicchio and S. L. Sondhi, Quantum adiabatic algorithm and scaling of gaps at first-order quantum phase transitions. *Phys. Rev. Lett.* **109**, 030502 (2012).
54. Q. Luo, J. Zhao and X. Wang, Intrinsic jump character of first-order quantum phase transitions. *Phys. Rev. B* **100**, 121111(R) (2019).
55. T. W. B. Kibble, Some implications of a cosmological phase transition. *Phys. Rep.* **67**, 183 (1980).
56. W. H. Zurek, Cosmological experiments in condensed matter systems. *Phys. Rep.* **276**, 177 (1996).
57. P. C. Hohenberg and B. I. Halperin, Theory of dynamic critical phenomena. *Rev. Mod. Phys.* **49**, 435 (1977).
58. M. Campostrini, J. Nespolo, A. Pelissetto and E. Vicari, Finite-size scaling at first-order quantum transitions of quantum Potts chains. *Phys. Rev. E* **91**, 052103 (2015).
59. M. Campostrini, A. Pelissetto and E. Vicari, Quantum transitions driven by one-bond defects in quantum Ising rings. *Phys. Rev. E* **91**, 042123 (2015).
60. A. Yuste, C. Cartwright, G. De Chiara and A. Sanpera, Entanglement scaling at first order quantum phase transitions. *New J. Phys.* **20**, 043006 (2018).
61. D. Rossini and E. Vicari, Ground-state fidelity at first-order quantum transitions. *Phys. Rev. E* **98**, 062137 (2018).
62. A. Pelissetto, D. Rossini and E. Vicari, Scaling properties of the dynamics at first-order quantum transitions when boundary conditions favor one of the two phases. *Phys. Rev. E* **102**, 012143 (2020).
63. M. Campostrini, A. Pelissetto and E. Vicari, Quantum Ising chains with boundary terms. *J. Stat. Mech.* P11015 (2015).
64. P. Pfeuty, The one-dimensional Ising model with a transverse field. *Ann. Phys.* **57**, 79 (1970).
65. T. W. Burkhardt and I. Guim, Finite-size scaling of the quantum Ising chain with periodic, free, and antiperiodic boundary conditions. *J. Phys. A: Math. Gen.* **18**, L33 (1985).
66. G. G. Cabrera and R. Jullien, Role of boundary conditions in the finite-size Ising model. *Phys. Rev. B* **35**, 7062 (1987).
67. A. Pelissetto, D. Rossini and E. Vicari, Finite-size scaling at first-order quantum transitions when boundary conditions favor one of the two phases. *Phys. Rev. E* **98**, 032124 (2018).
68. M. Campostrini, J. Nespolo, A. Pelissetto and E. Vicari, Scaling phenomena driven by inhomogeneous conditions at first-order quantum transitions. *Phys. Rev. E* **91**, 022108 (2015).
69. J. Sólyom and P. Pfeuty, Renormalization-group study of the Hamiltonian version of the Potts model. *Phys. Rev. B* **24**, 218 (1981).
70. F. Iglói and J. Sólyom, First-order transition for the (1+1)-dimensional $q \geq 4$ Potts model from finite lattice extrapolation. *J. Phys. C: Solid State Phys.* **16**, 2833 (1983).
71. R. B. Potts, Some generalized order-disorder transformations. *Math. Proc. Camb. Phil. Soc.* **48**, 106 (1952).
72. R. J. Baxter, Potts model at the critical temperature. *J. Phys. C: Solid State Phys.* **6**, L445 (1973); R. J. Baxter, H. N. V. Temperley and S. E. Ashley, Triangular Potts model at its transition temperature, and related models. *Proc. R. Soc. Lond. A* **538**, 535 (1978).

73. H. W. J. Blöte and M. P. Nightingale, Critical behaviour of the two-dimensional Potts model with a continuous number of states; A finite size scaling analysis. *Phys. A* **112**, 405 (1982).
74. F. Y. Wu, The Potts model. *Rev. Mod. Phys.* **54**, 235 (1982).
75. D. Kim, $1/q$-expansion for the magnetization discontinuity of Potts model in two dimensions. *Phys. Lett. A* **87**, 127 (1981).
76. R. J. Baxter, Magnetisation discontinuity of the two-dimensional Potts model. *J. Phys. A* **15**, 3329 (1982).
77. F. Iglói and E. Carlon, Boundary and bulk phase transitions in the two-dimensional Q-state Potts model ($Q > 4$). *Phys. Rev. B* **59**, 3783 (1999).
78. S.-J. Gu, Fidelity approach to quantum phase transitions. *Int. J. Mod. Phys. B* **24**, 4371 (2010).
79. D. Braun, G. Adesso, F. Benatti, R. Floreanini, U. Marzolino, M. W. Mitchell and S. Pirandola, Quantum-enhanced measurements without entanglement, *Rev. Mod. Phys.* **90**, 035006 (2018).
80. P. W. Anderson, Infrared catastrophe in Fermi gases with local scattering potentials. *Phys. Rev. Lett.* **18**, 1049 (1967).
81. P. Zanardi, M. G. A. Paris and L. Campos Venuti, Quantum criticality as a resource for quantum estimation. *Phys. Rev. A* **78**, 042105 (2008).
82. C. Invernizzi, M. Korbman, L. Campos Venuti and M. G. A. Paris, Optimal quantum estimation in spin systems at criticality. *Phys. Rev. A* **78**, 042106 (2008).
83. L. Campos Venuti and P. Zanardi, Quantum critical scaling of the geometric tensors. *Phys. Rev. Lett.* **99**, 095701 (2007).
84. M. H. S. Amin and V. Choi, First-order quantum phase transition in adiabatic quantum computation. *Phys. Rev. A* **80**, 062326 (2009).
85. A. P. Young, S. Knysh and V. N. Smelyanskiy, First order phase transition in the quantum adiabatic algorithm. *Phys. Rev. Lett.* **104**, 020502 (2010).
86. T. Jörg, F. Krzakala, G. Semerjian and F. Zamponi, First-order transitions and the performance of quantum algorithms in random optimization problems. *Phys. Rev. Lett.* **104**, 207206 (2010).
87. D. Nigro, D. Rossini and E. Vicari, Scaling properties of work fluctuations after quenches near quantum transitions. *J. Stat. Mech.* 023104 (2019).
88. D. Rossini and E. Vicari, Scaling of decoherence and energy flow in interacting quantum spin systems. *Phys. Rev. A* **99**, 052113 (2019).
89. A. Polkovnikov, K. Sengupta, A. Silva and M. Vengalattore, Colloquium: Nonequilibrium dynamics of closed interacting quantum systems. *Rev. Mod. Phys.* **83**, 863 (2011).
90. A. Chandran, A. Erez, S. S. Gubser and S. L. Sondhi, Kibble-Zurek problem: Universality and the scaling limit. *Phys. Rev. B* **86**, 064304 (2012).
91. F. Tarantelli and E. Vicari, Out-of-equilibrium dynamics arising from slow round-trip variations of Hamiltonian parameters across quantum and classical critical points. *Phys. Rev. B* **105**, 235124 (2022).
92. A. Pelissetto, D. Rossini and E. Vicari, Out-of-equilibrium dynamics driven by localized time-dependent perturbations at quantum phase transitions. *Phys. Rev. B* **97**, 094414 (2018).
93. L. D. Landau, On the theory of transfer of energy at collisions II. *Phys. Z. Sowjetunion* **2**, 46 (1932).
94. C. Zener, Non-adiabatic crossing of energy levels. *Proc. R. Soc. Lond. A* **137**, 696 (1932).

95. N. V. Vitanov and B. M. Garraway, Landau-Zener model: Effects of finite coupling duration. *Phys. Rev. A* **53**, 4288 (1996).
96. M. Abramowitz and I. A. Stegun, *Handbook of Mathematical Functions* (Dover, New York, 1964).
97. I. Bloch, Quantum coherence and entanglement with ultracold atoms in optical lattices. *Nature* **453**, 1016 (2008).
98. J. Simon, W. S. Bakr, R. Ma, M. E. Tai, P. M. Preiss and M. Greiner, Quantum simulation of antiferromagnetic spin chains in an optical lattice. *Nature* **472**, 307 (2011).
99. E. E. Edwards, S. Korenblit, K. Kim, R. Islam, M.-S. Chang, J. K. Freericks, G.-D. Lin, L.-M. Duan and C. Monroe, Quantum simulation and phase diagram of the transverse-field Ising model with three atomic spins. *Phys. Rev. B* **82**, 060412(R) (2010).
100. R. Islam, E. E. Edwards, K. Kim, S. Korenblit, C. Noh, H. Carmichael, G.-D. Lin, L.-M. Duan, C.-C. Joseph Wang, J. K. Freericks and C. Monroe, Onset of a quantum phase transition with a trapped ion quantum simulator, *Nat. Commun.* **2**, 377 (2011).
101. G.-D. Lin, C. Monroe and L.-M. Duan, Sharp Phase Transitions in a small frustrated network of trapped ion Spins. *Phys. Rev. Lett.* **106**, 230402 (2011).
102. K. Kim, S. Korenblit, R. Islam, E. E. Edwards, M.-S. Chang, C. Noh, H. Carmichael, G.-D. Lin, L.-M. Duan, C. C. Joseph Wang, J. K. Freericks and C. Monroe, Quantum simulation of the transverse Ising model with trapped ions. *New J. Phys.* **13**, 105003 (2011).
103. P. Richerme, Z.-X. Gong, A. Lee, C. Senko, J. Smith, M. Foss-Feig, S. Michalakis, A. V. Gorshkov and C. Monroe, Non-local propagation of correlations in long-range interacting quantum systems. *Nature* **511**, 198 (2014).
104. P. Jurcevic, B. P. Lanyon, P. Hauke, C. Hempel, P. Zoller, R. Blatt and C. F. Roos, Observation of entanglement propagation in a quantum many-body system. *Nature* **511**, 202 (2014).
105. S. Debnath, N. M. Linke, C. Figgatt, K. A. Landsman, K. Wright and C. Monroe, Demonstration of a small programmable quantum computer with atomic qubits. *Nature* **536**, 63 (2016).
106. H. Labuhn, D. Barredo, S. Ravets, S. de Leseleuc, T. Macri, T. Lahaye and A. Browaeys, Tunable two-dimensional arrays of single Rydberg atoms for realizing quantum Ising models. *Nature* **534**, 667 (2016).
107. A. Keesling, A. Omran, H. Levine, H. Bernien, H. Pichler, S. Choi, R. Samajdar, S. Schwartz, Pietro Silvi, S. Sachdev, P. Zoller, M. Endres, M. Greiner, V. Vuletić and M. D. Lukin, Quantum Kibble-Zurek mechanism and critical dynamics on a programmable Rydberg simulator. *Nature* **568**, 207 (2019).
108. A. Sinha, T. Chanda and J. Dziarmaga, Nonadiabatic dynamics across a first-order quantum phase transition: Quantized bubble nucleation. *Phys. Rev. B* **103**, L220302 (2021).
109. G. Di Meglio, D. Rossini and E. Vicari, Dissipative dynamics at first-order quantum transitions. *Phys. Rev. B* **102** (2020) 224302.
110. A. Nava and M. Fabrizio, Lindblad dissipative dynamics in the presence of phase coexistence. *Phys. Rev. B* **100**, 125102 (2020).
111. G. Lindblad, On the generators of quantum dynamical semigroups. *Commun. Math. Phys.* **48**, 119 (1976).
112. V. Gorini, A. Kossakowski and E. C. G. Sudarshan, Completely positive dynamical semigroups of N-level systems. *J. Math. Phys.* **17**, 821 (1976).

113. H.-P. Breuer and F. Petruccione, *The Theory of Open Quantum Systems* (Oxford University Press, New York, 2002).
114. A. Rivas and S. F. Huelga, *Open Quantum System: An Introduction* (Springer, New York, 2012).
115. M. E. Fisher, Critical phenomena. In M. S. Green (ed.) Varenna, Italy, July 27–August 8, 1970. Course 51, Proceedings of the International School of Physics "Enrico Fermi", (Academic Press, New York, 1971).
116. A. Aharony, Dependence of universal critical behavior on symmetry and range of interaction. In C. Domb and M. S. Green (eds.) Vol. 6, p. 357, *Phase Transitions and Critical Phenomena*, (Academic Press, London, 1976).
117. M. P. A. Fisher, P. B. Weichman, G. Grinstein and D. S. Fisher, Boson localization and the superfluid-insulator transition. *Phys. Rev. B* **40**, 546 (1989).
118. B. Schmittmann and R. K. P. Zia, Statistical mechanics of driven diffusive systems. In C. Domb and J. L. Lebowitz (eds.) Vol. 17, *Phase Transitions and Critical Phenomena*, (Academic Press, London, 1995).
119. V. Privman and J. Rudnick, Nonsymmetric first-order transitions: finite-size scaling and tests for infinite-range models. *J. Stat. Phys.* **60**, 551 (1990).
120. C. Bonati, M. D'Elia and E. Vicari, Universal scaling effects of a temperature gradient at first-order transitions. *Phys. Rev. E* **89**, 062132 (2014).
121. A. Pelissetto and E. Vicari, Off-equilibrium scaling behaviors driven by time-dependent external fields in three-dimensional $O(N)$ vector models. *Phys. Rev. E* **93**, 032141 (2016).
122. S. Scopa and S. Wald, Dynamical off-equilibrium scaling across magnetic first-order phase transitions. *J. Stat. Mech.* 113205 (2018).
123. K. Binder, K. Vollmayr, H.-P. Deutsch, J. D. Reger, M. Scheucher and D. P. Landau, Monte Carlo methods for first order phase transitions: some recent progress. *Int. J. Mod. Phys. C* **3**, 1025 (1992).
124. C. Borgs and J. Imbrie, A unified approach to phase diagrams in field theory and statistical mechanics. *Comm. Math. Phys.* **123**, 305 (1989).
125. S. Cabasino *et al.* (APE collaboration), The ape with a small jump. *Nucl. Phys. B (Proc. Suppl.)* **17**, 218 (1990).
126. M. Fukugita, H. Mino, M. Okawa and A. Ukawa, Resolving the order of phase transitions in Monte Carlo simulations. *J. Phys. A* **23**, L561 (1990).
127. J. F. McCarthy, Determination of the order of phase transitions in numerical simulations. *Phys. Rev. B* **41**, 9530 (1990).
128. A. Billoire, First order phase transitions of spin systems. *Nucl. Phys. B (Proc. Suppl.)* **42**, 21 (1995).
129. M. M. Tsypin and H. W. J. Blöte, Probability distribution of the order parameter for the three-dimensional Ising-model universality class: A high-precision Monte Carlo study. *Phys. Rev. E* **62**, 73 (2000).
130. A. Pelissetto and E. Vicari, Three-dimensional ferromagnetic CP^{N-1} models. *Phys. Rev. E* **100** (2019) 022122.
131. C. Bonati, A. Pelissetto and E. Vicari, Phase diagram, symmetry-breaking pattern, and critical behavior of three-dimensional lattice multiflavor scalar chromodynamics. *Phys. Rev. Lett.* **123**, 232002 (2019); Three-dimensional lattice multiflavor scalar chromodynamics: interplay between global and gauge symmetries. *Phys. Rev. D* **101**, 034505 (2020).
132. C. Bonati, A. Pelissetto and E. Vicari, Lattice Abelian-Higgs models with noncompact gauge field. *Phys. Rev. B* **103**, 085104 (2021).

133. C. Bonati, A. Franchi, A. Pelissetto and E. Vicari, Phase diagram and Higgs phases of 3D lattice SU(N_c) gauge theories with multiparameter scalar potentials. *Phys. Rev. E* **104**, 064111 (2021).
134. K. Binder, Finite size scaling analysis of Ising model block distribution functions. *Z. Phys. B* **43**, 119 (1981).
135. P. Calabrese, P. Parruccini, A. Pelissetto and E. Vicari, Critical behavior of O(2) $\otimes$ O(N) symmetric models. *Phys. Rev. B* **70**, 174439 (2004).
136. C. P. Enz (ed.), *Dynamical Critical Phenomena and Related Topics. Lecture Notes in Physics*, Vol. 104 (Springer, Berlin, 1979).
137. P. Calabrese and A. Gambassi, Ageing properties of critical systems. *J. Phys. A* **38**, R133 (2005).
138. R. Folk and G. Moser, Critical dynamics: A field-theoretical approach. *J. Phys. A* **39**, R207 (2006).
139. M. Creutz, *Quark, Gluons and Lattices* (Cambridge University Press, 1983).
140. P. Fontana, Scaling behavior of Ising systems at first-order transitions. *J. Stat. Mech.* 063206 (2019).
141. H. Panagopoulos and E. Vicari, Off-equilibrium scaling across a first-order transition. *Phys. Rev. E* **92**, 062107 (2015).
142. A. Pelissetto and E. Vicari, Dynamic finite-size scaling at first-order transitions. *Phys. Rev. E* **96**, 012125 (2017).
143. A. Pelissetto and E. Vicari, Dynamic off-equilibrium transition in systems slowly driven across thermal first-order transitions. *Phys. Rev. Lett.* **118**, 030602 (2017).
144. H. Panagopoulos, A. Pelissetto and E. Vicari, Dynamic scaling behavior at thermal first-order transitions in systems with disordered boundary conditions. *Phys. Rev. D* **98**, 074507 (2018).
145. A. Tröster and K. Binder, Microcanonical determination of the interface tension of flat and curved interfaces from Monte Carlo simulations. *J. Phys.: Condens. Matter* **24**, 284107 (2012).
146. J. L. Meunier and A. Morel, Condensation and metastability in the 2D Potts model. *Eur. Phys. J. B* **13**, 341 (2000).
147. E. S. Loscar, E. E. Ferrero, T. S. Grigera and S. A. Cannas, Nonequilibrium characterization of spinodal points using short time dynamics. *J. Chem. Phys.* **131**, 024120 (2009).
148. T. Nogawa, N. Ito and H. Watanabe, Static and dynamical aspects of the metastable states of first order transition systems. *Physics Procedia* **15**, 76 (2011).
149. M. Ibàñez Berganza, P. Coletti and A. Petri, Anomalous metastability in a temperature-driven transition. *Europhys. Lett.* **106**, 56001 (2014).
150. N. Liang and F. Zhong, Renormalization-group theory for cooling first-order phase transitions in Potts models. *Phys. Rev. E* **95** 032124 (2017).
151. F. Chippari, L. F. Cugliandolo and M. Picco, Low-temperature universal dynamics of the bidimensional Potts model in the large q limit. *J. Stat. Mech.* 093201 (2021).
152. F. Corberi, L. F. Cugliandolo, M. Esposito, O. Mazzarisi and M. Picco, How many phases nucleate in the bidimensional Potts model?. *J. Stat. Mech.* 073204 (2022).

https://doi.org/10.1142/9789811282386_0028

Chapter 28

Michael Fisher and Localization

Franz J. Wegner
Institute for Theoretical Physics,
Philosophenweg 19,
Ruprecht-Karls-University,
D-69120 Heidelberg, Germany

Michael Fisher presented many important contributions to critical phenomena and distributed it widely in review articles. His paper with Kenneth Wilson on the expansion of critical exponents in $4 - \epsilon$ dimensions gave the field of critical phenomena an enormous boost.

A few years later Alexander Polyakov presented his paper on the expansion around the lower critical dimension 2, which brought an important impetus for non-linear sigma-models. The Anderson localization transition due to disorder is described by such a model of supersymmetric matrices. It is applied to various dimensions and to ensembles with and without time-reversal invariance.

1. Critical Phenomena and Michael Fisher

Michael Fisher was one of the leading scientists in the development of critical phenomena. Cyril Domb describes[1] the first 16 years (1948–1966) of Michael's academic career at King's College in London, where he started as an undergraduate and ended as full professor. He excelled by his mathematical abilities. Domb's inaugural lecture 1955 *Statistical Physics and its problems* directed Michael to critical phenomena and related problems, which he started with enumeration of graphs and paths on lattices.[2]

Michael Fisher worked on high- and low-temperature expansions for various vector models in statistical mechanics. These expansions allow to estimate critical temperatures and critical exponents. He mastered various systems connected with critical behaviour. He presented vividly his knowledge in lectures and review articles. I mention only the References,[3–6] which gave him the title *Pope of Critical Phenomena.*

In 1966 Michael went to the Physics Department at Cornell University. Probably his most important paper was that with Ken Wilson.[7] In this paper they expanded the critical exponents for the Ising- and the XY-model in an ϵ-expansion in $d = 4-\epsilon$ dimensions, starting out from the recurrence relation derived by Wilson.[8]

This paper was seminal to many papers. Initially most of them were on the isotropic and anisotropic n-vector model, as can be seen for example in the review-articles in volume 6 of the series on *Phase Transitions and Critical Phenomena*,[9] the first volume of this series after the appearance of the Wilson-Fisher paper.

However, some problems, such as the localization problem, call for matrix models. This will be considered in the following Section.

2. Anderson Localization

2.1. *Introduction*

Phil Anderson argued in his 1958 paper[10] that disorder in a metal destroys diffusion close to the band tails, but it may still allow diffusion and conduction in the centre of the band. Thus at a certain Fermi-energy, called mobility edge, the system switches from insulating to conducting behaviour. The eigenstates in the insulating regime are localized, whereas they are extended in the conducting regime. David Thouless gives a historical survey[11] in the volume *50 years of Anderson localization.*[12] Interest in this phenomenon grew and attracted an increasing number of physicists, in particular Neville Mott[13,14] and coworkers and also David Thouless.[15] One of the problems was to distinguish, what is due to disorder and what is due to interactions like coupling to phonons or Coulomb interaction between electrons.

Here I consider only the localization problem due to disorder in the one-particle Hamiltonian as described by the Hamiltonian in Eq. (1). It became clear that the localization transition is a critical phenomenon.

One observation was that the amplitudes at the sites r occur pairwise. This became clear from the $1/n$-expansions of the corresponding n-orbital models (Oppermann and Wegner[16]), but also in the approach by Aharony and Imry,[17] and in the calculation by Harris and Lubensky,[18,19] although the later approaches did not come to valid conclusions. The $1/n$-expansion yields the leading terms of the critical exponents.

The transformation to the non-linear sigma-model was performed in the papers by Wegner[20] and by Schäfer and Wegner[21] in 1979 and 1980. There the replica trick was used: The calculation was performed for m replicas and finally the limit $m \to 0$ was taken. The result is a non-linear sigma-model of a matrix field Q, which obeys $Q^2 = 1$. The symmetry of Q is of hyperbolic nature. The correlations of the amplitudes of this model are expressed by correlations of elements of the Q-matrices. Efetov *et al.*[22] performed a similar calculation with the replica trick, however representing the amplitudes by fermionic variables. They obtained the renormalization in one-loop order for all three Dyson classes. The critical behaviour of the nonlinear sigma-models were analyzed with the renormalization group procedure of Polyakov[23] starting from the lower critical dimension two.

In 1979 the paper by Abrahams, Anderson, Licciardello, and Ramakrishnan[24] appeared with their renormalization approach for the conductance.

The replica trick may sometimes fail, see Ref.[25] However, in 1982 Efetov[26] introduced Grassmann variables in addition to the complex variables and thus avoided the replica trick and gave safer ground by this super-symmetric formulation of the non-linear sigma-model.

There are several review articles on this non-linear sigma-model and related topics. I mention only those by Efetov,[27,28] Verbaarschott *et al.*,[29] Belitz *et al.*,[30] Zirnbauer,[31–33] Guhr *et al.*,[34] Mirlin,[35] Evers *et al.*,[36] Abrahams in *Fifty years of Anderson Localization*,[12] and chapters 4 and 21-23 of Wegner.[39]

Subsection 2.2 follows with preliminaries. The tight-binding model is described. Then I explain the three symmetry classes according to Dyson.[40] Finally a few relations for anticommuting variables and for supermatrices are given.

In subsection 2.3 the non-linear sigma model for the tight-binding model is developed including the Hubbard-Stratonovich transformation.

In subsection 2.4 on the random matrix model we extend the models to n-orbital models. Then the technique will be used for the zero-dimensional case, the problem originally introduced by Wigner[41] as description for the statistics of nuclear levels. The correlations between the energy levels are given for the three Dyson ensembles.

In subsection 2.5 on hydrodynamics diffusion and conductivity are considered in the limit of small wave-vectors q and frequencies ω. We emphasize diffusons and the importance of cooperons in case of time-reversal invariance. For the Dyson-ensembles symmetry relations and β-functions are given both in powers of the inverse diffusion constant and in the inverse conductance. Remarks on density fluctuations and multifractality and the statistics of energy spacing are added.

A few other developments are mentioned in the final paragraph: All compact symmetric spaces correspond to non-linear sigma-models which adds to the three Wigner-Dyson classes three chiral classes and four Bogolubov-de Gennes classes. Exact proof of absence of diffusion for large disorder or low energies is mentioned. Further numerically determined exponents are listed. A few authors who worked on interacting disordered systems are cited.

2.2. *Preliminaries*

Tight-binding model I start out with the diffusive model, which is a tight-binding model with orbitals $|r\rangle$ at sites r and hopping matrix elements $f_{r,r'} + t_{r,r'}$

$$H = \sum_{r,r'} |r\rangle(f_{r,r'} + t_{r,r'})\langle r'| \tag{1}$$

The elements $f_{r,r'}$ are random matrix elements with average zero. For the sake of simplicity they are assumed to be Gaussian distributed

$$\overline{f_{r,r'}} = 0, \quad \overline{f_{r_1,r_1'} f_{r_2,r_2'}} = M_{r_1,r_1'} \delta_{r_1,r_2'} \delta_{r_2,r_1'} \tag{2}$$

$M_{r,r'}$ is assumed to be translational invariant,

$$M_{r,r'} = M_{r-r'} \tag{3}$$

In contrast t is assumed to be fixed, but also translational invariant,

$$t_{r,r'} = t_{r-r'}. \tag{4}$$

I will also consider the n-orbital model, where there are n orbitals at each lattice site.

Time-reversal invariance In accord with Dyson[40] we consider ensembles with three symmetries. The unitary ensemble applies to systems without time-reversal invariance. For $t_{r-r'} = 0$ it is also called phase-invariant ensemble, since it is invariant against transformations $|r\rangle \to \exp(\mathrm{i}\phi_r)|r\rangle$.

Disordered systems with time-reversal invariance are described by orthogonal and symplectic ensembles, where the first one neglects spin, and the second one takes spin-dependent hopping into account. The orthogonal ensemble is also called real-matrix ensemble, since the matrix elements f and t are real. The symplectic ensemble takes spin 1/2 into account. To preserve time-reversal invariance the matrix elements are real quaternions. This is described in Eqs. (108–111).

Grassmann variables Grassmann variables are anticommuting variables. Thus

$$\xi_1\xi_2 = -\xi_2\xi_1, \quad \xi_1^2 = 0 \tag{5}$$

for Grassmann variables ξ_1, ξ_2. Products of an even (odd) number of Grassmann variables and sums of these products are called even (odd) elements of the algebra.

These variables were introduced by Grassmann and later independently by Berezin.[42,43] For the algebra, analysis, and groups of these variables see Refs.[27,28,39,43,44]

Their usefulness becomes obvious from the Gaussian integrals

$$\int \exp\left(-\sum x_i^* A_{ij} x_j\right) \prod_i \frac{\mathrm{d}\Re x_i \mathrm{d}\Im x_i}{\pi} = (\det A)^{-1}, \tag{6}$$

$$\int \exp\left(-\sum \xi_i^* A_{ij} \xi_j\right) \prod_i \mathrm{d}\xi_i^* \mathrm{d}\xi_i = \det A \tag{7}$$

for complex x_i and Grassmannian ξ_i. The integrations in the first integral run from $-\infty$ to $+\infty$. The real parts of the eigenvalues of A have to be positive. The ξ and ξ^* have to be linearly independent in the second integral.

Supermatrices Let us denote the set of $(n+m) \times (n+m)$-matrices

$$K = \begin{pmatrix} a & \alpha \\ \beta & b \end{pmatrix} \tag{8}$$

by $\mathcal{M}(n,m)$, where a and α have n rows, β and b m rows, a and β n columns, and α and b have m columns. The matrices a and b consist of even elements, and α and β of odd elememts. a is called boson-boson matrix, b fermion-fermion matrix. For these matrices one defines the supertrace

$$\mathrm{str}(K) = \mathrm{tr}(a) - \mathrm{tr}(b) \tag{9}$$

It obeys for matrices $K, L \in \mathcal{M}(n,m)$

$$\mathrm{str}(KL) = \mathrm{str}(LK). \tag{10}$$

The superdeterminant is given by

$$\mathrm{sdet}(K) = \frac{\det(a)}{\det(b - \beta a^{-1}\alpha)} = \frac{\det(a - \alpha b^{-1}\beta)}{\det(b)}. \tag{11}$$

It obeys

$$\mathrm{sdet}(KL) = \mathrm{sdet}(K)\,\mathrm{sdet}(L). \tag{12}$$

One can express the superdeterminant by the supertrace

$$\mathrm{sdet}(K) = \exp(\mathrm{str}(\ln K)) \tag{13}$$

and its differential as

$$\mathrm{d}\,\mathrm{sdet}(K) = \mathrm{sdet}(K)\,\mathrm{str}(K^{-1}\mathrm{d}K). \tag{14}$$

2.3. *Tight-binding model*

Green's functions The one-particle Green's functions

$$G(r, r', z) = \langle r|\frac{1}{z - H}|r'\rangle \tag{15}$$

are obtained from integrals over time t

$$\mathrm{i}\int_{-\infty}^{0} \mathrm{d}t e^{(\mathrm{i}E+\eta-\mathrm{i}H)t} = \frac{1}{E - \mathrm{i}\eta - H} \quad \Im z < 0, \tag{16}$$

$$-\mathrm{i}\int_{0}^{+\infty} \mathrm{d}t e^{(\mathrm{i}E-\eta-\mathrm{i}H)t} = \frac{1}{E + \mathrm{i}\eta - H} \quad \Im z > 0. \tag{17}$$

The upper Green's function is called advanced, the lower one retarded. The density of states ρ at site r is obtained as the limit

$$\rho(r, E) = \lim_{\eta\to+0}(G(r, r, E - \mathrm{i}\eta) - G(r, r, E + \mathrm{i}\eta))/(2\pi\mathrm{i}). \tag{18}$$

The product of m Green's functions for the tight-binding Hamiltonian (1) reads

$$\prod_{p=1}^{m} G(r_p, r'_p, z_p) = \int \prod_{p=1}^{m} s_{p\mathrm{b}} x_p(r_p) x_p^*(r'_p)$$
$$\times \exp\left(- \sum_{p,r,r'} [s_{p\mathrm{b}} x_p^*(r)(z_p \delta_{r,r'} - f_{r,r'} - t_{r.r'}) x_p(r') \right.$$
$$\left. + s_{p\mathrm{f}} \xi_p^*(r)(z_p \delta_{r,r'} - f_{r,r'} - t_{r,r'}) \xi_p(r)] \right) \prod_{r,p} \left(\frac{s_{p\mathrm{f}}}{s_{p\mathrm{b}}} \mathrm{D}[S_p(r)] \right), \tag{19}$$

with the integrals over the complex variables x and the Grassmann variables ξ

$$(S_p^*(r) = (x_p^*(r), \xi_p^*(r)), \quad S_p(r) = \begin{pmatrix} x_p(r) \\ \xi_p(r) \end{pmatrix}, \tag{20}$$

$$\mathrm{D}[S_p(r)] = \int \frac{\mathrm{d}\Re x_p(r) \mathrm{d}\Im x_p(r)}{\pi} \mathrm{d}\xi_p^*(r) \mathrm{d}\xi_p(r). \tag{21}$$

Convergence requires the choice $s_{p\mathrm{b}} = -\mathrm{i}\Im z_p$ for the integration of x_p. We introduce

$$A_p(r, r') = s_{p\mathrm{b}} x_p^*(r) x_p(r') + s_{p\mathrm{f}} \xi_p^*(r) \xi_p(r') \tag{22}$$

and the functions

$$\mathcal{S}_{\mathrm{int}} = \sum_{p,r,r'} A_p(r, r')(f_{r,r'} + t_{r,r'}), \tag{23}$$

$$\mathcal{S}_{z,J} = \sum_{p,r} s_{\mathrm{b}p} z_p x_p^*(r) x_p(r) + \sum_{p,r} s_{\mathrm{f}p} z_p \xi_p^*(r) \xi_p(r)$$
$$- \sum_{p,p',r,r'} \sqrt{s_{p\mathrm{b}} s_{p'\mathrm{b}}} J_{p',p}(r', r) x_{p'}^*(r') x_p(r). \tag{24}$$

The contributions J are source terms. They allow the determination of Green's functions and products of Green's functions. The partition function

$$Z(J) = \int \exp(-\mathcal{S}_{\mathrm{int}} - \mathcal{S}_{z,J}) \prod_{p,r} \mathrm{D}[S_p(r)] \tag{25}$$

yields

$$\frac{\partial Z(J)}{Z(0) \partial J_{p,p}(r', r)} = G(r, r', z_p), \tag{26}$$

$$\frac{\partial^2 Z(J)}{Z(0)\partial J_{p_1,p_1}(r_1',r_1)\partial J_{p_2,p_2}(r_2',r_2)} = G(r_1,r_1',z_{p_1})G(r_2,r_2',z_{p_2}), \tag{27}$$

$$\frac{\partial^2 Z(J)}{Z(0)\partial J_{p_1,p_2}(r_1',r_2)\partial J_{p_2,p_1}(r_2',r_1)} = G(r_1,r_1',z_{p_1})G(r_2,r_2',z_{p_2}). \tag{28}$$

Ensemble average and Hubbard-Stratonovich transformation We separate

$$\mathcal{S}_{\text{int}} = \mathcal{S}_{\text{fluc}} + \mathcal{S}_{\text{t}}, \quad \mathcal{S}_{\text{fluc}} = \sum_{p,r,r} f_{r,r'} A_p(r,r'), \quad \mathcal{S}_{\text{t}} = \sum_{p,r,r'} t_{r.r'} A_p(r,r'). \tag{29}$$

The Gaussian distribution (2) of f yields the ensemble average

$$\overline{\exp(-\mathcal{S}_{\text{fluc}})} = \exp\left(\frac{1}{2}\sum_{p,p',r,r'} M_{r,r'} A_p(r,r') A_{p'}(r',r)\right) \tag{30}$$

The product $A_p A_{p'}$ contains products of four spin variables. They can be eliminated by means of the Hubbard-Stratonovich transformation.

For this purpose we introduce the inverse w of M,

$$(M^{-1})_{r,r'} = w_{r,r'}, \tag{31}$$

In order that the matrices M and w are positive definite, we require $w_{r,r'} \leq 0$ for $r \neq r'$. From now on we restrict ourselves to $m = 2$ energies z_1, z_2 and Green's functions and put them together

$$S(r) = \begin{pmatrix} x_{p_1}(r) \\ x_{p_2}(r) \\ \xi_{p_1}(r) \\ \xi_{p_2}(r) \end{pmatrix}. \tag{32}$$

and use

$$s = \text{diag}(s_{p_1\text{b}}, s_{p_2\text{b}}, s_{p_1\text{f}}, s_{p_2\text{f}}), \tag{33}$$

$$z = \text{diag}(z_{p_1}, z_{p_2}, z_{p_1}, z_{p_2}). \tag{34}$$

Then

$$A(r,r') = S^*(r) s S(r') \tag{35}$$

and

$$A(r,r')A(r',r) = \text{str}(B(r)B(r')) \tag{36}$$

holds with

$$B(r) = \sqrt{s} S(r) S^*(r) \sqrt{s} \in \mathcal{M}(2,2). \tag{37}$$

Hence we have

$$\overline{\exp(-\mathcal{S}_{\text{fluc}})} = \exp\left(\frac{1}{2}\sum_{r,r'} M_{r,r'} \operatorname{str}(B(r)B(r'))\right). \tag{38}$$

We introduce fields $R(r) \in \mathcal{M}(2,2)$, where two Green's functions are considered for energies $E \pm \omega/2$. Then also B belongs to $\mathcal{M}(2,2)$.

We observe

$$\begin{aligned}\sum_{r,r'} w_{r,r'} \operatorname{str}[(R(r) - \sum_{r'_1} M_{r,r'_1} B(r'_1)) \\ \times (R(r') - \sum_{r_1} M_{r',r_1} B(r_1))] \\ = \sum_{r,r'} w_{r,r'} \operatorname{str}(R(r)R(r')) - \sum_{r} \operatorname{str}(R(r)B(r)) \\ - \sum_{r'} \operatorname{str}(B(r')R(r')) + \sum_{r,r'} M_{r',r} \operatorname{str}(B(r')B(r)).\end{aligned} \tag{39}$$

One multiplies the expression (39) by $-1/2$, exponentiates it and integrates it over R. This integral yields one. One multiplies the integral by the expression for the average of the fluctuation average (38). Then one is left with

$$\begin{aligned}\overline{\exp(-\mathcal{S}_{\text{fluc}})} = \prod_r \int \mathrm{D}[R(r)] \exp\Bigg(-\frac{1}{2}\sum_{r,r'} w_{r,r'} \operatorname{str}(R(r)R(r')) \\ + \sum_r \operatorname{str}(B(r)R(r))\Bigg).\end{aligned} \tag{40}$$

Eqs. (39) and (40) constitute the Hubbard-Stratonovich transformation.[37,38] Since

$$\operatorname{str}(B(r)R(r))) = \sum_r \operatorname{str}(S^*(r)\sqrt{s}R(r)\sqrt{s}S(r)) \tag{41}$$

we obtain

$$\begin{aligned}Z(J) = \int \prod_r \left(\frac{s_{p_1\mathrm{f}} s_{p_2\mathrm{f}}}{s_{p_1\mathrm{b}} s_{p_2\mathrm{b}}} \mathrm{D}[S(r)]\right) \prod_r \mathrm{D}[R(r)] \exp\Bigg(\operatorname{str}\left[-\frac{1}{2}\sum_{r,r'} w_{r,r'} R(r)R(r')\right] \\ + \sum_{r,r'} \operatorname{str}[S^*(r)\sqrt{s}(-z\delta_{r,r'} + t_{r,r'} + J(r,r') + R(r)\delta_{r,r'})\sqrt{s}S(r')]\Bigg).\end{aligned} \tag{42}$$

Since the vectors S appear only bilinearly, integration over S can be performed and yields

$$Z(J) = \int \prod_r \mathrm{D}[R(r)] \exp\left(\mathrm{str}\left[-\frac{1}{2} \sum_{r,r'} w_{r,r'} R(r) R(r') \right] \right.$$
$$\times\ \mathrm{sdet}(R - z + t + J)^{-1}. \tag{43}$$

Fourier transform If the hopping matrix elements t vanish, then we can immediately go to the next step and determine the saddle point of the integrand of Eq. (43). However, with the hopping matrix elements t we have to perform a Fourier transform,

$$S(r) = \frac{1}{\sqrt{N}} \sum_q \mathrm{e}^{\mathrm{i}qr} \hat{S}(q), \quad S^*(r) = \frac{1}{\sqrt{N}} \sum_q \mathrm{e}^{-\mathrm{i}qr} \hat{S}^*(q). \tag{44}$$

and

$$\hat{t}_q = \frac{1}{N} \sum_{r,r'} \mathrm{e}^{\mathrm{i}q(r'-r)} t_{r,r'}, \quad \mathcal{S}_\mathrm{t} = \sum_{p,q} \hat{t}_q \hat{S}^*(q) \hat{S}(q) \tag{45}$$

with q confined to the first Brillouin zone. N is the number of lattice sites. We assume periodic boundary conditions. We indicate the Fourier transformed quantities by hats.

The corresponding transformation for (31) yields for $\hat{M}$ and $\hat{w}$

$$\hat{M}_q \hat{w}_q = 1. \tag{46}$$

The choice

$$\hat{R}_{q-q'} = \frac{1}{N} \sum_r \mathrm{e}^{\mathrm{i}(q'-q)r} R(r). \tag{47}$$

yields

$$\sum_{q,q'} \hat{S}^*(q) \sqrt{s} \hat{R}_{q-q'} \sqrt{s} \hat{S}(q') = \sum_r S^*(r) \sqrt{s} R(r) \sqrt{s} S(r) \tag{48}$$

for the expression (41).

Saddle point One obtains the saddle point of the integrand of $Z(0)$, Eq. (43) by setting $R(r) = R^{(0)}$ for all r. Then the saddle point equation reads

$$\hat{w}_0 R^{(0)} + \frac{1}{N} \sum_q \frac{1}{R^{(0)} - z + \hat{t}_q} = 0. \tag{49}$$

There are $N-1$ solutions along the real axis between the lowest and largest values of $z_p - \hat{t}_q$. Besides there is a pair of solutions outside the real axis, if z is inside the band. They are relevant for the saddle point and diagonal in p and ν,

$$R^{(0)}_{p\nu,p'\nu'}(r) = \delta_{p,p'}\delta_{\nu,\nu'}R^{\rm d}_p. \tag{50}$$

The one-particle Green's function is given within the saddle point approximation

$$\hat{G}^{(0)}(q,z_p) = \frac{1}{z_p - R^{\rm d}_p - \hat{t}_q}. \tag{51}$$

Second order and fluctuations Returning to Z in Eq. (43) we expand

$$\mathrm{sdet}(R - z + t + J) = \mathrm{sdet}(R - z + t)\,\mathrm{sdet}(1 - \mathcal{G}J), \quad \mathcal{G} := \frac{1}{z - R - t} \tag{52}$$

Then we obtain with $R = R^{(0)} + \delta R$

$$\mathcal{G} = G^{(0)} + G^{(0)}\delta R G^{(0)} + G^{(0)}\delta R G^{(0)}\delta R G^{(0)} + \cdots \tag{53}$$

and the one-particle Green's function up to order $(\delta R)^2$

$$\hat{G}_1(q,z_p) = \hat{G}^{(0)}(q,z_p) + (\hat{G}^{(0)}(q,z_p))^2 \sum_{q',p'\nu} \overline{\hat{G}^{(0)}(q',z_{p'})\delta\hat{R}_{q-q',p\mathrm{b},p'\nu}\delta\hat{R}_{q'-q,p'\nu,p\mathrm{b}}}, \tag{54}$$

where $\nu = 0$ stands for b and $\nu = 1$ for f.

The two-particle function yields

$$\begin{aligned}
&\hat{G}_2(q_1,q'_1,z_1;q_2,q'_2,z_2)\\
&= \overline{\hat{\mathcal{G}}_{1\mathrm{b},1\mathrm{b}}(q_1,q'_1)\hat{\mathcal{G}}_{2\mathrm{b},2\mathrm{b}}(q_2,q'_2)} + \overline{\hat{\mathcal{G}}_{1\mathrm{b},2\mathrm{b}}(q_1,q'_2)\hat{\mathcal{G}}_{2\mathrm{b},1\mathrm{b}}(q_2,q'_1)}\\
&= \delta_{q_1,q'_1}\hat{G}_1(q_1,z_1)\delta_{q_2,q'_2}\hat{G}_1(q_2,z_2)\\
&\quad + \hat{G}_1(q_1,z_1)\hat{G}_1(q'_1,z_1)\hat{G}_1(q_2,z_2)\hat{G}_1(q'_2,z_2)\\
&\quad \times (\overline{\delta\hat{R}_{q_1-q'_1,1\mathrm{b},1\mathrm{b}}\delta\hat{R}_{q_2-q'_2,2\mathrm{b},2\mathrm{b}}} + \overline{\delta\hat{R}_{q_1-q'_2,1\mathrm{b},2\mathrm{b}}\delta\hat{R}_{q_2-q'_1,2\mathrm{b},1\mathrm{b}}}).
\end{aligned} \tag{55}$$

The action up to second order in δR reads

$$\begin{aligned}
\mathcal{S}(R,0) &= \mathcal{S}(R^{(0)},0) + \frac{1}{2}\sum_{r,r'}\mathrm{str}(\delta R(r) w_{r,r'}\delta R(r')) - \frac{1}{2}\mathrm{str}(G^{(0)}\delta R G^{(0)}\delta R)\\
&= \frac{1}{2}N \sum_{q,p,\nu,p',\nu'} (-)^{\nu}(\hat{w}_q - \hat{\Pi}_q(z_{p'},z_p))\delta\hat{R}_{-q,p\nu,p'\nu'}\delta\hat{R}_{q,p'\nu',p\nu}.
\end{aligned} \tag{56}$$

with

$$\hat{\Pi}_q(z_{p'}, z_p) = \sum_{q'} \hat{G}^{(0)}(q+q', z_p)\hat{G}^{(0)}(q', z_{p'}). \tag{57}$$

This is the action in harmonic approximation in δR. It yields

$$\overline{\delta\hat{R}_{-q_1,p_1\nu_1,p_1',\nu_1'}\delta\hat{R}_{q_2,p_2'\nu_2',p_2\nu_2}} = \delta_{q_1,q_2}\delta_{p_1,p_2}\delta_{p_1',p_2'}\delta_{\nu_1,\nu_2}\delta_{\nu_1',\nu_2'}\frac{(-)^{\nu_1'}}{N}\hat{\Gamma}_{q_1}(z_{p_1'}, z_{p_1}) \tag{58}$$

with

$$\hat{\Gamma}_q(z_{p'}, z_p) = \frac{1}{\hat{w}_q - \hat{\Pi}_q(z_{p'}, z_p)}. \tag{59}$$

The two contributions $\nu = 0$ and $\nu = 1$ for the one-particle Green's function $\hat{G}_1$ cancel. Only the second fluctuation term to $\hat{G}_2$ contributes,

$$\begin{aligned}\hat{G}_2(q_1, q_1', z_1; q_2, q_2', z_2) &\\ = \delta_{q_1,q_1'}\hat{G}_1(q_1, z_1)\delta_{q_2,q_2'}\hat{G}_1(q_2, z_2) &\\ + \frac{\delta_{q_1+q_2,q_1'+q_2'}}{N}\hat{\Gamma}_{q_1-q_1'}(z_1, z_2)\hat{G}_1^{(0)}(q_1, z_1)\hat{G}_1^{(0)}(q_1', z_1)\hat{G}_1^{(0)}(q_2, z_2)\hat{G}_1^{(0)}(q_2', z_2).\end{aligned} \tag{60}$$

Nonlinear σ-model We choose now $z = E \pm \mathrm{i}0$ and obtain

$$R^{(0)}(E) = \mathrm{diag}(R_{\mathrm{A}}(E), R_{\mathrm{R}}(E), R_{\mathrm{A}}(E), R_{\mathrm{R}}(E)). \tag{61}$$

The density of states is given by Eq. (18). Thus

$$2\pi\rho = \hat{w}_0(R_{\mathrm{A}}(E) - R_{\mathrm{R}}(E)) \tag{62}$$

$$R = \Re R^{(0)}\mathbf{1}_4 + \frac{\pi\mathrm{i}\rho}{\hat{w}_0}\Lambda \tag{63}$$

with

$$\Lambda = \mathrm{diag}(1, -1, 1, -1). \tag{64}$$

Eq. (61) is not the general solution for the saddle point. This is given by

$$R(E, T) = \Re R^{(0)}\mathbf{1}_4 + \frac{\pi\mathrm{i}\rho}{\hat{w}_0}Q(T) \tag{65}$$

with

$$Q^2(T) = \mathbf{1}_4. \tag{66}$$

Then the saddle point Eq. (49) reads

$$\hat{w}_0\Re R^{(0)} + \pi\mathrm{i}\rho Q = \frac{1}{N}\sum_q \frac{E - \Re R^{(0)} - \hat{t}_q + \pi\mathrm{i}\rho Q/\hat{w}_0}{(E - \Re R^{(0)} - \hat{t}_q)^2 + (\pi\rho/\hat{w}_0)^2}. \tag{67}$$

The full matrix R is

$$R = \Re R^{(0)} \mathbf{1}_4 + (\pi \mathrm{i}\rho/\hat{w}_0)Q + P \tag{68}$$

The symmetry of R is the same as of B, since both are multiplied by R in Eq. (40),

$$B^\dagger = s^{-1} B s. \tag{69}$$

thus

$$Q^\dagger(T) = s^{-1} Q(T) s. \tag{70}$$

$$Q(T) = T \Lambda T^{-1}, \quad T s T^\dagger = s, \tag{71}$$

Q can be written as suggested by Efetov[28]

$$Q = U \begin{pmatrix} \lambda_1 & \mathrm{i}\mu_1 & 0 & 0 \\ \mathrm{i}\mu_1^* & -\lambda_1 & 0 & 0 \\ 0 & 0 & \lambda_2 & \mu_2 \\ 0 & 0 & \mu_2^* & -\lambda_2 \end{pmatrix} U^{-1} \tag{72}$$

with

$$\lambda_1 = \cosh\theta_1, \ \ \mu_1 = \sinh\theta_1 \exp(\mathrm{i}\phi_1), \tag{73}$$

$$\lambda_2 = \cos\theta_2, \ \ \mu_2 = \sin\theta_2 \exp(\mathrm{i}\phi_2). \tag{74}$$

The matrix T has the form of Q, but θ_i is replaced by $\theta_i/2$. The matrix U differs from the unit matrix only by anticommuting terms,

$$U^{\pm 1} = \begin{pmatrix} 1 - \frac{1}{2}\alpha^*\alpha & 0 & \mp\alpha^* & 0 \\ 0 & 1 + \frac{1}{2}\beta^*\beta & 0 & \mp\mathrm{i}\beta^* \\ \pm\alpha & 0 & 1 - \frac{1}{2}\alpha\alpha^* & 0 \\ 0 & \pm\mathrm{i}\beta & 0 & 1 + \frac{1}{2}\beta\beta^* \end{pmatrix}. \tag{75}$$

There exist several parametrizations for the matrix P. We give here the Schäfer-Wegner parametrization

$$P(r) = T_0 \begin{pmatrix} P^{\mathrm{A}}_{\mathrm{bb}}(r) & 0 & P^{\mathrm{A}}_{\mathrm{bf}}(r) & 0 \\ 0 & P^{\mathrm{R}}_{\mathrm{bb}}(r) & 0 & P^{\mathrm{R}}_{\mathrm{bf}}(r) \\ P^{\mathrm{A}}_{\mathrm{fb}}(r) & 0 & P^{\mathrm{A}}_{\mathrm{ff}}(r) & 0 \\ 0 & P^{\mathrm{R}}_{\mathrm{fb}}(r) & 0 & P^{\mathrm{R}}_{\mathrm{ff}}(r) \end{pmatrix} T_0^{-1} \tag{76}$$

with fixed T_0, which is of the form T. P_{bb} is hermitian and P_{ff} antihermitian.

Both energies advanced or retarted If in contrast to eq. (61) both energies z_p are advanced or retarted, then we may use that

$$K\left(z+\frac{\omega}{2}, z-\frac{\omega}{2}\right) = K(z,z) - \frac{\omega^2}{4}\frac{\mathrm{d}^2K(z_1,z_2)}{\mathrm{d}z_1\mathrm{d}z_2}|_{z_1=z_2=z} + \cdots \tag{77}$$

Thus only corrections in order ω^2 appear, which can be neglected for small ω.

Invariant measure One obtains for the invariant measure in the space of the Q-matrix

$$[\mathrm{D}Q] = \frac{1}{(\lambda_1-\lambda_2)^2}\mathrm{d}\lambda_1\mathrm{d}\lambda_2\mathrm{d}\phi_1\mathrm{d}\phi_2\mathrm{d}\alpha\mathrm{d}\alpha^*\mathrm{d}\beta\mathrm{d}\beta^* \tag{78}$$

In order to obtain convergence we had to choose different signs s_{Ab} and s_{Rb} for the bosonic integrations. We choose for the fermionic integrations $s_{\mathrm{Af}} = s_{\mathrm{Rf}}$. Then the singularity $1/(\lambda_1-\lambda_2)^2$ appears only at $\lambda_1 = \lambda_2$.

Then the λ-integrals run over the intervals $\lambda_1 = 1...\infty$ and $\lambda_2 = -1...+1$. Closer investigation shows that the singularity at $\lambda_1 = \lambda_2 = 1$ creates a contribution at these λs with $\alpha = \alpha^* = \beta = \beta^* = 0$.

2.4. *The random matrix model*

***n*-Orbital models** Investigation of the n-orbital model in the limit $n \to \infty$[16,45] allowed the determination of critical exponents at the mobility edge up to order $1/n^2$.

Compared to the model (1) an additional index indicated by a Greek letter running from 1 to n is introduced. Thus Eq. (1) is replaced by

$$H = \sum_{r\alpha,r'\beta} |r,\alpha\rangle(f_{r\alpha,r'\beta} + t_{r,r'}\delta_{\alpha,\beta})\langle r',\beta| \tag{79}$$

with Gaussian distributed elements f with average 0

$$\overline{f_{r\alpha,r'\alpha'}} = 0, \quad \overline{f_{r_1\alpha_1,r_1'\alpha'}f_{r_2\alpha_2,r_2'\alpha_2'}} = \frac{1}{n}M_{r-r'}\delta_{r_1,r_2'}\delta_{\alpha_1,\alpha_2'}\delta_{r_2,r_1'}\delta_{\alpha_2,\alpha_1'}. \tag{80}$$

One uses

$$A(r\alpha,r'\beta) = S^*(r\alpha)sS(r'\beta), \quad B(r) = \sum_\alpha \sqrt{s}S(r\alpha)S^*(r,\alpha)\sqrt{s} \tag{81}$$

and one has to replace M by M/n and w by nw. Finally one obtains

$$Z(J) = \left(\int\prod_r \mathrm{D}[R(r)]\right)\exp\left(\mathrm{str}\left[-\frac{n}{2}\sum_{r,r'} w_{r,r'}R(r)R(r')\right]\right.$$
$$\times\ \mathrm{sdet}(R-z+t+J)^{-n}. \tag{82}$$

instead of Eq. (43). Thus the action $\mathcal{S}$ is multiplied by n.

Random Matrix Theory. $d = 0$ The random matrix model has been considered by Wigner[41] as a zero-dimensional model. It is the model Eq. (79), but with only one site r and matrix $t = 0$. Thus the Hamiltonian reads

$$H = \sum_{\alpha,\beta} |\alpha\rangle f_{\alpha,\beta} \langle\beta|. \tag{83}$$

The elements f are Gaussian distributed with average 0 and width

$$\overline{f_{\alpha,\alpha'}} = 0, \quad \overline{f_{\alpha_1,\alpha_1'} f_{\alpha_2,\alpha_2'}} = \frac{1}{n} M \delta_{\alpha_1,\alpha_2'} \delta_{\alpha_2,\alpha_1'}. \tag{84}$$

Initially Wigner considered the orthogonal case (see below). R obeys for the unitary case

$$wR^{(0)} + \frac{1}{R^{(0)} - E} = 0 \tag{85}$$

with the solution

$$R^{(0)} = \frac{E}{2} \pm \mathrm{i}\sqrt{M - E^2/4}, \quad \rho(E) = \frac{w}{\pi}\sqrt{M - E^2/4}. \tag{86}$$

This is the well-known semi-circle law for the density of states, which also holds in the orthogonal and the symplectic case. The density ρ is so normalized that its integral over E yields unity.

The one- and two-particle Green's functions G and K obey

$$\overline{G(\alpha,\beta,z)} = \delta_{\alpha,\beta} G(z), \tag{87}$$

$$\overline{G(\alpha,\beta,z_1)G(\gamma,\delta,z_2)} = \delta_{\alpha,\beta}\delta_{\gamma,\delta}K(z_1,z_2) + \delta_{\alpha,\delta}\delta_{\beta,\gamma}K'(z_1,z_2), \tag{88}$$

$$\begin{aligned} n^2\tilde{K}(z_1,z_2) &= \sum_{\alpha,\beta} \overline{G(\alpha,\alpha,z_1)G(\beta,\beta,z_2)} \\ &= n^2 K(z_1,z_2) + nK'(z_1,z_2), \end{aligned} \tag{89}$$

$$n\tilde{K}'(z_1,z_2) = \sum_{\alpha,\beta} \overline{G(\alpha,\beta,z_1)G(\beta,\alpha,z_2)}, \tag{90}$$

$$\tilde{K}'(z_1,z_2) = K(z_1,z_2) + nK'(z_1,z_2) = \frac{G(z_2) - G(z_1)}{z_1 - z_2}. \tag{91}$$

The matrix elements of Q_{bb}, which are used to calculate the expectation values, see Eq. (72), read

$$Q_{\mathrm{bb}}^{\mathrm{RR}} = \lambda_1 + \alpha^*\alpha(\lambda_2 - \lambda_1), \tag{92}$$

$$Q_{\mathrm{bb}}^{\mathrm{AA}} = -\lambda_1 + \beta^*\beta(\lambda_2 - \lambda_1), \tag{93}$$

$$Q_{\mathrm{bb}}^{\mathrm{RA}} = \mathrm{i}\mu_1 \left(1 - \frac{1}{2}\alpha^*\alpha\right)\left(1 + \frac{1}{2}\beta^*\beta\right) + \mathrm{i}\mu_2^*\alpha^*\beta, \tag{94}$$

$$Q_{\mathrm{bb}}^{\mathrm{AR}} = \mathrm{i}\mu_1^* \left(1 - \frac{1}{2}\alpha^*\alpha\right)\left(1 + \frac{1}{2}\beta^*\beta\right) + \mathrm{i}\mu_2\beta^*\alpha. \tag{95}$$

Thus

$$G(z_1) = w\left(\frac{z_1}{2} - \mathrm{i}\sqrt{M - \frac{z_1^2}{4}}\right),\quad G(z_2) = w\left(\frac{z_2}{2} + \mathrm{i}\sqrt{M - \frac{z_2^2}{4}}\right). \tag{96}$$

$$\tilde{K}(z_1, z_1) = G^2(z_1),\ \ \tilde{K}(z_2, z_2) = G^2(z_2) \tag{97}$$

and

$$\tilde{K}(z_1, z_2) = G(z_1)G(z_2) + (\pi\rho)^2 \int_1^\infty \mathrm{d}\lambda_1 \int_{-1}^1 \mathrm{d}\lambda_2 \exp^{\mathrm{i}\pi f(\lambda_1 - \lambda_2)}, \tag{98}$$

$$\tilde{K}'(z_1, z_2) = (\pi\rho)^2 \int_1^\infty \mathrm{d}\lambda_1 \int_{-1}^1 \mathrm{d}\lambda_2 \frac{\lambda_1 + \lambda_2}{\lambda_1 - \lambda_2} \exp^{\mathrm{i}\pi f(\lambda_1 - \lambda_2)}. \tag{99}$$

The mean level spacing is $1/(n\rho)$. Thus

$$f = n\rho\omega. \tag{100}$$

h Hence

$$\tilde{K}(z_1, z_2) = G(z_1)G(z_2) + (\pi\rho)^2 \frac{2\mathrm{i}\sin(\pi f)}{(\pi f)^2} \mathrm{e}^{\mathrm{i}\pi f}, \tag{101}$$

$$\tilde{K}'(z_1, z_2) = (\pi\rho)^2 \frac{2\mathrm{i}}{\pi f}. \tag{102}$$

This allows us to calculate the correlations between the levels

$$\overline{\rho\left(E + \frac{\omega}{2}\right)\rho\left(E - \frac{\omega}{2}\right)} = \rho^2(E)C_{\mathrm{U}}(f), \tag{103}$$

$$C_{\mathrm{U}}(f) = 1 + \delta(f) - \frac{\sin^2(\pi f)}{f^2} = \delta(f) + \frac{(\pi f)^2}{3} + O(f^4). \tag{104}$$

The delta function describes the self-correlation. The self-repulsion yields the f^2 increase. Erdös[46] reports a general form for the level correlation of k levels.

Orthogonal Gaussian ensemble The matrix elements in this ensemble are real $f_{\alpha,\beta} = f_{\beta,\alpha} \in \mathbb{R}$ and Gaussian distributed with

$$\overline{f_{\alpha,\beta}} = 0,\quad \overline{f_{\alpha_1,\alpha_1'} f_{\alpha_2,\alpha_2'}} = \frac{1}{n} M(\delta_{\alpha_1,\alpha_2'}\delta_{\alpha_2,\alpha_1'} + \delta_{\alpha_1,\alpha_2}\delta_{\alpha_1',\alpha_2'}). \tag{105}$$

This model was introduced by Wigner as a stochastic model for nuclear levels.[47] The number of field components S and consequently also of R and Q have to be doubled. (The same applies for the symplectic case.)

One obtains for the level correlation

$$C_{\mathrm{O}}(f) = 1 + \delta(f) - \frac{\sin^2(\pi f)}{(\pi f)^2} - \left(\frac{\pi}{2}\mathrm{sign}(f) - \mathrm{Si}(\pi f)\right)\left(\frac{\cos(\pi f)}{\pi f} - \frac{\sin(\pi f)}{(\pi f)^2}\right)$$
$$= \delta(f) + \frac{\pi^2}{6}|f| + O(f^2). \tag{106}$$

with the sine integral

$$\mathrm{Si}(x) = \int_0^x \mathrm{d}y \frac{\sin(y)}{y}, \qquad \lim_{x\to\infty} \mathrm{Si}(x) = \frac{\pi}{2}. \tag{107}$$

The self-repulsion yields an increase proportional to $|f|$.

Symplectic Gaussian ensemble Spin-dependent time-reversal invariant ensemble

$$H = \sum_{\alpha,m,\beta,m'} |\alpha,m\rangle f_{\alpha,m,\beta,m'} \langle \beta, m'| \tag{108}$$

Time-reversal invariance of H requires $\Theta H = H\Theta$ with the time-reversal operator $\Theta = -\mathrm{i}\sigma^y K$, where K denotes complex conjugation in real space. Thus the matrices f are real quaternions,

$$\begin{pmatrix} f_{\uparrow\uparrow} & f_{\uparrow\downarrow} \\ f_{\downarrow\uparrow} & f_{\downarrow\downarrow} \end{pmatrix} = \epsilon \begin{pmatrix} f^*_{\uparrow\uparrow} & f^*_{\uparrow\downarrow} \\ f^*_{\downarrow\uparrow} & f^*_{\downarrow\downarrow} \end{pmatrix} \epsilon^{-1} = \begin{pmatrix} f^*_{\downarrow\downarrow} & -f^*_{\downarrow\uparrow} \\ -f^*_{\uparrow\downarrow} & f^*_{\uparrow\uparrow} \end{pmatrix} \in \mathbb{H}. \tag{109}$$

If $H\psi = E\psi$, then also $H\Theta\psi = E\Theta\psi$. The wave-functions ψ and $\Theta\psi$ are orthogonal. Thus all eigenstates are doubly degenerate (Kramers doublett).

$$\overline{f_{\alpha,m_1,\beta,m_2}} = 0, \tag{110}$$

$$\overline{f_{\alpha_1,m_1,\alpha'_1,m'_1} f_{\alpha_2,m_2,\alpha'_2,m'_2}} = \frac{1}{2n} M(\delta_{\alpha_1,\alpha_2}\delta_{\alpha'_1,\alpha'_2}\epsilon_{m_1,m_2}\epsilon_{m'_1,m'_2}$$
$$+ \delta_{\alpha_1,\alpha'_2}\delta_{\alpha'_1,\alpha_2}\epsilon_{m_1,m'_2}\epsilon_{m'_1,m_2}). \tag{111}$$

The level correlation is given by

$$C_{\mathrm{S}}(f) = 1 + \delta(f) - \frac{\sin^2(2\pi f)}{(2\pi f)^2} + \mathrm{Si}(2\pi f)\left(\frac{\cos(2\pi f)}{2\pi f} - \frac{\sin(2\pi f)}{(2\pi f)^2}\right)$$
$$= \delta(f) + \frac{(2\pi f)^4}{135} + O(f^6). \tag{112}$$

The self-repulsion is particularly strong in the symplectic case. The correlation increases proportional to f^4 for small f. Level correlations for the random matrix ensemble were calculated by Wigner,[41] Porter,[48] and Mehta[49,50] with different techniques.

2.5. *Hydrodynamics*

Diffusion For $q = 0$ one obtains

$$\hat{\Pi}_0(z_j, z_i) = \frac{1}{N}\sum_{q'} \frac{1}{R_i^{\mathrm{d}} - z_i + \hat{t}_{q'}} \frac{1}{R_j^{\mathrm{d}} - z_j + \hat{t}_{q'}} = \frac{\hat{w}_0(R_j^{\mathrm{d}} - R_i^{\mathrm{d}})}{R_j^{\mathrm{d}} - z_j - R_i^{\mathrm{d}} + z_i}, \quad (113)$$

$$\hat{w}_0 - \hat{\Pi}_0(z_j, z_i) = \frac{\hat{w}_0(z_i - z_j)}{R_j^{\mathrm{d}} - z_j - R_i^{\mathrm{d}} + z_i} \quad (114)$$

One obtains as $\omega \to 0$ for $\Im\omega > 0$

$$\hat{w}_0 - \hat{\Pi}_0\left(E + \frac{1}{2}\omega, E - \frac{1}{2}\omega\right) = \frac{-\mathrm{i}\omega\hat{w}_0}{2\pi\rho(E)}. \quad (115)$$

Thus a diffusion pole appears in the hydrodynamic limit

$$\hat{\Gamma}_q\left(E + \frac{1}{2}\omega, E - \frac{1}{2}\omega\right) = \frac{1}{\hat{w}_q - \hat{\Pi}_q\left(E + \frac{1}{2}\omega, E - \frac{1}{2}\omega\right)} = \frac{2\pi\rho(E)/\hat{w}_0^2}{-\mathrm{i}\omega + D^{(0)}q^2}. \quad (116)$$

where $\hat{w}_q$ and $\hat{\Pi}_q$ are expanded up to order q^2 to obtain $D^{(0)}$. $\hat{w}_q - \hat{\Pi}_q$ approaches zero as ω and q tend to zero. It is obtained by summing the bubbles Π,

$$\hat{w}_q\hat{\Gamma}_q = 1 + \hat{M}_q\hat{\Pi}_q + (\hat{M}_q\hat{\Pi}_q)^2 + (\hat{M}_q\hat{\Pi}_q)^3 + \cdots, \quad (117)$$

which describes multiple scattering of a particle-hole pair by disorder. It is called Diffuson.

Conductivity One obtains for the frequency dependent conductivity at temperature T from Kubo[51] and Greenwood[52]

$$\sigma_T(\omega) = \frac{1}{\omega}\int \mathrm{d}E\left(f_T\left(E - \frac{1}{2}\omega\right) - f_I\left(E + \frac{1}{2}\omega\right)\right)\sigma(\omega, E) \quad (118)$$

with the Fermi-function $f_T(E) = 1/(\mathrm{e}^{E/(k_{\mathrm{B}}T)} + 1)$ and Boltzmann constant k_{B}.

In the limit $T = 0$ and $\omega = 0$ one obtains the d.c. conductivity $\sigma(0, \mu)$ for the chemical potential μ. Evaluation of this limit yields

$$\sigma(0, \mu) = e^2\rho(\mu)D^{(0)}(\mu) \quad (119)$$

with the elementary electric charge e.

The action of the nonlinear sigma-model is given in the hydrodynamic limit by

$$\mathcal{S}(Q) = \frac{1}{2}\pi\rho^{(0)} \sum_r \mathrm{str}(-D^{(0)}(\nabla Q(r))^2 - 2\mathrm{i}\omega\Lambda Q(r)). \tag{120}$$

Note that $-(\nabla Q(r))^2$ is positive semi-definite, if expressed by the gradients of θ_i and ϕ_i in Eq. (72).

Orthogonal and Symplectic Case

In these time-reversal invariant cases one requires for the spinless (orthogonal) case real matrix elements $f_{r,r'} = f_{r',r} \in \mathbb{R}$ as in (105), $t_{r,r'} = t_{r',r} \in \mathbb{R}$. The matrix-elements f and t of the spin dependent (symplectic) case are quaternions, compare Eq. (109).

Around Eq. (117) we realized that in the unitary case two particles called Diffusons running in the opposite direction lead to the long-range behaviour both in time and space (small ω and small q). In the time-invariant case one finds besides these diffusons similar correlations for two particles running in the same direction, called Cooperons, as described by Vollhardt and Wölfle.[53,54] The two-particle Green's function reads now

$$\begin{aligned}\hat{G}_2(q_1, q_1', z_1; q_2, q_2', z_2) &= \delta_{q_1,q_1'}\hat{G}_1(q_1, z_1)\delta_{q_2,q_2'}\hat{G}_1(q_2, z_2)\\ &+ \frac{1}{N}\delta_{q_1+q_2-q_1'-q_2'}\hat{G}^{(0)}(q_1, z_1)\hat{G}^{(0)}(q_1', z_1)\\ &\times \hat{G}^{(0)}(q_2, z_2)\hat{G}^{(0)}(q_2', z_2)(\hat{\Gamma}_{q_1-q_2'}(z_1, z_2) + \hat{\Gamma}_{q_1+q_2}(z_1, z_2))\end{aligned} \tag{121}$$

with the Cooperon in the second term $\hat{\Gamma}$. This contribution was called Cooperon by Altshuler *et al.*[55,56] in their papers on magnetoresistance, since magnetic effects destroy the Cooperon.

Symmetry relations There are symmetry relations between the ensembles for various symmetries.

$$\mathcal{S}(Q) = \frac{\kappa}{t}\int \mathrm{d}^d r\, \mathrm{str}\left[\frac{1}{2}(\nabla Q(r))^2 + \mathrm{i}\frac{\omega}{D^{(0)}}\Lambda Q(r)\right]. \tag{122}$$

with

$$\begin{array}{lcl} \text{ensemble} & \kappa & 1/t \\ \text{unitary} & -1 & \frac{1}{2}\pi\rho^{(0)}D^{(0)} \\ \text{unitary} & +1 & \frac{1}{2}\pi\rho^{(0)}D^{(0)} \\ \text{orthogonal} & -1 & \frac{1}{4}\pi\rho^{(0)}D^{(0)} \\ \text{symplectic} & +1 & \frac{1}{2}\pi\rho^{(0)}D^{(0)} \end{array} \tag{123}$$

In a formal power expansion in t one obtains the relations

$$Z_{\rm O}(t) = Z_{\rm Sp}(-t), \quad Z_{\rm U}(t) = Z_{\rm U}(-t), \tag{124}$$

where one exchanges $\lambda_1 - 1$ and $1 - \lambda_2$ in an expansion in these quantities. These symmetries are not valid in general, but only for small t, where the missing contributions for $\lambda_2 < -1$ can be neglected. Corresponding relations for expectation values were found by Jüngling and Oppermann.[57,58]

There are two types of compact supergroups, the unitary one and the unitary-orthosymplectic one. Their elements U have the property $U^\dagger U = \mathbf{1}$ in the unitary case and a similar relation for the unitary-orthosymplectic case, where however the adjoint is differently defined, similarly to the timereversal operation. If one takes away all Grassmann variables, then only the boson-boson and the fermion-fermion block are left. In the unitary case both blocks are unitary matrices, in the unitary-orthosymplectic case one is orthogonal, the other one is unitary-symplectic. We have switched the boson-boson block to the pseudo-form. Instead one can express the action by the fermion-fermion block as Efetov *et al.*[22] did. These two possibilities yield the connections Eqs. (124).

β-function and conductance The corresponding β-functions for t read[60–64]

$$\tilde\beta(t) = \begin{cases} \epsilon t - t^2 - \frac{3}{4}\zeta(3)t^5 + \frac{27}{64}\zeta(4)t^6 + O(t^7) & \text{orthogonal ensemble} \\ \epsilon t - \frac{1}{2}t^3 - \frac{3}{8}t^5 + O(t^7) & \text{unitary ensemble} \\ \epsilon t + t^2 - \frac{3}{4}\zeta(3)t^5 - \frac{27}{64}\zeta(4)t^6 + O(t^7) & \text{symplectic ensemble} \end{cases} \tag{125}$$

The conductance g (inverse resistance) of a cylindric body of length L and cross-section L^{d-1} is

$$g = \sigma L^\epsilon \propto t^{-1}. \tag{126}$$

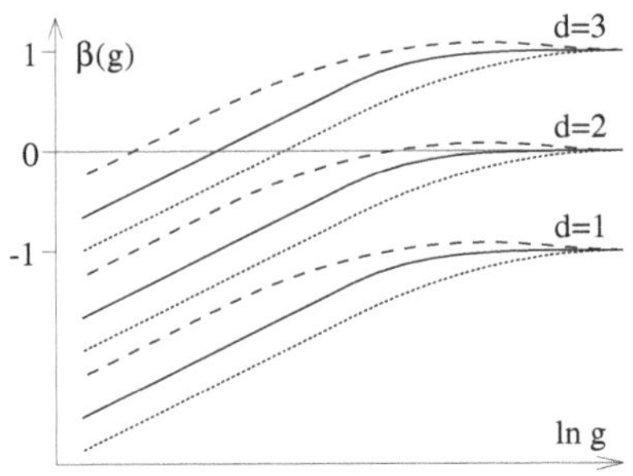

Fig. 1. Function $\beta(g)$ for dimensions $d = 1, 2, 3$ for the symplectic (upper dashed), unitary (middle full), and orthogonal (lower dotted line).

The conductance was introduced by Abrahams *et al.*[24] Its variation as function of L is given by

$$\frac{\mathrm{d}\ln g}{\mathrm{d}\ln L} = \beta(g) = t^{-1}\tilde{\beta}(t) = g\tilde{\beta}(1/g)$$
$$= \begin{cases} \epsilon - g^{-1} - \frac{3}{4}\zeta(3)g^{-4} + \frac{27}{64}\zeta(4)g^{-5} + O(g^{-6}) & \text{orthogonal ensemble} \\ \epsilon - \frac{1}{2}g^{-2} - \frac{3}{8}g^{-4} + O(g^{-6}) & \text{unitary ensemble} \\ \epsilon + g^{-1} - \frac{3}{4}\zeta(3)g^{-4} - \frac{27}{64}\zeta(t)g^{-5} + O(g^{-6}) & \text{symplectic ensemble} \end{cases} . \tag{127}$$

For one-dimensional wires $\beta(g)$ is always negative. Therefore g decays with increasing length, as long as inelastic scattering is negligible.

For two-dimensional films β approaches 0 as g tends to infinity. Then the conductance decays slowly for the orthogonal and unitary case, although finally it will decay. This is the region of weak localization. The conductance increases in the symplectic case. This is called weak anti-localization and appears when spin-orbit interactions become important. It was predicted by Hikami *et al.*[65]

Three dimensional bodies with sufficiently strong conductance at a microscopical scale will become typical conductors with $g \propto L$. The mobility edge is obtained, when $\beta(g) = 0$. Then the conductance does not vary with length L. If g starts out with negative $\beta(g)$, then the body is an insulator.

Density fluctuations and multifractality The amplitudes of the wave-functions show charateristic fluctuations. They are described by the inverse participation ratio P

$$P_q(E) = \overline{\sum_i |\psi_i(r)|^{2q}\delta(E-e_i)}/\rho(E) \tag{128}$$

in the region of the localized states and by the participation ratio

$$1/p_q(E) = \lim_{\eta\to 0} \overline{\left[\left[\sum_i |\psi_i(r)|^2\delta_\eta(E-e_i)\right]^q\right]}/\rho(E)^q \tag{129}$$

in the region of extended states. In the later region one uses the smeared δ-function

$$\delta_\eta(x) = \frac{\eta}{\pi(x^2+\eta^2)} = \frac{1}{2\pi\mathrm{i}}\left(\frac{1}{x-\mathrm{i}\eta} - \frac{1}{x+\mathrm{i}\eta}\right) \tag{130}$$

These quantities were introduced by Bell and Dean[66] and discussed by Thouless[67] and Wegner[68] for $q=2$. P_2 is realized by a localized state consisting of $1/P_2$ sites with $|\psi^2(r)| = P_2$, and p_2 by an extended state with Np_2 sites with $|\psi^2(r)| = 1/(Np_2)$.

As long as the localization length ξ is small in comparison to the size L of the system, one obtains

$$P_q(E) \propto \xi^{-\tau_q} \propto |E - E_c|^{\tau_q \nu}, \quad p_q(E) \propto \xi^{\Delta_q} \propto |E - E_c|^{-\Delta_q \nu}. \tag{131}$$

The exponents are related by

$$\tau_q = (d-1)q + \Delta_q. \tag{132}$$

Δ_q vanishes for $q = 0$ and $q = 1$,

$$\tau_0 = d, \quad \Delta_0 = 0, \quad \tau_1 = 0, \quad \Delta_1 = 0. \tag{133}$$

Δ obeys the symmetry relation

$$\Delta_q = \Delta_{1-q}. \tag{134}$$

This has been observed in $2+\epsilon$-expansions.[59,61,62] General results were derived by Mirlin *et al.*[69] and by Gruzberg *et al.*[70,71]

Statistics of energy spacing The statistics of energy spacing is in the region of localized states of Poisson type $\Delta N \propto N$, in the region of extended states of Wigner-Dyson type $\Delta N \propto \ln N$, and at the mobility edge of type $\Delta N \propto N^{1-1/(d\nu)}$ as found by Kravtsov *et al.*[72]

Final remarks This article cannot cover all the results on the Anderson transition. We mention only three other important developments.

Until now we considered only the three classes introduced by Dyson, called Wigner-Dyson classes. Altland and Zirnbauer[73] found that all compact symmetric spaces correspond to a class of nonlinear sigma-models. The seven classes besides the three Wigner-Dyson classes are known as three chiral classes und four Bogolubov-de Gennes classes. They differ in the statistics of the states close to the band center from the Wigner-Dyson classes. They are classified according to their behaviour under time-reversal invariance, particle hole symmetry and sublattice symmetry. The reader is referred to the papers by Zirnbauer,[31,33] the review by Evers aand Mirlin,[36] and the article by Chiu *et al.*[74] This leads also into the field of topological insulators and superconductors.

Fröhlich and Spencer[75] proved that there is no diffusion in the Anderson tight-binding model for sufficiently large disorder or at low energies.

Kramer, MacKinnon *et al.*[76,77] estimated critical exponents numerically by finite size scaling. They obtained for the exponent ν

$$\begin{array}{rl} 1.57 \pm 0.02 & d = 3 \text{ orthogonal symmetry} \\ 1.43 \pm 0.04 & d = 3 \text{ unitary symmetry} \\ 1.375 \pm 0.016 & d = 3 \text{ symplectic symmetry} \\ 2.73 \pm 0.02 & d = 2 \text{ symplectic symmetry} \\ 2.593 \pm 0.006 & \text{integer quantum Hall effect} \end{array} \tag{135}$$

Here we considered only systems with one-particle hamiltonians. When interaction between the particles, in particular the Coulomb interaction between electrons has to be considered, then the supersymmetry approach does no longer help and one has to return to the replica trick or to diagramatic methods. This has been done among others by Altshuler, Aronov and Lee,[78] by Finkel'stein,[79,80] and by Belitz and Kirkpatrick.[30]

Note added in proof Recently Martin Zirnbauer has published two papers,[81,82] in which he argues for a pure-point region between the absolute continuous metallic and the singular continuous insulating region. Indications for a different behaviour of the transition is given by the paper by Garcia-Mata[84] *et al.* focusing on the insulator side of the transition finding a Kosterlitz-Thouless type of transition and and by the paper by Sierant[83] *et al.* approaching the transition from the metallic side finding a non-Kosterlitz thouless transition. These renormalization group calculations are governed by two couplings. The suggested renormalization group flows are depicted in Fig. 1 in[81] and Figures 1 and 3 in.[82] This is in contrast to the $2+\epsilon$ expansion with only one coupling, which does not show a pure-point region.

References

1. C. Domb, Michael Fisher at King's College London, *Physica* **A177**, 1 (1991).
2. C. Domb and M. E. Fisher, Random walks with restricted reversals, *Proc. Cambridge Philos. Soc.* **54**, 48 (1958).
3. M. E. Fisher, The theory of equilibrium critical phenomena, *Rep. Prog. Phys.* **30**, 615 (1967).
4. M. E. Fisher, The renormalization group in the theory of critical behavior, *Rev. Mod. Phys.* **46**, 597 (1974).
5. M. E. Fisher, Renormalization group theory: Its basis and formulation in statistical physics, *Rev. Mod. Phys.* **70**, 653 (1998).
6. M. E. Fisher, *Renormalization Group Theory, the epsilon expansion, and Ken Wilson as I knew him*, in Baaquie, B.E., K. Huang, M.E. Peskin, and K.K. Phua, eds., *Ken Wilson Memorial Volume: Renormalization, Lattice Gauge Theory, The Operator Product Expansion and Quantum Fields*, World Scientific (2015) p. 43.
7. K. G. Wilson and M. E. Fisher, Critical exponents in 3.99 dimensions, *Phys. Rev. Lett.* **28**, 240 (1972).
8. K. G. Wilson, Renormalization group and Kadanoff scaling picture, *Phys. Rev.* **B4**, 3174 and 3184 (1971).
9. C. Domb and M. S. Green, (eds.) *Phase transitions and critical phenomena*, Academic Press vol. **6** (1975).
10. P. W. Anderson, Absence of diffusion in certain random lattices, *Phys. Rev.* **109**, 1492 (1958).
11. D. J. Thouless, *Anderson localization in the seventies and beyond*, in 50 years of Anderson localization,[12] p. 7 (2010).
12. E. Abrahams, (editor) *50 Years of Anderson localization*, World Scientific (2010).
13. N. F. Mott and W. D. Twose, The theory of impurity conduction *Adv. Phys.* **10**, 107 (1961).
14. N. F. Mott, Electrons in disordered structures, *Adv. Phys.* **16**, 49 (1967).

15. D. J. Thouless, Anderson's theory of localized states, *J. Phys.* **C3**, 1559 (1970).
16. R. Oppermann and F. J. Wegner, Disordered systems with n orbitals per site: $1/n$ expansion, *Z. Phys.* **B34**, 327 (1979).
17. A. Aharony and Y. Imry, The mobility edge as a spin glass problem, *J. Phys.* **C10**, 487 (1977).
18. A. B. Harris and T. C. Lubensky, Mean-field theory and ϵ expansion for Anderson localization, *Solid State Commun.* **34**, 343 (1980).
19. A. B. Harris and T. C. Lubensky, Mean-field theory and ϵ expansion for Anderson localization, *Phys. Rev.* **B23**, 2640 (1981).
20. F. Wegner, The mobility edge problem, continuous symmetry and a conjecture, *Z. Phys.* **B35**, 207 (1979).
21. L. Schäfer and F. Wegner, Disordered system with n orbitals per site: Lagrange formulation, hyperbolic symmetry, and goldstone modes, *Z. Phys.* **B38**, 113 (1980).
22. K. B. Efetov, A. I. Larkin and D. E. Khmelnitskii, Interaction of diffusion modes in the theory of localization, *Zh. Eksp. Teor. Fiz.* **79**, 1120 (1980); *Sov. Phys. JETP* **52**, 568 (1980).
23. A. M. Polyakov, Interaction of Goldstone particles in two dimensions, Applications to ferromagnets and massive Yang-Mills-fields, *Phys. Lett.* **B59**, 79 (1975).
24. E. Abrahams, P. W. Anderson, D. C. Licciardello and T. V. Ramakrishnan, Scaling theory of localization: Absence of quantum diffusion in two dimensions, *Phys. Rev. Lett.* **42**, 673 (1979).
25. J. J. M. Verbaarschott and M. R. Zirnbauer, Critique of the replica trick, *J. Phys.* **A17**, 1093 (1985).
26. K. B. Efetov, Supersymmetry method in localization theory, *Zh. Eksp. Teor. Fiz.* **82**, 872 (1982); *Sov. Phys. JETP* **55**, 514 (1982).
27. K. B. Efetov, Supersymmetry and theory of disordered metals, *Adv. Phys.* **32**, 53 (1983).
28. K. B. Efetov, *Supersymmetry in Disorder and Chaos*, Cambridge University Press, Cambridge (1997).
29. J. J. M. Verbaarschott, H. A. Weidenmüller and M. R. Zirnbauer, Grassmann integration in stochastic quantum physics: the case of compound nucleon scattering *Phys. Rep.* **129**, 367 (1985).
30. D. Belitz and T. R. Kirkpatrick, The Anderson-Mott transition, *Rev. Mod. Phys.* **66**, 261 (1994).
31. M. R. Zirnbauer, Riemannian symmetric superspaces and their origin in random matrix theory, *J. Math. Phys.* **37**, 4986 (1996).
32. M. R. Zirnbauer, *The supersymmetry method of random matrix theory*, Encyclopedia of mathematical physics (Elsevier, 2005).
33. M. R. Zirnbauer, *Symmetry classes in random matrix theory*, Encyclopedia of mathematical physics (Elsevier, 2005).
34. T. Guhr, A. Müller-Groeling and H. A. Weidenmüller, Random matrix theories in quantum physics: common concepts, *Phys. Rep.* **299**, 190 (1998).
35. A. D. Mirlin, Statistics of energy levels and eigenfunctions in disordered systems, *Phys. Rep.* **326**, 259 (2000).
36. F. Evers and A. D. Mirlin, Anderson transitions, *Rev. Mod. Phys.* **80**, 1355 (2008).
37. R. A. Stratonovich, On a method of calculating quantum distribution functions, *Dokl. Akad. Nauk SSSR* **115**, 1046 (1957); *Sov. Phys. Dokl.* **2**, 416 (1958).
38. B. Hubbard, Calculation of partition functions, *Phys. Rev. Lett.* **3**, 77 (1959).
39. F. Wegner, Supermathematics and its Applications in Statistical Physics, *Lect. Notes in Phys.* **920** Springer Berlin Heidelberg (2016).

40. F. J. Dyson, Statistical Theory of the energy levels of complex systems. *I. J. Math. Phys.* **3**, 140 (1962).
41. E. P. Wigner, On the distribution of the roots of certain symmetric matrices, *Ann. Math.* **67**, 325 (1958).
42. F. A. Berezin, *The method of second quantization*, (Academic Press, New York, 1966).
43. F. A. Berezin, *Introduction to Superanalysis*, (Reidel, Dordrecht, 1987).
44. V. Rittenberg and M. Scheunert, Elementary construction of graded Lie groups, *J. Math. Phys* **19**, 709 (1978).
45. F. Wegner, Disordered systems with n orbitals per site: $n = \infty$ limit, *Phys. Rev.* **B19**, 783 (1979).
46. L. Erdös, Universality of Wigner random matrices: a survey of recent results, arXiv1004.0861[math.ph]; *Russ. Math. Surv.* **66**, 507 (2011).
47. E. P. Wigner, On a class of analytic functions from the quantum theory of collisions, *Ann. Math.* **53**, 36 (1951).
48. C. E. Porter, *Statistical Theory of Spectra*, Academic, London, (1965).
49. M. L. Mehta, *Random matrices and the statistical theory of energy levels*, Academic Press, London, (1967).
50. M. L. Mehta, *Random matrices*, Academic Press, Boston, (1991).
51. R. Kubo, A general expression for the conductivity tensor, *Canad. J. Phys.* **34**, 1274 (1956).
52. D. A. Greenwood, The Boltzmann equation in the theory of electrical conduction in metals, *Proc. Phys. Soc. Lond.* **71**, 585 (1958).
53. D. Vollhardt and P. Wölfle, Anderson localization problem in $d \leq 2$ dimensions: A self-consistent diagrammatic theory, *Phys. Rev. Lett.* **45**, 482 (1980).
54. D. Vollhardt and P. Wölfle, Diagrammatic, self-consistent treatment of the Anderson localization problem in $d \leq 2$ dimensions, *Phys. Rev.* **B22**, 4666 (1980).
55. B. L. Altshuler, D. E. Khmelnitskii, A. I. Larkin and P. A. Lee, Magnetoresistance and Hall effect in a disordered two-dimensional electron gas, *Phys. Rev.* **B22**, 5142 (1980).
56. B. L. Altshuler, A. G. Aronov, A. I. Larkin and D. E. Khmelnitskii, Anomalous magneto-resistance in semiconductors, *Zh. Eksp. Teor. Fiz.* **81**, 768 (1981); *Sov. Phys. JETP* **54**, 411 (1981).
57. K. Jüngling and R. Oppermann, Random electronic models with spin-dependent hopping, *Phys. Lett.* **76A**, 449 (1980).
58. K. Jüngling and R. Oppermann, Effects of spin interactions in disordered electronic systems: loop expansions and exact relations among local gauge invariant models, *Z. Phys.* **B38**, 93 (1980).
59. F. Wegner, Inverse participation ratio in $2+\epsilon$ dimensions, *Z. Phys.* **B36**, 209 (1980).
60. E. Brezin, S. Hikami and J. Zinn-Justin, Generalized nonlinear sigma-models with gauge invariance, *Nucl. Phys.* **B165**, 528 (1980).
61. D. Hoef and F. Wegner, Calculation of anomalous dimensions for the nonlinear sigma model, *Nucl. Phys.* **B275** [FS17], 561 (1986).
62. F. Wegner, Anomalous dimensions for the nonlinear sigma-model in $2+\epsilon$ dimensions (II), *Nucl. Phys.* **B280** [FS18], 210 (1987).
63. W. Bernreuther and F. Wegner, Four-loop order β-function for two-dimensional nonlinear σ-models, *Phys. Rev. Lett.* **57**, 1383 (1986).
64. S. Hikami, Localization, Nonlinear σ model and string theory, *Prog. Theor. Phys. Supp.* **107**, 213 (1992).
65. S. Hikami, A. I. Larkin, Y. Nagaoka, Spin-orbit interaction and magnetoresistance in the two-dimensional random system, *Prog. Theor. Phys.* **63**, 707 (1980).

66. R. J. Bell and P. Dean, Atomic Vibrations in vitreous silica, *Disc. Faraday Soc.* **50**, 55 (1970).
67. D. J. Thouless, Electrons in disordered systems and the theory of localization, *D.J. Thouless, Phys. Repts.* **13C**, 93 (1974).
68. F. Wegner, Electrons in disordered systems. Scaling near the mobility edge, *Z. Phys.* **B25**, 327 (1976).
69. A. D. Mirlin, Y. V. Fyodorov, A. Mildenberger and F. Evers, Exact relations between multifractal exponents at the Anderson transition, *Phys. Rev. Lett.* **97**, 046803 (2006).
70. I. A. Gruzberg, A. W. W. Ludwig, A. D. Mirlin, and M. R. Zirnbauer, Symmetries of multifractal spectra and field theories of Anderson localization, *Phys. Rev. Lett.* **107**, 086403 (2011).
71. I. A. Gruzberg, A. D. Mirlin and M. R. Zirnbauer, Classification and symmetry properties of scaling dimensions of Anderson transitions, *Phys. Rev.* **B87**, 125144 (2013).
72. V. E. Kravtsov, I. V. Lerner, B. L. Altshuler and A. G. Aronov, Universal Spectral Correlations at the Mobility Edge, *Phys. Rev. Lett.* **72**, 888 (1994).
73. A. Altland and M. R. Zirnbauer, Nonstandard symmetry classes in mesoscopic non-superconducting hybrid structures, *Phys. Rev.* **B55**, 1142 (1997).
74. C. Chiu, Y. Teo, A. P. Schnyder and S. Ryu, Classification of topological quantum matter with symmetries, *Rev. Mod. Phys.* **88**, 035005 (2016).
75. J. Fröhlich and T. Spencer, Absence of diffusion in the Anderson tight binding model for large disorder or low energy, *Comm. Math. Phys.* **88**, 151 (1983).
76. B. Kramer and A. MacKinnon, Localization theory and experiment, *Rep. Prog. Phys.* **56**, 1469 (1993).
77. B. Kramer, A. MacKinnon, T. Ohtsuki and K. Slevin, *Finite size scaling analysis of the Anderson transition*, in[12] p. 347 (2010).
78. B. L. Altshuler, A. G. Aronov and P. A. Lee, Interaction effects in disordered Fermi systems in two dimensions, *Phys. Rev. Lett.* **44**, 1288 (1980).
79. A. M. Finkel'stein, The influence of Coulomb interaction on the properties of disordered metals, *Zh. Teor. Eksp. Fiz.* **84**, 168 (1993); *Sov. Phys. JETP* **57**, 97 (1993).
80. A. M. Finkel'stein, Weak localization and Coulomb interactions in disordered systems, *Z. Phys.* **B56**, 189 (1994).
81. J. Arenz and M. Zirnbauer, *Wegner model on a tree graph: U(1) symmetry breaking and a non-standard phase of disordered electronic matter*, arXiv2305.00243 v1.
82. M. Zirnbauer, *Wegner model in high dimension: U(1) symmetry breaking and a non-standard phase of disordered electronic matter I. One-replica theory*, arXiv2309. 17323 v1. I am indebted to Richard Bausch for informing me on this paper.
83. P. Sierant, M. Lewenstein and A. Scardiocchio, *Universality in Anderson localization on random graphs with varying connectivity*, arXiv2205.14614 v3
84. Garcia-Mata I., J. Martin, O. Giraud, B. Georgeot, R. Dibertrand, G. LeMarié, Critical properties of the Anderson transition on random graphs: Two-parameter scaling theory, Kosterlitz-Thouless type flow and many-body localization, *Phys. Rev.* **B106**, 214202 (2022)

https://doi.org/10.1142/9789811282386_0029

Chapter 29

Fermi Surface Renormalization Group

R. Shankar

Sloane Physics Lab, Yale University, 217 Prospect Street,
New Haven, CT 06520, USA

The low energy manifold for fermions at finite density is the Fermi surface. I describe renormalization group (RG) in which modes encountered on approaching the Fermi surface are systematically integrated out. The fixed point is described by strictly marginal coupling *functions* identified as the Landau parameters and marginally relevant coupling functions describing the BCS instability in various angular momentum channels.

1. How I Met Michael

I want to begin with a few words on how Michael entered my life. It was around 1978 and I had just moved to Yale to join its particle physics group. In the three previous years at Harvard, I was switching from S-matrix theory (in which I was trained by the high priest, Geoffrey Chew himself) to quantum field theory. Another switch followed when Polyakov, while visiting Yale, mentioned that the Zamolodchikovs had found a connection between factorizable S-matrices in $1 + 1$ dimensions and Baxter's eight-vertex model. It was shocking that an S-matrix in a relativistic field theory in the continuum could have any relation to the Baxter model on a lattice. From my collaboration with Witten, I was familiar with the S-matrices, but the Baxter model was a closed book to me. Since the Zamolodchikov connection had not appeared in print (and did not for some time), I decided to search for the interpolating steps. I wrote a short preprint on this and sent a copy to Michael since I knew he was a leader in statistical mechanics and I had relied on his numerous reviews to educate myself.[1] He asked me to give a talk in Joel Lebowitz's meeting at Rutgers. When I asked Joel how much time I had, he told me it was 60 minutes divided by the number of people wanting to speak, which turned out to be between 4 and 5 minutes. Though initially put off by this, over the years, I came to love the format: everybody got a shot, everybody got the same time and if you really had something to say, you could say it in 4–5 minutes, whereas it took considerably longer if you did not.

Based on this, Michael invited me to Cornell where I gave a blackboard talk which he liked. After that, I corresponded with him and even collaborated with him (and Rajiv Singh). Those who know him will not be surprised that the hardest obstacle to publication were not the referees but Michael's exacting standards. I was amazed that a person of his standing would get down into the weeds and not budge till everything was perfect. For this, for his civic sense and for the way he treated his correspondence and demanded clarity in the written and spoken word,[2] he became a role model for me. This is why my book on *Field Theory and Condensed Matter* was dedicated to him, Ben Widom and Ken Wilson, whose work on the RG had given me so much pleasure and profit.

Having imbibed the renormalization group (RG) Kool aid, I spent many years on its applications. I will now describe one that was quite different and whose novelty Michael appreciated.

2. RG for Non-Relativistic Fermions

To a man with a hammer, everything looks like a nail. Thus, in the late 1980s I was walking around looking for RG applications everywhere. At that time, the High-T_c stampede had just begun. Anderson[6] pointed out right away that the normal state of the superconductor was a non-Fermi liquid. This was a real possibility given that in one spatial dimension the slightest interaction kills the sharply defined quasi-particle, leaving behind just a cut, and only in three dimensions were Landau Fermi liquids known to exist. So which way will two dimensions go? While experimentally the high-T_c materials certainly looked non-Fermi, I wanted to know *if a Fermi liquid was at all possible in two dimensions for non-zero interactions.* I decided to use the RG because it would provide an unbiased way to settle the matter (unlike, say, Hartree–Fock). This suited me well since I did not know enough to have a bias.

The general strategy of the RG is to zero-in on the variables that control the asymptotic behavior and integrate away the rest. Usually, this corresponds to keeping just the long wavelength, low energy modes. In ϕ^4 theory, for example, the action is near the origin in Euclidean momentum space. One starts with a ball of radius Λ and lets $\Lambda \to 0$ and keeps track of the couplings. When the couplings are expanded as functions of the external momenta, only the constant quartic term u_0 survives (near four dimensions) and becomes the *coupling constant* of the theory.

In adapting the RG to fermions at finite density, I had to make some changes to suit the problem. The biggest difference was that the low energy states were not near the origin in momentum space, but near the Fermi surface. Let us focus on $d = 2$ which is easiest to explain and experimentally the most relevant. The region of interest is an annulus of thickness 2Λ, located symmetrically about the Fermi circle, as shown in Fig. 1. The RG consists of reducing Λ to zero. Since the low-energy region

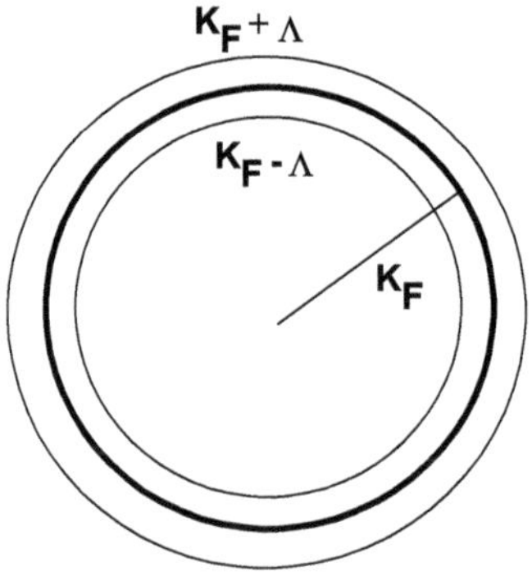

Fig. 1. The low-energy region for Fermions at finite density is an annulus of thickness 2Λ about the Fermi surface.

is a surface and not point, we will end up with *coupling functions* that vary over the surface. One can expect the flows of these functions to be complicated. What happened when I actually worked it out was a pleasant surprise. I will describe it in minimum detail now.

The Euclidean action for the free theory is

$$S_0 = \int_{-\infty}^{\infty} \int_{-\Lambda}^{\Lambda} \int_0^{2\pi} d\omega dk d\theta \, \bar{\psi}(\omega, k, \theta) \, [i\omega - k] \, \psi(\omega k, \theta), \tag{1}$$

where k is the deviation from the Fermi momentum K_F. Factors like K_F in the measure have been absorbed in the measure, and the Fermi velocity has been set to unity.

We see that the action is that of a Euclidean 1 + 1-dimensional field theory: ω is the frequency and k is the momentum. The angle θ is an internal isospin-like variable of which the energy is independent. (In the usual 1+1-dimensional theory,[7] the Fermi surface is a pair of disjoint points labeled left and right). Consequently, the only interaction that will survive is the quartic one, which will be marginal. Furthermore, its dependence on ω and k is irrelevant. However, it can depend on the external θ's:

$$u_0 = u_0(\theta_1, \ldots, \theta_4), \tag{2}$$

which is the new feature here.

Obviously, only three of the θ's are independent due to momentum conservation. However, because the momenta come not from the plane but the Fermi circle, only two of them, say, the incoming ones θ_1 and θ_2, can be specified freely, and the other two must be equal to them up to an irrelevant permutation. Thus, only forward-scattering amplitudes survive, as shown in the left half of Fig. 2.

A subtle point: the Fermions' momenta are not confined to the FS, but free to wiggle by an amount of order Λ. However, the coupling functions may be evaluated

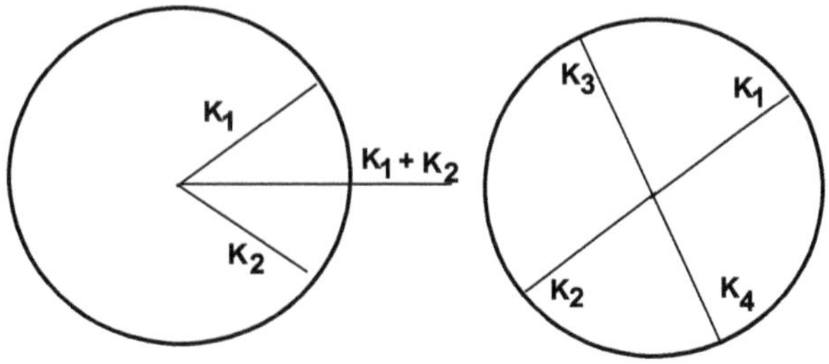

Fig. 2. Since the momenta come from the Fermi circle, the only way to conserve momentum is either (left) forward scattering or (right) Cooper scattering, where 1 and 2 are opposite as are 3 and 4, but in a different direction.

on the FS since their dependence on k is irrelevant. So, we have now

$$u_0 = u_0(\theta_1, \theta_2). \tag{3}$$

Finally, rotational invariance tells us that

$$u_0 = u_0(\theta_1, \theta_2) = u_0(\theta_1 - \theta_2) \equiv F(\theta), \tag{4}$$

and so, the asymptotic behavior of the fermions is controlled by a single function $u_0 \equiv F$. It is called the Landau parameter because he arrived at it in his own way decades ago without explicitly appealing to the RG.

There is an exception to the forward-scattering restriction shown in the right half of Fig. 2. If the incoming angles are exactly opposite, the final ones also have to be equal and opposite, but the final pair can be at any angle θ relative to the initial, leading to another function $V(\theta)$ that survives the RG and describes the scattering of Cooper pairs.

The fate of the marginal couplings has to be decided by a one-loop calculation. The diagrams for F and V appear in Fig. 3.

These diagrams have the same topology as ϕ^4 field theory, but behave differently. In the diagram labeled ZS (for zero-sound), there is no transfer of momentum since $\theta_1 = \theta_3$ and no transfer of ω because the external legs can be chosen to have $\omega = 0$. This means the two propagators in the loop have poles in the same half-plane in ω and give zero under ω-integration, for we may close the contour the other way. (It is well known that in the ZS channel, only particle–hole diagrams can survive the ω-integration.) In the ZS$'$ diagram with crossed outgoing momenta, the two external legs at the left vertex restrict the loop variables to a region of order $d\Lambda^2/K_F$, where a factor $d\Lambda$ is from the shell width and a factor $d\Lambda/K_F$ is from the angular restriction. Finally, in the BCS diagram, the two initial momenta again restrict the loop momenta to a region of size $d\Lambda^2/K_F$. So, $du/dt = 0$, where $dt = d\Lambda/\Lambda$. Later, I will argue that u does not flow to all orders.

The flow of the Cooper amplitude is similar with notable exception: the particle–particle bubble, with opposite ω and same energy (due to time-reversal invariance), makes a contribution of order $d\Lambda$ when the loop ω is integrated. Furthermore, the

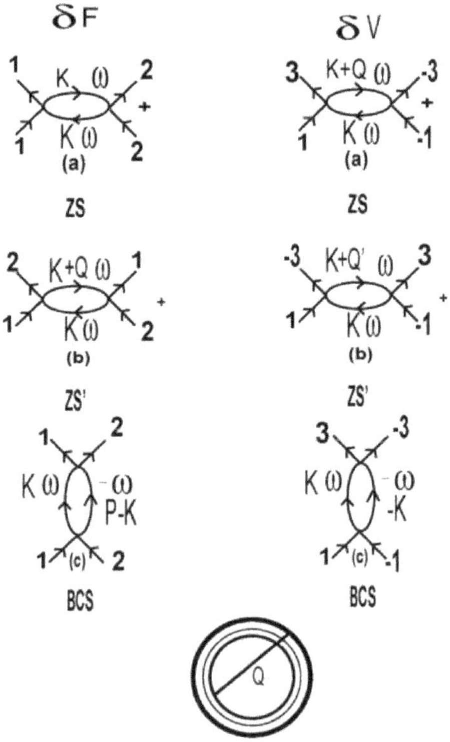

Fig. 3. The diagrams that contribute to the flow of F and V. Only ZS has a contribution of order $d\Lambda$ to δu, but the diagram is killed by ω-integration because the poles lie on the same half-plane. The BCS diagram contributes to δV to order $d\Lambda$ which survives the ω integration because the poles lie on opposite half-planes. The bottom figure shows that a momentum Q of order K_F determines the initial momentum on either shell being integrated up to a small multiplicity.

angle of the loop momenta can range freely over the FS as long as they too are opposite. This leads to the *functional flow equation*

$$\frac{dV(\theta_1 - \theta_3)}{dt} = - \int d\theta V(\theta_1 - \theta)\, V(\theta - \theta_3). \tag{5}$$

By expanding in terms of angular momentum eigenfunctions, we get an infinite number of flow equations for the coefficients V_m:

$$\frac{dV_m}{dt} = -V_m^2, \tag{6}$$

one for each coefficient. These equations tell us that if the potential in angular momentum channel m is repulsive, it will get renormalized down to zero (a result derived many years ago by Morel and Anderson,[8] while if it is attractive, it will run off, causing the BCS instability. This is the reason the V's are not a part of Landau theory, where the assumed adiabatic continuity requires that there be no phase transitions.

It was at this point that I tried to publish my results but ran into a brick wall. The main objection was that I had not computed any numbers. Finally, I got a break when I was asked to contribute to Michael Fisher's 60th birthday Festschrift in a special issue of Physica[9] in 1991. Since Michael knew of this work and had heartily encouraged me (as on many others occasions), I was delighted to publish my work and also honor him with a work he approved of.

For the next two years, I traveled a lot, describing my work. I discovered and resolved many subtleties I had initially overlooked. I learned a lot from my audience and clarified their issues. I summarized my findings in the *Reviews of Modern Physics* issue that came out in 1994.[13] John Wilkins insisted, as editor, that I should take all the space I needed to render the article accessible to a wide audience, especially graduate students. I needed no coaxing and jumped at that invitation. Here are some highlights from that review:

- In more than two spatial dimensions, Fermi liquids still behave like 1 + 1-dimensional theories, with θ getting replaced by more angles. The RG now allows non-forward scattering amplitudes, but their effect (say on lifetime) is small due to phase space. Only the forward scattering amplitudes enter the energetics.
- I was made to realize that the fixed point I had found described Landau's Fermi liquid theory.[10] In particular, the non-singular the fixed point coupling $F(\theta)$ was the Landau's parameter to which he was led by his deep intuition. Subsequently, his students[11] gave a diagrammatic proof of his results.

  Let us remember that getting the fixed point action is only half the task. One still needs to solve for the Greens functions of the cut-off theory. For these, one draws Feynman diagrams with vertices F and loop momenta that run in the range $-\Lambda \le k \le \Lambda$. (I ignore the Cooper channel following Landau.) Here, one finds a great simplification: the diagrams are given by a sum of iterated ZS bubbles that can be summed geometrically.

  For example, $\bar{\psi}\psi - \bar{\psi}\psi$ correlation function (with vanishing values of external frequency and momentum transfer and in the limit $\omega/q \to 0$) takes the form

$$\chi = \frac{\chi_0}{1 + F_0}, \tag{7}$$

  where F_0 is the angular average of $F(\theta)$ and χ_0 is the answer when $F = 0$. Note that the answer is not perturbative in F. Landau got this result by working with the ground-state energy as a functional of Fermi surface deformations. We get it by summing bubbles.

  The RG provides us with not just the ground-state energy as Landau theory does, but an effective Hamiltonian (operator) for all of low-energy physics. This operator problem can be solved using large N-techniques.

  The bubble sums should ring a bell with readers familiar with the $1/N$ expansion (say from the Gross–Neveu model[12]). I showed in my review that the small cut-off theory is a $1/N$ theory with $N \simeq K_F/\Lambda$. By dividing the annulus into

N regions of width 2Λ in the angular direction, one ends up with N patches, labeled by an index that runs from 1 to N. The kinematics is such that only diagrams where the patch index is freely summed survive in the large N limit. These turn out to be the ZS diagram (which does not contribute to the flow but does contribute to the iterated bubble in the computation of Green's functions) and the Cooper diagram that gives rise to the flow.

Since in the large N limit, the one-loop β-function for the fermion–fermion coupling is exact, it follows that the marginal nature of the Landau parameters F and the flow of V in Eq. (6) are both exact as $\Lambda \to 0$.

- If a Fermi surface is nested, that is, if there is a momentum Q that takes every point on the FS to another point on the FS, the system has infinite susceptibility to charge density wave (CDW) order at that Q. So, order sets in for arbitrarily small coupling, just as the BCS instability sets in for arbitrarily weak attraction. I showed that in this case, as $\Lambda \to 0$, the kinematics allows another marginal amplitude W besides F and V. At one loop, W flows to strong values, exhibiting the CDW instability. I computed the functional flow equation, but in the absence of rotational invariance, could not separate them into decoupled flows for individual angular momentum amplitudes.

  Thus, the RG seems to not only get the Fermi liquid as a fixed point, but also all its known instabilities. The only way a non-Fermi liquid can arise is if the entire diagrammatic approach fails at the outset due to strong interactions that invalidate any perturbative analysis that starts with free fermions.
- I laid out a road map for solving realistic problems by first integrating high-energy modes perturbatively, getting some effective couplings and then letting the RG run its course, mainly to see if a repulsive microscopic interaction could lead to pairing. Unlike Ken Wilson, I talked about it, but did not actually do this. Luckily, this was subsequently done by may others, as I will describe momentarily.

3. Contemporaneous and Subsequent Developments

After this rather personal description of events, I want to add a few words on related works. Benfatto and Gallavotti[14] considered this problem around the time I did. Their language is very different from mine, but the parallels (including the BCS flow) are readily visible. Joe Polchinski[15] discovered the marginal nature of the Fermi liquid couplings in his study of effective field theories. Feldman *et al.*[16] attacked the same problem more formally and later independently derived the $1/N$ expansion.[17]

One can ask how a non-Fermi liquid of the type Anderson[6] envisaged may emerge by asking how the above derivation could fail. One way is at strong coupling where the perturbative RG does not converge. Another option is to consider fermions coupled to gauge fields either in strongly correlated electrons (Polchinski[18]) or in the Fractional Quantum Hall problem at half-filling (Halperin, Lee and Read[19]).

Earlier, I had raised the question of whether a repulsive interaction can spawn superconductivity within the RG. This proposed attack has two stages: integrate high-energy modes perturbatively (to one loop say) to get an effective set of interactions and then run the flow as above starting with these as initial couplings. Among others, this was done by Zanchi and Schulz,[20] Halboth and Metzner,[21] Wang *et al.*,[22]), and Honerkamp *et al.*[23] More rigorous derivations of the ideas espoused here were derived, among others by Salmhofer[24]), though as of now, no small parameters besides $\Lambda/K_F = 1/N$ or weak coupling have emerged to justify approximations which inevitably follow in any practical calculation.

The $1/N$ expansion may be used to study problems where the interaction has frequency dependence, as in the BCS problem with retardation (Tsai *et al.*[25]).

It is possible to use the RG even if bosons are simultaneously present, provided the dynamical exponent $z = 1$ (Yamamoto and Si[26])).

The theory may be applied to nuclear matter or stars (Schwenk *et al.*[27]) as well as quarks at finite density (Rajagopal and Wilczek,[28] Alford *et al.*[29]).

One can also vary the power law of the force between fermions continuously to a controllable fixed point (Nayak and Wilczek[30]).

The theory may also be extended to quantum dots to address a problem with finite size, disorder and interactions (Murthy and Shankar[31]).

If the present approach is applied to fermions in d spatial dimensions, the scaling is done only for ω and $k = K - K_F$, since the other angles (like θ) on the Fermi surface behave like internal or isospin degrees of freedom. Thus, the Fermi liquid behaves like a $1+1$-dimensional field theory for all $d \geq 2$. A new generalization has been proposed (Senthil and Shankar[32]) to Fermi surfaces with more co-dimensions on which the energy depends linearly. For example, in three dimensions, a line Fermi surface has co-dimension 2 and behaves like a $2+1$-dimensional theory. It is also possible to vary these extra dimensions in an ε expansion and obtain new weak coupling fixed points. This idea has been advanced further by Lee.[33]

References

1. M. E. Fisher, in *Critical Phenomena.* In ed. F. W. J. Hahne, Lecture notes in Physics, **186**, Springer (1983).
2. M. E. Fisher, in *Coneceptual Foundatons of Quantum Field Theory,* p. 89, Ed. Tian Yu Cao, Cambridge University Press (1999).
3. L. P. Kadanoff, Scaling laws for Ising models near Tc. *Physics*, **2**, 263 (1966). See also Kadanoff, L.P., The application of renormalization group techniques to quarks and strings. *Rev.Mod. Phys.* **49**, 267–296 (1977).
4. K. G. Wilson, Renormalization Group and Critical Phenomena. I. Renormalization Group and the Kadanoff Scaling Picture, *Phys. Rev. B* **4**(9), 3174. See also Wilson, K. G., *Rev. Mod. Phys.*, 1975, The renormalization group: Critical phenomena and the Kondo problem, **47**, 773, (1971), and Wilson, K. G., and J. B. Kogut, 1974. *Physics Reports*, **12**, 7. *Phys. Rev. B* **4**, 3174 (1971).

5. Critical exponents in 3.99 dimensions — K. G. Wilson, and M. E. Fisher. *Phys. Rev. Lett.* **28**, 240–243, (1972). See also Fisher, M.E., The renormalization group in the theory of critical behavior. *Rev. Mod. Phys.* **4**, 597, 1974 and Fisher M. E., Renormalization group theory: Its basis and formulation in statistical physics. *Rev. Mod. Phys.* **70**, 653 (1998).
6. P. W. Anderson, Luttinger-liquid behavior of the normal metallic state of the 2D Hubbard model, *Phys. Rev. Lett.* **64**, 1839 (1990).
7. J. Solyom, The Fermi gas model of one-dimensional conductors. *Adv. Phys.* **28**, 201 (1979); Bourbonnais, C. and Caron L. G., *Int. J. Mod. Phys. B.* **5**, 1033 (1991).
8. P. Morel and P. W. Anderson, Calculation of the superconducting state parameters with retarded electron-phonon interaction. *Phys. Rev.* **125**, 1263 (1962).
9. R. Shankar, Renormalization group for interacting fermions in $d > 1$, *Physica, A* **177**, 530 (1991).
10. L. D. Landau, *Sov. Phys. JETP*, **3**, 920 (1956).
11. A. A. Abrikosov, L. P. Gorkov and I. E. Dzyaloshinski, *Methods of Quantum Field Theory in Statistical Mechanics*, Dover, New York (1963). Pines, D. and P. Nozieres, *The Theory of Quantum Liquids* (1966), Baym, G. and Pethick, C., *Landau Fermi Liquid Theory*, Wiley, New York (1991).
12. D. J. Gross and A. Neveu, Dynamical symmetry breaking in asymptotically free field theories. *Phys. Rev. D* **10**, 3235 (1974).
13. R. Shankar, Renormalization group approach to interacting fermions. *Rev. Mod. Phys.* **66**, 129 (1994).
14. G. Benfatto and G. Gallavotti, Renormalization-group approach to the theory of the Fermi surface. *Phys. Rev. B* **42**, 9967 (1990). Press. New York.
15. J. Polchinski, *Effective Field Theory and the Fermi Surface*, in Recent directions in particle theory, Ed. J. Harvey and J. Polchinski, TASI Lectures 1992, World Scientific, Singapore, 1993.
16. J. Feldman and E. Trubowitz, Perturbation theory for many fermion systems. *Helv. Phys. Acta*, **63**, 157 (1990).
17. J. Feldman, V. Magnan, Rivasseau and E. Trubowitz, An intrinsic 1/N expansion for many fermion systems. *Europhys. Lett.* **24**, 437 (1992).
18. J. Polchinski, Low energy dynamics of the spinon-gauge system. *Nucl. Phys. B* **422**, 617 (1993).
19. B. I. Halperin, P. A. Lee and N. Read, Theory of the half-filled Landau level. *Phys. Rev. B* **47**, 7312 (1993).
20. D. Zanchi and H. J. Schulz, Weakly correlated electrons on a square lattice: Renormalization-group theory. *Phys. Rev.* **61**, 13609 (2000).
21. C. J. Halboth and W. Metzner, Renormalization group analysis of the 2d Hubbard Model. *Phys. Rev. B* **61**, 7364 (2000).
22. Wang *et al.*, Functional renormalization group study of the pairing symmetry and pairing mechanism of the FeAs based high temperature superconductors. *Phys. Rev. Lett.* **102**, 047005 (2009).
23. C. Honerkamp, *et al.*, Breakdown of the Landau-Fermi liquid in two dimensions due to umklapp scattering. *Phys. Rev.* **63**, 035109 (2001); for a variation see Honerkamp *et al.*, Interaction flow method for many-fermion systems, *Phys. Rev. B* **70**, 235115 (2004).
24. M. Salmhofer, *Renormalization: An Introduction*, Springer, Heidelberg (1998).
25. S. W. Tsai, *et al.*, Renormalization group approach to superconductivity from weak to strong electron-phonon coupling. *Phil. Mag.* **86**, 2631 (2006).

26. J. Yamamoto and Q. Si, Renormalization group for mixed boson fermion systems, 2010. *Phys. Rev. B* **81**, 205106 (2010).
27. A. Schwenk, G. E. Brown, B. Friman, *Nucl. Phys. A* **713**, 191 (2003).
28. K. Rajagopal and F. Wilczeck, The condensed matter physics of QCD, arXiv.org/hep-ph0011333v2, 2000, in Festschrift in honor of B. L. Ioffe, *At the Frontier of Particle Physics/Handbook of QCD*, M. Shifman, ed., (World Scientific). v2.
29. G. A. Alford, A. Schmitt, K. Rajagopal and Schäfer, Color superconductivity in dense quark matter, *Rev. Mod. Phys.* **80**, 1455 (2008).
30. C. Nayak and F. Wilczek, Non-Fermi liquid fixed points in $2 + 1$ dimensions, *Nucl. Phys. B* **417**, 379 (1993).
31. G. Murthy and R. Shankar, Chaotic quantum dots with strongly correlated electrons. *Rev. Mod. Phys.* **80**, 379 (2008).
32. T. Senthil, Fermi surfaces in general codimension and a new controlled non-trivial fixed point. *Phys. Rev. Lett.* **102**, 046406 (2009).
33. S. S. Lee, Perturbative non-Fermi liquids from dimensional regularization. *Phys. Rev. B* **88**, 245106 (2013).

https://doi.org/10.1142/9789811282386_0030

Chapter 30

Critical Matter

L. Radzihovsky

Department of Physics, University of Colorado, Boulder, CO 80309, USA

I review a class of novel ordered states of "critical matter", that exhibit strongly fluctuating universal power-law orders, controlled by an infrared attractive, non-Gaussian fixed point. I will illustrate how RG methods pioneered by Wilson and Fisher can be used to deduce critical phenomenology of such critical phases, resembling that of a critical point of second-order phase transitions, but requiring no fine-tuning.

1. Introduction

Michael Fisher is a towering figure in theoretical physics, most notably recognized for his seminal development (in collaboration with Ken Wilson and building on the works of Kadanoff, Migdal, Widom, Larkin, Pokrovsky, and others) of the renormalization group (RG)[1–3] and its numerous early applications to critical phenomena of continuous phase transitions.[4] In these works, Michael transformed Wilson's deep ideas[2,3] into a practical and powerful theoretical tool for controlled calculations[1] (complementing more formal developments in quantum-field theory[5]) and demonstrated its power through numerous seminal applications to critical phenomena. With this, he elevated RG into a central calculation tool in modern theoretical physics.

Beginning with the earliest seminal analysis of the ϕ^4 field theory to describe criticality of the Ising paramagnet–ferromagnet transition, most applications of RG are used to treat enhanced fluctuations near a critical point of a continuous phase transition, where direct perturbation theory breaks down. In such applications, a critical behavior emerges after tuning a set of parameters, e.g., temperature, pressure, magnetic field, etc., to a (multi-) critical point in a phase diagram, as illustrated in Fig. 1. It is only then that fluctuations and non-linearities (interactions) become important (below the upper-critical dimension) and universal asymptotic critical behavior emerges, e.g., power-law, scale-free correlation functions and thermodynamic responses with universal exponents.[6]

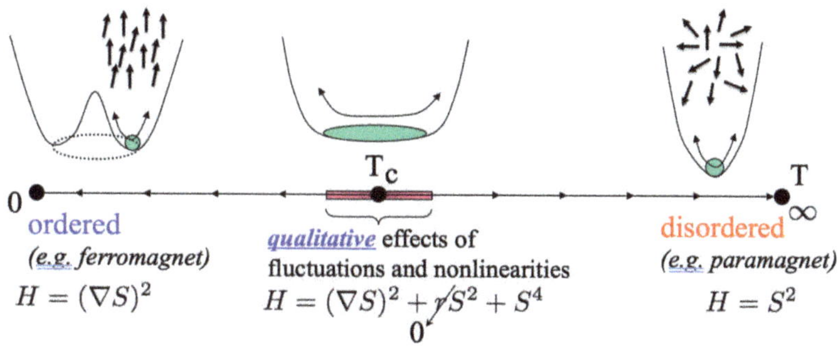

Fig. 1. Illustration of unimportance of fluctuations inside phases of conventional systems, where qualitative effects of thermal fluctuations are confined to a vicinity of a critical point.

In contrast to critical points, as illustrated in Fig. 1, phases of matter, by their very definition, are stable to weak generic perturbations that do not explicitly break their symmetry. They are thus characterized by infrared *attractive* fixed points and thereby require no fine-tuning, in contrast to a critical point that separates them. Fluctuations in ordered Landau phases of matter[8] (which is our focus here) are characterized by their Goldstone modes associated with the spontaneous breaking of a *continuous* symmetry — Landau phases that break *discrete* symmetries have no interesting fluctuations. Despite being gapless, generically within ordered phases, Goldstone-mode fluctuations are small, with finite root-mean-squared fluctuations, latter typically taken as the defining property of the ordered phase. In the RG parlance, this corresponds to typical phases that are controlled by a Gaussian attractive fixed point, as illustrated in Fig. 1. Thus, conventional Landau phases' description is effectively trivial, with fluctuations (above the lower-critical dimension, where the ordered phase is stable) only leading to small corrections to their mean-field description.

In this chapter, I instead focus on an exotic class of ordered phases of matter — "critical phases," e.g., smectics, cholesterics, columnar phases, membranes, elastomers, and many others as illustrated in Fig. 2, where in stark contrast to conventional phases, Goldstone-mode fluctuations are divergingly strong, interacting, and thus are controlled by a *non*-Gaussian infrared attractive fixed point, as illustrated in Fig. 3. As such, the resulting ordered critical phase, while breaking a continuous symmetry is highly non-trivial and strongly interacting, with little resemblance of its mean-field cartoon. It is characterized by universal critical exponents, but in striking contrast to a critical point requires no fine-tuning, protected by underlying spontaneously broken symmetries, with importance of strong fluctuations and non-linearities extending throughout the ordered phase. A generic ingredient of such critical phases (often associated with spontaneous breaking of *continuous* spatial symmetries) is that underlying symmetry enforces an exact *vanishing* of a subset of elastic moduli, resulting in particularly "soft" harmonic elasticity of the associated Goldstone modes, described by m-Lifshitz models.[9,51]

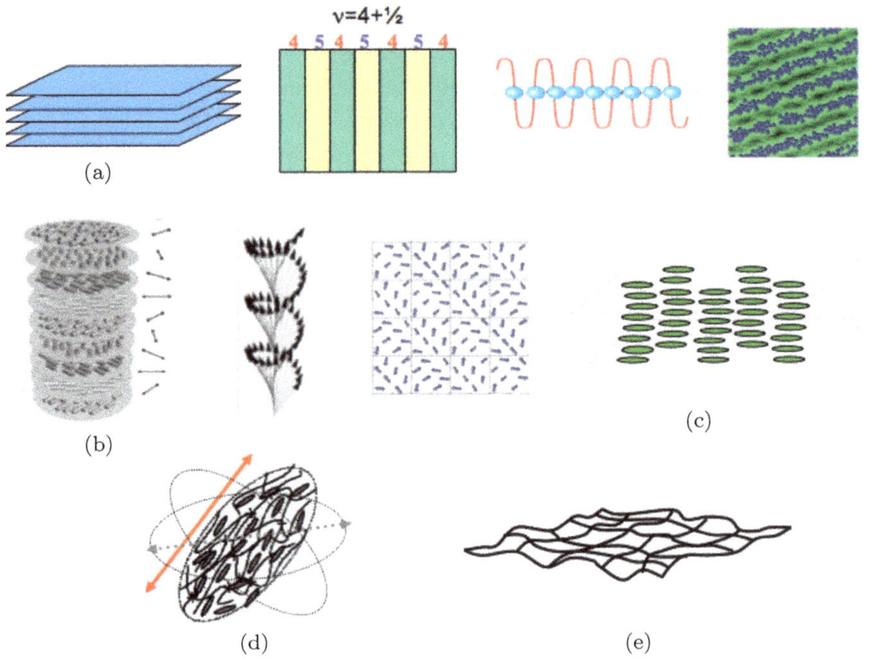

Fig. 2. Illustration of critical phases: (a) smectics realized as conventional liquid crystals,[4,10–14] 2D colloids,[15] quantum-Hall systems[16–22] and as striped "pair-density wave" FFLO[23,24] superconducting phases in degenerate atomic gases,[25–27] p-wave resonant,[28] spin–orbit coupled[29] and frustrated superfluids;[30] (b) cholesteric liquid crystals[4,10,31,32] and helical state of frustrated magnets;[32–36] (c) columnar liquid crystals and spontaneous vortex lattices;[37–42] (d) nematic elastomers;[43,44] (e) polymerized membranes.[45–50]

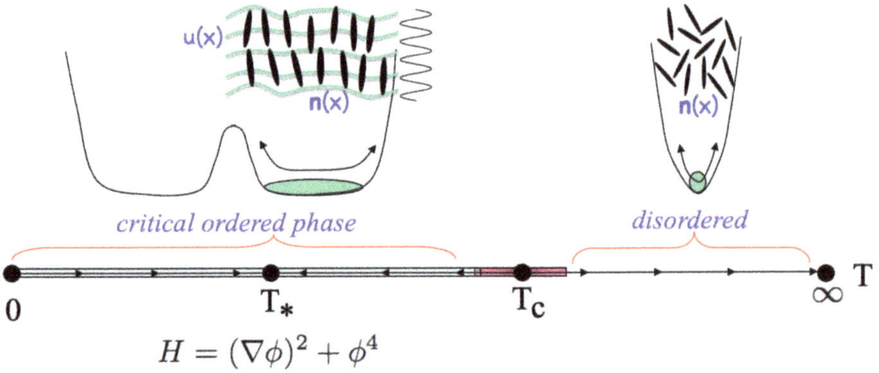

Fig. 3. Illustration of the importance of fluctuations inside ordered phases of "critical matter", in which even deep in the ordered phase, the quartic potential is missing the stabilizing quadratic contribution, resulting in divergent fluctuations akin to a critical point. The bottom of the figure is a schematic diagram of an RG flow, e.g., for a smectic state in $d < 3$ dimensions, illustrating that at low T, it is a "critical phase" displaying universal power-law phenomenology, controlled by a non-trivial infrared-stable fixed point.

In subsequent sections of this chapter, I illustrate the above phenomenology through a set of examples of critical phases, some of which are illustrated in Fig. 2. I will begin with a detailed discussion of a quintessential critical phases — uniaxially periodically modulated states that spontaneously break translational and rotational symmetry and are realized in numerous physical contexts, including smectic liquid crystals,[4,10,11] cholesterics,[32] helical magnets and bosons on frustrated lattices,[30,32–36] non-zero momentum superfluids such as Fulde–Ferrell–Larkin–Ovchinnikov (FFLO) superconductors,[25,26] p-wave resonantly paired bosons,[28] spin–orbit coupled bosons,[29] and quantum-Hall striped states.[16–21] In Section 3, I describe a columnar states, that appear in discotic liquid crystals[4,10,40] and in its line-crystal analogs, such as putative spontaneous vortex lattice in magnetic superconductors.[41,42] I analyze another rich example of a critical phase — a thermally fluctuating tensionless polymerized membrane in Section 4.[45–49] In Section 5, I discuss a nematic elastomer — a liquid crystal rubber, which mathematically is a three-dimensional amalgam of a smectic and columnar phases, displaying critical phenomenology throughout its phase.[43,44]

2. Uniaxially Periodic Critical States

There are a number of states of matter that *spontaneously* develop a periodic modulation along an arbitrarily chosen *single* axis. Strictly speaking, this can only take place in a system with underlying rotational invariance, though this can be exhibited over an extended intermediate regime when explicitly broken rotational symmetry couples only weakly, as, e.g., in low-density electronic systems.[16,17]

Beyond conventional liquid crystals, some of the novel systems include helical states of frustrated and spin–orbit coupled magnets and bosons,[29,30,33–35] and strongly correlated electronic and bosonic systems (quite surprising for isotropic and point constituents), such as FFLO[25] "striped" ("pair-density wave") superconductors,[20,27] finite momentum superfluids,[28–30] and a two-dimensional electron gas in the quantum Hall regime of half-filled Landau level.[16–22] The underlying unifying feature of these diverse sets of critical systems is the *spontaneously* broken rotational (in contrast to, e.g., density waves in crystalline solids) and translational symmetries. Based on these symmetries that they break, we collectively refer to this class of states as smectics akin to their conventional soft matter liquid crystal realizations, which we turn to next.

2.1. *Smectic liquid crystals*

Illustrated in Fig. 4, the most ubiquitous liquid crystal phases are the uniaxial nematic, that spontaneously breaks rotational symmetry of the parent isotropic fluid and the smectic state, a uniaxial one-dimensional density wave that further breaks translational symmetry along a single axis. The uniaxial nematic is characterized by a quadrupolar order parameter $Q_{ij} = S(\hat{n}_i\hat{n}_j - \frac{1}{3}\delta_{ij})$, with S the strength of

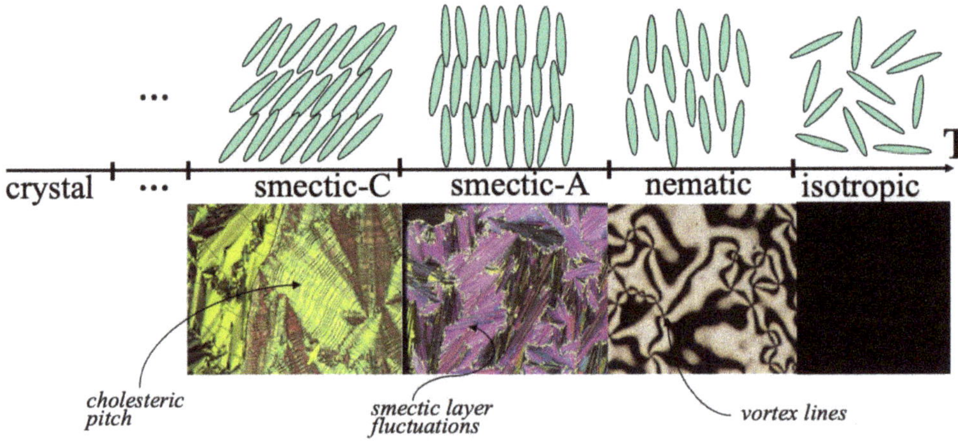

Fig. 4. Illustration of the most ubiquitous nematic (orientationally ordered uniaxial fluid), smectic-A and smectic-C (one-dimensional density wave with, respectively, isotropic and polar in-plane fluid orders) liquid crystal phases and their associated textures in cross-polarized microscopy (N.A. Clark laboratory).

the orientational order along the principle uniaxial axis $\hat{\mathbf{n}}$. A three-dimensional smectic-A is a periodic array of two-dimensional fluids, characterized by a uniaxial periodic density modulation with Fourier components that are integer multiples of the smectic ordering wavevector $\mathbf{q_0} = \hat{\mathbf{n}}2\pi/a$ (with a the layer spacing) parallel to the nematic director $\hat{\mathbf{n}}$; other type of smectics, e.g., smectic-C, where $\hat{\mathbf{n}}$ makes a non-zero angle with $\mathbf{q}_0$ are common.[4,10] The dominant lowest Fourier component $\psi(\mathbf{x}) \equiv \rho_{\mathbf{q}_0}$ can be taken as the local (complex scalar) order parameter which distinguishes the smectic-A from the nematic phase.[10] It is related to the molecular density $\rho(\mathbf{x})$ by

$$\rho(\mathbf{x}) = \mathrm{Re}[\rho_0 + \psi(\mathbf{x})e^{i\mathbf{q_0}\cdot\mathbf{x}}], \tag{1}$$

where ρ_0 is the mean density of the smectic, $\psi = \rho_{\mathbf{q}_0}$ is the Fourier amplitude of the density at wavevector $\mathbf{q}_0$, and Re is a real part.

As first introduced by de Gennes,[10] the effective Hamiltonian functional $H_{dG}[\psi, \hat{\mathbf{n}}]$, which describes the nematic-to-smectic-A (NA) transition at long length scales, is given by

$$H_{dG}[\psi, \hat{\mathbf{n}}] = \int d^d x \left[c|(\boldsymbol{\nabla} - iq_0\delta\hat{\mathbf{n}})\psi|^2 + t_0|\psi|^2 + \frac{1}{2}g_0|\psi|^4 \right] + H_F[\mathbf{n}], \tag{2}$$

where $t_0 \propto T - T_{NA}$ is the reduced temperature for transition at T_{NA},

$$\delta\hat{\mathbf{n}}(\mathbf{x}) \equiv \hat{\mathbf{n}}(\mathbf{x}) - \hat{\mathbf{n}}_0 = \delta\hat{\mathbf{n}}_\perp + \hat{\mathbf{n}}_0 \left(\sqrt{1 - \delta\hat{\mathbf{n}}_\perp^2} - 1 \right) \tag{3}$$

is the fluctuation of the local nematic director $\hat{\mathbf{n}}(\mathbf{x})$ away from its average value $\hat{\mathbf{n}}_0$, which I take to be $\hat{\mathbf{z}}$, and $H_F[\mathbf{n}]$ is the Frank effective Hamiltonian that describes the elasticity of the nematic director:

$$H_F[\hat{\mathbf{n}}] = \int d^d x \, \frac{1}{2} \left[K_s(\boldsymbol{\nabla} \cdot \hat{\mathbf{n}})^2 + K_t(\hat{\mathbf{n}} \cdot \boldsymbol{\nabla} \times \hat{\mathbf{n}})^2 + K_b(\hat{\mathbf{n}} \times \boldsymbol{\nabla} \times \hat{\mathbf{n}})^2 \right], \quad (4)$$

where K_s, K_t, and K_b are the bare elastic moduli for splay, twist and bend of the nematic director field, respectively.[4,10]

The "minimal" gauge-like coupling between $\mathbf{n}$ and ψ is enforced by the requirement of global rotational invariance.[10] It is important to emphasize, however, that although de Gennes Hamiltonian H_{dG} closely resembles the Ginzburg–Landau model of a superconductor and the Abelian–Higgs model, there are essential differences, which are discussed in the chapter by Tom Lubensky. The physical reality of the nematic director $\hat{\mathbf{n}}$ and the smectic order parameter ψ (in contrast to the gauge ambiguity in the definition of the vector potential and the superconducting order parameter) selects the liquid crystal gauge $\delta\hat{\mathbf{n}}_\perp \cdot \hat{\mathbf{n}}_0 = 0$ as the physical gauge in which $\hat{\mathbf{n}}$ and ψ are measured. The strict gauge invariance is also explicitly broken in H_{dG} by the splay term $K_s(\boldsymbol{\nabla} \cdot \hat{\mathbf{n}})^2$ of the Frank Hamiltonian (contrasting with the Maxwell action that involves purely gauge-invariant derivatives, e.g., $(\boldsymbol{\nabla} \times \mathbf{A})^2$), which, as we will see in the following, determines the smectic Goldstone-mode elasticity.

2.2. *Smectic elasticity*

2.2.1. *Smectic from a nematic liquid crystal fluid*

Within the ordered smectic phase, the fluctuations are conveniently described in terms of the magnitude and phase of the smectic order parameter ψ. It is easy to show that the fluctuations of the *magnitude* of ψ around the average value $|\psi_0| = \sqrt{t_0/g_0} =$ const. are "massive" and can therefore be safely integrated out of the partition function, leading to only *finite*, unimportant shifts in the effective elastic moduli. In contrast, the phase of ψ is a U(1) massless Goldstone mode, corresponding to spontaneously broken translational symmetry along $\mathbf{q}_0$. It is the low-energy phonon degree of freedom of the smectic state, describing local displacement of the smectic layers from its perfect periodic order. In accordance with this discussion, deep within the smectic phase, we can represent the smectic order parameter as

$$\psi(\mathbf{x}) = |\psi_0| e^{-iq_0 u(\mathbf{x})}, \quad (5)$$

safely neglecting fluctuations in the magnitude $|\psi_0|$ at temperature below the T_{NA}.

Using this low-temperature smectic description (5) inside the de Gennes Hamiltonian (2) and dropping constants, one finds

$$H[u, \delta\hat{\mathbf{n}}_\perp] = \int d^d x \left[\frac{B}{2}(\boldsymbol{\nabla}_\perp u + \delta\hat{\mathbf{n}}_\perp)^2 + \frac{B}{2}\left(\partial_z u - \frac{1}{2}\delta\hat{n}_\perp^2\right)^2 \right.$$
$$\left. + \frac{K_s}{2}(\boldsymbol{\nabla} \cdot \delta\hat{\mathbf{n}})^2 + \frac{K_t}{2}(\hat{\mathbf{z}} \cdot \boldsymbol{\nabla} \times \delta\hat{\mathbf{n}})^2 + \frac{K_b}{2}(\hat{\mathbf{z}} \times \boldsymbol{\nabla} \times \delta\hat{\mathbf{n}})^2 \right], \quad (6)$$

where $B = 2c|\psi_0|^2 q_0^2$ is the smectic compression modulus. I observe that the fluctuation mode $\nabla_\perp u + \delta\hat{\mathbf{n}}_\perp$ is "massive" and leads to an emergent Anderson–Higgs-like mechanism, a hallmark of gauge theories. As a consequence, after a simple Gaussian integration over $\delta\hat{\mathbf{n}}_\perp$, one finds that at long length scales, $\delta\hat{\mathbf{n}}_\perp$ fluctuations are constrained to follow $\nabla_\perp u$, corresponding to locking of the director $\hat{\mathbf{n}}$ to the smectic layer normal. The low-energy elastic smectic Hamiltonian is then obtained by the replacement

$$\delta\hat{\mathbf{n}}_\perp \to -\nabla_\perp u, \tag{7}$$

everywhere in Eq. (6) and Frank energy, Eq. (4). Valid in the long-wavelength limit and provided dislocations are confined, one thus obtains a non-linear elastic Goldstone-mode Hamiltonian of the smectic phase,

$$H_{\rm sm}[u] = \int d^d x \left[\frac{K}{2}(\nabla_\perp^2 u)^2 + \frac{B}{2}(\partial_z u - \frac{1}{2}(\nabla u)^2)^2\right], \tag{8}$$

where $K = K_s$ is the splay modulus.[4,11]

I observe that the harmonic elasticity of the smectic — $m = d-1$-Lifshitz model[9] — is highly anisotropic (at harmonic level with scaling $z \sim x_\perp^2$), displaying compression modulus B along $\mathbf{q}_0$ and higher-order *curvature* (Laplacian) modulus transverse to $\mathbf{q}_0$, with the "tension" (transverse gradient) modulus vanishing exactly. This is ensured by the underlying rotational symmetry, as discussed in the introduction to critical phases. As I will show in the forthcoming sections, as a consequence, smectic fluctuations are highly enhanced (relative to, e.g., its XY-model counterpart), resulting in importance of non-linear Goldstone modes' elasticity for $d \le 3$, which I therefore retained in (8). I also note that the compressional modulus B multiplies a square of the non-linear strain tensor, whose precise form is determined by the underlying invariance under arbitrary large rotation of smectic layers,

$$u_0(\mathbf{r}) = z(1-\cos\theta) + x\sin\theta, \tag{9}$$

and maintained under coarse-graining RG. I conclude this section by also noting that the aforementioned connection to the Anderson–Higgs mechanism elucidates why a smectic state is characterized by only a single Goldstone mode u, rather than three, despite partially breaking two rotational and one translational symmetries. In contrast, a charged superconductor is well known to have all its Goldstone modes "eaten" — fully gapped out by the Anderson–Higgs mechanism.

2.2.2. *Smectic from an isotropic fluid*

I now derive the non-linear smectic elasticity in a more basic, complementary way, starting instead with an isotropic fluid state.[11] I begin with a generic energy functional that captures system's tendency to develop a unidirectional wave at wavevector $\mathbf{q}_0$, with an arbitrary direction, and magnitude fixed at q_0,

$$\mathcal{H}_{\rm sm} = \frac{1}{2}J\left[(\nabla^2\rho)^2 - 2q_0^2(\nabla\rho)^2\right] + \frac{1}{2}t\rho^2 - w\rho^3 + v\rho^4 + \cdots, \tag{10}$$

where J, q_0, t, w, v are parameters of the isotropic fluid phase. Clearly, the first term is engineered so that dominant fluctuations and condensation are on a spherical surface at a non-zero wavevector with a magnitude q_0. Thus, I focus on the density at a wavevector $\mathbf{q}$, that for now is unrelated to q_0

$$\rho(\mathbf{x}) = \mathrm{Re}\left[\rho_q(\mathbf{x})e^{i\mathbf{q}\cdot\mathbf{x}}\right], \tag{11}$$

where $\rho_q(\mathbf{x})$ is a complex scalar. Without loss of generality, based on the discussion of the previous section, the order parameter $\rho_q(\mathbf{x}) = |\rho_q|e^{-iqu}$ is taken to have a (constant) magnitude $|\rho_q|$ and a phase $qu(\mathbf{x})$. Clearly, $u(\mathbf{x})$ is just a phonon displacement along $\mathbf{q}$. Substituting this expression for $\rho(\mathbf{x})$ and its gradients into $\mathcal{H}_{\mathrm{sm}}$, one finds

$$\begin{aligned}\mathcal{H}_{\mathrm{sm}} = J\rho_0^2\Bigg[\frac{1}{4}q^2(\nabla^2 u)^2 + \left(q\mathbf{q}\cdot\nabla u - \frac{1}{2}q^2(\nabla u)^2\right)^2 \\ + 4(q^2 - q_0^2)\left(q\mathbf{q}\cdot\nabla u - \frac{1}{2}q^2(\nabla u)^2\right)\Bigg] + \cdots,\end{aligned} \tag{12}$$

where constant parts as well as fast oscillating pieces were dropped as they average away after the spatial integration of the above energy density. Firstly, I observe that (as discussed on general grounds above) in a harmonic part linear gradient elasticity in u only appears for gradients *along* $\mathbf{q}$, namely $\mathbf{q}\cdot\nabla$, with elasticity transverse to $\mathbf{q}$ starting with a Laplacian form. Secondly, the elastic energy density is an expansion in a rotationally invariant strain tensor combination

$$u_{qq} = \hat{\mathbf{q}}\cdot\nabla u - \frac{1}{2}(\nabla u)^2, \tag{13}$$

whose non-linearities in u ensure that it is fully rotationally invariant even for large rotations. To see this (picking $\hat{\mathbf{q}} = \hat{\mathbf{z}}$), note that a rigid (distortion-free) rotation of $\mathbf{q}$ ($q_0\hat{\mathbf{z}} \to \mathbf{q} = q_0(\cos\theta\hat{\mathbf{z}} + \sin\theta\hat{\mathbf{x}})$ can be interpreted as a spatially linear "distortion" $u_0(\mathbf{x}) = z(1-\cos\theta) + x\sin\theta$, for which non-linear strain u_{qq} vanishes identically, thus, as required, corresponding to a vanishing energy. Thirdly, the last term in (12) vanishes for $|\mathbf{q}|$ picked to equal q_0, corresponding to energetically preferred choice of modulation wavevector.

Looking ahead, as one includes effects of fluctuations, the "bare" condition $q = q_0$ will need to be adjusted so as to eliminate the fluctuation-generated linear term in u_{qq} order by order, which amounts to an expansion in the non-linear strain u_{qq} around the correct (fluctuation-corrected) ground state (akin to fluctuation correction to T_c at a critical point). Finally, I note that the relation between the curvature modulus K of the Laplacian (first) term and the bulk modulus B gradient (second) term is not generic and can be relaxed to have distinct elastic constants, as can be seen if higher-order gradient terms are included in the original energy density, Eq. (10).

Choosing the coordinate system such that $\hat{\mathbf{z}}$ is aligned along $\mathbf{q}$, one finds that for $q = q_0$, the Goldstone-mode Hamiltonian (12) reduces to the standard smectic elastic energy density, (8), derived from the de Gennes model above.[4,10]

2.2.3. *Smectic as cholesteric and helical fluids*

Another beautiful example of a system that realizes a uniaxial periodic critical state is a cholesteric phase of chiral liquid crystals, as illustrated in Fig. 2(b). It is in fact the first liquid crystal phase discovered in cholesterol benzoate by Reinitzer in 1886.[4,10] Such orientational (as opposed to mass) density wave state is ubiquitous in nature and is equivalent to the helical co-planar spin-spiral state that is found in non-centrosymmetric magnets, such as MnSi, FeSi and many others[33] (neglecting crystal-field pinning that is always present due to spin–orbit interactions and explicit breaking of rotational symmetry.).

Following the analysis of Radzihovsky and Lubensky,[32] I discuss in the following the low-energy *non-linear* elasticity of a cholesteric, on length scales longer than its period, $a = 2\pi/q_0$. As first proposed by de Gennes, the latter is expected, based on symmetry and an explicit *harmonic* derivation by Lubensky,[10,31] to be identical to that of a smectic, i.e., that of a one-dimensional crystalline order, which spontaneously breaks the underlying translational and rotational symmetry, with constant-orientation "layers" transverse to the ordering wavevector, $\mathbf{q}_0$. Although the underlying chirality has only subtle surface effects within the cholesteric state, its effects become important in the nature of the phase transition in and out of the cholesteric critical state.

The simplest model of a chiral nematic liquid crystal (a cholesteric) is captured by the chiral Frank elastic free energy density,

$$\mathcal{H}_F^* = \frac{1}{2}K_s(\nabla\cdot\hat{n})^2 + \frac{1}{2}K_b(\hat{n}\times\nabla\times\hat{n})^2 + \frac{1}{2}K_t(\hat{n}\cdot\nabla\times\hat{n}+q_0)^2. \tag{14}$$

The broken chiral symmetry allows a chiral q_0 term, which leads to a twist of the nematic structure into a helix with a pitch $2\pi/q_0$ along a *spontaneously* chosen axis.

In a simplified isotropic limit of $K_s = K_b = K_t$, the chiral Frank energy density is then given by

$$\mathcal{H}_F^* = \frac{1}{2}K\left[(\partial_i\hat{n}_j)^2 + 2q_0\hat{n}\cdot\nabla\times\hat{n} + q_0^2\right] \tag{15}$$

$$= \frac{1}{2}K\left(\partial_i\hat{n}_j + q_0\epsilon_{ijk}\hat{n}_k\right)^2 - \frac{1}{2}Kq_0^2, \tag{16}$$

where after integration by parts, I utilized the identity

$$(\partial_i\hat{n}_j)^2 = (\nabla\cdot\hat{n})^2 + (\nabla\times\hat{n})^2 + \nabla\cdot[(\hat{n}\cdot\nabla)\hat{n} - \hat{n}\nabla\cdot\hat{n}]. \tag{17}$$

In the absence of topological defects, neglecting the boundary term, $\mathcal{H}_F^*$ is a sum of squares and is therefore minimized by a twist-only cholesteric state

$$\hat{n}(\mathbf{r}) = \hat{\mathbf{e}}_{10}\cos(\mathbf{q}\cdot\mathbf{r}) + \hat{\mathbf{e}}_{20}\sin(\mathbf{q}\cdot\mathbf{r}), \tag{18}$$

where $\hat{\mathbf{e}}_{10}, \hat{\mathbf{e}}_{20}, \hat{\mathbf{e}}_{30} \equiv \hat{\mathbf{e}}_{10}\times\hat{\mathbf{e}}_{20}$ form an orthonormal triad, with a constant twist

$$\hat{n}\cdot\nabla\times\hat{n} = -q_0. \tag{19}$$

I now derive the low-energy Goldstone-mode elasticity about the cholesteric helical ground state, which can be parameterized according to

$$\hat{n}(\mathbf{r}) = \hat{\mathbf{e}}_1(\mathbf{r})\cos(\mathbf{q}\cdot\mathbf{r} + \chi(\mathbf{r})) + \hat{\mathbf{e}}_2(\mathbf{r})\sin(\mathbf{q}\cdot\mathbf{r} + \chi(\mathbf{r})), \tag{20}$$

where fluctuations are captured by the spatially dependent orthonormal triad $\hat{\mathbf{e}}_1(\mathbf{r}), \hat{\mathbf{e}}_2(\mathbf{r}), \hat{\mathbf{e}}_3(\mathbf{r})$ and the helical phase

$$\chi(\mathbf{r}) = -q_0 u(\mathbf{r}), \tag{21}$$

which corresponds to the phonon field $u(\mathbf{r})$ of the constant-orientation layers.

The helical state breaks a group of three-dimensional translations and rotations $G = T_{x,y,z} \times O(3)$ of the isotropic fluid down to $H = T_x \times T_y \times U(1)$ (latter $U(1) = \text{diagonal}[T_z, O_z(2)]$). Thus, since $\dim[G/H = O(3)] = 3$, one may expect *three independent* Goldstone modes $\chi(\mathbf{r})$ and $\hat{\mathbf{e}}_3(\mathbf{r})$, corresponding to three degrees of freedom of the orthonormal triad; the azimuthal angle $\phi(\mathbf{r})$ defining the orientation of the $\hat{\mathbf{e}}_{1,2}$ around $\hat{\mathbf{e}}_3$ is not independent of $\chi(\mathbf{r})$ as it can be absorbed into it. The low-energy coset space is isomorphic to $S^1 \times S^2$, a ball (radius π) of the group manifold of SO(3). However, despite this standard counting, as I will show in the following and as expected on general grounds anticipated by de Gennes, the cholesteric state is characterized by a *single* smectic-like Goldstone mode $u(\mathbf{r})$.

To this end, substituting the form for $\hat{\mathbf{n}}(\mathbf{r})$ from Eq. (20) into the Frank energy of the chiral nematic (14), and defining effective connection gauge fields, $\mathbf{a}$ and $\mathbf{c}_{1,2}$,

$$\partial_i \hat{\mathbf{e}}_{1j} = a_i \hat{\mathbf{e}}_{2j} + c_{1i}\hat{\mathbf{e}}_{3j}, \tag{22a}$$

$$\partial_i \hat{\mathbf{e}}_{2j} = -a_i \hat{\mathbf{e}}_{2j} + c_{2i}\hat{\mathbf{e}}_{3j}, \tag{22b}$$

$$a_i = \hat{\mathbf{e}}_2 \cdot \partial_i \hat{\mathbf{e}}_1, \quad c_{1i} = -\hat{\mathbf{e}}_1 \cdot \partial_i \hat{\mathbf{e}}_3, \quad c_{2i} = -\hat{\mathbf{e}}_2 \cdot \partial_i \hat{\mathbf{e}}_3, \tag{22c}$$

one finds

$$\mathcal{H}_F^* = \frac{K}{2}(\nabla\chi + \mathbf{a} + \mathbf{q} - q_0\hat{\mathbf{e}}_3)^2 + \frac{K}{4}(\mathbf{c}_1 + q_0\hat{\mathbf{e}}_2)^2 + \frac{K}{4}(\mathbf{c}_2 - q_0\hat{\mathbf{e}}_1)^2. \tag{23}$$

In the above, I again neglected spatially oscillating and constant terms.

Taking $\mathbf{q} = q_0\hat{\mathbf{z}}$, with $\hat{\mathbf{z}}$ defining the helical axis (distinct from the normal to the helical plane, $\hat{\mathbf{e}}_3$) and noting the compatibility condition on effective flux or its equivalent vanishing of the Pontryagin density,

$$\nabla \times \mathbf{a} = \epsilon_{ij}\hat{\mathbf{e}}_3 \cdot \partial_i \hat{\mathbf{e}}_3 \times \partial_j \hat{\mathbf{e}}_3 = 0, \tag{24}$$

required by well-defined cholesteric layers, i.e., in the absence of dislocations and disclinations in the layer structure, allows one to take

$$\mathbf{a} = \boldsymbol{\nabla}\phi. \tag{25}$$

Under this condition, ϕ can be eliminated in favor of χ, i.e., $\chi + \phi \to \chi$ and in the laboratory coordinate system $\hat{\mathbf{x}}, \hat{\mathbf{y}}, \hat{\mathbf{z}}$, the fluctuations are characterized by the local helical frame described by χ and $\hat{\mathbf{e}}_3$, with

$$\hat{\mathbf{e}}_3 = \hat{\mathbf{e}}_{3\perp} + \hat{\mathbf{z}}\sqrt{1-\hat{\mathbf{e}}_{3\perp}^2} \approx \hat{\mathbf{e}}_{3\perp} + \hat{\mathbf{z}}\left(1 - \frac{1}{2}\hat{\mathbf{e}}_{3\perp}^2\right). \tag{26}$$

I thus obtain

$$\mathcal{H}_F^* = \frac{1}{2}(\nabla_\perp\chi - q_0\hat{\mathbf{e}}_{3\perp})^2 + \frac{1}{2}\left(\partial_z\chi + \frac{1}{2}q_0\hat{\mathbf{e}}_{3\perp}^2\right)^2 + \frac{1}{4}(\mathbf{c}_1^2 + \mathbf{c}_2^2)$$
$$+ \frac{q_0}{2}(\mathbf{c}_1 \cdot \hat{\mathbf{e}}_2 - \mathbf{c}_2 \cdot \hat{\mathbf{e}}_1). \tag{27}$$

From the minimization of the first term (or equivalently integrating out the independent $\hat{\mathbf{e}}_{3\perp}$ degree of freedom), I obtain an effective constraint

$$\nabla_\perp\chi = q_0\hat{\mathbf{e}}_{3\perp}, \tag{28}$$

which is an example of an emergent Higgs mechanism (akin to smectic liquid crystals discussed above) locking the cholesteric layers with the molecular frame orientation. With this constraint (valid at low energies), the effective cholesteric Hamiltonian reduces to

$$\mathcal{H}_F^* = \frac{1}{2}\left[\partial_z\chi + \frac{1}{2q_0}(\nabla_\perp\chi)^2\right]^2 + \frac{1}{4q_0^2}(\hat{\mathbf{e}}_{\alpha\gamma}\cdot\partial_\beta^\perp\partial_\gamma^\perp\chi)^2$$
$$+ \frac{q_0}{2}(\hat{\mathbf{e}}_{1\alpha}\hat{\mathbf{e}}_{2\beta}\partial_\beta\partial_\alpha\chi - \hat{\mathbf{e}}_{2\alpha}\hat{\mathbf{e}}_{1\beta}\partial_\beta\partial_\alpha\chi) \tag{29a}$$

$$= \frac{B}{2}\left[\partial_z u - \frac{1}{2}(\nabla u)^2\right]^2 + \frac{K}{2}(\nabla^2 u)^2, \tag{29b}$$

where compressional modulus $B = Kq_0^2$, $\hat{\mathbf{e}}_{\alpha,i}\hat{\mathbf{e}}_{\alpha,j} = \delta_{ij}$ and $\hat{\mathbf{e}}_{1,i}\hat{\mathbf{e}}_{2,j} - \hat{\mathbf{e}}_{2,i}\hat{\mathbf{e}}_{i,j} = \hat{\mathbf{e}}_{3,k}\epsilon_{ijk}$ were used and to eliminate the last term in the first line above, I used the condition of the single-valuedness of the phase field χ, i.e., well-defined cholesteric layers with no dislocations. However, this last condition can be violated by singular dislocations and thus distinguishes the cholesteric from a smectic near a phase transition out of the cholesteric state.

Thus, as advertised and expected on symmetry grounds, one indeed finds,[32] that cholesteric Goldstone-mode elasticity, even at the non-linear level, is identical to that of a conventional smectic — at harmonic level controlled by the $m = d-1$-Lifshitz model.[9] Thus, smectic critical state universal phenomenology, to be worked out in the following, will equally apply to the cholesteric state.

2.2.4. *Smectic as a helical bosonic superfluid*

I next discuss a quantum realization of a smectic state, namely via interacting superfluid bosons on a frustrated honeycomb lattice.[30] Other bosonic realizations also appear in p-wave Feshbach-resonant[28] and Rashba spin–orbit coupled[29] bosons, engineered in a controlled way in cold atom experiments[29,52] to condense at a non-zero momentum, $\mathbf{k}_0$.

I refer the reader to the original literature for these latter systems and here focus on bosons on a frustrated honeycomb lattice, modeled by a tight-binding dispersion with nearest t_1 and next-nearest (antiferromagnetic) $-t_2$ neighbor hopping amplitudes, with

$$H = -t_1 \sum_{\langle ij \rangle} a^\dagger_{i,1} a_{j,2} + t_2 \sum_{\langle\langle ij \rangle\rangle} (a^\dagger_{i,1} a_{j,1} + a^\dagger_{i,2} a_{j,2}) + h.c. + \frac{U}{2} \sum_i \sum_{s=1,2} n_{i,s}(n_{i,s} - 1). \tag{30}$$

For sufficiently frustrated hopping, with $t_2/t_1 > 1/6$ (that has been engineered in an optical lattice through Floquet techniques[52]), the above system displays a dispersion minimum at a non-zero momentum closed contour, as illustrated in Fig. 5.

Such non-zero dispersion, with a closed minimum contour, is approximately circular for $t_2/t_1 \approx 1/6 + \epsilon$ and can be faithfully modeled by a continuum field theory, encoded in a non-interacting Hamiltonian $H_0 = \int_{\mathbf{r}} \hat{\Psi}^\dagger_{\mathbf{r}} \hat{\varepsilon}_{\mathbf{r}} \hat{\Psi}_{\mathbf{r}}$ with

$$\hat{\varepsilon}_{\mathbf{r}} = J(-\nabla^2 - k_0^2)^2 + \varepsilon_0, \tag{31}$$

with a quartic dispersion

$$\varepsilon_{\mathbf{k}} = J(\mathbf{k}^2 - k_0^2)^2 + \varepsilon_0, \tag{32}$$

and minima lying on contour $\mathbf{k}^2 = k_0^2$. The corresponding Euclidean (imaginary-time τ) Lagrangian is given by

$$\mathcal{L} = \Psi^* \partial_\tau \Psi + J|\nabla^2 \Psi|^2 - 2Jk_0^2 |\nabla \Psi|^2 + (\tilde{\varepsilon}_0 - \mu)|\Psi|^2 + \frac{U}{2}|\Psi|^4. \tag{33}$$

For $\mu > \epsilon_0$, bosons condense into a non-zero-momentum superfluid state, the simplest version of which is a helical condensate at a single point $\mathbf{k}_0$ on the dispersion

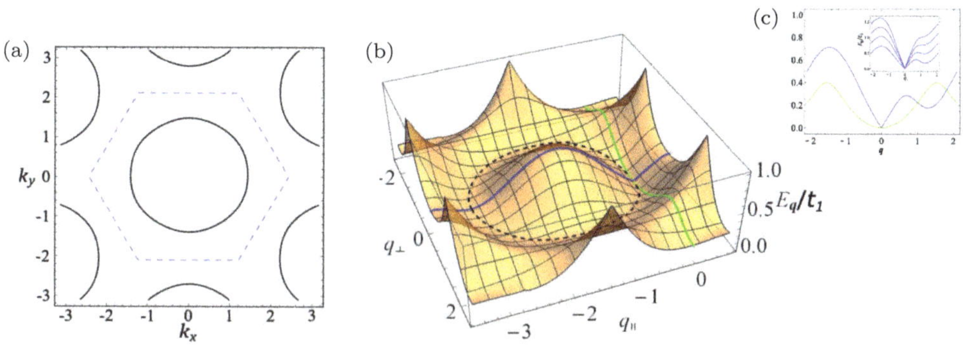

Fig. 5. (a) A black circular curve is a non-zero momentum $\mathbf{k}_0$ minimum in the non-interacting microscopic dispersion $\epsilon^-_{\mathbf{k}}$ (around the Γ point in the Brillouin zone, dashed blue curve) of bosons on a frustrated honeycomb lattice, (30). (b) The dispersion of the *interacting* helical superfluid, $E_{\mathbf{q}}$ around $\mathbf{q} = \mathbf{k} - \mathbf{k}_0 = 0$, with green (see inset) indicating quadratic $q_\perp^2$ form along $q_\perp$ and blue (see inset) $|q_\parallel|$ along the condensate momentum, $\mathbf{k}_0$, consistent with symmetry-expected smectic form, given by (35). The black-dashed contour indicates the degenerate minimum of the non-interacting band $\epsilon^-_{\mathbf{k}}$.

contour minimum — encoded in $\varepsilon_{\mathbf{k}}$, (32). In the density-phase representation, the state is characterized by

$$\Psi(\mathbf{r}) = \sqrt{n}e^{i\mathbf{k}_0\cdot\mathbf{r}+i\phi} = \sqrt{n_0 + \pi}e^{i\mathbf{k}_0\cdot\mathbf{r}+i\phi}, \tag{34}$$

with a non-zero condensate density (at mean-field level) $n_0 = (\mu - \varepsilon_0)/U$ and momentum k_0 for $\mu > \varepsilon_0$ (vanishing otherwise), and density and phase fluctuations, π and ϕ, respectively. To quadratic order, the helical superfluid is then described by a Goldstone-mode Lagrangian density, which, after integrating out density field π gives a zero-temperature quantum "columnar" Lagrangian – $m = d - 2$-Lifshitz model in d spacetime dimensions[9] — (to also appear in the subsequent sections)

$$\mathcal{L}_\phi \approx B_\tau(\partial_\tau\phi)^2 + B(\partial_\|\phi)^2 + K(\partial_\perp^2\phi)^2. \tag{35}$$

At non-zero T, (35) reduces to the classical smectic ($m = d - 1$-Lifshitz model[9]) Hamiltonian for ϕ discussed in the previous sections.

2.2.5. *Smectic as a Fulde–Ferrell–Larkin–Ovchinnikov superconductor*

A singlet superconductor, frustrated by a depairing Zeeman field has been predicted by Fulde and Ferrell[23] and by Larkin and Ovchinnikov[24] (FFLO) to pair at a non-zero momentum, thereby exhibiting a periodic modulation of the Cooper-pair amplitude — a pair-density wave (PDW).[25,27] Such FFLO state, in the absence of deleterious orbital field and pinning lattice effects was proposed[26,53,54] to be realizable in spin-imbalanced (polarized) Feshbach-resonant paired atomic superfluids. It is now believed to have been observed in an array of decoupled one-dimensional traps, with recent progress toward a higher-dimensional realization of the FFLO superfluid.[55]

While microscopic analysis is a bit involved,[25,53,54] its upshot is a derivation of an effective Landau theory of the form (33) for the pairing amplitude, with a harmonic dispersion ε_k, for intermediate Zeeman field (chemical potential for spin imbalance) $h_{c1} < h < h_{c2}$, displaying a minimum on non-zero momentum closed contour, akin to that of the helical superfluid (33), above. Fulde–Ferrell and Larkin–Ovchinnikov are two simplest states that minimize this Lagrangian. The FF state is characterized by a pairing amplitude,[23]

$$\Psi_{FF}(\mathbf{r}) = \Psi_{q_0}e^{i\mathbf{q}_0\cdot\mathbf{r}+i\phi}, \tag{36}$$

which is a plane wave with the momentum $\mathbf{q}_0$ and a single Goldstone mode ϕ, a local superconducting phase. The state carries a non-zero, uniform spontaneously directed supercurrent (the total current vanishes, due to the additional back-flow contribution from the quasi-particle current)

$$\mathbf{j}_{FF} = \frac{1}{m}|\Psi_{q_0}|^2(\mathbf{q}_0 + \nabla\phi), \tag{37}$$

and thereby breaks time-reversal and rotational symmetries, chosen spontaneously along $\mathbf{q}_0$, as well as the global gauge symmetry, corresponding to the total atom conservation. The low-energy Lagrangian for $\phi(\mathbf{r},\tau)$ takes the smectic (columnar at $T=0$) form, (35).

The LO state[24] is a time-reversal symmetric counterpart, which is a superposition of $\pm\mathbf{q}_0$ paired condensates, given by

$$\Psi_{LO}(\mathbf{r}) = \Psi_+(\mathbf{r})e^{i\mathbf{q}\cdot\mathbf{r}} + \Psi_-(\mathbf{r})e^{-i\mathbf{q}\cdot\mathbf{r}} \tag{38a}$$

$$= 2|\Psi_{q_0}|e^{i\frac{1}{2}(\phi_+ + \phi_-)}\cos\left[\mathbf{q}_0\cdot\mathbf{r} + \frac{1}{2}(\phi_+ - \phi_-)\right] \tag{38b}$$

$$= 2|\Psi_{q_0}|e^{i\phi}\cos[\mathbf{q}_0\cdot\mathbf{r} + \theta]. \tag{38c}$$

It is characterized by the superfluid phase ϕ and the phonon $u = \theta/q_0$ Goldstone modes, with a combination of the XY and smectic sectors in a classical Hamiltonian,

$$\mathcal{H}_{LO} = \frac{1}{2}K(\nabla^2 u)^2 + \frac{1}{2}B\left(\partial_\| u - \frac{1}{2}(\nabla u)^2\right)^2 + \frac{1}{2}\rho_s^\|(\partial_\|\phi)^2 + \frac{1}{2}\rho_s^\perp(\nabla_\perp\phi)^2, \tag{39}$$

whose zero-temperature quantum counterpart is a combination of the XY and "columnar" — $m = d-2$-Lifshitz model in d spacetime dimensions.[9] At non-zero temperature, one thus expects FF and LO states to display critical smectic phenomenology that I discuss in the following section.

2.2.6. *Smectic as quantum Hall stripes*

One other prominent example of a quantum smectic has been argued theoretically[18–21] to be realized in half-filled large N Landau level regime and observed experimentally[16,17] at such filling $\nu = N + 1/2$ through appearance of a highly anisotropic transport below 100 milli-Kelvin. At mean-field level, the state is described as a periodic array of alternating integer filling N and $N+1$ quantum Hall stripes, predicted to be the exact ground state in the $N \gg 1$ limit. Beyond the Hartree–Fock theory,[18,19] fluctuations of such a striped state are well described by chiral edge Luttinger liquids, corresponding to inter- (positions) and intra- (shape) stripe phonons of this quantum Hall smectic. Much of the analysis discussed in the following, importantly beyond its harmonic fluctuations, will also apply to this state at non-zero temperature leading to a critical smectic state. At zero temperature, it is described by "cholesteric"-like, i.e., $m = d-2$-Lifshitz model in d spacetime dimensions.[9]

2.3. *Non-zero T Gaussian fluctuation in a harmonic smectic*

To assess the extent of thermal fluctuations of the smectic Goldstone mode $u(\mathbf{r})$, I first analyze them within a harmonic approximation, neglecting elastic nonlinearities in H_{sm}.[4,56] In terms of the Fourier modes $u_{\mathbf{k}}$, the Hamiltonian decouples, reducing to

$$H_{sm} = \frac{1}{2}\int \frac{d^dk}{(2\pi)^d}\left(Kk_\perp^4 + Bk_z^2\right)|u_{\mathbf{k}}|^2, \tag{40}$$

thus allowing a straightforward computation of phonon correlation functions via standard Gaussian integrals or equivalently via equipartition. This gives mean-squared fluctuations

$$\langle u^2\rangle_0^T = \int_{L_\perp^{-1}}^{\Lambda_\perp} \frac{d^dk}{(2\pi)^d}\frac{T}{Bk_z^2 + Kk_\perp^4} \approx \begin{cases} \frac{T}{2\sqrt{BK}} C_{d-1} L_\perp^{3-d}, & d < 3, \\ \frac{T}{4\pi\sqrt{BK}} \ln q_0 L_\perp, & d = 3, \end{cases} \tag{41}$$

where I defined a constant $C_d = S_d/(2\pi)^d = 2\pi^{d/2}/[(2\pi)^d\Gamma(d/2)]$, with S_d a surface area of a d-dimensional sphere, and introduced an infrared cutoff by considering a system of finite extent $L_\perp \times L_z$, with L_z the length of the system along the ordering (z) axis and $L_\perp$ transverse to z. Unless it has a huge aspect ratio, such that $L_z \sim L_\perp^2/\lambda >> L_\perp$, any large system ($L_\perp, L_z >> \lambda$) will have $\lambda L_z \ll L_\perp^2$.

The key observation here is that the smectic phonons exhibit fluctuations that diverge, growing logarithmically in 3D and linearly in 2D with system size $L_\perp$; for $d > 3$, fluctuations are bounded. Thus, as a consequence of their "soft" elasticity (a vanishing $(\boldsymbol{\nabla}_\perp u)^2$ modulus) 3D smectics are akin to 2D XY systems, such as superfluid films and two-dimensional crystals,[8,57–60] exhibit a power-law quasi-long-range order in three dimensions.

The expression for the mean-squared phonon fluctuations in (41) leads the emergence of important crossover length scales $\xi_\perp, \xi_z$, related by

$$\xi_\perp = (\xi_z\sqrt{K/B})^{1/2} \equiv \sqrt{\xi_z\lambda}, \tag{42}$$

which characterize the finite-temperature smectic state. These are defined as scales $L_\perp, L_z$ at which phonon fluctuations are large, comparable to the smectic period $a = 2\pi/q_0$, namely, setting

$$\langle u^2\rangle_0^T \approx a^2 \tag{43}$$

in Eq. (41), one finds

$$\xi_\perp \approx \begin{cases} \frac{a^2\sqrt{BK}}{T} \sim \frac{K}{Tq_0}, & d = 2, \\ ae^{4\pi a^2\sqrt{BK}/T} \sim ae^{\frac{cK}{Tq_0}}, & d = 3, \end{cases} \tag{44}$$

where in the second form of the above expressions, I took the simplest approximation for the smectic anisotropy length $\lambda = \sqrt{K/B}$ to be $\lambda = a \sim 1/q_0$ and introduced $O(1)$ Lindemann constant c,[61] which depends on the somewhat arbitrary definition of "large" phonon root-mean-squared fluctuations.

The smectic connected correlation function

$$C_u(\mathbf{x}_\perp, z) = \langle[u(\mathbf{x}_\perp, z) - u(\mathbf{0}, 0)]^2\rangle_0 \tag{45}$$

is also straightforwardly worked out in 3D giving the logarithmic Caillé form[56]

$$C_u^{3d}(\mathbf{x}_\perp, z) = 2T\int \frac{d^2k_\perp dk_z}{(2\pi)^3}\frac{1-e^{i\mathbf{k}\cdot\mathbf{x}}}{Kk_\perp^4 + Bk_z^2} \equiv \frac{T}{2\pi\sqrt{KB}}\, g_T^{3d}\left(\frac{z\lambda}{x_\perp^2}, \frac{x_\perp}{a}\right) \tag{46a}$$

$$= \frac{T}{2\pi\sqrt{KB}}\left[\ln\left(\frac{x_\perp}{a}\right) - \frac{1}{2}\mathrm{Ei}\left(\frac{-x_\perp^2}{4\lambda|z|}\right)\right] \tag{46b}$$

$$\approx \frac{T}{2\pi\sqrt{KB}}\begin{cases}\ln\left(\frac{x_\perp}{a}\right), & x_\perp \gg \sqrt{\lambda|z|}\\ 1/2\ln\left(\frac{4\lambda z}{a^2}\right), & x_\perp \ll \sqrt{\lambda|z|}\end{cases}, \tag{46c}$$

where $\mathrm{Ei}(x)$ is the exponential-integral function. As indicated in the last form, in the asymptotic limits of $x_\perp \gg \sqrt{\lambda z}$ and $x_\perp \ll \sqrt{\lambda z}$, the above 3D correlation function reduces to a logarithmic growth with $x_\perp$ and z, respectively.

In 2D, one instead finds[62]

$$C_u^{2d}(x, z) = 2T\int \frac{dk_x dk_z}{(2\pi)^2}\frac{1-e^{i\mathbf{k}\cdot\mathbf{x}}}{Kk_x^4 + Bk_z^2} \equiv \frac{T}{2\pi\sqrt{KB}}\, g_T^{2d}\left(\frac{z\lambda}{x^2}, \frac{x}{a}\right) \tag{47a}$$

$$= \frac{2T}{B}\left[\left(\frac{|z|}{4\pi\lambda}\right)^{1/2} e^{-x^2/(4\lambda|z|)} + \frac{|x|}{4\lambda}\mathrm{erf}\left(\frac{|x|}{\sqrt{4\lambda|z|}}\right)\right] \tag{47b}$$

$$\approx \frac{2T}{B}\begin{cases}\left(\frac{|z|}{4\pi\lambda}\right)^{1/2}, & x \ll \sqrt{\lambda|z|},\\ \frac{|x|}{4\lambda}, & x \gg \sqrt{\lambda|z|},\end{cases} \tag{47c}$$

where $\mathrm{erf}(x)$ is the error function.

As a consequence of the above divergent phonon fluctuations, the smectic density-wave order parameter (11) vanishes in thermodynamic limit

$$\langle\rho(\mathbf{x})\rangle_0 = 2|\rho_q|\langle\cos[\mathbf{q}_0\cdot\mathbf{x} - qu(\mathbf{x})]\rangle_0 \tag{48a}$$

$$= 2|\rho_q|e^{-\frac{1}{2}q_0^2\langle u^2\rangle_0}\cos(\mathbf{q}_0\cdot\mathbf{x}) \tag{48b}$$

$$= 2\tilde{\rho}_q(L_\perp)\cos(\mathbf{q}_0\cdot\mathbf{x}), \tag{48c}$$

with the thermally suppressed order parameter amplitude given by

$$\tilde{\rho}_q(L_\perp) = |\rho_q|\begin{cases}e^{-L_\perp/\xi_\perp}, & d = 2\\ \left(\frac{a}{L_\perp}\right)^{\eta/2}, & d = 3\end{cases} \tag{49a}$$

$$\to 0, \quad \text{for } L_\perp \to \infty, \tag{49b}$$

where I used results for the phonon and phase fluctuations (41) and defined the Caillé exponent

$$\eta = \frac{q_0^2 T}{8\pi\sqrt{BK}}. \tag{50}$$

Thus, in qualitative contrast to its mean-field cartoon, at long scales (longer than $\xi_{\perp,z}$), the smectic state is actually characterized by a *uniform* mass density.

Since the average density is actually uniform, a better characterization of the smectic state is through the structure function, $S(\mathbf{q})$, a Fourier transform of the density correlation function, which in 3D is given by

$$S(\mathbf{q}) = \int d^3x \langle \delta\rho(\mathbf{x})\delta\rho(0)\rangle e^{-i\mathbf{q}\cdot\mathbf{x}} \tag{51a}$$

$$\approx \frac{1}{2}\sum_{q_n} |\rho_{q_n}|^2 \int_{\mathbf{x}} \langle e^{-iq_n(u(\mathbf{x})-u(0))}\rangle_0 e^{-i(\mathbf{q}-q_n\hat{z})\cdot\mathbf{x}} \tag{51b}$$

$$\approx \frac{1}{2}\sum_{n} \frac{|\rho_{q_n}|^2}{|q_z - nq_0|^{2-n^2\eta}}, \quad \text{for } d = 3, \tag{51c}$$

where I approximated phase and phonon fluctuations by Gaussian statistics (in 3D valid up to weak logarithmic corrections[11]). Thus, as anticipated, one finds that the logarithmically divergent 3D phonon fluctuations lead to a structure function, with highly anisotropic ($q_z \sim q_\perp^2/\lambda$) *quasi*-Bragg peaks replacing the true (δ-function) Bragg peaks characteristic of a true long-range periodic order.[4,10,63]

In two dimensions, smectic order is even more strongly suppressed by thermal fluctuations. The linear growth of the 2D phonon fluctuations leads to exponentially short-ranged correlations of the density, expected to result in dislocation unbinding at any non-zero temperature, thereby completely destroying smectic state in 2D.[62]

2.4. *Non-linear elasticity: Beyond Gaussian fluctuations*

2.4.1. *Perturbation theory*

As is clear from the analysis of the previous section, the restoration of the translational symmetry (a vanishing $\langle\rho_q(\mathbf{x})\rangle$, etc.) by thermal fluctuations is a robust prediction of the quadratic theory, which cannot be overturned by the left-out non-linearities. However, as discussed in Section 1, the instability of the smectic Gaussian fixed point (in $d < 3$) illustrated in Fig. 3 and the asymptotic long-scale form of the correlation functions computed within the harmonic approximation only extends out to the non-linear length scales $\xi_{\perp,z}^{NL}$. On longer scales, the divergently large smectic phonon fluctuations invalidate the neglect of phonon non-linearities,

$$\mathcal{H}_{\text{nonlinear}} = -\frac{1}{2}B(\partial_z u)(\nabla u)^2 + \frac{1}{8}B(\nabla u)^4. \tag{52}$$

These will necessarily qualitatively modify predictions (46c), (47c), and (51c) on scales longer than the crossover scales $\xi_{\perp,z}^{NL}$, which I compute next.

To see this, one uses a perturbative expansion in the non-linear operators (52) to assess the size of their contribution to, e.g., the free energy. Following the standard field-theoretic RG analysis, pioneered by Wilson and Fisher[1] (but here applied to the critical smectic *phase* rather than to a critical point), these can be accounted for as corrections to the compressional B and bend K elastic moduli, with the

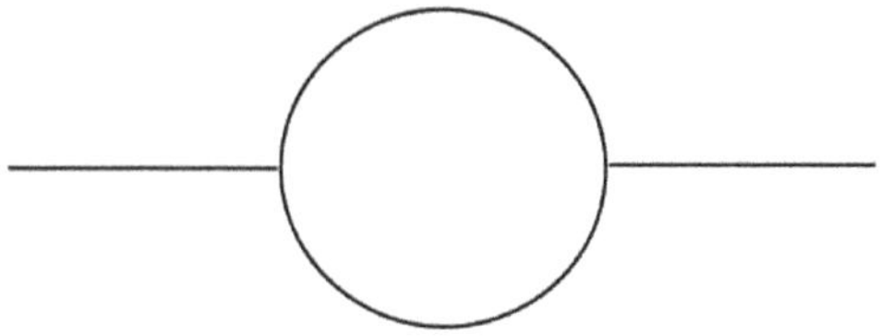

Fig. 6. Feynman graph that renormalizes the elastic moduli K, B of the smectic state.

leading softening (negative) contribution to δB, summarized graphically in Fig. 6, and given by

$$\delta B = -\frac{1}{2}TB^2 \int_{\mathbf{q}} q_\perp^4 G_u(\mathbf{q})^2 \tag{53a}$$

$$\approx -\frac{1}{2}TB^2 \int_{-\infty}^{\infty} \frac{dq_z}{2\pi} \int_{L_\perp^{-1}} \frac{d^{d-1}q_\perp}{(2\pi)^{d-1}} \frac{q_\perp^4}{(Kq_\perp^4 + Bq_z^2)^2}$$

$$\approx -\frac{1}{8}\frac{C_{d-1}T}{3-d}\left(\frac{B}{K^3}\right)^{1/2} L_\perp^{3-d} B. \tag{53b}$$

In the above, I used the smectic correlator, $G_u(\mathbf{q})$, focused on $d \leq 3$ (which allowed the dropping of the uv-cutoff (Λ) dependent part that vanishes for $\Lambda \to \infty$), and cut off the divergent contribution of the long-wavelength modes via the infrared cutoff $q_\perp > 1/L_\perp$ by considering a system of a finite extent $L_\perp$. Similar analysis gives a *positive* fluctuation correction to the bend modulus K, stiffening the undulation mode.[25]

Clearly, the anharmonicity becomes important when the fluctuation corrections to the elastic constants (e.g., δB above) become comparable to its bare microscopic value. As the following analysis demonstrates, the divergence of this correction as $L_\perp \to \infty$ signals the breakdown of the conventional harmonic elastic theory on length scales longer than a crossover scale

$$\xi_\perp^{NL} \approx \begin{cases} \frac{1}{T}\left(\frac{K^3}{B}\right)^{1/2}, & d = 2, \\ ae^{\frac{c}{T}\left(\frac{K^3}{B}\right)^{1/2}}, & d = 3, \end{cases} \tag{54}$$

which I define here as the value of $L_\perp$ at which $|\delta B(\xi_\perp^{NL})| = B$. Within the approximation of the smectic screening length $\lambda = a$, these non-linear crossover lengths reduce to the phonon disordering lengths (42), (44), defined by the Lindemann-like criterion.[61] Clearly, on scales longer than $\xi_{\perp,z}^{NL}$ (also a crossover scale from Gaussian to smectic fixed point in Fig. 1), the perturbative contributions of non-linearities dominate and therefore cannot be neglected. Their contributions are thus expected to qualitatively modify the harmonic predictions of the previous section.

2.4.2. *Renormalization-group analysis in $d = 3 - \epsilon$ dimensions*

To describe the physics beyond the crossover scales, $\xi^{NL}_{\perp,z}$, i.e., to make sense of the infrared divergent perturbation theory found in Eq. (53b), requires an RG analysis.[1–3] This was first performed in the context of conventional 3D smectic liquid crystals and Lifshitz points in a seminal work by Grinstein and Pelcovits (GP).[11] In the following, I complement GP's treatment with Wilson's momentum-shell RG analysis, extending it to an arbitrary dimension d, so as to connect to the behavior in 2D, which has an exact solution.[64]

To this end, I integrate (perturbatively in $\mathcal{H}_{\text{non-linear}}$) short-scale Goldstone modes in an infinitesimal cylindrical shell of wavevectors, $\Lambda e^{-\delta\ell} < q_\perp < \Lambda$ and $-\infty < q_z < \infty$ ($\delta\ell \ll 1$ is infinitesimal). The leading perturbative momentum-shell coarse-graining contributions come from terms found in direct perturbation theory above, but with the system size divergences controlled by the infinitesimal momentum shell. The thermodynamic averages can then be equivalently carried out with an effective coarse-grained Hamiltonian of the form (8), with momentum-dependent couplings obtained from integrating up their infinitesimal corrections, obtained via momentum-shell RG. For smectic moduli B and K, this gives

$$\delta B \approx -\frac{1}{8} g B \delta\ell, \quad \delta K \approx \frac{1}{16} g K \delta\ell, \tag{55}$$

where the dimensionless coupling is given by

$$g = C_{d-1}\Lambda_\perp^{d-3} T \left(\frac{B}{K^3}\right)^{1/2} \approx \frac{T}{2\pi}\left(\frac{B}{K^3}\right)^{1/2}, \tag{56}$$

and in the second form, I approximated g by its value in 3D. Equations (55) show that B is softened and K is stiffened by the non-linearities in the presence of thermal fluctuations, making the system effectively more isotropic, as one may expect on general physical grounds.

For convenience, I then anisotropically rescale the lengths and the remaining long-wavelength part of the fields $u^<(\mathbf{r})$ according to $r_\perp = r'_\perp e^{\delta\ell}$, $z = z' e^{\omega\delta\ell}$ and $u^<(\mathbf{r}) = e^{\phi\delta\ell} u'(\mathbf{r}')$, so as to restore the ultraviolet cutoff $\Lambda_\perp e^{-\delta\ell}$ back up to $\Lambda_\perp$. The underlying rotational invariance ensures that the non-linear fluctuation corrections preserve the rotationally invariant strain operator $(\partial_z u - \frac{1}{2}(\nabla_\perp u)^2)$, renormalizing it as a whole. It is therefore convenient (but not necessary) to choose the dimensional rescaling that also preserves this form. It is easy to see that this choice leads to

$$\phi = 2 - \omega. \tag{57}$$

The leading (one-loop) corrections to the effective coarse-grained and rescaled free energy functional can then be summarized by differential RG flows

$$\frac{dB(\ell)}{d\ell} = (d + 3 - 3\omega - \frac{1}{8} g(\ell)) B(\ell), \tag{58a}$$

$$\frac{dK(\ell)}{d\ell} = (d - 1 - \omega + \frac{1}{16} g(\ell)) K(\ell). \tag{58b}$$

From these, one readily obtains the flow of the dimensionless coupling $g(\ell)$

$$\frac{dg(\ell)}{d\ell} = (3-d)g - \frac{5}{32}g^2, \tag{59}$$

whose flow for $d < 3$ away from the $g = 0$ Gaussian fixed point (illustrated in Fig. 3) encodes the long-scale divergences found in the direct perturbation theory above. This resembles the Wilson–Fisher flow at the critical point just below $d = 4$,[1] but, as discussed in Section 1 requires no fine-tuning. As summarized in Fig. 3 for $d < 3$, the flow terminates at a non-zero fixed-point coupling $g_* = \frac{32}{5}\epsilon$ (with $\epsilon \equiv 3 - d$), which determines the non-trivial long-scale behavior of the smectic critical phase (see the following). As with treatments of critical points,[1,4] but here extending over the whole smectic phase, the RG procedure is quantitatively justified by the proximity to $d = 3$, i.e., controlled by the smallness of ϵ.

One can now use a standard matching calculation to determine the long-scale asymptotic form of the correlation functions on scales beyond $\xi^{\mathrm{NL}}_{\perp,z}$, namely, applying the above coarse-graining RG analysis to a computation of correlation functions allows one to relate its form at long length scales of interest (that because of infrared divergences is impossible to compute via a direct perturbation theory in non-linearities) to its counterpart at short scales, evaluated with coarse-grained couplings, $B(\ell)$, $K(\ell)$,.. In contrast to the former, the latter is readily computed via a perturbation theory, which because of shortness of the length scale is convergent. The result of this matching calculation to lowest order gives correlation functions from an effective Gaussian theory,

$$G_u(\mathbf{k}) \approx \frac{T}{B(\mathbf{k})k_z^2 + K(\mathbf{k})k_\perp^4}, \tag{60}$$

with moduli $B(\mathbf{k})$ and $K(\mathbf{k})$ that are singularly wavevector-dependent, determined by the solutions $B(\ell)$ and $K(\ell)$ of the RG flow equations (58a) and (58b), with initial conditions given by the microscopic values B and K.

2d analysis: In $d = 2$, at long scales, $g(\ell)$ flows to a non-trivial infrared stable fixed point $g_* = 32/5$, and the matching analysis predicts correlation functions characterized by anisotropic wavevector-dependent moduli

$$K(\mathbf{k}) = K\left(k_\perp \xi_\perp^{NL}\right)^{-\eta_K} f_K(k_z \xi_z^{NL}/(k_\perp \xi_\perp^{NL})^\zeta), \tag{61a}$$
$$\sim k_\perp^{-\eta_K},$$
$$B(\mathbf{k}) = B\left(k_\perp \xi_\perp^{NL}\right)^{\eta_B} f_B(k_z \xi_z^{NL}/(k_\perp \xi_\perp^{NL})^\zeta), \tag{61b}$$
$$\sim k_\perp^{\eta_B}.$$

Thus, on scales longer than $\xi^{\mathrm{NL}}_{\perp,z}$ these qualitatively modify the real-space correlation function asymptotics of the harmonic analysis in the previous section. In Eqs. (61), the universal anomalous exponents are given by

$$\begin{aligned}\eta_B &= \frac{1}{8}g_* = \frac{4}{5}\,\epsilon \\ &\approx \frac{4}{5}, \quad \text{for } d = 2, \end{aligned} \tag{62a}$$

$$\begin{aligned}\eta_K &= \frac{1}{16}g_* = \frac{2}{5}\,\epsilon \\ &\approx \frac{2}{5}, \quad \text{for } d = 2, \end{aligned} \tag{62b}$$

determining the $z - \mathbf{x}_\perp$ anisotropy exponent via (60) to be

$$\zeta \equiv 2 - (\eta_B + \eta_K)/2 \tag{63a}$$

$$= \frac{7}{5}, \quad \text{for } d = 2, \tag{63b}$$

as expected reduced by thermal fluctuations down from its harmonic value of 2. The $\mathbf{k}_\perp - k_z$ dependence of $B(\mathbf{k}), K(\mathbf{k})$ is determined by universal scaling functions, $f_B(x), f_K(x)$ that I will not compute here. The underlying rotational invariance gives an *exact* relation between the two anomalous $\eta_{B,K}$ exponents

$$3 - d = \frac{\eta_B}{2} + \frac{3}{2}\eta_K, \tag{64a}$$

$$1 = \frac{\eta_B}{2} + \frac{3}{2}\eta_K, \quad \text{for } d = 2, \tag{64b}$$

which is obviously satisfied by the anomalous exponents, Eqs. (62b) and (62a), computed here to first order in $\epsilon = 3 - d$.

Thus, as advertised, I find that at non-zero temperature, a 2D smectic state is highly non-trivial and qualitatively distinct from its mean-field perfectly periodic form. In addition to a vanishing density modulation and associated fluctuation-restored translational symmetry, it is characterized by universal non-local length-scale-dependent moduli, Eq. (61). Consequently, its Goldstone-mode theory and the associated correlations are not describable by a local harmonic field theory, which is an analytic expansion in local field operators. Instead, in 2D, on length scales beyond $\xi_{\perp,z}^{\mathrm{NL}}$ (but shorter than the expected dislocation unbinding length ξ_{disl}), thermal fluctuations and correlations of this smectic *critical* phase are controlled by a non-trivial fixed point, as illustrated in Fig. 3, characterized by *universal* anomalous exponents $\eta_{K,B}$ and scaling functions $f_{B,K}(x)$, defined above.

Above, I obtained this non-trivial structure from an RG analysis and estimated these exponents within a controlled but approximate ϵ-expansion. Remarkably, in 2D, an exact solution of this problem was discovered by Golubovic and Wang.[64]

It predicts critical phenomenology in a qualitatively agreement with the RG predictions above and gives exact exponents

$$\eta_B^{2d} = 1/2, \quad \eta_K^{2d} = 1/2, \quad \zeta^{2d} = 3/2, \tag{65}$$

derived by mapping onto the 1+1D KPZ equation.[65]

3D analysis: In $d = 3$, the nonlinear coupling $g(\ell)$ is marginally irrelevant, flowing to 0 at long scales. Despite this, as discovered by Larkin and Khmel'nitskii and by Michael Fisher *et al.*, in the context of the Ising model,[2,66,67] the marginal flow to the Gaussian fixed point is sufficiently slow (logarithmic in lengths) that (as usual at a marginal dimension[4]) its power law in ℓ dependence leads to a universal, asymptotically *exact* logarithmic wavevector dependence found in a smectic by Grinstein and Pelcovits,[11]

$$K(\mathbf{k}_\perp, k_z = 0) \sim K|1 + \frac{5g}{64\pi}\ln(1/k_\perp a)|^{2/5}, \tag{66a}$$

$$B(\mathbf{k}_\perp = 0, k_z) \sim B|1 + \frac{5g}{128\pi}\ln(\lambda/k_z a^2)|^{-4/5}. \tag{66b}$$

This translates into the smectic order parameter correlations given by

$$\langle \rho_q^*(\mathbf{x})\rho_q(0)\rangle \sim e^{-c_1(\ln z)^{6/5}}\cos(q_0 z), \tag{67}$$

with $2/5, 4/5$, and $6/5$ *exact* universal exponents and c_1 a non-universal constant.[11] Although these 3D anomalous effects are less dramatic and likely to be difficult to observe in practice, theoretically they are quite significant as they represent a qualitative breakdown of the mean-field and harmonic descriptions, which respectively ignore interactions and thermal fluctuations.

I conclude this section by noting that all of the above analysis is predicated on the validity of the purely elastic model, Eq. (8), which neglects topological defects, such as dislocations. If these unbind (as they undoubtedly do in 2D at any non-zero temperature[62]), then the above prediction only holds on scales shorter than the separation ξ_{disl} between these defects.

3. Biaxially Periodic Critical States: Spontaneous Line Crystals

Another class of critical states are those partially crystallized along *two* of the three axes. They can be dubbed as "*spontaneous* line crystals" as they are a periodic 2D crystalline array of 1D liquids, oriented along a *spontaneously* chosen axis, illustrated in Fig. 2(c), and in this sense dual to a uniaxially periodic critical state — a 1D periodic array of 2D liquids — the $m = d - 1$-Lifshitz model,[9] as discussed in Section 2. As I discuss in the following, its key feature is rotational symmetry-enforced vanishing of the tilt modulus — the $m = 1$-Lifshitz model,[9] which leads to its enhanced critical fluctuations.

3.1. *Columnar liquid crystal*

One example of a spontaneous line crystal is a ubiquitous "columnar" liquid crystal. It emerges from a nematic fluid composed of a high aspect ratio disk-shaped constituents, as illustrated in Fig. 2(c).[10] In this phase, disk-shaped molecules stack into one-dimensional fluid columns along a spontaneous nematic axis, which freeze into a 2D crystal (typically triangular lattice), retaining fluid order along the columns. Such discotic liquid crystal, forming a 2D crystal of 1D fluid columns compliments the smectic state of 1D periodic array of 2D fluids, as discussed in earlier sections.

Simple analysis similar to that of a smectic shows that for $d < 5/2$, the columnar liquid crystal ($m = 1$-Lifshitz model[9]) is a critical phase, though given this dimensional constraint not a very practical one for experimental realization. However, although I do not pursue the subject here, I note that far stronger distortion effects of quenched disorder (e.g., columnar state in aerogel[12–14]) lead to a critical columnar glass for $d < 7/2$, characterized by an infrared attractive fixed point controlled by disorder.[40]

The columnar state is characterized by a two-component phonon field $\mathbf{u} = (u_x, u_y)$, which are two Goldstone modes associated with translational symmetry-breaking in the plane transverse to the columns. Analogous to a smectic, the Goldstone modes associated with the rotational symmetry are gapped out by the emergent Higgs mechanism.[32]

Although one can derive the Goldstone-mode elasticity for the columnar state by following the approach analogous to that of a smectic, here I will simply write it down based on symmetry and experience with the smectic. To this end, I note that the columnar state exhibits two types of rigidities: the in-plane (transverse to the columnar axis $\hat{z}$) crystalline elasticity and bending elasticity of the columns. The former is characterized by shear and bulk elasticity of a 2D crystal. The latter is captured by the higher-derivative curvature (rather than tension) filament elasticity. Together, these give the energy density

$$\mathcal{H}_{col} = \frac{1}{2}\kappa(\partial_z^2 \mathbf{u})^2 + \frac{\lambda}{2}\, u_{\alpha\alpha}^2 + \mu\, u_{\alpha\beta}^2, \tag{68}$$

where κ is the curvature modulus, μ and λ are Lamé elastic moduli of 2D triangular lattice,[4] and (with the summation convention over repeated indices) the strain tensor is given by

$$u_{\alpha\beta} = \frac{1}{2}(\partial_\alpha u_\beta + \partial_\beta u_\alpha - \partial_\gamma u_\alpha \partial_\gamma u_\beta) \approx \frac{1}{2}(\partial_\alpha u_\beta + \partial_\beta u_\alpha - \partial_z u_\alpha \partial_z u_\beta). \tag{69}$$

The two phonons, $\mathbf{u}(x, y, z) = (u_x, u_y)$, are fields in a three-dimensional space, and in the second approximate form above, I only kept the most important strain non-linearities.

As in the discussion of the smectic elasticity, the columnar non-linear elasticity (68) is strictly constrained by symmetry. The linear z derivative in the harmonic terms is forbidden by the rotational invariance about the x-axes (broken spontaneously) at infinitesimal level corresponding to $\mathbf{u} \to \mathbf{u} + \theta z\hat{\mathbf{y}}$. This "softness" with respect to $\partial_z \mathbf{u}$ modes then requires one to keep corresponding non-linearities in the in-plane strain tensor, $u_{\alpha\beta}$ above, the form of which is dictated by the in-plane rotational invariance.

As in a smectic, here too, the analysis of quadratic phonon fluctuations leads to power-law divergence for $d < 5/2$, requiring inclusion of elastic non-linearities in (68) via (69). Account of these via an RG analysis very similar to that of the smectic in the previous section leads to a non-trivial infrared stable fixed point controlled by $\epsilon = 5/2 - d$, which characterizes the universal properties of the resulting columnar *critical phase* (as in Fig. 3). For technical details, I refer the reader to the original literature.[40]

3.2. *Spontaneous vortex lattice in magnetic superconductors*

Another putative realization of a biaxially periodic (line-crystal) critical state is a "spontaneous vortex lattice"' in a ferromagnetic supercondutor (FS). Rare-earth borocarbide materials are believed to be examples of the latter, exhibiting a rich phase diagram that includes superconductivity, antiferromagnetism, ferromagnetism and spiral magnetic order.[37–39] In particular, there is now ample experimental evidence that, at low temperatures, superconductivity and ferromagnetism competitively coexist in $ErNi_2B_2C$ compounds and perhaps in more recently discovered high-temperature superconductor $Sr_2Y\,Ru_{1-x}Cu_xO_6$.

For sufficiently strong ferromagnetism, such FSs have been predicted[39] to exhibit a *spontaneous* vortex (SV) state driven by the spontaneous magnetization rather than by an external magnetic field $\mathbf{H}$. It is clear based purely on symmetry arguments that for $\mathbf{H} = 0$ and in the absence of atomic crystal anisotropy, the elastic properties of the resulting SV solid differ dramatically and *qualitatively* from those of a conventional Abrikosov lattice. The key underlying difference is the aforementioned *vanishing* of the tilt modulus (68), which is guaranteed by the underlying rotational invariance. Although this invariance is broken by the magnetization, $\mathbf{M}$, the tilt modulus remains zero because this breaking is *spontaneous* (neglecting the ever-present crystal field anisotropy). This contrasts strongly with a conventional vortex solid, where the rotational symmetry is *explicitly* broken by the *applied* field $\mathbf{H}$ and lattice anisotropy.

Consequently, the elasticity of the SV crystal is identical to that of the columnar phase, described by (68) together with (69), neglecting crystalline anisotropy. In the presence of purely thermal fluctuations, elastic non-linearities remain irrelevant for such 3D SV state, since $d = 3 > 5/2$. However, as discovered and explored in Refs. 41 and 42, in the presence of ever-present quenched disorder, SV is a disorder-controlled (rather than temperature-controlled) critical glass state, treatable via RG perturbative in $\epsilon = 7/2 - d$, with many interesting properties, e.g., the unusually singular $B(H)$ relation.

4. Polymerized Membrane

A fluctuating, *tensionless* and therefore curvature-controlled (as opposed to tension-controlled) membrane is a strongly fluctuating system. Focus on such membranes was originally stimulated in part by ubiquitous biophysical realizations (e.g., liposomes, cellular membranes, cytoskeleton of a red blood cell, etc.),[45] but more recently found an ideal realization in fluctuating graphene sheets.[68] The latter, polymerized (solid) membrane, characterized by a non-zero in-plane shear rigidity, is a particularly fascinating case of a critical phase for its internal dimension $D < 4$.

As was first argued by Nelson and Peliti,[46] in addition to the rotationally invariant "crumpled" phase of linear polymers and liquid membranes, polymerized membranes exhibit a finite temperature "flat" phase. As I discuss in the following, this too is a *critical* phase characterized by universal power-law correlations (e.g., membrane roughness) and anomalous elasticity, with the state's very existence in two-dimensional membranes being a beautiful illustration of a phenomena of order-from-disorder. I note in passing that, as illustrated in Fig. 7, a non-self-avoiding — "phantom" anisotropic polymerized membrane was predicted[69,70] and observed in simulations[71] to exhibit an intermediate tubule phase, crumpled along one dimension and extended along the other, that, despite its one-dimensionality, is stable to thermal fluctuations.

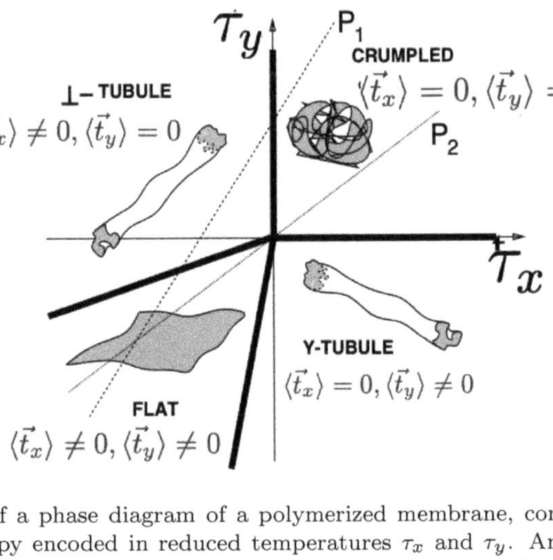

Fig. 7. Illustration of a phase diagram of a polymerized membrane, controlled by temperature and in-plane anisotropy encoded in reduced temperatures τ_x and τ_y. An isotropic membrane is described by $\tau_x = \tau_y \equiv \tau$, exhibiting temperature τ-driven transition along path P_2 from a crumpled phase (with vanishing tangent $\vec{t}_{x,y} = 0$ order parameters) to a "flat" phase with $\vec{t}_{x,y} \neq 0$. Flat phase spontaneously breaks the rotational $O(3)$ symmetry of the embedding space and is a critical phase that exhibits rich universal power-law phenomenology, controlled by a non-trivial infrared stable fixed point. A polymerized membrane with in-plane anisotropy is predicted to undergo a two-stage transition along P_1 from the crumpled-to-tubule phase (latter characterized by only one of the two tangent order parameters non-zero), followed by tubule-to-flat phase transition,[69,70] later observed in simulations.[71]

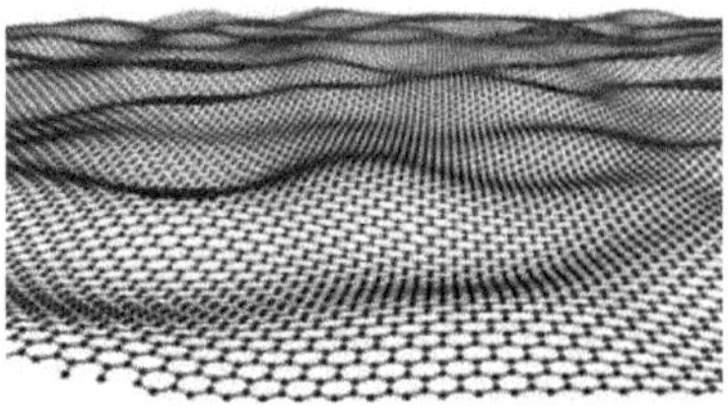

(a)

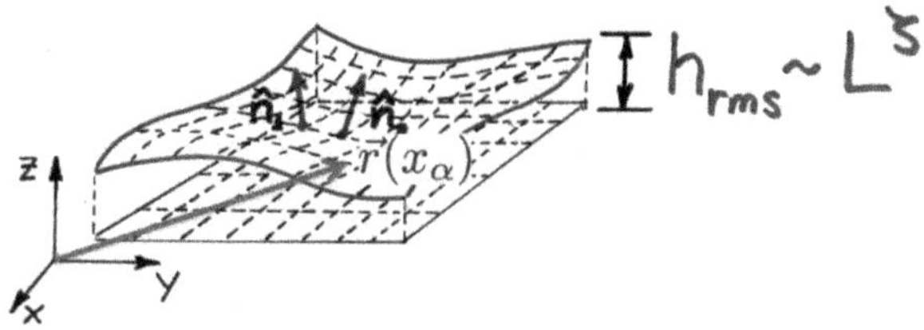

(b)

Fig. 8. Illustration of (a) a graphene sheet as a modern ideal realization of a thermally fluctuating polymerized membrane, predicted to exhibit rich anomalous elasticity as a critical phase, (b) flat phase of a polymerized membrane defining its normals, Monge gauge embedding and the definition of the roughness exponent, ζ.

A detailed derivation of the flat-phase Goldstone-modes elasticity is available[45,46] starting from the Landau theory of the crumpled phase in terms of the tangent vectors $\boldsymbol{\nabla}\vec{r}(\mathbf{x})$, where $\vec{r}(\mathbf{x})$ is the embedding of the D-dimensional membrane (parameterized by $\mathbf{x}$) in the d-dimensional space, $\vec{r}$ (with $D = 2$ and $d = 3$, the physical case). The flat phase parameterized by

$$\vec{r}(\mathbf{x}) = (t_0\mathbf{x} + \mathbf{u}(\mathbf{x}), \vec{h}(\mathbf{x})), \tag{70}$$

in the Monge gauge, spontaneously breaks the $O(d)$ symmetry of the crumpled phase down to $O(D)$, where t_0 scale factor is the effective order parameter of the flat phase, $\mathbf{u}(\mathbf{x})$ is the D-component in-plane phonon, and $\vec{h}(\mathbf{x})$ is the "height function" describing the membrane's transverse undulation into the embedding space where we have implicitly generalized to arbitrary codimension $d_c = d - D$.

The resulting energy functional is a sum of the bending and the in-plane elastic contributions:

$$H_{flat}[\vec{h}, \mathbf{u}] = \int d^D x \left[\frac{\kappa}{2}(\nabla^2\vec{h})^2 + \mu u_{\alpha\beta}^2 + \frac{\lambda}{2}u_{\alpha\alpha}^2\right], \tag{71}$$

where the non-linear strain tensor is

$$u_{\alpha\beta} = \frac{1}{2}(\partial_\alpha\vec{r}\cdot\partial_\beta\vec{r} - \delta_{\alpha\beta}) \approx \frac{1}{2}(\partial_\alpha u_\beta + \partial_\beta u_\alpha + \partial_\alpha\vec{h}\cdot\partial_\beta\vec{h}), \tag{72}$$

defined as the deviation of the embedding-induced metric $g_{\alpha\beta}$ from the flat metric, and in the second form, I neglected in-plane elastic non-linearities that are subdominant at long scales. A crucial feature of $H_{flat}[\vec{h}, \mathbf{u}]$ (71) (as with other critical phases) is the higher-order curvature elasticity of $\vec{h}$, with the surface tension modulus for $(\boldsymbol{\nabla}\vec{h})^2$ vanishing exactly, enforced by the underlying rotational invariance (of the embedding space) and tension-free membrane.

One can also integrate out the non-critical in-plane phonon field $\mathbf{u}$, that appears only harmonically, to obtain a convenient equivalent form, purely in terms of $\vec{h}$[46,47,50,72]

$$H_{flat}[\vec{h}] = \int d^D x \left[\frac{\kappa}{2}(\nabla^2\vec{h})^2 + \frac{1}{4d_c}(\partial_\alpha\vec{h}\cdot\partial_\beta\vec{h})R_{\alpha\beta,\gamma\delta}(\partial_\gamma\vec{h}\cdot\partial_\delta\vec{h})\right], \tag{73}$$

where, for convenience, I rescaled Lamé coefficients so that the quartic coupling is of order $1/d_c$. The four-point coupling fourth-rank tensor is given by

$$R_{\alpha\beta,\gamma\delta} = \frac{K - 2\mu}{2(D-1)} P^T_{\alpha\beta} P^T_{\gamma\delta} + \frac{\mu}{2}(P^T_{\alpha\gamma} P^T_{\beta\delta} + P^T_{\alpha\delta} P^T_{\beta\gamma}), \tag{74}$$

where $P^T_{\alpha\beta} = \delta_{\alpha\beta} - q_\alpha q_\beta / q^2$ is a transverse (to $\mathbf{q}$) projection operator. The convenience of this decomposition is that $K = 2\mu(2\mu + D\lambda)/(2\mu + \lambda)$ and μ moduli renormalize independently and multiplicatively.[49,72]

I note that this last h-only elastic form indeed reflects the general discussion of critical phases in Section 1, namely that they are described by an energy functional (of ϕ^4 form with $\vec{\phi}_\alpha \equiv \partial_\alpha \vec{h}$) enforced to be critical (i.e., missing the ϕ^2 term) by the underlying rotational symmetry, as illustrated in Fig. 3.

Applying Wilson–Fisher momentum-shell RG analysis to membrane's elastic non-linearities[46–49,72] one sees that for $D < 4$ (and arbitrary embedding dimension d), the membrane exhibits height undulations that diverge with its extent L. As for smectics and other critical phases discussed earlier, one can make sense of the associated divergent perturbation theory in elastic non-linearities by performing an RG analysis controlled by $\epsilon = 4 - D$[47] or equivalently $1/d_c$ (large embedding dimension, which is the analog of $1/N$ large N expansion of the $O(N)$ model's Wilson–Fisher critical point).[48] The RG approach for arbitrary d gives flows of the dimensionless non-linear coupling constants, $\hat{\mu}(\ell) = \mu/\kappa^2, \hat{\lambda}(\ell) = \lambda/\kappa^2$, after integrating out modes in a momentum shell $\Lambda e^{-\delta\ell} < q < \Lambda$ (as for a smectic in the previous section) and rescaling for convenience.

The dimensionless couplings flow to a tuning-free infrared attractive Aronovitz–Lubensky (AL) fixed point,[46–49] with two other (in addition to the Gaussian) fixed points that describe a marginally mechanically unstable membrane (of physical relevance to nematic elastomers of the following section). The resulting globally infrared stable AL fixed point controls the properties of the highly non-trivial "flat" phase, which is critical and power-law rough. A complementary approach of Le Doussal and Radzihovsky[49] is via a solution of self-consistent integral equations (SCSA) that build on the large embedding $1/d_c$ expansion, with virtue of being exact in three limits: (i) $D \to 4$ (ϵ expansion) for arbitrary d_c, (ii) $d_c \to \infty$ ($1/d_c$ expansion) and (iii) $d_c = 0$ for arbitrary D, thereby providing best quantitative predictions, as judged by numerics and experiments.

The "flat" critical phase is characterized by *universal anomalous* elasticity with length-scale-dependent elastic moduli

$$\kappa(q) \sim q^{-\eta}, \quad \mu(q) \sim \lambda(q) \sim q^{\eta_u}, \tag{75}$$

where for physical membranes $\eta \approx 0.82$ and the underlying embedding-space rotational invariance imposes exact relations,[46–49]

$$\eta_u = 4 - D - 2\eta_\kappa, \quad \zeta = (4 - D - \eta)/2. \tag{76}$$

These indicate that at long scales, $q \to 0$, thermal fluctuations stiffen the effective bending modulus $\kappa(q)$ and soften membrane's in-plane moduli, $\mu(q), \lambda(q)$, as one can easily test by playing around with a wrinkled piece of paper.

Based on these, the SCSA[49] predicts a *universal* and *negative* Poisson ratio,

$$\operatorname*{Lim}_{q \to 0} \sigma \equiv \frac{\lambda(q)}{2\mu(q) + (D-1)\lambda(q)} = -\frac{1}{3} \quad \text{for } D = 2, \tag{77}$$

which measures the ratio of in-plane compression of a membrane along an axis transverse to its strained direction. Its negative value indicates that such membrane actually expands transversely as it is being stretched. This approximate value of $-1/3$ compares extremely well with largest simulations[73] and higher-order computations. This amazing fluctuation-driven anomalous elasticity phenomenology, uncovered using systematic Wilson–Fisher RG, can be understood qualitatively by playing around with a roughened piece of paper.

Finally, I conclude by emphasizing that the universal anomalous elasticity is at the heart of stability of the 2D "flat" phase. To see this, observe that in contrast to the unstable Gaussian fixed point (describing fluctuations of a microscopic membrane), the "flat" *critical* phase at the AL fixed point, with a renormalized, momentum-dependent $\kappa(q)$, is characterized by root-mean-squared height undulations

$$h_{\text{rms}} = \sqrt{\langle h^2(\mathbf{x}) \rangle} = \left[\int_{L^{-1}} \frac{d^D q}{(2\pi)^d} \frac{T}{\kappa(q) q^4} \right]^{1/2} \sim L^{\zeta}, \tag{78}$$

with the roughness exponent $\zeta < 1$. Its best estimate for $D = 2, d = 3$[49] is

$$\zeta = 0.59$$

and has also been computed within ϵ and $1/d_c$ expansions.[47–49,74] I note that despite unbounded power-law height undulations (78), the *rotational* $O(d)$ symmetry of the embedding space remains broken since $L^{\zeta} \ll L \to \infty$, despite two-dimensionality ($D = 2$) of the polymerized membrane field theory. It thus circumvents a naive application of the Hohenberg–Mermin–Wagner–Coleman theorem.[58–60] The rigorous form of the latter does not *a priori* apply to the higher-derivative polymerized membrane field theory, and thus does not forbid above findings, e.g., stability of the 2D flat phase.[75] More physically, as can be seen from (78), the theorem's common heuristic application fails precisely because the state is controlled by a non-Gaussian fixed point with anomalous, momentum-dependent elastic moduli, specifically with $\eta > 0$. As a result, $h_{\text{rms}}(L)/L \sim L^{-\eta/2} \to 0$ indicates that the "flat" phase is stabilized (for non-zero positive $\eta > 0$ generated by thermal fluctuations) by the very fluctuations that attempt to destabilize it, a striking phenomenon known as order-from-disorder.

Finally, I note that random heterogeneity also ubiquitously appears in most membrane realizations (see, e.g., work on graphene[76]) and leads to a disorder-driven critical membrane phase, studied extensively starting with the work of Nelson and Radzihovsky,[77,78] with many that followed.[72,79,80]

5. Nematic Elastomer

A final example of a critical phase that I will briefly discuss is a nematic elastomer, namely rubber — a randomly crosslinked polymer network composed of mesogenic groups, as illustrated in Fig. 9. It exhibits a spontaneous transition to a nematic phase, thereby driving an accompanying spontaneous uniaxial distortion of the elastic matrix, as illustrated in Fig. 9.[43]

Even in the absence of fluctuations, bulk nematic elastomers were predicted[81] and later observed to display an array of fascinating phenomena,[43,82] the most striking of which is the vanishing of stress for a range of strain, applied transversely to the spontaneous nematic axis. This striking softness is generic, stemming from the spontaneous orientational symmetry-breaking by the nematic state, accompanied by the nemato-elastic Goldstone mode, that leads to the observed soft distortion and strain-induced director reorientation,[81,83] as illustrated in Fig. 10. This unique

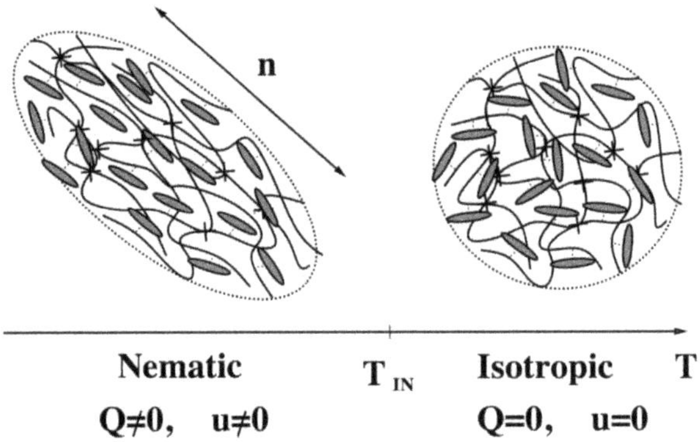

Fig. 9. Spontaneous uniaxial distortion of nematic elastomer driven through the isotropic–nematic transition.

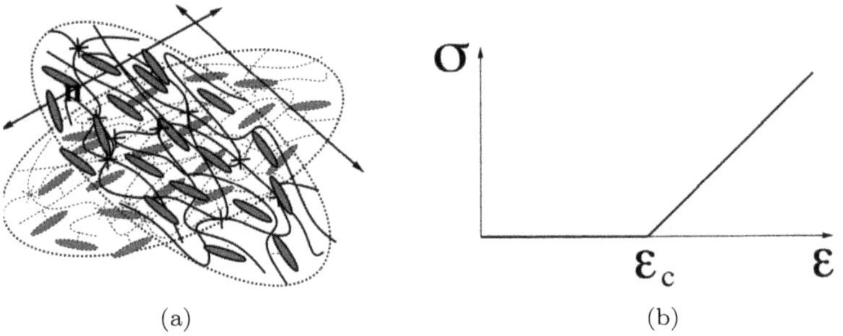

Fig. 10. (a) Simultaneous reorientation of the nematic director and of the uniaxial distortion is a low-energy nemato-elastic Goldstone mode of an ideal elastomer, which is responsible for its softness and (b) its resulting idealized flat (vanishing stress) stress–strain curve for a range of strains $0 < \varepsilon < \varepsilon_c$.[83]

elastic phenomenon is captured even at the harmonic approximation of the fully non-linear uniaxial Hamiltonian in Eq. (79),

$$\mathcal{H}_{\text{elast}} = \frac{1}{2}B_z\, w_{zz}^2 + \lambda_{z\perp}\, w_{zz}\, w_{\alpha\alpha} + \frac{1}{2}\,\lambda\, w_{\alpha\alpha}^2 + \mu\, w_{\alpha\beta}^2 + \frac{K}{2}\,(\nabla_\perp^2 u_z)^2, \qquad (79)$$

where akin to earlier examples, the components of the (rescaled) effective non-linear strain tensor $\underline{\underline{w}}$ are given by

$$w_{zz} = \partial_z u_z + \frac{1}{2}(\nabla_\perp u_z)^2, \qquad (80)$$

$$w_{\alpha\beta} = \frac{1}{2}(\partial_\alpha u_\beta + \partial_\beta u_\alpha - \partial_\alpha u_z \partial_\beta u_z). \qquad (81)$$

The underlying rotational symmetry of the parent isotropic phase guarantees the vanishing of one of the five elastic constants,[81,83] $\mu_{\perp z}$ (i.e., the $w_{z\alpha}^2$ is missing from (79)), that usually characterizes harmonic deformations of a three-dimensional uniaxial solid,[8] here with the uniaxial axis $\hat{z}$ chosen spontaneously. Interestingly, examining (81), w_{zz} and $w_{\alpha\beta}$ have the structure of, respectively, a uniaxially (e.g., smectic) and biaxially (e.g., columnar) periodic critical state. Thus, nematic elastomer is a three-dimensional amorphous solid that is an amalgam of a smectic and columnar liquid crystals.

Given this softness of the harmonic elasticity associated with symmetry-imposed vanishing of the $\mu_{z\perp}$ modulus, thermal and heterogeneity-driven fluctuations are divergent in 3D.[44,84] Concomitantly, the elastic terms $(\boldsymbol{\nabla}_\perp u_z)^2$ and $(\partial_z \mathbf{u}_\perp)^2$ associated with transverse distortions are missing from the quadratic part of the elastic energy. Thus, as in other critical matter discussed in earlier sections, one expects qualitative importance of above elastic non-linearities (81) in the presence of thermal fluctuations and network heterogeneity.[44,84]

Similar to their effects in smectic, columnar liquid crystals, polymerized membranes, and other critical matter discussed earlier, in bulk elastomers, thermal fluctuations (and network heterogeneity) lead to anomalous elasticity controlled by a tuning-free infrared attractive fixed point, with universal length-scale-dependent elastic moduli,[44,84]

$$K_{\text{eff}} \sim L^{\eta}, \quad \mu_{\text{eff}} \sim L^{-\eta_u}. \qquad (82)$$

The resulting critical state can be shown[44] to be characterized by a universal non-Hookean stress–strain relation

$$\sigma_{zz} \sim (\varepsilon_{zz})^{\delta} \qquad (83)$$

and a *negative* Poisson ratio for extension $\varepsilon_{xx} > 0$ transverse to the nematic axis

$$\varepsilon_{yy} = \frac{5}{7}\varepsilon_{xx}, \quad \varepsilon_{zz} = -\frac{12}{7}\varepsilon_{xx}. \qquad (84)$$

While considerable progress has been made in understanding these fascinating materials, many questions, particularly associated with network heterogeneity, remain open.

6. Summary and Conclusions

In this chapter, I discussed a novel class of ordered states of matter that I dub as "critical matter". The common crucial feature of these exotic states is that, as a result of the underlying symmetry broken in the ordered state, a subset of elastic moduli of the associated Goldstone modes vanish identically. As a result, for sufficiently low dimensions, such systems exhibit harmonic fluctuations that are divergent in the thermodynamic limit and require treatment of Goldstone modes' non-linearities throughout the ordered phase (not just near a critical point). Treating these within the RG analysis leads to an infrared stable fixed point that controls the resulting highly non-trivial, strongly interacting, and universal ordered state. The latter exhibits properties akin to that of a critical point, but extending over the entire phase, which I therefore naturally refer to as a "critical phase."

As prominent examples of such systems, I have discussed smectics (with many soft and hard matter realizations from paired superfluids (FFLO), spin–orbit coupled and frustrated bosons to quantum Hall stripes), cholesterics, helical magnets, columnar phases, putative spontaneous vortex lattices, polymerized membranes, nematic elastomers, and many others. While significant progress has been made in characterizing these systems, discovery of new systems (e.g., as quantum examples in physically accessible dimensions), understanding of effects of random heterogeneity and dynamics remain as challenging open problems.

I finish by noting that there are also critical phases (many appearing as counterparts of the ones discussed in this chapter), that I have not discussed here, where fluctuations are driven by quenched disorder rather than by temperature or quantum fluctuations. These exhibit critical states of glassy Goldstone modes and generically appear in dimensions higher than their thermal counterparts,[12–14,40–42,44,72,77–80,85] and are thus of even stronger experimental relevance.

Acknowledgments

The material presented in these lectures is based on the research done with a number of wonderful colleagues and friends, most notably David Nelson, John Toner, Pierre Le Doussal, Tom Lubensky, Xiangjun Xing and Tzu-Chi Hsieh. I am indebted to them for much of my insight into the material presented here. This research was supported by the Simons Investigator Award from the Simons Foundation and in part by the National Science Foundation under Grant No. NSF PHY-1748958. I thank Kavli Institute for Theoretical Physics for its hospitality during the preparation at this manuscript.

On a personal note, I have met and discussed physics with Michael Fisher at conferences, most often at the wonderful Rutgers Stat Mech meetings. I recall fondly Michael always sitting in the front row, sometimes flanked by Daniel and Matthew, strikingly paying close attention and asking probing questions even in the three-minute student talks and as attentively as in invited talks by distinguished senior faculty. I recall him always being kind and supportive in my own career. Although I have never had the pleasure of working with Michael, I have greatly benefited from his insight through his many seminal papers that influenced my own RG work (some described in this chapter). But perhaps more importantly, I indirectly learned physics from Michael through David Nelson, Daniel and Matthew, all of whom have strongly shaped my style, taste and approach to physics. And for that, I am eternally grateful.

References

1. K. G. Wilson and M. E. Fisher, Critical exponents in 3.99 dimensions. *Phys. Rev. Lett.* **28**(4), 240–243 (1972).
2. M. E. Fisher, The renormalization group in the theory of critical behavior. *Reviews of Modern Physics*, **46**(4), 597 (1974).
3. K. G. Wilson and J. Kogut, The renormalization group and the ϵ expansion. *Physics Reports* **12**, 78–199 (1974).
4. P. Chaikin and T. C. Lubensky, *Principles of Condensed Matter Physics* (Cambridge University Press, Cambridge 1995).
5. Jean Zinn-Justin, *Quantum Field Theory and Critical Phenomena*, 5th Edition, (International Series of Monographs on Physics).
6. In quantum phase transitions, temperature T introduces a finite size cutoff $\hbar/k_BT$ into the imaginary time. As a result critical behavior can be probed even in a fan of $\delta g(T)$) away from a quantum critical point.
7. S. Sachdev, *Quantum Phase Transitions* (Cambridge University Press, 1999).
8. L. D. Landau, In D. ter Haar (ed.) *Collected papers of L. D. Landau*, p. 209 (Gordon and Breach, New York, (1965); L. D. Landau and E. M. Lifshitz, *Statistical Physics* (Pergamon, London, 1969).
9. A generalized m-Lifshitz model, has elasticity with m "soft" (Laplacian) axes and $d-m$ complementary "hard" (gradient) axes. In this nomenclature, the conventional classical 3d smectic is the $m=2$-Lifshitz model, and 2+1d quantum smectic and 3d classical columnar liquid crystal are described by the $m=1$-Lifshitz model. Other generalizations include a nonscalar Goldstone-mode field as in e.g., tethered membranes 45,69,70,72 and nematic elastomers 44.
10. P. de Gennes and J. Prost, *The Physics of Liquid Crystals* (Clarendon Press, Oxford, 1993).
11. G. Grinstein and R. A. Pelcovits, *Phys. Rev. Lett.* **47**, 856 (1981).
12. L. Radzihovsky and J. Toner, Nematic–to–smectic-A transition in aerogel. *Phys. Rev. Lett.* **79**, 4214 (1997).
13. L. Radzihovsky and J. Toner, Smectic liquid crystals in random environments. *Phys. Rev. B* **60**, 206 (1999).
14. T. Bellini, L. Radzihovsky, J. Toner and N. A. Clark, Universality and scaling in the disordering of a smectic liquid crystal. *Sci.* **204**, 5544 (2001).
15. Matthew A. Glaser, Gregory M. Grason, Randall D. Kamien, A. Kosmrlj, Christian D. Santangelo and P. Ziherl, Soft spheres nake more mesophases. *Europhys. Lett.* **78**, 46004 (2007).
16. M. P. Lilly, K. B. Cooper, J. P. Eisenstein, L. N. Pfeiffer and K. W. West, Evidence for an anisotropic state of two-dimensional electrons in high Landau levels. *Phys. Rev. Lett.* **82**, 394 (1999).

17. R. R. Du, D. C. Tsui, H. L. Stormer, L. N. Pfeiffer, K. W. Baldwin and K. W. West, Strongly anisotropic transport in higher two-dimensional Landau levels. *Solid State Comm.* **109**, 389 (1999).
18. A. A. Koulakov, M. M. Fogler and B. I. Shklovskii, Charge density wave in two-dimensional electron liquid in weak magnetic field. *Phys. Rev. Lett.* **76**, 499 (1996).
19. R. Moessner and J. T. Chalker, Exact results for interacting electrons in high Landau levels. *Phys. Rev. B* **54**, 5006 (1996).
20. E. Fradkin and S. A. Kivelson, Liquid-crystal phases of quantum Hall systems. *Phys. Rev. B* **59**, 8065 (1999).
21. A. H. MacDonald and M. P. A. Fisher, Quantum theory of quantum Hall smectics. *Phys. Rev. B* **61**, 5724 (2000).
22. L. Radzihovsky and A. T. Dorsey, Theory of Quantum Hall Nematics. *Phys. Rev. Lett.* **88**, 216802 (2002).
23. P. Fulde and R. A. Ferrell, Superconductivity in a strong spin-exchange field. *Phys. Rev.* **135**, A550 (1964).
24. A. I. Larkin and Yu. N. Ovchinnikov, Nonuniform state of superconductors. *Sov. Phys. JETP* **20**, 762 (1965).
25. L. Radzihovsky and A. Vishwanath, Quantum liquid crystals in an imbalanced Fermi gas: fluctuations and fractional vortices in Larkin-Ovchinnikov states. *Phys. Rev. Lett.* **103**, 010404 (2009); L. Radzihovsky, Fluctuations and phase transitions in Larkin-Ovchinnikov liquid-crystal states of a population-imbalanced resonant Fermi gas. *Phys. Rev. A* **84**, 023611 (2011); L. Radzihovsky, *Quantum liquid-crystal order in resonant atomic gases* (invited review). *Physica C* **481**, 189–206 (2012).
26. Leo Radzihovsky and Daniel E. Sheehy, Imbalanced Feshbach-resonant Fermi gases. *Rep. Prog. Phys.* **73**, 076501 (2010).
27. D. F. Agterberg, J. S. Davis, S. D. Edkins, E. Fradkin, D. J. Van Harlingen, S. A. Kivelson, P. A. Lee, L. Radzihovsky, J. M. Tranquada and Y. Wang, The physics of pair-density waves: Cuprate superconductors and beyond. *Annu. Rev. Condens. Matter Phys.* **11**, 231 (2020).
28. L. Radzihovsky and S. Choi, P-wave resonant Bose gas: A finite-momentum spinor superfluid. *Phys. Rev. Lett.* **103**, 095302 (2009); *Phys. Rev. A* **84**, 043612 (2011).
29. Hui Zhai, Degenerate quantum gases with spin-orbit coupling: A review. *Rep. Prog. Phys.* **78**, 026001 (2015).
30. Tzu-Chi Hsieh, Han Ma and Leo Radzihovsky, Helical superfluid in a frustrated honeycomb Bose-Hubbard model. *Phys. Rev. A* **106**, 023321 (2022).
31. T. C. Lubensky, Low-temperature phase of infinite cholesterics. *Phys. Rev. Lett.* **29**, 206 (1972).
32. L. Radzihovsky and T. C. Lubensky, Nonlinear smectic elasticity of helical state in cholesteric liquid crystals and helimagnets. *Phys. Rev. E* **83**, 051701 (2011).
33. S. Muhlbauer, B. Binz, F. Jonietz, C. Peiderer, A. Rosch, A. Neubauer, R. Georgii and P. Boni, Skyrmion lattice in a chiral magnet. *Sci.* **323**, 915 (2009).
34. Doron Bergman, Jason Alicea, Emanuel Gull, Simon Trebst and Leon Balents, Order by disorder and spiral spin liquid in frustrated diamond lattice antiferromagnets. *Nat. Phys.* **3**, 487 (2007).
35. A. Mulder, R. Ganesh, L. Capriotti and A. Paramekanti, Spiral order by disorder and lattice nematic order in a frustrated Heisenberg antiferromagnet on the honeycomb lattice. *Phys. Rev. B* **81**, 214419 (2010).
36. Tzu-Chi Hsieh and Leo Radzihovsky, *O(N) smectic σ-model*, arXiv:2310.13046, to appear in Phys. Rev. B (2023).

37. P. Canfield, P. Gammel and D. Bishop, New magnetic superconductors: A toy box for solid-state physicists. *Phys. Today*, October (1998).
38. U. Yaron, P. L. Gammel, A. P. Ramirez, D. A. Huse, D. J. Bishop, A. I. Goldman, C. Stassis, P. C. Canfield, K. Mortensen and M. R. Eskildsen, Microscopic coexistence of magnetism and superconductivity in $ErNi_2B_2C$. *Nat.* **386**, 236 (1996).
39. H. S. Greenside, E. I. Blount and C. M. Varma, Possible coexisting superconducting and magnetic states. *Phys. Rev. Lett.* **46**, 49 (1981); E. I. Blount and C. M. Varma, *ibid* **42**, 1079 (1979); T. K. Ng and C. M. Varma, *ibid.* **78**, 330, *ibid.* 3745 (1997).
40. K. Saunders, L. Radzihovsky and J. Toner, A discotic disguised as a smectic: A hybrid columnar bragg glass. *Phys. Rev. Lett.* **85**, 4309 (2000).
41. A. M. Ettouhami, Karl Saunders, L. Radzihovsky and John Toner, 'Soft' anharmonic vortex glass in ferromagnetic superconductors. *Phys. Rev. Lett.* **87**, 27001 (2001).
42. A. M. Ettouhami, Karl Saunders, L. Radzihovsky and John Toner, Elasticity, fluctuations and vortex pinning in ferromagnetic superconductors: A columnar elastic glass. *Phys. Rev. B* **71**, 224506 (2005).
43. M. Warner and E. M. Terentjev, *Liquid Crystal Elastomers* (Oxford University Press, 2003).
44. X. Xing and L. Radzihovsky, Thermal fluctuations and anomalous elasticity of homogeneous nematic elastomers. *Europhys. Lett.* **61**, 769 (2003); Nonlinear elasticity, fluctuations and heterogeneity of nematic elastomers. *Ann. Phys.* **323**, 105–203 (2008). Universal elasticity and fluctuations of nematic gels. *Phys. Rev. Lett.* **90**, 168301 (2003).
45. For a review, and extensive references, see the articles in *Statistical Mechanics of Membranes and Interfaces*, 2nd edition, Jerusalem Winter School, D. R. Nelson, T. Piran and S. Weinberg (eds.) (World Scientific, Singapore, 1989).
46. D. R. Nelson and L. Peliti, Fluctuations in membranes with crystalline and hexatic order. *J. Phys. (Paris)* **48**, 1085 (1987).
47. J. A. Aronovitz and T. C. Lubensky, Fluctuations of solid membranes. *Phys. Rev. Lett.* **60**, 2634 (1988); J. A. Aronovitz, L. Golubovic and T. C. Lubensky, Fluctuations and lower critical dimensions of crystalline membranes. *J. Phys. (Paris)* **50**, 609 (1989).
48. F. David and E. Guitter, Crumpling transition in elastic membranes: Renormalization group treatment. *Europhys. Lett.* **5**, 709 (1988).
49. P. Le Doussal and L. Radzihovsky, Self-consistent theory of polymerized membranes. *Phys. Rev. Lett.* **69**, 1209 (1992).
50. L. Radzihovsky, *Statistical Mechanics and Geometry of Random Manifolds*, Ph.D. Thesis, Harvard University, (1993).
51. Another, distinct class of "self-organized" critical phases have been proposed in Per Bak, Chao Tang and Kurt Wiesenfeld, Self-organized criticality: An explanation of the 1/f noise. *Phys. Rev. Lett.* **59**, 381 (1987). We will not discuss such nonequilibrium dynamical systems.
52. J. McIver, Monika Aidelsburger and M. Claassen, Floquet engineering of quantum materials. *Nature Commun. Phys. Focus, 17* March (2022).
53. Daniel E. Sheehy and Leo Radzihovsky, BEC-BCS crossover in "magnetized" Feshbach-resonantly paired superfluids. *Phys. Rev. Lett.* **96**, 060401 (2006).
54. Daniel E. Sheehy and Leo Radzihovsky, BEC-BCS crossover, phase transitions and phase separation in polarized resonantly-paired superfluids. *Ann. Phys.* **322**, 1790 (2007).
55. M. C. Revelle, J. A. Fry, B. A. Olsen and R. G. Hulet, 1D to 3D Crossover of a spin-imbalanced fermi gas. *Phys. Rev. Lett.* **117**, 235301 (2016).

56. A. Caillé, Remarks on the scattering of x-rays by A-type smectics. *C. R. Acad. Sci. Ser. B* **274**, 891 (1972).
57. R. E. Peierls, *Helv. Phys. Acta Suppl.* **7**, 81 (1934).
58. P. C. Hohenberg, Existence of long-range order in one and two dimensions. *Phys. Rev.* **158**, 383 (1967).
59. N. D. Mermin and H. Wagner, Absence of ferromagnetism or antiferromagnetism in one- or two-dimensional isotropic heisenberg models. *Phys. Rev. Lett.* **17**, 1133 (1966).
60. S. Coleman, There are no Goldstone bosons in two dimensions. *Commun. Math. Phys.* **31**, 259 (1973).
61. F. Lindemann, About the calculation of molecular own frequencies. *Z. Phys.* **11**, 609 (1910).
62. J. Toner and D. R. Nelson, Smectic, cholesteric, and Rayleigh-Benard order in two dimensions. *Phys. Rev. B* **23**, 316 (1981).
63. J. Als-Nielsen, J. D. Litster, R. J. Brigeneau, M. Kaplan, C. R. Safinya, A. Lindegaard-Andersen and S. Mathiesen, Observation of algebraic decay of positional order in a smectic liquid crystal. *Phys. Rev. B* **22**, 312 (1980).
64. L. Golubovic and Z. Wang, Anharmonic elasticity of smectics A and the Kardar-Parisi-Zhang model. *Phys. Rev. Lett.* **69**, 2535 (1992).
65. M. Kardar, G. Parisi and Y.-C. Zhang, Dynamic scaling of growing interfaces. *Phys. Rev. Lett.* **56**, 889 (1986).
66. A. I. Larkin and D. E. Khmel'nitskii, Phase transition in uniaxial ferroelectrics. *Zh. Eksp. Teor. Fiz* **56**, 2087 (1969).
67. M. E. Fisher, S. K. Ma and B. G. Nickel, Critical exponents for long-range interactions. *Phys. Rev. Lett.* **29**, 917 (1972).
68. Geim A. K. and K. S. Novoselov, The rise of graphene. *Nat. Mater.* **6**, 183–191 (2007).
69. L. Radzihovsky and J. Toner, A new phase of tethered membranes: Tubules. *Phys. Rev. Lett.* **75**, 4752 (1995).
70. L. Radzihovsky and J. Toner, Elasticity, shape fluctuations and phase transitions in the new tubule phase of anisotropic tethered membranes. *Phys. Rev. E* **57**, 1832–1863 (1998).
71. M. Bowick, M. Falcioni and G. Thorleifsson, Numerical observation of a tubular phase in anisotropic membranes. *Phys. Rev. Lett.* **79**, 885 (1997).
72. P. L. Doussal and L. Radzihovsky, Anomalous elasticity, fluctuations and disorder in elastic membranes. *Ann. Phys.* **392**, 340–410 (2018).
73. M. Falcioni, M. Bowick, E. Guitter and G. Thorleifsson, The Poisson ratio of crystalline surfaces. *Europhys. Lett.* **38**, 67 (1997).
74. O. Coquand, D. Mouhanna and S. Teber, Flat phase of polymerized membranes at two-loop order. *Phys. Rev. E* **101**, 062104 (2020).
75. P. Le Doussal and L. Radzihovsky, *"Tattereed" membrane*, arXiv:2311.00752.
76. P. Y. Huang, *et al.*, Grains and grain boundaries in single-layer graphene atomic patchwork quilts. *Nature* **469**, 389 (2011).
77. D. R. Nelson and L. Radzihovsky, Polymerized membranes with quenched random internal disorder. *Europhys. Lett.*, **16**, 79 (1991).
78. L. Radzihovsky and D. R. Nelson, Statistical mechanics of randomly polymerized membranes. *Phys. Rev. A* **44**, 3525 (1991).
79. D. C. Morse and T. C. Lubensky, Curvature disorder in tethered membranes: A new flat phase at $T = 0$. *Phys. Rev. A* **46**, 1751 (1992). D. C. Morse, T. C. Lubensky and G. S. Grest, Quenched disorder in tethered membranes. *Phys. Rev. A* **45**, R2151 (1992).

80. Le Doussal and L. Radzihovsky, Flat glassy phases and wrinkling of polymerized membranes with long range disorder. *Phys. Rev. B* R**48**, 3548 (1993).
81. L. Golubovic and T. C. Lubensky, Nonlinear elasticity of amorphous solids. *Phys. Rev. Lett.* **63**, 1082 (1989).
82. M. Warner and E. M. Terentjev, Nematic elastomers — a new state of matter? *Prog. Polym. Sci.* **21**, 853(1996), and references therein; E. M. Terentjev, Liquid-crystalline elastomers. *J. Phys. Cond. Mat.* **11**, R239 (1999).
83. T. C. Lubensky, R. Mukhopadhyay, L. Radzihovsky and X. Xing, Symmetries and elasticity of nematic gels. *Phys. Rev. E* **66**, 011702 (2002).
84. O. Stenull and T. C. Lubensky, Phase transitions and soft elasticity of smectic elastomers. *Europhys. Lett.* **61**, 779 (2003).
85. P. L. Doussal and L. Radzihovsky, Thermal buckling transition of crystalline membranes in a field. *Phys. Rev. Lett.* **127**, 015702 (2021).

Finite Size Scaling, Monte Carlo

https://doi.org/10.1142/9789811282386_0031

Chapter 31

Multiparameter Universality and Intrinsic Diversity in Weakly Anisotropic Bulk and Confined Systems

V. Dohm

Institute for Theoretical Physics, RWTH Aachen University, 52056 Aachen, Germany

An overview of recent advances in the theory of critical phenomena in d-dimensional weakly anisotropic systems is given. On the basis of a generalized shear transformation between anisotropic and isotropic systems, exact and approximate results are discussed for bulk and confined systems in two and three dimensions, where conformal-field theory and the minimal renormalization without ε-expansion play a crucial role. Stimulation for this research comes from the seminal work by V. Privman and M.E. Fisher in 1984 in which the principle of two-scale-factor universality for bulk systems has been extended to finite systems. Based on this principle for isotropic systems, we predict the validity of multiparameter universality with up to $d(d+1)/2+1$ non-universal parameters in d-dimensional anisotropic bulk and confined systems with periodic boundary conditions (BC). The verification of multiparameter universality for confined anisotropic systems with realistic BC and the study of the intrinsic diversity of the critical behavior of magnetic materials, superconductors, liquid crystals, and solids with non-cubic symmetry are a major challenge to future research.

1. Introduction

The traditional theory of ordinary critical phenomena with short-range interactions[1–8] including conformal-field theory[9,10] as well as the more recent development of the functional renormalization group[11] have been primarily focused on the critical behavior in *spatially isotropic* systems, apart from anisotropic two-dimensional models.[12–24] Spatial anisotropy is a fundamental property that is omnipresent in condensed matter physics where it is the origin of a wide variety of non-universal effects. Substantial evidence has emerged over the last two decades[25–39] that part of this non-universal diversity persists even in the critical region as well as in the near-critical Goldstone regime of the so-called weakly anisotropic systems. This includes magnetic materials, superconductors, alloys, liquid crystals, compressible solids and solids with structural phase transitions. Recently, an unexpected complex form of self-similarity of anisotropy effects in finite weakly anisotropic systems with periodic boundary conditions (BC) has been found[38] near the instability where weak anisotropy breaks down.

Ordinary bulk critical phenomena can be divided into universality classes characterized by the dimension d and the symmetry properties of the ordered state.[1,5,7] As an example we consider $O(n)$-symmetric systems with short-range interactions and with an n-component order parameter. Within each (d, n) universality class, all systems have the same universal quantities (critical exponents, amplitude ratios, and scaling functions) which includes isotropic and weakly anisotropic systems since spatial anisotropy is only a marginal perturbation in the renormalization-group sense.[4,5,40–43] The principle of two-scale-factor universality (or hyperuniversality)[3–5,41,44–47] predicts that, once the universal quantities of a universality class are known, the asymptotic critical behavior of any particular system of this universality class is known completely provided that only two non-universal amplitudes are given. This principle was stated to be valid for all systems in a universality class.[5,48,49] Furthermore, it was asserted[6,9,43,50–54] that asymptotic isotropy can be restored in weakly anisotropic systems by a suitable anisotropic scale transformation and that universality can be restored,[52] reintroduced,[54] or repaired[5] in some cases. However, the question was left unanswered how the various non-universal effects due to non-cubic anisotropy in bulk[6,12–14,17,19–21,40,42] and confined[9,15,16,18,51–56] systems could be reconciled with the principle of two-scale-factor universality.

A systematic study of the effect of non-cubic spatial anisotropy on the critical behavior in bulk and confined systems was initiated[25] by introducing a non-diagonal anisotropy matrix $\mathbf{A}$ into the φ^4 field theory. An anisotropy-induced non-universality was discovered for the critical Casimir amplitude[57–59] and critical Binder cumulant ratio[5] which were the hallmarks of finite-size universality. These results demonstrated that restoring isotropy relates the critical behavior of anisotropic systems to that of isotropic systems but does not restore two-scale-factor universality. Shortly thereafter, the non-universality of the Binder cumulant ratio was confirmed by Monte Carlo simulations of an anisotropic Ising model.[27] The analytic results[25] were based on the exact large n limit and on the φ^4 theory in the minimal subtraction scheme in three dimensions.[60–62] The violation of two-scale-factor universality in weakly anisotropic bulk and confined systems was subsequently confirmed,[28–30,32–34,36,37] most recently by exact analytic results for the critical free energy and Casimir amplitude in two dimensions[38] based on conformal-field theory[10] as well as by Monte Carlo simulations.[39]

Weakly anisotropic systems do not have a unique single bulk correlation length but rather an angular-dependent correlation length[37] with d independent non-universal amplitudes in the d principal directions which are determined by $d(d-1)/2$ non-universal angles. This has led to the hypothesis[36] (first formulated[30] for bulk amplitude relations) that in weakly anisotropic bulk and confined systems, two-scale-factor universality is absent and is replaced by multiparameter universality which allows for up to $d(d+1)/2+1$ independent non-universal parameters. This implies a revised notion of a universality class: It must be divided into subclasses[30,31] of isotropic and weakly anisotropic systems with different scaling forms for the bulk correlation function and for the free energy of confined systems within the same universality class. These anisotropic scaling forms are governed by a reduced anisotropy matrix $\bar{\mathbf{A}}$ which has a universal structure in terms of principal correlation lengths and principal directions where the latter depend in a generically unknown way on the anisotropic interactions. So far, there exists no general proof for this hypothesis. In the following, we report on recent advances[36–38,63–65] of such a proof based on the validity of two-scale-factor universality in isotropic bulk[3,5,41,44] and confined[4,5] systems and on a generalized shear transformation[63] between anisotropic and isotropic systems. Among the $d(d+1)/2+1$ non-universal parameters of weakly anisotropic systems, there are only two parameters that can be determined by macroscopic thermodynamic measurements,[63] whereas the $d(d+1)/2-1$ independent parameters contained in $\bar{\mathbf{A}}$ cause a high degree of intrinsic diversity arising from the non-universal angular dependence of the critical correlations.

2. $O(n)$-Symmetric Models with Lattice Anisotropy

In order to discuss the issue of multiparameter universality, we analyze different types of $O(n)$-symmetric anisotropic models that belong to the same universality classes. As examples, we consider (i) the φ^4 model and (ii) the n-vector model.

(i) The φ^4 lattice Hamiltonian and the total free energy divided by k_BT are

$$H = v\left[\sum_{i=1}^{N}\left(\frac{r_0}{2}\varphi_i^2 + u_0(\varphi_i^2)^2\right) + \sum_{i,j=1}^{N}\frac{K_{i,j}}{2}(\varphi_i-\varphi_j)^2\right], \tag{1}$$

$$\mathcal{F}_{\text{tot}} = -\ln\left[\prod_{i=1}^{N}\frac{\int d^n\varphi_i}{v^{n(2-d)/(2d)}}\right]\exp(-H). \tag{2}$$

The variables $\varphi_i \equiv \varphi(\mathbf{x}_i)$ are n-component vectors on N lattice points $\mathbf{x}_i \equiv (x_{i1}, x_{i2}, \ldots, x_{id})$ of a d-dimensional Bravais lattice of volume $V = Nv$, where v is the volume of the primitive cell. We assume periodic BC. The large-distance anisotropy arising from the couplings $K_{i,j}$ is described by a dimensionless symmetric anisotropy matrix $\mathbf{A}$[25] with matrix elements[28]

$$A_{\alpha\beta} = N^{-1}\sum_{i,j=1}^{N}(x_{i\alpha}-x_{j\alpha})(x_{i\beta}-x_{j\beta})\,K_{i,j}. \tag{3}$$

The continuum version of this model in terms of the vector field $\varphi(\mathbf{x})$ has the Hamiltonian[25,30]

$$H_{\text{field}} = \int_V d^d x \left[\frac{r_0}{2}\varphi^2 + \sum_{\alpha,\beta=1}^{d} \frac{A_{\alpha\beta}}{2} \frac{\partial \varphi}{\partial x_\alpha} \frac{\partial \varphi}{\partial x_\beta} + u_0(\varphi^2)^2 \right]. \tag{4}$$

Weakly anisotropy systems have the same critical exponents as the isotropic system which requires[25] $\det \mathbf{A}(\{K_{i,j}\}) > 0$. We shall discuss the exact structure of the anisotropic bulk correlation function

$$G(\mathbf{x_i} - \mathbf{x_j}, t) = \lim_{V \to \infty} \left[< \varphi(\mathbf{x}_i) \cdot \varphi(\mathbf{x}_j) > - < \varphi >^2 \right] \tag{5}$$

in $d \geq 2$ dimensions. Near T_c, the decomposition

$$\mathcal{F}_{\text{tot}} = \mathcal{F}_s + \mathcal{F}_{ns} \tag{6}$$

into singular and non-singular parts is appropriate. We shall discuss the singular part $f_s = \mathcal{F}_s/V$ of the free energy density $f = \mathcal{F}_{\text{tot}}/V$ in a $d = 3$ block geometry[36] as well as the excess free energy

$$\mathcal{F}^{\text{ex}} = \mathcal{F}_{\text{tot}} - \mathcal{F}_b \tag{7}$$

and the critical free energy of the finite system at T_c

$$\mathcal{F}_c = \lim_{T \to T_c} \mathcal{F}^{\text{ex}} = \lim_{T \to T_c} \mathcal{F}_s \tag{8}$$

in rectangular and parallelogram geometries,[38,63,64] where $\mathcal{F}_b = V f_b$ is the bulk part of $\mathcal{F}_{\text{tot}}$ with $f_b = \lim_{V \to \infty} \mathcal{F}_{\text{tot}}/V$ being the bulk free-energy density.

(ii) The anisotropic $O(n)$-symmetric n-vector model is defined on the same lattice with the same BC as the φ^4 model. It has the Hamiltonian and total free energy (divided by $k_B T$)

$$H^{\text{sp}} = -\sum_{i,j} E_{i,j} S_i \cdot S_j, \tag{9}$$

$$\mathcal{F}^{\text{sp}}_{\text{tot}} = -\ln \left[\prod_{i=1}^{N} \int d^n S_i \right] \exp\left(-\beta H^{\text{sp}}\right), \tag{10}$$

with pair interactions $E_{i,j}$ and $\beta = 1/(k_B T)$. The continuous spin variables S_i are n-component vectors with a fixed length $S_i^2 = 1$. The definitions of $G^{\text{sp}}, \mathcal{F}^{\text{sp}}_s, \mathcal{F}^{\text{sp,ex}}$, and $\mathcal{F}^{\text{sp}}_c$ are analogous to (5)–(8). We shall discuss the exact structure of G^{sp} for $d \geq 2$ and exact results for $\mathcal{F}^{\text{sp,ex}}$ and $\mathcal{F}^{\text{sp}}_c$ in $d = 2$ dimensions.

For $n = 1, 2, 3, \infty$, the models (i) and (ii) belong to the Ising, XY, Heisenberg and spherical universality classes. In forthcoming papers,[63–65] it is shown that the hypothesis of multiparameter universality[36] is indeed valid for G, G^{sp} in $d \geq 2$

dimensions for general n and for $\mathcal{F}^{\rm ex}, \mathcal{F}^{\rm sp,ex}$ in the $(d = 2, n = 1)$ Ising universality class. Our strategy is to employ exact or approximate results for $G^{\rm iso}, G^{\rm sp,iso}$ and $\mathcal{F}^{\rm ex,iso}, \mathcal{F}^{\rm sp,ex,iso}$ of isotropic systems for which two-scale-factor universality[3–5,41] can be invoked and to derive the structure of $G, G^{\rm sp}$ and $\mathcal{F}^{\rm ex}, \mathcal{F}^{\rm sp,ex}$ of weakly anisotropic systems by means of non-universal inverse shear transformations.[63] An extension of this analysis to the Gaussian universality class is given in Refs. 63 and 64.

3. Isotropic Case: Two-Scale-Factor Universality

The couplings $K_{i,j}$ and $E_{i,j}$ can be chosen such that the bulk system has isotropic correlations in the large-distance scaling regime near T_c. The isotropic bulk correlation function has the established scaling form for general n above and at T_c $(+)$ and $n = 1$ below T_c $(-)$[4,37,63]

$$G^{\rm iso}(|\mathbf{x}|, t) = \frac{\Gamma_+^{\rm iso}(\xi_{0+}^{\rm iso})^{-2+\eta}}{|\mathbf{x}|^{d-2+\eta}}\, \Psi_\pm\left(\frac{|\mathbf{x}|}{\xi_\pm^{\rm iso}(t)}\right), \tag{11}$$

$$\xi_\pm^{\rm iso}(t) = \xi_{0\pm}^{\rm iso}\,|t|^{-\nu}, \quad t = (T - T_c)/T_c, \tag{12}$$

with the universal critical exponents η and ν, the universal scaling function $\Psi_\pm$, and the non-universal amplitudes $\Gamma_\pm^{\rm iso}$ and $\xi_{0\pm}^{\rm iso}$ of the susceptibility $\chi^{\rm iso}(t) = \Gamma_\pm^{\rm iso}|t|^{-\gamma}$ and of the correlation lengths, respectively. The latter are universally related as

$$\xi_{0+}^{\rm iso}/\xi_{0-}^{\rm iso} = X_\xi = \text{universal}. \tag{13}$$

The scaling function $\Psi_\pm$ depends on the universality class. It is known exactly for the $d = 2$ Ising universality class,[37] for the spherical universality class corresponding to the large n limit[63] and for the Gaussian universality class corresponding to the φ^4 Hamiltonians (1) and (4) with $u_0 = 0$.[63] We recall that the universal scaling regime is defined by the limit $|\mathbf{x}| \to \infty, \xi_\pm^{\rm iso}(t) \to \infty$ at fixed finite ratio $|\mathbf{x}|/\xi_\pm^{\rm iso}(t)$, whereas scaling and universality are not valid in the regime $|\mathbf{x}| \gg \tilde{a}, \xi_\pm^{\rm iso}(t) \gg \tilde{a}, |\mathbf{x}| \gg \xi_\pm^{\rm iso}(t)$ even arbitrarily close to T_c, where $\tilde{a}$ is some microscopic reference length.[30]

The singular part of the bulk free-energy density $f_{b,s,\pm}^{\rm iso}$ of isotropic systems has the asymptotic form[5,7,63]

$$f_{b,s,\pm}^{\rm iso}(t) = \begin{cases} A_\pm^{\rm iso}|t|^{d\nu} & \text{for } 2 < d < 4, \\ \dfrac{1}{2}A_\pm^{\rm iso}|t|^{2\nu}\ln|t| & \text{for } d = 2, \end{cases} \tag{14}$$

where the non-universal amplitudes $A_\pm^{\rm iso}$ have the universal ratio $A_+^{\rm iso}/A_-^{\rm iso}$. According to two-scale-factor universality for isotropic bulk systems,[3,5,41,44] these amplitudes are universally related to the correlation-length amplitude $\xi_{0+}^{\rm iso}$ through

$$\left(\xi_{0+}^{\rm iso}\right)^d A_+^{\rm iso} = Q_1, \tag{15}$$

with the universal constant Q_1. The amplitude $A_+^{\rm iso}$ enters the singular part of the specific heat $C_{b,s,+}^{\rm iso}(t) = -\partial^2 f_{b,s,+}^{\rm iso}(t)/\partial t^2$, thus the correlation length $\xi_{0+}^{\rm iso}$ is determined by the thermodynamic amplitude of the specific heat.[63] The important consequence of two-scale-factor universality for isotropic bulk systems is that all non-universal parameters of the isotropic bulk correlation function can be completely determined by the thermodynamic amplitudes of the bulk susceptibility and the bulk specific heat without involvement of measurements of the spatial dependence of the correlation function via scattering experiments. We shall show in Section 4 that weak anisotropy destroys this important feature due to the intrinsic non-universality of the anisotropic bulk correlation function.

The hypothesis of two-scale-factor universality for *finite* systems (with a large characteristic length L_0, a small ordering field h, and a small reduced temperature $|t|$) in a $L_1 \times L_2 \cdots \times L_d$ block geometry with the volume V has been formulated in terms of the singular part of the free-energy density in the Privman–Fisher scaling form as[4]

$$f_s(t, h, L_1, L_2, \ldots) = L_0^{-d}\, Y(C_1 t L_0^{1/\nu}, C_2 h L_0^{\Delta/\nu}; L_1/L_0, L_2/L_0, \ldots), \tag{16}$$

with universal critical exponents ν, Δ and, for given BC, with the universal scaling function Y. The shape ratios L_j/L_0 are supposed to approach finite constants as $L_0 \to \infty$. The two metric factors C_1 and C_2 are the only non-universal parameters entering (16). With the choice $L_0 = V^{1/d}$, we see that the Privman–Fisher scaling function Y has the meaning of the singular part of the free energy

$$\mathcal{F}_s = V f_s = Y(C_1 t V^{1/(d\nu)}, C_2 h V^{\Delta/(d\nu)}; L_1/V^{1/d}, L_2/V^{1/d}, \ldots). \tag{17}$$

In particular, we shall consider the critical free energy

$$\mathcal{F}_c = \lim_{t\to 0, h\to 0} \mathcal{F}_s = Y(0, 0; L_1/V^{1/d}, L_2/V^{1/d}, \ldots). \tag{18}$$

The constants C_1 and C_2 are universally related to the constants A_1 and A_2 of the singular part of the bulk free-energy density[4]

$$f_{b,s}(t, h) = \lim_{\{L_j\}\to\infty} f_s = A_1 |t|^{d\nu}\, W_{\pm}(A_2 h |t|^{-\Delta}), \tag{19}$$

which can be measured via the bulk specific heat and susceptibility. Due to the logarithmic factor in (14) for $d = 2$, it is necessary to decompose $f_s = f_{b,s} + f_s^{\mathrm{ex}}$ and to reformulate the scaling forms (16) and (17) for $d = 2$ only for f_s^{ex} and $\mathcal{F}_s^{\mathrm{ex}}$ (where we have omitted a possible surface contribution for nonperiodic BC).

The function Y has been asserted[5,48,49] to be the same for all systems in a given universality class in $d < 4$ dimensions. While the validity of the structure of (11)–(18) is well established for isotropic systems, the hypothesis of multiparameter universality[36] predicts weak anisotropy to cause significant changes of these structures. This is confirmed by recent advances[37–39,63–65] to be discussed in the following section.

4. Multiparameter Universality and Intrinsic Diversity of the Bulk Correlation Function

In this section, we present the exact scaling forms of the anisotropic bulk correlation functions of both the φ^4 and the n-vector models in $d \geq 2$ dimensions and show the validity of multiparameter universality.

4.1. *Anisotropic bulk correlation function of the φ^4 theory*

We first sketch the derivation of our result[36] for $G(\mathbf{x}, t)$ in the anisotropic φ^4 theory. It is possible to relate $G(\mathbf{x}, t)$ to the correlation function $G'(|\mathbf{x}'|, t)$ of the isotropic system. For this purpose, a shear transformation $\mathbf{x} \to \mathbf{x}'$ is performed by[25,63]

$$\mathbf{x}' = \boldsymbol{\lambda}^{-1/2}\mathbf{U}\mathbf{x}, \tag{20}$$

$$\varphi'(\mathbf{x}') = (\det \mathbf{A})^{1/4}\varphi(\mathbf{x}), \tag{21}$$

$$u_0' = (\det \mathbf{A})^{-1/2}u_0, \tag{22}$$

where $\mathbf{A}$ is given by (3). It consists of a rotation provided by the orthogonal matrix $\mathbf{U}$ and a subsequent spatial rescaling by the diagonal rescaling matrix $\boldsymbol{\lambda}$ such that the Hamiltonian (4) is brought into the isotropic form

$$H_{\text{field}} = H'_{\text{field}} = \int_{V'} d^d x' \left[\frac{r_0}{2}\varphi'(\mathbf{x}')^2 + \frac{1}{2}(\nabla'\varphi')^2 + u_0'(\varphi'^2)^2\right], \tag{23}$$

with the asymptotic bulk correlation function $G'(|\mathbf{x}'|, t)$ for $V' \to \infty$. The matrix elements of $\mathbf{U}(\{\mathbf{e}^{(\alpha)}\})$ are $U_{\alpha\beta} = e_\beta^{(\alpha)}$, which are determined by the d eigenvectors $\mathbf{e}^{(\alpha)}$ of the matrix $\mathbf{A}$, whose Cartesian components are denoted by $e_\beta^{(\alpha)}$. They satisfy the eigenvalue equation $\mathbf{A}\ \mathbf{e}^{(\alpha)} = \lambda_\alpha\ \mathbf{e}^{(\alpha)}$, $\alpha = 1, 2, \ldots, d$ with $\mathbf{e}^{(\alpha)} \cdot \mathbf{e}^{(\beta)} = \delta_{\alpha\beta}$. The matrix $\mathbf{U}$ diagonalizes the matrix $\mathbf{A}$ according to

$$\boldsymbol{\lambda} = \mathbf{U}\mathbf{A}\mathbf{U}^{-1}, \tag{24}$$

whose diagonal elements are the eigenvalues $\lambda_\alpha > 0$ of $\mathbf{A}$. The eigenvectors $\mathbf{e}^{(\alpha)}$ determine the directions of the principal axes of the large-distance correlations of the anisotropic system.[25,30,36] Due to the transformation of φ, (21), the anisotropic bulk correlation function (5) is transformed as

$$G'(|\mathbf{x}'|, t) = (\det \mathbf{A})^{1/2}G(\mathbf{x}, t). \tag{25}$$

The bulk susceptibility $\chi(t) = \Gamma_\pm|t|^{-\gamma} = \chi'(t) = \Gamma'_\pm|t|^{-\gamma}$ remains invariant, i.e., $\Gamma'_\pm = \Gamma_\pm$, as follows from the sum rules[30,63]:

$$\chi'(t) = \int d^d\mathbf{x}'\ G'(|\mathbf{x}'|, t) = \int d^d\mathbf{x}\ G(\mathbf{x}, t) = \chi(t). \tag{26}$$

From the isotropic scaling form (11), we have

$$G'(|\mathbf{x}'|,t) = \frac{\Gamma'_+(\xi'_{0+})^{-2+\eta}}{|\mathbf{x}'|^{d-2+\eta}} \Psi_\pm \left(\frac{|\mathbf{x}'|}{\xi'_\pm(t)}\right), \tag{27}$$

$$\xi'_\pm(t) = \xi'_{0\pm}\ |t|^{-\nu}, \tag{28}$$

where $\xi'_\pm(t)$ is the correlation length of the isotropic system (23). The amplitudes ξ'_{0+} and ξ'_{0-} are universally related by

$$\xi'_{0+}/\xi'_{0-} = X_\xi, \tag{29}$$

where X_ξ is the same universal constant as in (13). The principal correlation lengths of the anisotropic system above and below T_c have been determined as[25,30]

$$\xi^{(\alpha)}_\pm(t) = \xi^{(\alpha)}_{0\pm}|t|^{-\nu} = \lambda_\alpha{}^{1/2}\xi'_\pm(t). \tag{30}$$

As a consequence, their amplitude ratios are independent of the direction α and are universal quantities given by

$$\xi^{(\alpha)}_{0+}/\xi^{(\alpha)}_{0-} = \xi'_{0+}/\xi'_{0-} = X_\xi \quad \text{for each } \alpha, \tag{31}$$

where X_ξ is independent of α and the same universal constant as in (13).

It has been shown recently[36,63] that by inverting the shear transformation, one obtains from $G(\mathbf{x},t) = (\det \mathbf{A})^{-1/2}G'(|\mathbf{x}'|,t)$ the correlation function of the anisotropic φ^4 model for general $n \geq 1, T \geq T_c$ and for $n = 1, T < T_c$ in $2 \leq d < 4$ dimensions

$$G(\mathbf{x},t) = \frac{\Gamma_+(\bar\xi_{0+})^{-2+\eta}}{[\mathbf{x}\cdot\bar{\mathbf{A}}^{-1}\mathbf{x}]^{(d-2+\eta)/2}}\Psi_\pm\left(\frac{[\mathbf{x}\cdot\bar{\mathbf{A}}^{-1}\mathbf{x}]^{1/2}}{\bar\xi_\pm(t)}\right), \tag{32}$$

with the same universal scaling function $\Psi_\pm$ as for the isotropic system where

$$\bar\xi_\pm(t) = \left[\prod_{\alpha=1}^{d}\xi^{(\alpha)}_\pm(t)\right]^{1/d} = \bar\xi_{0\pm}|t|^{-\nu}, \quad \bar\xi_{0\pm} = \left[\prod_{\alpha=1}^{d}\xi^{(\alpha)}_{0\pm}\right]^{1/d} \tag{33}$$

is the mean correlation length and

$$\bar{\mathbf{A}} = \mathbf{A}/(\det\mathbf{A})^{1/d} = \mathbf{U}^{-1}\bar{\boldsymbol\lambda}\mathbf{U} \tag{34}$$

is the reduced anisotropy matrix with the diagonal reduced rescaling matrix

$$\bar{\boldsymbol\lambda} = \boldsymbol\lambda/(\det\boldsymbol\lambda)^{1/d}. \tag{35}$$

Its diagonal elements are expressed in terms of ratios of principal correlation lengths as

$$\bar\lambda_\alpha(\{\xi^{(\alpha)}_{0\pm}\}) = \prod_{\beta=1,\ \beta\neq\alpha}^{d}\left(\frac{\xi^{(\alpha)}_{0\pm}}{\xi^{(\beta)}_{0\pm}}\right)^{2/d} = \left(\frac{\xi^{(\alpha)}_{0\pm}}{\bar\xi_{0\pm}}\right)^2. \tag{36}$$

Here, the isotropic correlation length $\xi'_{0\pm}$ has been canceled. Thus, instead of expressing $\bar{\mathbf{A}}$ as a non-universal function of the couplings $K_{i,j}$, we obtain the structure

$$\bar{\mathbf{A}} = \bar{\mathbf{A}}\big(\{\xi_{0\pm}^{(\alpha)}, \mathbf{e}^{(\alpha)}\}\big) = \mathbf{U}(\{\mathbf{e}^{(\alpha)}\})^{-1}\bar{\boldsymbol{\lambda}}\big(\{\xi_{0\pm}^{(\alpha)}\}\big)\mathbf{U}(\{\mathbf{e}^{(\alpha)}\}) \tag{37}$$

in terms of ratios of principal correlation lengths. The appearance of the matrix $\bar{\mathbf{A}}$ in (32) violates two-scale-factor universality and introduces an intrinsic diversity of the correlation function.

4.2. *Anisotropic bulk correlation function of the n-vector model*

As a representative of weakly anisotropic systems other than the φ^4 model, we take the fixed-length n-vector Hamiltonian H^{sp}, (9), for general d and n. The special shear transformation (20)–(22) of the φ^4 theory is not applicable to the n-vector model (9) where it is unknown how to construct an appropriate continuum Hamiltonian with a large-distance anisotropy matrix as a function of the couplings $E_{i,j}$ that plays the same role as $\mathbf{A}$ in the anisotropic φ^4 theory. Instead, we have introduced[63] a generalized shear transformation that does not need an anisotropy matrix corresponding to $\mathbf{A}$ and that is applicable to all weakly anisotropic systems. Our only assumptions for the n-vector model (9) are (i) that there exist d principal unit vectors $\mathbf{e}^{\mathrm{sp}(\alpha)}$ in the directions of d principal axes together with d principal correlation lengths above and below T_c

$$\xi_{\pm}^{\mathrm{sp}(\alpha)}(t) = \xi_{0\pm}^{\mathrm{sp}(\alpha)}|t|^{-\nu} \tag{38}$$

along these directions where ν is the critical exponent of the isotropic n-vector model and (ii) that the amplitude ratios satisfy

$$\xi_{0+}^{\mathrm{sp}(\alpha)}/\xi_{0-}^{\mathrm{sp}(\alpha)} = X_\xi \quad \text{for each } \alpha, \tag{39}$$

where X_ξ is the same universal constant as in (13) for isotropic systems of the (d,n) universality class.

We introduce the orthogonal set of d *principal correlation vectors*

$$\boldsymbol{\xi}_{0\pm}^{\mathrm{sp}(\alpha)} = \xi_{0\pm}^{\mathrm{sp}(\alpha)}\mathbf{e}^{\mathrm{sp}(\alpha)}, \quad \alpha = 1, \dots, d. \tag{40}$$

We use a Cartesian coordinate system with orthogonal unit vectors $\boldsymbol{\epsilon}^{(\alpha)}$, $\alpha = 1, \dots, d$ along the d Cartesian axes. The shear transformation is achieved by a rotation by means of $\widehat{\mathbf{U}} = \mathbf{U}\big(\{\mathbf{e}^{\mathrm{sp}(\alpha)}\}\big)$ with matrix elements $\widehat{U}_{\alpha\beta} = e_\beta^{\mathrm{sp}(\alpha)}$, such that these vectors point along the direction of the Cartesian axes,

$$\widehat{\mathbf{U}}\mathbf{e}^{\mathrm{sp}(\alpha)} = \boldsymbol{\epsilon}^{(\alpha)}, \quad \widehat{\mathbf{U}}\boldsymbol{\xi}_{0\pm}^{\mathrm{sp}(\alpha)} = \xi_{0\pm}^{\mathrm{sp}(\alpha)}\boldsymbol{\epsilon}^{(\alpha)}, \tag{41}$$

and by a subsequent rescaling of their lengths by means of a diagonal matrix $\widehat{\boldsymbol{\lambda}}^{-1/2}$. This yields the d transformed correlation vectors

$$\widehat{\boldsymbol{\xi}}_{0\pm}^{(\alpha)} = \widehat{\boldsymbol{\lambda}}^{-1/2}\widehat{\mathbf{U}}\boldsymbol{\xi}_{0\pm}^{\mathrm{sp}(\alpha)} = \widehat{\lambda}_\alpha^{-1/2}\xi_{0\pm}^{\mathrm{sp}(\alpha)}\boldsymbol{\epsilon}^{(\alpha)}, \tag{42}$$

where $\widehat{\lambda}_\alpha$ are the diagonal elements of $\widehat{\boldsymbol{\lambda}}$. We note that the orthogonal matrix $\widehat{\mathbf{U}}$ contains $d(d-1)/2$ independent matrix elements which are needed to specify the directions $\mathbf{e}^{\mathrm{sp}(\alpha)}$ of the principal axes. These d mutually orthogonal axes can be described by $d(d-1)/2$ independent angles Ω_i, $i = 1, 2, \ldots, d(d-1)/2$, i.e., one angle Ω in two dimensions, three angles in three dimensions, etc.:

$$\mathbf{U}(\{\mathbf{e}^{\mathrm{sp}(\alpha)}\}) = \mathbf{U}(\Omega_1, \Omega_2, \Omega_3, \ldots). \tag{43}$$

This statement also holds for the φ^4 model. In order to obtain a system that is isotropic in the scaling region, it is necessary and sufficient that $\widehat{\lambda}_\alpha^{1/2}$ is proportional to $\xi_{0\pm}^{\mathrm{sp}(\alpha)}$. Thus, we may make the choice[63]

$$\widehat{\lambda}_\alpha^{1/2} = \xi_{0\pm}^{\mathrm{sp}(\alpha)}/\xi_{0\pm}^{\mathrm{iso}}, \tag{44}$$

where ξ_{0+}^{iso} and ξ_{0-}^{iso} are free parameters, together with the requirement[63] that

$$\xi_{0+}^{\mathrm{iso}}/\xi_{0-}^{\mathrm{iso}} = X_\xi = \text{universal}. \tag{45}$$

This requirement is necessary in order to comply with the fact implied by two-scale-factor universality that the ratio of the correlation lengths above and below T_c of any isotropic system near T_c must satisfy the universal relation (13). Equations (42)–(45) transform the d different principal correlation vectors (40) to the d vectors:

$$\widehat{\boldsymbol{\xi}}_{0\pm}^{(\alpha)} = \xi_{0\pm}^{\mathrm{iso}}\boldsymbol{\epsilon}^{(\alpha)}, \quad |\widehat{\boldsymbol{\xi}}_{0\pm}^{(\alpha)}| = \xi_{0\pm}^{\mathrm{iso}}, \quad \alpha = 1, \ldots, d, \tag{46}$$

with the angular-independent lengths ξ_{0+}^{iso} and ξ_{0-}^{iso} above and below T_c, respectively, thus representing an isotropic system.[63]

We apply this generalized shear transformation to the lattice vectors $\mathbf{x}$ of the anisotropic system

$$\widehat{\mathbf{x}} = \widehat{\boldsymbol{\lambda}}^{-1/2}\mathbf{U}\mathbf{x}. \tag{47}$$

This generates an n-vector model on a transformed lattice with lattice points $\widehat{\mathbf{x}}$ and with isotropic correlations characterized by the correlation length $\xi_{0\pm}^{\mathrm{iso}}$, for which two-scale-factor universality can be invoked. Our generalized shear transformation differs significantly from the special transformation (20)–(22) in that our transformation is a pure coordinate transformation without transforming the field φ and the coupling u_0. As a consequence, this transformation leaves the amplitude of the order-parameter correlation function G^{sp} invariant. Thus, the special transformation (25) is replaced by the invariance

$$G^{\mathrm{sp}}(\mathbf{x}, t) = \widehat{G}(|\widehat{\mathbf{x}}|, t) = \widehat{G}(|\widehat{\boldsymbol{\lambda}}^{-1/2}\mathbf{U}\mathbf{x}|, t), \tag{48}$$

where the transformed correlation function $\widehat{G}$ has the isotropic structure (11),

$$\widehat{G}(|\widehat{\mathbf{x}}|,t) = \frac{\widehat{\Gamma}_+(\xi_{0+}^{\rm iso})^{-2+\eta}}{|\widehat{\mathbf{x}}|^{d-2+\eta}}\,\Psi_\pm\left(\frac{|\widehat{\mathbf{x}}|}{\xi_\pm^{\rm iso}(t)}\right), \tag{49}$$

with $\widehat{\Gamma}_+ \equiv \Gamma_+^{\rm iso}$ and with the universal scaling function $\Psi_\pm$. The susceptibility $\chi^{\rm sp}(t) = \Gamma_\pm\,|t|^{-\gamma}$ of the anisotropic system, however, is not invariant under the generalized shear transformation but is transformed as[63]

$$\widehat{\chi}(t) = \int d^d\widehat{\mathbf{x}}\ \widehat{G}(\widehat{\mathbf{x}},t) = \widehat{\Gamma}_\pm\,|t|^{-\gamma} \tag{50}$$

$$= (\det\widehat{\boldsymbol{\lambda}})^{-1/2}\int d^d\mathbf{x}\ G^{\rm sp}(\mathbf{x},t) = \left(\xi_{0\pm}^{\rm iso}/\bar{\xi}_{0\pm}^{\rm sp}\right)^d\chi^{\rm sp}(t), \tag{51}$$

where $\bar{\xi}_{0\pm}^{\rm sp}$ is the amplitude of the mean correlation length $\bar{\xi}_\pm^{\rm sp}(t) = \bar{\xi}_{0\pm}^{\rm sp}|t|^{-\nu}$. This implies the amplitude relation $\widehat{\Gamma}_\pm = \left(\xi_{0\pm}^{\rm iso}/\bar{\xi}_{0\pm}^{\rm sp}\right)^d\,\Gamma_\pm^{\rm sp}$. It has been shown[63] that by inverting the shear transformation, one obtains from (48) the scaling form of the anisotropic bulk correlation function of the n-vector model for $2 \le d < 4$

$$G^{\rm sp}(\mathbf{x},t) = \frac{\Gamma_+^{\rm sp}\left(\bar{\xi}_{0+}^{\rm sp}\right)^{-2+\eta}}{\left[\mathbf{x}\cdot\left(\bar{\mathbf{A}}^{\rm sp}\right)^{-1}\mathbf{x}\right]^{(d-2+\eta)/2}}\Psi_\pm\left(\frac{\left[\mathbf{x}\cdot\left(\bar{\mathbf{A}}^{\rm sp}\right)^{-1}\mathbf{x}\right]^{1/2}}{\bar{\xi}_\pm^{\rm sp}(t)}\right), \tag{52}$$

with the same universal scaling function $\Psi_\pm$ as for isotropic systems in the (d,n) class. The reduced anisotropy matrix of the n-vector model is given by

$$\bar{\mathbf{A}}^{\rm sp} = \mathbf{U}^{-1}\bar{\boldsymbol{\lambda}}^{\rm sp}\mathbf{U} = \bar{\mathbf{A}}\left(\{\xi_{0\pm}^{{\rm sp}(\alpha)},\mathbf{e}^{{\rm sp}(\alpha)}\}\right) \tag{53}$$

$$= \mathbf{U}(\{\mathbf{e}^{{\rm sp}(\alpha)}\})^{-1}\bar{\boldsymbol{\lambda}}^{\rm sp}\left(\{\xi_{0\pm}^{{\rm sp}(\alpha)}\}\right)\mathbf{U}(\{\mathbf{e}^{{\rm sp}(\alpha)}\}), \tag{54}$$

with the diagonal reduced rescaling matrix

$$\bar{\boldsymbol{\lambda}}^{\rm sp} = \widehat{\boldsymbol{\lambda}}/\left(\det\widehat{\boldsymbol{\lambda}}\right)^{1/d}. \tag{55}$$

It has the diagonal elements

$$\bar{\lambda}_\alpha^{\rm sp} = \prod_{\beta=1,\,\beta\neq\alpha}^{d}\left(\frac{\xi_{0\pm}^{{\rm sp}(\alpha)}}{\xi_{0\pm}^{{\rm sp}(\beta)}}\right)^{2/d} = \left(\frac{\xi_{0\pm}^{{\rm sp}(\alpha)}}{\bar{\xi}_{0\pm}^{\rm sp}}\right)^2, \tag{56}$$

where the free parameter $\xi_{0\pm}^{\rm iso}$ has been canceled. Thus, $\bar{\mathbf{A}}^{\rm sp}$ has the same structure as the matrix $\bar{\mathbf{A}}$ in (37) for the φ^4 model, but with the arguments $\xi_{0\pm}^{(\alpha)}$ and $\mathbf{e}^{(\alpha)}$ replaced by $\xi_{0\pm}^{{\rm sp}(\alpha)}$ and $\mathbf{e}^{{\rm sp}(\alpha)}$. This can be extended to any weakly anisotropic system beyond the φ^4 theory without explicit knowledge of an anisotropy matrix

$\mathbf{A}$ as a function of the couplings. This proves the validity of multiparameter universality for the anisotropic bulk correlation function for general d and n with $d(d+1)/2+1$ non-universal parameters but a universal scaling function $\Psi_\pm$ and a universal structure of the reduced anisotropy matrix $\bar{\mathbf{A}}^{\rm sp}$. Here, we have not made any assumptions other than the validity of two-scale-factor universality for isotropic bulk systems and the existence of principal correlation lengths and principal axes for weakly anisotropic systems together with (39). Our derivation cannot, of course, make any prediction about the dependence of the non-universal parameters on the couplings $E_{i,j}$ of the n-vector model, unlike the dependence on the couplings $K_{i,j}$ within the φ^4 model.[36,37]

We have shown[63] that macroscopic measurements of the amplitude of specific heat of anisotropic systems can determine the mean correlation length $\bar{\xi}^{\rm sp}_{0+}$ appearing in the bulk correlation function (52). Together with macroscopic measurements of the amplitude $\Gamma^{\rm sp}_+$ of the susceptibility of anisotropic systems, one obtains the two amplitudes determining the overall amplitude of the bulk correlation function (52). By contrast, the $d(d+1)/2-1$ non-universal parameters contained in the reduced anisotropy matrix $\bar{\mathbf{A}}^{\rm sp}$ cannot be determined from macroscopic measurements but require scattering experiments that resolve the non-universal angular-dependent correlations. Thus, weak anisotropy destroys two-scale-factor universality and introduces a considerable intrinsic diversity of correlation functions in the asymptotic scaling regime with additional five non-universal parameters in three dimensions that are generically unknown[15,37,63] for the anisotropic n-vector model (9) and for real systems in most cases.

4.3. *Exact bulk correlation functions in two dimensions*

We apply our analysis to the anisotropic bulk correlation function of the ($d=2$, $n=1$) universality class with the critical exponents $\nu=1, \eta=1/4$. This includes both anisotropic scalar φ^4 models and anisotropic Ising models. In the isotropic case, two-scale-factor universality implies that all of these systems have the same universal structure of the correlation function given in (11)

$$G^{\rm iso}(|\mathbf{x}|,t) = \frac{\Gamma^{\rm iso}_+(\xi^{\rm iso}_{0+})^{-7/4}}{|\mathbf{x}|^{1/4}}\,\Psi_\pm\left(\frac{|\mathbf{x}|}{\xi^{\rm iso}_\pm(t)}\right), \tag{57}$$

$$\xi^{\rm iso}_\pm(t) = \xi^{\rm iso}_{0\pm}\,|t|^{-1}, \quad \xi^{\rm iso}_{0+}/\xi^{\rm iso}_{0-} = 2, \tag{58}$$

with two independent non-universal parameters $\Gamma^{\rm iso}_+, \xi^{\rm iso}_{0+}$. Exact correlation functions of the two-dimensional Ising model on several "isotropic lattices" are known for a long time,[12,20] but the universal part of the scaling function was left undetermined. To determine the universal scaling function $\Psi_\pm$, it suffices to consider the simplest non-trivial example. This is the standard Ising model with equal nearest-neighbor (NN) couplings on a square lattice.[12] Recently, we have identified[37] the universal scaling function $\Psi_\pm(y_\pm)$ of this isotropic Ising model.

In the anisotropic case, the two-dimensional φ^4 or Ising models have two additional non-universal parameters: the angle Ω or $\Omega^{\rm Is}$ describing the orientation of the two principal axes and the ratio $q = \xi_{0\pm}^{(1)}/\xi_{0\pm}^{(2)}$ or $q^{\rm Is} = \xi_{0\pm}^{\rm Is(1)}/\xi_{0\pm}^{\rm Is(2)}$ of the two principal correlation lengths. These parameters enter the rotation and reduced rescaling matrices

$$\mathbf{U}(\Omega) = \begin{pmatrix} \cos\,\Omega & \sin\,\Omega \\ -\sin\,\Omega & \cos\,\Omega \end{pmatrix}, \quad \bar{\boldsymbol{\lambda}}(q) = \begin{pmatrix} q & 0 \\ 0 & q^{-1} \end{pmatrix}, \tag{59}$$

which yield the reduced two-dimensional anisotropy matrix

$$\bar{\mathbf{A}}(q,\Omega) = \mathbf{U}(\Omega)^{-1}\bar{\boldsymbol{\lambda}}(q)\mathbf{U}(\Omega) = \begin{pmatrix} q\, c_\Omega^2 + q^{-1} s_\Omega^2 & (q - q^{-1})\, c_\Omega\, s_\Omega \\ (q - q^{-1})\, c_\Omega\, s_\Omega & q\, s_\Omega^2 + q^{-1}\, c_\Omega^2 \end{pmatrix} \tag{60}$$

of the anisotropic φ^4 models and the matrix

$$\bar{\mathbf{A}}^{\rm Is}(q^{\rm Is},\Omega^{\rm Is}) = \bar{\mathbf{A}}(q^{\rm Is},\Omega^{\rm Is}), \tag{61}$$

of the Ising models, respectively, with the abbreviations $c_\Omega \equiv \cos\Omega, s_\Omega \equiv \sin\Omega$. This leads to the exact anisotropic two-dimensional correlation functions

$$G(\mathbf{x},t) = \frac{\Gamma_+\,(\bar{\xi}_{0+})^{-7/4}}{\left[\mathbf{x}\cdot\left(\bar{\mathbf{A}}\right)^{-1}\mathbf{x}\right]^{1/8}}\,\Psi_\pm\left(\frac{[\mathbf{x}\cdot\left(\bar{\mathbf{A}}\right)^{-1}\mathbf{x}]^{1/2}}{\bar{\xi}_\pm(t)}\right) \tag{62}$$

and

$$G^{\rm Is}(\mathbf{x},t) = \frac{\Gamma_+^{\rm Is}\,(\bar{\xi}_{0+}^{\rm Is})^{-7/4}}{\left[\mathbf{x}\cdot\left(\bar{\mathbf{A}}^{\rm Is}\right)^{-1}\mathbf{x}\right]^{1/8}}\,\Psi_\pm\left(\frac{[\mathbf{x}\cdot\left(\bar{\mathbf{A}}^{\rm Is}\right)^{-1}\mathbf{x}]^{1/2}}{\bar{\xi}_\pm^{\rm Is}(t)}\right) \tag{63}$$

for the anisotropic φ^4 and Ising models, respectively. They satisfy the feature of multiparameter universality in that they have the same universal scaling function $\Psi_\pm$ and the same universal structure of the anisotropy matrix but depend on four independent non-universal parameters $\Gamma_+, \bar{\xi}_{0+}, q, \Omega$ and $\Gamma_+^{\rm Is}, \bar{\xi}_{0+}^{\rm Is}, q^{\rm Is}, \Omega^{\rm Is}$, respectively, thus violating two-scale-factor universality. We conclude that our result predicts the exact scaling form of the correlation function of all weakly anisotropic systems of the $(d = 2, n = 1)$ universality class.[63] This implies that the calculation of the correlation function of any two-dimensional weakly anisotropic system of the Ising universality class no longer requires a new calculation of a scaling function and of an anisotropy matrix but can be restricted to the much simpler task of determining four non-universal parameters, namely the amplitude of the susceptibility and the specific heat above T_c, the ratio of the two principal correlation lengths, and the angle Ω determining the orientation (angles Ω and $\Omega + \pi/2$) of the two principal axes. This constitutes a fundamental simplification in the analytic theory of two-dimensional anisotropic systems as well as in the analysis of numerical or experimental data of anisotropic correlation functions.

A rigorous confirmation for this conclusion comes from the analysis[37,63] of the exact correlation function $G^{\rm tr}(\mathbf{x},t)$[13] of the fully anisotropic triangular-lattice Ising model with the Hamiltonian

$$H^{\rm tr} = \sum_{j,k}[-E_1\sigma_{j,k}\sigma_{j,k+1} - E_2\sigma_{j,k}\sigma_{j+1,k} - E_3\sigma_{j,k}\sigma_{j+1,k+1}], \quad \sigma_{j,k}=\pm 1, \tag{64}$$

with horizontal, vertical, and diagonal couplings E_1, E_2, E_3 on a square lattice. The condition for a ferromagnetic critical point with weak anisotropy is

$$E_1+E_2>0, \quad E_1+E_3>0, \quad E_2+E_3>0. \tag{65}$$

The structure of $G^{\rm tr}(\mathbf{x},t)$ was found[37,63] to be in exact agreement with (61) and (63) and the exact non-universal parameters $\Gamma^{\rm tr}_+$, $\bar\xi^{\rm tr}_{0+}, q^{\rm tr}, \Omega^{\rm tr}$ were determined analytically[37,63,64] in this range. It would be worthwhile to verify the validity of multiparameter universality by analyzing the universality properties of the correlation functions of other two-dimensional anisotropic models.[17,19–21,23,66]

4.4. *Angular-dependent correlation vector*

We briefly discuss the notion of an angular-dependent correlation vector introduced recently[63] that is relevant to both bulk and confined systems with arbitrary boundary conditions. Since such vectors exist in all weakly anisotropic systems including φ^4 and Ising models, we suppress the superscript "Is" in the following. In a two-dimensional anisotropic system, it is of interest to consider a correlation length as a measure of the spatial range of the critical correlations in a certain spatial direction with an angle θ described by a unit vector

$$\mathbf{e}(\theta) = \begin{pmatrix}\cos\theta\\ \sin\theta\end{pmatrix}. \tag{66}$$

This can be done by introducing polar coordinates $\mathbf{x} = r\,\mathbf{e}(\theta)$ for the argument $\mathbf{x}$ of the anisotropic bulk correlation function $G(\mathbf{x},t)$. By rewriting the argument of the scaling function $\Psi_\pm$ in (63) as

$$\frac{[\mathbf{x}\cdot\bar{\mathbf{A}}(q,\Omega)^{-1}\mathbf{x}]^{1/2}}{\bar\xi_\pm(t)} = \frac{r}{\xi_\pm(t,\theta-\Omega,q)}, \tag{67}$$

we obtain from (60) the angular-dependent correlation length[37]

$$\xi_\pm(t,\theta-\Omega,q) = \xi_{0\pm}(\theta-\Omega,q)|t|^{-1}, \tag{68}$$

$$\xi_{0\pm}(\theta-\Omega,q) = \frac{\bar\xi_{0\pm}}{f(\theta-\Omega,q)}, \tag{69}$$

$$f(\theta-\Omega,q) = [q\sin^2(\theta-\Omega) + q^{-1}\cos^2(\theta-\Omega)]^{1/2}. \tag{70}$$

This yields the representation of the correlation function in polar coordinates

$$G(\mathbf{x},t) = \frac{\Gamma_+}{\left(\bar{\xi}_{0+}\right)^2}\left(\frac{\xi_{0\pm}(\theta-\Omega,q)}{r}\right)^{1/4}\Psi_\pm\left(\frac{r}{\xi_\pm(t,\theta-\Omega,q)}\right). \tag{71}$$

The usefulness of the correlation length $\xi_{0\pm}(\theta-\Omega,q)$ is its property of having two extrema[37] with respect to θ at the angles

$$\theta^{(1)} = \Omega, \quad \theta^{(2)} = \Omega + \pi/2, \tag{72}$$

which determine the principal directions, i.e.,

$$\xi_{0\pm}(0,q) = \bar{\xi}_{0\pm}q^{1/2} = \xi_{0\pm}^{(1)}, \tag{73}$$

$$\xi_{0\pm}(\pi/2,q) = \bar{\xi}_{0\pm}q^{-1/2} = \xi_{0\pm}^{(2)}. \tag{74}$$

Here, we extend the definition of the amplitude $\xi_{0\pm}(\theta-\Omega,q)$ to a definition of an *angular-dependent correlation vector*

$$\boldsymbol{\xi}_\pm(t,\theta,\Omega,q) = \boldsymbol{\xi}_{0\pm}(\theta,\Omega,q)|t|^{-\nu}, \tag{75}$$

$$\boldsymbol{\xi}_{0\pm}(\theta,\Omega,q) = \xi_{0\pm}(\theta-\Omega,q)\,\mathbf{e}(\theta), \tag{76}$$

which is oriented in the direction of the angle θ. This is a generalization of the principal correlation vectors (40) which are obtained from (69)–(76) for $\theta=\theta^{(1)}$ and $\theta=\theta^{(2)}$. It can be verified that the generalized shear transformation of Section 4.2 indeed transforms the angular-dependent correlation vector (76) to a vector with a rescaled *isotropic* length $\xi_{0\pm}^{\mathrm{iso}}$,

$$\widehat{\boldsymbol{\lambda}}^{-1/2}\mathbf{U}(\Omega)\,\boldsymbol{\xi}_{0\pm}(\theta,\Omega,q) = \xi_{0\pm}^{\mathrm{iso}}\,\mathbf{e}(\theta-\Omega,q), \tag{77}$$

$$\mathbf{e}(\theta-\Omega,q) = \frac{1}{\left[1+q^2\tan^2(\theta-\Omega)\right]^{1/2}}\begin{pmatrix}1\\ q\tan(\theta-\Omega)\end{pmatrix}, \quad |\mathbf{e}(\theta-\Omega,q)|=1, \tag{78}$$

with an orientation that depends on the relative angle $\theta-\Omega$.

The same angular-dependent representation can be employed in the shear transformation applied to any lattice point $\mathbf{x}$ of the anisotropic system with the length $r=|\mathbf{x}|$. For the generalized shear transformation (47), this yields the transformed vector

$$\widehat{\mathbf{x}} = \widehat{\boldsymbol{\lambda}}^{-1/2}\,\mathbf{U}(\Omega)\,\mathbf{x} = \frac{\xi_{0\pm}^{\mathrm{iso}}}{\xi_{0\pm}(\theta-\Omega,q)}\,|\mathbf{x}|\,\mathbf{e}(\theta-\Omega,q), \tag{79}$$

where the factor $\xi_{0\pm}^{\mathrm{iso}}/\xi_{0\pm}(\theta-\Omega,q)$ describes the amount of rescaling of the length. For the special shear transformation (20) of the φ^4 model, the corresponding representation of the transformed vector $\mathbf{x}'$ reads

$$\mathbf{x}' = \boldsymbol{\lambda}^{-1/2}\,\mathbf{U}(\Omega)\,\mathbf{x} = \frac{\xi'_{0\pm}}{\xi_{0\pm}(\theta-\Omega,q)}\,|\mathbf{x}|\,\mathbf{e}(\theta-\Omega,q). \tag{80}$$

The notion of an angular-dependent representation of correlation vectors and lattice points is applicable to both bulk and confined systems. This is of relevance to the application of the shear transformation to the boundaries of finite systems discussed in Section 5 and in Refs. [63–65]. The discussion given above is also applicable to the Gaussian model.[63]

5. Multiparameter Universality in Finite Anisotropic $d = 2$ Systems

In this section, we report on recent advances[38,39,63–65] in finite-size theory of weakly anisotropic systems with periodic BC in two dimensions.

5.1. *Exact critical free energy and Casimir amplitude on a rectangle*

Recently,[38] exact results have been derived for the critical free energy $\mathcal{F}_c$ and the ensuing critical Casimir amplitude X_c of the anisotropic φ^4 theory at T_c on a finite rectangle with periodic BC. These results were based on (i) the special shear transformation (20)–(22) of the anisotropic φ^4 model on a rectangle to an isotropic φ^4 model on a parallelogram, (ii) the exact critical free energy $\mathcal{F}_c^{\rm CFT}$ at T_c of conformal-field theory[10] for the isotropic Ising model on a parallelogram, and (iii) the universality of $\mathcal{F}_c^{\rm CFT}$ according to the Privman–Fisher hypothesis.[4] Corresponding predictions were presented[38] for the finite anisotropic triangular-lattice Ising model[13,37] (64) on the basis of the assumption that multiparameter universality[36] is valid for this model, but no proof was given. A quantitative test was performed[39] by high-precision Monte Carlo simulations for a special Ising model on a square with a diagonal anisotropic coupling and remarkable agreement was found with the predicted[38] amplitude $\mathcal{F}_c^{\rm Is}$. Here, we report on an analytic proof[63] for the validity of multiparameter universality for weakly anisotropic Ising models with the general Hamiltonian

$$H^{\rm Is} = -\sum_{i,j} E_{i,j}\sigma_i\sigma_j, \quad \sigma_i = \pm 1. \tag{81}$$

The strategy of this proof is (a) to apply of the generalized shear transformation (47) to the boundaries of the anisotropic Ising model on a rectangle which is transformed to an isotropic Ising model on a parallelogram, (b) to exploit the invariance of the critical free energy under this shear transformation,[63] (c) to employ the exact result of conformal-field theory[10] for the critical free energy of the isotropic system, and (d) to invert the shear transformation by substituting the transformation formulae.

We assume a finite $L_\parallel \times L_\perp$ rectangle spanned by the vectors $\mathbf{L}_\parallel = L_\parallel\,(1,0)$ and $\mathbf{L}_\perp = L_\perp\,(0,1)$ in the horizontal and vertical directions with the aspect ratio

$$\rho_{\rm rec} = L_\perp/L_\parallel. \tag{82}$$

We apply our generalized shear transformation (47) directly to the anisotropic Ising model on this rectangle in order to obtain an isotropic Ising model on a parallelogram. The transformation applied to the boundaries of the rectangle reads

$$\mathbf{L}_{p\|} = \widehat{\boldsymbol{\lambda}}^{-1/2}\mathbf{U}(\Omega^{\mathrm{Is}})\mathbf{L}_{\|}, \quad \mathbf{L}_{p\perp} = \widehat{\boldsymbol{\lambda}}^{-1/2}\mathbf{U}(\Omega^{\mathrm{Is}})\mathbf{L}_{\perp}, \tag{83}$$

where $\mathbf{U}(\Omega)$ is given by (59) and where $\widehat{\boldsymbol{\lambda}}$ has the diagonal elements $\lambda_1^{\mathrm{Is}} = [\xi_{0\pm}^{\mathrm{Is}(1)}/\xi_{0\pm}^{\mathrm{iso}}]^2, \lambda_2^{\mathrm{Is}} = [\xi_{0\pm}^{\mathrm{Is}(2)}/\xi_{0\pm}^{\mathrm{iso}}]^2$. This generates an isotropic Ising model with the bulk correlation-length amplitude $\xi_{0\pm}^{\mathrm{iso}}$ on a parallelogram spanned by the vectors $\mathbf{L}_{p\|}$ and $\mathbf{L}_{p\perp}$ with the transformed aspect ratio

$$\rho_{\mathrm{p}} = |\mathbf{L}_{p\perp}|/|\mathbf{L}_{p\|}|. \tag{84}$$

The angle α between the vectors $\mathbf{L}_{p\perp}$ and $\mathbf{L}_{p\|}$ is determined by

$$\cos\alpha = \frac{\mathbf{L}_{p\perp}\cdot\mathbf{L}_{p\|}}{|\mathbf{L}_{p\perp}||\mathbf{L}_{p\|}|}. \tag{85}$$

This yields the transformed aspect ratio[63]

$$\rho_{\mathrm{p}}(\rho_{\mathrm{rec}}, q^{\mathrm{Is}}, \Omega^{\mathrm{Is}}) = \rho_{\mathrm{rec}}\left[\frac{\tan^2\Omega^{\mathrm{Is}} + (q^{\mathrm{Is}})^2}{1+(q^{\mathrm{Is}})^2\tan^2\Omega^{\mathrm{Is}}}\right]^{1/2} = \rho_{\mathrm{rec}}\left[\frac{\bar{\mathbf{A}}(q^{\mathrm{Is}},\Omega^{\mathrm{Is}})_{11}}{\bar{\mathbf{A}}(q^{\mathrm{Is}},\Omega^{\mathrm{Is}})_{22}}\right]^{1/2} \tag{86}$$

and the angle α determined by

$$\cot\alpha(q^{\mathrm{Is}},\Omega^{\mathrm{Is}}) = [(q^{\mathrm{Is}})^{-1} - q^{\mathrm{Is}}]\cos\Omega^{\mathrm{Is}}\sin\Omega^{\mathrm{Is}} = -\bar{\mathbf{A}}(q^{\mathrm{Is}},\Omega^{\mathrm{Is}})_{12}. \tag{87}$$

No specific properties of the weakly anisotropic Ising model were needed in the derivation of (86) and (87), thus these relations have a universal structure. In Eqs. (7) and (8) of Ref. 38, the same formulae as (86) and (87), but with $q^{\mathrm{Is}}, \Omega^{\mathrm{Is}}$ replaced by q, Ω, were first obtained for the φ^4 model. These formulae were then adopted[38] for the Ising model by the substitution $q \to q^{\mathrm{Is}}, \Omega \to \Omega^{\mathrm{Is}}$ on the basis of the hypothesis of multiparameter universality. Here, we have provided an analytic justification for this substitution directly within the Ising model, without any assumption and without recourse to the φ^4 model, through our generalized shear transformation (47) and (83) for the Ising model. Thus, this transformation between an anisotropic and an isotropic Ising model specifies what was called "effective shear transformation" of the Ising model in Fig. 1 of Ref. 38.

The critical free energy of the transformed isotropic Ising model on the parallelogram is denoted by $\mathcal{F}_c^{\mathrm{Is,iso}}(\alpha, \rho_{\mathrm{p}})$ which, according to the Privman–Fisher hypothesis,[4] is a universal function of the geometric parameters α and ρ_{p}. The generalized shear transformation is a smooth coordinate transformation which leaves the singular part of the free energy invariant.[63] Denoting the critical free energy of the anisotropic Ising model on the rectangle by $\mathcal{F}_c^{\rm Is}$, we have the exact invariance relation

$$\mathcal{F}_c^{\rm Is} = \mathcal{F}_c^{\rm Is,iso}(\alpha, \rho_{\rm p}). \tag{88}$$

It is well known[10] that a parallelogram with periodic BC is topologically equivalent to a torus. As pointed out recently,[38] the exact result for $\mathcal{F}_c^{\rm Is,iso}(\alpha, \rho_{\rm p})$ can be taken directly from the known critical free energy $\mathcal{F}_c^{\rm CFT}(\tau)$ of conformal-field theory for the isotropic Ising model on a torus[10]:

$$\mathcal{F}_c^{\rm Is,iso}(\alpha, \rho_{\rm p}) = \mathcal{F}_c^{\rm CFT}(\tau) = -\ln Z^{\rm CFT}(\tau), \tag{89}$$

$$\tau(\alpha, \rho_{\rm p}) = {\rm Re}\,\tau + i\,{\rm Im}\,\tau = \tau_0 + i\,\tau_1 = \rho_{\rm p}\exp(i\,\alpha), \tag{90}$$

which is characterized by the complex torus modular parameter $\tau(\alpha, \rho_{\rm p})$. The τ-dependence of $\mathcal{F}_c^{\rm CFT}(\tau)$ is universal and is expressed in terms of Jacobi theta functions[10,38] $\theta_i(0|\tau) \equiv \theta_i(\tau)$ as

$$Z^{\rm CFT}(\tau) = \big(|\theta_2(\tau)| + |\theta_3(\tau)| + |\theta_4(\tau)|\big)/\big(2|\eta(\tau)|\big), \tag{91}$$

with $\eta(\tau) = \left[\frac{1}{2}\theta_2(\tau)\theta_3(\tau)\theta_4(\tau)\right]^{1/3}$. The crucial step is to transfer this exact information from the isotropic Ising model to the anisotropic Ising model by means of inverting the generalized shear transformation. This is achieved by defining the $(q^{\rm Is}, \Omega^{\rm Is})$-dependent torus parameter

$$\tau(\rho_{\rm rec}, q^{\rm Is}, \Omega^{\rm Is}) = \tau\big(\alpha(q^{\rm Is}, \Omega^{\rm Is}), \rho_{\rm p}(\rho_{\rm rec}, q^{\rm Is}, \Omega^{\rm Is})\big), \tag{92}$$

with $\alpha(q^{\rm Is}, \Omega^{\rm Is})$ and $\rho_{\rm p}(\rho_{\rm rec}, q^{\rm Is}, \Omega^{\rm Is})$ given by (87) and (86) and by substituting (92) into the isotropic formula (89). Using (88), we then obtain the exact result for the critical free energy and the Casimir amplitude $X_c^{\rm Is}$ of the anisotropic Ising model on the rectangle as

$$\mathcal{F}_c^{\rm Is}(\rho_{\rm rec}, q^{\rm Is}, \Omega^{\rm Is}) = -\ln Z^{\rm CFT}\big(\tau(\rho_{\rm rec}, q^{\rm Is}, \Omega^{\rm Is})\big), \tag{93}$$

$$X_c^{\rm Is}(\rho_{\rm rec}, q^{\rm Is}, \Omega^{\rm Is}) = -\rho_{\rm rec}^2\,\partial\mathcal{F}_c^{\rm Is}(\rho_{\rm rec}, q^{\rm Is}, \Omega^{\rm Is})/\partial\rho_{\rm rec}, \tag{94}$$

with two non-universal parameters $q^{\rm Is}, \Omega^{\rm Is}$. Going from (89) to (93) is equivalent to performing a non-universal inverse shear transformation from the isotropic to the anisotropic system. The universal exact structure of (93) and (94) holds for all weakly anisotropic systems in the $(d = 2, n = 1)$ Ising universality class and confirms the predictions[38] obtained on the basis of the hypothesis of multiparameter universality. This completes our proof.

We add the following comments. The difference between (89) and (93) is that the former relation contains a universal function of the geometric variables α and $\rho_{\rm p}$ in agreement with two-scale-factor universality,[4] whereas (93) contains a universal function reflecting multiparameter universality with the two non-universal anisotropy parameters $q^{\rm Is}$ and $\Omega^{\rm Is}$. The same structure was derived[38] for the φ^4 model without any assumption other than two-scale-factor universality for isotropic systems. This establishes the universality of the complex self-similar structures of the critical free energy and the Casimir amplitude discovered[38] for all weakly anisotropic systems with periodic BC in this universality class.

As noted already in the context of the anisotropic bulk correlation function, there exists an intrinsic diversity in finite weakly anisotropic systems as compared to isotropic systems: It arises from the non-universal anisotropy parameters q, Ω and $q^{\rm Is}, \Omega^{\rm Is}$ that do not exist in isotropic systems. Furthermore, there is a basic difference between $q(\{K_{i,j}\})$ and $\Omega(\{K_{i,j}\})$ of the φ^4 model (which are known exactly as functions of $K_{i,j}$[37]) and $q^{\rm Is}(\{E_{i,j}\})$ and $\Omega^{\rm Is}(\{E_{i,j}\})$ which are generically unknown[15,37] for the general Ising model (81). We conclude that weak anisotropy destroys the universality of the critical Casimir amplitude X_c of isotropic systems and makes X_c to become an unknown quantity for general anisotropic systems whose principal angles and correlation lengths are unknown. The triangular-lattice Ising model[13,37] (64) is a very rare example for a system other than the φ^4 model for which these parameters are known exactly as a function of the microscopic couplings.[37,63] It would be worthwhile to extend this knowledge to other anisotropic Ising models[17,19–21] whose correlation functions have been calculated but whose anisotropy parameters $q^{\rm Is}$ and $\Omega^{\rm Is}$ are as yet unknown.

5.2. *Exact excess free energy on a rectangle away from T_c*

A complete understanding of finite-size properties of weakly anisotropic two-dimensional systems requires to extend the theory to the whole scaling region above and below T_c. Exact results for the excess free energy in an anisotropic rectangular geometry away from T_c are available for the Gaussian model[30,36] and the Ising model[18] where in the latter case the result was presented in a non-universal form. A more general analysis of the universality properties away from T_c has been achieved in forthcoming papers.[64,65] Here, we report on some of the results of these papers.

We consider the anisotropic Ising model (81) on a square lattice at $T \neq T_c$ with the shape of the rectangle with the area

$$V_{\rm rec} = L_{\|} L_{\perp}. \tag{95}$$

The exact excess free energy of this model is determined via the same transformation formulae (86) and (87) as in the case $T = T_c$, but in addition one needs a t-dependent scaling variable. For $T \neq T_c$, it is the transformation of the area $V_{\rm rec}$ of the rectangle to the area

$$V^{\rm iso} \equiv V_{\rm p} = |\mathbf{L}_{p\|}|\, |\mathbf{L}_{p\perp}| \, \sin\alpha \tag{96}$$

of the isotropic parallelogram that will enter the t-dependent scaling variable, as we shall see in (107). The generalized shear transformation (47) yields the relation between these areas[63]:

$$V^{\rm iso} = (\det \widehat{\boldsymbol{\lambda}})^{-1/2}\, V_{\rm rec} = \big(\xi^{\rm iso}_{0\pm}/\bar{\xi}^{\rm Is}_{0\pm}\big)^2\, V_{\rm rec}, \tag{97}$$

where we have used (44). This implies the invariance of the area ratios

$$\frac{V^{\rm iso}}{\big(\xi^{\rm iso}_{0+}\big)^2} = \frac{V_{\rm rec}}{\big(\bar{\xi}^{\rm Is}_{0+}\big)^2}, \quad \frac{V^{\rm iso}}{\big(\xi^{\rm iso}_{0-}\big)^2} = \frac{V_{\rm rec}}{\big(\bar{\xi}^{\rm Is}_{0-}\big)^2}, \tag{98}$$

where $\big(\xi^{\rm iso}_{0\pm}\big)^2 = V^{\rm iso}_{{\rm cor},\pm}$ and $\big(\bar{\xi}^{\rm Is}_{0\pm}\big)^2 = \xi^{\rm Is(1)}_{0\pm}\xi^{\rm Is(2)}_{0\pm} = V_{\rm cor\pm}$ are the isotropic (spherical) and anisotropic (ellipsoidal) correlation areas above and below T_c, respectively.

For the transformed isotropic system, the Privman–Fisher scaling form (16) for the excess free-energy density $f^{\rm ex,iso} = \mathcal{F}^{\rm ex,iso}/V^{\rm iso}$ at $h = 0$ reads

$$f^{\rm ex,iso} = L_0^{-2}\, F^{\rm ex,iso}\big(C_1 t L_0, \alpha, \rho_{\rm p}\big), \tag{99}$$

where $F^{\rm ex,iso}$ is a universal function. Here, we have set $\nu = 1$ for the $d = 2$ Ising universality class. The metric factor C_1 and the characteristic lengths L_0 can be chosen as

$$C_1 = (\xi^{\rm iso}_{0+})^{-1}, \quad L_0 = (V^{\rm iso})^{1/2}. \tag{100}$$

Then we obtain

$$f^{\rm ex,iso} = (V^{\rm iso})^{-1}\, F^{\rm ex,iso}\big(\widetilde{x}(t), \alpha, \rho_{\rm p}\big), \tag{101}$$

with the scaling variable

$$\widetilde{x}(t) = t\big[V^{\rm iso}/(\xi^{\rm iso}_{0+})^2\big]^{1/2}. \tag{102}$$

This yields the excess free energy of the isotropic parallelogram

$$\mathcal{F}^{\rm ex,iso} = V^{\rm iso}\, f^{\rm ex,iso} = F^{\rm ex,iso}\big(\widetilde{x}(t), \alpha, \rho_{\rm p}\big). \tag{103}$$

As discussed above for the case $T = T_c$, the shape dependence can be parameterized in terms of the complex torus modular parameter $\tau(\alpha, \rho_{\rm p})$, (90), as given in (89),

$$F^{\rm ex,iso}(0, \alpha, \rho_{\rm p}) = \mathcal{F}_c^{\rm CFT}\big(\tau(\alpha, \rho_{\rm p})\big). \tag{104}$$

It is suggestive to conjecture that a description of the shape dependence in terms of τ is more generally valid for $\widetilde{x}(t) \neq 0$ such that the excess free energy (103) of isotropic systems on a parallelogram can be written in the scaling form

$$\mathcal{F}^{\rm ex,iso} = \mathcal{F}^{\rm ex,iso}(\widetilde{x}(t), \tau) \tag{105}$$

in the whole asymptotic scaling region above and below T_c. This is indeed the case as will be substantiated in the following. We denote the excess free energy of the anisotropic Ising model (81) on the rectangle by $\mathcal{F}^{\rm Is,ex}$. Since the shear transformation does not change the excess free energy,[63] we have the invariance relation

$$\mathcal{F}^{\mathrm{Is,ex}} = \mathcal{F}^{\mathrm{ex,iso}}(\widetilde{x}(t), \tau), \tag{106}$$

where $\widetilde{x}(t)$ and τ have to be expressed in terms of the parameters $q^{\mathrm{Is}}, \Omega^{\mathrm{Is}}, \rho_{\mathrm{rec}}$, $V_{\mathrm{rec}}/\left(\bar{\xi}_{0+}^{\mathrm{Is}}\right)^2$ of the anisotropic system. For τ, this has already been done in (92) by substituting the transformation formulae (86) and (87). For the scaling variable $\widetilde{x}(t)$, this is achieved by observing its invariance under the shear transformation, namely

$$\widetilde{x}(t) = t\left[V^{\mathrm{iso}}/(\xi_{0+}^{\mathrm{iso}})^2\right]^{1/2} = t\left[V_{\mathrm{rec}}/(\bar{\xi}_{0+}^{\mathrm{Is}})^2\right]^{1/2} \tag{107}$$

as follows from (102) and (98). Thus, $\widetilde{x}(t)$ can be written in two ways and can be used as a scaling variable for both the isotropic and anisotropic systems. An analogous result was obtained in three dimensions in Eq. (6.12) of Ref. 36.

Finally, it is necessary to determine the function $\mathcal{F}^{\mathrm{ex,iso}}(\widetilde{x}, \tau)$. According to the Privman–Fisher hypothesis[4] of two-scale-factor universality, this function is the same for all (isotropic) members of the two-dimensional Ising universality class. Thus, it suffices to calculate it for one isotropic system of this universality class which has been performed recently[65] for the triangular-lattice Ising model (64). The exact universal result for isotropic systems in the Ising universality class on a parallelogram in the entire scaling region above and below T_c reads[64,65]

$$\mathcal{F}^{\mathrm{ex,iso}}(\widetilde{x}, \tau) = -\ln\left\{\frac{1}{2}\left[e^{\mathcal{G}(\widetilde{x}^2,\tau,0,1/2)} + e^{\mathcal{G}(\widetilde{x}^2,\tau,1/2,0)} + e^{\mathcal{G}(\widetilde{x}^2,\tau,1/2,1/2)} - \frac{\widetilde{x}}{|\widetilde{x}|}e^{\mathcal{G}(\widetilde{x}^2,\tau,0,0)}\right]\right\}, \tag{108}$$

with the function

$$\mathcal{G}(\widetilde{x}^2, \tau, \Delta_1, \Delta_2) = -\sqrt{\frac{\widetilde{x}^2\tau_1}{8\pi^2}} \times \sum_{\substack{m,n=-\infty \\ (m,n)\neq(0,0)}}^{\infty} \frac{K_1\left(|m+n\tau|\sqrt{\frac{\widetilde{x}^2}{2\tau_1}}\right)}{|m+n\tau|} e^{-2\pi i(m\Delta_1 + n\Delta_2)}, \tag{109}$$

with $\tau_1 = \mathrm{Im}\, \tau > 0$, where

$$K_\mu(y) = \frac{1}{2}\left(\frac{y}{2}\right)^\mu \int_0^\infty \frac{dz}{z^{\mu+1}} \exp\left(-z - \frac{y^2}{4z}\right) \tag{110}$$

is the modified Bessel function of the second kind. The function (109) can be derived[64] from the Gaussian scaling function in Eqs. (A1) and (6.7) of Ref. 36 for the case of periodic boundary conditions ($\Delta_i = 0$) and from its extension to antiperiodic boundary conditions ($\Delta_i = 1/2$).

The only non-universal parameter of this result is the amplitude of the isotropic correlation length above T_c contained in the isotropic scaling variable $\widetilde{x}$, (102), with $\nu = 1$. This exact information on the isotropic system can be transferred to the anisotropic system by inverting the shear transformation, i.e., by substituting the transformation formulae (86), (87) and (107) into the invariance relation (106). Then we arrive at the exact universal scaling form of the excess free energy of the anisotropic Ising model (64) on the rectangle above, at, and below T_c[64]:

$$\mathcal{F}^{\mathrm{Is,ex}}(\rho_{\mathrm{rec}}, q^{\mathrm{Is}}, \Omega^{\mathrm{Is}}, tV_{\mathrm{rec}}^{1/2}/\bar{\xi}_{0+}^{\mathrm{Is}}) = \mathcal{F}^{\mathrm{ex,iso}}\big(tV_{\mathrm{rec}}^{1/2}/\bar{\xi}_{0+}^{\mathrm{Is}}, \tau(\rho_{\mathrm{rec}}, q^{\mathrm{Is}}, \Omega^{\mathrm{Is}})\big), \quad (111)$$

with three non-universal parameters $q^{\mathrm{Is}}, \Omega^{\mathrm{Is}}, \bar{\xi}_{0+}^{\mathrm{Is}}$, whereas two-scale-factor universality would allow only one non-universal parameter C_1 (at $h = 0$). No specific properties of the Ising model have been used in the derivation of (111). Thus, this result is universally valid for all systems in the $d = 2, n = 1$ Ising universality class including the scalar φ^4 model, in agreement with multiparameter universality, with three different non-universal parameters $q, \Omega, \bar{\xi}_{0+}$. The latter are known explicitly for the φ^4 model[37] as a function of the microscopic couplings $K_{i,j}$ and the lattice structure through the matrix elements (3) of $\mathbf{A}$ and the correlation length $\xi'_{0+}(u'_0)$ of the Hamiltonian (23). For the general anisotropic Ising model (81), however, the non-universal parameters $q^{\mathrm{Is}}, \Omega^{\mathrm{Is}}, \bar{\xi}_{0+}^{\mathrm{Is}}$ are generically unknown and need to be determined for each specific model under consideration. Plotting (111) in a contour plot in the $q^{\mathrm{Is}} - \Omega^{\mathrm{Is}}$ plane yields universal self-similar structures[64] for $t \neq 0$ similar to those found previously[38] for $t = 0$.

For the special case of the Ising model (64), one can also plot explicitly[64] the non-universal scaling functions $\mathcal{F}^{\mathrm{Is,ex}}[t, \rho_{\mathrm{rec}}, E_1/E_3, E_2/E_3]$ and $X^{\mathrm{Is}}[t, \rho_{\mathrm{rec}}, E_1/E_3, E_2/E_3]$ for $t \neq 0$ (compare Fig. 5 of Ref. 38 for $t = 0$). These functions are of interest for a comparison with the exact result of Nash and O'Connor[18] for $\mathcal{F}^{\mathrm{Is,ex}}$, who use a different non-universal representation. Exact agreement is found[64] with their result, and the exact equivalence of (108)–(110) with the different representation of Ref. 18 is proven analytically in Ref. 65.

5.3. *Exact excess free energy on a parallelogram*

The above results can be extended to anisotropic systems on an arbitrarily shaped parallelogram.[64] We consider the Ising model (81) an a parallelogram lattice with the shape of a parallelogram spanned by the vectors

$$\mathbf{L}_1 = L_1 \begin{pmatrix} 1 \\ 0 \end{pmatrix}, \quad \mathbf{L}_2 = L_2 \begin{pmatrix} \cos\vartheta \\ \sin\vartheta \end{pmatrix} \quad (112)$$

in the horizontal direction and in the direction of the angle ϑ, respectively. The aspect ratio ρ_{para} and the area V_{para} of the parallelogram are given by

$$\rho_{\mathrm{para}} = \frac{|\mathbf{L}_2|}{|\mathbf{L}_1|} = \frac{L_2}{L_1}, \quad V_{\mathrm{para}} = L_1 L_2 \sin\vartheta. \quad (113)$$

For $\vartheta = \pi/2$, this includes the special case of rectangular geometry at T_c discussed above. The generalized shear transformation applied to the boundaries of the parallelogram reads

$$\mathbf{L}_{p1} = \widehat{\boldsymbol{\lambda}}^{-1/2}\mathbf{U}(\Omega^{\rm Is})\mathbf{L}_1, \quad \mathbf{L}_{p2} = \widehat{\boldsymbol{\lambda}}^{-1/2}\mathbf{U}(\Omega^{\rm Is})\mathbf{L}_2. \tag{114}$$

This yields a changed isotropic parallelogram which has the transformed area:

$$V^{\rm iso}_{\rm para} = |\mathbf{L}_{p1}|\,|\mathbf{L}_{p2}|\,\sin\alpha_{\rm para} = (\det\widehat{\boldsymbol{\lambda}})^{-1/2}\,V_{\rm para} = \left(\xi^{\rm iso}_{0\pm}/\bar{\xi}^{\rm Is}_{0\pm}\right)^2 V_{\rm para}, \tag{115}$$

the transformed aspect ratio

$$[\widehat{\rho}_{\rm para}(\vartheta,\rho_{\rm para},q^{\rm Is},\Omega^{\rm Is})]^2 = \frac{|\mathbf{L}_{\rm p2}|^2}{|\mathbf{L}_{\rm p1}|^2} = \rho^2_{\rm para}\,\frac{\cos^2(\vartheta-\Omega^{\rm Is}) + (q^{\rm Is})^2\sin^2(\vartheta-\Omega^{\rm Is})}{\cos^2\Omega^{\rm Is} + (q^{\rm Is})^2\sin^2\Omega^{\rm Is}}, \tag{116}$$

and the transformed angle $\alpha_{\rm para}$ determined by

$$\cot\alpha_{\rm para}(\vartheta,q^{\rm Is},\Omega^{\rm Is}) = \frac{(q^{\rm Is})^{-1} - q^{\rm Is}\tan\Omega^{\rm Is}\,\tan(\vartheta-\Omega^{\rm Is})}{\tan\Omega^{\rm Is} + \tan(\vartheta-\Omega^{\rm Is})}. \tag{117}$$

The corresponding torus parameter is

$$\tau(\alpha_{\rm para},\widehat{\rho}_{\rm para}) = \tau_0 + i\,\tau_1 = \widehat{\rho}_{\rm para}\exp(i\,\alpha_{\rm para}) = \tau(\vartheta,\rho_{\rm para},q^{\rm Is},\Omega^{\rm Is}), \tag{118}$$

and the corresponding generalization of the scaling variable (107) is

$$\widehat{x}(t) = t\left[V^{\rm iso}_{para}/(\xi^{\rm iso}_{0+})^2\right]^{1/2} = t\left[V_{\rm para}/(\bar{\xi}^{\rm Is}_{0+})^2\right]^{1/2}. \tag{119}$$

Substituting these formulae into the invariance relation (106), we obtain the exact universal scaling form of the excess free energy of the anisotropic Ising model (81) on the parallelogram[64]

$$\mathcal{F}^{\rm Is,ex}(\vartheta,\rho_{\rm para},q^{\rm Is},\Omega^{\rm Is},tV^{1/2}_{\rm para}/\bar{\xi}^{\rm Is}_{0+}) = \mathcal{F}^{\rm ex,iso}\big(tV^{1/2}_{\rm para}/\bar{\xi}^{\rm Is}_{0+},\tau(\vartheta,\rho_{\rm para},q^{\rm Is},\Omega^{\rm Is})\big), \tag{120}$$

with three non-universal parameters $q^{\rm Is},\Omega^{\rm Is},\bar{\xi}^{\rm Is}_{0+}$, where $\mathcal{F}^{\rm ex,iso}(\widehat{x},\tau)$ is the same universal function as given in (108). For simplicity, we have not changed here the notation for the ϑ-dependent anisotropy parameters $q^{\rm Is},\Omega^{\rm Is},\bar{\xi}^{\rm Is}_{0+}$ of the Ising model on the anisotropic parallelogram which differ from those on the anisotropic rectangle because the parallelogram-lattice structure of the anisotropy parallelogram at $\vartheta \neq \pi/2$ differs from that of the square-lattice structure of the rectangle. The same comment applies to the non-universal parameters $q,\Omega,\bar{\xi}_{0+}$ of the φ^4 model. Universal self-similar structures are found,[64] and applications are given to the finite-size part of the specific heat and the Casimir force away from T_c. The explicit expressions for all non-universal parameters of the triangular lattice model (64) on a parallelogram lattice[64] demonstrate a considerable intrinsic diversity caused by weak anisotropy.

6. Field-Theoretic Approach to $d = 3$ Anisotropic Critical Behavior: Minimal Renormalization Without ε Expansion

The critical behavior of three-dimensional isotropic and weakly anisotropic systems with short-range interactions is significantly more complex than that of two-dimensional systems, and no exact analytic description is available in three dimensions, except in special cases, such as Gaussian models, spherical models, and in the large n limit. Here, we consider, for finite n, the field-theoretic renormalization-group approach[2,8,67] in the minimal renormalization scheme[68] at fixed dimension[36,60–62] applied to the φ^4 model in $2 < d < 4$ dimensions which is a powerful method to an analytic description of bulk and confined systems near criticality. This approach is applicable to both isotropic and weakly anisotropic systems since weak anisotropy does not generate new ultraviolet divergencies.[25,30] We restrict ourselves to a brief overview of those aspects that are relevant to finite-size effects in isotropic and weakly anisotropic three-dimensional systems. For an overview of the early history of the theory of finite-size scaling in isotropic systems, see Chapter 33 by Young.

6.1. *Lowest-mode separation approach to finite-size theory*

Ordinary perturbation theory for finite systems, in the sense of an expansion around mean-field theory, fails because of unphysical divergencies arising from the isolated lowest ($\mathbf{k} = \mathbf{0}$) mode at T_c and from the massless Goldstone modes at the coexistence line below T_c. A concept of separating the lowest mode from the higher modes was formulated within the $4 - \varepsilon$ and the $2 + \varepsilon$ expansions.[69,70]

We briefly sketch this lowest-mode separation approach based on the isotropic φ^4 Hamiltonian for general n

$$H^{\text{iso}}_{\text{field}} = \int_V d^d x \left[\frac{r_0}{2}\varphi(\mathbf{x})^2 + \frac{1}{2}(\nabla\varphi)^2 + u_0((\varphi)^2)^2\right], \tag{121}$$

with $r_0 = r_{0c} + a_0 t$, $a_0 > 0$, $u_0 > 0$. Pioneering work on finite-size calculations within this model has been performed previously[69,70] for a finite cubic volume $V = L^d$ with periodic BC. In the Fourier expansion,

$$\varphi(\mathbf{x}) = L^{-d} \sum_{\mathbf{k}} e^{i\mathbf{k}\cdot\mathbf{x}_j} \hat{\varphi}(\mathbf{k}), \tag{122}$$

the summations $\sum_{\mathbf{k}}$ run over the discrete vectors $\mathbf{k}$ of the first Brillouin zone of the reciprocal lattice. It was proposed to decompose $\varphi(\mathbf{x})$ as

$$\varphi(\mathbf{x}) = \Phi + \sigma(\mathbf{x}) \tag{123}$$

into the lowest (homogeneous) mode

$$\Phi = L^{-d}\hat{\varphi}(\mathbf{0}) = L^{-d} \int d^d x \varphi \tag{124}$$

and the higher inhomogeneous modes

$$\sigma(\mathbf{x}) = L^{-d} \sum_{\mathbf{k} \neq \mathbf{0}} e^{i\mathbf{k}\cdot\mathbf{x}_j} \hat{\varphi}(\mathbf{k}). \tag{125}$$

Correspondingly, the Hamiltonian is decomposed as

$$H_{\text{field}}^{\text{iso}} = H_0(\Phi^2) + \widetilde{H}(\Phi, \sigma), \tag{126}$$

with the zero-mode Hamiltonian

$$H_0(\Phi^2) = L^d \left(\frac{1}{2} r_0 \Phi^2 + u_0 \Phi^4 \right). \tag{127}$$

This yields the partition function

$$Z = \int d^n \Phi \exp\left\{ - \left[H_0(\Phi^2) + \Gamma(\Phi^2) \right] \right\}, \tag{128}$$

$$\Gamma(\Phi^2) = - \ln \int d^n \sigma \exp\left[-\widetilde{H}(\Phi, \sigma) \right]. \tag{129}$$

The main task of the theory is the calculation of the contribution of the higher modes arising from $\Gamma(\Phi^2)$. Some interesting results were obtained within a low-order renormalized treatment of $\Gamma(\Phi^2)$ based on the $\varepsilon = 4 - d$ expansion for $T \geq T_c$,[69,70] whereas the result for $n = 1$ below T_c turned out to be less satisfactory.[71] This has called for substantial extensions and methodological improvements in order to address several problems of finite-size theory:

(1) an improved perturbation approach capable of determining finite-size scaling functions below T_c not only for $n = 1$[30,71] but also for general $n > 1$ including the crossover from the low-temperature Goldstone regime to the critical region,[36,72]
(2) applications to the order-parameter distribution function above, at, and below T_c for general n[73] and in the presence of a finite external field h,[74]
(3) applications to the excess free energy and critical Casimir force in cubic[30,71] and block[35,36] geometries,
(4) a treatment of a finite-slab geometry with small but finite aspect ratio $\rho << 1$ in order to circumvent the unresolved problem of a dimensional crossover from d-dimensional bulk to $d - 1$-dimensional film critical behavior,[35,75]
(5) the extension to realistic Dirichlet BC[76] that are required to describe the observed critical Casimir force[77] near the superfluid transition in isotropic ^{4}He films of finite thickness near the minimum below bulk T_c,
(6) the prediction of a critical Casimir force scaling function for anisotropic superconducting films not only with idealized periodic BC[36,78] but also with realistic Dirichlet BC.

It is, of course, beyond the scope of this chapter to discuss all of these issues. Some general comments are given in the following section and the issue (6) is further discussed.

6.2. *Minimal renormalization without ε expansion in bulk and finite-size theory*

The description of the critical behavior requires perturbative calculations with respect to u_0 as well as multiplicative and additive renormalizations, followed by a mapping of the renormalized quantities from the critical to the non-critical region where perturbation theory is applicable. These renormalizations are the same in bulk and confined system since there exist no L-dependent ultraviolet divergencies,[67] and they are the same in isotropic and in weakly anisotropic systems after an appropriate shear transformation.[25,30] The perturbative calculations fall into two categories: those based on the Wilson–Fisher $\varepsilon = 4 - d$ expansion[79] or the $\varepsilon = d - 2$ expansion[80] and those that are carried out at fixed dimensions $d < 4$.[8,30,36,60–62,71,72,76] Within these two approaches, one can further distinguish between two types of renormalizations: the use of renormalization conditions[2] and the minimal subtraction scheme.[68]

The latter renormalization scheme is of particular simplicity and elegance, and has the advantage that the renormalization constants (Z-factors) attain the simplest form possible since they absorb just the pole terms $\propto \varepsilon^{-m}$. Furthermore, they are the same above and below T_c. It has been shown[60,61] that the minimal renormalization scheme can be combined with the perturbation theory at fixed dimension (see, e.g., applications in bulk theory up to four-loop[81] and five-loop[82] order for general n above and below T_c where spurious Goldstone singularities have been canceled). It has been demonstrated[36,72,73] that this is of particular importance in finite-size calculations below T_c for general n where a conventional $\varepsilon = 4 - d$ expansion would not be possible because of spurious Goldstone singularities for $n \geq 2$.

We substantiate this point in the context of the lowest-mode separation approach sketched above. The integrand of (128) plays the role of an order-parameter distribution function

$$P(\Phi^2) = Z^{-1} \int d^n\sigma \exp\big(- H^{\mathrm{iso}}_{\mathrm{field}}\big) = Z^{-1} \exp\big\{ - \big[H_0(\Phi^2) + \Gamma(\Phi^2)\big]\big\}, \qquad (130)$$

which is a physical quantity in its own right. The integration $\int d^n\sigma$ in (130) is to be performed over both longitudinal and transverse fluctuations of $\sigma(\mathbf{x}) = \sigma_L(\mathbf{x}) + \sigma_T(\mathbf{x})$. Great care must be taken in making approximations in the treatment of the transverse fluctuations σ_T in $\Gamma(\Phi^2)$ such that the exponential structure of $P(\Phi^2)$ is maintained. This means that appropriate approximations[36] should be made only for the quantity $\Gamma(\Phi^2)$ without further expanding $\exp\big[-\Gamma(\Phi^2)\big]$. This is in line with the minimal subtraction procedure at fixed d[36] which does not use the fixed-point value $u^* \sim O(\varepsilon)$ as smallness parameter. By contrast, the conventional ε expansion would destroy the exponential structure of $P(\Phi^2)$ and would produce spurious singularities below T_c arising from the transverse fluctuations.[36] An analytic calculation of $P(\Phi^2)$ for $n = 2$ and $n = 3$ within the lowest-mode-separation approach combined with the minimal renormalization at fixed dimension $d = 3$ in an isotropic cubic geometry has been shown to be in quantitative agreement with Monte Carlo data below T_c.[73]

6.3. *Anisotropic extension of the Privman–Fisher scaling form in three dimensions*

We report on the singular part of the free-energy density of the anisotropic φ^4 model in a finite $L_1 \times L_2 \times L_3$ block with periodic BC. By perturbation theory in the minimal renormalization scheme at fixed dimension $d = 3$, we have recently derived the anisotropic extension of the Privman–Fisher scaling form (16) as[36]

$$f_s^{\text{aniso}}(t, h, L_1, L_2, L_3) = L_0^{-3}\, Y\left(\hat{x}, \hat{x}_h; \frac{L_1}{L_0}, \frac{L_2}{L_0}, \frac{L_3}{L_0}, \bar{\mathbf{A}}\right), \tag{131}$$

with the scaling variables

$$\hat{x} = t\,(L_0/\bar{\xi}_{0+})^{1/\nu}, \quad \hat{x}_h = h\,(L_0/\bar{\xi}_c)^{\Delta/\nu}, \tag{132}$$

and with the reduced anisotropy matrix

$$\bar{\mathbf{A}} = \bar{\mathbf{A}}(\{\xi_{0\pm}^{(\alpha)}, \mathbf{e}^{(\alpha)}\}) = \bar{\mathbf{A}}\left(\frac{\xi_{0+}^{(1)}}{\bar{\xi}_{0+}}, \frac{\xi_{0+}^{(2)}}{\bar{\xi}_{0+}}, \frac{\xi_{0+}^{(3)}}{\bar{\xi}_{0+}}, \Omega_1, \Omega_2, \Omega_3\right). \tag{133}$$

Here, $\bar{\xi}_{0+}$ and $\bar{\xi}_c$ are the geometrical mean of the amplitudes $\xi_{0+}^{(\alpha)}$ and $\xi_c^{(\alpha)}$ of the three principal correlation lengths for $T > T_c$, $h = 0$ and for $T = T_c$, $h \neq 0$ of the anisotropic system. The three angles $\Omega_1, \Omega_2, \Omega_3$, determine the directions of the principal unit vectors $\mathbf{e}^{(\alpha)}, \alpha = 1, 2, 3$ of the three principal axes. The characteristic length L_0 may be chosen as $L_0 = V^{1/3}, V = L_1 L_2 L_3$. At $h = 0$, the anisotropic scaling function Y has been calculated above, at, and below T_c approximately for general finite n and exactly in the large n limit.[36] It describes the entire crossover from the low-temperature to the high-temperature regions including the crossover from Goldstone to the critical behavior.

In the isotropic case, we have $\bar{\mathbf{A}} = \mathbf{1}$, and (131) has the form

$$f_s^{\text{iso}}(t, h, L_1, L_2, L_3) = L_0^{-3}\, Y\left(\hat{x}^{\text{iso}}, \hat{x}_h^{\text{iso}}; \frac{L_1}{L_0}, \frac{L_2}{L_0}, \frac{L_3}{L_0}, \mathbf{1}\right), \tag{134}$$

with the scaling variables

$$\hat{x}^{\text{iso}} = t\,(L_0/\xi_{0+}^{\text{iso}})^{1/\nu}, \quad \hat{x}_h^{\text{iso}} = h\,(L_0/\xi_c^{\text{iso}})^{\Delta/\nu}, \tag{135}$$

where ξ_{0+}^{iso} and ξ_c^{iso} are the correlation lengths of the isotropic system. It has been shown[36] that $f_s^{\text{iso}}(t, 0, L_\parallel, L_\parallel, L)$ agrees well with available Monte Carlo data for the isotropic Ising ($n = 1$) and XY ($n = 2$) models in a $L_\parallel^2 \times L$ slab geometry with aspect ratios between $\rho = L/L_\parallel = 1/8$ and $\rho = 1$.

In the anisotropic case, the reduced anisotropy matrix depends on Ω_i and on the ratios of $\xi_{0+}^{(\alpha)}$. Only two of the three ratios $\xi_{0+}^{(\alpha)}/\bar{\xi}_{0+}$ are independent, thus $\bar{\mathbf{A}}$ depends on five independent non-universal anisotropy parameters. In addition, there are the two mean correlation lengths $\bar{\xi}_{0+}$ and $\bar{\xi}_c$ entering the scaling variables $\hat{x}$ and $\hat{x}_h$, thus there are seven independent non-universal parameters altogether rather than only two parameters in the isotropic case.

Our result[36] for the scaling function Y is valid for arbitrary Bravais lattices with an arbitrary anisotropy matrix $\mathbf{A}$ with $\det \mathbf{A} > 0$. It exhibits multiparameter universality within the class of weakly anisotropic three-dimensional φ^4 models for general n. There is essentially no doubt that this result is also universally valid for the n-vector model, with seven non-universal parameters in view of our proof of finite-size multiparameter universality for the two-dimensional Ising universality class and of bulk multiparameter universality for the (d, n) universality classes. Although Y and $\bar{\mathbf{A}}$ are universal functions of their arguments, weak anisotropy introduces a substantial degree of non-universality through the non-universal arguments both in experimental and theoretical analyses. In particular, the $d = 3$ finite-size amplitude at criticality

$$Y_c(\bar{\mathbf{A}}) = Y\big(0, 0; \rho_1, \rho_2, \rho_3, \bar{\mathbf{A}}\big), \tag{136}$$

the critical Binder cumulant ratio[5,30]

$$U_c(\bar{\mathbf{A}}) = \frac{1}{3}\left[\frac{\partial^4 Y(0, y; ; \rho_1, \rho_2, \rho_3, \bar{\mathbf{A}})/\partial y^4}{(\partial^2 Y(0, y; ; \rho_1, \rho_2, \rho_3, \bar{\mathbf{A}})/\partial y^2)^2}\right]_{y=0}, \tag{137}$$

and the ensuing critical Casimir amplitude are non-universal quantities. While $\bar{\xi}_{0+}$ and $\bar{\xi}_c$ can be determined by macroscopic thermodynamic measurements of the specific heat and the susceptibility, the remaining five non-universal parameters Ω_i and $\xi_{0+}^{(\alpha)}/\xi_{0+}^{(\beta)}$ require detailed scattering measurements to resolve the angular dependence of the anisotropic critical correlations. On the theoretical side, all seven anisotropy parameters are in principle calculable within the φ^4 theory from the matrix elements of the anisotropy matrix $\mathbf{A}$, but this is not easily possible within the n-vector model for which no large-distance anisotropy matrix $\mathbf{A}$ can be defined. The angles Ω_i of the principal axes are generically unknown functions of the couplings, and their determination in three dimensions for special cases is a non-trivial task as seen already from the research on anisotropic two-dimensional Ising models. Thus, unlike isotropic systems, weakly anisotropic systems exhibit a high degree of intrinsic diversity of critical finite-size behavior within the (d, n) universality classes even if the scaling function Y is known exactly. This affects all observable physical quantities that can be derived from the free-energy density, e.g., the specific heat, the susceptibility, and the critical Casimir force.

So far, only a few examples for the anisotropic scaling function[36] Y and of the ensuing Casimir force scaling function X have been discussed in some detail.[36,38] It is far beyond the scope of this chapter to enter a discussion of the complexity of three-dimensional self-similar structures to be expected on the basis of the recent exact analysis of the universal Casimir force scaling function[38] in the case of a planar anisotropy in the three-dimensional anisotropic φ^4 model with periodic BC in the large n limit. These complex structures exist not only at T_c but also far below T_c[38] and in the whole scaling region near T_c, as demonstrated in a forthcoming paper.[64]

6.4. *Critical Casimir forces in anisotropic superconductors*

In the past decades, there has been much interest in the critical Casimir force in confined systems near criticality which was first predicted by M.E. Fisher and P.G. De Gennes[83] in 1978. An experimental verification has been achieved so far only in isotropic systems, most prominently in superfluid ^{4}He films.[77] It has been pointed out by G. Williams[78] that a measurable critical Casimir force should also occur in superconducting films. Superconductors belong to the same universality class as superfluid ^{4}He and have the same (Dirichlet) BC but are anisotropic. Williams' scenario is the following: a superconducting film is connected to a bulk sample of the same material. He argues that below the bulk critical temperature, the film–bulk system can lower its free energy by a transfer of electrons (Cooper pairs) from the film to the bulk system which is analogous to helium atoms moving from the film to the superfluid bulk system. While in the helium system, this leads to a thinning of the film, the effect in the superconducting film–bulk system is a transfer of negative electrical charge from the film to the bulk system. Williams argues that this gives rise to an electrical potential difference which can be related to the free-energy difference (per unit area) between the film and the bulk and from which a Casimir force can be derived. He estimates the voltage difference to have a measurable magnitude.

The critical Casimir force is an observable only if the ordering degrees of freedom can enter and leave the system. Therefore, it has been claimed in the literature[84–87] that this force can be active only in isotropic fluids and that the issue of spatial anisotropy is not relevant in the context of the critical Casimir force. We argue that, unlike the localized degrees of freedom (magnetic moments) of the order parameter of an anisotropic magnetic material, the ordering degrees of freedom (Cooper pairs) of a superconductor are not localized at lattice points but play the role of an electrical superfluid in an anisotropic environment that can leave and enter the film connected to the bulk of the same material, as anticipated by Williams.[78] So far, no specific objection has been raised in the literature against this specific argumentation for superconductors, and in a comment[88] on Ref. 78, the measurability of the critical Casimir force in superconductors has not been questioned. Furthermore, we point to the largely unexplored area of thermodynamic Casimir forces in liquid crystals,[89] which exhibit a wide variety of spatial anisotropy and whose ordering degrees of freedom can leave and enter the system.

In closing, we note that experimental studies, Monte Carlo simulations, and further theoretical research are called for in view of the fact that at present no experimental or Monte Carlo data and no analytic predictions are available for the critical Casimir force in anisotropic systems with realistic BC. Also, theoretical efforts based on the functional renormalization group[11] applied to anisotropic confined systems could yield important contributions to this matter. Analytic results for the critical Casimir force in anisotropic films of finite thickness have been presented previously[36] for the case of periodic BC which refute earlier results,[78] where no anisotropy effect in anisotropic superconductors near T_c was found. An analytic renormalization-group study in three dimensions with realistic Dirichlet BC below T_c without adjustable parameters was performed,[76] which explains the depth and position of the deep minimum of the Casimir force scaling function observed in isotropic ^{4}He films[77] in the temperature regime $T_{c,\mathrm{film}} < T < T_c$ on a semi-quantitative level. It is conceivable that this study can be extended to anisotropic film systems with Dirichlet BC which could lead to quantitative predictions of the critical Casimir force in real superconductors below T_c.

7. Michael Fisher's Research and Comments on Weak Anisotropy

Michael Fisher was interested in lattice structure effects on the critical behavior as seen from his publications and from later private communication in spontaneous responses to our research.

In Table IV of the early review article,[1] he has listed the weakly anisotropic interaction types (1)–(4) of short spatial range with cubic symmetry, which exhibit a quadratic, i.e., isotropic momentum dependence at leading order with terms of cubic anisotropy at higher order. This table was later complemented in Ref. 30. In the work with Tarko[90] square, simple-cubic, and body-centered cubic Ising models were studied, which demonstrated the universality of several critical bulk amplitude relations that remained unaffected by the cubic weak anisotropy. For the anisotropic Ising model with non-cubic anisotropy, Aharony and Fisher[14] found non-universal ratios of correction amplitudes which in the isotropic case were known to be universal.[5] Furthermore, they pointed out[14] a non-universal anisotropy effect on the exact bulk correlation function[23] of the anisotropic two-dimensional Ising model. This non-universality was reanalyzed recently in the framework of multiparameter universality.[37,63,64]

In 2004 and 2008, Michael Fisher responded spontaneously to our research on weak anisotropy. With regard to the paper[25] by Xiaosong Chen and myself in 2004, he wrote that he felt it "important to have spelled out properly some of the quite subtle implications of lack of cubic symmetry". Then he added a "very mild critical comment" that we have "in some sense overemphasized the degree of non-universality." He argued that "there is a reasonable degree of restricted or modified universality in the sense that it is only certain moments of the anisotropy that play a role," and "the whole issue of universality boils down to how many parameters are needed." Thus, in some sense, he anticipated at an early stage the notion of multiparameter universality with $d(d+1)/2+1$ parameters that was introduced later[30] and was fully established recently.[36,37,63]

In response to my paper[30] in 2008, he commented on "your magnum opus on universality" and sovereignly admitted the "limitations of universality (especially!) in finite system relative to overoptimistic expectations." He added the "remark that it is the appreciation of the subtleties exhibited by your Figs. 1 and 2 that, many years ago, led David Jasnow and me to hesitate to give a mathematical definition of scaling that one might propose as a 'theorem' suitable for a proof!" See also the original letter of Michael Fisher in the Appendix.

The last time I met Michael Fisher was on a conference in Göttingen 2014 on the occasion of the 80th birthday of Guenter Ahlers. In a conversation with Michael Fisher (together with Pierre Hohenberg), he pointed to the large variety of crystal structures causing non-universal anisotropy effects near criticality, and Pierre Hohenberg completely agreed with our analysis of the consequences of weak anisotropy.

Acknowledgments

I thank F. Kischel and S. Wessel for the useful discussions and collaboration on Refs. 64, 65.

References

1. M. E. Fisher, The renormalization group in the theory of critical behavior. *Rev. Mod. Phys.* **46**, 597 (1974).
2. E. Brézin, J. C. Le Guillou and J. Zinn-Justin, Field theoretical approach to critical phenomena, in C. Domb and J. L. Lebowitz (eds.), *Phase Transitions and Critical Phenomena*, Vol. 6, p. 125 (Academic, New York, 1976).
3. P. C. Hohenberg, A. Aharony, B. I. Halperin and E. D. Siggia, Two-scale-factor universality and the renormalization group. *Phys. Rev. B* **13**, 2986 (1976).
4. V. Privman and M. E. Fisher, Universal critical amplitudes in finite-size scaling. *Phys. Rev. B* **30**, 322 (1984).
5. V. Privman, A. Aharony and P. C. Hohenberg, Universal critical-point amplitude relations. In C. Domb and J. L. Lebowitz (eds.), *Phase Transitions and Critical Phenomena*, Vol. 14, p. 1 (Academic, New York, 1991).
6. J. L. Cardy, *Scaling and Renormalization in Statistical Physics* (Cambridge University Press, 1996).
7. A. Pelissetto and E. Vicari, Critical phenomena and renormalization-group theory. *Phys. Rep.* **368**, 549 (2002).
8. J. Zinn-Justin, *Phase Transitions and Renormalization Group* (Oxford University Press, 2007).

9. J. L. Cardy, Conformal invariance. In C. Domb and J. L. Lebowitz (eds.), *Phase Transitions and Critical Phenomena*, Vol. 11, p. 55 (Academic, New York, 1987).
10. P. Di Francesco, P. Mathieu and D. Sénéchal, *Conformal Field Theory* (Springer, New York, 1997).
11. N. Dupuis, L. Canet, A. Eichhorn, W. Metzner, J. M. Pawlowski, M. Tissier and N. Wschebor, The nonperturbative functional renormalization group and its applications. *Phys. Rep.* **910**, 1 (2021).
12. T. T. Wu, B. M. McCoy, C. A. Tracy and E. Barouch, Spin-spin correlation function for the two-dimensional Ising model: Exact theory in the scaling region. *Phys. Rev. B* **13**, 316 (1976).
13. H. G. Vaidya, The spin-spin correlation functions and susceptibility amplitudes for the two-dimensional Ising model: Triangular lattice. *Phys. Lett. A* **57**, 1 (1976).
14. A. Aharony and M. E. Fisher, Universality in analytic corrections to scaling for planar Ising models. *Phys. Rev. Lett.* **45**, 679 (1980).
15. P. Nightingale and H. Blöte, The relation between amplitudes and critical exponents in finite-size scaling. *J. Phys. A* **16**, L 657 (1983).
16. D. Kim and P. A. Pearce, Scaling dimensions and conformal anomaly in anisotropic lattice spin models. *J. Phys. A* **20**, L 451 (1987).
17. H. Au-Yang and J. H. H. Perk, Critical correlations in a Z-invariant inhomogeneous Ising model. *Physica A* **144**, 44 (1987).
18. C. Nash and D. O'Connor, Modular invariance, lattice field theories, and finite size corrections. *Ann. Phys.* **273**, 72 (1999).
19. H. Au-Yang and J. H. H. Perk. In M. Kashiwara and T. Miwa (eds.), *MathPhys Odyssey 2001: Integrable Models and Beyond*, p. 23 (Birkäuser, Boston, 2002).
20. H. Au-Yang and J. H. H. Perk, Susceptibility calculations in periodic and quasiperiodic planar Ising models. *Physica A* **321**, 81 (2003).
21. J. H. H. Perk and H. Au-Yang, Some recent results on pair correlation functions and susceptibilities in exactly solvable models. *J. Phys.: Conf. Ser.* **42**, 231 (2006).
22. N. S. Izmailian, Finite-size effects for anisotropic 2D Ising model with various boundary conditions. *J. Phys. A* **45**, 494009 (2012).
23. B. M. McCoy and T. T. Wu, *The Two-dimensional Ising Model*, second edition (Dover, 2014), and references therein.
24. H. Hobrecht and A. Hucht, Anisotropic scaling of the two-dimensional Ising model I: the torus. *SciPost Phys.* **7**, 026 (2019).
25. X. S. Chen and V. Dohm, Nonuniversal finite-size scaling in anisotropic systems. *Phys. Rev. E* **70**, 056136 (2004).
26. V. Dohm, Phase Transitions. In R. G. Lerner and G. L. Trigg (eds.), *Encyclopedia of Physics*, Vol. 2, p. 1901 (Wiley-VCH Verlag, Weinheim, 2005).
27. W. Selke and L. N. Shchur, Critical Binder cumulant in two-dimensional anisotropic Ising models. *J. Phys. A* **38**, L 739 (2005).
28. V. Dohm, Anisotropy and restricted universality of critical phenomena. *J. Phys. A* **39**, L 259 (2006).
29. X. S. Chen and H. Y. Zhang, Finite-size scaling of the correlation length in anisotropic systems. *Int. J. Mod. Phys. B* **21**, 4212 (2007).
30. V. Dohm, Diversity of critical behavior within a universality class. *Phys. Rev. E* **77**, 061128 (2008).
31. V. Dohm, Kritisch, aber nichtuniversell. *Physik J.* **8**, 37 (2009).
32. W. Selke and L. N. Shchur, Critical Binder cumulant in a two-dimensional anisotropic Ising model with competing interactions. *Phys. Rev. E* **80**, 042104 (2009).

33. D. Dantchev and D. Grüneberg, Casimir force in $O(n)$ systems with a diffuse interface. *Phys. Rev. E* **79**, 041103 (2009).
34. B. Kastening and V. Dohm, Finite-size effects in film geometry with nonperiodic boundary conditions: Gaussian model and renormalization-group theory at fixed dimension. *Phys. Rev. E* **81**, 061106 (2010).
35. V. Dohm, Critical free energy and Casimir forces in rectangular geometries. *Phys. Rev. E* **84**, 021108 (2011).
36. V. Dohm, Crossover from low-temperature to high-temperature fluctuations: Universal and nonuniversal Casimir forces of isotropic and anisotropic systems. *Phys. Rev. E* **97**, 062128 (2018).
37. V. Dohm, Multiparameter universality and directional nonuniversality of exact anisotropic critical correlation functions of the two-dimensional Ising universality class. *Phys. Rev. E* **100**, 050101(R) (2019).
38. V. Dohm and S. Wessel, Exact critical Casimir amplitude of anisotropic systems from conformal field theory and self-similarity of finite-size scaling functions in $d \geq 2$ dimensions. *Phys. Rev. Lett.* **126**, 060601 (2021).
39. V. Dohm, S. Wessel, B. Kalthoff and W. Selke, Multiparameter universality and conformal field theory for anisotropic systems: test by Monte Carlo simulations. *J. Phys. A* **54**, 23LT01 (2021).
40. A. D. Bruce, On the critical behaviour of anisotropic systems. *J. Phys. C* **7**, 2089 (1974).
41. F. Wegner, The critical state, general aspects. in C. Domb and M. S. Green (eds.), *Phase Transitions and Critical Phenomena*, Vol. 6, p. 7 (Academic, New York, 1976).
42. A. Aharony, Dependence of universal critical behavior on symmetry and range of interactions. In C. Domb and M. S. Green (eds.), *Phase Transitions and Critical Phenomena*, Vol. 6, p. 357 (Academic, New York, 1976).
43. B. Schmittmann and R. K. P. Zia, Statistical mechanics of driven diffuse systems. In C. Domb and J. L. Lebowitz (eds.), *Phase Transitions and Critical Phenomena*, Vol. 17, p. 1 (Academic, New York, 1995).
44. D. Stauffer, M. Ferer and M. Wortis, Universality of second-oder phase transitions: The scale factor for the correlation length. *Phys. Rev. Lett.* **29**, 345 (1972).
45. A. Aharony, Two-scale-factor universality and the ε expansion. *Phys. Rev. B* **9**, 2107 (1974).
46. P. R. Gerber, Two scale factor universality in the spherical model. *J. Phys. A* **8**, 67 (1975).
47. V. Privman, Finite-size scaling theory. In V. Privman (ed.), *Finite Size Scaling and Numerical Simulation of Statistical Systems*, p. 1 (World Scientific, Singapore, 1990).
48. V. Privman, Universal size dependence of the free energy of finite systems near criticality. *Phys. Rev. B* **38**, 9261 (1988).
49. M. Henkel, Application of the Hamiltonian limit to critical phenomena, finite-size scaling and conformal invariance. In V. Privman (ed.), *Finite Size Scaling and Numerical Simulation of Statistical Systems*, p. 353 (World Scientific, Singapore, 1990).
50. J. L. Cardy, Critical behaviour at an edge. *J. Phys. A* **16**, 3617 (1983).
51. M. N. Barber, I. Peschel and P. A. Pearce, Magnetization at corners in two-dimensional Ising models. *J. Stat. Phys.* **37**, 497 (1984).
52. J. O. Indekeu, M. P. Nightingale and W. V. Wang, Finite-size interaction amplitudes and their universality: Exact, mean-field, and renormalization-group results. *Phys. Rev. B* **34**, 330 (1986).

53. K. Binder and J. S. Wang, Finite-size effects at critical points with anisotropic correlations: Phenomenological scaling theory and Monte Carlo simulations. *J. Stat. Phys.* **55**, 87 (1989).
54. M. P. Nightingale, Transfer matrices, phase transitions, and critical phenomena: numerical methods and applications. In V. Privman (ed.), *Finite Size Scaling and Numerical Simulation of Statistical Systems*, p. 287 (World Scientific, Singapore, 1990).
55. J. L. Cardy, Conformal invariance and finite-size scaling. In J. L. Cardy (ed.), *Finite-size scaling*, p. 325 (North Holland, Amsterdam, 1988).
56. M. A. Yurishchev, Hyperuniversality of a fully anisotropic three-dimensional Ising model. *Phys. Rev. B* **50**, 13533 (1994).
57. H. W. J. Blöte, J. L. Cardy and M. P. Nightingale, Conformal invariance, the central charge, and universal finite-size amplitudes at criticality. *Phys. Rev. Lett.* **56**, 742 (1986).
58. M. Krech, *The Casimir Effect in Critical Systems* (World Scientific, Singapore, 1994).
59. D. M. Dantchev and S. Dietrich, Critical Casimir effect: Exact results. *Phys. Rep.* **1005**, 1 (2023), and references therein.
60. V. Dohm, Nonuniversal critical phenomena along the lambda line of ^{4}He. I. Specific heat in three dimensions. *Z. Phys. B* **60**, 61 (1985).
61. R. Schloms and V. Dohm, Minimal renormalization without ε-expansion: Critical behavior in three dimensions. *Nucl. Phys. B* **328**, 639 (1989).
62. R. Schloms and V. Dohm, Minimal renormalization without ε-expansion: Critical behavior above and below T_c. *Phys. Rev. B* **42**, 6142 (1990).
63. V. Dohm, Multiparameter universality and intrinsic diversity of critical phenomena in weakly anisotropic systems, *Phys. Rev. E* **108**, 044149 (2023).
64. V. Dohm, F. Kischel and S. Wessel, Multiparameter universality of exact self-similar finite-size scaling functions at and away from criticality of anisotropic systems, manuscript in preparation.
65. F. Kischel, S. Wessel and V. Dohm, Exact free energy of the anisotropic triangular-lattice Ising model in parallelogram geometry, manuscript in preparation.
66. R. J. Baxter, *Exactly Solved Models in Statistical Mechanics* (Academic, London, 1982).
67. E. Brézin, An investigation of finite-size scaling. *J. Phys. (Paris) B* **43**, 15 (1982).
68. G. 't Hooft and M. Veltman, Regularization and renormalization of gauge fields. *Nucl. Phys. B* **44**, 189 (1972).
69. E. Brézin and J. Zinn-Justin, Finite size effects in phase transitions. *Nucl. Phys. B* **25 [FS] 14**, 867 (1985).
70. J. Rudnick, H. Guo and D. Jasnow, Finite-size scaling and the renormalization group. *J. Stat. Phys.* **41**, 353 (1985).
71. A. Esser, V. Dohm and X. S. Chen, Field theory of finite-size effects for systems with a one-component order parameter. *Physica A* **222**, 355 (1995).
72. V. Dohm, Crossover from Goldstone to critical fluctuations: Casimir forces in confined $O(n)$-symmetric systems. *Phys. Rev. Lett.* **110**, 107207 (2013).
73. X. S. Chen, V. Dohm and N. Schultka, Order-parameter distribution function of finite $O(n)$-symmetric systems. *Phys. Rev. Lett.* **77**, 3641 (1996).
74. X. S. Chen and V. Dohm, Renormalization-group theory of critical phenomena in confined systems: Order-parameter distribution function. *Int. J. Mod. Phys. B* **12**, 1277 (1998).
75. V. Dohm, Critical Casimir force in slab geometry with finite aspect ratio: Analytic calculation above and below T_c. *Europhys. Lett.* **86**, 20001 (2009).

76. V. Dohm, Pronounced minimum of the thermodynamic Casimir forces of $O(n)$ symmetric film systems: Analytic theory. *Phys. Rev. E* **90**, 030101(R) (2014).
77. R. Garcia and M. H. W. Chan, Critical fluctuation-induced thinning of ^{4}He films near the superfluid transition. *Phys. Rev. Lett.* **83**, 1187 (1999).
78. G. A. Williams, Vortex fluctuations in the critical Casimir effect of superfluid and superconducting films. *Phys. Rev. Lett.* **92**, 197003 (2004).
79. K. G. Wilson and M. E. Fisher, Critical exponents in 3.99 dimensions. *Phys. Rev. Lett.* **28**, 240 (1972).
80. E. Brézin and J. Zinn-Justin, Renormalization of the nonlinear σ model in $2 + \varepsilon$ dimensions — Application to the Heisenberg ferromagnets. *Phys. Rev. Lett.* **36**, 691 (1976).
81. M. Strösser and V. Dohm, Minimal renormalization without ε expansion: Four-loop free energy in three dimensions for general n above and below T_c. *Phys. Rev. E* **67**, 056115 (2003).
82. S. A. Larin, M. Mönnigmann, M. Strösser and V. Dohm, Five-loop additive renormalization in the φ^4 theory and amplitude functions of the minimally renormalized specific heat in three dimensions. *Phys. Rev. B* **58**, 3394 (1998).
83. M. E. Fisher and P. G. de Gennes, Phénomènes aux parois dans un mélange binaire critique. *C.R. Seances Acad. Sci. Paris Ser. B* **287**, 207 (1978).
84. F. P. Toldin and S. Dietrich, Critical Casimir forces and adsorption profiles in the presence of a chemically structured substrate. *J. Stat. Mech.* **P**11003 (2010).
85. H. W. Diehl and D. Grüneberg, Critical Casimir amplitudes for n-component φ^4 models with $O(n)$-symmetry breaking quadratic boundary terms. *Nucl. Phys. B* **822**, 517 (2009).
86. M. Burgsmüller, H. W. Diehl and M. A. Shpot, Fluctuation-induced forces in strongly anisotropic critical systems. *J. Stat. Mech.* **P**11020 (2010).
87. D. M. Dantchev and S. Dietrich, Critical Casimir effect: Exact results. *arXiv:* 2203.15050v1.
88. D. Dantchev, M. Krech and S. Dietrich, Thermodynamic Casimir force in models of ^{4}He films. *Phys. Rev. Lett.* **95**, 259701 (2005).
89. S. Singh, Phase transitions in liquid crystals. *Phys. Rep.* **324**, 107 (2000).
90. H. B. Tarko and M. E. Fisher, Theory of critical point scattering and correlations. III. The Ising model below T_c and in a field. *Phys. Rev. B* **11**, 1217 (1975).

Appendix: Letter from Michael Fisher

UNIVERSITY OF
MARYLAND
INSTITUTE FOR PHYSICAL SCIENCE AND TECHNOLOGY

College Park, Maryland 20742
301.405.4874 TEL 301.314.9363 FAX
301.314.9404 FAX

31 August 2008

Professor Volker Dohm
Institute Theoretical Physics
RWTM Aachen University
D-52056 Aachen, Germany

RE: Phys. Rev. E 77
061128 (2008)

Dear Volker:

I am writing to thank you for sending me a hard copy – great to have! – of your magnum opus on universality and its limitations in finite systems (especially!) relative to overoptimistic expectations. I value this article as much as a review as for the specific results; but it's very good to see the anisotropy matrix $\underset{\sim}{A}$ fully recognized as well as to see your revised interpretations of some of your earlier results with X.S. Chen (Have you thought of a Physics Report?)

You will allow me to also remark that it is the appreciation of the subtitles exhibited by your Figs. 1 and 2 that, many years ago, lead David Jasnow and me (in our infamous unpublished VOL. 4) to hesitate to give a mathematical definition of scaling that one might propose as a "theorem" suitable for proof!

All Best Wishes,

Michael E. Fisher

Cc: J.V. Sengers, IPST
Dr. Y. C. Kim, NIH
Prof. Walter Selke

https://doi.org/10.1142/9789811282386_0032

Chapter 32

Michael Fisher's Relationship to Monte Carlo

D.P. Landau* and W. Selke†

**The University of Georgia, Athens, GA 30602, USA*

†RWTH Aachen University, 52062 Aachen, Germany

1. Introduction

While Michael Fisher has often been viewed as a critic of Monte Carlo simulations, it is important to remember that his decades-long involvement with computers was substantial. From his early work on analog computing to his about 20 co-authored papers involving Monte Carlo simulations, his involvement in computational physics was manifest. In spite of his early criticism of Monte Carlo research, it is perhaps most accurate to say that, as with all of his works, he was a critic of poor Monte Carlo. The true meaning of the famous saying first attributed to John von Neumann (and upon occasion repeated by Michael Fisher), "The only good Monte Carlo is a dead Monte Carlo," was perhaps best explained by John Tukey; "This aphorism was coined to express the view that out of a well conducted Monte Carlo should come enough insight to allow us to use newly-developed or newly-chosen approximations to solve other cases of that particular complexity, thus needing to use Monte Carlo again only when we want to go still further or still deeper."[1]

In the following Sections, we choose a few examples of Michael Fisher's Monte Carlo work as well as other Monte Carlo studies that relate strongly to other research in statistical physics to which he has contributed fundamentally. The choice of research problems that will be included is personal, and there are numerous other important studies that could not be covered in the brief account presented here.

2. Monte Carlo Simulations Related to Michael Fisher's Fundamental Contributions

Michael Fisher is best known for his seminal contributions to the theory of phase transitions and critical phenomena. Indeed, he received numerous awards and honors mentioning his important ideas and discoveries. In this chapter, we primarily discuss the impact of some of his fundamental insights on Monte Carlo (MC) simulations related to critical phenomena.

2.1. *Finite-size scaling*

Phase transitions occur in the thermodynamic limit while MC simulations consider finite systems. In fact, consideration of finite-size effects is essential for analysis in the critical region. The crucial theoretical approach for dealing with this problem was introduced by Michael Fisher and Michael Barber, his doctoral student at that time.[2,3] Nowadays, their finite-size scaling (fss) ansatz provides a standard framework for the extraction of critical properties from finite system size MC data, see also articles by Volker Dohm and by Peter Young in this book.

Perhaps, the earliest validation of finite-size scaling theory for a "standard" model in statistical physics was presented by Kurt Binder's Monte Carlo study of Ising films with different numbers of layers n.[4] As shown in Fig. 1, the values for the order parameter (spontaneous magnetization) collapse onto a single curve when plotted in finite-size scaling form. Binder's simulations also confirm critical exponent spatial dimensional crossover with increasing film thickness. Finite-size scaling was further confirmed for the 2D Ising model via Monte Carlo simulations by David Landau (see Fig. 2).[5] These same Monte Carlo simulations were used to demonstrate finite-size scaling of the spin–spin correlation function $\Gamma(r, \epsilon, N)$, where r is the separation between spins, N is the length of a side of the lattice, and $\epsilon = |1 - T/T_c|$.

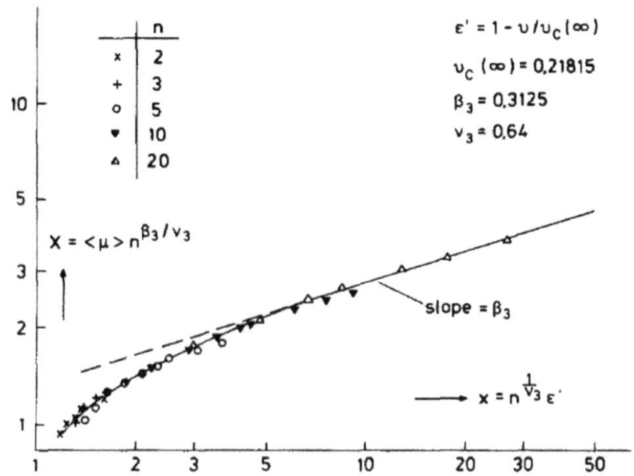

Fig. 1. Finite-size scaling function of the spontaneous magnetization for thin Ising films composed of n layers [After Ref. 4].

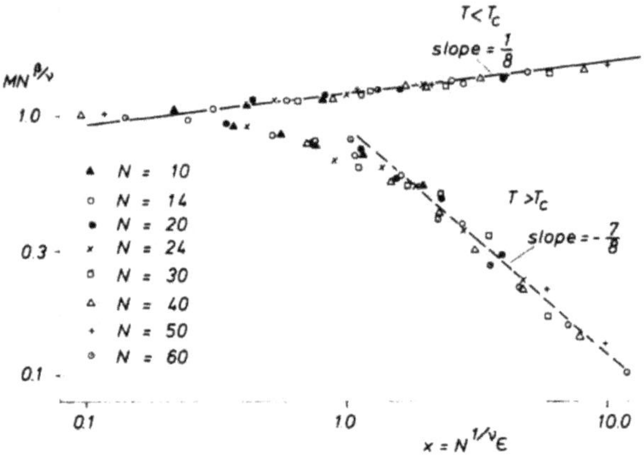

Fig. 2. Finite-size scaling function of the spontaneous magnetization for the $N \times N$ Ising models with periodic boundary conditions. Here, $\epsilon = |1 - T/T_c|$ [After Ref. 5].

2.2. *Critical behavior*

Apart from fss, Fisher contributed essentially to the foundations of the theory of phase transitions, beginning in the 1960s, by introducing critical exponents for the correlation length as well as clarifying relations between critical exponents, as has been discussed by Cyril Domb.[6] All these insights play an important role, for instance, in a large number of Monte Carlo simulations of the nearest-neighbor Ising model on the simple cubic lattice.

Building on the renormalization-group (RG) theory foundation, laid by Leo Kadanoff and Ken Wilson, Michael Fisher developed a picture of flows along a critical trajectory with the repeated application of a RG transformation, as shown in Fig. 3, from Ref. 7. This picture was clearly verified for the three-dimensional Ising model by Stuart Pawley and colleagues[8] using a Monte Carlo renormalization-group (MCRG) simulation that also showed that additional couplings were generated with repeated transformations in accordance with Fisher.[7] Martin Hasenbusch then exploited the possibility of using an "improved" Hamiltonian with additional couplings to determine accurately the critical exponents for the three-dimensional Ising model.[9] A higher resolution study of the nearest-neighbor model, complete with a careful finite-size analysis using Fisher's finite-size scaling theory and systems as large as one billion spins, was then conducted recently by Ferrenberg, Xu, and Landau.[10]

2.3. *Multicritical behavior*

Multicritical points may occur when lines of critical points converge and the nature of the phase transition changes at that point. Among Michael Fisher's seminal contributions to statistical physics was the application of RG theory to critical and multicritical phenomena. The 1970s was a time of tremendous interest in

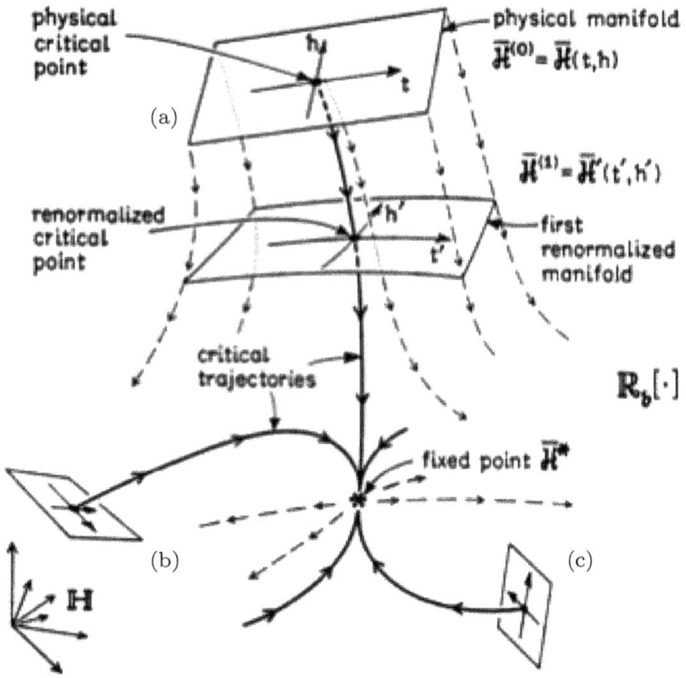

Fig. 3. A depiction of the space of Hamiltonians **H** showing initial or physical manifolds [labelled (a), (b), ...] and the flows induced by repeated application of a discrete RG transformation $\mathbf{R_b}$ with a spatial rescaling factor b. Critical trajectories are shown in bold: they all terminate, in the region of **H** shown here, at a fixed point $\mathbf{H}^*$. The full space might contain other non-trivial, critical fixed points, describing multicritical points and distinct critical-point universality classes (After Ref. 7).

such phenomena, and Robert Griffiths and Michael Fisher played important roles in introducing and naming such multicritical points, as has been discussed, for instance, in an article by Michael Kosterlitz, David Nelson, and Michael Fisher in 1976.[11] In that article, the three authors used RG to study the three-dimensional anisotropic Heisenberg antiferromagnet subject to uniform and staggered magnetic fields $(H_{||}, H_{||}^{+})$, respectively. In the $H_{||} - T$ plane, the low-T, low-$H_{||}$ ordered state is a two-sublattice antiferromagnet (AF) which is separated from the paramagnetic state (P) by a line of Ising-like, second-order transitions. Sweeping $H_{||}$ at low enough T produces a first-order transition to a canted "spin-flop" (SF) state which is also separated from the paramagnetic phase by a line of second-order transitions. The main focus of that article is on the possible existence of bicritical and tetracritical points in the phase diagrams as a function of temperature and these two fields, and it yielded the phase diagram prediction shown qualitatively in Fig. 4.

This RG study motivated careful Monte Carlo simulations by David Landau and Kurt Binder on these XXZ antiferromagnetic models.[12] Interestingly, topologies of the resulting phase diagram, see Fig. 5, showed some significant differences

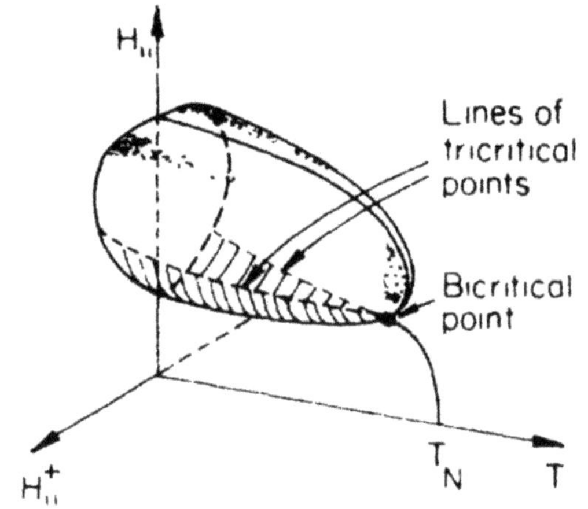

Fig. 4. Phase diagram with both uniform and staggered fields from RG predictions. The shaded lower portion of the "balloon" represents a region of first-order transitions which terminates in a line of tricritical points (After Ref. 11).

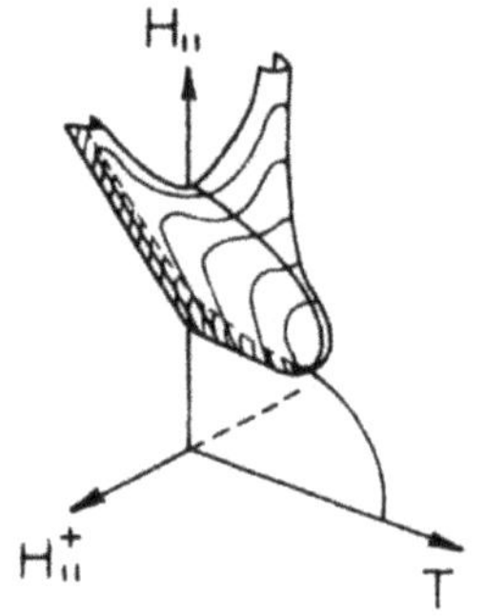

Fig. 5. Phase diagram with both uniform and staggered fields from Monte Carlo simulations (After Ref. 12).

compared to the analytic predictions. Indeed, the discrepancy has contributed to a long-lasting interest in multicritical points in those models using both analytic and simulational methods. For example, we refer to a recent publication by Amnon Aharony and Ora Entin-Wohlman about the nature of the bicritical point in anisotropic Heisenberg antiferromagnets.[13]

2.4. *Droplets and quantum chains*

The formation of equilibrium droplets at a first-order transition has been studied by Michael Fisher,[14] following earlier descriptions. The 'Fisher droplets' received much attention, especially within the context of Monte Carlo simulations. A few examples include applications in percolation theory,[15] in Ising models,[16] and in the development of cluster-flip algorithms.[17,18]

A further pioneering work of Fisher, which inspired simulations, is that on quantum anisotropic Ising–Heisenberg (XXZ) spin chains in a field. In a citation classic, Jill Bonner and Michael Fisher used exact machine calculations for rather short chains, to obtain results for various quantities, including dispersion laws, specific heats and susceptibilities.[19] Several related quantum Monte Carlo simulations have been performed, e.g., by Douglas Scalapino who published results on an array of such chains, using the world line method.[20] In the limit of vanishing couplings between the chains, results of Bonner and Fisher were recovered. In addition, evidence for a phase transition in the three-dimensional case was presented.

3. Publications Co-authored by Michael Fisher Using MC Simulations

Michael Fisher contributed to about 20 publications based largely on simulations. His major contribution was to interpret these numerical data and to relate them to analytical results.

The first Monte Carlo simulations co-authored by Michael Fisher were published in 1975. It was a joint work with the leading expert, at that time, on computational physics at Cornell University, Geoffrey Chester, and two, then junior scientists, Eytan Domany and Kin Mon.[21] The article deals with Ising models on a square lattice with nearest- and next-nearest-neighbor interactions as well as four-spin couplings, including generalized eight-vertex or Baxter models. Through the application of standard Monte Carlo techniques, the specific heat near criticality for rather small lattices was computed. Using finite-size scaling theory, findings were observed to agree, for various cases, with known exact analytic results.

In the following simulation, published in 1979 together with the Postdoc Walter Selke,[22] a model that had been rather neglected up to that time, the axial next-nearest-neighbor Ising or ANNNI model (a term introduced by Fisher[23]) on a simple cubic lattice, has been analyzed. In addition to nearest-neighbor ferromagnetic interactions, competing antiferromagnetic couplings to next-nearest-neighbor spins along one of the axes, e.g., the z-axis, are assumed. As a consequence, spatially modulated, commensurate or incommensurate structures occur. An example is seen in Fig. 6, showing the temperature dependence of the magnetization per layer at a fixed ratio of the competing interactions. At low temperatures, a simple commensurate structure is observed in the simulation, while at higher temperatures, the periodicity of an apparently sinusoidal magnetization pattern is changing. Obviously, a rather complicated phase diagram occurs, including a multicritical Lifshitz point. Michael Fisher showed great interest in these fascinating features. To establish fine details of the phase diagram, Fisher and Selke then performed low-temperature series expansions,[24] finding sequences of commensurate structures, resembling a devil's staircase. Later, they performed additional MC simulations, again using the Metropolis algorithm, for the ANNNI model on the square lattice.[25] In that

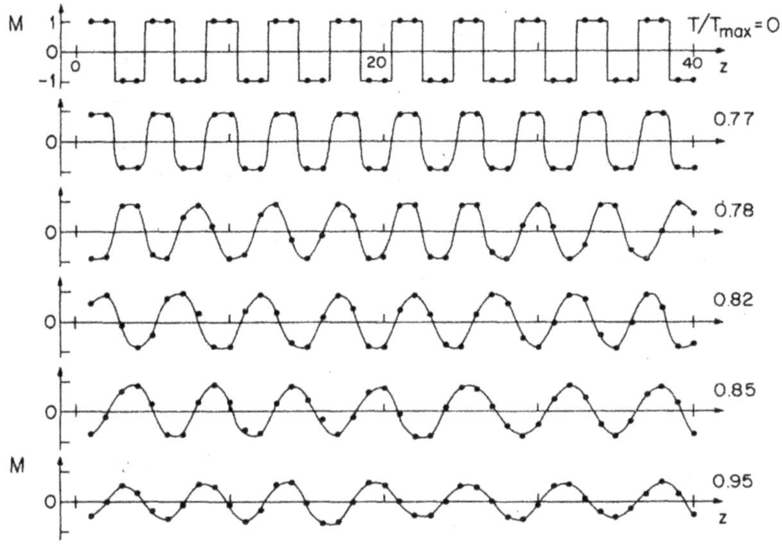

Fig. 6. Sequence of magnetization plots showing the squeezing effect in a $6 \times 6 \times 40$ system at $k_B T = 3.025 J_1$ where J_1 is the nearest-neighbor interaction in the z-direction. The evolution of the magnetization per layer, $M(x)$, in a single run between 180 and 260 MCS/S (averages are taken over 20 MCS/S) is shown (note that MCS/S means Monte Carlo steps per spin) (After Ref. 22).

case, the phase transition from the modulated phase to the disordered phase is expected to belong to the universality class of the two-dimensional XY model, i.e. to be of (Berezinskii)-Kosterlitz–Thouless type, with the unbinding of vortex pairs. Indeed, based on the simulated spin configurations, Michael Fisher identified the corresponding topological objects in the two-dimensional ANNNI model. Results on this and related models, in different spatial dimensions, obtained using various approaches, have been reviewed in Ref. 23.

The next set of Monte Carlo simulations, co-authored by Michael Fisher, has been performed in the years 1986 to 1992. These simulations dealt with properties of vesicles in two dimensions. The simulations started at Cornell University with the first publication on this topic appearing in 1987.[26] All three authors listed the Baker Laboratory of Cornell University as their home address, with Stanislas Leibler and Rajiv Singh being research associates. The simulations nicely confirmed, among others, predictions on various scaling functions and critical exponents for self-avoiding walks. Shortly later in 1987, Fisher moved to the University of Maryland where he then continued to work in this field both analytically and numerically. Using Metropolis Monte Carlo simulations as a primary tool, he studied various types of vesicles, including branched polymers and membranes. Sizes and shapes of the vesicles may change very markedly, leading to different regimes with typical scaling behavior and corresponding transformations or transitions. Interesting results appeared in several publications, e.g., Refs. 27–29. Among others, Carlos Camacho, who did his Ph.D. work at the University of Maryland, played an important role in these studies.

From 2000 until 2011, Michael Fisher co-authored more than 10 articles on Monte Carlo simulations about phase transitions in liquids. Grand canonical simulations were usually used to take into account important density fluctuations. In the beginning, numerical support was provided, especially, by Athanassios Panagiotopoulos and Gerrassimos Orkoulos, both originally from Greece, as well as Young Kim from Korea, and Eric Luijten from the Netherlands. All individuals had joined Fishers's team at the University of Maryland. Later Kurt Binder, one of the leading Monte Carlo experts in the world, participated in the project, together with his coworker Jürgen Horbach and with Subir Das. One may mention that Subir Das and Erik Luijten were both first postdocs of Binder in Mainz and then of Fisher at the University of Maryland. Some of the resulting publications on liquids, combining numerics and analytical insights, will be discussed shortly. In particular, critical properties of liquid–vapor transitions have been studied in the first stage of the project. Hereby, asymmetry effects for critical and scaling properties played a major role,[30–36] in comparing Monte Carlo data to established theories, like the Yang–Yang anomaly, and to modified analytic approaches. The study of phase transitions in such systems lacking clear symmetry is a challenging problem.

A concrete example of a study of the type described in the previous paragraph is the examination of asymmetric fluid criticality in which the second derivative of the chemical potential diverges as the critical point is approached from below.[35] Kim, Fisher, and Luijten (2003) used grand canonical Monte Carlo, together with a finite-size analysis, to identify and include pressure mixing effects for the hard-core square-well fluid and for the restricted primitive model (RPM) electrolyte. In Fig. 7 we show their high quality Monte Carlo data for the discontinuity of the density as the critical point is approached. While high resolution is achieved close to the critical point, finite size rounding begins to smear out the discontinuity and a finite size scaling study is needed. In fact, they performed such a study and found excellent scaling behavior from which the critical point could be accurately located and the coexistence curve could be precisely determined. Motivated by the simulations on fluids, screening properties of charge-symmetric electrolytes have been studied.[37] Another challenging class of liquids, within the context of phase transitions, are binary liquid mixtures. They have been investigated in joint collaboration of Michael Fisher, Kurt Binder, and coworkers.[38–40] Combining Monte Carlo simulations and molecular dynamics static and dynamic critical behavior of a symmetric binary Lennard-Jones mixture near its consolute point has been studied. Numerical results are found to agree with predictions by renormalization group and mode-coupling theories as well as experiments, provided finite-size corrections are taken into account properly.

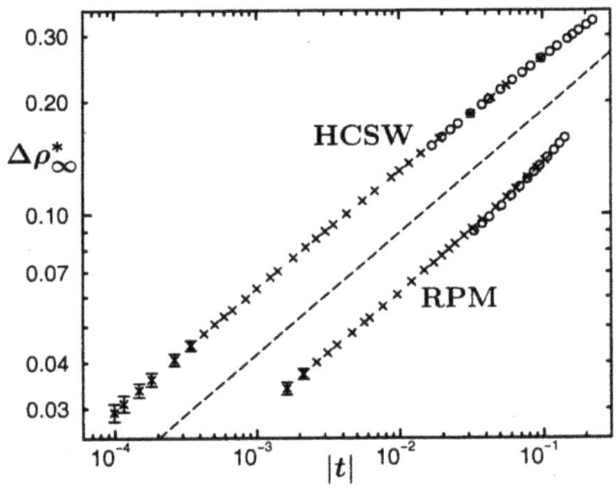

Fig. 7. A log–log plot of the reduced semi-density jump, $\rho_0^* = \frac{1}{2}[\rho^+(T) - \rho^-(T)]a^3$ vs $t = (|T - T_c/T_c|)$, where a is the hard- sphere diameter, for a HCSW fluid with interaction range $1.5a$ (and $\rho^* = 0.3067$) and for the RPM with $\rho^* = 0.079$. The circles report previous estimates for the RPM and HCSW fluid employing an equal-weight prescription. The dashed line has a slope $\beta_{Ising} = 0.326$ (After Ref. 35).

In summary, for many years, from 1975 to 2011, Michael Fisher coauthored articles on Monte Carlo simulations, comparing numerical results to theoretical predictions and experimental data.

4. Final Remarks and Acknowledgements

As a graduate student, DPL found, when performing experimental studies of thermodynamic behavior of an Ising-like antiferromagnet, $Dy_3Al_5O_{12}$, which orders at liquid Helium temperatures, that the applied field-induced transition appeared to be of higher than second order. His advisor, Werner Wolf, remarked that "Michael Fisher says it must be second order and Michael Fisher is always right." It was later shown that an applied field actually coupled to the staggered magnetization so that the experiment actually was conducted slightly out of the uniform field — temperature plane — and passed nearby the transition line but did not actually cross it. Hence, Michael Fisher was indeed right. WS expresses his appreciation for the excellent guidance provided by Michael Fisher as his host in 1978 and 1979 as well as numerous most useful correspondences and meetings in the following decades.

References

1. J. W. Tukey, Data analysis, computation and mathematics. *Quart. Appl. Math.* **30**, 51 (1972).
2. M. E. Fisher, The theory of critical point singularities. in Critical Phenomena, Proc. of 51st Enrico Fermi Summer School, Varenna, ed. by M.S. Green (Academic Press, New York, 1971).

3. M. E. Fisher and M. N. Barber, Scaling theory for finite-size effects in the critical region. *Phys. Rev. Lett.* **28**, 1516 (1972).
4. K. Binder, Monte Carlo study of thin magnetic Ising films. *Thin Solid Films* **20**, 367 (1974).
5. D. P. Landau, Finite-size behavior of the Ising square lattice. *Phys. Rev. B* **13**, 2997 (1976).
6. C. Domb, Michael Fisher at King's College London. *Physica A* **28**, 177 (1991).
7. M. E. Fisher, Renormalization group theory: Its basis and formulation in statistical physics. *Rev. Mod. Phys.* **70**, 653 (1998).
8. G. S. Pawley, R. H. Swendsen, D. J. Wallace and K. G. Wilson, Monte Carlo renormalization-group calculations of critical behavior in the simple-cubic Ising model. *Phys. Rev. B* **29**, 4030 (1984).
9. M. Hasenbusch, Finite size scaling study of lattice models in the three-dimensional Ising universality class. *Phys. Rev. B* **82**, 174433 (2010).
10. A. M. Ferrenberg, J. Xu and D. P. Landau, Pushing the limits of Monte Carlo simulations for the three-dimensional Ising model. *Phys. Rev. E* **97**, 043301 (2018).
11. J. M. Kosterlitz, D. R. Nelson and M. E. Fisher, Bicritical and tetracritical points in anisotropic antiferromagnetic systems. *Phys. Rev. B* **13**, 412 (1976).
12. D. P. Landau and K. Binder, Phase diagrams and multicritical behavior of a three-dimensional anisotropic Heisenberg antiferromagnet. *Phys. Rev. B* **17**, 2328 (1978).
13. A. Aharony and O. Entin-Wohlman, Puzzle of bicriticality in the XXZ antiferromagnet. *Phys. Rev. B* **106**, 094424 (2022).
14. M. E. Fisher, The theory of condensation and the critical point. *Physics Physique Fizika* **3**, 255 (1967).
15. D. Stauffer, Scaling theory of percolation clusters. *Phys. Rep.* **54**, 1 (1979).
16. A. Nussbaumer, E. Bittner, T. Neuhaus and W. Janke, Monte Carlo study of the evaporation/condensation transition of Ising droplets. *Europhys. Lett.* **75**, 716 (2006).
17. R. H. Swendsen and J.-S. Wang, Nonuniversal critical dynamics in Monte Carlo simulations. *Phys. Rev. Lett.* **58**, 86 (1987).
18. U. Wolff, Monte Carlo simulation of a lattice field theory as correlated percolation. *Nucl. Phys. B* **300**, 501 (1988).
19. J. C. Bonner and M. E. Fisher, Linear magnetic chains with anisotropic coupling. *Phys. Rev.* **135**, A640 (1964).
20. D. J. Scalapino, Simulations: A tool for studying quantum condensed matter systems. *J. Stat. Phys.* **43**, 757 (1986).
21. E. Domany, K. K. Mon, G. V. Chester and M. E. Fisher, Monte Carlo study of multicriticality in finite Baxter models. *Phys. Rev. B* **12**, 5025 (1975).
22. W. Selke and M. E. Fisher, Monte Carlo study of the spatially modulated phase in an Ising model. *Phys. Rev. B* **20**, 257 (1979).
23. W. Selke, The ANNNI model- Theoretical analysis and experimental application. *Phys. Rep* **170**, 213 (1988).
24. M. E. Fisher and W. Selke, Infinitely many commensurate phases in a simple ising model. *Phys. Rev. Lett.* **44**, 1502 (1980).
25. W. Selke and M. E. Fisher, Two-dimensional Ising models with competing interactions — A Monte Carlo study. *Z. Physik B* **40**, 71 (1980).
26. S. Leibler, R. R. P. Singh and M. E. Fisher, Thermodynamic behavior of two-dimensional vesicles. *Phys. Rev. Lett.* **59**, 1989 (1987).
27. A. C. Maggs, S. Leibler, M. E. Fisher and C. J. Camacho, Size of an inflated vesicle in two dimensions. *Phys. Rev. A* **42**, 691 (1990).

28. C. J. Camacho, M. E. Fisher and R. R. P. Singh, Semiflexible planar polymeric loops. *J. Chem. Phys.* **94**, 5693 (1991).
29. C. J. Camacho, M. E. Fisher and J. P. Straley, Rods of self-avoiding walks of trees in two dimensions. *Phys. Rev. A* **46**, 6300 (1992).
30. J. M. Romero-Enrique, G. Orkoulas, A. Z. Panagiotopoulos and M. E. Fisher, Coexistence and criticality in size-asymmetric hard-core electrolytes. *Phys. Rev. Lett.* **85**, 4558 (2000).
31. G. Orkoulas, M. E. Fisher and A. Z. Panagiotopoulos, Precise simulation of criticality in asymmetric fluids. *Phys. Rev. E* **63**, 051507 (2001).
32. A. Z. Panagiotopoulos and M. E. Fisher, Phase Transitions in 2.1 and 3.1 hard-core model electrolytes. *Phys. Rev. Lett.* **88**, 045701 (2002).
33. E. Luijten, M. E. Fisher and A. Z. Panagiotopoulos, Universality class of criticality in the restrictive primitive model electrolyte. *Phys. Rev. Lett.* **88**, 185701 (2002).
34. Y. C. Kim, M. E. Fisher and G. Orkoulas, Asymmetric fluid criticality. I. Scaling with pressure mixing. *Phys. Rev. E* **67**, 061506 (2003).
35. Y. C. Kim, M. E. Fisher and E. Luijten, Precise simulation of near-critical fluid coexistence. *Phys. Rev. Lett.* **91**, 065701 (2003).
36. Y. C. Kim, E. Luijten and M.E. Fisher, Screening in ionic systems: Simulations for the lebowitz length. *Phys. Rev. Lett.* **95**, 14570 (2005).
37. S. K. Das, Y. C. Kim and M. E. Fisher, When is a conductor not perfect? Sum rules fail under critical fluctuations. *Phys. Rev. Lett.* **107**, 215701 (2011).
38. S. K. Das, J. Horbach, K. Binder and M. E. Fisher, Static and dynamic critical behavior of a symmetrical binary fluid: A computer simulation. *J. Chem. Phys.* **125**, 024506 (2006).
39. S. K. Das, M. E. Fisher, J. V. Sengers, J. Horbach and K. Binder, Critical dynamics in a binary fluid: Simulations and finite-size scaling. *Phys. Rev. Lett.* **97**, 025702 (2006).
40. K. Binder, S. K. Das, M. E. Fisher, J. Horbach and J. V. Sengers, Interdiffusion in critical binary mixtures by molecular dynamics. *Diff. Fundam.* **6**, 10.1 (2007).

https://doi.org/10.1142/9789811282386_0033

Chapter 33

Finite-Size Scaling

A.P. Young

Physics Department,
University of California Santa Cruz,
Santa Cruz, CA 95060, USA
petery@ucsc.edu

I discuss the important work of Michael Fisher in developing the theory of finite-size scaling, which is now widely used in simulations of critical phenomena. A brief overview of the history of the subject and its formulation will be given. As an example of the power of FSS when combined with careful, large-scale Monte Carlo simulations, I briefly discuss the work which accurately determined the critical behavior of the three-dimensional Ising spin glass. Finite-size scaling breaks down above the upper critical dimension, which is $d = 4$ for most systems, because of a "dangerous irrelevant" variable, a term first coined by Michael. A brief discussion of how standard finite-size scaling has to be modified in this region will be given at the end.

1. My Debt to Michael Fisher

Although I was never a student or post-doc under Michael, and we never wrote a paper together, I nonetheless owe him a great debt of gratitude. I was a post-doc at Cornell for the academic year 1977–1978 and had just been informed that I was shortlisted for a lectureship at Imperial College, London, when I bumped into Michael in the elevator in Clark Hall. Normally, I would be invited for an interview, but Imperial didn't want to pay for my air fare from the US, so instead they asked me to send them my views on teaching. Unfortunately, at that time, I had never taught a class and so didn't have anything interesting to say on that topic. Michael, who went all the way through the British academic system from student to professor, emphasized the importance of the interview in the British system and told me that I should go if at all possible, even if I had to pay for the airfare myself. Fortunately, Freddie Laker had just started his low-cost airline, so last-minute tickets were available at a non-punitive cost from the East Coast to England and off I went. The interview must have gone well because I got a job offer. In fact, Imperial managed to create a second position for me in addition to the original position which they offered to David Pettifor. They would never have done this based on a non-descript statement from me on teaching. Hence, I owe my first academic job to Michael (with some additional help from Freddie Laker!).

2. Birth of Finite-Size Scaling

We know that a sharp phase transition can only take place in an infinite system, while for a finite system the change in behavior at a transition is "rounded out" in some way. One of Michael's many contributions to science was to pioneer a study[1,2] of how this rounding occurs in large but finite systems, a set of ideas which came to be known as finite-size scaling (FSS). The first articles on FSS are lectures by Michael in the proceedings of a summer school in Varenna[1] in 1971 and a paper with a student, Michael Barber,[2] in 1972. However, according to Barber (private communication), it is not clear if that part of the published lectures which is on FSS was actually presented orally at the school, and finite-size scaling theory may have first seen the light of day slightly later in 1971 at an IUPAP meeting in Chicago.

Nowadays, numerical studies of phase transitions are ubiquitous and a careful application of FSS to the results is necessary to arrive at accurate conclusions. However, in 1971, numerical simulations of phase transitions had not started, so the initial work of Michael on FSS was ahead of its time. In fact, the biggest application of FSS has been to Monte Carlo simulations, a technique which, at least by reputation, Michael did not like. However, I think that his criticism was really leveled against the many *careless* Monte Carlo papers in the early days, which did not check for equilibration and, in particular, did not extrapolate results to infinite system size using FSS rather than against the Monte Carlo method *per se*. Indeed, in later years, he was a coauthor on several high-quality Monte Carlo studies, such as his work with Erik Luijten and others on electrolytes.[3] See also similar comments in the chapter by Landau and Selke in this volume.

This chapter focuses on the history of FSS and its basic ideas. A useful reference for these aspects is the book by Cardy[4] which includes reprints of some of the early papers including Ref. 2.

3. The Formulation of Finite-Size Scaling

I will consider systems which are finite in all d dimensions, as in Monte Carlo simulations. One can also consider other geometries in which one or more dimensions are infinite, the most important case being *one* infinite dimension which corresponds to transfer matrix calculations and also to quantum systems at zero temperature. However, I will not consider that situation here. For simplicity, I assume that the system has the same size L in each of d dimensions, so the total size of the system is given by

$$N = L^d. \tag{1}$$

I will consider models for magnetic materials in which a "spin" lies on each site of a hypercubic regular lattice and interacts with its neighbors. Most commonly, a spin might be a scalar that takes values $S_i = \pm 1$, which is known as the Ising model, but sometimes classical m-component vectors are taken instead. The examples considered here will all be for Ising models.

As soon as we move away from the thermodynamic limit and discuss finite-size systems, we have to specify the boundary conditions. In simulations, one would like finite-size corrections to be as small as possible, so normally one considers period boundary conditions, where a site with coordinate $L + 1$ in one of the directions is identified with site 1 in that direction. This means that there is no surface. Here, I will mainly consider periodic boundary conditions, but in Section 6, I will include the effects of surfaces by choosing free boundaries, in which a site on the surface has no coupling going outside the surface. One can convert periodic boundary conditions to free boundary conditions simply by setting to zero those interactions which "wrap around" the system.

Even with periodic boundary conditions, one expects results to depend on the system size in the vicinity of a second-order phase transition because the correlation length ξ of an infinite system diverges like

$$\xi \sim t^{-\nu}, \tag{2}$$

where

$$t = \frac{T - T_c}{T_c} \tag{3}$$

is called the reduced temperature, T_c is the transition temperature, and ν is a critical exponent. When t is sufficiently small such that $\xi \gtrsim L$, results will depend on L. By contrast, well above T_c, where $L \gg \xi$, one expects properties of a system with periodic boundary conditions to differ by only exponentially small corrections of order $\exp(-L/\xi)$ from those of an infinite system. With free boundaries, where there is a surface, corrections will be much larger in this limit, typically of $1/L$ relative to the bulk values.

Let us consider some quantity X, say, which diverges at T_c as

$$X_\infty \approx A t^{-\lambda} \tag{4}$$

in the thermodynamic limit. A simple postulate for how X might vary on a finite system is

$$\frac{X_L(t)}{X_\infty(t)} = f\left(\frac{L}{\xi}\right) \tag{5}$$

for some function $f(x)$. One expects that f is universal in the renormalization group sense, i.e. it does not depend on the lattice structure or irrelevant variables. One does, however, expect that it will depend on the boundary conditions.

We can deduce the form of f in two limits:

(1) At a fixed t in the limit $L \to \infty$, we have $X_L \to X_\infty$, so

$$\lim_{x\to\infty} f(x) = 1. \tag{6}$$

(2) At L fixed and $t \to 0$, $X_L(t)$ is not singular because there can be no sharp transition in a finite system. This means that the behavior of $f(x)$ for $x \to 0$ should compensate for the divergence if X_∞, i.e.

$$f(x) = Cx^{\lambda/\nu} \quad \text{for } x \to 0. \tag{7}$$

In fact, $X_L(t)$ varies smoothly around $t = 0$ and one can incorporate this as well as have a single formula which works both above and below T_c by rewriting Eq. (5) as

$$X_L(t) = L^{\lambda/\nu}\, \tilde{X}(L^{1/\nu}\, t), \tag{8}$$

where, ignoring amplitudes like A in Eq. (4) for simplicity, one has

$$\tilde{X}(x) = x^{-\lambda} f(x^\nu). \tag{9}$$

Since $X_L(0)$ is non-zero and finite, we must have $\tilde{X}(0) = \text{const.}$, so Eq. (8) gives a prediction for the size dependence at T_c, namely

$$X_L(0) \sim L^{\lambda/\nu}. \tag{10}$$

Equation (8) is the basic postulate of finite-size scaling. For what quantities X would we typically apply it? Naturally, we would like a quantity with a strong divergence, i.e. λ is greater than zero and not too small. An "obvious" choice is the susceptibility χ, which diverges with an exponent γ which is exactly 7/4 in the two-dimensional Ising model and about 1.24 in the three-dimensional Ising model. However, determining χ is difficult in numerical simulations for the following reason. It is computed from spin correlations using the relation

$$\chi = \frac{1}{T}\frac{1}{N}\sum_{i,j} \left[\langle S_i S_j\rangle - \langle S_i\rangle\langle S_j\rangle\right] \tag{11}$$

$$= \frac{1}{T} N \left[\langle m^2\rangle - \langle m\rangle^2\right], \tag{12}$$

where

$$m = \frac{1}{N}\sum_{i=1}^{N} S_i \tag{13}$$

is the magnetization per site. However, for a finite system and in zero external magnetic field, strictly speaking, one has $\langle m\rangle = 0$. In practice, below T_c, the magnetization will very slowly fluctuate between the "up" state and the "down" state with a time that varies strongly with system size and temperature, and may be much longer than the time of the simulation. Hence, at least below T_c, it is useful to calculate only quantities which are invariant under the global symmetry of the model. Here, for an Ising model in zero field, this is inversion of all the spins.

Some proposed fixes for this problem are as follows:

(1) Replace the second term in Eq. (12) by $\langle |m| \rangle^2$. This gives the desirable result that the two terms in Eq. (12) tend to cancel below T_c, but unfortunately it does not give the correct susceptibility above T_c. For example, in the limit $T \to \infty$, one has $\chi = (1 - 2/\pi)/T$ instead of the correct result which is $1/T$.

(2) My preferred solution is to replace Eq. (11) by

$$\chi = \frac{1}{T}\left[\frac{1}{N}\sum_{i,j}\langle S_i S_j\rangle - \sum_i \langle S_i S_{i+\vec{L}/2}\rangle\right], \tag{14}$$

where $S_{i+\vec{L}/2}$ indicates the spin which is $L/2$ in every direction away from i, i.e. as far as possible from i. The second term is exponentially small in L/ξ above T_c, while below T_c it tends to cancel the first term. Unfortunately, this definition of χ in a finite system has not been adopted.

(3) Ignore the second term in Eqs. (11) and (12) completely. Typically, one also neglects the factor of $1/T$ since this varies only slightly in the vicinity of the transition. We will therefore define the correlation function

$$C = \frac{1}{N}\sum_{i,j}\langle S_i S_j\rangle = N\langle m^2\rangle. \tag{15}$$

Well above T_c, this is the susceptibility, apart from the factor of $1/T$, but below T_c, the dominant part varies as N times the magnetization squared, i.e. $N|t|^{2\beta}$, where β is the order parameter exponent defined by $\langle m\rangle \sim |t|^\beta$ for $t < 0$. For χ, this part is canceled by the second term in Eq. (12). From Eq. (8), we write

$$C = L^{2-\eta}\tilde{C}\left(L^{1/\nu}t\right), \tag{16}$$

where we used the scaling relation

$$\gamma = (2-\eta)\nu, \tag{17}$$

in which the exponent η characterizes the power-law decay of correlations with distance at the critical point. The scaling function $\tilde{C}(x)$ must have the following limiting behaviors:

$$\tilde{C}(x) \sim \begin{cases} x^{-\gamma}, & (x \to \infty) \\ \text{const.}, & (x = 0), \\ |x|^{2\beta}, & (x \to -\infty) \end{cases} \tag{18}$$

where in the last line, we used the following "hyperscaling" scaling relation:

$$\beta = \frac{1}{2}(d - 2 + \eta)\,\nu. \tag{19}$$

Hyperscaling relations are connections between exponents which also involve the space dimension d. I will discuss them in some detail in Section 6 when I consider finite-size scaling above the upper critical dimension.

Equations (15)–(19) describe the FSS of the second moment of the order parameter m. Following Binder,[5] it is of interest to consider other moments and, in particular, the whole distribution $P(m)$. In order that the second moment satisfies Eq. (16) and that the distribution is normalized, we must have

$$P(m) = L^{\beta/\nu}\tilde{P}\left(L^{\beta/\nu}m, L^{1/\nu}t\right). \tag{20}$$

The power of L in front of the scaling function $\tilde{P}$ has to be the same as the power of L multiplying m in the first argument of $\tilde{P}$ in order that the distribution is normalized, and this power has to be β/ν in order to get Eq. (16) (to see this, note Eq. (19)).

Binder[5] also pointed out that it is useful to look at ratios of moments such that the total power of m in the numerator and denominator are equal because then the power of L in front of the scaling function disappears. A simple and commonly studied case is the ratio of the fourth moment to the second moment squared,

$$g = \frac{1}{2}\left(3 - \frac{\langle m^4\rangle}{\langle m^2\rangle^2}\right), \tag{21}$$

which is known as the Binder ratio or Binder cumulant. The factors 3 and 1/2 are not essential but are often included so that g varies from 0 well above T_c (since fluctuations there are Gaussian, implying $\langle m^4\rangle = 3\langle m^2\rangle^2$) to 1 well below T_c (since $\langle m^4\rangle = \langle m^2\rangle^2$ in that region). The FSS behavior of g is

$$g = \tilde{g}\left(L^{1/\nu}t\right). \tag{22}$$

A critical point is characterized by two independent, universal, critical exponents, which can be taken to be ν and η, as well as the value of T_c. An advantage of using the Binder ratio to investigate a critical point, rather than, say, the correlation function C, is that one can determine each of these three quantities *separately* as follows:

(1) From Eq. (22), we see that g is independent of size at T_c, and so T_c can be determined from the temperature where the data for different sizes intersect, see Fig. 1 for example. If there are corrections to scaling, one has to extrapolate the intersection temperatures between sizes L and $L/2$, say, which we denote by $T^\star(L)$, to $L = \infty$, i.e. $T_c = \lim_{L\to\infty} T^\star(L)$. I will briefly discuss corrections to FSS in Section 5.

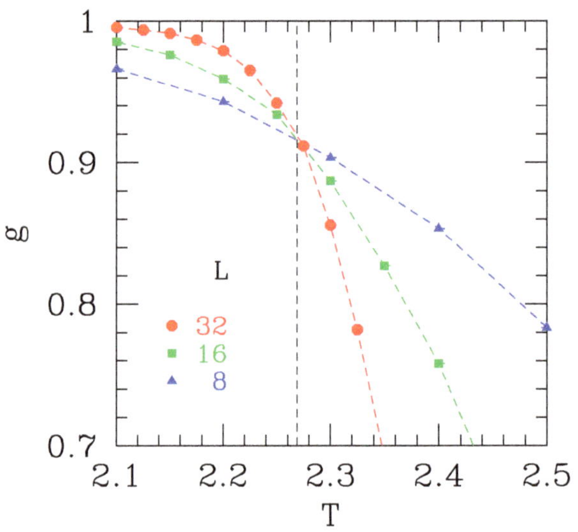

Fig. 1. Calculation of the Binder ratio defined in Eq. (21) for the two-dimensional Ising model on a square lattice with nearest-neighbor interactions. The dashed vertical line indicates the exact transition temperature $T_c = 2/\ln(\sqrt{2}+1) = 2.269\ldots$ The figure shows that T_c is the temperature where the data for g for different sizes intersect. Furthermore, the *value* of g at the critical (intersection) point is of interest because it is universal. From more detailed calculations, this value is found to be 0.916 to three decimal points. Universality means that the same value would occur for a triangular lattice, or a square lattice with second-neighbor interactions, for example, even though the transition temperatures would be different for these models. The value of $g(T_c)$ does, however, depend on the boundary conditions and is also different in different space dimensions.

(2) One then determines ν by requiring that a plot of g against $L^{1/\nu}(T - T_c)$ collapses on a single curve, see Eq. (22). Alternatively, one can determine ν from $(\partial g/\partial T)_{T_c} \sim L^{1/\nu}$.
(3) Finally, one determines η by requiring that a plot of $C/L^{2-\eta}$ against $L^{1/\nu}(T - T_c)$ collapses on a single curve according to Eq. (16). One can alternatively determine η from the size dependence of $\partial C/\partial T$ at T_c.

This is the best way to determine T_c and the exponents since they are obtained one at a time. Alternatively, if one just uses data for C, one fits several parameters at the same time, which leads to greater uncertainties because these parameters are correlated.

I have mentioned that the reason FSS is crucial to analyze numerical data at a second-order phase transition is that the correlation length diverges at T_c. It is interesting to investigate the correlation length itself, but the question then arises as to how to define it for a finite system. This can be done by generalizing the second moment of the order parameter in Eq. (15) to non-zero wavevector,

$$C(\mathbf{k}) = \frac{1}{N}\sum_{i,j} e^{i\mathbf{k}\cdot(\mathbf{R}_i - \mathbf{R}_j)}\,\langle S_i S_j\rangle. \tag{23}$$

Sufficiently far above T_c so that finite-size effects are unimportant, $C(\mathbf{k})$ follows the Ornstein-Zernicke form

$$C(\mathbf{k}) = \frac{C(0)}{1 + \xi^2 k^2 + \cdots} \tag{24}$$

at small k. The coefficient of k^2 is the square of the *bulk* correlation length ξ characterizing the exponential decay of the correlations with distance.

One can define the correlation length of a *finite* system, ξ_L, by slightly modifying this formula. As it stands, Eq. (24) is not a periodic function of $\mathbf{k}$, but it can be made so by the following replacement:

$$k^2 \to 2\sum_{\ell=1}^{d} (1 - \cos(k_\ell)) = \sum_{\ell=1}^{d} (2\sin(k_\ell/2))^2 \,, \tag{25}$$

where we have set the lattice spacing to be 1. Typically, we then extract the correlation length of the finite system by comparing $C(\mathbf{k})$ for the two smallest wavevectors, $\mathbf{k} = 0$ and $\mathbf{k} = \mathbf{k}_{\rm min} = (2\pi/L)(1, 0, \ldots, 0)$, which gives

$$\xi_L = \frac{1}{2\sin(\pi/L)} \left(\frac{C(0)}{C(\mathbf{k}_{\rm min})} - 1 \right)^{1/2} . \tag{26}$$

The ratio ξ_L/L is dimensionless and so should have a FSS form similar to that of the Binder ratio in Eq. (22), namely there is no power of L in front of the scaling function, i.e.

$$\frac{\xi_L}{L} = \tilde{\xi}(L^{1/\nu} t). \tag{27}$$

This means that data for ξ_L, like that for g, gives *directly* the value of T_c (ignoring corrections to FSS) since T_c is the temperature where the data for different sizes intersect, see Fig. 1 for an example of the Binder ratio. This property is very convenient, so dimensionless quantities like g and ξ_L/L play an important role in FSS analysis.

I already noted that the correlation function $C(0)$ is only the susceptibility above T_c but not below T_c because it does not include the subtracted term in Eq. (11). For the same reason, ξ_L is a correlation length only above T_c but not below it. However, it is a very convenient quantity to perform an FSS analysis at *all* temperatures.

One reason why the study of critical phenomena is particularly interesting is that many quantities are "universal", that is to say, they only depend on certain broad features of the problem, such as space dimensionality and the symmetry of the order parameter, but do not depend on microscopic details, such as the type of lattice and range of interactions (as long is they are not infinite range). Exponents are examples of universal quantities while the transition temperature depends on all the microscopic details and so is non-universal.

In addition, scaling functions, including finite-size scaling functions, have a degree of universality. Michael, together with his student Vladimir Privman,[6] clarified the situation for FSS functions. Suppose we take Eq. (8) for a general quantity X_L. There will be a non-universal amplitude (or metric factor) associated with the overall size of X_L and a non-universal amplitude in front of the reduced temperature in the scaling function, so we can write

$$X_L(t) = AL^{\lambda/\nu}\, \tilde{X}(BL^{1/\nu}t), \tag{28}$$

where A and B are non-universal. According to Privman and Fisher,[6] the resulting scaling function $\tilde{X}$ is universal, which means that it does not depend on microscopic details, but, since we are dealing with finite size effects, it does depend on the boundary conditions. If we include a magnetic field h, the scaling function has a second argument $CL^{y_H}h$. The amplitude C is non-universal, but the scaling function itself remains universal. Here, y_H is called the magnetic exponent and is related to exponents introduced earlier by

$$y_H = \frac{1}{2}(d + 2 - \eta). \tag{29}$$

The exponent $1/\nu$ which is the power of L multiplying t in the scaling functions is called the thermal exponent and labeled y_T, so

$$y_T = \frac{1}{\nu}. \tag{30}$$

The terminology of thermal and magnetic exponents will be discussed more in Section 5 when I discuss FSS from the point of view of the renormalization group in order to include corrections to the basic FSS expressions discussed so far.

An important application of FSS functions being universal occurs for dimensionless quantities such as the Binder ratio g or the reduced correlation length ξ_L/L. Taking g as an example, Eq. (22) doesn't have a non-universal overall scale factor like A in Eq. (28), multiplying the scaling function (since g is dimensionless), and so, we write

$$g = \tilde{g}(BL^{1/\nu}t) \tag{31}$$

with B non-universal and $\tilde{g}$ universal. Hence, the *value* of g at the intersection point (which is $T = T_c$) is a universal number, see Fig. 1. Consequently, if the intersection values for two different systems known to be in the same universality class are apparently different, it follows that the asymptotic scaling regime has *not* been reached and corrections to FSS are significant. This is a useful check. It should, however, be noted that in anisotropic systems there is less universality, see Refs. 19 and 20 and the article by V. Dohm in this volume.

4. Justification and Limit of Validity

The original proposal of FSS by Fisher and Barber[1,2] made plausible assumptions about the nature of finite-size effects. Subsequently, FSS was put on a firm footing by Brézin[7] who used renormalization group arguments to justify those assumptions.

In particular, Brézin used the field theoretical version of the renormalization group that he pioneered in collaboration with Le Guillou and Zinn-Justin.[8] The details are beyond the scope of this article, but essentially the RG equations obtained in Ref. 8 for an infinite system are shown in Ref. 7 to be not changed substantially when applied to finite-size systems. The resulting analysis leads to FSS results like Eq. (8). For more on the field theory approach to FSS, see the article by Dohm in this volume.

Brézin also noted that there is an implicit assumption that the FSS scaling functions are not singular and that this assumption breaks in dimensions above the "upper critical dimension" d_u, where critical exponents no longer depend on dimension and stick to their mean field values. For the most commonly studied magnetic models, one has $d_u = 4$. I will discuss how FSS has to be modified for $d > d_u$ in Section 6.

5. Corrections to Finite-Size Scaling

Finite-size scaling expressions, such as Eqs. (8), (16) and (22), are valid in the limit when the system size tends to infinity and the distance to the critical point tends to zero. In practice, these conditions are not met in numerical simulations, so there will be corrections to FSS whose form I will discuss briefly in this section. In some cases, such as the $d = 2$ Ising model for which some data are shown in Fig. 1, the corrections are small, but this is not always the case.

To include corrections, we need to adopt a renormalization group (RG) perspective. We consider a system on a lattice with lattice spacing a with a Hamiltonian $\mathcal{H}$ which has interactions $K_1, K_2, \ldots$, in which we have absorbed the factor of $1/(k_B T)$. In the RG, we perform a coarse graining of the system such that the new lattice spacing is $b\,a$ with $b > 1$ and the probability distribution of long wavelength fluctuations is unchanged. This distribution is governed by a new Hamiltonian $\mathcal{H}'$, of the same form as $\mathcal{H}$ but with new interactions $K'_1, K'_2, \ldots$.

If we are close to a second-order transition, the set of interactions will approach close to a fixed point $K^*_1, K^*_2, \ldots$. Near the fixed point is useful to consider linear combinations u_i of the deviations $K_\alpha - K^*_\alpha$ which transform simply, $u'_i = b^{y_i} u_i$, where y_i is an eigenvalue of the RG transformation linearized about the fixed point. If $y_i > 0$, then u_i is said to be a "relevant" variable, since iterating the RG transformation takes u_i away from the fixed point ($u_i = 0$), whereas if $y_i < 0$, then u_i is an "irrelevant" variable since u_i decreases to zero upon iterating.

For a standard second-order transition, one has to set the temperature to the critical temperature and the magnetic field to zero in order to be at the critical point, so there are two relevant operators: t the reduced temperature defined in Eq. (3) and the magnetic field h. One can show that the corresponding exponents, y_T and y_H, are related to other exponents by Eqs. (30) and (29). In the following discussion, I will include just the most important irrelevant operator, i.e. the one with the least negative eigenvalue and call it u. Following convention, I will write $y_u = -\omega$, where $\omega > 0$.

Under the RG transformation, quantities of interest are assumed to vary in the following way:

$$X(t, h, u, L) = b^{y_X}\, X(b^{y_T} y,\, b^{y_H} h,\, b^{-\omega} u,\, L/b), \tag{32}$$

where y_X is related to the exponent giving the divergence of X in the thermodynamic limit, as we shall see. If we fix $h = 0$, ignore the irrelevant variable, and setting $b = L$, we get

$$X_L(t) = L^{y_X} X(L^{y_T} t, 0, 0, 1) = L^{y_X} \tilde{X}(L^{y_T}\, t), \tag{33}$$

which agrees with the basic FSS expression in Eq. (8) with $y_X = \lambda/\nu = \lambda\, y_T$, where λ is the exponent for the divergence of X at the bulk critical point, see Eq. (4).

We are now in a position to characterize corrections to FSS. These are of two types:

(1) "singular" corrections to scaling, which come from expanding functions like the RHS of Eq. (32) in powers of the irrelevant variables;
(2) "analytic" corrections to scaling which arise because the temperature-like variable multiplying b^{y_T} in Eq. (32) is only precisely equal to the reduced temperature very close to T_c, and, more generally, t should be replaced by a temperature scaling variable g_T which can be expanded in powers of t and h consistent with the symmetry about $h = 0$, i.e.

$$g_T = t + a\, t^2 + b\, t^3 + c\, t h^2 + \cdots . \tag{34}$$

Similarly, the field should be replaced by a field scaling variable which can be expanded as follows:

$$g_H = h + d\, t h + e\, t^2 h + f\, h^3 + \cdots . \tag{35}$$

Thus, one needs to replace t by g_T and h by g_H in Eq. (32) and expand the function on the RHS.

Clearly, then, there are many corrections and these can be difficult to separate out in the numerics. However, if one correction dominates and one has high-quality data over a broad range of lattice sizes, then it should be possible to extract the form of this leading correction including the appropriate exponent.

To keep things simple here, I will just take an example where one singular correction dominates and will investigate how the intersection temperatures for the Binder ratio, $T^*(L)$, computed from sizes L and sL, say, vary with L. In the absence of corrections, $T^*(L) = T_c$ for all L whereas including corrections $T^*(L) \neq T_c$ but tends to T_c for $L \to \infty$. We would like to understand how the deviation $T^*(L) - T_c$ varies with L and be able to compute the true critical temperature T_c.

Including the leading singular correction, Eq. (22) is modified to

$$g(t,L) = \tilde{g}\left(L^{1/\nu}t\right) + L^{-\omega}\tilde{\tilde{g}}\left(L^{1/\nu}t\right). \tag{36}$$

We are interested in t and $1/L$ small, so we expand

$$g(t,L) \simeq \tilde{g}(0) + L^{1/\nu}\tilde{g}'(0)\,t + L^{-\omega}\tilde{\tilde{g}}(0). \tag{37}$$

We compute the reduced temperature t_L^* where sizes L and $s\,L$ intersect, i.e. where $g(t_L^*, s\,L) = g(t_L^*, L)$. A bit of algebra gives

$$\frac{T_L^* - T_c}{T_c} \equiv t_L^* = A_s L^{-\omega-1/\nu}, \tag{38}$$

where the amplitude is given by

$$A_s = \frac{(1-s^{-\omega})\,\tilde{\tilde{g}}(0)}{(s^{1/\nu}-1)\,\tilde{g}'(0)}. \tag{39}$$

Equation (38) shows that fitting the intersection temperatures as a function of size determines the combination of exponents $\omega + 1/\nu$. By also computing a second dimensionless quantity, one can actually determine ω and $1/\nu$ separately, see e.g. Refs. 9–11 for details.

One of the most challenging and long-standing problems which has been resolved satisfactorily (at least in my opinion) by Monte Carlo is the critical behavior of the three-dimensional Ising spin glass. This is a hard problem because slow dynamics limit the range of sizes which can be equilibrated, there are significant sample-to-sample fluctuations which have to be averaged over, and three dimensions is close to the lower critical dimension (below which fluctuations destroy the transition), whose numerical value is close to 2.5.[12] However, Refs. 10 and 11, have succeeded in getting an accurate estimate of T_c and the critical exponents including the correction exponent ω. In my view, these two papers are a real tour-de-force, combining massive simulations using the most efficient algorithms and, what is relevant here, a very detailed FSS analysis of the results, including nailing down the leading correction to scaling, without which the uncertainties in exponents would have been very large.

6. Finite-Size Scaling Above the Upper Critical Dimension

The simplest approximate theory to describe critical phenomena is the mean field theory (MFT), according to which the critical exponents I have mentioned so far have values

$$\gamma = 1, \qquad \beta = 1/2, \qquad \nu = 1/2, \qquad \eta = 0 \qquad \text{(MFT)}. \tag{40}$$

It is well established by the RG that MFT gives the exponents correctly for dimension d greater than the upper critical dimension $d = 4$. Whereas for $d < 4$, the exponents vary with d, for $d > 4$ they "stick" to their MF values.

Earlier, we have mentioned two scaling relations, Eqs. (17) and (19), relating different exponents. Equation (17) relating γ and η *is* satisfied by their MF values. However, Eq. (19), which relates β and ν, and also involves d, does *not* work for $d > 4$, though it is correct at the borderline dimension between MF and non-MF behavior, $d = 4$. Relations between exponents which involve d are called "hyperscaling" relations. Equation (19) and other hyperscaling relations are not valid in the mean field region $d > 4$. In this section, I will discuss the physics behind this violation and show how this physics also gives rise, for $d > 4$, to a form of FSS which is different from what we have discussed so far.

In RG treatments of critical phenomena, one starts with a "soft-spin" version of the problem, in which an initial coarse graining over the discrete "fixed length" spins has been carried out. This Hamiltonian, associated with the names of Ginzburg, Landau and Wilson, is, for scalar (Ising) spins,

$$\mathcal{H}_{GLW} = \int \left[\frac{1}{2}r(T)\phi^2 + \frac{1}{2}(\nabla\phi)^2 + \frac{1}{4}u\phi^4\right] d^d x, \tag{41}$$

where $\phi(x)$ is the "spin" variable and the partition function is given by $Z = \int D[\phi(x)] \exp(-\mathcal{H}_{GLW})$. The parameter $r(T)$ varies smoothly with temperature and goes negative at low T in order to induce an ordering in ϕ. In this formulation, MFT corresponds to taking $\phi(x)$ to be independent of x and minimizing the resulting function with respect to ϕ. The transition temperature is when $r = 0$, and at lower temperatures, one finds

$$\langle\phi\rangle = \left(\frac{|r|}{u}\right)^{1/2} \qquad \text{(MFT)}. \tag{42}$$

Since r varies smoothly with T, i.e. $r(T) \propto T - T_c$, we have $\langle\phi\rangle \propto (T_c - T)^{1/2}$, so $\beta = 1/2$, the MF value that I quoted in Eq. (40).

Renormalization group arguments show that for $d > 4$ the quartic coupling u is irrelevant, so its value iterates to zero under repeated RG transformations. However, the expression for $\langle\phi\rangle$ in Eq. (42) becomes singular in this limit, so we can't just set $u = 0$. This is important because, in order to derive the *standard* FSS result in Eq. (8), we set the irrelevant variable to zero in Eq. (33). However, for $d > 4$, we cannot do this, just as we cannot set $u = 0$ in Eq. (42), but must rather consider the singular nature of the scaling function $X(t, h, u, L)$ as u *tends* to 0. Irrelevant variables which give singularities if they are set to zero are said to be "dangerous" irrelevant variables, a term first coined by Michael Fisher.[13]

Hence, Eqs. (8), (16), (22), and (27), for example, are not valid for $d > 4$. The way these equations have to be modified was shown by Brézin and Zinn-Justin[14] and Binder *et al.*[15] The argument of the standard FSS functions is $L^{1/\nu}\, t$. If we put the MF value $\nu = 1/2$, this would be $L^2 t$. However, Refs. 14 and 15 show that the correct result involves the square root of the volume, i.e. $L^2 t$ is replaced by $L^{d/2}\, t$. The prefactor in front of the scaling function may also change in order to give the correct results for an infinite system. For $d > 4$, the expressions corresponding to Eqs. (8), (16), (22) and (27) are modified to

$$X_L = L^{d\lambda/2}\tilde{X}(L^{d/2}\,t), \quad C = L^{d/2}\tilde{C}(L^{d/2}\,t), \quad g = \tilde{g}(L^{d/2}\,t), \quad \xi_L = L^{d/4}\tilde{\xi}(L^{d/2}\,t). \tag{43}$$

I emphasize that these results are for periodic boundary conditions. There are some subtleties for free boundary conditions, which I will discuss in Section 6.1.

These results indicate that finite-size effects start when $L \simeq t^{-d/2}$, at which point $\xi_L \sim L^{d/4}$, which is larger than L. Normally, one imagines that finite-size effects begin when $\xi_L \sim L$, but for $d > 4$ and periodic boundary conditions, nothing happens at this point and it is only closer to T_c, when $\xi_L \gg L$ (for large L) that finite-size effects appear. This surprising result is presumably due to the following: (i) we have the periodic boundary conditions, which means there is no surface at a distance L, and (ii) the fixed point value of the coupling u in Eq. (41) is zero, so long wavelength fluctuations interact very weakly.

I emphasize that the reason we have the power $d/2$ rather than 2 in Eq. (43) ultimately comes from Eq. (42), namely that the order parameter $\langle\phi\rangle$ is singular for $u \to 0$. What form would FSS take if we look at some quantity which does not involve the order parameter (which is a $\mathbf{k} = 0$ quantity)? Wittmann and the author[16] investigated this question numerically for the five-dimensional Ising model. They found, for example, that computing the wave-vector-dependent correlation function $C(\mathbf{k})$ in Eq. (23) for, say, the smallest *non-zero* wavevector (which is orthogonal to the order parameter), the data collapsed with the *standard* FSS form with $\eta = 0, \nu = 1/2$. A natural hypothesis is that *all* quantities which do not involve the order parameter satisfy standard FSS expressions even for $d > 4$. This implies that the "true" susceptibility, as this author likes to define it in Eq. (14), would follow the standard FSS scaling form, since the contribution from the order parameter cancels between the two terms. To my knowledge, this has not been confirmed numerically.

6.1. *Free boundaries*

In the previous section, we noted that for $d > 4$ and with periodic boundary conditions, there is no change in behavior when $\xi_L \sim L$, but only when the correlation length has reached the larger value $\sim L^{d/4}$. While surprising, this result is nonetheless possible with periodic boundary conditions. However, if we have free boundary conditions, surely *something* must happen when $\xi \sim L$ because the correlations must be affected by the surfaces.

To our knowledge, the question of what happens with free boundary conditions for $d > 4$ was first addressed by Rudnick *et al.*[17] They showed that one needs to consider two exponents describing finite-size effects: a "shift" exponent and a "rounding" exponent.

I will describe shortly how one defines shift and rounding. But I note first that the possibility that one might need two exponents to describe finite-size effects was considered in the original papers by Fisher and Barber.[1,2] Part of the reason they considered this came from their studies of the spherical model in a finite geometry for which the shift was found to be bigger than the rounding. However, it was later realized that the spherical model in a finite geometry is unphysical.[18] Furthermore, it was shown by Brézin[7] that if hyperscaling is satisfied, then the standard formulation of FSS described in Section 3 is correct with just a single exponent describing the range of temperature over finite-size effects occur. However, when hyperscaling is violated for $d > 4$, it turns out that one *can* have different exponents for shift and rounding, at least for free boundary conditions, as we shall now see.

To clarify what is meant by shift and rounding, consider the data in Fig. 2 for the Binder ratio g in five dimensions for free boundary conditions, which is taken from Ref. 16. The transition temperature, which can be obtained most accurately from simulations with periodic boundary conditions, is $T_c = 0.877846$. We can define a characteristic temperature $T(L)$ for each size based on where the Binder ratio takes

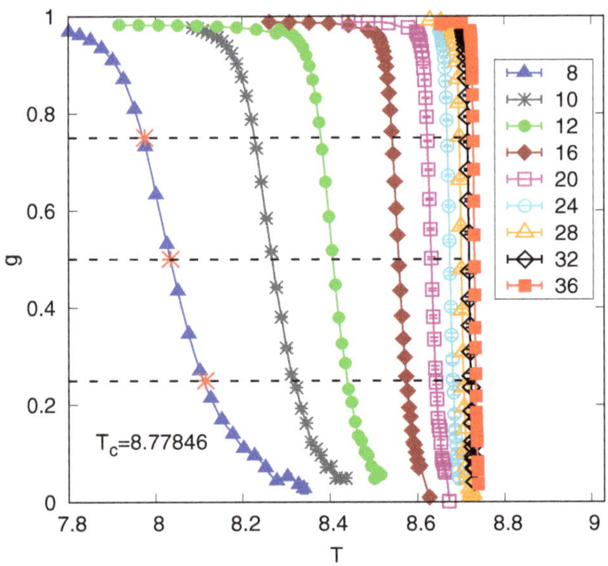

Fig. 2. The Binder ratio defined in Eq. (21) for the five-dimensional Ising model with free boundary conditions. The data is from Ref. 16. The region of finite-size rounding, t_L, is defined here, for each size, to be the range of temperature, where g has values 0.75 and 0.25. For example, for $L = 8$, we have $t_8 = 8.115 - 7.975 = 0.140$ (these points are indicated by crosses). The shift ΔT_L is defined here to be difference between $T_c (= 8.77846)$ and the value of T where $g = 0.5$. For $L = 8$, we find $g = 0.5$ at $T = 8.035$ (shown as a cross in the figure), so $\Delta T_8 = 8.77846 - 8.035 = 0.743$. Analysis of the data shows that the rounding tends to zero as $1/L^{5/2}$ while the shift tends to zero as $1/L^2$, which is bigger.

some convenient value, say, 0.5. The temperature shift will vary as

$$\Delta T_L \equiv T_c - T(L) = \frac{A}{L^\lambda}, \tag{44}$$

which defines the shift exponent λ. The rounding temperature can be defined for each size as the difference in temperatures, where g has two convenient values, say, 0.25 and 0.75, so

$$t_L \equiv T(g = 0.25) - T(g = 0.75) = \frac{B}{L^\mu}, \tag{45}$$

which defines the rounding exponent μ. For $L = 8$, the points where $g = 0.25, 0.5$ and 0.75 are indicated in the figure.

One can see by eye that the shift is greater than the rounding, and indeed, the exponents λ and μ turn out to be different. Rudnick *et al.*[17] showed analytically that $\lambda = 1/\nu = 2$ and $\mu = d/2 = 5/2$ and an analysis of the data in Fig. 2 agrees with these values.[16] Since the shift is bigger than the rounding, there are no intersections in the data and in fact[16] $g(T_c) \to 0$ for $L \to \infty$.

7. Conclusions

In this article, I have given a brief history of FSS, emphasizing the pioneering work of Michael Fisher, and have described the main aspects of the theory. To emphasize how crucial FSS has been to the study of critical phenomena, I mentioned state-of-the-art work[10,11] which solved the long-standing question of the critical behavior of the Ising spin glass in three dimensions. Finally, as a quirky afterthought, I discussed the rather surprising changes in FSS which occur for $d > 4$ where hyperscaling is violated. Interestingly, these changes arise from a "dangerous relevant variable", another concept which was first discussed by Michael.

Acknowledgments

I have benefited from discussions on finite-size scaling with many people, including Victor Martin-Mayor, Helmut Katzgraber, and Matt Wittmann. I should particularly mention the stimulating interactions on many topics, including finite-size scaling, with the late Kurt Binder. Finally, I thank Michael Barber for an interesting email exchange on the history of the subject while I was preparing this manuscript.

References

1. M. E. Fisher, The theory of critical point singularities. In M. S. Green (ed.), *Critical Phenomena, Proceedings of the 51st Enrico Fermi Summer School, Varenna*, p. 1 (Academic Press, New York, 1971).
2. M. E. Fisher and M. N. Barber, Scaling theory for finite-size effects in the critical region. *Phys. Rev. Lett.* **28**, 1516 (1972).

3. E. Luijten, M. E. Fisher and A. Z. Panagiotopoulos, Universality class of criticality in the restricted primitive model electrolyte. *Phys. Rev. Lett.* **88**, 18570 (2002).
4. J. Cardy, Ed., *Finite-Size Scaling* (Elsevier, New York, USA, 1988).
5. K. Binder, Finite size scaling analysis of Ising model block distribution functions. *Z. Phys. B.* **43**, 119 (1981).
6. V. Privman and M. E. Fisher, Universal critical amplitudes in finite-size scaling. *Phys. Rev. B.* **30**, 322 (1984).
7. E. Brézin, An investigation of finite size scaling. *J. Phys. (Paris).* **43**, 15 (1982).
8. E. Brézin, J. Le Guillou and J. Zinn-Justin, Field theory approach to critical phenomena. In eds. C. Domb and M. S. Green, *Phase Transitions and Critical Phenomena*, vol. 6, p. 127 (Academic Press, New York, 1976).
9. R. A. Baños, L. A. Fernandez, V. Martin-Mayor and A. P. Young, The correspondence between long-range and short-range spin glasses. *Phys. Rev. B.* **86**, 134416 (2012).
10. M. Hasenbusch, A. Pelissetto and E. Vicari, The critical behavior of three-dimensional Ising glass models. *Phys. Rev. B.* **78**, 214205 (2008).
11. M. Baity-Jesi *et al.*, Critical parameters of the three-dimensional Ising spin glass. *Phys. Rev. B.* **88**, 224416 (2013).
12. S. Boettcher, Stiffness of the Edwards-Anderson model in all dimensions. *Phys. Rev. Lett.* **95**, 197205 (2005).
13. M. E. Fisher. In J. Gunton and M. Green (eds.), *Renormalization Group in Critical Phenomena and Quantum Field Theory* (Temple University, Philadelphia, 1974).
14. E. Brézin and J. Zinn-Justin, Finite size effects in phase transitions. *Nucl. Phys. B.* **257**, 867 (1985).
15. K. Binder, M. Nauenberg, V. Privman and A. P. Young, Finite-size tests of hyperscaling. *Phys. Rev. B.* **31**, 1498 (1985).
16. M. Wittmann and A. P. Young, Finite-size scaling above the upper critical dimension. *Phys. Rev. E.* **90**, 062137 (2014).
17. J. Rudnick, G. Gaspari and V. Privman, Effect of boundary conditions on the critical behavior of a finite high-dimensional Ising model. *Phys. Rev. B.* **32**, 7594 (1985).
18. E. Brézin, Finite size scaling: Part II. *Ann. NY Acad. Sci.* **410**, 339 (1983).
19. X. S. Chen and V. Dohm, Non universal finite-size scaling in anisotropic systems, *Phys. Rev. E.* **70**, 056136 (2004).
20. V. Dohm and S. Wessel, Exact critical Casimir amplitude of anisotropic systems from conformal field theory and self-similarity of finite-size scaling functions in $d \geq 2$ dimensions, *Phys. Rev. Lett.* **126**, 060601 (2021).

Field Theory

https://doi.org/10.1142/9789811282386_0034

Chapter 34

Should We Worry about Renormalons in the ϵ-expansion?

E. Brézin

Laboratoire de Physique de l'Ecole normale supérieure, ENS, Université PSL, CNRS, Sorbonne Université, Université de Paris, F-75005 Paris,
brezin@lpt.ens.fr

Turning the divergent ϵ-expansion into a numerically sensible algorithm, relies on the knowledge of the behavior of the large order contributions. Two different pictures are known to compete there. The first one was based on Lipatov's instantons, which is known to deal with the multiplicity of Feynman diagrams which grows factorially at high orders. However, this was challenged by 't Hooft's renormalons, which pointed out that renormalization could yield a similar growth through one single diagram. We study here a well-known model the $O(N)$ model, in the large N-limit. The reason for returning to this familiar model, is that it deals with diagrams known to give renormalon effects. Through an explicit analytic result, we find no sign of a non-analyticity of perturbation theory due to these renormalons.

1. Introduction

A little more than 50 years ago, K. Wilson and M. Fisher introduced the ϵ-expansion in the celebrated article[1] *Critical Exponents in 3.99 Dimensions.* This article has had a considerable influence both in the area of critical phenomena and in quantum field theory in general. However, in practice, if the first two terms of the expansion provided often a reasonable approximation to the measured indices in dimension three, the situation deteriorated if one pushed the procedure to higher orders.[2] Clearly, the expansion looked divergent, it is believed to be at best asymptotic and in fact limited in its applicability. Of course, summation procedures, such as Borel transforms, Padé approximants, could be tried, but in the absence of further indication, it was a blind shot. Therefore, at the time, analytic calculations, and computer simulations, appeared to be limited in their ability of reproducing precision measurements by the renormalization-group approach.

The situation changed significantly in 1977 with Lipatov's approach to the characterization of large orders in perturbation theory.[3] His instanton method, which he developed for the field theory $g\phi^4$ in dimension four, predicted a large-order behaviour of the perturbation expansion of the various correlation functions, β-function, etc., of the form $g^k k!(-a)^k k^b c$ for large k, with calculable coefficients a, b, c. I shall recall in the following why this information, the explicit knowledge of a, b, c, is essential to extract a numerically sensible result from such a divergent series. With Le Guillou and Zinn-Justin,[4] we first checked these results in lower dimensions. For instance the one-dimensional quartic anharmonic oscillator had been extensively studied by Bender and Wu[5] who had found, on the basis of a WKB method, that indeed perturbation theory diverged with a $k!$ growth as above. We verified that the instanton method reproduced exactly what they had found. Extension to field theory in higher dimensions was similar[4] except that the instanton solution is not known in analytic form except in dimension four.[6] In dimensions lower than four the interpretation of these results is simple: At a given order, all Feynman diagrams have the same sign (we are considering an Euclidean field theory) and their number grows proportionally to $k!$. The situation is less transparent in dimension four since diagrams require counterterms and subtractions. This is the central point of the investigation that we try to focus on in this paper.

The knowledge of the large-order behaviour of the perturbative expansion was the source of considerable improvements in the ability to get a sensible answer from those rapidly divergent series.[8] In practice, it led to numerically convergent algorithms developed around Zinn-Justin and collaborators,[8,9] Borel transform of the series, followed by conformal mappings relying on the explicit knowledge of the above coefficients $a.b, c$, were among the techniques that they used, and examples, such as the anharmonic oscillator, revealed that the divergent perturbation series could lead, after those mappings, to results with a precision of many-digits.

In our 1977 paper,[4] we tried to carry the large-order knowledge of perturbative series to the ϵ-expansion. In most procedures, one has to perform a double expansion in g and ϵ. Since the fixed point g^* is of order ϵ, one has to consider at order k, terms of order $g^k, g^{k-1}\epsilon, g^{k-2}\epsilon^2, \ldots$. Fortunately, the well-known technique of *minimal subtraction* allows one to avoid this double expansion. In this renormalization scheme[12] by definition, the counter-terms contain only poles in ϵ without any finite part. As a result, one obtains the renormalization-group β and γ functions in dimension d from their four-dimensional counterpart

$$\beta_d(g) = -\epsilon g + \beta_4(g),$$
$$\gamma_d(g) = \gamma_4(g). \tag{1}$$

The fixed point g^* is then ϵ-expanded from

$$\epsilon g^* = \beta_4(g^*) \tag{2}$$

($\beta(g)$ is a series starting at order two) and the critical exponent η, which is (twice) the conformal anomaly of the ϕ-field, is then expanded in ϵ from

$$\eta = \gamma_4(g^*). \tag{3}$$

Therefore, the computation of the ϵ-expansion is reduced to four-dimensional calculations. Using the 4D large-order calculation from Lipatov's method, we obtained an estimate of the asymptotic orders in the ϵ-expansion,[4] which was again of the $k!$-type as before. Then Zinn-Justin and co-workers based a summation procedure on this large-order behaviour,[8,9] and the process looked nicely convergent: at least adding one more order in the ϵ-expansion improved the previous result instead of destroying it as the straight expansion does.

The confidence in this process was severely affected after 't Hooft's discovery of a phenomenon,[10] now called *renormalons,* which appears only in renormalizable theories such as ϕ^4 in dimension four (and not in lower dimensions). His argument was based on the fact that one single diagram at order k could be proportional to $k!$, whereas the $k!$, for $d < 4$, resulted from their multiplicity. In a renormalizable theory, a diagram, such as the (renormalized) bubble diagram, grows logarithmically for a large external momentum. A repeated insertion of such diagrams leads to an integral with a log at kth power, which gives after integration of a $k!$. Clearly, this shed doubt on Lipatov's estimate for large orders, although 't Hooft's argument didn't show explicitly that it was wrong. It was not really clear either how those renormalons affected quantitatively the actual calculation.

I have remained puzzled by this problem since then, wondering whether the ambitious RG work based on the ϵ-expansion, was simply an approximation which should not be pushed too far. I want to present here a simple well-known problem, namely the $O(N)$-model in the large N limit. The reason for returning to this familiar model is that it is clearly a candidate for showing up renormalon effects, whereas this time we do not need an instanton asymptotics since there are only a few diagrams of a given order. I do not pretend that it solves the instanton–renormalon competition, but I wonder on the basis of this calculation whether renormalons affect the perturbation series as badly as one could think.

2. The $O(N)$ Model in Dimension Four

This, half a century-old, model consists of an N-component order parameter ϕ_a, $a = 1, \ldots, N$, with an interaction invariant under $O(N)$.[7] This model has been studied by hundreds of authors, and elaborate techniques have allowed us to compute several orders in a $1/N$ expansion. We will limit ourselves here to the leading large-N terms for the RG functions $\beta(g)$ and $\gamma(g)$ in the minimal subtraction scheme, which allows us to make easy contact with the ϵ-expansion. The reason for our interest is that $\gamma(g)$ is of order $1/N$ and the expansion in powers of g of this leading term involves exactly the diagrams which have been identified as generating renormalon singularities in the perturbative expansion. With the help of previous results on critical indices, we will try to understand what the renormalons do to this expansion.

The model is given by an (Euclidean) action

$$S = \int d^4x \left[\frac{1}{2}\nabla\phi_a\nabla\phi_a + \frac{1}{2}m_0^2\phi^2 + \frac{g_0}{4!N}(\phi^2)^2\right], \tag{4}$$

which one could regularize by an ultraviolet cutoff. For the reasons mentioned above, we prefer here the dimensional regularization by going to $4-\epsilon$ dimensions and then renormalize minimally by a coupling constant renormalization Z_1 and a field rescaling $\phi = \sqrt{Z}\varphi$. We work in the massless theory (critical temperature) and the action in terms of the renormalized field reads (omitting mass counterterms)

$$S = \int d^dx \left[\frac{Z}{2}\nabla\varphi_a\nabla\varphi_a + \mu^\epsilon \frac{gZ_1}{4!N}(\varphi^2)^2\right]. \tag{5}$$

Varying μ at fixed bare theory, we obtain the standard renormalization-group functions of the Callan–Symanzik equation[11]

$$\epsilon g + \beta(g)\left(1 + g\frac{d}{dg}\ln(Z_1/Z^2)\right) = 0, \tag{6}$$

$$\gamma(g) = \beta(g)\frac{d}{dg}\ln(Z). \tag{7}$$

We are interested in computing the leading terms in a $1/N$ expansion of Z_1, which is $O(N^0)$ and Z since $Z = 1 + O(1/N)$.

2.1. *The β-function at order N^0*

The coupling constant, involving an explicit factor $1/N$ and the leading terms maximize diagrams with internal index loops which provide a compensating factor N. For the two-point function, the leading diagrams are of order $1/N$ and the field renormalization $Z = 1+O(1/N)$. For the four-point function, the leading diagrams consist of a string of "bubbles", all of order $1/N$, and this yields a vertex renormalization Z_1 of order N^0, which we now compute.

Let us begin with a bubble with external momentum p

$$B(p) = \int \frac{d^dq}{q^2(p-q)^2}, \tag{8}$$

where we use the convention of omitting the usual geometric factor $\frac{2\pi^{d/2}}{(2\pi)^d\Gamma(d/2)}$ included into a rescaling of g that will be implied; we take the scale factor μ as unit of momentum:

$B(p) =$ 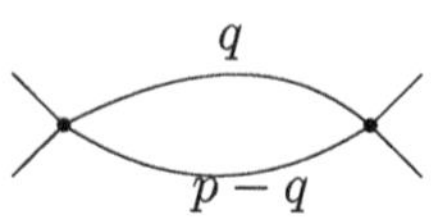

A standard calculation (using a Feynman parameter) yields

$$B(p) = a(\epsilon)\frac{p^{-\epsilon}}{\epsilon},$$
$$a(\epsilon) = \frac{1-\epsilon/2}{1-\epsilon}\,\frac{\Gamma^3(1-\epsilon/2)\Gamma(1+\epsilon/2)}{\Gamma(1-\epsilon)} = 1 + \frac{\epsilon}{2} + O(\epsilon^2), \tag{9}$$

i.e.,

$$B(p) = \frac{1}{\epsilon} + 1/2 - \ln p. \tag{10}$$

Then the four-point function, for external indices a, a, b, b, is given by a geometric series

$\Gamma^{(4)} =$ 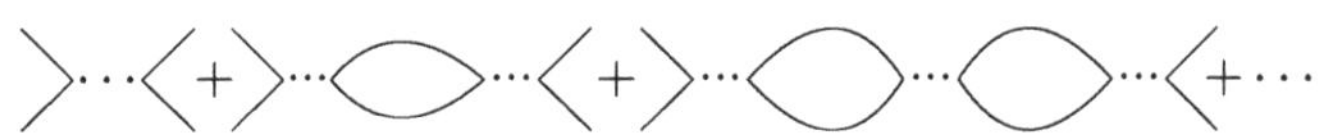

$$-\frac{1}{2}\Gamma^{(4)} = \frac{1}{N6}gZ_1 - \frac{g^2 Z_1^2}{N6^2}B(p) + \frac{g^3 Z_1^3}{N6^3}B^2(p) + \cdots$$
$$= \frac{1}{N}\frac{gZ_1/6}{1 + gZ_1/6\;B(p)} = \frac{1}{N}\frac{g/6}{1/Z_1 + g/6\;B(p)} + O(1/N^2). \tag{11}$$

Note that the Z factor omitted here would give a $1/N^2$ contribution. Then taking

$$1/Z_1 = 1 - \frac{g}{6\epsilon}, \tag{12}$$

which satisfies the minimal subtraction rule, we obtain a finite $\Gamma^{(4)}$ in the limit $\epsilon \to 0$, at order zero in $1/N$. The β-function follows immediately:

$$0 = \epsilon g + \beta(g)\left[1 + \frac{g}{6\epsilon - g}\right], \tag{13}$$

i.e.,

$$\beta(g) = -\epsilon g + \frac{1}{6}g^2 + O(1/N). \tag{14}$$

One can check this result which is valid to all orders in g, but zeroth order in $1/N$, in the literature. We copy from Zinn-Justin's book,[13] in which he used the minimal subtraction scheme:

$$\beta = -\epsilon g + \frac{N+8}{6N}g^2 - \frac{3N+14}{12N^2}g^3 + \cdots, \tag{15}$$

and higher terms in g are of order $1/N$, which agrees with (14) when N goes to infinity.

2.2. *The two-point function*

We are now considering the diagrams for the inverse two-point function $\Gamma^{(2)}(p)$:

$\Gamma^{(2)}(p) =$ 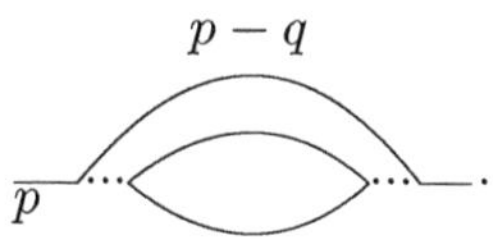

At order g^2, we have one inserted bubble diagram, thus

$$\Gamma^{(2)}(p) = Zp^2 - 2\frac{(gZ_1)^2}{N6^2}\int d^dq \frac{B(q)}{(p-q)^2} + O(g^3, 1/N^2). \tag{16}$$

The integral over q diverges as $-\frac{1}{8\epsilon}$, and we will compute explicitly the integral with a string of bubbles of arbitrary length hereafter. So, at order g^2,

$$Z = 1 - \frac{1}{144N}\frac{g^2}{\epsilon} + O(g^3, 1/N^2), \tag{17}$$

giving

$$\gamma(g) = \frac{g^2}{72N} + O(g^3, 1/N^2). \tag{18}$$

To all orders in g, the string of bubbles gives

$$\Gamma^{(2)}(p) = Zp^2 - \frac{2}{N}\sum_{k=1}\left(\frac{-gZ_1}{6}\right)^{k+1}\int d^dq \frac{B^k(q)}{(p-q)^2} + O(1/N^2). \tag{19}$$

We will sum the series later, but it is interesting to study the finite order k dealing thus with the integral

$$I_k = \int d^dq \frac{B^k(q)}{(p-q)^2} = \frac{a^k(\epsilon)}{\epsilon^k}\int d^dq \frac{q^{-k\epsilon}}{(p-q)^2}, \tag{20}$$

with $B(q)$ and $a(\epsilon)$ given in (9).

$I_k =$ 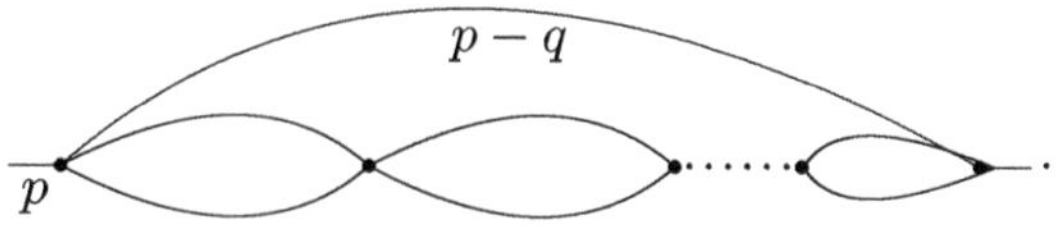

It is interesting to compute I_k explicitly to understand what happens at higher k's. Standard techniques give

$$I_k = -\frac{k}{4(k+1)}\frac{a^k p^{2-\epsilon(k+1)}}{\epsilon^k}\frac{\Gamma^2(1-\epsilon/2)\Gamma(1-\epsilon(k+1)/2)\Gamma(1+\epsilon(k+2)/2)}{\Gamma(1-\epsilon(k+2)/2)\Gamma(1+\epsilon k/2)}$$

$$\times \frac{(1-\epsilon/2)}{(1-\epsilon(k+2)/4)(1-\epsilon(k+2)/2)}$$

$$= -\frac{k}{4(k+1)} \frac{p^{2-\epsilon(k+1)}}{\epsilon^k} [1 + (5k/4+1)\epsilon + O(\epsilon^2)]. \tag{21}$$

The calculations from hereon are straightforward, we just have to expand in powers of ϵ the various explicit functions which appear in Eqs. (9) and (20) and chose Z to cancel all the poles in ϵ. For instance, at order g^3, if we take

$$Z = 1 - \frac{g^2}{N144\epsilon} - \frac{g^3}{N6^4\epsilon^2}(1-\epsilon/4) + O(g^4/N, 1/N^2), \tag{22}$$

we verify that $\Gamma^{(2)}$ is finite up to this order, as renormalization theory implies

$$\Gamma^{(2)}(p) = p^2 - \frac{g^2}{144N}p^2(2\ln p - 7/8) - \frac{9g^3}{6^4N}p^2\ln^2 p + O(g^4/N). \tag{23}$$

From (22), we obtain immediately at leading order $1/N$,

$$\gamma(g) = \frac{1}{72N}\left[g^2 - \frac{1}{24}g^3 + O(g^4)\right], \tag{24}$$

which agrees for N large with the result in[13]

$$\gamma(g) = \frac{N+2}{72N^2}g^2\left[1 - \frac{N+8}{24N}g + \frac{5(-N^2+18N+100)}{576N^2}g^2\right] + \cdots. \tag{25}$$

We can proceed in this fashion to all orders in g, but it is tedious. One can also sum the series (19), but it is not simple either to extract Z from the sum:

$$\Gamma^{(2)} = Zp^2 - \frac{g^2 Z_1}{18N}\int \frac{d^d q}{(p-q)^2}\frac{1}{\frac{1}{Z_1 B(q)} + g/6}. \tag{26}$$

Fortunately, previous results on the $1/N$ expansion of this model allow us to recover $\gamma(g)$, at order $1/N$, to all orders in g, as explained in the following section.

3. Where are the Renormalons?

If instead of dimensional regularization, we had stayed in four dimensions with an ultraviolet cutoff Λ, we could have computed the same diagrams The bubble diagram (8),

$$\int_\Lambda \frac{d^4 v}{(v-q)^2 v^2}, \tag{27}$$

behaves as $\ln q/\Lambda$ at small momentum. Inserted into the two point function, the k-th iterated bubble behaves like $\ln^k q/\Lambda$, and inserted in the two-point function, it yields the integral

$$\int_\Lambda d^4q \left[\frac{1}{(p-q)^2} - \frac{1}{q^2}\right] \ln^k q/\Lambda, \tag{28}$$

where we have explicited the zero-momentum subtraction of the massless theory (which automatically vanished in the minimal scheme). The resulting integral is proportional to p^2 and it yields an integral over q which is infra-red singular in the small p region. Taking $\ln q/\Lambda = -x$, the singular part is given by a power of a logarithmic singularity in p, with a coefficient which behaves for large x as $\int dxe^{-x}(-x)^k$, i.e., a factorial growth with alternating signs. This is the argument for a perturbation expansion exhibiting (infra-red) renormalons.[7]

Our goal is to compute the renormalized correlation functions, the scaling limit of the theory in which distances are much larger than the lattice spacing Λ^{-1}. We have seen in the previous section how complex the interplay between the diagrams and the counterterms is and this is not transparent in the above cutoff regularized theory. So, let us return to the results of the previous section within dimensional regularization. Looking back at the kth order, i.e., the integral (21), we see explicitly that the large k and small ϵ limits do not commute. The renormalization procedure is strictly defined as ϵ goes to zero first, canceling the poles in ϵ through the poles coming from Z and Z_1. It is only after we have removed those singular terms that we may examine the asymptotic behavior for large k. In the minimal subtraction scheme that we have followed here, this is done by canceling all the poles and multiple poles in ϵ occurring in the results such as (21) with the poles in Z and Z_1.

Could that produce the same -k! renormalon? Indeed, if we return to (21), the factor $\frac{1}{\epsilon^k}p^{2-\epsilon(k+1)}$ will end up expanded k times in powers of ϵ by the time all the subtractions manage to produce a finite four-dimensional theory (i.e, $\epsilon = 0$). This might yield a behavior of the two-point function at order k with a term $p^2[-(k+1)\ln p]^k$ and of course $(-k)^k \simeq (-a)^k k!$, but this is far from obvious, the algebra could produce a $1/k!$, which would kill this would-be renormalon. However, our goal here is to understand the large orders of the ϵ-expansion, and that relies on the expansion of the renormalization-group γ function. There, we will see that there is no room for a renormalon large-order behavior.

4. An Explicit Solution Through Earlier Results

Since the removal of the poles in ϵ is increasingly cumbersome when the order increases, fortunately we can call on previous results on the large-N limit to bypass this long algebra. In fact, there are better ways of dealing with the $1/N$ expansion, like adding to the action a Lagrange multiplier $\lambda(\varphi^2 - \psi)$, replacing the quartic term in φ by ψ^2 and tracing out the Gaussian φ's. The expansion around the saddle point of the resulting (λ, ψ) action yields the $1/N$ expansion.[7,13] The reason for not following this procedure here is that we needed to stick to the minimal subtraction scheme. Several terms of the $1/N$ expansion of the critical exponents have been computed for arbitrary dimensions, much more than what we need here. In particular, we find in[7] the critical exponent η at leading order and copy here the result:

$$\eta = \frac{1}{N}\left[\frac{\epsilon^2}{2}\frac{\Gamma(1-\epsilon)}{\Gamma^3(1-\epsilon/2)\Gamma(1+\epsilon/2)}\frac{1-\epsilon}{(1-\epsilon/4)(1-\epsilon/2)}\right] + O\left(\frac{1}{N^2}\right). \quad (29)$$

But we know that $\eta = \gamma(g^*)$ and $g^* = 6\epsilon + O(1/N)$. Therefore, we obtain the RG function $\gamma(g)$ by replacing ϵ by $g/6$ in η, providing the result to all orders in g:

$$\gamma(g) = \frac{g^2}{72N}\left[\frac{\Gamma(1-g/6)}{\Gamma^3(1-g/12)\Gamma(1+g/12)}\frac{1-g/6}{(1-g/24)(1-g/12)}\right] + O\left(\frac{1}{N^2}\right). \quad (30)$$

This is the result valid to all orders in g, first order in $1/N$, that we were looking for. One verifies easily that this, expanded in g, reproduces what we had found before at low order (24).

5. Concluding Remarks

- The result (30) is analytic in g in the neighbourhood of the origin: it is meromorphic in g with the closest singularity at $g = 12$. Had the large orders of the expansion in g be growing factorially, the result would not be analytic at $g = 0$. This is reminiscent of what is familiar in matrix models: the matrix integral is not analytic in g, the coefficient in $g\mathrm{TrM}^4$. However, if one considers the large N limit, and the successive terms of the $1/N$ expansion, every term of the expansion is analytic at $g = 0$. It has been argued by previous authors that renormalons are not present in φ_4^4[16]: the explicit calculation performed here confirms this position.
- The potentially dangerous renormalons do not show up in the final result.
- We have not shown that other correlation functions, other than the one we have computed here; we could not show renormalons, but it seems likely, in view of what we did, that they are simply absent at first order in $1/N$, and we are inclined to believe that this remains true to all orders in a $1/N$ expansion.
- We have not shown that renormalons would not show up at fixed N, but the argument in their favor being *a priori* operative, but finally absent, for the case that we have considered here, we see no reason to believe that they spoil the old result[4] on the large orders of the ϵ-expansion.
- The model that we have studied was considered as a candidate for infrared renormalons. We have nothing to say on potential UV renormalons as in gauge theories.

- Many contemporary scientists probably consider that the problem discussed here is obsolete; who needs the ϵ-expansion given the magnificent precision of the conformal bootstrap,[14] which defeats earlier methods in their accuracy at predicting critical exponents? However, I believe that there are many problems of interest in the scaling region for which the tools of conformal bootstrap are not (not yet?) available. For instance, the universal scaling equation of state, still relies on expansions: it was my first article (with Wallace and Wilson)[15] using the ϵ-expansion, 50 years ago!

Acknowledgement

I thank Giorgio Parisi for a discussion which led me to reconsider this ancient story. Discussions with Kay Wiese are gratefully acknowledged.

References

1. K. G. Wilson and M. E. Fisher, Critical Exponents in 3.99 Dimensions. *Phys. Rev. Lett.* **28**, 240 (1972).
2. E. Brézin, J. C. Le Guillou, J. Zinn-Justin, J. and B. G. Nickel, Higher order contributions to critical exponents. *Phys. Lett.* **44A** (1973) 227. This was only $O(\epsilon^3)$; nowadays five or six terms of the ϵ-expansion have been computed, confirming its numerical divergence.
3. L. N. Lipatov, Divergence of the perturbation-theory series and the quasi-classical theory. *Sov. Phys. JETP* **45**, 216 (1977).
4. E. Brézin, J.-C. Le Guillou and J. Zinn-Justin, Perturbation theory at large order; I and II. *Phys. Rev.* **D15**, 1544 (1977), *Phys. Rev.* **D15**, 1558 (1977).
5. C. M. Bender and T. T. Wu, Large order behavior of perturbation theory. *Phys. Rev. Lett.* **27**, 461 (1971); *Phys. Bev. D*7, 1620 (1973).
6. E. Brézin and G. Parisi, Critical exponents and large-order behavior of perturbation theory. *J. Stat. Phys.* **19**, 269 (1978).
7. M. Moshe and J. Zinn-Justin, Quantum field theory in the large N limit: A review. *Phys. Rep.* **385**, 69 (2003).
8. J. C. Le Guillou and J. Zinn-Justin, Critical exponents from field theory. *Phys. Rev. B* **21**, 3976 (1980).
9. R. Guida and J. Zinn-Justin, Critical exponents of the N-vector model. *J. Phys. A* **31**, 8103 (1998).
10. G.'t Hooft, Can We Make Sense Out of Quantum Chromodynamics?. *Subnucl. Ser.* **15**, 943 (1979).
11. C. G. Callan, Broken Scale Invariance in Scalar Field Theory. *Phys. Rev.* **D2**, 1541 (1970). K. Symanzik, Small distance behaviour in field theory and power counting. *Comm. Math. Phys.* **18**, 227 (1970).
12. G.'t Hooft, Dimensional regularization and the renormalization group. *Nucl. Phys. B* **61**, 455 (1973). S. Weinberg, New approach to the renormalization group. *Phys. Rev.* **D8**, 3497 (1973).
13. J. Zinn-Justin *Quantum Field Theory and Critical Phenomena* (Clarendon Press, Oxford, 1989).

14. S. El-Showk, M. F. Paulos, D. Poland, S.Rychkov and D. Simmons-Duffin, Solving the 3d Ising Model with the Conformal Bootstrap II. c-Minimization and Precise Critical Exponents and many subsequent work. *J. Stat. Phys.* **157**, 869 (2014).
15. E. Brézin, D. J. Wallace and K. G. Wilson, Feynman-graph expansion for the equation of state near the critical point. *Phys. Rev. Lett.* **29**, 591 (1972).
16. I. M. Suslov, *Sov. Phys. JETP* **100**, 1188 (2005). Divergent perturbation series.

https://doi.org/10.1142/9789811282386_0035

Chapter 35

Field-Theoretic Renormalization Group in Models of Critical Behavior, Turbulence and Non-equilibrium Dynamics

L. Adzhemyan,[*,†,**] N. Antonov,[*,†,††] M. Hnatič,[†,‡,§,‡‡] J. Honkonen,[¶,§§] P. Kakin,[*,¶¶] D. Kazakov,[†,‖‖] M. Komarova,[*,***] M. Kompaniets,[*,†,†††] A. Kudlis,[‖,‡‡‡] T. Lučivjanský,[‡,§§§] M. Nalimov[*,†,¶¶¶]

[*]*Department of Physics, St Petersburg State University, Universitetskaya nab. 7/9, St Petersburg, 199034, Russia*

[†]*N.N. Bogoliubov Laboratory of Theoretical Physics Joint Institute for Nuclear Research, Joliot-Curie 6 Dubna, Moscow region, 141980, Russia*

[‡]*Pavol Jozef Šafárik University in Košice, Šrobárova 2 04180 Košice, Slovakia*

[§]*Institute of Experimental Physics of Slovak Academy of Sciences Watsonova 47, 04001 Košice, Slovakia*

[¶]*Department of Physics, P.O. Box 64 FI-00014 University of Helsinki, Finland*

[‖]*Abrikosov Center for Theoretical Physics, MIPT, Institutskiy Pereulok 9, Dolgoprudny, Moscow Region, 141701, Russia*

[**]*l.adzhemyan@spbu.ru*
[††]*n.antonov@spbu.ru*
[‡‡]*hnatic@saske.sk*
[§§]*juha.honkonen@helsinki.fi*
[¶¶]*p.kakin@spbu.ru*
[‖‖]*kazakovd@theor.jinr.ru*
[***]*frommarina@mail.ru*
[‡‡‡]*andrewkudlis@gmail.com*
[§§§]*tlucivjansky@gmail.com*
[¶¶¶]*m.nalimov@spbu.ru*

Numerous physical systems of different nature exhibit singular behavior in the vicinity of their critical points: Various thermodynamic and correlation functions acquire scaling forms with apparently universal critical exponents. The theory of phase transitions and critical phenomena is one of the most interesting areas of modern theoretical physics. The crucial breakthrough came with the application of the renormalization group and ε-expansion by K. Wilson, M. Fisher and others. Most conventional critical systems (liquid–vapour, magnets, superfluid transition) belong to the universality class of the $O(n)$ symmetric ϕ^4 model and its descendants. There, the application of the field-theoretic renormalization group allows for high-order calculations of the critical exponents within various regular schemes: $4-\varepsilon$-, $2+\varepsilon$-, $1/n$- and τ-expansions with subsequent resummation of the resulting asymptotic series. Another direction of research is the application of renormalization group to more complex, dynamic and non-equilibrium systems such as growth phenomena, chemical reactions and turbulence. This chapter presents the results obtained by our informal international network (St. Petersburg, Dubna, Košice, Helsinki) and covers both directions.

1. Introduction

Numerous physical systems of very different nature exhibit interesting singular behavior in the vicinity of their critical points: Various thermodynamic and correlation functions acquire scaling forms with apparently universal critical exponents. The theory of phase transitions and critical phenomena is one of the most interesting areas of modern theoretical physics.

In 1971, K. Wilson applied the renormalization-group (RG) procedure (in the form of certain recursion relations) to the problem of critical behavior. Later, K. Wilson and M. Fisher identified a formal small parameter $\varepsilon = 4-d$, the difference between the upper critical dimension 4 (above which the mean-field theory works) and the physical spatial dimension d.[1] Then Wilson showed, on the basis of field-theoretic diagrammatic expansions, that ε can serve as an expansion parameter in a regular systematic perturbation scheme.[2]

Soon after that, J.C. Le Guillou, E. Brezin and J. Zinn-Justin revealed a close resemblance between the Wilsonian procedure and the RG developed in quantum-field theory (QFT) to meet the needs of relativistic particle physics.[3–6] Those works made it possible to apply the well-developed techniques of the QFT: functional methods (including Schwinger–Dyson equations, Ward identities, Legendre transformations of generating functionals), various methods of calculation of higher-order Feynman diagrams, renormalization theory and renormalization group (including composite operators), short-distance operator product expansions, instanton calculus and so on. This also allowed one to systematically improve the results of the ε-expansion by calculating of the higher orders with subsequent resummation of the corresponding asymptotic series.

More subtle characteristics of the critical state (scaling functions, amplitude ratios, surface and finite-size effects, decaying and ageing), as well as alternative expansions ($1/n$-, $2+\varepsilon$- and τ-expansions, expansions in regulators of analytic renormalization, RG at fixed d and so on) became accessible.

The area of applicability of the RG approach has become wider and now includes the dynamics of conventional equilibrium phase transitions (critical dynamics), driven diffusive systems, percolation and growth phenomena, polymer dilutes, roughening of fluctuating surfaces, self-organized criticality, turbulence and turbulent transfer; this is not an exhaustive list.

A more detailed discussion of the RG, its applications to various problems and bibliography can be found in Refs. 5–12.

2. Various Types of Expansions: Regular Expansions

2.1. ε: *special place*

In the work of Wilson and Fisher,[1,2] it was proposed to use the deviation of space dimension from the upper critical dimension 4 as an expansion parameter in calculations. The $4 - \varepsilon$ expansion in the RG approach has since then become both the most significant and the most widely used. It has a strictly defined dimensionless parameter of perturbation theory. This made it possible to improve the results obtained on the basis of mathematically sound procedures discussed in the following. Most importantly, within the ε-expansion, it is easy to understand the connection between the RG equation and UV renormalization. The common phrase "deviation from the upper critical dimension" leads to an erroneous interpretation of what is happening. Instead of the upper critical dimension, one should speak about the logarithmic theory, in which it is possible to clearly trace the connection between the UV and IR problems of the perturbation theory.[5] For logarithmic theories, the most effective renormalization scheme of minimal subtractions (MS) in dimensional regularization[13] was developed as well as simple methods for solving RG equations and calculating scaling functions clearly demonstrating which singularities of perturbation theory are relevant and how they are summed by RG.

2.2. *Double expansion*

The QFT renormalization with the use of the $\varepsilon = 4 - d$-expansion is based on the dimensional regularization put forward by 't Hooft and Veltman.[14] The Bogolyubov–Parasyuk theorem was analyzed by Speer[15] for the alternative analytic regularization, but it has been rarely used to renormalize diagrams of the perturbation theory. However, in several models of critical phenomena and stochastic dynamics, interactions or correlations of fluctuations are described as powerlike functions of momenta, which can be treated as propagators in analytic regularization. Deviations of the exponents of propagators from their logarithmic values (regulators) can be used as expansion parameters.

For instance, in stochastic field-theoretic models, UV-regulators often appear in exponents of powerlike correlation functions of random sources and can serve as expansion parameters in analytic renormalization. Such models do not require dimensional regularization in the physical space dimension $d = 3$. However, at dimensions $d = 2$ or $d = 4$, additional divergences occur, and when they are included in renormalization, an additional regulator is required and the double expansion comes into play. In the RG analysis of turbulence, the regulator was introduced either in space dimension in Refs. 16 and 16 or in the exponent of a non-local correlation function in Ref. 17 with the subsequent expansion in the respective regulator.

Comparison of the results of different expansions for the same model (e.g., $\varepsilon = 4 - d$, analytical and $1/n$) provides additional information about the scope of each of them. For instance, Sak pointed out[18] that in the limit of vanishing analytic regulator, the local counterpart of the long-range term must be taken into account. The effect to critical behavior of the interplay of short-range and long-range correlations of fluctuations of the local transition temperature in the Ginzburg–Landau model was analyzed by Weinrib and Halperin, who put forward the term "double expansion".[19] They applied Wilson's approach and replica trick, giving rise to effective interaction of the structure $\phi^2 \left(v + wk^{4-d-\delta}\right) \phi^2$.

A field-theoretic analysis of this interplay to all orders of perturbation theory was given by Honkonen and Nalimov,[20,21] who considered the action

$$S = \int d\mathbf{x} \left[-\frac{1}{2}\nabla\phi\nabla\phi - \frac{b}{2}\phi(-\nabla^2)^{1-\alpha}\phi - \frac{\lambda}{24}\phi^4 \right] \tag{1}$$

as well as the model of Weinrib and Halperin. Model (1) becomes logarithmic under the choice of the parameter α at an arbitrary d. For finite $\alpha > 0$, the local quadratic term in (1) is IR-irrelevant. When the parameter α is a regulator of UV-divergences, the renormalized model is obtained in the limit $\alpha \to 0$. The local term is generated in renormalization and must be included in the multiplicatively renormalizable action. Renormalization gives rise to an anomalous dimension of the local term competing with that of the non-local term. The $\varepsilon = 4 - d$-expansion in the local model was shown to be a consistent choice for describing the critical behavior of the standard ϕ^4 model. For models with a long-range order (arbitrary $\alpha > 0$), the crossover point is $\alpha = \eta$, where η is the Fisher exponent.

Following Weinrib and Halperin, the term "double expansion" usually refers to a field theory with both dimensional and analytic renormalization. Double expansion in this sense was used in the asymptotic analysis of various problems of random walk in quenched random fields with long-range correlations: random walk in isotropic vector field,[22] Lévy flights in constrained vector field,[23] vicious walks[24] to mention a few. Double expansion within the Weinrib–Halperin model[19] was used to study polymers in long-range disorder.[25]

In the RG analysis of turbulence with the non-local force correlation function used by De Dominicis and Martin,[17] additional divergences appeared in the limit $d \to 2$. It was pointed out by Honkonen and Nalimov that in a consistent renormalization, these divergences must be absorbed in a local term added to the action,[26] leading to a regular expansion in $d-2$ and the regulator of the correlation function.

The double expansion has allowed us to assess the effect of low-dimensional fluctuations on the scaling behavior: In the first two-loop calculation of the randomly forced Navier–Stokes equation, it was pointed out that explicit expressions of diagrams can yield large relative contributions when the diagrams are divergent in two dimensions.[27] Accounting for additional divergences as $d \to 2$ within the double expansion scheme was shown to significantly improve numerical results for non-universal quantities (the Kolmogorov constant in the first place).[28,29] It was also shown by explicit calculation that taking into account both local and non-local force correlations is imperative for consistent renormalization of the model. Within this approach, already the one-loop values of the Kolmogorov constant and the skewness factor were in reasonable agreement with the recommended experimental values.[30]

Double expansion was also used in the problem of a scalar[31] and transverse vector[32] advected by a synthetic velocity field. In the Kolmogorov scaling regime, the inverse Prandtl number, the Kolmogorov constant, and the Obukhov–Corrsin constant were calculated in the leading order in the double expansion for the stochastic Navier–Stokes equation in Ref. 33. The analysis was extended to structure functions in Ref. 34. The large-scale asymptotic behavior of the solution of randomly forced magnetohydrodynamic equations for space dimensions $d \geq 2$ with the use of the double expansion was consistently analyzed in Ref. 35. Here, the "principle of maximum divergences" was declared as a guide for constructing a renormalization procedure.

It should be noted that in many applications of the double expansion, it was assumed that the regulators are proportional to each other (the "ray scheme"). At two-loop and higher orders of perturbation theory, the ray scheme simplifies calculations significantly because the MS scheme can be used. However, in analytic renormalization, renormalized Green functions are analytic functions of all the regulators. In the ray scheme, renormalized Green functions are not analytic functions of regulators beyond one-loop approximation, which raises serious doubts about the consistency of the ray scheme. In the two-loop calculation for the stochastic Navier–Stokes equation, however, the values for the critical exponents did not depend on the renormalization scheme,[29] but there are indications that care should be taken in calculations beyond one loop.

2.3. *Critical-point $1/n$-expansion*

A useful alternative to the ε-expansion in calculating critical behavior is the $1/n$-expansion, where n is the number of components of the order parameter. In this expansion, the coefficients of the leading powers of $1/n$ can be calculated exactly as functions of the space dimension d. Expansion of these coefficients in $4-d$ and $d-2$ helps one to check the results of the ε-expansions and provides information on inferring these expansions. A detailed consideration of various ways of constructing a large-n expansion was given by Vasil'ev.[5]

The $1/n$ expansion appeared in the calculation of critical properties of the n-vector model[36] and the $O(n)$-symmetric n-component ϕ^4 model (1),[37] in the limit $n \to \infty$ equivalent to the exactly solvable spherical model.[38] The calculation of corrections to the spherical-model values of critical exponents of the n-vector model (whose critical behavior is equivalent to that of the non-linear σ model[5]) and ϕ^4 model at the beginning of 1970s was reviewed by Moshe and Zinn-Justin.[39]

Results in higher orders appeared slowly. In three dimensions, Abe calculated η_2 for the n-vector model.[40] The subscript refers to the coefficient of the $1/n$-expansion of the exponent: $\eta = \eta_0 + \eta_1/n + \eta_2/n^2 + \cdots$, etc. Okabe and Oku calculated β_2, γ_2 and ν_2[41,42] and later the same result with a different method was obtained by Kondor and Temesvári.[43] The first $O(1/n^2)$ result in an arbitrary space dimension was the calculation of η_2 by Symanzik (unpublished), which was first published by Kondor, Temesvári, and Herényi[44] and a bit later by Vasil'ev, Pis'mak and Honkonen.[45] The exponent ν at $O(1/n^2)$ in arbitrary space dimension[46] and the first third-order result η_3[47] were obtained by Vasil'ev, Pis'mak and Honkonen. The coefficient ω_2 (ω is the exponent of corrections to the leading scaling behavior, e.g., $\langle\phi(\mathbf{k})\phi(-\mathbf{k})\rangle \sim k^{-2+\eta}\,(1 + Ak^{\omega} + \cdots)$) was calculated in the non-linear σ model by Broadhurst, Gracey and Kreimer.[48]

Critical exponents beyond order $O(1/n)$ in an arbitrary space dimension were calculated with the use of various skeleton equations. These include the self-consistent screening approximation of Bray[49] and the conformal bootstrap due to Parisi and Peliti.[50] In both cases, the model is analyzed directly at the Wilson–Fisher critical point (massless propagators and conformal-invariant vertices,[51,52] all consisting of powerlike functions of spatial distances). This approach, amended by conformal-invariant integration (method of uniqueness) and novel transformations of massless master graphs proposed by Vasil'ev, Pis'mak and Honkonen[45–47] (dubbed the critical-point $1/n$-expansion in Ref. 53) allowed one to obtain the highest-order results in the $1/n$-expansion of the n-vector model and later in the n-component Gross–Neveu model.[54]

In the analysis of phase transitions in particle physics, the Gross–Neveu model with a $U(n)$-symmetric four-fermion interaction[55] is a benchmark model, whose critical exponents were calculated in the $d-2$-, $4-d$- and $1/n$-expansions.[5,6] For the $1/n$-expansion, the Euclidean action of n Dirac spinor fields is cast to the Yukawa form with the coupling constant $g \sim 1/n^5$:

$$S = \int d\mathbf{x}\left(\overline{\psi}\partial\!\!\!/\psi - \frac{1}{2g}\sigma^2 + \overline{\psi}\psi\sigma\right).$$

The Gross–Neveu model is not multiplicatively renormalizable in the $d-2$-expansion, but its critical behavior corresponds to a fixed point of the generic multiplicatively renormalizable four-fermion model; therefore, the critical exponents of the Gross–Neveu model are meaningful.[56] The critical exponents η, κ and ν (see Ref. 5 for definitions) in order $O(1/n)$ were calculated by Gat *et al.*[57] and others. The coefficient η_2 was calculated by Gracey,[58,59] κ_2 and η_3 were calculated by Vasil'ev *et al.*,[54] λ_2 by Vasil'ev and Stepanenko[60] and Gracey.[61,62] All results beyond the order $O(1/n)$ were obtained within the critical-point $1/n$-expansion approach.

Dimensional regularization and the scheme of minimal subtractions are the mainstream tools in the $4-d$- and $d-2$-expansions. The consistency of the critical-point $1/n$ expansion from the point of view of renormalization theory was demonstrated in a scheme based on analytic regularization put forward by Vasil'ev and Nalimov on the example of the non-linear σ model[63] and completed by Vasil'ev and Stepanenko.[64] In particular, within this approach, the two correction exponents ω of the two-charge CP^{n-1} model (whose critical behavior is equivalent to that of scalar electrodynamics) were calculated by Vasil'ev and Nalimov[65] and all basic exponents and mixing matrices of composite operators of dimensions 2 and 4 (determining four correction exponents ω) of the $n \times p$ matrix gauge-invariant σ model (whose critical behavior is equivalent to that of scalar chromodynamics) were calculated by Vasil'ev, Nalimov and Honkonen[66] at the order $O(1/n)$.

Recent results and application of the critical-point large n expansion to models of particle physics were reviewed by Gracey.[53]

2.4. *Fixed spatial dimension*

An alternative approach to the ε expansion, when the entire calculation is performed directly at the physical dimension of space was proposed by Parisi.[67] It turned out to be extremely effective, making it possible to obtain numerical estimates of critical exponents with very high accuracy significantly exceeding the outcome of the ε-expansion of that period.[68–70] The deviation from high-temperature expansions and lattice calculations was a few percent. The integrals were calculated numerically in the massive scheme, and the renormalization constants were found from the normalization conditions for the renormalized Green functions, which are those of the logarithmic model. Almost all renormalization constants turn out to be finite: They cease to be those carriers of divergences as in the ε-expansion. Moreover, the theory in three dimensions is superrenormalizable and it has only two divergent diagrams. The first multiloop calculations within this scheme were made in Ref. 71. A couple of years later, the same authors performed a more detailed numerical analysis of the series using knowledge of the asymptotic behavior.[72] They also presented a detailed report with the values of all six-loop diagrams for the four-point Green function and seven-loop diagrams for the two-point Green function[73]; an unprecedented calculation that has not yet lost its relevance. The report was a kind of bible for those who have studied the critical behavior of ϕ^4 models with different symmetries within the RG in the physical spatial dimension. They achieved this amazing length of expansions by cleverly collapsing internal loops in the diagrams (e.g., see Ref. 74 and references therein), which allowed them to lower the degree of integration. Nickel and Murray in 1991 carried out an improved calculation of the exponent associated with the anomalous field dimension within seven loops.[75]

However, seven loops for the β-function have never been obtained. Since then, great progress has not been observed. In Ref. 76, the authors described only ideas for the full automation of calculations, without which it is difficult to imagine extracting higher-order results (more than six loops). They demonstrated the problems researchers might face in order to complete a full seven-loop calculation.

Over the years, with the help of this technique, numerical estimates were obtained for critical exponents for many models of critical behavior,[68–70,77–79] to mention a few.

In the convex of narrative about expansion in a fixed spatial dimension, we would like to note another approach that was proposed by Dohm *et al.*, where they suggested its own version of renormalization in a fixed-dimensional space.[80–83]

3. Technical Features and New Approaches (Calculation Methods)

3.1. *Method of uniqueness*

To construct the $1/n$-expansion 2.3 of the critical non-linear σ model (or, equivalently, the ϕ^4 model), it is necessary to introduce an auxiliary field ψ to convert the interaction of the model to the form $\propto \int d\mathbf{x}\,\psi\phi^2$, where ϕ is the basic field. Due to scale invariance full (dressed) propagators in skeleton graphs of the model are of the form $D_\phi = A|\mathbf{x}|^{-2a}$ and $D_\psi = B|\mathbf{x}|^{-2b}$, where a and b are scaling dimensions of the fields. Conformal symmetry fixes the one-irreducible three-point function to a product of powers of separations of its coordinates.[84] Therefore, skeleton graphs of Green functions for the critical-point $1/n$-expansion contain three-point vertices only.

Calculations in the method of uniqueness are carried out in the position space and they are based on transformations,[45,46] which allow us to change exponents in the power-law propagators to arrive at combinations satisfying the condition $a+b+c=d$, in which case the "star-triangle" relation holds,[51,52] connecting the "unique" vertex and triangle ($\mathbf{x}_{ij} \equiv \mathbf{x}_i - \mathbf{x}_j$, $H(a) = \Gamma(d/2-a)/\Gamma(a)$, $H(a,b,c) = H(a)H(b)H(c)$; notation from[5]):

$$\int d\mathbf{x}_4 \frac{1}{|\mathbf{x}_{14}|^{2a}|\mathbf{x}_{24}|^{2b}|\mathbf{x}_{34}|^{2c}} = \frac{\pi^{d/2}H(a,b,c)}{|\mathbf{x}_{12}|^{2(d/2-c)}|\mathbf{x}_{13}|^{2(d/2-b)}|\mathbf{x}_{23}|^{2(d/2-a)}}. \tag{2}$$

The basic tool to construct these transformations is the convolution relation

$$\int d\mathbf{x}_3 \frac{1}{|\mathbf{x}_{13}|^{2a}|\mathbf{x}_{23}|^{2b}} = \frac{\pi^{d/2}H(a,b,d-a-b)}{|\mathbf{x}_{12}|^{2(a+b-d/2)}}, \tag{3}$$

which allows the manipulation of the propagator exponents. Using relations (2) and (3) together with conformal inversion and switching from position to momentum space, Vasil'ev, Pis'mak and Honkonen put forward a set of transformations of massless skeleton graphs to arrive at "unique" vertex integrals, giving rise to expressions in a closed form.[45,46] This powerful method of calculation of multiloop integrals in the critical-point $1/n$-expansion, dubbed "the method of uniqueness" by Kazakov,[85] made it possible to push calculations of static critical exponents beyond the order $O(1/n)$ in an arbitrary space dimension and was widely used in calculations in the $4-d$-expansion as well.[86–88] A detailed consideration of methods of calculating Feynman diagrams with massless propagators was given by Vasil'ev.[5]

Amended by the trick of dimensional transformation proposed by Derkachov, Honkonen and Pis'mak,[89] the method of uniqueness was used to calculate the three-loop contribution to the anomalous dimension of the diffusion coefficient in a model of diffusion in a long-range potential random field in an arbitrary space dimension.

The transformation group of a two-loop massless master graph put forward in Ref. 46 attracted considerable attention itself[90,91] and became an important tool in multiloop calculations of Feynman diagrams (for review, see Refs. 53,92).

3.2. *Higher-order asymptotics*

The computational success of the ε-expansion in calculating the critical exponents led to a natural question about the behavior of the resulting series. Based on the instanton analysis, Lipatov showed that the series of $4-\varepsilon$-expansions are asymptotic with zero radius of convergence.[93] This situation is common in theoretical physics; both the perturbation theory series in quantum electrodynamics and the special relativity corrections to the heat capacity of an ideal gas have a similar property. However, the expansion parameter in the case of real critical behavior is not small ($\varepsilon = 1$). Results of the ε-expansion usually cannot be trusted starting from the third-fourth order because the next order of the expansion produces the large contribution and then the previous. For the critical behavior of composite operators, sometimes even the first order is dubious. However, the key to solving this problem was also put forward in Ref. 93. For the series in ε, the asymptotic behavior of the form $F^{(N)} \approx N!a^N N^{b(F)} C(F)$ was proved. For these power series with "Lipatov" asymptotics, resummation according to the Borel–Leroy method is possible with purely physical assumptions of the properties of the analyzed quantities.[6,94] There are different schemes. Padé, Padé-Borel, conformal Borel, etc. Compared to using Padé, the conformal Borel approximant has the advantage of using more established information — the constants a, b from Lipatov's asymptote. In Ref. 95, by comparing different resummation schemes for the dynamic critical exponent in the dynamic model A, it was shown that although all resummation schemes give similar results, only schemes using the known Lipatov asymptotic parameters lead to intersecting error intervals.

In dimensional regularization using the MS renormalization scheme, the Lipatov asymptotics for the ϕ^4 theory was calculated.[96–98] An error found in Ref. 96 in the calculation of one contribution to constants of type $C(F)$ was corrected in Ref. 99.

Any additional information available can improve the results and accuracy of the resummation. It was proposed to use the strong coupling asymptotics to this end.[100–104]

The Lipatov asymptotics in the ϕ^4 theory and other statical equilibrium models was obtained on the basis of the found instanton solution of the classical equation of motion. In dynamic stochastic models with a Gibbs equilibrium limit, a similar problem was solved in Ref. 105–107, after which the dynamic critical index z for the $O(n)$ symmetric model A was resummed in Ref. 95 in four orders of ε-expansion for $n = 1, 2, 3$ and in Ref. 108 in five orders of ε expansion for $n = 1$. A finite radius of convergence of the perturbation series was found in some Kraichnan models.[109–111] But these models exhaust the class of models of critical behavior or stochastic dynamics with known asymptotes of higher orders. For the other stochastic and random-walk models mentioned here, Borel summability has not yet been proven.

4. Static Critical Phenomena

4.1. *The basic model of critical statics: ϕ^4 field theory with $O(n)$-symmetric order parameter*

Starting with the works of Wilson and Fisher,[1] the $O(n)$-symmetric model or the generalized Heisenberg model is the most studied in the theory of critical phenomena.[5,6,9,68] This model is truly rich in the description of various physical systems. For different dimensions of the order parameter, n, this model describes the behavior in the vicinity of a second-order phase transition for some polymers ($n = 0$), the liquid–vapour system and uniaxial ferromagnets ($n = 1$), the superfluid Bose–liquid and planar ferromagnets ($n = 2$), the Heisenberg (isotropic) ferromagnets ($n = 3$), etc.[112–117]

After the breakthrough work of Wilson and Fisher in the 1980–1990s, there was a rapid increase in accuracy due to the calculation of new contributions to the perturbation theory. The record for the $O(n)$-symmetric model at that time was five loops in the ε technique[87,118–122] and six-loop analysis in the physical spatial dimension ($d = 3$).[72,123] Over the next two decades, although $O(n)$-symmetric results were used for models with other symmetries, there was no serious development in the theory. The situation changed only after almost 20 years with new contributions to the perturbation theory.[104,124–128] The prerequisites for this breakthrough were the emergence of a whole set of new numerical methods and various implementations in special software. First, the sector decomposition technique,[129] which enabled one to process all integrable singularities in the calculation of Feynman diagrams. It allowed an independent check of five- and six-loop results also[104,130] in the framework of the ε-expansion and an extension of the calculation of the exponent η in a fixed space dimension to the six-loop order ($D = 2$).[131] Second, active use of integration by parts[46,132–134] and the R^* operation,[135–138] which was extensively used for computation of the six-loop contribution to the exponent η in the ε-expansion.[124]

Table 1. Comparison of numerical estimates of critical exponents ν and η for $O(n)$-symmetric universality classes in case of $n = 0, 1, 2, 3$, obtained by means of different theoretical approaches in $D = 3$. The notation of methods: 3D RG — renormaliation group approach in $D = 3$, CB — conformal bootstrap, ε^6 — six-loop ε-expansion, ε^7 — seven-loop ε-expansion, MC — Monte Carlo, NPRG — non-perturbative RG.

Method	Exp.	$n = 0$	$n = 1$	$n = 2$	$n = 3$
3D RG [143]	ν	0.5882(11)	0.6304(13)	0.6703(15)	0.7073(35)
	η	0.0284(25)	0.0335(25)	0.0354(25)	0.0355(25)
CB	ν	0.5877(12) [144]	0.62999(5) [145]	0.67175(10) [146]	0.7121(28) [147]
	η	0.0282(4) [144]	0.03631(3) [145]	0.03852(64) [147]	0.0386(12) [147]
MC	ν	0.5875970(4) [148]	0.62998(5) [149]	0.67169(7) [150]	0.71164(10) [151]
	η	0.031043(3) [152]	0.036284(40) [149]	0.03810(8) [150]	0.03784(5) [151]
ε^6 [104]	ν	0.5874(3)	0.6292(5)	0.6690(10)	0.7059(20)
	η	0.0310(7)	0.0362(6)	0.0380(6)	0.0378(5)
ε^7 [153]	ν	0.58770(17)	0.62977(22)	0.67076(38)	0.70906(18)
	η	0.03121(70)	0.03653(65)	0.03810(56)	0.03809(62)
NPRG	ν	—	0.63012(16) [154]	0.6716(6) [154]	0.7114(9) [154]
	η	—	0.0361(11) [154]	0.0380(13) [154]	0.0376(13) [154]

Third, the calculation of parametric integrals with hyperlogarithms[139–141] for the full six-loop calculation,[104,125] graphical functions[142] for the full seven-loop calculation,[126] and the recently obtained record-breaking eight-loop calculations for the anomalous field dimension.[128] Such calculations are truly bewitching.

The RG approach for this canonical model allowed us to enrich physics with high-precision values of critical exponents, the most basic characteristics of a system in the vicinity of the second-order phase transition. We would like to conclude this section by presenting the values of critical exponents obtained by various theoretical methods for the most popular universality classes: $n = 0, 1, 2, 3$ in case of three-dimensional space. They are presented in Table 1.

4.2. *Various symmetries for* ϕ^4

The Bogolyubov–Parasyuk approach makes it possible to generalize the counterterms of diagrams calculated for the $O(1)$-symmetric (scalar) model to any theory with non-trivial symmetry due to the factorization of tensor structures (see, e.g., Ref. 155). This implies that for all models with different symmetry, the results should be known up to seven loops. To date, however, the majority of such models were treated only within the six-loop approximation. Moreover, within six loops in Ref. 156, the authors managed to derive RG functions for a general ϕ^4-based theory with arbitrary symmetry. The multiple crosschecking procedures and independent calculations have confirmed correctness of the results obtained there. Thus, within this section, we list only the most important cases.

The most natural extension of the isotropic model is the model with an additional term invariant with respect to the cubic group of transformations. It is difficult to imagine a ferromagnet that would not have any anisotropy of its magnetic subsystem. A detailed study of the n-vector cubic model including the evaluation of critical exponents and n_c was carried out by many groups[77,79,157–184] having used both field-theoretic methods and lattice calculations. At the moment, the most advanced here is the six-loop analysis within the ε-expansion[185] and the 3D RG approach.[79] Speaking of lattice calculations, here the most accurate analysis was made in Ref. 186. We end with the results that have been obtained recently using the $1/n$ expansion[187] and the conformal bootstrap technique.[188] Let us note that the main question is the value of the critical dimension of the order parameter that determines the stable critical behavior in real cubic ferromagnets. The most relevant studies demonstrate that the cubic universality class is stable and the boundary dimension is less than three.[79,185] This conclusion can also be achieved by establishing the numerical difference in the values of the critical exponents of the Heisenberg and cubic universality classes.[186]

A special case worthy of separate consideration is the weakly disordered Ising model which, with the help of the replica trick, can be reduced to a cubic model with zero dimension of the order parameter ($n = 0$). The first steps in this field were laid by Harris and Lubensky,[189,190] Khmelnitskii,[191] and Grinstein and Luther.[192] It was proposed to consider an effective Hamiltonian describing critical thermodynamics of the randomly diluted m-vector model. A particular case is the randomly diluted Ising model ($m = 1$) which, in accordance with the Harris criterion,[193] possesses a set of critical exponents different from those of the pure system. In the model with impurities, within the lowest order of the perturbation theory, it is not possible to find a non-trivial fixed point describing random critical behavior. This is due to the degeneracy of the equations for the zeros of the β function. However, a solution was found[191,192]: instead of the usual ε-expansion, an $\sqrt{\varepsilon}$-expansion was proposed. Exponents and corresponding numerical estimates were obtained. Later, two-, three- and, many years later, five-loop corrections were found.[176,194–197] For a detailed description of the problems associated with the irregularity of the corresponding series and the numerical estimates obtained on their basis, see Ref. 198, in which the six-loop contribution to $\sqrt{\varepsilon}$ was obtained. In addition to the ε analysis, this model was treated within the RG in real space. Two-loop,[199,200] three-loop,[173,201,202] four-loop,[172,203] five-loop[77] and six-loop[204] RG contributions were calculated. With this technique, it is possible to obtain numerical estimates of exponents whose values agree much better with the results of the lattice calculation than in the case of the $\sqrt{\varepsilon}$ technique.

A $O(m) \times O(n)$-symmetric theory describes the critical behavior of frustrated spin systems with non-collinear and non-coplanar ordering.[205–213] Stacked triangular antiferromagnets and helical magnets are examples of such reasons for these frustrations can be the geometric features of the crystal lattice and also the nature of the interaction between the nearest and next-to-nearest spins. Understanding the critical properties of this model allows one to determine the type of phase transition that occurs in real materials. Specifically, it is uncertain whether the transition is continuous with a unique set of critical exponents belonging to the chiral universality class or a first-order transition. To this end, it is necessary to analyze the boundary dimensions of the order parameter or marginal dimensions, which play a key role here. The first attempts to analyze this model within the RG approach were made almost 50 years ago.[205,214,215] Later, in the case of physically interesting couples $\{m, n\}$, the results were obtained by different field-theoretic approaches: ε-expansion,[216–218] $1/n$-expansion,[217,219] 3D RG technique,[212,220] $2-\varepsilon$ machinery,[221,222] pseudo-ε-expansion,[218] "exact" (functional, non-perturbative) renormalization group (ERG).[223] They all testify to the fact that, for each m, some upper bound (marginal) value of n – n^+ – exists, beyond which the system undergoes a continuous phase transition with a specific set of critical exponents. For values of n below n^+, it is believed that the transition is of the first order. The record-breaking calculations in terms of perturbation theory order are six-loop calculations within both 3D RG[220] and ε-expansion.[224] There is also a conformal bootstrap analysis.[225] Regarding the stability of chiral universality class for different couples $\{m, n\}$, the situation here is quite contradictory — different approaches give opposite conclusions. A detailed overview of the situation is presented in Ref. 224.

Another model having $\mathrm{U}(n) \times \mathrm{U}(m)$-symmetry is connected with the description of the finite-temperature phase transition in QCD in the limit of vanishing quark masses with $n = m = N_f$ flavors and unbroken anomaly at the critical temperature.[115,226] The first attempt to analyze the model for $m = n$ within the RG approach was made in Refs. 226–228, where one-loop ε expansions were obtained. A stable fixed point was found in light QCD. The analysis in the five-loop approximation was performed in Ref. 229. The authors considered the most general case $m = n$ with a detailed description of the features of the model. These numerical estimates were refined almost 20 years later in Ref. 230, where a six-loop analysis was performed within the ε formalism. Along with the ε technique, this model was processed using RG in real space, where the series of the same — six-loop — length were obtained[231] and reexamined in Ref. 232.

In addition, two models deserve to be pointed out. One of them describes the multicritical behavior in frustrated spin systems with non-collinear ordering. The most accurate analysis was carried out within the ε-expansion technique in the five-loop approximation.[233] The other model describes the multicritical behavior of a system with two competing order parameters, each of which has $O(n)$ symmetry.[234] Of those published to date, the most advanced here is the five-loop analysis over ε.

We would like to finish this section with a recently published work,[156] where the results were obtained for general scalar field theory within the six-loop approximation. This work covers all available data obtained by means of the ε-expansion for ϕ^4 theories with different symmetries.

5. Dynamic Critical Phenomena

Models of critical dynamics were originally classified by Halperin and Hohenberg.[235] The classification includes several models; we will focus on three of them: A, E, and F that describe important universality classes.[236,237]

5.1. *Model A: Main results*

The phenomenon of critical slowing down in systems with a non-conserved order parameter and energy is marked by the critical dynamic exponent z, which relates the correlation length ξ to the typical time scale of fluctuations τ as $\tau \propto \xi^z$. The order parameter, the correlation length and typical time of fluctuations increase at criticality due to strong fluctuations in the system approaching the critical point. To explore this phenomenon, one could examine the model A of critical dynamics. This model is the simplest dynamic extension of the ϕ^4 field theory and is capable of describing the critical slowing down of systems with non-conserved order parameter and energy. In the following, we briefly give the results for the Ising universality class, which corresponds to the one-component model A of critical dynamics. For a review of an arbitrary dimension of the order parameter, see, e.g., Ref. 238.

The model has been analyzed using various field-theoretic methods as well as within the lattice calculations. From a large number of published papers (see references in Ref. 238) we would like to point out the most up-to-date and accurate calculation of the critical exponent z, which was implemented in the work.[239] For the three-dimensional case, the following estimate was obtained: $z = 2.0245(15)$.

Let us recall the results in the highest orders of the perturbation theory within the RG approach. First, the analysis of model A of critical dynamics, which was carried out with the help of RG in the physical dimension of space in the four-loop approximation.[240] Using advanced resummation techniques for the 2D and 3D one-component model A for z, the authors were able to infer the following numbers, 2.0842(39) and 2.0237(55), respectively.[241]

The most accurate results to date were obtained in the ε-expansion. This was achieved by taking into account the five-loop order implemented in Ref. 242. Using the diagram reduction technique, the authors were able to significantly cut the number of calculated diagrams: from 1025 to 201. Calculation of the diagrams is much more complicated than in statics. To this end, the decomposition technique[243] tested in the sector of static problems was adapted. Using advanced resummation techniques, the authors extracted the following value for the three-dimensional one-component model A: $z = 2.0235(8)$.

Let alone the RG in the physical dimension of space and the ε-expansion, the model A was analyzed using the functional RG. The most accurate and up-to-date numbers were obtained in Refs. 244 and 245. For example, in Ref. 244 the authors obtained 2.025 for a three-dimensional one-component model, which is in good agreement with the numbers extracted by alternative methods.

5.2. *Superfluidity and models E/F*

The phenomenological models E and F have long been the main candidates for describing the critical dynamics of a Bose gas. According to the standard classification,[235] they contain a set of fields suitable for description of essential hydrodynamic flows: a nonconserved order parameter and a conserved density. Model F seems more representative, but the data on the experimental and theoretical determination of the critical index α for the specific heat show[5] that the corresponding systems probably degenerate to a simpler model E in the critical region. However, it remains unclear which of the two non-trivial fixed points of the RG equations for model E determines the real critical behavior of the boson gas.

Models E and F do not explain superfluid properties of the Bose gas, e.g., the critical viscosity exponent, closely related to the correct allowance for dissipation in the hydrodynamic equations. Recently, it has been proposed to study the stability of the E and F models against the influence of the hydrodynamic velocity field[246–248] and perturbations by density waves.[249] The RG equations turned out to be sensitive to the hydrodynamic modes, which significantly affect the critical behavior and the exponents. Taking into account the compressibility of the medium has led to an unexpected conclusion: Models E and F are unstable to the density waves, and their IR behavior boils down to that of the complex model A, which has a single non-trivial IR stable fixed point of the RG equations.

5.3. *Real-time Green functions*

In Refs. 249–252, a new model for calculation of the dynamic critical exponent z for the λ transition was proposed. The model was based on the action (in the unit system with $\hbar = 1$)

$$S = \int_C d\mathbf{x} \int dt \left[\psi^+ \left(i\partial_t + \frac{\Delta}{2m} + \mu\right)\psi - \frac{g}{2}(\psi^+\psi)^2\right] \tag{4}$$

for quantum boson particles with local repulsive interaction. Here, $\psi(\mathbf{x}, t)$, $\psi^+(\mathbf{x}, t)$ are complex conjugated fields, m_0 denotes the particle mass, μ denotes the chemical potential, g denotes the coupling constant, and Δ stands for the Laplace operator. The formalism of time-dependent Green functions at finite temperature was used. Thus, the evolution parameter t belongs to the Schwinger–Keldysh contour in the complex t plane.[253,254] An effective IR field-theory model was put forward on the basis of a change of variables corresponding to the change from the standard Green functions to the retarded, advanced and Keldysh Green functions in the original paper of Keldysh.[254]

The resulting effective field theory with only IR-relevant terms retained and a suitable temporal regularization introduced turned out to be multiplicatively renormalizable.[250] The results of the three-loop RG calculation in the proposed effective model[255] support the conjecture that the dynamics of the superfluid phase transition is described by a model that belongs to the same universality class as the stochastic model A.

6. Turbulence and Turbulent Transport Phenomena

6.1. *Turbulence, multiscaling and RG*

Theoretical understanding of turbulence is often referred to as the last unsolved problem of classical physics. An important representative example is the fully developed (homogeneous, isotropic, inertial-range) hydrodynamic turbulence.[256–258] The quantities of interest are, in particular, the equal-time structure functions

$$S_n(r) = \langle [\theta(\mathbf{x}) - \theta(\mathbf{x}')]^n \rangle = (Wr)^{n/3} \, f_n(r/l, r/L), \quad r \equiv |\mathbf{x} - \mathbf{x}'|. \tag{5}$$

Here, the field $\theta = \theta(t, x)$ can be a component of the velocity field or some other relevant scalar field in the problem: temperature or entropy of the fluid, concentration of an impurity, etc. The angle brackets $\langle \cdots \rangle$ denote the ensemble averaging, and the time argument common to all the quantities is omitted. The second relation follows from dimensionality reasons, where W is the mean energy dissipation rate, l is the internal (Kolmogorov, dissipation, UV) length, L is the external (outer, integral, IR) turbulence scale, and f_n are certain scaling functions.

One of the most interesting problems is the behavior of the structure functions (5) in the inertial range $l \ll r \ll L$, where both the pumping of energy and its dissipation are irrelevant. According to the celebrated Kolmogorov–Obukhov (K41) phenomenological theory,[256–258] the structure functions in the inertial range are independent of both the external L and internal l scales (the first and second Kolmogorov hypotheses, respectively) and, therefore, are determined by the single parameter W. Then dimensionality considerations determine the functions (5), apart from numerical coefficients, in the form

$$S_n(r) \simeq (Wr)^{n/3}. \tag{6}$$

The relation $S_3(r) = -(4/5)(Wr)$, based on the energy conservation law, is considered as a firm theoretical result.

The most remarkable features of the fully developed turbulence are related to the phenomenon of intermittency related to strong fluctuations of the energy flux W. It leads to deviations from the predictions of the K41 theory. Such deviations, referred to as anomalous scaling, manifest themselves in a singular dependence of the correlation functions on the integral scale L. For the functions (5), they are conventionally written in the form

$$S_n(r) \simeq (Wr)^{n/3} \, (r/L)^{q_n}, \tag{7}$$

where the anomalous exponents q_n are certain non-trivial functions of n ($q_3 = 0$).

Within the framework of numerous semi-heuristic models, these exponents are related to statistical properties of the local dissipation rate, the characteristics of non-trivial spatial formations (vortex filaments) and so on, see Refs. 256–258. Such models are only loosely related to underlying hydrodynamics equations, involve arbitrary adjusting parameters and, therefore, cannot be considered to be the basis for the construction of a systematic perturbation theory in a certain expansion parameter; see, e.g., the remark in Ref. 259. Thus, serious doubts remain about the universality of anomalous exponents and the very existence of deviations from the K41 theory.

In Refs. 260 and 261, Adzhemyan *et al.* showed that the second Kolmogorov hypothesis (independence of UV scale) can be justified by the standard RG techniques. It is equivalent to the representation $S_n(r) \simeq (Wr)^n f_n(r/L)$, where the form of scaling functions $f_n(r/L)$ is not determined by the RG equations.

The issues of the first Kolmogorov hypothesis and the anomalous scaling (7) lie beyond the area of expertise of the RG in the narrow sense of the word: They can be studied within the general RG framework with the help of more advanced techniques: inclusion of composite fields ('operators' in field-theoretic terms) and short-distance operator-product expansions (OPE). This gives the following representation:

$$f_n(r/L) = \sum_F C_{nF}\,(r/L)^{-\Delta_F}, \tag{8}$$

where C_{nF} are the coefficients regular in (r/L), the summation is implied over all possible composite operators F allowed by the symmetry and Δ_F are their scaling dimensions. The key role is played by composite operators with negative scaling dimensions, $\Delta_F < 0$, termed 'dangerous' in Refs. 260 and 261. Such operators are absent in standard models of critical behavior like ϕ^4.

The typical models of developed turbulence are described to be the equations of stochastic hydrodynamics for the relevant fields (velocity, temperature, density of pollutants etc.) with random stirring noises (forces). The role of the RG expansion parameters (analogs of $\varepsilon = 4 - d$ in the ϕ^4 model) is played by various exponents in the correlation functions of these noises; see Refs. 260 and 261.

Dangerous composite operators in the Navier–Stokes model occur only for finite values of ε and cannot be reliably identified within the ε expansion. Moreover, these operators (provided they exist) enter into the OPE in the form of infinite families with the spectrum of scaling dimensions unbounded from below (for the proof, see, e.g., Appendix A in Ref. 262), and the analysis of the small-(r/L) behavior implies the summation of their contributions in representations like (8). Thus, for Galilean invariant quantities, like the equal-time structure functions (5), one has to go beyond the ε-expansion to find scaling dimensions of all Galilean invariant composite operators and to perform summation of their contributions in representations (8). This is clearly an extremely difficult task that requires considerable improvement of the present techniques.

However, a similar problem was solved for different-time correlation functions, where the most singular contributions in representations (8) were identified with the powers of the velocity field. Using the Ward identities for Galilean symmetry, the authors of Ref. 260 found their scaling dimensions exactly, summed up their contributions and established the substantial resulting dependence of the correlation functions on L for $(r/L) \to 0$. From physics viewpoint, this corresponds to the well-known phenomenon of sweeping of the small-scale vortices by the large-scale ones.[263,264] Therefore, this effect is neither ignored nor overlooked within the RG formalism, as it was sometimes suspected,[265,266] but the basic perturbative RG should be combined with more advanced methods like OPE and exact relations like exact Schwinger–Dyson and Ward functional equations.

Another optimistic example is provided by numerous models of passively advected scalar and vector (magnetic) fields, where the anomalous exponents in representations like (7) can be identified with dimensions of individual operators; see the following section.

6.2. *Turbulent advection of scalar fields*

Both natural and numerical experiments suggest that the deviation from the classical K41 theory is even more pronounced for passively advected scalar fields than for the velocity field itself; see, e.g. Refs. 267–271 and the literature cited therein. At the same time, the problem of passive advection appears easier tractable theoretically: even simplified models describing the advection by a 'synthetic' velocity field with a given statistics reproduce many of the anomalous features of genuine turbulent heat or mass transport observed in experiments.

Probably, the most important progress in the subject, achieved in the last decade of the 20th century, was related to a simplified model of the fully developed turbulence known as Kraichnan's rapid-change model. The model, which dates back to classical studies of Batchelor, Obukhov, Kraichnan and Kazantsev, describes a scalar quantity passively advected by the Gaussian velocity field. The pair correlation function is taken in the form $\langle v_i v_j \rangle \simeq \delta(t-t')k^{-d-\varepsilon}$, where k is the wavenumber and ε is an arbitrary parameter. For the first time, the existence of anomalous scaling was established on the basis of a microscopic model[272] and the corresponding anomalous exponents were derived within controlled approximations[273,274] and a regular perturbation scheme,[275] namely it was shown that the inertial-range structure functions in such a model exhibit anomalous scaling behavior:

$$S_{2n}(r) \simeq r^{n(2-\varepsilon)} \, (r/L)^{\Delta_n}, \tag{9}$$

with negative anomalous exponents Δ_n, whose first terms of the expansion in ε[273] and $1/d$[274] have the forms

$$\Delta_n = -2n(n-1)\varepsilon/(d+2) + O(\varepsilon^2) = -2n(n-1)\varepsilon/d + O(1/d^2). \tag{10}$$

The relation $\Delta_1 = 0$ is exact. Another quantity of interest is the local dissipation rate of scalar fluctuations, $E(x) = \kappa_0 \partial_i \theta(x) \partial_i \theta(x)$, where κ_0 is the diffusivity coefficient. The equal-time correlation functions of its powers in the inertial range have the forms[273,274]

$$\langle E^n(x) E^p(x') \rangle \simeq r^{-\Delta_n - \Delta_p} (mr)^{\Delta_{n+p}}, \tag{11}$$

with the same exponents Δ_n from (10). These results were obtained within the 'zero-mode' approach developed in Refs. 273, 274; a detailed discussion can be found in Ref. 271.

The second systematic approach, proposed in Ref. 275, is based on the field-theoretic RG and operator product expansion (OPE). The solution proceeds in two main stages. In the first stage, the multiplicative renormalizability of the corresponding field-theoretic model is demonstrated and the differential RG equations for its correlation functions are obtained. The asymptotic behavior of the latter on their UV argument (r/l) for $r \gg l$ and any fixed (r/L) is given by IR-stable fixed points of those equations. It involves some 'scaling functions' of the IR argument (r/L), whose form is not determined by the RG equations. In the second stage, their behavior at $r \ll L$ is found from the OPE within the framework of the general solution of the RG equations. There, the crucial role is played by the critical dimensions of various composite fields (composite operators in the QFT terminology), which give rise to an infinite family of independent scaling exponents and hence to multiscaling. Of course, both these stages have long been known in the RG theory of critical behavior, where the OPE is used in the analysis of the small-(r/L) form of the scaling functions; see, e.g., Refs. 6 and 5 and references therein. The distinguishing feature, specific to models of turbulence, is the existence of composite operators with negative critical dimensions. Such operators are sometimes termed 'dangerous' because their contributions to the OPE diverge as $(r/L) \to 0$.

The OPE and the concept of dangerous operators in the stochastic hydrodynamics were introduced and developed in Ref. 260; a detailed discussion of the NS case can be found in Refs. 276 and 261 and Chapter 6.[5]

The RG analysis of Ref. 275 showed that dangerous operators are indeed present in the rapid-change model and that their dimensions can be calculated systematically within a regular perturbation expansion similar to the ε-expansion of the critical exponents. Owing to the linearity of the original stochastic equations in the passive field, only a finite number of dangerous operators can contribute to any given structure function, which allows one to identify the corresponding anomalous exponent with the critical dimension of an individual composite operator. The actual calculations were performed to the second[275] and third[277] orders in ε (two-loop and three-loop approximations, respectively). Generalizations to the cases of compressible[278,279] and anisotropic[280] velocity ensembles and the vector advected field[281,282] were obtained.

The two approaches complement each other nicely: the zero-mode technique allows for exact (non-perturbative) solutions for the anomalous exponents related to the second-order correlation functions,[274,283,284] while the RG approach forms the basis for systematic perturbative calculations of the higher-order anomalous exponents.[275,277–279] In the cases of anisotropic velocity ensembles or/and passively advected vector fields, as well as advection of extended objects, where the calculations become rather involved, all the existing results for higher-order correlation functions were derived only by means of the RG approach ε.[280–282]

However, the main advantage of the RG+OPE approach is the possibility to consider more realistic velocity ensembles and include finite correlation time and non-Gaussianity; see, e.g., Ref. 31,285,286.

6.3. *First-principles intermittency, decaying spectra*

A characteristic feature of strongly developed stationary turbulence at high Reynolds numbers $Re \gg 1$ is the presence of a wide range of spatial and temporal scales, in which the statistical correlations of the velocity field have a universal power-law behavior with the values of the critical exponents predicted by the phenomenological Kolmogorov theory and their possible corrections. It is known as the inertial interval, and, for example, for spatial scales, it is limited from below by the Kolmogorov dissipation length l and from above by the external turbulence scale $L = Re^{3/4} l$ (obviously, $L \gg l$).

In the energy-containing region with scales of the order of the turbulent length L, the experimental data show that such universality is not observed. However, the study of this area is also very interesting from the point of view of studying decaying turbulence, when the source of energy injection into the turbulent system, present to compensate for energy losses due to dissipation, is turned off. It is clear that in this case the stationarity disappears and the inertial interval gradually narrows with a simultaneous decrease in the outer length L according to a power law in time.

It turns out that using the results of the RG and the ε-expansion to find the kinetic energy spectrum and the transfer energy function, this complex phenomenon can be studied by solving the energy balance equation. In Ref. 287, on the basis of the energy balance equation, an integro-differential non-linear equation for the kinetic energy spectrum was derived and its solution was found in the energy-containing region as a function of the dimensionless variable kL with the correct Kolmogorov asymptotics in the inertial interval, where k is the modulus of the wave vector. At the same time, a power law was found for the temporal decrease of the external turbulent length L, which is in good agreement with the experimental data and predictions of the Kolmogorov theory.

If already known results, obtained in the experiment behind the grid (see, e.g., Ref. 257), are scaled in terms of the dimensionless variable kL, then the non-universal jumpy data obtained at different distances from the grid collapse, fitting onto one smooth curve,[288] which is described with very good accuracy by the obtained solution in Ref. 287. From the solution for the longitudinal spectrum of the kinetic energy, the value of the experimentally measured Kolmogorov constant C_k is also determined. The found value $C_k \approx 2.07$ agrees well with the experimental value ≈ 2.01.

We also mentioned the works,[289,290] in which an alternative model for describing the developed turbulence based on the principle of maximum randomness was proposed. The RG analysis of this model was carried out and the Kolmogorov scaling was obtained. Based on this model, we also managed to obtain the turbulence damping spectrum.[291]

6.4. *Magnetic hydrodynamics*

The developed hydrodynamic turbulence is the most important open complex system described by the field-theoretic model brought about by the stochastic Navier–Stokes equation.[292–294] A closely related problem is the turbulence generated by the stochastic magnetohydrodynamics (MHD). In the leading order of the ε-expansion (one-loop approximation), the stochastic MHD equations were first analyzed within RG with the aid of Wilson's recurrence relations by Fournier *et al.*.[295] An equivalent field-theoretic model with a generic matrix of noise correlators for any space dimension $d > 2$ was formulated by Adzhemyan, Vasil'ev and Hnatič.[296] In both papers, two IR-stable fixed RG points — kinetic and magnetic — were found in the one-loop approximation (leading order in ε). The former controls the scaling behavior of statistical correlations of fluctuations of the velocity field and the magnetic field in a large-scale region of space and time. The one-loop expressions for the critical exponents are exact with the Kolmogorov values attained for the physical values of the model parameters. This result confirms the statement about the equipartition of kinetic and magnetic energy in the inertial interval formulated by Kraichnan.[297] On the contrary, the critical exponents for the magnetic fixed point are given by non-trivial expansions in ε.

In the three-charge MHD model, charges have zero values at some fixed points. This complicates the RG analysis of asymptotic solutions because the amplitudes of the leading terms can vanish. The detailed analysis in Refs. 295 and 298 revealed that the corrections can lead to non-universal scaling behavior, i.e., the universality of solutions of the RG equations can be violated. As noted in Ref. 5, the RG analysis of the large-scale asymptotics for a magnetic fixed point is of significant methodological importance since it shows that in complex systems with multiple charges not only universal scaling but also more complex non-universal behavior of statistical correlations can emerge. Thus, the RG approach is more generic than the idea of critical scaling.

The value of another relevant quantity — the Prandtl number — was calculated in Refs. 295 and 296. It is essential in the theory of turbulent dynamo[299] (see the following). An analysis of the influence of anisotropy, compressibility, and mirror symmetry-breaking was carried out in Ref. 298 and 300–303. The principal conclusions are as follows: Symmetry-breaking does not violate the main conclusions of the early analysis.[295,296] However, in simplified MHD models (e.g., Kraichnan–Kazantsev model and its descendants) with prescribed statistical properties of velocity, scaling functions determining the intermittent behavior of the magnetic field can be calculated. The critical exponents are not universal because they explicitly depend on the parameters of anisotropy and helicity.[301,302] The two-loop calculation[302] confirmed the stability of the kinetic fixed point and made it possible to determine the Prandtl number with greater accuracy up to linear order in ε and also to obtain its dependence on the degree of mirror symmetry-breaking α. Violation of mirror symmetry gives rise to exponential instability of magnetic fluctuations in the IR region that ultimately cause the formation of a large-scale average magnetic field (see, e.g. Ref. 304). The turbulent dynamo mechanism describes the transformation of the kinetic energy of turbulent motion into magnetic energy. Within the field-theoretic model of MHD, the turbulent dynamo was explained by the mechanism of spontaneous symmetry-breaking, which results in the generation of a uniform magnetic field $\mathbf{B}$,[300,302,303] whose value $|\mathbf{B}| = 16\nu\sqrt{\rho}/3\sqrt{\pi Pr\, l}$ was found in the one-loop approximation. Here, ν is kinematic viscosity , ρ is the density, Pr is the Prandtl number, and l is the Kolmogorov dissipation length. In the leading order, the magnitude of the magnetic field does not depend on the parameter of helicity. Thus, mirror symmetry-breaking is only a trigger to the generation of instabilities, while the spontaneous field can exist in a medium with unbroken reflection symmetry. The dynamo mechanism is accompanied by linear-in-time deformations of Alfvén waves, which is reminiscent of the appearance of Goldstone modes in the form of spin waves in ferromagnets.

7. Non-Equilibrium Phenomena and Stochastic Equations

Critical behavior and phase transitions in systems far from thermodynamic equilibrium have attracted constant attention over decades; see, e.g., Refs. 305–317 and the literature cited therein. Examples are provided by kinetic roughening of surfaces or interfaces,[306–308] open driven dissipative systems,[309] models with self-organized criticality,[310] transitions between fluctuating and absorbing states,[311,312] reaction–diffusion problems,[313] dynamic and directed percolation processes,[314] random walks and diffusion in random (e.g., porous or turbulent) media[315] as well as many others. In comparison with the equilibrium case, the non-equilibrium critical behavior is much richer and more diverse; it is also sometimes rather unusual but much less understood.

It has long been realized that the behavior of a real system near its critical point is extremely sensitive to external disturbances, geometry of the experimental setup, gravity, presence of impurities and so on; see, e.g., the monograph[318] for the general discussion and references therein. "Ideal" critical behavior of an infinite stationary system can be obscured by finite-size effects, finite time of evolution (ageing) and so on. What is more, some disturbances (randomly distributed impurities or turbulent mixing) can produce completely new types of critical behavior with rich and rather exotic properties, like, e.g., expansion in $\varepsilon^{1/2}$ rather than in ε.[319–322] Those issues become especially important for non-equilibrium phase transitions because the ideal conditions of a "pure" stationary critical state can hardly be achieved in real chemical or biological systems, and the effects of various disturbances can never be completely excluded. In particular, intrinsic turbulence effects cannot be avoided in chemical catalytic reactions or forest fires. One can also speculate that atmospheric turbulence can play an important role in spreading infectious diseases by flying insects or birds. Investigation of the effects of various kinds of deterministic or chaotic flows (laminar shear flows, turbulent convection and so on) on the behavior of critical fluids (like binary liquid mixtures near their consolution points) has shown that the flow can destroy the usual critical behavior, typical of the ϕ^4 or Reggeon models. It can change to the mean-field behavior or, under some conditions, to more complex behavior described by new non-equilibrium universality classes.[319,323–329]

7.1. *Reaction–diffusion systems*

There are many physical systems that are not in thermal equilibrium. This can be caused either by external or even various intrinsic mechanisms.[330,331] A broad class of such systems encompasses reaction–diffusion models, which find their applications in a plethora of models in physics, biology, growth processes, ecology, sociology, etc. They can be viewed as stochastic systems in which interacting particles evolve according to prescribed probabilistic rules.[331] The irreversible character of interactions and strong fluctuation effects, especially in low space dimensions, cause intriguing behavior in which emergent collective behavior appears, and it is possible to employ continuum approximation. Then sophisticated field-theoretic methods can be used to study universal behavior. The RG method is a convenient tool to analyze and characterize macroscopic scaling behavior.

The simplest systems in this class are random walks (diffusion) in random environments (quenched random fields) in which the particle number is conserved. First applications of the RG to these problems were carried out by Luck[332] and Fisher[333] for diffusion in an isotropic random field with local correlations. In a constrained random field the problem was analyzed by Fisher *et al.*[334] and Kravtsov *et al.*[335] The effect of long-range correlations of random advection on diffusion were analyzed in Refs. 22, 336 and 337. A perturbative proof of the triviality of the β function in the special case of a potential advection field was given by Honkonen, Pis'mak and Vasil'ev,[338] and the anomalous dimension of the diffusion coefficient in two dimensions was calculated exactly by Honkonen and Pis'mak.[339] A three-loop calculation of the anomalous dimension of the long-range correlated potential was carried out by Derkachov *et al.*[89,340]

A related field-theoretic model[341,342] of a "true" self-avoiding random walk was shown to be, in general, nonrenormalizable except for a few special cases, for which a two-loop RG analysis was carried out by Derkachov *et al.*[340,343]

Among reaction–diffusion models, a paradigmatic model is a unimolecular binary annihilation process (BAP) and it plays a paramount role as many of its features are also present in more complicated reaction schemes. The dynamics of this process can be effectively described by means of the Doi–Peliti approach.[344–347] The resulting model was successfully renormalized by Peliti[348] to all orders in the loop expansion. RG analysis was further elaborated by Lee[349] where, in particular, the decay exponent was calculated. This required a non-trivial summation of all orders in the initial density of reacting particles.

A typical chemical reaction takes place in a fluid environment. Thermal fluctuations of this underlying medium exert an additional influence on the reacting particles. Hence, it is of interest to investigate this influence on BAP itself. In order to model the effects of the ambient environment on the reaction process, various approaches have been undertaken. Most of the RG analyses of the effect of random drift on the BAP in the framework of the Doi approach have been carried out for the case of a quenched random drift field. Potential random drift with long-range[350,351] and short-range correlations[352] as well as turbulent flow (i.e. quenched solenoidal random field) with potential disorder[353,354] have been studied. A more realistic description of a turbulent flow time-dependent velocity field would be more appropriate. In Ref. 354, dynamic disorder with a given Gaussian distribution was considered, whereas the most ambitious approach is based on a velocity field generated by the stochastic Navier–Stokes equation.[355] The model for velocity fluctuations encompasses both thermal fluctuations and a turbulent velocity field with the Kolmogorov scaling behavior.[294] The actual RG analysis was performed within a double expansion scheme in which two small deviation parameters Δ and ε are introduced.[29] The former is a deviation from the upper critical dimension of the system, whereas the latter describes the power-law behavior of random force correlations in the momentum representation.[294]

The RG analysis[355,356] reveals the existence of altogether five asymptotic regimes. Interestingly, it turns out that a fixed point with diffusion movement becomes unstable only with respect to the inclusion of velocity fluctuations.

RG analysis at two-loop approximation[356] allows further derivation of a renormalized fluctuation-amended rate equation with one-loop corrections taken into account, which again is a step forward in the accuracy of the description of the system and opens up the possibility of analyzing direct effects on the number density of velocity fluctuations that do not enter explicitly in the leading-order mean-field rate equation. This non-linear integro-differential equation was solved iteratively in the framework of the double (ε, Δ)-expansion, and the scaling function for the mean number density has been calculated for the IR-stable regimes. This equation allows research into heterogeneous systems while taking into account the effect of density and advecting velocity fluctuations.

7.2. *Percolation*

Initially, DP was developed[357] to model the spreading of a fluid through an irregular porous medium. However, it finds its applications in research areas as diverse as high energy physics,[358] population dynamics,[359] reaction–diffusion problems,[330] or transition from laminar to turbulent flow.[360]

It is believed that the model falls under the universality class of directed percolation once certain conditions are met, which are known as the Janssen and Grassberger conjecture.[361,362] Although, this DP hypothesis has not been proven rigorously yet, there is a number of numerical support in its favor.[311,312] Despite a lot of effort put into a theoretical analysis of DP, even its simplest formulation in $(1 + 1)$ dimensions remains unsolvable exactly.[311]

Universal properties of the DP problem are captured by the corresponding field-theoretic action,[358,361] which can be treated by RG methods. A one-loop calculation was performed in the context of the Reggeon field theory,[363] whereas two-loop calculations were carried out by Janssen[361] and in high-energy physics by Bronzan and Dash.[364] The technical details associated with the analytical two-loop expression for the Feynman graphs can be found in Ref. 365 and 314. Recently, it has become possible to extend the existing two-loop results and put forward a semi-analytical technique that allows extracting a three-loop prediction on critical exponents.[366–368]

7.3. *Interfaces, growth, self-organized criticality*

Among many instances of phenomena in systems far from equilibrium, the random growth and surface roughening occupy a special place that justifies consistent attention they receive from scientific community. The sheer width of the range of examples simultaneously provides technological relevance and teases possible (if far-fetched) paradigmatic implications. The most prominent examples include the deposition of a substance on the surface and the growth of the corresponding phase boundary, propagation of flame and smoke fronts, propagation of solidification fronts, growth of vicinal surfaces, molecular beam epitaxy and many others; see Refs. 306–308,369 and references therein.

Self-organized criticality is another important concept in the study of non-equilibrium dynamics. While the classic sandpile models nominally describe the evolution of the surface, the mechanism behind "self-tuning" of those systems to their critical states is different from the one causing surface roughening.[310] Self-organized criticality is already believed to manifest itself in various dissipative open systems (biological, neural, social, etc.) while the growing sophistication of data collection and analysis gives hope that the existence of self-organized criticality can be established beyond doubt very soon.

By using semi-phenomenological models that describe the evolution of coarse-grained fields, one can study both surface roughening and self-organized criticality.

Investigation of turbulence and random motion effects on surface growth and systems with self-organized criticality revealed many intriguing features. White-in-time (columnar) quenched random noise together with the environment motion lead to the emergence of a new non-linearity (interaction) in the equation for the surface growth. What is interesting, both the noise and the motion are not sufficient by themselves to cause the emergence.[370] It is possible to study models of surface growth with an infinite number of coupling constants and obtain the exact result for critical exponents from them.[371,372] Moreover, these models can inspire non-perturbative functional RG analysis that can uncover further interesting features.[373] Coupling of non-linear stochastic equations (such as the running sandpile model of self-organized criticality) to the Navier–Stokes equation for viscous incompressible fluid introduces several dimensionless ratios of viscosity coefficients from both equations. Such ratios serve as additional coupling constants and transform the expected picture of RG flows and IR attractive fixed points of RG equations into a complex pattern that implies lose of universality.[374] Dimensionless parameters can appear for other reasons as well (e.g., after the introduction of finite correlation time velocity field ensemble[375]). In one interesting case, placing an anisotropic system into isotropic environment resulted in regime of a critical behavior that corresponds to a generalized scaling (in the spirit of G. Stell and M. Fisher) for systems with different characteristic scales or different scaling laws.[376]

8. Conclusion

The application of the RG techniques to critical phenomena, pioneered by K.G. Wilson and M. Fisher, provided the firm base for the theoretical description of of phase transitions and critical phenomena. One of the most fruitful concepts is the RG and ε-expansion. Further development of these ideas led to significant progress in the subject: Various classical and quantum phase transitions, equilibrium and strongly non-equilibrium dynamic phenomena, quantum systems at high temperatures, pressures and external fields can be studied within the general RG framework. Recent higher-order calculations in the ϕ^3 and ϕ^4 models and applications to numerous non-equilibrium systems show that the RG approach to critical behavior remains a very interesting and still open area of modern theoretical physics.

Acknowledgments

The authors are grateful to our teachers A.N. Vasil'ev and A.I. Sokolov, who are no longer with us.

We also would like to thank our dear friends and colleagues S. É. Derkachov, N. M. Gulitskiy, G. Kalagov, A. Lanotte, N. M. Lebedev, A. Mazzino, L. Mižišin P. Muratore Ginanneschi, Y. M. Pis'mak and M. M. Tumakova (Kostenko) for fruitful collaboration and the numerous inspiring discussions. The authors are also indebted to Amnon Aharony for offering to contribute a chapter to this edition.

P.I.K. was supported by the Foundation for the Advancement of Theoretical Physics and Mathematics "BASIS" (project 22-1-3-33-1).

M.H. and T.L. were supported by the Ministry of Education, Science, Research and Sport of the Slovak Republic (VEGA Grant No. 1/0535/21).

References

1. K. G. Wilson and M. E. Fisher, Critical exponents in 3.99 dimensions. *Phys. Rev. Lett.* **28**, 240 (1972).
2. K. G. Wilson and J. B. Kogut, The renormalization group and the ε expansion. *Phys. Rep.* **12**, 75 (1974).
3. E. Brézin, J. Le Guillou and J. Zinn-Justin, *Field theoretical Approach to Critical Phenomena.* vol. 6, *Phase Transitions and Critical Phenomena* (Academic Press, London, 1976).
4. D. J. Amit, *Field Theory, Renormalization Group, and Critical Phenomena* (McGraw-Hill, New York, 1978).
5. A. N. Vasil'ev, *Quantum Field Renormalization Group in Critical Behavior Theory and Stochastic Dynamics* (Chapman & Hall/CRC, Apr. 2004). ISBN 978-0-415-31002-4. originally published in Russian in 1998 by St. Petersburg Institute of Nuclear Physics Press; translated by Patricia A. de Forcrand-Millard.
6. J. Zinn-Justin, *Quantum Field Theory and Critical Phenomena* (Clarendon Press, Oxford, 1989, 1993, 2002).
7. A. Z. Patashinskii and V. L. Pokrovskii, *Fluctuation Theory of Phase Transitions* (Pergamon Press, Oxford, 1979).
8. S.-K. Ma, *Modern Theory Of Critical Phenomena* (Routledge, New York, 2001), 1st ed. edition. doi: 10.4324/9780429498886.
9. H. Kleinert and V. Schulte-Frohlinde, *Critical Properties of ϕ^4-Theories* (World Scientific, Singapore, 2001).
10. J. Cardy, *Scaling and Renormalization in Statistical Physics.* Cambridge Lecture Notes in Physics (Cambridge University Press, 1996). doi: 10.1017/CBO9781316036440.
11. N. Goldenfeld, *Lectures on Phase Transitions and the Renormalization Groups* (1st edn.) (ImprintCRC Press, Boca Raton, 1992). doi: 10.1201/9780429493492.
12. N. N. Bogoliubov and D. V. Shirkov, *Introduction to the Theory of Quantized Fields* (John Wiley & Sons Inc, 1980).
13. A. A. Vladimirov, D. I. Kazakov and O. V. Tarasov, Calculation of critical exponents by quantum field theory methods. *Zh. Eksp. Teor. Fiz.* **77**, 1035 (1979).
14. G. 't Hooft and M. Veltman, Regularization and renormalization of gauge fields. *Nucl. Phys. B* **44**, 189 (1972).
15. E. R. Speer, *Generalized Feynman Amplitudes. (AM-62)* (Princeton University Press, 1969).
16. D. Forster, D. R. Nelson and M. J. Stephen, Long-time tails and the large-eddy behavior of a randomly stirred fluid. *Phys. Rev. Lett.* **36**, 867 (1976).

17. C. DeDominicis and P. C. Martin, Energy spectra of certain randomly-stirred fluids. *Phys. Rev. A* **19**, 419 (1979).
18. J. Sak, Recursion relations and fixed points for ferromagnets with long-range interactions. *Phys. Rev. B* **8**, 281 (1973).
19. A. Weinrib and B. I. Halperin, Critical phenomena in systems with long-range-correlated quenched disorder. *Phys. Rev. B* **27**, 413 (1983).
20. J. Honkonen and M. Y. Nalimov, Crossover between field theories with short-range and long-range exchange or correlations. *J. Phys. A* **22**, 751 (1989).
21. J. Honkonen, Critical behaviour of the long-range $(\phi^2)^2$ model in the short-range limit. *J. Phys. A* **23**, 825 (1990).
22. Z. S. Gevorkian and Y. E. Lozovik, Classical diffusion in random fields with long-range correlations. *J. Phys. A* **20**, L659 (1987).
23. J. Honkonen, Stochastic processes with stable distributions in random environments. *Phys. Rev. E* **53**, 327 (1996).
24. I. Goncharenko and A. Gopinathan, Vicious walks with long-range interactions. *Phys. Rev. E* **82**, 011126 (2010).
25. V. Blavats'ka, C. v. Ferber and Y. Holovatch, Polymers in long-range-correlated disorder. *Phys. Rev. E* **64**, 041102 (2001).
26. J. Honkonen and M. Y. Nalimov, Two-parameter expansion in the renormalization-group analysis of turbulence. *Z. Phys. B* **99**, 297 (1995).
27. L. T. Adzhemyan, N. V. Antonov, M. V. Kompaniets and A. N. Vasil'ev, Renormalization-group approach to the stochastic Navier–Stokes equation: Two-loop approximation. *Int. J. Mod. Phys. B* **17**, 2137 (2003).
28. L. T. Adzhemyan, J. Honkonen, M. V. Kompaniets and A. N. Vasil'ev, Improved ε expansion for three-dimensional turbulence: Summation of nearest dimensional singularities. *Phys. Rev. E* **68**, 055302 (2003).
29. L. T. Adzhemyan, J. Honkonen, M. V. Kompaniets and A. N. Vasil'ev, Improved ϵ expansion for three-dimensional turbulence: Two-loop renormalization near two dimensions. *Phys. Rev. E* **71**, 036305 (2005).
30. L. T. Adzhemyan, M. Hnatich and J. Honkonen, Improved ε expansion in the theory of turbulence: summation of nearest singularities by inclusion of an infrared irrelevant operator. *Eur. Phys. J. B* **73**, 275 (2010).
31. N. V. Antonov, Anomalous scaling regimes of a passive scalar advected by the synthetic velocity field. *Phys. Rev. E* **60**, 6691 (1999).
32. N. V. Antonov, M. Hnatich, J. Honkonen and M. Jurčišin, Turbulence with pressure: Anomalous scaling of a passive vector field. *Phys. Rev. E* **68**, 046306 (2003).
33. M. Hnatich, J. Honkonen, D. Horvath and R. Semancik, Advection of a passive scalar near two dimensions. *Phys. Rev. E* **59**, 4112 (1999).
34. A. V. Gladyshev, E. Jurčišinová, M. Jurčišin, R. Remecký and P. Zalom, Anomalous scaling of a passive scalar field near two dimensions. *Phys. Rev. E* **86**, 036302 (2012).
35. M. Hnatich, J. Honkonen and M. Jurcisin, Stochastic magnetohydrodynamic turbulence in space dimensions $d \geq 2$. *Phys. Rev. E* **64**, 056411 (2001).
36. H. E. Stanley, Dependence of critical properties on dimensionality of spins. *Phys. Rev. Lett.* **20**, 589 (1968).
37. K. G. Wilson, Quantum field — theory models in less than 4 dimensions. *Phys. Rev. D* **7**, 2911 (1973).
38. T. H. Berlin and M. Kac, The spherical model of a ferromagnet. *Phys. Rev.* **86**, 821 (1952).
39. M. Moshe and J. Zinn-Justin, Quantum field theory in the large N limit: A review. *Phys. Rep.* **385**, 69 (2003).

40. R. Abe, Critical exponent η up to $1/n^2$ for the three-dimensional system with short-range interaction. *Prog. Theor. Exp. Phys.* **49**, 1877 (1973).
41. Y. Okabe and M. Oku, $1/n$ Expansion up to order $1/n^2$. II: Critical exponent β for $d = 3$. *Prog. Theor. Exp. Phys.* **60**, 1277 (1978).
42. Y. Okabe and M. Oku, $1/n$ Expansion up to order $1/n^2$. III: Critical exponents γ and ν for $d = 3$. *Prog. Theor. Exp. Phys.* **60**, 1287 (1978).
43. I. Kondor and T. Temesvári, Calculation of critical exponents to $o(\frac{1}{n^2})$. *Phys. Rev. B* **21**, 260 (1980).
44. I. Kondor, T. Temesvári and L. Herényi, Resummation of the $\frac{1}{n}$ expansion through a self-consistent approach. *Phys. Rev. B* **22**, 1451 (1980).
45. A. N. Vasil'ev, Y. M. Pis'mak and J. R. Honkonen, Simple method of calculating the critical indices in the $1/n$ expansion. *Theor. Math. Phys.* **46**, 104 (1981).
46. A. N. Vasil'ev, Y. M. Pis'mak and J. R. Honkonen, $1/n$ Expansion: Calculation of the exponents η and ν in the order $1/n^2$ for arbitrary number of dimensions. *Theor. Math. Phys.* **47**, 465 (1981).
47. A. N. Vasil'ev, Y. M. Pis'mak and J. R. Honkonen, $1/n$ Expansion: Calculation of the exponent ν in the order $1/n^3$ by the conformal bootstrap method. *Theor. Math. Phys.* **50**, 127 (1982).
48. D. Broadhurst, J. Gracey and D. Kreimer, Beyond the triangle and uniqueness relations: Non-zeta counterterms at large n from positive knots. *Z. Phys. C* **75**, 559 (1997).
49. A. J. Bray, Self-consistent screening calculation of the critical exponent η. *Phys. Rev. Lett.* **32**, 1413 (1974).
50. G. Parisi and L. Peliti, Calculation of critical indices. *Lett. nuovo cimento.* **2**, 627 (1971).
51. M. D'eramo, G. Parisi and L. Peliti, Theoretical predictions for critical exponents at the λ-point of bose liquids. *Lett. nuovo cimento.* **2**, 878 (1971).
52. K. Symanzik, On calculations in conformal invariant field theories. *Lett. nuovo cimento.* **3**, 734 (1972).
53. J. A. Gracey, Large n_f quantum field theory. *Int. J. Mod. Phys. A* **33**, 1830032 (2018).
54. A. N. Vasil'ev, S. E. Derkachev, N. A. Kivel' and A. S. Stepanenko, The $1/n$ expansion in the Gross-Neveu model: Conformal bootstrap calculation of the index η in order $1/n^3$. *Theor. Math. Phys.* **94**, 127 (1993).
55. D. J. Gross and A. Neveu, Dynamical symmetry breaking in asymptotically free field theories. *Phys. Rev. D* **10** 3235 (1974).
56. A. N. Vasil'ev, M. I. Vyazovskii, S. E. Derkachev and N. A. Kivel', On the equivalence of renormalizations in standard and dimensional regularizations of $2D$ four-fermion interactions. *Theor. Math. Phys.* **107**, 441 (1996).
57. G. Gat, A. Kovner, B. Rosenstein and B. J. Warr, Four-fermi interaction in $(2+1)$ dimensions beyond leading order in $1/n$. *Phys. Lett. B* **240**, 158 (1990).
58. J. A. Gracey, Calculation of exponent η to $O(1/N^2)$ in the $\mathcal{O}(n)$ Gross Neveu model. *Int. J. Mod. Phys. A* **06**, 395 (1991).
59. J. A. Gracey, Erratum: Calculation of exponent η to $O(1/N^2)$ in the $\mathcal{O}(n)$ Gross Neveu model. *Int. J. Mod. Phys. A* **06**, 2755 (1991).
60. A. N. Vasil'ev and A. S. Stepanenko, The $1/n$ expansion in the Gross-Neveu model: Conformal bootstrap calculation of the exponent $1/\nu$ to the order $1/n^2$. *Theor. Math. Phys.* **97**, 1349 (1993).
61. J. A. Gracey, Computation of $\beta'(g_c)$ at $O(1/N^2)$ in the $\mathcal{O}(n)$ Gross-Neveu model in arbitrary dimensions. *Int. J. Mod. Phys. A* **09**, 567 (1994).

62. J. A. Gracey, Computation of critical exponent η at $O(1/N^3)$ in the four-Fermi model in arbitrary dimensions. *Int. J. Mod. Phys. A* **09**, 727 (1994).
63. A. N. Vasil'ev and M. Y. Nalimov, Analog of dimensional regularization for calculation of the renormalization-group functions in the $1/n$ expansion for arbitrary dimension of space. *Theor. Math. Phys.* **55**, 423 (1983).
64. A. N. Vasil'ev and A. S. Stepanenko, A method of calculating the critical dimensions of composite operators in the massless nonlinear σ model. *Theor. Math. Phys.* **95**, 471 (1993).
65. A. N. Vasil'ev and M. Y. Nalimov, The CP^{N-1} model: Calculation of anomalous dimensions and the mixing matrices in the order $1/N$. *Theor. Math. Phys.* **56**, 643 (1983).
66. A. N. Vasil'ev, M. Y. Nalimov and J. R. Honkonen, $1/N$ expansion: Calculation of anomalous dimensions and mixing matrices in the order $1/N$ for $N \times p$ matrix gauge-invariant σ model. *Theor. Math. Phys.* **58**, 111 (1984).
67. G. Parisi, Field-theoretic approach to second-order phase transitions in two- and three-dimensional systems. *J. Stat. Phys.* **23**, 49 (1980).
68. J. C. Le Guillou and J. Zinn-Justin, Critical exponents for the n-vector model in three dimensions from field theory. *Phys. Rev. Lett.* **39**, 95 (1977).
69. R. Guida and J. Zinn-Justin, Critical exponents of the N-vector model. *J. Phys. A* **31**, 8103 (1998).
70. S. A. Antonenko and A. I. Sokolov, Critical exponents for a three-dimensional $\mathcal{O}(n)$-symmetric model with $n > 3$. *Phys. Rev. E* **51**, 1894 (1995).
71. G. A. Baker, B. G. Nickel, M. S. Green and D. I. Meiron, Ising-model critical indices in three dimensions from the callan-symanzik equation. *Phys. Rev. Lett.* **36**, 1351 (1976).
72. G. A. Baker, B. G. Nickel and D. I. Meiron, Critical indices from perturbation analysis of the callan-symanzik equation. *Phys. Rev. B* **17**, 1365 (1978).
73. B. G. Nickel, D. I. Meiron and G. A. Baker. Compilation of 2-pt. and 4-pt. graphs for continuous spin. unpublished University of Guelph report (1977), available on http://users.physik.fu-berlin.de/~kleinert/nickel/.
74. B. G. Nickel, Evaluation of simple feynman graphs. *J. Math. Phys.* **19**, 542 (1978).
75. D. B. Murray and B. G. Nickel. Compilation of 2-pt. and 4-pt. graphs for continuous spin. unpublished, University of Guelph report (1991).
76. R. Guida and P. Ribeca, Towards a fully automated computation of RG functions for the three-dimensional $\mathcal{O}(n)$ vector model: parametrizing amplitudes. *J. Stat. Mech. Theory Exp.* **2006**, P02007 (2006).
77. D. V. Pakhnin and A. I. Sokolov, Five-loop renormalization-group expansions for the three-dimensional n-vector cubic model and critical exponents for impure ising systems. *Phys. Rev. B* **61**, 15130 (2000).
78. P. Calabrese, P. Parruccini and A. I. Sokolov, Chiral phase transitions: Focus driven critical behavior in systems with planar and vector ordering. *Phys. Rev. B* **66**, 180403 (2002).
79. J. Manuel Carmona, A. Pelissetto and E. Vicari, n-component ginzburg-landau hamiltonian with cubic anisotropy: A six-loop study. *Phys. Rev. B* **61**, 15136 (2000).
80. V. Dohm, Nonuniversal critical phenomena along the lambda line of ^{4}He. *Z. Phys. B Cond. Matt.* **60**, 61 (1985).
81. R. Schloms and V. Dohm, Minimal renormalization without ϵ expansion: Critical behavior above and below t_c. *Phys. Rev. B* **42**, 6142 (1990).

82. S. S. Burnett, M. Strösser and V. Dohm, Minimal renormalization without ϵ expansion: Amplitude functions for $o(n)$ symmetric systems in three dimensions below t_c. *Nuclear Physics B* **504**, 665 (1997).
83. S. A. Larin, M. Mönnigmann, M. Strösser and V. Dohm, Five-loop additive renormalization in the φ^4 theory and amplitude functions of the minimally renormalized specific heat in three dimensions. *Phys. Rev. B* **58**, 3394 (1998).
84. A. Polyakov, Conformal symmetry of critical fluctuations. *JETP Lett.* **12**, 381 (1970).
85. D. I. Kazakov, Calculation of Feynman diagrams by the "Uniqueness" method. *Theor. Math. Phys.* **58**, 223 (1984).
86. N. I. Usyukina, Calculation of many-loop diagrams of perturbation theory. *Theor. Math. Phys.* **54**, 78 (1983).
87. D. I. Kazakov, The method of uniqueness, a new powerful technique for multiloop calculations. *Phys. Lett. B* **133**, 406 (1983).
88. D. I. Kazakov, Many-loop calculations: The uniqueness method and functional equations. *Theor. Math. Phys.* **62**, 84 (1985).
89. S. E. Derkachov, J. Honkonen and Y. M. Pis'mak, Three-loop calculation of the random walk problem: An application of dimensional transformation and the uniqueness method. *J. Phys. A* **23**, 5563 (1990).
90. S. G. Gorishnii and A. P. Isaev, An approach to the calculation of many-loop massless Feynman integrals. *Theor. Math. Phys.* **62**, 232 (1985).
91. D. J. Broadhurst, Exploiting the 1, 440-fold symmetry of the master two-loop diagram. *Z. Phys. C* **32**, 249 (1986).
92. A. G. Grozin, Massless two-loop self-energy diagram: Historical review. *Int. J. Mod. Phys. A* **27**, 1230018 (2012).
93. L. N. Lipatov, Divergence of the perturbation-theory series and the quasi-classical theory. *Sov. Phys. JETP* **45**, 216 (1977). translation of Zh. Eksp. Teor. Fiz. 72, 411 (1977).
94. D. I. Kazakov and D. V. Shirkov, Asymptotic series of quantum field theory and their summation. *Fortschritte der Physik.* **28**, 465 (1980).
95. D. I. Kazakov and D. V. Shirkov, Borel resummation of the ε- expansion of the dynamical exponent z in model a of the $\phi^4(\mathcal{O}(n))$ theory. *Theor. Math. Phys.* **159**, 496 (2009).
96. M. Komarova and M. Nalimov, Asymptotic behaviour of renormalization constants in higher orders of the perturbations expansion for the $(4-\varepsilon)$ dimensionally regularized $\mathcal{O}(n)$-symmetric ϕ^4 theory. *Theor. Math. Phys.* **126**, 339 (2001).
97. M. Komarova and M. Nalimov, Asymptotic behaviour of higher-order perturbations: scaling functions of the $\mathcal{O}(n)$-symmetric ϕ^4 theory in the $(4-\varepsilon)$-expansion. *Theor. Math. Phys.* **129**, 1631 (2001).
98. M. Komarova and M. Nalimov, Large — oder asymptotic terms in perturbation theory: the first $(4-\varepsilon)$ — expansion correction to the renormalization constants in the $\mathcal{O}(n)$ — symmetric theory. *Theor. Math. Phys.* **143**, 211 (2005).
99. A. J. McKane, Perturbation expansions at large order: Results for scalar field theories revisited. *J. Phys. A* **52**, 055401 (2019).
100. D. I. Kazakov, O. V. Tarasov and D. V. Širkov, Analytic continuation of the results of perturbation theory for the model $g\phi^4$ to the region $g \gtrsim 1$. *Theor. Math. Phys.* **38**, 9 (1979).
101. Y. Kubyshin, Corrections to the asymptotic formula for high orders of perturbation theory. *Theor. Math. Phys.* **57**, 1196 (1984).

102. Y. Kubyshin, Corrections to the asymptotic formula for high orders of perturbation theory. *Theor. Math. Phys.* **57**, 363 (1983).
103. M. Kompaniets, Prediction of the higher-order terms based on borel resummation with conformal mapping. *J. Phys. Conf. Series* **762**, 012075 (2016).
104. M. V. Kompaniets and E. Panzer, Minimally subtracted six-loop renormalization of $\mathcal{O}(n)$-symmetric ϕ^4 theory and critical exponents. *Phys. Rev. D* **96**, 036016 (2017).
105. J. Honkonen, M. V. Komarova and M. Y. Nalimov, Large-order asymptotes for dynamic models near equilibrium. *Nucl. Phys. B* **707**, 493 (2005).
106. J. Honkonen, M. V. Komarova and M. Y. Nalimov, Instantons for dynamic models from B to H. *Nucl. Phys. B* **714**, 292 (2005).
107. A. Andreanov, J. Honkonen, M. Komarova and M. Y. Nalimov, Large-order asymptotes for dynamic models. *J. Phys. A* **39**, 7815 (2006).
108. L. T. Adzhemyan, E. Ivanova, M. Kompaniets and S. Y. Vorobyeva, Diagram reduction in problem of critical dynamics of ferromagnets: 4-loop approximation. *J. Phys. A* **51**, 155003 (2018).
109. A. Y. Andreanov, M. V. Komarova and N. M. Yu., Large-order asymptotes of the quantum-field expansion for the kraichnan model of passive scalar advection. *J. Phys. A* **39**, 7801 (2006).
110. M. V. Komarova, I. S. I. S. Kremnev and M. Y. Nalimov, Family of instantons of the kraichnan model with a frozen velocity field. *Theor.Math.Phys.* **158**, 167 (2009).
111. M. V. Komarova, I. S. I.S. Kremnev and M. Y. Nalimov, Convergence of perturbation series for renormalization constants in kraichnan model with "frozen" velocity field. *Eur. Phys. J* **71**, 1646 (2011).
112. N. D. Mermin and C. Stare, Ginzburg-landau approach to $l \neq 0$ pairing. *Phys. Rev. Lett.* **30**, 1135 (1973).
113. A. I. Sokolov, Phase diagram of superfluid He_3. *JETP* **51**, 998 (1980).
114. J. A. Sauls and J. W. Serene, ${}^3p_2$ pairing near the transition temperature in neutron-star matter. *Phys. Rev. D* **17**, 1524 (1978).
115. F. Wilczek, Application of the renormalization group to a second-order QCD phase transition. *Int. J. Mod. Phys. A* **7**, 3911 (1992).
116. K. Rajagopal and F. Wilczek, Static and dynamic critical phenomena at a second order QCD phase transition. *Nucl. Phys. B* **399**, 395 (1993).
117. A. I. Sokolov, Phase transformations in a superfluid neutron liquid. *JETP* **49**, 1137 (1980).
118. K. G. Chetyrkin, S. G. Gorishny, S. A. Larin and F. V. Tkachov, Five-loop renormalization group calculations in the $g\phi^4$ theory. *Phys. Lett. B* **132**, 351 (1983).
119. K. G. Chetyrkin, A. L. Kataev and F. V. Tkachov, Five-loop calculations in the $g\phi^4$ model and the critical index η. *Phys. Lett. B* **99**, 147 (1981).
120. K. G. Chetyrkin, A. L. Kataev and F. V. Tkachov, Errata. *Phys. Lett. B* **101**, 457 (1981).
121. J. Le Guillou and J. Zinn-Justin, Accurate critical exponents from the ε-expansion. *J. Physique Lett.* **46**, 137 (1985).
122. H. Kleinert, J. Neu, N. Schulte-Frohlinde, K. G. Chetyrkin and S. A. Larin, Five-loop renormalization group functions of $\mathcal{O}(n)$-symmetric $\varphi 4$-theory and ε-expansions of critical exponents up to ε^5. *Phys. Lett. B* **272**, 39 (1991).
123. E. V. Orlov and A. I. Sokolov, Critical thermodynamics of two-dimensional systems in the five-loop renormalization-group approximation. *Phys. Solid State* **42**, 2151 (2000).

124. D. V. Batkovich, M. V. Kompaniets and K. G. Chetyrkin, Six loop analytical calculation of the field anomalous dimension and the critical exponent η in $\mathcal{O}(n)$-symmetric φ^4 model. *Nucl. Phys. B* **906**, 147 (2016).
125. M. V. Kompaniets and E. Panzer. Renormalization group functions of ϕ^4 theory in the ms-scheme to six loops. In *Loops and Legs in Quantum Field Theory Leipzig, Germany, April 24–29*, p. PoS(LL2016)038. Proceedings of Science, (2016). PoS(LL2016)038.
126. O. Schnetz, Numbers and functions in quantum field theory. *Phys. Rev. D* **97**, 085018 (2018).
127. M. Serone, G. Spada and G. Villadoro, $\lambda\phi^4$ Theory I: The symmetric phase beyond NNNNNNNNLO. *JHEP* **8**, 148 (2018).
128. O. Schnetz, Seven loops ϕ^4. *arXiv:2212.03663 [hep-th]*. (2022).
129. T. Binoth and G. Heinrich, An automatized algorithm to compute infrared divergent multiloop integrals. *Nucl. Phys. B* **585**, 741 (2000).
130. L. T. Adzhemyan and M. V. Kompaniets, Five-loop numerical evaluation of critical exponents of the φ^4 theory. *J. Phys. Conf. Ser.* **523**, 012049 (2014).
131. L. T. Adzhemyan, Y. V. Kirienko and M. Kompaniets, Critical exponent η in 2d $\mathcal{O}(n)$ — symmetric φ^4-model up to 6 loops. *arXiv:1602.02324*. (2016).
132. K. G. Chetyrkin and F. V. Tkachov, Integration by parts: The algorithm to calculate β-functions in 4 loops. *Nucl. Phys. B* **192**, 159 (1981).
133. P. A. Baikov and K. G. Chetyrkin, Four loop massless propagators: An algebraic evaluation of all master integrals. *Nucl. Phys. B* **837**, 186 (2010).
134. R. N. Lee, A. V. Smirnov and V. A. Smirnov, Master integrals for four-loop massless propagators up to weight twelve. *Nucl. Phys. B* **856**, 95 (2012).
135. K. G. Chetyrkin and F. V. Tkachov, Infrared r-operation and ultraviolet counterterms in the MS-scheme. *Phys. Lett. B* **114**, 340 (1982).
136. K. G. Chetyrkin and V. A. Smirnov, R* -operation corrected. *Phys. Lett. B* **144**, 419 (1984).
137. K. G. Chetyrkin, Combinatorics of $\mathbf{R}-$, $\mathbf{R}^{-1}$, and $\mathbf{R}^*$-operations and asymptotic expansions of feynman integrals in the limit of large momenta and masses. *arXiv:1701.08627*. (2017).
138. D. V. Batkovich and M. V. Kompaniets, Toolbox for multiloop feynman diagrams calculations using r^* operation. *J. Phys. Conf. Ser.* **608**, 012068 (2015).
139. F. C. S. Brown, The massless higher-loop two-point function. *Commun. Math. Phys.* **287**, 925 (2009).
140. E. Panzer, Algorithms for the symbolic integration of hyperlogarithms with applications to Feynman integrals. *Comput. Phys. Commun.* **188**, 148 (2015).
141. F. C. S. Brown and D. Kreimer, Angles, scales and parametric renormalization. *Lett. Math. Phys.* **103**, 933 (2013).
142. O. Schnetz, Graphical functions and single-valued multiple polylogarithms. *arXiv:1302. 6445*. (2013).
143. R. Guida and J. Zinn-Justin. *J. Phys. A* **31**, 8103 (1998).
144. H. Shimada and S. Hikami, Fractal dimensions of self-avoiding walks and ising high-temperature graphs in 3d conformal bootstrap. *J. Stat. Phys.* **165**(6), 1006 (2016).
145. S. El-Showk, M. F. Paulos, D. Poland, S. Rychkov, D. Simmons-Duffin and A. Vichi, Solving the 3d ising model with the conformal bootstrap II. c-minimization and precise critical exponents. *J. Stat. Phys.* **157**, 869 (2014).

146. S. M. Chester, W. Landry, J. Liu, D. Poland, D. Simmons-Duffin, N. Su and A. Vichi. Carving out OPE space and precise o(2) model critical exponents. *J. High Energy Phys.* **2020** (2020).
147. F. Kos, D. Poland, D. Simmons-Duffin and A. Vichi, Precision islands in the ising and o(n) models. *J. High Energy Phys.* **2016**, 36 (2016).
148. N. Clisby, Scale-free monte carlo method for calculating the critical exponent γ of self-avoiding walks. *J. Phys. A* **50**, 264003 (2017).
149. M. Hasenbusch, Restoring isotropy in a three-dimensional lattice model: The ising universality class. *Phys. Rev. B* **104**, 014426 (2021).
150. M. Hasenbusch, Monte carlo study of an improved clock model in three dimensions. *Phys. Rev. B* **100**, 224517 (2019).
151. M. Hasenbusch, Monte carlo study of a generalized icosahedral model on the simple cubic lattice. *Phys. Rev. B* **102**, 024406 (2020).
152. N. Clisby and B. Dünweg, High-precision estimate of the hydrodynamic radius for self-avoiding walks. *Phys. Rev. E* **94**, 052102 (2016).
153. A. M. Shalaby, Critical exponents of the $O(n)$-symmetric ϕ^4 model from the ε^7 hypergeometric-meijer resummation. *Eur. Phys. J. C* **81**, 1 (2021).
154. G. De Polsi, I. Balog, M. Tissier and N. Wschebor, Precision calculation of critical exponents in the o(n) universality classes with the nonperturbative renormalization group. *Phys. Rev. E* **101**, 042113 (2020).
155. N. V. Antonov, M. V. Kompaniets and N. M. Lebedev, Critical behaviour of the $\mathcal{O}(n) - \varphi^4$ model with an antisymmetric tensor order parameter. *J. Phys. A* **46**, 405002 (2013).
156. A. Bednyakov and A. Pikelner, Six-loop beta functions in general scalar theory. *J. High Energy Phys.* **2021** (2021).
157. A. Aharony, Critical behavior of anisotropic cubic systems. *Phys. Rev. B* **8**, 4270 (1973).
158. D. J. Wallace, Critical behaviour of anisotropic cubic systems. *J. Phys. C* **6**, 1390 (1973).
159. I. J. Ketley and D. J. Wallace, A modified epsilon expansion for a hamiltonian with cubic point-group symmetry. *J. Phys. A* **6**, 1667 (1973).
160. D. R. Nelson, J. M. Kosterlitz and M. E. Fisher, Renormalization-group analysis of bicritical and tetracritical points. *Phys. Rev. Lett.* **33**, 813–817 (1974).
161. E. Breźin, J. C. Le Guillou and J. Zinn-Justin, Discussion of critical phenomena for general n-vector models. *Phys. Rev. B* **10**, 893 (1974).
162. I. F. Lyuksyutov and V. L. Pokrovskii, First-order phase transitions in systems with cubic anisotropy. *JETP Lett.* **21**, 9 (1975).
163. T. Nattermann and S. Trimper, Critical behaviour and cubic anisotropy. *J. Phys. A* **8**, 2000 (1975).
164. T. Nattermann, On the decay of anisotropy approaching the critical point. *J. Phys. C* **9**, 3337 (1976).
165. A. L. Korzhenevskii, Connection between the properties of the system of renormalization-group equations and the symmetry of the hamiltonian in the phase-transition problem. *Sov. Phys. JETP* **44**, 751 (1976).
166. M. C. Yalabik and A. Houghton, Approximate renormalization group calculation of cubic crossover. *Phys. Lett.* **61A**, 1 (1977).
167. A. I. Sokolov, Phase-transition in 3-dimensional model-effect of cubic anisotropy. *Sov. Phys. Solid State* **19**, 433 (1977).

168. M. Ferer, J. Van Dyke and W. Camp, Effect of a cubic crystal field on the critical behavior of a 3d model with Heisenberg exchange coupling: A high-temperature series investigation. *Phys. Rev. B* **23**, 2367 (1981).
169. K. E. Newman and E. K. Riedel, Cubic n-vector model and randomly dilute ising model in general dimensions. *Phys. Rev. B* **25**, 264 (1982).
170. I. O. Maier and A. I. Sokolov, Critical-behavior of cubic-crystals under structure phase-transitions. *Izv. Akad. Nauk SSSR, Ser. Fiz.* **51**, 2103 (1987).
171. I. O. Maier and A. I. Sokolov, Is a cubic crystal "isotropic" in the critical point? *Ferroelectrics Lett. Sect.* **9**, 95 (1988).
172. I. O. Mayer, A. I. Sokolov and B. N. Shalayev, Critical exponents for cubic and impure uniaxial crystals: most accurate (?) theoretical values. *Ferroelectrics* **95**, 93 (1989).
173. N. A. Shpot, Critical behavior of the mn-component field model in three dimensions II. Three-loop results. *Phys. Lett. A* **142**, 474 (1989).
174. H. Kleinert and S. Thoms, Large-order behavior of a two-coupling-constant φ^4 theory with cubic anisotropy. *Phys. Rev. D* **52**, 5926 (1995).
175. H. Kleinert and V. Schulte-Frohlinde, Exact five-loop renormalization group functions of θ^4-theory with $\mathcal{O}(N)$-symmetric and cubic interactions. Critical exponents up to ε^5. *Phys. Lett. B* **342**, 284 (1995).
176. B. N. Shalaev, S. A. Antonenko and A. I. Sokolov, Five-loop ε for random ising model and marginal spin dimensionality for cubic systems. *Phys. Lett. A* **230**, 105 (1997).
177. H. Kleinert, S. Thoms and V. Schulte-Frohlinde, Stability of a three-dimensional cubic fixed point in the two-coupling-constant φ^4 theory. *Phys. Rev. B* **56**, 14428 (1997).
178. M. Caselle and M. Hasenbusch, The stability of the $\mathcal{O}(n)$ invariant fixed point in three dimensions. *J. Phys. A* **31**, 4603 (1998).
179. K. B. Varnashev, Stability of a cubic fixed point in three dimensions: Critical exponents for generic n. *Phys. Rev. B* **61**, 14660 (2000).
180. R. Folk, Y. Holovatch and T. Yavors'kii, Effective and asymptotic critical exponents of a weakly diluted quenched ising model: Three-dimensional approach versus $\sqrt{\varepsilon}$ expansion. *Phys. Rev. B* **61**, 15114 (2000).
181. D. V. Pakhnin and A. I. Sokolov, Renormalization group and nonlinear susceptibilities of cubic ferromagnets at criticality. *Phys. Rev. B* **64**, 094407 (2001).
182. M. Tissier, D. Mouhanna, J. Vidal and B. Delamotte, Randomly dilute ising model: A nonperturbative approach. *Phys. Rev. B* **65**, 140402 (2002).
183. M. Hasenbusch and E. Vicary, Anisotropic perturbations in three-dimensional $\mathcal{O}(N)$-symmetric vector models. *Phys. Rev. B* **84**, 125136 (2011).
184. A. Kudlis and A. I. Sokolov, Anisotropy of a cubic ferromagnet at criticality. *Phys. Rev. E* **94**, 042107 (2016).
185. L. T. Adzhemyan, E. V. Ivanova, M. V. Kompaniets, A. Kudlis and A. I. Sokolov, Six-loop ε expansion study of three-dimensional n-vector model with cubic anisotropy. *Nucl. Phys. B* **940**, 332 (2019).
186. M. Hasenbusch, Cubic fixed point in three dimensions: Monte carlo simulations of the ϕ^4 model on the simple cubic lattice. *Phys. Rev. B* **107**, 024409 (2023).
187. D. J. Binder, The cubic fixed point at large n. *J. High Energy Phys.* **2021**(9) (2021).
188. S. M. Chester, W. Landry, J. Liu, D. Poland, D. Simmons-Duffin, N. Su and A. Vichi, Bootstrapping Heisenberg magnets and their cubic instability. *Phys. Rev. D* **104**, 105013 (2021).

189. A. B. Harris and T. C. Lubensky, Renormalization-group approach to the critical behavior of random-spin models. *Phys. Rev. Lett.* **33**, 1540 (1974).
190. T. C. Lubensky, Critical properties of random-spin models from the ε expansion. *Phys. Rev. B* **11**, 3573 (1975).
191. D. E. Khmelnitskii, Second-order phase transition in inhomogeneous bodies. *Sov. Phys. JETP* **41**, 981 (1975).
192. G. Grinstein and A. Luther, Application of the renormalization group to phase transitions in disordered systems. *Phys. Rev. B* **13**, 1329 (1976).
193. A. B. Harris, Effect of random defects on the critical behaviour of ising models. *J. Phys. C* **7**, 1671 (1974).
194. A. Aharony, New singularities in the critical behavior of random ising models at marginal dimensionalities. *Phys. Rev. B* **13**, 2092 (1976).
195. C. Jayaprakash and H. J. Katz, Higher-order corrections to the $\epsilon^{\frac{1}{2}}$ expansion of the critical behavior of the random ising system. *Phys. Rev. B* **16**, 3987 (1977).
196. B. N. Shalaev, The phase transition in a weakly disordered uniaxial ferromagnet. *Sov. Phys. JETP* **46**, 1204 (1977).
197. R. Folk, Y. Holovatch and T. Yavors'kii, The correction-to-scaling exponent in dilute systems. *JETP Lett.* **69**, 747 (1999).
198. M. V. Kompaniets, A. Kudlis and A. I. Sokolov, Critical behavior of the weakly disordered ising model: Six-loop $\sqrt{\varepsilon}$ expansion study. *Phys. Rev. E* **103**, 022134 (2021).
199. G. Jug, Critical behavior of disordered spin systems in two and three dimensions. *Phys. Rev. B* **27**, 609 (1983).
200. Y. Holovatch and M. Shpot, Critical exponents of random ising-like systems in general dimensions. *J. Stat. Phys.* **66**, 867 (1992).
201. A. I. Sokolov and B. N. Shalaev, On critical behavior of Ising model with impurities. *Sov. Phys. Solid State* **23**, 1200 (1981).
202. I. O. Maier and A. I. Sokolov, Critical exponents of impure ising model. *Sov. Phys. Solid State* **26**, 2076 (1984).
203. I. O. Mayer, Critical exponents of the dilute Ising model from four-loop expansions. *J. Phys. A* **22**, 2815 (1989).
204. A. Pelissetto and E. Vicari, Randomly dilute spin models: A six-loop field-theoretic study. *Phys. Rev. B* **62**, 6393 (2000).
205. T. Garel and P. Pfeuty, Commensurability effects on the critical behaviour of systems with helical ordering. *J. Phys. C: Solid State Phys.* **9**, 245 (1976).
206. H. Kawamura, Phase transitions in Heisenberg antiferromagnets on triangular and layered-triangular lattices. *J. Appl. Phys.* **61**, 3590 (1987).
207. D. R. T. Jones, A. Love and M. A. Moore, Phase transitions in superfluid He_3. *J. Phys. C: Solid State Phys.* **9**, 743 (1976).
208. D. Bailin, A. Love and M. A. Moore, *J. Phys. C: Solid State Phys.* **10**, 1159 (1977).
209. H. Kawamura, Renormalization-group approach to the frustrated Heisenberg antiferromagnet on the layered-triangular lattice. *J. Phys. Soc. Jpn.* **55**, 2157 (1986).
210. H. Kawamura, Renormalization-group analysis of chiral transitions. *Phys. Rev. B* **38**, 4916 (1988).
211. H. Kawamura, Erratum: Renormalization-group analysis of chiral transitions. *Phys. Rev. B* **42**, 2610 (1990).
212. P. Calabrese, P. Parruccini, A. Pelissetto and E. Vicari. *Phys. Rev. B* **70**, 174439 (2004).
213. M. De Prato, A. Pelissetto and E. Vicari, Normal-to-planar superfluid transition in ^3He. *Phys. Rev. B* **70**, 214519 (2004).

214. P. Bak and D. Mukamel, Physical realizations of $n \geq 4$-component vector models. III. phase transitions in Cr, Eu, Mns_2, Ho, Dy, and Tb. *Phys. Rev. B* **13**, 5086 (1976).
215. Z. Barak and M. B. Walker, First-order phase transitions in Tb, Dy, and Ho. *Phys. Rev. B* **25**, 1969 (1982).
216. H. Kawamura, Generalized chiral universality. *J. Phys. Soc. Jpn.* **59**, 2305 (1990).
217. A. Pelissetto, P. Rossi and E. Vicari, Large-n critical behavior of $\mathrm{O}(n)\times\mathrm{O}(m)$ spin models. *Nucl. Phys. B* **607**, 605 (2001).
218. P. Calabrese and P. Parruccini, Five-loop ε expansion for $\mathrm{O}(n)\times\mathrm{O}(m)$ spin models. *Nucl. Phys. B* **679**, 568 (2004).
219. J. Gracey, Critical exponent ω at $\mathrm{O}(1/n)$ in $\mathrm{O}(n)\times\mathrm{O}(m)$ spin models. *Nucl. Phys. B* **644**, 433 (2002).
220. P. Parruccini, Critical behavior of frustrated spin systems with nonplanar orderings. *Phys. Rev. B* **68**, 104415 (2003).
221. P. Azaria, B. Delamotte and T. Jolicoeur, Nonuniversality in helical and canted-spin systems. *Phys. Rev. Lett.* **64**, 3175 (1990).
222. P. Azaria, B. Delamotte, F. Delduc and T. Jolicoeur, A renormalization-group study of helimagnets in $D = 2 + \varepsilon$ dimensions. *Nucl. Phys. B* **408**, 485 (1993).
223. M. Tissier, D. Mouhanna and B. Delamotte, Nonperturbative approach of the principal chiral model between two and four dimensions. *Phys. Rev. B* **61**, 15327 (2000).
224. M. Kompaniets, A. Kudlis and A. Sokolov, Six-loop ε expansion study of three-dimensional $\mathcal{O}(n) \times \mathcal{O}(m)$ spin models. *Nucl. Phys. B* **950**, 114874 (2020).
225. Y. Nakayama and T. Ohtsuki, Bootstrapping phase transitions in qcd and frustrated spin systems. *Phys. Rev. D* **91**, 021901 (2015).
226. R. D. Pisarski and F. Wilczek, Remarks on the chiral phase transition in chromodynamics. *Phys. Rev. D* **29**, 338 (1984).
227. R. D. Pisarski and D. L. Stein, Critical behavior of linear Φ^4 models with $g \times G'$ symmetry. *Phys. Rev. B* **23**, 3549 (1981).
228. A. J. Paterson, Coleman-Weinberg symmetry breaking in the chiral $\mathrm{SU}(n)\times\mathrm{SU}(n)$ linear σ model. *Nucl. Phys. B* **190**, 188 (1981).
229. P. Calabrese and P. Parruccini, Five-loop ε expansion for $\mathrm{U}(n)\times\mathrm{U}(m)$ models: finite-temperature phase transition in light QCD. *J. High Energy Phys.* **2004**, 018 (2004).
230. L. Adzhemyan, E. Ivanova, M. Kompaniets, A. Kudlis and A. Sokolov, Six-loop ε expansion of three-dimensional $\mathrm{U}(n)\times\mathrm{U}(m)$ models. *Nucl. Phys. B* **975**, 115680 (2022).
231. A. Butti, A. Pelissetto and E. Vicari, On the nature of the finite-temperature transition in QCD. *J. High Energy Phys.* **2003**, 029 (2003).
232. A. Pelissetto and E. Vicari, Relevance of the axial anomaly at the finite-temperature chiral transition in QCD. *Phys. Rev. D* **88**, 105018 (2013).
233. P. Calabrese, A. Pelissetto and E. Vicari, Multicritical behavior in frustrated spin systems with non-collinear order. *Nucl. Phys. B* **709**, 550 (2005).
234. P. Calabrese, A. Pelissetto and E. Vicari, Multicritical phenomena in $\mathrm{O}(n_1) \bigoplus \mathrm{O}(n_2)$-symmetric theories. *Phys. Rev. B* **67**, 054505 (2003).
235. P. C. Hohenberg and B. I. Halperin, Theory of dynamic critical phenomena. *Rev. Modern Phys.* **49**, 435 (1977).
236. P. Calabrese and A. Gambassi, Ageing properties of critical systems. *J. Phys. A* **38**, R133 (2005).
237. R. Folk and G. Moser, Critical dynamics: a field-theoretical approach. *J. Phys. A* **39**, R207 (2006).

238. L. Adzhemyan, D. Evdokimov, M. Hnatič, E. Ivanova, M. Kompaniets, A. Kudlis and D. Zakharov, Model a of critical dynamics: 5-loop ε expansion study. *Phys. A: Stat. Mech. Appl.* **600**, 127530 (2022).
239. M. Hasenbusch, Dynamic critical exponent z of the three-dimensional ising universality class: Monte carlo simulations of the improved blume-capel model. *Phys. Rev. E* **101**, 022126 (2020).
240. V. V. Prudnikov, A. V. Ivanov and A. A. Fedorenko, Critical dynamics of spin systems in the four-loop approximation. *J. Exp. Theor. Phys.* **66**, 835 (1997).
241. A. Krinitsyn, V. V. Prudnikov and P. V. Prudnikov, Calculations of the dynamical critical exponent using the asymptotic series summation method. *Theor. Math. Phys.* **147**, 561 (2006).
242. L. Adzhemyan, D. Evdokimov, M. Hnatič, E. Ivanova, M. Kompaniets, A. Kudlis and D. Zakharov, The dynamic critical exponent z for $2d$ and $3d$ ising models from five-loop ε expansion. *Phys. Lett. A* **425**, 127870 (2022).
243. T. Binoth and G. Heinrich, Numerical evaluation of multi-loop integrals by sector decomposition. *Nucl. Phys. B* **680**, 375 (2004).
244. D. Mesterházy, J. H. Stockemer and Y. Tanizaki, From quantum to classical dynamics: The relativistic $\mathcal{O}(n)$ model in the framework of the real-time functional renormalization group. *Phys. Rev. D* **92**, 076001 (2015).
245. C. Duclut and B. Delamotte, Frequency regulators for the nonperturbative renormalization group: A general study and the model a as a benchmark. *Phys. Rev. E* **95**, 012107 (2017).
246. M. Dančo, M. Hnatič, M. V. Komarova, T. c. v. Lučivjanský and M. Y. Nalimov, Superfluid phase transition with activated velocity fluctuations: Renormalization group approach. *Phys. Rev. E* **93**, 012109 (2016).
247. M. Komarova, D. M. Krasnov and M. Nalimov, Bose condensation: The viscosity critical dimension and developed turbulence. *Theor. Math. Phys.* **169**, 89 (2011).
248. M. Dančo, M. Hnatič, M. V. Komarova, D. M. Krasnov, T. Lučivjanský, L. Mizisin and M. Y. Nalimov, Influence of hydrodynamic fluctuations on the phase transition in the e and f models of critical dynamics. *Theor. Math. Phys.* **176**, 888 (2013).
249. Y. A. Zhavoronkov, M. V. Komarova, Y. G. Molotkov, M. Y. Nalimov and J. Honkonen, Critical dynamics of the phase transition to the superfluid state. *Theor. Math. Phys.* **200**, 1237 (2019).
250. J. Honkonen, M. V. Komarova, Y. G. Molotkov and M. Y. Nalimov, Effective large-scale model of boson gas from microscopic theory. *Nucl. Phys. B* **939**, 105 (2019).
251. J. Honkonen, M. V. Komarova, Y. G. Molotkov and M. Y. Nalimov, Kinetic Theory of Boson Gas. *Theor. Math. Phys.* **200**, 1360 (2019).
252. J. Honkonen, M. V. Komarova, Yu. G. Molotkov, M. Yu. Nalimov and Yu. A. Zhavoronkov, Renormalization group in non-relativistic quantum statistics. *EPJ Web Conf.* **226**, 01005 (2020).
253. J. Schwinger, Brownian motion of a quantum oscillator. *J. Math. Phys.* **2**, 407 (1961).
254. L. V. Keldysh, Diagram technique for nonequilibrium processes. *Sov. Phys. JETP* **20**, 1018 (1965).
255. J. Honkonen, M. Komarova, Y. Molotkov, M. Nalimov and A. Trenogin, Critical dynamics of the superfluid phase transition: Multiloop calculation of the microscopic model. *Phys. Rev. E* **106**, 014126 (2022).
256. U. Frisch, *Turbulence: The Legacy of A.N. Kolmogorov.* (Cambridge University Press, Cambridge, MA, 1995).
257. A. S. Monin and A. M. Yaglom, *Statistical Fluid Mechanics*, Vol. 2 (MIT Press, Cambridge, MA, 1975).

258. W. D. McComb, *Physics of Fluid Turbulence* (Clarendon, Oxford, 1990).
259. P. L. Sulem, J. D. Fournier and U. Frisch, Fully developed turbulence and renormalization group. *Lecture Notes Phys.* **104**, 321 (1979).
260. L. T. Adzhemyan, N. V. Antonov and A. N. Vasil'ev, Infrared divergences and renormalization group in the theory of fully developed turbulence. *Sov. Phys. JETP* **68**, 733 (1989).
261. L. T. Adzhemyan, N. V. Antonov and A. N. Vasiliev, *The Field Theoretic Renormalization Group in Fully Developed Turbulence* (Gordon and Breach, London, 1999).
262. N. V. Antonov and M. M. Kostenko, Anomalous scaling in magnetohydrodynamic turbulence: Effects of anisotropy and compressibility in the kinematic approximation. *Phys. Rev. E* **92**, 053013 (2015).
263. R. H. Kraichnan, Kolmogorov's hypotheses and eulerian turbulence theory. *Phys. Fluids* **7**, 1723 (1964).
264. R. H. Kraichnan, Lagrangian-history closure approximation for turbulence. *Phys. Fluids* **8**, 575 (1965).
265. S. Chen and R. H. Kraichnan, Sweeping decorrelation in isotropic turbulence. *Phys. Fluids A* **1**, 2019 (1989).
266. S. L. Woodruff, A similarity solution for the direct interaction approximation and its relationship to renormalization-group analyses of turbulence. *Phys. Fluids* **6**, 3051 (1994).
267. R. A. Antonia, E. J. Hopfinger, Y. Gagne and F. Anselmet, Temperature structure functions in turbulent shear flows. *Phys. Rev. A* **30**, 2704 (1984).
268. K. R. Sreenivasan and R. A. Antonia, The phenomenology of small-scale turbulence. *Annu. Rev. Fluid Mech.* **29**, 435 (1997).
269. M. Holzer and E. D. Siggia, Turbulent mixing of a passive scalar. *Phys. Fluids* **6**, 1820 (1994).
270. L. Mydlarski and Z. Warhaft, Passive scalar statistics in high-Pèclet-number grid turbulence. *Journal of Fluid Mechanics* **358**, 135 (1998).
271. G. Falkovich, K. Gawędzki and M. Vergassola, Particles and fields in fluid turbulence. *Rev. Mod. Phys.* **73**, 913 (2001).
272. R. H. Kraichnan, Anomalous scaling of a randomly advected passive scalar. *Phys. Rev. Lett.* **72**, 1016 (1994).
273. K. Gawędzki and A. Kupiainen, Anomalous scaling of the passive scalar. *Phys. Rev. Lett.* **75**, 3834 (1995).
274. M. Chertkov, G. Falkovich, I. Kolokolov and V. Lebedev, Normal and anomalous scaling of the fourth-order correlation function of a randomly advected passive scalar. *Phys. Rev. E* **52**, 4924 (1995).
275. L. T. Adzhemyan, N. V. Antonov and A. N. Vasil'ev, Renormalization group, operator product expansion, and anomalous scaling in a model of advected passive scalar. *Phys. Rev. E* **58**, 1823 (1998).
276. L. T. Adzhemyan, N. V. Antonov and A. N. Vasiliev, Quantum field renormalization group in the theory of fully developed turbulence. *Phys-Usp.* **39**, 1193 (1996).
277. L. T. Adzhemyan, N. V. Antonov, V. A. Barinov, Y. S. Kabrits and A. N. Vasil'ev, Anomalous exponents to order ε^3 in the rapid-change model of passive scalar advection. *Phys. Rev. E* **63**, 025303 (2001).
278. L. T. Adzhemyan and N. V. Antonov, Renormalization group and anomalous scaling in a simple model of passive scalar advection in compressible flow. *Phys. Rev. E* **58**, 7381 (1998).

279. N. V. Antonov and J. Honkonen, Anomalous scaling in two models of passive scalar advection: Effects of anisotropy and compressibility. *Phys. Rev. E* **63**, 036302 (2001).
280. L. T. Adzhemyan, N. V. Antonov, M. Hnatich and S. V. Novikov, Anomalous scaling of a passive scalar in the presence of strong anisotropy. *Phys. Rev. E* **63**, 016309 (2000).
281. N. V. Antonov, J. Honkonen, A. Mazzino and P. Muratore-Ginanneschi, Manifestation of anisotropy persistence in the hierarchies of magnetohydrodynamical scaling exponents. *Phys. Rev. E* **62**, R5891 (2000).
282. L. T. Adzhemyan, N. V. Antonov and A. V. Runov, Anomalous scaling, nonlocality, and anisotropy in a model of the passively advected vector field. *Phys. Rev. E* **64**, 046310 (2001).
283. M. Vergassola, Anomalous scaling for passively advected magnetic fields. *Phys. Rev. E* **53**, R3021 (1996).
284. I. Arad, L. Biferale and I. Procaccia, Nonperturbative spectrum of anomalous scaling exponents in the anisotropic sectors of passively advected magnetic fields. *Phys. Rev. E* **61**, 2654 (2000).
285. L. T. Adzhemyan, N. V. Antonov and J. Honkonen, Anomalous scaling of a passive scalar advected by the turbulent velocity field with finite correlation time: Two-loop approximation. *Phys. Rev. E* **66**, 036313 (2002).
286. L. T. Adzhemyan, N. V. Antonov, J. Honkonen and T. L. Kim, Anomalous scaling of a passive scalar advected by the navier-stokes velocity field: Two-loop approximation. *Phys. Rev. E* **71**, 016303 (2005).
287. L. T. Adzhemyan, M. Hnatich, D. Horvath and M. Stehlik, Calculation of spectra of turbulence in energy-containing and inertial ranges. *Phys. Rev. E* **58**, 4511 (1998).
288. W. K. George, The decay of homogeneous isotropic turbulence. *Phys. Fluids. A: Fluid Dynamics* **4**, 1492 (1992).
289. L. T. Adzhemyan and M. Y. Nalimov, The principle of maximal randomness in the statistical theory of fully developed turbulence 1. The isotropic homogeneous turbulence. *Theor. Math. Phys.* **91**, 532 (1992).
290. L. T. Adzhemyan and M. Y. Nalimov, The principle of maximal randomness in the statistical theory of fully developed turbulence 2. The isotropic decaying turbulence. *Theor. Math. Phys.* **96**, 872 (1993).
291. L. T. Adzhemyan, S. V. Borisonok and M. Y. Nalimov, Calculation of the spectra for developed decaying turbulence in the energy-containing and inertial regions. *Theor. Math. Phys.* **106**, 341 (1996).
292. H. K. Janssen, On a lagrangean for classical field dynamics and renormalization group calculations of dynamical critical properties. *Z. Phys. B* **23**, 377 (1976).
293. C. de Dominicis, Technics of field renormalization and dynamics of critical phenomena. *J. Phys. Colloq. France* **37**, C1.247 (1976).
294. L. T. Adzhemyan, A. N. Vasil'ev and Y. M. Pis'mak, Renormalization group approach in the turbulence theory. *Theor. Math. Phys.* **57**, 268 (1983).
295. J. D. Fournier, P. L. Sulem and A. Pouquet, Infrared properties of forced magnetohydrodynamic turbulence. *J. Phys. A* **15**, 1393 (1982).
296. L. T. Adzhemyan, A. N. Vasil'ev and M. Hnatich, Quantum-field renormalization group in the theory of turbulence: Magnetohydrodynamics. *Theor. Math. Phys.* **64**, 777 (1985).
297. R. H. Kraichnan, Inertial-range spectrum of hydromagnetic turbulence. *Phys. Fluids* **8**, 1385 (1965).

298. L. Ts. Adzhemyan, M. Hnatich, D. Horváth and M. Stehlik, Infrared properties of an anisotropically driven MHD turbulence. *Int. J. Mod. Phys. B* **09**, 3401 (1995).
299. H. K. Moffatt, *Magnetic Field Generation in Electrically Conducting Fluids*, p. 353 (Cambridge, England, Cambridge University Press, 1978).
300. L. T. Adzhemyan, A. N. Vasil'ev and M. Hnatich, Turbulent dynamo as spontaneous symmetry breaking. *Theor. Math. Phys.* **72**, 940 (1987).
301. M. Hnatich, J. Honkonen, M. Jurcisin, A. Mazzino and S. Sprinc, Anomalous scaling of passively advected magnetic field in the presence of strong anisotropy. *Phys. Rev. E* **71**, 066312 (2005).
302. M. Hnatič and P. Zalom, Helical turbulent Prandtl number in the A model of passive vector advection. *Phys. Rev. E* **94**, 053113 (2016).
303. M. Hnatič, J. Honkonen and T. Lučivjanský, Symmetry breaking in stochastic dynamics and turbulence. *Symmetry* **11** (2019).
304. F. Krause and K. H. Rädler, *Mean-field Magnetohydrodynamics and Dynamo Theory* (Pergamon Press, 1980). ISBN 9780080250410.
305. H. Krug, J. Spohn, *Solids Far from Equilibrium* (Cambridge University Press Cambridge, New York, 1991). ISBN 052141170.
306. T. Halpin-Healy and Y.-C. Zhang, Kinetic roughening phenomena, stochastic growth, directed polymers and all that. aspects of multidisciplinary statistical mechanics. *Phys. Rep.* **254**, 215 (1995).
307. H. S. A.-L. Barabási, *Fractal Concepts in Surface Growth* (Cambridge University Press, New York, 1995).
308. H. Yan, D. A. Kessler and L. M. Sander, Kinetic roughening in surface growth. *MRS Online Proceedings Library (OPL)* **278**, 237 (1992).
309. B. Schmittmann and R. Zia. Statistical mechanics of driven diffusive systems. In eds. B. Schmittmann and R. Zia, *Statistical Mechanics of Driven Diffusive System*, vol. 17, *Phase Transitions and Critical Phenomena*, p. 3. Academic Press (1995).
310. G. Pruessner, *Self-Organized Criticality: Theory, Models and Characterisation* (Cambridge University Press, Cambridge, MA, USA, 2012).
311. H. Hinrichsen, Non-equilibrium critical phenomena and phase transitions into absorbing states. *Adv. Phys.* **49**, 815 (2000).
312. M. Henkel, H. Hinrichsen and S. Lübeck, *Non-Equilibrium Phase Transitions*, Vol. I: *Absorbing Phase Transitions* (Springer, Dordrecht, The Netherlands, 2008).
313. M. Hnatič, J. Honkonen and T. Lučivjanský, Advanced field-theoretical methods in stochastic dynamics and theory of developed turbulence. *Acta Phys. Slovaca* **66**, 69 (2016).
314. H.-K. Janssen and U. C. Täuber, The field theory approach to percolation processes. *Ann. Phys.* **315**, 147, (2005). Special Issue.
315. J.-P. Bouchaud and A. Georges, Anomalous diffusion in disordered media: Statistical mechanisms, models and physical applications. *Phys. Rep.* **195**, 127 (1990).
316. M. Cross and H. Greenside, *Pattern Formation and Dynamics in Nonequilibrium Systems* (Cambridge University Press, Cambridge, UK, 2009).
317. M. Pietruszka, Non-equilibrium phase transition at a critical point of human blood. *Sci. Rep.* **11**, 22398 (2021).
318. D. Y. Ivanov, *Critical Behaviour of Non-Idealized Systems* (Fizmatlit, Moscow, 2003).
319. G. Satten and D. Ronis, Critical phenomena in randomly stirred fluids: Correlation functions, equation of motion, and crossover behavior. *Phys. Rev. A* **33**, 3415 (1986).

320. D. E. Khmelnitskii, Second-order phase transition in inhomogeneous bodies. *Sov. Phys. JETP* **41**, 981 (1975).
321. B. N. Shalaev, The phase transition in a weakly disordered uniaxial ferromagnet. *Sov. Phys. JETP* **46**, 1204 (1977).
322. H.-K. Janssen, K. Oerding and E. Sengespeick, On the crossover to universal criticality in dilute ising systems. *J. Phys. A: Math. Gen.* **28**, 6073 (1995).
323. A. Onuki and K. Kawasaki, Critical phenomena of classical fluids under flow. I: Mean field approximation. *Prog. Theor. Exp. Phys.* **63**, 122 (1980).
324. J. A. Aronovitz and D. R. Nelson, Turbulence in phase-separating binary mixtures. *Phys. Rev. A* **29**, 2012 (1984).
325. N. V. Antonov, M. Hnatich and J. Honkonen, Effects of mixing and stirring on the critical behaviour. *J. Phys. A* **39**, 7867 (2006).
326. N. V. Antonov and A. S. Kapustin, Effects of turbulent mixing on critical behaviour in the presence of compressibility: Renormalization group analysis of two models. *J. Phys. A* **43**, 405001 (2010).
327. D. Beysens, M. Gbadamassi and L. Boyer, Light-scattering study of a critical mixture with shear flow. *Phys. Rev. Lett.* **43**, 1253 (1979).
328. N. V. Antonov and A. A. Ignatieva, Critical behaviour of a fluid in a random shear flow: Renormalization group analysis of a simplified model. *J. Phys. A* **39**, 13593 (2006).
329. N. V. Antonov, V. I. Iglovikov and A. S. Kapustin, Effects of turbulent mixing on the nonequilibrium critical behaviour. *J. Phys. A* **42**, 135001 (2009).
330. G. Ódor, Universality classes in nonequilibrium lattice systems. *Rev. Mod. Phys.* **76**, 663 (2004).
331. U. C. Täuber. *Critical Dynamics: A Field Theory Approach to Equilibrium and Non-equilibrium Scaling Behavior* (Cambridge University Press, 2014).
332. J. M. Luck, Diffusion in a random medium: A renormalization group approach. *Nucl. Phys. B* **225**, 169 (1983).
333. D. S. Fisher, Random walks in random environments. *Phys. Rev. A* **30**, 960 (1984).
334. D. S. Fisher, D. Friedan, Z. Qiu, S. J. Shenker and S. H. Shenker, Random walks in two-dimensional random environments with constrained drift forces. *Phys. Rev. A* **31**, 3841 (1985).
335. V. E. Kravtsov, I. V. Lerner and V. I. Yudson, Random walks in media with constrained disorder. *J. Phys. A* **18**, L703 (1985).
336. J. Bouchaud, A. Comtet, A. Georges and P. Le Doussal, Anomalous diffusion in random media of any dimensionality. *J. Phys. France* **48**, 1445 (1987).
337. J. Honkonen and E. Karjalainen, Diffusion in a random medium with long-range correlations. *J. Phys. A* **21**, 4217 (1988).
338. J. Honkonen, Y. M. Pis'mak and A. N. Vasil'ev, Zero beta function for a model of diffusion in potential random field. *J. Phys. A* **21**, L835 (1988).
339. J. Honkonen and Y. M. Pis'mak, Exact calculation of the anomalous dimension of the diffusion coefficient for a model of a random walk in a random potential. *J. Phys. A* **22**, L899 (1989).
340. S. E. Derkachov, J. Honkonen and A. N. Vasil'ev, Renormalisation of random walks with memory. *J. Phys. A* **23**, 2479 (1990).
341. D. J. Amit, G. Parisi and L. Peliti, Asymptotic behavior of the "true" self-avoiding walk. *Phys. Rev. B* **27**, 1635 (1983).
342. L. Peliti and Y.-C. Zhang, Renormalisation of the long-range "true" self-avoiding walk. *J. Phys. A* **18**, L709 (1985).

343. S. E. Derkachov, J. Honkonen, E. Karjalainen and A. N. Vasil'ev, Renormalisability and renormalisation of the "true" self-avoiding random walks. *J. Phys. A* **22**, L385 (1989).
344. M. Doi, Second quantization representation for classical many-particle system. *J. Phys. A* **9**, 1465 (1976).
345. M. Doi, Stochastic theory of diffusion-controlled reaction. *J. Phys. A* **9**, 1479 (1976).
346. L. Peliti, Path integral approach to birth-death processes on a lattice. *J. Phys. France* **46**, 1469 (1985).
347. U. C. Täuber, M. Howard and B. P. Vollmayr-Lee, Applications of field-theoretic renormalization group methods to reaction–diffusion problems. *J. Phys. A* **38**, R79 (2005).
348. L. Peliti, Renormalisation of fluctuation effects in the $A+A \to A$ reaction. *J. Phys. A* **19**, L365 (1986).
349. B. P. Lee, Renormalization group calculation for the reaction $kA \to \emptyset$. *J. Phys. A* **27**, 2633 (1994).
350. J.-M. Park and M. W. Deem, Disorder-induced anomalous kinetics in the $A+A \to \emptyset$ reaction. *Phys. Rev. E* **57**, 3618 (1998).
351. W. J. Chung and M. W. Deem, Numerical observation of disorder-induced anomalous kinetics in the $A+A \to \emptyset$ reaction. *Phys. A: Stat. Mech. Appl.* **265**, 486 (1999).
352. M. J. E. Richardson and J. Cardy, The reaction process $A+A \to \emptyset$ in Sinai disorder. *J. Phys. A* **32**, 4035 (1999).
353. M. W. Deem and J.-M. Park, Reactive turbulent flow in low-dimensional, disordered media. *Phys. Rev. E* **58**, 3223 (1998).
354. N. le Tran, J.-M. Park and M. W. Deem, "superfast" reaction in turbulent flow with potential disorder. *J. Phys. A* **32**, 1407 (1999).
355. M. Hnatich and J. Honkonen, Velocity-fluctuation-induced anomalous kinetics of the $A+A \to \emptyset$ reaction. *Phys. Rev. E* **61**, 3904 (2000).
356. M. Hnatič, J. Honkonen and T. Lucivjanský, Two-loop calculation of anomalous kinetics of the reaction $A+A \to \emptyset$ in randomly stirred fluid. *Eur. Phys. J. B* **86**, 214 (2013).
357. S. R. Broadbent and J. M. Hammersley. Percolation processes: I. Crystals and mazes. In *Mathematical Proceedings of the Cambridge Philosophical Society*, Vol. 53, p. 629. Cambridge University Press (1957).
358. J. L. Cardy and R. L. Sugar, Directed percolation and Reggeon field theory. *J. Phys. A* **13**, L423 (1980).
359. U. C. Täuber, Population oscillations in spatial stochastic lotka-volterra models: A field-theoretic perturbational analysis. *J. Phys. A* **45**, 405002 (2012).
360. M. Sano and K. Tamai, A universal transition to turbulence in channel flow. *Nature Physics* **12**, 249 (2016).
361. H.-K. Janssen, On the nonequilibrium phase transition in reaction-diffusion systems with an absorbing stationary state. *Z. Phys. B* **42**, 151 (1981).
362. P. Grassberger, On phase transitions in Schlögl's second model. *Z. Phys. B* **47**, 365 (1982).
363. H. D. I. Abarbanel, J. B. Bronzan, A. Schwimmer and R. L. Sugar, Intercept of the Pomeron. *Phys. Rev. D* **14**, 632 (1976).
364. J. B. Bronzan and J. W. Dash, Higher-order ϵ terms in reggeon field theory. *Phys. Rev. D* **10**, 4208 (1974).
365. H.-K. Janssen, Directed percolation with colors and flavors. *J. Stat. Phys.* **103**, 801 (2001).

366. L. Ts. Adzhemyan, M. Hnatič, M. Kompaniets, T. Lučivjanský and L. Mižišin, Numerical calculation of scaling exponents of percolation process in the framework of renormalization group approach. *EPJ Web Conf.* **108**, 02005 (2016).
367. L. Ts. Adzhemyan, M. Hnatič, M. Kompaniets, T. Lučivjanský and L. Mižišin, Directed percolation: Calculation of Feynman diagrams in the three-loop approximation. *EPJ Web Conf.* **173**, 02001 (2018).
368. L. Ts. Adzhemyan, M. Hnatič, M. Kompaniets, T. Lučivjanský and L. Mižišin. Renormalization group approach of directed percolation: Three-loop approximation. In eds. C. H. Skiadas and I. Lubashevsky, *11th Chaotic Modeling and Simulation International Conference*, p. 195, Cham (2019). Springer International Publishing. ISBN 978-3-030-15297-0.
369. T. Song and H. Xia, Kinetic roughening and nontrivial scaling in the Kardar–Parisi–Zhang growth with long-range temporal correlations. *J. Stat. Mech.* **2021**, 073203 (2021).
370. P. I. Kakin, M. A. Reiter, M. M. Tumakova, N. M. Gulitskiy and N. V. Antonov, Stirred Kardar-Parisi-Zhang equation with quenched random noise: Emergence of induced nonlinearity. *Universe* **8** (2022).
371. N. V. Antonov and P. I. Kakin, Scaling in erosion of landscapes: renormalization group analysis of a model with turbulent mixing. *J. Phys. A* **50**, 085002 (2017).
372. N. V. Antonov, A. A. Babakin and P. I. Kakin, Strongly nonlinear diffusion in turbulent environment: A problem with infinitely many couplings. *Universe* **8** (2022).
373. C. Duclut and B. Delamotte, Nonuniversality in the erosion of tilted landscapes. *Phys. Rev. E* **96**, 012149 (2017).
374. N. Antonov, P. Kakin, N. Lebedev and A. Luchin, Renormalization group analysis of a self-organized critical system: Intrinsic anisotropy vs random environment. *arXiv preprint arXiv:2212.01913 (cond-mat.stat-mech).* (2022).
375. N. V. Antonov, N. M. Gulitskiy, P. I. Kakin and V. D. Serov, Effects of turbulent environment and random noise on self-organized critical behavior: Universality versus nonuniversality. *Phys. Rev. E* **103**, 042106 (2021).
376. N. V. Antonov, N. M. Gulitskiy, P. I. Kakin and M. N. Semeikin, Dimensional transmutation and nonconventional scaling behavior in a model of self-organized criticality. *Int. J. Mod. Phys. A* **37**, 2240022 (2022).

https://doi.org/10.1142/9789811282386_0036

Chapter 36

Critical Phenomena and the Bootstrap

D. Poland and D. Simmons-Duffin

1. Introduction

We would like to begin by looking back to Michael Fisher's 1965 lectures "The Nature of Critical Points,"[1] which gave a beautiful and surprisingly modern summary of the state of the field. At the time, a number of different fluid, magnetic, and lattice systems were known to exhibit critical points, and serious attempts were being made to extract their critical exponents and compare them to theoretical calculations. The state of knowledge at the time is concisely summarized in table given in Fig. 1.

While the exact solution of the 2D Ising critical point was known, the 3D Ising critical point was poorly understood[a] and even less was known about the 3D Heisenberg critical point.

There was hope that these fixed points could admit an exact solution, and various conjectures were made about the possible exact values of the exponents. For example, the β exponent for the 3D Ising critical point, characterizing the power-law behavior of the order parameter (the fluid density or magnetization), was conjectured to be $\beta = 5/16 = 0.3125$. Much discussion in Fisher's lectures was also given to the possibility of $\beta = 1/3$, as potentially indicated by various experimental results. Similarly, the γ exponent, characterizing the power-law behavior of the susceptibility, was conjectured to be $\gamma = 5/4 = 1.25$.

Thanks to the subsequent development of the ϵ-expansion,[2] improved lattice simulations, and more recently the conformal bootstrap, we now know that these conjectures were too optimistic. As we will discuss, the bootstrap approach in fact gives rigorous[b] upper and lower bounds on these exponents and has revealed that

[a] To quote the lectures: "What rigorous results have been obtained for the Ising models (a) in three dimensions? (b) in a non-zero magnetic field? The answer to (a) is that essentially nothing is known!"
[b] With a suitable definition of rigor.

Temperatures	$T \leqq T_C$			$T=T_C$	$T \geqq T_C$	
Exponent	α'	β	γ'	δ	α	γ
Classical Theory	$0_{\text{discont.}}$	$\frac{1}{2}$	1	3	$0_{\text{discont.}}$	1
Ising d = 2	$0_{\text{log.}}$	$\frac{1}{8}$	$1\frac{3}{4}$	15	0_{log}	$1\frac{3}{4}$
Ising d = 3	$\geqq 0$	$\cong\frac{5}{16}$	$\cong 1\frac{1}{4}$(?)	$\cong 5.2$	$\cong 0.2$	$1\frac{1}{4}$
Heisenberg d = 3	?	?	?	?	$\geqq 0$?	$\cong 1\frac{1}{3}$
Experiment Magnets	$\cong 0_{\text{log}}$?	0.33	$\geqq 1.22$	$\cong 4.22$	$\geqq 0.1$?	$\cong 1.35$
Experiment Fluids	$\cong 0_{\text{log}}$	0.33-0.36	$\geqq 1.18$	$\cong 4.2$	$\geqq 0.1$?	>1.1(?)

Fig. 1. Table of critical exponents reproduced from Ref. 1.

the actual values lie in the ranges $\beta = 0.326419(3)$ and $\gamma = 1.237075(10)$. The bootstrap approach has moreover helped to reveal an extremely rich theoretical structure underlying these critical points that we expect will be exciting to study for decades to come. We review these developments in the following section.

2. CFT Review

A crucial insight due to[3] was the observation that critical points are governed by conformal field theories which possess an enhanced symmetry group $\mathrm{SO}(d+1,1)$. This symmetry group is generated by the usual translations P^μ and rotations $M^{\mu\nu}$, but also includes dilatations D (scale transformations) and special conformal transformations K^μ. These symmetries can be used to tightly constrain the observables of the critical points, including correlation functions of local operators $\mathcal{O}_i(x)$. Such operators are characterized by their behavior under the different conformal transformations, and one can focus on primary operators (those annihilated by K^μ) which are labeled by their scaling dimension Δ_i (the eigenvalue of D) and a spin representation.

A key element of the bootstrap approach is to focus on the algebra of local operators, governed by the operator product expansion (OPE):

$$\mathcal{O}_i(x)\mathcal{O}_j(y) = \sum_k \lambda^a_{ijk} C^a(x-y,\partial_y)\mathcal{O}_k(y). \tag{1}$$

In a conformal field theory, the OPE is a convergent expansion that allows one to reduce n-point correlators to a sum of $n-1$-point correlators. Amazingly, conformal symmetry completely fixes the form of the differential operators $C^a(x-y,\partial_y)$ in terms of the scaling dimensions and spins of the operators and a finite choice of tensor structures (labeled by a). Thus, one is left with a discrete set of data characterizing the theory, captured by the OPE coefficients λ^a_{ijk} and scaling dimensions Δ_i.

Via the OPE, these data determine the correlation functions of local operators. For example, in the 3D Ising CFT, there is a "spin" field $\sigma(x)$ and its two-point correlation function displays a power-law behavior dictated by its scaling dimension:

$$\langle \sigma(x_1)\sigma(x_2)\rangle = \frac{1}{x_{12}^{2\Delta_\sigma}}, \tag{2}$$

where $x_{12}^2 \equiv |x_1 - x_2|^2$.

In the language of critical phenomena, one typically focuses on a small collection of critical exponents that are important in practical observables like the heat capacity and susceptibility. These critical exponents are related to scaling dimensions, e.g., the exponent η is given by $\Delta_\sigma = \frac{1}{2} + \frac{\eta}{2}$. However, from the perspective of CFT, the scaling dimension Δ_σ is more fundamental. It is part of an infinite set of data that is tightly constrained by the OPE (1). For example, there are an infinite number of other operators in the $\sigma \times \sigma$ OPE alone, which we'll write schematically as

$$\sigma \times \sigma \sim \mathbf{1} + \epsilon + \epsilon' + T^{\mu\nu} + \cdots, \tag{3}$$

where the leading scalar ϵ with dimension Δ_ϵ is the "energy" operator, ϵ' with dimension $\Delta_{\epsilon'}$ is the next subleading scalar, $T^{\mu\nu}$ is the spin-2 stress-energy tensor, and the "$\cdots$" denote an infinite tower of other contributions. By studying all of these data together, one finds novel connections that would otherwise be hidden.

Let us pause to review why critical exponents are related to scaling dimensions. In the language of effective field theory, this arises from a "matching" calculation between a microscopic ultraviolet (UV) theory and an infrared (IR) effective theory given by the deformed CFT. For example, when the IR theory is the 3D Ising CFT, the IR effective action takes the form

$$S = S_{\mathrm{CFT}} + t\int d^3x\, \epsilon(x) + \cdots, \tag{4}$$

where S_{CFT} is the action of the CFT, t is an IR coupling constant, and "$\cdots$" indicates an infinite sum of irrelevant operators. (For simplicity, we work at zero applied magnetic field.) The matching between the UV and IR theory is analytic in the couplings of both theories, and thus, we have $t \propto (T - T_c) + O((T - T_c)^2)$, where T_c is defined as the temperature of the UV theory where t vanishes. Once we have the effective action (4), the rest is dimensional analysis. The parameter t has length dimension $-\Delta_t = -(3 - \Delta_\epsilon)$. Assuming we are close enough to the critical point that the irrelevant operators can be ignored, this is the only dimensionful quantity in the IR theory and it follows that the correlation length must have the form $\xi \propto t^{-1/(3-\Delta_\epsilon)}$. From this result, we read off the critical exponent $\nu = 1/(3 - \Delta_\epsilon)$.

Other exponents can be computed in a similar fashion. In d-dimensions, the relations to conventional critical exponents take the following form:

$$\alpha = \frac{d-2\Delta_\epsilon}{d-\Delta_\epsilon}, \qquad \beta = \frac{\Delta_\sigma}{d-\Delta_\epsilon}, \qquad \gamma = \frac{d-2\Delta_\sigma}{d-\Delta_\epsilon},$$
$$\delta = \frac{d-\Delta_\sigma}{\Delta_\sigma}, \qquad \eta = 2\Delta_\sigma - d + 2, \qquad \nu = \frac{1}{d-\Delta_\epsilon},$$
$$\omega = \Delta_{\epsilon'} - d. \tag{5}$$

While these are all given in terms of Δ_σ, Δ_ϵ, and $\Delta_{\epsilon'}$, we emphasize again that the full set of CFT data contains an infinite number of scaling dimensions, and there are correspondingly an infinite number of "subleading" critical exponents which could in principle be defined.

The bootstrap approach aims to constrain this data by writing down consistency conditions. The simplest such conditions arise from a single four-point correlation function like $\langle\sigma(x_1)\sigma(x_2)\sigma(x_3)\sigma(x_4)\rangle$. This can be expanded using the OPE in different ways (called "channels"):

$$\begin{aligned}\langle\sigma(x_1)\sigma(x_2)\sigma(x_3)\sigma(x_4)\rangle &= \frac{1}{x_{12}^{2\Delta_\sigma}x_{34}^{2\Delta_\sigma}}\sum_i \lambda_{\sigma\sigma i}^2 g_{\Delta_i,\ell_i}(z,\bar{z})\\ &= \frac{1}{x_{23}^{2\Delta_\sigma}x_{14}^{2\Delta_\sigma}}\sum_i \lambda_{\sigma\sigma i}^2 g_{\Delta_i,\ell_i}(1-z,1-\bar{z}).\end{aligned} \tag{6}$$

In the first line, we took the $\sigma(x_1)\times\sigma(x_2)$ OPE, while in the second line, we took the $\sigma(x_2)\times\sigma(x_3)$ OPE. The functions $g_{\Delta,\ell}(z,\bar{z})$ are called conformal blocks and depend on two conformally invariant cross-ratios, here captured by the complex variable z and its conjugate $\bar{z}$.[c]

Significant effort in the last decade has gone into developing methods for calculating conformal blocks in general dimensions and for arbitrary external operators. For our purposes, the salient features are that the blocks are fully determined by conformal symmetry in terms of the scaling dimensions and spins of the operators involved. Moreover, both of the conformal block expansions in (6) are expected to converge as long as one keeps z away from branch cuts along $[1,+\infty)$ and $(-\infty,0]$. In numerical analysis of these equations, one often expands around the symmetric point $z=\bar{z}=1/2$, corresponding to all four operators being arranged in a square.

The resulting "crossing" equation (6) can be written as a sum rule that must be satisfied by any consistent CFT:

$$0 = \sum_i \lambda_{\sigma\sigma i}^2 F^{\Delta_\sigma}_{\Delta_i,\ell_i}(z,\bar{z}), \tag{7}$$

where $F^{\Delta_\sigma}_{\Delta_i,\ell_i}(z,\bar{z})$ is a difference of conformal blocks in the two different channels. Moreover, in unitary (or reflection positive) CFTs, the coefficients must be positive

[c]These are related to the positions via $z\bar{z} = \frac{x_{12}^2x_{34}^2}{x_{13}^2x_{24}^2}$ and $(1-z)(1-\bar{z}) = \frac{x_{23}^2x_{14}^2}{x_{13}^2x_{24}^2}$.

and all scaling dimensions are known to satisfy a general spin-dependent lower bound $\Delta \geq \Delta_\ell^{\min}$, where $\Delta_0^{\min} = d/2 - 1$ and $\Delta_{\ell>0}^{\min} = \ell + d - 2$.

The sum rule (7), involving an infinite number of *a priori* unknown operator dimensions and OPE coefficients, has been well known for decades. But due to its complexity, it has proven extremely difficult to find solutions. Currently, no non-trivial exact solutions for CFTs with $d > 2$ have been constructed. However, an important breakthrough occurred in Ref. 4, which took a different perspective. Instead of trying to solve the equation, the authors of Ref. 4 showed how to prove properties of its solutions by exploiting positivity of the coefficients $\lambda_{\sigma\sigma i}^2$ (which follows from unitarity of the underlying CFT). Their central idea was to act on (7) with a linear functional satisfying certain positivity constraints, whose existence determines whether the sum rule can be satisfied.

In practice, the search for such a functional is most efficient to perform numerically. It can only be done after introducing certain truncations to make the problem finite: choosing a finite-dimensional basis for the linear functional (typically by taking derivatives up to some order), introducing a rational approximation for the conformal blocks, and imposing positivity constraints over a finite set of spins. However, once these choices are made, the problem can be reduced to an optimization problem involving matrices of polynomials, which can be solved using well-known semidefinite programming methods. A detailed review of this approach can be found in Ref. 5.

As an example, one can start by imposing a hypothetical condition on the CFT spectrum, e.g., that the first scalar operator appearing the OPE has a dimension $\Delta_\epsilon \geq \Delta_\epsilon^{\min}$. If one can find a functional that is positive when acting on all the terms in (7), then one concludes that a unitary CFT cannot exist subject to the assumptions that were made. By searching over different possibilities for $\Delta_\epsilon^{\min}$ and looking for the boundary between disallowed values and possibly allowed values, one can obtain an upper bound on Δ_ϵ. This upper bound is essentially rigorous.[d] The bound can be improved by increasing the search space for the functional and including more crossing relations. However, the logic is such that it can only become stronger and never weaker.

3. Bootstrap Results

3.1. *Ising model*

The numerical method described above was first applied to 4D CFTs in Ref. 5 and later adapted to 3D CFTs in Refs. 6, 7. Surprisingly, the resulting bound on Δ_ϵ as a function of Δ_σ in 3D showed a prominent "kink" near $(\Delta_\sigma, \Delta_\epsilon) \sim (0.518, 1.412)$,

---

[d] There are tiny numerical errors induced by the approximation to the conformal blocks and issues related to numerical precision of the computation. However, these can be made arbitrarily small at the expense of additional computation time — in practice, they are negligible.

whose location coincides with estimates of the scaling dimensions in the 3D Ising CFT from other methods.

For example, the ϵ-expansion gives as its first few terms

$$\Delta_\sigma = 1 - \frac{\epsilon}{2} + \frac{1}{108}\epsilon^2 + \cdots , \quad (8)$$

$$\Delta_\epsilon = 2 - \epsilon + \frac{\epsilon}{3} + \frac{19}{162}\epsilon^2 + \cdots . \quad (9)$$

These expansions can be computed using the usual perturbative methods, but more recently, it was understood that the leading terms can also be elegantly reproduced using CFT constraints on three-point functions,[8] by using weakly broken higher-spin symmetry,[9] or through the analytic bootstrap.[10,11] These expansions are now known up to $O(\epsilon^8)$ for Δ_σ and $O(\epsilon^7)$ for Δ_ϵ. An excellent summary of these results along with a list of references can be found in Ref. 12. Predictions from the ϵ-expansion for 3D can be made by applying resummation methods, e.g., a Borel resummation of the ϵ-expansion up to $O(\epsilon^6)$[13,14] has yielded

$$\Delta_\sigma^{\epsilon-\text{exp}} = 0.5181(3), \quad (10)$$

$$\Delta_\epsilon^{\epsilon-\text{exp}} = 1.4108(12), \quad (11)$$

in very close agreement with the kink location in Fig. 2. The physical CFT appears to live not only on the boundary of the allowed region from crossing symmetry and unitarity but also at a corner!

One can even attempt to compute numerical bootstrap bounds in fractional dimensions. The computation itself is technically straightforward (at least for scalar correlators) since the spacetime dimension simply enters as a parameter in the conformal blocks that can be varied continuously. Fractional dimensional bounds

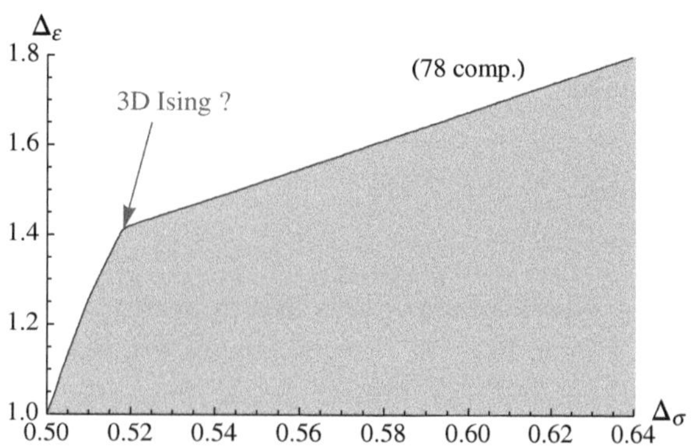

Fig. 2. Upper bound on the scaling dimension of ϵ as a function of the scaling dimension of σ obtained from the conformal bootstrap applied to the $\langle\sigma\sigma\sigma\sigma\rangle$ four-point function in three dimensions, first computed in Ref. 6. Plot reprinted with permission from Ref. 7.

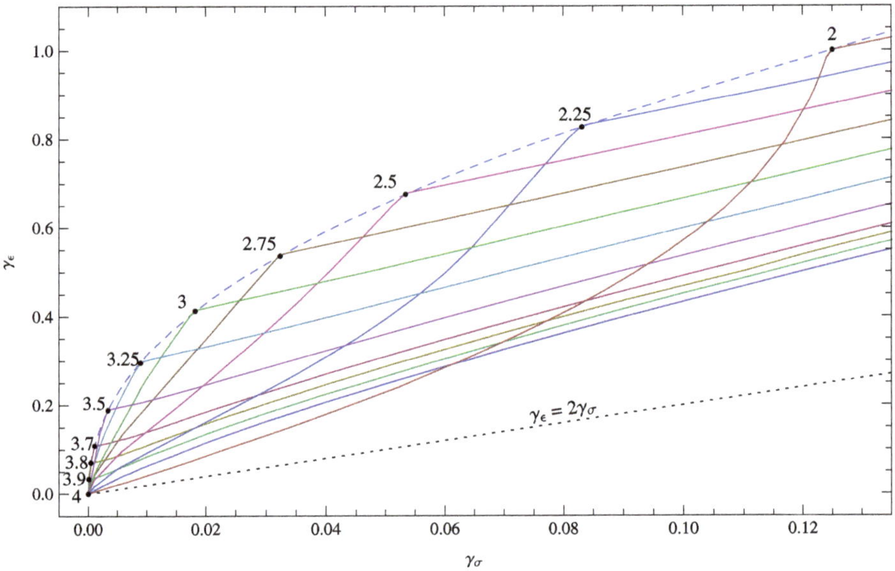

Fig. 3. Upper bounds on the anomalous dimension of ϵ as a function of the anomalous dimension of σ obtained from the conformal bootstrap applied to the $\langle\sigma\sigma\sigma\sigma\rangle$ four-point function in $2 \leq d \leq 4$ dimensions. Here, the anomalous dimension is related to the scaling dimension as $\gamma_{\mathcal{O}} = \Delta_{\mathcal{O}} - (d-2)/2$. Reprinted with permission from Ref. 15. Copyright 2014 © the American Physical Society.

were first explored in Ref. 15 and recently improved and extended in Refs. 14, 16. This leads to a sequence of kinks, shown in Fig. 3, which interpolate between the 4D free scalar CFT, the 3D Ising CFT, and the 2D Ising CFT. The evolution of these kinks match very well with the predictions of the ϵ-expansion.

However, there are subtleties in interpreting these bounds in fractional dimensions. It was shown in Ref. 17 that the Wilson–Fisher CFT in fractional dimensions actually displays violations of unitarity, which is a central assumption in numerical bootstrap bounds. In practice, the effects of these violations of unitarity are expected to be small, since they first appear in operators with large scaling dimension $\Delta \sim 23$. Hence, it is possible that the bounds in Fig. 3 are correct in some approximate sense, but this has not been carefully quantified.

The bootstrap approach can be pushed further by considering systems of correlation functions. By combining the crossing relations for $\langle\sigma\sigma\sigma\sigma\rangle$ with $\langle\sigma\sigma\epsilon\epsilon\rangle$ and $\langle\epsilon\epsilon\epsilon\epsilon\rangle$, Kos *et al.*[18] was able to obtain an isolated region in parameter space which contains the 3D Ising CFT. To do so only requires mild physical input: the assumption of a $\mathbb{Z}_2$ symmetry and that σ is the only relevant $\mathbb{Z}_2$ odd operator and ϵ is the only relevant $\mathbb{Z}_2$ even operator. Then by using a computing cluster to explore a very high-dimensional functional search space (and additionally scanning over possible values of the OPE coefficient ratio $\frac{\lambda_{\sigma\sigma\epsilon}}{\lambda_{\epsilon\epsilon\epsilon}}$), one can obtain the very precise island shown in Fig. 4.[19]

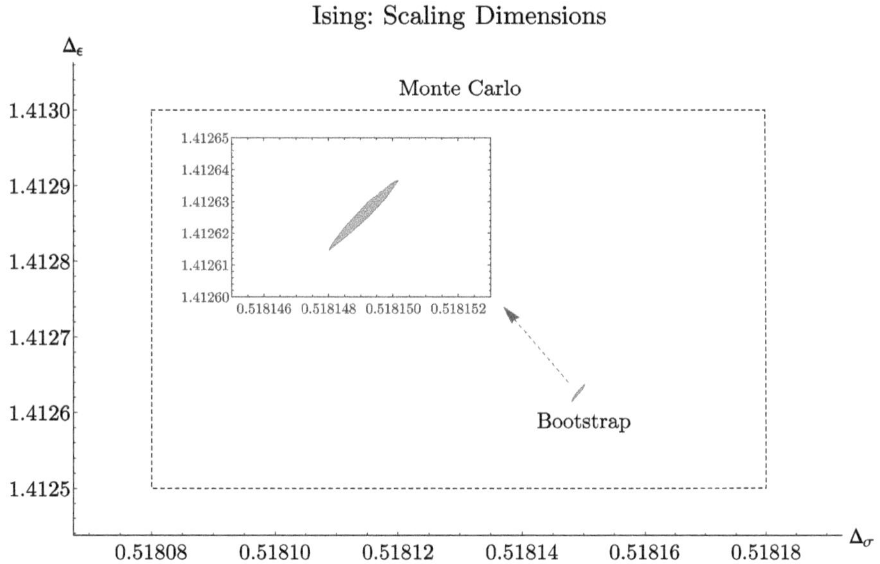

Fig. 4. Conformal bootstrap island obtained from the system of four-point functions containing $\{\sigma, \epsilon\}$ after assuming $\Delta_{\sigma'} \geq 3$ and $\Delta_{\epsilon'} \geq 3$, reprinted from Ref. 19. In the plot, the results are compared to the Monte Carlo simulations performed in Ref. 20.

The resulting scaling dimensions are

$$\Delta_\sigma^{\mathrm{CB}} = 0.5181489(10), \tag{12}$$

$$\Delta_\epsilon^{\mathrm{CB}} = 1.412625(10). \tag{13}$$

At present, the most precise Monte Carlo simulations have been performed by Hasenbusch[21] using a generalized Blume–Capel model on a cubic lattice. By comparison, this study gives the values

$$\Delta_\sigma^{\mathrm{MC}} = 0.518142(20), \tag{14}$$

$$\Delta_\epsilon^{\mathrm{MC}} = 1.41265(13), \tag{15}$$

with estimated errors roughly an order of magnitude larger than the (rigorous) bootstrap error bars. However, one can see that there is an impressive agreement between the two methods!

In the above discussion, we focused on the leading scaling dimensions, but a variety of both rigorous bounds and non-rigorous estimates have been computed using bootstrap methods for other CFT data, including OPE coefficients like $\lambda_{\sigma\sigma\epsilon}$ and $\lambda_{\epsilon\epsilon\epsilon}$, the dimensions of subleading scalar operators like $\Delta_{\epsilon'}$ and $\Delta_{\epsilon''}$, and data connected to the lowest "Regge" trajectories of spinning operators. A comprehensive review of various determinations of this data can be found in Ref. 12.

Here, we would like to highlight that a non-rigorous approach called the "extremal functional method" (EFM) can be used to extract the full set of CFT data associated with an approximate solution of the crossing equations which extremizes some quantity (e.g., an OPE coefficient).[22,23] This provides a wealth of data (often hundreds of operators across many spins) which can be compared with calculations using other approaches. In particular, it has been extremely fruitful to compare EFM data extracted from the numerical bootstrap with computations of the same data using various analytical methods.

3.2. *O(N) models*

The 3D Ising CFT is the first of a sequence of models of N interacting scalar fields ϕ_i called the $O(N)$ vector models. Of these, the 3D $O(2)$ model (sometimes called the XY model), the 3D $O(3)$ model (sometimes called the Heisenberg model), and also the large N limit of the $O(N)$ model have attracted the most scientific interest. The $O(2)$ and $O(3)$ models make appearances in a variety of critical systems, but perhaps, the most well known are the application of the $O(2)$ model to describe the λ-transition in superfluid Helium and the application of the $O(3)$ model to describe the Curie transition in isotropic magnets.

In fact, the $O(2)$ model has been at the center of a controversy. The most precise measurements of this model were done by the Lambda-point experiment in the 1990s[24] which measured the heat capacity of ^{4}He across the transition. Analysis of their data produced a determination of the α exponent, which in the language of CFT translates into a measurement of the scaling dimension $\Delta_s = \frac{3}{2}\frac{1-\alpha}{1-\alpha/2} = 3 - \frac{1}{\nu}$, where s is the leading $O(2)$-singlet scalar:

$$\Delta_s^{\text{exp}} = 1.50946(22). \tag{16}$$

On the other hand, Monte Carlo simulations have consistently produced determinations that show a disagreement with this value well outside of the error bars, e.g., the recent computations of Hasenbusch[25] give

$$\begin{aligned} \Delta_s^{\text{MC}} &= 1.51122(15), \\ \Delta_\phi^{\text{MC}} &= 0.519050(40). \end{aligned} \tag{17}$$

Determinations have also been made of the scaling dimension of the two-index tensor t^{ij}, connected to anisotropic perturbations of the model, e.g., Hasenbusch and Vicari[26] give

$$\Delta_t^{\text{MC}} = 1.2361(11). \tag{18}$$

The $\sim 8\sigma$ disagreement in the determinations of Δ_s can be seen in Fig. 5, where we also show the most precise computations performed using the conformal bootstrap from Ref. 27:

$$\begin{aligned} \Delta_s^{\text{CB}} &= 1.51136(22), \\ \Delta_\phi^{\text{CB}} &= 0.519088(22), \\ \Delta_t^{\text{CB}} &= 1.23629(11). \end{aligned} \tag{19}$$

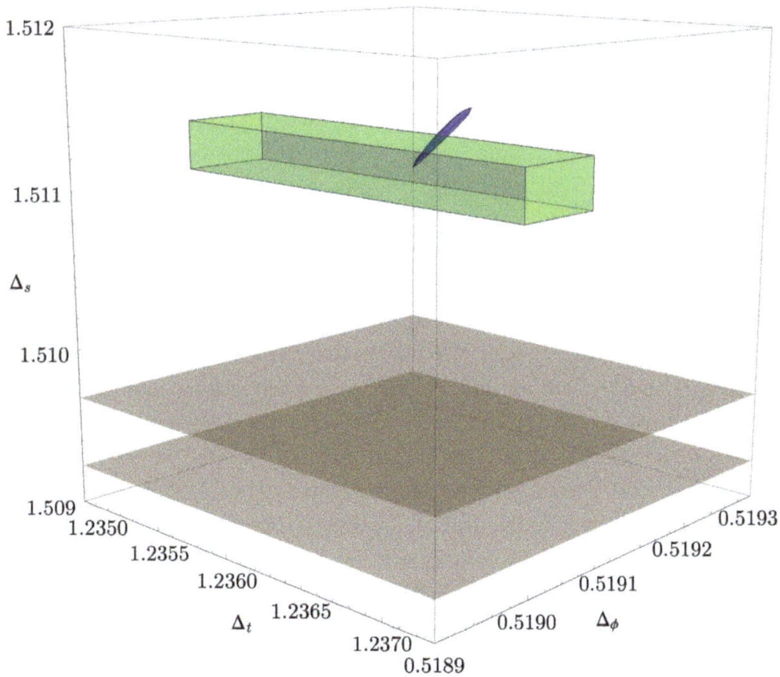

Fig. 5. Conformal bootstrap determinations of Δ_ϕ, Δ_s, and Δ_t in the 3D $O(2)$ model (blue) compared with Monte Carlo simulations (green box) and experiment (brown planes). Plot reprinted from Ref. 27.

The bootstrap islands show excellent agreement with results from Monte Carlo simulations, while the ultimate origin of the disagreement on the experimental side remains to be understood.

These bootstrap computations involved several new ingredients. First, they required the introduction of information about the global symmetry, disentangling operator representations and the associated tensor structures. Second, the most precise computations involved the crossing relations for all four-point functions containing three operators: the $O(2)$ vector ϕ, the $O(2)$ singlet s, and the $O(2)$ symmetric tensor t. Ultimately, this means that there is a six-dimensional parameter space to map out involving the three scaling dimensions $\{\Delta_\phi, \Delta_s, \Delta_t\}$ as well as the three ratios of their OPE coefficients $\left\{\frac{\lambda_{\phi\phi s}}{\lambda_{sss}}, \frac{\lambda_{tts}}{\lambda_{sss}}, \frac{\lambda_{\phi\phi t}}{\lambda_{sss}}\right\}$. The plot of Fig. 5 is actually a 3D projection of this full six-dimensional region.

The work of Ref. 27 introduced various new search strategies for mapping out this space, including Delaunay triangulation methods for searching through dimension space and a "cutting surface" algorithm for efficiently searching through OPE space. Also crucial to the success of this computation were improvements to the software SDPB used for solving the underlying optimization problems.[28,29] This computation serves to highlight that searching through high-dimensional spaces is a necessary element of precise bootstrap computations and it is important to develop methods that scale well as the size of the search space increases.

Similar methods were applied to the 3D $O(3)$ model in Ref. 30, again producing precise determinations of the leading scaling dimensions:

$$\begin{aligned} \Delta_\phi^{\text{CB}} &= 0.518936(67), \\ \Delta_s^{\text{CB}} &= 1.59488(81), \\ \Delta_t^{\text{CB}} &= 1.20954(32). \end{aligned} \tag{20}$$

These can be compared against the recent Monte Carlo simulations of Hasenbusch[31] (along with that of[26] for the anisotropic perturbation Δ_t):

$$\begin{aligned} \Delta_\phi^{\text{MC}} &= 0.518920(25), \\ \Delta_s^{\text{MC}} &= 1.5948(2), \\ \Delta_t^{\text{MC}} &= 1.2094(3). \end{aligned} \tag{21}$$

In this case, the Monte Carlo simulations have achieved a slightly higher (though non-rigorous) precision, but again, there is a satisfying and impressive agreement. Moreover, the bootstrap has also determined the leading OPE coefficient data which are poorly understood from other methods.

An interesting question in the $O(3)$ model has been whether it is stable or unstable to cubic anisotropy. This decades-old question is related to whether the leading four-index symmetric tensor operator t_4^{ijkl} in the CFT is relevant or irrelevant. If it is relevant and allowed by the microscopic symmetries, then it can induce a flow to the nearby fixed point with cubic symmetry rather than $O(3)$ symmetry. A variety of attempts have been made over the years to either estimate the scaling dimension of this operator in the $O(3)$ model or to estimate the critical value of N in the $O(N)$ model at which it transitions from relevance to irrelevance, e.g., the critical value of N has been computed in the ϵ-expansion[32–34]:

$$N_{\text{crit}} = 4 - 2\epsilon + 2.58847559\epsilon^2 - 5.87431189\epsilon^3 + 16.82703902\epsilon^4 + \cdots, \tag{22}$$

which has been resummed to $N_{\text{crit}} \approx 2.89(2)$ or $2.89(4)$, depending on the method, implying that $N = 3$ is on the unstable side.

More directly, the Monte Carlo study[26] produced the estimate

$$\Delta_{t_4}^{\text{MC}} = 2.987(4), \tag{23}$$

again implying that it is a barely relevant operator.

This question can also be elegantly probed using bootstrap methods. In the study of Chester *et al.*,[30] an algorithm called "tiptop" was developed to find the

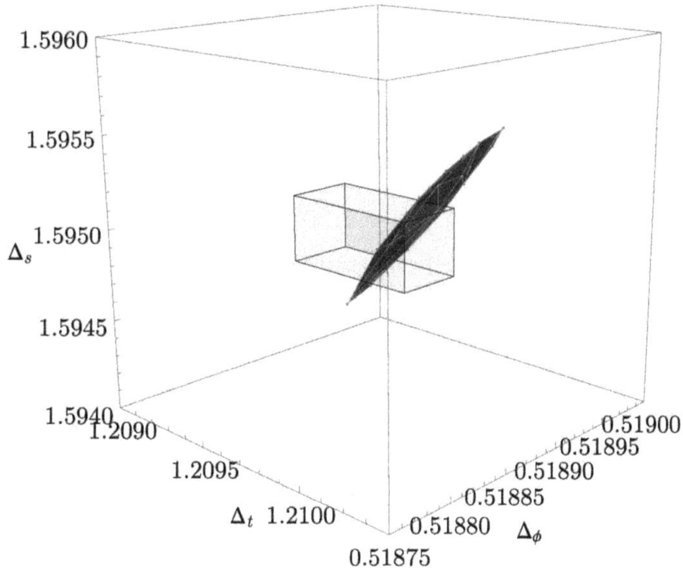

Fig. 6. Conformal bootstrap determinations of Δ_ϕ, Δ_s, and Δ_t (black) in the 3d $O(3)$ model compared with results from Monte Carlo simulations (green). Plot reprinted from Ref. 30.

maximum value of Δ_{t_4} over the allowed region of Fig. 6. The conclusion was a rigorous upper bound

$$\Delta_{t_4} < 2.99056, \tag{24}$$

conclusively establishing the instability of 3D Heisenberg magnets. It is still an open (but low-hanging) problem to compute a lower bound on Δ_{t_4} using the bootstrap.

3.3. *Other CFTs*

The Ising and O(N) models described above are just the tip of the iceberg in a vast ocean of CFTs. There are many other known CFTs connected to critical phenomena in 3D, which include theories of interacting fermions (Gross–Neveu–Yukawa fixed points), theories of gauge fields coupled to fermions (3d QED or QCD) or scalars (CP^n models, connected to deconfined critical points). These theories all constitute important targets for the bootstrap, not to mention their supersymmetric extensions and the many fascinating CFTs in 2D, 4D, and even higher dimensions. A more comprehensive discussion of targets for the numerical bootstrap program can be found in Refs. 5, 35. Ultimately, we hope to fully classify all CFTs with a small number of degrees of freedom.

Here, let us highlight that bootstrap techniques have recently produced small islands in parameter space for the O(N)-invariant Gross–Neveu–Yukawa models containing a scalar coupled to N Majorana fermions.[36] Particularly strong constraints have been placed on the $N = 1$ model which possesses 3D $\mathcal{N} = 1$ supersymmetry.[37]

However, there are many extensions of these models, including chiral GNY models (with $O(N/2)_L \times O(N/2)_R$ symmetry) and variants with multiple scalars such as the XY and Heisenberg GNY models. We believe that these theories are good candidates to be numerically "solved" by the bootstrap in the near future.

More generally, some theories (such as models of interacting scalars and fermions) seem to be relatively easy to isolate using the bootstrap, while others (such as conformal gauge theories) seem to be much more difficult to isolate. For example, some recent attempts targeting 3D QED can be found in Ref. 38–40, where it was found that one must make some very aggressive (and not fully justified) assumptions in order to obtain an island in the space of CFT data. Perhaps, this is simply because we have yet to consider the right system of correlators for these theories.

In any case, it is an interesting task to understand precisely why some theories are easier to bootstrap than others and how this can be overcome. At some level, it is related to how the dynamics of the theory are encoded in its spectrum — in the Ising CFT, the perturbative equation of motion imposes a large gap in the $\mathbb{Z}_2$ odd spectrum (due to the equation of motion $\partial^2\sigma \sim \sigma^3$) and it is easy to probe this with the bootstrap. In conformal gauge theories, the underlying dynamics are reflected in a more intricate pattern in the spectrum of gauge-invariant operators, and the optimal way to isolate their solutions is less clear; see Ref. 41 for a recent discussion. With that said, there is hope that scaling up the bootstrap to larger systems of mixed correlators, particularly those involving spinning operators such as currents and the stress tensor, will help get around these issues.

3.4. *Scaling up the bootstrap*

Where is the bootstrap heading in the coming years? At least on the numerical side, the field is still young and we expect that algorithmic improvements could still yield orders of magnitude improvements in computation time. These could come in many forms: better bases for the functionals, hybrid algorithms incorporating partial analytical solutions,[42] better search strategies through high-dimensional spaces, more efficient starting points for solutions,[43,44] and more.

As the bootstrap is scaled up, there are interesting choices to be made about the best way to add new constraints. In the current framework, one typically considers a finite set of four-point functions (say, all four-point functions containing the operators σ and ϵ) and then Taylor expands their crossing relations up to some derivative order Λ. The resulting bounds get stronger as one increases Λ, but they also get stronger as one adds additional four-point functions. What is the best strategy?

In practice, we have computed many bounds where increasing Λ is initially very effective, but eventually one sees diminishing returns. On the other hand, adding additional external operators causes significant additional complexity — it increases the number of parameters that must be searched through, it adds many new 4-point functions that must be included in order to maintain positivity, it increases the sizes of matrices (leading to larger memory requirements), and so on. Ultimately, with fixed computational power, one cannot go to as large a value of Λ if one adds new external operators. However, in some examples (such as the O(N) models described above), this tradeoff has been worth it and has led to orders of magnitude improvement. So, it seems that the future is likely in going to larger and larger systems of correlators.

The big challenge with this future is that the constraints are highly non-linear in the external data. As one adds more and more non-linear parameters, one must think hard about the best way to search through this space. One recent idea, called the "navigator" method, involves the introduction of a function which is positive outside the allowed region and negative inside of it.[45] By performing gradient descent on this function, one can find an allowed island in a high-dimensional search space, even starting from a random disallowed point. The navigator can also be used to efficiently extremize a given scaling dimension or OPE coefficient across a multi-dimensional island. This strategy seems to offer significant improvements for high-dimensional search spaces, particularly when "hotstarting" by initializing each optimization run using the saved output from a previous one. On the other hand, this approach comes with dangers such as becoming trapped in local minima or the gradient descent behaving poorly if the navigator function is not sufficiently smooth. "Navigating" these issues will be an important topic for future work.

One can gain some further intuition by connecting to the analytical bootstrap, as reviewed recently in, e.g., the work of Hartman *et al.*[46] It has been established that all CFTs are expected to have trajectories of operators called multi-twist operators whose existence can be established analytically at large spin. Schematically, these are composite operators like $\sigma\partial^{\mu_1}\ldots\partial^{\mu_\ell}\partial^{2n}\sigma$ built out of two or more fundamental fields and derivatives. The leading double-twist trajectories can be seen in the extremal functional solutions of the numerical bootstrap (see, e.g., Ref. 47), while higher-twist trajectories are typically not seen. An example of these are the operators built out of two stress tensors like $T^{\mu\nu}\partial^{\mu_1}\partial^{\mu_\ell}T^{\rho\sigma}$, which are properly considered as quadruple-twist operators since $T^{\mu\nu} \sim \sigma\partial^{\mu}\partial^{\nu}\sigma$. We expect that going to systems of correlators containing both σ and $T^{\mu\nu}$, we will more effectively probe these operators, allowing one to increase numerical precision of the solution.

An alternate way to probe multi-twist operators is to consider higher-point functions, such as $\langle\sigma\sigma\epsilon\sigma\sigma\rangle$ and $\langle\sigma\sigma\sigma\sigma\sigma\sigma\rangle$. In principle, these contain the same information as four-point functions of arbitrary spinning operators, but it is possible that higher-point functions give a more efficient way to probe the multi-twist spectrum. There has been recent progress on the use of these objects in the analytic bootstrap,[48–50] as well as preliminary results from the numerical bootstrap.[51] We expect that the future of the numerical bootstrap will also involve exciting new studies of the constraints that can be extracted from higher-point functions.

Acknowledgements

DP is supported by the US DOE grant DE-SC00-17660 and Simons Foundation grant 488651 (Simons Collaboration on the Nonperturbative Bootstrap). DSD is supported by the Simons Foundation grant 488657 (Simons Collaboration on the Non-perturbative Bootstrap) and a DOE Early Career Award under grant no. DE-SC0019085.

References

1. M. E. Fisher, *The Nature of Critical Points.* (University of Colorado Press, 1965).
2. K. G. Wilson and M. E. Fisher, Critical exponents in 3.99 dimensions. *Phys. Rev. Lett.* **28**, 240–243 (1972). Doi:10.1103/PhysRevLett.28.240.
3. Polyakov, A. M. Conformal symmetry of critical fluctuations. *JETP Lett.* **12**, 381–383 (1970).
4. R. Rattazzi, V. S. Rychkov, E. Tonni and A. Vichi, Bounding scalar operator dimensions in 4D CFT. *JHEP* **12**, 031 (2008). Doi:10.1088/1126-6708/2008/12/031, arXiv:0807.0004 [hep-th].
5. D. Poland, S. Rychkov and A. Vichi, The conformal bootstrap: Theory, numerical techniques and applications. *Rev. Mod. Phys.* **91**, 015002 (2019). Doi:10.1103/RevModPhys.91.015002, arXiv:1805.04405 [hep-th].
6. S. El-Showk, M. F. Paulos, D. Poland, S. Rychkov, D. Simmons-Duffin and A. Vichi, Solving the 3D Ising model with the conformal bootstrap. *Phys. Rev. D* **86**, 025022 (2012). Doi:10.1103/PhysRevD.86.025022, arXiv:1203.6064 [hep-th].
7. S. El-Showk, M. F. Paulos, D. Poland, S. Rychkov, D. Simmons-Duffin and A. Vichi, Solving the 3d Ising model with the conformal bootstrap II. c-Minimization and precise critical exponents. *J. Stat. Phys.* **157**, 869 (2014b). Doi:10.1007/s10955-014-1042-7, arXiv:1403.4545 [hep-th].
8. S. Rychkov and Z. M. Tan, The ϵ-expansion from conformal field theory. *J. Phys. A* **48**(29), 29FT01 (2015). Doi:10.1088/1751-8113/48/29/29FT01, arXiv:1505.00963 [hep-th].
9. L. F. Alday, Solving CFTs with weakly broken higher spin symmetry. *JHEP* **10**, 161 (2017). Doi:10.1007/JHEP10(2017)161, arXiv:1612.00696 [hep-th].
10. R. Gopakumar, A. Kaviraj, K. Sen and A. Sinha, A Mellin space approach to the conformal bootstrap. *JHEP* **05**, 027 (2017). Doi:10.1007/JHEP05(2017)027, arXiv:1611.08407 [hep-th].
11. L. F. Alday, J. Henriksson and M. van Loon, Taming the ϵ-expansion with large spin perturbation theory. *JHEP* **07**, 131 (2018). Doi:10.1007/JHEP07(2018)131, arXiv:1712.02314 [hep-th].
12. J. Henriksson, The critical O(N) CFT: Methods and conformal data. *Phys. Rept.* **1002**, 1–72 (2023). Doi:10.1016/j.physrep.2022.12.002, arXiv:2201.09520 [hep-th].
13. M. V. Kompaniets and E. Panzer, Minimally subtracted six loop renormalization of $O(n)$-symmetric ϕ^4 theory and critical exponents. *Phys. Rev. D* **96**(3), 036016 (2017). Doi:10.1103/PhysRevD.96.036016, arXiv:1705.06483 [hep-th].
14. C. Bonanno, A. Cappelli, M. Kompaniets, S. Okuda and K. J. Wiese, Benchmarking the Ising Universality Class in $3 \leq d < 4$ dimensions (2022). arXiv:2210.03051 [hep-th].
15. S. El-Showk, M. Paulos, D. Poland, S. Rychkov, D. Simmons-Duffin and A. Vichi, Conformal field theories in fractional dimensions. *Phys. Rev. Lett.* **112**, 141601 (2014a). Doi:10.1103/PhysRevLett.112.141601, arXiv:1309.5089 [hep-th].

16. J. Henriksson, S. R. Kousvos and M. Reehorst, Spectrum continuity and level repulsion: the Ising CFT from infinitesimal to finite ε. *JHEP* **02**, 218 (2023). Doi:10.1007/JHEP02(2023)218, arXiv:2207.10118 [hep-th].
17. M. Hogervorst, S. Rychkov and B. C. van Rees, Unitarity violation at the Wilson-Fisher fixed point in 4-ϵ dimensions. *Phys. Rev. D* **93**(12), 125025 (2016). Doi:10.1103/PhysRevD.93.125025, arXiv:1512.00013 [hep-th].
18. F. Kos, D. Poland and D. Simmons-Duffin, Bootstrapping mixed correlators in the 3D Ising model. *JHEP* **11**, 109 (2014). Doi:10.1007/JHEP11(2014)109, arXiv:1406.4858 [hep-th].
19. F. Kos, D. Poland, D. Simmons-Duffin and A. Vichi, Precision islands in the Ising and $O(N)$ models. *JHEP* **08**, 036 (2016). Doi:10.1007/JHEP08(2016)036, arXiv:1603.04436 [hep-th].
20. M. Hasenbusch, Finite size scaling study of lattice models in the three-dimensional ising universality class. *Physical Review B* **82**, 17 (2010). Doi:10.1103/physrevb.82.174433, https://doi.org/10.1103%2Fphysrevb.82.174433.
21. M. Hasenbusch, Restoring isotropy in a three-dimensional lattice model: The Ising universality class. *Phys. Rev. B* **104**(1), 014426 (2021). Doi:10.1103/PhysRevB.104.014426, arXiv:2105.09781 [cond-mat.stat-mech].
22. D. Poland and D. Simmons-Duffin, Bounds on 4D conformal and superconformal field theories. *JHEP* **05**, 017 (2011). Doi:10.1007/JHEP05(2011)017, arXiv:1009.2087 [hep-th].
23. S. El-Showk and M. F. Paulos, Bootstrapping conformal field theories with the extremal functional method. *Phys. Rev. Lett.* **111**(24), 241601 (2013). Doi:10.1103/PhysRevLett.111.241601, arXiv:1211.2810 [hep-th].
24. J. A. Lipa, J. A. Nissen, D. A. Stricker, D. R. Swanson and T. C. P. Chui, Specific heat of liquid helium in zero gravity very near the lambda point. *Phys. Rev. B* **68**, 174518 (2003). Doi:10.1103/PhysRevB.68.174518, https://link.aps.org/doi/10.1103/PhysRevB.68.174518.
25. M. Hasenbusch, Monte Carlo study of an improved clock model in three dimensions. *Phys. Rev. B* **100**(22), 224517 (2019). Doi:10.1103/PhysRevB.100.224517, arXiv:1910.05916 [cond-mat.stat-mech].
26. M. Hasenbusch and E. Vicari, Anisotropic perturbations in three-dimensional o(n)-symmetric vector models. *Phys. Rev. B* **84**, 125136 (2011). Doi:10.1103/PhysRevB.84.125136, https://link.aps.org/doi/10.1103/PhysRevB.84.125136.
27. S. M. Chester, W. Landry, J. Liu, D. Poland, D. Simmons-Duffin, N. Su and A. Vichi, Carving out OPE space and precise $O(2)$ model critical exponents. *JHEP* **06**, 142 (2020). Doi:10.1007/JHEP06(2020)142, arXiv:1912.03324 [hep-th].
28. D. Simmons-Duffin, A semidefinite program solver for the conformal bootstrap. *JHEP* **06**, 174 (2015). Doi:10.1007/JHEP06(2015)174, arXiv:1502.02033 [hep-th].
29. W. Landry and D. Simmons-Duffin, Scaling the semidefinite program solver SDPB. (2019). arXiv:1909.09745 [hep-th].
30. S. M. Chester, W. Landry, J. Liu, D. Poland, D. Simmons-Duffin, N. Su and A. Vichi, Bootstrapping Heisenberg magnets and their cubic instability. *Phys. Rev. D* **104**(10), 105013 (2021). Doi:10.1103/PhysRevD.104.105013, arXiv:2011.14647 [hep-th].
31. M. Hasenbusch, Monte Carlo study of a generalized icosahedral model on the simple cubic lattice. *Phys. Rev. B* **102**(2), 024406 (2020). Doi:10.1103/PhysRevB.102.024406, arXiv:2005.04448 [cond-mat.stat-mech].

32. H. Kleinert and V. Schulte-Frohlinde, Exact five-loop renormalization group functions of θ^4-theory with o(n)-symmetric and cubic interactions. critical exponents up to ϵ^5. *Phys. Lett. B* **342**(1), 284–296 (1995). Doi:https://doi.org/10.1016/0370-2693(94)01377-O, https://www.sciencedirect.com/science/article/pii/037026939401377O.
33. K. B. Varnashev, Stability of a cubic fixed point in three-dimensions: Critical exponents for generic N. *Phys. Rev. B* **61**, 14660 (2000). Doi:10.1103/PhysRevB.61.14660, arXiv:cond-mat/9909087.
34. J. M. Carmona, A. Pelissetto and E. Vicari, The N component Ginzburg-Landau Hamiltonian with cubic anisotropy: A Six loop study. *Phys. Rev. B* **61**, 15136–15151 (2000). Doi:10.1103/PhysRevB.61.15136, arXiv:cond-mat/9912115.
35. D. Poland and D. Simmons-Duffin, Snowmass white paper: The numerical conformal bootstrap. (2022). In *2022 Snowmass Summer Study*, arXiv:2203.08117 [hep-th].
36. R. S. Erramilli, L. V. Iliesiu, P. Kravchuk, A. Liu, D. Poland and D. Simmons-Duffin, The Gross-Neveu-Yukawa archipelago. *JHEP* **02**, 036 (2023). Doi:10.1007/JHEP02(2023)036, arXiv:2210.02492 [hep-th].
37. A. Atanasov, A. Hillman, D. Poland, J. Rong and N. Su, Precision bootstrap for the $\mathcal{N} = 1$ super-Ising model. *JHEP* **08**, 136 (2022). Doi:10.1007/JHEP08(2022)136, arXiv:2201.02206 [hep-th].
38. He, Y.-C., J. Rong and N. Su, Conformal bootstrap bounds for the $U(1)$ Dirac spin liquid and $N = 7$ Stiefel liquid. *SciPost Phys.* **13**(2), 014 (2022). Doi:10.21468/SciPostPhys.13.2.014, arXiv:2107.14637 [cond-mat.str-el].
39. S. Albayrak, R. S. Erramilli, Z. Li, D. Poland and Y. Xin, Bootstrapping $N_f = 4$ conformal QED_3. *Phys. Rev. D* **105**(8), 085008 (2022), Doi:10.1103/PhysRevD.105.085008, arXiv:2112.02106 [hep-th].
40. Y.-C. He, J. Rong, N. Su and A. Vichi, Non-Abelian currents bootstrap. (2023). arXiv:2302.11585 [hep-th].
41. Y.-C. He, J. Rong and N. Su, A roadmap for bootstrapping critical gauge theories: Decoupling operators of conformal field theories in $d > 2$ dimensions. *SciPost Phys.* **11**, 111 (2021). Doi:10.21468/SciPostPhys.11.6.111, arXiv:2101.07262 [hep-th].
42. N. Su, The hybrid bootstrap. (2022). arXiv:2202.07607 [hep-th].
43. M. Go and Y. Tachikawa, Autoboot: A generator of bootstrap equations with global symmetry. *JHEP* **06**, 084 (2019). Doi:10.1007/JHEP06(2019)084, arXiv:1903.10522 [hep-th].
44. N. Afkhami-Jeddi, T. Hartman and A. Tajdini, Fast conformal bootstrap and constraints on 3d gravity. *JHEP* **05**, 087 (2019). Doi:10.1007/JHEP05(2019)087, arXiv:1903. 06272 [hep-th].
45. M. Reehorst, S. Rychkov, D. Simmons-Duffin, B. Sirois, N. Su and B. van Rees, Navigator function for the conformal bootstrap. *SciPost Phys.* **11**, 072 (2021). Doi:10.21468/SciPostPhys.11.3.072, arXiv:2104.09518 [hep-th].
46. T. Hartman, D. Mazac, D. Simmons-Duffin and A. Zhiboedov, Snowmass white paper: The analytic conformal bootstrap. In *Snowmass 2021*, (2022). arXiv:2202. 11012 [hep-th].
47. D. Simmons-Duffin, The lightcone bootstrap and the spectrum of the 3d Ising CFT. *JHEP* **03**, 086 (2017). Doi:10.1007/JHEP03(2017)086, arXiv:1612.08471 [hep-th].
48. C. Bercini, V. Gonçalves and P. Vieira, Light-cone bootstrap of higher point functions and Wilson loop duality. *Phys. Rev. Lett.* **126**(12), 121603 (2021). Doi:10.1103/PhysRevLett.126.121603, arXiv:2008.10407 [hep-th].

49. A. Kaviraj, J. A. Mann, L. Quintavalle and V. Schomerus, Multipoint lightcone bootstrap from differential equations. (2022). arXiv:2212.10578 [hep-th].
50. M. S. Costa, V. Goncalves, A. Salgarkar and J. Vilas Boas, Conformal multi-Regge theory. (2023). arXiv:2305.10394 [hep-th].
51. D. Poland, V. Prilepina and P. Tadić, The five-point bootstrap. (2023). arXiv:2305.08914 [hep-th].

https://doi.org/10.1142/9789811282386_0037

Chapter 37

Quantum Statistical Mechanics of the Sachdev–Ye–Kitaev Model and Charged Black Holes

Subir Sachdev

Department of Physics, Harvard University, Cambridge MA 02138, USA

The first example of a quantum many body system not expected to have any quasiparticle excitations was the Wilson-Fisher conformal field theory. The absence of quasiparticles can be established in the compressible, metallic state of the Sachdev-Ye-Kitaev model of fermions with random interactions. The solvability of the latter model has enabled numerous computations of the non-quasiparticle dynamics of chaotic many-body states, such as those expected to describe quantum black holes. This chapter reviews thermodynamic properties of the SYK model, and describes how they have led to an understanding of the universal structure of the low energy density of states of charged black holes without low energy supersymmetry.

1. Introduction

The early work of Michael Fisher on critical points in fluid and magnetic systems[1] pioneered a line of thought which ultimately transformed the fields of condensed matter physics and high energy particle physics. I recall the strong impression his articles made on me as a young student in the early 80's. The idea of 'universality' seemed almost magical: why should an Ising model of magnetic spins on a lattice have the same critical exponents as the liquid-gas critical point of carbon dioxide? The combination of mathematical elegance and experimental relevance was unbeatable, and convinced me to become a condensed matter theorist. I was fortunate to learn more about these topics from my Ph.D. advisor David Nelson, one of Michael Fisher's Ph.D. advisees, and so become part of Michael Fisher's remarkable academic tree (see Fig. 1).

The ideas of universality and universal scaling functions led to the development of the renormalization group, and of renormalization group fixed points, by Fisher, Kadanoff, Wilson, and others. These developments greatly broadened the impact of Michael Fisher's ideas. The Wilson-Fisher fixed point[2] provided the first example of a conformal field theory which was strongly interacting. Earlier examples, such

Fig. 1. Michael Fisher on the occasion of the Lee Historical Lecture at Harvard in 2017. Shown here with 5 generations in the Fisher academic tree: Michael Fisher, David R. Nelson, the author, T. Senthil, and Chong Wang.

as the Bethe ansatz for the antiferromagnetic Heisenberg spin chain[3] and Onsager's solution of the two-dimensional Ising model on the square lattice,[4] provided fixed points which ultimately permitted a free field description, although the relationship between the free fields and the microscopic degrees of freedom was quite non-trivial.

In a more recent language, the Wilson-Fisher fixed point provides the first example of a many-body quantum system which is not expected to have any quasiparticle excitations *i.e.* the two-spatial-dimensional Ising model in a critical transverse field is a quantum system which has no particle-like description of its excitations, even asymptotically. One indication of the absence of quasiparticles is that the retarded dynamic susceptibility of the critical Ising model at momentum k and frequency ω obeys[5]

$$\chi(k,\omega) \sim \frac{1}{[c^2k^2 - (\omega + i0^+)^2]^{1-\eta/2}}, \qquad (1)$$

with η the anomalous dimension exponent introduced by Fisher and Burford.[6] A non-zero η implies that the spectral density of excitations has a branch cut for $|\omega| > ck$, and there is no pole associated with a quasiparticle. This point of view is developed in my first book,[7] and should be contrasted from that invariably taken in quantum field theory monographs, where the asymptotic scattering states are reduced to free particles via a LSZ reduction. I should note that the non-quasiparticle property of the Wilson-Fisher fixed point has not been proven, although it is highly likely: it is not ruled out that some exotic and non-local quasi-particle basis may be discovered in the future, leading to poles in the associated non-local operator.

Michael Fisher expressed the strongly interacting property by the concept of 'hyperscaling':[8] for a quantum critical system in d spatial dimensions with dynamic critical exponent z hyperscaling implies that the entropy density $\mathcal{S}$ has the following dependence on temperature

$$\mathcal{S}(T \to 0) = a_1 T^{d/z} \tag{2}$$

for some constant a_1. For a system with relativistic invariance with a velocity c at low energies, we have $z = 1$, and we obtain a length scale $\hbar c/(k_B T)$, and then the dimensional constant a_2 in

$$\mathcal{S}(T \to 0) = a_2 k_B (k_B T/(\hbar c))^d \tag{3}$$

becomes a universal number,[8] dependent only upon the universality class of the critical theory. The number a_2 is closely connected to the 'central charge' of conformal field theories.

Another consequence of hyperscaling, and the associated absence of quasiparticles, is that the dynamics of such systems is 'Planckian', an idea finding much resonance in studies of strongly correlated electrons.[9] The relaxational and dissipational dynamics of the quantum-critical fluid is controlled only by the Planckian time $\hbar/k_B T$,[5,10] and is entirely independent of the energy scale of the interactions between the microscopic degrees of freedom. This appears remarkable when viewed from the lens of the Landau theory of quasiparticles, where the relaxational dynamics is determined by collisions between quasiparticles, and the collision rate is proportional to the square of the interaction strength via Fermi's Golden Rule.

A fresh perspective on the dynamics of systems without quasiparticles has recently appeared in form of out-of-time-order correlators (OTOCs).[11] These relate the absence of quasiparticles to the rapid growth of quantum many-body chaos. The Wilson-Fisher fixed point is also expected to become chaotic in a time of order $\hbar/(k_B T)$,[12] and such a time is the shortest possible,[7] as has been proven for a class of models with many local degrees of freedom.[11]

This chapter will describe perhaps the simplest system without quasiparticle excitations: the Sachdev-Ye-Kitaev (SYK) model of fermions with random interactions. The simplicity has allowed proof of the absence of quasiparticles, and the computation of numerous properties of non-quasiparticle dynamics that are intractable for the Wilson-Fisher theory. Moreover, the SYK model provides a realization of the low energy dynamics of black holes with a near-horizon AdS_2 geometry,[13] and this has led to new insights on the nature of the Bekenstein-Hawking entropy of black holes which resolve long-standing open questions.[14,15] The SYK model has also led to significant advances in the theory of strange metals in correlated electron compounds.[16,17] Only the black hole connection is described in this chapter, and the connection to strange metals is described in a companion article.[18] The presentation is designed for condensed matter theorists, in alignment with Michael Fisher's vision of crossing inter-disciplinary boundaries.

The SYK model has no spatial structure, and should be considered as a critical quantum system in spatial dimension $d = 0$. Naively extending the results (2,3) to $d = 0$, we conclude that the $T = 0$ entropy of the SYK model should be a universal constant. This naive expectation is essentially correct, but we do have to carefully consider the orders of limits of $N \to \infty$ (where N is the number of fermion flavors) and $T \to 0$, and the influence of irrelevant operators. For a SYK model with fermion density $\mathcal{Q}$ ($0 \leq \mathcal{Q} \leq 1$), we have for the total entropy S in the limit $N \to \infty$ followed by $T \to 0$

$$\frac{S(T,\mathcal{Q})}{k_B} = N(s_0 + \gamma\, k_B T) - \frac{3}{2}\ln\left(\frac{U}{k_B T}\right) - \frac{\ln N}{2} + \mathcal{O}(1/N). \tag{4}$$

The leading result in (4) is the Ns_0 term,[19] and this is the analog of the scaling result in (3) in $d = 0$. Here s_0 is a universal function of the charge $\mathcal{Q}$; for half-filling with $\mathcal{Q} = 1/2$, we have

$$s_0 = \frac{\mathcal{G}}{\pi} + \frac{\ln(2)}{4} = 0.4648476991708051\ldots, \tag{5}$$

where $\mathcal{G}$ is Catalan's constant. So this term indicates a non-zero entropy density in the limit $T \to 0$. However, this term does not imply an exponentially large ground state degeneracy; determining the ground state degeneracy requires consideration of the limit $T \to 0$ followed by $N \to \infty$, and we will describe the structure of this limit in Section 2.

The first correction to Ns_0 in (4) is the term proportional to $\gamma \sim 1/U$, where U is the root-mean-square strength of the interactions. This correction is due to the leading irrelevant operator, associated with the 'Schwarzian', which will play a crucial role in the analysis in Section 2.[20–24]

Finite N fluctuation corrections have also been computed, and these require an exact path integral over the effective action of the Schwarzian operator. Such a path integral leads to the $(3/2)\ln(T)$ correction in (4). The key property of this correction is that it diverges as $T \to 0$, indicating a breakdown of the $1/N$ expansion in this limit. Such a breakdown is a characteristic feature of '*dangerously irrelevant*' operators, a concept introduced by Michael Fisher.[25] By comparing the γ term with the $(3/2)\ln(T)$, we see that the latter becomes larger at a temperature scale $T \sim e^{-N}$, and so the Schwarzian operator is dangerously irrelevant below this exponentially small T. Section 2 will discuss the $\ln N$ term[23,26] in (4), and the nature of the spectrum of the SYK model at low energy scales.

Let us now turn to the quantum theory of charged black holes. There are a variety of reasons for suspecting that there is a connection between this seemingly unrelated system and the SYK model:[13]

- It has long been known that black holes exhibit 'Planckian' dynamics;[27] more specifically, the ring-down time τ_r of quasi-normal modes of black holes, when expressed in terms of the black hole Hawking temperature T,[28] is of the order of $\hbar/(k_B T)$.

- The AdS/CFT correspondence of string theory[29] describes a duality between conformal field theories in $d+1$ spacetime dimensions and quantum gravity in AdS_{d+2} space for $d \geq 1$. For the $d=0$ case, it was pointed out[13] that conformal structure of correlators in SYK model was identical to correlators in the AdS_2 horizon of charged black holes.[30]
- Charged black holes have a non-zero entropy as $T \to 0$ at fixed black hole charge Q. This matches the behavior of the large N limit of the SYK model in (4). Hawking's famous result for black hole entropy is[28]

$$S = \frac{\mathcal{A}(T)c^3}{4\hbar G}, \tag{6}$$

where $\mathcal{A}(T)$ is the area of the horizon at a temperature T. For the case of a black hole with total charge Q in a $3+1$ dimensional spacetime which is asymptotically Minkowski, $\mathcal{A}_0 \equiv \mathcal{A}(T \to 0) = 2GQ^2/c^4$ is non-zero, and Hawking's result for the low T entropy is[30–32]

$$\frac{S(T,Q)}{k_B} = \frac{c^3}{4\hbar G}\left(\mathcal{A}_0 + 2\sqrt{\pi}\mathcal{A}_0^{3/2}\frac{k_B T}{\hbar c} + \mathcal{O}(T^2)\right). \tag{7}$$

Note the perfect match with the T dependence of the first 2 terms in (4). As we will review in Section 3, this match extends also to finite N corrections, which leads to a common many-body density of states. Here, we view (7) as obtained in a small G expansion, corresponding to the large N expansion in (4).

Hawking's result for black hole entropy[28] was revolutionary, and raised many questions. Foremost among them was that it did not identify any quantum degrees of freedom whose many-body density of states, $D(E)$, was the exponential of the entropy. The first identification[33] of possible degrees of freedom appeared using string theory for charged black holes which had supersymmetry at the lowest energies. This theory had an exponentially large degeneracy of stringy states in the zero energy ground state, and it was established that

$$D(E) = e^{S(T\to 0,Q)/k_B}\,\delta(E) + \cdots \tag{8}$$

for supersymmetric black holes, but the form of $D(E)$ for non-zero E was not described.

Here we will review work building on the SYK model which has obtained the form of $D(E)$ for a generic charged black hole without supersymmetry; see Fig. 2, which summarizes the main results. The $D(E)$ of such black holes does *not* have an exponentially large degeneracy of the ground state. Indeed, there are no large degeneracies (apart from order unity degeneracies required by symmetries), but the energy levels have an exponentially small spacing. Supersymmetric generalizations of the SYK model have also been studied,[35,38] and these do have the a $D(E)$ of the

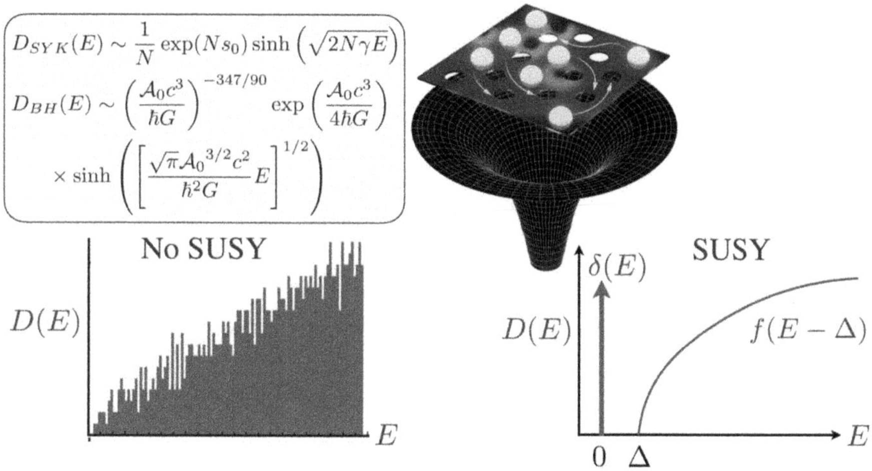

Fig. 2. Comparison of the many-body densities of states of SYK models ($D_{SYK}(E)$) and charged black holes ($D_{BH}(E)$) with and without supersymmetry (SUSY). $D_{BH}(E)$ is for a charged black holes in $3+1$ dimensional asymptotically Minkowski space without SUSY,[34] where $\mathcal{A}_0$ is the area of the horizon at $T = 0$, see (66); $D_{SYK}(E)$ is from (32).[26] The left plot is for Majorana SYK, as in Fig. 5. Both black holes and SYK models with sufficient low energy SUSY have an energy gap Δ, above a delta function with a co-efficient as in (8).[24,35–37] Black holes and SYK models without SUSY do not have delta function, nor a gap, but an exponentially dense spacing of levels down to $E = 0$; their $D(E)$ vanishes as $E \to 0$, but with an exponentially large pre-factor, and the latter is sufficient to yield a zero temperature entropy.

form in (8), confirming that the exponentially large degeneracy is a special feature of low energy supersymmetry. This connection to supersymmetric SYK models has also allowed determination of the non-zero E contributions to the $D(E)$ of supersymmetric black holes in (8), and such computations show that there is a gap in the many-body density of states above the zero energy delta function.[24,36,37]

A generic thermodynamic system with an extensive entropy has an energy level spacing which is exponentially small in its size (N) at an extensive energy above the ground state. Here, we are interested in the behavior of non-supersymmetric systems as the energy density becomes sub-extensive. For a conventional quantum system, the low energy excitations are quasiparticles, and hence the lowest energy level spacing is of order $1/N^a$ (for some positive constant a), implying $D(E) \sim N^a$ at small E. In this perspective, the remarkable feature of the SYK model, and of charged black holes, is that the energy level spacing remains of order e^{-N} down to very low energies, and this is linked to the non-zero $T \to 0$ limits of the first two terms in (4) and (8). Indeed, this exponentially small energy level spacing establishes the absence of quasiparticles (although the converse need not be true): the number of quasiparticle states is $\sim N$, and it is not possible to combine them to obtain an exponentially large number of states with a non-extensive energy. The wavefunctions of the low energy states change completely between successive states which happen to have energies exponentially close to each other, a signal of their chaotic nature. In contrast, in quasiparticle systems, successive states are very similar to each other, and differ only by the motion of a few quasiparticles.

In Section 2, we will further describe the finite N corrections in (4), and determine their consequences for $D(E)$ at the lowest energies. Remarkably, similar corrections apply also to the black hole result in (7), and lead to similar results for the $D(E)$ of generic charged black holes, which will be presented in Section 3. Section 4 will present brief discussion of the $D(E)$ of the Wilson-Fisher theory placed on a sphere, and discuss its relationship to the SYK model and black holes.

2. The SYK Model

The Hamiltonian of a version of a SYK model is illustrated in Fig. 3. We take a system with fermions c_i, $i = 1 \ldots N$ states. Depending upon physical realizations, the label i could be position or an orbital, and it is best to just think of it as an abstract label of a fermionic qubit with the two states $|0\rangle$ and $c_i^\dagger|0\rangle$. We now place $\mathcal{Q}N$ fermions in these states, so that a density $\mathcal{Q} \approx 1/2$ is occupied, as shown in Fig. 3. The quantum dynamics is restricted to *only* have a 'collision' term between the fermions, analogous to the right-hand-side of the Boltzmann equation. However, in stark contrast to the Boltzmann equation, we will not make the assumption of statistically independent collisions, and will account for quantum interference between successive collisions: this is the key to building up a many-body state with non-trivial entanglement. So a collision in which fermions move from sites i and j to sites k and ℓ is characterized not by a probability, but by a quantum amplitude $U_{ij;k\ell}$, which is a complex number.

The model so defined has a Hilbert space of order 2^N states, and a Hamiltonian determined by order N^4 numbers $U_{ij;k\ell}$. Determining the spectrum or dynamics of such a Hamiltonian for large N seems like an impossibly formidable task. But if we

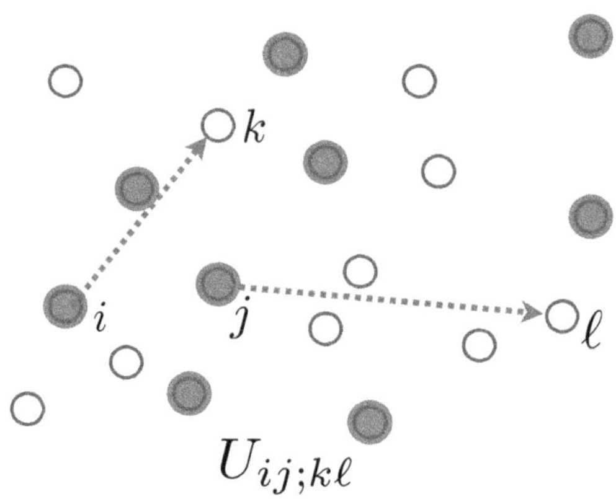

Fig. 3. The SYK model: fermions undergo the transition ('collision') shown with quantum amplitude $U_{ij;k\ell}$.

now make the assumption that the $U_{ij;k\ell}$ are statistically independent random numbers, remarkable progress is possible. Note that we are not considering an ensemble of SYK models with different $U_{ij;k\ell}$, but a single fixed set of $U_{ij;k\ell}$. Most physical properties of this model are self-averaging at large N, and so as a technical tool, we can rapidly obtain them by computations on an ensemble of random $U_{ij;k\ell}$. In any case, the analytic results we now describe have been checked by numerical computations on a computer for a fixed set of $U_{ij;k\ell}$. We recall that even for the Boltzmann equation, there was an ensemble average over the initial positions and momenta of the molecules that was implicitly performed.

Specifically, the Hamiltonian in a chemical potential μ is[39–41]

$$\mathcal{H} = \frac{1}{(2N)^{3/2}} \sum_{i,j,k,\ell=1}^{N} U_{ij;k\ell}\, c_i^\dagger c_j^\dagger c_k c_\ell - \mu \sum_i c_i^\dagger c_i \tag{9}$$

$$c_i c_j + c_j c_i = 0, \quad c_i c_j^\dagger + c_j^\dagger c_i = \delta_{ij} \tag{10}$$

$$\mathcal{Q} = \frac{1}{N} \sum_i c_i^\dagger c_i; \quad [\mathcal{H}, \mathcal{Q}] = 0; \quad 0 \le \mathcal{Q} \le 1, \tag{11}$$

and its large N limit is most simply taken graphically, order-by-order in $U_{ij;k\ell}$, and averaging over $U_{ij;k\ell}$ as independent random variables with $\overline{U_{ij;k\ell}} = 0$ and $\overline{|U_{ij;k\ell}|^2} = U^2$. In the large N limit, only the graph for the self energy in Fig. 4 survives, and the on-site fermion Green's function is given by the solution of the following equations

$$G(i\omega) = \frac{1}{i\omega + \mu - \Sigma(i\omega)}, \quad \Sigma(\tau) = -U^2 G^2(\tau) G(-\tau)$$
$$G(\tau = 0^-) = \mathcal{Q}, \tag{12}$$

where τ is imaginary time, and ω is imaginary frequency.

For general ω and T, Eqs. (12) have to be solved numerically. But an exact analytic solution is possible in the limit $\omega, T \ll U$. At $T = 0$, the asymptotic forms

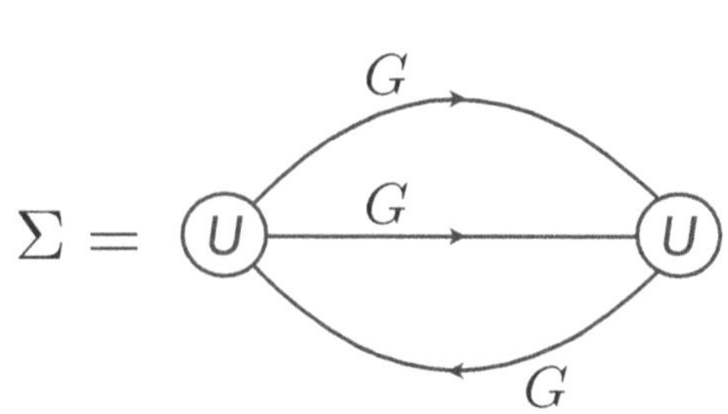

Fig. 4. Self-energy for the fermions of $\mathcal{H}$ in (9) in the limit of large N. The intermediate Green's functions are fully renormalized.

can be obtained straightforwardly[39]

$$G(i\omega) \sim -i\mathrm{sgn}(\omega)|\omega|^{-1/2}, \quad \Sigma(i\omega) - \Sigma(0) \sim -i\mathrm{sgn}(\omega)|\omega|^{1/2}, \tag{13}$$

and a more complete analysis allows one to obtain the exact form at non-zero T ($\hbar = k_B = 1$)[42]

$$G(\omega) = \frac{-iCe^{-i\theta}}{(2\pi T)^{1/2}} \frac{\Gamma\left(\frac{1}{4} - \frac{i\omega}{2\pi T} + i\mathcal{E}\right)}{\Gamma\left(\frac{3}{4} - \frac{i\omega}{2\pi T} + i\mathcal{E}\right)} \quad |\omega|, T \ll U. \tag{14}$$

Here is $\mathcal{E}$ is a dimensionless number which characterizes the particle-hole asymmetry of the spectral function; both $\mathcal{E}$ and the pre-factor C are determined by an angle $-\pi/4 < \theta < \pi/4$

$$e^{2\pi\mathcal{E}} = \frac{\sin(\pi/4 + \theta)}{\sin(\pi/4 - \theta)}, \quad C = \left(\frac{\pi}{U^2 \cos(2\theta)}\right)^{1/4}, \tag{15}$$

and the value of θ is determined by a Luttinger relation to the density $\mathcal{Q}$

$$\mathcal{Q} = \frac{1}{2} - \frac{\theta}{\pi} - \frac{\sin(2\theta)}{4}. \tag{16}$$

The striking property of (14) is its conformally invariant form, which shows that at low energies the characteristic frequency scale is determined solely by the Planckian frequency scale $k_B T/\hbar$ and is independent of the value of U, as discussed in Section 1.

2.1. *G-Σ theory*

We analyze fluctuations about the large N solution (14) by the path integral method.[22,23]

After introducing replicas $a = 1 \ldots n$, and integrating out the disorder, the Grassman path integral of the Hamiltonian $\mathcal{H}$ in (9) can be written as

$$\mathcal{Z} = \int \mathcal{D}c_{ia}(\tau) \exp\left[-\sum_{ia} \int_0^\beta d\tau\, c_{ia}^\dagger \left(\frac{\partial}{\partial\tau} - \mu\right) c_{ia} - \frac{U^2}{4N^3} \sum_{ab} \int_0^\beta d\tau d\tau' \left|\sum_i c_{ia}^\dagger(\tau) c_{ib}(\tau')\right|^4\right]. \tag{17}$$

For simplicity, we neglect the replica indices, and introduce the identity

$$1 = \int \mathcal{D}G(\tau_1,\tau_2)\mathcal{D}\Sigma(\tau_1,\tau_2) \exp\left[-N \int_0^\beta d\tau_1 d\tau_2 \Sigma(\tau_1,\tau_2)\left(G(\tau_2,\tau_1) + \frac{1}{N}\sum_i c_i(\tau_2)c_i^\dagger(\tau_1)\right)\right]. \tag{18}$$

Then the partition function can be written as a path integral with an action S analogous to a Luttinger-Ward functional

$$\begin{aligned} \mathcal{Z} &= \int \mathcal{D}G(\tau_1,\tau_2)\mathcal{D}\Sigma(\tau_1,\tau_2)\exp(-NI) \\ I &= \ln\det\left[\delta(\tau_1-\tau_2)(\partial_{\tau_1}+\mu)-\Sigma(\tau_1,\tau_2)\right] \\ &\quad + \int d\tau_1 d\tau_2 \left[\Sigma(\tau_1,\tau_2)G(\tau_2,\tau_1)+(U^2/2)G^2(\tau_2,\tau_1)G^2(\tau_1,\tau_2)\right] \end{aligned} \tag{19}$$

This is a 'G-Σ' theory,[19,22,23,41] written as a path integral over bi-local in time functions $G(\tau_1,\tau_2$ and $\Sigma(\tau_1,\tau_2)$, in contrast to the path integral over N fermionic fields in (17). The saddle point of (19) yields the equations (12) for the Green's function $G(\tau_1-\tau)$, and the self energy $\Sigma(\tau_1-\tau_2)$.

2.2. *Emergent symmetries*

To analyze fluctuations of (19) about its saddle point, we need a better understanding of its symmetries, as this will help motivate an effective action for the lowest energy fluctuations. At frequencies $\ll U$, the time derivative in the determinant is less important, and without it the path integral is invariant under the time reparameterization $(f(\sigma))$ and gauge $(\phi(\sigma))$ transformations

$$\begin{aligned} \tau &= f(\sigma) \\ G(\tau_1,\tau_2) &= [f'(\sigma_1)f'(\sigma_2)]^{-1/4}e^{-i\phi(\sigma_1)+i\phi(\sigma_2)}\,G(\sigma_1,\sigma_2) \\ \Sigma(\tau_1,\tau_2) &= [f'(\sigma_1)f'(\sigma_2)]^{-3/4}e^{-i\phi(\sigma_1)+i\phi(\sigma_2)}\,\Sigma(\sigma_1,\sigma_2) \end{aligned} \tag{20}$$

where $f(\sigma)$ and $\phi(\sigma)$ are arbitrary functions.

We also need a better understanding of the symmetries of the saddle-point solutions of (19), which we denote as $G_s(\tau_1-\tau_2)$ and $\Sigma_s(\tau_1-\tau_2)$. The solutions in (13) written in imaginary time are

$$\begin{aligned} G_s(\tau_1-\tau_2) &\sim (\tau_1-\tau_2)^{-1/2} \\ \Sigma_s(\tau_1-\tau_2) &\sim (\tau_1-\tau_2)^{-3/2}. \end{aligned} \tag{21}$$

The saddle point will be invariant under a reparamaterization $f(\tau)$ when choosing $G(\tau_1,\tau_2)=G_s(\tau_1-\tau_2)$ leads to a transformed $\widetilde{G}(\sigma_1,\sigma_2)=G_s(\sigma_1-\sigma_2)$ (and similarly for Σ). It turns out this is true for (21) only for the SL(2, R) transformations under which

$$f(\tau)=\frac{a\tau+b}{c\tau+d}, \quad ad-bc=1. \tag{22}$$

So the (approximate) reparametrization symmetry is spontaneously broken down to SL(2, R) by the saddle point.

Let us also note the extension of (22) to $T > 0$. The $T > 0$ solution in (14) is the Fourier transform of

$$G(\tau_1 - \tau_2) = -A\frac{e^{-2\pi\mathcal{E}T(\tau_1-\tau_2)}}{\sqrt{1+e^{-4\pi\mathcal{E}}}}\left(\frac{T}{\sin(\pi T(\tau_1-\tau_2))}\right)^{2\Delta}. \tag{23}$$

This is invariant under PSL(2, R) transformations which map the thermal circle onto itself, and an associated gauge transformation

$$\frac{\tan(\pi T f(\tau))}{\pi T} = \frac{a\frac{\tan(\pi T\tau)}{\pi T} + b}{c\frac{\tan(\pi T\tau)}{\pi T} + d}, \quad ad - bc = 1,$$
$$-i\phi(\tau) = -i\phi_0 + 2\pi\mathcal{E}T(\tau - f(\tau)). \tag{24}$$

Indeed, we can derive the $T > 0$ form (23) for (21) by using (24).[41]

2.3. *From the SYK model to the Schwarzian action*

We now return to the path integral in (19). We focus on the vicinity of the saddle point, G_s, Σ_s, and the low energy "Nambu-Goldstone" modes associated with breaking time reparameterization and U(1) gauge symmetries by writing[22,23]

$$G(\tau_1, \tau_2) = [f'(\tau_1)f'(\tau_2)]^{1/4} G_s(f(\tau_1) - f(\tau_2))e^{i\phi(\tau_1)-i\phi(\tau_2)} \tag{25}$$

(and similarly for Σ). Then the path integral (19) is approximated by

$$\mathcal{Z} = \int \mathcal{D}f(\tau)\mathcal{D}\phi(\tau)e^{-E_0/T+Ns_0-NI_{\text{eff}}[f,\phi]}, \tag{26}$$

where $E_0 \propto N$ is the ground state energy. Demanding that S_{eff} is invariant under the symmetries in (24), we obtain the following effective action in a gradient expansion (this can be viewed as 'non-linear sigma model' associating with the breaking of time reparameterizations to SL(2,R))

$$I_{\text{eff}}[f,\phi] = \frac{K}{2}\int_0^{1/T} d\tau(\partial_\tau\phi + i(2\pi\mathcal{E}T)\partial_\tau\epsilon)^2 - \frac{\gamma}{4\pi^2}\int_0^{1/T} d\tau\,\{\tan(\pi T(\tau+\epsilon(\tau))), \tau\}, \tag{27}$$

where $f(\tau) \equiv \tau + \epsilon(\tau)$, the couplings K, γ, and $\mathcal{E}$ can be related to thermodynamic derivatives. Specifically, the coupling γ is the same as that appearing in (4). We have used the Schwarzian

$$\{g, \tau\} \equiv \frac{g'''}{g'} - \frac{3}{2}\left(\frac{g''}{g'}\right)^2, \tag{28}$$

which is obtain from the requirement that effective action at $T = 0$ obey

$$I_{\text{eff}}\left[f(\tau) = \frac{a\tau + b}{c\tau + d}, \phi(\tau) = 0\right] = 0. \tag{29}$$

2.4. *Many-body density of states*

It now remains to evaluate the path integral in (26). Remarkably, this path integral can be evaluated exactly.[24] We will not present the details here, and refer the reader to the literature.[20–24,26] The final result is best expressed in terms of the many-body density of states, related to the partition function by a Laplace transform

$$\mathcal{Z}(T,\mathcal{Q}) = \int_{0^-}^{\infty} D(E,\mathcal{Q}) e^{-E/T}, \tag{30}$$

while the entropy is given by

$$S(T,\mathcal{Q}) = \frac{\partial}{\partial T}(T \ln \mathcal{Z}(T,\mathcal{Q}))\bigg|_{\mathcal{Q}}. \tag{31}$$

This procedure yields the main result for the SYK model (see Fig. 5)

$$D(E,\mathcal{Q}) \sim \frac{1}{N} \exp(N s_0) \sinh(\sqrt{2N\gamma E}). \tag{32}$$

It should be noted that the $1/N$ pre-factor does not follow from the low energy theory in (26), but requires an analysis of the full partition function in (19);[26] this applies also to the $\ln N$ term in (4). The remaining terms in (32) are more universal, and rely mainly on the SL(2,R) symmetry of (27). The result for the entropy in (4) now follow from (30) and (32). The thermodynamic limit of the entropy $\lim_{N\to\infty} S(T)/N$ in (4) yields the microcanonical entropy

$$S(E)/k_B = N s_0 + \sqrt{2N\gamma E}, \tag{33}$$

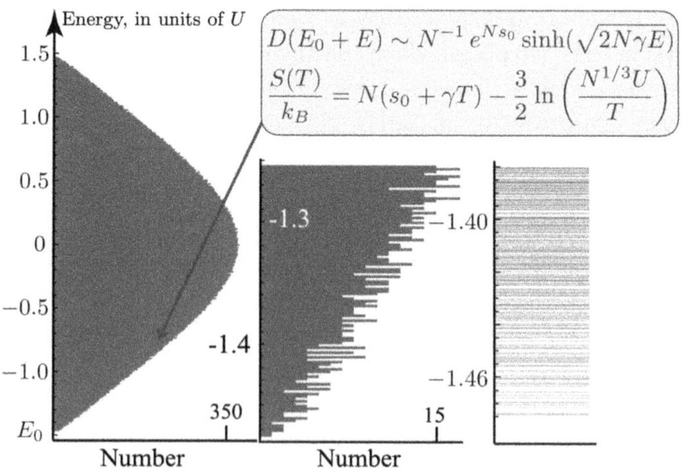

Fig. 5. Plot of the 65536 many-body eigenvalues of a $N = 32$ Majorana SYK Hamiltonian (credit: G. Tarnopolsky); however, the analytical results quoted here are for the SYK model with complex fermions which has a similar spectrum. The coarse-grained low energy and low temperature behavior is described by (32) and (4).

and this connects to the extensive E limit of (32) after using Boltzmann's formula $D(E) \sim \exp(S(E)/k_B)$. We summarize these results, and compare with numerics on the SYK model in Fig. 5 and Fig. 2.

3. Charged Black Holes

This section will begin by reviewing Hawking's computation of the entropy of a Schwarzschild black hole with no net charge[43] in Section 3.1. This is extended to charged black holes in Section 3.2, and subsequent subsections discuss reduction of the near-horizon theory of the charged black hole to the Schwarzian theory.

3.1. *Gibbons-Hawking computation of black hole entropy*

The Gibbons-Hawking method proceeds by simply evaluating the action of classical Einstein gravity in Euclidean time. Planck's constant appears only via the requirement that the temporal period is $\hbar/(k_B T)$. In Euclidean time, we will see below that the spacetime geometry along the radial and temporal directions is a 'cigar' which closes off at the horizon, as shown in Fig. 6. Consequently, the computation is performed entirely outside the black hole, and we sidestep our ignorance of the black hole interior. The black hole entropy is then seen to be a consequence of entanglement around the horizon, as shown in Fig. 7.

The Einstein action for gravity in $3+1$ dimensions is

$$I_E = \int d^4x\sqrt{g}\left[-\frac{1}{2\kappa^2}\mathcal{R}_4\right], \quad \mathcal{Z} = \int \mathcal{D}g\exp(-I_E), \tag{34}$$

where $\kappa^2 = 8\pi G$ is the gravitational constant, $\mathcal{R}_4$ is the Ricci scalar. The Schwarzschild solution of the saddle-point equations is

$$ds^2 = V(r)d\tau^2 + r^2 d\Omega_2^2 + \frac{dr^2}{V(r)} \tag{35}$$

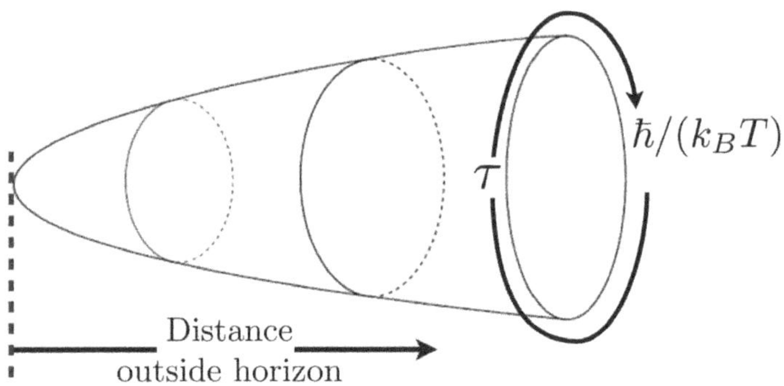

Fig. 6. Cigar geometry of a Schwarzschild black hole in Euclidean time.

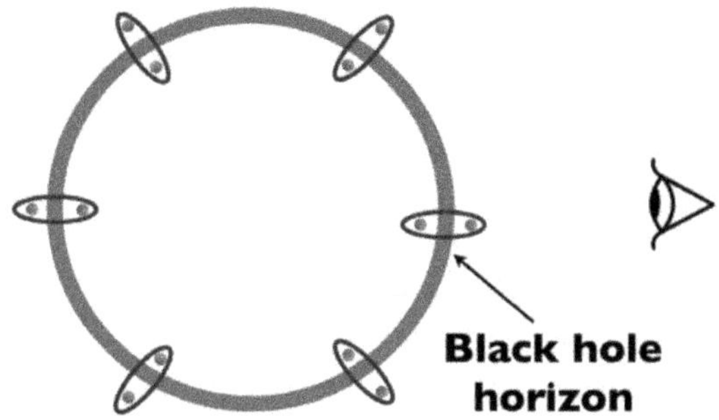

Fig. 7. Observation of entanglement across the horizon.

where $d\Omega_2^2$ is the metric of the 2-sphere, and

$$V(r) = 1 - \frac{m}{r}. \tag{36}$$

The gravitational mass of the black hole is $M = 2Gm$. The black hole horizon is at $r = r_0$ where $V(r_0) = 0$; so

$$r_0 = m. \tag{37}$$

The $T > 0$ quantum partition function is obtained in a spacetime which is periodic as a function of τ with period $\hbar/(k_B T)$, as in Fig. 6. We have to ensure that there is no singularity at the horizon r_0 where $V(r_0) = 0$. Let us change radial co-ordinates to y, where $r = r_0 + y^2$. Then for small y

$$ds^2 = \frac{4}{V'(r_0)} \left[\frac{(V'(r_0))^2}{4} y^2 d\tau^2 + dy^2 \right] + r_0^2 d\Omega_2^2 = \frac{4}{V'(r_0)} \left[y^2 d\theta^2 + dy^2 \right] + r_0^2 d\Omega_2^2 \tag{38}$$

The expression in the square brackets is the metric of the flat plane in polar co-ordinates, with radial co-ordinate y and angular co-ordinate $\theta = V'(r_0)\tau/2$. Smoothness requires periodicity in θ with period 2π, and so

$$4\pi T = V'(r_0) = \frac{1}{m}. \tag{39}$$

The free energy $\beta F = I_E$, where $\beta = 1/T$. So the entropy is (as in (31))

$$S = -\frac{\partial F}{\partial T} = \left(\beta \frac{\partial}{\partial \beta} - 1 \right) I_E \tag{40}$$

However, the metric is τ-independent, and the only explicit dependence of the action is via $I_E = \beta H$. Such an action implies $S = 0$.

Gibbons and Hawking[43] argued that a proper evaluation of the gravity action requires a boundary term, I_{GH}, so that the gravity path integral in (34) is replaced by

$$I_{\rm grav} = I_E + I_{GH}, \quad I_{GH} = \int_\partial d^3x \sqrt{g_b} \left[-\frac{1}{\kappa^2} \mathcal{K}_3 \right], \quad \mathcal{Z} = \int \mathcal{D}g \exp(-I_{\rm grav}), \tag{41}$$

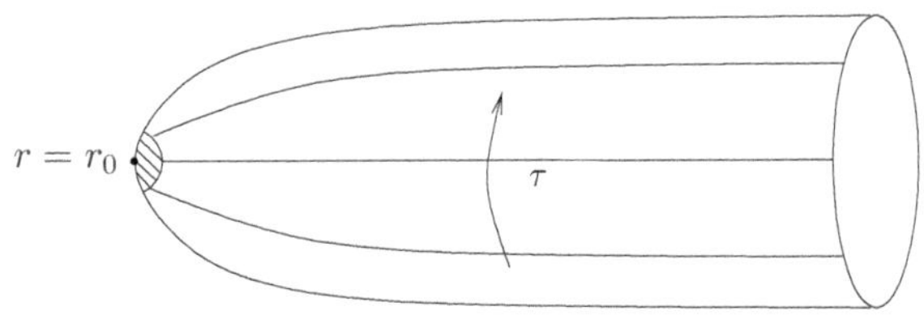

Fig. 8. The action I_{grav} is evaluated in the shaded region at the end of the cigar (figure adapted from[44]).

where $\mathcal{K}_3$ is the extrinsic scalar curvature of the 3-dimensional boundary of space-time. I_{GH} is the Gibbons-Hawking boundary term, deduced by the requirement that the Euler-Lagrange equations of I_{grav} co-incide with the Einstein equations, with no additional boundary terms. Because the bulk action does not contribute any entropy, we can move the boundary from asymptotically far region to the vicinity of horizon, as shown in Fig. 8.[44] The entire contribution to the entropy then comes from I_{GH} evaluated in the vicinity of the co-ordinate singularity at $r = r_0$. We evaluate I_{GH} by using the identity

$$\int_\partial d^3x\sqrt{g_b}\,\mathcal{K}_3 = \frac{\partial}{\partial n}\int_\partial d^3x\sqrt{g_b} \tag{42}$$

where n is the Gaussian normal co-ordinate of the boundary. Evaluating at $y = \epsilon$, we have

$$\int_\partial d^3x\sqrt{g_b} = 2\pi\epsilon\mathcal{A} \tag{43}$$

where $\mathcal{A} = 4\pi r_0^2$ is the area of the horizon. Combining everything, and noting that the contribution of I_E in the shaded region of Fig. 8 vanishes in the limit $\epsilon \to 0$, we have the famous result of Hawking

$$S = \frac{2\pi\mathcal{A}}{\kappa^2} = \frac{\mathcal{A}}{4G}. \tag{44}$$

3.2. *Hawking entropy of a charged black hole*

For a charged black hole, we need the Einstein-Maxwell theory of g and a U(1) gauge flux $F = dA$, and simply need to compute the area of the horizon in this modified geometry. The Einstein-Maxwell action is

$$I_{EM} = \int d^4x\sqrt{g}\left[-\frac{1}{2\kappa^2}\mathcal{R}_4 + \frac{1}{4g_F^2}F^2\right], \quad \mathcal{Z}_Q = \int \mathcal{D}g\mathcal{D}A\exp(-I_{EM} - I_{GH}). \tag{45}$$

The saddle-point equations now yield a solution with[30,31]

$$V(r) = 1 + \frac{\Theta^2}{r^2} - \frac{m}{r}; \quad A_\tau = i\mu\left(1 - \frac{r_0}{r}\right)$$
$$\Theta = \frac{\kappa r_0}{\sqrt{2}g_F}\mu; \quad Q = \frac{4\pi\mu r_0}{g_F^2}; \quad S = \frac{2\pi\mathcal{A}}{\kappa^2}, \tag{46}$$

where Q is the total charge, the chemical potential is μ, and as before the horizon is where $V(r_0) = 0$, the temperature $T = V'(r_0)/(4\pi)$, and $\mathcal{A} = 4\pi r_0^2$. This defines a two parameter family of charged black hole solutions of I_{EM} determined by T and Q.

Now we take the limit $T \to 0$ at fixed Q. Then we find the remarkable feature that the horizon radius remains finite

$$R_h \equiv r_0(T \to 0, Q) = \frac{Q\kappa g_F}{4\pi} \tag{47}$$

In this limit, entropy becomes

$$S(T \to 0, Q) = \frac{4\pi R_h^2}{G} + \gamma T, \quad \gamma \equiv \frac{4\pi^2 R_h^3}{G}, \tag{48}$$

corresponding to the result in (7).

For the near-horizon metric, it is useful to introduce the co-ordinate ζ

$$r = R_h + \frac{R_h^2}{\zeta} \tag{49}$$

so that the horizon at $T = 0$ is at $\zeta = \infty$. Then in the near-horizon regime $R_h \ll \zeta < \infty$ the $T = 0$ metric is

$$ds^2 = R_h^2 \frac{d\tau^2 + d\zeta^2}{\zeta^2} + R_h^2 d\Omega_2^2 \tag{50}$$

The spacetime in (50) is $AdS_2 \times S^2$, as illustrated in Fig. 9. The dominant low energy excitations involve the AdS_2 component, and so the near-horizon metric is effectively $1 + 1$ dimensional.

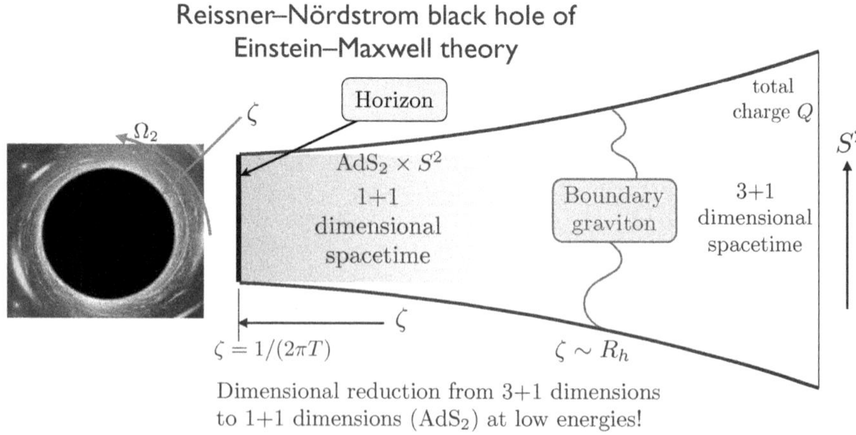

Fig. 9. Schematic of a charged (Reissner-Nördstrom) black hole.

3.3. *AdS₂ and its symmetries*

The AdS_2 metric

$$ds^2 = \frac{d\tau^2 + d\zeta^2}{\zeta^2} \tag{51}$$

is invariant under isometries which are SL(2,R) transformations. The reader can verify that the co-ordinate change

$$\tau' + i\zeta' = \frac{a(\tau + i\zeta) + b}{c(\tau + i\zeta) + d}, \quad ad - bc = 1, \tag{52}$$

with a,b,c,d real, leaves the AdS_2 metric invariant. This is a crucial fact, and establishes a connection to the SL(2,R) invariance of the SYK model in (22).

As for the SYK model, we can map the $T = 0$ metric to $T > 0$ metric. The co-ordinate transformation

$$\begin{aligned} \zeta &= \frac{1}{\cosh(2\pi T\rho) - \sinh(2\pi T\rho)\cos(2\pi T\hat{\tau})}, \\ \tau &= \frac{\sinh(2\pi T\rho)\sin(2\pi T\hat{\tau})}{\cosh(2\pi T\rho) - \sinh(2\pi T\rho)\cos(2\pi T\hat{\tau})} \end{aligned} \tag{53}$$

maps the metric to

$$ds^2 = 4\pi^2 T^2 \left[d\rho^2 + \sinh^2(2\pi T\rho)d\hat{\tau}^2\right], \tag{54}$$

which is illustrated in Fig. 10. The $\hat{\tau}$ directions is periodic with period $1/T$.

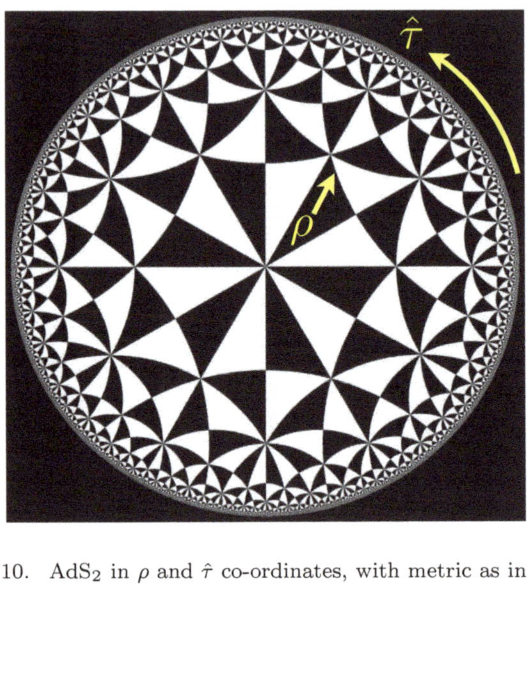

Fig. 10. AdS_2 in ρ and $\hat{\tau}$ co-ordinates, with metric as in (54).

3.4. *From Einstein-Maxwell theory in 3 + 1 dimensions to the Schwarzian action*

We have shown above how a universal AdS$_2$ metric emerges in the near-horizon region of a charged black hole in $3+1$ dimensions (a similar mapping applies in $d+2$ dimensions with $d \geq 2$). This subsection will carry out the same mapping at the level of the action, and so allow us to go beyond the saddle-point approximation at low energies.

3.4.1. *From I_{EM} in $3+1$ dimensions to $I_{EM,2}$ in $1+1$ dimensions.*

This dimensional reduction is carried out the by simply taking all fields dependent only upon the radial co-ordinate r and imaginary time τ. We make the metric ansatz[45]

$$ds^2 = \frac{ds_2^2}{\Phi(\zeta,\tau)} + [\Phi(\zeta,\tau)]^2 \, d\Omega_2^2 \tag{55}$$

where ds_2^2 is an arbitrary metric in the (ζ,τ) spacetime, and Φ is a scalar field in the (ζ,τ) spacetime. Here we do not discuss the dimensional reduction of the U(1) gauge field A, which has been described elsewhere.[32,45]

3.4.2. *JT gravity as the low energy limit of $I_{EM,2}$*

We take the low energy limit of $I_{EM,2}$ by mapping it to a near-horizon theory, I_{JT}, in a $1+1$ dimensional spacetime with a boundary.[46–50] The low energy theory on the (ζ,τ) spacetime involves a metric h, and a scalar field Φ_1 given by

$$\lim_{\zeta\to\infty} [\Phi(\zeta,\tau)]^2 = R_h^2 + \Phi_1(\zeta,\tau), \tag{56}$$

obeying the action

$$I_{JT} = -\frac{2\pi \mathcal{A}_0}{\kappa^2} + \int d^2x\sqrt{h}\left[-\frac{2\pi}{\kappa^2}\,\Phi_1\left(\mathcal{R}_2 + \frac{2}{R_h^3}\right)\right] - \frac{4\pi}{\kappa^2}\int_\partial dx\sqrt{h_b}\Phi_1\,\mathcal{K}_1 \tag{57}$$

where $\mathcal{A}_0 = 4\pi R_h^2$ is the area of the horizon at $T=0$, and $\mathcal{K}_1$ is the extrinsic curvature of the one-dimensional boundary $\zeta \to 0$ where

$$h_{\tau\tau}(\zeta\to 0) = \frac{R_h^3}{\zeta^2}, \quad \Phi_1(\zeta\to 0) = \frac{2R_h^3}{\zeta} \tag{58}$$

3.4.3. *Fluctuations about the AdS$_2$ saddle point of I_{JT}*

Einstein gravity in $1+1$ dimensions has no graviton, and is 'pure gauge'. In the JT-gravity theory with boundary, there is a remnant degree of freedom which is a boundary graviton, as illustrated in Fig. 9. The action for this boundary graviton turns out to be the advertised Schwarzian theory: consequently, the partition function of the $1+1$ dimensional JT gravity theory can be evaluated exactly (here we are ignoring the gauge field path integral, which is subdominant at fixed Q[26,51]).

The partition function of JT gravity is

$$\mathcal{Z}_Q = \int \mathcal{D}h\mathcal{D}\Phi_1 \exp\left(-I_{JT}\right) \tag{59}$$

The action is linear in Φ_1, and the integral over Φ_1 yields a constraint $\mathcal{R}_2 = -2/R_h^3$ *i.e.* the metric h is rigidly AdS_2. The only dynamical degree of freedom in JT gravity is a time reparameterization along the boundary $\tau \to f(\tau)$.[46,48] To ensure that the bulk metric obeys its boundary condition, we also have to make the spatial co-ordinate ζ a function of τ, so we map $(\tau, \zeta) \to (f(\tau), \zeta(\tau))$. Then the metric obeys its boundary condition provided $\zeta(\tau)$ is related to $f(\tau)$ by (here ζ_b is a small constant whose value cancels in the final result)

$$\zeta(\tau) = \zeta_b f'(\tau) + \zeta_b^3 \frac{\left[f''(\tau)\right]^2}{2f'(\tau)} + \mathcal{O}(\zeta_b^4) \tag{60}$$

Finally, we evaluate I_{GH} along this boundary curve. In this manner we obtain the action

$$I_{1,\text{eff}}[f] = -\frac{2\pi\mathcal{A}_0}{\kappa^2} - \frac{\gamma}{4\pi^2}\int d\tau\,\{f(\tau),\tau\}, \quad \{f,\tau\} \equiv \frac{f'''}{f'} - \frac{3}{2}\left(\frac{f''}{f'}\right)^2 \tag{61}$$

where $\gamma = 32\pi^3 R_h^3/\kappa^2$ is precisely the linear-T co-efficient in the black hole entropy in (48).

After a conformal map to finite temperature (and ignoring the contribution of the gauge field fluctuation), we can write the low energy partition function of a $3+1$-dimensional black hole with charge $Q = 4\pi R_h/(\kappa g_F)$, as a path integral over a single field $f(\tau)$ in one time dimension:

$$\mathcal{Z}_Q = \exp\left(\frac{2\pi\mathcal{A}_0}{\kappa^2}\right)\int \frac{\mathcal{D}f}{||\mathrm{SL(2,R)}||}\exp\left(\frac{\gamma}{4\pi^2}\int_0^{1/T} d\tau\,\{\tan(\pi T f(\tau)),\tau\}\right) \tag{62}$$

where $\gamma = 32\pi^3 R_h^3/\kappa^2$, $\mathcal{A}_0 = 4\pi R_h^2$, and $f(\tau)$ is a monotonic function of τ obeying

$$f(\tau + 1/T) = f(\tau) + 1/T. \tag{63}$$

We divide by the (infinite) volume of the SL(2,R) group because

$$\{f,\tau\} = \left\{\frac{af+b}{cf+d},\tau\right\} \tag{64}$$

where a, b, c, d are constants with $ad-bc = 1$. This is precisely the theory obtained in Section 2.3 for the SYK model. The integral over the phase field ϕ is not present in (62), but it also appears in the same form after including the dimensional reduction of the gauge field A.[32,45]

3.5. *Many-body density of states*

With the mapping to the Schwarzian in (62), we can now immediately apply the results of Section 2.4 to the charged black hole. The final results for the entropy and many-body density of states are[34]

$$\begin{aligned}\frac{S(T,Q)}{k_B} &= \frac{c^3}{4\hbar G}\left(\mathcal{A}_0 + 2\sqrt{\pi}\mathcal{A}_0^{3/2}\frac{k_B T}{\hbar c}\right) \\ &\quad - \frac{3}{2}\ln\left(\frac{(\hbar c^5/G)^{1/2}}{k_B T}\right) - \frac{559}{180}\ln\left(\frac{\mathcal{A}_0 c^3}{\hbar G}\right) + \cdots . \end{aligned} \tag{65}$$

$$D(E,Q) \sim \left(\frac{\mathcal{A}_0 c^3}{\hbar G}\right)^{-347/90} \exp\left(\frac{\mathcal{A}_0 c^3}{4\hbar G}\right) \sinh\left(\left[\sqrt{\pi}\mathcal{A}_0^{3/2}\frac{c^3}{\hbar G}\frac{E}{\hbar c}\right]^{1/2}\right). \tag{66}$$

These results should be compared to (4) and (32) for the SYK model, as in Fig. 2. For the entropy, the first two terms are the saddle-point contributions, which are proportional to $1/G$ and N respectively. The $(3/2)\ln T$ terms are identical between (65) and (4), and arise from the Schwarzian path integral. Finally, the $\ln(\mathcal{A}_0)$ term in (65) is the analog of the $\ln N$ term in (4), and these do have different co-efficients, reflecting their sensitivity to microscopics. The $\ln N$ term relies on the fact that (9) contains only 4-fermion terms,[23,26] while the $\ln \mathcal{A}_0$ term has a co-efficient dependent upon the number of massless fields in the quantum gravity theory under consideration.[34] The $\ln \mathcal{A}_0$ term was partly computed in earlier work.[52,53]

The many-body density of states in (66) is connected to the entropy in (65) via (30), and the $\ln \mathcal{A}_0$ term in the entropy determines the power of $\mathcal{A}_0$ in the pre-factor. However, the E dependence is identical to that in (32).

4. Wilson-Fisher Conformal Field Theory

It is instructive to compare the above results for the many-body density of states of the SYK model and charged black holes with those of the Wilson-Fisher conformal field theory. To this end, we place the Wilson Fisher theory in $2+1$ dimensions on the two-dimensional surface of a sphere of radius R. Via the state-operator correspondence of conformal field theory, there is a one-to-one correspondence between the energy eigenvalues E_n and the scaling dimensions of operators Δ_n:

$$E_n = \frac{\hbar c \Delta_n}{R} \tag{67}$$

where c is the velocity of 'light', and the ground state has zero energy. The lowest energy levels correspond to operators originally studied by Wilson and Fisher,[2] and they have Δ_n and their spacings of order unity. So for $E \sim \hbar c/R$, we have the many-body density of states $D(E) \sim R/(\hbar c)$. This small value of $D(E)$ at low E is quite different from the SYK model and charged black holes.

Let us now consider larger energies with $E \gg \hbar c/R$. At sufficiently large energies, we expect chaotic behavior with eigenstate thermalization, and so the system should be characterized locally by a flat space theory at a temperature T. From the entropy density in (3) for $d = 2$, we obtain the total entropy

$$S(T) = (4\pi R^2)\, a_2 k_B (k_B T/(\hbar c))^2. \tag{68}$$

We can now use familiar thermodynamic identities ($S = -\partial F/\partial T$, $F = E - TS$) to obtain the energy

$$E(T) = \frac{2T}{3} S(T), \tag{69}$$

and then convert to the microcanonical ensemble

$$S(E) = k_B (9\pi a_2)^{1/3} \left(\frac{ER}{\hbar c}\right)^{2/3}. \tag{70}$$

Finally, from Boltzmann's relation we obtain the many-body density of states

$$D(E) \sim \exp\left[(9\pi a_2)^{1/3} \left(\frac{ER}{\hbar c}\right)^{2/3}\right]. \tag{71}$$

This shows that the spacing between energy levels for $E \gg \hbar c/R$ is exponentially small in $E^{2/3}$, and (71) should be compared with the exponentials of $E^{1/2}$ in (32) for the SYK model, and in (66) for charged black holes. These large dimension operators of the Wilson-Fisher theory correspond to neutral black hole states, although their gravity theory is surely more complicated than Einstein gravity. It would be interesting to study such black hole states numerically by the methods of.[54]

We can now see that a significant advantage of the SYK model was that its black holes states extended all the way down to the ground state, and this was an important reason for its simplicity. In contrast, the Wilson-Fisher theory is black hole-like only above some threshold energy, and the low-lying states are not exponentially dense.

When carried out for a conformal field theory in d spatial dimensions, the above argument yields $D(E) \sim \exp\left[bE^{d/(d+1)}\right]$, for some constant b, for black hole states.[55,56] This does not match the SYK model $D(E) \sim \exp\left[bE^{1/2}\right]$ at $d = 0$ because of the important role of the Schwarzian-induced breaking of conformal symmetry: the state-operator correspondence does not apply to the SYK model.

5. Discussion

The main results for the common structure of the many-body densities of states of non-supersymmetric charged black holes and the SYK model have been obtained in Sections 2 and 3, and were summarized in Fig. 2. This connection shows that it is possible to realize Hawking's black hole entropy in a manner consistent with the standard quantum mechanics of many-body systems without supersymmetry. The match is at the level of the density of states coarse-grained over a few level spacings, and extends also to multi-point correlators of the density of states.[57]

The precise energy levels of the SYK model in Fig. 5, of course, depend upon the particular realization chosen for the couplings $U_{ij;k\ell}$. We do not expect any such instance of the SYK model to described the ultimate microstructure of a realistic black hole. But numerous universal features of realistic black holes are indeed captured by the SYK model,[58] after coarse-graining over an exponentially small level spacing.

The reader may find the analogy with single-particle quantum chaos instructive. A particle moving in a quantum billiard defines a problem with no randomness in the Hamiltonian, analogous to a black hole. Nevertheless, the energy levels are chaotic, and their statistical properties are equivalent to those of a random matrix ensemble.[59,60] For the black hole, the SYK model defines a random matrix in the many-body Fock space of size $2^N \times 2^N$ using of order N^4 random numbers: so the SYK matrix is quite sparse and has significant structure, and is very far from a fully random matrix with 2^{2N} independent elements.

Finally, we note that the SYK model was originally introduced[39] as a toy model of the strange metal state of the cuprates. This connection has also undergone rapid development in recent years, and has led to realistic and universal model of strange metal behavior,[17] as reviewed in a companion article.[18]

Acknowledgements

This article is based on lectures presented at the College de France, Paris in May-June 2022. I am grateful to Antoine Georges for his warm hospitality, and for many discussions and collaborations over the years. I also thank Matthew Heydeman, Luca Iliesiu, Upamanyu Moitra, Sameer Murthy, Leo Radzihovsky, Sandeep Trivedi, G. Joaquin Turiaci, and Spenta Wadia for valuable discussions. This research was supported by the U.S. National Science Foundation grant No. DMR-2002850, and by the Simons Collaboration on Ultra-Quantum Matter which is a grant from the Simons Foundation (651440, S.S.).

References

1. M. E. Fisher, *The Nature of Critical Points*. University of Colorado Press (1965), http://personal.rhul.ac.uk/UHAP/027/PH4211/PH4211_files/Fisher-1965.pdf.
2. K. G. Wilson and M. E. Fisher, Critical exponents in 3.99 dimensions, *Phys. Rev. Lett.* **28**, pp. 240–243 (1972), https://link.aps.org/doi/10.1103/PhysRevLett.28.240.
3. H. Bethe, Zur theorie der metalle, *Zeitschrift für Physik* **71**, 3, pp. 205–226 (1931), ISBN 0044-3328, https://doi.org/10.1007/BF01341708.
4. L. Onsager, Crystal Statistics. I. A Two-Dimensional Model with an Order-Disorder Transition, *Phys. Rev.* **65**, pp. 117–149 (1944), https://link.aps.org/doi/10.1103/PhysRev.65.117.
5. S. Sachdev and J. Ye, Universal quantum-critical dynamics of two-dimensional antiferromagnets, *Phys. Rev. Lett.* **69**, 16, pp. 2411–2414 (1992), doi:10.1103/PhysRevLett.69.2411, arXiv:cond-mat/9204001 [cond-mat].

6. M. E. Fisher and R. J. Burford, Theory of critical-point scattering and correlations. i. the ising model, *Phys. Rev.* **156**, pp. 583–622 (1967), doi:10.1103/PhysRev.156.583, https://link.aps.org/doi/10.1103/PhysRev.156.583.
7. S. Sachdev, *Quantum Phase Transitions.* Cambridge University Press, Cambridge, UK (1999), ISBN 9780521514682, https://www.cambridge.org/9780521514682.
8. M. E. Fisher and P. G. de Gennes, Phénomènes aux parois dans un mélange binaire critique, *Comptes Rendus Acad. Sc. Paris B* **287**, p. 207 (1978), https://gallica.bnf.fr/ark:/12148/bpt6k62353730/f61.item.
9. S. A. Hartnoll and A. P. Mackenzie, Colloquium: Planckian dissipation in metals, *Rev. Mod. Phys.* **94**, 4, p. 041002 (2022), doi:10.1103/RevModPhys.94.041002, arXiv:2107.07802 [cond-mat.str-el].
10. S. Chakravarty, B. I. Halperin and D. R. Nelson, Two-dimensional quantum Heisenberg antiferromagnet at low temperatures, *Phys. Rev. B* **39**, pp. 2344–2371 (1989), doi:10.1103/PhysRevB.39.2344, https://link.aps.org/doi/10.1103/PhysRevB.39.2344.
11. J. Maldacena, S. H. Shenker and D. Stanford, A bound on chaos, *JHEP* **08**, p. 106 (2016), doi:10.1007/JHEP08(2016)106, arXiv:1503.01409 [hep-th].
12. D. Chowdhury and B. Swingle, Onset of many-body chaos in the $O(N)$ model, *Phys. Rev. D* **96**, 6, p. 065005 (2017), doi:10.1103/PhysRevD.96.065005, arXiv:1703.02545 [cond-mat.str-el].
13. S. Sachdev, Holographic metals and the fractionalized Fermi liquid, *Phys. Rev. Lett.* **105**, p. 151602 (2010), doi:10.1103/PhysRevLett.105.151602, arXiv:1006.3794 [hep-th].
14. J. Preskill, P. Schwarz, A. Shapere, S. Trivedi and F. Wilczek, Limitations on the statistical description of black holes, *Modern Physics Letters A* **06**, 26, pp. 2353–2361 (1991), https://doi.org/10.1142/S0217732391002773.
15. J. M. Maldacena and A. Strominger, Universal low-energy dynamics for rotating black holes, *Phys. Rev. D* **56**, pp. 4975–4983 (1997), doi:10.1103/PhysRevD.56.4975, arXiv:hep-th/9702015.
16. D. Chowdhury, A. Georges, O. Parcollet and S. Sachdev, Sachdev-Ye-Kitaev models and beyond: Window into non-Fermi liquids, *Rev. Mod. Phys.* **94**, 3, p. 035004 (2022), doi:10.1103/RevModPhys.94.035004, arXiv:2109.05037 [cond-mat.str-el].
17. A. A. Patel, H. Guo, I. Esterlis and S. Sachdev, Universal theory of strange metals from spatially random interactions, *Science* **381**, p. 790 (2023), doi:10.1126/science.abq6011, arXiv:2203.04990 [cond-mat.str-el].
18. S. Sachdev, Strange Metals and Black Holes: Insights From the Sachdev-Ye-Kitaev Model, in *Oxford Research Encyclopedia of Physics.* Oxford University Press (December 2023), doi:10.1093/acrefore/9780190871994.013.48, arXiv:2305.01001 [cond-mat.str-el].
19. A. Georges, O. Parcollet and S. Sachdev, Quantum fluctuations of a nearly critical Heisenberg spin glass, *Phys. Rev. B* **63**, 13, 134406 (2001), doi:10.1103/PhysRevB.63.134406, arXiv:cond-mat/0009388 [cond-mat.str-el].
20. J. S. Cotler, G. Gur-Ari, M. Hanada, J. Polchinski, P. Saad, S. H. Shenker, D. Stanford, A. Streicher and M. Tezuka, Black Holes and Random Matrices, *JHEP* **05**, p. 118 (2017), doi:10.1007/JHEP05(2017)118, arXiv:1611.04650 [hep-th], [Erratum: JHEP 09, 002 (2018)].
21. D. Bagrets, A. Altland and A. Kamenev, Power-law out of time order correlation functions in the SYK model, *Nucl. Phys. B* **921**, pp. 727–752 (2017), doi:10.1016/j.nuclphysb.2017.06.012, arXiv:1702.08902 [cond-mat.str-el].

22. J. Maldacena and D. Stanford, Remarks on the Sachdev-Ye-Kitaev model, *Phys. Rev. D* **94**, 10, p. 106002 (2016), doi:10.1103/PhysRevD.94.106002, arXiv:1604.07818 [hep-th].
23. A. Kitaev and S. J. Suh, The soft mode in the Sachdev-Ye-Kitaev model and its gravity dual, *Journal of High Energy Physics* **05**, p. 183 (2018), doi:10.1007/JHEP05(2018)183, arXiv:1711.08467 [hep-th].
24. D. Stanford and E. Witten, Fermionic Localization of the Schwarzian Theory, *Journal of High Energy Physics* **10**, p. 008 (2017), doi:10.1007/JHEP10(2017)008, arXiv:1703.04612 [hep-th].
25. M. E. Fisher, *Renormalization group in critical phenomena and quantum field theory, J. Gunton and M. S. Green eds.* Temple University, Philadelphia (1973).
26. Y. Gu, A. Kitaev, S. Sachdev and G. Tarnopolsky, Notes on the complex Sachdev-Ye-Kitaev model, *Journal of High Energy Physics* **02**, p. 157 (2020), doi:10.1007/JHEP02(2020)157, arXiv:1910.14099 [hep-th].
27. C. V. Vishveshwara, Scattering of Gravitational Radiation by a Schwarzschild Black-hole, *Nature* **227**, 5261, pp. 936–938 (1970), ISBN 1476-4687, https://doi.org/10.1038/227936a0.
28. S. W. Hawking, Particle Creation by Black Holes, *Commun. Math. Phys.* **43**, pp. 199–220 (1975), https://doi.org/10.1007/BF02345020.
29. J. M. Maldacena, The Large N limit of superconformal field theories and supergravity, *Adv. Theor. Math. Phys.* **2**, pp. 231–252 (1998), doi:10.1023/A:1026654312961, arXiv:hep-th/9711200.
30. T. Faulkner, H. Liu, J. McGreevy and D. Vegh, Emergent quantum criticality, Fermi surfaces, and AdS_2, *Phys. Rev. D* **83**, p. 125002 (2011), doi:10.1103/PhysRevD.83.125002, arXiv:0907.2694 [hep-th].
31. R. Emparan, C. V. Johnson and R. C. Myers, Surface terms as counterterms in the AdS/CFT correspondence, *Phys. Rev. D* **60**, p. 104001 (1999), doi:10.1103/PhysRevD.60.104001, arXiv:hep-th/9903238 [hep-th].
32. S. Sachdev, Universal low temperature theory of charged black holes with AdS_2 horizons, *J. Math. Phys.* **60**, 5, p. 052303 (2019), doi:10.1063/1.5092726, arXiv:1902.04078 [hep-th].
33. A. Strominger and C. Vafa, Microscopic origin of the Bekenstein-Hawking entropy, *Phys. Lett. B* **379**, pp. 99–104 (1996), doi:10.1016/0370-2693(96)00345-0, arXiv:hep-th/9601029.
34. L. V. Iliesiu, S. Murthy and G. J. Turiaci, Revisiting the Logarithmic Corrections to the Black Hole Entropy, (2022), arXiv:2209.13608 [hep-th].
35. W. Fu, D. Gaiotto, J. Maldacena and S. Sachdev, Supersymmetric Sachdev-Ye-Kitaev models, *Phys. Rev. D* **95**, 2, p. 026009 (2017), doi:10.1103/PhysRevD.95.026009, arXiv:1610.08917 [hep-th], [Addendum: Phys.Rev.D 95, 069904 (2017)].
36. M. Heydeman, L. V. Iliesiu, G. J. Turiaci and W. Zhao, The statistical mechanics of near-BPS black holes, *J. Phys. A* **55**, 1, p. 014004 (2022), doi:10.1088/1751-8121/ac3be9, arXiv:2011.01953 [hep-th].
37. L. V. Iliesiu, S. Murthy and G. J. Turiaci, Black hole microstate counting from the gravitational path integral, (2022), arXiv:2209.13602 [hep-th].
38. M. Heydeman, G. J. Turiaci and W. Zhao, Phases of $\mathcal{N} = 2$ Sachdev-Ye-Kitaev models, *JHEP* **01**, p. 098 (2023), doi:10.1007/JHEP01(2023)098, arXiv:2206.14900 [hep-th].
39. S. Sachdev and J. Ye, Gapless spin-fluid ground state in a random quantum Heisenberg magnet, *Phys. Rev. Lett.* **70**, 21, pp. 3339–3342 (1993), doi:10.1103/PhysRevLett.70.3339, arXiv:cond-mat/9212030 [cond-mat].

40. A. Kitaev, A simple model of quantum holography, talk given at kitp program: entanglement in strongly-correlated quantum matter, *KITP, University of California, Santa Barbara, April 2015* (2015).
41. S. Sachdev, Bekenstein-Hawking Entropy and Strange Metals, *Phys. Rev. X* **5**, 4, p. 041025 (2015), doi:10.1103/PhysRevX.5.041025, arXiv:1506.05111 [hep-th].
42. O. Parcollet and A. Georges, Non-Fermi-liquid regime of a doped Mott insulator, *Phys. Rev. B* **59**, 8, pp. 5341–5360 (1999), doi:10.1103/PhysRevB.59.5341, arXiv:cond-mat/9806119 [cond-mat.str-el].
43. G. W. Gibbons and S. W. Hawking, Action integrals and partition functions in quantum gravity, *Phys. Rev. D* **15**, pp. 2752–2756 (1977), https://doi.org/10.1103/PhysRevD.15.2752.
44. S. F. Ross, Black hole thermodynamics, (2005), arXiv:hep-th/0502195.
45. U. Moitra, S. P. Trivedi and V. Vishal, Extremal and near-extremal black holes and near-CFT_1, *Journal of High Energy Physics* **07**, p. 055 (2019), doi:10.1007/JHEP07(2019)055, arXiv:1808.08239 [hep-th].
46. A. Almheiri and J. Polchinski, Models of AdS_2 backreaction and holography, *JHEP* **11**, p. 014 (2015), doi:10.1007/JHEP11(2015)014, arXiv:1402.6334 [hep-th].
47. A. Almheiri and B. Kang, Conformal Symmetry Breaking and Thermodynamics of Near-Extremal Black Holes, *JHEP* **10**, p. 052 (2016), doi:10.1007/JHEP10(2016)052, arXiv:1606.04108 [hep-th].
48. J. Maldacena, D. Stanford and Z. Yang, Conformal symmetry and its breaking in two dimensional Nearly Anti-de-Sitter space, *PTEP* **2016**, 12, p. 12C104 (2016), doi: 10.1093/ptep/ptw124, arXiv:1606.01857 [hep-th].
49. S. Choi and F. Larsen, Effective Field Theory of Quantum Black Holes, (2021), arXiv:2108.04028 [hep-th].
50. S. Choi and F. Larsen, AdS_2 Holography and Effective QFT, (2023), arXiv:2302.13917 [hep-th].
51. L. V. Iliesiu and G. J. Turiaci, The statistical mechanics of near-extremal black holes, *Journal of High Energy Physics* **05**, p. 145 (2021), doi:10.1007/JHEP05(2021)145, arXiv:2003.02860 [hep-th].
52. A. Sen, Logarithmic Corrections to $\mathcal{N} = 2$ Black Hole Entropy: An Infrared Window into the Microstates, *Gen. Rel. Grav.* **44**, 5, pp. 1207–1266 (2012), doi: 10.1007/s10714-012-1336-5, arXiv:1108.3842 [hep-th].
53. N. Banerjee and M. Saha, Revisiting leading quantum corrections to near extremal black hole thermodynamics, *JHEP* **07**, p. 010 (2023), doi:10.1007/JHEP07(2023)010, arXiv:2303.12415 [hep-th].
54. W. Zhu, C. Han, E. Huffman, J. S. Hofmann and Y.-C. He, Uncovering Conformal Symmetry in the 3D Ising Transition: State-Operator Correspondence from a Quantum Fuzzy Sphere Regularization, *Phys. Rev. X* **13**, 2, p. 021009 (2023), doi: 10.1103/PhysRevX.13.021009, arXiv:2210.13482 [cond-mat.stat-mech].
55. E. Shaghoulian, Modular forms and a generalized Cardy formula in higher dimensions, *Phys. Rev. D* **93**, 12, p. 126005 (2016), doi:10.1103/PhysRevD.93.126005, arXiv:1508.02728 [hep-th].
56. A. Belin, J. de Boer, J. Kruthoff, B. Michel, E. Shaghoulian and M. Shyani, Universality of sparse $d > 2$ conformal field theory at large N, *JHEP* **03**, p. 067 (2017), doi:10.1007/JHEP03(2017)067, arXiv:1610.06186 [hep-th].
57. P. Saad, S. H. Shenker and D. Stanford, JT gravity as a matrix integral, (2019), arXiv:1903.11115 [hep-th].

58. R. Bousso, X. Dong, N. Engelhardt, T. Faulkner, T. Hartman, S. H. Shenker and D. Stanford, Snowmass White Paper: Quantum Aspects of Black Holes and the Emergence of Spacetime, (2022), arXiv:2201.03096 [hep-th].
59. O. Bohigas, M. J. Giannoni and C. Schmit, Characterization of chaotic quantum spectra and universality of level fluctuation laws, *Phys. Rev. Lett.* **52**, pp. 1–4 (1984), doi:10.1103/PhysRevLett.52.1, https://link.aps.org/doi/10.1103/PhysRevLett.52.1.
60. S. Müller, S. Heusler, A. Altland, P. Braun and F. Haake, Periodic-orbit theory of universal level correlations in quantum chaos, *New Journal of Physics* **11**, 10, 103025 (2009), doi:10.1088/1367-2630/11/10/103025, arXiv:0906.1960 [nlin.CD].

Soft Matter, Complex Matter and Biological Physics

https://doi.org/10.1142/9789811282386_0038

Chapter 38

Topological Transformations of Biomembranes and Vesicles *via* Fission and Fusion

R. Lipowsky
Max Planck Institute of Colloids and Interfaces,
Science Park Golm, 14424 Potsdam, Germany

The cells of our body are compartmentalized by biomembranes and vesicles, which consist of molecular bilayers with a thickness of a few nanometers and represent two-dimensional fluids. Due to their fluidity, the bilayer membranes can easily remodel their composition, shape, and topology. Here, we focus on their topology and transformations between different topologies *via* fission and fusion processes. In general, these topological transformations can be characterized by changes in the Euler characteristic and in the topological genus. Fission processes proceed via the closure and cleavage of membrane necks as recently demonstrated for giant unilamellar vesicles (GUVs) and for unilamellar nanovesicles assembled *in silico*. Neck cleavage is controlled by constriction forces that compress the neck and provide a general physical mechanism for membrane fission. Fusion processes proceed via the adhesion of two membranes and by the formation of a fusion pore, which has the same shape as a membrane neck. In fact, when two membranes fuse, they undergo the same sequence of shapes as during fission but in reversed order. However, the 'local surgery,' necessary to form the fusion pore, involves alternative molecular pathways as reviewed here for tension-induced fusion. Multispherical vesicles, doped with membrane proteins that drive homotypic fusion, can be used to form high genus vesicles, which are accessible to experimental studies.

1. Introduction

Our body contains a huge number of biomembranes that enclose our cells and most intracellular organelles. These membranes are fluid and create a flexible architecture that partitions each cell into separate liquid compartments. This membrane architecture is rather dynamic and continuously remodeled. First, the flexibility of these membranes allows them to easily adapt both their local composition and their shape to different cues in their surroundings. Second, the membrane compartments can change their topology via fusion and fission processes. Thus, if we looked at the intracellular membrane compartments with nanoscopic resolution, we would see heavy traffic of vesicles between different organelles, as displayed in Fig. 1.[1] This

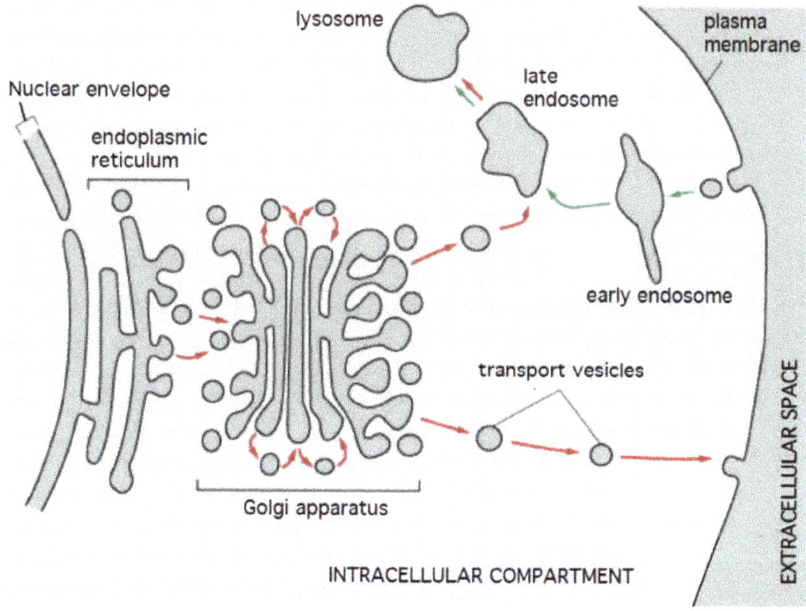

Fig. 1. Schematic view of vesicular traffic within eukaryotic cells. Vesicles are formed by budding and fission from donor membranes, transported across the cell, and eventually integrated by adhesion and fusion into acceptor membranes. The red arrows indicate the outward secretory pathway, the green arrows the inward endocytic pathway[1] [with permission from W.W. Norton and Company].

traffic involves both the production of new vesicles via budding and fission from a donor membrane as well as the uptake of such vesicles by acceptor membranes via adhesion and fusion.

In Fig. 1, the different membrane-bound organelles are displayed in a highly schematic manner. In particular, only the inner part of the endoplasmic reticulum (ER) is shown, which is known as the perinuclear region and is located close to the nuclear envelope.[2] Another particularly fascinating part of the ER is its peripheral region, which forms a network of membrane nanotubes that extend throughout the whole cell, as shown by the light microscopy image in Fig. 2(a). The network consists of irregular polygons connected by three-way junctions with contact angles close to 120°.[3,4]

In spite of its complex architecture, the whole nanotubular network of the ER is formed by a single membrane which encloses a continuous nanofluidic network of water channels.[2] As a consequence, the ER membrane creates a bicontinuous structure that partitions the intracellular space into two separate, interpenetrating subcompartments, the lumen of the ER network and the surrounding cytosol. In fact, the ER membrane itself provides two additional quasi-two-dimensional subcompartments corresponding to the two leaflets of the lipid–protein bilayer,

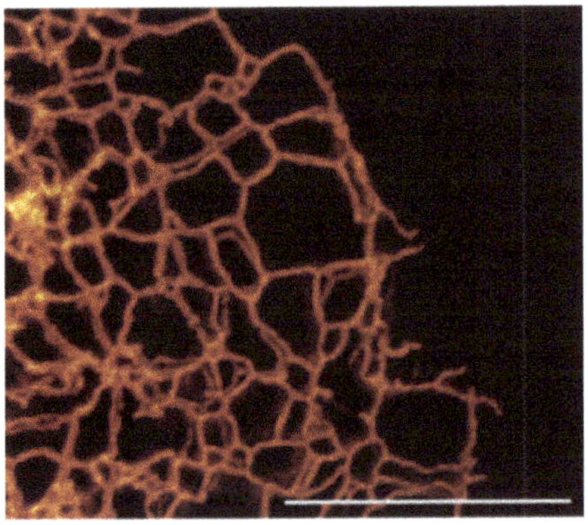

(a)

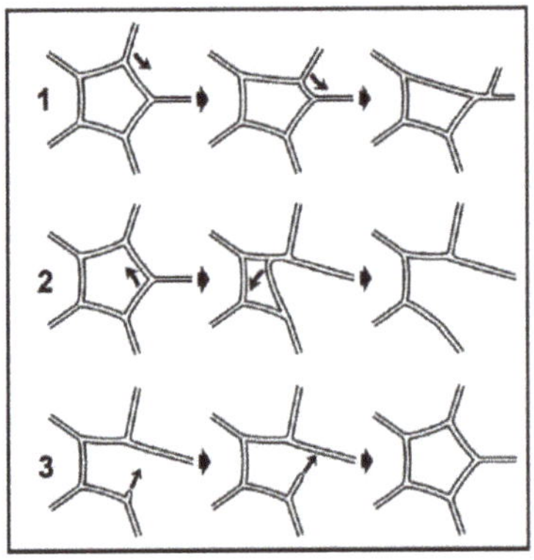

(b)

Fig. 2. (a) Nanotubular network of the peripheral endoplasmic reticulum (ER) with a very high topological genus, which is equal to the number of irregular polygons formed by the membrane nanotubes. Scale bar: $10\,\mu m^2$ [with permission from Jennifer Lipponcott-Schwartz]. (b) Dynamic processes of sliding, ring closure, and tubule branching as shown in rows 1, 2, and 3. Ring closure removes one polygon from the network, thereby reducing the genus of the ER membrane, whereas tubule branching increases this genus by creating a new polygon[5] [with permission from International Review of Cytology, 2001].

which can accommodate membrane-bound molecules and processes. All four subcompartments are fluid, which implies that widely separated regions of the cell can communicate with each other both via the ER membrane and via the ER lumen.

From a topological point of view, the peripheral ER is rather fascinating because of its very large topological genus, corresponding to the huge number of irregular polygons formed by the ER membrane. The genus of the peripheral ER is quite stable over extended periods of time but is occasionally remodeled by ring closure and tubule branching, see Fig. 2(b), Refs. 4,5. Ring closure deletes one polygon from the network, thereby decreasing the topological genus of the ER. Branching, on the other hand, creates one additional polygon, thereby increasing the topological genus. *In vivo*, these remodeling processes are controlled, at least to some extent, by interactions of the ER membranes with the cytoskeleton. On the other hand, *in-vitro* reconstitution of the peripheral ER has demonstrated that the nanotubular networks are also formed in the absence of any cytoskeletal elements.

In cell biology, a lot of effort has been devoted to identify the different membrane proteins that are involved in the topologial transformations of cellular and intracellular membranes. These proteins perturb the membranes locally, thereby inducing their fission and fusion. In this chapter, we use a coarse-grained description for the molecular details of this 'local surgery' and focus on the membrane geometry of the remodeling processes and the membrane elasticity which includes both their curvature elasticity and membrane tension.

Furthermore, we illustrate the geometric and membrane-elastic features for two types of synthetic membrane compartments: giant unilamellar vesicles (GUVs) as studied by light microscopy and nanovesicles as visualized by molecular dynamics simulations.

This chapter is organized as follows. We first introduce, in Section 2, the Euler characteristic and the topological genus in order to obtain a general classification of the membrane topology and its topological transformations. Fission processes which reduce the genus and increase the Euler characteristic will be addressed in Section 3, where we emphasize the formation of membrane necks for out- and in-budded vesicles as well as the constriction forces compressing these necks. Section 4 describes three recent examples for fission processes that drive the division of GUVs, as observed by light microscopy, as well as the division of nanovesicles and their endocytosis of condensate droplets, as studied in molecular dynamics simulations. Fusion processes, which increase the genus and reduce the Euler characteristic, are considered in Section 5, focusing on the relatively simple mechanism of tension-induced fusion, which exhibits several fusion pathways. Section 6 is devoted to the fusion and fission of multispherical vesicles, which are useful in order to study higher genus vesicles, similar to the ER. Finally, Section 7 contains a summary and an outlook on related topics that provide challenges for future studies.

2. Topology of Closed Membrane Surfaces

The topology of a closed membrane surface can be characterized in a global manner, using the Euler characteristic χ or the topological genus $\mathfrak{g}$, which are two integer numbers. Obtaining the precise values of these numbers may be quite difficult, as one can envisage by looking at the reticular network in Fig. 2(a). On the other hand, we can also focus on the 'local surgery' of the membrane by individual fission and fusion events, which change χ by ± 2 and $\mathfrak{g}$ by ± 1. In the following sections, we first characterize topologically simple membrane surfaces with small χ- and $\mathfrak{g}$-values in a global manner and subsequently look at the changes of χ and $\mathfrak{g}$ arising from the 'local surgery' of these surfaces.

2.1. *Euler characteristic and topological genus*

In general, the topology of any closed surface in three dimensions can be characterized by the integer Euler characteristic χ. This number can be obtained from any partitioning of the surface into a discrete set of surface segments.[6] One widely applied discretization method is triangulation, but one may also use a mesh of smooth curves that are embedded in the surface. Any discretization involves the surface segments themselves, usually called 'faces,' the edges between neighboring faces, and the vertices at which several edges come together. Counting the number of faces, F, the number of edges, E, and the number of vertices, V, the Euler

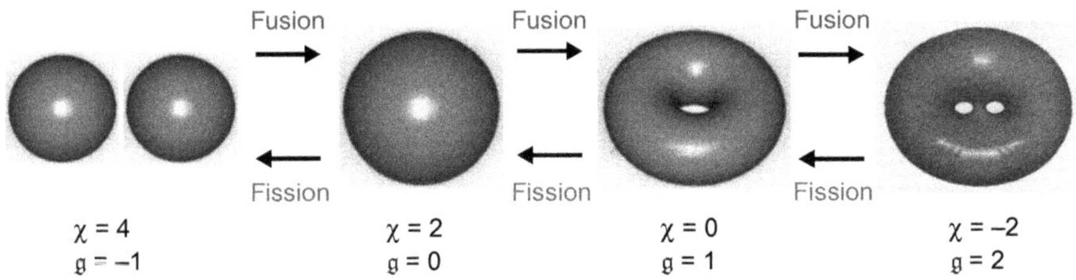

Fig. 3. Closed membrane surfaces with different topologies, distinguished by their Euler characteristic χ and their topological genus $\mathfrak{g}$: Two spherical vesicles are characterized by $\chi = 4$ and $\mathfrak{g} = -1$; a single spherical vesicle by $\chi = 2$ and $\mathfrak{g} = 0$; a torus (or doughnut shape) by $\chi = 0$ and $\mathfrak{g} = 1$; and a double torus (or button shape) by $\chi = -2$ and $\mathfrak{g} = 2$. Membrane fusion decreases the Euler characteristic χ and increases the topological genus $\mathfrak{g}$, whereas membrane fission increases χ and decreases $\mathfrak{g}$. Additional fission events lead to more spherical vesicles with a large and positive χ. Additional fusion events lead to a membrane surface with a large number of handles, corresponding to a large and positive $\mathfrak{g}$, see the reticular network in Fig. 2(a). The formation of membrane surfaces with $\mathfrak{g} \geq 1$ involves self-adhesion of the membrane.

characteristic χ is obtained via

$$\chi = F - E + V. \tag{1}$$

The three numbers F, E, and V depend on the chosen partitioning of the surface, as one can easily see by using different polyhedra to discretize a sphere. In contrast, the Euler characteristic χ itself is independent of this partitioning and defines the topology of the surface. Thus, we obtain $\chi = 2$ for a tetrahedron, a cube or an icosahedron as well as for any other partitioning of a spherical surface.

Some examples for closed membrane surfaces with different χ-values are displayed in Fig. 3. The torus or doughnut surface has the Euler characteristic $\chi = 0$, the double-torus or button surface has $\chi = -2$, the sphere has $\chi = 2$, and the combined surface of two spheres has $\chi = 4$. Likewise, the topology of N_{ve} spherical vesicles is characterized by $\chi = 2N_{ve}$, which extends the sequence of surface topologies in Fig. 3 to the left. Furthermore, a membrane surface, which forms $\mathfrak{g}$ handles or loops, has $\chi = 2 - 2\mathfrak{g}$, which extends the surface topologies in Fig. 3 to the right.[a]

The number $\mathfrak{g}$ of handles or loops provides another, equivalent characterization of a closed surface, the so-called topological genus $\mathfrak{g}$, which satisfies

$$\mathfrak{g} = 1 - \tfrac{1}{2}\chi \quad \text{and} \quad \Delta\mathfrak{g} = -\tfrac{1}{2}\,\Delta\chi, \tag{2}$$

where $\Delta\mathfrak{g} \equiv \mathfrak{g}_2 - \mathfrak{g}_1$ and $\Delta\chi \equiv \chi_2 - \chi_1$ is the difference between the genera of two surfaces 1 and 2 and between their Euler characteristics. For a sphere and a torus, the topological genus is equal to $\mathfrak{g} = 0$ and $\mathfrak{g} = 1$, respectively. The topological genus $\mathfrak{g}$ can be deduced directly from the global surface geometry by counting

[a]Handles or loops are sometimes called 'pores,' but, for biomembranes, a 'pore' is a water channel across the bilayer membrane.

the number of handles or loops formed by the surface without any partitioning or discretization of this surface. Thus, when we look at the nanotubular network of the peripheral endoplasmic reticulum in Fig. 2(a), we can immediately conclude that this network has a very large and positive genus $\mathfrak{g}$ and, thus, a very large and negative Euler characteristic χ.

2.2. *Local topology changes by fission and fusion*

By definition, a topological transformation of a vesicle or another closed membrane compartment implies that this compartment changes its topology and, thus, its Euler characteristic and topological genus. In the living cell, the membrane-bound organelles continuously undergo such topological transformations by membrane fission and membrane fusion, as illustrated in Fig. 1. Likewise, the simple topologies in Fig. 3 can be transformed into each other by individual fusion and fission processes. These individual events proceed via some 'local surgery' which can be characterized without knowing the global shape and topology of the membrane compartment.

A single, local *fission* event as depicted in Figs. 4(a) and (b) leads to an increase of the Euler characteristic by $\Delta\chi = +2$ and to a decrease of the topological genus by $\Delta\mathfrak{g} = -\frac{1}{2}\Delta\chi = -1$. On the other hand, a single, local *fusion* event leads to a decrease of the Euler characteristic by $\Delta\chi = -2$ and to an increase of the genus by $\Delta\mathfrak{g} = +1$, see Figs. 4(c) and (d). These positive and negative values of $\Delta\chi$ and $\Delta\mathfrak{g}$ for single fission and fusion events are completely general and apply to any shape and topology of the membrane compartment as long as this compartment forms a closed surface without bilayer pores or bilayer edges both before and after such an event. Therefore, computing the Euler characteristic before and after a topological transformation provides a simple and general method to distinguish fission and fusion processes which may not be obvious otherwise.

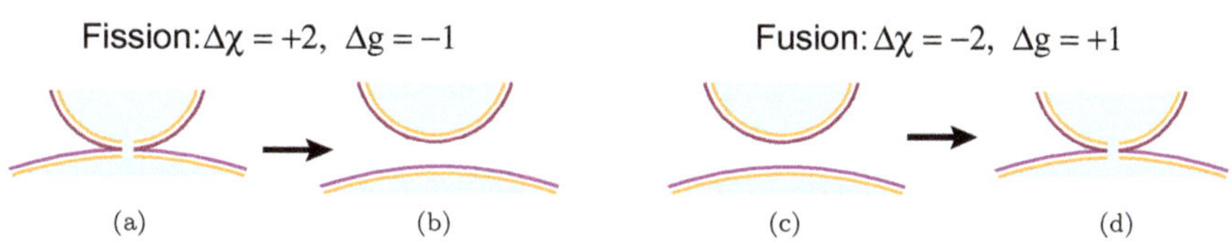

Fig. 4. 'Local surgery' of membrane topology by single fission and fusion events: (a) A single fission event proceeds via the formation and cleavage of a membrane neck. (b) After neck cleavage, the Euler characteristic χ and the topological genus $\mathfrak{g}$ of the membrane compartment have been changed by $\Delta\chi = +2$ and $\Delta\mathfrak{g} = -1$. (c) A single fusion event proceeds via the adhesion of two membrane segments and the formation of a fusion pore. (d) After the fusion pore has been formed, the Euler characteristic and the genus of the membrane compartment have been changed by $\Delta\chi = -2$ and $\Delta\mathfrak{g} = +1$. These changes of χ and $\mathfrak{g}$ apply irrespective of the overall shape and global topology of the compartment. Biomembranes are molecular bilayers with two leaflets, which typically have a certain transbilayer asymmetry as indicated by the two colors, purple and orange, for the outer and inner leaflets. The light blue aqueous solution in contact with the inner leaflet (orange) represents the lumen of the compartment.

3. Fission via Budding and Neck Cleavage

A single fission event as displayed in Figs. 4(a) and (b) proceeds in several steps. First, the mother or donor membrane forms a bud which is still connected to the donor membrane by a narrow or closed membrane neck. Second, the bilayer membrane is cleaved across this neck, thereby forming two bilayer edges in contact with the aqueous solutions. Third, these bilayer edges are resealed, which leads to a donor membrane with reduced membrane area and to a newly formed daughter vesicle. In order to obtain additional insight into the formation and stability of the membrane neck, we first consider the simplest case as provided by the fission of a vesicle with genus $\mathfrak{g} = 0$ which has the topology of a sphere. For genus-0 vesicles, the fission process has been studied in some detail, both by analytical theory, experiment, and simulations.[7–10] The two substeps, formation and cleavage of a membrane neck will now be described in more detail.

3.1. *Two budding geometries for genus-0 vesicles*

For genus-0 vesicles, the formation of a membrane bud can proceed in two different ways, as shown in Fig. 5. First, the vesicle membrane may form an out-bud, as depicted in Fig. 5(a). The out-bud represents a membrane deformation toward the exterior space, which is still connected to the mother vesicle by a closed membrane neck. Second, the vesicle membrane can form an in-bud, as in Fig. 5(c). The in-bud

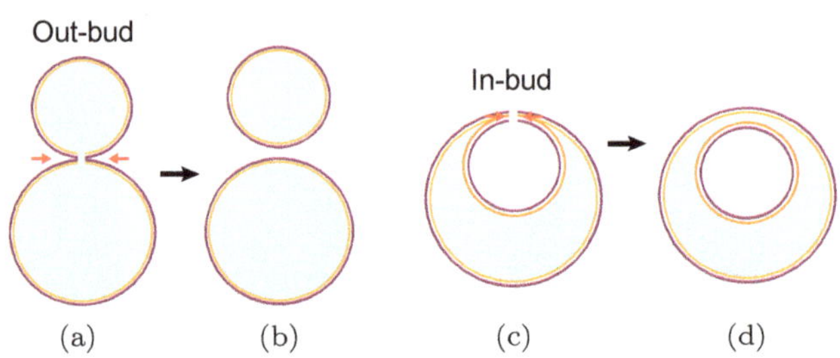

Fig. 5. Fission of genus-0 vesicles by the formation and cleavage of membrane necks:: (a) Out-budded vesicle, which is formed when the membrane has a sufficiently large positive spontaneous curvature. The closed membrane neck between the out-bud and the vesicle is compressed by a curvature-dependent constriction force (red arrows). (b) The constriction force can cleave the neck, thereby dividing the out-budded vesicle into two daughter vesicles, which both have the same transbilayer asymmetry as the original genus-0 vesicle before division. The out-budded vesicle in (a) has Euler characteristic $\chi = 2$ and genus $\mathfrak{g} = 0$, the two daughter vesicles in (b) have $\chi = 4$ and $\mathfrak{g} = -1$. (c) In-budded vesicle, which is formed when the membrane has a sufficiently large negative spontaneous curvature. The closed membrane neck between the in-bud and the vesicle is again compressed by a curvature-dependent constriction force (red arrows). (d) Neck cleavage now leads to two nested daughter vesicles with opposite transbilayer asymmetries. The in-budded vesicle in (c) has $\chi = 2$ and $\mathfrak{g} = 0$, the two nested daughter vesicles in (d) are characterized by $\chi = 4$ and $\mathfrak{g} = -1$.

points toward the interior vesicle compartment, again connected to the mother vesicle by a closed membrane neck.[b]

The formation of an out-bud and the subsequent cleavage of the membrane neck leads to two daughter vesicles that are 'mutually disjoint' in the sense that they enclose two disjoint spatial regions. Furthermore, the two daughter vesicles arising from an out-bud have the same bilayer asymmetry as the mother vesicle. In contrast, the formation of an in-bud and the subsequent neck fission leads to two daughter vesicles that are 'nested' in the sense that the spatial region enclosed by the larger daughter vesicle includes the smaller one. In addition, the two daughter vesicles have opposite bilayer asymmetries as indicated in Fig. 5(d).

Different views of closed membrane necks: Closed membrane necks look quite different when we look at them with nanoscale and mesoscale resolution. These two different views are displayed in Fig. 6. In panels a and b of this figure, we see a closed membrane neck at the nanoscale, as observed in molecular dynamics simulations. The lipid bilayer has a thickness of about 4 nm and forms a neck with an hourglass shape. The neck is axisymmetric and has a circular waistline. For a closed neck, the outer radius of this waistline is equal to twice the bilayer thickness whereas the inner radius vanishes. For the examples shown in Figs. 6(a) and (b), the vesicle size is about 36 nm, which is the radius of the spherical vesicles that we would obtain from the dumbbell-shaped vesicles by osmotic inflation. Inspection of the simulation snapshots shows that the hourglass-shaped neck is highly curved in the sense that its contour curvature is large and negative, whereas its other principal

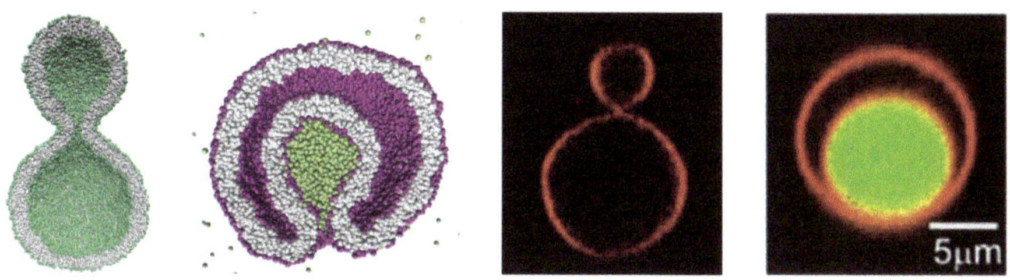

(a) (b) (c) (d)

Fig. 6. Closed membrane necks on nanoscopic and micrometer scales: (a,b) Hourglass shapes of closed necks on the nano scale, visualized *via* molecular dynamics simulations of nanovesicles with an out-bud in (a)[12] and an in-bud in (b).[10] (c,d) On the micrometer scale, closed necks appear to be pointlike as observed by fluorescence microscopy of giant unilamellar vesicles (GUVs) with fluorescently labeled lipids (red) with an out-bud in (c)[13] and an in-bud in (d).[14] In both (b) and (d), the membrane segment of the in-bud encloses a condensate droplet (green).

[b]In Ref. 11, the neck of an out-bud is viewed as an 'interior neck' because it is filled with the interior solution. Likewise, the neck of an in-bud is viewed as an 'exterior neck' because it is filled with exterior solution.

curvature is equal to $+1/R_{\text{ne}}$, which diverges as the neck radius R_{ne} vanishes during neck closure.

At the micrometer scale, the view of the membrane neck as obtained *via* light microscopy is quite different, see panels c and d of Fig. 6. Indeed, when observed under the light microscope, the hourglass-shaped membrane segment is no longer visible but is replaced by the touching point of the two membrane segments connected by the neck. On the one hand, the pointlike neck is a direct consequence of the limited optical resolution. On the other hand, one can characterize this pointlike neck locally in terms of two important properties. First, during neck closure, the two principal curvatures diverge, but the mean curvature attains a finite limit. Second, this mean curvature of the neck is directly related to the curvature radii of the two adjacent membrane segments which one can read off from the optical images. To derive these two features, we now make a short detour into the theory of curvature elasticity.

3.2. *Curvature elasticity of biomembranes*

Spontaneous curvature model: In order to obtain a quantitative description of vesicle shapes with closed membrane necks, we consider the spontaneous curvature model.[15–17] This model characterizes the membrane surface in terms of its (local) mean curvature M and its (local) Gaussian curvature G and describes the elastic curvature energy E_{cu} of the membrane by an area integral, as given by

$$E_{\text{cu}} = \int \mathrm{d}A \, \left[2\kappa(M-m)^2 + \kappa_G G\right] . \tag{3}$$

This expression contains three parameters: the spontaneous (or preferred) curvature m, the bending rigidity κ, and the Gaussian curvature modulus κ_G. The spontaneous curvature m provides a quantitative measure for the transbilayer asymmetry between the two leaflets of the bilayer membrane.

For a closed vesicle without membrane edges or pores, the Gauss–Bonnet theorem of differential geometry implies that the area integral over the Gaussian curvature is proportional to the Euler characteristic χ, as defined by Eq. (1), and equal to[6]

$$\int \mathrm{d}A \, G = 2\pi\chi = 2\pi(2 - 2\mathfrak{g}), \tag{4}$$

with the topological genus $\mathfrak{g} = 1 - \frac{1}{2}\,\chi$, as in Eq. (2). For a budding process, which does not involve a topological transformation, the constant energy term proportional to the Gaussian curvature modulus κ_G can be ignored and the curvature energy reduces to the elastic bending energy

$$E_{\text{be}} = \int \mathrm{d}A \, 2\kappa(M-m)^2, \tag{5}$$

which becomes small when the mean curvature M is close to the spontaneous curvature m. In order to determine the shape of a vesicle, the bending energy E_{be} has to be supplemented by constraints on the membrane area A and the vesicle volume V. At constant temperature, the membrane area A is constant, reflecting the ultralow solubility of the lipid molecules and membrane proteins. Likewise, the volume of the vesicle is conserved for constant pressure difference

$$\Delta P = P_{\text{ex}} - P_{\text{in}} \tag{6}$$

between the pressures P_{ex} and P_{in} of the exterior and interior solutions, which requires constant osmotic conditions. We are then led to minimize the vesicle's shape energy[16–18]

$$F_{\text{ve}} = \Delta P\, V + \Sigma A + E_{\text{be}} = \Delta P\, V + \Sigma A + 2\,\kappa \int \mathrm{d}A\,(M-m)^2 \tag{7}$$

and to treat the pressure difference ΔP and the membrane tension Σ as Lagrange multipliers, which are used to perform the constrained minimization of the bending energy for a prescribed vesicle volume V and a prescribed membrane area A. Several recent studies have demonstrated that the shapes of giant vesicles calculated in this manner agree quantitatively with the experimentally observed shapes,[8,13] provided the vesicle membrane contains a molecular component such as cholesterol that undergoes frequent flip-flops from one leaflet to the other. Otherwise, one has to take area-difference elasticity[19–21] into account, which leads to a spatially non-local term in the curvature energy.

Dimensionless shape parameters: In order to reduce the number of parameters, it is convenient to define the vesicle size R_{ve} in terms of the membrane area A via

$$R_{\text{ve}} \equiv \sqrt{A/(4\pi)} \tag{8}$$

and to use R_{ve} as the basic length scale. For a closed membrane compartment with volume V and membrane area A, the shape of this compartment depends primarily on two parameters;[16] the rescaled volume

$$v \equiv \frac{V}{\frac{4\pi}{3} R_{\text{ve}}^3} = 6\sqrt{\pi}\frac{V}{A^{3/2}}, \tag{9}$$

which satisfies $0 \leq v \leq 1$, and the rescaled spontaneous curvature

$$\bar{m} \equiv m R_{\text{ve}}, \tag{10}$$

which can be positive or negative. More precisely, for each point $(\bar{m}, v)$, one finds a discrete set of energy branches that differ in their bending energy. For simplicity, we have tacitly assumed here that the membrane is laterally uniform with a uniform molecular composition which implies laterally uniform curvature-elastic parameters.

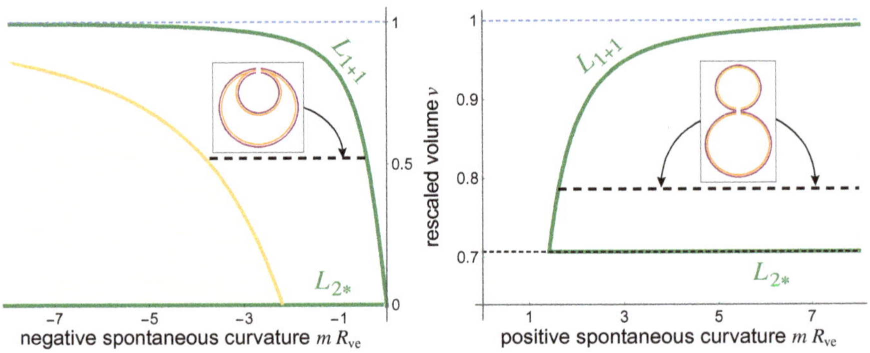

(a) (b)

Fig. 7. Stability regimes for (1+1)-multispheres with closed membrane necks: (a) For negative spontaneous curvature, stable (1+1)-shapes have in-buds, see inset and Fig. 5(c). (b) For positive spontaneous curvature, stable (1+1)-shapes exhibit out-buds, see inset and Fig. 5(a). Both stability regimes (light green) are bounded by two lines of limit shapes, L_{1+1} and L_{2*} (dark green). In (a), the stability regime is also bounded by the orange line, at which the small in-budded sphere undergoes a sphere-prolate instability. Within each stability regime, the multispherical shapes are independent of the spontaneous curvature $\bar{m}$ but depend only on the rescaled volume v as defined by Eq. (9). Thus, the (1+1)-multispherical shapes remain unchanged when $\bar{m}$ is varied for fixed v (horizontal dashed lines). Note the different range of v-values in (a) and (b).

3.3. *Stability regimes for closed membrane necks*

Closed membrane necks as displayed in Figs. 5(a) and (c) exhibit large stability regimes within the two-dimensional morphology diagram, defined by the two shape parameters v and $\bar{m}$ in Eqs. (9) and (10). These stability regimes are displayed in Fig. 7 for multispherical shapes consisting of one small and one large sphere, the so-called (1+1)-multispheres. In general, vesicles can form $(N_l + N_s)$-multispheres, consisting of N_l large and N_s small spheres.[13,22] The multispheres represent limit shapes in the sense that they can be reached from shapes with open necks by the closure of these necks. Therefore, the multispherical shapes have the same genus, $\mathfrak{g} = 0$, as a sphere.

For (1+1)-multispheres, we have to distinguish two stability regimes depending on the sign of the spontaneous curvature. For *negative* spontaneous curvatures, the vesicles form in-buds with closed membrane necks as displayed in Fig. 5(c). These in-budded shapes are stable within the light green stability regime of Fig. 7(a). On the other hand, out-budded shapes as in Fig. 5(a) form stably closed necks in the light green stability regime of Fig. 7(b), which applies to *positive* spontaneous curvature.

3.4. *Local stability conditions for closed necks*

The stability of a closed membrane neck depends on the mean curvatures M_a and M_b of the two membrane segments adjacent to the neck. Inspection of Fig. 5 shows that the mean curvature of the mother vesicle, which we denote by M_a, is always positive. The mean curvature of the bud will be denoted by M_b. For an out- and in-bud, M_b is positive and negative, respectively. If the out- and in-budded vesicles are (1+1)-multispheres as in Fig. 5, the mean curvatures can be expressed in terms of the sphere radii R_a and R_b with $M_a = 1/R_a$ and $M_b = \pm 1/R_b$. The two mean curvatures of the membrane segments adjacent to the membrane neck define the neck curvature[7,13]

$$M_{\rm ne} \equiv \frac{1}{2}(M_a + M_b) = \frac{1}{2}\left(\frac{1}{R_a} \pm \frac{1}{R_b}\right), \tag{11}$$

where the first equality applies to the closed neck of any vesicle shape, whereas the second equality applies to (1+1)-multispheres with the plus and minus sign for out- and in-buds. For an out-budded vesicle shape as in Fig. 5(a), both M_a and M_b are positive, which implies a positive neck curvature $M_{\rm ne}$. For an in-budded vesicle shape as in Fig. 5(c), the membrane segment of the mother vesicle has a positive mean curvature $M_a > 0$, whereas the membrane segment of the in-bud has a negative mean curvature $M_b < 0$ with $|M_b| > M_a$. The latter inequality is necessary in order to accommodate the in-bud within the mother vesicle without intersections between the two membrane segments. As a consequence, the neck curvature $M_{\rm ne}$ is negative for an in-budded vesicle.

A closed membrane neck with positive neck curvature $M_{\rm ne}$ is stable if the spontaneous curvature m is *positive* and sufficiently large with[7,17]

$$m \geq M_{\rm ne} > 0 \quad \text{(stably closed neck for positive } m\text{)}. \tag{12}$$

The limiting case with $m = M_{\rm ne} > 0$ corresponds to the green line of limit shapes L_{1+1} in Fig. 7(b), which provides one boundary for the green shaded stability regime of (1 + 1)-multispheres with positive spontaneous curvature. The second boundary of this stability regime, denoted by L_{2*} in Fig. 7(b), corresponds to the limit shapes of two equally sized spheres. The line L_{2*} is a direct consequence of the geometry for (1+1)-spheres with out-buds. Indeed, this geometry implies that the rescaled volume v cannot be smaller than $v_2 = 1/\sqrt{2} = 0.7071$. Each line of limit shapes separates the stability region of the (1+1)-multispherical shapes from a region of the morphology diagram, in which the vesicles have a smoothly curved shape with open necks.

For a closed neck with negative neck curvature $M_{\rm ne}$, the closed membrane neck is stable if the spontaneous curvature m is *negative* and has a sufficiently large absolute value with[11,17]

$$m \leq M_{\rm ne} \leq 0 \quad \text{(stably closed neck for negative } m\text{)}. \tag{13}$$

The limiting case with $m = M_{\rm ne} < 0$ corresponds to the green line of limit shapes L_{1+1} in Fig. 7(a), which provides one boundary for the green shaded stability regime of (1 + 1)-multispheres with negative spontaneous curvature. The second boundary, denoted by L_{2*} in Fig. 7(a), corresponds to the line of limit shapes L_{2*}, for which the in-bud has the same size as the mother vesicle. The line L_{1+1} again separates the stability regime of the $(1+1)$-multispheres from smoothly curved vesicle shapes with open necks. For negative spontaneous curvatures, the limit shapes L_{2*} of two equally sized spheres are located at $v = v_2 = 0$ which implies that one cannot cross this line to reach some smoothly curved vesicle shapes.

The stability relations for stably closed necks in Eqs. (12) and (13) were originally obtained from numerical calculations for axially symmetric vesicle shapes approaching the limit shapes L_{1+1}.[16] For out-budded vesicles, the stability relations were also obtained from a certain parametrization of the shape in terms of piecewise constant-mean-curvature surfaces.[23] The stability relations for (1+1)-multispheres have been generalized to vesicles with two intramembrane domains[24] and non-axisymmetric vesicle shapes.[25] The limit shapes L_{2*} of two equally sized spheres follow from a detailed analysis of the multisphere geometry.[7] In addition, the stability relations for closed necks lead to constriction forces acting on these necks. These constriction forces increase with the absolute value of the spontaneous curvature and are enhanced both by the line tension of domain boundaries and by the complete engulfment of adhesive nanoparticles, see Section 3.7.

3.5. *Geometric relation between volume and neck curvature*

As mentioned, the neck curvature M_{ne} is equal to the spontaneous curvature m along the two lines of limit shapes L_{1+1} in the two panels of Fig. 7. For these two lines, one can obtain an explicit expression for the volume v as a function of the spontaneous curvature m in the $(\bar{m}, v)$ morphology diagram.[16,17] Substituting the rescaled spontaneous curvature $\bar{m}$ by the rescaled neck curvature

$$\bar{M}_{\text{ne}} \equiv M_{\text{ne}} R_{\text{ve}} = \frac{1}{2}\left(\frac{R_{\text{ve}}}{R_a} \pm \frac{R_{\text{ve}}}{R_b}\right), \tag{14}$$

one obtains the explicit relationship

$$v = v_{1+1}(x) \equiv -\frac{1}{4x^3} + \left(1 - \frac{1}{2x^2}\right)\sqrt{1 + \frac{1}{4x^2}} \quad \text{with} \quad x \equiv \bar{M}_{\text{ne}} \tag{15}$$

between the rescaled volume v and the rescaled neck curvature $\bar{M}_{\text{ne}}$. The purely geometric relationship in Eq. (15) applies both to the first branch of limit shapes L_{1+1} with negative spontaneous curvature in Fig. 7(a) and to the second branch of L_{1+1} with positive spontaneous curvature in Fig. 7(b). Furthermore, inspection of panels a and b of Fig. 7 shows that the volume $v = v_{1+1}(\bar{M}_{\text{ne}})$ as given by the relationship in Eq. (15) increases monotonically for increasing positive $\bar{M}_{\text{ne}} \geq \sqrt{2}$ and for decreasing negative $\bar{M}_{\text{ne}} \leq 0$. This monotonicity implies that the functional relation in Eq. (15) can be inverted to obtain the neck curvature $\bar{M}_{\text{ne}}$ as a function of v for both branches of limit shapes L_{1+1}.

3.6. *Emergence of closed neck conditions*

Next, we discuss the closure of the membrane neck in more detail, starting from the hourglass-shaped neck in panels a and b of Fig. 6, as observed on the nanoscale. In these simulation snapshots, the waistline (wl) of the neck forms a circle with radius $R_{\rm ne}$. Along this waistline, the neck is characterized by two principal curvatures, the negative contour curvature $C_{1,\rm wl} < 0$ perpendicular to the waistline and the positive principal curvature $C_{2,\rm wl} = 1/R_{\rm ne} > 0$ parallel to the waistline. When the neck closes, the neck radius goes to zero and the principal curvature $C_{2,\rm wl}$ diverges. However, the mean curvature

$$M_{\rm wl} = \frac{1}{2}\left(C_{1,\rm wl} + C_{2,\rm wl}\right)$$

remains finite and satisfies the asymptotic equality

$$M_{\rm wl} \approx M_{\rm ne} \equiv \frac{1}{2}\left(M_a + M_b\right) \tag{16}$$

in the limit of small $R_{\rm ne}$,[26] with the mean curvature $M_{\rm ne}$ of the closed neck determined by the mean curvatures M_a and M_b of the two membrane segments, a and b, adjacent to the neck. Thus, as the neck closes, the positive singular contribution from the second principal curvature $C_{2,\rm wl} = 1/R_{\rm ne} > 0$ is canceled by a negative contribution arising from the contour curvature $C_{1,\rm wl}$. Therefore, neck closure involves a singular catenoid with vanishing mean curvature that does not contribute to the mean curvature of the membrane neck. After the cancelation of the two singular contributions, we are left with the neck curvature $M_{\rm ne}$ as given by Eq. (16), which is positive for out-buds with $M_a > 0$ and $M_b > 0$ but negative for in-buds with $M_a > 0$ and $M_b < -M_a < 0$.

3.7. *Constriction forces at closed membrane necks*

Curvature-induced constriction forces: As mentioned, the shape of a (1+1)-multisphere depends only on the rescaled volume v and, thus, does not change when we vary the spontaneous curvature $\bar{m}$ within the stability regimes, see horizontal dashed lines in Fig. 6. However, this variation leads to a change in the constriction force acting at the membrane neck. Indeed, the closed membrane necks displayed in panels a and c of Fig. 5 can experience significant constriction forces, f, that increase with the absolute value of the spontaneous curvature. These constriction forces have the form[7,8]

$$f \approx 8\pi\kappa(m - M_{\rm ne}) \quad \text{for } 0 < M_{\rm ne} \leq m, \tag{17}$$

$$f \approx 8\pi\kappa(M_{\rm ne} - m) \quad \text{for } m \leq M_{\rm ne} \leq 0, \tag{18}$$

where the asymptotic equalities are valid to leading order in the neck radius $R_{\rm ne}$. These constriction forces vanish along the line of limit shapes, L_{1+1}, for which the neck curvature $M_{\rm ne}$ is equal to the spontaneous curvature m. Furthermore, as we vary the spontaneous curvature, the constriction force is proportional to the excess curvature $m - M_{\text{ne}}$ for $m > 0$ and to the excess curvature $M_{\text{ne}} - m$ for $m < 0$.

The constriction forces in Eqs. (17) and (18) can be obtained in a simple and direct manner starting from the bending energy E_{be} of a dumbbell shape with an open neck of radius R_{ne}, which has the form[7]

$$E_{\text{be}}(R_{\text{ne}}) \approx E_{\text{be}}(0) \pm 8\pi\kappa(m - M_{\text{ne}})R_{\text{ne}} \tag{19}$$

up to first order in R_{ne} where the plus and minus sign applies to positive and negative spontaneous curvature, respectively. The constriction force f compressing the membrane neck is then given by

$$f = \partial E_{\text{be}}/\partial R_{\text{ne}}, \tag{20}$$

which leads to Eqs. (17) and (18) depending on the sign in Eq. (19). In both cases, the force depends only on local properties of the membrane segments close to the neck. Therefore, the constriction forces in Eqs. (17) and (18) are not restricted to the necks of out- and in-buds but apply to any neck with $m \geq M_{\text{ne}} > 0$ and to any neck with $m \leq M_{\text{ne}} \leq 0$, respectively. In particular, these constriction forces apply to the necks of multispheres to be considered in Section 6.

When we look at a closed neck with molecular resolution as in panels a and b of Fig. 6, we see that the membrane neck is formed by two bilayer segments in close contact and that the neck diameter is about twice the bilayer thickness. For an open neck, the two bilayer segments are fully hydrated and each bilayer has a thickness of 4 to 5 nm, depending on the chain length of the lipid molecules. This bilayer thickness will persist until neck closure with $m = M_{\text{ne}}$ and vanishing constriction force, $f = 0$, but will start to decrease for $m > M_{\text{ne}}$ when the two segments are pushed together by a positive constriction force, $f > 0$. In the latter case, the constriction force is balanced by an elastic counter force arising from the locally compressed bilayer segments.

Magnitude of curvature-induced constriction forces: As an example, let us consider a GUV enclosed by a lipid membrane with positive spontaneous curvature and bending rigidity $\kappa = 48\, k_{\text{B}}T$, as studied experimentally in Ref. 8, see Section 4.1. The corresponding curvature-induced constriction force f is depicted in Fig. 8(a) and plotted in Fig. 8(b) as a function of the excess curvature $m - M_{\text{ne}}$. For comparison, this plot also includes literature values for constriction forces, as generated by specialized protein complexes *in vivo*. Inspection of Fig. 8(b) reveals that the constriction force acting on the closed neck of the giant vesicle in Fig. 8(a) lies within the range of constriction forces generated by proteins *in vivo*.

Constriction forces enhanced by line tension of domain boundaries: The curvature-induced constriction force is enhanced if the neck contains a domain

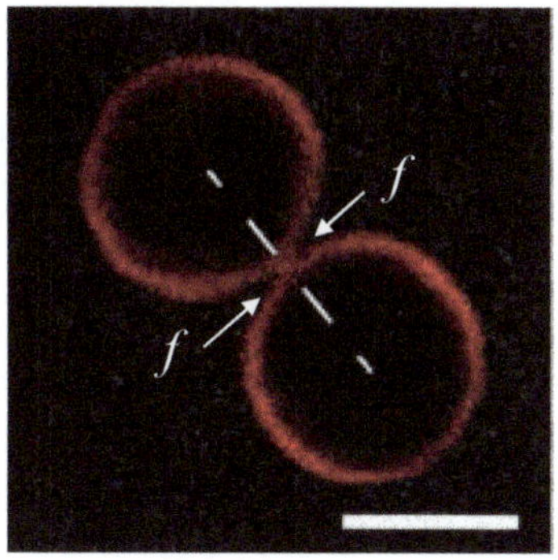

(a)

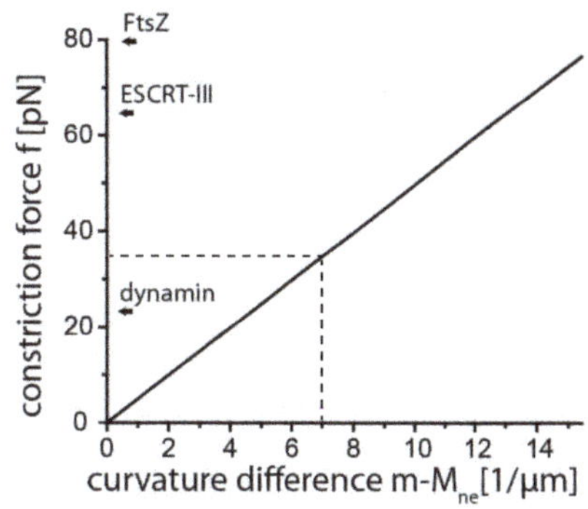

(b)

Fig. 8. Curvature-induced constriction force f at closed membrane neck: (a) Confocal image of a GUV membrane (red) that consists of two equally sized spheres connected by a closed neck. The spontaneous curvature is generated by His-tagged GFP which binds to the membrane from the exterior solution. The constriction force f (white arrows) acts to compress this neck. Scale bar: $5\,\mu$m. (b) Constriction force f as a function of the excess curvature $m - M_{\rm ne}$. The straight line corresponds to Eq. (17) with $\kappa = 48\,k_{\rm B}T$. The dashed lines represent the values of $m - M_{\rm ne}$ and f for the membrane neck in (a). For comparison, the plot also includes literature values for the constriction forces as generated by the specialized protein complexes of dynamin,[27] ESCRT-III,[28] and FtsZ.[29] Thus, the constriction force in (a) lies within the range of force values found *in vivo*.[8]

boundary between two membrane domains a and b that differ in their molecular composition. The line tension $\lambda_{\rm db}$ of the domain boundary must be positive to ensure that the domains are thermodynamically stable. If the two membrane domains have comparable curvature-elastic properties, the constriction force is given by[11]

$$f \approx 8\pi\kappa(m - M_{\rm ne}) + 2\pi\lambda_{\rm db} \quad \text{for } 0 < M_{\rm ne} \leq m + \lambda/(4\kappa), \tag{21}$$

$$f \approx 8\pi\kappa(M_{\rm ne} - m) + 2\pi\lambda_{\rm db} \quad \text{for } 0 \geq M_{\rm ne} \geq m - \lambda/(4\kappa). \tag{22}$$

Therefore, the line tension $\lambda_{\rm db}$ of the domain boundary enhances the constriction force f at the closed neck for both out-budded and in-budded vesicles. In both cases, the neck curvature $M_{\rm ne}$ is again a purely geometric quantity, compare Section 3.5, whereas the spontaneous curvature m and the line tension $\lambda_{\rm db}$ are material parameters.

In general, the two membrane domains a and b may have different curvature-elastic properties. The corresponding constriction forces are obtained from Eqs. (21) and (22) by replacing the terms $\pm 8\pi\kappa(m - M_{\rm ne})$ by $\pm 4\pi[\kappa_a(m_a - M_a) + \kappa_b(m_b - M_b)]$ while the line tension terms remain unchanged.[11]

Constriction forces enhanced by adhesion: The constriction forces are also enhanced by the engulfment of adhesive nanoparticles. The competition between the adhesion energy per unit area, W, and the bending rigidity κ is encoded in the adhesion length[30,31]

$$R_W = \sqrt{2\kappa/|W|}. \tag{23}$$

If the adhesive particle is completely engulfed by an out-bud, which initiates the process of particle exocytosis, the constriction force f has the form[32]

$$f \approx 8\pi\kappa(m - M_{\rm ne}) + \frac{4\pi\kappa}{R_W} = 8\pi\kappa\left(m - M_{\rm ne} + \frac{1}{2R_W}\right), \tag{24}$$

which implies

$$f \geq 0 \quad \text{for} \quad 0 < M_{\rm ne} \leq m + \frac{1}{2R_W} = m + \sqrt{\frac{|W|}{8\kappa}}. \tag{25}$$

On the other hand, if the adhesive particle is completely engulfed by an in-bud, the first step during particle endocytosis, the constriction force is given by[32]

$$f \approx 8\pi\kappa(M_{\rm ne} - m) + \frac{4\pi\kappa}{R_W} = 8\pi\kappa\left(M_{\rm ne} - m + \frac{1}{2R_W}\right), \tag{26}$$

which implies

$$f \geq 0 \quad \text{for} \quad 0 \geq M_{\rm ne} \geq m - \frac{1}{2R_W} = m - \sqrt{\frac{|W|}{8\kappa}}. \tag{27}$$

Therefore, the complete engulfment of an adhesive nanoparticle leads to an increase of the constriction force f at the closed neck by

$$\Delta f = \frac{4\pi\kappa}{R_W} = 2\pi\sqrt{2\kappa|W|} \tag{28}$$

for both out-budded and in-budded vesicles. This adhesion-induced enhancement of the constriction force agrees with the intuitive notion that it will be more difficult to open the neck when it is stabilized by adhesion.

The adhesion-induced enhancement of the constriction force has been derived for the engulfment of spherical nanoparticles, which are solid, have a fixed shape, and cannot be deformed by the membrane. Strictly speaking, the latter constraint does not apply to the engulfment of condensate droplets. On the other hand, when these droplets are completely engulfed, as in Figs. 6(b) and 6(d), they usually attain a spherical shape. In this case, the adhesion-induced enhancement of the constriction force is expected to apply to liquid droplets as well, provided we identify the parameter W with the affinity contrast between the two aqueous phases.[33]

Large constriction forces drive fission: We now consider an out-budded or in-budded vesicle with a closed membrane neck and increase the absolute value of the spontaneous curvature $\bar{m}$, thereby moving along one of the horizontal dashed lines in Fig. 7. When we increase the absolute value $|\bar{m}|$, the shape of the (1+1)-vesicle remains unchanged, but the constriction forces as described by Eqs. (17)–(24) increase. A sufficiently large constriction force acting on an individual membrane neck leads to the fission of this neck and to the division of the vesicle. Such fission processes have been recently observed for several biomimetic membrane systems, as will be described in Section 4.

3.8. *Free energy landscape for fission*

The fission process involves the closure and subsequent cleavage of a membrane neck, as shown schematically in Figs. 9(a)–(c), where we display the fission of a small nanovesicle in order to explicitly resolve the lipid bilayer. The neck cleavage has to overcome a free energy barrier arising from the local disruption of the bilayer structure. On the supramolecular scale, one may envisage this barrier for neck fission to arise from a cut across the bilayer and the creation of two ring-like hydrophobic bilayer edges across the neck. The 'healing' of these bilayer edges then leads to two separate daughter vesicles. Furthermore, the energy barrier associated with fission is proportional to the total length of the bilayer edges, which is minimal if the bilayer is cut across the neck.

A cut across the neck creates two bilayer pores with a diameter that is comparable to the size of the closed membrane neck before fission. The resulting free

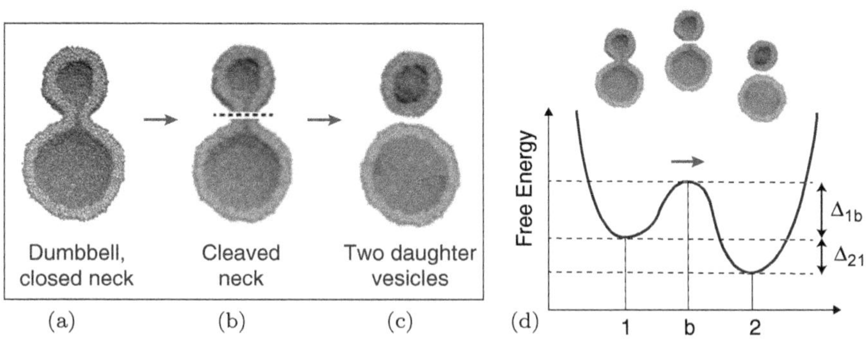

Fig. 9. Fission of small nanovesicle and corresponding free energy landscape: (a–c) Schematic fission process which starts from a dumbbell with a closed neck in (a), proceeds via the cleavage of this neck in (b), corresponding to the shortest possible cut (broken line) across the vesicle membrane, and ends up with two separate daughter vesicles in (c). The cut leads to two circular membrane edges which have a positive line tension and determine the free energy barrier. (d) Schematic free energy landscape versus an abstract reaction coordinate, with the free energy barrier provided by the cleaved neck. For the fission process to be 'downhill' (exergonic) in free energy, the free energy difference Δ_{21} between the two vesicle state 2 and the dumbbell state 1 must be negative. For $\Delta_{21} < 0$, the fission process can proceed without being coupled to another process such as nucleotide hydrolysis, but the fission rate is determined by the positive free energy barrier Δ_{1b}.

energy barrier is governed by the edges of these two pores and the associated edge energy, which is equal to the edge tension λ_{ed} times the combined circumference of the two pores. To lower this barrier by a significant amount, the constriction force f in Eq. (17) must perform mechanical work that is comparable to the edge energy.[7,17] One then finds that the neck undergoes fission if the spontaneous curvature m exceeds the threshold value $m_* \equiv \lambda/(2\kappa)$. This threshold value is reduced if the neck contains a domain boundary between two intramembrane domains, see Eqs. (21) and (22), or if the membrane segment adjacent to the neck contains an adhesive particle or droplet, see Eqs. (24) and (26).

The process displayed in Fig. 9 represents a topological transformation from a single vesicle to two separate daughter vesicles. Both states have essentially the same bending energy because the bending energy of the closed neck is proportional to its area which is much smaller than the total membrane area of the vesicle. However, the two states have different topologies which implies that the Gaussian curvature term in Eq. (3) makes a different contribution to the one- and two-vesicle states. The latter contribution is equal to $4\pi\kappa_G$ for the one-vesicle state and to $8\pi\kappa_G$ for the two-vesicle state, where κ_G is the Gaussian curvature modulus. Therefore, the difference in free energy, $G_2 - G_1$ between the two- and the one-vesicle state is equal to $4\pi\kappa_G$.

Both experimental studies[34,35] and computer simulations[36] indicate that the Gaussian curvature modulus is negative with $\kappa_G \simeq -\kappa$. For the lipid membranes studied here, we then obtain the estimate $\kappa_G \simeq -48 k_B T$ which leads to the free energy difference $G_2 - G_1 \simeq -603\, k_B T$. Therefore, neck fission and GUV division is a strongly exergonic process and can, in principle, occur spontaneously. However, the rate with which this process proceeds is governed by the free energy barrier that separates the one-vesicle from the two-vesicle state.

3.9. *Contact lines with positive and negative line tensions*

In the previous sections, we implicitly assumed that the closure of a membrane neck proceeds in an axisymmetric manner. This assumption applies to the neck closure of GUV membranes with a uniform composition as in Fig. 8. The same assumption also applies to domain-induced budding of vesicle membranes, for which the membrane neck contains the domain boundary between two different membrane domains, a and b. As mentioned, the line tension λ_{db} of such a domain boundary is always positive, which implies that this boundary prefers to attain a circular shape. On the other hand, the assumption about axisymmetry is less obvious for the engulfment of condensate droplets because the line tension of the contact line between membrane and droplets is not necessarily positive but can be negative as well.

Indeed, strongly non-axisymmetric membrane necks have been observed in molecular dynamics simulations for the engulfment of condensate droplets by planar[37] and by vesicle[10] bilayers. In both cases, the contact line between droplet and membrane becomes strongly non-circular, which reveals a *negative* line tension. As a consequence, the membrane neck close to the contact line attains an unusual, tight-lipped shape. Such a tight-lipped neck has also been observed for the membrane engulfment of solid and undeformable nanoparticles by planar bilayers.[38] A tight-lipped membrane neck prevents the fission of the neck and, thus, the division of the vesicle.

For a contact line between condensate droplet and membrane, the sign of the line tension is controlled by the initial stress asymmetry $\Delta\Sigma = \Sigma_1 - \Sigma_2$ between the two leaflet tensions, Σ_1 and Σ_2, of the two leaflets, as measured in the absence of the droplet.[10,38] A relatively small stress asymmetry leads to a negative line tension and a tight-lipped membrane neck, which does not undergo fission, whereas a sufficiently large stress asymmetry leads to a positive line tension, which ensures that neck closure proceeds in an axisymmetric manner and that the closed neck undergoes fission.[10,38]

4. Fission of Biomimetic Membranes

4.1. *Division of giant unilamellar vesicles*

A particularly amazing property of living cells is their ability to divide into two daughter cells. During cell division, the plasma membrane undergoes a series of morphological transformations which ultimately lead to membrane fission. Analogous remodeling processes have been induced by low densities of proteins bound to the membranes of GUVs.[8] When these proteins are fluorescently labeled and bound to the membranes via histidine-tags and anchor lipids, an unprecedented control over the spontaneous curvature has been achieved. In fact, the spontaneous curvature can be directly controlled by the molar concentration of the histidine-tagged proteins in the exterior solution. Fine-tuning of this curvature leads to dumbbell-shaped vesicles with closed membrane necks as in Fig. 8(a). Since the proteins bind to the outer leaflet of the GUV membranes, the induced spontaneous curvature is positive and generates a significant constriction force, see Fig. 8(b), which is sufficient to cleave the neck and divide the GUV, as illustrated in Fig. 10 using green fluorescent protein (GFP). These experiments confirm the curvature-induced constriction force, as described by Eq. (17) in a quantitative manner.

The division of GUVs as illustrated in Fig. 10 involves two key parameters: the (rescaled) spontaneous curvature $\bar{m}$ and the (rescaled) vesicle volume v. The spontaneous curvature is controlled by adding histidine-tagged proteins to the exterior solution of the GUVs. The vesicle volume can be directly controlled by the osmotic conditions. For a sufficiently small volume, a dumbbell shape is reached as in Fig. 10(a). Increasing the spontaneous curvature of the GUV membrane by adding GFP to the exterior solution then leads to the two-sphere shape with a closed

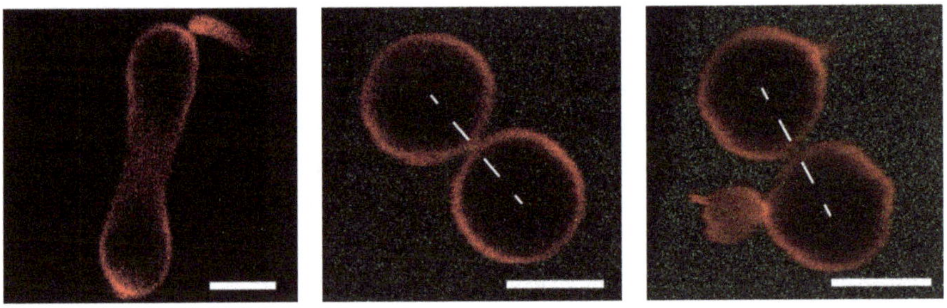

(a) (b) (c)

Fig. 10. Curvature-induced division of a GUV: (a) Dumbbell-shaped vesicle obtained by osmotic deflation. (b) When green fluorescent protein (GFP) is added to the exterior solution, the vesicle attains a two-sphere shape with a closed neck. The GFP binds to anchor lipids in the outer leaflet of the GUV membrane, generating a positive spontaneous curvature as well as a substantial constriction force compressing the neck. (c) Increasing the spontaneous curvature by adding more GFP, the membrane neck is cleaved and the GUV is divided into two daughter vesicles, even though the GFP concentration is still very low and only 39 nM. All scale bars: 5 μm.[8]

neck, as displayed in Fig. 10(b). This shape is close to the line of limit shapes L_{2*} in Fig. 7(b). Adding more GTP, we move along this line toward larger values of the spontaneous curvature, eventually reaching a sufficiently large constriction force that cleaves the neck. It is important to note that neck fission and vesicle division are obtained for very low GTP concentrations in the nanomolar regime. The membrane neck in Fig. 10(b), for example, was cleaved for a GTP concentration of only 39 nM.

Comparison with alternative experimental approaches: Other experimental approaches have been pursued in order to achieve the division of lipid vesicles.[26,39–44] In what follows, these alternative approaches will be briefly summarized. First, PEG-dextran solutions within a GUV undergo aqueous phase separation into a PEG-rich and a dextran-rich droplet. The GUV membrane enclosing these two droplets can form a closed membrane neck in between the two droplets, thereby partitioning the membrane into two segments in contact with the PEG-rich and dextran-rich phase.[39] However, these two membrane segments remained connected by a membrane nanotube (or 'tether'), which prevented their genuine division. Another approach was based on the reconstitution of the bacterial division machinery provided by FtsZ proteins. The formation of relatively large rings of these proteins within GUVs has been observed, albeit with rather low frequency.[40] For about 1.2% of the GUVs, protein rings were observed to induce progressive constrictions of the GUVs, in some cases leading to two subcompartments connected by closed membrane necks. However, the subsequent division into two separate membrane compartments has not been reported. Furthermore, the buffer used in this latter approach involved two nucleotides, ATP and GTP. In the absence of GTP, no progressive constrictions of GUVs could be detected.

In Ref. 41, extruded vesicles with a diameter of 200 nm were exposed to histidine-tagged GFP. The vesicle membranes contained 20 mol% anchor lipids and were exposed to a GFP solution concentration of 5 and 20 µM. As a consequence, the anchor-lipid mole fraction was at least 20 times higher and the GFP solution concentration was at least 128 times larger compared to the GUV system described by us in Ref. 8 and displayed in Fig. 10. In fact, it was concluded in Ref. 41 that crowding of membrane-bound GFP is a prerequisite for successful vesicle division. In contrast, our results clearly demonstrate that vesicle division can be achieved by membrane-bound proteins in the dilute regime, in which the separation of the membrane-bound proteins is much larger than their lateral size.

A microfluidic-based approach was used in Ref. 42 to mechanically split GUVs into two daughter vesicles. The GUVs were formed from double-emulsion droplets and subsequently flowed against the sharp edge of a wedge-shaped microfluidic junction. The resulting division process competed primarily with two alternative outcomes, bursting of the GUV and futile division attempts with no splitting (called 'snaking' in Ref. 42). As a consequence, the probability for division was observed to depend strongly on the size of the GUV and follow a bell-shaped curve with a maximum at a GUV diameter of about 6 µm. For the latter size, the division probability was about 0.38. Both for smaller and for larger sizes, the division probability decreased rapidly to zero: For smaller sizes, futile division attempts became the typical outcomes, whereas larger GUVs were destroyed by bursting. In contrast, our approach (Figs. 8 and 10) should be size-independent because the curvature-induced constriction force as given by Eq. (17) compresses the neck locally and does not involve the size of the GUV.

GUVs enclosing Min proteins undergo shape oscillations,[43] during which the vesicles change their neck size in a recurrent manner.[26] These oscillations are driven by the Min proteins that bind to and unbind from the vesicle membrane,[45] with the detachment being controlled by the ATP hydrolysis. The observed shape oscillations can be understood in terms of a spontaneous curvature that undergoes cyclic changes in time.[26] During each oscillation cycle, the vesicle shape is transformed from a symmetric dumbbell with two equally sized subcompartments to an asymmetric dumbbell with two subcompartments that differ in their size, followed by the reverse, symmetry-restoring transformation.

As previously mentioned, see Eq. (21), the constriction force compressing the membrane neck can be enhanced by the line tension λ_{db} of a domain-boundary between two membrane domains when this boundary is located at the membrane neck. Thus, the presence of such a domain boundary should facilitate the fission of the neck. The latter process may have been observed in Ref. 44, but the published optical images make it difficult to decide whether or not the two membrane segments remain connected by a membrane nanotube as in Ref. 39.

4.2. *Division of unilamellar nanovesicles*

Division via large transbilayer asymmetry: In molecular dynamics simulations, the transbilayer asymmetry of vesicle bilayers can be directly controlled by the number of lipids, $N_{\rm il}$ and $N_{\rm ol}$, which are assembled in the inner and outer leaflets of the vesicle bilayer. Instead of the lipid numbers, which are extensive variables, one can consider the leaflet tensions $\Sigma_{\rm il}$ and $\Sigma_{\rm ol}$ of the inner and outer leaflets, which represent intensive variables. The sum of the two leaflet tensions is equal to the bilayer tension, $\Sigma = \Sigma_{\rm ol} + \Sigma_{\rm il}$. The stress asymmetry between the two leaflets of the vesicle bilayer is given by[10,38,46]

$$\Delta\Sigma = \Sigma_{\rm ol} - \Sigma_{\rm il}, \tag{29}$$

which vanishes for symmetric bilayer states with $\Sigma_{\rm ol} = \Sigma_{\rm il}$. On the other hand, the leaflets of tensionless bilayers with vanishing bilayer tension, $\Sigma = \Sigma_{\rm ol} + \Sigma_{\rm il} = 0$, experience opposite leaflet tensions, $\Sigma_{\rm il} = -\Sigma_{\rm ol}$, corresponding to one stretched and one compressed leaflet. One particularly important bilayer state is provided by the reference state with tensionless leaflets and $\Sigma_{\rm ol} = \Sigma_{\rm il} = 0$.

In our first nanovesicle simulations,[12] we focused on nanovesicles, which were assembled from a fixed total number of $N_{\rm ol} + N_{\rm il} = 10100$ lipids in both leaflets, and varied the lipid number $N_{\rm ol}$ in the outer leaflet from $N_{\rm ol} = 5700$ to $N_{\rm ol} = 6300$. These vesicles have a radius of $22\,d$, where $d \simeq 0.8\,{\rm nm}$ is the basic length scale used to parametrize the coarse-grained molecular model. For $N_{\rm ol} = 5700$ and $N_{\rm il} = 4400$, the spherical vesicles with tensionless bilayers experience the stress asymmetry $\Delta\Sigma = 1.69\,\gamma$, with the tension scale $\gamma \equiv k_{\rm B}T/d^2$, which has the value $\gamma \simeq 6.4\,{\rm mN/m}$ at room temperature. Starting with such a spherical vesicle, a reduction of the vesicle volume creates an in-budded vesicle with an open neck.[12] When the lipid number $N_{\rm ol}$ in the outer leaflet is further reduced to $N_{\rm ol} = 5500$, the stress asymmetry of the tensionless bilayer is increased to $\Delta\Sigma = 2.67\,\gamma$. For the latter stress asymmetry, the reduction of the vesicle volume leads to an in-bud with a closed neck that undergoes fission, see Fig. 11.

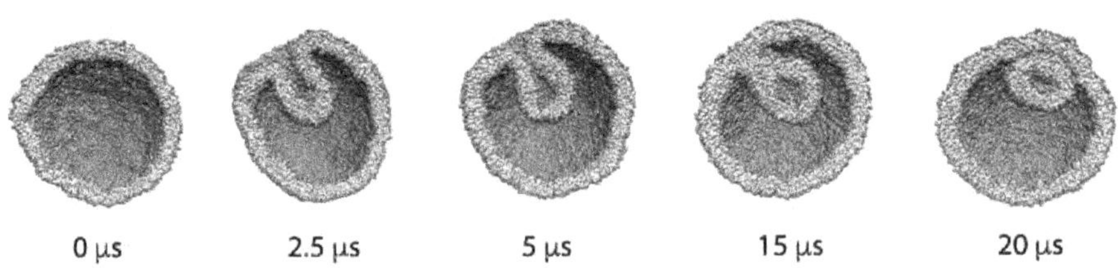

Fig. 11. Time-lapse snapshots for the division of a nanovesicle with $N_{\rm ol} = 5500$ lipids in the outer and $N_{\rm il} = 4600$ lipids in the inner leaflet. Until time $t = 0\,\mu{\rm s}$, the vesicle has a spherical shape with a certain volume, as determined by the number of water molecules within the interior vesicle compartment. At $t = 0\,\mu{\rm s}$, the vesicle volume is reduced by 20%. After this volume reduction, the vesicle develops an in-bud with a membrane neck that is closed at $t = 5\,\mu{\rm s}$. At about $t = 15\,\mu{\rm s}$, the neck undergoes fission, thereby generating an intraluminal daughter vesicle that adheres to the inner leaflet of the larger daughter vesicle.[38]

Nanovesicle division via solute adsorption: Budding and fission of nanovesicles can also be induced by the adsorption of small solutes onto the vesicle membranes.[9] We used an off-lattice model for a binary mixture consisting of water and solute beads and studied this model by molecular dynamics simulations. In Ref. 9, we focused on low solute concentrations and on the one-phase region of the binary mixture, in which the solutes formed adsorption layers on the outer leaflet of the vesicle membranes. This adsorption generated a significant spontaneous curvature that could be used to change the morphology of the vesicles.[c] In addition to the solute concentration, we identified the solvent conditions as a second key parameter for these processes. By definition, good solvent conditions imply that the solution remains spatially uniform for all solute concentrations. A poor solvent, on the other hand, leads to a certain range of solute concentrations in which the solution undergoes liquid–liquid phase separation. Examples for this kind of phase behavior are provided by aqueous two-phase systems,[47–49] by water-in-water emulsions,[50] and by biomolecular condensates.[51,52]

For *good* solvent conditions, the budding of a nanovesicle can be controlled by reducing the vesicle volume for constant solute concentration or by increasing the solute concentration for constant vesicle volume. After the budding process has been completed, the budded vesicle consists of two membrane subcompartments which are connected by a closed membrane neck. The budding process is reversible, as can be demonstrated explicitly by reopening the closed neck via a decrease of the solute concentration. For *poor* solvent conditions, on the other hand, we observed two unexpected morphological transformations of the nanovesicles, as shown by the simulation snapshots in Fig. 12. Close to the binodal line, at which the aqueous solution undergoes phase separation, the vesicle exhibits recurrent shape changes with closed and open membrane necks, see the simulation snapshots in Fig. 12(a), which cover a time period of about 35 μs and are reminiscent of the flickering fusion pores (kiss-and-run) that have been observed for synaptic vesicles.[53–55]

As we approach the binodal line even closer, the recurrent shape changes are truncated by the fission of the membrane neck which leads to the division of the nanovesicle into two daughter vesicles, see the simulation snapshots in Fig. 12(b), which cover a relatively short time period of less than a microsecond and show the last neck closure and the subsequent fission event. Inspection of Fig. 12(b) and the associated time-lapse movie reveals that the underlying fission mechanism is provided by the solute-mediated adhesion of the two membrane segments close to the neck, which leads to a growing, non-axisymmetric contact area between these two segments. In this way, our simulations reveal a nanoscale mechanism for the

[c]In Ref. 9, the solute molecules were contained in the exterior solution and their adsorption onto the outer vesicle leaflet generated a positive spontaneous curvature. A negative spontaneous curvature is obtained when the solutes are adsorbed from the interior solution onto the inner vesicle leaflet.

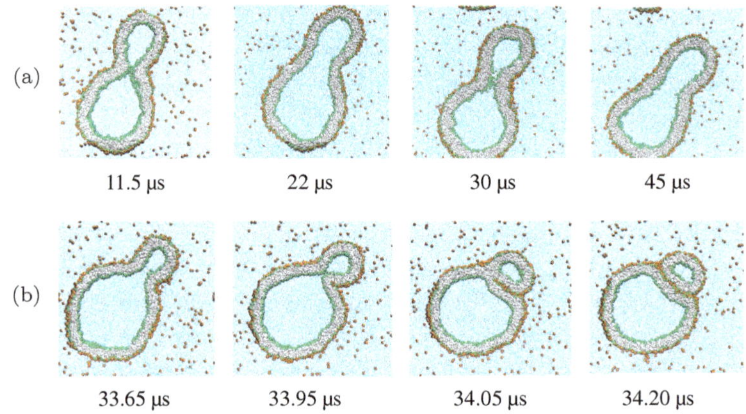

Fig. 12. Unusual morphological transitions of nanovesicles exposed to small solutes (orange dots) in the exterior aqueous solution (blue), for low solute concentrations Φ and poor solvent conditions. The nanovesicles have a size of about 36 nm. The cross-sections of the vesicle membranes depict lipid bilayers with green head groups and grey hydrocarbon chains. The solutes form adsorption layers (orange–green) at the outer leaflets of the bilayers: (a) Time series of budded nanovesicle with recurrent shape changes between dumbbells with open and closed necks for solute mole fraction $\Phi = 0.025$. (b) Division of nanovesicle by fission of the membrane neck for $\Phi = 0.026$, which leads to two daughter vesicles that adhere via the adsorbed solutes. Both mole fractions belong to the one-phase region but are located close to the binodal line.[9]

budding and fission of nanovesicles, a mechanism that arises from the interplay between membrane elasticity and solute-mediated membrane adhesion.

Other simulation studies of nanovesicle fission: Fission of nanovesicles has also been observed in other simulation studies. When a solid nanoparticle is completely engulfed by the vesicle membrane, a neck is formed which may undergo fission, as observed in Brownian dynamics simulations without explicit water.[56] This process is governed by the adhesion between the membrane and the nanoparticle, which leads to the adhesion-dependent constriction force described by Eq. (24).[31,32] Fission has also been observed in coarse-grained molecular simulations of bilayers with two lipid components[57,58] and in dissipative particle dynamics of monolayers assembled from triblock copolymers.[59] The nanovesicle in Ref. 57 was formed by an asymmetric bilayer exposed to two different types of water beads in the interior and exterior compartment. The associated spontaneous curvature generates a constriction force that can be sufficiently large to cleave the neck, in analogy to the experimentally observed behavior of giant vesicles[8] displayed in Figs. 8 and 10. The latter mechanism also applies to the fission of one-component monolayers of asymmetric block copolymers.[59] The two-component bilayers and monolayers in Refs. 58 and 59 consisted of two membrane domains with the domain boundary being located within the membrane neck of the dumbbell-shaped vesicles. In such a situation, the cleavage of the neck depends both on the spontaneous curvature and on the line tension $\lambda_{\rm db}$ of the domain boundary, see Eq. (21). Furthermore, using coarse-grained molecular models, the cleavage of membrane necks by protein helices mimicking dynamin has also been studied by simulations in Ref. 60.

4.3. *Endocytosis of condensate droplets*

Condensate droplets are formed in aqueous solutions of macromolecules that undergo phase separation into two liquid phases.[33] A well-studied example are solutions of the two polymers PEG and dextran which have been used for a long time in biochemical analysis and biotechnology.[47,50] More recently, phase separation has also been observed in living cells where it leads to membrane-less or droplet-like organelles. In the latter case, the condensate droplets are enriched in certain types of proteins. Examples for these biomolecular condensates include germ P-bodies,[51,61] nucleoli,[62] and stress granules.[63] These biomolecular condensates are believed to form via liquid–liquid phase separation in the cytoplasm[51,52] and can be reconstituted in vitro.[64–67] They are enriched in certain types of proteins that have intrinsically disordered domains and interact via multivalent macromolecular interactions.[52,66–69]

The adhesion of condensate droplets to biomembranes and vesicles leads to different adhesion and wetting morphologies as well as wetting transitions as we change the molecular composition or the temperature.[33,48] Aqueous phase separation of PEG-dextran solutions within GUVs was first reported by Christine Keating and coworkers.[70] Particularly fascinating remodeling processes of membranes interacting with condensate droplets are the formation of membrane nanotubes,[71–73] the formation of two spherical membrane segments that enclose two different condensate droplets,[39] and the complete engulfment of the droplets by the membranes.[14] Complete engulfment of a dextran-rich condensate droplet by a GUV membrane leads to a closed membrane neck, as displayed in Fig. 6(d). So far, the fission of such necks has not been observed experimentally.

However, neck fission and endocytosis of condensate droplets have been observed in molecular dynamics simulations,[10] in which the formation of condensate droplets was studied using the same off-lattice model for a binary water–solute mixture as in Fig. 12, but now focussing on the two-phase coexistence region of this mixture. The endocytic process starts with the spreading of the droplet onto the membrane, followed by partial or complete engulfment of the droplet. One example for partial engulfment is shown in Fig. 13, which was obtained for solute mole fraction $\Phi_{\rm S} = 0.004$ and solubility $\zeta = 25/70 = 0.36$, which is located within the two-phase coexistence region of the solute–water phase diagram. Initially, both the nanodroplet and the nanovesicle are fully immersed in the liquid phase α, as shown in Fig. 13(a). When the droplet gets into contact with the vesicle membrane, a small contact area is formed as in Fig. 13(b). After this onset of adhesion, the vesicle membrane starts to engulf the membrane. This process continues by pulling

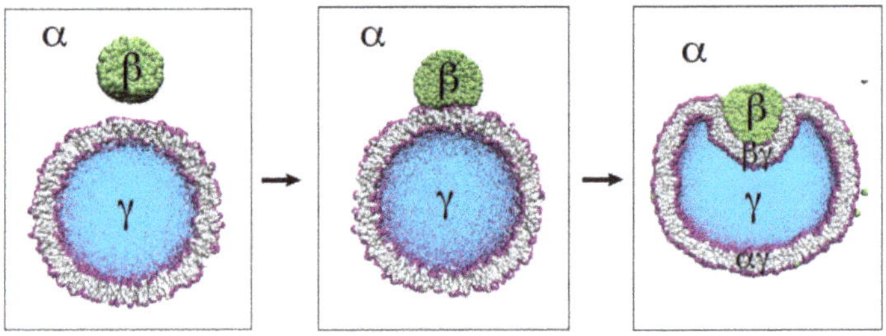

(a) (b) (c)

Fig. 13. Partial engulfment of a condensate droplet (green) by the lipid bilayer (purple–grey) of a nanovesicle, as observed in molecular dynamics simulations.[10] The vesicle encloses the aqueous solution γ (blue). Both the nanodroplet and the nanovesicle are immersed in the aqueous bulk phase α (white): (a) Initially, the droplet is well separated from the vesicle which implies that the outer leaflet of the bilayer is only in contact with the α phase. (b) When the droplet is attracted toward the vesicle, it spreads onto the lipid bilayer, thereby forming an increasing contact area with the vesicle membrane. (c) Partial engulfment of the droplet by the membrane after the vesicle–droplet couple has relaxed to a new stable state. The contact area between bilayer and β droplet defines the $\beta\gamma$ segment of the bilayer membrane, whereas the rest of the bilayer represents the $\alpha\gamma$ segment still in contact with the α phase. Vesicle and droplet have a diameter of 37 nm and 11.2 nm, respectively.

out membrane area from the thermally excited undulations, thereby increasing the bilayer tension in the membrane. Eventually, a new stable morphology, corresponding to a partially engulfed droplet, is reached, as shown in Fig. 13(c).

The contact line between the condensate droplet and the lipid membrane partitions this membrane into two segments: the $\alpha\gamma$ segment exposed to the aqueous bulk phase α and the $\beta\gamma$ segment in contact with the β droplet, see Fig. 13(c). The membrane area of the $\beta\gamma$ segment represents the contact area between membrane and droplet. When we now reduce the volume of the nanovesicle by osmotic deflation, the contact area further increases following three possible pathways.[10] On the one hand, a droplet of volume V cannot be completely engulfed by the vesicle membrane if the area A of this membrane is too small and satisfies $(A/2)^3 < 36\pi V^2$ as follows from the isoperimetric inequality. On the other hand, complete engulfment is possible for $(A/2)^3 \geq 36\pi V^2$ and can then proceed in an axisymmetric or non-axisymmetric manner, depending on the line tension of the contact line between membrane and droplet.

As previously discussed in Section 3.9, the contact line tension λ_{co} can be positive or negative. In fact, the sign of λ_{co} is determined by the stress asymmetry between the two leaflets of the bilayer.[10] The line tension λ_{co} is negative if the bilayer is symmetric or only weakly asymmetric in the absence of the droplets and positive if the bilayer exhibits a sufficiently large stress asymmetry.[10,33] A positive line tension leads to a circular contact line and to an axisymmetric membrane neck that becomes closed and then undergoes fission. One example for this axisymmetric process of complete engulfment and endocytosis is displayed in Fig. 14. In contrast, a negative

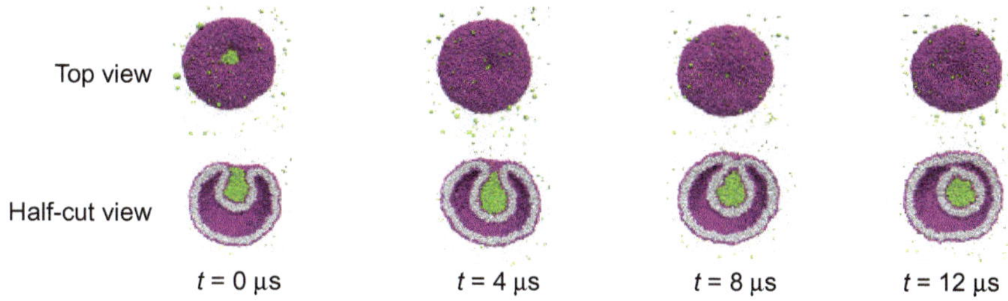

Fig. 14. Endocytosis of a nanodroplet (green) via complete axisymmetric engulfment by the vesicle membrane (purple–grey) followed by the division of the vesicle into two nested daughter vesicles: The nanovesicle is formed by a lipid bilayer with $N_{\rm ol} = 5500$ lipids in the outer leaflet and $N_{\rm il} = 4600$ lipids in the inner one. The vesicle has a size of 37 nm and the droplet a diameter of 11.2 nm. The displayed time lapse starts with a reduction of the vesicle volume at time $t = 0$, which is kept constant for all later times. The membrane neck closes at $t = 4\,\mu$s and undergoes fission at $t = 9\,\mu$s, generating a small intraluminal vesicle around the droplet. The latter morphology remains unchanged for at least $30\,\mu$s.[10]

contact line tension leads to a strongly non-circular contact line and to a tight-lipped shape of the membrane neck, which prevents the fission of this neck.

5. Fusion via Adhesion and Pore Formation

We now address the process of membrane fusion, focussing again on geometric and membrane-elastic aspects. As far as the Euler characteristic χ and the topological genus $\mathfrak{g}$ are concerned, each individual fusion process represents the reverse of a fission process (Figs. 3 and 4). However, the energetics of fusion and fission is quite different. Since the fission process is exergonic or 'downhill' in free energy, as shown in Sections 3 and 4, the fusion process must be endergonic or 'uphill' in free energy. In the cell, fusion processes are often coupled to the hydrolysis of ATP or GTP. From a physical point of view, the presumably simplest mechanism to drive membrane fusion is via the mechanical tension acting within one of the membranes in their pre-fusion state.

5.1. *Membrane necks created by fusion*

Let us start with the fusion of two genus-0 vesicles, as depicted in panels a and c of Fig. 15. In Fig. 15(a), the two vesicles enclose two mutually disjoint spatial regions and could have been formed by the fission process in Figs. 5(a) and (b). The fusion of these mutually disjoint vesicles creates an out-budded vesicle with a narrow fusion neck as in Fig. 15(b). In Fig. 15(c), the larger vesicle encloses the smaller one, which could be the result of the fission process in Figs. 5(c) and (d). The fusion of these nested vesicles creates the in-budded vesicle with a narrow membrane neck as in Fig. 15(d). In the literature on membrane fusion, the membrane neck is usually denoted as the 'fusion pore.' The term 'fusion neck' is used here to emphasize the intimate relation to the fission processes in Fig. 5.

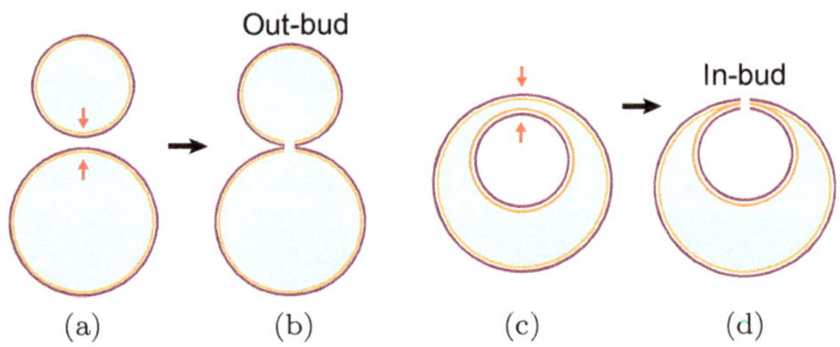

Fig. 15. Formation of membrane necks via fusion starting from two genus-0 vesicles with Euler characteristic $\chi = 4$ and genus $\mathfrak{g} = -1$ that fuse into a single genus-0 vesicle with $\chi = 2$ and $\mathfrak{g} = 0$: (a) Two vesicles, which enclose two mutually disjoint spatial regions prior to fusion, adhere with their outer leaflets as indicated by the two red arrows. (b) Fusion then creates an out-budded genus-0 vesicle with a narrow membrane neck, reversing the two fission substeps in panels a and b of Fig. 5. (c) When the two vesicles are nested prior to fusion, the outer leaflet of the smaller one adheres to the inner leaflet of the larger one (red arrows). (d) Fusion now creates an in-budded genus-0 vesicle with a narrow membrane neck, reversing the two fission substeps in panels c and d of Fig. 5.

5.2. *Pathways of tension-induced fusion*

Experimentally, tension-induced fusion was originally studied by osmotic swelling of vesicles that adhere to a planar bilayer.[74] The molecular mechanisms underlying tension-induced fusion have been visualized and elucidated by molecular dynamics simulations.[75–77] More recently, several experimental studies provided additional evidence for tension-induced fusion,[78–81] which can be strongly enhanced by electrostatic interactions between oppositely charged membranes.[79] In the simulations, the close proximity of two lipid bilayers can lead to a variety of alternative outcomes such as stable hemifused bilayers and bilayer rupture, but fusion events were always observed within a certain range of positive bilayer tensions. Furthermore, the simulations revealed three different fusion pathways, as described in the following.

Non-axisymmetric fusion pores after hemifusion: In our first set of simulations,[76,82] we studied the fusion of a planar bilayer with a nanovesicle, for which we observed an unusual pathway, as illustrated by the time lapse snapshots in Fig. 16. In order to visualize this pathway, the lipids in the planar bilayer have red head groups and green tails, whereas the lipids in the vesicle have orange head groups and yellow tails. The observed fusion pathway proceeds via three distinct substeps. After the vesicle and the planar bilayer have come into contact at time $t = 78.5$ ns, the vesicle starts to spread onto the bilayer, thereby forming a highly curved membrane segment along the contact line with the planar bilayer. The lipids along the contact line then undergo interbilayer flips from the vesicle to the planar bilayer, see the yellow lipid tails that appear in the lower leaflet of the planar bilayer. These flips lead to the intermixing and partial fusion of the two bilayers, followed by the nucleation of a bean-shaped hemifusion diaphragm at about $t = 1177$ ns. The bean

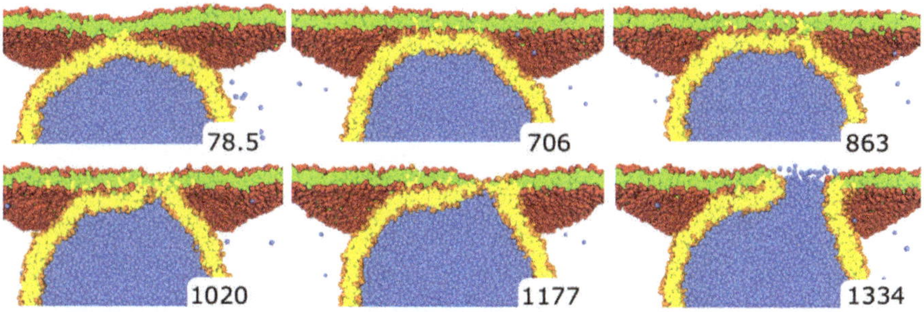

Fig. 16. Fusion of a nanovesicle with a planar bilayer as observed in molecular dynamics simulations: The vesicle consists of 6869 lipids (orange heads, yellow chains) and has a diameter of 28 nm, while the planar membrane contains 6911 lipids (red heads, green chains). The nanovesicle has a diameter of 28 nm. The water beads originally inside the vesicle are blue, those outside are not shown for clarity (white background). The figure displays six time lapse snapshots which illustrate the development of the fusion process from 78.5 ns, when the first contact between the two membranes occurs and lipid tails start to undergo interbilayer flips until the opening of the fusion pore after 1334 ns.[76]

shape of the diaphragm can be clearly seen when one looks at sections across the planar bilayer, which are perpendicular to the side view in Fig. 16. Finally, this diaphragm ruptured close to the contact line at $t = 1334\,\text{ns}$, thereby opening up a strongly asymmetric fusion pore that provides a water channel between the interior vesicle solution (blue) and the water compartment (white) above the planar bilayer in Fig. 16.

Time scales of fusion substeps: The three substeps for the formation of an asymmetric fusion pore are governed by three tension-dependent time scales, as displayed in Fig. 17, where these time scales are plotted as functions of the (projected) area per lipid, A, in the planar bilayer. This area per lipid increases with increasing bilayer tension. The time scale $\langle t_\alpha \rangle$ is the average time it takes from the initial contact between the two membranes until the first interbilayer flip. The time scale $\langle t_\beta \rangle$ is the average reordering time between the first interbilayer flip and the nucleation of the hemifused patch. The average fusion time is denoted by $\langle t_{\text{fu}} \rangle$ and the width of the fusion time distribution by Δt_{fu}.

All time scales that are plotted in Fig. 17 decay exponentially with increasing area per lipid, A, and thus with increasing bilayer tension. Such a behavior indicates that the free energy barriers that determine these time scales are reduced by the bilayer tension in a linear fashion. For tensionless bilayers, the free energy barrier is found to be about $9\,k_{\text{B}}T$ for the interbilayer flips independent of the vesicle size, whereas the barrier for the nucleation of the hemifused diaphragm is about $12\,k_{\text{B}}T$ for the 14-nm vesicles in Fig. 17(a) and about $16\,k_{\text{B}}T$ for the 28-nm vesicles in Fig. 17(b). The rupture time for the hemifused diaphragm was observed to be always between 150 and 300 ns, irrespective of the bilayer tension.

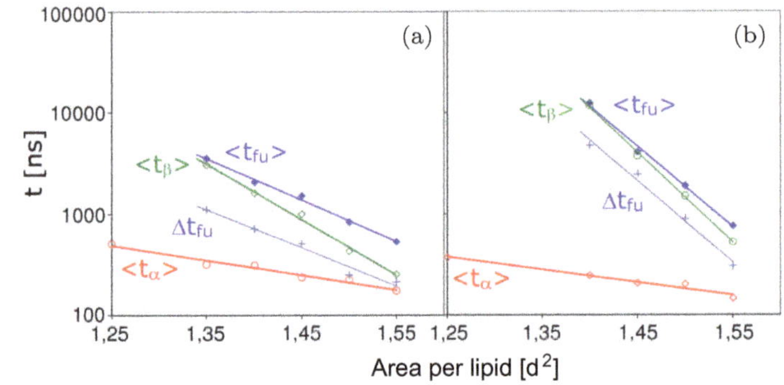

Fig. 17. Different time scales for the fusion between nanovesicle and planar bilayer: The diameter of the vesicles in (a) and (b) is equal to 14 and 28 nm, respectively. The time scales are plotted semi-logarithmically versus the area per lipid within the planar bilayer, given in units of d^2 where d is the basic length scale of the molecular model. For the force parameters used here, the area per lipid is proportional to the bilayer tension. The different time scales are the average first flipping time $\langle t_\alpha \rangle$ (red) until the first lipid flip from the vesicle bilayer to the planar bilayer; the average reordering time $\langle t_\beta \rangle$ (green), between the first interbilayer flip and the nucleation of a hemifused patch; the average fusion time $\langle t_{fu} \rangle$ (blue); and the width Δt_{fu} of the fusion time distribution (blue). These time scales are obtained from more than 140 statistically independent simulations. A comparison of the two sets of data in (a) and (b) shows that the average first flipping time $\langle t_\alpha \rangle$ is independent of the vesicle size, whereas the average reordering time $\langle t_\beta \rangle$ increases with the size of the vesicle.[76]

Two alternative fusion pathways: We also studied the fusion of two nanovesicles, using a somewhat different set of force parameters for the coarse-grained molecular model which made the lipid bilayers more rigid.[77,83] As a result, we observed, depending on the magnitude of the bilayer tensions, two alternative fusion pathways, as displayed in Figs. 18 and 19. For relatively low tensions, one of the two vesicles develops a bilayer pore within the contact area, see Fig. 18(a). This pore grows in size and leads to the two-vesicle geometry, as displayed in Figs. 18(b). At first sight, the latter geometry resembles a hemifused state, but the diaphragm between the two vesicles consists almost exclusively of lipids from the unruptured vesicle, in contrast to a hemifusion diaphragm, which would contain a mixture of lipids from the outer leaflets of both vesicles.

For relatively high bilayer tension, the molecular model with the stiffer lipid tails leads to the fusion pathway shown in Fig. 19. In this case, the local adhesive contact between the two vesicles leads to a relatively small contact area and to fast disruption of both bilayers. In this high-tension pathway, the fusion pore is formed without an intermediate diaphragm, see Fig. 19(b).

Fusion of tensionless bilayers with stretched outer leaflets: All pathways of tension-induced fusion, discussed so far and displayed in Figs. 16–19, are driven by sufficiently large bilayer tensions, $\Sigma = \Sigma_{ol} + \Sigma_{il} > 0$, within the two bilayers to be fused. The positive bilayer tensions act to stretch all four leaflets of the two

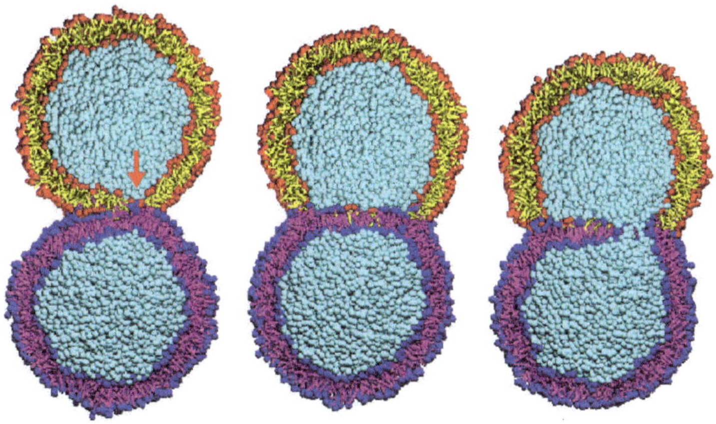

(a) (b) (c)

Fig. 18. Low-tension pathway for two nanovesicles with more rigid bilayers: (a) The two nanovesicles adhere and form a small contact area. The lipids in the upper vesicle have red head groups and yellow tails, whereas those in the lower vesicle have blue head groups and purple tails. The adhesion generates a substantial bilayer tension, which leads to the poration of the upper bilayer close to the boundary of the contact area (red arrow). This bilayer pore grows until the upper bilayer has completely retracted from the contact area in (b). The latter geometry is different from a hemifused state because the diaphragm between the two vesicles consists almost exclusively of blue-purple lipids from the unporated vesicle bilayer. Finally, an asymmetric fusion pore is formed in (c) after the blue–purple diaphragm has ruptured.[77]

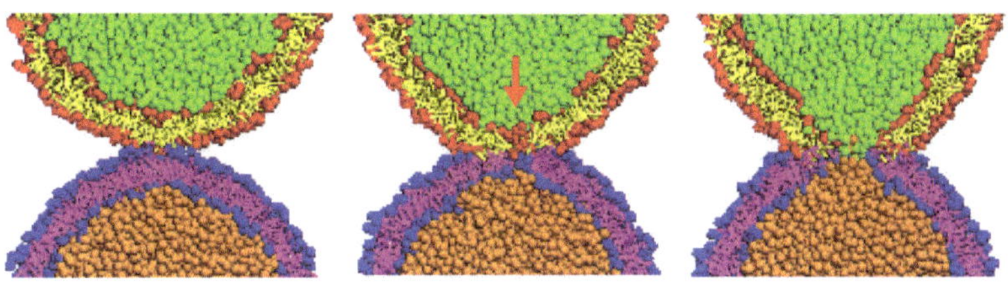

(a) (b) (c)

Fig. 19. High-tension pathway for two nanovesicles with more rigid bilayers: (a) local adhesive contact between the two nanovesicles; (b) disruption of the two adhering bilayers (red arrow); and (c) opening of a fusion pore which provides a connection between the two water compartments (orange and green) enclosed by the two vesicles.[77]

bilayers. However, in order to obtain vesicle fusion, it is not necessary to stretch all leaflets of the two bilayers. Indeed, it is sufficient to stretch only the two outer leaflets of the two bilayers by sufficiently large outer leaflet tensions $\Sigma_{\mathrm{ol}} > 0$.[38] As a consequence, it is possible to obtain fusion of vesicles, which are bounded by *tensionless* bilayers with $\Sigma = \Sigma_{\mathrm{ol}} + \Sigma_{\mathrm{il}} = 0$, provided the outer leaflet tensions Σ_{ol} are sufficiently large and positive. One example for such a fusion event is displayed in Fig. 20.

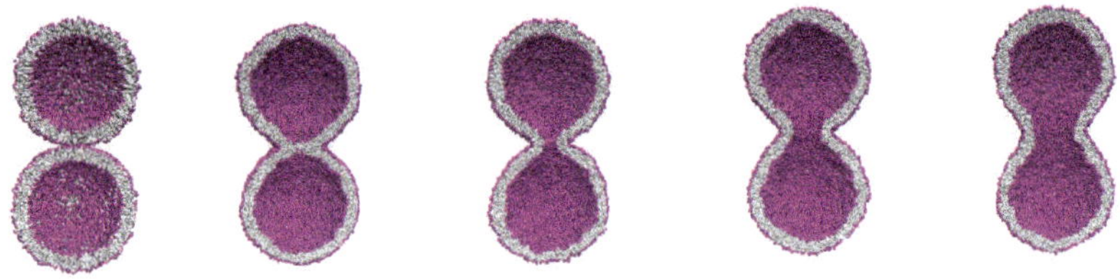

Fig. 20. Time-lapse snapshots of two identical vesicles with tensionless bilayers, which fuse when their outer leaflets are stretched by a sufficiently large leaflet tension $\Sigma_{ol} > 0$. Before they come into contact, each vesicle bilayer is spherical and contains $N_{ol} = 5600$ lipids in its outer leaflet as well as $N_{il} = 4500$ lipids in its inner leaflet. The outer leaflets are stretched by the leaflet tension $\Sigma_{ol} = +1.02\, k_B T/d^2$, whereas the inner leaflets are compressed by the inner leaflet tension $\Sigma_{il} = -1.02\, k_B T/d^2$. Note that both vesicle bilayers are tensionless with bilayer tensions $\Sigma = \Sigma_{ol} + \Sigma_{il} = 0$. At time $t = 0\,\mu$s, the two vesicles are brought into contact and promptly undergo fusion within $0.3\,\mu$s.[38]

5.3. *Insights from fusion simulations*

First of all, our simulations revealed that even the presumably simplest driving force for fusion as provided by bilayer tension can lead to several distinct pathways. Second, these simulations demonstrated that the fusion process is stochastic in nature and that the distinct substeps are governed by different probability distributions,[76,77,82] as illustrated by the different time scales in Fig. 16. In hindsight, the stochastic nature of the fusion process is quite obvious because the process has to overcome one or several free energy barriers. However, the large majority of fusion simulations ignored this aspect because it is computationally quite expensive, even for coarse-grained molecular dynamics. Furthermore, our simulations revealed two more specific molecular features, interbilayer flips of splayed lipids and non-axisymmetric fusion pores.

Interbilayer flips of splayed lipids: For a splayed lipid, the two tails point in opposite directions, as originally proposed in Ref. 84. When a lipid has such a splayed conformation, it can bridge the water gap between the two adjacent bilayers in their pre-fusion state by inserting one tail into each bilayer. Some experimental evidence for this bridging conformation of splayed lipids has been obtained by fluorescence spectroscopy[85] and by using lipids with a azobenze group that can undergo light-induced trans/cis isomerization.[86] Furthermore, splayed lipids have also been observed in other simulation studies.[76,87–92]

Non-axisymmetric fusion pores: To some extent, our fusion simulations were motivated by the so-called stalk model[93–95] which assumes that the fusion process proceeds via axisymmetric membrane shapes, i.e., via axisymmetric intermediates and axisymmetric fusion pores. In contrast, the simulations snapshots displayed in Figs. 16 and 18 clearly demonstrate that the fusion process can involve both strongly non-axisymmetric intermediates and non-axisymmetric fusion pores. Non-axisymmetric fusion pores have now been observed by many other groups both in experimental and simulation studies.[96–101]

5.4. *Mechanisms of protein-mediated fusion*

Membrane fusion of cellular membranes involves the formation of protein complexes anchored to the membranes in their pre-fusion state. Three types of fusion mechanisms can be distinguished, depending on how these proteins mediate the adhesion between the membranes. The simplest fusion mechanism is observed for viral fusion proteins that are only anchored to the membrane envelope of the virus. The viral fusion protein inserts into the acceptor membrane of the host cell, a process that does not seem to require additional proteins anchored to the acceptor membrane. One example for such a viral fusion protein is provided by hemagglutinin of influenza viruses,[102] another by the Spike protein of corona viruses, which contains a fusion peptide sequence within its S2 subunit.[103,104]

A second type of protein-mediated fusion mechanism is used in homotypic fusion of intracellular membrane segments that enclose the endoplasmic reticulum or belong to the outer membranes of two mitochondria. In this case, the apposing membranes are pulled together by forming trans-dimers of identical membrane proteins anchored to both membranes. These proteins have a GTPase domain that hydrolyzes GTP. Examples are proteins from the atlastin family, which drive GTP-dependent fusion of the endoplasmic reticulum membranes in multicellular animals,[105–107] as well as analogous membrane GTPases such as Seyp1 and RHD3, which catalyze membrane fusion in yeast[108] and plant[109] cells. Likewise, membrane GTPases from the mitofusin family drive the homotypic fusion of the outer membranes of two mitochondria.[110–112]

A third type of multiprotein complex is used along the outward secretory pathway in eukaryotic cells (red arrows in Fig. 1). This pathway, which involves the heterotypic fusion of different organelles and vesicles, is dominated by proteins from the SNARE family.[113–115] For SNARE-mediated fusion, each of the fusing membranes contributes a different set of membrane proteins to the multiprotein complex that mediates their adhesion.

6. Fusion and Fission of Multispherical Vesicles

In Figs. 5–7, we described $(1+1)$-multispheres, which represent the simplest examples for stable multispherical shapes. In fact, the theory of curvature elasticity predicts stable multispheres consisting of an arbitrary number of (punctured) spheres connected by closed membrane necks.[7,22] This prediction is a consequence of two general curvature-elastic features: (i) the shape equation for spherical segments is quadratic in the mean curvature of these segments and (ii) the stability condition for closed membrane necks, as given by Eqs. (12) and (13). The quadratic nature of the shape equation implies that only up to two different sphere radii can coexist on the same vesicle. Likewise, the local nature of the stability conditions for closed membrane necks as given by Eqs. (12) and (13) ensures that these conditions can be applied to the necks of any multispherical shape.

6.1. *Multispherical shapes with more than two spheres*

Experimentally, a striking variety of multispherical shapes of GUVs has been observed when the GUVs were exposed to asymmetric solutions of two simple sugars, sucrose and glucose.[13,22,116] When the interior solution contained only sucrose and the exterior solution primarily glucose, the vesicle membranes acquired a positive spontaneous curvature of about $1/\mu m$. In general, the observed multispheres consist of N_l large spheres with radius R_l and N_s small spheres with radius R_s. In addition, special multispheres consisting of N equally sized spheres were also observed. These shapes correspond to the limiting case, in which the large-sphere-radius R_l and the small-sphere-radius R_s become equal, transforming the $(N_l + N_s)$-multisphere into an N-multisphere with $N = N_l + N_s$. Several examples of N-multispheres are displayed in Fig. 21.

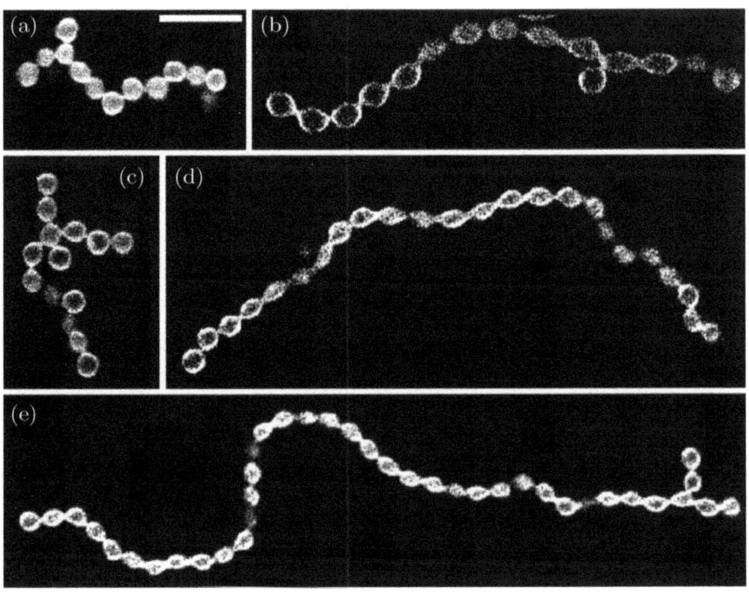

Fig. 21. Multispherical shapes of giant unilamellar vesicles with positive spontaneous curvature, consisting of N equally sized spheres, as observed by light microscopy.[13] The sphere number N is equal to $N = 14$ in (a), $N = 15$ in (b) and (c), $N = 24$ in (d), and $N = 39$ in (e). All (N_*)-multispheres provide examples for constant-mean-curvature surfaces. Furthermore, apart from (d), all N-multispheres displayed here exhibit multispherical junctions. Each junction consists of one central sphere that is connected to more than two neighboring spheres. Scale bar in (a) is $10\,\mu m$ and applies to all panels. In spite of their complex morphology, all multispherical shapes displayed in this figure have the same genus-0 topology as a single sphere.

It is interesting to note that the N-multispheres represent constant-mean-curvature (CMC) surfaces.[22] For a long time, the only examples for freely suspended CMC surfaces with non-zero mean curvature were provided by the unduloids of Delaunay,[117] which provide a one-parameter family of tubular shapes that interpolate smoothly between multispherical tubes consisting of equally sized (and punctured) spheres and cylindrical tubes. More recently, additional CMC surfaces have been constructed by perturbing a cluster of identical spheres that touch each other.[118–122] One example are triunduloids[121,122] that consist of three unduloidal arms connected by a central core. These triunduloids provide smoothly curved shapes with a low bending energy for the three-way junctions observed in the endoplasmic reticulum, see Fig. 2(a) and Ref. 123.

In the research field of CMC surfaces, the physical systems used to motivate these surfaces are the shapes of soap films and liquid droplets. However, when the initial cluster of identical and touching spheres is viewed as a cluster of liquid droplets, the resulting CMC surface is not stable. Indeed, the cluster will either fall apart and then form many small droplets or it will coalesce into one large droplet that will eventually attain the shape of a single sphere. In contrast, when the initial cluster of droplets is enclosed by a membrane, this membrane can lead to a stable multispherical shape as predicted theoretically[7] and observed experimentally,[13] see examples in Fig. 21.

6.2. *Formation of higher genus vesicles by fusion*

The multispherical shapes in Fig. 21 consist of spheres with a radius of the order of 1 μm. This size is convenient in order to observe the multispherical shapes by conventional light microscopy. One can reduce the size of the spherical vesicles to a diameter below 100 nm by the preparation of liposomes and proteoliposomes. Furthermore, proteoliposomes that contain a membrane GTPase such as atlastin fuse with each other and form long nanotubular structures after the addition of GTP.[105,106,124,125] The diameter of these nanotubes has been estimated by electron microscopy to be below 50 nm.[124]

Homotypic fusion of proteoliposomes: To obtain some insight into the formation of nanotubes from proteoliposomes, we now consider N proteoliposomes that undergo homotypic fusion to form an N-multisphere. In the absence of GTP, the proteoliposome are taken to be relaxed in the sense that the mean curvature of their membrane is close to the membrane's spontaneous curvature m, which implies that the radius R of the proteoliposomes is close to $1/m$ and that their bending energy in Eq. (5) is close to zero. The dimensionless shape parameters v and $\bar{m}$, as defined in Eqs. (9) and (10), are then equal to $v = v_1 = 1$ and $\bar{m} = \bar{m}_1 = 1$ for each proteoliposomes. When we add GTP, the proteoliposomes start to fuse to form multispherical shapes. More precisely, the fusion of N proteoliposomes leads a multisphere with the shape parameters $v = v_N \equiv 1/\sqrt{N}$ and $\bar{m} = \bar{m}_N \equiv \sqrt{N}$, provided both the total membrane area and the total vesicle volume are conserved during the fusion process. For $N = 2$, the shape parameters $v = v_2$ and $\bar{m} = \bar{m}_2$ correspond to the corner point of the stability regime in Fig. 7(b), at which the two lines of limit shapes, L_{1+1} and L_{2*}, meet. For $N > 2$, the shape parameters $v = v_N$ and $\bar{m} = \bar{m}_N$ define analogous corner points of the corresponding stability regimes.[22]

The nanotube shapes can be resolved by super-resolution microscopy, which has not been reported, so far, for the nanotubes reconstituted from proteoliposomes. However, structured illumination microscopy has been applied to reticular nanotubes *in vivo*,[2,126] which revealed that these shapes resemble unduloids. In the $(\bar{m}, v)$ morphology diagram, unduloids with minimal bending energy are found for $\bar{m} < \bar{m}_N$ and for $v < v_N$. Such unduloid shapes will form *via* the fusion of proteoliposomes when some membrane proteins get lost during the fusion process, thereby reducing the spontaneous curvature $\bar{m}$, or if the fusion process is leaky and the interior volume of the proteoliposomes is not conserved during the process.

Homotypic fusion of multispherical tubes: Now, let us assume that we have obtained a multispherical tube consisting of N equally sized spheres. One example with $N = 18$ is displayed in Fig. 22. The tubular membrane contains membrane proteins that can hydrolyze GTP, thereby driving homotypic fusion between different membrane segments. In Fig. 22(a), we start from an unbranched multispherical tube with topological genus $\mathfrak{g} = 0$ in the absence of GTP. When we add GTP, the two terminal spheres can fuse, which leads to a multispherical tube with $\mathfrak{g} = 1$ and no junctions, as shown in Fig. 22(b). In fact, one of the terminal spheres may fuse with any other sphere of the multispherical tube, again producing a multisphere with $\mathfrak{g} = 1$, which now involves a three-way junction. A somewhat more complex shape evolution is depicted in Fig. 22(c)–(e). We now start, in the absence of GTP,

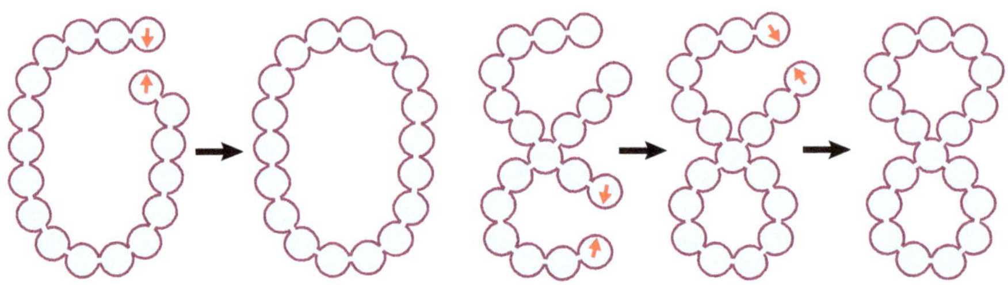

(a) (b) (c) (d) (e)

Fig. 22. Self-adhesion and fusion of multispherical tubes with $N = 18$ equally sized spheres: (a) Genus-0 multisphere with 18 spheres connected by 17 membrane necks. Fusion of the two terminal ends (red arrows) creates the genus-1 multisphere in (b) with 18 spheres and 18 necks. (c) Genus-0 multisphere with 18 spheres, 17 necks, and one four-way junction, which is formed by the sphere with four membrane necks. Fusion of the two terminal spheres in the lower toroidal ring (red arrows) leads to the genus-1 shape in (d) with 18 spheres and 18 necks. (d) Fusion of the two terminal spheres in the upper toroidal ring (red arrows) creates the genus-2 shape in (e) with 18 spheres and 19 necks.

with a genus-0 multisphere that consists of a four-way junction with four linear tube segments. After the addition of GTP, homotypic fusion of the two terminal spheres below the junction now leads to the genus-1 multisphere in Fig. 22(d). One additional fusion event between the two terminal spheres above the junction then leads to the genus-2 multisphere in Fig. 22(e).

Adding more multispherical tubes to the small networks displayed in Fig. 22 as well as GTP, the homotypic fusion between different membrane segments will create tubular networks with a large genus, which resemble the ER networks in Fig. 2. However, these networks can exhibit junctions with more than three necks. One example is the network in Fig. 22(c), which involves a four-way junction, consisting of a central sphere that is connected to four neighboring spheres *via* four membrane necks. It has been argued elsewhere that the predominance of three-way junctions in the ER arises from a significant bilayer tension that acts to stretch the membrane nanotubes.[22,123]

6.3. *Fission of higher genus vesicles*

Finally, we illustrate the fission of higher genus vesicles by looking at the cleavage of the multispherical tubes in panels b and e of Fig. 22. These membrane tubes exhibit many membrane necks, each of which is governed by the stability relation in Eq. (12) and compressed by the constriction force in Eq. (17). Thus, in order to cleave one of these necks, we may simply increase the spontaneous curvature m, e.g., by binding His-tagged GFP to the outer leaflets of the tube membranes, a procedure that was successfully used for the genus-0 dumbbell vesicles in Figs. 8 and 10. After one of the necks in Fig. 22 has been cleaved, we obtain the multispherical shapes in Fig. 23.

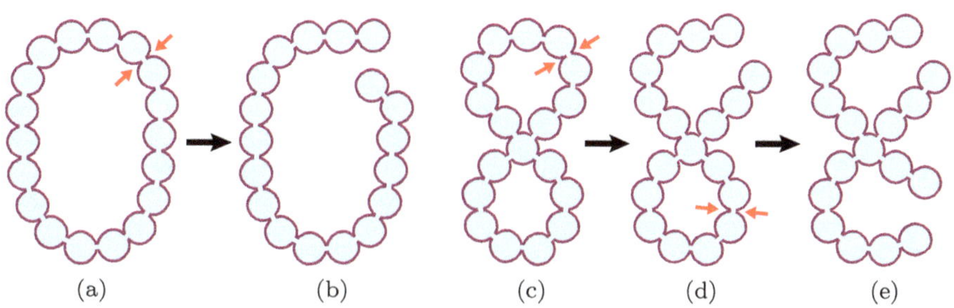

Fig. 23. Fission of genus-1 and genus-2 vesicles, illustrated for multispheres consisting of $N = 18$ equally sized spheres: (a) Genus-1 multisphere with 18 spheres connected by 18 membrane necks. Cleavage of one neck (red arrows) leads to the genus-0 multisphere in (b) with 18 spheres and 17 necks. This fission process reverses the fusion process in panels a and b of Fig. 22. (c) Genus-2 multisphere with 18 spheres connected by 19 membrane necks. Cleavage of one neck in the upper toroidal ring (red arrows) leads to the genus-1 multisphere in (d) with 18 spheres and 18 necks. (d) Another neck cleavage in the lower toroidal ring (red arrows) leads to the genus-0 multisphere in (e) with 18 spheres and 17 necks.

For a laterally uniform membrane, each neck is subject to the same constriction force which implies that the fission time is governed by the same dwell time distribution. This degeneracy is lifted if the neck contains a domain boundary, separating two distinct intramembrane domains, because the line tension λ_{db} of the domain boundary enhances the constriction force, as described by Eq. (21). Likewise, if one of the spheres engulfs an adhesive particle or condensate droplet, the necks connected to this sphere will experience the enhanced constriction force described by Eq. (24).

7. Summary and Outlook

In this chapter, recent insights about the topological transformations of biomembranes and vesicles have been described in order to further elucidate the physical mechanisms for these processes. These transformations can be characterized in a very general manner by changes in the Euler characteristic χ and in the topological genus $\mathfrak{g}$ (Figs. 3 and 4). We then addressed fission processes which are controlled by constriction forces that compress closed membrane necks. For laterally uniform membranes, these forces are described by Eqs. (17) and (18) for the necks of out- and in-buds, respectively. The constriction forces are enhanced by the line tension λ_{db} of a domain boundary, residing in the neck, see Eqs. (21) and (22), as well as by adhesive particles or droplets engulfed by the membranes, see Eqs. (24) and (26). A sufficiently large constriction force leads to the fission of the membrane neck, as observed for the division of GUVs (Figs. 8 and 10), for the division of nanovesicles (Figs. 11 and 12b), and for the endocytosis of condensate droplets by nanovesicles (Fig. 14).

Topological transformations by fusion were discussed in Section 5. On the one hand, it was emphasized that the fusion process involves the same membrane shapes as fission, but in reversed order, as one concludes from a comparison of Fig. 15 with Fig 5. On the other hand, the 'local surgery' of the adhering bilayers can proceed via different intermediates, as illustrated by Figs. 16–20 for tension-induced fusion, which represents the simplest driving force for membrane fusion. Likewise, protein-mediated fusion processes can involve different intermediate conformations of the proteins, as briefly discussed in Section 5.4, where we distinguish viral fusion from homotypic and heterotypic fusion.

Finally, Section 6 was devoted to multispherical vesicles, as predicted by the theory of curvature elasticity[7,22] and experimentally observed for GUVs exposed to aqueous solutions of simple sugars.[13,116] We focused on multispherical tubes consisting of N equally sized spheres, as displayed in Figs. 21–23. So far, all experimentally observed multispheres were genus-0 vesicles and had the same topology as a single sphere (Fig. 21). We then performed a 'Gedankenexperiment' by doping the vesicle membranes in Fig. 21 with membrane proteins that are capable of homotypic fusion in the presence of GTP. These fusion processes were illustrated by some simple examples in Fig. 22, corresponding to the formation of genus-1 and genus-2 multispheres. The fission of these vesicles as displayed in Fig. 23 can be accomplished by binding His-tagged GFP to the outer leaflets of the multispherical membranes, a method that was previously used for the division of (1+1)-vesicles (Figs. 8 and 10).

In the context of synthetic biology, one long-term challenge is to mimick the vesicular traffic in eukaryotic cells, as schematically shown in Fig. 1. The vesicles are generated by the formation of buds from donor membranes and by the fission of the resulting membrane necks (Fig. 4(a)). These vesicles move across the cell and are eventually reintegrated into different membrane-bound organelles by adhesion to the acceptor membranes of these organelles and by the subsequent fusion between vesicle and acceptor membrane (Fig. 4(b)). In a steady state, the two processes balance each other and lead to an average membrane area that is approximately constant. Such a balance should hold on time scales that are small compared to the mean lifespan (or turnover time) of the cells. The latter time scale depends on the cell type and varies over many orders of magnitude but the mean lifespan of human cells always exceeds a couple of days.[127]

In order to obtain a steady state between fission and fusion processes, one can start from GUVs exposed to condensate droplets in the exterior solution. A reduction of the GUV volume *via* osmotic deflation will lead to the endocytic uptake of the droplets by the GUVs, thereby generating intraluminal vesicles as in Fig. 14. The fusion of these intraluminal vesicles with the GUV membrane can be triggered by adding caged GTP[128] to the interior solution of the GUV and by doping the GUV membranes with atlastin or another membrane-bound GTPase for homotypic fusion (Section 5.4). When the GTP is uncaged by exposure to light and hydrolyzed by the membrane-bound atlastin, the intraluminal vesicles will fuse with the GUV membrane, thereby releasing the condensate droplets back to the exterior solution. The latter process will be facilitated by increasing the bilayer tension within the GUV membrane via osmotic inflation, which also leads to an increase in the GUV volume. One can then start another cycle of fission and fusion by osmotic deflation and inflation where the GTP remains caged during deflation but becomes uncaged during inflation.

Abbreviations

ATP: Adenosine triphosphate
CMC: constant-mean-curvature
ER: Endoplasmic reticulum
GFP: Green fluorescent protein
GTP: Guanosine triphosphate
GTPase: Enzyme that hydrolyzes GTP
GUV: Giant unilamellar vesicle

His-tag: Histidine-tag
PEG: Polyethylene glycol

Acknowledgments

I thank Rikhia Ghosh for her exploratory vesicle simulations, which led to the time lapse snapshots in Figs. 11 and 20, as well as all my collaborators for enjoyable interactions. This paper is dedicated to the memory of Michael E. Fisher, with whom I had very stimulating and fruitful interactions for many years, in particular from 1984 until 1986, when I was a postdoc in his group at Cornell University. We published several joint papers on wetting phenomena. Such phenomena have now reemerged as important aspects of condensate droplets in living cells.

References

1. B. Alberts, D. Bray, A. Johnson, J. Lewis, M. Raff, K. Roberts and P. Walter, *Essential Cell Biology: An Introduction to the Molecular Biology of the Cell* (Garland, New York, 1998).
2. C. J. Obara, A. S. Moore and J. Lippincott-Schwartz, Structural diversity within the endoplasmic reticulum — From the microscale to the nanoscale. *Cold Spring Harb. Perspect. Biol.* doi: 10.1101/cshperspect.a041259, 2022.
3. M. Terasaki, L. B. Chen and K. Fujiwara, Microtubules and the endoplasmic reticulum are highly interdependent structures. *J. Cell. Biol.* **103**, 1557–1568 (1986).
4. C. Lee and L. B. Chen, Dynamic behavior of endoplasmic reticulum in living cells. *Cell* **54**, 37–46 (1988).
5. O. Baumann and B. Walz, Endoplasmic reticulum of animal cells and its organization into structural and functional domains. *Int. Rev. Cyt.* **205** (2001).
6. M. P. do Carmo, *Differential Geometry of Curves and Surfaces* (Prentice-Hall, Englewood Cliffs, NJ, USA, 1976).
7. R. Lipowsky, Understanding and controlling the morphological complexity of biomembranes. In *Advances in Biomembranes and Lipid Self-Assembly*, Vol. 30 Chapter 3, pp. 105–155 (Elsevier, London, UK, 2019).
8. J. Steinkühler, R. L. Knorr, Z. Zhao, T. Bhatia, S. Bartelt, S. Wegner, R. Dimova and R. Lipowsky, Controlled division of cell-sized vesicles by low densities of membrane-bound proteins. *Nat. Commun.* **11**, 905 (2020).
9. R. Ghosh, V. Satarifard, A. Grafmüller and R. Lipowsky, Budding and Fission of Nanovesicles induced by membrane adsorption of small solutes. *ACS Nano* **15**, 7237–7248 (2021).
10. R. Ghosh, V. Satarifard and R. Lipowsky, Different pathways for engulfment and endocytosis of droplets by nanovesicles. *Nat. Commun.* **14**, 615 (2023).
11. R. Lipowsky, Remodeling of membrane shape and topology by curvature elasticity and membrane tension. *Adv. Biol.* **6**, 2101020 (2022).
12. R. Ghosh, V. Satarifard, A. Grafmüller and R. Lipowsky, Spherical nanovesicles transform into a multitude of nonspherical shapes. *Nano Lett.* **19**, 7703–7711 (2019).
13. T. Bhatia, S. Christ, J. Steinkühler, R. Dimova and R. Lipowsky, Simple sugars shape giant vesicles into multispheres with many membrane necks. *Soft Matt.* **16**, 1246–1258 (2020).

14. Yanhong Li, H. Kusumaatmaja, R. Lipowsky and R. Dimova, Wetting-induced budding of vesicles in contact with several aqueous phases. *J. Phys. Chem. B* **116**, 1819–1823 (2012).
15. W. Helfrich, Elastic properties of lipid bilayers: Theory and possible experiments. *Z. Naturforsch.* **28c**, 693–703 (1973).
16. U. Seifert, K. Berndl and R. Lipowsky, Shape transformations of vesicles: Phase diagram for spontaneous curvature and bilayer coupling model. *Phys. Rev. A* **44**, 1182–1202 (1991).
17. R. Lipowsky, Understanding giant vesicles: A theoretical perspective. In R. Dimova and C. Marques (eds.), *The Giant Vesicle Book*, Chapter 5, pp. 73–168 (Taylor & Francis, 2020).
18. H. J. Deuling and W. Helfrich, The curvature elasticity of fluid membranes: A catalogue of vesicle shapes. *J. Physique* **37**, 1335–1345 (1976).
19. U. Seifert, L. Miao, H.-G. Döbereiner and M. Wortis, Budding transition for bilayer fluid vesicles. In R. Lipowsky, D. Richter and K. Kremer (eds.), *The Structure and Conformation of Amphiphilic Membranes*, Springer Proceedings in Physic, pp. 93–96 (Springer-Verlag, 1992).
20. H.-G. Döbereiner, E. Evans, M. Kraus, U. Seifert and M. Wortis, Mapping vesicle shapes into the phase diagram: A comparison of experiment and theory. *Phys. Rev. E* **55**(4), 4458–4474 (1997).
21. S. Svetina and B. Žekš, Shape behavior of lipid vesicles as the basis of some cellular processes. *Anat. Rec.* **268**, 215–225 (2002).
22. R. Lipowsky, Multispherical shapes of vesicles highlight the curvature elasticity of biomembranes. *Adv. Colloid Interface Sci.* **301**, 102613 (2022).
23. B. Fourcade, L. Miao, M. Rao, M. Wortis and R. K. P. Zia, Scaling analysis of narrow necks in curvature models of fluid lipid–bilayer vesicles. *Phys. Rev. E* **49**, 5276–5286 (1994).
24. F. Jülicher and R. Lipowsky, Domain-induced budding of vesicles. *Phys. Rev. Lett.* **70**, 2964–2967 (1993).
25. P. Yang, Q. Du and Z. C. Tu, General neck condition for the limit shape of budding vesicles. *Phys. Rev. E* **95**, 042403 (2017).
26. S. Christ, T. Litschel, P. Schwille and R. Lipowsky, Active shape oscillations of giant vesicles with cyclic closure and opening of membrane necks. *Soft Matt.* **17**, 319–330 (2021).
27. A. Roux, G. Koster, M. Lenz, B. Sorre, J.-B. Manneville, P. Nassoy and P. Bassereau, Membrane curvature controls dynamin polymerization. *PNAS* **107**(9), 4242–4146 (2010).
28. J. Schoeneberg, M. R. Pavlin, S. Yan, M. Righini, I.-H. Lee, L.-A. Carlson, A. H. Bahrami, D. H. Goldman, X. Ren, G. Hummer, C. Bustamante and J. H. Hurley, ATP-dependent force generation and membrane scission by ESCRT-III and Vps4. *Science* **362**, 1423–1428 (2018).
29. J. Xiao and E. D. Goley, Redefining the roles of the FtsZ-ring in bacterial cytokinesis. *Curr. Opin. Microbiol.* **34**, 90–96 (2016).
30. U. Seifert and R. Lipowsky, Adhesion of vesicles. *Phys. Rev. A* **42**, 4768–4771 (1990).
31. J. Agudo-Canalejo and R. Lipowsky, Critical particle sizes for the engulfment of nanoparticles by membranes and vesicles with bilayer asymmetry. *ACS Nano* **9**, 3704–3720 (2015).
32. J. Agudo-Canalejo and R. Lipowsky, Stabilization of membrane necks by adhesive particles, substrate surfaces, and constriction forces. *Soft Matt.* **12**, 8155–8166 (2016).

33. R. Lipowsky, Remodeling of biomembranes and vesicles by adhesion of condensate droplets. *Membranes* **13**, 223 (2023).
34. A. Derzhanski, A. G. Petrov and M. D. Mitov, Molecular asymmetry and Saddle-Splay elasticity in lipid bilayers. *Ann. Phys.* **3**, 297 (1978).
35. S. Lorenzen, R.-M. Servuss and W. Helfrich, Elastic torques about membrane edges: A study of pierced egg lecithin vesicles. *Biophys. J.* **50**, 565–572 (1986).
36. Mingyang Hu, J. J. Briguglio and M. Deserno, Determining the Gaussian Curvature Modulus of Lipid Membranes in Simulations. *Biophys. J.* **102**(6), 1403–1410 (2012).
37. V. Satarifard, A. Grafmüller and R. Lipowsky, Nanodroplets at membranes create tight-lipped membrane necks via negative line tension. *ACS Nano* **12**, 12424–12435 (2018).
38. R. Lipowsky, R. Ghosh, V. Satarifard, A. Sreekumari, M. Zamaletdinov, B. Różycki, M. Miettinen and A. Grafmüller, Leaflet tensions control the spatio-temporal remodeling of lipid bilayers and nanovesicles. *Biomolecules* **13**, 926 (2023).
39. M. Andes-Koback and C. D. Keating, Complete budding and asymmetric division of primitive model cells to produce daughter vesicles with different interior and membrane compositions. *JACS* **133**(24), 9545–9555 (2011).
40. M. Osawa and H. P. Erickson, Liposome division by a simple bacterial division machinery. *PNAS* **110**(27), 11000–1004 (2013).
41. W. T. Snead, C. C. Hayden, A. K. Gadok, C. Zhao, E. M. Lafer, P. Rangamani and J. C. Stachowiak, Membrane fission by protein crowding. *Proc. Nat. Acad. Sci. USA* **114**, E3258–E3267 (2017).
42. S. Deshpande, W. K. Spoelstra, M. van Doorn J. Kerssemakers and C. Dekker, Mechanical Division of Cell-Sized Liposomes. *ACS Nano* **12**, 2560–2568 (2018).
43. T. Litschel, B. Ramm, R. Maas, M Heymann and P. Schwille, Beating Vesicles: Encapsulated Protein Oscillations Cause Dynamic Membrane Deformations. *Angew. Chemie Int. Ed.* **57**, 16286–16290 (2018).
44. Y. Dreher, K. Jahnke, E. Bobkova, J. P. Spatz and K. Göpfrich, Division and regrowth of phase-separated giant unilamellar vesicles. *Angew. Chemie Int. Ed.* **60**, 10661–10669 (2021).
45. Z. Hu, E. P. Gogol and J. Lutkenhaus, Dynamic assembly of MinD on phospholipid vesicles regulated by ATP and MinE. *Proc. Natl. Acad. Sci.* **99**, 6761–6766 (2002).
46. A. Sreekumari and R. Lipowsky, Large stress asymmetries of lipid bilayers and nanovesicles generate lipid flip-flops and bilayer instabilities. *Soft Matter* **18**, 6066–6078 (2022).
47. P. A. Albertsson, *Partition of Cell Particles and Macromolecules: Separation and Purification of Biomolecules, Cell Organelles Membranes, and Cells in Aqueous Polymer Two-Phase Systems and Their Use in Biochemical Analysis and Biotechnology*, (3rd edn.) (Wiley, 1986).
48. Yanhong Li, R. Lipowsky and R. Dimova, Transition from complete to partial wetting within membrane compartments. *JACS* **130**, 12252–12253 (2008).
49. R. Dimova and R. Lipowsky, Giant vesicles exposed to aqueous two-phase systems: Membrane wetting, budding processes, and spontaneous tubulation. *Adv. Mater. Interf.* **4**, 1600451 (2017).
50. J. Esquena, Water-in-water (W/W) emulsions. *Curr. Opin. Colloid Interface Sci.* **25**, 109–119 (2016).
51. C. P. Brangwynne, C. R. Eckmann, D. S. Courson, A. Rybarska, C. Hoege, J. Gharakhani, F. Jülicher and A. A. Hyman, Germline P granules are liquid droplets that localize by controlled dissolution/condensation. *Science* **324**, 1729–1732 (2009).

52. S. F. Banani, H. O. Lee, A. A. Hyman and M. K. Rosen, Biomolecular condensates: Organizers of cellular biochemistry. *Nature Rev.: Mol. Cell Biol.* **18**, 285–295 (2017).
53. R. Fesce, F. Grohovaz, F. Valtorta and J. Meldolesi, Neurotransmitter release: Fusion or 'Kiss-and-run'? *Trends Cell Biol.* **4**(1–4) (1994).
54. R. G. W Staal, E. V. Mosharov and D. Sulzer, Dopamine neurons release transmitter via a flickering fusion pore. *Nat. Neurosci.* **7**, 341–346 (2004).
55. R. M. Wightman and C. L. Haynes, Synaptic vesicles really do kiss and run. *Nat. Neurosci.* **7**, 321–322 (2004).
56. H. Noguchi and M. Takasu, Adhesion of nanoparticles to vesicles: A Brownian dynamics simulation. *Biophys. J.* **83**, 299–308 (2002).
57. A. J. Markvoort, P. Spijker, A. F. Smeijers, K. Pieterse, R. A. van Santen and P. A. J. Hilbers, Vesicle deformation by draining: Geometrical and topological shape changes. *J. Phys. Chem. B* **113**, 8731–8737 (2009).
58. N. Urakami, T. Jimbo, Y. Sakuma and M. Imai, Molecular mechanism of vesicle division induced by coupling between lipid geometry and membrane curvatures. *Soft Matt.* **14**, 3018–3027 (2018).
59. X. Li, Y. Liu, L. Wang, M. Deng and H. Liang, Fusion and fission pathways of vesicles from amphiphilic triblock copolymers: A dissipative particle dynamics simulation study. *Phys. Chem. Chem. Phys.* **11**, 4051–4059 (2009).
60. M. Pannuzzo, Z. A. McDargh and M. Deserno, The role of scaffold reshaping and disassembly in dynamin driven membrane fission. *eLife* **7**, e39441 (2018).
61. A. W. Fritsch, A. F. Diaz-Delgadillob, O. Adame-Arana, C. Hoege, M. Mittasch, M. Kreysing, M. Leaver, A. A. Hyman, F. Jülicher and C. A. Weber, Local thermodynamics govern formation and dissolution of Caenorhabditis elegans P granule condensates. *PNAS* **118**(37), e2102772118 (2021).
62. J. Berry, S. C. Weber, N. Vaidya, M. Haatajaa and C. P. Brangwynne, RNA transcription modulates phase transition-driven nuclear body assembly. *Proc. Natl. Acad. Sci. USA* **112**, E5237–E5245 (2015).
63. S. Jain, J. R. Wheeler, R. W. Walters, A. Agrawal, A. Barsic and R. Parker, ATPase-Modulated Stress Granules Contain a Diverse Proteome and Substructure. *Cell* **164**, 487–498 (2016).
64. P. Li, S. Banjade, H.-C. Cheng, S. Kim, B. Chen, L. Guo, M. Llaguno, J. V. Hollingsworth, D. S. King, S. F. Banani, P. S. Russo, Q.-X. Jiang, B. T. Nixon and M. K. Rosen, Phase transitions in the assembly of multivalent signalling proteins. *Nature* **483**, 336–341 (2012).
65. A. Patel, H. O. Lee, L. Jawerth, S. Maharana, M. Jahnel, M. Y. Hein, S. Stoynov, J. Mahamid, S. Saha, T. M. Franzmann, S. Grill, D. Drechsel, A. A. Hyman and S. Alberti, A liquid-to-solid phase transition of the ALS protein FUS accelerated by disease mutation. *Cell* **162**, 1066–1077 (2015).
66. Y. Lin, D. S. W. Protter, M. K. Rosen and R. Parker, Formation and maturation of phase-separated liquid droplets by RNA-binding proteins. *Mol. Cell* **60**, 208–219 (2015).
67. A. Molliex, J. Temirov, J. Lee, H. J. Kim, T. Mittag and J. P. Taylor, Phase separation by low complexity domains promotes stress granule assembly and drives pathological fibrillization. *Cell* **163**, 123–133 (2015).
68. T. W. Han, M. Kato, S. Xie, L. C. Wu, H. Mirzaei, J. Pei, M. Chen, Y. Xie, J. Allen, G. Xiao and S. L. McKnight, Cell-free formation of RNA granules: Bound RNAs identify features and components of cellular assemblies. *Cell* **149**, 768–779 (2012).
69. L. Guo and J. Shorter, It's raining liquids: RNA tunes viscoelasticity and dynamics of membraneless organelles. *Mol. Cell* **60**, 189–192 (2015).

70. M. R. Helfrich, L. K. Mangeney-Slavin, M. S. Long, K. Y. Djoko and C. D. Keating, Aqueous phase separation in giant vesicles. *J. Am. Chem. Soc. (JACS)* **124**, 13374–13375 (2002).
71. Yanhong Li, R. Lipowsky and R. Dimova, Membrane nanotubes induced by aqueous phase separation and stabilized by spontaneous curvature. *Proc. Nat. Acad, Sci. USA* **108**, 4731–4736 (2011).
72. R. Lipowsky, Spontaneous tubulation of membranes and vesicles reveals membrane tension generated by spontaneous curvature. *Faraday Discuss.* **161**, 305–331 (2013).
73. Yonggang Liu, J. Agudo-Canalejo, A. Grafmüller, R. Dimova and R. Lipowsky, Patterns of flexible nanotubes formed by liquid-ordered and liquid-disordered membranes. *ACS Nano* **10**, 463–474 (2016).
74. J. Zimmerberg A. Finkelstein and F. S. Cohen, Osmotic swelling of vesicles: Its role in the fusion of vesicles with planar phospholipid bilayer membranes and its possible role in exocytosis. *Anu. Rev. Physiol.* **48**, 163–174 (1986).
75. J. Shillcock and R. Lipowsky, Tension-induced fusion of bilayer membranes and vesicles. *Nat. Materials* **4**, 225–228 (2005).
76. A. Grafmüller, J. Shillcock and R. Lipowsky, Pathway of membrane fusion with two tension-dependent energy barriers. *Phys. Rev. Lett.* **98**, 218101 (2007).
77. Lianghui Gao, R. Lipowsky and J. C. Shillcock. Tension-induced vesicle fusion: Pathways and pore dynamics. *Soft Matt.* **4**, 1208–1214 (2008).
78. T.-T. Kliesch, J. Dietz, L. Turco, P. Halder, E. Polo, Marco Tarantola, R. Jahn and A. Janshoff, Membrane tension increases fusion efficiency of model membranes in the presence of SNAREs. *Sci. Rep.* **7**, 12070 (2017).
79. R. B. Lira, T. Robinson, R. Dimova and K. A. Riske, Highly efficient protein-free membrane fusion: A giant vesicle study. *Biophys. J.* **116**, 79–91 (2019).
80. I. Ivanov, R. B. Lira, T.-Y. D. Tang, T. Franzmann, A. Klosin, L. C. da Silva, A. A. Hyman, K. Landfester, R. Lipowsky, K. Sundmacher and R. Dimova, Directed growth of biomimetic microcompartments. *Adv. Biosyst.* **3**, 1800314 (2019).
81. S. Deshpande, S. Wunnava, D. Hueting and C. Dekker, Membrane tension–mediated growth of liposomes. *Small* **15**, 1902898 (2019).
82. A. Grafmüller, J. C. Shillcock and R. Lipowsky, The fusion of membranes and vesicles — Pathway and energy barriers from Dissipative Particle Dynamics. *Biophys. J.* **96**, 2658–2675 (2009).
83. Lianghui Gao, J. Shillcock and R. Lipowsky, Improved dissipative particle dynamics simulations of lipid bilayers. *J. Chem. Phys.* **126**, 015101 (2007).
84. P. K. J. Kinnunen, Fusion of lipid bilayers: A model involving mechanistic connection to Hn phase forming lipids. *Chem. Phys. Lipids* **63**, 251–258 (1992).
85. J. M. Holopainen, J. Y. A. Lehtonen and P. K. J. Kinnunen, Evidence for the extended phospholipid conformation in membrane fusion and hemifusion. *Biophys. J.* **76**, 2111–2120 (1999).
86. H. A. Scheidt, K. Kolocaj, D. B. Konrad, J. A. Frank, D. Trauner, D. Langosch and D. Huster, Light-induced lipid mixing implies a causal role of lipid splay in membrane fusion. *BBA Biomembr.* **1862**, 183438 (2020).
87. M. J. Stevens, J. H. Hoh and T. B. Woolf, Insights into the molecular mechanism of membrane fusion from simulation: Evidence for the association of splayed tails. *Phys. Rev. Lett.* **91**(18), 188102 (2003).
88. K. Pieterse A. F. Smeijers, A. J. Markvoort and A. J. Hilbers, A detailed look at vesicle fusion. *J. Phys. Chem. B* **110**, 13212–13219 (2006).

89. Y. G. Smirnova, S. J. Marrink, R. Lipowsky and V. Knecht, Solvent-exposed tails as prestalk transition states for membrane fusion at low hydration. *JACS* **132**, 6710–6718 (2010).
90. D. Mirjanian, A. N. Dickey, J. H. Hoh, T. B. Woolf and M. J. Stevens, Splaying of aliphatic tails plays a central role in barrier crossing during liposome fusion. *J. Phys. Chem. B* **114**, 11061–11068 (2010).
91. M. A. Tahir, R. C. Van Lehn, S. H. Choi and A. Alexander-Katz, Solvent-exposed lipid tail protrusions depend on lipid membrane composition and curvature. *Biochim. Biophys. Acta* **1858**, 1207–1215 (2016).
92. J. M. Gardner and C. F. Abrams, Lipid flip-flop vs. lateral diffusion in the relaxation of hemifusion diaphragms. *BBA Biomembr.* **1860**, 1452–1459 (2018).
93. D. P. Siegel, Energetics of intermediates in membrane fusion: Comparison of stalk and inverted micellar intermediate mechanisms. *Biophys. J.* **65**, 2124–2140 (1993).
94. L. K. Tamm, J. Crane and V. Kiessling, Membrane fusion: A structural perspective on the interplay of lipids and proteins. *Curr. Opin. Struct. Biol.* **13**, 453–466 (2003).
95. L. V. Chernomordik and M. M. Kozlov, Mechanics of membrane fusion. *Nat. Struct. Mol. Biol.* **15**(7), 675–683 (2008).
96. J. Nikolaus, M. Stoeckl, D. Langosch, R. Volkmer and A. Herrmann, Direct visualization of large and protein-free hemifusion diaphragms. *Biophys. J.* **98**, 1192–1199 (2010).
97. H. J. Risselada, Y. Smirnova and H. Grubmüller, Free energy landscape of rim-pore expansion in membrane fusion. *Biophys. J.* **107**, 2287–2295 (2014).
98. T. Brandt, L. Cavellini, W. Kuehlbrandt and M. M. Cohen, A mitofusin-dependent docking ring complex triggers mitochondrial fusion in vitro. *eLife* **5**, e14618 (2016).
99. M. D'Agostino, H. J. Risselada, L. J. Endter, V. Comte-Miserez and A. Mayer, SNARE-mediated membrane fusion arrests at pore expansion to regulate the volume of an organelle. *EMBO J.* **37**, **e99193** (2018).
100. E. M. Blokhuis, M. D'Agostino, A. Mayer and H. J. Risselada, Fusion pores live on the edge. *J. Phys. Chem. Lett.* **11**, 1204–1208 (2020).
101. H. J. Risselada and H. Grubmüller. How proteins open fusion pores: Insights from molecular simulations. *Eur. Biophys. J.* **50**, 279–293 (2021).
102. S. Boonstra, J. S. Blijleven, W. H. Roos, P. R. Onck, E. van der Giessen and A. M. van Oijen, Hemagglutinin-mediated membrane fusion: A biophysical perspective. *Annu. Rev. Biophys.* **47**, 153–173 (2018).
103. F. Li, Structure, function, and evolution of coronavirus spike proteins. *Annu. Rev. Virol.* **3**, 237–261 (2016).
104. C. B. Jackson, M. Farzan, B. Chen and H. Choe, Mechanisms of SARS- CoV-2 entry into cells. *Nat. Rev. Mol. Cell Biol.* **23**, 3–20 (2022).
105. G. Orso, D. Pendin, S. Liu, J. Tosetto, T. J. Moss, J. E. Faust, M. Micaroni, A. Egorova, A. Martinuzzi, J. A. McNew and A. Daga, Homotypic fusion of ER membranes requires the dynamin-like GTPase Atlastin. *Nature* **460**, 978–983 (2009).
106. J. Hu, Y. Shibata, P.-P. Zhu, C. Voss, N. Rismanchi, W. A. Prinz, T. A. Rapoport and C. Blackstone, A class of dynamin-like GTPases involved in the generation of the tubular ER network. *Cell* **138**, 549–561 (2009).
107. N. Wang and T. A. Rapoport, Reconstituting the reticular ER network — mechanistic implications and open questions. *J. Cell Sci.* **132**, jcs227611 (2019).
108. K. Anwar, R. W. Klemm, A. Condon, K. N. Severin, M. Zhang, R. Ghorlando, J. Hu, T. A. Rapoport and W. A. Prinz, The dynamin-like GTPase Sey1p mediates homotypic ER fusion in S. cerevisiae. *J. Cell Biol.* **197**(2), 209–217 (2012).

109. M. Zhang and J. Hu, Homotypic fusion of endoplasmic reticulum membranes in plant cells. *Front. Plant Sci.* **4**, Article 514 (2013).
110. D. C. Chan, Mitochondrial fusion and fission in mammals. *Annu. Rev. Cell Dev. Biol.* **22**, 79–99 (2006).
111. L. Tilokani, S. Nagashima, V. Paupe and J. Prudent, Mitochondrial dynamics: Overview of molecular mechanisms. *Essays Biochem* **62**, 341–360 (2018).
112. A. Green, T. Hossain and D. M. Eckmann, Mitochondrial dynamics involves molecular and mechanical events in motility, fusion and fission. *Front. Cell Dev. Biol.* **10**, 1010232 (2022).
113. R. Jahn and R. H. Scheller, SNAREs — Engines for membrane fusion. *Nat. Rev. Mol. Cell Biol.* **7**, 631–643 (2006).
114. S. Koike and R. Jahn, SNARE proteins: Zip codes in vesicle targeting? *Biochem. J.* **479**, 273–288 (2022).
115. D. Mion, L. Bunel, P. Heo and F. Pincet, The beginning and the end of SNARE-induced membrane fusion. *FEBS Open Bio.* **12**, 1958–1979 (2022).
116. A. Nowbagh, A. Deshwal, M. Kadu, A. Chaudhuri, S. Maiti, R. Lipowsky and T. Bhatia, Generation of bilayer asymmetry and membrane curvature by the sugar-cleaving enzyme invertase. *ChemSystemsChem* **5**, e202200027 (2022).
117. C. Delaunay, Sur la surface de révolution dont la courbure moyenne est constante. *J. Math. Pures et Appl. Sér. 1* **6**, 309–320 (1841).
118. N. Kapouleas, Complete constant mean curvature surfaces in euclidean three-space. *Ann. Math.* **131**, 239–330 (1990).
119. N. J. Korevaar, R. Kusner and B. Solomon, The structure of complete embedded surfaces with constant mean curvature. *J. Differ. Geom.* **30**, 465–503 (1989).
120. N. Korevaar and R. Kusner, The global structure of constant mean curvature surfaces. *lnvent. Math.* **114**, 311–332 (1993).
121. K. Grosse-Brauckmann and K. Polthier, Constant mean curvature surfaces derived from Delaunay's and Wente's examples. In H. C. Hege and K. Polthier (eds.), *Visualization and Mathematics*, p. 386 (Springer Verlag, 1997).
122. K. Grosse-Brauckmann, R. B. Kusner and J. M. Sullivan, Triunduloids: Embedded constant mean curvature surfaces with three ends and genus zero. *Journal fur die reine und angewandte Mathematik* **564**, 35–61 (2003).
123. R. Lipowsky, S. Pramanik, A. S. Benk, J. Spatz and R. Dimova, Elucidating the Morphology of the Endoplasmic Reticulum: Puzzles and Perspectives. *ACS Nano* **17**, 11957–11968 (2023).
124. R. E. Powers, S. Wang, T. Y. Liu and T. A. Rapoport, Reconstruction of the tubular endoplasmic reticulum network with purified components. *Nature* **543**, 257–272 (2017).
125. M. A. Betancourt-Solis, T. Desai and J. A. McNew, The atlastin membrane anchor forms an intramembrane hairpin that does not span the phospholipid bilayer. *J. Biol. Chem.* **293**(48) 18514–18524 (2018).
126. D. Holcman, P. Parutto, J. E. Chambers, M. Fantham, L. J. Young, S. J. Marciniak, C. F. Kaminski, D. Ron and E. Avezov, Single particle trajectories reveal active endoplasmic reticulum luminal flow. *Nat. Cell Biol.* **20**, 1118–1125 (2018).
127. R. Sender and R. Milo, The distribution of cellular turnover in the human body. *Nat. Med.* **27**, 45–48 (2021).
128. C. Kötting and K. Gerwert, What vibrations tell us about GTPases. *Biol. Chem.* **396**, 131–144 (2015).

https://doi.org/10.1142/9789811282386_0039

Chapter 39

Universal Scaling Function Ansatz for Finite Temperature Jamming

S.A. Ridout, J.P. Sethna, and A.J. Liu

Department of Physics and Astronomy, University of Pennsylvania, Philadelphia, PA 19104 USA
Department of Physics, Emory University, Atlanta, GA 30322, USA
LASSP, Cornell University, Ithaca, NY 14850, USA

We cast a non-zero-temperature analysis of the jamming transition[1] into the framework of a scaling ansatz. We show that four distinct regimes for scaling exponents of thermodynamic derivatives of the free energy, such as pressure, bulk and shear moduli, can be consolidated by introducing a universal scaling function with two branches. Both the original analysis and the scaling theory assume that the system always resides in a single basis in the energy landscape. The two branches are separated by a line $T^*(\Delta\phi)$ in the $T-\Delta\phi$ plane, where $\Delta\phi = \phi - \phi_c^\Lambda$ is the deviation of the packing fraction from its critical, jamming value, ϕ_c^Λ, for that basin. The branch for $T < T^*(\Delta\phi)$ reduces at $T = 0$ to an earlier scaling ansatz[2] that is restricted to $T = 0$, $\Delta\phi \geq 0$, while the branch for $T > T^*(\Delta\phi)$ reproduces exponents observed for thermal hard spheres. In contrast to the usual scenario for critical phenomena, the two branches are characterized by different exponents. We suggest that this unusual feature can be resolved by the existence of a dangerous irrelevant variable u, which can appear to modify exponents if the leading $u = 0$ term is sufficiently small in the regime described by one of the two branches of the scaling function.

Like crystallization, jamming provides a paradigm for how particulate systems become rigid. For ideal spheres that increasingly repel each other as they overlap and do not interact if they do not overlap, the onset of rigidity with increasing packing fraction at zero temperature T is discontinuous for the most stable possible state, the perfect FCC crystal, and critical for the least stable possible state, the jammed state at the jamming transition at a packing fraction ϕ_c.[3,4] While crystallization is a straightforward first-order transition, the extent to which the jamming transition can be described within the framework of critical phenomena is still unclear. Certainly, many quantities exhibit power-law scaling, but are those scalings consistent with universal scaling functions with associated scaling relations among exponents?

In contrast to normal critical transitions, in which a diverging correlation length corresponds to fluctuations that diverge at the critical point, the jamming transition displays diverging correlations of fluctuations that *vanish* at the transition.[5] Moreover, the jamming transition point depends on the history of the material — the protocol with which one increases the density and decreases the temperature to reach jamming. It is encouraging, however, that a universal scaling ansatz[2] successfully described the zero-temperature elastic properties. Moreover, a jamming-like transition in a diluted spring network[6] has been cast into a universal scaling form (see Supplement[6]) and since turned into a general linear response theory.[7,8]

The situation at non-zero temperature $\tilde{T}$ is more challenging because the systems are evolving with time. Work by De Giuli *et al.*[1] has bypassed this difficulty by assuming that at $\tilde{T} > 0$, the system always resides in a given basin in the energy landscape, whose jamming transition is at ϕ_c^Λ. That is, they describe a system near jamming using an effective contact network with contacts between particles that collide. This description is valid on sufficiently short time scales where this contact network does not change, analogous to methods developed to study stresses in experiments measuring colloidal particle positions.[9] While this assumption must break down at long time scales, it is reasonable to ask whether the scaling ansatz can be extended to non-zero temperatures within this approximation. We note that the zero-temperature scaling ansatz makes a similar approximation because the protocol and sample-to-sample fluctuations of the critical jamming packing fraction ϕ_c are swept under the rug by constructing a scaling ansatz in terms of $\Delta\phi = \phi - \phi_c^\Lambda$, where ϕ_c^Λ can vary among configurations $\{\Lambda\}$.

The theoretical arguments of De Giuli *et al.*[1] apply to systems of N soft, frictionless spheres in a volume V at temperature $\tilde{T}$. The packing fraction is $\phi = \frac{1}{V}\sum_i V_i$, where V_i is the d-dimensional volume of particle i. The spheres interact via pairwise harmonic repulsions when they overlap, with a spring stiffness k. We introduce the dimensionless free energy, pressure, shear stress, bulk modulus and shear modulus using k with appropriate factors of D_{avg}, the average particle diameter, so that F = free energy$/kD_{\text{avg}}^2$, p = pressure $* D_{\text{avg}}/k$, etc. Note that we are departing from the usual custom of using temperature T to construct a dimensionless free energy; this is important because we specifically interested in describing behavior at $T = 0$ as well as $T > 0$. Instead, we introduce the dimensionless temperature $T = \tilde{T}/kD_{\text{avg}}^2$. The $T = 0$ scaling ansatz[2] is expressed in terms of these dimensionless quantities for the same reason.

The marginal stability arguments and effective medium theory of De Giuli *et al.*[1] suggest that there are four different scaling regimes in the $T - \phi$ plane,[1] Fig. 1 (top), only one of which is described by the existing scaling ansatz.[2] The predicted

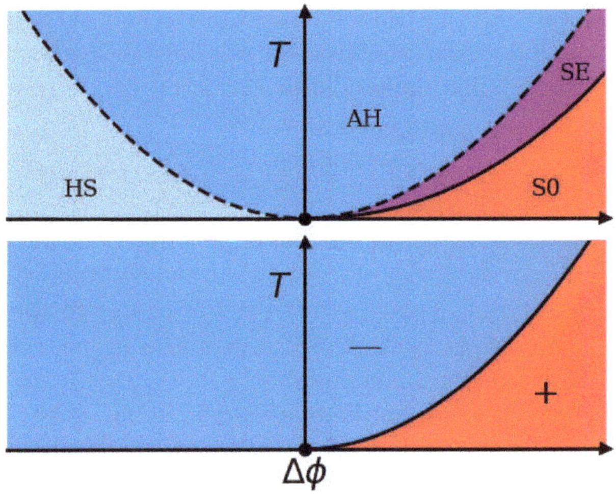

Fig. 1. Top: The phase diagram of De Giuli *et al.*[1] Different colors, separated by lines, indicate four distinct scaling regimes, which are not necessarily separated by sharp phase transitions. Bottom: The phase diagram corresponding to the scaling ansatz Eq. (1). The − branch reproduces the scaling exponents of three of De Giuli *et al.*'s scaling regimes. The phase boundary lies along a constant ratio $T^*/\phi^{\zeta_+/\delta_{\phi,+}} = C^*$ in the notation of Eq. (1) (see footnote a).

exponents in the different regimes are consistent with mean-field solutions for soft spheres[10] and the perceptron model,[11] although these later investigations showed only a crossover, not a sharp transition — perhaps the smearing expected by the evolving landscape at finite temperatures.

The existence of four distinct scaling regimes all terminating at the same critical point could suggest rather exotic criticality. Here, we show that the results of De Giuli *et al.*[1] are captured by a scaling function in a manner similar to that of ordinary criticality, with two branches (Fig. 1, bottom) separated by a continuous phase transition. This scaling function fully describe all of the scaling behaviors seen near the jamming transition in the full space described by the jamming phase diagram, namely temperature T, shear strain ϵ, and packing fraction $\Delta\phi = \phi - \phi_c$, where ϕ_c is sample-dependent. Unlike in ordinary criticality, the two branches are characterized by different exponents. We suggest that the difference in exponents can result from the existence of a dangerous irrelevant variable u, which can appear to modify exponents if the leading $u = 0$ term is small in the regime described by one of the two branches of the scaling function.

Each state is also characterized by average number of interacting neighbors per particle (the contact number, Z), which satisfies $Z \geq Z_{\min}$ where $Z_{\min} = 2d - (2d-2)/N$ approaches the isostatic value $Z_{\text{iso}} \equiv 2d$ at the jamming transition in the thermodynamic limit[3,12]; we define $\Delta Z = Z - Z_{\min}$. The scaling ansatz of Ref. 2 was written for the elastic energy with the excess coordination beyond isostaticity, ΔZ, as the control parameter analogous to reduced temperature t in the Ising model. We will follow suit, constructing a scaling ansatz for the dimensionless free energy F instead of the dimensionless elastic energy E in order to describe behavior at $T \geq 0$:

$$F(\Delta Z, \Delta\phi, \varepsilon, N, T) = \Delta Z^{\zeta_\pm} \mathcal{F}_\pm \left(\frac{\Delta\phi}{\Delta Z^{\delta_{\phi,\pm}}}, \frac{\varepsilon}{\Delta Z^{\delta_{\varepsilon,\pm}}}, N\Delta Z^{\psi_\pm}, \frac{T}{\Delta Z^{\zeta_\pm}} \right), \quad (1)$$

where $\pm$ denote the two branches of the scaling function and the scaling exponents for the two branches, $\zeta_\pm$, $\delta_{\phi,\pm}$, $\delta_{\varepsilon,\pm}$ and $\psi_\pm$, are to be determined. Equation (1) is set up so that the leading singular part of the free energy in the thermodynamic limit is proportional to $\Delta Z^{\zeta_\pm}$. Note that $\zeta_\pm$ also appear in the temperature scaling; this follows simply from dimensional analysis.[2]

Note that the finite-size scaling is expressed in terms of the total number of particles, N, instead of the system length, $L \sim N^{1/d}$. We use N instead of L because of arguments[2,12–14] that the upper critical dimension for jamming is $d_u = 2$. Finite-size effects for $d \geq d_u$ should scale with N with the exponent $\psi = d_u\nu$, where ν is the mean-field correlation length exponent.[15]

The excess packing fraction $\Delta\phi$ and shear strain ε in Eq. (1) represent components of the same strain tensor (compression and shear, respectively) but are allowed to scale differently. Note that we assume systems are prepared isotropically so that different shear directions are statistically equivalent; this is why we represent them with a single ε. For systems prepared by shearing or by applying other loads, a more complicated formulation of the scaling ansatz is needed. For the $T = 0$, $\phi > \phi_c$ case, Ref. 16 points the way toward such a formulation for shear-jammed systems.

DeGiuli *et al.*[1] identified four regimes in the $T - \phi$ plane (Fig. 1, top): (1) The regime "S0" in Ref. 1 corresponds to low T at $\phi > \phi_c$. The exponents in this regime are those for $T = 0$ soft spheres, described by the scaling ansatz of Ref. 2 (see Table 1 of Ref. 1 for a list of exponents in each regime). (2) A regime at somewhat higher temperatures at $\phi > \phi_c$, called "SE" in Ref. 1, or "entropic soft spheres." In this regime, the pressure is controlled by overlaps, but the vibrational properties are strongly affected by thermal collisions. (3) An anharmonic regime at packing fractions ϕ close to ϕ_c at non-zero T, called "AH" in Ref. 1. In this regime, the pressure and other properties are purely entropic and controlled only by T. (4) A regime at $\phi < \phi_c$ and lower T, extending down to $T = 0$, called "HS" in Ref. 1 because it is described by the exponents near the jamming transition for thermal hard spheres. We will use the predicted exponents of De Giuli *et al.*[1] in each of the four regimes to determine the exponents introduced in the scaling ansatz and will show that they are fully consistent with the ansatz.

At $T = 0$, $\phi > \phi_c$, Eq. (1) reduces to the scaling ansatz already proposed in Ref. 2, with $\mathcal{F}_+ = \mathcal{E}_0$ from Eq. (1) of Ref. 2. Therefore, the exponents for the positive branch of the scaling function can be read off from Table 1 of Ref. 2: $\zeta_+ = 4$,

$\delta_{\phi,+} = 2$, $\delta_{\varepsilon,+} = 3/2$, $\psi_+ = 1$. These exponents yield exponents for the pressure, bulk and shear moduli, etc. that are in agreement with those listed in the S0 regime in Ref. 1; the exponents in this regime have been established for a long time.[3] As in Ref. 2, $\psi_+ = d_u \nu_T$, where ν_T is the exponent for the transverse length scale, ℓ_T. This length scale characterizes the wavelength of transverse phonons at the boson peak frequency, ω^*[17,18]: $\ell_T \sim c_T/\omega^*$, where the speed of transverse sound, c_T, scales as the square root of the shear modulus, G. Note that the boson peak frequency corresponds to the Ioffe–Regel crossover frequency where the phonon wavelength becomes comparable to the mean-free path between phonon scattering events.[19] Thus, ℓ_T also corresponds to the phonon mean-free path at ω^*. Equivalently, ℓ_T corresponds to the length scale of correlations of polarization vectors in vibrational modes at ω^*[1,20] (their $\ell_S(\omega^*)$). It is straightforward to show that $\ell_T \sim \sqrt{G}/\omega^*$ has the same scaling as $\ell_S(\omega^*)$ in Table 1 of Ref. 1 in all four regimes S0, SE, AH and HS.

The next step is to calculate the exponents for the negative branch of the scaling function. This can be done by comparing to exponents in the SE, AH or HS regimes of Ref. 1. We will use the SE exponents and then show that the resulting exponents for the negative branch are fully consistent with the exponents for the AH and HS regimes, as well, so that the negative branch describes all three regimes.

We first consider the finite-size scaling behavior. The free energy must be continuous at all $T > 0$ for any ϕ for any system size N. This implies that we must have equality of the finite-size scaling exponents for both branches: $\psi_- = \psi_+ = 1$. This implies a correlation length exponent of $\nu = \psi/d_u = 1/2$, just as in the S0 regime. From Table 1 in Ref. 1, we see that the length scale ℓ_T ($\ell_S(\omega^*)$ in their notation) scales as $(T/\Delta\phi)^{\frac{a-1}{4}}$, where $a = (1-\theta_e)/(3+\theta_e)$ and $\theta_e \approx 0.42$ is the exponent characterizing the low force tail of the force distribution in the mean-field theory,[21,22] as well as in low dimensions when bucklers are excluded.[23] Note that ΔZ scales as $(T/p)^{2b}$, where $p \sim \Delta\phi$ in this regime, where $b = (1+\theta_e)/(3+\theta_e)$. Thus, $\ell_T \sim \Delta Z^{-1/2}$, so that $\nu_T = 1/2 = \nu$. Thus, the relevant length scale controlling finite-size effects in both the S0 and SE regimes is the transverse length scale ℓ_T.

The remaining exponents can be obtained by differentiating Eq. (1) and comparing with the scalings of Table 1 of Ref. 1. For example, by differentiating F with respect to $\Delta\phi$, we find that the pressure satisfies $p \sim \Delta Z^{\zeta_- - \delta_{\phi,-}}$. But from Table 1, we see that $p \sim \Delta\phi \sim \Delta Z^{\delta_{\phi,-}}$, so $\zeta_- = 2\delta_{\phi,-}$. But we also have $\Delta Z \sim (T/p)^{2b} \sim \Delta Z^{2b\zeta_-}/\Delta Z^{2b(\zeta_- - \delta_{\phi,-})} \sim \Delta Z^{2b\delta_{\phi,-}}$. Thus, $\delta_{\phi,-} = 1/(2b) = (3+\theta_e)/(1+\theta_e)$ and $\zeta_- = 2(3+\theta_e)/(1+\theta_e)$. Similarly, $\delta_{\varepsilon,-} = (2+\theta_e)/(1+\theta_e)$.

It is straightforward to show that if one adopts the exponents from Table 1 of Ref. 1 in the hard- sphere (HS) regime, one obtains the same values of the exponents ψ_-, ζ_-, $\delta_{\phi,-}$ and $\delta_{\varepsilon,-}$. Moreover, by adopting these exponents and applying the

Table 1. Values of critical exponents for the two branches (+ and −) of the scaling ansatz of Eq. (1).

	+	−
ζ	4	$2(3+\theta_e)/(1+\theta_e)$
δ_ϕ	2	$(3+\theta_e)/(1+\theta_e)$
δ_ε	3/2	$(2+\theta_e)/(1+\theta_e)$
ψ	1	1

scaling ansatz in Eq. (1) at $\Delta Z = 0$, $T > 0$, where the behavior must be analytic, one recovers the exponents for the anharmonic (AH) regime of Table 1 of Ref. 1. Thus, three of the four regimes, HS, AH and SE, of Fig. 1 in Ref. 1 (the blue and purple regions) all correspond to the negative branch of the scaling function with identical exponents (for details, see Appendix A.2). The remaining regime (S0) is described by the positive branch of the scaling function. In every regime, the length scale controlling finite-size effects is the transverse length scale ℓ_T ($\ell_S(\omega^*)$ in the notation of Ref. 1).

The free energy must be continuous everywhere, so the two branches of the scaling function must yield the same free energy along a curve $T^*(\Delta\phi, \varepsilon)$ for every system size N. Here, we restrict ourselves to the case where $\varepsilon = 0$ and $N \to \infty$ to compare directly to the prediction of Ref. 1. Clearly, we must have

$$\Delta Z^{\zeta_+} \mathcal{F}_+^\infty \left(\frac{\Delta\phi}{\Delta Z^{\delta_{\phi,+}}}, \frac{T^*}{\Delta Z^{\zeta_+}} \right) = \Delta Z^{\zeta_-} \mathcal{F}_-^\infty \left(\frac{\Delta\phi}{\Delta Z^{\delta_{\phi,-}}}, \frac{T^*}{\Delta Z^{\zeta_-}} \right). \tag{2}$$

Eliminating ΔZ from the equation (see Appendix A.1), we obtain

$$T^* \tilde{\mathcal{F}}_+^\infty \left(\frac{\Delta\phi^{\zeta_+/\delta_{\phi,+}}}{T^*} \right) = T^* \tilde{\mathcal{F}}_-^\infty \left(\frac{\Delta\phi^{\zeta_-/\delta_{\phi,-}}}{T^*}, \right). \tag{3}$$

Remarkably, we note that $\zeta_+/\delta_{\phi,+} = \zeta_-/\delta_{\phi,-} = 2$. If these ratios were not equal for the two branches of the scaling function, continuity of F would not be possible. Equation (3) implies that $T^* \sim \Delta\phi^2$. This agrees with earlier predictions[2,24,25] but differs from the results of Ref. 1, who found $T^* \sim \Delta\phi^{(2-a)/(1-a)}$; our arguments reproduce their scalings without introducing this anomalous temperature scale.

The resulting phase diagram in the $T - \phi$-plane is as depicted in Fig. 1. The exponents are summarized in Table 1. The red regime is described by the positive branch of the scaling function while the blue regime is described by the negative branch. The curve $T^* \sim \Delta\phi$ describes the boundary between the two branches. The free energy is continuous at T^*, but its derivatives are not necessarily continuous in the limit $N \to \infty$, so $T^*(\Delta\phi)$ is a line of phase transitions. This can be checked by calculating derivatives of the functions $\tilde{F}_\pm$ for different N and evaluating them at T^*.

For jammed soft spheres, Goodrich *et al.*[2] argued for a scaling relation connecting shear stress to the pressure:

$$s^2 N/p^2 = \mathcal{S}_+(Np^{1/2}). \tag{4}$$

This holds for the positive branch of the scaling function. A consequence of the scaling ansatz is that for the negative branch, there is anomalous scaling of shear-stress fluctuations so that

$$s^2 N/p^x = \mathcal{S}_-(Np^{2b}), \tag{5}$$

at fixed $T/\Delta\phi^2$ or T/p^2, where $x = \frac{7+\theta}{3+\theta} \approx 2.17$ and $2b = (1+\theta_e)/(3+\theta_e)$. Since at fixed p, T the correlation length is finite, the scaling function $\mathcal{S}_-(X)$ must approach a constant as $X \to \infty$.

As shown in Appendix A.3, for small $|\Delta\phi|$, the scaling ansatz predicts the anomalous scaling

$$s^2 \propto \frac{T^{x/2}}{N} \approx \frac{T^{1.09}}{N}. \tag{6}$$

This is a slightly stronger scaling than equipartition for a fixed spring network, consistent with the thermal stiffening of the effective contacts.

We note that the critical exponents in our scaling function differ on the two sides of the transition. There is a longstanding tradition in statistical mechanics of allowing for this possibility, but almost always the exponents are the same — as follows from the linearized flow of the renormalization group. (The exponents are ratios of the eigenvalues of the flow at a fixed point and are thus equal on both sides of the fixed point.) Different sets of exponents for the two branches would seem to imply a highly unusual form of criticality.

It is possible, however, to generate this apparent difference of exponents through an alternative scenario, which, though somewhat contrived, has the advantage of being consistent with the traditional renormalization group picture. Consider the effects of an irrelevant variable u, which we incorporate by adding another argument $X_u = u\Delta Z^{\zeta_- - \zeta_+}$ to the new scaling function $\mathcal{F}_u$. Here, u is irrelevant because $\zeta_- > \zeta_+$. Since u is irrelevant, near the transition, we may expand the scaling function around $u = 0$. We do so in an unorthodox way, evaluating the derivative not at zero (as in a Taylor expansion) but at the current value $X_u = u\Delta Z^{\zeta_- - \zeta_+}$:

$$\begin{aligned} F(\Delta Z, \Delta\phi, \varepsilon, T) &= \Delta Z^{\zeta_+} \mathcal{F}_u\left(\frac{\Delta\phi}{\Delta Z^{\delta_{\phi,+}}}, \frac{\varepsilon}{\Delta Z^{\delta_{\varepsilon,+}}}, \frac{T}{\Delta Z^{\zeta_+}}, u\Delta Z^{\zeta_- - \zeta_+}\right) \\ &\approx \Delta Z^{\zeta_+}\left(\mathcal{F}_u\left(X_\phi^+, X_\varepsilon^+, X_T^+, 0\right) + X_u \mathcal{F}_u^{[0,0,0,1]}\left(X_\phi^+, X_\varepsilon^+, X_T^+, X_u\right)\right), \end{aligned} \tag{7}$$

where the invariant scaling combinations $X_x^{+/-}$ are those valid in the plus/minus region. The first term $\mathcal{F}_u(X_\phi^+, X_\varepsilon^+, X_T^+, 0)$ equals $\mathcal{F}_+(X_\phi^+, X_\varepsilon^+, X_T^+)$, the traditional universal scaling function dominant as $\Delta Z \to 0$. But suppose that this dominant term is zero in the minus region[a] (as the scaling form for the Ising magnetization is zero for $T > T_c$). Then the free energy in the minus region will be given by the second term, subdominant by a factor $\Delta Z^{\zeta_- - \zeta_+}$, as desired. To get the scaling form of Eq. (1) for the minus region, this derivative must take the form (Appendix A.4)

$$\begin{aligned} u\mathcal{F}_u^{[0,0,0,1]}\left(X_\phi^+, X_\varepsilon^+, X_T^+, X_u\right) &= \mathcal{F}_-\left(X_\phi^-, X_\varepsilon^-, X_T^-\right) \\ &= \mathcal{F}_-\left(u^{1/2}X_\phi^+/X_u^{1/2}, u^{1/4}X_\varepsilon^+/X_u^{1/4}, uX_T^+/X_u\right) \end{aligned} \tag{8}$$

for the particular value of u for this system. This leads to

$$\begin{aligned} F(Z,\phi,\varepsilon,T) &= \Delta Z^{\zeta_+}\mathcal{F}\left(\frac{\Delta\phi}{\Delta Z^{\delta_{\phi,+}}}, \frac{\varepsilon}{\Delta Z^{\delta_{\varepsilon,+}}}, \frac{T}{\Delta Z^{\zeta_+}}\right) \\ &\quad + \Delta Z^{\zeta_-}\mathcal{F}_-\left(\frac{\Delta\phi}{\Delta Z^{\delta_{\phi,-}}}, \frac{\varepsilon}{\Delta Z^{\delta_{\varepsilon,-}}}, \frac{T}{\Delta Z^{\zeta_-}}\right) \end{aligned} \tag{9}$$

as desired.

One prediction of this picture is that it should be possible to find an irrelevant variable that, when tuned to a large value, induces corrections to the ordinary zero-T soft sphere scaling that realize the hard-sphere exponents. Some changes to the potential are irrelevant. Changes to the preparation protocol of states are also irrelevant, and it would be interesting to explore whether they could be responsible for these predicted corrections to scaling.

In summary, a scaling theory can be constructed for the jamming transition as a function of the variables encapsulated by the jamming phase diagram, namely packing fraction, shear strain and temperature. We find that the known behaviors are captured by two branches of a scaling function for the free energy. The two branches join together at a temperature $T^* \sim \Delta\phi^2$.

Recall that temperature T is dimensionless: $T = \tilde{T}/kD_{\mathrm{avg}}^2$ where $\tilde{T}$ is the dimensional temperature. We can interpret the +-branch of the scaling function, corresponding to $T < T^*$, as the regime in which the behavior is dominated by the particle stiffness k or the interaction energy. On the other hand, the −-branch, where $T > T^*$, is dominated by temperature, $\tilde{T}$, or thermal collisions.

We note that the arguments presented here can be generalized to a family of potentials described by the exponent α:

$$U(r_{ij}) = \frac{U_0}{\alpha}\left(1 - \frac{r_{ij}}{R_i + R_j}\right)^\alpha \Theta\left(1 - \frac{r_{ij}}{R_i + R_j}\right), \tag{10}$$

where r_{ij} is the distance between the centers of particles i and j and R_i and R_j are the radii of the particles. Here, $\Theta(x)$ is the Heaviside step function and U_0 is the

[a]The plus region is $X_\phi > 0$, $X_T < C^* X_\phi^{\zeta/\delta_\phi} = C^* X_\phi^2$, see Fig. 1.

interaction energy scale. Following earlier work,[19] it is useful to define an effective spring constant, $k_{\text{eff}} = U_0(\alpha - 1)/D_{\text{avg}}^2 \Delta Z^{2(\alpha-2)}$, where D_{avg} is the average particle diameter. If we use k_{eff} to construct dimensionless free energy, temperature, etc., as discussed before Eq. (1), then the results are generally applicable for any $\alpha > 1$.

The limit $\alpha = 1$, where $k_{\text{eff}} = 0$, is singular. There the de-dimensionalization of scaling variables using k_{eff} clearly fails. In this case, scaling exponents are predicted to be zero and simulations and mean-field calculations by Sclocchi, Urbani, and Franz[26–28] show that logarithmic corrections are dominant. It would be interesting to see if the scaling ansatz applies to these logarithmic corrections as well.

In our formulation of the scaling ansatz, we used ΔZ as a control variable, following the approach of Goodrich *et al.*[2] This choice is appealing because it treats $\Delta\phi$ and ε — compressive and shear strain — in the same way. However, Goodrich *et al.*[2] pointed out that we could alternatively choose $\Delta\phi$ as the control variable. This would lead to the scaling ansatz:

$$F(\Delta Z, \Delta\phi, \varepsilon, N, T) = \Delta Z^{\zeta_\pm} \mathcal{F}_\pm\left(\frac{\Delta\phi}{\Delta Z^{\delta_{\phi,\pm}}}, \frac{\varepsilon}{|\Delta\phi|^{\delta_{\varepsilon,\pm}/\delta_{\phi,\pm}}}, N|\Delta\phi|^{\psi_\pm/\delta_{\phi,\pm}}, \frac{T}{|\Delta\phi|^{\zeta_\pm/\delta_{\phi,\pm}}}\right). \quad (11)$$

Note, however, that this form requires using $|\Delta\phi|$ to scale ϵ, N and T, which causes problems because it introduces non-analyticity at $\Delta\phi = 0$ even at $T > 0$. It would seem, therefore, that ΔZ is a better choice for the control variable. The discovery of a correlation length that characterizes ΔZ fluctuations and diverges at the jamming transition[5] suggests still another choice for the control variable, namely the pressure, p, which is conjugate to $\Delta\phi$. So far, the length scale characterizing contact number fluctuations is the only diverging correlation length that has been identified for the jamming transition, although there are multiple diverging length scales have been identified that are not associated with correlation functions. It could be argued that ℓ_T is a diverging correlation length because it corresponds to the length scale for correlations in normal modes of vibrations at the boson peak frequency ω^*. The length scale $\ell_S(\omega^*)$, however, involves the boson peak frequency ω^* and does not correspond to the correlation length of any of the variables that appear in the scaling ansatz. We note that the ΔZ correlation length (which is associated with ΔZ, one of the variables in the scaling ansatz) does not appear to scale the same way as $\ell_S(\omega^*)$ (which controls finite-size scaling). This lack of correspondence between a correlation length and the length scale characterizing finite-size effects, while unusual, appears to be characteristic of other systems with sharp global transitions but configuration-dependent critical densities, as originally identified in the depinning of charge-density waves.[29–31]

While it is normal for a correlation length to diverge at a critical point, it is decidedly abnormal for the fluctuations to vanish at long length scales — usually they diverge. Contact number fluctuations must vanish at long length scales because the average contact number at the transition must be $Z = 2d$, as dictated by isostaticity. As a result, contact number fluctuations are hyperuniform,[5,13] implying that there can be no square-gradient expansion — no Ginzburg–Landau theory — for jamming. It is not clear how to construct a renormalization group for a system that is hyperuniform at long length scales. The existence of other systems that exhibit hyperuniformity at transitions, such as systems with transitions that separate absorbing states from non-absorbing ones,[32] suggests that the challenge may not be restricted to the jamming transition. It is interesting and non-trivial that despite being highly unusual in exhibiting hyperuniformity, the jamming transition can nevertheless be described by a scaling ansatz.

It should be noted, however, that the vanishing of ΔZ fluctuations may be specific to systems prepared at fixed pressure (pressure control). A recent study at fixed $\Delta\phi$ (phi-control) suggests that ΔZ fluctuations might diverge instead of vanish and that even the finite-size scaling exponent is different in that case, at least for ΔZ and energy fluctuations.[33] It would be interesting to see whether a different scaling ansatz can be constructed for that case, and how the ansatzes for pressure- and ϕ-control might be related.

Finally, we must again raise the caveat that the arguments of De Giuli *et al.*[1] and our scaling ansatz are predicated on the approximation that the system is restricted to a single basin in the energy landscape. At $T > 0$, we know that this is not true — given enough time, the system will explore multiple basins in the landscape, each with a minimum corresponding to a different value of the jamming critical packing fraction, ϕ_c. It is likely that thermal fluctuations that drive the system over energy barriers will destroy the line of phase transitions at the boundary separating the two branches of the scaling function. In other words, glass physics may well smear the phase transition line into a crossover. However, the jamming transition and the line that emanates from it have such unusual properties in the canon of criticality and are so important to our understanding of disordered solids that it is crucial to across whether the line is singular or only demarcates a crossover.

Acknowledgments

We thank G. Biroli, S. Franz, and F. Zamponi for the useful discussions, the Simons Foundation for supporting this work through the "Cracking the glass problem" collaboration (#454945, SAR and AJL), the NSF (DMR-1719490 to JPS), and the 2019 Beg Rohu Summer School, where this work was primarily done, as well as the Center for Computational Biology at the Flatiron Institute, where the manuscript was primarily written. This chapter is dedicated to the memory of Michael E. Fisher, the thesis advisor of Andrea J. Liu and colleague of J. P. Sethna. Liu, in particular, is grateful for having had the opportunity to absorb scaling theory and criticality directly from Fisher. We hope that this contribution helps to demonstrate the lasting value of looking at the world through the lens of critical phenomena.

Appendix

A.1. *Converting between ΔZ and $\Delta\phi$ in the scaling ansatz*

As in the zero-temperature case,[2] the scaling ansatz has the unusual feature that the variables in it are not all independent: at fixed $\Delta\phi, T, N$, and preparation protocol, the coordination ΔZ is fixed by some equation of state.

In addition to the scaling ansatz, we assume a scaling equation of state

$$\frac{\Delta\phi}{\Delta Z^{\delta_{\phi,\pm}}} = f\left(\frac{\varepsilon}{\Delta Z^{\delta_{\varepsilon,\pm}}}, N\Delta Z^{\psi_\pm}, \frac{T}{\Delta Z^{\zeta_\pm}}\right). \tag{A.1}$$

If this is assumed, then it is possible to eliminate ΔZ from the scaling ansatz, in order to obtain Eq. (3) in the main text or make comparisons to De Giuli *et al.*'s scaling predictions (which are expressed in terms of $\Delta\phi$ rather than ΔZ).

This equation of state can be inverted to yield

$$\Delta Z^{\delta_{\phi,\pm}} = |\Delta\phi|\, g_\pm\left(\frac{\varepsilon}{|\Delta\phi|^{\delta_{\varepsilon,\pm}/\delta_{\phi,\pm}}}, N\,|\Delta\phi|^{\psi_\pm/\delta_{\phi,\pm}}, \frac{T}{|\Delta\phi|^{\zeta_\pm/\delta_{\phi,\pm}}}\right). \tag{A.2}$$

Here, the branches $f_\pm$ correspond to the sign of $\Delta\phi$ rather than the regions $-, +$ of the phase diagram; if we wished to completely eliminate ΔZ from the scaling ansatz in all regions simultaneously, it may introduce a nonanlyticity at $\Delta\phi = 0$. However, this is not a problem if we only wish to eliminate ΔZ from an equation with the sign of $\Delta\phi$ fixed.

Subject to this restriction, this allows us to eliminate ΔZ from any equation of the original scaling variables:

$$h\left(\frac{\Delta\phi}{\Delta Z^{\delta_{\phi,\pm}}}, \frac{\varepsilon}{\Delta Z^{\delta_{\varepsilon,\pm}}}, N\Delta Z^{\psi_\pm}, \frac{T}{\Delta Z^{\zeta_\pm}}\right)$$
$$= \tilde{h}_\pm\left(\frac{\varepsilon}{|\Delta\phi|^{\delta_{\varepsilon,\pm}/\delta_{\phi,\pm}}}, N\,|\Delta\phi|^{\psi_\pm/\delta_{\phi,\pm}}, \frac{T}{|\Delta\phi|^{\zeta_\pm/\delta_{\phi,\pm}}}\right). \tag{A.3}$$

Restricting to $\Delta\phi > 0$, we may thus eliminate ΔZ to obtain Eq. (3) in the main text and as necessary in Appendix A.2 to verify that the scaling ansatz reproduces the scaling regimes of De Giuli *et al.*[1]

A.2. *Deriving the De Giuli et al. exponents from the scaling ansatz*

Here, we show in more detail how to obtain the various scalings described by De Giuli *et al.*[1] from the scaling ansatz.

We begin with the scaling behaviour of the pressure. Start with the ansatz of Eq. (1) and take $N \to \infty$ and $\epsilon = 0$. Taking a derivative with respect to the first argument yields the scaling ansatz for the pressure,

$$P = \Delta Z^{\zeta_\pm - \delta_{\phi,\pm}} \mathcal{P}_\pm\left(\frac{\Delta\phi}{\Delta Z^{\delta_{\phi,\pm}}}, \frac{T}{\Delta Z^{\zeta_\pm}}\right). \tag{A.4}$$

Eliminating ΔZ as above gives

$$P = |\Delta\phi|^{\zeta_\pm/\delta_{\phi,\pm} - 1}\, \mathcal{P}_\pm\left(\frac{T}{|\Delta\phi|^{\zeta_\pm/\delta_{\phi,\pm}}}\right). \tag{A.5}$$

First, we study the "anharmonic" limit $T \gg \Delta\phi^2$, in which De Giuli *et al.* predict $P \propto \sqrt{T}$. We set $\Delta\phi \to 0$ in the scaling function at fixed T and require that the behavior of the scaling function is not singular in $\Delta\phi$. This requires that the asymptotic behavior of the scaling function has the right power law to balance all powers of $\Delta\phi$, i.e.,

$$P \propto |\Delta\phi|^{\zeta_-/\delta_{\phi,-} - 1}\left(\frac{T}{|\Delta\phi|^{\zeta_-/\delta_{\phi,-}}}\right)^{\frac{\zeta_- - \delta_{\phi,-}}{\zeta_-}} \tag{A.6}$$

$$= T^{\frac{\zeta_- - \delta_{\phi,-}}{\zeta_-}}. \tag{A.7}$$

Thus, consistency with De Giuli *et al.* requires that

$$\zeta_- = 2\delta_{\phi,-}. \tag{A.8}$$

Note that ζ_+ and $\delta_{\phi,+}$ are already known to satisfy the same requirement.

In the hard-sphere limit ($\Delta\phi < 0, T \to 0$), dimensional analysis requires that $P \propto T$. Thus, we extract the asymptotic equation of state

$$P \propto |\Delta\phi|^{\zeta_-/\delta_{\phi,-} - 1}\left(\frac{T}{|\Delta\phi|^{\zeta_-/\delta_{\phi,-}}}\right) \propto \frac{T}{|\Delta\phi|}. \tag{A.9}$$

Thus, the hard-sphere pressure scaling is automatically satisfied.

The pressure scaling in the regimes that De Giuli *et al.* call "soft-zero-T" and "soft-entropic" are obtained by taking $T \to 0$ for positive $\Delta\phi$; this recovers the previous zero-temperature scaling ansatz, giving $P \propto \Delta\phi$ in agreement with their results.

Now, we must verify the scaling of the shear modulus G. At zero strain, it obeys the scaling form

$$G = \Delta Z^{\zeta_\pm - 2\delta_{\epsilon,\pm}} \mathcal{G}_\pm\left(\frac{\Delta\phi}{\Delta Z^{\delta_{\phi,\pm}}}, \frac{T}{\Delta Z^{\zeta_\pm}}\right). \tag{A.10}$$

The anharmonic and hard-sphere limits are treated identically to the case of the pressure, demanding a finite value at zero $\Delta\phi$ in the anharmonic case while requiring $G \propto T$ in the hard-sphere case. Matching the anharmonic limit for G with De Giuli *et al.* requires

$$\frac{\delta_{\epsilon,-}}{\zeta_-} = \frac{1}{4}\frac{4+2\theta_e}{3+\theta_e}, \tag{A.11}$$

while the hard-sphere limit requires

$$\frac{\delta_{\epsilon,-}}{\delta_{\phi,-}} = \frac{1}{2}\frac{4+2\theta_e}{3+\theta_e}. \tag{A.12}$$

Note that these equations are consistent with Eq. (A.8), and are both satisfied by the exponents in Table 1.

The "soft-entropic" regime of De Giuli *et al.* is less obvious, but still consistent with the scaling ansatz in the $-$ branch. After eliminating ΔZ, we have

$$G = |\Delta\phi|^{2-2\delta_{\epsilon,-}/\delta_{\phi,-}}\, \mathcal{G}_-\left(\frac{T}{\Delta\phi^2}\right) \tag{A.13}$$

$$= T^{1-2\delta_{\epsilon,-}/\delta_{\phi,-}}\, |\Delta\phi|^{2\delta_{\epsilon,-}/\delta_{\phi,-}}\, g_-\left(\frac{T}{\Delta\phi^2}\right). \tag{A.14}$$

If the function g_- approaches a finite limit, this matches the "soft-entropic" scaling as long as

$$\frac{\delta_{\epsilon,-}}{\delta_{\phi,-}} = \frac{1}{2}\frac{4+2\theta_e}{3+\theta_e}, \tag{A.15}$$

which shows consistency with the exponents in the hard-sphere and anharmonic regimes.

To see the "soft-entropic" scaling in our picture, we must move along the phase boundary, $T = C^*\Delta\phi^2$, since our picture does not have the anomalous temperature scale T^* of De Giuli *et al.*; g_- is constant along this trajectory as required.

A.3. *Shear-stress fluctuations at zero* $\Delta\phi$

At finite T, in the limit of small $\Delta\phi$ the scaling form for the shear-stress fluctuations s^2 is

$$s^2 = |\Delta\phi|^{2(\zeta_- - \delta_{\epsilon,-})/\delta_{\phi,-}}\, \mathcal{S}_-\left(N\,|\Delta\phi|^{\psi_-/\delta_{\phi,-}}, \frac{T}{|\Delta\phi|^2}\right). \tag{A.16}$$

A finite correlation length requires $1/\sqrt{N}$ fluctuations of shear stress for large N:

$$s^2 = \frac{1}{N}\,|\Delta\phi|^{(2\zeta_- - 2\delta_{\epsilon,-} - \psi_-)/\delta_{\phi,-}}\, \tilde{\mathcal{S}}_-\left(\frac{T}{|\Delta\phi|^2}\right). \tag{A.17}$$

Finally, at finite T, the fluctuations cannot diverge or vanish as $\Delta\phi \to 0$. Imposing the required power law to cancel all factors of $\Delta\phi$ yields

$$s^2 N \propto T^{(2\zeta_- - 2\delta_{\epsilon,-} - \psi_-)/2\delta_{\phi,-}} = T^{\frac{7+\theta_e}{6+2\theta_e}}. \tag{A.18}$$

A.4. *The necessary form of the small-u scaling function*

Note that

$$\zeta_- - \zeta_+ = \frac{2 - 2\theta_e}{1 + \theta_e}, \tag{A.19}$$

$$\delta_{\phi,-} - \delta_{\phi,+} = \frac{1 - \theta_e}{1 + \theta_e} = \frac{1}{2}(\zeta_- - \zeta_+), \tag{A.20}$$

$$\delta_{\varepsilon,-} - \delta_{\varepsilon,+} = \frac{1 - \theta_e}{2 + 2\theta_e} = \frac{1}{4}(\zeta_- - \zeta_+). \tag{A.21}$$

Thus, the $-$ exponents are recovered if the correction to scaling is, for small u, a function of X_T/X_u, $X_\phi/X_u^{1/2}$, and $X_\phi/X_u^{1/4}$.

References

1. E. Degiuli, E. Lerner and M. Wyart, Theory of the jamming transition at finite temperature. *J. Chem. Phys.* **142**(16), 164503 (2015).
2. C. P. Goodrich, A. J. Liu and J. P. Sethna, Scaling ansatz for the jamming transition. *Proc. Nat. Acad. Sci.* **113**(35), 9745–9750 (2016).
3. C. S. O'Hern, L. E. Silbert, A. J. Liu and S. R. Nagel, Jamming at zero temperature and zero applied stress: The epitome of disorder. *Phys. Rev. E.* **68**(1), 011306 (2003).
4. C. P. Goodrich, S. Dagois-Bohy, B. P. Tighe, M. van Hecke, A. J. Liu and S. R. Nagel, Jamming in finite systems: Stability, anisotropy, fluctuations, and scaling. *Phys. Rev. E* **90**(2), 022138 (Aug., 2014).
5. D. Hexner, A. J. Liu and S. R. Nagel, Two diverging length scales in the structure of jammed packings. *Phys. Rev. Lett.* **121**(11), 115501 (2018).
6. D. B. Liarte, O. Stenull, X. M. Mao and T. C. Lubensky, Elasticity of randomly diluted honeycomb and diamond lattices with bending forces. *J. Phys.-Cond. Matt.* **28**(16), 165402 (2016). ISSN 0953-8984. doi: 10.1088/0953-8984/28/16/165402. URL ⟨GotoISI⟩://WOS:000373457300005.
7. D. B. Liarte, S. J. Thornton, E. Schwen, I. Cohen, D. Chowdhury and J. P. Sethna, Universal scaling for disordered viscoelastic matter near the onset of rigidity. *Phys. Rev. E* **106**, L052601 (November, 2022). doi: 10.1103/PhysRevE.106.L052601. URL https://link.aps.org/doi/10.1103/PhysRevE.106.L052601.
8. S. J. Thornton, D. B. Liarte and J. P. Sethna. Work in progress (2023).
9. N. Y. C. Lin, M. Bierbaum, P. Schall, J. P. Sethna and I. Cohen, Measuring nonlinear stresses generated by defects in 3d colloidal crystals, *Nat. Mater.* **15**, 1172–1176 (2016). doi: 10.1038/nmat4715.
10. C. Scalliet, L. Berthier and F. Zamponi, Marginally stable phases in mean-field structural glasses. *Phys. Rev. E* **99**(1), 012107 (2019).
11. S. Franz, Personal communication (2023).
12. C. P. Goodrich, A. J. Liu and S. R. Nagel, Finite-size scaling at the jamming transition. *Phys. Rev. Lett.* **109**(9), 095704 (2012).
13. D. Hexner, P. Urbani and F. Zamponi, Can a large packing be assembled from smaller ones?, *Phys. Rev. Lett.* **123**(6), 068003 (2019).
14. H. Ikeda, Jamming below upper critical dimension. *Phys. Rev. Lett.* **125**(3), 038001 (2020).

15. K. Binder, M. Nauenberg, V. Privman and A. Young, Finite-size tests of hyperscaling. *Phys. Rev. B* **31**(3), 1498–1502 (1985).
16. M. Baity-Jesi, C. P. Goodrich, A. J. Liu, S. R. Nagel and J. P. Sethna, Emergent so (3) symmetry of the frictionless shear jamming transition. *J. Stat. Phys.* **167**, 735–748, (2017).
17. L. E. Silbert, A. J. Liu and S. R. Nagel, Vibrations and diverging length scales near the unjamming transition. *Phys. Rev. Lett.* **95**, 098301 (2005).
18. A. J. Liu and S. R. Nagel, The jamming transition and the marginally jammed solid. *Annu. Rev. Condens. Matter Phys.* **1**(1), 347–369 (2010).
19. V. Vitelli, N. Xu, M. Wyart, A. J. Liu and S. R. Nagel, Heat transport in model jammed solids. *Phys. Rev. E* **81**, (February, 2010).
20. E. Lerner, E. DeGiuli, G. During and M. Wyart, Breakdown of continuum elasticity in amorphous solids. *arXiv* (December, 2013).
21. P. Charbonneau, J. Kurchan, G. Parisi, P. Urbani and F. Zamponi, Exact theory of dense amorphous hard spheres in high dimension. III. The full replica symmetry breaking solution. *J. Stat. Mech. Theory Exp.* **2014**(10), P10009 (2014).
22. P. Charbonneau, J. Kurchan, G. Parisi, P. Urbani and F. Zamponi, Fractal free energy landscapes in structural glasses. *Nat. Commun.* **5**(1), 3725 (2014).
23. P. Charbonneau, E. I. Corwin, G. Parisi and F. Zamponi, Jamming criticality revealed by removing localized buckling excitations. *Phys. Rev. Lett.* **114**(12), 1–5 (2015). ISSN 10797114. doi: 10.1103/PhysRevLett.114.125504.
24. Z. Zhang, N. Xu, D. T. N. Chen, P. J. Yunker, A. M. Alsayed, K. B. Aptowicz, P. Habdas, A. J. Liu, S. R. Nagel and A. G. Yodh, Thermal vestige of the zero-temperature jamming transition. *Nature* **459**(7244), 230–233 (May, 2009).
25. A. Ikeda, L. Berthier and G. Biroli, Dynamic criticality at the jamming transition. *J. Chem. Phys.* **138**(12), 12A507 (2013).
26. S. Franz, A. Sclocchi and P. Urbani, Critical jammed phase of the linear perceptron. *Phys. Rev. Lett.* **123**(11), 115702 (2019).
27. S. Franz, A. Sclocchi and P. Urbani, Critical energy landscape of linear soft spheres. *SciPost Phys.* **9**, 012 (2020). doi: 10.21468/SciPostPhys.9.1.012.
28. S. Franz, A. Sclocchi and P. Urbani, Surfing on minima of isostatic landscapes: avalanches and unjamming transition. *J. Stat. Mech. Theory Exp.* **2021**(2), 023208 (February, 2021). doi: 10.1088/1742-5468/abdc16. URL https://dx.doi.org/10.1088/1742-5468/abdc16.
29. C. R. Myers and J. P. Sethna, Collective dynamics in a model of sliding charge-density waves. I. Critical behavior. *Phys. Rev. B* **47**, 11171–11193 (May, 1993). doi: 10.1103/PhysRevB.47.11171.
30. C. R. Myers and J. P. Sethna, Collective dynamics in a model of sliding charge-density waves. II. Finite-size effects. *Phys. Rev. B* **47**, 11194–11203 (May, 1993). doi: 10.1103/PhysRevB.47.11194.
31. A. A. Middleton and D. S. Fisher, Critical-behavior of charge-density waves below threshold — Numerical and scaling analysis. *Phys. Rev. B* **47**(7), 3530–3552 (1993).
32. D. Hexner and D. Levine, Hyperuniformity of critical absorbing states. *Phys. Rev. Lett.* **114**(11), 110602 (2015).
33. H. Ikeda, Control parameter dependence of fluctuations near jamming, The Journal of Chemical Physics. 158(5), 056101 (02, 2023). ISSN 0021-9606. doi: 10.1063/5.0127064. URL https://doi.org/10.1063/5.0127064

https://doi.org/10.1142/9789811282386_0040

Chapter 40

How I Met Michael Fisher and Started Working in Biophysics

A.B. Kolomeisky

Department of Chemistry, Department of Physics and Astronomy,
Department of Chemical and Biomolecular Engineering,
Center for Theoretical Biological Physics,
Rice University, 6100 Main Street, Houston TX, 77005-1892, USA
tolya@rice.edu

I present my personal recollections of working together with Michael Fisher and what I learned from him. He stimulated the shift of my research interests toward the direction of theoretical biophysics. More specifically, together we started to investigate the mechanisms of motor proteins, biological molecular machines and cytoskeleton protein filaments. A very brief review of some of these topics is also presented.

1. First Meetings with Michael Fisher

I was a graduate student at the Department of Chemistry at Cornell University from 1994 till 1998, working in the research group of Prof. Ben Widom. It was the time when I first heard about Michael Fisher and his research. In 1996, I took a graduate chemistry class in statistical mechanics that was taught by Ben Widom. In this class, Ben spent a significant time explaining critical phenomena and renormalization-group theory analysis. He frequently emphasized that the central role in this probably the most fundamental development in statistical mechanics in 20th century was played by Michael Fisher. I was even more impressed by the beauty and the logic of these theoretical concepts because Ben Widom added to his scientific explanations the real stories of how these developments occurred which he witnessed himself. One of these stories, I recall, was that nobody could understand the original ideas about the renormalization that Ken Wilson was trying to introduce. Only Michael Fisher was able to understand, significantly extend and explain them to the overall scientific community, eventually leading to a large breakthrough that influenced physics and chemistry. All these exciting events happened at Cornell University, including the classroom where we were taking our statistical mechanics class!

So, I became very excited when I learned that Michael Fisher is coming to Cornell Chemistry in the Fall of 1997 as a Baker Lecturer. Those lecturers were chosen among the best scientists in the world, and they usually presented several lectures devoted to one specific topic while staying for some time at the Department of Chemistry at Cornell. Michael gave a set of lectures titled "Understanding Criticality in Electrolytes and Other Fluids." It was a huge success! The room where he presented his lectures was always full, and I had to come 30 minutes before to make sure that I have a good seat. I have never seen in my life such an emotional speaker with such a clear scientific message that even I, a chemistry graduate student, could easily comprehend. The fact that Michael Fisher is a genius was clear to me. One non-scientific thing that I immediately recall from his lectures was the simultaneous use of two overhead projectors that I have not seen before in seminars given by various speakers at Cornell. I found it very convenient since I could follow the logic of presentation and I could always come back to the previous statement if I was lost.

I was graduating in 1998 and started searching for postdoctoral positions. Ben Widom recommended me to apply to Michael Fisher. I thought that this was a very long shot and my chances are low. Not only because almost all people who were postdocs with Michael before were physicists, but also because he was considered as the most famous scientist working in the field of statistical physics. The smartest young people were trying to get into his group, which was always quite small. Although I had a decent mathematical training during my education at Moscow State University in Russia, I always considered my physics background as relatively weak. At the same time, I recalled that the topic of Baker Lectures given by Michael at Cornell was somehow familiar to me. Electrolytes are frequently utilized and studied in physical chemistry. So, I applied. To my great surprise, Michael chose me and I was very excited to come to Maryland to work with him on the investigation of the critical phenomena in electrolyte systems.

2. Starting a Postdoctoral Work with Michael Fisher at University of Maryland

In the Summer of 1998, I came to University of Maryland at College Park to work as a postdoctoral research associate in the group of Michael Fisher at the Institute of Physical Science and Technology (IPST). My first task was to develop a lattice version of Debye–Huckel theory of electrolytes in various dimensions. This topic was related to investigating the critical phenomena in the systems with long-range interactions that Michael was doing at the time. While I started working on this project, Michael asked me to give a seminar at IPST. Apparently, this was a tradition for all new incoming postdocs working with different advisers at IPST.

I decided to talk about biological molecular motors, a topic that started to become popular in the scientific field at the end of 1990s. It was already known

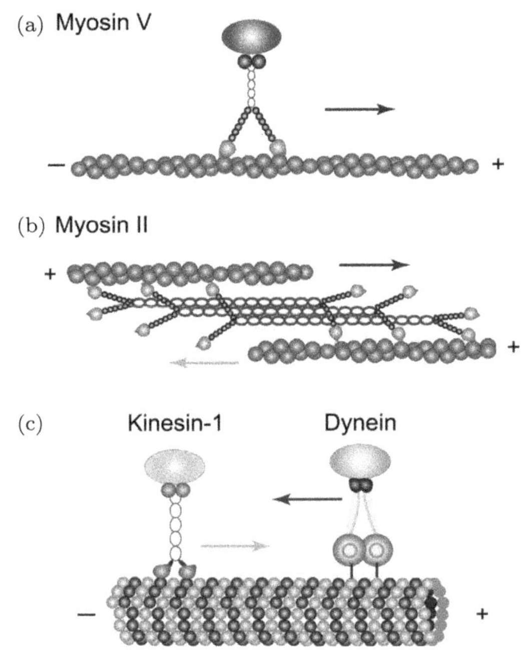

Fig. 1. Examples of several classes of most known motor proteins: (a) myosin-V motors that step along actin filaments; (b) a group of myosin-II motor proteins that can move along several actin filaments; and (c) kinesin and dynein motor proteins that move in opposite direction along microtubules. This figure is reproduced with permission from Ref. 10.

that there are several classes of enzymatic molecules that preferentially move in specific directions along the cytoskeleton filaments.[1–3] They catalyze the hydrolysis of adenosine triphosphate (ATP), and the energy released during this chemical reaction is utilized to support their mechanical motion. Some of the most prominent motor proteins are schematically illustrated in Fig. 1.[3,9,10]

A protein kinesin that travels along the microtubules probably was the best known and investigated example of motor proteins at the time. The striking development was also that the first single-molecule experiments were already able to visualize some aspects of the dynamics of kinesins in *in vitro* conditions.[4] In addition, some theoretical investigations on the mechanisms of motor proteins just came out.[5,6] It was also suggested theoretically that the mechanism of motor proteins can be described as "ratchets."[5] The main idea of this approach is that the moving motor protein molecule can be found in one of several chemical states. In each state, it experiences different potentials that are periodic but asymmetric within a period. Transitions between different motor protein's states were modeled as stochastic switching between different potentials, while inside each potential, the molecular motors exhibited a random Brownian motion. One of the simplest ratchet models with only two states (switching between two potentials) is illustrated schematically in Fig. 2.

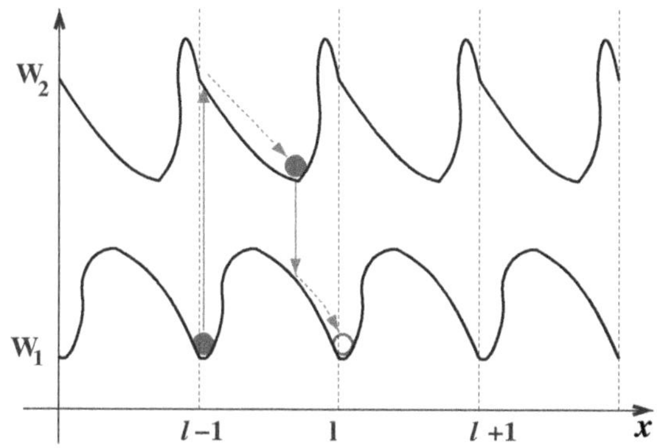

Fig. 2. A schematic view of the motion of motor proteins according to the ratchet models. The molecular motor moves as Brownian particle in the potential, and it might stochastically switch between different states that correspond to different potentials. The potentials are periodic but asymmetric. The figure is reproduced with permission from Ref. 10.

In 1997, my Ph.D. adviser Ben Widom went to Paris where he learned about motor proteins and ratchet models that have been proposed to describe their complex dynamic behavior.[5] Ben returned from this trip with a very simple model which he was able to solve analytically at all times, obtaining a number of important dynamic properties of motor proteins, including the drift velocity V, the diffusion coefficient D and the effective force F that drives the motor proteins. This model assumed only one intermediate state in the catalytic cycle of kinesin motor proteins, similar to what is shown in Fig. 2. When he presented this work at Cornell at our group meeting, I immediately realized a connection of his model with a more general set of one-dimensional periodic hopping models for single random-walk particles that were introduced and fully investigated by Bernard Derrida.[7] The model proposed by Ben Widom corresponded to the dynamic motion of particles on lattices with a period of $N = 2$, while Derrida obtained results in the stationary state limit for periods of arbitrary size. Based on this connection, we presented a paper in which the properties of molecular motors using the simplified ratchet model were evaluated using direct calculations and Derrida's approach.[8] I was very proud of this work and decided to present it at IPST seminar.

My seminar on modeling of biological molecular motors was a complete failure! I have been asked by so many questions that I could not answer, and Michael Fisher was the one who asked the great majority of them. He skillfully attacked my scientific arguments and completely destroyed them. I have never felt so sad since I realized how much I do not know. I was thinking that Michael now understood how weak I am and he would stop my postdoctoral position in his group.

Next morning, Michael invited me to his office and I went there feeling that this is my last day as a postdoc. To my big surprise, Michael was very happy! This was one of the happiest moments that I ever remember him in our communications. He explained to me that, although my presentation was a disaster, it stimulated him to move toward a very new scientific area, which he desired for some time, and he was very excited about the opportunity to investigate these problems. He told me that there is a lot of work for us to do, but we can work together to achieve high goals. It was emphasized that my experience as chemist is a plus in this endeavor because biological systems to a large degree specified by biochemical reactions and biologists utilize the language similar to chemists. This was the beginning of our joint work in biophysics that completely changed my scientific life!

We started to read a lot of biological literature to educate ourselves. It was not an easy task, but we frequently discussed the papers together. Michael was excellent in choosing the important things and neglecting the less relevant information, which was not easy for many biological papers. I tried to learn this from him, but I am not sure that I succeeded. We also went to American Biophysical Society meetings that took place every Spring in different US cities. Listening to talks and talking with experts helped us a lot. I am also recalling a scene when we presented our first results as a poster in one of Biophysical Society meetings. Typically, posters were presented by graduate students or postdocs. It was an amazing picture to see Michael standing in front of our poster and being surrounded by young people with other posters. Nobody knew us in this field at the beginning and we were totally neglected. I was very upset, but Michael was never moved at all. It was enough for him to start any conversation with his clear British accent and people listened to him attentively. Slowly, people in the field started to pay attention to what we were saying. I am confident that they were able to see how brilliant Michael is and how his ideas provide new physical–chemical insights into some of these complex biological problems.

3. Working with Michael Fisher on Various Biophysical Problems

Now, let me briefly discuss some biophysics-oriented scientific projects that I did with Michael Fisher while I worked as a postdoc in his group. As a first step in the direction of investigating biophysical problems, Michael decided that we have to correct the issues that were encountered in the simplest ratchet models that I presented in my seminar at IPST. One of them was an estimation of the driving force of the motor protein that was based on Einstein relation,[8]

$$F = k_B T \frac{V}{D}, \tag{1}$$

where V and D are the velocity and the diffusion coefficient of the molecular motor. Several single-molecule *in vitro* experiments that observed the dynamics of individual motor proteins have already appeared.[4,11–15] Using the measured values for velocities and diffusion constants for kinesins, one could estimate the force using the Einstein relation leading to $F \simeq 1$pN. However, experiments indicated that forces up to 7pN were need to stop the motion of this biological molecular motor. Also, from theoretical point of view, the Einstein relation should work only for equilibrium systems or for systems that are not far away from the equilibrium while the motor proteins due to large energy consumption from hydrolysis of ATP molecules were definitely far away from the equilibrium. This indicated that the application of the Einstein relation to estimate the forces of biological molecular motors might not be valid at all.

We suggested that to understand better the forces that the motor protein can exert, it is better to utilize a traditional chemical-kinetic description of molecular motors as illustrated in Fig. 3.[16] In the simplest version, the motion of the motor protein between neighboring binding sites on the cytoskeleton filament, which is coupled to a hydrolysis of one ATP molecule, can be viewed as a set of N sequential chemical states. From the intermediate state j, the motor protein can move to the state $j+1$ with a rate u_j while the backward transition to the state $j-1$ is taking place with a rate w_j ($j = 0, 1, \ldots, N-1$). Then from the statistical mechanics point of view, the molecular transformations described by the chemical-kinetic scheme in Fig. 3 correspond to hopping of a single particle along the periodic (with a period N) asymmetric lattice. This allowed us to utilize the results of Derrida to obtain explicit expressions for dynamic properties of molecular motors.[7,16] We argued that a dimensionless parameter,

$$\Gamma = \prod_{j=0}^{N-1} \left(\frac{u_j}{w_j}\right), \tag{2}$$

plays a critical role in understanding the dynamics of motor proteins that follow the kinetic scheme in Fig. 3. From the chemical point of view, this quantity is just an equilibrium constant for the effective chemical reaction to go from the binding site l to the binding site $l+1$ via N consecutive chemical transitions, providing a connection to the free-energy change in this process,

$$\Gamma = \exp\left(\frac{\Delta G}{k_B T}\right). \tag{3}$$

This allowed us to evaluate the maximal effective force that the motor protein can exert since the motor moved by the distance d,

$$F = \frac{k_B T}{d} \ln \Gamma. \tag{4}$$

Using experimentally obtained results for dynamic proprieties of motor proteins and fitting them to the discrete-state stochastic models led us to the evaluation of

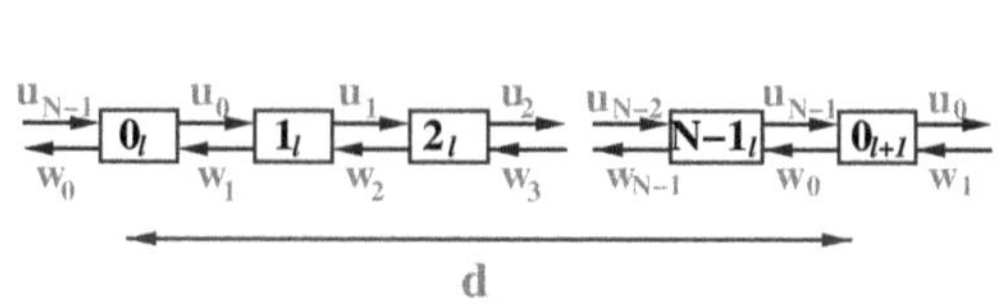

Fig. 3. Chemical-kinetic scheme for the simplest discrete-state stochastic model of motor proteins. The distance between neighboring binding sites l and $l+1$ is equal to d. The motor protein preferentially moves to the right. The forward and backward transition rates are u_j and w_j, respectively, with $j = 0, 1, 2, \ldots, N-1$.

individual transition rates.[17] Substituting these rates into Eq. (4), the agreement with experimental stall forces for motor proteins has been observed, supporting our theoretical arguments.

Single-molecule experimental measurements of motor proteins dynamics utilized the force (produced mainly via laser optical trap setups) to probe the microscopic picture of their functioning. Together with Michael, we proposed an explicit approach to take the effect of external forces into account. More specifically, it was argued that transition rates between chemical states can be modified using Kramer's view of chemical rates,[16,17]

$$u_j(F) = u_j(0) \exp\left(-\frac{\theta_j^+ F d}{k_B T}\right), \quad w_j(F) = w_j(0) \exp\left(+\frac{\theta_j^- F d}{k_B T}\right), \tag{5}$$

where θ_j^+ and θ_j^- are load distribution factors (with $\sum_{j=0}^{N-1}(\theta_j^+ + \theta_j^-) = 1$) that reflect how the external forces influence individual transition between intermediate chemical states. They contain an important molecular information about the underlying free-energy landscapes for the motor proteins dynamics. This approach allowed us to fully describe the experimental observations on kinesins and myosins V motor proteins,[17,18] providing new theoretical predictions that were later also confirmed.[3,9]

Motivated by the single-molecule experiments and theoretical studies on motor proteins, our original approach was later extended to take into account the possibility of more complex chemical pathways for molecular motors.[19,20] More specifically, we obtained exact analytic expressions for the drift velocities V and the diffusion constants D of single particles in the discrete-state stochastic models that included parallel pathways, branched states and irreversible dissociations. Our theoretical analysis indicated that some topological features of chemical pathways might have a strong effect on dynamic properties of molecular motors. Furthermore, we extended our theoretical calculations for the systems where some of the chemical states cannot be observed experimentally requiring to develop the effective stochastic models with transitions between states deviating from exponential waiting times.[21] We derived exact results for dynamic properties of motor proteins with general waiting-time distributions for transitions between states.

4. Lessons I Learned from Michael Fisher

Working with Michael Fisher was the biggest gift in my life! It stimulated my shift in the direction of theoretical biophysics where I could apply most efficiently my knowledge of chemistry and statistical physics. But there are a lot of other things that Michael taught me. I think it is important to share them.

1. Michael emphasized to me many times that each research paper must be a separate story. It could be short or long story, but it must be logically complete. It is not good to publish a paper that is not finished as well

as to mix several stories in the same paper (with the exception of review articles).

2. When someone is trying to analyze complex phenomena, the simplest model, preferably analytical, that can account for all available data and can predict future results that can be tested experimentally is the most efficient and practical approach. There is no point in making the model too complex if the simpler picture works well. All models are imperfect, but to stimulate the scientific progress, we must first fully account the observed results in the simplest possible way. Only when new observations that contradict the existing theoretical description come, one should start looking for more advanced models.
3. Computer simulations are critical for the modern research. But they are still far from being satisfactory, and even when they would become perfect, at best, they will never generate new knowledge and, thus, will never be able to substitute analytical models. Computer simulations is the tool to test the proposed scientific ideas. They depend on many factors that are beyond our control (force fields, details of calculations, thermal baths, time of simulations, etc.), and it is dangerous to fully believe them. Thus, anyone who is saying that we do not need theoretical scientists that develop analytically solvable models is absolutely wrong. Even the approximate models might provide a better physical picture of underlying microscopic processes than the most advance computer simulations. I guess, we can extend this lesson to machine-learning methods that recently became very popular and fashionable.
4. In developing theoretical model for biological processes, it is important to make sure that they do not violate fundamental laws of chemistry and physics. As one example, I remember Michael's arguments that ALL chemical reactions are reversible, and it is not correct to neglect the transition rates even if they are very small, especially considering the energetics of the biological processes.
5. If the developed theoretical model contains no mathematical errors but still disagrees with experimental observations, it is important not to try change the model so that it can fit the data even if some parameters are non-physical. You have to be honest, and experiments might also be wrong. I am recalling our work with Michael Fisher on modeling of force–velocity relations for growing microtubules.[22] It analyzed the experimental data on microtubule growth where the previous analysis determined that one of the rates is negative, which can never happen in chemistry.

These are the lessons that I am trying to follow in my research and I teach my students and postdocs. For me, Michael Fisher will always be the example of an ideal scientist and an ideal human being, a status I will never be able to reach, but I will try. I miss Michael very much!

References

1. H. Lodish, *et al.*, *Molecular Cell Biology* (Macmillan, 2008).
2. B. Alberts, *Molecular Biology of the Cell* (WW Norton and Company, 2017).
3. A. B. Kolomeisky, *Motor Proteins and Molecular Motors* (CRC Press, 2015).
4. M. Schnitzer and S. M. Block, *Nature* **388**, 386–390 (1997).
5. J. Prost, J. F. Chauwin, L. Peliti and A. Ajdari, A., *Phys. Rev. Lett.* **72**, 2652–2655 (1994).
6. I. Derenyi and T. Viscek, *Proc. Natl. Acad. Sci. USA* **93**, 6775–6779 (1996).
7. B. Derrida, *J. Stat. Phys.* **31**, 433–450 (1983).
8. A. B. Kolomeisky and B. Widom, *J. Stat. Phys.* **93**, 633–645 (1998).
9. A. B. Kolomeisky and M. E. Fisher, *Ann. Rev. Phys. Chem.* **58**, 675–695 (2007).
10. A. B. Kolomeisky, *J. Phys. Condens. Matter* **25**, 463101 (2013).
11. K. Svoboda and S. M. Block, *Cell* **77**, 773–784 (1994).
12. K. Svoboda, P. P. Mitra and S. M. Block, *Proc. Natl. Acad. Sci. USA* **91**, 11782–11786 (1994).
13. H. Kojima, E. Muto, H. Higuchi and T. Yanagida, *Biophys. J.* **73**, 2012–2022 (1997).
14. H. Higuchi, E. Muto, Y. Inoue and T. Yanagida, *Proc. Natl. Acad. Sci. USA* **94**, 4395–4400 (1997).
15. M. D. Wang, M. J. Schnitzer, H. Yin, R. Landick, J. Gelles and S. M. Block, *Science* **282**, 902–907.
16. M. E. Fisher and A. B. Kolomeisky, *Proc. Natl. Acad. Sci. USA* **96**, 6597–6602 (1999).
17. M. E. Fisher and A. B. Kolomeisky, *Proc. Natl. Acad. Sci. USA* **98**, 7748–7753 (2001).
18. A. B. Kolomeisky and M. E. Fisher, *Biophys. J.* **84**, 1642–1650 (2003).
19. A. B. Kolomeisky and M. E. Fisher, *Physica A* **279**, 1–20 (2000).
20. A. B. Kolomeisky, *J. Chem. Phys.* **115**, 7253–7259 (2001).
21. A. B. Kolomeisky and M. E. Fisher, *J. Chem. Phys.* **113**, 10867–10877 (2000).
22. A. B. Kolomeisky and M. E. Fisher, *Biophys. J.* **80**, 149–154 (2001).

https://doi.org/10.1142/9789811282386_0041

Chapter 41

Topological Defects in Living Matter

J.M. Yeomans
The Rudolf Peierls Centre for Theoretical Physics,
Clarendon Laboratory, Parks Road, Oxford, OX1 3PU, UK
julia.yeomans@physics.ox.ac.uk

It is a great pleasure and privilege to contribute to a volume of papers in honor of Professor Michael E. Fisher. Michael gave a colloquium in Oxford around 1978 and I was sufficiently enthralled to dare to talk to him afterward. Perhaps, I asked the right questions because I had the good fortune to have the opportunity to pursue a postdoc at Cornell. Michael was a superb postdoc advisor, he expected high standards and did so by example. It was an exciting time and a stimulating place to do science.

Later, in his career, Michael became interested in what are now termed active systems, in particular motor proteins.[1] He was also much enthused by David Mermin's work on topological defects.[2] Here, I examine the two themes together and review recent works on nematic defects in living systems, particularly active, self-propelled defects.

1. Active Nematics

Long, thin molecules or colloids with apolar symmetry can order in a nematic phase which has long-range orientational order, but no long-range positional order (Fig. 1(a)).[4] A corresponding order parameter is the director field **n** which is a headless vector. In a passive liquid crystal, described by the usual laws of equilibrium thermodynamics, small distortions in the nematic ordering are quickly relaxed by elastic forces. However, more complex, topological, defects which are singularities in the ordering can also occur. In two dimensions, these are characterised by a topological charge

$$m = \frac{1}{2\pi} \oint d\theta, \tag{1}$$

where the integral is taken around the defect core and θ is the polar angle of the director field. The director configurations around the most common two-dimensional defects, $m = 1/2$ and $m = -1/2$, are shown in Fig. 1(b). In passive nematics, defects of opposite change attract and annihilate, thus reducing elastic energy and restoring nematic order.

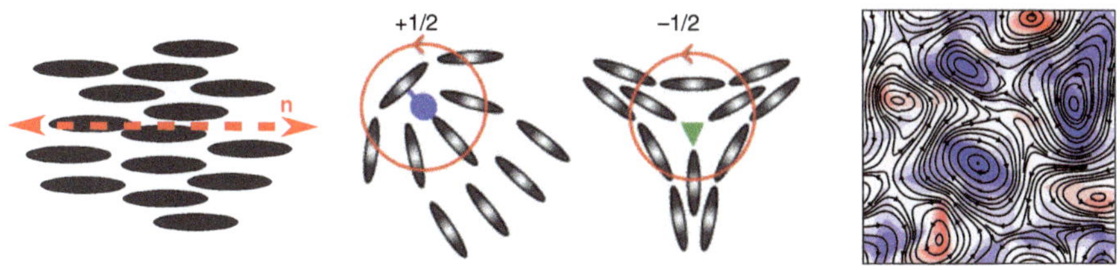

(a) (b) (c)

Fig. 1. Schematic representation of (a) nematic alignment along the director **n**, (b) $\pm 1/2$ topological defects, (c) snapshot of active turbulence characterised by chaotic flows and strong vorticity. Black lines are streamlines and the colormap shows vorticity normalized to its maximum value, varying from $+1$ (red) for anticlockwise to -1 (blue) for clockwise. Adapted from Ref. 3.

Active nematics continuously take energy from their surroundings which is then dissipated as viscosity or friction. Hence, they remain out of thermodynamic equilibrium.[3] Examples of dense active nematics include swarms of swimming or crawling bacteria, microtubule-motor protein suspensions and aggregates of eukaryotic cells. Indeed, quoting Schrödinger,[5] all living matter "evades the decay to equilibrium" and hence provides, often very complex, examples of non-equilibrium statistical physics.

In wet active nematics, each nematogen acts as a force dipole producing stresses which lead to flows whenever there is a distortion in the order-parameter field. Small fluctuations in the nematic order do not relax, but rather are enhanced by the flows. This drives the ordered nematic unstable and the system achieves a dynamical steady state, termed active turbulence, which is characterised by chaotic flows and high vorticity (Fig. 1(c)).

The input of energy, and the destruction of long-ranged nematic ordering, means that topological defects in active nematics can be created as well as annihilated. Thus, active turbulence is characterised by a gas of topological defects which are continually created and destroyed in oppositely charged pairs. The director field around defects is strongly distorted and hence leads to flows. The polar symmetry of $+1/2$ defects means that the flows are unbalanced, and therefore, these defects are self-propelled. If the active dipolar forces act outward along the long axis of the nematic particles, they are called extensile and the defects move toward their heads; if the forces act inward, the nematogens are contractile and the $+1/2$ defects move toward their tails.

In recent years, several biological systems have been reported to show the dynamics of active turbulence with associated short-range nematic order and topological defects. Moreover, in some instances, the relevance of the defects to biological function has been identified.

2. Bacteria

Myxococcus xanthus is a rod-shaped bacterium that lives in dense colonies in the soil. The bacteria live as a thin sheet, coating surfaces, but when nutrients are scarce, they form dome-shaped structures called fruiting bodies. The first step in this process is the sequential formation of layers of motile bacteria on top of one another. Copenhagen *et al.*[6] showed that new layers are preferentially initiated at the positions of $+1/2$ defects, whereas holes in the layers preferentially form at $-1/2$ defects. By modelling the bacteria as an extensile active nematic with anisotropic friction, they explain their observations by showing that cells tend to move toward $+1/2$ and away from $-1/2$ defects.

Meacock *et al.*[7] tell the *Hare and the Tortoise* story of the impact of topological defects on mixtures of two strains of *Pseudomonas aeruginosa*. This pathogen lives in dense colonies on surfaces and expands its range through twitching motility driven by grappling-hook-like appendages called type IV pili. The experiments addressed a mixture of wild-type bacteria and Δ*pilH* cells, which were genetically modified to have more pili, and hence to move faster, than the wild type. Surprisingly, the slower, wild-type colonies were able to expand more efficiently, an effect explained by invoking active liquid crystal physics. When defects with topological charge $+1/2$ approach each other, the resultant flow fields cause the fast-moving mutant cells to rotate to point vertically and form stable, vertical clusters which subsequently trap more fast cells, a fate avoided by the slow cells.

Other authors have considered colonies that expand as a result of cell division rather than bacterial motility. The dividing cells push outward along their long axis, and hence, the dividing colony behaves like an extensile active nematic.[8] There are interesting differences between two sets of results presented in the literature.

Dell'Arciprete *et al.*[10] studied the growth of small colonies of rod-shaped *E. coli* bacteria plated on a glass slide and growing just under the surface of soft agarose. They found that the bacteria formed small nematic domains, of typically about 10 cells, and attributed this to break up of nematic order due to the active instability (see Fig. 2). $\pm 1/2$ defects were created in the bulk of the colony, with the $+1/2$ defects moving toward their heads as expected in extensile flows. There was strong tangential alignment of the bacteria at the colony boundaries which is also an expected consequence of the flows due to extensile activity. Further insights into a similar system, from experiment and agent-based modelling, are given in Ref. 11.

By contrast when dividing *E. coli* expand inward to fill holes in a dense layer, the cells tend to align radially and hence form a $+1$ defect as they fill the hole.[12] The authors argue that the radial alignment is driven by flow-induced torques. There is no break-up into nematic microdomains which suggests that, in these experiments, the length scale of the active instability is larger than the hole size. One explanation for the difference in active length scale is the different substrate, these

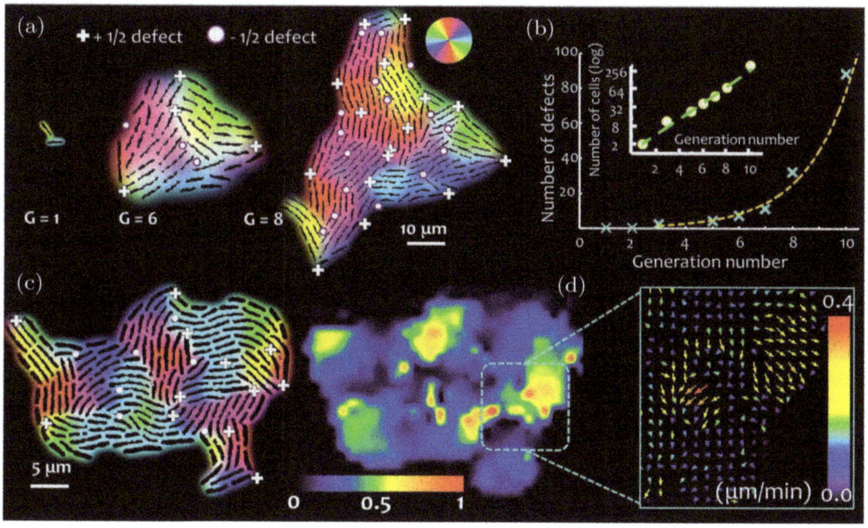

Fig. 2. Topological defects in a growing bacterial colony: (a) Time lapse snapshots of a growing colony of non-motile *E. coli* bacteria. $G \sim 45$ min is the generation time (the number of cell doubling events). As the colony expands, topological defects emerge spontaneously, either at the periphery or within the bulk of the colony. (b) The total number of topological defects increases approximately exponentially with time (represented here by the generation number, G), and approximately linearly with the number of cells in the colony. (c) Cell divisions drive active flows with high vorticity. The relative strength of the flows is indicated by the heat map scale bar. (d) Active flow color coded by flow speed. Stronger flows are observed in regions with a higher density of topological defects. Figure taken from Ref. 9.

colonies were grown on an agarose surface and so were subject to different frictional forces. The authors speculate that the flow-driven alignment that leads to the radial ordering is stronger for longer bacteria and may give them a competitive advantage in colonizing substrates. The contrasts between the results in Refs. 10 and 12 suggest interesting questions about the differences between inward and outward colony growth, the relative roles of flow-driven alignment and active instabilities in determining bacterial ordering, the competition between radial ordering in the bulk and tangential ordering at surfaces, and the impact of this physics on mixed colonies.

Dhar *et al.*[13] extend these ideas to the single to multilayer transition in growing colonies. They show that in regions where the monolayer is under sufficient in-plane stress, normally close to the centre of a colony, cell divisions can lead to extrusions. In their experiments, all extrusion events were localized close to topological defects, although not every defect triggered an extrusion. A possible explanation for this behavior is that it is a consequence of the additional stresses in the vicinity of both $+1/2$ and $-1/2$ defects. Note the differences to the first example in this section. In the case of motile bacteria, cell flows toward $+1/2$ and away from $-1/2$ topological defects lead, respectively, to the formation or destruction of layers. Here, the authors report layer formation at both $\pm 1/2$ defects, with just a small increase in the probability of extrusion at the $+1/2$'s.

3. Eukaryotic Cells

The most obvious place to look for topological defects in eukaryotic cells is motile, spindle-shaped cells. Active turbulence has been observed in human bronchial epithelial cell (HBEC) cultures where, between confluence and jamming, the properties of the flow vortices and topological defects are well described by the theory of extensile active nematics.[14]

Duclos *et al.* studied mouse embryo fibroblasts (NIH 3T3) plated on fibronectin-coated glass substrates.[16] As the cells reach confluence, their motility combines with steric repulsion to align them into nematic domains, but with the ordering disturbed by $\pm 1/2$ topological defects. The $+1/2$ defects self-propel in the direction of their tails, characteristic of contractile activity. $+1/2$ and $-1/2$ defects annihilate, but the cells are in a regime where there is insufficient active energy to create new defects. As the cells divide their velocity decreases, the system jams and defect positioning becomes controlled by elasticity rather than activity. A clear demonstration of this is that, when confined in mesoscopic circular domains, the system evolves toward two identical $+1/2$ defects facing each other. The authors suggest that high cell–substrate friction is responsible for the suppression of active turbulence in the fibroblast layer.

Sustained active extensile turbulence is however seen in fibroblasts that move more rapidly: human fibrosarcoma cells (HT1080). Yashunsky *et al.* show that close to an interface, this cell line exhibits a spontaneous edge current with broken left–right symmetry, while in the bulk, the cell flow remains chaotic.[17] The current is driven by an array of $+1/2$ topological defects at the boundary, with its direction determined by cell chirality. Cancer-associated fibroblasts are considered to be an important factor in tumor development, and an active matter perspective may provide a way to help understand their ordering and dynamics.

Another link of topological defects to cancer biology is suggested in Ref. 18. The mesothelium is a single layer of epithelial cells which forms the lining of several body cavities. The authors show that high aspect ratio mesothelial cells form patterns very similar to fibroblasts, with topological defects separating nematic domains of different orientations. They show that cells tend to move toward $+1/2$ defects and away from $-1/2$ defects, leading to higher cell densities at the former and lower cell densities at the latter. Mesothelial clearance is the process by which cancer cells can invade the mesothelium to invade underlying tissue. The authors suggest that this is reduced around defect sites.

Single murine neural progenitor cells can easily change their shape and elongate. In dense layers, they extend and align, moving along their long axis, with 180° changes in direction. Kawaguchi *et al.* showed that the cells form nematic domains interspersed with topological defects.[19] Here also, cells tend to accumulate at $+1/2$ defects, leading to distinct mounds, and to reduce in density around $-1/2$ defects. Similarities are flagged to the dynamics of cells in the rostral migratory stream along which neuronal precursor cells that originate in the subventricular zone of the brain migrate to reach the main olfactory bulb.

Perhaps, more surprisingly, topological defects have also been observed in Madin Darby canine kidney (MDCK) epithelial cells which are, on average, isotropic.[15] When confluent, the cells elongate in a correlated way to give short-ranged nematic order which can support $\pm 1/2$ topological defects. Saw *et al.* establish a link between apoptosis (cell death and ejection from the layer) and the position of, particularly, $+1/2$ defects. This is explained in terms of the enhanced stress at defect sites which promotes the motion of yes-associated protein (YAP) from the cytoplasm to the nucleus, a mechanotransduction pathway implicated in cell death. These experiments are illustrated in Fig. 3.

The motile topological defects in Saw *et al.* were identified as extensile. To understand this further, Balasubramanian *et al.* showed that the weakening of inter-cellular contacts, by knocking out E-cadherin juntions, could change the behavior of defects in an MDCK layer from extensile to contractile.[20] Other work suggests that stresses at topological defects can initiate hole formation in monolayers on soft substrates.[21] There is still much to be done to understand the extent to which the simple theories of active nematics, and the concepts of extensility and contractility, which are clear when describing bacterial flow fields, apply to nematogens, such as eukaryotic cells, that can be easily deformed.

4. Morphogenesis

Topological defects can also drive changes in shape. Much of the work on this has been motivated by compelling experiments arguing that the regeneration of *Hydra* is related to defect patterning.[22] *Hydra* is a marine polyp a few mm in length when fully grown (Fig. 4). It has a head at one end, a foot at the other, and an array of tentacles. The body comprises a double layer of cells which contain arrays of parallel supracellular actin fibers. These are aligned along the body axis in the outer layer and perpendicular to the body axis in the inner layer. Myosin motors move along the fibres exerting contractile forces. Treating the direction of the fibres in the outer layer as a nematic field, the head and foot regions correspond to $+1$ topological defects, and each tentacle has a $+1$ defect at its tip and two $-1/2$ defects at opposite sides of its base.

Hydra can regenerate from a small, flat excised tissue fragment, and Maroudas-Sacks *et al.* followed the dynamics of the actin-defined nematic field during this process.[22] At early times, as the tissue folds, and the supercellular fibres re-organize, $+1/2$ and $+1$ defects predominate. The $+1/2$'s are motile and tend to merge to form $+1$'s or to be eliminated in encounters with $-1/2$'s. The authors show that

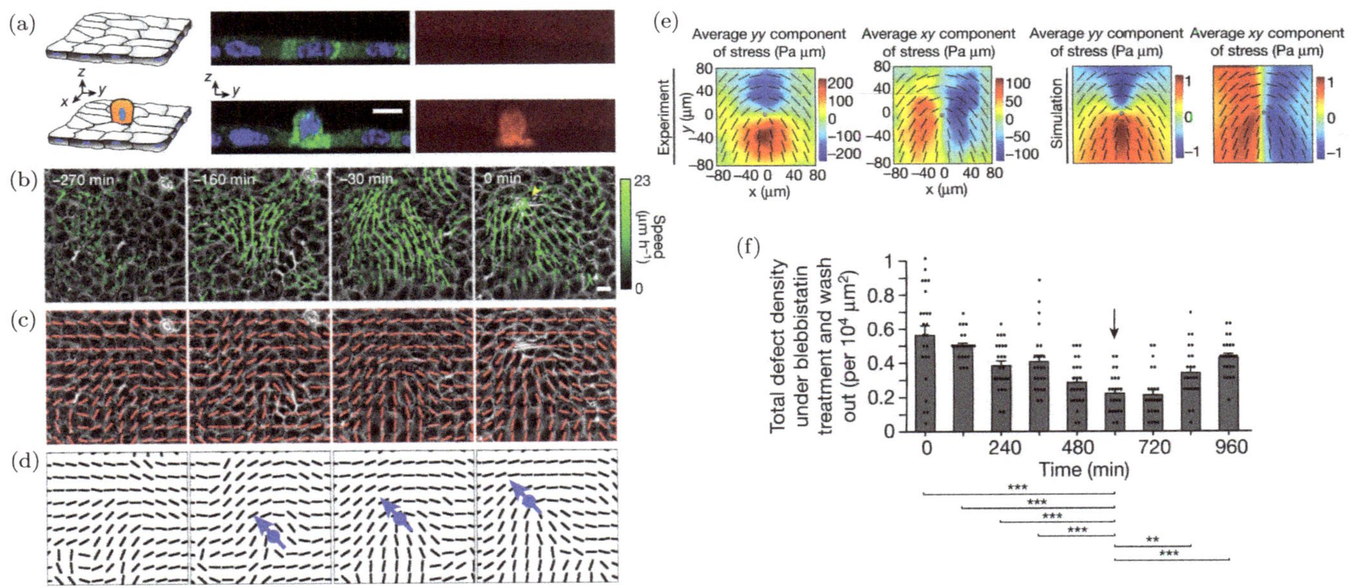

Fig. 3. Topological defects and cell extrusion in MDCK monolayers: (a) Left: diagrams of confluent monolayer (top) and extruding cell (bottom). Middle: side view confocal image of confluent MDCK monolayer (top) and extruding cell (bottom). Green, actin; blue, nucleus. Right: corresponding images of activation of caspase-3 signal (a marker for cell death). (b) Phase-contrast images showing monolayer dynamics before extrusion (yellow arrowhead) at $t = 0$ min, overlaid with velocity field vectors. Length of vectors is proportional to their magnitude. Colour code: bright (dark) green for high (low) speeds. (c, d) Corresponding images overlaid with red lines (in (c)), and represented as black lines (in (d)) showing average local orientation of cells. The group of cells moving toward the extrusion forms a $+1/2$ defect. (e) Map of average yy- and xy-components of stress around a $+1/2$ defect in experiments (left pair of plots), compared with simulations of an extensile, active nematic liquid crystal (right pair of plots). Colour code is positive for tensile stress and negative for compressive stress. Overlaid black lines in panels show representative nematic directors, grey circle denotes defect core. (f) Time evolution of (total) defect areal density under 50 μm blebbistatin treatment and washout (arrow). Adding blebbistatin reduces activity and defects annihilate but are not created. Washout restores activity and the defect number recovers. Adapted from Ref. 15.

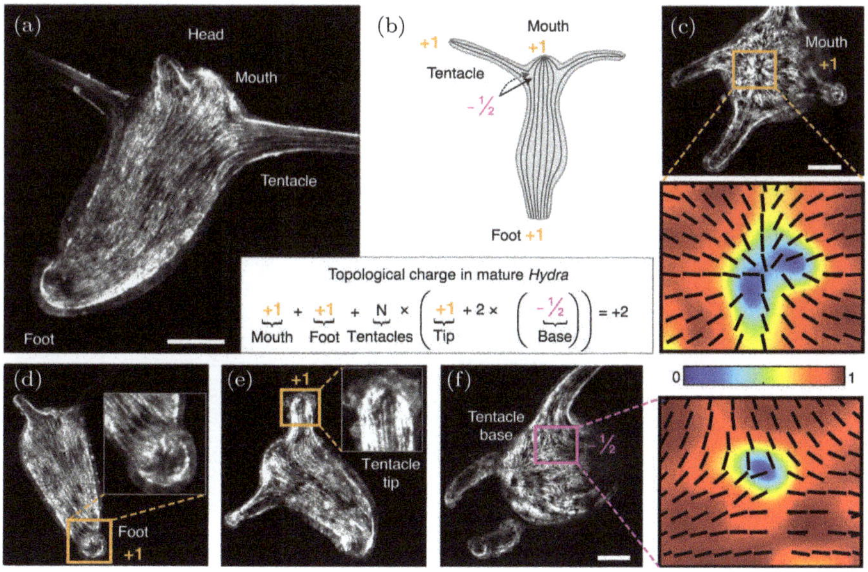

Fig. 4. Topological defects in Hydra: (a) image and (b) schematic of the nematic actin fibre organization in the ectoderm of a small mature *Hydra* indicating topological defects; (c) +1 defect at the tip of the head; (d) +1 defect at the apex of the foot; (e) +1 defect at the tip of a tentacle (f) −1/2 defect at the base of a tentacle. The zoomed maps in (c) and (f) depict the actin fibre orientation (black lines) and the order parameter magnitude. Figure adapted from Ref. 22.

an early +1 defect becomes the site of head formation, whereas two +1/2 defects merge in the vicinity of the emerging foot. Tentacle formation typically involves the later appearance of a +1 defect at the tip of the growing tentacle, together with two −1/2 defects at its base. These results strongly suggest the importance of mechanical signaling coupling to biochemical fields to drive *Hydra* morphogenesis.

Guillamat *et al.* confined an elongated cell-line, C2C12 myoblasts, to a disc with diameter of order the nematic correlation length, thus stabilizing a +1 topological defect at the centre of the disc.[23] At low density, the cells self-organised into persistently rotating spirals. As the density increased, they reorganised to align radially and they ceased to rotate. Flows were inward toward the centre of the colony where a mound formed. If the cells were allowed to differentiate, they did so preferentially in the centre of the mound, suggesting a link between mechanical stresses and differentiation. When differentiation was inhibited, the cellular mounds grew further in height, forming cylindrical protrusions of up to hundreds of micrometers, with collective cell rotation around the protrusion axis. A theoretical discussion of the active instability that leads to a protrusion around +1 defects is given in Ref. 24.

Combining ideas from mechanobiology and active matter physics is leading to new ways to think about cell motility and morphogenesis. It is becoming increasingly apparent that mechanical signalling plays an important role in many biological changes. As localised hotspots of stress, topological defects may drive, and can help interrogate, this coupling. See Refs. 9, 25, 26 for other reviews on these topics.

We are just at the beginning. Active turbulence is well established in two-dimensional confluent cell layers and in well-defined bacterial assays. But what is the role of activity in three dimensions, *in vivo*, and in the complex natural environments of biofilms and tissues?

References

1. A. B. Kolomeisky and M. E. Fisher, Molecular motors: A theorist's perspective. *Ann. Rev. Phys. Chem.* **58**, 675 (2007).
2. N. D. Mermin, The topological theory of defects in ordered acedia. *Rev. Mod. Phys.* **51**, 591 (1979).
3. A. Doostmohammadi, J. Ignés-Mullol, J. M. Yeomans and F. Sagués, Active nematics. *Nat. Comms.* **9**, 3246 (2018).
4. P. G. de Gennes and J. Prost, *The Physics of Liquid Crystals.* (Oxford University Press, 1995).
5. E. Schrödinger, *What Is Life? The Physical Aspect of the Living Cell* (Cambridge University Press, 1944).
6. K. Copenhagen, R. Alert, N. Wingreen and J. Shaevitz, Topological defects promote layer formation in *Myxococcus Xanthus* colonies, *Nat. Phys.* **17**, 211 (2021).
7. O. J. Meacock, A. Doostmohammadi, K. R. Foster, J. M. Yeomans and W. M. Durham, Bacteria solve the problem of crowding by moving slowly. *Nat. Phys.* **17**, 205 (2021).
8. A. Doostmohammadi, S. P. Thampi, T. B. Saw, C. T. Lim, B. Ladoux and J. M. Yeomans, Celebrating Soft Matter's 10th anniversary: Cell division: a source of active stress in cellular monolayers. *Soft Matter* **11**, 7328 (2015).
9. A. Sengupta, Microbial active matter: A topological framework. *Fron. Phys.* **8**, 184 (2020).
10. D. Dell'Arciprete, M. L. Blow, A. T. Brown, F. D. C. Farrell, J. S. Lintuvuori, A. F. McVey, D. Marenduzzo and W. C. K. Poon, A growing bacterial colony in two dimensions as an active nematic. *Nat. Comms.* **9**, 4190 (2018).
11. R. Los, D. van Holthe tot Echten, G. Nordemann, M. Wehrens, S. J. Tans and T. Idema, Defect dynamics in growing bacterial colonies, arXiv:2003.10509v2 (2023).
12. M. Basaran, Y. I. Yaman, T. C. Yuece, R. Vetter and A. Kocabas, Large-scale orientational order in bacterial colonies during inward growth. *Elife* **11**, e72187 (2022).
13. J. Dhar, A. L. P. Thai, A. Ghoshal, L. Giomi and A. Sengupta, Self-regulation of phenotypic noise synchronizes emergent organization and active transport in confluent microbial environments. *Nat. Phys.* **18**, 945 (2022). ISSN 1745-2473.
14. C. Blanch-Mercader, V. Yashunsky, S. Garcia, G. Duclos, L. Giomi and P. Silberzan, Turbulent dynamics of epithelial cell cultures. *Phys. Rev. Lett.* **120**, 208101 (2018).
15. T. B. Saw, A. Doostmohammadi, V. Nier, L. Kocgozlu, S. Thampi, Y. Toyama, P. Marcq, C. T. Lim, J. M. Yeomans and B. Ladoux, Topological defects in epithelia govern cell death and extrusion. *Nature* **544**, 212 (2017).
16. G. Duclos, C. Erlenkämper, J. F. Joanny and P. Silberzan, Topological defects in confined populations of spindle-shaped cells. *Nat. Phys.* **13**, 58 (2017).
17. V. Yashunsky, D. J. G. Pearce, C. Blanch-Mercader, F. Ascione, P. Silberzan and L. Giomi, Chiral edge current in nematic cell monolayers. *Phys. Rev. X.* **12**, 041017 (2022).

18. J. Zhang, N. Yang, P. K. Kreeger and J. Notbohm, Topological defects in the mesothelium suppress ovarian cancer cell clearance. *APL Bioeng* **5**, 036103 (2021).
19. K. Kawaguchi, R. Kageyama and M. Sano, Topological defects control collective dynamics in neural progenitor cell cultures. *Nature* **545**, 327 (2017).
20. L. Balasubramaniam, A. Doostmohammadi, T. B. Saw, G. Narayana, R. Mueller, T. Dang, M. Thomas, S. Gupta, S. Sonam, A. S. Yap, Y. Toyama, R.-M. Mège, J. M. Yeomans and B. Ladoux, Investigating the nature of active forces in tissues reveals how contractile cells can form extensile monolayers. *Nat. Mat.* **20**, 1156 (2021).
21. S. Sonam, L. Balasubramaniam, S.-Z. Lin, Y. Ivan, I. Pi-Jauma, C. Jebane, M. Karnat, Y. Toyama, P. Marcq, J. Prost, R.-M. Mège, J.-F. Rupprecht and B. Ladoux, Mechanical stress driven by rigidity sensing governs epithelial stability. *Nat. Phys.* **19**, 132 (2023).
22. Y. Maroudas-Sacks, L. Garion, L. Shani-Zerbib, A. Livshits, E. Braun and K. Keren, Topological defects in the nematic order of actin fibers as organization centers of *Hydra* morphogenesis. *Nat. Phys.* **17**, 251 (2021).
23. P. Guillamat, C. Blanch-Mercader, G. Pernollet, K. Kruse and A. Roux, Integer topological defects organize stresses driving tissue morphogenesis. *Nat. Mat.* **21**, 588 (2022).
24. L. A. Hoffmann, L. N. Carenza, J. Eckert and L. Giomi, Theory of defect-mediated morphogenesis. *Sci. Adv.* **8**, eabk2712 (2022).
25. A. Ardaseva and A. Doostmohammadi, Topological defects in biological matter. *Nat. Rev. Phys.* **4**, 354 (2022).
26. A. Doostmohammadi and B. Ladoux, Physics of liquid crystals in cell biology. *Trends Cell Biol.* **32**, 140 (2022).

https://doi.org/10.1142/9789811282386_0042

Chapter 42

Critical Casimir Forces in Soft Matter

A. Gambassi[*,§] and S. Dietrich[†,‡,¶]

[*] *SISSA — International School for Advanced Studies and INFN, via Bonomea 265, 34136 Trieste, Italy*

[†] *Max-Planck Institute for Intelligent Systems, Heisenbergstr. 3, 70569 Stuttgart, Germany*

[‡] *Institut für Theoretische Physik IV, Universität Stuttgart, Pfaffenwaldring 57, 70569 Stuttgart, Germany*

[¶] *dietrich@is.mpg.de*

[§] *gambassi@sissa.it*

In this chapter, we review recent advances in the theoretical, numerical, and experimental studies of critical Casimir forces in soft matter, with particular emphasis on their relevance for the structures of colloidal suspensions and on their dynamics. Distinct from other interactions which act in soft matter, such as electrostatic and van der Waals forces, critical Casimir forces are effective interactions characterized by the possibility to control reversibly their strength via minute temperature changes, while their attractive or repulsive character is conveniently determined via surface treatments or by structuring the involved surfaces. These features make critical Casimir forces excellent candidates for controlling the equilibrium and dynamical properties of individual colloids or colloidal dispersions as well as for possible applications in micro-mechanical systems. In the past 25 years, a number of theoretical and experimental studies have been devoted to investigate these forces primarily under thermal equilibrium conditions, while their dynamical and non-equilibrium behavior is a largely unexplored subject open for future investigations.

1. Introduction

Critical Casimir forces (CCFs) emerge at the surfaces of objects immersed in a fluid medium whenever they confine the fluctuations of the order parameter (OP) of a continuous phase transition occurring within that medium. Accordingly, these forces are deeply related to phase transitions and critical phenomena in statistical systems due to the effects of spatial confinements on them,[1] i.e., two topics which are part of the scientific legacy of Michael E. Fisher, as he gave fundamental contributions to both. This is the reason why we deem it natural to present this review in a book dedicated to his memory.

The very same idea of the emergence of peculiar solvation forces in a critical medium — which, later on, have been called *critical Casimir forces* — can be traced back to a seminal paper[3] written in 1978 by Fisher and Pierre-Gilles de Gennes. These forces owe their name to the fact that they are the thermodynamicly analogous ones related to the celebrated Casimir effect of quantum electrodynamics,[4] in which the relevant fluctuations are those of quantum nature of the electromagnetic fields in vacuum, while the confining surfaces are provided by perfectly conducting metallic plates. As in the latter case, the dependence of CCFs on the relevant physical constants and geometrical parameters characterizing the system is largely universal in a sense we shall specify in the following.

The magnitude of CCFs can be estimated on the basis of dimensional analysis: As they are essentially determined by thermal fluctuations of an OP, their typical energy scale is set by the thermal energy $k_B T$ at temperature T. In turn, this is the scale typically involved in the physics of mesoscopic particles such as *colloids*, having radii in the micrometer and sub-micrometer range. Therefore, CCFs are expected to give rise to relevant effects in soft matter, which primarily involves colloids dispersed in a liquid medium. For this reason, the present contribution focuses on CCFs in soft matter, with particular emphasis on colloidal suspensions.

Chapter 15 discusses in detail yet another relevant manifestation of CCFs on the behavior of fluids at the microscopic scale, e.g., the way in which they affect the thickness of complete wetting films. From the investigations reviewed here, it clearly emerges that, different from the typical interactions in soft matter,[5,6] CCFs offer a large degree of tunability which is already finding interesting practical applications.

This review is organized as follows. In Section 2, we briefly present (following Ref. 7) the basic notions of statistical physics of spatially confined systems and discuss the effects of introducing physical boundaries on their behavior, explaining if and how CCFs emerge. As anticipated, these forces are characterized by universal scaling functions, the analytical and numerical determination of which are presented in Section 3. While most of these predictions refer to critical media confined to the film geometry, in Section 4, we discuss how this knowledge can be used to study the case in which curved or inhomogeneous surfaces are involved, e.g., spherical and non-spherical colloids or patterned substrates of experimental relevance. Increasing the complexity of the system further, we consider many-body effects of these forces as well as their subtle interplay with other forces which are usually at play in soft matter, such as electrostatic interactions in the presence of electrolytes. In Section 5, we review how the critical Casimir effect has been detected experimentally, starting from what is arguably its earliest evidence,[8] dating back to 1985, to the first indirect, quantitatively reliable measurement[9] in 1999 (see also Chapter 15), and finally to the direct measurement[10] in 2008. We also provide an overview of the experimental investigations of CCFs in the presence of patterned substrates as well as in more complex solvents, i.e., if in addition van der Waals and electrostatic interactions are present. Following this overview of the equilibrium features of CCFs, we review our current theoretical, numerical, and experimental understanding of their dynamical behavior which, beyond the cases of specific aspects or toy models, is still a largely unexplored subject. In Section 7, we provide our perspective on this subject, mentioning some possible further applications of CCFs in soft matter.

A general introduction to CCFs can be found in the early Refs. 11 and 12 and in the subsequent short reviews in Refs. 7 and 13. More recently, some specific aspects of CCFs have been reviewed, including their role in determining the phase behavior of colloidal suspensions,[14] establishing exact analytical results,[15] and measurements involving optical trapping.[16] Accordingly, our focus here is primarily on aspects which have not yet been covered in the above references. Moreover, due to length constraints, the available literature will not be reviewed completely, but pertinent studies will be at least quoted, as guidance for the interested reader.

2. Statistical Systems Under Confinement and the Critical Casimir Effect

The physical properties of soft matter are determined by various effective interactions, which depend on its structure at various length scales and on its thermodynamic conditions. For example, in colloidal suspensions, a prominent role is played by screened electrostatic, hydrodynamic, and depletion forces. As anticipated in Section 1, thermal fluctuations can also generate effective forces[3] in analogy to the celebrated Casimir effect in quantum electrodynamics.[4] In particular, the latter effective forces are heuristically expected to become particularly relevant whenever the thermal fluctuations are pronounced and spatially correlated in such a way that they influence both qualitatively and quantitatively the thermodynamic and structural properties of the fluctuating medium. This is known to be the case upon approaching a continuous (i.e., second-order) phase transition of the medium, close to which the relevant thermal fluctuations are those of a spatially varying OP $\phi(\boldsymbol{x})$ associated with the phase transition.[13]

2.1. *Critical phenomena and universality*

Upon approaching the critical point of a continuous phase transition, the thermal fluctuations of the order parameter $\phi(\boldsymbol{x})$ become spatially correlated across a typical distance ξ. Due to the emergence of a collective behavior, this correlation length actually diverges algebraically upon approaching the phase transition. Accordingly, close to a critical point, ξ eventually becomes the dominant length scale. It is significantly larger than the microscopic scales characterizing the molecular structure of the medium (and, possibly, larger than the scales of other effective interactions at work in the system). In this case, the free energy $\mathscr{F}$ of the medium itself receives a relevant non-analytic contribution due to the thermal fluctuations of $\phi(\boldsymbol{x})$. The physical nature of the quantity playing the role of the OP depends on the specific critical point of interest: In the case of the normal-to-superfluid transition of liquid ^{4}He, ϕ is identified with the quantum wavefunction of the superfluid, whereas for the demixing transition of a classical binary liquid mixture — which we primarily consider here — ϕ is proportional to the difference $c_A(\boldsymbol{x}) - c_B(\boldsymbol{x})$ between the local concentrations $c_{A,B}$ of the two species A and B forming the mixture. The thermal fluctuations of ϕ are controlled by an effective Hamiltonian $\mathscr{H}[\phi]$ which determines their probability $\propto \mathrm{e}^{-\mathscr{H}[\phi]}$ and thus the effective free energy $\mathscr{F} = -k_BT \ln \int [\mathrm{d}\phi] \mathrm{e}^{-\mathscr{H}[\phi]}$ of the medium. In principle, the effective Hamiltonian $\mathscr{H}[\phi]$ can be derived by coarse-graining the microscopic Hamiltonian of the system. Therefore, it is expected to depend on the structure of the medium at various length scales and on its specific material and microscopic properties. However, this generally prohibitive task turns out to be unnecessary upon approaching the critical point because only few of these features are relevant and thus eventually crucial in determining the thermodynamic behavior at mesoscopic length scales. Accordingly, the form of $\mathscr{H}[\phi]$ can be drastically simplified and only its dependence on a few gross features of the medium, such as the ranges and the symmetries of the microscopic interactions, and on general properties of the order parameter, e.g., the number of its components and the nature of the broken symmetry, matter. After this reduction, statistical systems, which differ significantly at the microscopic level, eventually display a common singular behavior upon approaching a critical point — a fact encoded in the notion of *universality* and universality classes within critical phenomena. This means, for example, that upon approaching the temperature T_c of the critical point from the homogeneous, i.e., disordered phase, the correlation length ξ as a function of temperature T increases algebraically as $\xi \simeq \xi_0^+ |T/T_c - 1|^{-\nu}$, with a critical exponent ν and a typical molecular scale ξ_0^+ of the medium. Whereas T_c and ξ_0^+ depend on the material properties of the medium undergoing the transition, the exponent ν does not and is therefore universal. For example, ν takes the same value for all binary liquid mixtures undergoing a second-order demixing transition, and therefore, it is universal.

If an object, such as a plate or a particle, is immersed into the fluid medium, the OP and its thermal fluctuations are affected by the presence of the additional interaction between the molecules of the object and those of the medium. In particular, the resulting perturbation is expected to extend into the medium by a typical distance around the object of the order of the correlation length ξ. As in the case of bulk properties, it turns out that, upon approaching the critical point, the microscopic details of the interaction between the object and the medium are largely irrelevant for determining the behavior of this inhomogeneous system at mesoscopic scales. This observation leads naturally to an extension of the notion of universality to the case of surface critical behavior and therefore to the introduction of the so-called *surface universality classes*.[17,18] In fact, for the practical purposes of a coarse-grained description, the presence of such objects results into suitable effective boundary conditions (BCs) for the OP ϕ at their surfaces. In particular, the interaction with the object can induce a local enhancement of ϕ upon approaching its surface, such that close to it either positive or negative values of ϕ are favored. These adsorption preferences effectively lead to the BC which is usually denoted by (+) or (−), respectively. Alternatively, there might be no preferential adsorption, leading to the Dirichlet (O) BC $\phi = 0$ at the surface. In the presence of enhanced interactions among the ordering degrees of freedom near the surface, one can also encounter effective Neumann BCs which, beyond mean-field theory, eventually lead to a multicritical, so-called surface-bulk (SB) surface transition.[17–19] The generic inequality of the interactions between the molecules of the confining solid wall and those two types of molecules forming the two components of a classical binary liquid mixture results in the (+) or the (−) BC. Concerning quantum fluids, the necessary continuity of the wavefunction causes generically ϕ (e.g., for the superfluid transition in ^{4}He) to vanish at the surface of the immersed object. However, exceptions may emerge[20] upon confining quantum mixtures, such as those of ^{3}He and ^{4}He.

2.2. *Spatial confinement and the critical Casimir effect*

If a second object is immersed into the fluid medium, it interacts locally with ϕ and its fluctuations, which are possibly affected by the presence of the first object. In particular, this is expected to be the case whenever the distance L of closest approach between the two objects is comparable to or smaller than the correlation length ξ, such that the regions of the medium, within which each surface separately affects its surrounding, overlap. If, on the contrary, $L \gg \xi$, the second object interacts with a medium which is not significantly perturbed by the first one, yielding a negligible effective interaction between the objects, the range of which is therefore set by ξ. Similarly, immersing a third object, sufficiently close to the previous two, will also affect their mutual interaction, leading to non-additive many-body effects, which will be discussed in Section 4.3. This effective interaction, mediated by the OP and its near-critical fluctuations — known as the *critical Casimir effect* — inherits the universality of the critical behaviors in the bulk and at the surfaces. Accordingly, the associated force depends only on the bulk and surface universality classes, the latter being characterized by the effective BC imposed on the OP at the surfaces as well as on the shape of the boundaries (see Section 4). The very thermodynamic nature of the critical Casimir effect can be understood by considering the grand canonical free energy Ω_V of a fluid medium in thermal equilibrium at temperature T, which, in the film geometry, is confined by two parallel surfaces of large area A, separated by a distance L, and with total volume $V = AL$. Upon increasing L, the free energy naturally decomposes as

$$\frac{\Omega_V(L,T)}{A} = -Lp_b(T) + \Omega_s(T) + \frac{k_B T}{L^2} X_{ex}(L/\xi) + \text{corrections}, \qquad (1)$$

where we assume $L, \xi \gg \xi_0^+$, and $A \to \infty$. Here, we focus on the non-analytic contribution Ω_V to the actual free energy of the system, which also contains a non-singular background term due to the non-critical degrees of freedom and fluctuations within the system. The latter typically result into other effective interactions (usually referred to as *solvation forces*) which also emerge as described in the following. Referring to Ω_V, the fact that the dependence of X_{ex} on T occurs only via the correlation length $\xi(T)$ is actually a consequence of finite-size scaling in critical phenomena.[1,12,21–23] The first term on the r.h.s. of Eq. (1) corresponds to the free energy $-ALp_b$ which a volume $V = AL$ of the medium contained inside the film would have in the bulk, i.e., in the absence of confining surfaces; accordingly, p_b is the bulk pressure. The second term corresponds to the additional total surface free energy $A\Omega_s(T)$, with $\Omega_s(T) = \Omega_{s,1}(T) + \Omega_{s,2}(T)$ due to the independent (i.e., for $L \to \infty$) presence of the two opposing and confining surfaces within the medium, facing each other and contributing with $\Omega_{s,1}$ and $\Omega_{s,2}$. The third term corresponds to the interaction free energy (also referred to as excess free energy) due to the simultaneous presence of the two plates, the temperature dependence of which is encoded solely in the dimensionless, *universal* scaling function X_{ex}, along with its dependence on the effective BC. To be concrete, here and in the following, we consider the experimentally relevant case of statistical systems in spatial dimension $d = 3$, although CCFs have been investigated in various spatial dimensions, especially within the field-theoretical context. As a consequence of the dependence of Ω_V on L and of the decomposition in Eq. (1), each of the two confining surfaces is subject to a force

$$F_V = -\partial\Omega_V/\partial L = Ap_b(T) + F_{\text{Cas}} + \text{corrections}, \tag{2}$$

pointing from inside the film toward its outside, where F_{Cas} is the CCF

$$F_{\text{Cas}} = -A\,\frac{\partial}{\partial L}\left[\frac{k_{\text{B}}T}{L^2}X_{\text{ex}}(L/\xi)\right] \equiv A\frac{k_{\text{B}}T}{L^3}X_{\text{Cas}}(L/\xi), \tag{3}$$

which vanishes as $L \to \infty$. Consider now the case of a confining surface of large but finite lateral extension, which is completely surrounded by the fluctuating medium. The force F_{in} exerted on this wall from inside the film toward its outside is given by $F_{\text{in}} = F_V(L)$, while the force F_{out} from outside toward the film inside and due to the unconfined medium outside the film is given by $F_{\text{out}} = F_V(L \to \infty)$. Accordingly, using also Eq. (2), the *total* force acting on the wall is $F_{\text{in}} - F_{\text{out}} = F_{\text{Cas}} + \text{corrections}$, i.e., its leading term as L increases is actually provided by the CCF. The analogies and differences between the expression in Eq. (3) for the CCF and the corresponding one for the Casimir effect in quantum electrodynamics are discussed, e.g., in Ref. 13. As anticipated, the range of F_{Cas} is set by the correlation length ξ. The scaling function X_{Cas} is universal in the sense that it depends only on the gross features of the critical behavior of the system (i.e., the bulk universality class) and on the kind of effective BCs imposed by the two plates on the OP ϕ, i.e., $(+,+)$ or $(+,-)$ for a binary liquid mixture of classical fluids and (O,O) for pure ^{4}He. Analogously to the cases of bulk systems and surfaces, the emerging universality allows one to predict the behavior of the scaling function upon approaching a critical point in terms of suitably simplified models, which act as representatives of the corresponding universality class and thus retain all the relevant gross features of the physical system under study. In fact, numerical simulations of lattice models and analytic studies of continuum theories for the OP field ϕ both in the bulk[24–26] and at surfaces[18] have provided significant insight into the critical behavior of a variety of different physical systems and yield remarkably accurate theoretical predictions both for exponents and scaling functions (see, e.g., Ref. 27). For example, the bulk critical properties of a demixing classical binary liquid mixture (as well as those of ferromagnetic/paramagnetic transitions) are captured on a lattice by the Ising model (with $\mathbb{Z}_2$ symmetry) of one-component spins, whereas those of the normal–superfluid transition of ^{4}He are captured by the XY model (with $O(2)$ symmetry) of two-component spins. Alternatively, these two cases can be studied via the Landau–Ginzburg–Wilson effective Hamiltonian $\mathscr{H}[\phi]$ of an n-component field ϕ with a $(\phi^2)^2$ self-interaction, $O(n)$ symmetry, for $n = 1$ and $n = 2$, respectively. In Section 3, we discuss how universality has been exploited, via analytical and numerical techniques, in order to determine the scaling function X_{Cas} in Eq. (3).

The universal scaling functions X_{ex} and X_{Cas} in Eqs. (1) and (3), respectively, depend on the scaling variable $y = L/\xi$. As a function of the temperature T, y may assume the same value at two temperatures, one above and one below T_c, which correspond to the same correlation length ξ. In order to distinguish these two cases, sometimes the equivalent scaling variable $x = (T/T_c - 1)(L/\xi_0^+)^{1/\nu}$ is considered.

Before discussing in the next sections the characteristic features of CCFs, we emphasize that these forces are fluctuating quantities themselves because, as explained above, they arise from confining the thermal fluctuations of an order parameter. While their mean value, given in Eq. (3), is characterized by a large degree of universality, their variance depends on the microscopic details of the system and is generically large compared to the average.[28] This raises the question of how this fact affects the numerical and experimental determination of the scaling function X_{Cas}, which is addressed in Ref. 29.

3. Scaling Function: Renormalization Group and Numerical Studies

Similar to what occurs for critical phenomena in the bulk, universality allows one to determine the scaling function X_{Cas} of the CCF in the cases mentioned above by studying the most convenient representative model belonging to the universality class under investigation. In particular, as a first step, one has to consider this model for the slab geometry, with macroscopically large transverse area A, finite thickness L, and with BCs which realize the surface universality classes of interest. For example, upon approaching the demixing transition of a classical binary liquid mixture, X_{Cas} can be determined by studying the Ising model (or any other spin model within the same universality class) in the slab geometry with fixed spins at the boundaries, pointing either in the same direction or in the opposite one, such as to realize effectively $(+,+)$ or $(+,-)$ BCs. Equivalently, this model can be investigated by considering the Landau–Ginzburg–Wilson effective Hamiltonian $\mathscr{H}[\phi]$ for a scalar OP ϕ within a slab with suitable surface fields at the boundaries. Similarly, X_{Cas} for the superfluid transition in ^{4}He can be determined by studying the XY model in a slab with unconstrained spins at the boundaries, which eventually realize (O,O) BC. Equivalently, this problem can be investigated by considering $\mathscr{H}[\phi]$ for a two-component order parameter, within a slab geometry, and with Dirichlet BCs.

In view of the illustration in Section 5 of some experimental evidences for the occurrence of CCFs, the previous discussion primarily focused on a medium undergoing a continuous phase transition which belongs to the universality class of either the normal–superfluid transition in ^{4}He (XY universality class) or the demixing transition of classical binary liquid mixtures (Ising universality class). In particular, we considered only those effective BCs, i.e., surface universality classes, which generically characterize the possible experimental realizations of these classes. As a rule of thumb, here we anticipate that the CCF turns out to be attractive for equal BCs $[(+,+),\ (O,O)]$ and repulsive for unequal BCs $[(+,-)]$, independent of the actual bulk universality class of the phase transition under investigation.

In this section, we review the available knowledge about X_{Cas}. In particular, in Section 3.1, we consider various analytical studies — primarily based on a field-theoretical approach, while a number of exact predictions based on lattice models in two bulk spatial dimensions ($d = 2$) or based on exactly solvable models in general d have been reviewed recently in Ref. 15. In Section 3.2, instead, we focus on the predictions which were obtained on the basis of Monte Carlo (MC) simulations of lattice models and which provide quantitatively accurate estimates for X_{Cas}, which are generically out of reach of analytical approaches, especially in the experimentally relevant case $d = 3$.

3.1. *Analytical approaches*

Early theoretical investigations of the critical Casimir effect in the film geometry were based primarily on field-theoretical approaches[30–32] which build upon previous analyses based on the minimization of a Landau–Ginzburg–Wilson functional for $\mathscr{H}[\phi]$.[3,31] In particular, these studies can be cast into the general framework in which one considers a field theory in a bounded domain in the presence of contributions localized at the boundaries. As in the bulk, and in view of universality, the form of these terms can be deduced from symmetry arguments and from power counting. For the simple but important case of a scalar OP ϕ, one finds that the effective free energy to be considered is given by

$$\mathscr{H}[\phi] = \int_{\mathscr{V}} \mathrm{d}^d x \left\{ \frac{1}{2}(\nabla\phi)^2 + \frac{r}{2}\phi^2 + \frac{u}{4!}\left(\phi^2\right)^2 \right\} + \sum_{i=1}^{2} \int_{\mathscr{A}_i} \mathrm{d}^{d-1} x_\parallel \left\{ \frac{c_i}{2}\phi^2 - h_i\phi \right\}, \quad (4)$$

where the confined volume corresponds to $\mathscr{V} = [0, L] \times \mathscr{A}$ with $|\mathscr{A}| = |\mathscr{A}_1| = |\mathscr{A}_2| = A$ and with the same shape as $\mathscr{A}_i$, while the boundary contributions to the action are collected in the second term in which $\boldsymbol{x} = (x_\perp, \boldsymbol{x}_\parallel)$ with $x_\perp = 0$ for $\boldsymbol{x}_\parallel \in \mathscr{A}_1$ and $x_\perp = L$ for $\boldsymbol{x}_\parallel \in \mathscr{A}_2$. In addition to the usual couplings in the bulk, which are controlled by the parameters r (proportional to the actual reduced temperature $(T - T_c)/T_c$ of the system) and $u > 0$, the so-called *surface enhancements* $c_{1,2}$ and the surface fields $h_{1,2}$ appear at the boundaries. Eventually, they determine the behavior of $\phi(\boldsymbol{x})$ upon approaching the boundaries, i.e., for $x_\perp \to 0$ or $x_\perp \to L$, and thus, they generate the effective BCs (see Refs. 18 and 33). In particular, in the absence of symmetry-breaking terms at the boundaries, i.e., for $h_i = 0$, the renormalization-group fixed-point values c_i^* of interest of the coupling c_i turns out to be $c_i^* = +\infty$. This results (as anticipated above) into Dirichlet BCs for the (fluctuating) field $\phi(\boldsymbol{x})$ such that the mean value $\langle\phi(\boldsymbol{x})\rangle$ vanishes upon approaching the boundary with a universal behavior, which requires the introduction of a novel critical *surface* exponent β_1 in addition to the bulk ones. If, instead, $h_i \neq 0$, the symmetry-breaking at the boundary implies, at bulk criticality and at the renormalization-group fixed point, an enhancement of $\langle\phi(\boldsymbol{x})\rangle$ of the form $\langle\phi(\boldsymbol{x})\rangle \sim x_\perp^{-\beta/\nu}$ upon approaching the boundary, where here $x_\perp$ is the spatial distance from the nearest boundary. The sign of h_i equals the sign of the enhancement near the surface and thus renders the $(+)$ or the $(-)$ effective BC. Depending on the choices made for $c_{1,2}$ and $h_{1,2}$, it is possible to realize various combinations of effective BCs at the two surfaces of the film.

As mentioned above, the functional in Eq. (4) and generalizations thereof (e.g., accounting for an n-component order parameter, which requires some care in specifying the boundary terms) form the basis for a number of theoretical investigations of CCFs of either analytic or numerical character (see the following).

3.1.1. *Mean-field approximation*

Concerning the analysis of the free energy Ω_V of a system, described by $\mathscr{H}[\phi]$ and confined to a finite volume V, the simplest corresponding approximation consists of neglecting the statistical fluctuations in calculating the partition function $\int[\mathrm{d}\phi]\mathrm{e}^{-\mathscr{H}[\phi]}$. This implies that the free energy Ω_V is then given by $\Omega_V = \mathscr{H}[\bar{\phi}]$, where $\bar{\phi}(\boldsymbol{x})$ is the L-dependent configuration which minimizes $\mathscr{H}[\phi]$ and which satisfies the BCs at the confining walls $\mathscr{A}_{1,2}$. The dependence on L of this mean-field free energy Ω_V of the confined system provides access to the mean-field CCF and the corresponding scaling function $X_{\mathrm{ex}}(L/\xi)$ according to Eqs. (1) and (2), in which — consistent with the mean-field approximation — one has $\xi \propto |r|^{-1/2}$. This approach requires the knowledge of the OP configuration $\bar{\phi}(\boldsymbol{x})$ in the whole film in order to be able to determine the critical Casimir pressure on the confining walls, i.e., on a well-localized part. In certain cases (involving both analytical and numerical calculations), this may constitute a serious drawback which can be circumvented by resorting to the notion of the stress tensor. In fact, the stress tensor $T_{ij}(\boldsymbol{x})$ (with i and j running over the d components of the position vector) is a suitable space-dependent quantity which is constructed on the basis of the (fluctuating) OP field $\phi(\boldsymbol{x})$ and its spatial derivatives, such that the expectation value of its components renders the pressure locally exerted at position $\boldsymbol{x}$ on the boundary. In fact, the corresponding force $\mathrm{d}\mathbf{f}$ acting on a small surface element with lateral extension $\mathrm{d}A$, centered at point $\boldsymbol{x}$, and belonging to the boundary of the system, is given by

$$\mathrm{d}f_i = \mathrm{d}A \sum_{j=1}^{d} \langle T_{ij}(\boldsymbol{x})\rangle \hat{n}_j, \tag{5}$$

where $\hat{\mathbf{n}}$ is the unit normal vector pointing toward the portion of space occupied by the field. Accordingly, in order to obtain the pressure locally exerted around a certain point at the boundary, it is sufficient to know $\phi(\boldsymbol{x})$ (and its fluctuations) right at that boundary. In addition, under thermal equilibrium conditions, mechanical equilibrium within the system requires that the stress tensor is conserved and therefore $\langle \partial_j T_{ij}\rangle = 0$. (This follows also formally from the fact that T_{ij} can be viewed as a conserved Noether current associated with the lateral translational invariance.) The explicit expression of T_{ij} in terms of $\phi(\boldsymbol{x})$ follows from Noether's theorem (see, e.g., Ref. 11). Within mean-field theory, it is given by $T_{ij} = [\delta\mathscr{H}/\delta(\partial_i\phi)]\partial_j\phi - \delta_{i,j}\mathscr{H}$ in terms of $\mathscr{H}$. With this, the expectation value in Eq. (5) can be calculated as for any local quantity. In particular, within the mean-field approximation discussed here, $\langle T_{ij}\rangle$ coincides with the value which T_{ij} assumes for the given mean-field profile $\bar{\phi}(\boldsymbol{x})$. The conservation of T_{ij} actually allows one to calculate the pressure on the boundaries of the film by considering the value which $\bar{\phi}(\boldsymbol{x})$ has on a suitably displaced boundary. This fact is useful especially for numerical calculations as well as analytic ones, whenever $\bar{\phi}(\boldsymbol{x})$ is singular upon approaching the boundaries (as in the case of $(\pm)$ BCs). In particular, the determination of the CCF for films via T_{ij} can be easily generalized to more complex geometries and topographically or chemically inhomogeneous surfaces, which will be discussed in Section 4.2.

The expression of $X_{\mathrm{ex}}(L/\xi)$, derived as discussed above within the mean-field approximation, generically depends on the values of the bulk coupling constant u and the surface couplings (c_i, h_i), which appear in Eq. (4) and which are left undetermined within the mean-field approximation. Accordingly, in order to derive a quantitative estimate of X_{ex} to be compared with MC simulation or experimental data, two strategies are possible: (i) improve the mean-field estimate by tuning the various constants to their renormalization-group fixed-point values, which are independently known from the studies of bulk and surface criticality (see, e.g., Refs. 27 and 33); however, this approach generically renders inaccurate values.

(ii) Alternatively, one can normalize the mean-field prediction for $X_{\rm ex}(L/\xi)$ with the critical Casimir amplitude $X_{\rm ex}(0)$ and use this *ratio* in order to estimate the scaling function reduced this way.

Among the available mean-field predictions (see Ref. 15 for a comprehensive review) obtained as described above, we briefly discuss here those which turned out to be relevant for a qualitative understanding of the CCF in experimentally accessible systems, such as critical binary liquid mixtures and superfluids.

In particular, in Ref. 34, the effective Hamiltonian in Eq. (4) was considered for films of thickness L with $(+,+)$ and $(+,-)$ BCs. This is relevant for describing the fluctuation-induced force arising close to the consolute point of a binary liquid mixture. One of the two corresponding species might be preferentially adsorbed at the surfaces of the film. An analytic expression in terms of Jacobi elliptic functions was derived for the mean-field profile $\bar{\phi}(x_\perp)$ as a function of the coordinate $x_\perp$ across the film and of the reduced temperature r. These expressions were used to calculate the value of T_{ij} at the confining walls of the film and thus the CCF, which acts on each of them. The important qualitative features of the resulting scaling function $X_{\rm Cas}$ — confirmed by quantitative, accurate estimates based on MC simulations and by exact results in spatial dimension $d = 2$ (see, e.g., Ref. 15) — are the following: (a) for $(\pm,\pm)$ BCs, the force is *attractive* for all (positive and negative) values of the scaling variable x (see the end of Section 2.2) and it attains its maximum strength below T_c; (b) for $(\pm,\mp)$ BCs, the force is *repulsive* for all values of the scaling variable x and it attains its maximum strength above T_c; (c) for a fixed value of x, the attraction for $(\pm,\pm)$ BCs is generically weaker than the repulsion for $(\pm,\mp)$ BCs. In all cases, the force vanishes exponentially for $L \gg \xi$. Here, we have assumed that the fluctuating medium in the bulk is in its disordered phase for temperatures $T > T_c$, although also the opposite case can apply to binary liquid mixtures with a lower critical point. As discussed in Section 4.2, in the case of chemically patterned substrates or of weak surface fields $h_{1,2}$ in Eq. (4), within this mean-field approximation, the CCF loses some of these qualitative features.

The effective Hamiltonian in Eq. (4), generalized to a two-component OP $\phi(x)$, belongs to the same universality class as the XY lattice model, and therefore, it describes the transition from normal to superfluid ^{4}He. In this case, as anticipated above, the natural BCs at the surfaces of the film are Dirichlet ones because of the vanishing of the superfluid OP outside the film. In the actual realization of the film, as discussed in detail in Chapter 15, the fluid film is actually confined between a metallic substrate and its vapor, within which the OP vanishes. The mean-field analyses of this effective Hamiltonian — which was done both within the field-theoretical approach discussed here[35,36] and by considering suitable lattice models[36] — revealed that the resulting mean-field CCF takes the scaling form in Eq. (3) with the appropriate mean-field critical exponents. The scaling function exhibits the following qualitative features: (a) it vanishes above the bulk superfluid transition temperature T_λ, i.e., for $x > 0$ (see the end of Section 2.2 with $T_c \mapsto T_\lambda$),

(b) it becomes attractive for $x < 0$, i.e., $X_{\mathrm{Cas}} < 0$, (c) upon decreasing $x < 0$ (i.e., reducing ξ) X_{Cas} develops a sharp and pronounced minimum, (d) followed by an exponential decay as a function of $|x|^{\nu} \sim L/\xi$. Some of these qualitative features are confirmed by MC simulations (see Section 3.2) and are borne out experimentally (see Chapter 15). The presence of a pronounced minimum belongs to those features of X_{Cas} which are valid beyond mean-field theory, while the fluctuations cause X_{Cas} to become non-zero for $T > T_{\lambda}$ and to decay exponentially upon increasing $T \gg T_{\lambda}$, whereas it attains a non-zero value[35] for $T \ll T_{\lambda}$ with, *inter alia*, a contribution due to capillary-wavelike surface fluctuations[37] if the film is bounded on one side by a fluid–vapor interface.

While typically the mean-field predictions mentioned above are modified by statistical fluctuations below the upper critical dimensionality $d_c = 4$ of the transition under consideration, there are experimentally relevant instances in which this is not the case and the mean-field predictions turn out to be (up to logarithmic corrections) quantitatively reliable. This holds, e.g., for the tricritical end point,[32] which emerges in the phase diagram of ^{3}He–^{4}He mixtures when the line of the continuous normal-to-superfluid transitions joins the critical point of demixing at the top of the two-phase coexistence region between a ^{4}He-rich superfluid phase and a ^{3}He-rich normal phase. As discussed also in Chapter 15, the CCFs, which arise at this tricritical point, were investigated experimentally,[38–45] leading also to a quantitative determination[41] of X_{Cas}. (As it will be mentioned in the following, X_{Cas} near the tricritical point actually depends on two scaling variables.) Interestingly enough, the force turned out to be *repulsive*, a fact which was theoretically rationalized via a suitable microscopic model[20] (the vectoralized Blume–Emery–Griffiths model) in terms of the emergence of $(O, +)$ BCs for that specific experimental setting. In order to describe the tricritical behavior, the bulk contribution in Eq. (4) has to be modified as to include the term $v(\phi^2)^3/6!$, where the mean-field tricritical point is accessed by a suitable tuning, such that $u = 0$. (This is physically achieved by tuning the ^{4}He–^{3}He concentration to its critical value.) Correspondingly, the tricritical Casimir force can be calculated as explained above within the mean-field approximation,[20,36] yielding predictions for X_{Cas} in very good agreement with the available experimental data.

In addition to the functional in Eq. (4), one can also consider effective functionals to be minimized (as in the case of the mean-field approach described above) which are constructed in such a way that the values of the bulk critical exponents are close to those determined numerically in the physically relevant spatial dimension $d = 3$. This approach, introduced by M.E. Fisher and collaborators in Refs. 46 and 47, can also be successfully used in order to estimate X_{ex} as done, e.g., in Refs. 48–50 for the Ising universality class in a slab in $d = 3$ with symmetry-breaking BCs.

3.1.2. *Effects of fluctuations*

The approaches discussed above, i.e., determining either the finite-size free energy or the expectation value of the stress tensor at a suitable surface, can be readily generalized to the case in which statistical fluctuations around $\bar{\phi}(\boldsymbol{x})$ are considered. In both cases, accounting for fluctuations requires a suitable field-theoretical renormalization of the original effective Hamiltonian $\mathscr{H}[\phi]$ which, however, proceeds as in the bulk, i.e., in the absence of confinement.[51]

Within this field-theoretical framework (as well as in the cases of exact solutions of lattice models and the corresponding MC simulations), it is actually rather natural to consider this way also BCs, such as, *inter alia*, periodic and antiperiodic ones, or to analyze surface multicriticality. However, these specific characteristics are of limited interest for the description of directly accessible experimental systems. Yet, these kinds of studies are highly valuable in order to rationalize the results of certain pertinent Monte Carlo simulations[52] of lattice models and for highlighting possible limitations[53,54] of some of the analytical approaches mentioned above. These systems allow one also to study crossover phenomena[55,56] and the role of surface universality classes in determining the attractive versus the repulsive character of the associated CCF. These field-theoretical investigations extend in several respects certain exact results available in $d = 1$[15,57] and $d = 2$[15,57–62] for various bulk universality classes. These results are based either on the exact solution of suitable lattice models or on the consequences of conformal invariance at the critical point.

Rather generically, if the bulk symmetry of the unconfined model is broken explicitly by the BCs (as, e.g., in the case of surfaces with ($\pm$) BCs and zero bulk field) or if the temperature is within the range of values for which such a symmetry is spontaneously broken, a non-trivial mean-field OP profile $\bar{\phi}(\boldsymbol{x})$ emerges. In this case, analytical progress beyond mean-field theory[34–36] is significantly impeded and requires a considerable effort.[34] Accordingly, except for those cases in which the mean-field or Gaussian approximations are quantitatively reliable, such as at the tricritical point of ^{4}He–^{3}He mixtures[36] discussed above or for understanding the low-temperature fluctuations[37] of ^{4}He, a quantitatively reliable analysis of the experimentally relevant case of confined systems in $d = 3$ has to rely on MC simulations of suitable, so-called improved, lattice models (see Section 3.2).

As an illustration of a field-theoretical calculation of X_{Cas} in an experimentally relevant context, we consider the case of the superfluid transition described by Eq. (4) with a two-component OP and with (O, O) BCs. As mentioned in Section 3.1.1, the mean-field scaling function X_{Cas} vanishes for $T > T_\lambda$. Therefore, the contribution to the force is determined only by statistical fluctuations beyond the mean-field approximation. The corresponding prediction can be read off by specializing the expressions presented in Ref. 51. It follows from the calculation of the renormalized one-loop free energy for a variety of pairs of BCs with $\bar{\phi} = 0$ (i.e., periodic, antiperiodic, Dirichlet, Neumann). This accounts for Gaussian fluctuations. In particular, it turns out that (see Eq. (6.6) in Ref. 51)

$$X_{\mathrm{Cas}}^{(O,O)}(y_+) = -\frac{2}{(4\pi)^{(d-1)/2}} \frac{y_+^d}{\Gamma((d+1)/2)} \int_1^\infty \mathrm{d}x \, \frac{(x^2-1)^{(d-1)/2}}{\mathrm{e}^{2xy_+} - 1}, \tag{6}$$

in terms of the scaling variable $y_+ = L/\xi_+$, where ξ_+ is the so-called second-moment correlation length above the critical temperature, while d is the spatial dimension of the confined system. This one-loop result is exact for $d = 4$ because the only relevant fluctuations are the Gaussian ones considered here. The value of the scaling function at the critical point, i.e., for $y_+ = 0$, is the so-called *critical Casimir amplitude* $\Delta^{(O,O)}$, which was calculated in Ref. 51 up to two-loop order (see Eq. (5.16) therein) where the lowest-order perturbative correction $\propto u$ was consistently evaluated at the Wilson–Fisher fixed-point $u = u^*$ for the present case of a two-component OP with $u^* = 3\varepsilon/10 + O(\varepsilon^2)$; $\varepsilon = 4 - d$ is the deviation from the upper critical dimensionality $d_u = 4$. The scaling function in Eq. (6) corresponds to an attractive force, the modulus of which decreases from the value $|\Delta^{(O,O)}|$ at $y_+ = 0$ and which vanishes exponentially upon increasing y_+. Interestingly, universality implies that the scaling function in Eq. (6), calculated within the Gaussian approximation of Eq. (4), is actually the same[63] as the one of the CCF which arises in a confined ideal Bose gas in the grand canonical ensemble upon approaching the Bose–Einstein condensation point for vanishing chemical potential $\mu = 0$; similarly, the same applies in the presence of interactions. This can be shown explicitly by a direct analytic calculation[64] of the grand canonical potential of the confined gas, based on elementary statistical mechanics[65] (i.e., without resorting to the field-theoretical model in Eq. (4)) and by taking into account the quantization of the free particle wavevectors due to the Dirichlet BCs at the walls. From this grand potential, one can infer the finite-size contribution according to Eq. (1) and thus the scaling function of the CCF in terms of the correlation length $\xi_+(\mu \to 0^-) \propto (-\mu)^{-1/2}$. Finally, we mention that significant analytical insight into the emergence of CCFs in confined statistical systems has been provided by the solution of the exactly solvable models reviewed in Ref. 15, e.g., the spherical model[66–70] with various BCs.

3.2. *Numerical approaches*

As anticipated above, exact predictions for the scaling function of the CCF are available in $d = 2$ (based on the solutions of either lattice models or field theories due to conformal invariance as well as for various BCs and geometries of the boundaries) or in general dimensions d for simplified models which actually do not belong to the universality classes of experimentally accessible systems. Instead, predictions for the corresponding universality classes in $d = 3$ can be obtained via a perturbative expansion (as discussed in Section 3.1), a derivative expansion,[71,72] or within effective models,[48–50] which necessarily involve approximations of various nature. This implies that, in general, *quantitative* predictions for X_{Cas} are beyond the reach of these approaches, while they can be obtained efficiently by MC simulations of lattice models which belong to the universality classes of interest or, to a lesser extent, by molecular dynamics simulations.

In particular, X_{Cas} for the demixing transition of a classical binary liquid mixture can be determined by simulating a three-dimensional slab of the lattice Ising model (or improvements thereof, see the following) with fixed spins at the boundaries, So as to realize the physically motivated symmetric $(\pm,\pm)$ or antisymmetric $(\mp,\pm)$ BCs.[73–76] Similarly, for the superfluid transition in ^{4}He, X_{Cas} can be conveniently determined by simulating a slab of the XY model (or improvements thereof, see the following) with free BCs, which realize (O,O) BCs.[73,74,77–80]

However, extracting the CCF from MC simulations requires the development of suitable approaches. In fact, simulations of lattice models in thermal equilibrium are particularly efficient for determining the expectations values of *local* quantities, which can be expressed in terms of a small number of local degrees of freedom on the lattice. Unfortunately, the CCF acting on a surface cannot be expressed directly in such a form, given that the natural observable for achieving this, i.e., the stress tensor discussed in Section 3.1.1 for a field theory, cannot be generalized to lattice systems (and discrete degrees of freedom) apart from very special cases.[81]

Accordingly, in order to determine the CCF, one has to first calculate numerically the free energy $\Omega(\beta, L, A)$ of a film of transverse area A, thickness L, and in equilibrium at temperature β^{-1}. (Compared to the notation introduced in Section 2.2 for the finite-size free energy Ω_V, we omit hereafter the indication of the subscript V in order the simplify the notation.) Then, from the dependence of $\Omega(\beta, L, A)$ on L, one infers the force according to Eqs. (1) and (3). With this approach, one still faces the problem that $\Omega(\beta, L, A)$ cannot be directly expressed as the statistical average of a suitable local quantity, which lends itself to numerical simulation. However, its variation $\Delta\Omega$ as a function of L (e.g., $\Delta\Omega(\beta, L, A) = \Omega(\beta, L, A) - \Omega(\beta, L-1, A)$ or analogous, discrete estimates of the derivative with respect to L, measured in units of the lattice spacing), can be actually expressed[73,77,78,82] as an integral over an auxiliary, dimensionless parameter λ which the statistical average of a suitable local quantity depends upon. In particular, depending on the physical significance of λ and of that local quantity, three general strategies have been considered:

(i) $\lambda \in [0, 1]$ controls the continuous interpolation between the Hamiltonians H_1 and H_2 of the system consisting, respectively, of the model in the film geometry with thickness L and of the model in the film with thickness $L-1$ plus an isolated layer of "thickness" 1 (see, e.g., Fig. 1 of Ref. 73), such that both H_1 and H_2 involve the same number of degrees of freedom. The free energy difference $\Delta\Omega$ can then be expressed as $\int_0^1 \mathrm{d}\lambda \langle H_2 - H_1\rangle_{H_{\text{cr}}(\lambda)}$, where the expectation value $\langle \cdots \rangle_{H_{\text{cr}}(\lambda)}$ is the canonical one at the temperature β^{-1} with the interpolating crossover Hamiltonian $H_{\rm cr}(\lambda) = (1-\lambda)H_1+\lambda H_2$. This strategy was introduced in Ref. 73 and used in order to investigate various bulk and surface universality classes,[74,83] also in the presence of variable[84] or random[85,86] surface fields. It was also extended in order to calculate the CCF in the presence of bulk ordering fields.[87] This approach was also used for investigating the CCF emerging with patterned substrates[88–91] as well as with a more complex shape of one of the confining surfaces,[92,93] as discussed further in the following. Alternatively, a similar approach consists of considering an interpolation between H_1 and H_2 which is more sophisticated compared to $H_{\rm cr}(\lambda)$. For example, one starts from H_1 corresponding to a film of thickness $L-1$ and adds one by one the degrees of freedom which are necessary in order to add a complete layer in order to form the film of thickness $L-1$ (see Ref. 78). In this case, $\Delta\Omega$ can be expressed as $\sum_{i=1}^{N} \ln\langle\exp[-\beta(H^{(i+1)} - H^{(i)})]\rangle_{H^{(i+1)}}$ where $\{H^{(1)}, H^{(2)}, \ldots, H^{(N)}\}$ is the sequence of Hamiltonians involved in the interpolation, and the canonical expectation value $\langle\cdots\rangle_{H^{(i+1)}}$ is taken with the Hamiltonian $H^{(i+1)}$ (see Ref. 78 for further details).

(ii) In the second strategy, λ is identified with the inverse temperature β, and the thermodynamic relationship[65] $\mathscr{U} = \partial(\beta\Omega)/\partial\beta$ between the (finite-size) free energy Ω and the (finite-size) internal energy $\mathscr{U}$ is used in order to calculate $\Delta\Omega$. (Here, it is assumed that the remaining thermodynamic parameters of the system are kept constant.) In fact, based on the numerical determination of $\Delta\mathscr{U} \equiv \mathscr{U}(\beta, L, A) - \mathscr{U}(\beta, L-1, A)$ for a suitable set of values of temperatures, it is possible to provide an accurate numerical estimate of the integral of $\Delta\mathscr{U}$ over β, which is required in order to invert the previous relationship between $\mathscr{U}$ and Ω and therefore between $\Delta\mathscr{U}$ and $\Delta\Omega$. In turn, $\Delta\mathscr{U}$ and $\mathscr{U}$, from which $\Delta\Omega$ is derived, can be directly expressed as the canonical expectation value of the Hamiltonian H_1 of the lattice system, which is amenable to MC simulations. This approach was proposed in Ref. 77 and applied for studying various bulk and surface universality classes[75,79,80,94] as well as crossover phenomena in the latter.[56] Similarly, this approach was also used beyond the film geometry.[95,96]

(iii) In a magnetic model (e.g., the Ising model as representative of the corresponding universality class), λ is identified with the external magnetic field h. The free energy $\Omega(\beta, h, L, A)$ of a film of thickness L, transverse surface area A, at inverse temperature β, and in the presence of such a non-vanishing field h, is obtained from that $\Omega(\beta, h_0, L, A)$ of a reference state with field h_0 via thermodynamic integration: $\Omega(\beta, h, L, A) = \Omega(\beta, h_0, L, A) + \int_{h_0}^{h} {\rm d}h' M(\beta, h', L, A)$, where $M(\beta, h', L, A)$ is the magnetization of the system which is amenable to Monte Carlo determination. The value of $|h_0|$ can be chosen to be sufficiently large so that the corresponding correlation length is small, whereby $\Omega(\beta, h_0, L, A)$ coincides with the free energy which a volume $L \times A$ of the system would have in the *bulk*. Thus, the critical Casimir contribution is encoded in the integral, together with a surface contribution. Eventually, the scaling function of the free energy can be extracted from the values of a suitable estimate of these integrals. This approach was first used in Refs. 82 and 97 for obtaining the scaling function of the CCF also in the presence of a bulk field, both in the film geometry[82] or for two spheres[97] (see the following).

An alternative approach for determining the CCF, which does not require a computation of $\Delta\Omega$ and which is, in spirit, close to what is pursued in certain experimental settings (see Section 5), was proposed and used in Ref. 98. In particular, one considers a wall in $d = 2$ (i.e., a line of the lattice where the degrees of freedom are fixed according to the BCs) which is immersed in a critical medium confined to a strip, with boundaries parallel to the wall and certain BCs. The wall separates the systems into two adjacent sub-strips to which the medium is confined. Therefore, the wall is subject to a bulk pressure and a CCF from both sides. The rigid wall is assumed to be mobile, such that its position h across the strip is allowed to fluctuate. Accordingly, the MC simulation of the whole system includes also moves in which the wall interchanges its position with one of the two neighboring lines of degrees of freedom (such as the spins for the Ising model considered in Ref. 98), with suitable flipping rates which satisfy detailed balance. As a consequence, the statistical distribution $P(h)$ of h, which can be measured numerically, is determined by the potential $\Phi(h)$ of the total force acting on the wall via the relation $P(h) \propto \exp\{-\beta\Phi(h)\}$; thus, $\Phi(h)$ follows from $P(h)$. The latter contains the contribution of the CCF, which the medium confined to each sub-strip exerts on the wall, and, in principle, it allows one to infer the scaling function of the force.[98] This approach was also extended to the case involving a spherical particle.[98]

The numerical determination of X_{Cas} is also affected by finite-size corrections to the leading finite-size scaling behavior of various origin. Accordingly, it is important to devise strategies which reduce them. In this respect, it is particularly convenient to focus on lattice models, which belong to the universality class of interest, but which have been modified by the introduction of irrelevant terms in their Hamiltonian, which allow one to choose them so as to minimize corrections to scaling and thus corrections to finite-size scaling. These so-called improved models — widely used for an accurate determination of bulk critical exponents[27] — have been also used to study the CCF for the Ising universality class, represented by a suitable Blume–Capel model,[56,75,86,88–91,94,96] and the XY universality class, represented by Eq. (4) for a two-component OP and a suitable value[79,80] u^* of the coupling constant u.

Concerning the Ising universality class, an alternative to the various approaches mentioned above consists of performing a molecular dynamic simulation of a model binary fluid confined within a slab and of determining the pressure which the fluid exerts on the confining walls in the semi-grand-canonical ensemble.[99] Upon approaching the second-order demixing transition, one observes the emergence of a contribution to the pressure on the walls, which is due to the CCF and which, in principle, can be used to extract the corresponding scaling function. However, in comparison with MC simulations, this method still suffers from sizeable finite-size corrections because the system sizes which can be actually simulated are rather limited by the high computational effort. Similarly, the number of possible BCs which can be realized is reduced. However, this approach, if developed further, could provide access to the non-equilibrium properties of the CCF (see Section 6).

The CCF can be studied, *inter alia*, also numerically by exploiting the well-known relationship between the equilibrium behavior of quantum and classical statistical models.[72] This connection implies that a two-dimensional *quantum* system with surface area A in thermal equilibrium and at the inverse temperature β has the same partition function (and thus the same statistical properties) as the corresponding *classical* model in a three-dimensional system of size $A \times \beta$ and periodic BCs along the additional dimension of linear extension β. The temperature T of the corresponding classical system is determined solely by the parameters entering the Hamiltonian of the two-dimensional quantum system and is not related to the temperature β^{-1} of the original quantum system. Accordingly, reversing the argument, the free energy Ω (derived from the partition function) of a three-dimensional classical film of thickness L (and transverse area A) depends on L in the same way as the free energy of the corresponding two-dimensional quantum system (of area A) does on the inverse temperature $\beta = L$. (For simplicity, here we use the units, such that $h = c = 1$.) The CCF can then be extracted from the thermodynamic relationship $\partial\Omega/\partial L = T\,\mathscr{U}_q(\beta = L)$, where $\mathscr{U}_q(\beta)$ is the internal energy of the two-dimensional *q*uantum system in equilibrium and at the inverse temperature β. In turn, the latter can be determined via a variety of approaches developed for studying quantum systems and quantum phase transitions, encompassing analytical and numerical approaches. From the perspective of numerical studies, which we are concerned with in this section, we mention the possibility to use quantum Monte Carlo methods in $d = 2$, which turn out to confirm the results of classical MC simulations for Ising films in $d = 3$.[72] However, beyond its conceptual appeal, this approach seems to be limited to the case of periodic BCs at the surfaces of the film, which unfortunately lacks experimental relevance.

Beyond the homogeneous film geometry: In the above overview of the numerical investigation of CCFs, we have focused primarily on the film geometry with homogeneous preferential adsorption along its long axis. However, as we shall discuss in Section 4, experimental settings often involve curved surfaces (such as those of colloidal particles) and chemically patterned substrates. These cases have also been studied numerically by using the various approaches mentioned above, possibly accompanied by the development of tailored algorithms.

In particular, the case in which one of the two confining surfaces of the film is chemically patterned with a single chemical step[91] or with a periodic pattern of alternating adsorption preferences[88–90] (realizing alternatively $(+)$ and $(-)$ BCs) was considered, with the other surfaces imposing $(+)$ or O BCs. For an isolated step, the line contribution to the CCF was determined as the necessary information needed to investigate a pattern with well-separated steps. For a generic pattern, in addition to the scaling variable $y \equiv L/\xi$ in Eq. (3), or equivalently to x (see end of Section 2.2), the scaling function X_{Cas} depends additionally on scaling variables related to the patterning, i.e., $S_{\pm}/L$ where S_+ and S_- are the widths of the stripes with (+) and (−) BC, respectively (with $S_+ = S_-$ in Refs. 88–90). Depending on the BCs at the homogeneous surfaces, a change of the force from attractive to repulsive may occur upon varying x with the other scaling variables kept fixed. If, instead, the medium is confined by a surface with a certain preferential adsorption, say (+), and another one with a quenched, random local adsorption preference which can be either (+) or (−), one observes a crossover between an attractive case $(+,+)$ and a repulsive one $(+,-)$, with a crossover function which depends on the disorder strength via a suitable scaling variable.[86,100]

The numerical investigation of CCFs beyond the film geometry can be carried out on the basis of the approaches mentioned above, although one might need to generalize standard algorithms, such as the cluster algorithm,[95,96,101] or to consider lattices with adaptive spacing.[102] In particular, the various geometrical arrangements, which were investigated numerically, include in $d = 2$, circular particles[98] or elongated ones ("needle") within a strip,[93] cubic particles within a film in $d = 3$,[92] spheres close to a plane,[95] two spheres[97] in the bulk of the system, and more circles in $d = 2$.[101] All aforementioned studies focused on the Ising universality class with (±) BCs at the involved surfaces.

4. From Films to Colloids

As discussed in the previous section, the emergence of CCFs was originally studied both theoretically and experimentally (see, c.f., Section 5 and Chapter 15) in the film geometry. However, in view of possible applications to soft matter as well as for a direct measurement of these forces, it turned out to be useful to consider geometries involving spherical particles, such as colloids: In fact, the thermal energy scale, which controls the strength of CCFs, is the same as one of the interactions which are typically at play in soft matter; accordingly, they are expected to have a measurable impact on the observed behavior of, e.g., colloidal suspensions.

The change of the geometry of the confining surfaces poses significant challenges for the theoretical study of the critical Casimir effect via the analytical approaches discussed in Section 3.1. Hardly any analytic progress can be made as soon as the BCs are imposed on non-planar surfaces, and the geometry is more complex than that of films. Within mean-field theory (see Section 3.1.1), some advances are still possible concerning the numerical minimization of the functional in Eq. (4) with modified boundaries and the determination of the CCF via the stress tensor. This was carried out, for example, in the case of a spherical[103] or ellipsoidal[104] particle approaching a flat and homogeneous surface, as well as in the cases of films in which one or both confining surfaces have a sawtooth cross-section[105] or one is a crenelated surface.[106] Similarly, within the mean-field approximation, it was possible to study the case in which a cylindrical colloidal particle approaches the soft and therefore responsive interface separating the two liquid phases in one of which the colloid is immersed.[107] As an exception, analytical and reliable predictions can also be formulated beyond the film geometry for systems which are exactly at their critical point (i.e., formally $\xi = \infty$). For such systems, conformal invariance (see, e.g., Ref. 59) allows one to map geometries with non-planar boundaries (specifically spheres) into the film one. This enables one to calculate analytically the forces acting on the former in terms of those acting on films.[108] In $d = 2$, this approach can be extended further to boundaries with a modulated adsorption preference,[109] which are also discussed, c.f., in Section 4.2. As we mentioned at the end of Section 3.2, the CCF acting on non-planar boundaries can be studied via MC simulations. Nonetheless, the complexity of the problem rises due to the increased number of scaling variables the scaling function eventually depends on.

In fact, in addition to the mesoscopic lengthscales which are relevant in the film geometry, i.e., the correlation length ξ and the surface-to-surface distance L (i.e., the film thickness), the presence of a spherical particle introduces its radius R as the sole parameter describing the geometry of the surface. Accordingly, one generically expects that the CCF F_{Cas}, which acts on a sphere and which is due to the effective confinement of the medium when the sphere approaches a plane at a closest distance L, depends additionally on the scaling variable $\Delta \equiv L/R$, such that Eq. (3) generalizes to[103]

$$F_{\text{Cas}} = \frac{k_B T}{R} X_{\circ|}(y = L/\xi, \Delta = L/R). \tag{7}$$

In the case, for example, of two spherical colloidal particles with different radii R_1 and R_2, the CCF will have a scaling form as the one above, but with an additional dependence on the dimensionless scaling variable R_1/R_2. An analogous extension is expected, e.g., when the particle has a spheroidal shape and it approaches a planar surface or another (spheroidal) particle.

The scaling function $X_{\circ|}$ of two variables, which appears in Eq. (7), as well as the dependence of the scaling function X_{Cas} in Eq. (3) on the additional scaling variables related to possible patterns of the film surfaces (see the end of Section 3.2), can in principle both be determined by numerical simulations of suitable lattice models, as anticipated in Section 3.2. However, this might be computationally demanding due to the increased number of variables involved. Accordingly, it is often useful to consider first some relevant limits. In particular, the case of small particles, formally corresponding to the limit $\Delta \to \infty$, can be addressed, on the basis of the knowledge of the critical behavior in the presence of boundaries,[18,33] via the so-called *small-sphere expansion*.[103] This approach can also be extended to non-spherical small particles such as dumbbells and needles[93,110] and also to chemically inhomogeneous (Janus) particles in $d = 2$.[111] The opposite limit $\Delta \to 0$, instead, can be addressed on the basis of the available information concerning the CCF acting between parallel plates. This is accomplished by using the so-called *Derjaguin approximation*, also known as *proximity force approximation*,[5,6,112] which assumes pairwise additivity of the fluctuation-induced forces. Since most of the available experimental data fulfil the limit $\Delta \ll 1$, here we focus primarily on this case.

4.1. *Pairwise additivity*

Within the Derjaguin approximation, two non-planar surfaces, which approach each other, are completely decomposed into a set of pairs of opposing and parallel surface elements of infinitesimal area $\mathrm{d}A$; the two elements of such a pair belonging to the two different surfaces are separated by a distance $L(\mathrm{d}A)$. Each of these pairs is assumed to provide a contribution $\mathrm{d}F$ to the force. The latter is the CCF acting on a surface of extension $\mathrm{d}A$, which belongs to the confining surfaces of a film of thickness $L(\mathrm{d}A)$, given by Eq. (3). The total force F_{Cas} is calculated as the integral over all these contributions under the assumption of pairwise additivity. Accordingly, the knowledge of the scaling function X_{Cas} for two macroscopic parallel plates allows one to predict the force acting in the presence of non-planar boundaries, such as those formed by a spherical particle approaching a flat surface, another spherical particle, or a topographically patterned but chemically homogeneous substrate. As it will be discussed in more detail in Section 4.2, the Derjaguin approximation allows one to also study the cases of surfaces with spatially varying preferential adsorption as well as of non-spherical particles. In all these instances, one can take advantage of the quantitatively accurate determination of the CCF inside homogeneous films with the appropriate BCs, obtained, e.g., via MC simulations.

For the concrete and experimentally relevant case of a spherical particle approaching a planar surface, the Derjaguin approximation is expected to be valid when the radius R of the sphere is significantly larger than the distance L of closest approach between the particle and the surface. In Eq. (7), this corresponds to the limit $\Delta = R/L \to 0$. Within this approximation, it turns out that $X_{\circ|}(y, \Delta \to 0) \to \Delta^{-2} X_{\circ|}^{(\mathrm{DA})}(y)$, where $X_{\circ|}^{(\mathrm{DA})}(y)$ can be expressed in terms of a certain elementary integral of the scaling function $X_{\mathrm{Cas}}(y)$ for the CCF in the film geometry (see Eq. (3)). The potential $\Phi_{\mathrm{Cas}}(L)$ of the resulting CCF, which is typically the quantity accessible to experiments (see Section 5), turns out to be given, within the same approximation, by

$$\frac{\Phi_{\mathrm{Cas}}(L)}{k_{\mathrm{B}}T} = \frac{R}{L}\vartheta_{\circ|}(L/\xi). \tag{8}$$

The scaling function $\vartheta_{\circ|}$ follows from an integration of $X_{\circ|}^{(\mathrm{DA})}(y)$ and inherits the universal features of $X_{\circ|}(y, \Delta)$ (with $\Phi_{\mathrm{Cas}}(L \to \infty) = 0$). Analogously, in the case of spheroidal particles, which are characterized by a set $\{R_i\}$ of radii, one expects the Derjaguin approximation to become more accurate for sufficiently small ratios $\{\Delta_i \equiv R_i/L\}$. In particular, in Refs. 10 and 113, the Derjaguin approximation was used for predicting $\vartheta_{\circ|}(x)$ on the basis of the MC simulation data for the scaling function in the film geometry (see Section 3.2), which in turn was found to be in agreement with the corresponding experimental data. The qualitative features, which emerge in the case of homogeneous preferential adsorption at the sphere and at the planar surface of the type ($\pm$), are the same as those for the corresponding BCs in the film geometry. This implies that the force is attractive for equal BCs and repulsive otherwise, while a richer behavior is expected for colloids exposed to patterned substrates, which are discussed in Section 4.2.

The range of validity of the Derjaguin approximation can be tested in various ways, for example, by deriving, within the same approach, the scaling function of the CCF in the film geometry and the one in the sphere–plate geometry, $X_{\circ|}(y, \Delta)$. The former is then used to calculate $X_{\circ|}^{(\mathrm{DA})}(y)$, which is then compared with $\Delta^2 \times X_{\circ|}(y, \Delta)$ for decreasing values of Δ. This comparison, based on the results of MC simulations for the Ising universality class in $d = 3$,[95] shows both for $(+,+)$ and $(+,-)$ BCs that the Derjaguin approximation is actually accurate even up to $\Delta \lesssim 0.5$ for all values of the scaling variable $x = (T/T_c - 1)(L/\xi_0^+)^{1/\nu}$ in the case of $(+,+)$ BCs. Instead, for $(+,-)$ BCs, the accuracy rapidly deteriorates for $x < 0$. A similar comparison can be carried out within mean-field theory, i.e., by using the mean-field predictions for the scaling function in the sphere–plane geometry and those for the film geometry. This comparison was studied for an ellipsoidal particle approaching a plane,[104] where the ensuing *critical Casimir torque* for a non-spherical particle was investigated in detail. In particular, it turned out that sizeable deviations appear between the two approaches upon increasing the aspect ratio of the ellipsoidal particle. On the other hand, for the case of a spherical colloid and a substrate with chemical patterns, as discussed in the following section, the suitably extended[114] Derjaguin approximation within the mean-field approximation turned out to provide rather accurate predictions as long as the typical lengthscale of the patterns is larger than the geometric mean of the particle radius and its distance of closest approach to the substrate.[115] However, beyond the aforementioned cases, a detailed investigation of the validity of this approximation is still missing. The Derjaguin approximation was also used in order to predict the CCF acting within the film geometry with one crenelated surface[106] on the basis of the mean-field predictions for a film with two flat confining surfaces. A suitable modification of the approximation turned out to provide predictions which are in very good agreement with those of a direct numerical calculation within the mean-field approximation.

In passing, we mention that the results of the experimental investigations of the CCFs available in the literature so far (see Section 5) agree with the predictions derived on the basis of the Derjaguin approximation, which is therefore confirmed within the present experimental accuracy and range of experimental parameters.

4.2. *From homogeneous to patterned surfaces*

The possibility of altering locally the adsorption preference of a material via suitable surface chemical treatments, which, for example, turn it from being hydrophobic to hydrophilic, can be used in order to realize surfaces with varying adsorption preferences for the two components of a binary liquid mixture (one of which is typically water). Accordingly, the effective BCs for the corresponding OP ϕ can be spatially modulated between the two cases (+) and (−) discussed above, which realizes chemically patterned substrates. For the film geometry, involving two confining surfaces with a periodically alternating adsorption preference, the emergence of a *lateral* component of the CCF, in addition to the one perpendicular to the surfaces, has been studied within mean-field approximation[116] or MC simulations (see Section 3.2). Such a component also emerges in the experimentally relevant case in which colloidal particles with a certain adsorption preference are exposed to these substrates with spatially modulated adsorption preferences. As a consequence, colloidal particles, while diffusing, tend to localize at those portions of the surface which have the same adsorption preference while they are repelled from those parts of the surface with the opposite preference. Similarly, spherical colloids subject to inhomogeneous surface treatments and colloidal clusters obtained by assembling colloids with different adsorption preferences realize effectively (anisotropic) Janus particles, which are subject to a critical Casimir *torque*, even in the presence of a homogeneous substrate. By combining these various possibilities, one can actually control the spatial distribution and orientation of colloidal particles to a degree which can be reversibly controlled by temperature: Upon moving away from the critical point, these critical Casimir interactions become generically negligible and the corresponding spatially modulated force vanishes.

Motivated by actual experiments (see Section 5.1), the case of a single, homogeneous, and spherical colloid with a certain adsorption preference in front of a substrate, which exhibits a chemical step between the same and the opposite adsorption preference, was considered in Ref. 114. The force acting on the colloid, as a function of the position of its centre in this geometrical arrangement, was determined both within mean-field theory and by using the Derjaguin approximation discussed in Section 4.1. The latter takes into account that the various pairs of infinitesimal flat surfaces dA, into which the system is effectively decomposed, realize distinct BCs depending on which part of the substrate they belong to. In particular, the available Monte Carlo data for the force acting within the film geometry were used in this Derjaguin approximation. Due to the intrinsic additivity of this approximation, the previous step case can be used to predict the force also in the presence of an isolated, say (+), stripe in an otherwise (−) substrate, as well as in the case of periodic lateral patterns. As we also discuss in Section 5.1, in the former case, it turns out that the probability distribution of the position of a single colloid close to the striped substrate, which, *inter alia*, is due to the action of CCFs, is even able to reveal details of this stripe, such as possible irregular boundaries, which would be otherwise unaccessible.[114,117]

Rather generically, if a colloid is sufficiently close to a patterned substrate, their interaction is dominated by the contribution stemming from the area of the substrate and of the colloid located around the corresponding points of closest approach. Accordingly, it might be repulsive or attractive depending on the local adsorption preferences on the substrate, provided the colloid surface is chemically homogeneous. On the other hand, if the distance between the particle and the substrate increases, the force follows from an average of the contributions stemming from a larger area around the points of closest approach. Accordingly, its character might change from attractive to repulsive or vice versa. The fact that within the film geometry, the CCF is approximately four times stronger in the case of repulsion than in the case of attraction (see Section 3) actually helps to facilitate such a change of character of the force. For a suitably designed substrate, one is able to find a point of *stable levitation*. This one is controlled by the CCF only, and therefore, it is extremely sensitive to changes of temperature.[115] These investigations were reviewed in Ref. 7.

In the presence of non-spherical colloids, either homogeneous or inhomogeneous as far as the preferential adsorption is concerned, the interaction with the patterned substrate may involve the emergence of critical Casimir torques which tend to align the colloidal particle with possible stripes on the substrate which they are exposed to. The case of cylindrical homogeneous or Janus-like colloids was investigated in Ref. 118 using both the Derjaguin approximation and the mean-field theory. This demonstrates that, in principle, these critical Casimir interactions can be exploited in order to achieve a well-defined and reversible alignment of both chemically homogeneous and Janus cylinders.

The self-assembly and phase behavior of colloidal suspensions, which consist of patchy or Janus particles, are currently under intensive theoretical and experimental investigation. This calls for a deep understanding of the possible effective interactions at play in these systems. As illustrated above, CCFs stand out in this respect because they provide both attractive and repulsive, reversible, and temperature-controlled interactions, which can be engineered via surface chemical treatments. In particular, in Ref. 119, the interaction between a single Janus particle and a laterally homogeneous substrate or a substrate with a chemical step was studied within the mean-field approximation and a suitably modified Derjaguin approximation. Due to the pairwise additivity of this approximation, it can be used to calculate the effective force between two Janus cylinders and between two Janus spheres, depending on their relative distances and orientations. Similarly, the effective pair interaction between two identical spherical colloids — carrying a circular patch of arbitrary size on their surfaces, characterized by an adsorption preference different from that on the rest of the surface — can be predicted within the same approximation and can be compared with the available experimental results.[120] In addition, a generic relationship between the scaling function of the inter-particle potential and that of the critical Casimir torque emerges, even beyond the aforementioned specific case.[120]

4.3. *Many particles: Beyond pairwise additivity and many-body effects*

Effective interactions in physics are generically expected to display many-body effects: In the presence of many particles, the total force acting on one of them is *not* simply given by the sum of its pairwise interactions with the other particles in the system, contrary to what occurs, e.g., for the gravitational interaction or the electrostatic force in vacuum. Among the various effective interactions relevant for soft matter,[5] which *a priori* are not pairwise additive, also CCFs are expected to exhibit many-body effects. They are out of reach of any theoretical investigation based on the Derjaguin approximation discussed in Section 4.1. Thus, their theoretical studies, as currently available, rely primarily on either mean-field approaches[121,122] or numerical simulations.[95] (The experimental investigation of three-body effects is discussed in, c.f., Section 5.2.)

In particular, many-body effects can be extracted and quantified by subtracting from the force acting on a particle in a certain configuration, the force which would act on it assuming pairwise additivity, starting from the simplest case of three particles immersed in the medium. In the case of two spheres facing a planar and homogeneous flat substrate immersed in a near-critical binary liquid mixture, which are studied within the mean-field approximation, three-body effects appear both for the normal and the lateral CCF acting on one of the colloids.[121] Depending on the relative geometrical arrangement of the two colloids and the substrate, as well as on their BCs and the deviation from the critical temperature of the solvent, a number of interesting features emerge, including a change of sign of the force. As expected heuristically, the many-body contribution is quantitatively more relevant close to criticality. It increases upon decreasing the colloid–colloid and colloid–substrate distances, and it may reach 25% of the total force. In order to investigate the role of the geometric configuration of the involved bodies for the many-body effects, the case of three parallel cylindrical colloids was also considered,[122] with the focus on the change of the interaction between two of them due to the presence of a third colloid. Depending on the temperature and the relative positions of the three cylinders, the same interesting features emerged as those mentioned above. This indicates that they are of generic character.

As in the cases discussed in Section 3, a quantitative estimate of CCFs beyond mean-field theory can be obtained via MC simulations of representative models for the universality class of interest. Referring to the case of binary liquid mixtures, the Ising model was simulated in $d = 2$, where the colloids correspond to regions of the lattice with a fixed spin orientation. With a suitably developed cluster algorithm (see Section 3.2), the three-body critical Casimir interaction potential was determined.[101] In particular, the case is considered in which two particles (each with the shape of disks) touch each other. The potential is determined as a function of the position of the centre of the third particle. All three particles are of the same size.

This highlights the relevance of three-body effects and the large degree of non-additivity of these forces. In fact, it turns out that the interaction at sufficiently large distances between the two disks in contact, each of radius R, and the third disk is almost identical to the interaction between the third disk and an effective disk of radius R_{eff} in place of the first two. Interestingly enough, in $d = 2$ and at $T = T_c$, this kind of spatial configuration (with two disks at contact) is also amenable to an almost exact solution. This can be obtained by using conformal invariance and a suitable conformal transformation,[101] which confirms the numerical results of the corresponding MC simulations, predicting, *inter alia*, $R_{\text{eff}} = \pi R/2$.

Beyond the cases mentioned above, many-body effects have also been investigated in geometrical arrangements which might be relevant for predicting the behavior of colloidal dispersions under confinement. This setting consists of a pair of spherical colloids which are immersed in a medium confined by two parallel walls forming a narrow slit. This involves a total of four different bodies (two colloids plus two walls).[123] In the absence of many-body effects (and thus in the case of pairwise additivity) and if the colloids keep an equal distance from the walls, the component of the force acting on the colloids, which is parallel to the surfaces, would not be affected by the presence of the slit confinement. However, the analysis of this configuration within mean-field theory and MC simulations shows that the effective interactions are strongly non-additive and that the interaction between the two colloids under confinement can be a few times stronger than the corresponding sum of the effective pair potentials. This fact influences, *inter alia*, the phase behavior of confined colloids, which acquires a dependence on the slit width.

Many-body effects are expected to be relevant in dense colloidal dispersions or if aggregation occurs — also due to the critical Casimir interaction. The latter might play a role in also determining the structure of the resulting aggregates or the very phase diagram of the dispersion. This was numerically shown to be the case, e.g., in $d = 2$.[124] However, beyond what is discussed above and in Section 5.2, a detailed theoretical or experimental knowledge of the CCFs-induced many-body effects is still lacking. The available experimental data can often be rationalized qualitatively by assuming pairwise additivity of the critical Casimir interactions. In particular, various aspects of the phase behavior of colloidal dispersions in near-critical solvents, which result from the interplay between electrostatic inter-particle repulsion and the pairwise critical Casimir attraction (or resulting from the presence of electrolytes as solvents, see also Section 4.4), were studied experimentally in Refs. 125–130 and theoretically (analytically or numerically) in Refs. 125,131,132. This holds also for a binary mixture of colloids with opposite preferential adsorption.[133] A recent review of these studies is provided in Ref. 14, so that here we do not discuss them further. More recently, in view of possible applications, the theoretical and experimental studies of colloidal dispersions, which also interact via CCFs, have primarily focused on the possibility to harness these interactions. This is done so as to control the structures of colloids of various kinds resulting from aggregation[134] or from the deposition on a substrate,[135,136] e.g., quantum dots as in Ref. 134, patchy particles,[137] or colloidal assemblies in the form of "superballs."[138] The phase behavior and the structure of colloidal dispersions, in which an attractive CCF competes with a repulsive magnetic dipolar interaction, has also been explored theoretically[139,140] in $d = 2$.

4.4. *Critical Casimir effect in the presence of electrolytes*

The CCF discussed here, between, say, colloidal particles, is one the various effective interactions which might act in soft matter,[5] such as depletion, electrostatic, and van der Waals interactions.[6] Different from these other interactions, which occur rather generically, CCFs emerge as a consequence of the onset of a collective behavior at critical points. Accordingly, they are particularly relevant only in the neighborhood of certain points in the phase diagram of the solvent, although they might be continuously related to other solvation forces, such as depletion forces.[141,142] This is theoretically embodied by the fact that CCFs follow from the contribution Ω_V (see Eq. (1)) to the actual total effective free energy of the confined medium, which is a *non-analytic* function of the thermodynamic distance from the critical point. The other interactions, instead, contribute to the total effective free energy with terms which are characterized by a smooth dependence on temperature. Due to their rather different physical origins, the various effective interactions are largely independent of each other in the sense that they provide additive contributions to the effective interaction between, say, two colloidal particles. This is clearly an *approximation*: In fact, e.g., upon approaching the critical point and depending on the effective BCs, adsorption profiles might develop close to the confining surfaces, rendering the confined medium inhomogeneous. This inhomogeneity generically implies a spatial modulation of the frequency-dependent dielectric function $\varepsilon(\omega)$ of the medium at distances close to these surfaces. In turn, this might affect both the possible electrostatic and the van der Waals interactions between them,[6] which also depend on $\varepsilon(\omega)$ (see, e.g., the discussion in Ref. 113).

However, most of the experimental data available so far (see Sec. 5) could be interpreted theoretically by neglecting the interplay between these forces, with the sole exception of the experiment in Ref. 143, which involved a critical binary liquid mixture in the presence of a suitable concentration of electrolytes provided by certain salts. It was observed that the interaction between a charged substrate and a colloid resulting from the simultaneous action of an expected *repulsive* CCF and a *repulsive* electrostatic force might be, depending on the temperature of the medium, *attractive* within a certain range of distances. This unexpected observation called for a better theoretical understanding (i.e., beyond the Deby–Hückel theory[5]) of the electrostatic interaction occurring between surfaces immersed in a critical solvent and of the subtle coupling between the spatial distribution of the ions in the solvent and the OP of the binary liquid mixture. Within a simplified lattice model for such a mixture,[144] one considers the fact that the space-dependent number densities $a^{-3}\varphi(\boldsymbol{x})$ and $a^{-3}(1-\varphi(\boldsymbol{x}))$ of the two components of the solvent (taken, for simplicity, to be identical and of volume a^3) are generically coupled to the ionic number densities $a^{-3}\varrho_+(\boldsymbol{x})$ and $a^{-3}\varrho_-(\boldsymbol{x})$ of the cations and anions, respectively. In turn, this implies a coupling between the OP $\phi(\boldsymbol{x}) = \varphi(\boldsymbol{x}) - \varphi_c$ of the demixing phase transition of the solvent (occurring at the bulk critical density φ_c) and the ionic densities, which generates[145] potentials $V_\pm(\varphi(\boldsymbol{x}))$ acting on $\varrho_\pm(\boldsymbol{x})$. $V_\pm$ accounts for the difference of the bulk solvation free energies of the cations and anions in solvents with $\varphi = 1$ and $\varphi = 0$, i.e., in the two pure phases of the mixture (typically water and lutidine). This corresponds to the fact that the anions and cations of the salt have different solubilities in the two pure fluid phases formed by the binary liquid mixture. Accordingly, the spatial distributions $\varrho_\pm(\boldsymbol{x})$ of the ions are influenced by the spatial variation of the local concentration $\phi(\boldsymbol{x})$ of the binary liquid mixture. The analysis of the grand potential of this simple, yet rich, model was carried out within mean-field theory. It predicted that *attraction* between two like-charged walls with opposite preferential adsorption can indeed occur, depending on the ratio ξ/κ^{-1} between the correlation length ξ of the critical fluctuations of ϕ and the Debye screening length κ^{-1} of the electrolyte. This attraction occurs rather generically for $\xi/\kappa^{-1} \simeq 1$, while the expected repulsion is found both for $\xi/\kappa^{-1} \ll 1$, which corresponds to the electrostatic repulsion in the almost homogeneous medium, and for $\xi/\kappa^{-1} \gg 1$, in which case the relevant long-distance interaction is provided by the repulsive CCF. This interplay between electrostatic and CCFs does not lead to any qualitative changes in the case of critical Casimir attraction and agrees with the experimental observations. Similar conclusions were reached by considering various ion–solvent couplings[146,147] and via a Landau–Ginzburg theory such as the one in Eq. (4), but extended in order to account for the electrostatic phenomena. The Landau–Ginzburg theory leads to a modified Debye–Hückel theory for the ion concentration close to a wall and in the presence of critical adsorption.[148] This predicts that the interaction between charged walls[149,150] or colloidal particles (see Ref. 151 and references therein) changes as described above (see, c.f., the end of Section 5.3).

We considered above the interplay between CCFs and electrostatics, motivated by corresponding experimental observations (see also Section 5.3). However, a spatial variation of ϕ is generically accompanied by a variation of the number density of the fluid solvent. This is expected to affect the total contribution of (i) the fluid–fluid interaction and (ii) the fluid–wall interactions to the effective free energy Ω_V and therefore to the total force. This is particularly the case when these interactions are long-ranged as in the case, e.g., of dispersion forces and therefore there is no clear scale separation between their range, the correlation length ξ of the critical solvent, and the film thickness L. As discussed above for the electrostatic interaction, it turns out that for thick films sufficiently close to criticality, the universal feature of the CCF prevails in the corresponding solvation force. However, away from the critical point, a rich behavior might emerge also in connection with the occurrence of capillary condensation, as discussed in Refs. 152–155, which the reader is referred to for further details.

5. Measuring Critical Casimir Forces

In order to have experimental access to CCFs, the relevant fluctuating degrees of freedom have to be able to enter and leave the space between the confining surfaces. Accordingly, phenomena such as magnetostriction and the associated stress should not be considered as a manifestation of CCFs discussed here: The former are indeed due to the variation of the magnetic exchange interaction as function of the lattice constant (which shrinks as a consequence of the interaction) and therefore due to a change of the bulk free energy density, which is completely independent of the presence of confining surfaces. On the other hand, in principle, ferrofluids[156] could allow for an observation of CCFs because, in their fluid phase, the colloids which carry the magnetic moments can enter or leave a confined geometry. However, so far, these systems have not been used in experiments.

In the following, we briefly review the various experimental investigations of CCFs, focussing particularly on those which allowed their quantitative measurement.

The occurrence of CCFs among colloidal particles was invoked[103] as one of the possible explanations of the early observation of their reversible aggregation[8,157] in the vicinity of the demixing (critical) point of the solvent.[14] However, the available experimental data allowed neither a conclusion in this respect nor a quantitative determination of the acting force. The first attempts to measure *indirectly* the CCFs focused on the film geometry, which is naturally realized by controllable complete wetting films[2,32,158,159] of quantum or classical fluids with a critical point within the fluid phase, as described in Chapter 15. In fact, a detailed study[32] of the dependence of the thickness of such wetting films on the thermodynamic parameters which control the system (primarily the undersaturation and the temperature) allows an *indirect* experimental determination of F_{Cas} and the associated scaling function X_{Cas} in Eq. (3). Following this theoretical suggestion, this kind of experimental approach successfully provided the first quantitative experimental evidence of the occurrence of CCFs in wetting films — of pure liquid ^{4}He close to its superfluid phase transition[9,160] and of a classical binary liquid mixture,[161] see Chapter 15.

As expected on theoretical grounds — see Section 3 — in the case of ^{4}He, the attractive force leads to a thinning of wetting films. With classical binary liquid mixtures instead, it is possible to realize both $(+,-)$ and $(+,+)$ effective BCs via a suitable choice of their two components and of the substrate on which wetting occurs. For example, a mixture of methylcyclohexane and perfluoromethylcyclohexane between its vapor phase and an SiO_2 substrate[161] or, correspondingly, heptane and methanol on an Si substrate[162] correspond to $(+,-)$ BCs and lead to a thickening of the wetting film. In the latter case, replacing heptane with nonane turns out to change the BCs from $(+,-)$ to $(+,+)$.[162] Correspondingly, one observes experimentally[162] a thinning of the wetting film, which can lead to its collapse if the attractive CCF overwhelms its dispersion counterpart in the effective interface potential. These resulting very thin films no longer probe the universal CCF. In this respect, the observation of Ref. 162 suggests that an attractive CCF is at play, but it does not allow a reliable quantitative determination of such a force, as done in Ref. 161 for the repulsive case. The qualitative and quantitative features of the scaling functions X_{Cas}, determined experimentally by studying the wetting films mentioned above as well as those formed by ^{3}He–^{4}He mixtures,[39–41] turned out to be in remarkable agreement with the theoretical predictions.[20,36] We note that the non-universal amplitude ξ_0^+, which normalizes the scaling variable $x = (T/T_c - 1)(L/\xi_0^+)^{1/\nu}$ of the scaling function X_{Cas}, takes different values for the numerical and the experimental data. In the aforementioned cases, ξ_0^+ can be inferred independently from the temperature dependence of the corresponding bulk correlation length ξ, which is measured both in the experiments and in the MC simulations. With these normalizations, there are no adjustable parameters left, a fact which makes the agreement between the theoretical and the experimental data rather remarkable.

Additional indirect evidences for the occurrence of CCFs were experimentally sought in the phenomenon of *critical-point wetting* of a binary liquid mixture (see Refs. 38 and 43 for a discussion). Indeed, the contact angle θ formed by the two liquid phases of a binary liquid mixture with a third phase, e.g., a solid, is expected to vanish upon approaching the critical demixing point, realizing complete wetting. However, if a long-ranged attractive critical Casimir force contributes to the disjoining pressure Π (i.e., minus the derivative of the effective interface potential with respect to the film thickness),[42,44] which determines θ, this angle may not vanish at T_c (partial wetting). An accurate determination of the temperature dependence of θ can provide quantitative information about this force. Preliminary experiments[42,44] on wetting of ^{3}He–^{4}He mixtures suggested that the relation $\theta(T = T_c) \neq 0$ was indeed fulfilled (see also Ref. 45). However, as in the case of the experiment with wetting films, a rather detailed knowledge of the non-critical background forces is necessary in order to be able to extract reliably a possible contribution of the CCF from the experimental data. A subsequent experimental study[43] of θ for the same system unfortunately reached the conclusion that part of the preliminary evidences mentioned above was probably due to optical artefacts in the course of complete wetting. Furthermore, within the experimental conditions of these studies, a possible contribution due to CCFs would be anyhow rather small compared with the contribution due to dispersion forces.

The experimental approaches mentioned so far present inherent difficulties associated with their *indirect* nature: In particular, the determination of the CCF hinges on observing phenomena in which universal and non-universal features are intertwined. These difficulties can be overcome by devising approaches which allow the *direct* measurement of the CCF or, equivalently, of the associated potential. This can be achieved by considering the force acting effectively not on an interface (as in the case of the wetting films mentioned above) but on a spherical particle, which is immersed in the considered fluid and approaches the wall of the container of the critical fluid or another particle. This can be studied via a variety of optical methods (see Ref. 16 for a short overview) which will be discussed in the following section.

5.1. *A single colloid close to a homogeneous or a patterned substrate*

In the case of a spherical particle of radius R approaching a flat surface, both the wall and the sphere impose effective BC onto the OP ϕ. Therefore, as discussed in Section 4, a CCF F_C acts on the sphere. If the distance z of closest approach of the sphere from the homogeneous wall is much smaller than R, the potential $\Phi_{\mathrm{Cas}}(z)$ of the CCF takes the form given in Eq. (8). Within the Derjaguin approximation, the function $\vartheta_{|\circ}(y)$ can be expressed in terms of $X_{\mathrm{Cas}}(y)$, as discussed in Section 4.1. These functions share their qualitative and universal properties. Within this geometrical setting, the experimental challenge is to detect the onset of the contribution F_{Cas} to the total force acting on the sphere as the critical point of the surrounding fluid is approached. In order for the effects due to F_{Cas} to be detectable for such a sphere floating in the mixture, F_{Cas} has to be comparable in magnitude with the typical forces at play. This suggests the use of colloids, i.e., micrometer-sized particles, dispersed in classical binary liquid mixtures. Indeed, the typical energy scale of the interaction within this system is provided by $k_{\mathrm{B}}T$, i.e., the same scale as the one for $\Phi_{\mathrm{Cas}}(z)$ in Eq. (8). The necessary sensitivity for the measurement of the force could be achieved by using total internal reflection microscopy[16,163] (TIRM), which is actually capable to measure forces with femto-Newton force resolution. With this technique, it was possible[10,113] to study the onset of CCFs acting on a single colloid near a wall as the temperature T of a water–lutidine mixture, used as the solvent, is increased toward its (lower) critical value T_c at fixed lutidine critical mass fraction. Figure 1 shows the experimental results and their comparison with the available theoretical predictions.

With reference to the water–lutidine mixture, "+" and "−" indicate the preferred adsorption of lutidine and water, respectively. Accordingly, a hydrophilic colloid (with $R \simeq 1\,\mu\mathrm{m}$) close to a hydrophilic wall realizes $(-,-)$ BCs. In this case, the measured potential $\Phi(z)$ in Fig. 1(a) for $T_c - T = 0.3\,\mathrm{K}$, i.e., sufficiently far from the critical point, results only from the combined action of buoyancy and of the electrostatic repulsion between the colloid and the wall. (In fact, as discussed in Ref. 113, dispersion forces turn out to be negligible in this experiment.) This repulsion is due to the fact that surfaces in contact with highly polar fluids

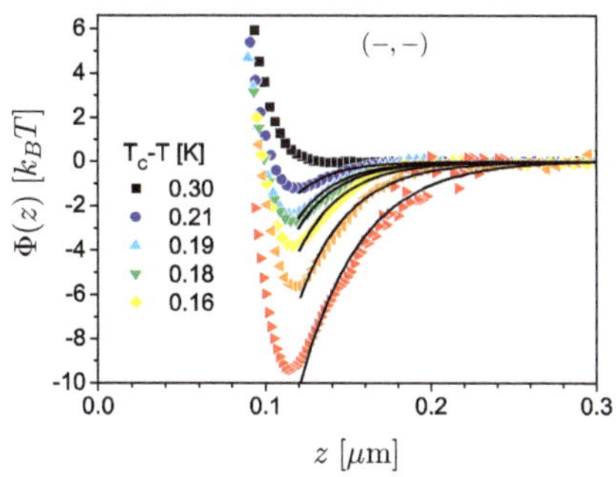

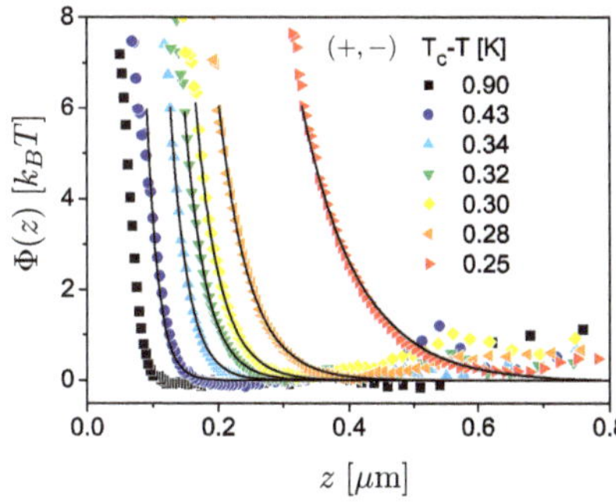

Fig. 1. Potential $\Phi(z)$ (symbols) of the forces acting on a colloid immersed in a water–lutidine mixture and at surface-to-surface distance z from an hydrophilic wall for various temperatures T of the mixture at the critical concentration.[10,113] Upon increasing T toward the critical value $T_c \simeq 34°$C, an attractive CCF is observed with a hydrophilic particle [$(-,-)$ BC, panel (a)], whereas with a hydrophobic particle [$(+,-)$ BC, panel (b)], the force is repulsive. The solid lines in (a) and (b) correspond to the theoretical predictions for the contribution $\Phi_{\mathrm{Cas}}(z)$ (see Eq. (8)) of the CCF to the total potential $\Phi(z)$. The value $\nu \simeq 0.63$ of the critical exponent of the correlation length $\xi(T)$ is fixed by the Ising universality class, while the non-universal amplitude $\xi_0^+ \simeq 0.2\,$nm was independently determined by light-scattering experiments for this mixture (see Refs. 10, 113 for details). Accordingly, there are no adjustable parameters. (Plots are taken from Ref. 10.)

acquire a surface charge after releasing a negligible amount of ions into the mixture. Buoyancy, instead, is characterized by a well-known linearly increasing potential which has been subtracted from all the data presented in Fig. 1. Upon approaching the critical point, an increasingly deep potential well develops as a consequence of the attractive CCF which provides a negative contribution $\Phi_{\mathrm{Cas}}(z)$ to the potential of the remaining forces. As expected theoretically, the spatial range ξ of $\Phi_{\mathrm{Cas}}(z)$ increases for $T \to T_c$. The set of measured potentials $\Phi(z)$ can be compared with the theoretical prediction in Eq. (8) for $\Phi_{\mathrm{Cas}}(z)$ — within the range of distances z for which the (non-critical) electrostatic interaction (basically given by the curve corresponding to $T_c - T = 0.3\,$K) is expected to be negligible. Within this range, the theoretical predictions are reported as solid lines in Fig. 1(a) and indeed they are in excellent agreement with the experimental data. In the case of a hydrophobic colloid (with $R \simeq 2\,\mu$m), the corresponding BCs change from $(-,-)$ to $(-,+)$ and, on theoretical grounds, repulsion is expected. Indeed, far below the critical point, the potential Φ in Fig. 1(b) for $T_c - T = 0.9\,$K consists of the electrostatic repulsion only. Upon approaching the critical point, the repulsive part of the potential curves shifts toward larger values of z due to the fact that, as expected, increasingly strong repulsive Casimir forces act on the particle and yield a positive contribution $\Phi_{\mathrm{Cas}}(z)$ to the total potential $\Phi(z)$. The corresponding theoretical predictions in Eq. (8) for $\Phi_{\mathrm{Cas}}(z)$ are reported as solid lines in Fig. 1(b). Within the region where the non-critical contribution to the interaction is negligible, the agreement with the experimental data for $\Phi(z)$ is remarkable. If, via a suitable chemical surface treatment, the adsorption preference of the silica wall is changed from being hydrophilic (−) to hydrophobic (+), attraction is recovered, in agreement with the theoretical expectations and predictions.[113] In contrast to the smooth and reversible onset of the CCF, in mixtures with a lutidine mass fraction x smaller than the critical one x_c and with $(+,+)$ BC or with $x > x_c$ and $(-,-)$ BCs, one observes the abrupt formation of a potential well upon approaching the phase transition line.[10,113] This fact can be interpreted in terms of the onset of liquid bridge formation between the particle and the wall. This occurs only if the mixture is poor in the component preferred by both the colloid and the wall. This tells that surface contributions to the free energy become sufficiently important so that the system minimizes its interfacial energy via bridge formation.

The *universal* character of the CCF was confirmed, at least in the attractive case, by experiments[164] which used an aqueous critical micellar solution of the non-ionic surfactant $C_{12}E_5$ instead of the standard molecular binary liquid mixture of water and lutidine primarily considered in experiments. Different from the latter, at the concentrations explored in the experiment, the surfactant forms wormlike micelles which are approximately 10 times larger than the molecules involved in the molecular liquid. In spite of the rather different microscopic structure of these two solvents, it turned out[164] that their confinement gives rise to the same CCFs on a colloid opposing a substrate or on two opposing colloids, both in agreement with the theoretical predictions. Differences between the interactions measured for these two solvents emerge only at small inter-particle distances, where other, possibly different, non-universal contributions come into play.

The experiments briefly presented above demonstrated that the CCFs can be strong enough to influence significantly the behavior of soft matter at the sub-micrometer scale. Compared with the various interactions typically acting in these systems, such as electrostatic, dispersion, and entropic ones, the CCF is characterized by an uncommonly high degree of tunability with respect to strength, range, and sign. All these features can be harnessed in order to serve dedicated purposes, such as for colloidal suspensions. In order to explore these possibilities and to have full experimental control of CCFs, a variety of systems involving colloidal particles has been investigated by means of various experimental techniques. They range from light scattering to digital video microscopy and surface plasmon spectroscopy.

Inhomogeneous surfaces: In particular, chemically patterned substrates can be used to produce a lateral modulation of the CCF. This allows one to control *reversibly* the spatial distribution and the formation of structures in a dilute colloidal suspension with a near-critical solvent.[165] These experimental studies, together with the corresponding theoretical analyses,[114,115] have revealed, *inter alia*, that the sensitivity of the resulting force with respect to details of the imprinted patterns can even be exploited in order to infer their fine geometrical features.[117] Suitable surface treatments allow one to realize substrates which are characterized by preferential adsorption for one of the components of a binary liquid mixture which varies gradually along a certain direction.[166] These substrates have been used for investigating the CCF in the presence of variable boundary conditions, giving rise to crossover phenomena. These experimental results, reviewed in Ref. 7, offer possible applications toward actively controlling the rheology and diffusion of colloidal suspensions which are exposed to such engineered substrates.

5.2. *More than two colloids and many-body effects*

The theoretical predictions discussed in Section 4.3 showed that the expected many-body effects in the critical Casimir interaction exhibit a rather complex dependence on the relative positions of the involved bodies, their geometrical shape, and the temperature of the medium; moreover, they might be quantitatively rather significant. However, it has been unclear whether they could be detected in actual experiments involving colloids. This issue was addressed in Ref. 167 by studying the interaction between two trapped colloidal particles in the presence or in the absence of a third particle close to them. The spatial positions of these three fluctuating colloids were controlled via holographic optical tweezers.[16] These positions (actually projected onto the horizontal plane identified by the beam waists of the three optical tweezers) were measured via digital videomicroscopy, such as to determine the statistical distribution of the thermally fluctuating, inter-particle projected distance between two particles forming one of three possible pairs. In the absence of the third particle (i.e., if it is kept sufficiently far from the other two), this distribution was found to be well fitted by that one expected on the basis of the theoretical predictions for the inter-particle critical Casimir potential (within the Derjaguin approximation, see Section 4.1) and for the electrostatic repulsion. This comparison allows one to determine the material parameters of the system and therefore characterize, e.g., the electrostatic repulsion as well as the stiffness of the optical trapping. When the third particle is brought close to the previous two, it affects the distribution of the inter-particle, projected distance between the first two particles. This distribution was compared with the one expected due to assuming pairwise additivity of the interaction between the three pairs of particles and by using the pair potentials determined previously. Discrepancies were observed beyond the statistical and systematic uncertainties. This demonstrated the occurrence of genuine many-body effects, particularly within the range of distances where CCFs provide the dominant contribution to the interaction compared, e.g., to the electrostatic repulsion.

As anticipated in Section 4.3, besides many-body effects, the influence of the CCF on the physical properties of colloidal suspensions is significant. This is confirmed by a number of theoretical and experimental studies briefly mentioned in that section and reviewed in Ref. 14, which the reader is referred to for further details. Here, we only mention that upon changing the temperature-dependent range of the critical Casimir interaction between two colloidal particles in a near-critical solvent, it is possible to induce and access experimentally a liquid–gas phase separation of the colloidal fluid and possibly the associated critical point. This was observed for solvents consisting of a molecular binary liquid mixture[128,133] or of an aqueous solution of globular solid micelles.[141,142] Moreover, as a convenient way to engineer and control the total interaction energy between colloids in suspension, the CCF can be used in order to study static and dynamic phenomena. They depend on the features of this interaction, such as the formation of structures resulting from aggregation,[126,129] which may change in zero gravity.

5.3. *Interplay with other forces*

In Section 4.4, we anticipated that CCFs in soft matter usually act together with forces of different nature, such as electrostatic, van der Waals, or depletion forces. Accordingly, it is natural to explore the extent to which their corresponding effects simply add or if these various forces might influence each other. In most of the cases of experimental relevance and in view of the currently available experimental accuracy, it turns out that this possible (and somehow expected) mutual influence is quantitatively negligible within the range of parameters studied so far. A notable exception was provided by the investigations of CCFs emerging in binary liquid mixtures with added salt.[125,143] In fact, in a sufficiently dilute colloidal suspension, one expects that the only relevant interactions are provided by the electrostatic repulsion and the critical Casimir attraction, the spatial ranges of which are controlled by the Debye screening length κ^{-1} and the correlation length ξ, respectively. Depending on the ratio of these two lengths and thus on the relationship between temperature and ion concentration in solution, one expects the resulting total potential to be capable to drive aggregation or not. The investigation of this phenomenon in the ξ–κ^{-1} plane[125] (see also Refs. 131 and 168) motivated a better understanding of the interaction between a (charged) colloid and a surface in the presence of an electrolyte. The corresponding potential was experimentally determined via TIRM (see Section 5.1 and Ref. 143). As anticipated in Section 4.4, the attractive force, observed for *similar* preferential adsorption at the colloid and the surface, was qualitatively not affected by the presence of the salt. The phenomenon observed in the presence of *opposite* preferential adsorption (leading, in the absence of salt, to a repulsive CCF) was richer. In fact, starting at a temperature T far from T_c, one first observes the expected electrostatic repulsion (as in Fig. 1(b)). Then, upon approaching T_c, an *attractive* contribution to the total potential develops before this gradually fades and a predominantly *repulsive* component eventually emerges close to criticality, as expected in the absence of salt and in line with the concept that asymptotically for $T \to T_c$, the *universal* scaling function should be recovered. This experimental observation motivated the theoretical investigations discussed in Section 4.4, which eventually turned out to explain qualitatively the phenomenon. A quantitative analysis, instead, is still lacking, as it would require a careful characterization of the material properties of the system (such as the surface charges), which crucially determine the non-universal interactions within this system.

6. Dynamics

The previous sections focussed on the *stationary* properties of CCFs occurring *in thermal equilibrium*. However, these forces are expected to also manifest themselves in *non-stationary* and possibly in *out-of-equilibrium* conditions, such as those generated by sudden perturbations of the system due to temperature changes,[169–171] the action of external forces,[172] or due to the non-equilibrium nature of the stationary state in which the confined fluctuations occur.[173–175] Considering non-equilibrium conditions or time-dependent properties in equilibrium poses a conceptual challenge to the approaches discussed in Section 3. In fact, while the equilibrium expectation value of the force can be derived from the effective free energy Ω or, alternatively, from the expectation value of the stress tensor T_{ij}, only the latter option is viable out of equilibrium,[169] taking into account dependence of the OP field $\phi(\boldsymbol{x},t)$ on time t. (Note, however, that the necessity of this choice has been debated.[176–178]) Accordingly, the dynamical behavior of CCFs can be predicted on the basis of the dynamics of the OP field $\phi(\boldsymbol{x},t)$. For the sake of brevity, here, we shall not present in detail the well-established theory of dynamical critical behavior.[179,180] Instead, we provide the theoretical prediction and the experimental evidence that universality emerges also in this case, in the sense that, on the long time scale, the collective dynamical behavior of the system is determined by the gross features of its dynamics, such as symmetries and possible conservation laws, and is largely independent of the microscopic details of the dynamics. In analogy to what occurs in equilibrium, this is due to the fact that the relevant timescale τ of the dynamics also displays an algebraic singularity as the critical point is approached, i.e., $\tau \sim \xi^z$, where z is the so-called *dynamic critical exponent*, which characterizes the dynamic universality class. In particular, from the theoretical point of view, the dynamics of the OP field is generically expressed in terms of stochastic evolution equations with thermal white noise, possibly involving additional fields beyond the OP itself. If the system under investigation is in thermal equilibrium, these stochastic differential equations satisfy the fluctuation–dissipation theorem, and the equilibrium universality class might split into various dynamic universality classes, depending on possible conservation laws.

6.1. *Analytical and numerical studies*

Among the simplest possible instances of *equilibrium* dynamics for a scalar OP field $\phi(\boldsymbol{x},t)$ are the so-called models A and B of Ref. 179, according to which

$$\partial_t \phi(\boldsymbol{x},t) = -D(-\nabla^2)^a \frac{\delta \mathscr{H}}{\delta \phi(\boldsymbol{x},t)} + \zeta(\boldsymbol{x},t), \tag{9}$$

where $\mathscr{H}$ is the Landau–Ginzburg effective free energy given in Eq. (4) (omitting, in general, the boundary terms), $\zeta(\boldsymbol{x},t)$ is a Gaussian white noise with $\langle \zeta(\boldsymbol{x},t)\rangle = 0$ and correlation function $\langle \zeta(\boldsymbol{x},t)\zeta(\boldsymbol{x}',t')\rangle = 2Dk_\mathrm{B}T(-\nabla^2)^a \delta(\boldsymbol{x}-\boldsymbol{x}')\delta(t-t')$, where

$a = 0$ corresponds to model A and $a = 1$ to model B. In the latter case, the OP ϕ is locally conserved in the sense that $\partial_t \phi = -\boldsymbol{\nabla} \cdot \boldsymbol{J}$ for a suitable stochastic current $\boldsymbol{J}$, as it occurs if ϕ is the OP of a binary liquid mixture. However, we note that the actual dynamics of these mixtures is significantly more complicated because one has to also account for the coupling of ϕ to other slow modes, such as the fluid velocity. This is encoded in the so-called model H of Ref. 179, briefly discussed further in the following.

As in the case of the stationary properties, the dynamics is affected by the presence of boundaries (see, e.g., Ref. 18). The stochastic evolution of $\phi(\boldsymbol{x}, t)$ (and possibly additional conserved quantities) has to be supplemented by the BCs discussed in Section 2.1. This possibly leads to a further splitting of the dynamic universality class depending on the presence or absence of locally conserved dynamics at the boundaries.[33] Similarly, the dynamics in the presence of spatial confinement can be theoretically investigated by considering the simultaneous presence of various boundaries. However, up to now, analytical progress has been possible only for the aforementioned simplest cases of dynamics and for the film geometry. In this respect, we note that the dynamics of CCFs is much less explored than their stationary properties in spite of the fact that these dynamical properties are within experimental reach and might be investigated in the near future.

Film geometry: In particular, the dynamics of a scalar Gaussian field $\phi(\boldsymbol{x}, t)$ (corresponding to $u = 0$ in Eq. (4)), confined within a film with ordinary BCs, was considered in Ref. 172. If $\phi(\boldsymbol{x}, t)$ is locally perturbed inside the film, the resulting critical Casimir pressures, acting on the two confining surfaces, become time- and space-dependent. These two pressures are no longer necessarily equal as they are in thermal equilibrium. Within model A, the associated scaling function was determined after switching off the perturbation. This analysis was extended[169] to the case of different BCs at the two confining surfaces (Dirichlet/Neumann) and for a global quench of the temperature of the medium, i.e., for a sudden change from a very high value to the critical point. In both cases, the strength of the force on the two surfaces turns out to be different while, as expected, it approaches the same value as the relaxation toward an equilibrium condition occurs. In an analogous setting, the case of a Gaussian field with short-ranged spatial correlations (modeled by neglecting the term $\propto (\nabla\phi)^2$ in Eq. (4)), conserved dynamics (model B), and no-flux condition $\boldsymbol{J} = \boldsymbol{0}$ at the boundaries was considered in Ref. 170. In particular, upon quenching the strength of the noise (i.e., the temperature) from an initial state with $\phi = 0$ (and therefore a vanishing fluctuation-induced force), the emergence of a transient, long-ranged dynamical effective Casimir-like force also away from criticality was observed. This is the case because the emergence of these long-ranged correlations is due to the conservation law of the dynamics. However, this force disappears in the stationary state, which is attained long after the quench because the correlation length of the fluctuations within the medium was assumed to be vanishingly small. The case of inclusions embedded in the fluctuating field and the possible realization of the aforementioned quench in active matter (see also the following) were discussed. The interplay between the long-ranged correlations generated out of equilibrium, in accordance with model B, and those emerging when the system is poised at its critical point, becomes particularly rich after a quench to this point. In Ref. 171, the resulting non-equilibrium critical Casimir effect in the film geometry was investigated for a Gaussian field with non-symmetry-breaking BCs. In particular, it turns out that complex transient regimes arise, including correlation functions and forces with oscillatory evolution, depending on the conditions in which the medium is prepared ahead of the quench to the critical point and depending on the specific choice of the BCs.

Alternatively, a non-equilibrium condition can be generated by modulating externally, as a function of time t, the distance $L(t)$ between a mobile plate and a fixed one. The two plates confine a scalar, fluctuating medium ϕ to a film with Dirichlet BCs. As a consequence, the force acting on the fixed plate is also modulated and its dynamics and universal form were determined analytically for a critical, scalar Gaussian field with model A dynamics and for a harmonic dependence of $L(t)$ on t under the assumption of a small modulation amplitude compared to the average distance between the boundaries.[181] In this limit, the resulting force also depends harmonically on time, with the same angular frequency ω_0 as $L(t)$ does and with a ω_0-dependent amplitude and phase shift. In addition, this force displays two distinct contributions, one generated by diffusion of stress within the medium and another one is related to resonant dissipation in the cavity (see Ref. 181 for details).

The theoretical investigations mentioned above considered dynamical evolutions in a confined geometry with BCs which do not break the $\phi \leftrightarrow -\phi$ symmetry of the Gaussian model. The choice of the latter properties are primarily motivated by the search for analytical solutions. However, as extensively discussed in the previous sections, in many cases of experimental interest, the boundaries are responsible for critical adsorption, which breaks such a symmetry and can be effectively represented by introducing surface fields, such as in Eq. (4). As a consequence of this choice, $\langle\phi(\boldsymbol{x},t)\rangle$ exhibits a non-trivial evolution, which can be conveniently studied within a mean-field approximation of the evolution equations. In Ref. 182, the case of a scalar OP was studied with the effective free energy $\mathscr{H}$ given by Eq. (4), with model B dynamics, and with no-flux BCs at the $(+,+)$ boundaries of a film. The dynamics of the film was considered after an instantaneous quench from a homogeneous, high-temperature state with $\langle\phi(\boldsymbol{x},t)\rangle = 0$ to T_c or to still supercritical but lower temperatures. The evolution of the CCF acting on the confining surfaces was determined together with the corresponding dynamic scaling functions. It turns out that such a force contains a non-equilibrium contribution which, depending on the parameters of the system, might have a non-monotonic dependence on time, while being generically *repulsive*. It actually turns out[183] that even the attractive or repulsive character of the CCF in *equilibrium* depends on whether the fluctuations of the field occur within a grand canonical or canonical ensemble. The former case is the one discussed in all previous sections. In particular, for equal BCs, the force is attractive within the grand canonical ensemble, while it is repulsive in the canonical case.[183] This is naturally realized as a consequence of the conserved dynamics.

Colloidal particles: Beyond the instances discussed above with fixed or slightly displacing boundaries, one might be interested in cases in which these surfaces move according to the very same effect of the CCF acting on them, as it occurs, e.g., for colloidal particles. Imposing BCs of the kind discussed in Section 2.1 on moving, non-planar boundaries turns out to be challenging. Thus, it is convenient to think of the particles as localized, compact parts of the medium with properties (e.g., compressibility or viscosity), which differ from the rest of the medium.[170,184] Within these parts, a symmetry-breaking field is applied instead of imposing BCs.[185–187] This portion of the medium then moves, e.g., according to an overdamped Langevin dynamics, which takes into account the total force acting on each particle due to its interaction with the (remaining part) of the medium and which, in thermal equilibrium, satisfies the fluctuation–dissipation theorem. Rather generically, even in the simplest case, in which the fluctuating field is Gaussian and the particle acts on the medium only as an external field, the effective dynamics of such a single particle turns out to be described by a Langevin equation with a *non-linear* memory kernel. This determines the effective, time-dependent friction and the correlations of its fluctuations,[185] and it spans an increasingly wide time interval as the critical point of the medium is approached. It causes, for example, an algebraic — instead of exponential — long-time relaxation of the position of a harmonically trapped particle when initially displaced from the centre of the trap, independent of the dynamics of the OP field corresponding to model A or B.[186]

The simple representation of a colloidal particle as outlined above allows one to investigate[187] the evolution of the effective interaction between two particles when they interact with the same Gaussian field. This is the case for the experiments discussed further in the following, in which two optically trapped colloids are in contact with a near-critical medium. In particular, one can consider the case that the position of one of the two particles is periodically modulated in time with, say, a harmonic function. This might induce (depending on the type of dynamics and whether the medium is critical or not) a non-harmonic motion of the other particle, which is kept away from the first one by a quadratic confining potential. At late times, the field-mediated interaction synchronizes the motion of these two particles, and the nonlinear response of the second particle can be determined as a function of the driving frequency of the first one.[187] The response can be compared with the adiabatic approximation which assumes an instantaneous relaxation of the OP to its equilibrium configuration appropriate for the actual arrangement of the particles. This comparison highlights the effects of *retardation* in the propagation of the force due to the slow critical dynamics of the medium.

Beyond the simplified models discussed above, the investigation of the evolution of CCFs, which is relevant for the critical dynamics of actual physical systems, has to rely on numerical calculations, even within the mean-field approximation. This is due to the increasing complexity of the evolution equations for more realistic models. For example, in the case of binary liquid mixtures, in addition to the OP ϕ, one also has to account for the velocity field $\boldsymbol{v}$ of the mixture. The dynamics, which captures the critical properties of ϕ and $\boldsymbol{v}$, corresponds to the so-called model H,[179] which includes hydrodynamics. Within the mean-field approximation, one considers the evolution equation without noise, in which ϕ is locally conserved, i.e., it obeys the continuity equation with the current $\boldsymbol{J} = \phi\boldsymbol{v} - D\boldsymbol{\nabla}\delta\mathscr{H}/\delta\phi$, where $\mathscr{H}$ is the effective free energy given in Eq. (4). The boundary terms are modified, as described further in the following. The local velocity $\boldsymbol{v}$ of the medium (which will eventually also encompass the particles according to the idea discussed above) obeys the Navier–Stokes equation, $\rho(\partial_t + \boldsymbol{v}\cdot\nabla)\boldsymbol{v} = \boldsymbol{f} - \boldsymbol{\nabla}\cdot\overleftrightarrow{\boldsymbol{\sigma}}$, with $\overleftrightarrow{\boldsymbol{\sigma}} = p\overleftrightarrow{\boldsymbol{I}} - \eta[(\nabla\boldsymbol{v})^{\dagger} + \nabla\boldsymbol{v}]$, where $\overleftrightarrow{\boldsymbol{I}}$ is the unit tensor, ρ the mass density of the liquid (which is assumed to be spatially constant), and η its viscosity. The hydrostatic pressure p facilitates to satisfy the incompressibility condition $\nabla\cdot\boldsymbol{v} = 0$, and $\boldsymbol{f}(\boldsymbol{x})$ is the force density $\boldsymbol{f} = -\phi\boldsymbol{\nabla}\delta\mathscr{H}/\delta\phi$. According to the approach outlined above, the presence of N particles labeled by the index $i \in \{1,\ldots,N\}$ at positions $\{\boldsymbol{x}_i\}$ in the medium is represented by a spatial modulation $\chi(\boldsymbol{x}-\boldsymbol{x}_i)$ of its local properties, where $\chi(\boldsymbol{x})$ is a smooth function which vanishes for $|\boldsymbol{x}|$ larger than the particle radius while it rapidly attains 1 upon decreasing $|\boldsymbol{x}|$ below that radius. This function determines the boundary contribution to $\mathscr{H}$ in the form $\sum_{i=1}^{N}\int \mathrm{d}^d x[c_i\chi(\boldsymbol{x}-\boldsymbol{x}_i)\phi^2(\boldsymbol{x}) - h_i|\boldsymbol{\nabla}\chi(\boldsymbol{x}-\boldsymbol{x}_i)|\phi(\boldsymbol{x})]$, where $c_i > 0$ plays the role of the surface enhancement and h_i that of the surface field. We note that this expression generalizes the boundary term appearing in Eq. (4). In addition, χ controls the spatially varying viscosity $\eta(\boldsymbol{x}) = \eta_l + (\eta_c - \eta_l)\sum_{i=1}^{N}\chi(\boldsymbol{x}-\boldsymbol{x}_i)$, where η_l is the viscosity of the *l*iquid and $\eta_c \gg \eta_l$ the (common) viscosity of the *c*olloidal particles. The particles are assumed to be a region of space filled by a highly viscous fluid. (From this approach, the notion of "fluid particle dynamics" earns its name; see, e.g., Ref. 184). Similarly, the diffusivity D acquires a space dependence, $D(\boldsymbol{x}) = D[1 - \sum_{i=1}^{N}\chi(\boldsymbol{x}-\boldsymbol{x}_i)]$, which suppresses diffusion within the particles, while $\boldsymbol{f}$ (see above) receives a contribution $-\mathscr{V}^{-1}\sum_{i=1}^{N}\chi(\boldsymbol{x}-\boldsymbol{x}_i)\boldsymbol{\nabla}_{\boldsymbol{x}_i}\mathscr{H}$, where $\mathscr{V} = \int \mathrm{d}^d x\,\chi(\boldsymbol{x})$ is the effective volume of each particle. The resulting evolution equations for $\phi(\boldsymbol{x},t)$ and $\boldsymbol{v}(\boldsymbol{x},t)$ are solved numerically[184] together with the dynamics of the centre $\boldsymbol{x}_i$ of the particle i, which is determined by its instantaneous velocity $\boldsymbol{V}_i(t) = \int \mathrm{d}^d x\,\boldsymbol{v}(\boldsymbol{x},t)\chi(\boldsymbol{x}-\boldsymbol{x}_i)/\mathscr{V}$. In addition to the terms discussed above, a mechanical potential $U(\boldsymbol{x}_1,\ldots,\boldsymbol{x}_N)$ might be introduced in $\mathscr{H}$ in order to account for possible direct interactions among the particles.

Based on the approach described above, the authors of Ref. 184 investigated the temperature dependence of the drag force on a single particle steadily moving in a near-critical fluid mixture. They also studied the relative motion of two particles due to the non-equilibrium CCF acting on them. Due to the slow dynamics of the OP and depending on the Péclet number of the particle motion, significant deviations of the instantaneous force acting on the colloids from its equilibrium value at the instantaneous separation between the two particles are observed. These retardation effects reduce the strength of the force during particle aggregation.

Before concluding this part, we mention that various simplifications of the model H described above, which are related to the velocity field $\boldsymbol{v}$, were considered in the literature. They allowed one to determine in semi-analytic form, e.g., the drag coefficient of a solid spherical colloid in the near-critical binary liquid mixture (see Ref. 188 and references therein) or the amplitude of the fluctuations of the position of such a colloid when it is confined by an harmonic trap (see Ref. 189 and references therein). In the latter case, it would be rewarding to investigate how such an amplitude emerges from the equilibrium distribution of the particle displacement, which is expected to be determined solely by the total potential of the mechanical forces acting on the particle.

6.2. *Experiments*

The dynamical aspects of CCFs are experimentally still largely unexplored. This is primarily due to the fact that the relaxation times τ within experimental reach are still rather short compared to the time scales at which the experimental measurements are performed. Accordingly, as we shall discuss in the following, the available experimental evidences can still be interpreted in terms of the occurrence of equilibrium CCFs, at least within the currently available experimental accuracy. No retardation effects have been observed so far. However, the possibility of using these forces also for controlling the dynamics of colloidal particles was demonstrated.[190,191]

Critical Casimir forces can be used to synchronize the motion of two identical colloidal particles which are immersed in a binary liquid mixture and are trapped by two optical tweezers separated by a certain distance. Upon moving periodically the centre of one of the two traps and therefore by moving one of the two colloids, it was shown[190] that one can induce the motion of the other colloid, in a way which depends on the thermodynamic distance from criticality. If the mixture is far from being critical and thus the correlation length ξ is smaller than the typical inter-particle distance ℓ, the motion of one colloid does not actually affect the other, as expected. This also demonstrates that the hydrodynamic interactions are negligible in this setting.[190] The situation changes when, upon approaching criticality, ξ becomes comparable to ℓ and one colloid senses the presence of the other in terms of the action of CCFs. Eventually, their motion synchronizes and the average ℓ decreases because of the occurrence of these attractive forces.

Alternatively, CCFs can also be used to influence the dynamics of diffusing colloidal particles in the absence of optical trapping.[191] In particular, by using blinking optical tweezers, which can be periodically switched on and off, the dynamics of two colloidal particles in a near-critical binary liquid mixture was experimentally observed during diffusion. This way one can highlight the effect of their mutual interaction possibly due to CCFs. By studying the position of the particles when both traps are switched on, one can determine the non-universal parameters (e.g., trap stiffnesses, Debye screening length, critical temperature, etc.), which control, *inter alia*, the colloidal interaction in thermal equilibrium and which therefore fix its form. This interaction potential can then be used to predict, via numerical (Langevin or Monte Carlo) simulations, the statistical properties of the dynamics of the particles in the absence of the trapping. Then these predictions can be compared with the actual experimental data for the relative velocity of the two particles in order to highlight the possible emergence of retardation effects. However, as anticipated above, no significant discrepancy was found upon this comparison. A similar conclusion was reached on the basis of the data for the distribution of the first passage time, i.e., of the first time at which the inter-particle distance reaches a certain threshold value when they are released from an initial distance larger or smaller than this threshold. Depending on the temperature, one clearly observes that the emergence of the (attractive) critical Casimir interaction between the particles influences the distribution of these first passage times.[191]

7. Perspectives

In the previous sections, we focused on Casimir-like effects induced by the confinement of the critical fluctuations, which characterize a continuous phase transition occurring in a liquid medium at thermal equilibrium. Before concluding, we mention analogous instances in which Casimir-like forces emerge because of the fluctuations of other relevant collective fields. These include, for example, fluctuations in liquid crystals (see, e.g., Refs. 192 and 193 and references therein), in polymers,[194] of the heights of interfaces[195,196] and membranes,[197–200] in complex fluids,[201] and in granular materials,[173,202] as well as fluctuations associated with non-equilibrium phase transitions[174] or with transitions to non-equilibrium steady states.[203–205] More recently, motivated by the increasing interest in active matter, this subject has also been studied in the present context.[175,206] In particular, an early numerical study of an ensemble of run-and-tumble particles in $d = 2$, confined by parallel lines of *finite* length, revealed the emergence of an attractive force, the magnitude of which increases upon increasing the run length of the particles, but decreases *exponentially* upon increasing the distance L between the lines. This attraction may be turned into a repulsion via varying L and the line length. In order to highlight the effects associated with confining an emergent collective behavior within these systems, the phenomenon of flocking in active matter has been considered[175] in $d = 2$. In fact, a system of interacting self-propelled particles is known to undergo a flocking transition, characterized by the emergence of a well-defined flocking direction of collective motion (see Ref. 175 and references therein). In the presence of two reflecting or partly repelling parallel walls, the flocking direction is parallel to them. An attractive Casimir-like effect emerges on the walls, which decays *algebraically* upon increasing their separation, while displaying a certain degree of universality. However, a systematic understanding of the various features of such forces caused by the long-ranged correlations of the fluctuations in active matter systems is still lacking. Therefore, this stands as an interesting problem for future investigations.

In this chapter, we discussed the emergence of CCFs in soft matter, starting from very basic theoretical ideas up to reporting the first experimental evidences in wetting films, and, most importantly, in colloidal suspensions. These studies brought to light the following peculiar and useful features of CCFs: (i) they can be reversibly switched on and off by controlling the temperature of the solvent; (ii) their strength and spatial direction can be controlled by suitable treatments of the *surfaces* of the involved bodies; (iii) they exhibit a universal character, and therefore, they are largely independent of the actual material properties of the involved bodies. These features are rather unique to the CCFs, especially when compared with other forces acting in soft matter, such as electrostatic, depletion, van der Waals, hydrodynamics forces, etc. Most of them sensitively depend on the microscopic properties of the involved materials, and they can hardly be tuned, as far as their strength, range, and spatial orientation are concerned, by varying an external parameter. In this respect, CCFs may find a number of possible applications in controlling and manipulating soft matter. In fact, this is the scientific direction into which most of the corresponding research has been developing recently. For example, the various studies mentioned in Section 4.3 focussed on exploiting CCFs in order to control the assembly of colloidal dispersions and the possible structures of aggregates emerging from the assembly, with an eye on concrete, technical applications. In addition, a recent experimental study[207] demonstrated that these forces provide a generic method for size-selective nanoparticle purification/separation. The aggregation they cause in a heterogeneous colloidal dispersion depends on the relative particle sizes. The larger colloids aggregate while the smaller ones remain in solution. This allowed the authors to propose a novel approach which turned out to be very efficient in purifying commercial silica nanoparticles in a water–lutidine mixture. These investigations open the door to a number of possible applications of CCFs to nanotechnologies. In this respect, it is also of high importance to understand how these forces can be used to counteract those which are usually at play in electromechanical devices at the micrometer or nanometer scale, i.e., MEMS or NEMS, respectively. Among the latter forces, the most prominent is the attractive Casimir–Lifshitz force, usually responsible for the undesired stiction of metallic parts. While in the experiments with colloidal particles described in Sections 5 and 6.2, such a van-der-Waals force turned out to be negligible, this might not always be the case. This possibility motivated the experimental study presented in Ref. 208.

It involves a microscopic gold flake suspended above a flat gold-coated substrate and immersed in a critical water–lutidine binary liquid mixture. It turns out that the attractive Casimir–Lifshitz force acting on the flake, which, on its own, cannot be easily made repulsive, can be conveniently counterbalanced by the repulsive critical Casimir force. This offers an active control of the total force exerted on the flake. Beyond the cases discussed here and in Sections 4.4 and 5.3, the investigation of the interplay of CCFs with other interactions in devices and in soft matter has still to reach a satisfactory level of understanding and quantitative accuracy. This stands as another interesting open problem for future studies.

For example, it has been realized only recently[209] that in previous theoretical studies of the wetting transition in systems with short-ranged interactions, a low-temperature Casimir contribution had been overlooked in the derivation of the interface potential describing the effective interaction between the free interface of the wetting film and a substrate. This contribution leads to a number of qualitative and quantitative changes compared to long-standing predictions. This also resolves a controversy concerning the comparison between corresponding theoretical predictions and available numerical simulations.

In addition to some of the aforementioned open problems and directions for future research as discussed in Section 6, the dynamics of CCFs in and out of thermal equilibrium pose a theoretical and experimental challenge and are a largely unexplored subject beyond simplistic models. Accordingly, any progress in this direction, e.g., by accounting for the dynamics of fluctuations beyond mean-field theory or by considering more realistic geometries for their confinement, would be highly valuable also in view of promising experimental access to these phenomena.

The scientific activity reviewed in this contribution was largely motivated by the inspiring work of Michael E. Fisher and it enjoyed his support and encouragement in the course of the years. The wealth and the variety of theoretical and experimental results, which originated from these studies, are therefore part of his scientific legacy and stand as a tribute to his outstanding scientific oeuvre.

References

1. M. E. Fisher and M. N. Barber, *Phys. Rev. Lett.* **28**, 1516 (1972).
2. S. Dietrich, In C. Domb and J. L, Lebowitz (eds.), *Phase Transitions and Critical Phenomena*, Vol. 12, P. 1. (Academic, New York, 1988).
3. M. E. Fisher and P. G. de Gennes, *C. R. Seances Acad. Sci. Paris Ser. B* **287**, 207 (1978).
4. H. B. Casimir, *Proc. K. Ned. Akad. Wet.* **51**, 793 (1948).
5. J. N. Israelachvili, *Intermolecular and Surface Forces* (Academic, London, 2011).
6. V. A. Parsegian, *Van der Waals Forces* (Cambridge University Press, 2006).
7. A. Gambassi and S. Dietrich, *Soft Matter* **7**, 1247 (2011).
8. D. Beysens and D. Estève, *Phys. Rev. Lett.* **54**, 2123 (1985).
9. R. Garcia and M. H. W. Chan, *Phys. Rev. Lett.* **83**, 1187 (1999).

10. C. Hertlein, L. Helden, A. Gambassi, S. Dietrich and C. Bechinger, *Nature* **451**, 172 (2008).
11. M. Krech, *The Casimir Effect in Critical Systems* (World Scientific, Singapore, 1994).
12. J. G. Brankov, D. M. Dantchev and N. S. Tonchev, *The Theory of Critical Phenomena in Finite-Size Systems — Scaling and Quantum Effects* (World Scientific, Singapore, 2000).
13. A. Gambassi, *J. Phys.: Conf. Ser.* **161**, 012037 (2009).
14. A. Maciołek and S. Dietrich, *Rev. Mod. Phys.* **90**, 045001 (2018).
15. D. M. Dantchev and S. Dietrich, *Phys. Rep.* **1005**, 1 (2023).
16. A. Callegari, A. Magazzù, A. Gambassi and G. Volpe, *Eur. Phys. J. Plus* **136**, 213 (2021).
17. K. Binder, In C. Domb and J. L. Lebowitz (eds.), *Phase Transitions and Critical Phenomena* Vol. 8, P. 1, (Academic, London, 1983).
18. H. W. Diehl, In C. Domb and J. L. Lebowitz (eds.), *Phase Transitions and Critical Phenomena*, Vol. 10, P. 76, (Academic, New York, 1986).
19. H. W. Diehl, *Eur. Phys. J. B* **93**, 195 (2020).
20. A. Maciołek and S. Dietrich, *EPL* **74**, 22 (2006).
21. E. Brézin, *J. Physique* **43**, 15 (1982).
22. V. Privman and M. E. Fisher, *J. Phys. A: Math. Gen.* **16**, L295 (1983).
23. J. L. Cardy, (ed.), *Finite-Size Scaling* (North-Holland, Amsterdam, 1988).
24. G. Parisi, *Statistical Field Theory* (Addison Wesley, New York, 1988).
25. J. Zinn-Justin, *Quantum Field Theory and Critical Phenomena* (Clarendon, Oxford, 2002).
26. D. P. Landau and K. Binder, *A Guide to Monte Carlo Simulations in Statistical Physics* (Cambridge University Press, London, 2005).
27. A. Pelissetto and E. Vicari, *Phys. Rep.* **368**, 549 (2002).
28. D. Bartolo, A. Ajdari, J.-B. Fournier and R. Golestanian, *Phys. Rev. Lett.* **89**, 230601 (2002).
29. M. Gross, A. Gambassi and S. Dietrich, *Phys. Rev. E* **103**, 062118 (2021).
30. K. Symanzik, *Nucl. Phys. B* **190**, 1 (1981).
31. J. O. Indekeu, M. P. Nightingale and W. V. Wang, *Phys. Rev. B* **34**, 330 (1986).
32. M. Krech and S. Dietrich, *Phys. Rev. A* **46**, 1922 (1992).
33. H. W. Diehl, *Int. J. Mod. Phys. B* **11**, 3503 (1997).
34. M. Krech, *Phys. Rev. E* **56**, 1642 (1997).
35. R. Zandi, A. Shackell, J. Rudnick, M. Kardar and L. P. Chayes, *Phys. Rev. E* **76**, 030601 (2007).
36. A. Maciołek, A. Gambassi and S. Dietrich, *Phys. Rev. E* **76**, 031124 (2007).
37. R. Zandi, J. Rudnick and M. Kardar, *Phys. Rev. Lett.* **93**, 155302 (2004).
38. S. Balibar and R. Ishiguro, *Pramana* **53**, 1 (1999).
39. R. Garcia and M. H. W. Chan, *J. Low. Temp. Phys.* **121**, 495 (2000).
40. R. Garcia and M. H. W. Chan, *Physica B* **280**, 55 (2000).
41. R. Garcia and M. H. W. Chan, *Phys. Rev. Lett.* **88**, 086101 (2002).
42. T. Ueno, S. Balibar, T. Mizusaki, F. Caupin, M. Fechner and E. Rolley, *J. Low. Temp. Phys.* **130**, 543 (2003).
43. R. Ishiguro and S. Balibar, *J. Low. Temp. Phys.* **140**, 29 (2005).
44. T. Ueno, S. Balibar, T. Mizusaki, F. Caupin and E. Rolley, *Phys. Rev. Lett.* **90**, 116102 (2003).
45. T. Ueno and S. Balibar, *J. Low. Temp. Phys.* **134**, 507 (2004).
46. M. E. Fisher and P. J. Upton, *Phys. Rev. Lett.* **65**, 2402 (1990).
47. M. E. Fisher and P. J. Upton, *Phys. Rev. Lett.* **65**, 3405 (1990).

48. Z. Borjan and P. J. Upton, *Phys. Rev. Lett.* **81**, 4911 (1998).
49. Z. Borjan and P. J. Upton, *Phys. Rev. Lett.* **101**, 125702 (2008).
50. P. J. Upton and Z. Borjan, *Phys. Rev. B* **88**, 155418 (2013).
51. M. Krech and S. Dietrich, *Phys. Rev. A* **46**, 1886 (1992).
52. M. Krech and D. P. Landau, *Phys. Rev. E* **53**, 4414 (1996).
53. H. W. Diehl, D. Grüneberg and M. A. Shpot, *EPL* **75**, 241 (2006).
54. D. Grüneberg and H. W. Diehl, *Phys. Rev. E* **77**, 115409 (2008).
55. F. M. Schmidt and H. W. Diehl, *Phys. Rev. Lett.* **101**, 100601 (2008).
56. M. Hasenbusch, *Phys. Rev. B* **83**, 134425 (2011).
57. J. Rudnick, R. Zandi, A. Shackell and D. Abraham, *Phys. Rev. E* **82**, 041118 (2010).
58. J. L. Cardy, *Nucl. Phys. B* **275**, 200 (1986).
59. H. W. J. Blöte, J. L. Cardy and M. P. Nightingale, *Phys. Rev. Lett.* **56**, 742 (1986).
60. I. Affleck, *Phys. Rev. Lett.* **56**, 746 (1986).
61. R. Evans and J. Stecki, *Phys. Rev. B* **49**, 8842 (1994).
62. D. B. Abraham and A. Maciołek, *Phys. Rev. Lett.* **105**, 055701 (2010).
63. A. Gambassi and S. Dietrich, *EPL* **74**, 754 (2006).
64. P. A. Martin and V. A. Zagrebnov, *EPL* **73**, 15 (2006).
65. K. Huang, *Statistical Mechanics* (2nd edn.) (John Wiley & Sons, New York, 1987).
66. D. Dantchev, *J. Stat. Phys.* **73**, 267 (1993).
67. D. Dantchev, *Phys. Rev. E* **53**, 2104 (1996).
68. D. M. Dantchev, *Phys. Rev. E* **58**, 1455 (1998).
69. H. Chamati and D. M. Dantchev, *Phys. Rev. E* **70**, 066106 (2004).
70. D. Dantchev and D. Grüneberg, *Phys. Rev. E* **79**, 041103 (2009).
71. P. Jakubczyk and M. Napiórkowski, *Phys. Rev. B* **87**, 165439 (2013).
72. A. Rançon, L.-P. Henry, F. Rose, D. L. Cardozo, N. Dupuis, P. C. W. Holdsworth and T. Roscilde, *Phys. Rev. B* **94**, 140506 (2016).
73. O. Vasilyev, A. Gambassi, A. Maciołek and S. Dietrich, *EPL* **80**, 60009 (2007).
74. O. Vasilyev, A. Gambassi, A. Maciołek and S. Dietrich, *Phys. Rev. E* **79**, 041142 (2009).
75. H. Hasenbusch, *Phys. Rev. B* **82**, 104425 (2010).
76. M. Hasenbusch, *Phys. Rev. E* **91**, 022110 (2015).
77. A. Hucht, *Phys. Rev. Lett.* **99**, 185301 (2007).
78. M. Hasenbusch, *Phys. Rev. E* **80**, 061120 (2009).
79. M. Hasenbusch, *Phys. Rev. B* **81**, 165412 (2010).
80. M. Hasenbusch, *J. Stat. Mech.: Theor. Exp.* P07031 (2009).
81. D. Dantchev and M. Krech, *Phys. Rev. E* **69**, 046119 (2004).
82. D. Lopes Cardozo, H. Jacquin and P. C. W. Holdsworth, *Phys. Rev. B* **90**, 184413 (2014).
83. C. M. Rohwer, A. Squarcini, O. Vasilyev, S. Dietrich and M. Gross, *Phys. Rev. E* **99**, 062103 (2019).
84. O. Vasilyev, A. Maciołek and S. Dietrich, *Phys. Rev. E* **84**, 041605 (2011).
85. A. Maciołek, O. Vasilyev, V. Dotsenko and S. Dietrich, *Phys. Rev. E* **91**, 032408 (2015).
86. F. Parisen Toldin, *Phys. Rev. E* **91**, 032105 (2015).
87. O. A. Vasilyev and S. Dietrich, *EPL* **104**, 60002 (2014).
88. F. Parisen Toldin and S. Dietrich, *J. Stat. Mech.: Theor. Exp.* P11003 (2010).
89. F. Parisen Toldin and S. Dietrich, In K. A. Milton and M. Bordag (eds.), *Proceedings of the Ninth Conference on Quantum Field Theory Under the Influence of External Conditions (QFEXT09)* (World Scientific, Singapore, 2010), p. 355.

90. F. Parisen Toldin, M. Tröndle and S. Dietrich, *Phys. Rev. E* **88**, 052110 (2013).
91. F. Parisen Toldin, M. Tröndle and S. Dietrich, *J. Phys.: Condens. Matter* **27**, 214010 (2015).
92. O. Vasilyev and A. Maciołek, *J. Non-Cryst. Solids* **407**, 376 (2015).
93. O. A. Vasilyev, E. Eisenriegler and S. Dietrich, *Phys. Rev. E* **88**, 012137 (2013).
94. M. Hasenbusch, *Phys. Rev. B* **85**, 174421 (2012).
95. M. Hasenbusch, *Phys. Rev. E* **87**, 022130 (2013).
96. M. Hasenbusch, *Phys. Rev. E* **91**, 022110 (2015).
97. O. A. Vasilyev, *Phys. Rev. E* **90**, 012138 (2014).
98. H. Hobrecht and A. Hucht, *EPL* **106**, 56005 (2014).
99. F. Puosi, D. L. Cardozo, S. Ciliberto and P. C. W. Holdsworth, *Phys. Rev. E* **94**, 040102 (2016).
100. A. Maciołek, O. A. Vasilyev, V. Dotsenko and S. Dietrich, *J. Stat. Mech.: Theor. Exp.* 113203 (2017).
101. H. Hobrecht and A. Hucht, *Phys. Rev. E* **92**, 042315 (2015).
102. M. Hasenbusch, *Phys. Rev. E* **91**, 033304 (2015).
103. A. Hanke, F. Schlesener, E. Eisenriegler and S. Dietrich, *Phys. Rev. Lett.* **81**, 1885 (1998).
104. S. Kondrat, L. Harnau and S. Dietrich, *J. Chem. Phys.* **131**, 204902 (2009).
105. M. Tröndle, L. Harnau and S. Dietrich, *J. Chem. Phys.* **129**, 124716 (2008).
106. M. Tröndle, L. Harnau and S. Dietrich, *J. Phys.: Condens. Matter* **27**, 214006 (2015).
107. A. D. Law, L. Harnau, M. Tröndle and S. Dietrich, *J. Chem. Phys.* **141**, 134704 (2014).
108. T. W. Burkhardt and E. Eisenriegler, *Phys. Rev. Lett.* **74**, 3189 (1995).
109. J. Dubail, R. Santachiara and T. Emig, *J. Stat. Mech.: Theor. Exp.*, 033201 (2017).
110. E. Eisenriegler, *J. Chem. Phys.* **121**, 3299 (2004).
111. A. Squarcini, A. Maciołek, E. Eisenriegler and S. Dietrich, *J. Stat. Mech.: Theor. Exp.*, 043208 (2020).
112. B. Derjaguin, *Kolloid Z.* **69**, 155 (1934).
113. A. Gambassi, A. Maciołek, C. Hertlein, U. Nellen, L. Helden, C. Bechinger and S. Dietrich, *Phys. Rev. E* **80**, 061143 (2009).
114. M. Tröndle, S. Kondat, A. Gambassi, L. Harnau and S. Dietrich, *EPL* **88**, 40004 (2009).
115. M. Tröndle, S. Kondat, A. Gambassi, L. Harnau and S. Dietrich, *J. Chem. Phys.* **133**, 074702 (2010).
116. M. Sprenger, F. Schlesener and S. Dietrich, *J. Chem. Phys.* **124**, 134703 (2006).
117. M. Tröndle, O. Zvyagolskaya, A. Gambassi, D. Vogt, L. Harnau, C. Bechinger and S. Dietrich, *Mol. Phys.* **109**, 1169 (2011).
118. M. Labbé-Laurent, M. Tröndle, L. Harnau and S. Dietrich, *Soft Matter* **10**, 2270 (2014).
119. M. Labbé-Laurent and S. Dietrich, *Soft Matter* **12**, 6621 (2016).
120. N. Farahmand Bafi, P. Nowakowski and S. Dietrich, *J. Chem. Phys.* **152**, 114902 (2020).
121. T. G. Mattos, L. Harnau and S. Dietrich, *J. Chem. Phys.* **138**, 074704 (2013).
122. T. Mattos, L. Harnau and S. Dietrich, *Phys. Rev. E* **91**, 042304 (2015).
123. O. A. Vasilyev, S. Dietrich and S. Kondrat, *Soft Matter* **14**, 586 (2018).
124. J. R. Edison, N. Tasios, S. Belli, R. Evans, R. van Roij and M. Dijkstra, *Phys. Rev. Lett.* **114**, 038301 (2015).

125. D. Bonn, J. Otwinowski, S. Sacanna, H. Guo, G. Wegdam and P. Schall, *Phys. Rev. Lett.* **103**, 156101 (2009).
126. S. J. Veen, O. Antoniuk, B. Weber, M. A. C. Potenza, S. Mazzoni, P. Schall and G. H. Wegdam, *Phys. Rev. Lett.* **109**, 248302 (2012).
127. M. T. Dang, A. V. Verde, V. D. Nguyen, P. G. Bolhuis and P. Schall, *J. Chem. Phys.* **139**, 094903 (2013).
128. V. D. Nguyen, S. Faber, Z. Hu, G. H. Wegdam and P. Schall, *Nat. Commun.* **4**, 1584 (2013).
129. M. A. C. Potenza, A. Manca, S. J. Veen, B. Weber, S. Mazzoni, P. Schall and G. H. Wegdam, *EPL* **106**, 68005 (2014).
130. P. B. Shelke, V. D. Nguyen, A. V. Limaye and P. Schall, *Adv. Mater.* **25**, 1499 (2013).
131. A. Gambassi and S. Dietrich, *Phys. Rev. Lett.* **105**, 059601 (2010).
132. T. F. Mohry, A. Maciołek and S. Dietrich, *J. Chem. Phys.* **136**, 224902 (2012).
133. O. Zvyagolskaya, A. J. Archer and C. Bechinger, *EPL* **96**, 28005 (2011).
134. E. Marino, D. M. Balazs, R. W. Crisp, D. Hermida-Merino, M. A. Loi, T. E. Kodger and P. Schall, *J. Phys. Chem. C* **123**, 13451 (2019).
135. O. A. Vasilyev, E. Marino, B. B. Kluft, P. Schall and S. Kondrat, *Nanoscale* **13**, 6475 (2021).
136. E. Marino, O. A. Vasilyev, B. B. Kluft, M. J. B. Stroink, S. Kondrat and S. P., *Nanoscale Horiz.* **6**, 751 (2021).
137. H. J. Jonas, S. G. Stuij, P. Schall and P. G. Bolhuis, *J. Chem. Phys.* **155**, 034902 (2021).
138. C. L. Kennedy, D. Sayasilpi, P. Schall and J.-M. Meijer, *J. Phys.: Condens. Matter* **34**, 214005 (2022).
139. K. Marolt, M. Zimmermann and R. Roth, *Phys. Rev. E* **100**, 052602 (2019).
140. K. Marolt and R. Roth, *Phys. Rev. E* **102**, 042608 (2020).
141. S. Buzzaccaro, J. Colombo, A. Parola and R. Piazza, *Phys. Rev. Lett.* **105**, 198301 (2010).
142. R. Piazza, S. Buzzaccaro, A. Parola and J. Colombo, *J. Phys.: Condens. Matter* **23**, 194114 (2011).
143. U. Nellen, J. Dietrich, L. Helden, S. Chodankar, K. Nygård, J. F. van der Veen and C. Bechinger, *Soft Matter* **7**, 5360 (2011).
144. M. Bier, A. Gambassi, M. Oettel and S. Dietrich, *EPL* **95**, 60001 (2011).
145. M. Bier, A. Gambassi and S. Dietrich, *J. Chem. Phys.* **137**, 034504 (2012).
146. S. Samin and Y. Tsori, *EPL* **95**, 36002 (2011).
147. S. Samin and Y. Tsori, *J. Chem. Phys.* **136**, 154908 (2012).
148. A. Ciach and A. Maciołek, *Phys. Rev. E* **81**, 041127 (2010).
149. F. Pousaneh and A. Ciach, *J. Phys.: Condens. Matter* **23**, 412101 (2011).
150. F. Pousaneh, A. Ciach and A. Maciołek, *Soft Matter* **8**, 7567 (2012).
151. R. Okamoto and A. Onuki, *Phys. Rev. E* **84**, 051401 (2011).
152. A. Maciołek, A. Drzewiński and P. Bryk, *J. Chem. Phys.* **120**, 1921 (2004).
153. D. Dantchev, F. Schlesener and S. Dietrich, *Phys. Rev. E* **76**, 011121 (2007).
154. D. Dantchev, J. Rudnick and M. Barmatz, *Phys. Rev. E* **75**, 011121 (2007).
155. D. Dantchev, J. Rudnick and M. Barmatz, *Phys. Rev. E* **80**, 031119 (2009).
156. B. Groh and S. Dietrich, *Phys. Rev. E* **50**, 3814 (1994).
157. T. Narayanan, A. Kumar, E. S. R. Gopal, D. Beysens, P. Guenoun and G. Zalczer, *Phys. Rev. E* **48**, 1989 (1993).
158. M. P. Nightingale and J. O. Indekeu, *Phys. Rev. Lett.* **54**, 1824 (1985).
159. J. O. Indekeu, *J. Chem. Soc. Faraday Trans. II* **82**, 1835 (1986).

160. A. Ganshin, S. Scheidemantel, R. Garcia and M. H. W. Chan, *Phys. Rev. Lett.* **97**, 075301 (2006).
161. M. Fukuto, Y. F. Yano and P. S. Pershan, *Phys. Rev. Lett.* **94**, 135702 (2005).
162. S. Rafaï, D. Bonn and J. Meunier, *Physica A* **386**, 31 (2007).
163. D. C. Prieve, *Adv. Colloid. Interf. Sci.* **82**, 93 (1999).
164. L. Helden, T. Knippenberg, L. Tian, A. Archambault, F. Ginot and C. Bechinger, *Soft Matter* **17**, 2737 (2021).
165. F. Soyka, O. Zvyagolskaya, C. Hertlein, L. Helden and C. Bechinger, *Phys. Rev. Lett.* **101**, 208301 (2007).
166. U. Nellen, L. Helden and C. Bechinger, *EPL* **88**, 26001 (2009).
167. S. Paladugu, A. Callegari, Y. Tuna, L. Barth, S. Dietrich, A. Gambassi and G. Volpe, *Nat. Commun.* **7**, 11403 (2016).
168. D. Bonn, G. Wegdam and P. Schall, Bonn, Wegdam and Schall reply, *Phys. Rev. Lett.* **105**, 059602 (2010).
169. A. Gambassi, *Eur. Phys. J. B* **64**, 379 (2008).
170. C. M. Rohwer, M. Kardar and M. Krüger, *Phys. Rev. Lett.* **118**, 015702 (2017).
171. M. Gross, C. M. Rohwer and S. Dietrich, *Phys. Rev. E* **100**, 012114 (2019).
172. A. Gambassi and S. Dietrich, *J. Stat. Phys.* **123**, 929 (2006).
173. C. Cattuto, R. Brito, U. M. B. Marconi, F. Nori and R. Soto, *Phys. Rev. Lett.* **96**, 178001 (2006).
174. R. Brito, U. M. B. Marconi and R. Soto, *Phys. Rev. E* **76**, 011113 (2007).
175. G. Fava, A. Gambassi and F. Ginelli, *arXiv:2211.02644* (2022).
176. D. S. Dean and A. Gopinathan, *J. Stat. Mech.: Theor. Exp.* L08001 (2009).
177. D. S. Dean and A. Gopinathan, *Phys. Rev. E* **81**, 041126 (2010).
178. M. Krüger, A. Solon, V. Demery, C. M. Rohwer and D. S. Dean, *J. Chem. Phys.* **148**, 084503 (2018).
179. P. C. Hohenberg and B. I. Halperin, *Rev. Mod. Phys.* **49**, 435 (1977).
180. U. C. Täuber, *Critical Dynamics: A Field Theory Approach to Equilibrium and Nonequilibrium Scaling Behavior* (Cambridge University Press, Cambridge, 2014).
181. A. Hanke, *PLoS ONE* **8**, e53228 (2013).
182. M. Gross, A. Gambassi and S. Dietrich, *Phys. Rev. E* **98**, 032103 (2018).
183. M. Gross, O. Vasilyev, A. Gambassi and S. Dietrich, *Phys. Rev. E* **94**, 022103 (2016).
184. A. Furukawa, A. Gambassi, S. Dietrich and H. Tanaka, *Phys. Rev. Lett.* **111**, 055701 (2013).
185. U. Basu, V. Démery and A. Gambassi, *SciPost Phys.* **13**, 078 (2022).
186. D. Venturelli, F. Ferraro and A. Gambassi, *Phys. Rev. E* **105**, 054125 (2022).
187. D. Venturelli and A. Gambassi, *Phys. Rev. E* **106**, 044112 (2022).
188. S. Yabunaka and Y. Fujitani, *J. Fluid Mech.* **886**, A2 (2020).
189. Y. Fujitani, *J. Phys. Soc. Jpn.* **85**, 044401 (2016); see also erratum, ibid., **86**, 098001 (2017).
190. I. A. Martínez, C. Devailly, A. Petrosyan and S. Ciliberto, *Entropy* **19**, 77 (2017).
191. A. Magazzù, A. Callegari, J. P. Staforelli, A. Gambassi, S. Dietrich and G. Volpe, *Soft Matter* **15**, 2152 (2019).
192. P. Ziherl, R. Podgornik and S. Žumer, *Chem. Phys. Lett.* **295**, 99 (1998).
193. F. Karimi Pour Haddadan and S. Dietrich, *Phys. Rev. E* **73**, 051708 (2006).
194. N. Uchida, *Phys. Rev. Lett.* **87**, 216101 (2001).
195. M. Oettel, A. Domínguez and S. Dietrich, *Phys. Rev. E* **71**, 051401 (2005).
196. H. Lehle, M. Oettel and S. Dietrich, *EPL* **75**, 174 (2006).
197. A.-F. Bitbol, P. G. Dommersnes and J.-B. Fournier, *Phys. Rev. E* **81**, 050903 (2010).

198. A.-F. Bitbol, K. S. Ronia and J.-B. Fournier, *EPL* **96**, 40013 (2011).
199. B. B. Machta, S. L. Veatch and J. P. Sethna, *Phys. Rev. Lett.* **109**, 138101 (2012).
200. J. Benet, F. Paillusson and H. Kusumaatmaja, *Phys. Chem. Chem. Phys.* **19**, 24188 (2017).
201. N. Gnan, E. Zaccarelli and F. Sciortino, *Nat. Commun.* **5**, 3267 (2014).
202. R. Brito, R. Soto and U. M. B. Marconi, *Granular Matter* **10**, 29 (2007).
203. T. R. Kirkpatrick, J. M. Ortiz de Zárate and J. V. Sengers, *Phys. Rev. Lett.* **110**, 235902 (2013).
204. A. Aminov, Y. Kafri and M. Kardar, *Phys. Rev. Lett.* **114**, 230602 (2015).
205. T. R. Kirkpatrick, J. M. Ortiz de Zárate and J. V. Sengers, *Phys. Rev. Lett.* **115**, 035901 (2015).
206. D. Ray, C. Reichhardt and C. J. O. Reichhardt, *Phys. Rev. E* **90**, 013019 (2014).
207. H. Guo, G. Stan and Y. Liu, *Soft Matter* **14**, 1311 (2018).
208. F. Schmidt, A. Callegari, A. Daddi-Moussa-Ider, B. Munkhbat, R. Verre, T. Shegai, M. Käll, H. Löwen, A. Gambassi and G. Volpe, *Nat. Phys.* **19**, 271 (2023).
209. A. Squarcini, J. M. Romero-Enrique and A. O. Parry, *Phys. Rev. Lett.* **128**, 195701 (2022).

https://doi.org/10.1142/9789811282386_bmatter

Michael E. Fisher — Publications Until 2017

We provide here a selected list of Fisher's publications aggregated throughout his long, illustrious career. This list was edited by Fisher himself and by his assistant. We thank Anne Suplee from UMD for sending it to us.

Fisher's self-portrait.

A. BOOKS, LECTURE SERIES, THESES

1. "The Solution of Problems in Theoretical Physics by Electronic Analogue Methods," Ph.D. Thesis (London, 1957), pp. 314.
2. "Analogue Computing at Ultra-High Speed: an Experimental and Theoretical Study" (with D.M. MacKay) Chapman and Hall (London, 1962); John Wiley (New York, 1962) pp. xv. + 395.

3. "The Nature of Critical Points," lectures delivered at the Summer Institute for Theoretical Physics, University of Colorado, Boulder, 1964; "Lectures in Theoretical Physics," Vol. VIIC, pp. 1–159 (Ed. W.E. Brittin, The University of Colorado Press, 1965).
4. "Природа Критических Точек," M.III. Гитерман 1968, Translation of "The Nature of Critical Points" by M.Sh. Giterman, pp. 1–221 (M.I.R., Moscow, 1968).
5. "Excursions in the Land of Statistical Physics: Selected Reviews by Michael E. Fisher."
   World Scientific Publishing Co. Pte. Ltd. (Singapore, 2016) pp. xi + 401.
6. "Wolf prize in Physics." (Edited by Tsvi Piran, Hebrew University)
   [Michael Ellis Fisher, pp. 2–74] World Scientific Publishing co. Pte. Ltd. (Singapore, 2016) pp. xiii + 1158.

B. SCIENTIFIC PAPERS

1. Slide rule versus bead frame, J. Royal Air Force Technical College, 3, 53–60 (Oct. 1954).
2. The matrix approach to filters and transmission lines, Part I, Electronic Engineering 27, 198–204 (1955).
3. The matrix approach to filters and transmission lines, Part II, Electronic Engineering 27, 258–263 (1955).
4. Higher order differences in the analogue solution of partial differential equations, Journées Inter- nationales de Calcul Analogique — Sept. 55 — Proc. Internat. Conf. Analog. Computation (Brussels, Sept. 1955) pp. 208–213.
5. Higher order differences in the analogue solution of partial differential equations, J. Assoc. Comp. Mach. 3, 325–347 (1956).
6. On iterative processes and functional equations (C. Domb and M.E.F.) Proc. Camb. Phil. Soc. 52, 652–662 (1956).
7. On the continuous solution of integral equations by an electronic analogue, Proc. Camb. Phil. Soc. 53, 162–174 (1957).
8. Optimum design of quarter-squares multipliers with segmented characteristics, J. Sci. Insts. 34, 312–316 (1957).
9. Avoiding the need for dividing units in setting up differential analyzers, J. Sci. Insts. 34, 334–335 (1957).
10. A wide band analogue multiplier using crystal diodes and its application to the study of a non-linear differential equation, Electronic Engineering 29, 580–585 (1957).
11. On random walks with restricted reversals (C. Domb and M.E.F.) Proc. Camb. Phil. Soc. 54, 48–59 (1958).
12. Proposed methods for the analog solution of Fredholm's integral equation, J. Assoc. Comp. Mach. 5, 357–369 (1958).

13. Stability and convergence limitations on the use of analogue computers with resistance network analogues, Brit. J. Appl. Phys. 9, 288–291 (1958).
14. Chain configurations of polymers and polyelectrolytes, J. Chem. Phys. 28, 756–761 (1958).
15. The excluded-volume problem — Remarks on a paper by H.N.V. Temperley, Discussions of the Faraday Soc. No. 25, 1958, "Configurations and Interactions of Macromolecules and Liquid Crystals" (Aberdeen Univ. Press, Aberdeen) 200–2.
16. On the stabilization of matrices and the convergence of linear iterative processes (M.E.F. and A.T. Fuller) Proc. Camb. Phil. Soc. 54, 417–425 (1958).
17. The susceptibility of the Ising model of an antiferromagnet (M.E.F. and M.F. Sykes) Phys. Rev. Lett. 1, 321–2 (1958).
18. Transformations of Ising models, Phys. Rev. 113, 969–981 (1959).
19. Excluded volume problem and the Ising model of ferromagnetism (M.E.F. and M.F. Sykes) Phys. Rev. 114, 45–58 (1959).
20. The susceptibility of the plane Ising model, Physica 25, 521–524 (1959).
21. Lattice statistics in a magnetic field I. A two-dimensional super-exchange antiferromagnet, Proc. Roy. Soc. A 254, 66–85 (1960).
22. On a non-linear differential equation for the zero-point energies of the rare gas solids (M.E.F. and I.J. Zucker) Proc. Camb. Phil. Soc. 57, 107–114 (1961).
23. Limitations due to noise, stability and component tolerance on the solution of partial differential equations by differential analyzers, J. Elec. and Control 8, 113–126 (1960).
REPRINTED in Simulation 3, No. 5, 45–52 (1964).
24. Lattice statistics in a magnetic field II. Order and correlations of a two-dimensional super-exchange antiferromagnet, Proc. Roy. Soc. A 256, 502–513 (1960).
25. The perpendicular susceptibility of an anisotropic antiferromagnet, Physica 26, 618–622, 1028(1960).
26. Association problem in statistical mechanics — A critique of the treatment of H.S. Green and R. Leipnik (M.E.F. and H.N.V. Temperley) Rev. Mod. Phys. 32, 1029–1031 (1960).
27. Configuration and free energy of a polymer molecule with solvent interactions (M.E.F. and B.J. Hiley) J. Chem. Phys. 34, 1253–67 (1961).
28. Some cluster size and percolation problems (M.E.F. and J.W. Essam) J. Math. Phys. 2, 609–619 (1961).
29. Critical probabilities for cluster size and percolation problems, J. Math. Phys. 2, 620–627 (1961).
30. Dimer problem in statistical mechanics — An exact result (H.N.V. Temperley and M.E.F.) Phil. Mag. 6, 1061–63 (1961).
31. Statistical mechanics of dimers on a plane lattice, Phys. Rev. 124, 1664–1672 (1961).

32. Solution of a combinatorial problem — Intermediate statistics, Amer. J. Phys. 30, 49–51 (1962).
33. On the theory of critical point density fluctuations, Physica 28, 172–180 (1962).
34. Antiferromagnetic susceptibilities of the plane square and honeycomb Ising lattices (M.F. Sykes and M.E.F.) Physica 28, 919–938 (1962).
35. Antiferromagnetic susceptibilities of the simple cubic and body-centered cubic Ising lattices (M.E.F. and M.F. Sykes) Physica 28, 939–956 (1962).
36. The entropy of an antiferromagnet in a magnetic field (J.C. Bonner and M.E.F.) Proc. Phys. Soc. 80, 508–515 (1962).
37. Relation between the specific heat and susceptibility of an antiferromagnet, Phil. Mag. 7, 1731–1743 (1962).
38. Perpendicular susceptibility of the Ising model, J. Math Phys. 4, 124–135 (1963).
39. Lattice statistics — A review and an exact isotherm for a plane lattice gas, J. Math. Phys. 4, 278–286 (1963).
40. Padé approximant studies of the lattice gas and Ising ferromagnet below the critical point (J.W. Essam and M.E.F.) J. Chem. Phys. 38, 802–812 (1963).
40A. Free energy, correlations and phase transitions — A new approach to statistical mechanics [abstract] Proc. 2nd Eastern Theoret. Physics Conf., Chapel Hill, North Carolina, October 1963.
41. Statistical mechanics of dimers on a plane lattice II. Dimer correlations and monomers (M.E.F. and J. Stephenson) Phys. Rev. 132, 1411–1431 (1963).
42. Ising model and self-avoiding walks on hypercubical lattices and "high-density" expansions (M.E.F. and D.S. Gaunt) Phys. Rev. 133, A224–A239 (1964).
43. Magnetism in one-dimensional systems — The Heisenberg model for infinite spin, Amer. J. Phys. 32, 343–346 (1964).
44. Linear magnetic chains with anisotropic coupling (J.C. Bonner and M.E.F.) Phys. Rev. 135, A640–A658 (1964).
    REPRINTED in:
    (i) "Mathematical Physics in One-Dimension," Eds. E.H. Lieb and D.C. Mattis (Academic Press, New York 1966) pp. 487–505;
    (ii) "The Many-Body Problem: An encyclopedia of exactly solved models in one dimension," Ed. D.C. Mattis (World Scientific, Singapore, 1993) pp. 717–735.
45. Correlation functions and the critical region of simple fluids, J. Math Phys. 5, 944–962 (1964).
    REPRINTED in:
    (i) "The Equilibrium Theory of Classical Fluids," Eds. H.L. Frisch and J.L. Lebowitz (W.B. Benjamin Inc., N.Y., 1964) pp. III-75-93;
    (ii) "Series of selected Papers in Physics," No. 49 (Phys. Soc. Japan, Tokyo, 1972) pp. 66–84.

46. Cluster size and percolation theory, Proc. I.B.M. Scientific Computing Symposium on Combinatorial Problems, held March 16–18, 1964 (I.B.M., N.Y., 1966) Chap. 11, pp. 179–198.
47. The free energy of a macroscopic system, Arch. Ratl. Mech. Anal. 17, 377–410 (1964).
48. Deviations from van der Waals behaviour on the critical isobar, J. Chem. Phys. 41, 1877–1878 (1964).
49. Specific heat of a gas near the critical point, Phys. Rev. 136, A1599-A1604 (1964).
50. Detailed magnetic behaviour of nickel near its Curie point (J.S. Kouvel and M.E.F.) Phys. Rev. 136, A1626–A1632 (1964).
51. The critical behaviour of ferromagnets, Proc. Internat. Conf. Magnetism, Nottingham, 7–11 September 1964 (Phys. Soc., London, 1965) pp. 79–84.
52. Critical isotherm of a ferromagnet and of a fluid (D.S. Gaunt, M.E.F., M.F. Sykes, and J.W. Essam) Phys. Rev. Lett. 13, 713–715 (1964).
53. Bounds for the derivatives of the free energy and the pressure of a hard-core system near close packing, J. Chem. Phys. 42, 3852–3856 (1965).
54. Hard-sphere lattice gases I. Plane-square lattice (D.S. Gaunt and M.E.F.) J. Chem. Phys. 43, 2840–2863 (1965).
55. Correlation functions and the coexistence of phases, J. Math. Phys. 6, 1643–1653 (1965).
56. The stability of many-particle systems (M.E.F. and D. Ruelle) J. Math. Phys. 7, 260–270 (1966).
57. The Theory of Condensation and the Critical Point, Physics 3, 255–283 (1967). Expanded version of The Theory of Condensation, Proc. Centennial Conf. on Phase Transformation, University of Kentucky, 18–20 March 1965 (Univ. of Kentucky, unpublished).
58. (a) Theory of critical fluctuations and singularities, in "Critical Phenomena," Proc. Conf. "Phenomena in the Neighborhood of Critical Points," N.B.S., Washington D.C., 5–8 April, 1965, Eds. M.S. Green and J.V. Sengers (N.B.S. Misc. Publ. 273, 1966) pp. 108–115.
    (b) Notes, definitions and formulas for critical point singularities, *ibid*, pp. 21–26.
59. The shape of a self-avoiding walk or polymer chain, J. Chem. Phys. 44, 616–622 (1966).
60. On hearing the shape of a drum, J. Comb. Theory 1, 105–125 (1966).
61. Effect of excluded volume on phase transitions in bio-polymers, J. Chem. Phys. 45, 1469–1473 (1966).
    REPRINTED in "Theory of Helix-Coil Transitions in Biopolymers," Eds. D. Poland and H.A. Scheraga (Academic Press, N.Y. 1970) pp. 739–743.
62. Quantum corrections to critical point behavior, Phys. Rev. Lett. 16, 11–14 (1966).

63. On the dimer solution of planar Ising models, J. Math. Phys. 7, 1776–1781 (1966).
64. Theory of critical point scattering and correlations I. The Ising model (M.E.F. and R.J. Burford) Phys. Rev. 156, 583–622 (1967).
65. Magnetic critical point exponents — Their interrelations and meaning, J. Appl. Phys. 38, 981–990 (1967).
66. The theory of equilibrium critical phenomena, Rep. Prog. Phys. 30, 615–731 (1967).
67. Critical temperatures of anisotropic Ising lattices I. Lower bounds (C.-Y. Weng, R.B. Griffiths and M.E.F.) Phys. Rev. 162, 475–479 (1967).
68. Critical temperatures of anisotropic Ising lattices II. General Upper Bounds, Phys. Rev. 162, 480–485 (1967).
69. Interfacial, boundary and size effects at critical points (M.E.F. and A.E. Ferdinand) Phys. Rev. Lett. 19, 169–172 (1967).
70. Intrinsic critical velocity of a superfluid (J.S. Langer and M.E.F.) Phys. Rev. Lett. 19, 560–563 (1967).
    REPRODUCED (on CD-ROM) in "The Physical Review: The first hundred years," Ed. H.H. Stroke (Amer. Inst. Phys. Press, New York, 1995).
71. Resistive anomalies at magnetic critical points (M.E.F. and J.S. Langer) Phys. Rev. Lett. 20, 665–668 (1968).
72. Renormalization of critical exponents by hidden variables, Phys. Rev. 176, 257–272 (1968).
73. The decay of superflow in helium, Proc. Conf. on "Fluctuations in Superconductors," Asilomar, California, March 13–15, 1968, Eds. W.S. Goree and F. Chilton (Stanford Research Institute, Menlo Park, 1968) pp. 357–380.
74. Asymptotic behavior of Toeplitz matrices and determinants (R.E. Hartwig and M.E.F.) Arch. Ratl. Mech. Anal. 32, 190–225 (1969).
75. Toeplitz determinants: Some applications, theorems and conjectures (M.E.F. and R.E. Hartwig) Adv. Chem. Phys. 15, 333–353 (1968).
76. Decay of correlation in linear systems (M.E.F. and B. Widom) J. Chem. Phys. 50, 3756–3772 (1969).
77. (a) Rigorous inequalities for critical point correlation exponents, Phys. Rev. 180, 594–600 (1969).
    (b) Exponent inequalities for critical point spin correlation functions [abstract] J. Appl. Phys. 40, 1278 (1969).
78. Bounded and inhomogeneous Ising models I. Specific heat anomaly of a finite lattice (A.E. Ferdinand and M.E.F.) Phys. Rev. 185, 832–846 (1969).
    REPRINTED in "Conformal Invariance and Applications to Statistical Mechanics," C. Itzykson, H. Salem and J.-B. Zuber (World Scientific, Singapore, 1988) pp. 289–303.
79. Mobile-electron Ising ferromagnets [abstract] (M.E.F. and P.E. Scesney) J. Appl. Phys. 40, 1554 (1969).

80. (a) Aspects of equilibrium critical phenomena, J. Phys. Soc. Japan, Suppl. 26, 87–93 (1969) [Proc. I.U.P.A.P. Conf. on Statistical Mechanics, Kyoto, September 1968].
   (b) A restatement of Flory's theory of excluded volume in polymers, *ibid* pp. 44–55 [discussion remarks].
81. Phase transitions and critical phenomena, in "Contemporary Physics," Vol. I, Trieste Symposium 1968 [Proc. International Symposium, 7–28 June 1968] (International Atomic Energy Agency, Vienna, 1969) pp. 19–46.
82. Absence of anomalous averages in systems of finite non-zero thickness or cross section (G.V. Chester, M.E.F. and N.D. Mermin) Phys. Rev. 185, 760–762 (1969).
83. Light scattering and pseudospinodal curves: The isobutyric-acid-water system in the critical region (B. Chu, F.J. Schoenes and M.E.F.) Phys. Rev. 185, 219–226 (1969).
84. Phase transitions in one-dimensional cluster-interaction fluids IA. Thermodynamics (M.E.F. and B.U. Felderhof) Ann. Phys.(N.Y.) 58, 176–216 (1970).
85. *idem* IB. Critical behavior (M.E.F. and B.U. Felderhof) Ann. Phys.(N.Y.) 58, 217–267 (1970).
86. Phase transitions in one-dimensional classical fluids with many-body interactions, "Systèmes a un Nombre Infini de Degrés de Liberté," Gif-sur-Yvette, France, 7–10 May 1969, Colloq. Internat. du C.N.R.S. No. 181 (Editions du C.N.R.S., Paris 1970) pp. 87–103.
87. Broken symmetry and decay of order in restricted dimensionality (D. Jasnow and M.E.F.) Phys. Rev. Lett. 23, 286–288 (1969).
88. Phase transitions in one-dimensional cluster-interaction fluids II. Simple logarithmic model (B.U. Felderhof and M.E.F.) Ann. Phys. (N.Y.) 58, 268–280 (1970).
88 ˆ*idem* III. Correlation functions (B.U. Felderhof) Ann. Phys.(N.Y.) 58, 281–300 (1970).
89. Some basic definitions in graph theory (J.W. Essam and M.E.F.) Rev. Mod. Phys. 42, 271–288 (1970).
90. Asymptotic free energy of a system with periodic boundary conditions (M.E.F. and J.L. Lebowitz) Commun. Math. Phys. 19, 251–272 (1970).
91. Visibility of critical-exponent renormalization (M.E.F. and P.E. Scesney) Phys. Rev. A 2, 825–835 (1970).
92. Decay of order in isotropic systems of restricted dimensionality I. Bose superfluids (D. Jasnow and M.E.F.) Phys. Rev. B 3, 895–907 (1971).
93. *idem* II. Spin systems (M.E.F. and D. Jasnow) Phys. Rev. B 3, 907–924 (1971).
94. Sums of inverse powers of sines: Solution to Problem 69–14 [SIAM Review 11, 621 (1969)] SIAM Review 13, 116–119 (1971) [Note corrections].

REPRINTED in "Problems in Applied Mathematics: Selections from SIAM Review," Ed. M.S. Klamkin (SIAM, Philadelphia, 1990) pp. 157–160.

95. Critical behavior in the one-dimensional cluster-interaction models, Proc. Midwest Conf. Theoret. Phys., University of Notre Dame, 3–4 April 1970 (National Science Foundation, Washington, 1970) pp. 50–60.
96. Zeros of the partition function for the Heisenberg, ferro-electric and general Ising models (M. Suzuki and M.E.F.) J. Math Phys. 12, 235–246 (1971).
97. Phase transitions, symmetry and dimensionality [Chicago Solid State Colloq. Series] Essays in Physics, Vol. 4 (Academic Press, London, 1972) pp. 43–89.
98. Behavior of two-point correlation functions at high temperatures (W.J. Camp and M.E.F.) Phys. Rev. Lett. 26, 73–77 (1971).
99. Behavior of two-point correlation functions near and on a phase boundary (M.E.F. and W.J. Camp) Phys. Rev. Lett. 26, 565–568 (1971).
100. (a) Critical behavior in spherical models of finite thickness (M.E.F., G.A.T. Allan and M.N. Barber) [summary] I.U.P.A.P. Conference on Statistical Mechanics, Chicago, March-April 1971.
     (b) Critical behavior of an ideal Bose gas of finite thickness (M.E.F. and M.N. Barber) [abstract] I.U.P.A.P. Conference on Statistical Mechanics, Chicago, March-April 1971.
     (c) Finite size effects in the spherical model and ideal Bose gas (M.E.F., M.N. Barber, and G.A.T. Allan) [abstract] Bull. Am. Phys. Soc. 16, 394 (1971).
101. Excitons in Mott-Hubbard insulators (S. Doniach, B.J. Roulet and M.E.F.) Phys. Rev. Lett. 27, 262–265 (1971).
102. (a) Theory of critical-point scattering and correlations II. Heisenberg models (D.S. Ritchie and M.E.F.) Phys. Rev. B 5, 2668–2692 (1972).
     (b) Critical scattering in the Heisenberg model (D.S. Ritchie and M.E.F.)[abstract] Amer. Inst. Phys. Conf. Proc. No. 5, p. 1250 (Amer. Inst. Phys., New York, 1972).
103. Remnant functions (M.E.F. and M.N. Barber) Arch. Ratl. Mech. Anal. 47, 205–236 (1972).
104. The Theory of Critical Point Singularities, pp. 1–99 in "Critical Phenomena," Proc. 1970 Enrico Fermi International School of Physics, Course No. 51, Varenna, Italy, Ed. M.S. Green (Academic Press for the Italian Physical Society, New York, 1971).

     I. Phase transitions and the order parameter
     II. Critical exponents and homogeneity
     III. Scaling, power laws and analyticity
     IV. The droplet model: its extensions and implications
     V. Finite size and boundary effects

TRANSLATED by M.Sh. Giterman and published by MIR (Moscow, 1973) in Устойчивость и фазовые переходы ("Stability and Phase Transitions") pp. 245–369: "Theory of Critical Point Singularities" from "Critical Phenomena".

105. Critical exponents in 3.99 dimensions (K.G. Wilson and M.E.F.) Phys. Rev. Lett. 28, 240–243 (1972).
REPRINTED in: "Series of Selected Papers in Physics," No. 78, "Critical Phenomena II," Eds. S. Hikami and Y. Iwasaki (Phys. Soc. Japan, Tokyo, 1982) pp. 17–20;
(ii) "The Physical Review: The first hundred years," Ed. H.H. Stroke (Amer. Inst. Phys. Press, New York, 1995).

106. (a) Estimation of spectra from moments — Application to the Hubbard model (M.E.F. and W.J. Camp) Phys. Rev. B 5, 3730–3737 (1972).
(b) *idem* [abstract] Bull. Amer. Phys. Soc. 17, 79 (1972).

107. Spectra, moments, and lattice walks for Hubbard magnetic insulators (B.J. Roulet, M.E.F. and S. Doniach) Phys. Rev. B 7, 403–420 (1973).

108. Ferromagnetic Heisenberg lattice films (D.S. Ritchie and M.E.F.) Amer. Inst. Phys. Conf. Proc. No. 5, Magnetism and Magnetic Materials, 1971 (Amer. Inst. Phys., 1972) pp. 1245–1249.

109. Critical behavior of the anisotropic *n*-vector model (M.E.F. and P. Pfeuty) Phys. Rev. B 6, 1889–1891 (1972).

110. On discontinuity of the pressure, Commun. Math. Phys. 26, 6–16 (1972).

111. Decay of order in classical many-body systems I. Introduction and formal theory (W.J. Camp and M.E.F.) Phys. Rev. B 6, 946–959 (1972).

111 ˆ*idem* II. Ising model at high temperatures (W.J. Camp) Phys. Rev. B 6, 960–979 (1972);.

111 ˆˆ*idem* III. Ising model at low temperatures (W.J. Camp) Phys. Rev. B 7, 3187–3203 (1973).

112. Scaling theory for finite-size effects in the critical region (M.E.F. and M.N. Barber) Phys. Rev. Lett. 28, 1516–1519 (1972).
REPRINTED in "Finite-Size Scaling," J.L. Cardy, Ed.(North Holland, Amsterdam, 1988) Paper 1.1, pp. 8–11.
REPRODUCED (on CD-ROM) in "The Physical Review: The first hundred years," Ed. H.H. Stroke (Amer. Inst. Phys. Press, New York, 1995).

113. Critical phenomena in systems of finite thickness I. The spherical model (M.N. Barber and M.E.F.) Ann. Phys. (N.Y.) 77, 1–78 (1973).

114. Finite-size and surface effects in Heisenberg films (D.S. Ritchie and M.E.F.) Phys. Rev. B 7, 480–494 (1973).

115. Critical exponents for long-range interactions (M.E.F., S.-K. Ma and B.G. Nickel) Phys. Rev. Lett. 29, 917–920 (1972).

116. Quantum lattice gas and the existence of a supersolid (K.-S. Liu and M.E.F.) J. Low Temp. Phys. 10, 655–683 (1973).
117. Critical phenomena in systems of finite thickness III. Specific heat of an ideal boson film (M.N. Barber and M.E.F.) Phys. Rev. A 8, 1124–35 (1973).
118. Critical phenomena — Series expansions and their analysis [summary] in "Padé Approximants and their Applications," Ed. P.R. Graves-Morris (Academic Press, 1973) pp. 159–162.
119. Crossover exponent φ for spin systems (P. Pfeuty, M.E.F., and D. Jasnow) Amer. Inst. Phys. Conf. Proc. No. 10, "Magnetism and Magnetic Materials," 1972 (Amer. Inst. Phys., 1973), pp. 817–821.
120. Helicity Modulus, superfluidity and scaling in isotropic systems (M.E.F., M.N. Barber, and D.M. Jasnow) Phys. Rev. A 8, 1111–1124 (1973).
121. Critical behavior of magnets with dipolar interactions I. Renormalization group near four dimensions (A. Aharony and M.E.F.) Phys. Rev. B 8, 3323–3341 (1973).
121 ˆ*idem* II-V. (A. Aharony) Phys. Rev. B 8, 3342–48, 3349–57, 3358–62, 3363–70 (1973).
122. Critical point phenomena — The role of series expansions, Proc. Int. Conf. Padé Approximants, "Continued Fractions and Related Topics" (Univ. Colorado, Boulder, June 1972) Rocky Mountain J. Math. 4, 181–201 (1974).
123. Dipolar interactions at ferromagnetic critical points (M.E.F. and A. Aharony) Phys. Rev. Lett. 30, 559–562 (1973).
124. Three-state Potts model and anomalous tricritical points (J.P. Straley and M.E.F.) J. Phys. A 6, 1310–1326 (1973).
125. Classical, n-component spin systems or fields with negative even integral n, Phys. Rev. Lett. 30, 679–681 (1973).
126. Discussion remarks in Proc. Symp. "Statistical and Probilistic Problems in Metallurgy," Ed. W.L. Nicholson, Suppl. Adv. Appl. Prob. (Dec. 1972) pp. 46, 67–8, 144–5, 216–220, etc.
127. Theory of critical point scattering and correlations III. The Ising model below T_c and in a field (H.B. Tarko and M.E.F.) Phys. Rev. B 11, 1217–1253 (1975). As regards the estimates in Table XIV, see Footnote 25 in [328], below.
128. Critical phenomena in films and surfaces [Proc. Conf. Thin Film Phenomena, 15–16 March 1973, IBM San Jose] J. Vacuum Sci. Techn. 10, 665–673 (1973).
129. (a) Renormalization group, exponents, and scaling [summary, tables, and discussion] pp. 10–20;
(b) Crossover effects and operator expansions, [summary and discussion], pp. 65–72;
(c) Discussion contributions, pp.2,5–9,22–26,30–35,38–41, etc.; in "Renormalization Group in Critical Phenomena and Quantum Field Theory: Proceedings of a Conference," Eds. J.D. Gunton and M.S. Green (Temple University Conference Center, Pennsylvania, 29–31 May 1974).

130. General scaling theory for critical points, Proc. Nobel Symp. XXIV held in Aspenäsgården, Lerum, Sweden, 12–16 June 1973, "Collective Properties of Physical Systems," Eds. B. Lundqvist and S. Lundqvist (Academic Press, New York, 1974) pp. 16–37.
131. Critical scattering in a field and below T_c (H.B. Tarko and M.E.F.) Phys. Rev. Lett. 31, 926–930 (1973).
132. (a) Scaling function for critical scattering (M.E.F. and A. Aharony) Phys. Rev. Lett. 31, 1238–1241, 1537 (1973).
(b) *idem* [abstract] I.U.P.A.P. van der Waals Centennial Conference on Statistical Mechanics, Amsterdam, 27–31 August 1973, Eds. C. Prins and M. Blumendal (North-Holland Publ. Co. 1973).
133. (a) Soluble renormalization groups and scaling fields for low-dimensional spin systems (D.R. Nelson and M.E.F.) Ann. Phys. (N.Y.) 91, 226–274 (1975).
(b) Exact renormalization groups for one-dimensional spin systems [summary] Amer. Inst. Phys. Conf. Proc. No. 18, Magnetism and Magnetic Materials, 1973 (Amer. Inst. Phys., N.Y. 1974) pp. 888–890.
134. Crossover scaling functions for exchange anisotropy (P. Pfeuty, D. Jasnow and M.E.F.) Phys. Rev. B 10, 2088–2112 (1974).
135. The renormalization group in the theory of magnetism, Proc. Internat. Conf. Magnetism, Moscow, 22–28 August 1973, Vol. I, pp. 51–79 (NAUKA, Moscow, 1973).
136. Scaling function for two-point correlations I. Expansion near four dimensions (M.E.F. and A. Aharony) Phys. Rev. B 10, 2818–2833 (1974).
136 ˆ*idem* II. Expansion to order $1/n$(A. Aharony) Phys. Rev. B 10, 2834–44 (1974).
137. Renormalization-group analysis of metamagnetic tricritical behavior (D.R. Nelson and M.E.F.) Phys. Rev. B 11, 1030–1039 (1975).
138. Spin flop, supersolids, and bicritical and tetracritical points (M.E.F. and D.R. Nelson) Phys. Rev. Lett. 32, 1350–53 (1974).
139. Renormalization-group analysis of bicritical and tetracritical points (D.R. Nelson, J.M. Kosterlitz and M.E.F.) Phys. Rev. Lett. 33, 813–817 (1974). REPRINTED in "Selected Papers in Physics," No. 78, "Critical Phenomena II," Eds. S. Hikami and Y. Iwasaki (Phys. Soc. Japan, Tokyo, 1982) pp. 51–55.
140. Gonzalo's scaling function for the equation of state (T.W. Capehart and M.E.F.) Phys. Rev. B 11, 1262–1263 (1975).
141. Tests of strong scaling in the three-dimensional Ising model (M.E.F. and H.B. Tarko) Phys. Rev. B. 11, 1131–1133 (1975).
142. Critical temperatures of classical n-vector models on hypercubic lattices (P.R. Gerber and M.E.F.) Phys. Rev. B 10, 4697–4703 (1974).
143. The renormalization group in the theory of critical behavior, Rev. Mod. Phys. 46, 597–616 (1974).

144. Susceptibility scaling functions for ferromagnetic Ising films (T.W. Capehart and M.E.F.) Phys. Rev. B 13, 5021–5038 (1976).
145. Universality of magnetic tricritical points (M.E.F. and D.R. Nelson) Phys. Rev. B 12, 263–266 (1975).
146. Bounded and inhomogeneous Ising models II. Specific heat scaling function for a strip (H. Au-Yang and M.E.F.) Phys. Rev. B 11, 3469–3487 (1975).
147. Bicritical and tetracritical points in anisotropic antiferromagnetic systems (J.M. Kosterlitz, D.R. Nelson and M.E.F.) Phys. Rev. B 13, 412–432 (1976).
148. Theory of multicritical transitions and the spin-flop bicritical point, Amer. Inst. Phys. Conf. Proc. No 24, "Magnetism and Magnetic Materials," 1974 (Amer. Inst. Phys., N.Y., 1975) pp. 273–280.
149. The renormalization group and its application to critical phenomena [summary] "Statistical Physics," Bose Memorial Symposium, Bangalore, July 1974, Eds. N. Mukunda, A.K. Rajagopal, and K.P. Sinha, Suppl. J. Ind. Inst. Sci., Bangalore, June 1975, pp. 51–55.
150. Critical temperatures of continuous spin models and the free energy of a polymer (P.R. Gerber and M.E.F.) J. Chem. Phys. 63, 4941–4946 (1975).
151. Scaling axes and the spin-flop bicritical phase boundaries, Phys. Rev. Lett. 34, 1634–1638 (1975).
152. Self-interacting walks, random spin systems and the zero-component limit (D. Jasnow and M.E.F.) Phys. Rev. B 13, 1112–1118 (1976).
153. Monte Carlo study of multicriticality in finite Baxter models (E. Domany, K.K. Mon, G.V. Chester and M.E.F.) Phys. Rev. B 12, 5025–5033 (1975).
154. Scaling theory of nonlinear critical relaxation (M.E.F. and Z. Rácz) Phys. Rev. B 13, 5039–5041 (1976).
155. (a) Regularly spaced defects in Ising models (M.E.F. and H. Au-Yang) J. Phys. C: Solid St. Phys., 8, L418-L421 (1975).
(b) Critical effects of regularly spaced defects (M.E.F. and H. Au-Yang) [abstract] Amer. Inst. Phys. Conf. Proc. No. 29, Magnetism and Magnetic Materials 1975 (Amer. Inst. Phys., 1976) p. 490.
156. Bounded and inhomogeneous Ising models III. Regularly spaced defects (H. Au-Yang, M.E.F. and A.E. Ferdinand) Phys. Rev. B 13, 1238–1265 (1976).
156 ˆ*idem* IV. Specific-heat amplitude for regular defects (H. Au-Yang) Phys. Rev. B 13, 1266–1271 (1976).
157. Crossover behavior of the specific heat and nonordering susceptibility of the anisotropic Heisenberg ferromagnet (P.R. Gerber and M.E.F.) Phys. Rev. B 13, 5042–5053 (1976).
158. Equations of state for bicritical points II. Ising-like ordered phase (E. Domany, D.R. Nelson and M.E.F.) Phys. Rev. B 15, 3493–3509 (1977).
159. Magnetization of cubic ferromagnets and the three-component Potts model (D. Mukamel, M.E.F. and E. Domany) Phys. Rev. Lett. 37, 565–568 (1976).

160. Novel two-variable approximants for studying magnetic multicritical behavior, Proc. Int. Conf. on Magnetism, Amsterdam, 1976, 5E-1; Physica 86–88, 590–592 (1977).
161. Critical behavior of cubic and tetragonal ferromagnets in a field (D. Mukamel, E. Domany and M.E.F.) Proc. Int. Conf. on Magnetism, Amsterdam, 1976, 7E-1; Physica 86–88, 572–574 (1977).
162. (a) Destruction of first-order transitions by symmetry-breaking fields (E. Domany, D. Mukamel and M.E.F.) Phys. Rev. B 15, 5432–5441 (1977).
(b) Symmetry-breaking fields and destruction of first-order transitions (D. Mukamel, E. Domany and M.E.F.) Proc. Int. Conf. Magnetism, Ann. Israel Phys. Soc. 2, 451–454 (1978).
163. Equations of state for bicritical points III. Cubic anisotropy and tetracriticality (E. Domany and M.E.F.) Phys. Rev. B 15, 3510–3521 (1977).
164. Series expansion approximants for singular functions of many variables, in "Statistical Mechanics and Statistical Methods in Theory and Application" – Proc. Symposium in honor of E.W. Montroll, Ed. U. Landman (Plenum Press, N.Y., 1977) pp. 3–32.
165. Critical points and multicritical points [abstract] in "Statistical Mechanics and Statistical Methods in Theory and Application" – Proc. Symposium in honor of E.W. Montroll, Ed. U. Landman (Plenum Press, N.Y., 1977) pp. 1–2.
166. Wall and boundary free energies I. Ferromagnetic scalar spin systems (M.E.F. and G. Caginalp) Commun. Math. Phys. 56, 11–56 (1977).
167. Partial differential approximants for multicritical singularities (M.E.F. and R.M. Kerr) Phys. Rev. Lett. 39, 667–670 (1977).
168. Lars Onsager 1903–1976 (H.C. Longuet-Higgins and M.E.F.) Bio. Mem. Fell. Roy. Soc. 24, 445–471 (1978).
REPRINTED (with revisions) in:
(i) Bio. Mem. N.A.S. (U.S.A.), Vol.60, pp. 182–232(1991);
(ii) J. Stat. Phys. 78, 605–640 (1995);
(iii) "The Collected Works of Lars Onsager," Eds. P.C.Hemmer, H. Holden, and S. Kjelstrup Ratke (World Scientific, Singapore, 1996) pp. 9–34.
169. Scaling functions for quantum crossover (I.D. Lawrie and M.E.F.) J. Appl. Phys. 49, 1353–55 (1978).
170. Tricritical scaling in the spherical model limit (S. Sarbach and M.E.F.) J. Appl. Phys. 49, 1350–52 1978).
171. The Yang-Lee edge singularity in spherical models (D.A. Kurtze and M.E.F.) J. Stat. Phys. 19, 205–218 (1978).
172. Wall and boundary free energies II. General domains and complete boundaries (G. Caginalp and M.E.F.) Commun. Math. Phys. 65, 247–280 (1979).
172 ˆ*idem* III. Correlation decay and vector spin systems (G. Caginalp), Commun. Math. Phys, 76, 149–163 (1980).

173. Tricriticality and the failure of scaling in the many-component limit (S. Sarbach and M.E.F.) Phys. Rev. B 18, 2350–2363 (1978).

174. The Yang-Lee edge singularity and φ^3 field theory, Phys. Rev. Lett. 40, 1610–1613 (1978).

175. (a) Nonuniversality of tricritical behavior (M.E.F. and S. Sarbach) Phys. Rev. Lett. 41, 1127–1130 (1978).
 (b) Nonuniversality in magnetic tricriticality (S. Sarbach and M.E.F.) J. Appl. Phys. 50, 1802–1803 (1979).

176. Phénomenes aux parois dans un mélange binaire critique (M.E.F. and P.-G. de Gennes) C.R. Acad. Sc. Paris, Ser. B 287, 207–209 (1978).

177. Inhomogeneous differential approximants for power series (M.E. Fisher and H. Au-Yang) J. Phys. A 12, 1677–1692 (1979); 13, 1517 (1980).

178. Yang-Lee edge singularity in the hierarchial model (G.A. Baker, Jr., M.E.F. and P. Moussa) Phys. Rev. Lett. 42, 615–618 (1979).

179. Yang-Lee edge singularities at high temperatures (D.A. Kurtze and M.E.F.) Phys. Rev. B 20, 2785–2796 (1979).

180. Monte-Carlo study of the spatially modulated phase in an Ising model (W. Selke and M.E.F.) Phys. Rev. B 20, 257–265 (1979).

181. Tricritical coexistence in three dimensions: The multi-component limit (S. Sarbach and M.E.F.) Phys. Rev. B 20, 2797–2817 (1979).

182. Spatially modulated phases in Ising models with competing interactions (W. Selke and M.E.F.) Proc. Int. Conf. Magnetism, Münich, Sept. 1979: J. Mag. Magn. Matls. 15–18, 403–404 (1980).

183. Wall effects in critical systems: Scaling in Ising model strips (H. Au-Yang and M.E.F.) Phys. Rev. B 21, 3956–3970 (1980).

184. Critical phenomena in statistical mechanics — Aspects of renormalization group theory [Summary of 1979 Cargèse Lectures], "Bifurcation Phenomena in Mathematical Physics and Related Topics," Eds. C. Bardos and D. Bessis (Reidel Publ. Co., Dordrecht, 1980) pp. 61–68.

185. (a) Critical wall perturbations and a local free energy functional (M.E.F. and H. Au-Yang) Physica 101A, 255–264 (1980).
 (b) *idem* [abstract] Bull.Amer.Phys.Soc. 25, No.3, 277(1980).

186. Infinitely many commensurate phases in a simple Ising model (M.E.F. and W. Selke) Phys. Rev. Lett. 44, 1502–1505; 45, E148 (1980).
 REPRODUCED (on CD-ROM) in "The Physical Review: The first hundred years," Ed. H.H. Stroke (Amer. Inst. Phys. Press, New York, 1995).

187. Universality tests at Heisenberg bicritical points (M.E.F., J.-H. Chen, and H. Au-Yang) J. Phys. C 13, L459–464 (1980); corrig. location unknown.

188. The states of matter — A theoretical perspective, Proc. Welch Foundation Conf. XXIII, "Modern Structural Methods," Houston, 1979, Ed. W.O. Milligan (R.A. Welch Foundation, Houston, 1980) pp. 74–145; and discussion contributions, pp. 146–175.

189. Cubic fields, bicritical crossover, the spherical and van der Waals limits (P. Seglar and M.E.F.) J. Phys. C 13, 6613–6625 (1980).
190. Two-dimensional Ising models with competing interactions — A Monte Carlo study (W. Selke and M.E.F.) Z. Physik. B — Condensed Matter 40, 71–77 (1980).
191. Discussion remarks in "Order and Fluctuations in Equilibrium and Nonequilibrium Statistical Mechanics:" XVII International Solvay Conference on Physics, Eds. G. Nicolis, G. Dewel and J.W. Turner (John Wiley and Sons, Inc., New York 1981) pp. 14–22; see also pp. 32, 99–100, 105, and 362.
192. This Week's Citation Classic: Fisher, M.E., The theory of equilibrium critical phenomena, Current Contents, P.C. & E.S., 20(46) 18 (1980).
193. Universality in analytic corrections to scaling for planar Ising models (A. Aharony and M.E.F.) Phys. Rev. Lett. 45, 679–682, 1044 (1980).
194. Low-temperature analysis of the axial next-nearest neighbour Ising model near its multiphase point (M.E.F. and W. Selke) Phil. Trans. Roy. Soc. 302, 1–44 (1981).
195. Yang-Lee edge behavior in one-dimensional systems, Prog. Theoret. Phys. Suppl. 69, 14–29 (1980).
196. (a) Multiphase behavior and modulated ordering in soluble Ising models (D.A. Huse, M.E.F. and J.M. Yeomans) Phys. Rev. B 23, 180–185 (1981).
(b) Exactly soluble Ising models exhibiting multiphase points (D.A. Huse, J.M. Yeomans and M.E.F.) J. Appl. Phys. 52, 2028–30 (1981).
197. Simple Ising models still thrive! A review of some recent progress, Physica 106A, 28–47 (1981).
198. Three-component model and tricritical points: A renormalization group study I. Two dimensions (M. Kaufman, R.B. Griffiths, J.M. Yeomans and M.E.F.) Phys. Rev. B 23, 3448–3459 (1981).
199. Three-component model and tricritical points: A renormalization group study II. General dimensions and the three-phase monohedron (J.M. Yeomans and M.E.F.) Phys. Rev. B 24, 2825–2840 (1981).
200. An infinity of commensurate phases in a simple Ising system: The ANNNI model, J. Appl. Phys. 52, 2014–18 (1981).
201. Critical point correlations of the Yvon-Born-Green equation (G.L. Jones, J.J. Kozak, E. Lee, S. Fishman and M.E.F.) Phys. Rev. Lett. 46, 795–798, 1350 (1981).
202. Specific heats of classical spin systems and inhomogeneous differential approximants (J.-H. Chen and M.E.F.) J. Phys. A 14, 2553–2566 (1981).
203. Critical point scaling in the Percus-Yevick equation (S. Fishman and M.E.F.) Physica 106A, 1–13 (1981).
204. Bicriticality and partial differential approximants (M.E.F. and J.-H. Chen) in "Phase Transitions: Cargèse 1980," Eds. M. Levy, J.C. Le Guillou, and J. Zinn-Justin (Plenum Publ. Corp., New York, 1982) pp. 169–216.

205. Many commensurate phases in the chiral Potts model or asymmetric clock models (J.M. Yeomans and M.E.F.) J. Phys. C: Solid State Phys. 14, L835–L839 (1981).
206. Critical scattering and integral equations for fluids (M.E.F. and S. Fishman) Phys. Rev. Lett. 47, 421–423 (1981).
207. (a) Scaling theory for the criticality of fluids between plates (M.E.F. and H. Nakanishi) J. Chem. Phys. 75, 5857–5863 (1981).
    (b) Fluid criticality near walls and between plates [abstract and summary] Int. J. Quant. Chem.: Quantum Chem. Symp. 16, 237–240 (1982).
208. Analysis of the multiphase region in the three-state chiral clock model (J.M. Yeomans and M.E.F.) Physica 127A, 1–37 (1984).
209. Scaling, universality, and renormalization group theory (Lecture notes prepared with the assistance of A.G. Every) in Lect. Notes in Physics, Vol. 186, "Critical Phenomena," Proc. Summer School at the University of Stellenbosch, January 1982, Ed. F.J.W. Hahne (Springer-Verlag, Berlin, 1983) pp. 1–139.
210. Wall wandering and the dimensionality dependence of the commensurate-incommensurate transition (M.E.F. and D.S. Fisher) Phys. Rev. B 25, 3192–3198 (1982).
211. Unbiased estimation of corrections to scaling by partial differential approximants (J.-H. Chen, M.E.F. and B.G. Nickel) Phys. Rev. Lett 48, 630–634 (1982).
212. Partial differential approximants for multivariable power series I. Definitions and faithfulness (M.E.F. and D.F. Styer) Proc. Roy. Soc. A 384, 259–298 (1982).
213. Partial differential approximants for multivariable power series II. Invariance properties (D.F. Styer and M.E.F.) Proc. Roy. Soc. A 388, 75–102 (1983).
213 ˆ*idem* III. Enumeration of invariance transformations (D.F. Styer) Proc. Roy. Soc. A 390, 321–339 (1983).
214. Melting, order, flows, mappings and chaos (M.E.F. and D.A. Huse) "Melting, Localization and Chaos," Proc. Ninth Midwest Solid State Theory Symposium, Argonne, Nov. 1981, Eds. R. Kalia and P. Vashishta (Elsevier Sci. Publ. Co., New York, 1982) pp. 259–293.
215. Scaling for first-order transitions in thermodynamic and finite systems (M.E.F. and A.N. Berker) Phys. Rev. B 26, 2507–2513 (1982).
216. The decoupling point of the axial next-nearest-neighbour Ising model and marginal crossover (D.A. Huse and M.E.F.) J. Phys. C: Solid State Phys. 15, L585–L595 (1982).
217. Nonlinear scaling fields and corrections to scaling near criticality (A. Aharony and M.E.F.) Phys. Rev. B 27, 4394–4400 (1983).
218. Domain walls and the melting of commensurate surface phases (D.A. Huse and M.E.F.) Phys. Rev. Lett. 49, 793–796 (1982).

219. Criticality in the Yvon-Born-Green and similar integral equations (M.E.F. and S. Fishman) J. Chem. Phys. 78, 4227–4244 (1983).
220. Critical phenomena in fluid films: Critical-temperature-shifts (H. Nakanishi and M.E.F.) J. Phys. C, 16, L95–L97 (1983).
221. Multicriticality of wetting, prewetting and surface transitions (H. Nakanishi and M.E.F.) Phys. Rev. Lett. 49, 1565–1568 (1982).
222. Critical point shifts in films (H. Nakanishi and M.E.F.) J. Chem. Phys. 78, 3279–3293 (1983).
223. Surface tension variation in multiphase fluid systems (D.J. Klinger, M.E.F. and B. Widom) J. Phys. Chem. 87, 2841–2845 (1983).
224. Commensurate melting, domain walls, and dislocations (D.A. Huse and M.E.F.) Phys. Rev. B 29, 239–270 (1984).
225. Melting and wetting transitions in the three-state chiral clock model (D.A. Huse, A.M. Szpilka, and M.E.F.) Physica 121A, 363–398 (1983).
226. Continuum fluids with a discontinuity in the pressure (G.W. Milton and M.E.F.) J. Stat. Phys. 32, 413–438 (1983).
227. Critical behavior of a three-dimensional dimer model (S.M. Bhattacharjee, J.F. Nagle, D.A. Huse and M.E.F.) J. Stat. Phys. 32, 361–374 (1983).
228. Convergence of finite-size scaling renormalization techniques (V. Privman and M.E.F.) J. Phys. A: Math. Gen. 16, L295–L301 (1983).
REPRINTED in "Finite-Size Scaling," Ed., J.L. Cardy (North Holland, Amsterdam, 1988) Paper 1.11, pp. 149–181.
229. Finite-size effects at first order transitions (V. Privman and M.E.F.) J. Stat. Phys. 33, 385–417 (1983).
REPRINTED in "Finite-Size Scaling," Ed., J.L. Cardy (North Holland, Amsterdam, 1988) Paper 2.8, pp. 292–298.
230. (a) The melting of commensurate phases and new universality classes [abstract](M.E.F. and D.A. Huse) J. Phys. Soc. Japan 52, Suppl. (1983) p. 31.
(b) Transition to "New Type of Ordered Phase": Four illustrations From the summary talk by M.E.F., *ibid,* pp.00,94,172,234.
231. Lectures on multicritical behavior in "Multicritical Phenomena," Proc. NATO Advanced Study Institute, Geilo, Norway, 10–21 April 1983, Eds. R. Pynn and A. Skjeltorp (Plenum Publ. Corp., New York, 1984) [summaries]:-

(a) Multicriticality: A theoretical introduction, pp. 1–5;
(b) A plenitude of commensurate phases in simple models, pp. 233–235;
(c) Commensurate melting and domain walls in surface phases, pp. 289–291.

232. Wetting transitions near bulk triple points (R. Pandit and M.E.F.) Phys. Rev. Lett. 51, 1772–5 (1983).
233. Nonlinear extensions of square gradient theory for fluid pair correlations (D.J. Klinger, M.E.F. and S. Fishman) J. Chem. Phys. 80, 3392–98 (1984).

234. Walks, walls, wetting and melting, J. Stat. Phys. 34, 667–729 (1984).
235. Universal critical amplitudes in finite-size scaling (V. Privman and M.E.F.) Phys. Rev. B 30, 322–327 (1984).
236. Nonlinearities in differential equations for near-critical pair correlation functions (D.J. Klinger and M.E.F.) Phys. Rev. Lett. 52, 400 (1984).
237. Partial differential approximants and the elucidation of multisingularities (D.F. Styer and M.E.F.) in Lect. Notes in Math. Vol. 1105, "Rational Approximation and Interpolation," [Proc. Conf., Tampa, 1983] Eds. P.R. Graves-Morris, E.B. Saff and R.S. Varga (Springer-Verlag, Heidelberg, 1984) pp. 313–330.
238. The winding angle of planar self-avoiding walks (M.E.F., V. Privman and S. Redner) J. Phys. A: Math. Gen. 17, L569–L578 (1984).
239. Critical behavior of a dilute interacting Bose fluid (M. Rasolt, M.J. Stephen, M.E.F. and P.B. Weichman) Phys. Rev. Lett. 53, 798–801 (1984).
240. Corrections to scaling and crossover in two-dimensional Ising-like systems (M. Barma and M.E.F.) Phys. Rev. Lett. 53, 1935–38 (1984).
241. Finite-size rounding of first-order transitions (V. Privman and M.E.F.) J. Appl. Phys. 57, 3327–28 (1985).
242. Two-dimensional Ising-like systems: Corrections to scaling in the Klauder and double Gaussian models (M. Barma and M.E.F.) Phys. Rev. B 31, 5954–75 (1985).
243. Walks, walls and ordering in low dimensions, in "Fundamental Problems in Statistical Mechanics VI," Proc. Summer School, Trondheim, Norway, June 1984, Ed. E.G.D. Cohen (Elsevier Science Publ., B.V., Amsterdam, 1985) pp. 1–50.
244. First-order transitions breaking O(n) symmetry: Finite-size scaling (M.E.F. and V. Privman) Phys. Rev. B 32, 447–464 (1985).
245. First-order transitions in spherical models: Finite-size scaling (M.E.F. and V. Privman) Commun. Math. Phys. 103, 527–548 (1986).
246. (a) Critical Phenomena: Some recent theoretical developments, J. Appl. Phys. 57, 3265–67 (1985).
(b) Phase transitions and criticality: Finite-size effects, conformal covariance, and other current problems, Proc. ICM'85, J. Mag. Magn. Matls. 54–57, 646–648 (1986).
247. The validity of hyperscaling in three dimensions for scalar spin systems (M.E.F. and J.-H. Chen) J. Phys. (Paris) 46, 1645–54 (1985).
248. Spin-wave corrections to single-domain behavior in finite ferromagnetic systems (V. Privman and M.E.F.) Proc. ICM'85, J. Mag. Magn. Matls. 54–57, 663–4 (1986).
249. Barma and Fisher Respond [to Comment "Universality among scalar spin systems" by G.A. Baker, Jr. and J.D. Johnson] (M.E.F. and M. Barma) Phys. Rev. Lett. 54, 2462 (1985).

250. Criticality and superfluidity in a dilute Bose fluid (P.B. Weichman, M. Rasolt, M.E.F. and M.J. Stephen) Phys. Rev. B 33, 4632–4663 (1986).
251. Wetting in random systems (R. Lipowsky and M.E.F.) Phys. Rev. Lett. 56, 472–475 (1986).
252. Spatial symmetries and critical phenomena, Proc. Fifth Phillip Morris Science Symposium, 1985, "Natural Products Research: The Impact of Scientific Advances," Ed. C.G. Lunsford (Phillip Morris, Inc., New York, 1987) pp. 17–60.
253. Finite-size effects in the spherical model of ferro-magnetism: Zero-field susceptibility under antiperiodic boundary conditions (S. Singh, R.K. Pathria and M.E.F.) Phys. Rev. B 33, 6415–6422 (1986).
254. The location of renormalization-group fixed points (M.E.F. and M. Randeria) Phys. Rev. Lett. 56, 2332 (1986).
255. (a) Interface wandering in adsorbed and bulk phases, pure and impure, J. Chem. Soc., Faraday Trans. 2, 82, 1569–1603 (1986) (Faraday Symp. 20).
(b) General discussions [Pure-to-random crossover] *loc. cit.* pp. 1818–19.
256. Classifying first-order phase transitions (M.E.F. and G.W. Milton) Physica 138A, 22–54 (1986).
257. Multisingularity and scaling in partial differential approximants. I. (M. Randeria and M.E.F.) Proc. Roy. Soc. A 419, 181–203 (1988).
258. (a) Domain-wall interactions and spatially modulated phases (A.M. Szpilka and M.E.F.) Phys. Rev. Lett. 57, 1044–47 (1986).
(b) Domain-wall interactions and high-order commensurate phases [abstract] (A.M. Szpilka and M.E.F.) Proc. STATPHYS-16, Boston University, August 1986.
259. (a) Helium in Vycor, constrained randomness and the Harris criterion (P.B. Weichman and M.E.F.) Phys. Rev. B 34, 7652–7665 (1986).
(b) Dilute Bose fluids, helium in Vycor and constrained randomness [abstract](P.B. Weichman and M.E.F.) Bull. Amer. Phys. Soc. 31 (4) 782 (1986).
260. Long-range crossover and "nonuniversal" exponents in micellar solutions, Phys. Rev. Lett. 57, 1911–14 (1986).
261. Unusual bifurcation of renormalization group fixed points for interfacial transitions (R. Lipowsky and M.E.F.) Phys. Rev. Lett 57, 2411–14 (1986).
262. Domain wall interactions I. General features and phase diagrams for spatially modulated phases (M.E.F. and A.M. Szpilka) Phys. Rev. B 36, 644–666 (1987).
263. Scaling regimes and functional renormalization for wetting transitions (R. Lipowsky and M.E.F.) Phys. Rev. B 36, 2126–2141 (1987).

264. Domain wall interactions II. High-order phases in the axial next-nearest-neighbor Ising model (M.E.F. and A.M. Szpilka) Phys. Rev. B 36, 5343–5362 (1987).

265. Domain wall interactions III. High-order phases in the three-state chiral clock model (A.M. Szpilka and M.E.F.) Phys. Rev. B 36, 5363–5376 (1987).

266. (a) Random coupling crossover in Ising ferromagnets (R.R.P. Singh and M.E.F.) Phys. Rev. B 37, 1980–85 (1988).
   (b) Bond randomness at Ising criticality (R.R.P. Singh and M.E.F.) Proc. MMM Conf. 1987, J. Appl. Phys. 63, 3082 (1988).

267. Diffusion from an entrance to an exit, IBM J. Res. Develop. 32, 76–81 (1988).

268. Disordered systems which escape the bound $\nu \geq 2/d$ (R.R.P. Singh and M.E.F.) Phys. Rev. Lett 60, 548 (1988).

269. Short-range Ising spin-glasses in general dimensions (R.R.P. Singh and M.E.F.) Proc. MMM Conf. 1987, J. Appl. Phys. 63, 3994–96 (1988).

270. (a) Thermodynamic behavior of two-dimensional vesicles (S. Leibler, R.R.P. Singh and M.E.F.) Phys. Rev. Lett. 59, 1989–92 (1987).
   REPRINTED (with revisions) in:
   (i) J.A. Blackman and J. Tagüeña, Eds., "Disorder in Condensed Matter Physics" (Oxford Univ. Press, 1990) Chap. 30, pp. 389–396;
   (ii) F. Ramos-Gómez, Ed., "Proc. Fifth Mexican School on Statistical Physics," Oaxtepec, 1989 (World Scientific Pub., Singapore, 1991) pp. 41–56.

270. (b) Membranes and vesicles: Their statistical mechanics in two dimensions, Proc. 3rd Univ. California Conf. Stat. Mech., Ed. C. Garrod (North Holland Publ. Co., 1988) Nucl. Phys. B (Proc. Suppl.) 5A, 165–167 (1988).
   (c) The statistical mechanics of two-dimensional vesicles, J. Math. Chem. 4, 395–399 (1990).

271. Condensed matter physics: Does quantum mechanics matter? in "Niels Bohr: Physics and the World," Eds. H. Feshbach, T. Matsui and A. Oleson (Harwood Academic Publ., Chur, 1988) pp. 65–115.

272. (a) Phases and phase transitions in less than three dimensions [summary], in "Physics in a Technological World," Proc. XIX General Assembly, I.U.P.A.P., Ed. A.P. French (Amer. Inst. Phys., New York, 1988), pp. 177–183.
   (b) Phase transitions and fluctuations in less than three dimension, in "Frontiers of Physics," Proc. Landau Memorial Conf., Tel Aviv, June 1988, Eds. E. Gotsman, Y. Ne'eman and A. Voronel (Pergamon Press, Oxford, 1990) pp. 195–205.

273. Critical points, large-dimensionality expansions, and the Ising spin glass (M.E.F. and R.R.P. Singh) in "Disorder in Physical Systems," Eds. G. Grimmett and D.J.A. Welsh (Oxford Univ. Press, 1990) pp. 87–111.

274. Finite-size effects in surface tension: Thermodynamics and the Gaussian interface model (M.P. Gelfand and M.E.F.) Proc. 10th Symp. Thermophysical Props. 1988, Eds. A. Cezairliyan and J.V. Sengers, Int. J. Thermophys. 9, 713–727 (1988).
275. Double Summations (Solution to Problem 87–109 by D. Sommers) SIAM Review 30, 320–322 (1988).
276. The reunions of three dissimilar vicious walkers (M.E.F. and M.P. Gelfand) J. Stat. Phys. 53, 175–189 (1988); 55, 472 (1989)[E].
277. Wetting in a two-dimensional random-bond Ising model (M. Huang, M.E.F. and R. Lipowsky) Phys. Rev. B 39, 2632–2639 (1989).
278. The three-dimensional Ising model revisited numerically (A.J. Liu and M.E.F.) Physica A 156, 35–76 (1989).
279. Surface tension of the three-dimensional Ising model: A low-temperature series analysis (L.J. Shaw and M.E.F.) Phys. Rev. A 39, 2189–93 (1989).
280. The spin-1/2antiferromagnetic XXZ chain: New results and insights (R.R.P. Singh, M.E.F. and R. Shankar) Phys. Rev. B 39, 2562–2567 (1989).
281. Interfaces: Fluctuations, interactions and related transitions, in "Statistical Mechanics of Membranes and Surfaces," Eds. D.R. Nelson, T. Piran, and S. Weinberg (World Scientific Publ., Singapore, 1989) pp. 19–44.
REPRINTED (with corrections and additions) in an Extended Second Edition (World Scientific, Singapore, 2004) pp. 19–47.
282. (a) The universal critical adsorption profile from optical experiments (A.J. Liu and M.E.F.) Phys. Rev. A 40, 7202–7221 (1989).
(b) Theoretical analysis of critical adsorption experiments [abstract] (A.J. Liu and M.E.F.) Bull. Amer. Phys. Soc., 34 (3) 650 (1989).
283. The shapes of bowed interfaces in the two-dimensional Ising model (L.-F. Ko and M.E.F.) J. Stat. Phys. 58, 249–264 (1990).
284. Fractal and nonfractal shapes in two-dimensional vesicles, Physica D 38, 112–118 (1989); Proc. Conf. "Fractals in Physics," 1–4 October 1989, Vence, France, Eds. A. Aharony and J. Feder (North Holland Publ. Co., 1989).
285. On the corrections to scaling in three-dimensional Ising models (A.J. Liu and M.E.F.) J. Stat. Phys. 58, 431–442(1990).
286. (a) Phases and phase diagrams: Gibbs' legacy today, in "Proceedings of the Gibbs Symposium, Yale University, May 1989," Eds. D.G. Caldi and G.D. Mostow (Amer. Math. Soc., Rhode Island, 1990) pp. 39–72.
(b) Phases and phase diagrams: Gibbs' legacy today [summary] Physica A 163, 15–16 (1990).
287. Tunable fractal shapes in self-avoiding polygons and planar vesicles (C.J. Camacho and M.E.F.) Phys. Rev. Lett. 65, 9–12 (1990).
288. Finite-size effects in fluid interfaces (M.P. Gelfand and M.E.F.) Physica A 166, 1–74 (1990).

289. Size of an inflated vesicle in two dimensions (A.C. Maggs, S. Leibler, M.E.F. and C.J. Camacho) Phys. Rev. A 42, 691–695 (1990).
290. Low-dimensional quantum antiferromagnets: Criticality and series expansions at zero temperature [extended abstract] Proc. Conf. "Frontiers of Condensed Matter Physics," Physica A 168, 22 (1990).
291. Interfaces, membranes: rough, smooth and interacting [summary] Proc. Conf. "Nonlinear Science: The Next Decade," Physica D 51, 498–500 (1991).
292. Universality and interfaces at critical end points (M.E.F. and P.J. Upton) Phys. Rev. Lett. 65, 2402–2405 (1990).
293. Critical endpoints, interfaces, and walls, Proc. Amer. Chem. Soc. "Symposium on Surfaces and Interfaces," Physica A 172, 77–86 (1991).
294. Fluid interface tensions near critical end points (M.E.F. and P.J. Upton) Phys. Rev. Lett. 65, 3405–8 (1990).
295. Phase boundaries near critical end points I. Thermo-dynamics and universality (M.E.F. and M.C. Barbosa) Phys. Rev. B 43, 11177–184 (1991).
296. Phase boundaries near critical end points II. General spherical models (M.C. Barbosa and M.E.F.) Phys. Rev. B 43, 10635–646 (1991).
297. Semiflexible planar polymeric loops (C.J. Camacho, M.E.F. and R.R.P. Singh) J. Chem. Phys. 94, 5693–5700 (1991).
298. Two-dimensional lattice vesicles and polygons (M.E.F., A.J. Guttmann and S.G. Whittington) J. Phys. A: Math. Gen. 24, 3095–3106 (1991).
299. Effective potentials, constraints, and critical wetting theory (M.E.F. and A.J. Jin) Phys. Rev. B 44, 1430–33 (1991).
300. Simulations of planar vesicles and their transitions (C.J. Camacho and M.E.F.) in "Computer Simulation Studies in Condensed Matter Physics IV," Eds. D.P. Landau, K.K. Mon and H.B. Schuttler (Springer Verlag, Berlin, 1993) pp. 189–193.
301. Effective interface Hamiltonians for short-range critical wetting (A.J. Jin and M.E.F.) Phys. Rev. B 47, 7365–7388 (1993).
302. Microcanonical density functionals for critical systems: An exact one-dimensional example (L.V. Mikheev and M.E.F.) J. Stat. Phys. 66, 1225–1244 (1992).
303. The critical behavior of model electrolytes, Comment on: *J. Chem. Phys. 93, 8405 (1990)*, J. Chem. Phys. 96, 3352–54 (1992); corrected version has an abstract: pp. 3352–55.
304. The isothermal binodal curves near a critical endpoint (Y.C. Kim, M.E.F. and M.C. Barbosa) J. Chem. Phys. 115, 933–950 (2001).
305. Interfacial stiffness and the wetting parameter: The simple cubic Ising model (M.E.F. and H. Wen) Phys. Rev. Lett. 68, 3654 (1992).
306. Rods to self-avoiding walks to trees in two dimensions (C.J. Camacho, M.E.F. and J.P. Straley) Phys. Rev. A 46, 6300–6310 (1992).

307. (a) Is short-range "critical" wetting a first-order transition? (M.E.F. and A.J. Jin) Phys. Rev. Lett. 69, 792–5 (1992).
     (b) An interfacial-stiffness instability in critical wetting [abstract] (A.J. Jin and M.E.F.) Bull. Amer. Phys. Soc. 37 (2) 952 (1992) G10 8.
308. (a) Surface tension of helium at the superfluid critical end-point (L.V. Mikheev and M.E.F.) J. Low Temp. Phys. 90, 119–138 (1993).
     (b) The λ-point singularity of the liquid helium surface tension [abstract] (M.E.F. and L.V. Mikheev) Bull. Amer. Phys. Soc. 37 (2) 952 (1992) G107.
309. Exact variational analysis of layered planar Ising models (L.V. Mikheev and M.E.F.) Phys. Rev. Lett. 70, 186–189 (1993).
310. Stiffness instability in short-range critical wetting (A.J. Jin and M.E.F.) Phys. Rev. B 48, 2642–2658 (1993).
311. Two-dimensional layered Ising models: Exact variational formulation and analysis (L.V. Mikheev and M.E.F.) Phys. Rev. B 49, 378–402 (1994).
311 ˆReentrant dimension crossover in planar ising superlattices (L.V. Mikheev), J. Stat. Phys. 78, 79–101.
312. Criticality in ionic fluids: Debye-Hückel, Bjerrum and beyond (M.E.F. and Y. Levin) Phys. Rev. Lett. 71, 3826–3829 (1993).
313. The story of Coulombic criticality, J. Stat. Phys. 75, 1–36 (1994).
314. On the stiffness of an interface near a wall (M.E.F., A.J. Jin and A.O. Parry) Ber. Bunsenges. Phys. Chem. 98, 357–361 (1994).
315. Cavity forces and criticality in electrolytes (X.-J. Li, Y. Levin and M.E.F.) Europhys. Lett. 26, 683–688 (1994).
316. The interaction of ions in an ionic medium (M.E.F., Y. Levin and X.-J. Li) J. Chem. Phys. 101, 2273–82 (1994).
317. Coulombic criticality in general dimensions (Y. Levin, X.-J. Li and M.E.F.) Phys. Rev. Lett. 73, 2716–19 (1994); 75 [E] 3374 (1995).
318. On the absence of intermediate phases in the two-dimensional Coulomb gas (M.E.F., X.-J. Li and Y. Levin) J. Stat. Phys. 79, 1–11; 81 [E] 865 (1995).
319. Criticality in the hard-sphere ionic fluid (Y. Levin and M.E.F.) Physica A 225, 164–220 (1996).
320. On the critical polynomial of the simple cubic Ising model, J. Phys. A: Math. Gen. 28, 6323–33 (1995).
321. The universal repulsive-core singularity and Yang-Lee edge criticality (S.-N. Lai and M.E.F.) J. Chem. Phys. 103, 8144–55 (1995).
322. Universal surface-tension and critical isotherm amplitude ratios in three dimensions (S.-Y. Zinn and M.E.F.) Physica A 226, 168–180 (1996).
323. Foreword: About the author and the subject, to "The Critical Point" by C. Domb (Taylor and Francis, London, 1996), pp. xiii–xviii.
324. (a) Density fluctuations in an electrolyte from generalized Debye-Hückel theory (B.P. Lee and M.E.F.) Phys. Rev. Lett. 76, 2906–10 (1996).

(b) Inhomogeneous Debye-Hückel theory and density fluctuations near Coulombic criticality [abstract] (B.P. Lee and M.E.F.) Bull. Amer. Phys. Soc. 41 (1) 377 1996) I20 1.

325. (a) Ginzburg criterion for Coulombic criticality (M.E.F. and B.P. Lee) Phys. Rev. Lett. 77, 3561–64 (1996).

(b) The Ginzburg criterion for ionic criticality according to Debye-Hückel based theories [abstract] (M.E.F. and B.P. Lee) Bull. Amer. Phys. Soc. 41 (1) 377 (1996) I20 2.

326. (a) Dipolar-ion pairs and Debye-Hückel theory: The charging process and reciprocity [abstract](B.P. Lee and M.E.F.) Bull. Amer. Phys. Soc. 41 (1) 356 (1996) I7 11.

(b) Generalized Debye-Hückel theory and ionic criticality (B.P. Lee and M.E.F.) [in preparation].

327. Criticality in Gaussian-molecule mixtures (S.-N. Lai and M.E.F.) Molec. Phys. 88, 1373–1397 (1996).

328. Renormalized coupling constants and related amplitude ratios for Ising systems (S.-Y. Zinn, S.-N. Lai, and M.E.F.) Phys. Rev. E 54, 1176–1182 (1996).

329. (a) Coulombic criticality in general dimensions: Progress and challenges [extended abstract] Proc. Hayashibara Forum '95: "Coherent Approaches to Fluctuations," Eds. M. Suzuki and N. Kawashima (World Scientific Publ. Co., Singapore, 1996) pp. 2–5.

(b) The nature of criticality in ionic fluids, Proc. Third Liquid Matter Conference, J. Phys.: Condens. Matter 8, 9103–9109 (1996).

(c) Criticality in hard-core ionic systems: Progress and challenges [extended abstract] Proc. Inauguration Conf. Asia Pacific Center for Theoretical Physics, "Current Topics in Physics," Eds. Y.M. Cho, J.B. Hong and C.N. Yang (World Scientific Publ. Co., Singapore, 1998) pp. 100–105.

330. The superfluid transition in a dilute Bose gas: Experiments and theory [abstract] Proc. Inauguration Conf. Asia Pacific Center for Theoretical Physics, "Current Topics in Physics," Eds. Y.M. Cho, J.B. Hong and C.N. Yang (World Scientific Publ. Co., Singapore, 1998) p. 175.

331. Right and wrong near critical endpoints (M.E.F. and Y.C. Kim) J. Chem. Phys. 117, 779–787 (2002).

332. (a) Prewetting transitions in a near-critical metallic vapor (V.F. Kozhevnikov, D.I. Arnold, S.P. Naurzakov and M.E.F.) Phys. Rev. Lett. 78, 1735–1738 (1997).

(b) Prewetting phenomena in mercury vapor (V.F. Kozhevnikov, D.I. Arnold, S.P. Naurzakov, and M.E.F.) Fluid Phase Equilibria, 150–151, 625–632 (1998).

333. Renormalization group theory: Its basis and formulation in statistical physics, Rev. Mod. Phys. 70, 653–681 (1998).

REPRINTED (with revisions) in "Conceptual Foundations of Quantum Field Theory," Ed., T.Y. Cao (Cambridge Univ. Press, 1998) Part IV, Chap. 8, pp. 89–135.

334. (a) Critique of primitive model electrolyte theories (D.M. Zuckerman, M.E.F. and B.P. Lee) Phys. Rev. E 56, 6569–6580 (1997).
(b) Critique of elcctrolyte theories using thermodynamic bounds (M.E.F., D.M. Zuckerman and B.P. Lee) in "Strongly Coupled Coulomb Systems," Proc. Conf. held in Boston College, August 1997, Eds. G.J. Kalman, J.M. Rommel and K.B. Blagoev (Plenum Press, New York 1998) pp. 415–418.

335. (a) Charge oscillations in Debye-Hückel theory (B.P. Lee and M.E.F.) Europhys. Lett. 39, 611–616 (1997).
(b) Charge oscillations in electrolytes from generalized Debye-Hückel theory [abstract] (B.P. Lee and M.E.F.) Bull. Amer. Phys. Soc. 42 (1), 725 (1997) Q16 9.

336. (a) Diverging correlation lengths in electrolytes: Exact results at low densities (S. Bekiranov and M.E.F) Phys. Rev. E 59, 492–511 (1999).
(b) Exact density correlation length for electrolytes at low densities [abstract] (S. Bekiranov and M.E.F.) Bull. Amer. Phys. Soc. 42 (1) 725 (1997) Q16 8.

337. Electrolyte criticality and generalized Debye-Hückel theory (M.E.F., B.P. Lee and S. Bekiranov) in "Strongly Coupled Coulomb Systems," Proc. Conf. held in Boston College, August 1997, Eds. G.J. Kalman, J.M. Rommel and K. B. Blagoev (Plenum Publ. Corp., 1998) pp. 33–41.

338. Exact thermodynamic formulation of chemical association (M.E.F. and D.M. Zuckerman) J. Chem. Phys. 109, 7961–81 (1998).

339. Chemical association via exact thermodynamic formulations (M.E.F. and D.M. Zuckerman) Chem. Phys. Lett. 293, 461–468 (1998).

340. (a) Fluctuations in electrolytes: The Lebowitz and other correlation lengths (S. Bekiranov and M.E.F.) Phys. Rev. Lett. 81, 5836–39 (1998).
(b) Charge and density fluctuations in electrolytes: The Lebowitz and other correlation lengths (M.E.F. and S. Bekiranov) Proc. STATPHYS 20, Physica A 263, 466–476 (1999).
(c) Correlation lengths in electrolytes: Exact results and sensible approximations [abstract] Proc. 1999 Int. Conf. on "Strongly Coupled Coulomb Systems," Saint-Malo, France, 4–10 September 1999, Eds. C. Deutsch, B. Jancovici, and M.-M. Gombert (EDP Sciences, Les Ulis, France, 2000) J. Physique IV 10, Pr5–17 (2000).

341. The shape of the van der Waals loop and universal critical amplitude ratios (M.E.F. and S.-Y. Zinn) J. Phys. A: Math. Gen. 31, L629-L635 (1998).

342. Understanding criticality: Simple fluids and ionic fluids [summary] Proc. NATO Advanced Study Institute, "New Approaches to New and Old Problems in Liquid State Theory," Messina, Sicily, Eds. C. Caccamo et al. (Kluwer Academic Publ., Dordrecht, 1999) pp. 3–8.
343. Some views from forty years as a statistical mechanician [retrospective outline] Proc. STATPHYS 20, Physica A 263, 554–556 (1999).
344. Trigonometric models for scaling behavior near criticality (M.E.F., S.-Y. Zinn and P.J. Upton) Phys. Rev. B 59, 14533–545 (1999); Erratum: *ibid* 64, 149901(E) (2001).
345. Crossover scales at the critical points of fluids with electrostatic interactions (A.G. Moreira, M.M. Telo da Gama and M.E.F.) J. Chem. Phys. 110, 10058–66 (1999).
346. Scaling for interfacial tensions near critical endpoints (S.-Y. Zinn and M.E.F.) [arXiv: cond-mat/0410673] Phys. Rev. E 71, 011601:1–17 (2005).
347. The force exerted by a molecular motor (M.E.F. and A.B. Kolomeisky) Proc. Natl. Acad. Sci. USA, 96, 6597–6602 (1999).
348. Understanding criticality: Magnetism as the key [abstract] invited talk in "The History of Magnetism" session, American Physical Society Centennial Meeting, Bull. Amer. Phys. Soc. 44(1) Part II, 1400 (1999) UA01 4.
349. Discussion remarks on quantum field theory, invited contributions in "Conceptual Foundations of Quantum Field Theory," Ed., T.Y. Cao (Cambridge University Press, 1999) pp. 72–73, 86–88, 135, 165, 231, 250, 266–267, 269–272, 274–275, 278–286, 377, 383.
350. Identity of the universal repulsive-core singularity with Yang-Lee edge criticality (Y. Park and M.E.F.) Phys. Rev. E 60, 6323–28 (1999).
351. Molecular motors and the forces they exert (M.E.F. and A.B. Kolomeisky) Physica A 274, 241–266 (1999) and in "Applications of Statistical Physics," Proc. NATO Adv. Res. Workshop, Budapest, May 1999, Eds. A. Gadomski, J. Kertész, H.E. Stanley and N. Vandewalle (Elsevier, Amsterdam, 1999) pp. 241–266.
352. Criticality and crossover in accessible regimes (G. Orkoulas, A.Z. Panagiotopoulos, and M.E.F.) Phys. Rev. E 61 5930–5939 (2000).
353. Periodic sequential kinetic models with jumping, branching, and deaths (A.B. Kolomeisky and M.E.F.) Physica A 279, 1–20 (2000); Erratum: *ibid* 284, 496 (2000).
354. High resolution study of fluid criticality (G. Orkoulas, M.E.F. and A.Z. Panagiotopoulos) in "Computer Simulation Studies in Condensed Matter Physics XIII," Eds. D.P. Landau, S.P. Lewis and H.B. Schüttler (Springer Verlag, Heidelberg, Berlin, 2000) pp. 167–171.
355. (a) Simple mechanochemistry describes the dynamics of kinesin molecules (M.E.F. and A.B. Kolomeisky) Proc. Natl. Acad. Sci. USA 98, 7748–53 (2001).

(b) Describing the dynamics of single kinesin molecules [abstract] (M.E.F. and A.B. Kolomeisky) Biophys. J. 80, 513a (2001) 2202-Plat.
(c) Motor proteins: Observations and theory (M.E.F. and A.B. Kolomeisky) [abstract for an invited talk] Bull. Amer. Phys. Soc. 46 (1) 1104 (2001) X21 1.

356. The Yang-Yang anomaly in fluid criticality: Experiment and scaling theory (M.E.F. and G. Orkoulas) Phys. Rev. Lett. 85, 696–699 (2000).
357. (a) The Yang-Yang anomaly in fluid criticality: An exactly soluble model [abstract] (C.A. Cerdeiriña, G. Orkoulas and M.E.F.) XV Congress of Statistical Physics, Royal Spanish Physical Society, Salamanca (27–29 March 2008).
(b) Exactly soluble fluid models exhibiting Yang-Yang anomalies at criticality [abstract and poster] (C.A. Cerdeiriña, G. Orkoulas and M.E.F.) Proc. 7th Liquid Matter Conf. (Lund, Sweden, June 2008).
(c) See [420] below.
358. The Yang-Yang relation and the specific heats of propane and carbon dioxide (G. Orkoulas, M.E.F. and C. Üstün) J. Chem Phys. 113, 7530–45 (2000).
359. (a) Force-velocity relation for growing microtubules (A.B. Kolomeisky and M.E.F.) Biophys. J. 80, 149–154 (2001).
(b) The growth of microtubules against an external force [abstract] (A.B. Kolomeisky and M.E.F.) Biophys. J. 80, 514a (2001) 2203-Plat.
360. Extended kinetic models with waiting-time distributions: Exact results (A.B. Kolomeisky and M.E.F.) J. Chem. Phys. 113, 10867–877 (2000).
361. Coexistence and criticality in size-asymmetric hard-core electrolytes (J.M. Romero-Enrique, G. Orkoulas, A.Z. Panagiotopoulos and M.E.F.) Phys. Rev. Lett. 85, 455861 (2000).
362. Asymmetric primitive-model electrolytes: Debye-Hückel theory, criticality and energy bounds (D.M. Zuckerman, M.E.F. and S. Bekiranov) Phys. Rev. E. 64, 011206:1–13 (2001).
363. (a) The heat capacity of the restricted primitive model electrolyte (E. Luijten, M.E.F. and A.Z. Panagiotopoulos) J. Chem. Phys. 114, 5468–71 (2001).
(b) Criticality and charge fluctuations in the restricted primitive model electrolyte [abstract] (E. Luijten, M.E.F. and A.Z. Panagiotopoulos) Bull. Amer. Phys. Soc. 46 (1) 71 (2001) A11 4.
364. Precise simulation of criticality in asymmetric fluids (G. Orkoulas, M.E.F. and A.Z. Panagiotopoulos) Phys. Rev. E 63, 051507:1–14 (2001).
365. The critical locus of a simple fluid with added salt (Y.C. Kim and M.E.F.) J. Phys. Chem. B 105, 11785–95 (2001).
366. Phase transitions in 2:1 and 3:1 hard-core model electrolytes (A.Z. Panagiotopoulos and M.E.F.) Phys. Rev. Lett. 88, 045701:1–4 (2002).
367. (a) Universality class of criticality in the restricted primitive model electrolyte (E. Luijten, M.E.F. and A. Z. Panagiotopoulos) Phys. Rev. Lett. 88, 185701:1–4 (2002).

(b) How to simulate fluid criticality: The simplest ionic model has Ising behavior but the proof is not so obvious! Proc. Internat. Conf. Theoretical Physics: TH-2002, UNESCO, Paris, 22–27 July 2002, Ann. Henri Poincaré 4, 413–416 (2002).

(c) Charge fluctuations and criticality in the restricted primitive model electrolyte [abstract] (E. Luijten, M.E.F. and A.Z. Panagiotopoulos) in "STATPHYS 21 Conference Abstracts," Eds, D. López, M. Barbosa and A. Robledo (IUPAP, Cancun, Mexico, 2001) p. 66.

368. Lattice models of ionic systems (V. Kobolev, A.B. Kolomeisky and M.E.F.) J. Chem. Phys. 116, 7589–7598 (2002).

369. Criticality in charge asymmetric ionic fluids (J.-N. Aqua, S. Banerjee and M.E.F.) [arXiv:cond-mat/0410692] Phys. Rev. E 72, 041501:1–25(2005).

370. Motion of kinesin on a microtubule [abstract] (M.E.F. and A.B. Kolomeisky) Biophys. J. 82, 62a (2002) 305-Pos.

371. (a) A simple kinetic model describes the processivity myosin-V (A.B. Kolomeisky and M.E.F.) Biophys. J. 84, 1642–1650 (2003).

(b) A simple stochastic model can explain the motility of myosin V molecules [abstract] (A.B. Kolomeisky and M.E.F.) Biophys. J. 82, 15a (2002) 71-Plat.

372. Screening in ionic systems: Simulations for the Lebowitz length (Y.C. Kim, E. Luijten and M.E.F.) Phys. Rev. Lett. 95, 145701:1–4 (2005).

373. Asymmetric fluid criticality I. Scaling with pressure mixing (Y.C. Kim, M.E.F. and G. Orkoulas) Phys. Rev. E 67, 061506:1–21 (2003).

374. Asymmetric fluid criticality II. Finite-size scaling for simulations (Y.C. Kim and M.E.F.) [arXiv:condmat/ 0306331] Phys. Rev. E 68, 041506:1–23 (2003).

375. First-order transitions from singly peaked distributions, Special issue: Physica A 389, 2873–79 (2010).

376. (a) Precise simulation of near-critical fluid coexistence (Y.C. Kim, M.E.F. and E. Luijten) [arXiv:cond-mat/ 0304032] Phys. Rev. Lett. 91, 065701:1–4 (2003).

(b) Criticality, coexistence and screening in electrolytes: High-resolution simulations [abstract] (M.E.F. and Y.C. Kim) program book for "15th Symposium on Thermophysical Properties," NIST, Boulder, Colorado, June 2003, p. 133.

377. Discretization dependence of criticality in model fluids: a hard-core electrolyte (Y.C. Kim and M.E.F.) [arXiv:cond-mat/0402275] Phys. Rev. Lett. 92, 185703:1–4 (2004).

378. Fluid critical points from simulations: the Bruce-Wilding method and Yang-Yang anomalies (Y.C. Kim and M.E.F.) [arXiv:cond-mat/0310247] J. Phys. Chem. B 108, 6750–6759 (2004).

379. Ionic criticality: an exactly soluble model (J.-N. Aqua and M.E.F.) [arXiv:cond-mat/0311491] Phys. Rev. Lett. 92, 135702:1–4 (2004).

380. (a) Reflections on the beguiling but wayward spherical model [Spanish title: "Reflexiones sobre el Modelo Esférico, Hechicero pero Caprichoso"] in "Current Topics in Physics," Proc. Symposium at the Universidad Nacional Autónoma de Mexico, 17 June 2003, in honor of Sir Roger J. Elliott, Eds. R.A. Barrio and K.K. Kaski (Imperial College Press, London, 2005) Chap. 1, pp. 3–31.
(b) The beguiling but wayward spherical model: Criticality, fluctuations and screening (M.E.F. and J.-N. Aqua) Rev. Mod. Phys. [in preparation].
381. (a) Vectorial loading of processive motor proteins: Implementing a landscape picture (Y.C. Kim and M.E.F.) [arXiv:cond-mat/0506185] J. Phys.: Condens. Matt. 17, S3821–S3838 (2005).
(b) Vectorial loading of processive motor proteins: Understanding kinesin [abstract] (M.E.F. and Y.C. Kim) Biophys. J. 86, 527a-528a (2004) 2738-Plat.
(c) Vectorial loading of processive motor proteins: Microtubule buckling experiment revisited [abstract] Bull. Amer. Phys. Soc., March 2006, [D29.12].
382. Charge and density fluctuations lock horns: ionic criticality with power-law forces (J.-N. Aqua and M.E.F.) J. Phys. A: Math. Gen. 37, L241–L248 (2004).
383. Convergence of fine-lattice discretization for near-critical fluids (S. Moghaddam, Y.C. Kim and M.E.F.) [arXiv:cond-mat/0502169] J. Phys. Chem. B 109, 6824–6837(2005).
384. Fluid coexistence close to criticality: Scaling algorithms for precise simulation (Y.C. Kim and M.E.F.) [arXiv: cond-mat/0411736] Comp. Phys. Commun. 169, 295–300 (2005).
384 ˆYang-Yang anomalies and coexistence curve diameters: Simulation of asymmetric fluids (Y.C. Kim) [arXiv:cond-mat/0503480] Phys. Rev. E 71, 051501:1–16 (2005).
385. Interfaces: Fluctuations, interactions and related transitions, Chap. 2 in "Statistical Mechanics of Membranes and Surfaces" (Second Edition, enlarged and revised) Eds. D. Nelson, T. Piran and S. Weinberg (World Scientific Publ., Singapore, 2004) pp. 19–47: see [281].
386. How multivalency controls ionic criticality (M.E.F., J.N. Aqua and S. Banerjee) [arXiv:cond-mat/0507077] Phys. Rev. Lett. 95, 135701:1–4 (2005).
387. Interfacial tensions near critical endpoints: Experimental checks of EdGF theory (S.-Y. Zinn and M.E.F.) [arXiv:cond-mat/05043857] Molec. Phys. 103, 2927–2942 (2005): special issue in honor of B. Widom.
388. Singular coexistence-curve diameters: Experiments and simulations (Y.C. Kim and M.E.F.) [arXiv:cond-mat/ 0507369] Chem. Phys. Lett. 414, 185–192 (2005).
389. (a) Kinesin crouches to sprint but resists pushing (M.E.F. and Y.C. Kim) Proc. Natl. Acad. Sci. USA 102, 16209–14 (2005).

(b) Kinesin binds ATP, crouches and swings 8 nm; But rebuffs help [abstract] (Y.C. Kim and M.E.F.) Biophys. J. 88, 649a (2005) 3189-Pos.

(c) Kinesin crouches before sprinting and resists forward and leftward loading [abstract] (Y.C. Kim and M.E.F.) Proc. 228th ACS National Meeting, "Biophysical Chemistry and Novel Imaging of Single Molecules and Single Cells," Division of Physical Chemistry, Philadelphia, PA (25 August 2004.)

390. Universality of ionic criticality: Size- and charge-asymmetric electrolytes (Y.C. Kim, M.E.F. and A.Z. Panagiotopoulos) Phys. Rev. Lett. 95, 195703:1–4 (2005).

391. (a) Static and dynamic critical behavior of a symmetrical binary fluid: A computer simulation (S.K. Das, J. Horbach, K. Binder, M.E.F. and J.V. Sengers) [arXiv: cond-mat/0603587 22 Mar 2006] J. Chem. Phys. 125, 024506:1–11 (2006).

(b) Interdiffusion in Critical Binary Mixtures by Molecular Dynamics Simulation (K. Binder, S.K. Das, M.E.F., J. Horbach, and J.V. Sengers) in "Diffusion Fundamentals II; L'Aquila 2007," Eds. S. Brandani, C. Chmelik, J. Kärger and R. Volpe (Leipziger Universitätsverlag, Leipzig, 2007) pp. 120–131.

392. Critical dynamics in a binary fluid: Simulations and finite-size scaling (S.K. Das, M.E.F., J.V. Sengers, J. Horbach and K. Binder) Phys. Rev. Lett. 97, 025702:1–4 (2006).

393. Molecular Motors: A theorist's perspective (A.B. Kolomeisky and M.E.F.) Ann. Rev. Phys. Chem. 58, 675–695 (2007).

394. (a) Backstepping, hidden substeps, and conditional dwell times in molecular motors (D. Tsygankov, M. Lindén and M.E.F.) [arXiv:q-bio BM/0611051] Phys. Rev. E 75, 021909:1–16 (2007).

(b) Back-stepping, dwell times, and hidden substeps in molecular motors [abstract] (D. Tsygankov, M. Lindén and M.E.F.) Biophys. J. 90 (Jan. 2007) 496a: 2369-Pos.

395. (a) Charge fluctuations and correlation lengths in finite electrolytes (Y.C. Kim and M.E.F.) Phys. Rev. E 77, 051502:1–7 (2008).

(b) Charge Fluctuations and correlations in Finite electrolytes, contributed talk (Y.C. Kim and M.E.F.) American Physical Society Meeting, Baltimore, about 16 March 2006.

396. (a) Mechanoenzymes under superstall and large assisting loads reveal structural features (D. Tsygankov and M.E.F.) Proc. Natl. Acad. Sci. USA 104, 19321–326 (2007).

(b) Superstall and assisting-load velocities of motor proteins can reveal mechano-chemical structure [abstract] (D. Tsygankov and M.E.F.) Biophys. J. 90 (Jan. 2007) 497a: 2371-Pos.

397. Simulating critical dynamics in liquid mixtures: Short-range and long-range contributions (S.K. Das, J.V. Sengers and M.E.F) J. Chem. Phys. 127, 144506:1–5 (2007).

398. Kinetic models for mechanoenzymes: Structural aspects under large loads (D. Tsygankov and M.E.F) J. Chem. Phys. 128, 015102:1–12 (2008).
399. Boltzmann Award Session; Remarks on the Boltzmann Medalists (*Kurt Binder* and *Giovanni Gallavotti*) Proc. STATPHYS 23, Euro. Phys. J. B 64, 301, 303–5 (2008).
400. (a) Velocity statistics distinguish quantum turbulence from classical turbulence (M.S. Paoletti, M.E.F., K.R. Sreenivasan and D.P. Lathrop) Phys. Rev. Lett. 101, 154501:1–4 (2008); designated an Editor's Selection and selected for a Viewpoint in *Physics Today.*
    (b) Quantum turbulence [abstract] (D.P. Lathrop, M.S. Paoletti, M.E.F., K.R. Sreenivasan) Bull. Amer. Phys. Soc. 54 (19) ET 1, 132 (2009).
    (c) Velocity statistics in superfluid and classical turbulence [abstract](K.R. Sreenivasan, D.A. Donzis, M.E.F., D.P. Lathrop, M.S. Paoletti, and P.K. Young) Bull. Amer. Phys. Soc. 54 (19) ET 2, 132 (2009).
401. (a) Comment on a recent conjectured solution of the three-dimensional Ising Model (F.Y. Wu, B.M. McCoy, M.E.F. and L. Chayes) Phil. Mag. 88, 3093–95 (2008).
    (b) Rejoinder to the Response to 'Comment on a recent conjectured solution of the three-dimensional Ising Model' (F.Y. Wu, B.M. McCoy, M.E.F. and L. Chayes) Phil. Mag. 88, 3103 (2008).
    (c) Erratum for 'Comment' and 'Rejoinder,'(F.Y.Wu *et al.*) Phil. Mag. 89, 195 (2009).
402. Criticality in multicomponent spherical models: Results and cautions (J.-N. Aqua and M.E.F) Phys. Rev. E 79, 011118:1–13(2009).
403. Critical charge and density coupling in ionic spherical models (J.-N. Aqua and M.E.F.) [in preparation].
404. Reconnection dynamics for quantized vortices (M.S. Paoletti, M.E.F. and D.P. Lathrop) Physica D 239, 1367–77 (2010).

404 ^'Search & Discovery': "*Filming vortex lines reconnecting in a turbulent superfluid*," B. Schwarzschild, Physics Today (July 2010) pp. 12–14.

405. (a) Atoms and Ions; Universality, singularity and particularity: On Boltzmann's vision a century later [abstract] p. 351,in "*Statistical Physics, High Energy, Condensed Matter and Mathematical Physics*," Proc. Conf. in Honor of C.N. Yang's 85th Birthday, Eds. M.L. Ge, C.H. Oh and K.K. Phua (World Scientific, Singapore (2008): PHOTO p. *xxiv.*
    (b) Discussion Remarks following lecture by Kerson Huang and a question by C.N. Yang, *ibid*, pp. 350–351.
406. Dynamics of the tug-of-war model for cellular transport (Y.-X. Zhang and M.E.F.) Phys. Rev. E 82, 011923:1–14 (2010).
407. For Professor Thomas Erber in recognition of his 80th Birthday, in "*Doing Physics*: *Festschrift for Tom Erber*," Ed. P.W. Johnson (Illinois Inst. Tech. Press, Chicago, IL, 2010) pp. 169–172. [Also pattern on the book's cover.]

408. Superfluid transitions: Proximity eases confinement, News & Views, Nature Phys. 6, 483–4 (2010). Comment on:

408ˆ "*Coupling and proximity effects in the superfluid transition in ^{4}He dots*," J.K. Perron, *et al.* Nature Phys. 6, 499–502 (2010).

409. (a) Measuring the limping of processive motor proteins (Y. Zhang and M.E.F.) J. Stat. Phys. 142, 1218–1251 (2011).
(b) Limping factors for motor proteins [abstract] (Y. Zhang and M.E.F.) Biophys. J. 100, 120a, 652-Pos (2011).

410. (a) Initial conditions for reconnection calculations of quantized vortices [abstract] (C. Rorai, D.P. Lathrop, M.E.F. and K.R. Sreenivasan) Bull. Amer. Phys. Soc. 55 (16) 327 (2010) M.
(b) Numerical investigations of reconnection of quantized vortices (C. Rorai, M.E.F., D.P. Lathrop, K.R. Sreenivasan and R.M. Kerr) Bull. Amer. Phys. Soc. 56 (18)DFD (2011) D.21.2.

411. (a) Entrevista con Michael E. Fisher (interview conducted in English by J.M. Ortiz de Zárate and translated by him into Spanish and edited by M.E.F. in both languages) Revista Iberoamericana de Física 6(1), 60–63 (2010).
(b) Interview with Michael E. Fisher [conducted by José M. Ortiz de Zárate] Europhys. News 42(1), 14–16 (2010).

412. When is a conductor not perfect? Sum rules fail under critical fluctuations (S.K. Das, Y.C. Kim and M.E.F.) Phys. Rev. Lett. 107, 215701:1–4(2011).

413. Cyril Domb: A personal view and appreciation [invited preface for special issue of the journal] J. Stat. Phys. 145, 510–517 (2011); 146, 883 (2012).

414. Quantized vortex reconnection: Fixed points and initial conditions (D.P. Meichle, C. Rorai, M.E.F. and D.P. Lathrop) Phys. Rev. B 86, 014509:1–4(2012).

415. Near critical electrolytes: Are the charge-charge sum rules obeyed? (S.K. Das, Y.C. Kim and M.E.F) J. Chem. Phys. 137, 074902:1–12(2012).

416. Propagating and annihilating vortex dipoles in the Gross-Pitaevskii equation (C.R. Rorai, K.R. Sreenivasan and M.E.F.) Phys. Rev. B 88, 134522:1–10(2013).

417. Criticality in alternating layered Ising models. I. Effects of connectivity and proximity (H. Au-Yang and M.E.F.) Phys. Rev. E 88, 032147:1–12(2013).

417ˆ Criticality in alternating layered Ising models. II. Exact scaling theory (H. Au-Yang) Phys. Rev. E 88, 032148:1–8 (2013).

418. (a) Renormalization group theory, the epsilon expansion, and Ken Wilson as I knew him, in *K. G. Wilson Memorial Volume: Renormalization, Lattice Gauge Theory, the Operator Product Expansion and Quantum Fields*, Eds. B.E. Baaquie, K. Huang, M.E. Peskin and K.K. Phua (World Scientific Publishers, Singapore, 2015);
Chap. 3, pp. 43–82.

(b) Renormalization group theory, the epsilon expansion, and Ken Wilson as I knew him, Int. J. Mod. Phys. B **29**, 1530006:1–39 (2015).

419. Comments concerning the Ising model and two letters by N.H. March (M.E.F. and J.H.H Perk) Phys. Lett. A **380**, 1339–40 (2016).

420. Soluble model fluids with complete scaling and Yang-Yang features (C.A. Cerdeiriña, G. Orkoulas, and M.E.F) Phys. Rev. Lett. 116, 040601:1–5(2016).

420 ˆCompressible cell gas models for asymmetric fluid criticality (C.A. Cerdeiriña and G. Orkoulas) Phys. Rev. E 95, 032105:1–15 (2017).

421. (a) Statistical Physics in the Oeuvre of Chen Ning Yang, Proc. Conf. 60 Years of Yang-Mills Gauge Field Theories: *CN Yang's Contributions to Physics*, Eds. L. Brink and K.K. Phua (World Scientific Publ., Singapore, 2015) Chap. 10, pp. 167–198.
(b) Statistical Physics in the Oeuvre of Chen Ning Yang [Invited review] Int. J. Modern Phys. B **29**, 1530013: 1–31(2015).
(c) Statistical Physics in the Oeuvre of Chen Ning Yang, Asia Pacific Newsletter 5, 22–30 (2016).

422. Michael Ellis Fisher: pp. 2–74, Commentary, pp. 6–13 (with Papers [45] and [384] and List of Publications in Wolf Prize in Physics, Ed. T. Piran (World Scientific Publ. Co., Singapore, 20

C. BOOK REVIEWS, (omitting contributions to Math. Reviews, 1966–67, etc.)

1. "Journal of Electronics," Vol. 1, No. 1, in Endeavour, February 1956.
2. L. Eisenbud and E.P. Wigner, "Nuclear Structure," in Atomics, January 1959.
3. L.D. Landau and E.M. Lifschitz, "Quantum Mechanics," in Science Progress, 1959.
4. E. Bodewig, "Matrix Calculus," 2nd Edition, in Proc. Phys. Soc. 74, 1959.
5. P. and T. Ehrenfest, "The Conceptual Foundations of the Statistical Approach in Mechanics," in Bull. Inst. Phys., 1960.
6. J.L. Powell and B. Craseman, "Quantum Mechanics," in Science Progress, 1961.
7. G.G. Hall, "Matrices and Tensors," in Proc. Phys. Soc., November 1963.
8. H.S.W. Massey and H. Kestelman, "Ancillary Mathematics," 2nd Edition, in Proc. Phys. Soc. 85, 1304 (1965).
9. J. de Boer and G.E. Uhlenbeck, Eds., "Studies in Statistical Mechanics. III," in Phil. Mag. 10 (1965).
10. D. Mattis, "The Theory of Magnetism," in Proc. Phys. Soc. 85 (1965).
11. "A consideration on the solution of partial differential equations by analog computer," T. Miura and J. Iwata, Simulation 6 (2) 105–108 (February 1966) in IEEE Trans. Electronic Computers (1966).

12. "Simulation of Field Problems using General Purpose Resistance Network Analogues," Jacques J. Vidal, Simulation 7, 190–203, October 1966, in IEEE Trans. on Computers c-17, 98 (1968).
13. E.G.D. Cohen, Ed., "Fundamental Problems in Statistical Mechanics. II," in Science, 21 February 1969.

C. BOOK REVIEWS, etc.

14. B.W. Roos, "Analytic Functions and Distributions in Physics and Engineering," in Physics Today 23, December(12) 52 (1970).
15. C. Domb and M.S. Green, Eds., "Phase Transitions and Critical Phenomena," Vols. 1 and 2, in Science 182, 576 (9 November 1973).
16. H. Haken and M. Wagner, Eds., "Cooperative Phenomena," in American Scientist (1974).
17. Remarks on M.S. Green in "Perspectives in Statistical Physics," Vol. 9 in Studies in Statistical Mechanics, Ed. H.J. Raveche (North Holland, Amsterdam, 1980).
18. "What's Mathematical Physics to Physics? Some examples from past, present, and future," Extract in Appendix to A. Bohm, Physica 124A, 103–114 (1984).
19. J. Mennier, D. Langevin and N. Boccara, Eds., "Physics of Amphiphilic Layers," in Applied Optics 28 (12) 2180 (15 June 1989).
20. Yu A. Izyumov and Yu N. Skryabin, "Statistical Mechanics of Magnetically Ordered Systems" (Consultants Bureau, N.Y., 1988) in Physics Today (January 1990) pp. 75–76.
21. "Mathematical Rigor in Physics: To What End?" Report on a lecture by M.E.F. in Summary from the Symposium "*Modern Perspectives of Mathematics: Mathematics as a Consumer Good, Mathematics in Academia*," Math. Sciences Institute, Cornell University, 29–31 March (1990) (M.S.I., Cornell Univ., Ithaca, N.Y., 1990) pp. 24–25.
22. J. Willard Gibbs and His Legacy: A Double Centennial, APS March Meeting, Austin, 3 March 2003 (Report by S.G. Brush and M.E.F.) Newsletter of the Forum on the History of Physics of The American Physical Society 9(1) 6–7 (Fall 2003).

D. UNPUBLISHED REPORTS, POSTERS, etc.

1. Discrete transforms and some of their applications, pp. 40 (1957).
2. Formalisms of thermodynamics and statistical mechanics, pp. 11 (Chapter 2 of a planned article on phase transitions) (1962).
3. Notes on the Ising model with long-range forces, pp. 11 (March 1963).
4. Comments on "Excluded volume effect in polymer chains" (with M.E. Sykes) pp. 5 (May 1963).
5. The Van Hove Theorem (July 1963).

6. Generalised Microcanonical Ensemble: Sketch of a direct and more general proof (July 1966).
7. Some Basic Definitions in Graph Theory (with J.W. Essam) (1967). Published in revised form in 1970: see [89].
8. On the transition entropy of K.D.P. (with J.F. Nagle) (September 1967).
9. The dimer problem and related lattice Combinatorics. Notes prepared (with aid) for Summer School on Statistics, Univ. North Carolina, June 1969. [Held on MEF's Dimer file.]
10. Finite-size effects in the spherical model (with G.A.T. Allan) (April 1970).
11. MIT Summer School Notes: Renormalization Group and Critical Phenomena (June 1972).
12. MIT Notes as above [revised] (June 1974).
13. Concerning research on tricritical and multicritical points at Cornell University [for MSC and NSF] (June 1974).
14. Theory of Critical Correlations in the Critical Region (with D. Jasnow) Chaps. 1–7: intended as Vol. IV in "Phase Transitions and Critical Phenomena," Eds. C. Domb and M.S. Green (Academic Press, Inc.).

D. UNPUBLISHED REPORTS, POSTERS, etc.

15. Spatially Modulated Phases in Ising Models with Competing Interactions (with Walter Selke) for Workshop on "The Contribution of Neutron Scattering to the Study of Incommensurable Modulations," Inst. Laue-Langevin, Grenoble, 13–14 October 1980.
16. Proof that Scaling cannot be derived from Thermodynamics (intended as an Appendix in H.E. Stanley's "Critical Phenomena," 2nd Edition) February 1981.
17. Some Progress in the Study of Wetting Phenomena (April 1984) for Condensed Matter Theory Program, N.S.F.
18. Helium in Vycor: Criticality in a Dilute Bose Fluid (P.B. Weichman and M.E.F.) Informal Proceedings of the Fifth Oregon Conference on Liquid Helium, 1–3 February 1987.
19. Critical Adsorption Profiles from Reflectivity Data (with Andrea J. Liu) September 1987. Circulated informally; formed a basis for the first part of a paper published in 1989: see [282].
20. Comment on work by A.L. Kholodenko and A.L. Beyerlein concerning the relation between electrolyte models and the spherical model (12 June 1991) [circulated informally].
21. On the Possibility of Convection in the 3M-PVTOS Experiment (by Sheng-Nan Lai, under supervision of M.E.F.) 26 March 1993. [Circulated informally and included in report to A.F.O.S.R. for NASA.]
22. Beyond the Ising Model: Understanding Criticality in Continuum Systems, extended abstract for the International Symposium on *New Developments in*

Statistical Physics to celebrate the 60th birthday of Professor Masuo Suzuki, 25–26 March 1997, Univ. of Tokyo. [Circulated in book of Abstracts for Meeting.]

23. Charge Fluctuations in Electrolytes: Exact results at low densities (S. Bekiranov and M.E.F.) Abstract for a (late) contributed poster at the American Physical Society 1998 March Meeting.

D. UNPUBLISHED REPORTS, POSTERS, etc., Continued

24. P.W. Debye and E. Hückel: "Zur theorie die Electrolyte," Physik. Zeit. 24 (1923): A translation of pages 192–193 (on basis of assistance from D. Brill and K. Vollmayr-Lee) August 1998.
25. The dynamics of single kinesin molecules described by simple stochastic models (M.E.F. and A.B. Kolomeisky) [summary with references and figures] March 2000.
26. Criticality in simple fluids: The Yang-Yang anomaly and other novelties [summary] Howard Reiss Symposium at the International Conference on Nucleation and Atmospheric Aerosols, University of Missouri-Rolla, August 2000.
27. The Yang-Yang Anomaly in Fluid Criticality: Exactly Soluble Models (R.T. Willis and M.E.F.) Poster for "Thermo-2005" Mid-Atlantic Meeting on Thermodynamics, University of Maryland, April 2005.
28. ARCC Workshop Statement for "*Phase Transitions in Physics, Computer Science, Combinatoriecs and Probability Theory*," American Institute of Mathematics, Palo Alto, California, 15 July 2006.
29. Definition of Thermodynamic Phases and Phase Transitions (M.E.F. with Charles Radin) set on American Institute of Mathematics (AIM) website, December 2006.
30. Backstepping, Dwell Times, and Hidden Substeps in Molecular Motors (D. Tsygankov, M. Lindén, and M.E.F.) Poster (with Abstract) for the 51st Annual Meeting of the Biophysical Society in Baltimore, March 2007.
31. Superstall and Assisting-Load Velocities of Motor Proteins can reveal Mechano-Chemical Structure (D. Tsygankov and M.E.F.) Poster (with Abstract) for the 51st Annual Meeting of the Biophysical Society in Baltimore, March 2007.
32. Limping Factors for Motor Proteins (Y.-X. Zhang and M.E.F.) Poster (with Abstract) for the 55th Annual Meeting of the Biophysical Society in Baltimore, March 2011.

D. UNPUBLISHED REPORTS, POSTERS, etc., Continued

33. Vortex Dipole Dynamics in the Gross-Pitaevskii Equation (C. Rorai, D. P. Lathrop and M.E.F.) [abstract with references] "Dynamics Days 2012," Baltimore, 4–7 January 2012.

https://doi.org/10.1142/9789811282386_bmatter

Index

$1/N$ expansion, 510
$1/n$-expansion, 253, 635
$O(N)$ vector models, 683
β-function, 509
ϵ-expansion, 249, 290, 311, 347, 348, 350, 358, 387, 388, 395, 398, 408, 619–621, 626–628, 680, 685
λ-transition, 683
ϕ^3 interaction, 285
$\sqrt{\varepsilon}$-expansion, 574–576, 632, 642
^{4}He film, 575, 579, 580
2D Ising critical point, 675
3D $O(2)$ model, 683
3D $O(3)$ model, 683
3D Heisenberg critical point, 675
3D Ising CFT, 677, 680
3D Ising critical point, 675
3D QED, 687

active instabilities, 798
active nematics, 795–797, 799, 800
active turbulence, 796, 799, 803
AdS_2, 708–710
Aharony, Amnon, 22, 24, 34, 62
Ahlers, Guenter, 251, 256
amplitude ratios, 253
"analytic" corrections, 609
Andereck–Patton theory, 402
Anderson Localization, 478, 479
Anderson, Phil W., 477–479, 497
Anderson–Higgs mechanism, 519
angle-like coexistence, 126, 131, 132
anisotropic perturbations, 683
anisotropy, 551,
  cubic, 580, 685
  non-cubic, 552, 580
  triangular, 566, 573
  weak, 551, 554, 559, 562–564
anisotropy matrix, 552, 553, 558, 561–563, 577
anomalous dimension, 314, 355–357
anomalous spin-Hall state, 351
antiferromagnet, 590, 591, 595
Ashcroft, Neil W., 41, 42
Aspen, 386
asymptotic, 305, 307, 308, 310–312, 316
autocorrelation function, 75, 81
avoided crossing, 137

bacteria, 796–798
Bardeen, Cooper and Schrieffer, 307
Bardeen, J., 306
Barenblatt, G. I., 317
BCS, 307, 308, 319
BCS instability, 507
Bell Laboratories, 62
Benfatto, Giuseppe, 509
Berezin, F. A., 480
beyond the bulk, 405
biaxially periodic critical states, 534
bicritical, 261, 590, 591
bicritical point, 27
binary liquid mixtures, 208
binary mixture film, 197
Binder cumulant ratio, 552, 578
Binder parameter, 463, 464
Binder ratio, 604, 606, 607
Binder, Kurt, 62
biological molecular motors, 786
Bishop, David, 62

black hole, 695, 705, 712, 713
Bogoliubov, 306–308, 311, 319
Bogoliubov, N. N., 306
Boltzmann equation, 699
Boltzmann Medal, 28
bootstrap condition, 299
boundary conditions, 564, 601
  antiperiodic, 571
  arbitrary, 564
  Dirichlet, 575
  nonperiodic, 556
  periodic, 552, 553, 566, 569, 580
Brézin, Édouard, 253
branch cut, 694
branched polymers, 292
Brownian motion, 787
Bruce, Alastair D., 22, 251

capillary wave fluctuations, 409
capillary waves, 410
CDW instability, 509
chaotic, 695, 699, 712
charge-density wave, 351, 353
Chemical Physics Seminar, 40
chiral-Heisenberg universality class, 356
cigar, 705
cluster expansion, 293
coarsening length, 78, 80
coexistence curve, 408
Combescot, Monique, 22
commensurate, 592
complex order parameter ψ, 386
condensed matter physics, 303–306, 319
conductance, 478, 479, 495, 496
  β-function, 479, 495
conformal blocks, 678
conformal bootstrap, 357, 675, 680, 682
conformal field theory, 297, 551, 566, 568, 676
constraint, 408
continuous symmetry, 407
Cooper, L., 306
Cooperon, 479, 494
Cornell, 386, 599
corrections to scaling, 253, 608
correlation function, 250, 253
  Fisher–Burford, 123
  two-spin, 250, 253
correlation length, 121, 140, 354, 405, 552, 601, 770
  amplitude ratio, 555, 558
  angular-dependent, 552, 564
  concentration fluctuations, 140
  critical fluctuations, 121
  isotropic, 555, 558, 560
  mean, 558, 561, 578
  polymer-chain, 125, 132, 140
  principal, 553, 558
correlation length of a *finite* system, 606
correlation vector, 559, 564–566
Coulomb gauge, 395
coupling functions, 505
critical Casimir force, 411, 575, 579, 580
critical dimension, 255, 266
  lower, 266
  upper, 255, 266
critical dynamics, 123, 632
critical exponents, 266, 631, 675, 769, 771, 772, 774–776
  $2+\epsilon$ expansion, 497
  $4-\epsilon$ expansion, 477
  effective, 266
  from finite size scaling, 497
  scaling relations, 770
critical line, 27
critical matter, 513, 515
critical phase, 515, 543
critical phenomena, 70, 120, 587, 588, 631, 675, 785
critical phenomena in electrolyte systems, 786
critical point, 675, 771
critical point Casimir effect, 197
critical wall perturbations, 405
critical-point universality, 119
criticality, 121, 135
  Ising-like, 135
  logarithmic corrections, 121, 134
criticality in electrolytes, 786
crossover, 575, 577, 771, 778
  dimensional, 575
  from Goldtstone to critical region, 575
  from low- to high-temperatures, 577
crossover theory, 290
cubic symmetry, 251
Curie transition, 683
cusp singularity, 71, 73, 77, 78, 84

cutoff of the order-parameter fluctuation, 197, 199
cytoskeleton filament, 787, 790

d-wave superconductors, 347, 349
dangerous relevant variable, 614
dangerously irrelevant, 696
Dasgupta, Chandan, 391, 392
Dasgupta–Halperin dual model, 402
de Gennes and Brochard theory, 137
de Gennes model, 386
de Gennes, Pierre Gilles, 197, 385
de Gennes–Sanchez scaling relations, 125, 126
Debye–Huckel theory of electrolytes, 786
degree of polymerization, 120, 121
diffusion, 478, 479, 493, 497
diffusion coefficient, 788
diffusive relaxation mode, 137
dilute polymer solutions, 311
dilution, 259
dimensional reduction, 711
dimensional regularization, 633
dimensionality shift, 255, 267, 268
dipolar interactions, 250, 251, 255, 257
    Heisenberg, 251
dipolar system, 250, 256
    critical behavior, 250
Dirac fermions, 347, 349–351
directed branched polymers, 294
directed percolation, 315
Dirichlet
    boundary conditions, 579, 580
Dirichlet–Dirichlet boundary conditions, 199
dislocations, 391
dissipation scale, 304
DNA denaturation, 89, 90, 97
double expansion, 634
drift velocity, 788
dynamic coupling, 136
dynamic critical phenomena, 61
dynamic properties of molecular motors, 791
dynamical critical exponent, 354
Dyson, Freeman, 40
Dzyaloshinskii, Lifshitz, and Pitaevskii theory, 199

Edward, Sir Samuel Frederick, 311
Efetov, Kostas B., 478, 479, 488, 495
effective field theory, 318, 411
effective Hamiltonian, 508
effective medium theory, 770
eigenstate thermalization, 712
Einstein gravity, 705
Einstein relation, 789
Einstein-Maxwell, 707
electrolytes, 786
emergence, 309, 317
emergent Higgs mechanism, 523, 535
energetics of the biological processes, 792
entropy, 705, 712, 713
epithelial cells, 800
equation of state, 253, 387
equilibrium constant, 790
equilibrium director, 386
eukaryotic cells, 796, 799, 800
excluded volume problem, 311
exclusion interaction, 98
experimental data, 402
extremal functional method, 682
Eyink, G. L., 313

factorization of the S matrix, 299
Fermi surface, 503, 504
Feynman diagrams, 632
finite-size scaling (FSS), 200, 206, 405, 411, 444, 451, 461, 588, 600, 772–774, 777
    out-of-equilibrium, 454, 456, 466, 467
first-order critical point, 96
first-order transition, 387
Fisher, Daniel, 63
Fisher, Sorrel, 28, 246
five-dimensional Ising model, 612, 613
fixed point, 263, 389, 508
    biconical, 263
    cubic, 263
    decoupled, 263
    isotropic, 263
Flory exponent, 98
Flory–Huggins theory, 126–128, 138
fluctuating surface, 75, 77, 82
fluctuation–dissipation relation, 314, 315
fluctuation-dominated clustering, 75, 82
fluctuation-dominated order, 71, 79, 83

fluctuation-driven anomalous elasticity, 540
FOCT (First-order classical transitions), 439, 459, 465, 467
FOQT (First-order quantum transitions), 439, 440, 450, 453
force of the motor protein, 789
force–velocity relations for growing microtubules, 792
four-point correlation function, 678
fractal dimensionality, 254
fractals, 254
fractional quantum hall effects, 304
Frank constants, 399
Frank elastic constants, 392
free boundary conditions, 601, 612
free energy, 128–130, 139, 769, 770, 772–774, 776
    density-dependent, 128
    field-dependent, 128
    polymer-chain, 129, 130, 139
fully developed turbulence, 646
functional flow equation, 507

Gallavotti, Giovanni, 509
Garland, C. W., 402
Garland–Nounesis, 402
gauge field $\mathcal{A}$, 386
gauge-field fluctuation, 387
gauge models, 386
gauged order parameter Ψ, 397
Gaussian and Heisenberg points, 390
generalized Widom crossover, 134
geometry, 554
    block, 554, 556, 575
    cubic, 575, 577
    film, 575, 579, 580
    parallelogram, 554, 566, 572
    rectangle, 566
    rectangular, 554
    slab, 575, 578
Georgi, H., 318
Gibbons-Hawking, 707
Ginsburg-Landau-Wilson, 23
Ginzburg criterion, 133, 140
Ginzburg number, 120, 133
Ginzburg parameter, 388
Ginzburg, Landau and Wilson, 611
glass, 778
Goldstone mode, 207, 514, 551, 574
    singularities, 576
    crossover to critical region, 575, 577
graphene, 347, 349
Grassmann, Hermann, 479, 480, 482, 495
Griffiths, Robert B., 39
Gross–Neveu model, 349, 357
Gross–Neveu–Yukawa (GNY) field theories, 347, 358
Gross–Neveu–Yukawa models, 686
Group theory, 264

Halperin, Bert I., 385, 386, 391, 392
hard-core square-well, 594
Harvard, 63
heat capacity, 683
Heisenberg, 590, 591
Heisenberg model, 683
Helfrich, Wolfgang, 391
helicity, 405
helicity modulus, 407
high temperature series, 267
high-temperature superconductors, 303
"HLM" paper, 386
Hohenberg, Pierre, 39, 44, 61, 62, 245
honeycomb lattice, 347, 349
horizon, 706
Hubbard model, 347, 349, 358
Hubbard–Stratonovich transformation, 286, 479, 483, 484
hybrid transitions, 90
Hydra, 800, 802
hydrolysis of adenosine triphosphate, 787
hyperscaling, 604, 611, 614, 695
hyperscaling relation, 256, 268

IDSI model, 91, 96
imaginary magnetic field, 282
Imry, Yoseph, 245, 266
Imry-Ma, 267
    domains, 267
incommensurate, 592
incomplete similarity, 317
infinite-randomness fixed point, 365
instanton, 619–621
interdisciplinary research, 305
interfaces, 405
interfacial profile, 408
intermittency, 314, 317

invariance relation, 568, 571–573
  of area ratio, 570
  of correlation function, 560
  of scaling variable, 571
irrelevant operator, 608
irrelevant variable
  dangerous, 775, 776
  dangerous irrelevant variable, 771
  irrelevant variable, dangerous, 769
Ising, 588, 589, 591, 592
Ising model, 310, 441, 562–564, 571–573, 580, 601
  classical, 459, 467
  gap, 442
  quantum, 441, 455
Ising model with long-range interactions, 71, 75, 79
Ising model, RG
  2D normal form with resonance, 418
  4D normal form, 418
  droplet singularity, 424
isostaticity, 771, 772, 778
isotropic gauge-transformed model, 398
isotropic-to-nematic (I–N) transition, 388

jamming transition, 90, 769–771, 777, 778
Josephson exponent relation, 408
JT gravity, 710

Kac, Marc, 33
Kadanoff, Leo P., 22, 40, 41, 313, 316, 319, 385
Kawasaki functions, 124
Kekule bond-density wave, 351, 353
Kibble–Zurek, 456, 467
kinesin, 787
Kolmogorov, 317
Kolmogorov theory, 650
Kolmogorov, A. N., 313
Kosterlitz, J. M., 26, 391
Kosterlitz–Thouless, 363, 369

Lake Como, 385
Lake Como School of Advanced Studies, 385
lambda line, 130
Lambda-point experiment, 683
laminar flow, 314
Landau expansion, 127, 128
Landau parameter, 506
Landau, Lev, 23
Landau–de Gennes free energy, 392
Landau–Ginzburg functional, 127, 128
Landau-Zener effect, 137
large order, 619–621, 626, 627
laser optical trap, 791
lattice
  triangular, 564
lattice gas, 283
lattice superconductor model (LSM), 391
lattice Villain model, 387
Lebowitz, Joel, 62
Lee, David, 21
level correlation, 491–493
levels of description, 306, 309, 310, 315, 317, 318
Lifson, Shneior, 34
light-scattering, 121, 122, 124, 125, 138
  dynamic, 124, 125, 138
  static, 125
limit of metastability, 388
Lipatov asymptotics, 639
liquid crystal, 551, 580
logarithmic corrections, 256, 410, 777
London penetration depth, 388
long-range interaction, 91, 89, 90
long-range order, 72, 73, 76, 84
Lorentz, Hendrik A., 44

M–SC theory, 389
M–SC transition, 400
Müller, K. Alexander, 251
Ma, Shang-keng, 61, 252, 266, 386
MacMillan ratio, 402
magnetic exponent, 607
magnetic field, 607
magnetohydrodynamics, 651
Mandelbrot, Benoit, 254
many-body localization, 363
marginal dimension, 132, 134
marginal stability, 770
marginal variable, 395
massive scheme, 637
matrix-tree theorem, 293
mean field theory, 610
measurements, 553, 578
  scattering, 556, 562, 578
  thermodynamic, 553, 556, 578, 562

Mermin, David, 21
Mermin, Dorothy, 40, 42, 45
method of uniqueness, 638
Michael and Sorrel, 386
microtubule-motor protein suspensions, 796
microtubules, 787
minimal model, 297, 306–309, 316, 317
minimal subtraction, 620, 621, 623, 626, 627
minimal subtractions, 633
mixed-order transition, 79, 89, 90, 96, 99
mobility edge, 478, 489, 496, 497
mode-coupling theory, 123
Model A of critical dynamics, 644
momentum-shell RG, 286
Monte Carlo, 587–595, 600, 610
Monte Carlo simulation, 387, 391, 682–685
morphogenesis, 800, 802
motor proteins, 787
Mott insulator, 349
Mott, Neville F., 478
multicritical, 589–592
multicritical points, 261
multifractality, 479, 496
multiparameter universality, 553, 562, 580, 581
  bulk correlation function, 557, 562–564
  critical Casimir amplitude, 568
  critical free energy, 566, 568
  excess free energy, 569, 572

n-orbital model, 478–480, 489
  Oppermann Wegner, 478
n-vector model, 554, 557, 559, 560, 562, 578
n-vector system, 412
N–SA transition, 391, 392, 402
Néel, 353
Néel order, 349
Navier–Stokes equations, 315
Nelson, David R., 21, 26, 63
nematic (N), 386
nematic-to-isotropic (NI) transition, 402
nemato-elastic Goldstone mode, 541
Nobel Prize, 20
non-equilibrium critical point, 315
non-equilibrium phase transitions, 653
non-Fermi liquid, 504
non-linear σ-model, 477–479
  Schäfer Wegner, 478, 488
non-linear elasticity, 529
non-symmetrical boundary conditions, 208
normal-metal to the superconducting state, 388
Nounesis, G., 402

Onsager Prize, 28
Onsager, Lars, 33
Oono, Y., 320
OPE coefficients, 677, 682
operator product expansion, 649, 676
order parameter, 129, 131, 603
  polymer chain, 129, 131
order-parameter distribution function, 575, 576
ordering field, 131
  polymer chain, 131
Ornstein–Zernike, 123
  correlation function, 123
Orsay branch of the University of Paris, 385
Osheroff, Douglas, 21
osmotic susceptibility, 122
out-of-time-order, 695

Parisi-Sourlas supersymmetry, 292
passive particles, 75, 76, 83
Penrose, Oliver, 33
percolation, 254, 655
percolation transitions, 90
period boundary conditions, 601
periodic boundary conditions, 612
periodic critical states, 516
periodic hopping models, 788
phase diagram, 771
phase fluctuation within the superfluid film, 207
phase transitions, 70, 75, 587–589, 592–594, 632
physical correlation functions, 397
physical gauge, 393, 398
pistachio trees, 310
Planckian, 695, 701
Platzman, Phil, 63
Polchinski, Joseph, 509
Polyakov, Alexander M., 477

polymer chain, 137
    correlation length, 137
polymer solution theory, 313
polymer solutions, 312
polymerized membrane, 537
Porod law, 73, 74, 82–84
port, 35
Potts model, 449, 456, 462, 468
    qantum, 449
predator–prey, 315, 316
principal angles, 563, 565, 569
principal axes, 557, 559, 560, 563, 577, 578
principal directions, 552, 553, 565
Privman–Fisher, 556, 566, 567, 570, 577
    hypothesis, 566, 567, 571
    scaling form, 556, 570, 577
    scaling function, 556
pseudopotential, 311

quantum anomalous Hall state, 351, 357
quantum chaos, 714
quantum field theories, 318
Quantum Fisher information, 451
quantum fluctuations, 261
quasicrystals, 303
quasiparticles, 694, 698
Quenches, 454
    classical, 466
    quantum, 454, 467

radius of gyration, 121, 140
random fields, 266
random matrices, 479, 489, 490, 493
Random-field Ising model
    2D normal form and scaling, 419
random-singlet state, 364
ratchets, 787
reaction–diffusion systems, 653
relevant operator, 608
renormalizability, 288, 318
renormalization at the physical dimension, 637
renormalization group, 20, 98, 123, 254, 305, 311, 478, 498, 551, 574, 580, 589, 594, 607, 632, 775
    epsilon expansion, 123
    field-theoretic, 574
    functional, 551, 580
    marginal perturbation, 552
    minimal renormalization, 552, 574, 576
    real space, 254
    renormalization at fixed dimension, 552, 574
renormalization-group theory, 785
renormalized-field theory, 287
renormalon, 619, 621, 625–627
replicas, 259
repulsive Casimir force, 208
repulsive gas singularity, 285, 290
response function formalism, 295
restricted primitive, 594
Reynolds number, 304
RG fixed points, 25
Ricci scalar, 705
Richardson, Robert, 21
Rohrer, Heinrich, 252
"rounding" exponent, 612
Royal Medal, 28
Rubin, Mort, 385
Rutgers, 62

Sachdev-Ye-Kitaev model (SYK), 695, 703, 712
scalar electrodynamics, 348
scaling ansatz, 769, 770, 772, 774, 777, 779
scaling dimensions, 677, 682
scaling function, 253, 772
    collapse of, 206
    universality of, 253
scaling function, universal, 769
scaling relation, 603
scaling variable, 570, 571, 577
    anisotropic system, 571, 573
    isotropic system, 570, 577
Schrödinger, Erwin, 796
Schrieffer, R., 306
Schwarzian, 696, 703, 712
Schwarzschild solution, 705
second-order inverted-xy-like, 392
second-order transition from a dual theory, 391
sector decomposition technique, 640
self-averaging, 270
self-avoiding walk singularity, 130
self-organized criticality, 655
self-portrait, 37
self-similar, 552, 569, 572, 573, 579

Sengers, Jan, 62
shear transformation, 553, 555, 557, 560, 565, 566
    effective, 567
    generalized, 560, 567, 573
    inverse, 555, 566, 568, 572
    inverting, 558
    special, 559, 560, 565
"shift" exponent, 612
Siggia, Eric, 61, 62
sine-Gordon transformation, 295
single-molecule experiments, 787
"singular" corrections, 609
SL(2,R), 702, 704, 709, 711
sliding particles, 71
smectic Goldstone mode, 526
smectic-A (SA), 386
specific heat, 555, 556, 562, 563, 573, 578
specific-heat exponent α, 391
spin anisotropy, 257
spin glass, 610, 614
spontaneous line crystals, 534
spontaneous magnetization, 588, 589
Static Critical Phenomena, 640
Stephen, Michael, 247
strange metal, 714
superconducting fixed points, 390
superconductivity, 306, 308, 319
superconductor, 551, 579, 580
superfluid, 309
superfluid density, 407
superfluid or the lambda transition, 197
superfluidity, 306
superrenormalizable theory, 637
supersymmetry, 267, 498, 697
susceptibility, 555–557, 562, 578, 602
Sykes, Martin, 33

tetracritical, 261
tetracritical point, 27
the Landau–Ginzburg model for superconductors, 386
the LC energies and correlation functions, 392
thermal exponent, 607
thermodynamic limit, 408
theta temperature, 127
theta-point, 120
Thouless effect, 90, 94
Thouless, David J., 20, 391, 478, 496
TIDSI model, 80, 91, 96, 98, 99
time reparameterization, 711
time-reversal, 352
time-reversal symmetry, 351, 357
topological defects, 795–800, 803
topological scaling laws, 318
torus modular parameter, 568, 570, 573
transitions in superconductors, 386
tricritical, 591
tricritical behavior, 121
    theta-point, 121
tricritical phase transition, 130, 131
tricritical point in ^{3}He–^{4}He mixtures, 207
tricritical points, 264
tricriticality, 408
triple line, 409
triple point, 262, 264
turbulence, 304, 312–315, 317
two-dimensional Ising model, 605
two-point correlation function, 677
two-scale-factor universality, 552, 555, 562, 571, 572
    bulk correlation function, 562
    bulk systems, 555
    finite systems, 556
type-I SCs, 388
type-II systems, 388

universal crossover behavior, 135
universal scaling function, 253, 770–772, 774–776
universal scaling functions, 312
universality, 606
universality class, 120, 121, 552–555, 578, 579
    (d,n), 552, 562, 578
    XY, 554, 578
    Gaussian, 555
    Heisenberg, 554
    Ising, 554, 555, 572, 578
    Ising-model, 120, 121
    spherical, 554, 555
universality class, 62
    revised notion, 553
    subclasses, 553

upper critical dimension, 287, 348, 610, 772

Villain model, 391
viscoelastic mode, 136, 137

Wagner, Herbert, 39, 40
weakly interacting Bose gas, 306
Weinberg, S., 320
Widom crossover, 133, 135
Widom scaling, 312
Widom variable, 132, 135
Widom, Benjamin, 22, 34, 36, 39, 40, 45, 250, 312, 504
Wigner, Eugene P., 479, 490, 492, 493
Wilson, Kenneth G., 21, 22, 39, 41, 250, 311, 385, 477, 504
Wilson–Fisher, 385
Wilson–Fisher CFT, 681
Wilson–Fisher expansion, 347
Wilson–Fisher paper, 61, 249
Wilson-Fisher fixed points, 347
Wolf Prize, 28

X-scattering, 398
XY model, 683

Yang–Lee edge, 281
  one dimension, 284
Yang–Lee theorem, 282
Yeshiva University, 62

Zirnbauer, Martin R., 479, 497, 498
Zurek, Woicjech, 42

Milton Keynes UK
Ingram Content Group UK Ltd.
UKHW051625051024
449147UK00004B/47

9 789811 28237